D0161142

Engineering Design

and

Design For Manufacturing

A Structured Approach

Text and Reference for Mechanical Engineers

John R. Dixon
Professor Emeritus
Corrado Poli
Professor and Head
Department of
Mechanical Engineering
University of Massachusetts
Amherst

Field Stone Publishers
Conway, Massachusetts, U.S.A.

Use At Your Own Risk

Every effort has been made to avoid errors in the preparation of this book. However, errors of both omission and commission are inevitable in a work of this size and nature. Readers and users of this book are therefore cautioned to verify by their own tests, checks, and methods any results or designs based on the data, methods, recommendations, and calculations presented herein. The authors and publisher cannot guarantee, and do not assume responsibility for, the efficacy or safety of designs or design methods based on the information contained in the book.

Thank You

The authors and publisher thank the copyright and trademark owners who have given their permission for use of copyrighted material and trademarks. We apologize for any errors or omissions by us in obtaining permissions or in giving proper credit. Such errors or omissions are unintentional and will be corrected at the first opportunity after the error or omission is brought to our attention. Please contact the publisher.

First Edition — *Second Printing*

Published By

Field Stone Publishers
331 Field Hill Road
Conway, Massachusetts 01341, U.S.A

ISBN 0-9645272-0-0

To Roe and Ann

who had to endure our absence while we “worked on the book”, and put up with our presence afterwards. Their love, patience, and good humor supported us through the long, hard times.

Acknowledgments

The authors of a technical book stand on the shoulders of others, and the debt that we owe is especially great. Chapter references list many of these people, but that is too implicit an acknowledgment of their contribution. Their work provided a foundation of knowledge and principles on which we could build.

There are too many shoulders for us to make special mention of them all, but we must mention a few. Geoff Boothroyd's work in design for assembly not only pioneered the field of DFA, but also opened the door to all the "design-for's". Gerhart Pahl and Wolfgang Beitz opened the door to a structured approach. Doug Wilde opened wide the door to optimization. Genichi Taguchi, Nam Suh, Hiroyuki Yoshikawa, and Tetsuo Tomiyama opened our minds to whole new points of view. Erskine Crossley added to our understanding of function. With their research, Herb Voelcker and Ari Requicha built a foundation for CAD. The classic machine design authors — Spotts, Shigley, Juvinall and Marshek, to name only a few — gave us huge amounts of useful, well organized knowledge on which to build. George Dieter's comprehensive design book was a special help. The work of the engineering design theory and methodology (DTM) research community — too many to list — has been invaluable. We owe much, too, to the insights of the British writers — Stuart Pugh, Nigel Cross, D. J. Lynch, Michael French, and Gordon Glegg. The professional societies and trade associations (e.g., ASME, ANSI, ASM, Society of the Plastics Industry, Spring Manufacturers Institute, Aluminum Association) have also been wonderfully helpful.

Our graduate students did much of the research that led to the book. In the UMass Mechanical Design Automation (MDA) Lab, these bright, enthusiastic people included: Richard Anderson, John Cunningham, Michael Duffey, Ajit Fathailall, Michael Guenette, Adele Howe, Rohinton Irani, Chris Jones, Vidu Kulkarni, Gene Libardi, Steven Luby, Ken Meunier, Bill Myers, Eric Nielsen, Mark Orelup, David Rosen, Qiang Sun, Mohan Vaghul, Ralph Verilli, Rick Welsh, Sally Wright, and Zhu Bo.

Graduate students contributed also to the research in DFM, and we give special thanks to Jiten Divgi, Sheng-Ming Kuo, Ricardo Fernandez, Lee Fredette, Ferruccio Fenoglio, Juan Escodero, Shyam Shanmugasundaram, Pratip Dastidar, Prashant Mahajan and Shrinivasan Chandrasrkar.

Faculty colleagues in the MDA Lab include Byung Kim and George Zinsmeister. Paul Cohen of the UMass Computer Science Department was always creative, enthusiastic, and knowledgeable. The contribution of Mel Simmons from the GE Research Laboratory is immeasurable; his insights, deep understanding, patience, questions, creativity, and good spirits often got us through. And we express our appreciation to our colleagues J. Edward Sunderland and Robert J. Graves who participated in the DFM work.

The contributions of the MDA Laboratory owe much to the financial and technical support of NSF, UMass, GE Plastics, GE Research Labs, GE Aircraft Engines, Polaroid, Xerox, Ford, Digital Equipment, Smith and Wesson, and Becton-Dickinson. But these firms contributed much more than money. The interest and advice of their people — like Barry Bebb, Hugh MacKenzie, Jim Mason, Ken Huebner, Bob Phillips, Mike Poccia, Gerry Trantina, Peter Will, and others — were very important and very special. J. B. Jones and Charles Hoover, and the National Research Council Committee on Engineering Design, taught us much. Peter Dalquist also provided valuable information and insight.

Industry sponsors of the DFM Laboratory included: Cambridge Tool and Manufacturing Co., Colonial Machine Co., Hobson and Motzer Inc., Jada Precision Plastics, K. F. Bassler Company, Kennedy Die Castings, Larson Tool and Stamping Company, Leicester Die and Tool Inc., Metropolitan Machine and Stamping Co., Newton New Haven Company, Sajar Plastics, Thomas Smith Company Inc., Tremont Tool and Gage, and Woldring Plastic Mold Technology. In addition, support from the Massachusetts Centers of Excellence Corporation is greatly appreciated.

Undergraduate students have been a tremendous help. Students in ME313 and ME375 commented on early drafts and pioneered the projects. Cedric Savineau, Mark Ouillette and Tom Gajewski were especially helpful. And we give special recognition to ME313 Design Teams A and G: thanks especially to Mark Manasas, Greg Batt, Kurt Shea, Ian Morrison, Scott Malenchini, Susan Hieb, Wells Hodous, Alan Jones, Chris Lukomski, and Julie Rodrigues.

Dorothy Kent's cheerful and indefatigable conversion of messy hand sketches into neat CAD representations was remarkable. Moreover, she was always there for going to the library, making copies, copy editing, and providing moral support through real and imagined emergencies. Without Dorothy, there would be no book, and the book there is would be much diminished.

To the anonymous reviewers who found ways to criticize and encourage at the same time, we offer special appreciation. And we owe much to Luis Longo and "V.E." Hu, who read and corrected the manuscript in detail.

Our spouses, children and grandchildren have paid a price. To Roe, Randy and Karen, Nicholas and Sarah, Linda and Stephen, and to Ann, Peter and Pam, Chuck and Debbie, and Kim, we apologize for our absences, both physical and mental.

Finally, to those who contributed but we have failed to acknowledge here, we also apologize. There just isn't space for you all, but we do sincerely appreciate your help. To the extent that the book is correct and useful, we owe much to all of you; to the extent that it is not, we are solely responsible.

John R. Dixon and Corrado Poli

Chapter Outline

Table of Contents

Front Matter

Part I - First Things

Prologue A Journey of a Thousand Miles

Chapter 1 The Proper Study of Mankind

Chapter 2 Best Practices of Product Realization

Chapter 3 Manufacturing for Designers

Part II - Engineering Conceptual Design

Chapter 4 Introduction to Part II: From Marketing Concept to Physical Concept

Chapter 5 - Formulating the Problem: The Engineering Design Specification

Chapter 6 - Generating Conceptual Alternatives

Chapter 7 - Evaluation and Redesign of Engineering Concepts

Chapter 8 - Completing the Conceptual Design of Special Purpose Parts

Chapter 9 - Selecting Materials and Processes

Part III - Configuration Design of Parts

Chapter 10 Formulating the Problem and Generating Altenatives

Chapter 11 Evaluating Part Configurations for Manufacturability: Injection Molding and Die Casting

Chapter 12 Evaluating Part Configurations for Manufacturability:Stamping.

Chapter 13 - Overall Evaluation and Redesign of Part Configurations

Part IV - Parametric Design Of Components

Chapter 14 - Introduction to Parametric Design

Chapter 15 Evaluating Parametric Designs for Manufacturability: Injection Molding and Die Casting

Chapter 18 - Introduction to Optimization

Chapter 19 Introduction to Taguchi Methods™

Chapter 20 Guided Iteration Applied to Common Mechanical Components

Chapter 21 Guided Iteration Applied to Thermal-Fluid Components

Part V Pandemic Topics

Chapter 22 Communications

Chapter 23 - Introduction to Engineering Economics

Chapter 24 - Patents, Liability, and Ethics

End Matter

Note to Instructors

> "Experience without theory teaches nothing."
>
> W. Edwards Deming [1]*
> *Out of the Crisis (1982)*

NI.1 Intended Use

This book is intended as a text for engineering students at the junior level or higher, and as continuing education and reference for engineering design practitioners in industry. Certain chapters, however, can be used in sophomore and freshman courses.

NI.2 A Structured Approach

The book takes a structured approach to both design and design for manufacturing (DFM), including a careful vocabulary for describing both designed objects and design processes. Structure is important for students and entry level practitioners so they can better observe and understand the processes within which they are working. Structure helps new designers see the forest among the trees, branches and leaves, and it ultimately helps engineers advance the profession of design.

The structure: After engineering design is placed in its context within the product realization process, it is described in three overlapping stages: engineering conceptual design, configuration design (of parts), and parametric design. Each stage is defined in terms of the information describing and related to the designed object that is to be determined and decided upon in that stage.

Structure is also provided by adherence to a single problem solving methodology called *guided iteration.* Guided iteration is presented as the synthesis process by which each of the three stages of engineering design is approached and resolved. Steps in the guided iteration process are:

- Formulating the problem,
- Generating alternative solutions,
- Evaluating the alternatives, and if none is acceptable,
- Redesigning, *guided by* the results of the evaluations.

In the text, specific methods are presented for each of these four steps in each of the three engineering design stages. A table of the these methods is shown in Table FM.1 on the next page.

Communications, an introduction to engineering economics, and design-related general issues (liability, patents, and ethics) are discussed in separate chapters.

NI.3 How to Use this Book

This book is intended for use in any one of three optional ways, and there also are a number of chapters students can read on their own.

Option A

The book can be used to support a one semester design course at the junior or senior level. In this case, selections from the following chapters may be included:

Chapters 1 through 10 inclusive;
Chapters 13 and 14;
Chapters 17 through 24 inclusive.

Depending on what is selected for inclusion, such a course can either supplement or replace a traditional course in machine design.

It is strongly recommended that this course include a limited scope term-long design project as described below. This project is intended to complement and reinforce the material in the text. It is not, however, intended as a substitute for a more comprehensive, more complex, project in a "capstone" course.

Option B

The book can also be used to support a one-semester course in *design for manufacturing* (or manufacturing for design) at the junior or senior level. For this purpose, the following chapters are recommended:

Chapters 3, 9, 11, 12, 15, 16, and 22.

It is strongly recommended that this course include a design for manufacturing (DFM) project as described below.

Such a course could be taught independently of a design course as described above in Option A, or the two courses could be taught simultaneously.

Option C

The book can also be used to support a two-semester course in design *and* design for manufacturing, including both the limited scope design project and the design for man-

---

* Numbers in brackets refer to references at the close of each chapter. This quoatation from *Out of the Crisis* is by permission of the M.I.T. Center for Advanced Engineering Study, Cambridge, MA.

STEPS IN THE GUIDED ITERATION PROCESS	STAGES IN THE DESIGN PROCESS: Engineering Conceptual Design	Configuration Design of Parts	Parametric Design
Formulation of Problem	Quality Function Deployment and "House of Quality" Engineering Design Specification	The Part Configuration Requirements Sketch	Identification of: Design Variables, Performance Parameters, Evaluation Criteria, and Analysis Procedures
Generation of Alternatives	Survey of Alternative Physical Principles Function First Decomposition Creative Techniques: Brainstorming, Synectics, Morphological Studies	Qualitative Physical Reasoning Re: Functionality and Failure Modes Qualitative DFM Reasoning and DFM "Advisors" Re: Tooling	Initial Trial Design Qualitative DFM Reasoning and DFM "Advisors" Re: Processing
Evaluation of Alternatives	Quality Function Deployment and "House of Quality" Pugh's Method Dominic's Method Pahl and Beitz Method Qualitative DFM Reasoning and DFM "Advisors" Re: Manufacturability	Qualitative Physical Reasoning Re: Functionality and Failure Modes Pugh's Method Dominic's Method Qualitative DFM Reasoning and DFM "Advisors" Re: Tooling Quantitative DFM Analyses Re: Tooling	Optimization Taguchi Methods™ for Robust Design Guided Iteration Using Dominic's Method Quantitative DFM Analyses Re: Processing
Guided Redesign	Guided by Results of the Evaluation	Guided by Results of the Evaluation	Guided by Results of the Evaluation

Table FM.1 Survey of Structure and Methods Described in This Book

ufacturing projects described below. For this purpose, all chapters may be included.

This two semester course could also logically and profitably be followed by a capstone course with a larger, more complex design project having design for manufacturing implications.

We recommend Option C because it provides the smoothest and most obvious integration of the design and DFM material. We have, however, made every effort to organize and write the book so that all three options listed above are feasible. We have also tried to write so that teachers in any option will have flexibility in selecting and re-ordering the material according to their own needs and preferences. Some individual chapters (e.g., 2, 3, 9, 22, 23, 24) can be used separately.

To conserve class time, there are several chapters (e.g., Chapters 1, 2, 22, 23, and 24) that average students can very comfortably read and understand independently. Indeed, some of these chapters can be handled well by sophomores and freshmen. In any case, for reasonable efficiency, students should read all the chapters themselves before class lectures and discussion; there is simply too much material for effective class coverage unless students have done the reading first.

Students in at least a half dozen courses have used the early drafts with good results. Tthese students report that they can successfully read and understand the material. Therefore, we strongly recommend minimal use of class time for specific lecturing on the text material; class time is far more valuable used for answering student questions, providing examples and counter-examples, and elaborating on the material from instructors' own backgrounds.

Integration of design and DFM begins with presentation of qualitative (DFM) guidelines in Chapter 3, and more is added in the chapters that concentrate on the design process. For example, students learn in configuration design of parts that reducing the number of parts in a product or assembly is usually the most important cost reducer, and that, for example, undercuts are relatively expensive to manufacture. Several qualitative DFM "Advisors" have been developed to assist with the presentation of the DFM guidelines for injection molding, die casting, stamping, extrusion, forging, and (to limitied extent) machining. These qualitative methods will suffice to sensitize design students to the issues involved in DFM, and they provide the information needed forearly "first cut" manufacturability evaluations of prospective designs.

For more accurate and complete DFM evaluations (e.g., how much more cost will a proposed undercut add?), quantitative methods are needed. These are presented in chapters 11, 12, 15, and 16 for the most common processes: injection molding, die casting, and stamping. The DFM analysis methods presented in these chapters inherently require considerably more attention to geometric detail, more than required by conceptual or configuration design. We have tried to smooth the transitions between the more abstract early design chapters and the more detailed DFM chapters, but there remains an unavoidable rapid change in the nature of the material. However, if students are to do quantitative DFM evaluations, they simply must take the time and trouble to learn to do them. Thus such transitions in level of detail simply go with the territory if one wants to perform quantitative DFM evaluations in a concurrent fashion.

Though at first glance the part coding systems that are the basis for these analyses appear cumbersome (there are a lot of new terms), they are in fact easy to learn. The reward is not only the ability to estimate relative tooling and processing costs, but also a good sense of design issues that drive part costs.

Another point where the level of abstraction changes rather suddenly is in Chapter 9, which deals with materials and process selection. Again, we have tried to put the more detailed material in its context, and to smooth the transition. There are, however, literally thousands of materials and hundreds of processes, and thus tens of thousands of combinations. By organizing materials hierarchically, and by approaching the selection processes in a design-oriented manner, we have done our best to make this particular forest visible among its many trees. But there still is no doubt that there is much more detailed material to deal with in Chapter 9 than in the conceptual design chapters that precede it and the configuration design chapters that follow.

NI.4 About Projects

In teaching from the preliminary drafts of the text, the authors have supplemented the reading and class discussions with projects as described below. We strongly recommend that projects of some type be included. The ones described below are examples of projects that the authors have used successfully. We assume, however, that most instructors will opt for projects of their own that will work better for them because they are their own.

Supplementary Design Project

For a supplementary design project (for either Option A or C), a project problem should be found that begins with a description of customer needs and wishes, together with certain presumed company capabilities and limitations. The problem's solution should ideally involve fairly distinct conceptual and configuration design stages, including both assembly and part configurations. Parametric design will always be involved, of course, but there may not be time in a course or two to do this phase thoroughly in the time frame of a course. That is okay; students have more difficulty and less experience with the conceptual and configuration stages, so the project helps most at these earlier design stages.

The project solution should involve designing a product or product line, but it cannot be a very complex product. There isn't time. Thus, there should be only a few parts. Only one or two of these parts need be selected for configuration design, and only one (perhaps only approximate) for parametric design.

When option A is being used, though any manufacturing process might be involved, we have found that aluminum extrusion is a good choice. Qualitative manufacturability issues in this domain are easily comprehended by students. Extensive data are available from major manufacturers. There is also relevant material in the text. Moreover, since extrusions are two dimensional, drawings are less time consuming for students to produce. Injection molding is the second choice for the project domain because its qualitative DFM guidelines are also fairly easy to follow.

We recommend having students read chapters 22 and 24 on their own at the beginning of the project. To get the project started quickly, some teachers may want to assign Chapter 5 ahead of Chapters 3 and/or 4. Chapters 1 and 2 require, at most, only one class period each if the students will do the reading; they should not be skipped, however.

Assigning students in teams with five or six members has worked well. Generally we feel it is best for the students to make up their own teams, though teachers may have to make some adjustments for numerical balance. Teams can select their own "weak managers" as discussed in Chapter 2.

Instructors must stress in class again and again the importance of teamwork, including such simple issues as attending all meetings, showing up on time, doing what was agreed upon, listening, contributing fully in good faith, being available for between-meeting communications, and so on. (Electronic mail works well for internal team communications.) The importance of teamwork can be reinforced by grading policy. See the example design project assignments below for one way to handle grading that stresses teamwork.

In our opinion, teachers need not micro-manage the students. If possible, good students (undergraduate or graduate level) who previously have taken the course can be used as teaching assistants to handle most student questions on the projects. To conserve instructor time, it should be required that these assistants be consulted before students are permitted to ask for help or advice from the instructor. In any case, help or advice from the course instructor should be provided only in normal office hours, or by special appointment.

Also, in our opinion, class time should not be used for the project-related questions unless the questions also connect to the text material. They will often connect, however, since the project work generally parallels fairly closely the flow of the textual material (i.e., conceptual, configuration, and parametric design).

Three written and one (final) oral project report are logically required as described in the sample assignments below. Students will be anxious for feedback on the early written reports, and this feedback must be given quickly and comprehensively. Though students will ask for grades, it is not necessary or even desirable that the interim reports be graded. The feedback given should be exclusively on the quality and technical merits of the design, and on the report.

Two sample design project assignments that have been used successfully are reproduced below. All students are given the assignments on the first day of the course even though teams have not yet been formed. The first of these example projects is followed through the text showing sample results at the end of the conceptual, configuration, and parametric stages. Sample transparencies from a final oral report are also shown in Chapter 22. However, these projects can only be considered as illustrative of a type. As noted above, each instructor will no doubt have his or her own knowledge on which to base other projects, better because of the instructor's first hand knowledge.

NI.5 Example Design Project A

The Assignment

(Note: This problem statement is repeated for students in Chapter 1, Section 1.8, and examples of student work provided at a number of stages throughout the book. The project is inspired by a product line of Bosch Automation Products, Buchanan, MI, and we sincerely appreciate the firm's cooperation.)

The company for which you are working (call it, say, Manufacturers Automation Supply, Inc., or MASI) as consultants or employees is a successful world manufacturer of consumer and industrial products. At the firm's plants there exist modern, efficient facilities for injection molding, die casting, plastic and aluminum extrusion, sheet metal stamping, and other common manufacturing processes. The firm has a reputation for high quality, reliable products.

MASI has a very effective marketing capability for its line of industrial machine tools and products used by other manufacturers who do both manual and sophisticated automated assembly of a wide variety of products. For example, MASI makes and markets pneumatic tools of many kinds, parts' bins, chairs and stools for work stations, small presses, parts feeding and orienting equipment, and electronic controls to support assembly processes. Indeed, MASIs customers can purchase just about everything needed to set up an automated assembly line except the tables, benches, and other structures to support the automation tools and equipment. That is, MASI's current product line requires customers to set everything up on the floor — or go to a competitor for the assembly line tables, etc. But when customers go to the competitors for tables, they also get exposed to the competitors' other products as well.

We therefore can well imagine that MASI's sales and marketing groups have found from discussions among themselves and with customers that there is a market opportunity for MASI to offer a line of work tables and benches, machine underframes, shelving for work stations, and the like. Since customers have a variety of needs, and these customers are always changing their facilities to keep up with their changing markets, the marketing concept is to offer a standardized or modularized system of some kind that will be "versatile, flexible, universally applicable, and readily reusable".

Other stated marketing requirements are: the structures should be "stable against torsional and bending stresses"; easily "extendible and convertible"; permit "extensive design possibilities", and be "easily erected and modified."

The company (represented by the Instructor) wants to review progress in the form of two page (maximum) reports as follows:

- *Engineering Design Specification at about 4 to 6 weeks,*
- *Conceptual Design (2 Page Maximum) at about 5 to 9 weeks, and*
- *Configuration Design of Selected Parts at about 7 to 11 weeks .*

A final report (10 pages maximum) is due at about 12 to 13 weeks. Questions will be answered in class or via e-mail.

NI.6 Example Design Project B

This is a second example project using aluminum extrusions. The assignment is:

Imagine that you are employed by a design consulting firm that has been retained by a large, diversified manufacturing corporation. Your job is to create a design that responds to a perceived market need for a new packaging system to compete with cardboard and wood containers for certain types of product shipping.

The corporation's marketing group has determined that it would be desirable for the packaging system to have the following characteristics:

- *fire retardant*
- *reusable for at least 50 round trips*
- *recyclable material*
- *size flexibility*
- *low maintenance in reuse*
- *compact storage when not in use, and small volume for empty return shipments*
- *easily cleaned and disinfected*
- *superior vertical stacking strength in up to 16 feet stacks*
- *temperatures in service: -20 to 120 F (A test is required on a 16 foot stack at 120 F for 7 days)*
- *cost competitive on a per trip basis with cardboard*
- *accommodates palletization*
- *superior resistance to normal shipping mishandling (e.g., drops, punctures, etc.). Containers dropable without damage from forklifts up to 12 inches. On a flat drop, the assumption is that forces are 75 Gsvertically and 15 Gslaterally into the side walls.*
- *superior resistance to vibrations during shipping on trucks and trains.*

The products that are to be contained and shipped include bulk raw materials, small loose piece parts and sub-assemblies, and military and commercial hardware. Density of loads will range up to 40 lbs/ft.^{3} Trailer trucks can handle total heights up to 100" and widths up to 96". Pallets are 55" high.

The client has facilities for and expertise in injection molding, die casting, aluminum extrusion, plastic extrusion, and stamping. Reports are required as in Project A.

NI.7 Comments to Instructors Regarding Design Projects

Examples of student reports for Project A at each of the suggested review points listed are included in the book.

Each team makes a ten minute oral, illustrated presentation during the last week of the course.

It is important to get the project started as quickly as possible. Students have difficulty at first getting the Engineering Design Specification from qualitative terms (e.g., "It must support a heavy load.") to quantitative ("It must support 400 pounds."). Thus Chapter 5 is important to them. They also need Chapter 3 as soon as possible. The project will help motivate the reading of both these chapters.

In Project A, some students seem to want to design the bench *tops* instead of just a basic structural framework system leaving the top to be some standard material that customers can choose based on their special needs. They need to be told not to design tops, but only to design to accommodate customers' tops of various types.

In both projects, it is helpful to hand out information on aluminum extrusions very early to supplement the text. For example, it does not say anywherein the text that aluminum extrusions are rather easily cut with a radial arm saw with an appropriate blade. Also, students are not generally familiar with the complexity of cross sections that is possible with extrusions. Though giving them this information sends a strong hint, not every team will necessarily develop a concept around aluminum extrusions. If an alternative concept seems feasible, instructors can allow the group to continue with it. However, no problems have been encountered by forcing teams to switch to aluminum extrusions after the conceptual stage if the initial alternatives are not likely to work out well. A switch should have no impact on the grade. This is one reason for not grading the interim reports.

The project grade has, in our classes, constituted 33% to 40% of the course grade for each student. At the end, each team is given a single grade for the project, and each member of the team shares that grade equally. This team grade counts for one-half of each student's project grade (i.e., 17% to 20% of the course grade). For the other half, each team member rates his or her teammates confidentially on their "participation as a team member". One way to do this that has worked well is to ask each student to distribute 100 points among his or her teammates based on the teamwork criteria. The points can then be converted to a grade in some reasonable manner.

These projects are not easy for the students. They will test most students' knowledge and work habits rather severely. But they are still doable with reasonable success. They involve a lot of work and effective organization on the part of individual students, and they absolutely require teamwork. Students must simultaneously grapple with learning and implementing the concepts in the book, with their interpersonal relations on the teams, and with the technical and design problems directly associated with the project. The student tendency to put things off until deadlines are imminent just won't do; it is therefore important that instructors insure that team members are in constant communication (e-mail is best) and have frequent, regular team meetings.

NI.8 Supplementary Design for Manufacturability Project

For a supplementary (DFM) project the authors have successfully used the following approach.

Students teamed in pairs select a consumer product such as a telephone, hair dryer, audio or video tape, beard trimmer, etc. Or, students can select some sub-assembly of a consumer product, such as a carburetor, alternator, etc. The assembly or sub-assembly must have a potential annual production volume of at least 50,000. We insist on a reasonably 'high' production volume so that students will be analyzing products which are produced using today's most common mass production techniques. Since students will be required to produce an exploded assembly drawing of the product and individual piece part drawings, it is suggested that they select products which appear reasonable to draw.

The product or sub-assembly should contain about 15 parts, more or less. If there are less than about 15 parts, the variety of parts is insufficient to make the project interesting. If there are more than about 20 parts, the project becomes too burdensome for the time available.

Following selection of their product, each team is required to produce an exploded assembly drawing of it. Then they are asked to estimate, using the methods presented in Chapter 3, the time required to assemble the product manually. They are also asked to redesign the product in order to reduce assembly time, hence assembly costs, and to estimate the total savings that can be achieved with their redesign.

Following this design for assembly analysis, the teams are next required to create a detailed part drawing for each injection molded, die cast, and stamped part, both in the original product and in the redesigned product. Then using the qualitative methods presented in Chapter 3, they can estimate approximate relative costs to produce each of these parts and compare the original designs with the redesigns.

As the course continues, and the more detailed DFM methods in Chapters 11, 12, 15, and 16 are encountered, the students repeat their relative part cost analyses and savings estimates using these methods.

While it is true that the DFM chapters can be taught without the use of a supplementary DFM project, our experience is clear and strong that student interest in manufacturing and how parts are made increases significantly when a project is included.

Though students are left essentially on their own to carry out the project, we have found that (human nature being the way human nature is) it is best if they are provided with a schedule. A schedule we have used is as follows:

Week 1: Search for product to be analyzed, read Chapter 22, and practice making isometric drawings, orthographic projections, and sectional views using AutoCAD or similar computer drawing package.

Week 2: Finalize selection of product.

Week 3: Work on assembly drawing. See Figure FM.1- for a sample. Number and label each part. For fasteners, indicate the method of fastening (press, snap, screw, etc.). Also indicate if the fastener is easily aligned and inserted. For non-fasteners, indicate if the part is easily aligned and inserted and the manufacturing method used to produce the part. For all parts indicate whether they are easy to grasp and manipulate with one hand using no holding tools.

Week 4: Turn in assembly drawing.

Week 5: Make corrections to drawing and resubmit with an analysis of the approximate time required to manually assemble the product. Make redesign suggestions to reduce assembly costs and determine the percent saving achievable. If, in an effort to reduce the number of parts, a part is being combined with one or more other parts, clearly state this on the part drawing after the part name. Be certain to indicate what the new part number will be on the assembly drawing for the redesigned assembly. If parts, such as screws, are being eliminated altogether and the fastening is to be accomplished using press/snap fits, be certain to clearly state this on the drawing after the part name.

Week 6: Submit assembly drawing for redesigned product. (See Figure FM.2 for a sample drawing). If screws have been eliminated from the original design and if the necessary fastening of parts is being accomplished by the use of press or snap fits be certain to indicate, in a separate drawing if necessary, how the snap fit features are being incorporated in the parts. Begin work on piece part drawings for all injection molded and die cast parts.

Week 7: Complete piece part drawings for injection molded and die cast parts. (See Figure FM.3 for a sample drawing.) Indicate on the drawing the overall part size, direction of mold closure and presence and location of undercuts. Obtain an estimate of the relative die costs for each part. Provide redesign suggestions.

Week 8: Turn in progress report with preliminary results comparing original design with redesign. Make preliminary recommendations.

Week 9: Begin work on piece part drawing for all stamped parts. Be certain to indicate overall part size, location of the primary plate, the presence of side-action features, etc. Obtain an estimate of the relative die costs for each part. Make redesign suggestions.

Week 10: Correct part drawings for injection molded and die cast parts. Add sufficient detail (size of ribs, bosses, wall thickness, etc.) so that an estimate of the relative processing cost of each part can be obtained.

Week 11: Correct the part drawing for all stamped parts. Estimate overall relative part cost.

Week 12: Work preparing final report and presentation.

Week 13: Complete project report and prepare presentation.

Week 14: Turn in final report and make final presentation.

Reference

[1] Deming, W. Edwards. *Out of the Crisis*, M.I.T. Center for Advanced Engineering Study, Cambridge, MA, 1982.

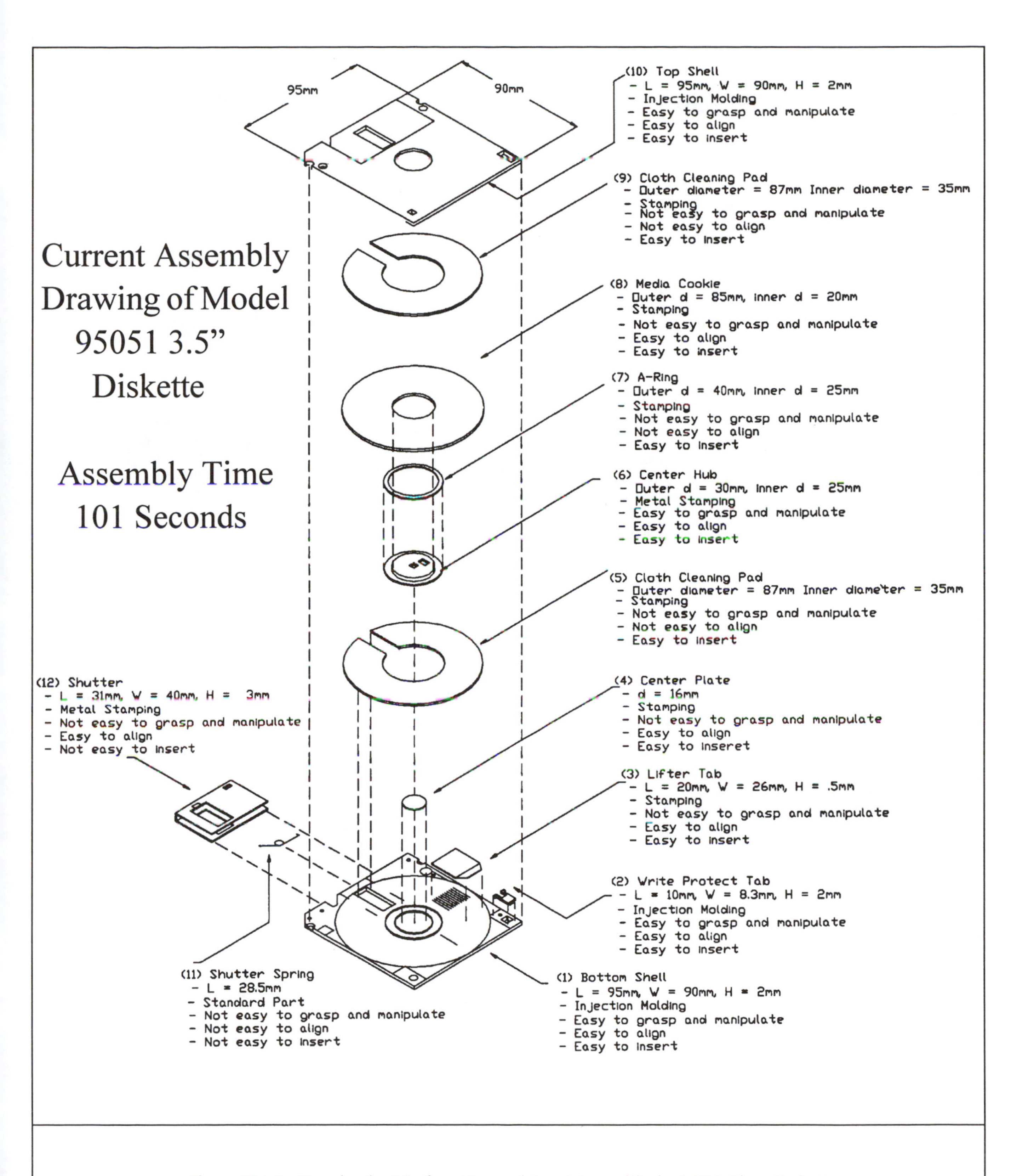

Figure FM.1 Drawing by Matthew Roy and Jean Maranville for DFM Class Project

Redesign Assembly Drawing of Model 95051 3.5" Diskette - Assembly Time 67 Seconds

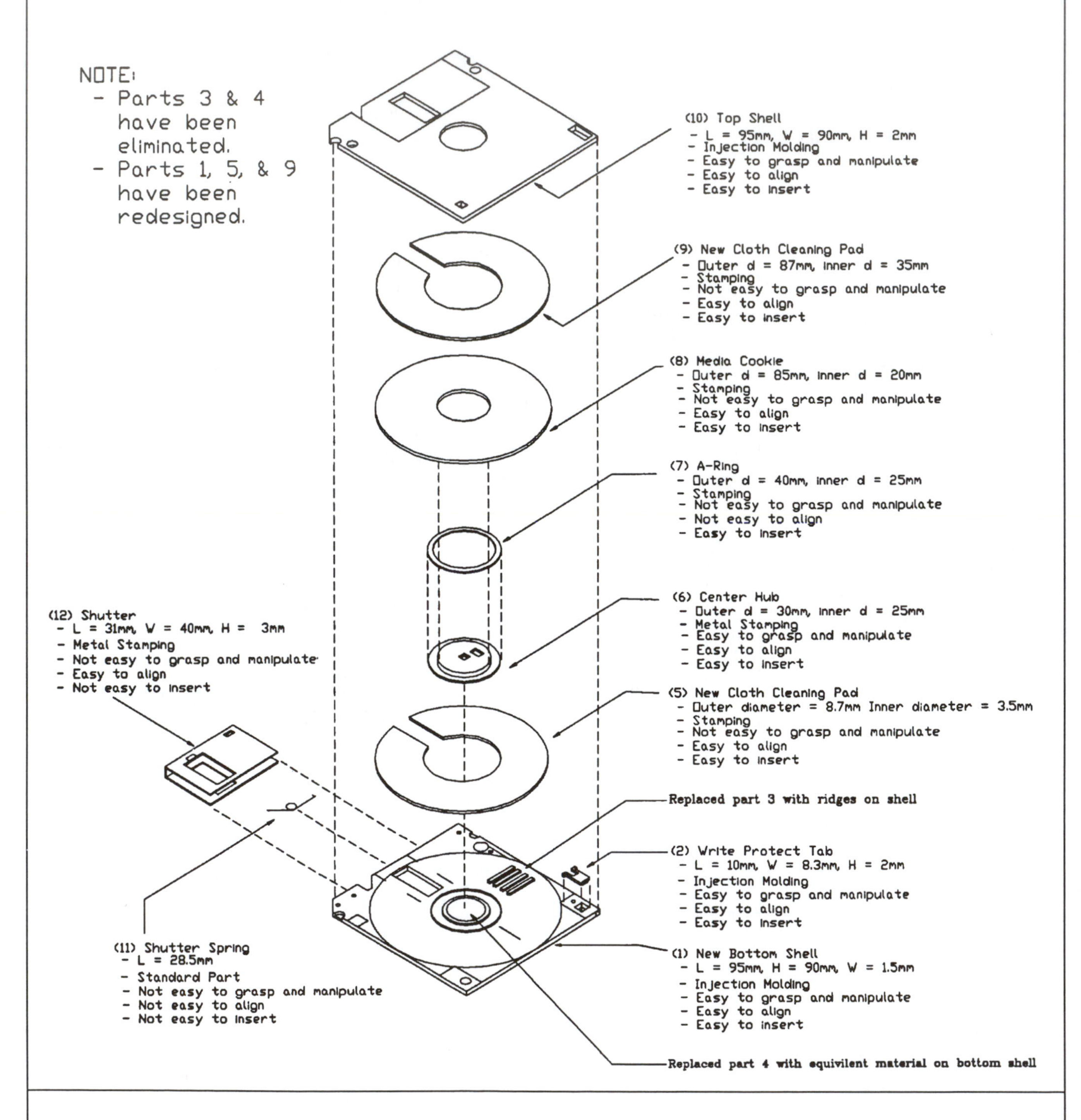

Figure FM.2 Drawing by Jean Maranville and Matthew Roy for DFMClass Project

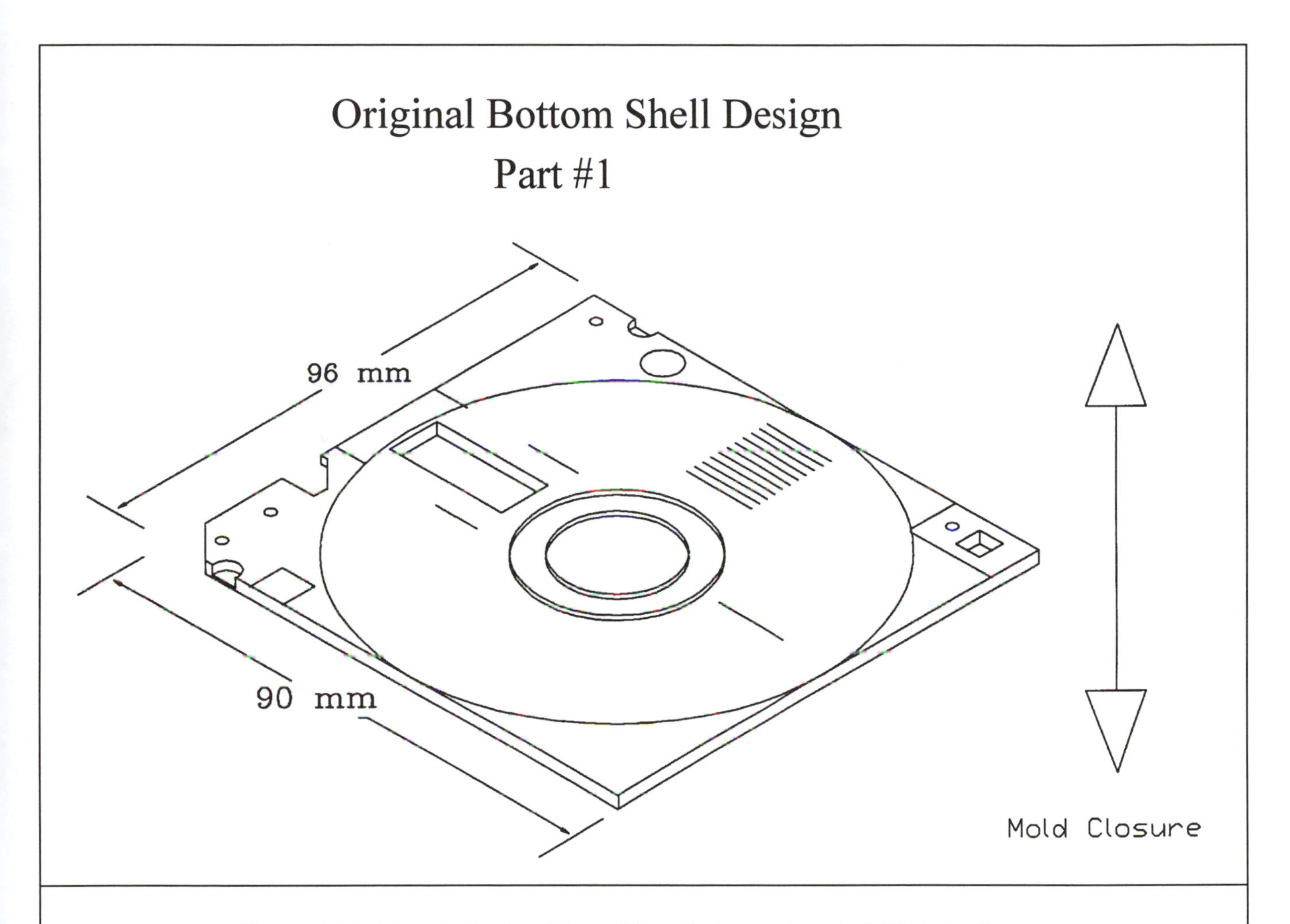

Figure FM.3 Drawing by Jean Maranville and Matthew Roy for DFM Class Project

This page is intentionally blank.

Part I First Things

Prologue

"A Journey of a Thousand Miles Must Begin With a Single Step"

Lao-tzu
The Way of Lao-tzu (600 B.C.)

To take the first step on a journey through the complex landscape of engineering design takes courage that can only come from a love of the subject, and from convictions about the best path to follow. It is the wonderful but excruciating nature of engineering design that every practitioner, teacher, researcher, and student loves the subject in his or her own way, and has his or her own convictions about how to describe it, how to do it, how to teach it, and how to learn it. And we, of course, have our convictions as well:

> The practice of engineering design and design for manufacturing (DFM) should be integrated at every stage of the design process. This being our conviction, we have presented the latest DFM knowledge throughout the engineering design story.

> Engineering design and DFM are intellectual, knowledge-based activities. Therefore, to do them well and fast requires extensive *know*-what, *know*-how, *know*-when, and *know*-why. Effective and efficient designers must not only have knowledge of and access to a great deal of factual information, but also knowledge of many methods (including when to use them), knowledge of how these methods relate to the process as a whole, and knowledge of the fundamentals upon which the methods are based. Therefore, we have included many methods, and how to perform them. And since the design process as a whole is greater than the sum of the methods used, we have placed the methods in their context within the design process.

> If design and DFM are knowledge based, then it follows that students and new practitioners will learn best if provided with an intellectual structure — that is, a language, logic, organization, and overall methodology — within which they can place the knowledge and methods that they encounter, and which they can use to develop an understanding and vision of the overall process. Such a structure is a secure starting place, and becomes a benchmark to be modified as designers develop their own understanding based on continued study and experience. Therefore, we have taken a structured approach to describing the design process. There are, in our view, the following well defined stages of engineering design, and the book is structured around them:

- Conceptual,
- Configuration, and
- Parametric.

> Students and new practitioners should be provided also with a structured approach to design problem solving. *Guided iteration* is the common problem solving methodology used in this book at every stage of the process. The steps in the guided iteration process are:

- Formulation of the Problem,
- Generation of Alternative Solutions,
- Evaluation of the Alternatives, and if none is acceptable,
- Redesign — *Guided by* the results of the evaluations.

> At each of these stages, there are methods available for implementing each step of the guided iteration process, and for performing DFM. The methods described in this book are summarized in Figure P1 on page FM-11 (seven pages back). There are other methods available that we have not included, and no doubt new and better methods will be developed. But those included here will give students and new practitioners a place to begin.

> Projects are an important part of the learning process *when the projects are related explicitly to knowledge and methods*. Thus, using actual projects, we illustrate how the intellectual structure and methods are employed.

> Communications (graphic, written, and oral), engineering economics, and certain legal (patents, liability) and ethical issues have special importance in design practice. Thus we have included chapters on these subjects.

We hope you enjoy and profit from the journey into and through engineering design and DFM that we describe in the pages that follow. Then we hope you go beyond our passion and convictions to develop your own.

"All this will not be finished in the first one hundred days. Nor will it be finished in the first one thousand days, nor perhaps in our lifetime on this planet. But let us begin."

John F. Kennedy
Inaugural Address (1961)

Chapter One

The Proper Study of Mankind

"The Proper Study of Mankind Is the Science of Design."

Herbert Simon [1]*
Sciences of the Artificial (1969)

1.1 Design, Engineering Design, and the Product Realization Process

Certainly design is the proper study of engineers. Design is where it all comes together for us. When we design, all our knowledge of physics, chemistry, mathematics, statics, materials, strength of materials, thermodynamics, fluids, heat transfer, instrumentation, analysis, and computers comes together. That's not all: history, psychology, literature, economics, sociology and philosophy are also a part of design. And that is still not all: design involves the whole person: character, creativity, values, and human understanding.

Best of all, design is fun. When we design, we not only get to apply all the science and engineering science we have worked so hard to learn, we also get to be creative, to express our individuality, and to apply our personal values and judgment. The reality of design is especially fun: if our design is good enough, it will get built, sold for a profit, and used by satisfied customers. And there is also a sense of idealism in design: if we design the right thing and design it well, we will help make a better world for people.

Design is exciting. It is exciting to see our designs work. It is exciting to beat the competition to the market with a product that is higher in quality and lower in cost. On the other hand, when our design does not work, that is exciting too — but not the kind of excitement we prefer.

Design is important. It is a major factor in the quality and cost of products, and hence is critical to competitiveness and success in a manufacturing business. Unless competitive quality has first been designed in by the designers, it cannot be built in during production, or inspected-in after production.

Design is important to a nation, too. This is not just because of military technology; probably a more important reason today is that competitive manufacturing is a critical ingredient in the economic security of modern industrialized nations. Successful manufacturing of well designed products creates wealth and provides good jobs. Design is especially beneficial when new products are conceived, developed, and brought to market based on new science and new technologies. And it is also important that there be continual improvement in the quality and value of existing products.

Design is also important to the survival and improvement of all human civilization, every bit as important as law, politics, medicine, and philosophy. People doing design conceived the water supply and sanitation systems that made cities possible, the equipment that made mining possible, the water wheels and engines and power plants that provide useful energy, efficient farm implements, the transportation equipment, radios and television, computers, and magnetic resonant imaging (MRI) systems.

Of course, not everything that has been designed and built by humans has had a 100% positive effect on human society, and there are those who argue that designers using technology and science to design things sometimes do more harm than good. Some environmental problems, for example, result from energy conversion plants, automobiles, and certain manufacturing processes.

* Numbers in brackets refer to References at the end of each chapter. The quotation here is by permission of the M.I.T Press, Cambridge, MA.

But even if we could, few of us would go back to a more primitive society. We are here now; we have a science- and technology-based civilization. We like the longer and healthier life span, the increased leisure, the reduction in back-breaking labor, the communications technology, and the mobility. Now we must do the best we can as engineers to use our education and our talents to design things that make our civilization better and safer.

Therefore, we study design so that we can do it better. There is a great deal to be learned about design, and more still to be discovered from research. The future depends on it. It is especially true for engineers that "the proper study of mankind is the science of design" [1].

Engineering Design. In this book, we are focussing on the engineering design of functional products and machines, including their parts and sub-assemblies. Engineering design transforms a relatively vague marketing goal, human need, or problem into a set of drawings (or CAD data) and other information sufficient to manufacture a product or machine that will help meet the goal, satisfy the need or solve the problem — and make a profit.

Of course, functional products and machines are only a fraction of the things that are designed by people. Buildings, roads, bridges, and communication systems are also designed. Gardens and parks are designed, furniture is designed, building interiors are designed, book covers are designed, clothes are designed, and so are curricula, design processes, commencement ceremonies, and fireworks displays. In this book, however, we deal only with engineering design — primarily of products and machines and their parts — though it is possible that some of the methods presented here are applicable to other areas of design as well.

The Product Realization Process. Products and machines are designed and manufactured in a complex set of inter-related activities that many firms call their product realization process, a process that involves the entire firm. See Figure 1.1. Engineering design is a part — a critically important part, but just a part — of the product realization process. As illustrated in the figure, engineering design fits approximately in the middle of the product realization process, roughly in between marketing and manufacturing.

As Figure 1.1 shows, engineering design is not an isolated part of the product realization process, but is intimately related to many other activities. A lot of what goes on in engineering design influences, and is influenced by, other parts of the larger product realization process. Thus, we must always think of engineering design as taking place within the context of a whole company.

1.2 About This Book

1.2.1 A Fantasy

This book is not a novel, but it nevertheless has a plot. Because the design process is complicated, the plot is surprisingly complex for a technical book. And not only is there a main plot, there are also a number of sub-plots that connect up to the main story in various ways. Keeping all this structure clear helps to understand how it all fits together into a whole. The little fantasy below is just for fun, but some readers have found it helpful for seeing the bigger picture.

> "The debt we owe to the play of imagination is incalculable."
>
> Carl Jung
>
> *Psychological Types (1923)*

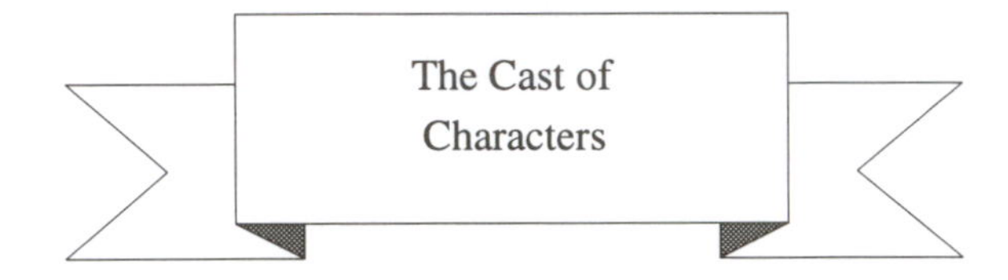

The main character and heroine in this story is Engineering Design. In the story she is always endeavoring to design better products quickly in order simultaneously to make a better world for people as well as a bigger profit for the firm. The character Engineering Design in this book must cope with three major kinds of problems:

Engineering conceptual design problems,

Configuration design problems (for parts), and

Parametric design problems.

Fortunately she is able to deal successfully with all three problems using a single problem solving process called guided iteration.

The country or environment within which this story of Engineering Design takes place is called the Product Realization Process, and our heroine Engineering Design must always be aware of this environment. We describe something about this territory of Product Realization in Chapter 2.

A major supporting character in the story is Design for Manufacturing, nicknamed DFM. DFM makes life more complicated for Engineering Design, but he also helps Engineering Design find better solutions to her problems. Working together, Engineering Design and DFM create products that are more easily and inexpensively manufactured.

To understand the characters in a novel, readers have to understand the characters' roots; that is, to know "where they are coming from". DFM, for example, comes from the physics and technology of a whole gang of Manufacturing Processes. Thus we take time in the book (especially in Chapter 3) to explain enough about the nature of these processes so that readers can better understand why DFM says and does the things he does.

There is another gang of supporting characters in this story called Materials. This is a huge gang with many different kinds of members, each with a different personality. Since Engineering Design has not yet figured out how to make something from nothing, she needs this Materials gang to solve her problems. Fortunately, the whole gang is not

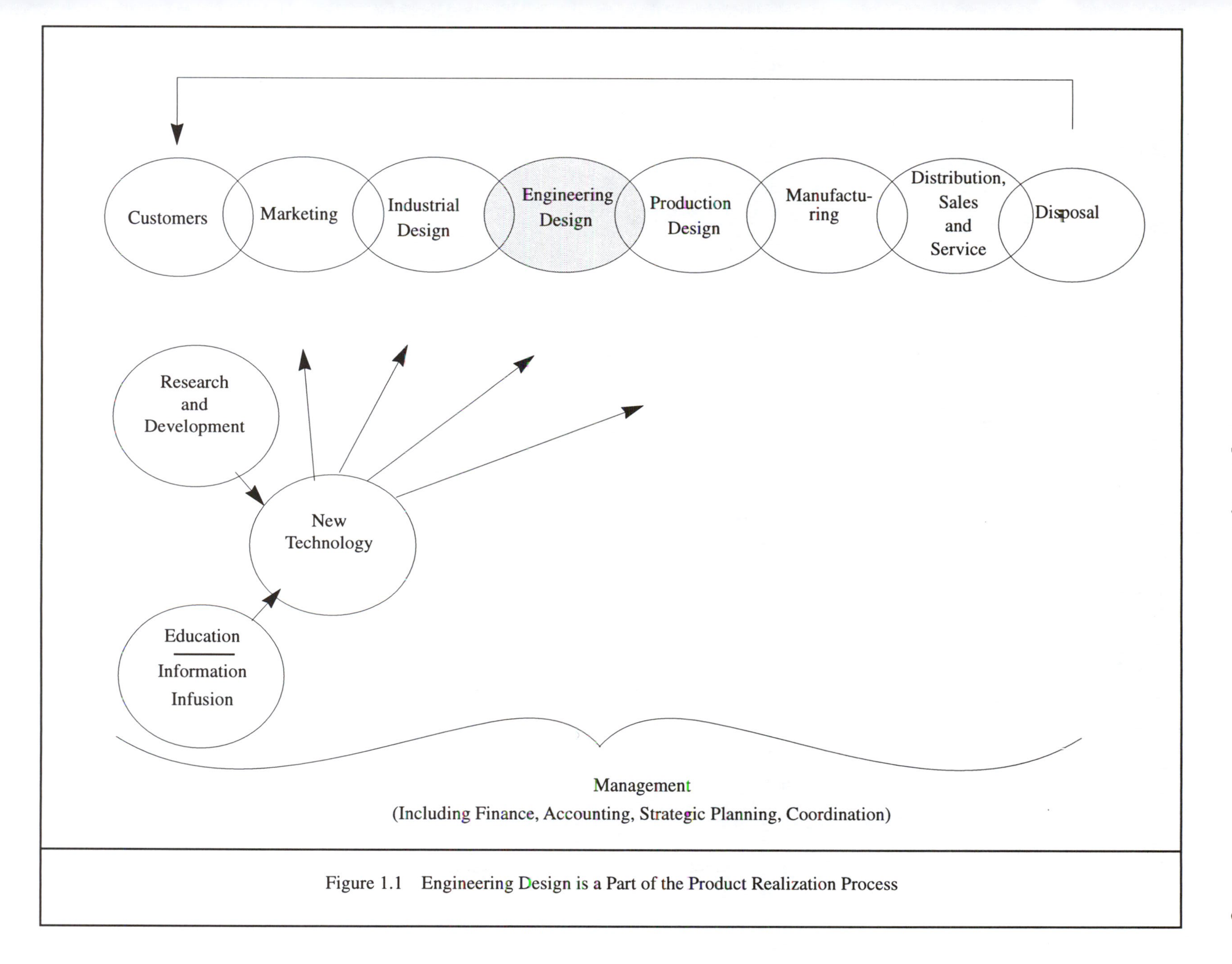

Figure 1.1 Engineering Design is a Part of the Product Realization Process

needed in every problem, but just one member for each part being designed. Of course, this means that Engineering Design has the constantly recurring problem of how to decide which member of the Materials' gang to select. The choice is made more complicated because, while some of the Materials' gang members get along very well with some of the Manufacturing Processes, others don't get along very well at all. How Engineering Design deals with selecting compatible Materials and Manufacturing Processes is discussed in Chapter 9.

There are also other supporting characters in this story. Communications is a wonderful, helpful character who likes to show everyone how to get along better and to exchange information more accurately and efficiently. She is discussed in Chapter 22. Economics is a meticulous, single-minded character with only one thing on his mind: money. But that one thing is a very important issue that Engineering Design must always be concerned about. This ever-present character is described in Chapter 23. Liability, Patents, and Ethics is an interesting, helpful, multi-faceted character who is a kind of legal and philosophical mentor to Engineering Design. He is discussed in Chapter 24.

How does our heroine use the guided iteration process to solve the conceptual design problem? How does she make use of DFM's advice in solving part configuration design problems? What methods can she use to solve parametric design problems? How can she sort through all those manufacturing process and material combinations to get the best one? How should she communicate the results to others in the process? What about those liability and patents issues; how do they effect her work? And how does she integrate it all together?

Sorry, there is not yet a movie; to get all the juicy details of this fascinating story, you'll just have to read the book and listen to your Instructors.

1.2.2 Back to Reality: The Design of the Book

The book is designed in five parts as described below:

Part I - First Things.

This first Part of the book (Chapters 1, 2, and 3), has four main goals:

Goal 1. To give you an overview; that is, the "big picture" or the main story line so you can always see the forest among the trees and branches, and the main story line among the sub-plots. Goal 1 is covered in this chapter.

Goal 2. To define a vocabulary of basic terms and concepts that will be used throughout the book. This is done so that we (the authors) can more effectively communicate with you. These definitions are also presented in this chapter.

Goal 3. To describe the product realization process so readers can understand the context within which engineering design takes place. This is done primarily in Chapter 2.

Goal 4. To describe the basic physical nature and technology of the more commonly used manufacturing processes, and to introduce some design for manufacturability (DFM) guidelines. Goal 4 is the subject of Chapter 3.

Part II - Engineering Conceptual Design

Part II (Chapters 4 through 9) describes how to apply the guided iteration methodology to engineering conceptual design. Before the engineering conceptual design stage, we have only information about the general need or goals that the product or machine is to fulfill. In engineering conceptual design, this vague goal is converted into an initial physical representation. The physical principles by which the designed object will work are established, and a physical embodiment of the design is created. Few details, however, are established at this stage.

Chapter 9 (the last chapter in Part II) focuses on the preliminary selection of materials and manufacturing processes for parts, a complicated topic because there are literally thousands of materials and hundreds of processes, and thus hundreds of thousands of material-process combinations. We obviously cannot consider every combination in this book so we have organized the subject hierarchically. We have also integrated material-process selection as smoothly as possible into the three stages of the engineering design process. Despite these efforts, however, there is no doubt that Chapter 9 contains information at a more detailed level than in Chapters 1 through 8. Readers should be prepared for this more detailed sub-plot when they get to Chapter 9; it really is essential to the main design story line. Parts made of the wrong materials have been responsible for many product design failures.

Part III - Configuration Design of Parts

Parts are made up of features like walls, holes, ribs, and the like. To configure a part means to determine what features will be present, and how those features are to be connected. How to perform this design task using guided iteration is the subject of Part III (Chapters 10 through 13). No dimensions are established at the configuration stage, though of course approximate sizes are generally obvious.

Chapters 11 and 12 in Part III introduce quantitative DFM methods that can be applied to parts at the configuration stage. Methods are presented for injection molded, die cast, and stamped parts.

Part IV - Parametric Design

After parts are configured, the parts and the assemblies to which they belong can be given dimensions, tolerances, and exact material specs. This is called parametric design, and it is the subject of Part IV (Chapters 14 through 21). Several methods are discussed: guided iteration, optimization, and design for robustness using the Taguchi Method™.

Chapters 15 and 16 deal with quantitative DFM methods for injection molded, die cast, and stamped parts when dimensions and tolerances are involved.

Part V - Pandemic Topics

Pandemic topics are those that apply everywhere throughout the product realization process. The final section of the book (Chapters 22, 23, and 24) cover three such topics: communications (Chapter 22); economic issues (Chapter 23); and liability, patent, and ethical issues (Chapter 24). Instead of leaving these chapters to be read last, readers may find it helpful to read one or more of them in conjunction with other parts of the book. For example, Chapter 22 may be helpful in preparing written and oral project reports.

1.3 Definitions Used in This Book

> "When I use a word," Humpty Dumpty said, in a rather scornful tone, "it means exactly what I choose it to mean -- neither more nor less."
>
> "The question is", said Alice, "whether you can make words mean so many different things."
>
> "The question is", said Humpty Dumpty, "which is to be master — that's all."
>
> Lewis Carroll
> *Through the Looking Glass (1872)*

1.3.1 Designed Objects or Artifacts

If we wish to master a subject, then we must master its language. That means having a clear set of definitions for the commonly used terms. Thus we begin to introduce some of these definitions for design in this Section.

A designed object or artifact is any material object that has been or is to be manufactured from information prepared for that purpose.

A design (noun) is information. The information may be in one or more of several forms: words, graphics, electronic data, and/or others. It may be partial or complete. It ranges from a small amount of highly abstract information early in the design process to very large amount of very detailed information at the end.

A design provides many different kinds of information about a designed object. For example, it may include, but is not limited to, information about size and shape, function, materials, marketing, simulated performance, manufacturing processes, tolerances, and more. Indeed, any and all information relevant to a designed object is a part of its design.

We will occasionally refer to the state of information about a designed object. The state of information is all the information known about the designed object at a given time.

Scope of the Book. We are concerned in this book with the design of functional objects as distinguished from objects designed and built exclusively or primarily for aesthetic purposes. Moreover, we consider the design of only those functional designed objects that are commonly called products or machines, including their parts (often called piece-parts) and sub-assemblies. This is a field often referred to as discrete part and assembly design, or sometimes mechanical design — though products and machines are increasingly complex systems of electronic, optical, software, thermal, and fluid components as well as strictly mechanical components. Our scope, therefore, includes but is considerably broader than the sub-field traditionally called machine design — which generally refers to the selection and analysis of standard mechanical elements such as beams, springs, gears, bolts, shafts, etc.

Here are some examples of discrete part and assembly designed objects whose parts, sub-assemblies, and assemblies are within our scope: paper clips, automobiles, appliances, injection-molding machines, cameras, the shutter in a camera, nuts and bolts, fans, the cover on a computer terminal, springs, heat exchangers, pumps, electric or manual drills, light bulbs, lamps, milling machines, file cabinets, pencils, hinges........ and so on.

We are not concerned in this book with functional designed objects such as bridges, buildings, roads, and the like. Nor are we concerned with the design and manufacture of materials, chemicals or textiles per se. The latter are continuous as distinguished from discrete products. Moreover, we do not cover the design or planning of systems — collections of products, machines, and people — as in a power plant or manufacturing facilities. And we do not consider the design of computers or electronic components and systems, though our functional designed objects may include them as components or sub-systems. Note, however, that the equipment and machines used to build bridges, to make chemicals, or to manufacture electronic products (e.g., construction equipment, bull dozers, and manufacturing equipment like heat exchangers and milling machines) are all functional objects within our scope.

Some of the principles and practices presented in this book may well apply, at least in part, to the design of objects outside the scope of discrete parts and assemblies. However, the field of discrete parts and assemblies is itself large and complex, and sufficiently important economically and intellectually so that it warrants a special focus. Thus no attempt has been made to consider extensions to any other types of designed objects.

This book is primarily intended as an introduction to the subject of engineering design of functional artifacts for students, recent graduates, and entry level practitioners. As such, the book cannot and does not reflect all the practical complexity of product design in a modern industrial setting. We do, however, attempt to present the basic concepts and practices in a structured way that will enable readers to better understand and cope with the complex realties of practice. Thus we describe basic engineering design problem types (conceptual, configuration, and parametric), a basic problem solving methodology (guided iteration), and a number of the more basic techniques employed to implement the methodology. We expect this structured approach to give readers the knowledge required to begin to contribute, the understanding required to continue learning, and the confidence required to seek the counsel and advice of those who know more.

1.3.2 Functional Designed Objects

We distinguish among the following types of functional designed objects, though not all of them are mutually exclusive: parts, assemblies, sub-assemblies, components, products and machines.

Parts. A *part* is a designed object that has no assembly operations in its manufacture. Joining operations (like welding and gluing) are considered assembly operations for the purposes of this definition. Parts may be made by a sequence of manufacturing processes (e.g., casting followed by milling), but parts are not assembled.

Parts are either *standard* or *special purpose*. A standard part is a member of a class of parts that has a generic function and is manufactured routinely without reference to its use in any particular product. Examples of standard parts are screws, bolts, rivets, jar tops, buttons, most beams, gears, springs, and washers. Tooling for standard parts is usually on hand and ready for use by manufacturers. Often standard parts themselves are carried in stock by manufacturers, distributors, or vendors. Standard parts are most often selected by designers from catalogs, often with help from vendors.

Special purpose parts are designed and manufactured for a specific purpose in a specific product or product line rather than for a generic purpose in several different products. Special purpose parts that are incorporated into the subassemblies and assemblies of products and machines are often referred to as piece parts. Special purpose parts that stand alone as products (e.g., paper clips, styrofoam cups) are referred to as single part products.

Even though screws, springs, gears, and the like, are generally manufactured as standard parts, a special or unique screw, spring, gear, etc., that is specially designed and manufactured for a special rather than general purpose is considered a special purpose part. This is not often done, however, because it is usually less expensive to use an available standard part if one will serve the purpose.

Assemblies and Sub-Assemblies. An assembly is a collection of two or more parts. A sub-assembly is an assembly that is included within another assembly or sub-assembly.

A standard module or standard assembly is an assembly or sub-assembly which — like a standard part — has a generic function and is manufactured routinely for general use or for inclusion in other assemblies or sub-assemblies. Examples of standard modules are electric motors, electronic power supplies or amplifiers, heat exchangers, pumps, gear boxes, v-belt drive systems, batteries, light bulbs, switches, and thermostats. Standard modules, like standard parts, are generally selected from catalogs.

Components. The term component in this book is a generic term that includes special purpose parts, standard parts, and standard assemblies or modules. In other words, only special purpose assemblies and sub-assemblies are not considered to be components.

Products and Machines. A product is a functional designed object that is made to be sold and/or used as a unit. Products that are marketed through retailing to the general public are called consumer products. Many manufactured products are designed for and sold to other businesses for use in their business; this is sometimes called the trade (or commercial, or industrial) market. For example, a manufacturer may buy a pump to circulate cooling water to a machine tool they have purchased. In addition, there are products, including especially standard parts and standard modules, that are sold to other manufacturers for use in products being manufactured; this is called the original equipment manufacturer (or OEM) market. An example is the purchase of a small motor for use in an electric fan. Trade marketing is usually done through a system of regional manufacturer's representatives and/or distributors.

A machine is a product whose function is to contribute to the manufacture of products and other machines.

1.3.3 Manufacturing, Design, and Design for Manufacturing (DFM)

There is unfortunate confusion created by different uses of the word *manufacturing*. Sometimes the word is used to refer to the entire product realization process; that is, to the entire spectrum of product-related activities in a firm that makes products for sale, including marketing (e.g., customer desires), design, production, sales, etc. This entire process is sometimes referred to as "Big-M manufacturing".

But the word manufacturing is also used as a synonym for production; that is, to refer only to the portion of the product realization process that involves the actual physical processing of materials and the assembly of parts. This is sometimes referred to as "Little-m manufacturing".

We will use the little-m meaning for manufacturing in this book. That is, manufacturing here consists of physical processes that modify materials in form, state, or properties. Thus in this book, manufacturing and production have the same meaning. When we wish to refer to Big-M manufacturing, we will call it the product realization process.

Design (as in a design process) is the series of activities by which the information known and recorded about a designed object is added to, refined (i.e., made more detailed), modified, or made more or less certain. The process of design changes the state of information that exists about a designed object. During successful design, the amount of information available about the designed object increases, and it becomes less abstract. In other words, as design proceeds the information becomes more complete and more detailed until finally there is sufficient information to perform manufacturing.

Design, therefore, is a process that modifies the information we have about an artifact or designed object, whereas manufacturing (i.e., production) modifies its physical state.

A *design problem* is created when there is a desire for a change in the state of information about a designed object. In other words, a design problem exists when there is a desire to generate more (or better) information about the designed object, when we want to develop a new (but presently unknown) state of information. For a simple example, we may know from the present state of information that a designed object is to be a beam, and we desire to know whether it is to be an I-beam, a box beam, an angle beam, or some other shape. Our desire to know more about the designed object defines a new design problem, in this example, determining the beam's shape. Later, once we know the shape (say it is to be an I-beam), another design problem is defined when we want also to know the dimensions. There are many kinds of design problems defined by the present and desired future states of information.

A *design method* is a procedure or prescription for how to solve a design problem. Usually methods are associated with particular problem types. For example, a method useful in finding dimensions for an I-beam is optimization.

A method generally applicable to a number of problem types is called (in this book) a *methodology*. The primary methodology discussed in this book is guided iteration; it is used to solve all of the major design problem types. The guided iteration methodology involves the following basic steps:

1. Formulating the problem,
2. Generating alternative solutions,
3. Evaluating alternatives, and if none is acceptable,
4. Redesigning, *guided by* the results of the evaluations.

Guided iteration is described in more detail in Section 1.4 below.

Design for manufacturing (DFM) is any aspect of the design process in which the issues involved in manufacturing the designed object are considered explicitly with a view to influencing the design. Examples of DFM include considerations of tooling costs or time required, processing costs or controllability, assembly time or costs, human concerns during manufacturing (e.g., worker safety or quality of work required), availability of materials or equipment, and so on. Design for manufacturing occurs — or should occur — throughout the design process.

1.3.4 The Product Realization Process

Product realization is the set of cognitive and physical processes by which new and modified products are conceived, designed, produced, brought to market, serviced, and disposed of. In other words, product realization is the entire "cradle to grave" cycle of all aspects of a product.

Product realization includes determining customers' needs, relating those needs to company strategies and products, developing the product's marketing concept, developing engineering specifications, designing both the product and the production tools and processes, operating those processes to make the product, and distributing, selling, repairing, and finally recycling or disposing of the product and the production facilities.

Product realization also includes those management, communication, and decision making processes that organize and integrate all of the above, including marketing, finance, strategic planning, design (industrial, engineering, detail, and production), manufacturing, accounting, research and development, distribution and sales, service, and legal operations.

We distinguish in this book among several overlapping stages of the product realization process including product development, industrial design, engineering design, and production design. These are defined in the paragraphs below. Figure 1.1 also provides a supporting illustration for the definitions of these terms.

Product development is the portion of the product realization process from inception to the point of manufacturing or production. Though product development does not, by this definition, include activities beyond the beginning of production, it does require the use of feedback from all the various downstream product realization activities for use in designing, evaluating, redesigning parts and assemblies, and planning production. This feedback includes information about manufacturing issues — that is, information about design for manufacturing. Thus product development (including product improvement and redevelopment) is an on-going activity even after production has begun.

1.3.5 Industrial (or Product) Design

The process of *industrial design* (sometimes the phrase *product design* is used) creates the first broadly functional description of a product together with its essential visual conception. Artistic renderings of proposed new products are made, and almost always physical models of one kind or another are developed. The models are often only very rough, non-functional ones showing external form, color, and texture only, but some models at this stage may also have a few moving parts.

There is great variability in the way industrial design is organized and utilized within different firms. In one firm, for example, a very small number of industrial designers are employed only to generate new or revised product concepts. These industrial designers are essentially a separate department but keep in close communication with colleagues in both marketing and engineering design. When a product concept has been approved by management for further development, then outside consultants in industrial design are brought in to work with the in-house industrial designers. The firm's engineering designers and manufacturing people are also consulted to help refine and complete the industrial design phase. Marketing remains involved during this phase.

In another firm, no industrial designers per se are employed. In this firm, initial product concepts are developed by creative engineering designers working with marketing. Once a concept has approval, then outside industrial design consultants are brought in to perform the work described above.

An example of the kinds of issues that must be resolved by cooperation among industrial and engineering designers at this early stage is determination of the basic size and shape of the product. Industrial designers will have aesthetics, company image, and style in mind in creating a proposed size and shape for a product. Engineering designers, on the other hand, will be concerned with how to get all the required functional parts into the (usually) small space proposed. Another issue requiring cooperation may be choices of materials for those parts that can be seen or handled by consumers. And of course, both design engineers and manufacturing engineers are concerned about how the product is to be made within the required cost and time constraints.

It is helpful to use the phrase "product marketing concept" to describe the results of industrial design. Sometimes the product marketing concept is written into a formal (or at least informal) product marketing specification. Whether written or not, the information available about the product after industrial design includes any and all information about the product that is essential to its marketing. Thus it will include qualitative statements of the marketing rationale and the in-use function of an artifact as perceived by potential customers. It can also include any special in-use features the product is to have, and it will certainly include any visual or other physical characteristics that are considered essential to marketing the product.

The product marketing concept or specification should, however, contain as little information as possible about engineering design and manufacturing. This is to allow as much freedom as possible to the engineering design and production phases that follow. Such a policy is called "least commitment", and it is a good policy at all stages of product realization. The idea is to allow as much freedom as possible for downstream decisions so that designers are free to develop the best possible solutions unconstrained by unnecessary commitments made at previous stages.

Some engineers who like things to be quite precise will fret over the fact that the definition of product marketing concept, and any associated product marketing specification, is fuzzy. However, the activity of industrial design that produces the product marketing concept is creative, involves aesthetics and psychology, and just naturally produces somewhat imprecise results. Industrial design is, nevertheless, extremely important to the product realization process. Engineers should respect it, and learn to live and work cooperatively with it, frustrating though it may be on occasion. As we shall see, one of the first tasks of engineering design is to convert the fuzzy goals of the marketing concept into much more concrete requirements.

The use of the terms "industrial design" or "product design" to describe this phase of product development is misleading. Both terms are nevertheless in customary use, and there is no hope of changing the custom. Logically, the whole product development process, including engineering design, could be accurately called product design. More accurate terms for industrial or product design would be *preliminary product concept design* or *marketing concept design.*

1.3.6 Engineering Design

> "Scientists study the world as it is, engineers create the world that never has been."
>
> Theodore von Karman

Engineering design generally follows but overlaps industrial design, as illustrated in Figure 1.1. We consider engineering design to consist of four roughly sequential but also overlapping stages or sub-processes, each corresponding to a design problem type:

- Engineering conceptual design,
- Configuration design of parts,
- Parametric design, and
- Detail design.

The first three of these basic engineering design problem types are introduced very briefly below. Then in Chapters 4 through 21, methods for solving each type using the guided iteration methodology are presented. Detail design is not covered in this book.

Engineering Conceptual Design. Several variations of the engineering conceptual design problem are encountered in the course of engineering design. The variations are slight and depend on whether the object being designed is

(1) a new product (usually an assembly),

(2) a sub-assembly within a product, or

(3) a part within a product or sub-assembly.

The desired state of information that is to result from the engineering conceptual design process is called the *physical concept.* It includes information about the physical principles by which the object will function. In addition, in the case of products and their sub-assemblies, the physical concept also includes identification of the principal functional sub-assemblies and components of which the product will be composed, including particular functions and couplings within the product. By "coupling", we mean their important inter-relationships such as physical connections, or the sharing of energy or other resources. More is said about the nature of the information in a physical concept in Chapters 4 and 6.

Engineering conceptual design problems, like all major engineering design problem types, are solved in this book by guided iteration. Thus the first step is problem formulation. In engineering conceptual design, this means preparation of an Engineering Design Specification. This Specification, the details of which are covered in Chapter 5, records the product's quantitative functional requirements as well as specific information on requirements for such factors as weight, cost, size, required reliability, etc.

Generating alternatives in engineering conceptual design (Chapter 6) requires a creative process called conceptual decomposition, or just decomposition. We might, for example, decompose a wheelbarrow into a wheel sub-assembly, the tub, and the carrying handle sub-assembly. The subsidiary sub-assemblies and components that make up a product or sub-assembly are created during the decomposition process for a product or other sub-assembly. There are many options open to designers in this creative decomposition process, and many opportunities for innovation. And the decomposition process is very important because the physical concepts chosen have a tremendous impact on the final cost and quality of the designed object.

For evaluating competing conceptual solutions, a number of methods are available. They are discussed in Chapter 7, together with redesign considerations.

The conceptual design of parts is a bit different from the issues in design of products and sub-assemblies, and so they are discussed separately in Chapter 8.

An activity that permeates the engineering design process is the selection of the materials of which the parts are to be made, and the processes by which they are to be manufactured. Because there are so many possibilities, these problems of choice are complex. Moreover, materials, processes, and the functional requirements of parts must all be compatible. Though most often it is only necessary at the conceptual design stage to select the broad class of material and process (e.g., plastic or thermoplastic injection molded, or aluminum die cast), we cover the subject of materials and process selection rather fully in Chapter 9.

Configuration Design of Parts and Components. During the conceptual design of products and their sub-assemblies, a number of components (that is, standard modules, standard parts, and special purpose parts) are created as concepts. In the case of standard modules and standard parts, configuration design involves identifying and selecting their type or class. For example, if a standard module is a pump, then configuration design involves deciding whether it is to be centrifugal pump, a reciprocating pump, a peristaltic pump, or some other type. Another example: if a standard part is to be a spring, then configuration design involves deciding whether it is to be a helical spring, a leaf spring, a beam spring, or other type.

For special purpose parts, configuration design includes determining the geometric features (e.g., walls, holes, ribs, intersections, etc.) and how these features are connected or related physically; that is, how the features are arranged or configured to make up the whole part. In the case of a beam, for example, configuration design makes the choice between I-beam or box beam and all the other possible beam cross section configurations. The features in an I-beam are the walls (called flanges and web), and they are configured as in the letter “I”.

At the configuration design stage, exact dimensions are not decided, though approximate sizes are generally quite obvious from the requirements of the Engineering Design Specification.

More information about material classes and manufacturing processes may be added at the configuration stage if the information is relevant to evaluating the configuration during the guided iteration process. For example, it may be necessary for evaluation to know that a high strength, engineering plastic is to be used, or that an aluminum extrusion is to be heat treated to a high level of strength. However, this more detailed information should be generated only if it is really needed; least commitment is the basic policy.

Configuration design of special purpose parts begins with problem formulation usually in the form of a part requirements' sketch. Once the requirements are established in a sketch, then alternative configurations are sketched and evaluated with important help from qualitative reasoning based on physical principles. This is all discussed in Chapters 10 and 13. Methods for considering the design for manufacturing (DFM) of parts at the configuration stage are the subject of Chapters 11 and 12 in this book.

Some designers use the word *configuration* to refer not only to the arrangement of the features of a part (walls, holes, intersections, etc.) but also to what we have called the decomposition of products and sub-assemblies. That is, they refer to the “configuration of an assembly”, meaning the way the various components, and sub-assemblies are arranged and connected. We use the term both ways in this book; usually the context is clear.

Parametric Design. In the parametric design stage, the initial state of information is the configuration. In other words, just about everything is known about the designed object except its exact dimensions and tolerances. It may also be in some cases that the exact material choice is also unknown, though the basic class of material will usually be included in the configuration information. The goal of parametric design, therefore, is to add the dimensions and any other specific information needed for functionality and manufacturability. The specific material is also selected if it has not previously been designated.

In the spirit of least commitment design, parametric design need not and should not specify information to any degree of precision not actually required by the Specifications.

Methods for parametric design, including design for manufacturing considerations, are described in Chapters 14 through 21 in this book. Methods for considering the design for manufacturing (DFM) of parts at the parametric stage are the subject of Chapters 15 and 16 in this book.

Detail Design. Detail design supplies any remaining dimensions, tolerances, and material information needed to describe the designed object fully and accurately in preparation for manufacturing. As every manufacturing engineer knows, in a large complex product, the result seldom provides all the dimensions and tolerances that exist. The rest of the details, if needed, are established in the manufacturing design phase. In this book, we do not consider the detailed design phase, but this does not imply that it is not important.

1.3.7 Production Design

Production design overlaps with detailed design. It involves finishing any of the design details left undone by detailed design, design of the tooling, planning the manufacturing process, planning for quality control, “ramping up”

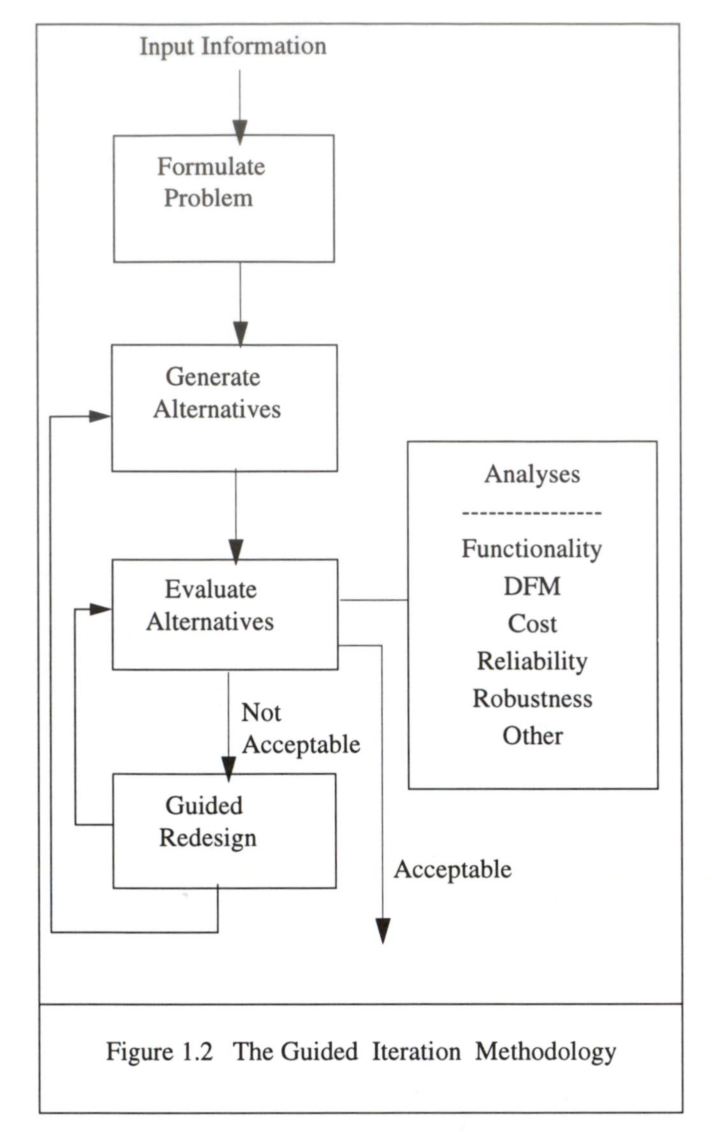

Figure 1.2 The Guided Iteration Methodology

the actual production, planning for quality and process control, and supporting the initial production runs. It is a substantial design task, though it is usually referred to as a part of the overall manufacturing or production process.

1.4 Overview of the Guided Iteration Methodology

In this book, we deal primarily with the first three stages of engineering design: engineering conceptual design, configuration design of parts, and parametric design. For each of these stages, we present a single problem-solving methodology called guided iteration. Though the guided iteration methodology does not absolutely insure optimum solutions, if employed with diligence and intelligence, it will produce at least satisfactory results, and often excellent results. At the parametric stage, we also present alternate methodologies (optimization and Taguchi Methods™) that can be used to advantage in certain situations. Throughout, we stress the importance of design for manufacturability, and present methods for doing it.

As noted above and illustrated in Figures 1.2 and 1.3, the guided iteration methodology involves the following basic steps: formulating the problem, generating alternative solutions, evaluating alternatives, and, if none is acceptable, redesigning, *guided by* the results of the evaluations. Each of these steps is discussed below in more detail.

Formulating the Problem. Problem formulation includes, first of all, identifying the type of problem that must be solved to advance the design. Then, assuming the problem is engineering conceptual design, configuration design, or parametric design, formulation includes: obtaining the initial information that describes and poses the problem; refining, completing, and checking that information (since it will seldom be complete, and it will sometimes even be inaccurate); and formulating the problem into a standard structure for solution by known methods and available tools.

Generating Alternative Solutions. Generating alternative solutions involves: the use of experience (i.e., memory, both broad and deep, of both facts and strategies); the use of creative idea generating methods; the use of physical principles and qualitative reasoning; and the ability to find information. Generating viable alternatives also requires that engineering designers have a basic qualitative understanding of manufacturing processes.

Evaluating the Alternatives. Evaluation of alternatives is supported heavily by engineering analysis procedures for predictions of performance, by methods for cost estimation, and by design for manufacturing analyses.

It is extremely important to distinguish clearly between (a) analyses and (b) the evaluations based on analyses. For example, analyses can determine that the deflection of a beam will be, say, 0.003 inches, or that the cost of tooling for a part will be, say, $8,500. Evaluation, on the other hand, determines whether or not that deflection and/or that tooling cost is good or bad in terms of the requirements of the design. Without the quantitative information provided by good analyses, good evaluation and hence good design is impossible. However, it is important also to understand that bad designs can be developed even when there are good analyses. There is much more to design than analysis. Indeed, analysis serves design, not the reverse.

Evaluation also includes deciding whether a design is acceptable. A prospective design solution is acceptable when it is perceived as "good enough" to be incorporated into the design, thus allowing the design or product realization process to move on to the next stage. Since moving on involves new commitments of time, money, and other resources, deciding to accept a proposed solution at any stage is an important decision. The final quality and cost of the designed object is increasingly fixed by each acceptability decision. Deciding whether to accept a proposed design solution or to continue to invest time and money in efforts to make improvements (or find new alternatives) is a constantly recurring part of the engineering design process.

Guided Redesign. When proposed solutions are not acceptable, redesign makes use of — is guided by — physical principles, knowledge of manufacturability issues, and

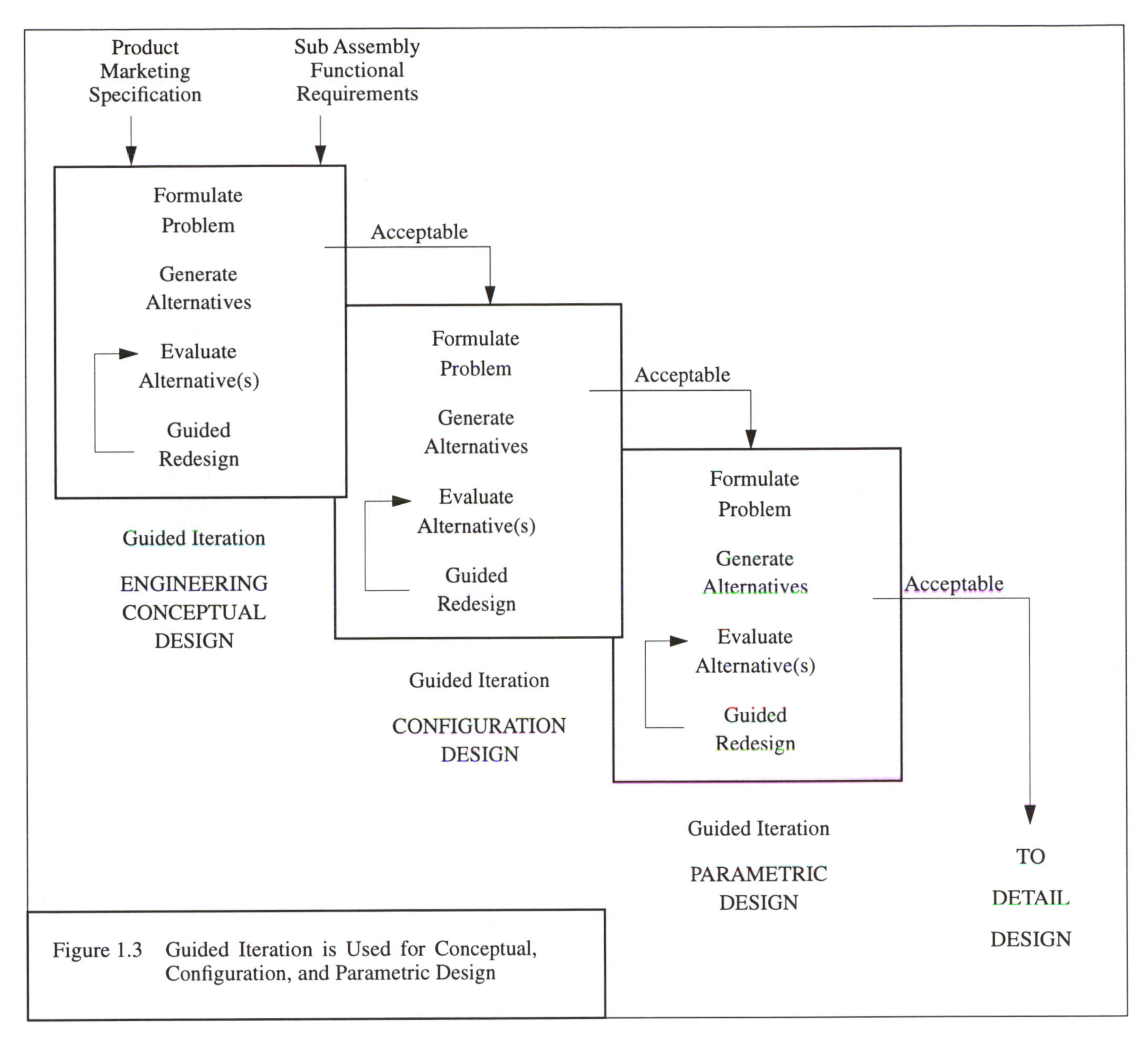

Figure 1.3 Guided Iteration is Used for Conceptual, Configuration, and Parametric Design

the results of previous evaluations to improve the existing design. When redesign fails, the evaluation results also help to guide a search for a new alternative. In most engineering design problems, the majority of designers' time and energy, however, is spent in guided redesign.

As described above, the basic steps for each design stage are always formulation, generation, evaluation, and guided redesign. There are, however, different techniques, methods and procedures available for implementing these steps at the conceptual, configuration, and parametric design stages.

Though it is not illustrated in Figure 1.3, if no satisfactory solution can be found at one stage, designers must backtrack to a previous stage (i.e., from parametric back to configuration, or from configuration back to conceptual). Since such backtracking is costly, especially of time, every effort should be made to minimize it. This is accomplished (in part) by generating a full set of alternatives at each stage, and by evaluating them thoroughly to get the best possible solution before moving forward. Sometimes in practice two competing solutions are carried forward when it is not clear which will work out best at the subsequent stage.

Students and others new to engineering design frequently resist using the basic process of guided iteration. Iteration apparently recalls the "trial and error" problems that everyone hated in early arithmetic classes. But guided iteration is not simply random trial and error. That would be like searching blindly for a needle in a haystack. In guided iteration, the path to a solution is effectively pointed out by the results of previous evaluations, by qualitative physical reasoning, and by knowledge of manufacturing processes.

1.5 Analysis Vis-a-Vis Design

One reason students resist using guided iteration is that they want to find a way to solve engineering design problems in a direct, linear way as is done in engineering science and analysis courses. In those courses, problems are solved by application of a physical principle or principles to an idealized model of an actual physical situation. Usually the principles are applied in the form of an equation or set of equations to be solved for the desired result. However, though solving sets of equations is often the way problems found in engineering science and analysis texts are solved, this is not the way design problems are solved. Instead, design problems are solved by guided iteration. Anyway, at the engineering conceptual and configuration stages, where many critical design decisions are made, there are very few numbers and essentially no equations. But even at the parametric stage of design where there are lots of numbers and equations, it is still the case that guided iteration, in one form or another, is generally required.

There is, to be sure, an important role for engineering analysis in the design process. Especially at the parametric stage, the results of engineering analyses are used in the evaluation of trial designs, and the data provided by engineering and other analyses are extremely important to design evaluation and redesign. Indeed, it is impossible to produce superior designs without superior analytical information.

But there is much, much more to engineering design than engineering analysis. Evaluation in design also involves important issues not based in engineering science and analysis — e.g., cost, customer expectations, time issues and, of course, manufacturability.

Just as there is a basic engineering design methodology — guided iteration — there is also a basic engineering analysis methodology. It involves defining the problem in quantitative terms, creating an idealized model to represent the real physical situation, making proper assumptions, applying the appropriate physical principles, performing the computations, and checking the results. This basic engineering analysis process is discussed and illustrated in texts by Ver Planck and Teare [2] and by Dixon [3].

1.6 Codes and Standards

There are certain types of design problems, especially when public health or safety are involved, whose solution methods are dictated by legal or professional society Codes or Standards. Most often in these special cases, the configurations and the equations for determining the design variable values are mandated by some time tested, empirical procedure. There are, for example, legal Codes that dictate the procedures by which the materials and dimensions of pressure vessels and certain structural members must be selected.

If legal Codes or Standards apply, their procedures must, of course, be used. In such cases, the guided iteration process applies only to the point in the process at which the Code procedure is employed. After that point, the design process is dictated by the Code.

1.7 Tolerances

A tolerance is a designer-specified allowed variation in a dimension or other geometric characteristic of a part. Proper tolerances are crucial to the proper functioning of products. But also, a common cause of excessive manufacturing cost is the specification by designers of too many tolerances and/or tighter than necessary tolerances.

On part drawings, simple dimensional tolerances are usually attached to dimensions as shown in Figures 1.4 and 1.5. In later chapters, we discuss tolerances in more detail at both configuration and parametric stages of design.

The issues of tolerances are at the intersection between the requirements of functionality and capabilities of manufacturing processes. Mass-produced parts cannot all be produced to any exact dimensions specified by designers. In a production run, regardless of the process used, there will always be variations in dimensions from nominal. Some of the reasons are: tools and dies wear, processing conditions change slightly during production, and raw materials vary in composition and purity. Modern methods for controlling processes are achieving ever more consistent and accurate dimensions, but variations of some frequency and magnitude are inevitable.

Designers can, in order to get the functionality required, limit the range of dimensional variations in those parts that reach the assembly line. However, the more strict these limitations, and the more dimensions subject to special tolerance limitations, the more expensive the part will be to produce.

Fortunately, the functionality of parts seldom, if ever, requires that all or even most dimensions of parts be controlled tightly. To achieve the desired functionality (and other requirements) of a part, a few dimensions and other geometric characteristics (e.g., straightness or flatness) may require quite accurate control, while others can be allowed to vary to a greater degree. Thoughtful design can result in parts and assemblies configured so that the number of characteristics requiring critical control is minimal. Also, those dimensions that do require critical control may be of a type or at a location where they are more easily controlled during manufacturing. We call designer attention to these matters by the term DFT — that is, Design for Tolerances.

Every manufacturing process has what are most often called standard or commercial tolerances. These are the tolerance levels that can be produced with the normal attention paid to process control and inspection. Though standard tolerance values are not always completely and accurately defined — and available in print for designers — they are well known to manufacturing process engineers. There is no need to guess; designers can go visit their friendly manufacturing colleague.

Parts designed so that there are no tolerance requirements tighter than standard will be the least expensive to produce. Moreover, and this is an important point for designers, the *number* of tolerances that must be critically controlled, whether standard or tighter, is crucial to ease and

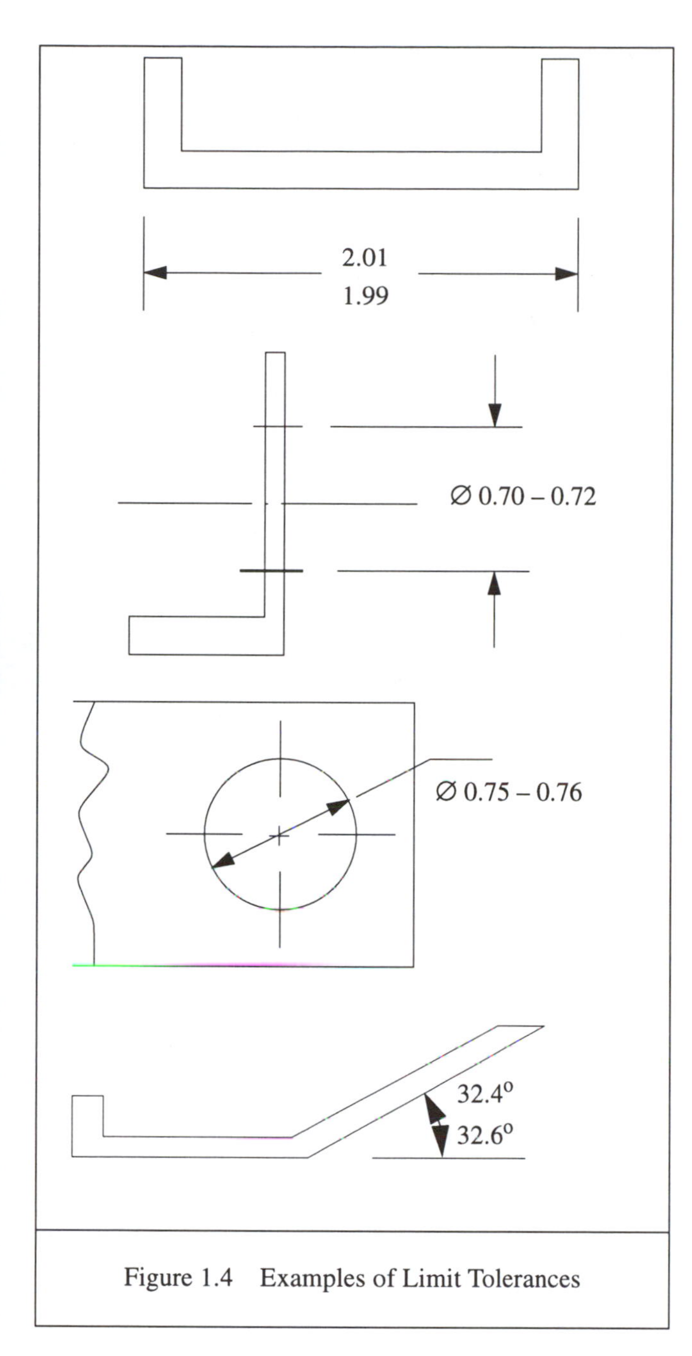

Figure 1.4 Examples of Limit Tolerances

cost of manufacturing. Controlling one or two critical dimensions, unless they are of an especially difficult type or extremely tight, is often relatively easy to do if the other dimensions of the part do not need special control.

1.8 Example Design Project A

(Note: This is an example semester project to be done by teams of five or six students. It is revisited at various points throughout the book and referred to as Example Design Project A.)

The company for which you are working (call it, say, Manufacturers' Automation Supply, Inc., or MASI) as consultants or employees is a very successful major world manufacturer of both consumer and industrial products. At the firm's various plants there exist modern, efficient facilities for injection molding, die casting, plastic and aluminum extrusion, sheet metal stamping, and other common manufacturing processes. The firm has a worldwide reputation for high quality, reliable products.

MASI has a very effective marketing capability for its line of industrial machine tools and products used by other manufacturers who do both manual and sophisticated automated assembly of a wide variety of products. For example, MASI makes and markets pneumatic tools of many kinds, parts' bins, chairs and stools for work stations, small presses, parts' feeding and orienting equipment, and electronic controls to support assembly processes. Indeed, MASI's customers can purchase just about everything needed to set up an automated assembly line except the tables, benches, and other structures needed to support the automation tools and equipment. That is, MASI's current product line requires customers to set everything up on the floor — or go to a competitor for the assembly line tables, etc. But when customers go to the competitors for tables, they also get exposed to the competitors' other products as well.

We therefore can well imagine that MASI's sales and marketing groups have found from discussions among themselves and with customers that there is a market opportunity for MASI to offer a line of work tables and benches, machine underframes, shelving for work stations, and the like. Since customers have a wide variety of needs, and since these customers are also always changing their facilities to keep up with their own changing markets, the marketing concept is to offer a standardized or modularized system of some kind that will be "versatile, flexible, universally applicable, and readily reusable."

Other stated requirements are: the structures should be "stable against torsional and bending stresses"; easily

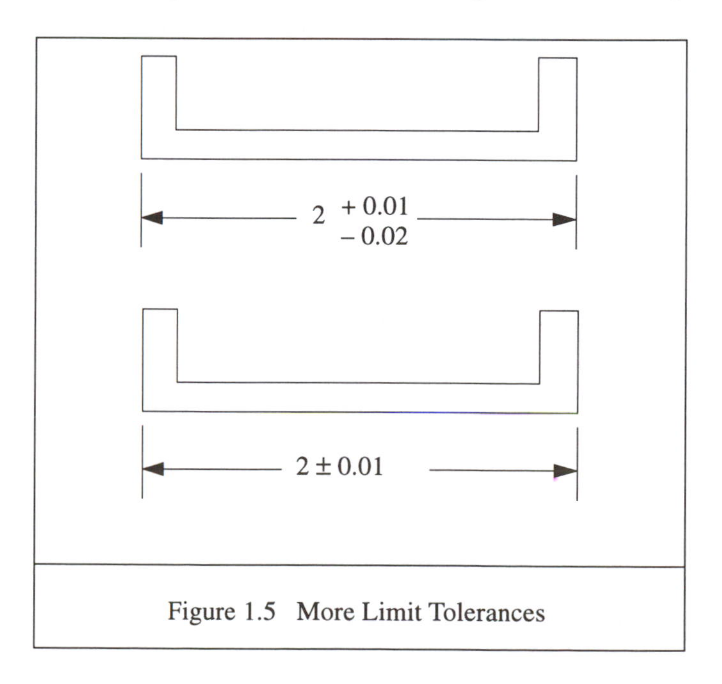

Figure 1.5 More Limit Tolerances

"extendible and convertible"; permit "extensive design possibilities", and be "easily erected and modified."

1.9 Summary and Preview

Summary. Our journey has begun. You have been briefly introduced to the main characters of the story — Engineering Design, Design for Manufacturing (DFM), Manufacturing Processes, and Materials — and you know the basic story line — that the three main stages of engineering design — engineering conceptual design, part configuration design, and parametric design, all use guided iteration as a problem solving methodology, and all involve DFM. You have a general idea of the guided iteration process: formulating the problem, generating alternatives, evaluating alternatives, and redesigning guided by the evaluations until an acceptable solution has evolved.

Preview. You know from this beginning that engineering design is a part of the Product Realization Process. So far, however, we have said very little about the rest of that process. Therefore in Chapter 2 we describe product realization more fully, tell you about some of the more successful ways of doing it (that is, its *best practices*), and outline briefly how three successful firms do it.

Recommended Supplementary Reading

There are many good tales of Engineering Design. Therefore, at the conclusion of most chapters, we list some books that relate closely to the material in the chapter. There are so many good books, it is impossible to list them all. In these lists, we have stressed the more recent books, but we have also included older books that are either still unique today, or that qualify as a kind of "best of their breed". We have no doubt omitted, either through oversight or the practical need to limit the length, some very good works. We apologize to the authors for these omissions. Some books could well be listed in more than one of our chapters. In this case, we have generally listed them only once.

We encourage readers to read some of these other books. Different and broader perspectives on engineering can only make you a better designer. Moreover, our lists should be only a starting place; bear in mind that we may well have missed books that you would find helpful. Moreover, new books are coming out all the time.

Engineering Design, Pahl, G. and Beitz, W. Edited by Ken Wallace, The Design Council, London, England, 1984.

The Design of Design, The Science of Design, The Development of Design, The Selection of Design, Four books by Gordon Glegg, Cambridge University Press, 1973-1981.

Design: Serving the Needs of Man, George C. Beakley and Ernest G. Chilton, MacMillan, 1974.

The Engineering Design Process, David Ullman, McGraw-Hill, 1992.

The Principles of Design, Nam P. Suh, Oxford University Press, 1990.

Fundamentals of Machine Component Design, Robert C. Juvinall and Kurt M. Marshek, Wiley, 1991.

Design Engineering: Inventiveness, Analysis, and Decision Making, John Dixon, McGraw-Hill, 1963.

The Engineering Design Process, A, Ernatz and J. C. Jones, Wiley, 1993.

Engineering Design, George Dieter, McGraw-Hill, 1983.

Engineering Design Methods, Nigel Cross, Wiley, 1989

Product Design and Development, Karl T. Ulrich and Steven D. Eppinger, McGraw-Hill, 1995.

Product Design for Manufacture and Assembly, G. Boothroyd, P. Dewhurst, and W. Knight, Marcel Dekker, Inc., New York, 1994.

References

[1] Simon, H. A. *The Sciences of the Artificial*, M.I.T. Press, Cambridge, MA, 1969. Quotation by permission of the publisher.

[2] Ver Planck, D. W. and Teare, B. R. *Engineering Analysis*, Wiley, New York, 1954.

[3] Dixon, J. R. *Design Engineering: Inventiveness, Analysis, and Decision Making*, McGraw-Hill, New York, 1966.

Problems

1.1 What is the smallest functional designed object you have heard of? The largest?

1.2 In a bicycle (or thermostat, or other product of your choice), give an example of a standard part, a special purpose part, a special purpose sub-assembly, a standard module, and a component.

1.3 Have you ever used the basic guided iteration process to solve a design problem? If so, describe your experience noting how you implemented each of the steps.

1.4 Have you ever used the basic guided iteration process to solve a problem other than a design problem? If so, describe your experience noting how you implemented each of the steps.

1.4 Go to your library and get a copy of the ASME Boiler Code (or some other available Code), and read enough of it to get the idea of what a Code is like.

1.5 Go to your library and find at least one of the books on the Recommended Supplementary Reading list, and read the introductory chapter. Report back to the class on how it is the same and how it is different from this chapter.

1.6 In Figure 1.1, identify the stages that constitute "production development" as defined in this chapter.

Chapter Two

Best Practices of Product Realization

"Those who join the [quality and productivity] revolution will become the survivors, while those who prefer the status quo will become the casualties."

Peter C. Reid [1]*
Well Made in America: Lessons From Harley Davidson on Being The Best (1990)

2.1 Introduction

Let us be among those who join the revolution.

In Chapter 1, we presented a vocabulary for the study of engineering design, and introduced the basic steps in the guided iteration process. In the course of the definitions for terms, we briefly described the product realization process. Readers may wish to review Section 1.3.4 and Figure 1.1.

In this chapter we expand the discussion of the product realization process. The purpose is to give engineering designers a more comprehensive understanding of the context within which their work takes place. We discuss the criteria for success of product realization: high quality, low cost products tailored to customers' needs, and brought quickly to market. We describe a number of the practices — called *best practices* — used by successful firms to achieve these goals. We close the chapter with brief outlines of the product realization processes from three successful companies.

Some readers will think this chapter is more about management than about engineering design. To some extent, this is true. However, just as an effective basketball player must understand the team's game plan, so too an effective engineering designer must understand the whole product realization strategy in his or her firm. One reason is that you need to understand the situation from your manager's perspective. Besides, many design engineers will one day become managers in the product realization process.

* Quotation by permission of the publisher, McGraw-Hill, Book Company, New York.

2.2 The Goals: Quality, Cost, Speed, and Flexibility

The overriding goal of any product realization process is to produce a product consistent with a firm's strategy, and one that meets customers' needs and desires responsibly (with concern for efficiency, safety, society, and the environment). The product must also generate a profit if the company is to continue paying employees, being a good citizen, and developing new and improved products. All this means that designers must develop products with a suitable (and competitive) combination of:

Quality,
Cost,
Time-to-market, and
Marketing flexibility.

Traditionally, cost has been considered of paramount importance, and no one would argue that cost is unimportant. However, in the 1980s, especially as a result of Japanese examples, quality — defined very broadly — became equally or possibly more important. Unfortunately in many firms, quality was rather narrowly viewed as exclusively related to *manufacturing or production* instead of to design and the rest of the process as well. This limited view was (is still?) a shortcoming of managers at all levels in some firms. It results in a serious error because it ignores the fact that many factors determining quality are largely decided in the design stages, especially the early design stages.

Now, not only cost and quality (broadly interpreted to include performance, reliability, environmental issues, safety, and others) but also time-to-market and market flexibility have also been recognized as critical to competitiveness. Companies that can get their new products and product improvements into the marketplace quicker often get a larger market share than competitors. Also, by going through the product development cycle more often, quicker companies learn more about the product, processes, and technology involved. They also learn more about how to improve their own organization and performance.

It used to be a firmly held belief that shortening the product development cycle would surely mean lower quality and/or higher cost products. However, this conventional wisdom is not necessarily correct. Shortening the time allowed for development may, in fact, improve the quality *and* lower the cost of the product developed [2, 3, 4].

How? Shorter times encourage engineers and others to "get it right the first time." The old attitude was "If it doesn't work, we can fix it later." This attitude cannot be tolerated in a rapid development process. Usually there isn't time or money later to fix it *right*, so some sort of 'kludge' is worked out, which results, then, in a design that is not the best possible one. (We are indebted to Dr. Barry Bebb, Barry Bebb and Associates, San Diego, CA., for this insight.)

Another goal of design and manufacturing systems that is becoming prominent is *marketing flexibility*. This means designing and manufacturing products tailored for individual customers, or at least for relatively small groups of customers. Though such flexibility has been talked about for some time, little progress has been made to date. Achieving real flexibility, however, may well require design and manufacturing techniques not yet conceived. The sub-field of *design for marketing flexibility* may yet be born.

A special factor in achieving flexibility in a global marketplace has to do with providing inexpensive ways to customize products for the cultural differences that exist throughout the world. In Germany, for example, stainless steel tubs in washing machines are considered highest quality; in the United States, fiberglass tubs are considered better. For customer acceptance, products often have to have slightly different features in different countries.

> "Engineering design...is the fundamental determinant of both the speed and cost with which new and improved products are brought to market, and of the quality and performance of those products. Design excellence is thus the primary means by which a firm can improve its profitably and competitiveness."
>
> *Improving Engineering Design (1991)*
> National Research Council [5]

To meet such variable marketing requirements, it is sometimes possible to design a product with a basic platform of common parts upon which variations can be inexpensively added for different international markets. If done correctly, this strategy can save a great deal of money, and add a great deal of flexibility compared with having a vastly different product for every market.

2.3 Some Best Practices of Product Realization

A number of specific practices — sometimes called *best practices* — have evolved in certain highly successful manufacturing firms. These are the practices that have contributed to developing and maintaining these firms' competitiveness in world markets. We therefore review and briefly discuss a number of these practices. Our emphasis is on those that relate closely to engineering design and design for manufacturing. The purpose is to further acquaint engineering design students and new practitioners with how engineering design is done and how it fits into the total industrial setting. Later in the book, we cover in much more detail how to employ many of these practices in the context of actual design problems.

The best practices described in this chapter have evolved over time, and will no doubt continue to evolve. As presented here, they are fairly general, and so their specific implementation in individual firms can take somewhat different forms. Nevertheless, the practices described are fundamental enough that, though specifics may vary, the general ideas are very likely to remain important for some time to come.

2.3.1 An *Explicit* On-going Mechanism for Generation, Evaluation, and Selection of Ideas for New and Revised Products

There are three primary sources of ideas for new or revised products in firms: customers, employees, and new technology. Before we look at how these can be utilized, consider the following related point of view:

In some (especially older) manufacturing firms there is a strong cultural bent towards the belief that the business of the firm is to *manufacture things*, specifically the things that constitute their current product or product line. However, the business of such a firm is never really a particular product or product line. Rather, their business is the *service* that these products perform for customers. For example, the business of a company manufacturing pencils is not making pencils; it is providing the service that pencils provide their users. That is, the business of such a firm is to provide the service that enables people to record their thoughts and other information onto hard copy.

A firm manufacturing pencils may never decide to

manufacture ball point pens or word processors or speech recognition systems, but at least to view their business as a service provider reveals who their competitors are (not just other pencil manufacturers) and who they may be in the future. It also gives the firm an incentive for inventing the next popular thought recording product — even if it is only a better pencil. It is thus important for manufacturing firms to determine and become conscious of what service their products perform for customers, and what the customer values about that service and the way it is provided.

Now let us look at the three most common sources of ideas for new and improved products:

Customers. Competitive manufacturing businesses require constant feedback from the customers who buy, sell, repair, or use the company's products. Getting such feedback cannot be left to chance or to mail surveys, or even to formal complaint records, though such mechanisms are certainly a part of the process. There are marketing professionals who know how to get the information. But in addition, if a design engineer is looking for positive new ideas as well as for shortcomings of current products, then he or she must get out personally and talk to the customers.

A famous example is the Japanese auto maker who sent design engineers to talk to American shoppers in super market parking lots discovering, among other things, that making it easy to get grocery bags into and out of car trunks was a convenience valued by many potential customers.

A successful major U.S. firm now requires every manager to talk to at least one actual customer *in person* every week.

A personal example: One of the authors once bought a car in which all the knobs on the various controls fell off within a month. When this annoyance was brought to the attention of the dealer, it turned out that this problem had been a characteristic of this particular automobile for at least seven or eight years! If the designers in that company had been listening to their customers, they would at least have supplied pliers with their cars so that owners could turn on the lights, adjust the choke, etc! Of course, that brand car was not purchased again by this customer.

The message is this: if you are a design engineer, then take time regularly to get out of your office and talk to the people outside the company who buy, sell, use, repair, and even dispose of the products you design. You may — no, you surely *will* — learn something extremely valuable for yourself or your company.

Design and product development engineers must also get out and talk to their "customers" *within* the company. These include not only marketing people, but also manufacturing engineers, analysts, draftsmen, accountants, machinists, production workers, and others.

Employees. Employees in the factory, shops, and offices are also an extremely valuable source of new ideas for products and product improvements. Good practice requires that there must be a believable, rewarding (financially), well understood, and low threshold (i.e., easy) mechanism for employees to get their new product, product improvement, and process improvement ideas heard and seriously considered.

Employees must also get rapid feedback on what happens to their ideas (good and bad), and why.

New Technology. Keeping abreast of new technologies and methodologies in materials, manufacturing, design, engineering, and management is another important source of ideas for new and improved products. Later in this chapter, we describe the need for an explicit mechanism in every firm for the infusion of new information about new technologies and methodologies of all kinds. Coupling new technological information with the search for new or improved product ideas is an essential part of the product development process that is not, strictly speaking, engineering design as we have defined it. But it is important for engineering designers to be able to contribute for best results within a company.

However, don't get carried away with a new technology (as some zealous engineers tend to do) and leave customers and the marketing and manufacturing people out of the process! This happened in one company where a new product idea based on new material technologies was developed privately within the engineering design group. After considerable time and expense, a prototype of the proposed new product was proudly unveiled by the engineers at a management meeting where marketing people heard of it for the first time. Probably needless to say, bad feelings happened. Whether the product idea was good or not, or whether it might have led to a good new product or not, will never be known! In a best practice product realization process, you've *got* to involve *everyone* right from the very start — and keep them *all* involved *all* along the way. It was excellent that the engineers developed a new product idea using a new technology, but they should have taken it immediately to marketing to find out how it could be used to make something customers might actually want.

A lively program for considering and evaluating ideas from all these three sources should be established. Regular meetings of an open-minded high level group responsible for sorting ideas submitted informally on a single sheet of paper can suffice. Ideas selected by this group get rewarded and sent to a marketing-engineering-design-manufacturing group for additional (but still preliminary) evaluation of their marketing potential, cost prospects, technological risk, and relationship to company strategy. Reports of this study then get presented to a top management group. Those selected move further into the product development process, and their originators get additional rewards. Only a very few ideas will finally make it into production, of course, but the constant, explicit search for and serious consideration of new or improved product and process ideas is a basic 'best practice' in top competitive firms.

Pugh's Method of Concept Evaluation. In evaluating new ideas, a best practice method has been developed by Professor Stuart Pugh has proved very successful [7]. The method is explained and illustrated in this book in Chapters 6 and 7.

2.3.2 Cross-Functional Teams

Sequential Versus Concurrent Processes. The old concept of product development was of a sequential-iterative process. Ideas were conceived somehow, *then* approved by management, *then* designed, *then* "engineered" (i.e., analyzed for performance), *then* detailed, and *then* sent to manufacturing for process planning, tool design, production, and quality control. Finally the resulting products were shipped, stored, distributed, sold, used, repaired, and disposed of. *Then* the next generation of products was considered. When something went wrong somewhere in this sequential operation, backtracking was required to fix the problem. Usually such fixes had to be patch jobs, because significant changes would cause great delays and expense.

Often, if *anyone* had overall responsibility for this sequential process, it was either someone at so low a level as to be merely a coordinator, or someone at so high (and abstract) a management level as to be merely a nominal director with no real knowledge of the process, technical control or decision making power. To a great extent, each step was independent from the rest. While there were "manufacturing reviews" during various stages, these generally could influence only details of the design. Engineering and detailed design were concerned mostly with function and fit, and manufacturing produced whatever resulted as best they could, usually very well considering the lack of attention to DFM in the design.

As is by now well recognized and publicized, there are several serious problems with this sequential process. The major one is that decisions made early in the process have *great* effects on subsequent stages, but do not properly take the downstream consequences into account. For example, certain early design decisions may affect, say, the cost of assembly or tooling, but do not appropriately take assembly or tooling costs into account.

Another problem with the linear process is that the product tends to lose its conceptual integrity and character because it is under the control of no single person or unified group. It looks like it was designed by a committee, or a bunch of committees. A third problem, as noted, is that the sequential process is often slow.

As a result of these shortcomings, it is common now to replace the sequential process with a *concurrent* or simultaneous process. Concurrency is widely believed to result in better designs, easier to manufacture products, and products more attuned to the marketplace and to a company's strategy. As with most ideas, however, stating the concept is easier than implementing it. For most small and moderately sized products, a method of implementing concurrency that has had success is to use *cross-functional* (or multi-functional) teams. These are teams consisting of people from, say, marketing, finance or accounting, manufacturing, engineering, product service, and design who have responsibility for the product development from the outset and continuing throughout the entire life cycle of the product. Such teams implement concurrency naturally because they include people with the various interests that must be considered.

There is very little known in general about how large cross-functional teams can or should be. For relatively small projects, probably there should be not more than five or six members. Larger projects may get done with teams of teams.

There is a great deal known and reported in the literature about how teams work in general, and what factors contribute to their success or failure. See, for example, [7]. We do not cover this subject in detail in this book. It is worth mentioning here, however, that members of cross-functional concurrent product development teams generally get selected on the basis of their technical expertise (to balance the technical knowledge of the team), but the teams no doubt function well or poorly depending more on the personal qualities of the individuals chosen, and their degree of personal identification with the project goals. It is probably better to have a team that works well together than it is to have all the technical bases fully covered. Temporary consultants can be called in by the team as needed for a specific technical issue.

Teams should be given a strong identity, and team members should have a clear personal stake in the success of the project. Beyond being given a strong and capable team leader, teams should not be micro-managed. Of course, they must have the resources (e.g., time, money, staff, management support) needed to do their task well and efficiently.

It is not really possible to include experts of all kinds on a cross-functional product development team. Major areas from manufacturing, engineering, and design can be included, and possibly a few others as appropriate. Thus the team members have responsibility for outreach to get the input needed from others whose interests may be more specialized. To insure that this indeed happens, clearly defined checkpoints must be established in the product development process where a representative for every possible downstream issue is required to review the design. The danger here is an overdose of bureaucracy which eventually becomes routine and meaningless, consuming time but contributing nothing. As noted above, review is not nearly as effective as regular participation, but teams must be kept to a reasonable size.

Team Leadership Styles. Whether or not a cross-functional team is developed to implement a product realization project, there must be some kind of program leadership. The selection of a team leader or program manager is extremely important. So is the manner in which the program manager's role is defined. Professors Hayes, Wheelwright, and Clark of the Harvard Business School, in their book entitled *Dynamic Manufacturing* [8] describe three types of team organization and management styles: (1) a "lightweight" program manager, (2) a "heavyweight" program manager, and (3) "tiger teams" (defined in the next paragraph), with either a lightweight or heavyweight manager.

In the first two of these, members remain in their respective regular functional organizations within the firm (e.g., engineering, manufacturing), but they are also assigned part time to the product development program. (At any given time, they may be assigned to several programs.) Team members thus have two (or more) bosses. In the case of a "tiger team"', on the other hand, the functional team members are temporarily reassigned full time to the product development program.

A lightweight manager is one whose role is best described as a coordinator or facilitator. He or she lacks great authority, and may even be out-ranked and out-experienced by some members of the team. A lightweight manager may also be out-ranked by the regular (i.e., functional organization) bosses of the team members.

A heavyweight manager, on the other hand, is usually a senior member of the firm, with considerable technical experience and ability, who is expected to take a strong leadership role in the program. He or she will generally out-rank the functional department managers from which the team members are supplied. Heavyweight managers have not only the authority of rank, but also the authority that comes from knowledge and successful experience. That is, they tend to be top technical people.

While any of these organization types can work well, studies point to advantages for the heavyweight manager mode (whether the team is or is not a tiger team). The reason is simple: a lightweight manager is in a very difficult position. He or she has responsibility for things over which no control can be exercised. Since the concept is that all team members must feel fully responsible for the success of the program, it *should* be possible for a lightweight manager to get the participation and support of all the team. But "should"' does not always work out well in practice. Of course, if a firm does not have a person qualified by ability, knowledge, or rank for the heavyweight role (often the case), then there is no choice but to use the lightweight model. Then it is especially critical that all team members feel very close personal and professional identity (i.e.,"ownership") with the program.

2.3.3 Focus on Quality

It is recognized throughout the most competitive companies that quality is crucial to competitiveness, and that *quality cannot be built-into or inspected-into a product unless it is first designed-in*. It is also recognized that time to market is a critical factor in profitability, and development times can be significantly shortened through appropriate management and engineering design approaches (e.g., concurrent design and design for manufacturing). Finally, competitive firms know that quality, time-to-market, and cost are all of a piece. None should or need be sacrificed for the other.

The most competitive firms tend to have established metrics (i.e., measurements) which indicate their performance regarding quality, cost, and time-to-market. One way to help establish such metrics is through *competitive benchmarking*. This involves looking in detail at the products and processes (both design and manufacturing processes) of one's very best competitors. Competing products can be purchased, taken apart, and analyzed for cost, performance, and manufacturability. Out of this process, metrics can be established for a company's own products and processes, and performance can be measured against these metrics.

A management philosophy (not discussed in this book) called Total Quality Management (TQM) has been conceived to focus the attention of all members of a firm on the critical importance of quality to all aspects of the firm's operation.

2.3.4 Concurrent Engineering, Design for X, and Design for Manufacturing (DFM)

Concurrent design attempts to organize the product realization process so as to have as much information and knowledge about *all* the issues in a product's life available at *all* stages of the design process. Sometimes this is also referred to as *Design for 'X',* where 'X' stands for the customer, for robustness, manufacturing (including tooling, assembly, processing, etc.), the environment, safety, reliability, inspectability, maintenance and service, shipping, disposability — and all the other issues in the life cycle of the designed object and its production [10]. The idea is that we must design from the beginning of the process with information about and constant and simultaneous concern for *all* the X's.

Implementing a concurrent engineering strategy is most often done by using cross-functional teams to involve *everyone* concerned with the entire life cycle of the product in the product realization process. Involving people from downstream processes early in the process here does not mean just occasional design reviews after critical decisions have essentially been made; it means *real* involvement in the discussions and decision making all along the way. The teams may not be all-inclusive in their membership, but they must be sure that any interests not represented on the team are heard from in a real and timely fashion.

It is not easy to implement concurrent design in a firm that has a history of doing things sequentially, but it can be done. It is partly a management problem, and partly a problem of technical knowledge. It is also partly a so-called cultural problem, which just means that people and organizations have a vested interest in the status quo, and hence are reluctant to change. It is, of course, the responsibility of management to conceive, organize, and manage change successfully.

In this section, we introduce several of the better known and most widely used best practice design for X methodologies. There are many others. And it is certain that these practices will be improved upon and that others will be developed.

Design for the Customer

Quality Function Deployment (QFD) [9] is a method for relating directly the specific needs and desires of customers to the features and engineering characteristics of a product being designed, evaluated or re-designed. A technique for implementing QFD, called the *House of Quality* [9], is generally used to perform the product or design evaluation and to guide the redesign for improved customer satisfaction.

Quality Function Deployment and the House of Quality are discussed in more detail and used in Chapters 5 and 7.

Design for Assembly

Design for Assembly (DFA) [11] is a method for evaluating qualitatively and quantitatively the time and hence cost of handling and inserting parts in a product or machine. The methods that have been developed can guide designers to design more easily assembled parts, and the methods also can assist designers and managers alike in evaluating and comparing existing or proposed designs. A key principle is that the number of parts should be as small as possible. Design for assembly is discussed again in Chapter 3.

Design for Manufacturing

It used to be the case in many firms that the design department and the manufacturing department seldom communicated, except when designers tossed their designs "over the wall" to manufacturing.

Design for Manufacturing is a generic term that refers to a whole set of methods relating design features and parameters of piece parts to the ease and cost of manufacturing them by specific processes such as injection molding, machining, die casting, extrusion, stamping, and the like. Thus there are *design for injection molding* methods for guiding the design and evaluating qualitatively and quantitatively the cost of tooling and processing for parts to be made by injection molding. And there are *design for die casting* methods that guide the design and allow evaluation of the design of parts that are to be made by die casting. And so on.

To speed up the process of getting a product into the market place, there are times when a firm will take some calculated risks by beginning the development of the production process (e.g., tool and die design and construction) *before* the design of the product is complete in every detail. This, of course, requires a close and trusting relationship among the designers and manufacturing people.

In this book, design for manufacturing (especially for injection molding, die casting, and stamping) is discussed in Chapters 3, 11, 12, 15, and 16.

Design for Robustness

Robustness refers to how well a product performs for its customers in the presence of variations in the manufacturing process, in the presence of variations in the environmental conditions in which it is used, and in the presence of the variations in the product as it wears in service. Obviously it is desirable that a product's performance be affected as little as possible by these inevitable variations (often called "noises"). Products whose performance is insensitive to these noises are said to be robust.

A popular method of designing for robustness was developed in Japan in the 1970's by Dr. Genichi Taguchi, a Japanese engineer with a special interest in statistical methods [12, 13, 14]. Taguchi's Method™* of designing for robustness is based on a somewhat mechanical use of the fundamental ideas of the field of *design of experiments* [15]. His methods contain a hidden (though reasonable) assumption about the function used to define and measure robustness quantitatively. For these reasons, the Taguchi Method™ has come in for criticism from the statistics and statistical quality control research communities [16]. Nevertheless, Taguchi's Method™ has a very good record of practical success and is relatively easy to learn. We present it in this book in Chapter 19, together with additional references to its critics and to other methods.

> "To furnish the means of acquiring knowledge is...the greatest benefit that can be conferred on mankind."
>
> John Quincy Adams (1846)
> *When Establishing the Smithsonian Institution*

2.3.5 An Explicit Program of Knowledge Infusion and New Learning

A key ingredient to providing a foundation for continuous competitiveness is new knowledge. A continuous influx of new knowledge in six basic areas is needed in every firm that is designing and manufacturing products:

- Materials,
- Manufacturing processes,
- Design methods and tools,
- Engineering analysis methods and tools,
- Management methods and tools, and
- Key company specific core technologies.

The meaning of the first five of these topic areas is obvious. The sixth — company core technologies — refers to those special but basic technical fields that relate directly to a firm's business. For example, one core technology for a company making home oil heating equipment is combustion. Another is sheet metal forming.

World competitive product development requires a continuous influx of new knowledge. (The Japanese know this and practice it; they are constantly scouring the world for new knowledge.) Every firm needs an aggressive mechanism to insure that new knowledge is received and distributed in a timely and broadly based way. Design and product development engineers must be a part of this process. It is a part of every engineer's job to stay abreast of his or her field. This means reading journals, attending technical and trade conferences, talking to vendors about new technologies, and exchanging information with others in the company.

It is not enough for a firm simply to adopt (i.e., copy) the world's best *current* product development practices. New knowledge will keep on being generated. The toughest competitors will keep on getting better. Establishing a permanent system of new knowledge infusion is a necessary part of continuous improvement in competitiveness [17].

---

* Taguchi's Method" is a trademark of the American Suppliers Institute, Dearborn, MI.

2.3.6 Attention to Physical Prototyping Policies

Prototypes can be either physical hardware, or they may be computational simulations of product performance done on the computer. In this section, we are referring to physical prototypes.

One specific issue related especially to product development time is the number and nature of physical prototypes planned in the product realization process. Reducing the number of planned prototypes (e.g., from four to three, or better, from three to two, or still better, from two to one) will save a great deal of time. The reason is that design engineers, knowing ideas can or will get tested in prototypes, are prone to take risks in their initial designs. But *the product realization process is not the time to take risks.* Risky ideas should be developed and tested in the laboratory *before* they are incorporated into product development programs, not tested for the first time in prototypes during an active program.

Ideas should be carefully evaluated when they are proposed, not later when they are already built into hardware, even prototypes. The reason is that untested ideas built into prototypes tend to get embedded. That is, once in, they are hard to get out. Instead, they get patched up and made to work one way or another, but they don't get fully corrected. And if they must get properly corrected, the resulting major changes take even more time to implement [18].

Along with reducing the number of prototypes goes defining much more carefully for all involved exactly what all the purposes are of each prototype.

The prototype shop in one company hand fitted a part for a prototype so that the product functioned beautifully. However, when the product went into production, the hand fitting, which turned out to be crucial to proper function, could not be done at a reasonable cost. The product's failure rate was outrageous. Some people thought the purpose of the prototype was to test only the functionality of the design concept; others independently thought it was also to test productivity. There was no communication, so the lack of agreement on the purpose of the prototype caused a disaster.

Speed of prototyping is important to reduce time-to-market. There are (as this is written) three major so-called *rapid* physical prototyping technologies in use for piece parts [19]. These are: (1) photo-polymer solidification (or stereolithography), (2) computer numerically controlled (CNC) machining, and (3) room temperature vulcanized rubber molding. As described in [20], the issues in selecting one of these are: appearance, strength and stiffness, dimensional accuracy, surface finish, feature definition, other material properties, and speed.

Though expensive when only one, or a very few, parts are needed, the CNC method is rated excellent on all criteria except speed; there must be a CAD model and the programming must also be done. The vulcanized rubber molding is slow, but provides good results in surface finish, features, and dimensionality. It is, however, only fair in other respects. Stereolithography has received a great deal of attention because of its speed (given a solid model representation). It has good feature definition and dimensional accuracy, but in terms of surface finish, strength, and appearance, it is rated only fair to poor. Thus different methods are best for different purposes.

Methods of physical prototyping are improving rapidly, and new methods are under development (e.g., making dies quickly by arc spray distribution) and these methods will be improved upon. This is a good example of an area in which designers must keep up to date through reading, attending technical meetings, and contact with colleagues and vendors.

2.3.7 Strategic Use of Computational Prototyping and Simulations

Modern computational methods, employing computers, make it possible to reduce or even eliminate more expensive and more time consuming physical prototyping. These methods also enable quality and performance testing that is often even more thorough than can be done with actual prototypes. Computer-aided design, solid modelling, finite element methods, and many kinds of simulation programs are used by best practice firms to improve quality and reduce design and development time.

A word of caution, however: the computer-based technologies of computational prototyping (i.e., the computers and software) cannot, by themselves, improve the product realization process. There must be people who know how to use them properly and to advantage. And most importantly, the technologies must be integrated strategically into the entire product realization process; they can't be effective as add-ons expected to fix an otherwise poor set of practices.

2.3.8 Exacting Control of Processes

The old idea of quality assurance was to inspect parts and assemblies *after* they had been produced. The new best practice is to control processes so rigorously that inspection is unnecessary.

A story has it that a U.S. firm ordered 1,000 sub-assemblies for their product from a Japanese manufacturer. When the order was placed, the president of the U.S. firm stressed to the President of the Japanese firm that 980 of the assemblies must be within specification. When the sub-assemblies were delivered, a large truck came with 980 good ones on board. The Japanese firm's president came along also with a small box containing 20 other out-of-spec sub-assemblies which he delivered in person. To the U.S. company president he said: "Here are the 20 bad sub-assemblies that you wanted. We made them special for you, though we can't imagine what you want with them".

An important method for controlling processes is called Statistical Process Control (SPC) [21]. This is another technique — like the Taguchi methods — that was first put into common use in Japan. In this case, however, the prime mover was an American, Edward Deming [22], who found the Japanese receptive to his ideas where U.S. industry had not been.

SPC is a method based on statistical reasoning and computations that enables a process to be monitored on line (i.e., in real time) to insure that it is within the control limits set for it. All processes, of course, involve a certain amount of random variability even when they are under proper control. What SPC does is make it possible to determine when the variability is within normal limits of randomness and when it is beginning to move outside the limits of normal variation; that is, out of control. Beyond this point, we do not discuss SPC in this book.

2.3.9 Intimate Involvement of Vendors

Vendors are manufacturers or distributors who supply piece parts and sub-assemblies to other manufacturers. There may well be dozens, hundreds, or even thousands of vendors involved in the manufacture of certain products and machines. Older practice was to prepare specifications which vendors must meet with their products, and which were used to obtain competitive bids from several competing vendors.

Newer practice (pioneered again by the Japanese) is to employ only a single vendor and to involve that vendor in the design of the product, especially, of course as it relates to the parts and sub-assemblies to be supplied by the vendor. (This is an aspect of concurrency not mentioned above). It is also a newer practice, then, for the manufacturing processes of the vendor to be closely monitored by the parent firm so that (referring to the example above) there are not 20 — or *any* — bad parts mixed in randomly with the 980 good ones.

2.3.10 Strategic Use of Computer Technology

Until after World War Two (i.e., about 1945 or so), many manufacturers did not use drawings for the manufacture of parts. Instead, actual reference parts were kept essentially "on file" to be used as a standard or model when more parts were needed. Today, the ability of computer-aided-systems and solid modelers to serve the design and manufacturing processes has increased dramatically. More needs to be done to make the computer a more useful tool, but it is already extremely helpful. Indeed, it has become essential in many best practice firms. And there is hardly any way that firms can remain competitive in the future without making full use of the computer's ability to support design and manufacturing activities.

2.3.11 New Accounting Procedure

The vast majority of the potential value of a product is added before manufacturing ever begins. To understand that this is so, note that most of the *information* needed to manufacture the product is contained in the design. Note, too, that if the design itself were to be sold as a product, its value would not be determined by how much the design cost to develop, but by how much net profit a purchaser could expect to realize by producing and marketing the product as it is designed.

> "Corporate management accounting systems are inadequate for today's environment."
>
> H. T. Johnson and R. S. Kaplan [23]*
> *Relevance Lost (1989)*

This view of the value of a product design is not, however, reflected in the accounting systems currently used in most (if not all) manufacturing firms. Instead product development is viewed as a cost or expense instead of as an investment that forms the basis for future profits.

The correct viewpoint is easier to see in the case of a design change. Suppose a change is proposed that results in, say, reduced manufacturing costs or increased sales. For example, in one well known firm, a concurrent design team redesigned a product line so that the parts involved were reduced from nearly 3,000 to under 1,000. The resulting savings in assembly, inventory, distribution, and overhead were tremendous, but could not be accounted for in such a way that the design group could get any direct credit. Management, looking at the cost of the group, was therefore critical of the design group's performance (no revolutionary new products) even though the savings to the company in the design changes were no doubt in the hundreds of millions of dollars annually.

This is not to say that value is not also added in actual production of parts and products, but there is a strong tendency to view the value added there as the difference between the final product's worth and the cost of materials, labor, and overhead, with the preceding design process being either a cost or part of overhead. But it is to a very great extent the value of the *information contained in the design* that determines the value added of the products produced.

If product realization is to be properly managed, then ultimately there will have to be accounting systems developed that reflect the value of the information produced during design. Otherwise there will never be proper investment by firms in design. In the book *Relevance Lost* [23], published by Harvard Business School Press, Boston, H. Thomas Johnson and Robert Kaplan describe the shortcomings of current cost and financial accounting systems, and describe potential new management accounting systems. To the author's knowledge, however, practical systems along these lines are not well established (if they are at all), but this is surely an issue for firms who would be extraordinary, long term competitors.

2.3.12 Research and Development in Product and Process Development

It is not possible to get ahead of competitors by copying them. It is not even possible to catch up by copying. The

* The quote is reprinted by permission from *Relevance Lost*, by H. Thomas Johnson and Robert S. Kaplan, Harvard Business School Press, Boston, MA, 1989, page xi.

best and toughest competitors keep improving. To get ahead, a company must do better than its best competitors are doing, and thus develop even more effective methods. The knowledge infusion program described above will enable a firm to keep up with what others are doing, and with new technical and management research and development results. But if firms are to get ahead of competing firms that employ extraordinary product development processes, then these firms are going to have to be privy to new research and development results (i.e., to new knowledge) related to product development *before* those results become common knowledge worldwide. This requires a program of research, development, and continuous improvement in their own design and manufacturing processes.

2.3.13 Management and Financial Issues

The involvement of management and the influence of financial issues have not been included in the above discussion, but they are ever-present issues. A product realization program's expected cost and profitability are periodically evaluated, and resources revised for better or worse depending on a variety of circumstances, some of which may have nothing to do with the value of the program itself. The timing and formality of these decision points varies from company to company, but they are always prominently there.

Another product realization activity, though not connected with a specific product in most cases, is on-going education in relevant or basic science, new design and manufacturing technologies, and materials. Many firms are increasingly also providing basic education and skills training as support for their product realization process.

2.4 Three Examples of Successful Product Realization Processes

In this section, the product realization processes followed by several successful United States manufacturing firms are outlined. The descriptions here are not detailed, so they do not fully reveal the great complexity and difficulty of actual implementation. Nevertheless, they are useful as an introduction to help put engineering design activities in perspective. Readers may wish to compare these firms' actual steps with the product realization process outlined in Chapter 1. The descriptions here for Polaroid and Hewlett-Packard are adapted from Reference [5]. Carrier Corporation kindly granted permission for the description of their process.

Polaroid Corporation

Polaroid Corporation makes instant cameras, film, and many other industrial and commercial products related to chemicals and electronic imaging. Polaroid is especially concerned that the entire corporation is informed about and involved in product development. Management emphasizes also that the steps shown below must be followed more or less in sequence.

1. Explore the business, marketing, and technical opportunities.
2. Define the customers' needs while continuously improving the product development process.
3. Define the long-range, customer-focused product line strategy and define the system's architecture for the family of future products.
4. Clearly and fully define the product performance specification with the product development team (manufacturing, marketing, finance, engineering, led by the program manager).
5. Insure the product definition requirs no inventions.
6. With clearly defined and agreed-to product specifications up front, there should be no performance specification changes during the design process (other than ones critical to customer needs).
7. Establish a benchmark process containing goals and driven by the need for continuous improvement.
8. In parallel, continuously develop core technology building blocks for future products.
9. Design the first system layout with computer-aided-engineering and design-for-manufacturing tools from the start, using multi-disciplined, professionally trained engineers and designers.
10. Concurrent with the product design process, design the manufacturing process.
11. Develop an information process for tracking world-class engineering design practices and share generic processes with universities and other U.S. companies.

Hewlett-Packard

Hewlett-Packard makes computers and a wide range of very high quality industrial laboratory, instrumentation, and test equipment. At Hewlett-Packard, fully half of the company's profits derive from products introduced in the most recent three years. This obviously places a great premium on new product development in the firm.

The company has a highly structured process called a *Phase Review Process*. There are seven phases; the endpoint of each phase is a well defined set of information as shown in the parentheses:

Phase 0. Requirements/Plan (System requirements)
Phase 1. Study/Define (System specifications)
Phase 2. Specify/Design (Functional plans)
Phase 3. Develop/Test (Functional verification)
Phase 4. User Test/Ramp Up (Functional sign-off and field preparedness)
Phase 5. Enhance/Support (Financial return verification)
Phase 6. Maturity (Plans for discontinuance)

Other management commitments are also tied to the phases. For example, funding for a specific system development is committed at the end of Phase 1, and funds for customer support are committed at the end of Phase 5.

Carrier Corporation

Carrier Corporation is the world leader in indoor climate control equipment and systems. Their planning recognizes that the product realization process involves several layers of detail, and thus they have a rather formal New Product Introduction Process that stresses concurrent engineering concepts.

At the top level, a Product Strategy Committee is very active in the formation of a project, and serves as a kind of supporting "watchdog" thereafter. For example, this high level management committee may review such steps as the following in the course of a project:

(1) A business opportunity proposal, including definition of customer requirements, the market, business opportunities, etc.;
(2) A Project Plan;
(3) Several formal, structured program reviews as the design proceeds;
(4) A Capital Appropriation just before parametric design begins;
(5) Several more program reviews prior to production; and
(6) A review and audit after one year of production.

Each program is managed on a day-to-day basis by a project team empowered to make key decisions except for approval of capital funds. Teams consist of representatives from all the functional groups (e.g., engineering, manufacturing, marketing, sales, logistics, purchasing). A simultaneous (or concurrent) engineering approach is used to ensure rapid and early consideration of the needs and requirements of all aspects of the product over its life cycle. Reports are made to the Product Strategy Committee or to Quality Review Boards as called for on the project schedule.

For the Project Team itself, there are fully sixty-one (!) milestones during the process that are explicitly defined in terms of the key requirements to be fulfilled. These milestones involve engineering, finance, manufacturing, purchasing, product engineering, marketing, service, sales, reliability and quality control, and the legal department. Explicit risk statements are required of the Project Team whenever new funds are committed.

Of course, every firm has its own more or less unique product realization process. The three described are, however, reasonably illustrative of both the commonalities and the differences.

> "...it is wretched taste to be gratified with mediocrity when the excellent lies before us."
>
> Isaac D'Isreali
> *Curiosities of Literature((1834)*

Lists of steps, activities, and decisions, however, do not do justice to the difficulty and complexity of the task. Moreover, two firms could have the same process outlines, but have very different success in the marketplace depending on how well the people involved implement the plans. The idea here is to show student readers that engineering designers must be prepared to work in the context of a complex organization, with other people, with many complex issues, and with understanding of the total purposes of the firm.

2.5 Summary and Preview

Summary. In this chapter, some requirements of competitive excellence in product realization have been outlined. A number of the practices used by the more competitive firms to achieve these goals have been described briefly. Such a list can never be complete, and the practices are always being improved. However, the most widely used current practices are included. Some of these practices are included in the remainder of this book.

Preview. In the next chapter, we describe a number of common manufacturing processes and the associated qualitative DFM guidelines. Of course, whole books are written that describe each of these process in much more detail. Our purpose is limited, however; we need only to present sufficient information for a reasonable understanding of the rationale for the DFM guidelines. Thus we can meet our purposes with a single chapter, though it is a rather long one.

We want to stress that Chapter 3 is important to the rest of the story of Engineering Design that follows. You will be able to use the DFM guidelines more comprehensively and appropriately if you know their origins in the physical nature and technology of the manufacturing processes.

Recommended Supplementary Reading

Improving Engineering Design: Designing for Competitive Advantage, National Academy Press, Washington, D.C.

Developing Products in Half the Time: Smith, P.G. and Reinertsen, D.G., Van Nostrand Reinhold, New York, 1991.

Compressing the Product Development Cycle: From Research to Marketplace, Slade, B. N., American Management Association, 135 W. 50th Street, New York, 1993.

Out of the Crisis, Deming, W.E., M.I.T. Center for Advanced Study, Cambridge, MA 02139, 1982.

Concurrent Design of Products and Processes, Nevins, J. L. and Whitney, D. E., McGraw-Hill, New York, 1989.

Well Made in America: Lessons from Harley-Davidson on Being the Best, Reid, P. C., McGraw-Hill, New York, 1990.

Design for Success, Rouse, W. B., Wiley, New York, 1991.

The Goal: A Process of On-going Improvement: Goldratt, E. M. and Cox, J.,North River Press, Inc., Croton-on-Hudson, New York, 1948.

Revolutionizing Product Development: Wheelwright, S. C. and Clark, K., The Free Press, New York, 1992.

Concurrent Engineering: The Product Development Environment for the 1990's, Carter, D. E. and Baker, S. B., Addison-Wesley, Reading, MA, 1992.

Creative Product Evolvement: Joseph Liston and Paul E. Stanley, Balt Publishers, Lafayette, Indiana, 1964.

Product Design for Manufacture and Assembly, G. Boothroyd, P. Dewhurst, and W. Knight, Marcal Dekker, New York, 1994.

Problems

2.1 Select a product you have used recently. As a customer, what would like to tell the project design team?

2.2 What service or services do the following products provide for their customers: stapler, bicycle, wrist watch, bookcase, you name it?

2.3 Speech recognition is becoming a viable new technology. List at least three products which may change *significantly* with the addition of speech recognition capability. One, for example, is the computer keyboard.

2.4 In your first assignment in industry to a product development cross-functional team, would you prefer to have a heavy weight or light weight manager? Why?

2.5 Concurrency means (among other things) designing the product and its manufacturing process simultaneously. To what extent are the following activities concurrent in this sense: building design and construction, making a cake, writing a book, doing an abstract painting, you name it?

2.6 What is a core technology in each of the following industries: automobiles, refrigerators, lathes, toys?

2.7 Name some activities other than product manufacturing where exacting control of the process is required because even a very occasional bad outcome is unacceptable. How are the processes controlled?

2.8 Discuss possible conflicts of interest between engineering designers and people in: marketing; accounting; industrial design; manufacturing; customer service; and....

2.9 Why does Polaroid's PRP say, "No inventions"?

2.10 Is the Hewlett-Packard PRP conducive to concurrent design? How is it concurrent, and how not?

2.11 In Item 6 of Polaroid's PRP, it says "...other than ones critical to customer needs". What trouble might this cause?

2.12 Consider and discuss what is meant by the "quality" of a designed object. Of a manufacturing process. Of a product development process. How would establish a way to measure these terms quantitatively?

2.13 Would you say that the Challenger disaster (when the o-ring failed leading to an explosion) was a design problem, a manufacturing problem, or a management problem? Explain.

2.14 Repeat 2.13 but consider instead the loss of market share of the U.S. automobile industry or the loss of the U.S. consumer electronics market.

2.15 Suppose that a product in mass production (1,000,000 per year) costs $1.00 and is sold by the manufacturer to distributors for $1.25. Now suppose that a design (or manufacturing) engineer makes a design (or manufacturing) change that enables the same product to be made for $0.96. What is the value of the *information* that constitutes the design (or manufacturing) change? Now discuss the fact that in most firms, current accounting procedures treat the cost of creating or modifying a design as an expense (bad: detracts from profits) rather than as an investment (good: may lead to increased profits).

References

[1] Reid, P. C. *Well Made in America: Lessons from Harley-Davidson on Being the Best*, McGraw-Hill, New York, 1988. Quote by permission of the publisher.

[2] Smith, P. G. and Reinertsen, D. G. *Developing Products in Half the Time*, Van Nostrand Reinhold, New York, 1991.

[3] Charney, C. *Time to Market: Reducing Product Lead Time*, Society of Manufacturing Engineers, Dearborn, MI, 1991.

[4] Slade, B. N. *Compressing the Product Development Cycle: From Research to Marketplace,* American Management Association, 135 W. 50th Street, New York, 1993.

[5] National Research Council, *Improving Engineering Design: Designing for Competitive Advantage*, National Academy Press, Washington, D.C., 1991.

[6] Pugh, S. *Total Design: Integrating Methods for Successful Product Engineering*, Addison-Wesley, Reading, MA, 1991.

[7] Ross, R. S. *Small Groups in Organizational Settings*, Prentice-Hall, Englewood Cliffs, NJ, 1989.

[8] Hayes, R. H., Wheelwright, S. C., and Clark K. B. *Dynamic Manufacturing: Creating the Learning Organization,* The Free Press, New York, 1988.

[9] Hauser, D. R. and Clausing, D. "The House of Quality." *Harvard Business Review,* May-June, 1988.

[10] Gatenby, D. A. "Design for 'X'(DFX) and CAD/CAE", *Proceedings of the 3rd International Conference on Design for Manufacturability and Assembly*, Newport, RI, June 6-8, 1988.

[11] Boothroyd, G. and Dewhurst, P. *Product Design for Assembly*, Boothroyd Dewhurst, Inc., Wakefield, RI, 1987.

[12] Taguchi, G. "Off-line and On-line Quality Control Systems", International Conference on Quality Control, Tokyo, Japan, 1978.

[13] Taguchi, G. *The Development of Quality Engineering*, The American Supplier Institute, Vol.1, No. 1, Fall, 1988.

[14] Taguchi, G. and Clausing, D. "Robust Quality", *Harvard Business Review,* January-February, 1990.

[15] Box, G., Hunter, W. G., and Hunter, J. *Statistics for Experimenters - An Introduction to Design, Data Analysis, and Model Building*, Wiley, New York, 1978.

[16] Box, G. and Bisgaard, S. "Statistical Tools for Improving Designs", *Mechanical Engineering*, January, 1988.

[17] Dixon, J. R. "Information Infusion Is Strategic Management", *Information Strategy,* Auerbach Publishers, New York, Fall, 1992.

[18] We are indebted to H. Barry Bebb, Barry Bebb and Associates, San Diego, CA. for this insight.

[19] Ashley, S. "Rapid Prototyping Systems", *Mechanical Engineering*, April, 1991.

[20] Wall, M. B., Ulrich, K., and Flowers, W. C. "Making Sense of Prototyping Technologies for Product Design", DE Vol. 31, Design Theory and Methodology, ASME, April, 1991.

[21] Galezian, R. *Process Control: Statistical Principles and Tools*, Quality Alert Institute, 257 Park Avenue South, New York, 1991.

[22] Deming, W.E.,. *Out of the Crisis*, M.I.T. Center for Advanced Study, Cambridge, MA, 1982.

[23] Johnson, H. T. and Kaplan, R. *Relevance Lost: The Rise and Fall of Management Accounting*, Harvard Business School Press, Boston, MA, 1989. Quotation by permission of the publisher.

Chapter Three

Manufacturing For Designers

"If the people don't want to come out to the park, nobody is going to stop them."

Yogi Berra

3.1 Manufacturing for Designers (MFD)

And if designers don't want to use design for manufacturing (DFM) principles and methods, nobody can stop them either. However, if *you* want to learn about the methods and use them intelligently, then it is first necessary to understand the basics of manufacturing processes at least to the level discussed in this chapter. We also present in this chapter several qualitative DFM "Advisors" that are easy to use and can help guide designers to more manufacturable designs early in the design process. Later, in Chapters 11, 12, 15, and 16, more detailed quantitative methods are presented for selected processes.

3.1.1 A True Story About MFD

A young manufacturing engineer has confided that when he began working for a company specializing in the manufacture of professional quality hand tools, he believed that manufacturing operations cranked out products routinely with unerringly high efficiency. After a few weeks on the job, however, he discovered that this was not generally true. Instead, there was a constant stream of problems. In this company, the problems most often were assembly difficulties, probably caused in large part by designers who had never attempted to assemble any of the parts and products which they had designed.

The young engineer also had believed, prior to beginning work, that engineering expertise would have optimized designs from both functionality and manufacturing points of view. Unfortunately, he found that this was not generally true either. Instead he found that most designers knew little about manufacturing and made little, if any, effort to design products so that they could also be easy to manufacture.

In addition, he had once believed that products went from the drawing board, to the assembly line, to the consumer in a straight forward methodical manner. Instead he found that the product followed a confusing, indirect path from one group to the next and back again, throughout the factory until it was eventually completed and staggered out of the plant to the consumer. Some of the groups failed to cooperate with each other, and each group tended to believe it was some other group's responsibility to resolve difficulties encountered.

One of the stumbling blocks to more efficient operation the engineer observed was the linear mode in which this company operated. In this mode, the product designer was satisfied by the production of a prototype which worked well; that is, when function and fit were accomplished. Following the development of the prototype, it was the job of manufacturing to select the production methods to be used, to design the tooling and fixturing required, and to develop the production process plans. Usually, this was the first time that manufacturing people would see the design. Manufacturing would then gloomily examine the complex geometry proposed, the severe tolerance requirements, and the costly materials being specified. While redesign recommendations were usually made that would significantly reduce the manufacturing cost of the designs, time to market considerations, coupled with the fact that many dollars had already been spent on the design, usually resulted in the rejection, by the design group, of the most significant redesign suggestions made by manufacturing.

The young engineer's initial attempts to influence product designs so that they would be easier to manufacture were ignored. However, as time passed, attitudes in the company began to change prompted by increased foreign competition and by the growing interest of a design group leader in design for manufacturing (DFM) concepts. When this team leader became interested, several designers in the group followed suit. Today, all new products in this firm are designed with DFM in mind.

3.1.2 About This Chapter

> "Now let us summarize the steps that must be taken to do integrated design right:
>
> 1. Treat the product design and the development of the manufacturing process in a unified, integrated manner."
>
> Bernard N. Slade [26]*
> *Compressing the Product Development Cycle (1990)*

At all stages of engineering design, designers must be concerned not just with functionality but also with manufacturability. Products must perform their function, but they must also be designed so as not to exceed the capabilities of the processes by which they will be manufactured. In addition, the cost of tooling and the time required for fabrication of tooling must be considered. Processing time and costs are also crucial *design* issues.

At the same time, the capabilities of processes should not be underestimated or underutilized. If a process can easily handle certain kinds of complexity, then that capability should be used to full advantage to avoid the need for more parts or to sacrifice product quality. In other words, the guideline that *the fewer parts, the better* means that the remarkable ability of processes like injection molding, casting, extrusion, and stamping to produce complex shapes should be *fully* used to reduce the total number of parts.

The point made in the above paragraph deserves emphasis. Processes have limitations that cannot be exceeded. However, if the part count can be reduced, parts do not necessarily have to be easy to produce. That is, a design with fewer more difficult parts will be less costly than one with more parts, even if the parts are simpler to produce. Thus designers must understand what *can* be done as well as what cannot be done by the various manufacturing processes. (We are indebted to Dr. Geoffrey Boothroyd for pointing out the importance of this point.)

Manufacturability concerns are important at *all* stages of engineering design, not just at the final, detailed stage. At the conceptual stage, concepts must be evaluated for their manufacturability, and feasible basic materials and manufacturing processes for parts must be selected. At the configuration stage, the sensitivity of tooling and die costs of many manufacturing processes to certain features and characteristics of parts must be considered. The number of parts must be minimized. Manufacturing is also important at the parametric stage where processing costs are often sensitive to the exact dimensions of parts and features (e.g., wall thicknesses or rib height).

The principles and practices of DFM follow directly from the nature and physics of the manufacturing processes. Thus, to accomplish full utilization of processes without exceeding their limits, designers must understand manufacturing processes very well — and also have good communications with manufacturing experts.

In this chapter, we introduce and survey a number of the most common manufacturing processes. Our purpose is to give designers a qualitative physical understanding of these processes to help support their design decisions and their communications with manufacturing engineers. With each process description, we include a discussion of design for manufacturing (DFM) issues, and present qualitative DFM guidelines intended for use early in the design process. In later chapters, more detailed DFM information and methods are presented to enable designers to perform more quantitative and detailed evaluations of manufacturability.

Figure 3.1 lists many of the existing manufacturing processes. It is not possible for us in this introductory design book to describe every one of these processes. Fortunately, some processes are much more widely used than others, so we emphasize these. Excellent books exist that cover the entire spectrum. [1, 2, 3].

Most consumer products consist of both standard components and special purpose parts. The vast majority of the special purpose parts are thin-walled parts produced by polymer processing, die casting, and stamping. Thus, in this chapter (and throughout the book) our discussion emphasizes these processes. This covers about 90% of the special purpose parts found in typical consumer products. We also include here an introduction to the process of aluminum extrusion because it makes a convenient two-dimensional domain in which to illustrate a number of design principles. Assembly and design for assembly (DFA) are included since almost all products are finally assembled. Machining and forging processes are also described and DFM guidelines included, though in less detail than for the more common processes.

3.2 Polymer Processing

3.2.1 The Processes

A large number of polymer processing techniques exist; among the most common are injection molding, compression molding, transfer molding, extrusion and extrusion blow molding.

Injection molding, compression molding, and transfer molding are capable of economical production of complex parts (with significant levels of geometric detail) and simple parts (with little detail). Extrusion is limited to the production of long parts with uniform cross-section. Extrusion blow

*Reprinted with permission of the publisher from *Compressing the Product Development Cycle: From Research to Marketplace*, Copyright 1993 Bernard N. Slade. Published by AMACOM, a Division of the American Management Association.

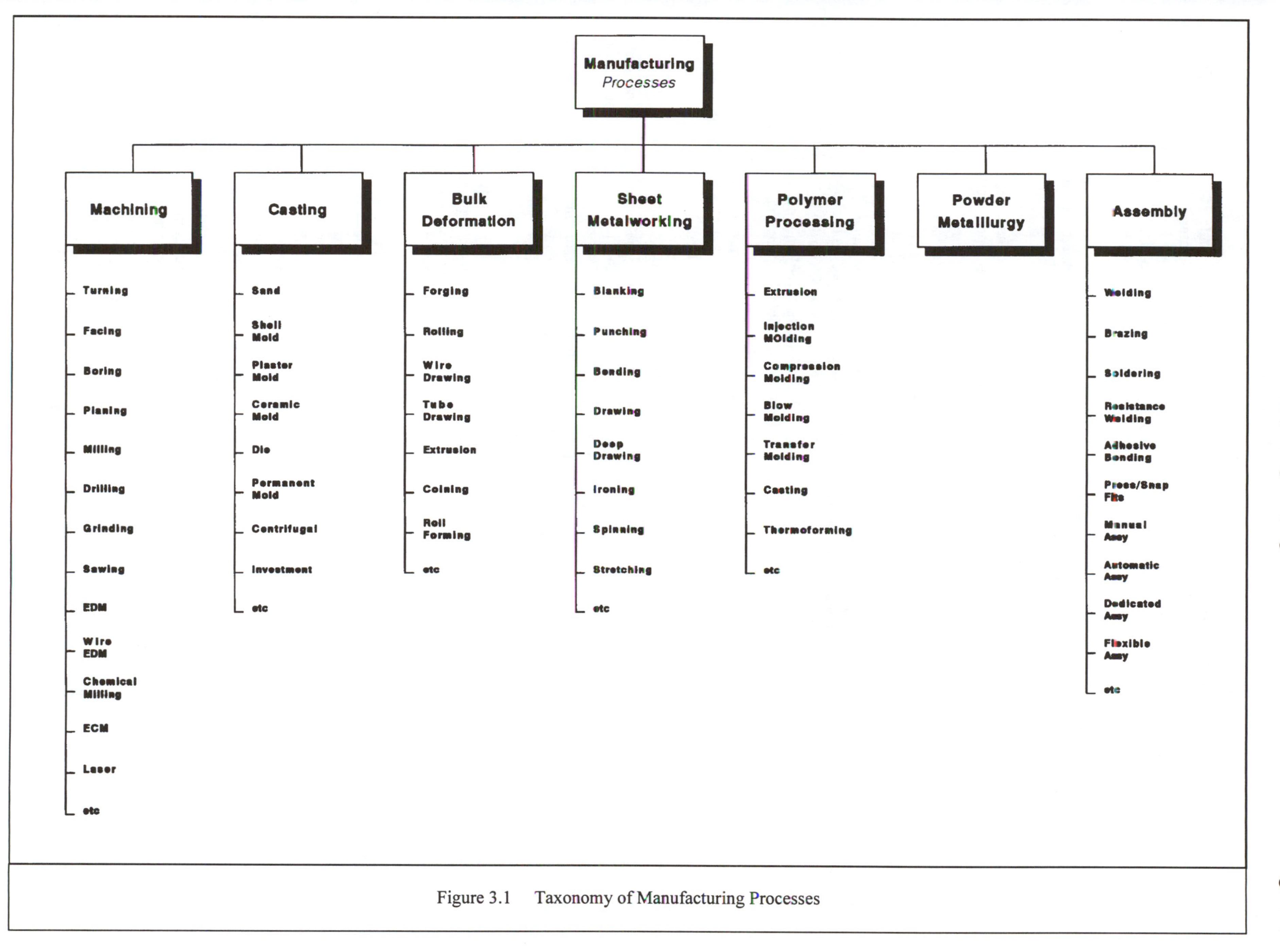

Figure 3.1 Taxonomy of Manufacturing Processes

molding is confined primarily to relatively simple hollow objects such as bottle containers.

Each of these processes is described in more detail later in this Section.

3.2.2 Materials Used in Polymer Processing

There are literally hundreds — maybe thousands — of polymeric materials available for processing, and more will continue to be developed. In general, these materials fall into two broad classes: thermoplastics and thermosets. Some polymers are available in both thermoplastics and thermosets formulations.

Thermoplastic materials can be repeatedly softened by heating and hardened by cooling, and are formed into parts primarily by injection molding, extrusion, and blow molding.

For common product applications, most parts made by injection molding are thermoplastic materials. Examples are gears, cams, pistons, rollers, valves, fan blades, rotors, washing machine agitators, knobs, handles, camera cases, battery cases, telephone and flashlight cases, sports helmets, luggage shells, housings and components for business machines, power tools and small appliances.

Thermoplastics are divided into two classes: crystalline and amorphous. *Crystalline* thermoplastics have a relatively narrow melting range. They are opaque, have good fatigue and wear resistance, high but predicable shrinkage, and relatively high melt temperatures and melt viscosities. Reinforcement of crystalline polymers with glass fibers or other materials improves their strength significantly. (Such reinforced plastics are often called composites.) Examples of crystalline plastics include acetal, nylon, polyethylene and polypropylene (PP).

Amorphous thermoplastics melt over a broader temperature range, are transparent, have less shrinkage, but relatively poor wear and fatigue resistance. The use of reinforcing fibers does not significantly improve the strength of amorphous thermoplastics at high temperatures. Examples of amorphous materials are ABS, polystyrene, and polycarbonate.

It should be noted that while amorphous polymers have no crystallinity, no polymer is more than about 90 percent crystalline. Thus, many thermoplastic polymers exhibit a mixture of amorphous as well as crystalline properties.

The small number of thermoplastics noted above include some that are called commodity or general purpose plastics (polystyrene, polyethylene and polypropylene) as well as engineering thermoplastics (ABS, polycarbonate, acetal and nylon-6). The largest proportion of thermoplastics used are commodity plastics used to produce film, sheet, tubes, toys and such throw away articles as bottles and food packaging. Compared to the commodity plastics, engineering thermoplastics are capable of supporting higher loads, for longer periods of time, and at higher temperatures.

Thermoset materials are polymeric materials that transform permanently on heating and cannot be remelted. Thermosets are formed primarily by compression molding and transfer molding. Parts made of thermoset materials can be subjected to higher temperatures without creeping, tend to have a harder surface, and are more rigid than thermoplastic parts made by injection molding. For this reason, parts used at higher temperatures (molded fryer pan housings, electrical connections, etc.), or parts which may be subjected to harsher environments (automotive carburetor spacers, automatic transmission thrust washers, etc.) are made of thermoset materials. Thus such parts are formed by compression or transfer molding.

Typical thermosets include phenolics, urea-formaldehyde, epoxies, polyesters and polyurethanes.

Thermoplastics and thermosets can both be combined with one or more additives (colorants, flame retardants, lubricants, heat or light stabilizers), fillers or reinforcements (glass fibers, hollow glass spheres), or with other polymers to form a blend or alloy to increase dimensional stability and improve their mechanical properties. Some of the commodity plastics, such as polypropylene, are reclassified as engineering plastics when they are reinforced with glass fibers.

With all the basic polymer materials available, together with all the possible combinations of fillers and additives, there is a dizzying array of possibilities for designers to choose from. There are also pitfalls as not all the properties of all these combinations are well known. Consultation with a polymer experts is well advised!

3.2.3 Injection Molding

In injection molding, thermoplastic pellets are melted, and the melt injected under high pressure (approximately 10,000 psi or about 70 MPa) into a mold. There the molten plastic takes on the shape of the mold, cools, solidifies, shrinks, and is ejected. Figure 3.2(a) shows an injection molding machine along with a mold used to form a simple box-shaped part.

Molds are generally made in two parts: (1) the cavity half gives a concave part its external shape, and (2) the core half gives such a part its internal shape. As the geometry of a part becomes complex, molds of course increase in complexity — and hence in cost.

As a part cools, it shrinks onto the core. Therefore, an ejector system is needed to push the part off. Because the fixed (cavity) half of the mold contains the "plumbing system" — elements called runners, sprues, and gates — used to transfer the melt to the mold, the ejector system is usually in the core (moving) half of the mold. The ejector system generally consists of pins which are used to push the part off the core. Careful examination of most injection molded parts will reveal the ejection pins marks — slight circular depressions about 3/16 inches in diameter. If a satisfactory flat part surface area does not exist to accommodate a pin, then a blade may be used to press against a narrow rib or part edge to eject the part.

A through-hole feature in the vertical wall such as the one shown in Figure 3.3(a) is referred to as an *external undercut*. To produce it requires a relatively costly side core to form the hole; the core is made to slide out of the way to per-

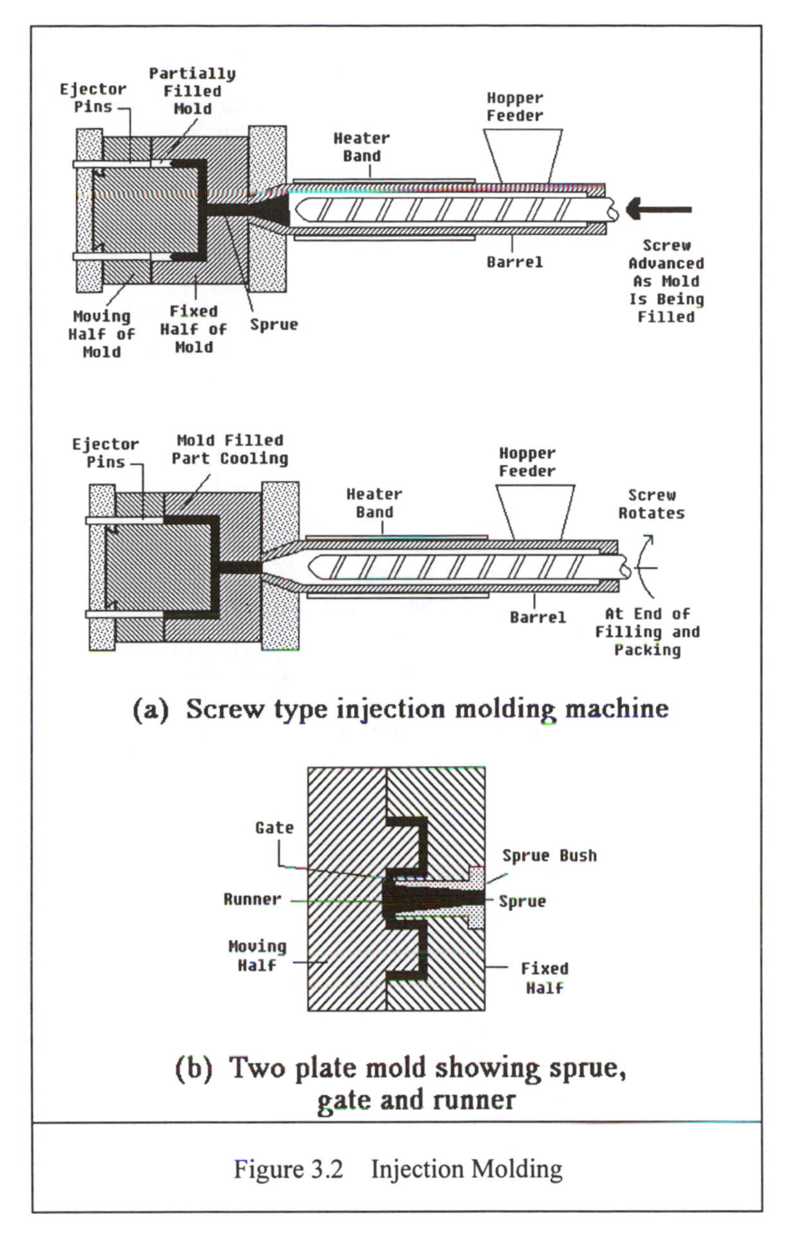

(a) Screw type injection molding machine

(b) Two plate mold showing sprue, gate and runner

Figure 3.2 Injection Molding

mit ejection of the part. In general, external undercuts are features that will, without special provisions, prevent the part from being extracted from the cavity half of the die.

An *internal undercut,* such as the one caused by the projection which exists on the inner wall of the boxed shaped part shown in Figure 3.3(b), is one which prevents the core mold half of the part from being extracted. In general, internal undercuts require even more costly molds than external undercuts.

Undercuts and their effect on tooling and processing costs for injection molded parts are discussed in more detail in Chapters 11 (for tooling) and 15 (for processing).

Per part processing time (or cycle time) for an injection molded part is primarily dependent on the time required for solidification, which accounts for about 70% of total cycle time. Solidification time in turn depends primarily on the thickness of the thickest wall. Typical solidification times for thermoplastic parts range from 15 seconds to about 60 seconds. Other part features that also influence cycle time are discussed in Chapter 15.

3.2.4 Compression Molding

Compression molding for forming thermoset materials uses molds similar to those for injection molding. The mold (Figure 3.4), mounted on a hydraulic press, is heated (by steam, electricity, or hot oil) to the required temperature. A slug of material, called a charge, is placed in the heated cavity where it softens and becomes plastic. The mold is then closed and the slug is subjected to pressures from 350 kPa (50 psi) to 80,000 kPa (12,000 psi), forcing the slug to take the shape of the mold.

The mold remains closed under pressure until the part hardens (cures). The mold is then opened, the part removed, and the cycle repeated.

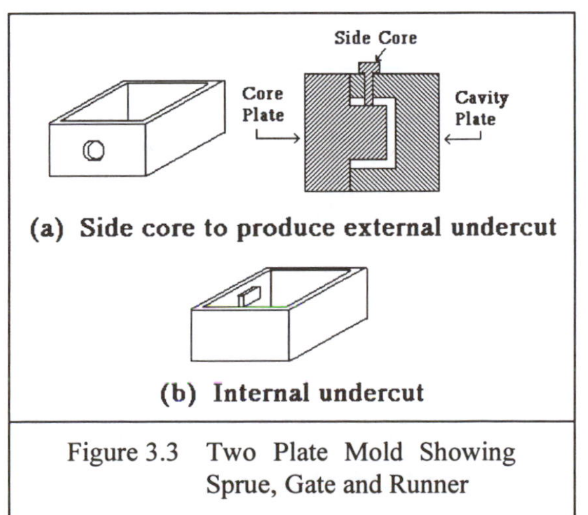

(a) Side core to produce external undercut

(b) Internal undercut

Figure 3.3 Two Plate Mold Showing Sprue, Gate and Runner

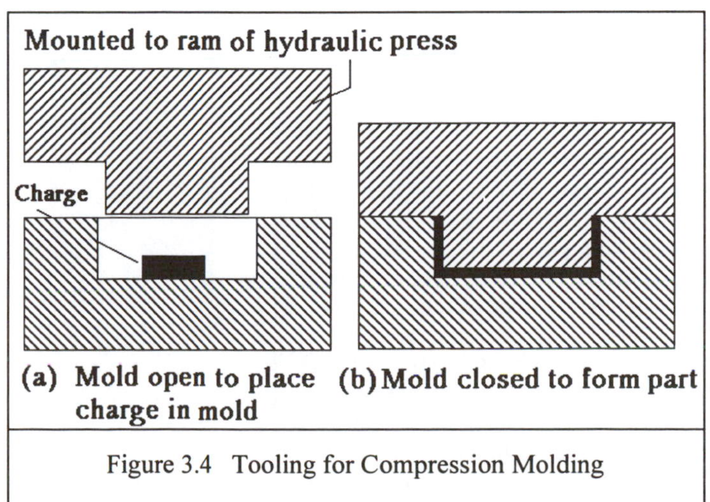

(a) Mold open to place charge in mold

(b) Mold closed to form part

Figure 3.4 Tooling for Compression Molding

The cure time can be as low as 20 seconds for small, thin-walled parts, from 1-3 minutes for larger parts, and as long as 24 hours for massive, thick-walled parts such as an aerospace rocket nozzle.

Compression molded parts may have external undercuts, but as in injection molding, undercuts increase tooling cost and should be avoided if possible. Compression molds, however, are somewhat simpler than injection molds since the compression molds do not need a sprue, runner and gates to feed and distribute the melt.

3.2.5 Transfer Molding

The main difference between compression molding and transfer molding is in the mold. In a transfer mold (Figure 3.5), the upper portion of the mold contains a pot where a slug (or charge) is placed, heated and melted. After the charge is melted, the mold is closed forcing the liquid resin through a sprue into the lower portion of the mold. The melt then takes the shape of the mold, hardens, and is removed.

3.2.6 Extrusion

In plastic extrusion, thermoplastic pellets are placed into a hopper which feeds a long cylinder (called a barrel) containing a rotating screw. The screw transports the pellets into a heated portion of the barrel where the pellets are melted and mixed. The resulting melt is then forced through a die hole of desired shape to form long parts of uniform cross section (Figure 3.6) such as tubes, rods, molding, sheets and other regular or irregular profiles. Short parts with uniform cross sectional shapes can also be produced by injection molding as well as extrusion, but the longer the part, the more advantageous the use of extrusion.

Shapes formed by extrusion can be subjected to post processing techniques by passing them through rollers (Figure 3.7) or stationary blades or formers that modify the shape of the (still hot and soft) extrusion. Sheets, for example, can be embossed to form patterns on them.

3.2.7 Extrusion Blow Molding

Extrusion blow molding is used to form hollow thermoplastic objects (especially bottles and containers). The process (Figure 3.8) takes a thin walled tube called a *parison* formed by extrusion, entraps it between two halves of a larger diameter mold, and then expands it by blowing air (at about 100 psi) into the tube forcing the parison out against the mold. The outside of the thin walled part takes the shape of the inside of the mold. By controlling variations in the parison thickness along its length, the wall thickness of the final part can be approximately controlled.

In addition to bottles and containers, blow molding is used to form such shapes as balls, light-weight baseball bats, dolls and animal toys. Although items like carrying cases for instruments and tools, large drums, ducts and automobile glove compartments can also be made by blow molding, this process is not usually used to produce such "engineering" type parts. No further consideration is given to blow molding in this book. For more on this subject we recommend references [4] and [5].

3.2.8 Other Polymer Processes

We have described only the most commonly used polymer processing techniques. Others exist — examples are calendering, foam processing, and rotational molding — but these processes are used for specialized or low production runs. For example, calendering is used to produce film and sheeting, foam processing for disposable cups and food containers, and rotational molding for battery cases and for very large parts.

For the production of shallow-shaped components, such as bus panels, boats, camper tops, lighting panels, trays, door and furniture panels, etc., a process called thermoforming is often used. Thermoforming involves heating to soften a thermoplastic sheet, clamping it over a mold, and then drawing and forcing it (via air pressure or vacuum) so that it takes the shape of the mold. References [4] and [5] contain detailed description of these, as well as some other, polymer processing techniques.

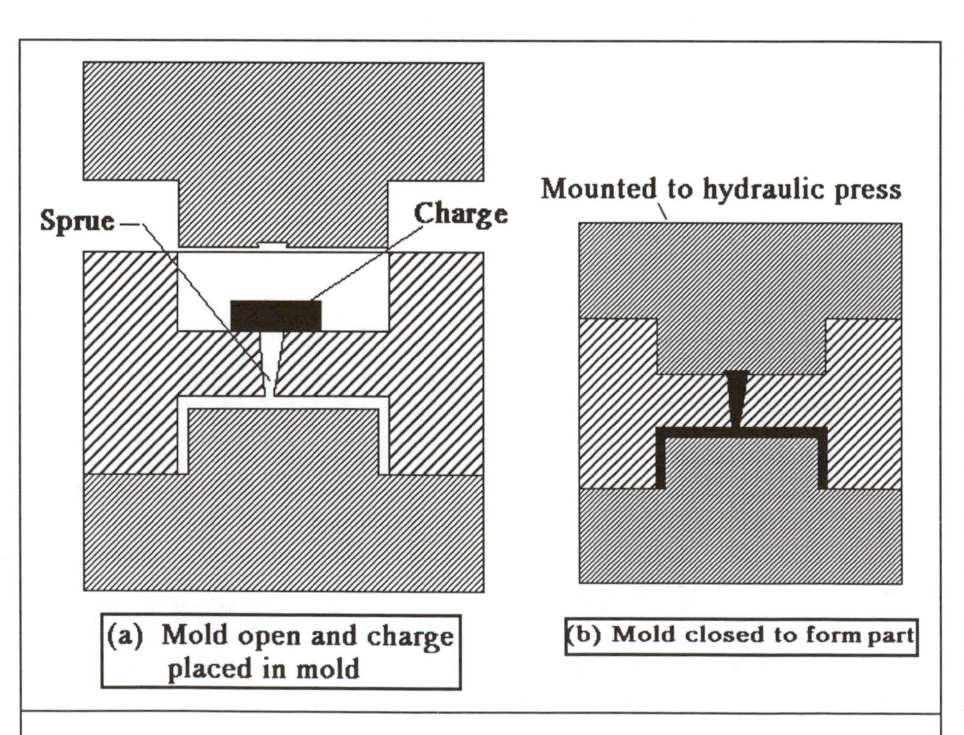

Figure 3.5 Tooling for Transfer Molding

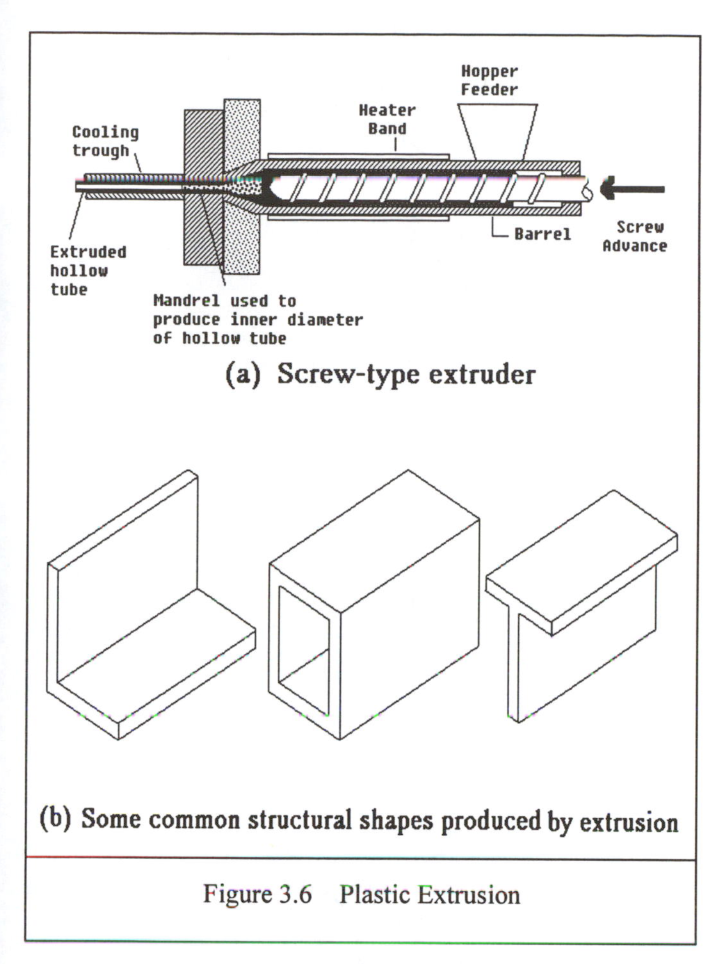

Figure 3.6 Plastic Extrusion

3.3 Metal Casting Processes

3.3.1 Introduction

As with polymer processing, there are also a number of metal casting processes. While there are distinct differences among these processes, there are also many common characteristics. For example, in all casting processes, a metal alloy is melted and then poured or forced into a mold where it is allowed to solidify before it is removed from the mold.

During solidification, most metals shrink (grey cast iron is an exception) so molds must be made slightly oversize in order to accommodate the shrinkage and still achieve the desired final dimensions.

The most common metal casting processes are sand casting, investment casting, and die casting. These are described in Sections 3.3.2 and 3.3.3. Other casting processes are briefly described in Section 3.3.4. The tolerances and surface finishes achievable are also different. For more information on casting processes and technology, References [1] and [6] should be consulted.

3.3.2 Sand Castings

Sand casting (Figure 3.9) is a process in which a sand mold is formed by packing sand around a wood or metal pattern which has the same external shape as the part to be cast. The pattern typically comes in two halves; each half is placed in a molding box with a top half (called a cope) and a bottom half (called a drag). After sand is packed around the pattern, holes used to pour the molten metal into the mold, and vents used to allow the escape of gases from the melt, are formed in the sand. Then the pattern is removed.

If the casting is to be a shape such as a thin walled cylinder, a sand core is placed in position so that the melt can not fill what is to be the open portion of the casting (e.g., the inside of the cylinder). The cope half of the mold box is then placed on top of the drag half. The melt is then poured, and the casting left to solidify. Once the casting has cooled, the sand mold is destroyed and the casting removed.

Sand castings are typically used to produce large parts such as machine-tool bases and components, structures, large housings, engine blocks, transmission cases, connecting rods, and other large components which, because of their size, cannot be cast by other processes. While almost any metal that can be melted can be sand cast, sand castings have (as a result of the sand mold) a grainy surface with large dimensional variations. Thus, sand castings often require local finish machining operations in order to obtain the necessary surface finishes and dimensional tolerances. Sand castings are not generally used for the production of parts which require high production volumes. They are sometimes used to produce prototype parts.

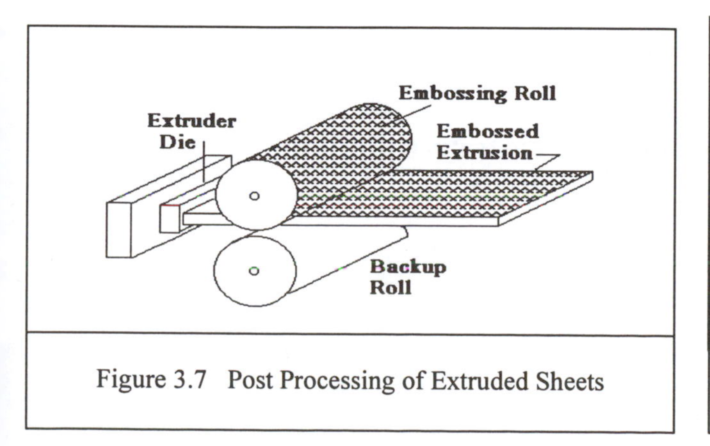

Figure 3.7 Post Processing of Extruded Sheets

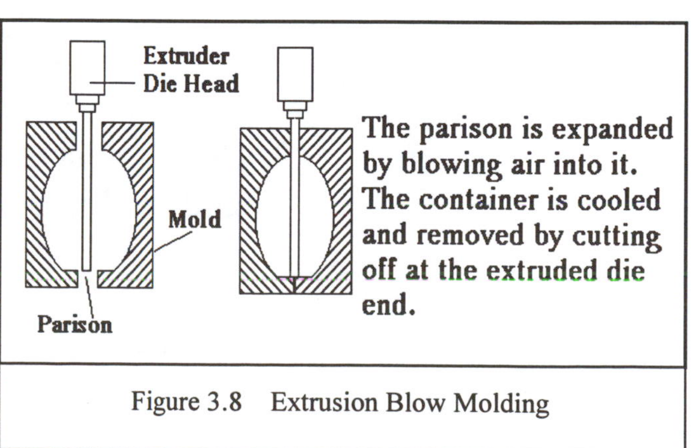

Figure 3.8 Extrusion Blow Molding

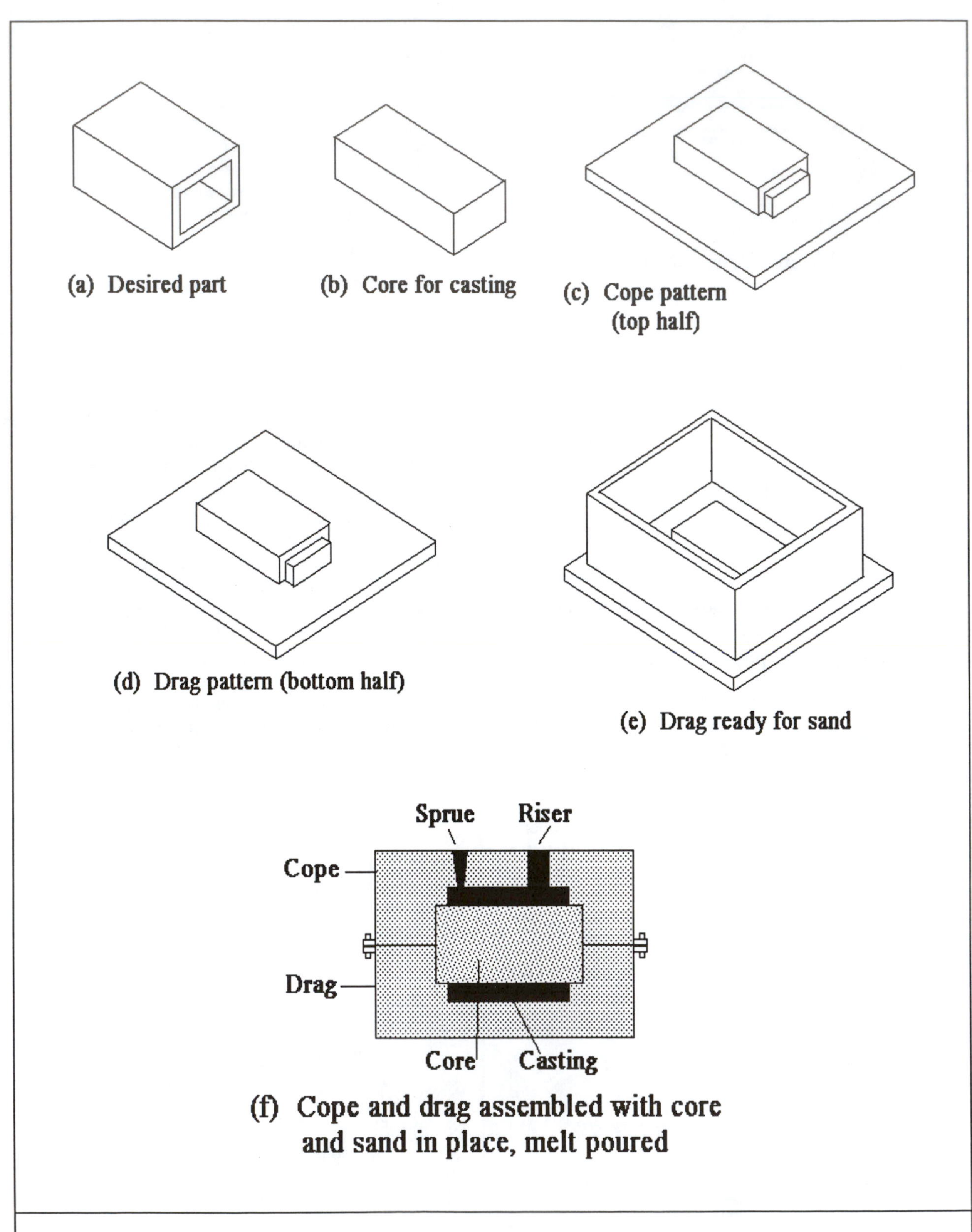

Figure 3.9 Sand Casting Process

3.3.3 Investment and Die Casting

Investment casting and die casting can produce parts of similar geometric shapes and size. In general, features that are difficult to die cast (e.g., undercuts) are also difficult to investment cast.

Investment casting is typically used when low production volumes are expected (e.g., less than 10,000 pieces), whereas die casting tends to be used when high production volumes are expected.

Investment cast parts can be made of steel, including stainless steel. Die casting, because of the fact that steel dies are used to produce the castings, must be restricted to metals with relatively low melting temperatures — primarily zinc and aluminum.

Investment Casting (Lost Wax Process). In investment casting (Figure 3.10), a metal die or mold is made by either machining or casting. The more complicated the shape (with undercuts, for example), the more costly the die.

After the mold is formed, wax is injected to form a pattern. The external shape of the wax pattern resembles the internal shape of the mold. The wax pattern is removed from the mold and coated with a slurry such as plaster of Paris or a very fine silica in several layers. Several wax patterns can be attached to the runners, gates, etc. that will feed and distribute the molten metal. The completed mold is placed in an oven and the wax removed by melting and evaporation.

To make parts, the mold cavity is filled with molten metal which is allowed to solidify. When the part has cooled, the mold is destroyed and the part removed. The tolerances and surface finishes achievable by investment casting are such that machining is not generally required.

Die Casting. Like injection molding, in die casting a melt is injected under pressure into a metal mold. The melt then cools and solidifies conforming to the internal shape of the mold. As in injection molding, as the part geometry becomes more complex, the cost of the mold increases. Also, as the wall thickness increases, the cycle time required also increases. Because of the difficulty of flash removal, internal undercuts are not generally die cast. Nevertheless, both injection molding and die casting can economically produce parts of great complexity.

There are two types of die casting machines: a hot chamber machine (Figure 3.11a) and a cold chamber machine (Figure 3.11b). In a hot chamber machine, the injection mechanism is submerged in the molten metal. Because the plunger is submerged in the molten metal, only alloys such as zinc, tin, and lead (which do not chemically attack or erode the submerged injection system) can be used. Aluminum and copper alloys are not suitable for hot chamber machines.

When the die is opened and the plunger retracted, the

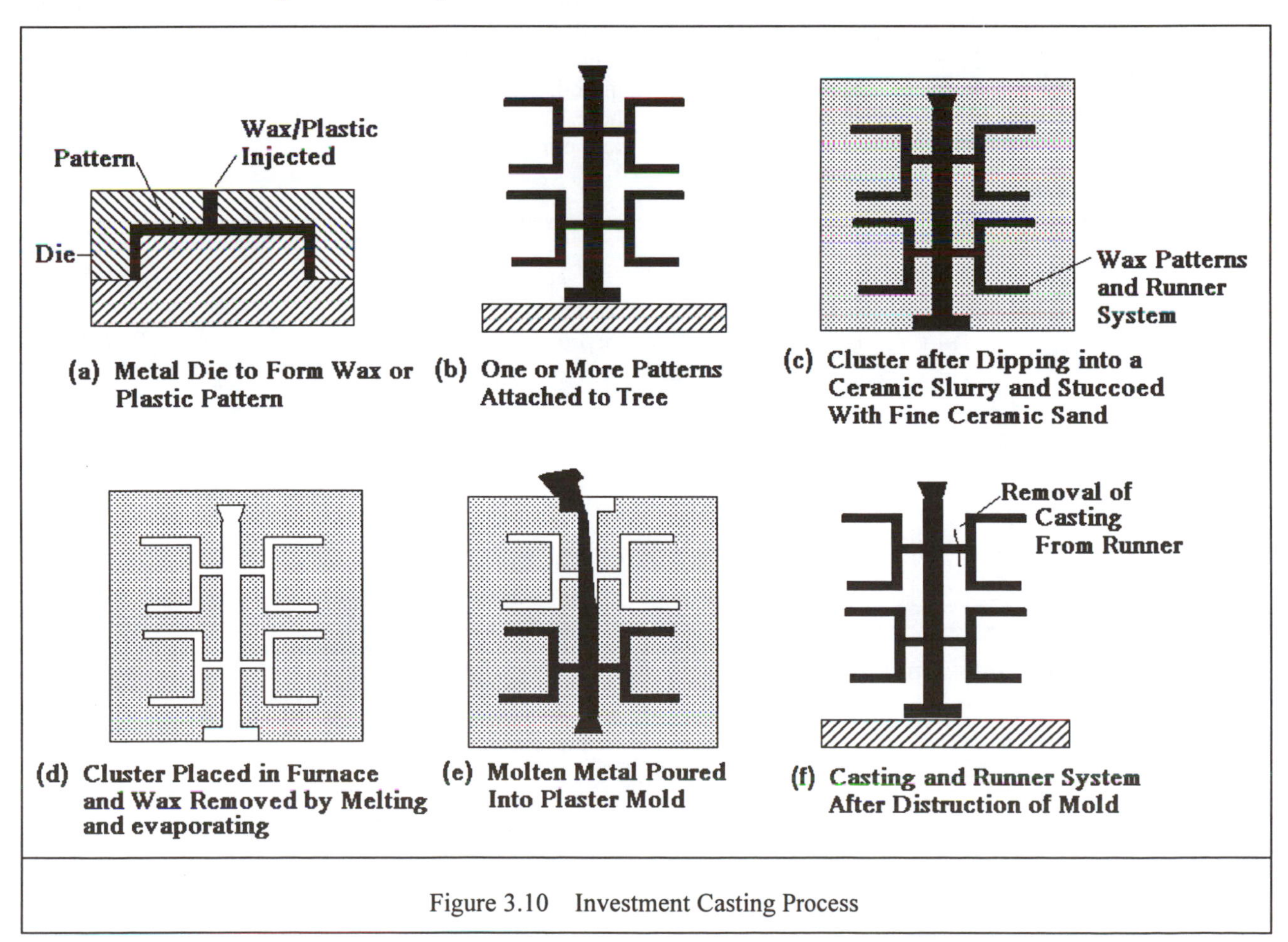

Figure 3.10 Investment Casting Process

molten metal flows into the pressure chamber (gooseneck). After the mold (die) is closed, the hydraulic cylinder is actuated and the plunger forces the melt into the die at pressures between 14 and 28 MPa (2,000 - 4,000 psi). After the melt solidifies, the die is opened, the part ejected, and the cycle repeated.

Because the higher temperatures used in casting aluminum and copper alloys shorten the life of hot chamber machines, cold chamber machines are often used (Figure 3.11b). In a cold chamber machine, molten metal from a separate holding furnace is ladled into the cold chamber sleeve after the mold is closed. The melt is then forced into the mold, and after solidification the mold is opened and the part ejected. Injection pressures in this type of machine usually range from 17 to 41 MPa (2,500 - 6,000 psi). Pressures as high as 138 MPa (20,000 psi) are possible.

Since the molds used in die casting are made of steel, only metals with relatively low melting points can be die cast. The vast majority of castings are made of either zinc alloys or aluminum alloys. Zinc alloys are used for most ornamental or decorative items; aluminum alloys are used for most nondecorative items.

3.3.4 Other Casting Processes

Die casting, investment casting and sand casting are the most commonly used casting processes. However, other casting processes, such as centrifugal casting, permanent mold casting, plaster mold casting, shell mold casting, and ceramic casting are also used.

In centrifugal casting, molten metal is poured into a mold revolving about a horizontal or vertical axis. Horizontal centrifugal casting is used to produce rotationally symmetric parts, such as pipes, tubes, bushings, etc. Vertical centrifugal casting is used to produce symmetrical as well as nonsymmetrical parts. However, since only a reasonable amount of imbalance can be tolerated for a nonsymmetrical part, the most common shapes produced are cylinders and rotationally symmetric flanged parts. Centrifugal casting of metal produces a finer grain structure and thinner ribs and webs than can be achieved in ordinary static mold casting.

In permanent mold casting, also referred to as gravity die casting, molten metal flows by gravity into a reusable permanent mold made of two or more parts. The tolerances and surface finishes achievable by this process are not as good as those obtainable by "pressure" die casting. Gravity die casting accounts for less than 5% of all die castings produced.

In plaster mold casting, molds are made by coating a pattern with plaster and allowing it to harden. The pattern is then removed and the plaster mold baked.

Ceramic molding is similar to plaster molding. In ceramic molding a fine-grain slurry is poured over the pattern and allowed to set chemically.

Shell mold casting is a process in which an expendable mold is formed by pouring resin-coated sand onto a heated pattern. The sand bonds to form a hardened shell corresponding to the outer shape of the pattern. Two shell halves are put together to form the single use mold. This process is used for small parts requiring a finer tolerance than is obtainable via sand casting. If better tolerances and surface finishes are required, then investment casting and pressure die casting are necessary. Details concerning each of these processes can be found in [1][4].

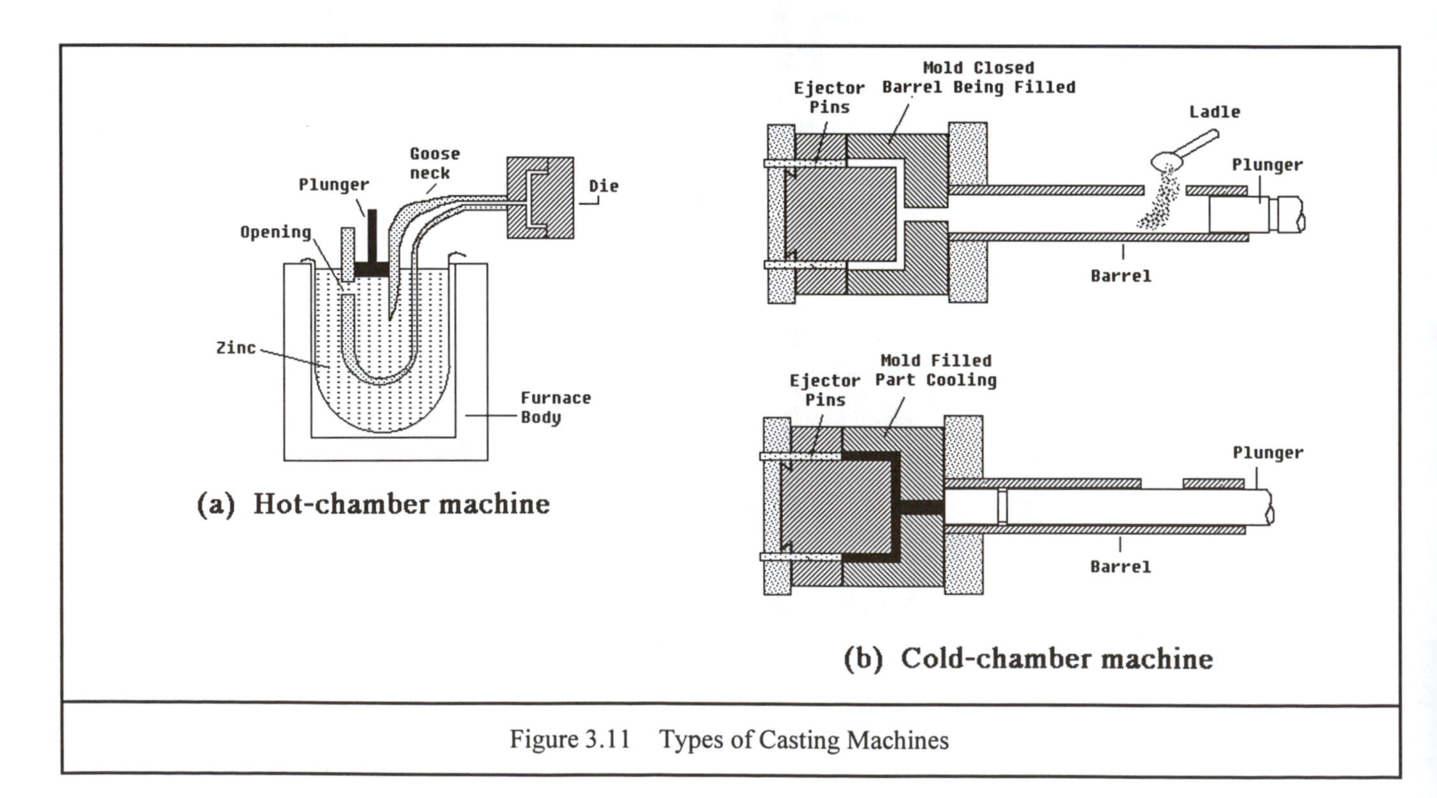

Figure 3.11 Types of Casting Machines

3.4 DFM Guidelines for Injection Molding, Compression Molding, Transfer Molding, and Die Casting

3.4.1 Qualitative DFM Guidelines

Injection molding, die casting, compression molding and transfer molding are all *internal flow processes, followed by cooling and solidification, followed by ejection from the mold.* That is, in each of these processes, a liquid (plastic resin or molten metal) flows into and fills a die cavity. Then the liquid is cooled to form a solid, and finally the part is ejected. The physical nature of these processes — flow, cooling to solidify, and ejection — provides the basis for a number of the qualitative DFM guidelines or rules of thumb that have been established. Parts should ideally be designed so that: (a) the flow can be smooth and fill the cavity evenly, (b) cooling, and hence solidification, can be rapid to shorten cycle time and uniform to reduce warpage, and (c) ejection can be accomplished with as little tooling complexity as possible.

To design parts properly for these manufacturing processes, designers must at least understand the meaning of (1) *mold closure direction* and (2) *parting surface*. Dies are made in two parts forming a cavity that is very close to the shape of the part. (The cavity may be slightly different from the part to allow for shrinkage and warping.) There is a closure direction for the die halves, and a parting "surface" (not necessarily planar) created where the die sections meet. The location of the parting surface, the direction of closure, and the design of the part must be considered simultaneously to provide for ejection of the part after solidification.

Knowing the mold closure direction enables designers to recognize and possibly avoid designing unnecessary undercuts. A fairly easy way to identify a potential undercut is to consider the shadows that would be created on the part from a light shining in the mold closure direction. If a part casts shadows onto itself, then the feature causing the shadow is an undercut. This is discussed again in Chapter 11.

Figures 3.12(a) and 3.12(b) illustrate how the choice of mold closure direction and location of the parting surface influence design and in particular tool design and tool cost.

With knowledge of the mold closure direction and location of the parting surface — and keeping in mind that the material should flow smoothly into and through the mold, solidify rapidly and uniformly, and then be easily ejected — designers can understand and make good use of the following DFM guidelines:

1. In designing parts to be made by injection molding, die casting, compression molding and transfer molding, designers must decide — *as a part of their design* — the direction of mold closure and the location of the parting surface. Though these decisions are tentative, and advice should be sought from a manufacturing expert, it is really impossible to do much design for manufacturing in these processes without considering the mold closure direction and parting surface location.

2. An easy to manufacture part must be easily ejected from the die, and dies will be less expensive if they do not require special moving parts (such as side cores) that must be activated in order to allow parts to be ejected. Since undercuts require side cores, parts without undercuts are less costly to mold and cast. Some examples of undercuts are shown in Figure 3.3. With knowledge of the mold closure direction and parting surface, designers can make tentative decisions about location(s) of features (holes, projections, etc.) in order to avoid undercuts wherever possible.

3. Because of the need for resin or metal to flow through the die cavity, parts that provide relatively smooth and easy internal flow paths with low flow resistance are desirable. For example, sharp corners and sudden changes or large differences in wall thickness should be avoided because they both create flow problems. Such features also make uniform cooling difficult.

4. Thick walls or heavy sections will slow the cooling process. This is especially true with plastic molding processes since plastic is a poor thermal conductor. Thus, parts with no thick walls or other thick sections are less costly to produce.

5. In addition, every effort should be made to design parts of uniform, or nearly uniform, wall thickness. If there are both thick and thin sections in a part, solidification may proceed unevenly causing difficult to control internal stresses and warping. Remember, too, that solidification time, and hence total cycle time, is largely determined by the thickest section.

6. We do not discuss gate location in this book except in this paragraph. However, in large or complex parts, two or more gates may be required through which resin or metal will flow in two or more streams into the mold. There will therefore be fusion lines in the part where the streams meet inside the mold. The line of fusion may be a weak region, and it may also be visible. Therefore, designers who suspect that multiple gates may be needed for a part should discuss these issues with manufacturing experts as early as possible in the design process. With proper design and planning, the location of the fusion lines can usually be controlled as needed for appearance and functionality.

These DFM rules are not absolute, rigorous rules. Note, for example, how the molded-in very thin hinge in a computer disk carrying case (Figure 3.13) violates the general thrust of the fifth rule above. Reference [8] contains an interesting discussion of how to design these hinges so that failure does not occur. If there are designs that have great advantage for function or marketing, then those designs can be given special consideration. Manufacturing engineers can sometimes solve the problems that may be associated with highly desirable functional but difficult to manufacture designs at a cost low enough to justify the benefit. However, relatively easy to manufacture designs should always be sought. More often than not, a design can be found that will be *both* efficient from a functional viewpoint *and* relatively easy to manufacture.

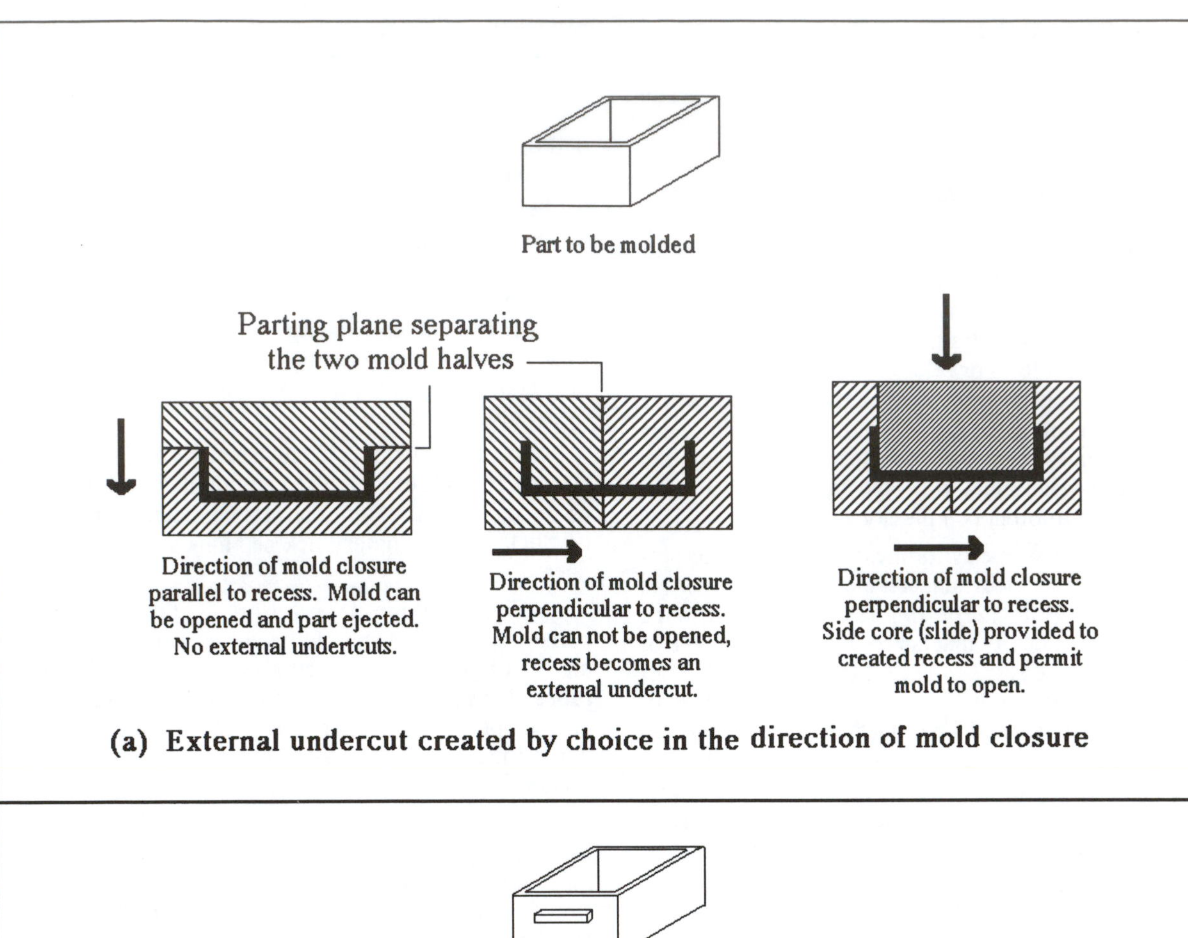

(a) External undercut created by choice in the direction of mold closure

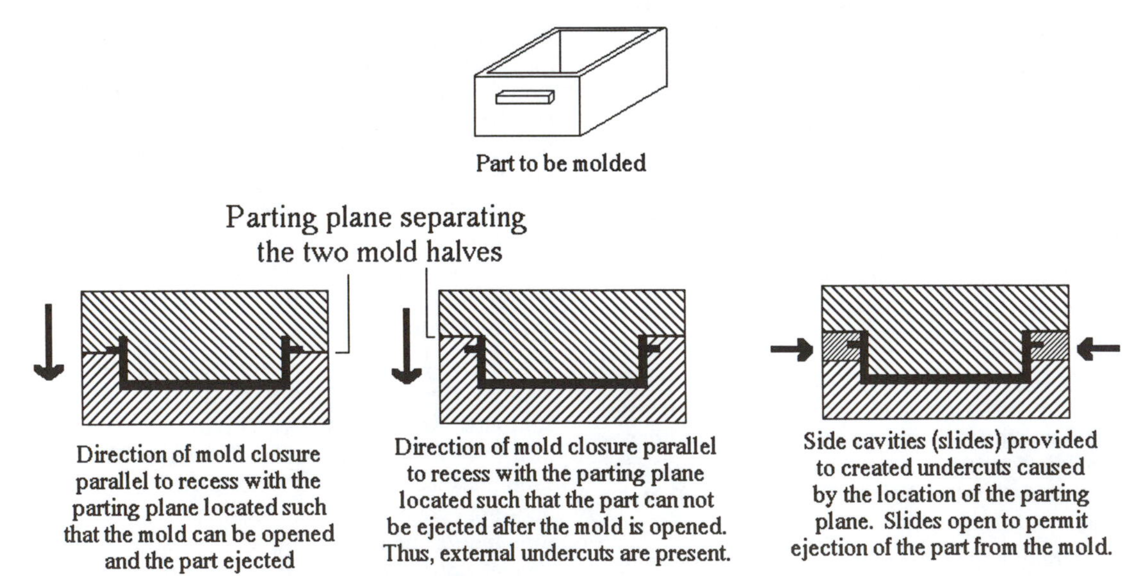

(b) External undercuts created by location of the parting planes between the die halves.

Figure 3.12 Mold Closure Direction and Parting Surface Influence the Design of Parts

3.4.2 Preliminary DFM Advisors

In later chapters (Chapters 11, 12, 15 and 16) we will present detailed methodologies for evaluating the manufacturability of injection moldings, die castings and stampings at both the configuration and parametric stages of design. It is useful at this time, however, to present a condensed, less detailed version of these methods in order to assist designers in performing quick evaluations of the manufacturability of a design during the early stages of design.

Figures 3.14 and 3.15 are called the *Injection Molding DFM Advisor* and the *Die Casting DFM Advisor,* respectively. Note that the origin of the matrices is in the upper left corner.

The wall thickness of the thickest section of a part to be evaluated is listed across the top of the *Advisor* increasing in the horizontal direction. Processing costs thus increase in the horizontal direction within the matrix. Also along the top, but beneath the values for wall thickness, the numbers in the small boxes indicate an approximate processing cycle time relative to the time to process a 1 mm (0.04") thick washer. These cycle times apply to all parts found in the same vertical column.

Part characteristics such as undercuts and basic shape are indicated along the vertical axis of the matrices, with part complexity increasing in the downward direction. Within the matrix, the small rectangular inset boxes indicate the dollar cost of tooling relative to the tooling costs for a simple, flat washer. The numbers show that tooling costs increase in the downward vertical direction.

In other words, as a part design moves from the top left corner to the bottom right corner, tooling and processing costs increase.

The shading in the figures provides a qualitative indication of combined tooling and processing costs expressed as Good, Fair, Poor, or Most Costly. Examples of what parts might look like in the various design categories are shown in the right hand side of Figures 3.14 and 3.15. A discussion of the knowledge and data acquisition procedures used to produce these figures is contained in [7].

As an example of the use of these Advisors, consider an injection molded part that is more or less box shaped with a maximum wall thickness of 2.75 mm, and has features creating three external undercuts. The *Injection Molding DFM Advisor* indicates an approximate processing time relative to a 1 mm flat washer of 2.4. That is, it will have 2.4 times more processing cost than the washer. The tooling cost relative to a flat washer for this part is about 2.2. That is, tooling will cost about 2.2 times more than tooling for the washer.

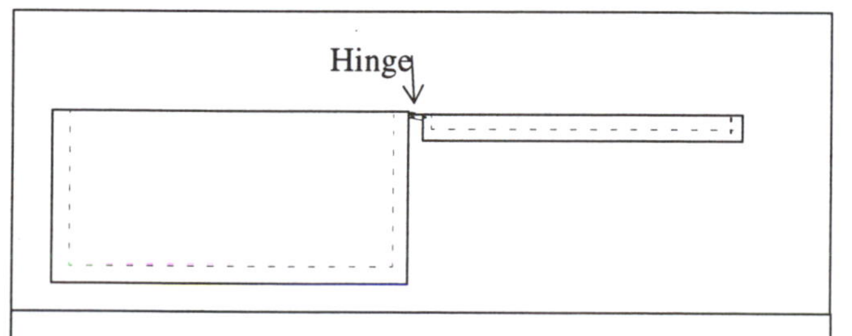

Figure 3.13 The Living Hinge Concept on One Type of Computer Disk Carrying Case

The shading in the box where the example is found indicates a relatively Poor design from the viewpoint of manufacturability. Though the design may be not be changeable, the Poor rating suggests that some effort be made to do so. Note that eliminating even one external undercut would cut tooling costs by about 10 percent, and changes the evaluation to Fair.

3.5 Stamping

3.5.1 Metals and Processes

From your previous study of materials you are probably aware that properties of metal alloys are accompanied by such terms as *cast* and *wrought.* Cast metals are those metals which have been formed by casting. Wrought metals are those which have been formed by rolling (Figure 3.16a), drawing (Figure 3.16b), extruding (Figure 3.16c), or forging (Figure 3.16d).

The previous section discussed various casting processes. In this section we will be discussing a process called stamping in which wrought metals, formed by rolling, are used to produce thin walled parts. In Sections 3.6 and 3.7 we will discuss forging and extruding.

More detailed information concerning rolling and drawing can be found in [1].

3.5.2 The Stamping Process

Stamping is a process in which thin walled metal parts (less than about 6.25 mm or 0.25 inches) are shaped by means of punches and dies driven by mechanical or hydraulic presses. Examples of parts made by stamping are can openers, fan blades, pulleys, ash trays, razor blades, buckles, kitchen utensils, cans, bottle caps, range tops, etc. Stamping is also used to produce a wide variety of parts for machines, power tools, appliances, automobiles, hardware, office equipment, electrical equipment, and clothing.

Stamping can be divided into two broad categories of press operations: (1) shearing or cutting, and (2) bending or forming.

Shearing is carried out by cutting a sheet that has been placed between a sharp punch and sharp die (Figure 3.17). The die imposes a rapid, high shearing stress to the sheet. Shearing can be used for several purposes: (a) to produce blanks, which are then subjected to further shearing, bending and/or forming; (b) to produce features such as holes, slots, notches, perforations, and lances; (c) to separate and remove a part from a strip of sheet metal; and (d) to trim and slit a part for size control. To be certain that shearing, and not drawing or bending, occurs, the clearance between the die and punch must be less than the sheet thickness.

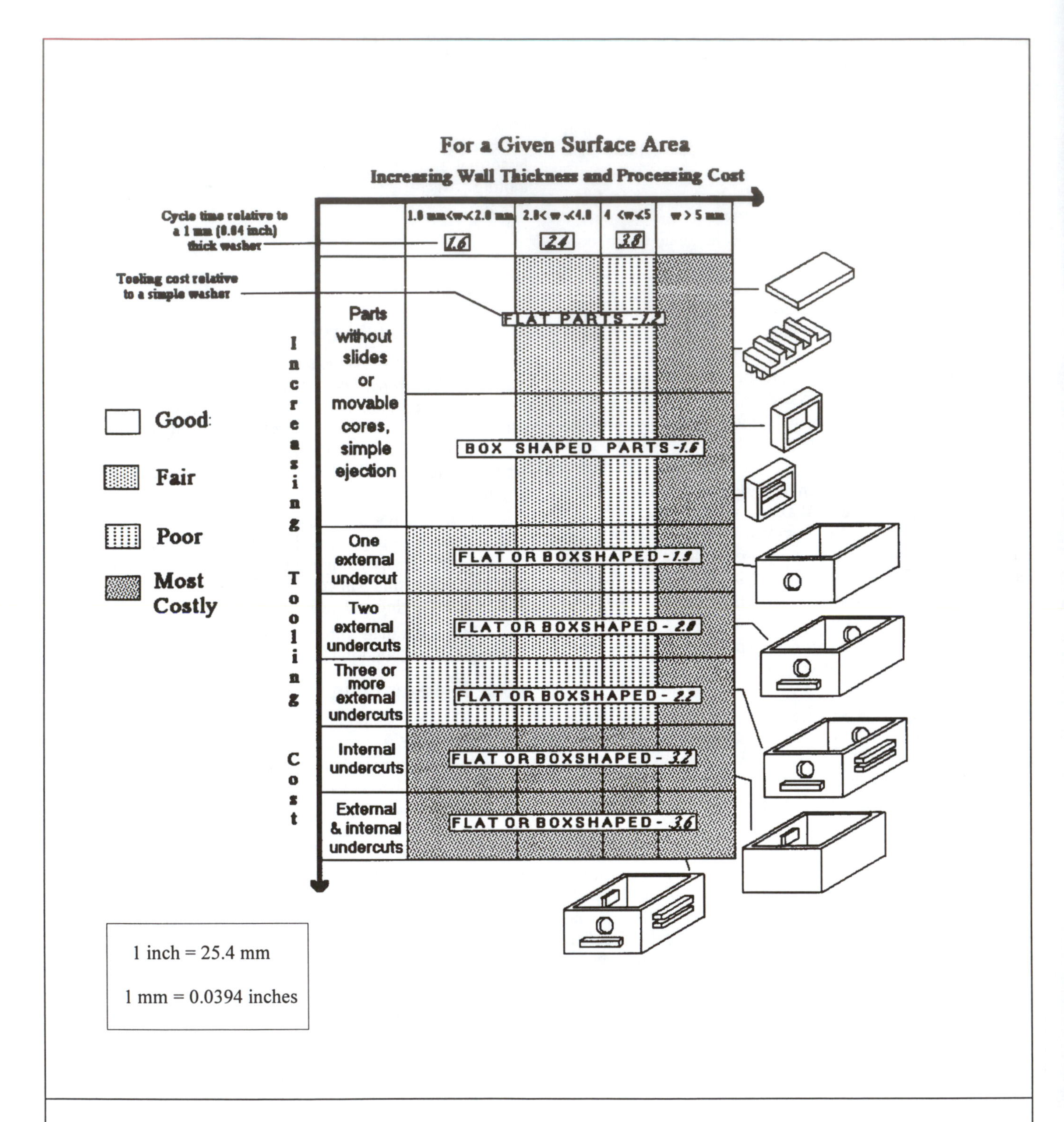

Figure 3.14 Injection Molding DFM Advisor

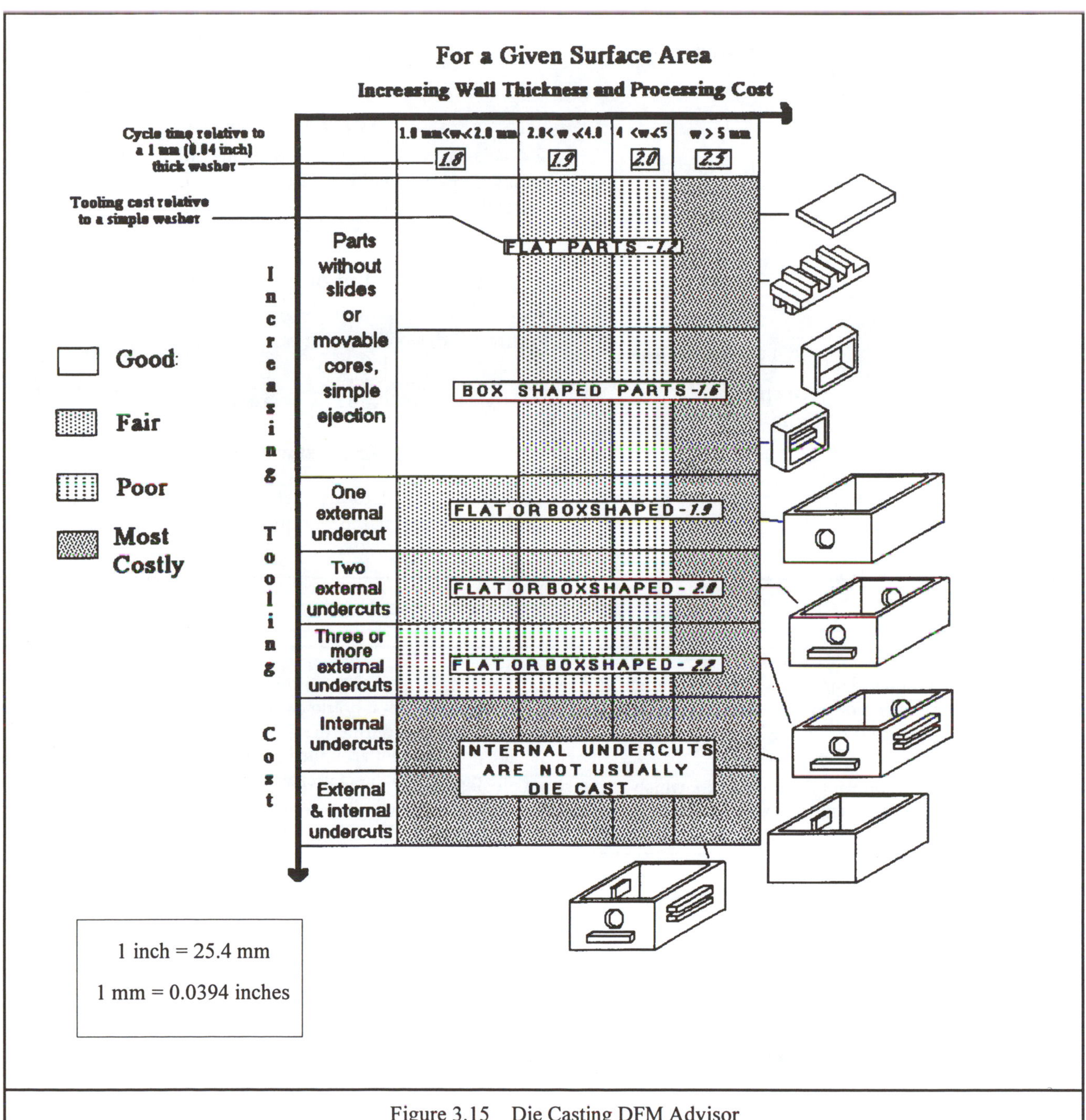

Figure 3.15 Die Casting DFM Advisor

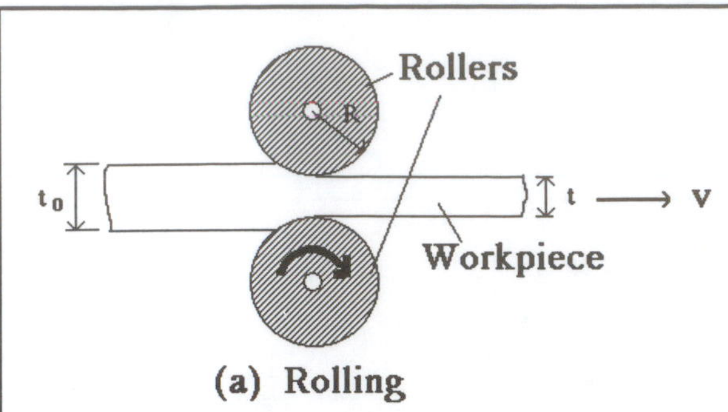

(a) Rolling

Rolling can produce large reductions in thickness of sheets, and can form metal strips into profiles. Large reductions are usually done hot; cold rolling produces better finish and tolerances.

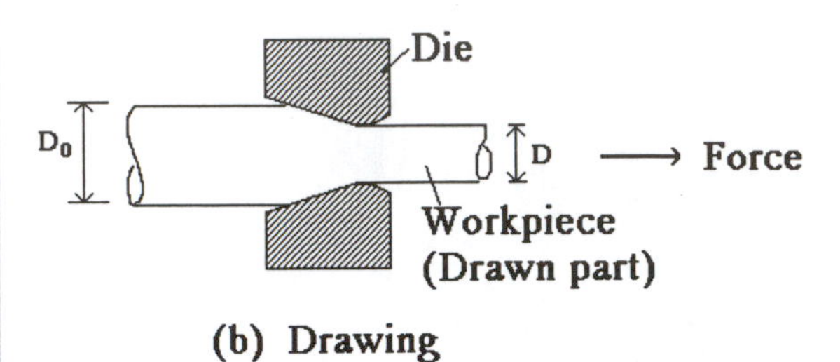

(b) Drawing

Drawing reduces the diameter of a wire, bar, or tube by pulling it through a die.

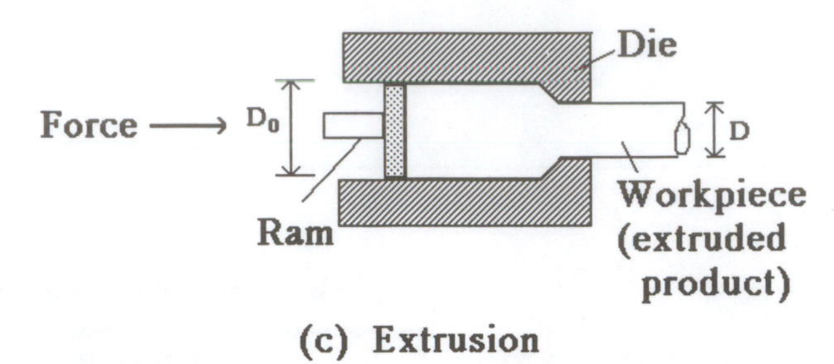

(c) Extrusion

Extrusion converts a cast or wrought billet into a long prismatic part of uniform cross section by pushing the metal through a die.

Dies mounted on press or hammer

Die Closure

Upper Die

Lower Die

(d) Impression die forging

Forging plastically deforms metal, usually hot, into desired shapes by compressing the workpiece between two dies.

Figure 3.16 Bulk Deformation Processes [1][2]

Bending is done using a matched punch and die set (Figure 3.18a) or, as is more commonly the case, by using a descending punch to fold or "wipe" the workpiece over the edge of the die (Figure 3.18b).

Drawing is a forming operation in which a punch draws a thin flat sheet into a die to form a recessed cup or box.

Punch Plate

(holds punches)

Piercing Punch

(pierces holes into strip)

Strip

Scrap

Die Block

Figure 3.17 The Shearing Process

Drawing is used to form such products as pots and pans, beverage containers, bottle caps, fire extinguishers, etc. To prevent shearing during the drawing process, the punch and dies are provided with rounded corners, and the clearance between the punch and die is greater than the sheet thickness.

Embossing is a forming operation used to form beads, ribs and lettering by use of shallow indentations (draws).

While other sheet metal forming operations (such as spinning, ironing, and others) exist, from the point of view of providing features to a consumer product component, the above stamping operations are the ones most commonly encountered.

3.5.3 Stamping Dies

A stamping die is a collection of parts mounted on a press used to produce a part. While each die is custom-made to produce a given part, many die components are identical or similar. In this section we concentrate on dies; in the next section we say more about presses. Most stampings require a series of operations. These operations can be carried out using a single die, called a *progressive die*, or on *compound* or *combination* dies. Progressive dies are divided into sections (called stations) that perform multiple operations with each

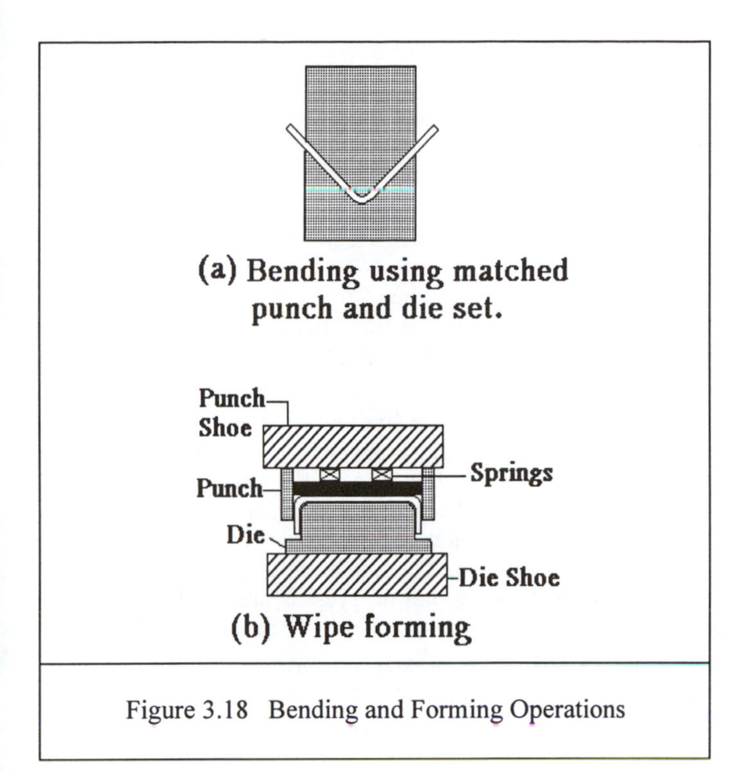

Figure 3.18 Bending and Forming Operations

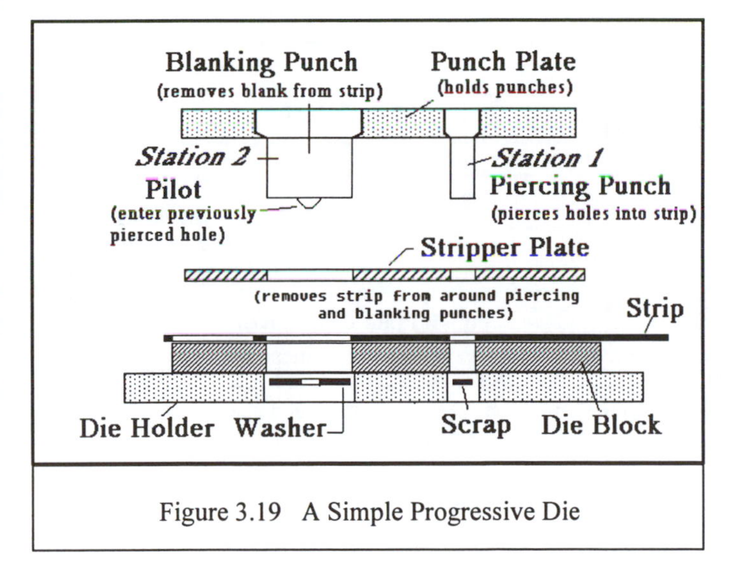

Figure 3.19 A Simple Progressive Die

stroke of the press (Figure 3.19). Compound or combination dies perform two or more operations (one inside of the other) with each stroke of the press. Figure 3.19 shows a two station progressive die used for producing a washer; Figure 3.20 shows a compound die used to produce the same washer. If a part cannot be produced using either a single or a combination die, then the part can be produced using a series of presses containing compound or combination dies. In this case, a partially completed stamping is transferred from workstation (press and die combination) to workstation having one or two operations performed on it at each station.

Most stamping operations carried out on parts less than 200 mm in size use *progressive dies.* Figure 3.19 shows a two station progressive die to produce a flat washer. Figure 3.20 shows a compound die used to produce the same washer. If a part cannot be produced using either a single or a combination die, then it can be produced using a series of presses containing compound or combination dies. In this case, the first workstation pierces the inner hole; the second workstation performs a blanking operation. Large parts are not usually produced on progressive dies because the size of the die becomes exorbitantly large.

Compound dies are used primarily for flat blanks. The process essentially simultaneously performs piercing of the inside holes and features together with blanking of the outside periphery.

Figure 3.21 shows a typical die set consisting of a stationary lower plate (called the die holder) to which the dies are attached, a movable upper plate to which the punches are attached (called the punch holder), and two sets of precisely fitted guide posts and guide bushings. Such units can be commercially purchased in various sizes and shapes.

Figure 3.22 shows a simplified version of a progressive die assembly including the die set, punches, punch plate (which holds the punches and is attached to the punch holder), dies (in this case in the form of a die block) and a stripper plate for producing a link.

When the part geometry is such that the workpiece can not remain attached to the strip, transfer dies are used. With transfer dies, a blank is transferred from station to station by equipment built into or mounted on the press itself.

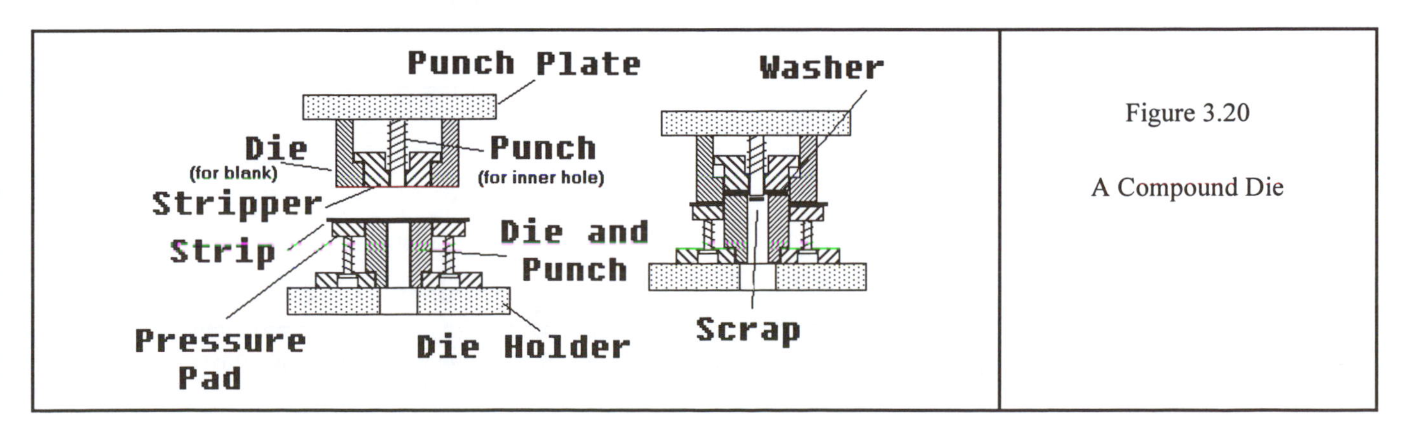

Figure 3.20

A Compound Die

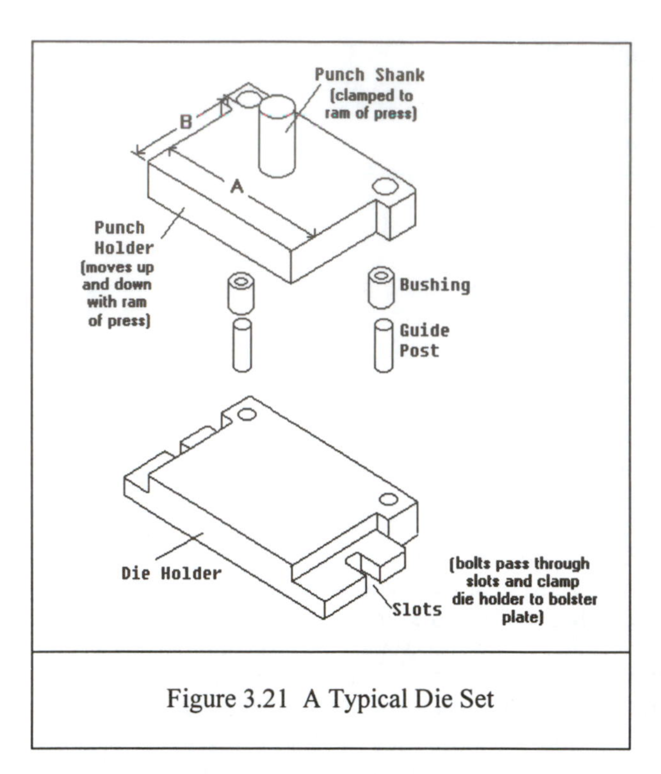

Figure 3.21 A Typical Die Set

As an example, imagine the link shown in Figure 3.22 with all its edges curled. In this case the curls can not be formed with the part attached to a strip. Hence, precut flat blanks which have been formed in another press, and which conform to the peripheral geometry of the part, are hand loaded into a magazine. A set of transfer arms removes the part from the magazine and places it into the first station of the die. The part is then transferred from station to station as successive operation are performed.

3.5.4 Stamping Presses

In order to appreciate the relative importance of tooling and processing costs for a stamped part, it is necessary for designers to understand the basic operations of the presses used.

All stamping presses contain a frame and bed, a ram or slide, a drive for the ram, and a power source and transmission. Power is supplied either mechanically or hydraulically [1].

Figure 3.23 shows a single-action (one ram) mechanical, open (C-frame) press. The bed, which forms the lower part of the frame, supports a bolster plate which in turn is used to support the die. T-slots in the top surface of the bolster plate are used when clamping the die to the plate. The ram holds the punch and, for a mechanical press, has a fixed stroke.

Two basic frame types are available for mechanical presses: straight-sided presses with closed frames and four posts (Figure 3.24), and gap or C-frames (Figure 3.23). Straight-sided presses are stiffer and deflect less than gap frame presses, but the dies are accessible only from the front or rear.

In most mechanical presses a flywheel is used to store the energy required for the operation. The flywheel runs continuously, engaged by a clutch only when a press stroke is needed. Slider crank mechanisms, gears, eccentric shafts and toggle mechanisms are used to convert the rotary motion of the flywheel to the straight line motion of the ram.

The source of power for hydraulic presses is a pressurized fluid, usually oil, acting against one or more pistons. Unlike mechanical presses, whose driving force varies with the length of the stroke, the full force of the hydraulically driven pistons can be provided over the entire length of the stroke. Thus, for operations such as deep drawing where the maximum force is required at the beginning of the draw, hydraulic presses are used. The stroke length and force of hydraulic presses can be easily changed.

However, hydraulic presses are slow compared to mechanical presses. A 150 ton press can easily operate at 40 strokes per minute. An equivalent hydraulic press rated at 100 tons can operate at only 12 strokes per minute.

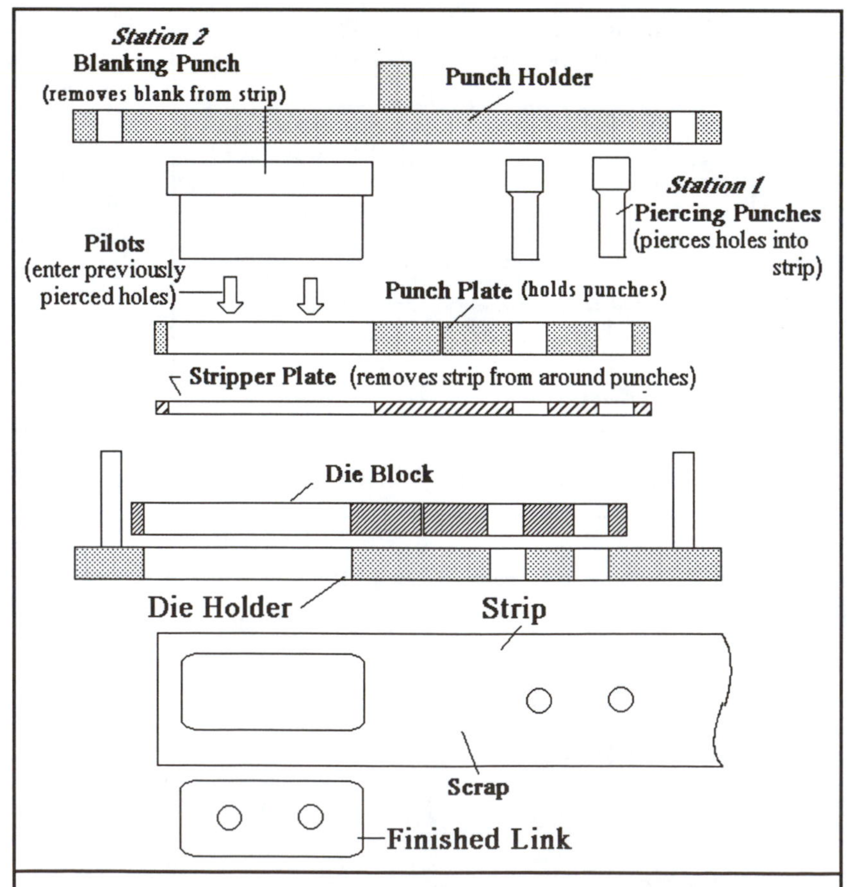

Figure 3.22 Die Assembly Producing a Link With a Progressive Die

Some presses are single action type, having only one ram, while others are double or triple action. A double action press has a ram within a ram. The outer ram descends first to seat the material. The majority of hydraulic presses used in drawing operations are the double action type.

Although hydraulic presses are less expensive than mechanical types, most stamping operations are carried out on high speed mechanical presses because they are much faster (24 strokes/min to 700 strokes/min). The cycle time per part is very low (less than 2.5 seconds), a value much less than the cycle time in injection molding or die casting (10s - 60 s). Consequently, the proportion of part cost due to processing is essentially negligible. At low and medium production volumes (10,000 - 100,000), most of the cost of a stamped part is due primarily to tooling. At very high production volumes (2,000,000), material cost is the dominating cost factor.

3.5.5 DFM Guidelines for Stamped Parts

Because of the short cycle times per part, tooling and material are generally the dominating cost factors for stampings. For example, at production volumes less than 100,000, the proportion of part cost due to tooling is about 75%, while the proportion of part cost due to processing is less than 2%. Even at production volumes of about 2,000,000, the proportion of part cost due to processing still remains low (less than 5%). At such large production volumes, the greatest proportion of part cost (about 75%) is due to material cost.

For large production volumes, designing parts to reduce the amount of scrap can be important. However, in most cases, it is the manufacturing process designer who designs the process (often referred to as developing the "strip layout" for the part) to achieve minimum scrap. Therefore, the discussion below is primarily about the relationship between part design and tooling cost.

Tooling cost is a function of tool construction cost and tool material costs. Both of these are functions of the size of the tool required. The size of the tool depends on:

Number of Features. In general, tools wear at different rates. That is, a punch used to produce a hole wears at a rate different from a punch used to produce an extruded hole or a tab. Thus, these tools must be removed and re-ground at different times. Hence, to facilitate their independent removal and maintenance from the die set, holes and extruded holes are not produced at the same station; instead one station is used to produce the holes and another station to produce the extruded holes. Obviously, as the number of distinct features increases, the number of die stations increases and both die construction costs and die material costs increase. Therefore:

The number of distinct features in a part should be kept to a minimum.

Spacing Between Features. If features are spaced closely together (less than three sheet thicknesses), then there may be insufficient clearance for the punches. Even if space permits, however, the die sections become thin making them susceptible to breakage, and punch breakage can occur due to metal deformation around any closely spaced features during piercing. As a result, two stations are required for each type of closely spaced feature; each station creates alternating features. Therefore:

Avoid closely spaced features when possible.

Narrow Cutouts and Narrow Projections. A link with a wide projection and a link with a wide cutout are shown in Figure 3.25(a); a link with a narrow projection is shown in Figure 3.25(b); and a link with a narrow cutout is shown in Figure 3.25(c). As in the case of a plain link (Figure 3.22), a link containing either a wide cutout or a wide projection can be blanked out (separated from the strip) at a single station. If a narrow projection is present, however, then to separate the link from the strip at one station would require a blanking punch with a narrow cutout along with a die containing a narrow projection. In this situation the narrow section of the die would be easily susceptible to damage. In addition, damage to the narrow groove in the punch would require replacement

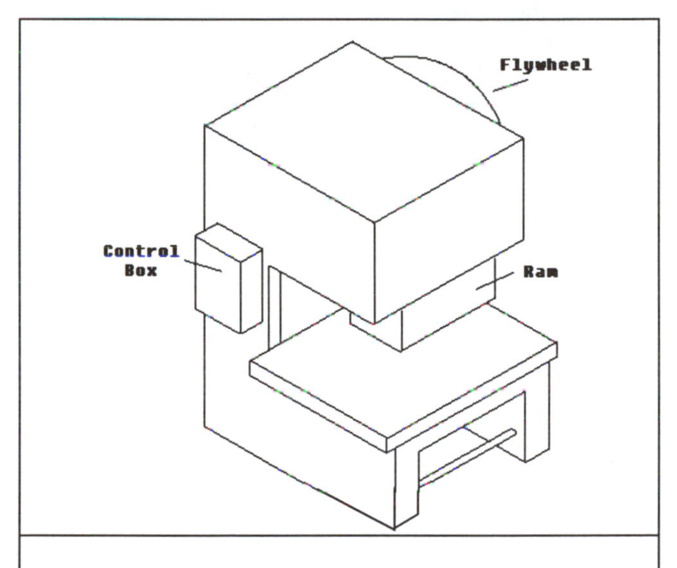

Figure 3.23 A Single-action Mechanical Press

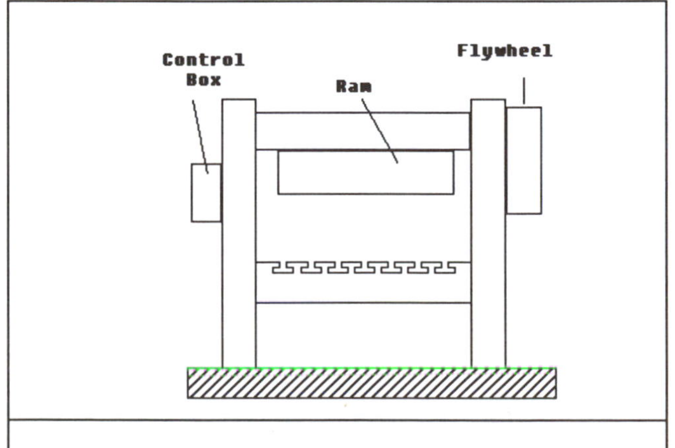

Figure 3.24 A Straight Sided Mechanical Press

of the entire blanking punch. For this reason the part is separated from the strip by first creating a notch to the left of the projection at one station [Figure 3.25(b)] and then creating the projection at the following station. Finally, the part is separated from the strip at a third station. Thus, in this case the addition of the narrow cutout to the link results in the addition of at least one and sometimes two additional stations.

In the case of the link with a narrow cutout, use of a single blanking tool requires the use of a narrow projection which is easily damaged. Damage to this tool requires replacement of the entire blanking tool. Thus, the narrow cutout is created at one station using a narrow punch, and the link is then separated from the strip at the following station. Thus:

The use of narrow cutouts and narrow projections should be avoided.

Bend Stages and Bend Directions. A U-shaped part, with both bends in the same direction, will have both bends created at the same time at one die station. An equivalent Z-shaped part, with bends in opposite directions, will require that each bend be separately created at two different die stations. Thus, bends in opposite directions create increased tool construction and die material costs. Therefore:

The number of bend stages in a part should be kept to a minimum.

In general, to keep the number of bend stages to a minimum, the number of bend axes should be minimized. A precise method for determining bend stages is described Chapter 12.

Overbends. To create a bend angle greater than 90° requires two die stations. At the first station the part is bent through 90°, while at the next station the part is bent past 90°. Clearly, then, parts with a bend angle greater than 90° require more costly tooling. Thus:

Bend angles greater than 90° should be avoided whenever possible.

Side-Action Features. Features, such as holes, whose shape and/or location from a bend line must be accurately located, must be created after bending. Once again, this requires that one or more additional die stations be added, thereby increasing tooling cost. Also, because the bends themselves cannot be closely controlled, they make poor reference points for important or close tolerances. Therefore:

Side-action features should be avoided or kept to a minimum.

In general, to keep side-action features to a minimum, the tolerances of feature dimensions that must be referred to bend lines should be generous.

The Stamping DFM Advisor

The effects of these rules on tooling costs are approximately summarized in the *Stamping DFM Advisor* shown in Figures 3.26a and 3.26b. Like Figures 3.14 and 3.15, these figures are condensations of more detailed matrices which appear later in Chapter 12. Reference [7] contains a discussion of the knowledge and data acquisition procedures used to produce these figures.

Figure 3.26(a) gives the relative tooling cost (i.e., with respect to a flat washer) for parts with localized features likes holes and ribs. Figure 3.26 (b) gives the relative tooling costs for parts with bends. The values in Figure 3.26b are added to those in Figure 3.26a in order to obtain the total relative tooling cost of the par

In Figure 3.26(a), the vertical axis categorizes the amount of die detail present (i.e., numbers of distinct features

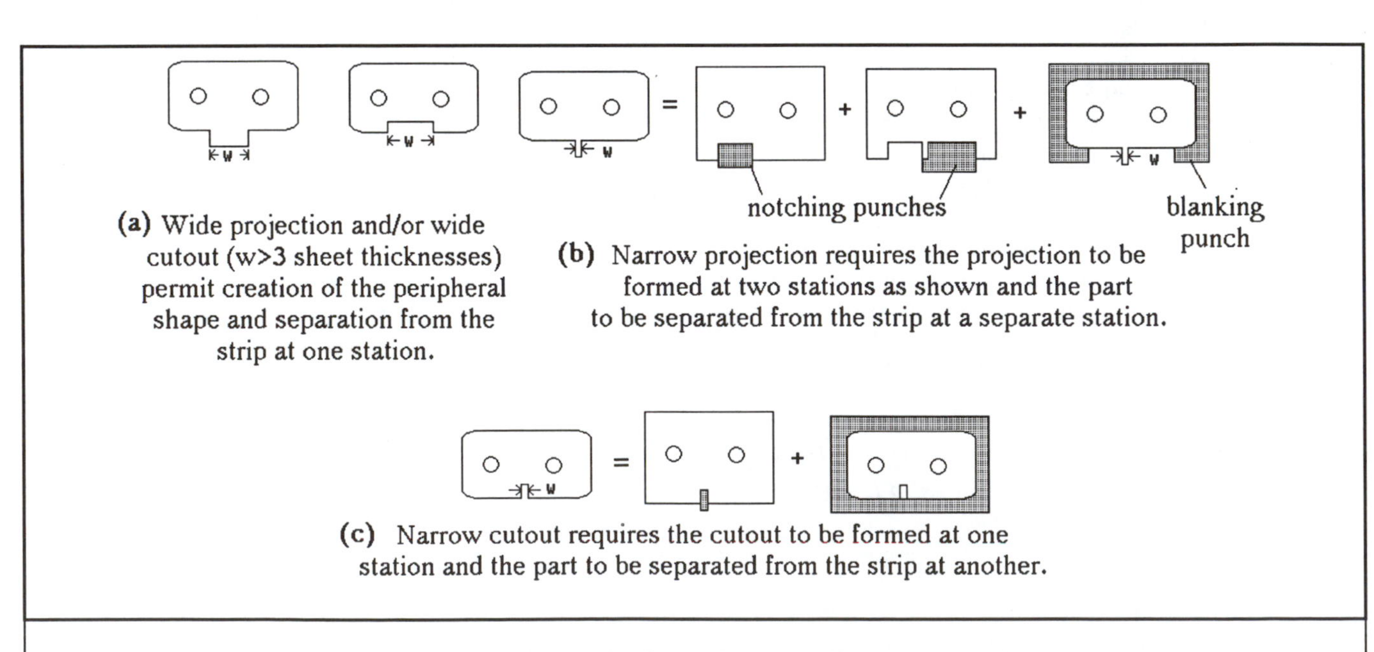

Figure 3.25 A Link with a Wide Projection and a Link with a Narrow Projection

Tooling cost relative to a simple washer

	Parts with no closely spaced features, no narrow cutouts, and no narrow projections	Parts with one closely spaced feature type and/or narrow cutouts and/or narrow projections	Parts with more than one closely spaced feature type with or without narrow cutouts and or narrow projections
Parts with low die detail	1.0	1.3	1.6
Parts with medium die detail	2.3	2.6	2.9
Parts with high die detail	3.0	3.3	3.5

Good

Fair

Poor

Most Costly

(a) Parts with holes, ribs, and other types of localized features

	Parts with all bend angles less than 90°	Parts with bend angles between 90° and 105°	Parts with bend angles greater than 105°
Parts with one bend stage and no side-action features	0.8	1.4	1.9
Parts with two bend stages and no side-action features	1.3	1.9	2.4
Parts with more than two bend stages and no side-action features	1.9	2.5	3.0
Parts with bends and side-action features	1+x	1+x	1+x

x-indicates the value for a similar part without side action features as obtained from the first three rows of this figure

(b) Parts with bends

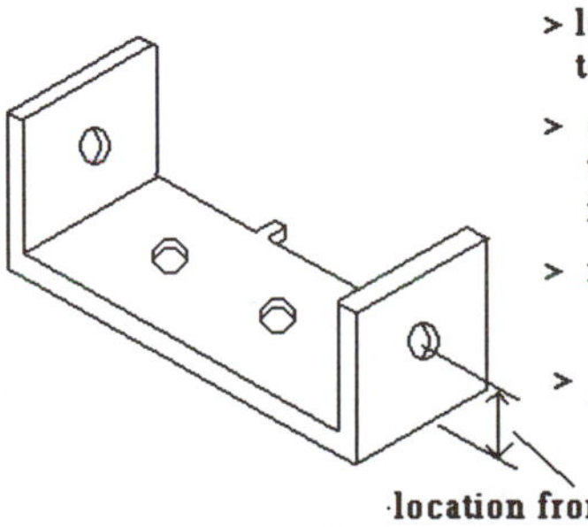

> low die detail; one narrow projection; relative tooling cost due to localized features = 1.3

> one bend stage; side action features; relative tooling cost due to bends and side action features = 1+x

> x = relative tooling cost without side-action features =0.8

> Total Relative Tooling Cost = 3.1

location from bend line is critical

(c) Example

Figure 3.26

Stamping DFM Advisor

present in the part) while the horizontal axis is used to give an indication of narrow cutouts, narrow projections and closely spaced features.

The vertical axis in Figure 3.26b ranks designs according to the number of bends present as well as the presence of side-action features. The horizontal axis is used to indicate whether or not overbends (bend angles greater than 90°) are present.

The most easily manufacturable designs have no more than one bend stage (i.e., parts with one bend, or parts with two or more bends having all bend lines in the same horizontal plane), with bend angles less than or equal to 90°, and with only a handful of distinct feature types not closely spaced.

As an example, consider the stamped part shown in Figure 3.26(c). Assume that the locations of the holes on the vertical faces of the part are critical with respect to the bend lines; that is, that they must be created after the sides are bent. In this case, then, they are side-action features.

A precise method for determining the die detail of a part is contained in Chapter 12, but here let us assume that any part which contains fewer than three distinct feature types is a part with "low" die detail. Let us also assume that if the part has more than four distinct feature types, it is considered to have "high" die detail. Obviously, all other parts are considered to have "medium" die detail. Neglecting the side-action features for the part shown in Figure 3.26 (c), we see that the part contains two identical holes. That is only one distinct feature type; thus, the part is considered to have low die detail. Since the part has one narrow cutout, the cost of the tooling required prior to the bending operations is 1.3 times the cost of the tooling required to stamp a simple washer.

While there are two bends here, both of these bends are about parallel axes and are wiped in the same direction. Thus, these bends can be created at the same station, so the number of bend stages is one. Since the bend angles are not greater than 90°, and the part has side action features, the tooling cost required to create these features relative to the tooling cost for a stamped washer is (1 + x). The value of x in this case is the value contained in row one column one of Figure 3.26 (b), which pertains to a similar part with bends less than 90° but without the side-action features. (Although a bend angle of 90° is indicated in Figure 3.26, it is assumed here that the bend angle does not have to be exactly 90°). In this case, x is 0.8. Thus, the tooling cost to stamp this part relative to the tooling cost required to stamp a simple washer is the sum of 1.3 and 1.8 or 3.1.

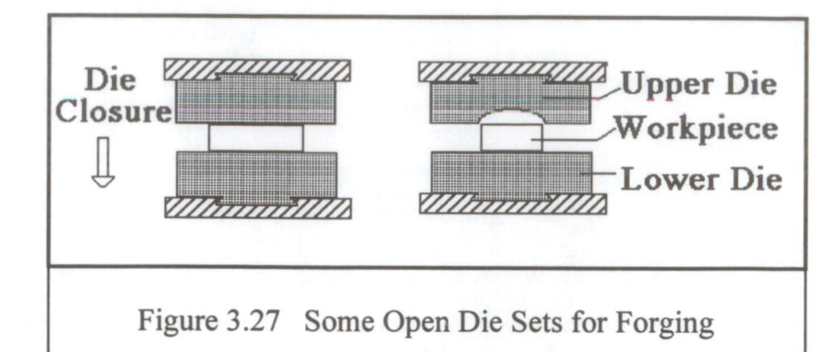

Figure 3.27 Some Open Die Sets for Forging

3.6 Forging

Forging is a bulk deformation process in which a part is shaped by squeezing (with a mechanical or hydraulic press) or hammering (with a gravity or power assisted hammer) a hot workpiece between two die halves attached to a press or hammer. Because the workpiece is plastically deformed, its cast structure is refined; the grains or fibers align in the direction of flow. This directional alignment of the fibers makes forgings stronger and more ductile than castings, and enables forgings to have greater resistance to shock and fatigue. Thus forging is used to produce some of the most critically stressed parts found in aircraft, automobiles and tools.

3.6.1 Types of Forgings

There are two broad categories of forgings: open die and closed die. Closed die forgings are also referred to as *impression forgings* since the forging dies partially enclose the workpiece material and restrict the flow of metal.

Open-die forging is a process where a hot metal workpiece is squeezed or hammered between flat, circular or v-shaped dies. The workpiece is not enclosed so the metal flow is not completely restricted. Figure 3.27 shows an example of dies used for open-die forging.

Open-die forgings are used for parts too large to be produced in closed dies, when the quantity involved is too small to justify closed dies, or when a short lead time exists.

While complex shapes can be produced by the use of open dies, their production requires skilled operators and is time consuming. Therefore, most open die forgings are restricted to the production of bars and shafts (round, square,

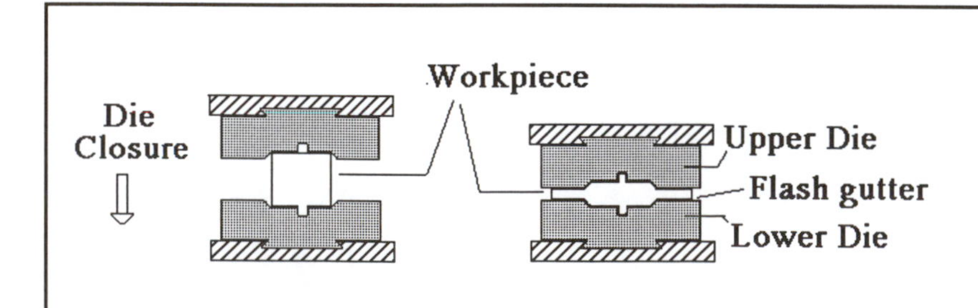

Figure 3.28 An Example of a Closed Die Forging Using a Parallel Flash Gutter.

Other gutter designs permit the dies to touch upon closure. The type of gutter used is a function of the equipment and the workpiece material [9].

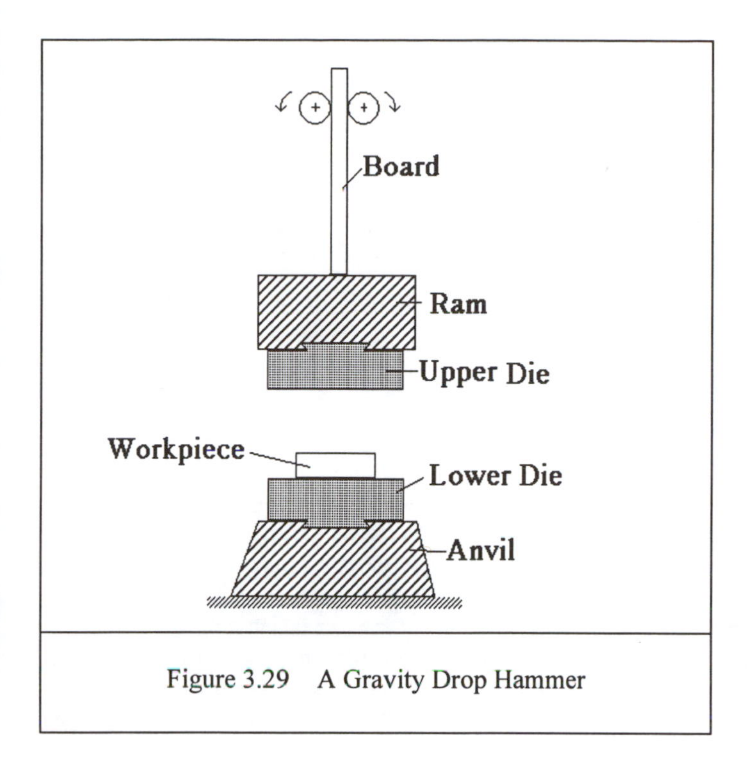

Figure 3.29 A Gravity Drop Hammer

rectangular, or hexagonal cross-sections) and flat pancake type parts. While similarly shaped shafts and bars may be available from off-the-shelf rolled stock, forged shafts are used when superior mechanical properties are required.

In *closed-die forging* the workpiece is squeezed or hammered between one or a series of dies enclosing the workpiece on all sides. Often a series of dies is used because the metal is difficult to form in a single stage.

Closed-die forgings are sub-divided into blocker-type, conventional forgings, and precision (low draft) forgings.

Blocker type forgings only roughly approximate the general shape of the final part with generous contours, large radii, large draft angles, and liberal finish tolerances. A blocker type forging requires considerable machining final dimensions, tolerances, and surface finishes.

Conventional forgings constitute the majority of forgings produced. They more accurately achieve final dimensions and with closer tolerances and smaller radii. To produce this more accurate part, more dies are required, and conventional forgings are often pre-formed using blocker dies prior to conventional forming.

Conventional forgings have portions of the part machined while other portions remain "as forged". Both blocker type and conventional forgings are produced in dies allowing for excess material ("flash") to escape from the die's cavity. See Figure 3.28.

3.6.2 Forging Machines

The two main types of forging machines used in commercial forging plants in the United States are *hammers* and *presses*.

Hammers. Hammers are either gravity type or power assisted. With a gravity drop hammer, the upper die is attached to a ram and is raised by either a board, belt or air (Figure 3.29). It then is allowed to freely fall to strike the workpiece. In power assisted drop hammers, air or steam is used against a piston to supplement the force of gravity during the downward stroke. The energy used to deform the workpiece is obtained from the kinetic energy of the moving ram and die. Because hammers are energy restricted machines, multiple blows (usually three) are required at each stage (i.e., for each die) during the forging process.

In addition to regular drop hammers, there are also counterblow hammers. In a counterblow hammer, two rams are activated simultaneously and driven toward each other. The two die halves strike the workpiece at some midway point between the two rams. Vertical counterblow hammers are used in order to avoid the need for the heavy anvil and foundation weights required with gravity drop hammers.

Horizontal counterblow hammers (also called impacters) also exist. In this case, two rams of equal weight are driven toward each other by compressed air in a horizontal plane. Horizontal counterblow hammers are used for the automatic forging of workpieces by utilizing a transfer unit to move the partially completed forging from die station to die station (Figure 3.30).

Presses. Forging presses can be either mechanically driven or hydraulically driven. While mechanical presses contain the same general components as stamping presses, forging presses tend to be stiffer and more robust. Unlike the blow delivered by a hammer, a mechanical press squeezes the metal between the two die halves; one squeeze is used for each stage (die) of the process. While hammers are energy restricted, mechanical presses are stroke limited. The largest force is delivered at bottom dead center.

The general components of a hydraulic forging press are similar to those of hydraulic presses used for stamping. Hydraulic presses are load limited since the maximum load can be delivered at any position within the press stroke range. Multiple ram hydraulic presses are available, and can be used to forge cavities which are equivalent to external undercuts found in die castings.

Presses tend to be more expensive than hammers. For this reason, hammers are used whenever possible. However, presses can produce all of the types of forgings produced by hammers and, in addition, can forge some low ductile alloys which might fragment under hammer blows. In addition, for rate-sensitive materials, hydraulic presses are preferred since the load can be slowly applied.

3.6.3 Factors Influencing Design for Forging

Materials. Forging difficulty is a function of both part material and part shape. These two factors are strongly interrelated; thus, a shape that is relatively easy to forge in one material (aluminum, for example) may be difficult or impossible to forge in another (say a nickel based superalloy).

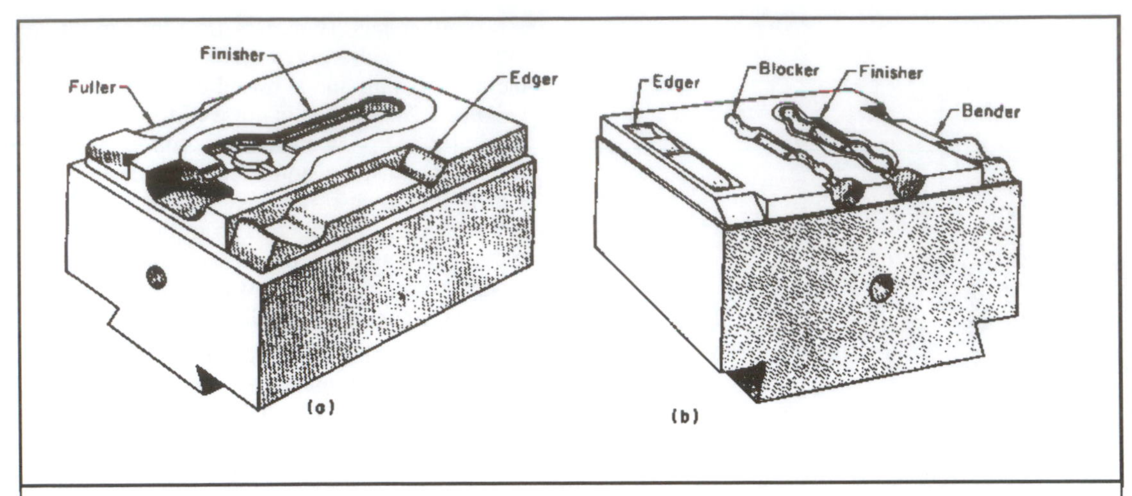

Figure 3.30 A Closed Die with Several Die Stations. (From *Metals Handbook*, Ninth Edition, Vol. 14, Forming and Forging International, 1988. With permission of the American Society of Metals.)

Table 3.1, taken from [10], contains a partial list of materials and alloys that can be forged. The materials have been divided into six groups, and have been ranked in general order of forging difficulty. Details concerning the actual material alloys allocated to each group together with recommendations for forging conditions, etc., can be found in [10]. Materials not generally forged (e.g., cast iron) have not been included.

Shapes. There are basically three types of forged parts produced: compact (or chunky) parts in which the length, width and height of the part are approximately equal; flat (disk like) parts; and long parts where the part length is significantly greater than the part width.

Compact shapes can be made from billets or blanks, and they have a simple forging sequence. In general, they can be formed using one blocker die set and one conventional

MATERIAL	COMMENTS
Light Alloys (Aluminum and magnesium)	High ductility; readily forged into precise, intricate shapes; low forging pressures; can be forged on both presses and hammers, but presses preferred when deformation is severe.
Copper and Copper Alloys (Brass, bronze, etc.)	Readily forged into intricate shapes; requires generally less pressure than equivalent shapes in low carbon steels; presses preferred.
Carbon and Alloy Steels	Most widely forged materials; readily forged into wide variety using conventional methods and standard equipment; hammers and presses both used.
Martensitic and Ferritic Stainless Steels, Maraging Steels and Tool Steels	Forged by conventional methods, but require higher pressures than carbon or low alloy steels; hammers and presses both used.
Austenitic Stainless and Austenitic Nickel Alloys of Iron	More difficult to forge than carbon and alloy steels; requires greater pressure; hammers preferred to presses.
Titanium and Titanium Alloys; Iron -, Cobalt-, and Nickel-based Superalloys; Refractory Metals; Beryllium	Forging pressures increase rapidly with decreasing temperature; both hammers and presses used.

Table 3.1 Materials for Forged Parts

die set. Thus, if formed on a press, two operations are required If formed using a hammer, generally the workpiece will be struck three times with each pair of dies. A significant proportion of parts in this category contain external undercuts which require multiple action forging machines.

Flat parts are also generally produced from billets or blanks, and upsetting (counterflow) type material flow predominate. Flat parts are usually produced using two or three die sets. For simple flat parts with uniformly thick walls, a blocker die is used to distribute the material properly and then an impression die is used to finish the forging.

Long parts are produced directly from bar stock and in general require elongation and drawing stages prior to the impression forging sequence. For example, in the case of long slender parts with two or more heavy sections separated by light sections, some preliminary preforming operations are used to thin down the metal in the center section and provide more mass at the ends for later operations. In general, three or four die sets are required to produce the final forged part.

From this discussion it should be apparent that the basic shape of a forged part affects the number of dies required, and hence the tooling costs.

As in die casting and injection molding, die costs increase if the part requires a non-planar parting surface, and if multiple action dies are required to produce external undercuts. Unlike injection molding and die casting, however, die costs are also a function of the part material as noted above. Material which is difficult to move requires more pre-forming stages, and decreases die life.

While part shape, part complexity and part material affect the number of dies or die stages required (hence die cost), as the number of dies or die stages increase, the sequence of operations necessary for processing the part also increases; hence, processing costs increase, too. In addition, multiple-action machines are more costly, so parts with external undercuts are also more costly to process.

3.6.4 Design for Manufacturing Guidelines For Forged Parts

The nature of forging process in which solid metal is squeezed and moved within a die set to form a part leads to the following broad DFM guidelines:

1. Because all the pre-forming operations required to forge a part result in long cycle times, and because the robustness required of the dies, hammers and presses result in high die and equipment cost, forging is an expensive operation compared to stamping and die casting. Thus, *if possible, forging should be avoided.*

Of course, there are times when functionality dictates a forged part, or when other processes are even more costly. In these cases:

2. *Select materials which are relatively easy to deform.* These materials will require fewer dies, shorten the processing cycle, and require a smaller hammer or press.

3. Because of the need for the metal to deform, *part shapes that provide smooth and easy external flow paths are desirable.* Thus, corners with generous radii are desirable. In addition, tall thin projections should be avoided since such projections require large forces (hence large presses and/or hammers), more pre-forming stages (hence more dies), cause rapid die wear, and result in increased processing cycle time.

4. For ease of producibility, *ribs should be widely spaced.* Spacing between longitudinal ribs should be greater than the rib height; spacing between radial ribs should be greater than 30 degrees. Closely spaced ribs can result in greater die wear and in increase in the number of dies required to produce the part.

Internal undercuts, and external undercuts caused by projections, must be avoided since they are impossible to form by the movement of solid metal. External undercuts which are the result of holes should be avoided since they increase both die costs and processing costs.

A more quantitative guide to design for forging complete with classification systems and relative cost models can be found in [10, 11, 12, 13].

3.7 Aluminum Extrusion

3.7.1 The Process

Like forging, aluminum extrusion is a solid metal flow process, but the process is very different. In aluminum extrusion, a round aluminum billet is heated to 700-900 F; the billet remains solid. It is then inserted into the cylindrical cavity of a large 'container' and, by means of a hydraulically powered ram, the metal is forced out through a die hole or holes of the desired shape (Figure 3.31).

After extrusion, the metal cools slowly, is given a slight mechanical stretching, and then cut to desired lengths. Almost always, extrusions are given additional heat treating to add strength and hardness, and sometimes a surface treatment called anodizing is done to improve appearance and provide better weatherability.

Aluminum extrusions can be formed into an amazingly wide variety of shapes, and anodized into many attractive colors. Examples of extrusions are found in residential window casements, commercial storefront structures, picture

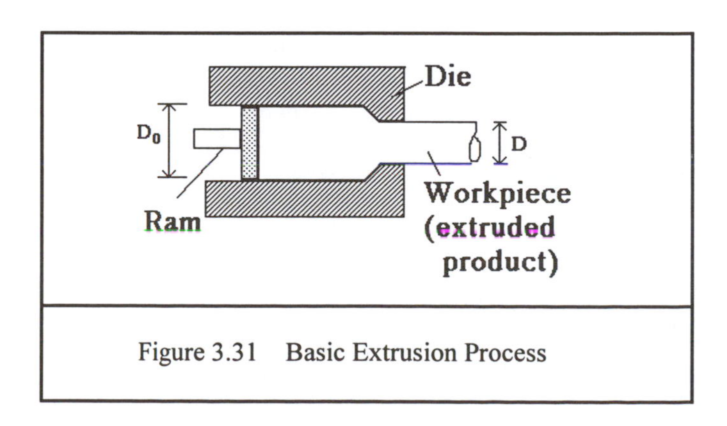

Figure 3.31 Basic Extrusion Process

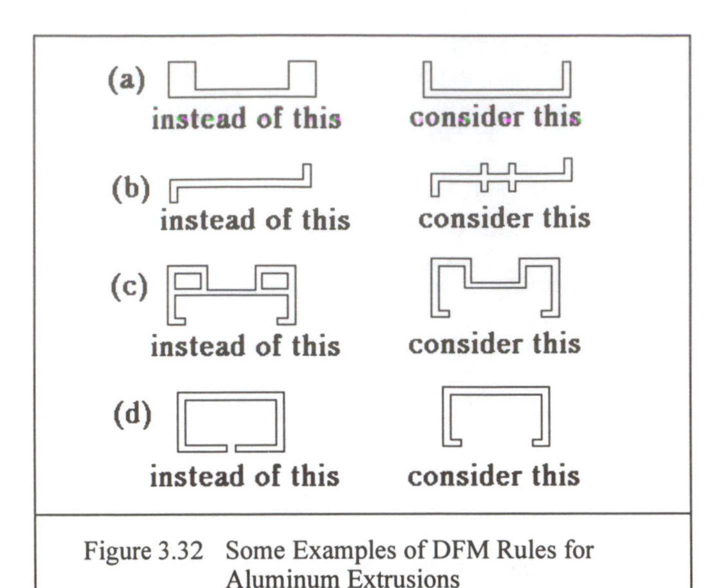

Figure 3.32 Some Examples of DFM Rules for Aluminum Extrusions

frames, chalk trays, and many standard structural shapes.

There are a number of aluminum alloys (most containing small amounts of magnesium, silicon, and other elements) that can be readily extruded and then heat treated [1]. Extrusion dies for aluminum are relatively inexpensive. If there is to be a reasonable quantity of production, die costs are usually essentially negligible on a per pound basis.

3.7.2 Qualitative Reasoning on Design for Manufacturing Aluminum Extrusions

Most DFM considerations for aluminum extrusions evolve from the difficulty of forcing the metal to flow *uniformly* from a large round billet through small complex die openings. This leads to the following "rules" that can help guide the design of more easily extrudable aluminum parts:

1) *Sections with both thick and thin sections are to be avoided.* Metal tends to flow faster where thicker sections occur, giving rise to distortions in the extruded shape. Use designs that will function with all the walls as uniform in thickness as possible. Avoid designs that require slugs of material (Figure 3.32a).

2) *Long, thin wall sections should be avoided*, since such shapes are difficult to keep straight and flat. If such sections are absolutely necessary, then the addition of ribs to the walls will help distribute the flow evenly (Figure 3.32b).

3) *Hollow sections are quite feasible, though they cost about 10% more per pound produced.* The added cost is often compensated for in the added torsional stiffness that the hollow shape provides. It is best if hollow sections can have a longitudinal plane of symmetry (Figure 3.32c).

4) *"Semi-hollow" features should be avoided.* A semi-hollow feature is one which requires the die to contain a very thin — and hence relatively weak — neck. Figure 3.32d shows the applications of this rule.

3.8 Machining

3.8.1 The Process and the Tools

Machining produces parts by removing material in the form of small chips from a solid workpiece using a single or multiple-edged cutting tool. Since the process removes material already paid for, machining is not an economical process, and is not generally used to produce special purpose parts for consumer products. However, it is often used to improve the tolerances or local surface finish of parts made via other processes (e.g., sand casting, forging, etc.) There are several kinds of machine tools, though not all of them will be discussed here. Among the most common are:

Lathes. Lathes are used primarily for the production of cylindrical or conical exterior and interior surfaces, via (see Figure 3.33) turning, facing, boring, and drilling. Lathes are also used for the production of screw threads. In a lathe, the workpiece is rotated while the cutting tool is moved ("fed") into the workpiece in a direction parallel and/or perpendicular to the axis of rotation of the workpiece.

Vertical and Horizontal Boring Machines. These machines are used in place of lathes for the machining of large workpieces. Boring machines can be used to perform turning, facing, as well as boring, and they are also used to form grooves, and for increasing the diameters of existing holes.

Vertical and Horizontal Milling Machines. Milling machines (Figure 3.34) are used to form slots, pockets, recesses, holes, etc. In this case the cutting tool is rotated, and the workpiece is fed.

Planing and Shaping Machines. These machines are used for reducing the thickness of blocks and plates, and for "squaring up" blocks and plates. Shapers are also used to machine notches and keyways, and to a form flat surface on parts formed by processes such as casting and forging. In the shaping machine shown in Figure 3.35, the workpiece is clamped to the table and the cutting tool moves horizontally. On the forward stroke, the cutting tool removes metal. During the return, when no metal is removed, the worktable is fed to the right in preparation for the cutting tool.

Surface and Cylindrical Grinding Machines. In a surface grinding machine the workpiece is fixed to the table which reciprocates longitudinally and is fed laterally. A grinding wheel is fixed to a rotating horizontal spindle and grinds the workpiece as the table reciprocates and is fed. Surface grinding machines are used primarily for improving the tolerance and surface finish of flat surfaces. Cylindrical grinding machines are also used to improve the tolerances and surface finishes of cylindrical surfaces. In this case both the workpiece and the grinding wheel rotate.

Electrical Discharge Machines (EDM). Electrical discharge machining (EDM) is a process of removing metal by means of an electrical discharge spark. The principle elements of an EDM machine are illustrated in Figure 3.36.

There are two types of EDM machines: "solid" and "wire"'. In the solid type, a tool (usually graphite but sometimes copper or brass) together with the workpiece itself are

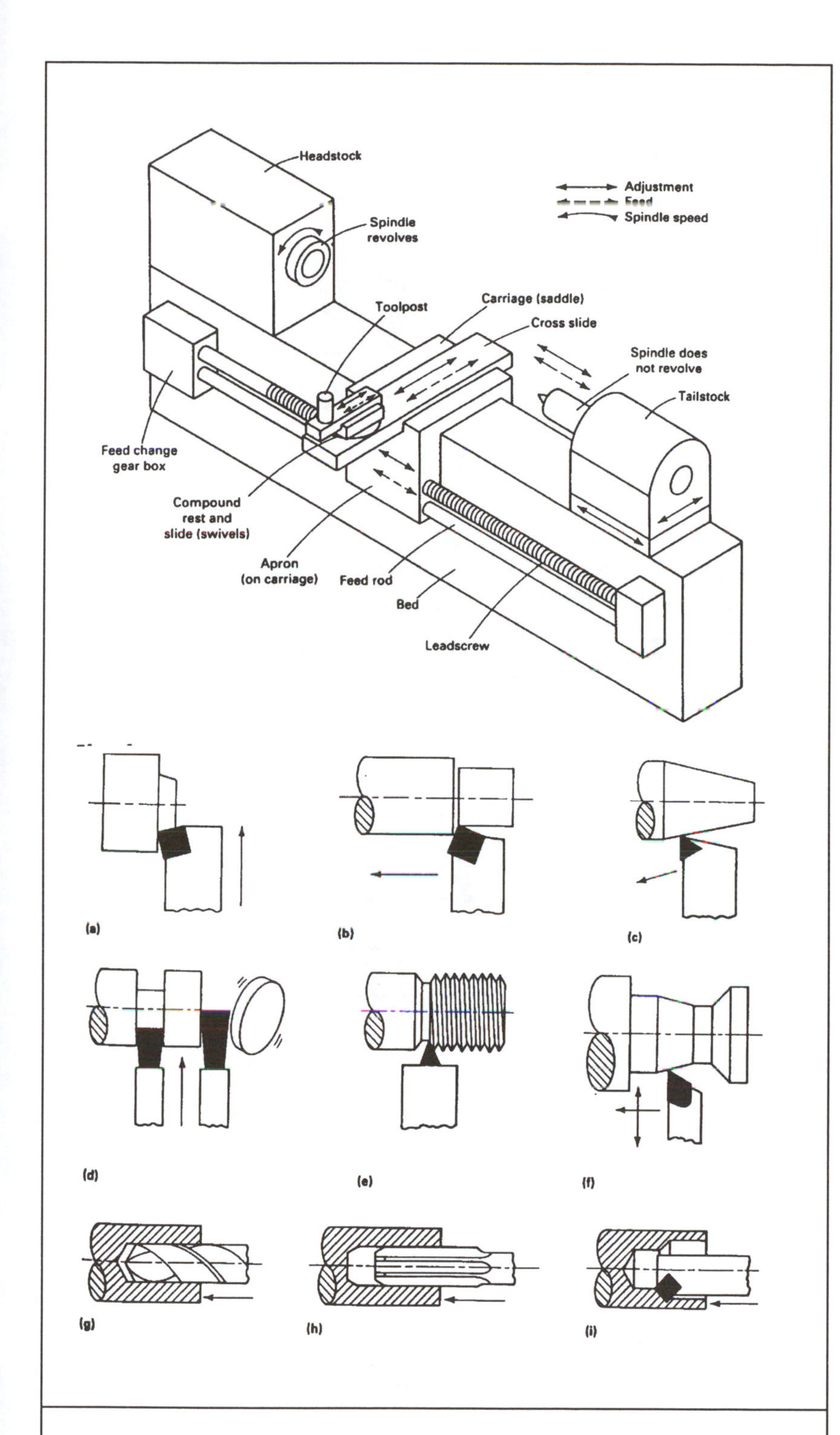

Figure 3.33 Principal Components and Movements of a Lathe and the Basic Operations that Can Be Performed. (a) Facing; (b) Straight turning; (c) Taper turning; (d) Grooving and cutoff; (e) Threading; (f) Tracer turning; (g) Drilling; (h) Reaming; (i) Boring. (Ref. *Metals Handbook*, Ninth Edition, Vol. 16, Machining, ASM International, 1989, with permission of the American Society of Metals.)

connected to a dc power source. The workpiece is placed in a tank filled with a dielectric fluid, and the tool is fed into the workpiece. When the potential between the two is sufficiently high, a spark is created which removes a small amount of material from the workpiece.

The solid EDM process is used to machine narrow slots, small holes, and complicated shapes. It can be used on any material that conducts electricity and is not significantly affected by the material hardness or strength. The EDM process is much slower than traditional machining as described above and leaves a "pitted" surface finish which may require further grinding or hand finishing. For these reasons, EDM is avoided when possible.

The wire EDM process is a variation of this process and is often used to produce stamping dies and punches. For a detailed description of each of these processes, as well as other machining processes, see [2].

3.8.2 Qualitative Guidelines on Design for Machining

In machining, the fact that metal already paid for is removed by use of a sharp cutting tool leads to the following design guidelines:

a) *If possible, avoid machining.* If the desired geometry can be produced by another process such as casting, molding, stamping, etc., the cost will almost certainly be lower (unless, of course, only a few parts are to made). However, if machining cannot be avoided, then following the DFM guidelines will help keep the cost down.

b) *Specify the most liberal tolerances and surface finishes.* Most machining operations are performed in two operations, a roughing operation and a finishing operation. The roughing operation is used to remove large quantities of metal without special regard to tolerances or surface finishes. The finish cut is performed to provide the necessary tolerances and surface finishes. Since surface roughness is directly related to the rate of feed used — large feeds produce high values of surface roughness (rough surfaces) and low feeds produce low values of surface roughness (smooth surfaces) — and low feeds result in longer machining times — surface finishes better than those absolutely necessary

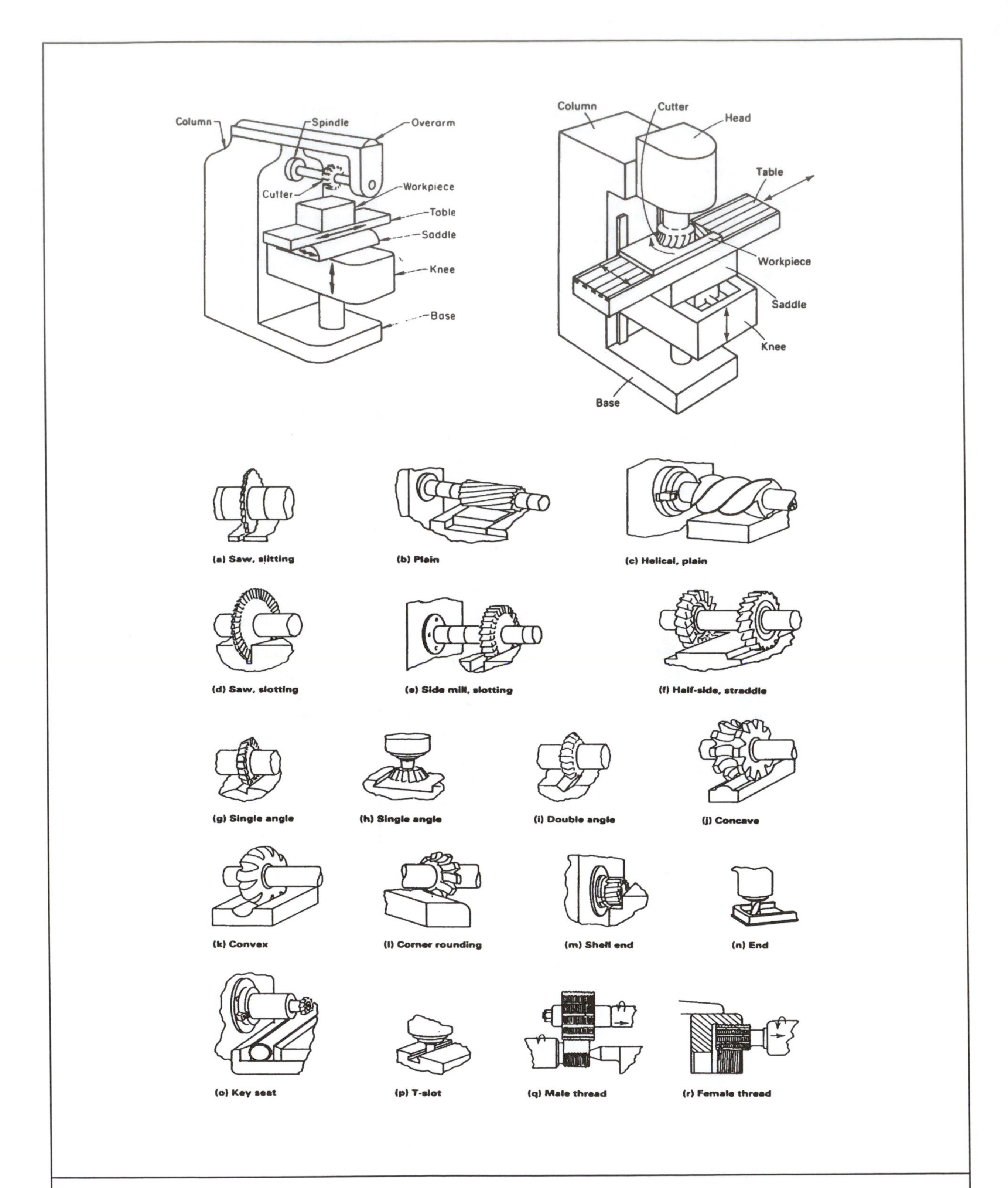

Figure 3.34 Principal Components and Movements of Horizontal and Vertical Milling Machines and the Basic Operations Performed On Them. (*Metals Handbook*, Ninth Edition, Volume 16, Machining, ASM International, 1989, with permission of the American Society of Metals.)

for functional purposes should not be specified.

c) *For turning operations on a lathe, avoid designs that require sharp internal corners.* Corner radii equal to the tool nose-radius of the cutting tool should be specified. Sharp internal corners call for either the use of sharp tools which more easily break and/or additional operations.

d) *For planing and milling operations, avoid sharp internal corners, radiused external corners and slot widths and shapes other than those available using standard off-the-shelf cutters.* Internal corners should be specified so that they are equal to the cutter radius of milling cutters and the tool radius of planing cutters. Since sharp external corners are a natural result of these processes, such corners should be specified.While these general machining guidelines apply across the board, a more detailed list of features which can be readily provided on machined components can be found in [15].

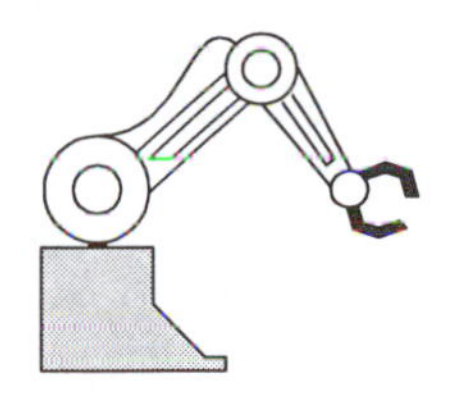

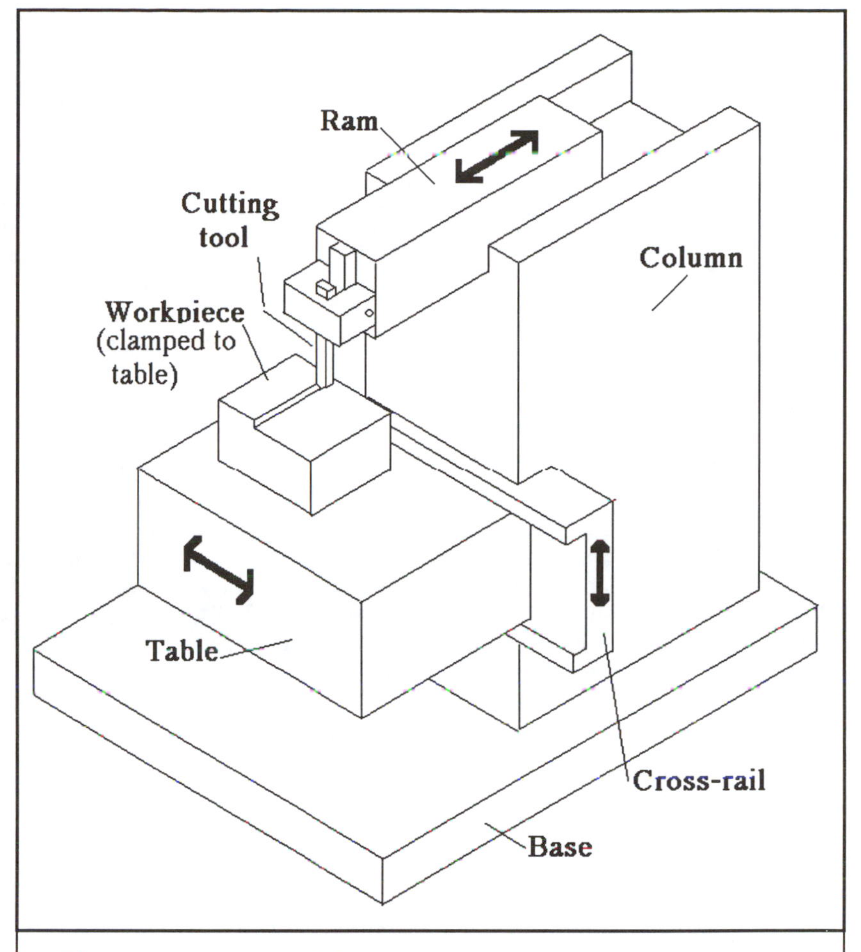

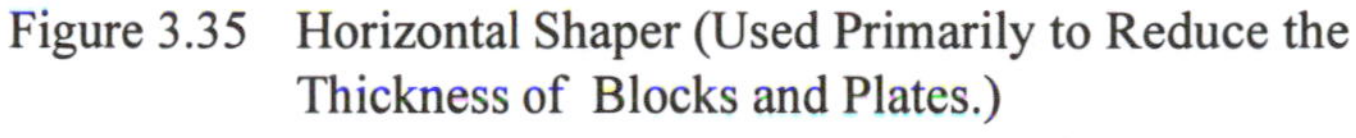
Figure 3.35 Horizontal Shaper (Used Primarily to Reduce the Thickness of Blocks and Plates.)

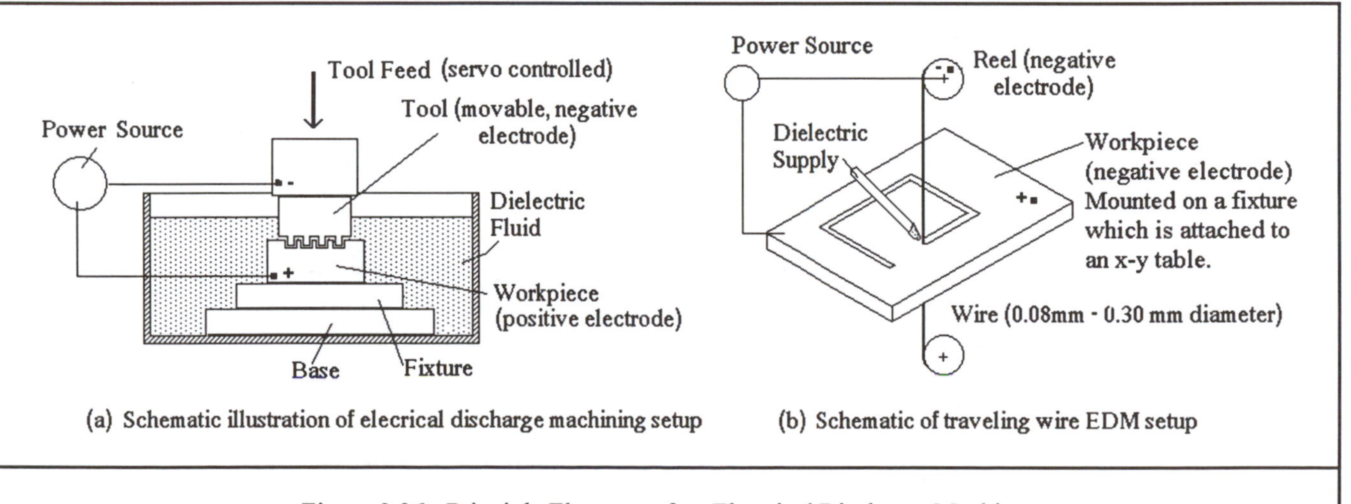

Figure 3.36 Principle Elements of an Electrical Discharge Machine

3.9 Assembly

3.9.1 Assembly Processes

Note: Material on assembly in this book is based on data and other information extracted primarily from references [16] through [23]. Readers are urged to consult these original references for more detailed information and analysis. We especially recommend Reference [16], "Product Design for Assembly" by G. Boothroyd and P. Dewhurst.

The manufacturing process of assembly is generally thought of as consisting of two distinct operations: *handling* followed by *insertion*. Both handling and insertion can be done either manually or automatically.

In *manual handling*, a human assembly operator stationed at a workbench reaches and grasps a part from a bin, and then transports, orients and pre-positions the part for insertion.

In *automatic handling*, parts are generally emptied into a parts feeder, such as a vibratory bowl feeder (Figure 3.37), which contains suitable orienting devices (Figure 3.38) so that only correctly oriented parts exit the feeder in preparation for insertion. Feedtracks are then used to transport the correctly oriented parts from the feeder to an automatic workhead. Escapement devices release the parts to the workhead.

In *manual insertion*, the human assembly operator places or fastens the part(s) together manually. Although power tools may be used, the process is still essentially one of manual insertion under human control.

When *automatic insertion* is used, automatic workheads, pick-and-place mechanisms, and robots are utilized.

3.9.2 Qualitative Guidelines on DFA

Reduce the Part Count

As pointed out by Boothroyd [16] and [17], the two main factors affecting the assembly cost of a product are (a) the *number of parts* contained in the assembly and (b) the *ease with which the parts can be handled* (transported, oriented and prepositioned) *and inserted* (placed or fastened into the assembly).

It is obvious that if one product has 50 component parts and if an alternative version of the same product has only 10 parts, then the one with the fewer number of parts will usually cost less to assemble. Moreover, the total cost of the parts themselves will also be less. Thus, the *best* method available for reducing assembly costs, as well as part costs, is to *reduce the number of parts* in the assembly. There is a caveat: if the complex parts are going to require longer time to produce, then this might in some cases have also to be considered [24].

Part reduction can be accomplished either by the outright elimination of individual component parts (eliminating screws and washers and using a press or snap fit to fasten two components, for example) or by combining several component parts into a single, perhaps injection molded, part. See Section 3.10 below.

Reduce the Manual Handling Time

Once a designer has reduced the number of parts contained in an assembly to its "minimum", the remaining parts must be designed so that they are easy to handle and insert. A part is easy to handle manually if (a) it is easy to grasp and manipulate with one hand without grasping tools, (b) it is both end-to-end symmetric (as defined below) as well as rotationally symmetric, and (c) its size and thickness are such that grasping tools and/or optical magnification are not required. Each of these points is discussed briefly below. For more details, see also [16] through [23].

Tangling and Nesting. Parts which nest or tangle (vending cups, helical springs, etc.) are difficult to grasp singly and manipulate with one hand, and they present obvious handling difficulties. In addition, parts which are sticky (a part coated with grease or an adhesive), sharp (razor blade), fragile (glass), slippery (ball bearing coated with light oil), or

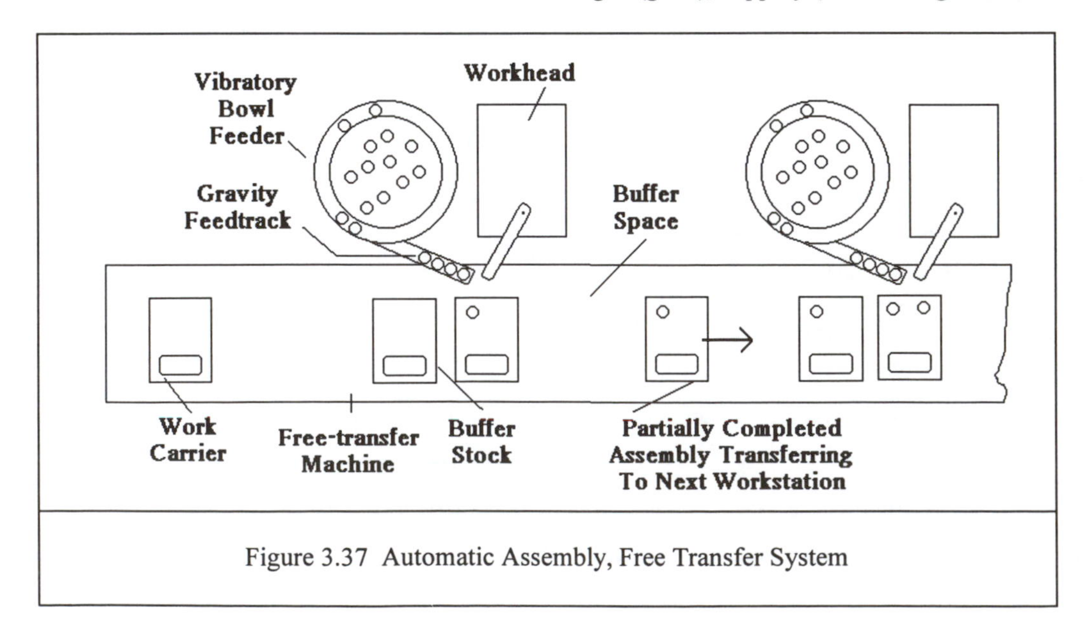

Figure 3.37 Automatic Assembly, Free Transfer System

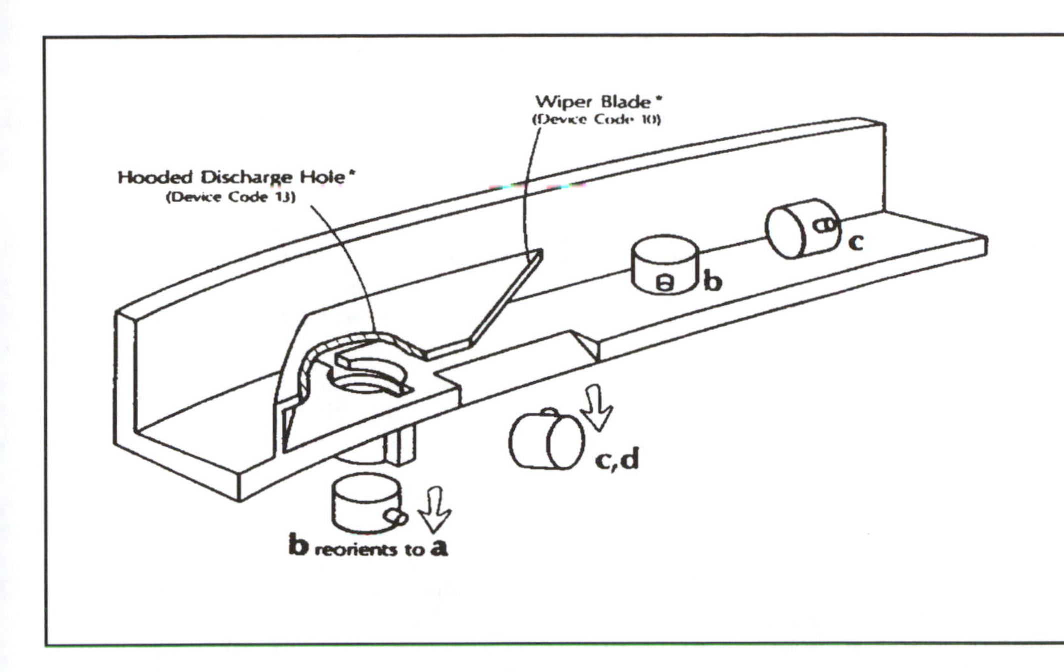

Figure 3.38

Example of Tooling for Automatic Handling.

From "Handbook of Feeding and Orienting Techniques for Small Parts," G. Boothroyd, C. Poli and L. Murch, Mechanical Engineering Department, University of Massachusetts Amherst.

flexible (belts, gaskets, etc.) are also difficult to grasp and manipulate with one hand.

Symmetry. To facilitate handling, parts should be designed with symmetry in mind. Parts with end-to-end symmetry (that is, parts that do not require end-to-end orientation prior to insertion) require less handling time than parts without such symmetry. A screw is an example of a part which does not have end-to-end symmetry, while a washer is an example of a part that does have end-to-end symmetry. A screw must be oriented so it can be inserted shank end first, while in the case of a washer either end (or side) may be inserted first.

Parts that have complete rotational symmetry (that is parts that do not require orientation about the axis of insertion such as a screw or washer) take less time to handle than parts that do not have any rotational symmetry — a house key for example. Parts with no symmetry take more time to orient than parts with some symmetry. For example all car keys lack end-to-end symmetry (i.e., only one end of the key can be inserted into the key hole), but some car keys have 180° rotational symmetry (i.e., either serrated edge can be aligned with the key hole) while others have no rotational symmetry (only one serrated edge exists and it must be properly aligned with the key hole). Many of us have probably experienced the fact that the key with no symmetry takes more time to properly orient than the key with 180° rotational symmetry.

Facilitate Automatic Handling

Parts can be easily handled automatically if they can be (a) easily fed (e.g., in a bowl feeder), and (b) easily oriented [16, 17, 22]. In general, parts that are difficult to grasp and manipulate manually are also difficult to feed in a feeder. Parts that have both end-to-end symmetry as well as rotational symmetry are also more easily and economically oriented in a bowl feeder than parts that do not have any symmetry.

A general rule of thumb [17] is: *if the part can not be made symmetrical, then accentuate its asymmetry.* That is, avoid making it almost symmetrical. A complete guide for designing parts for automatic handling is in [16] and [23].

Design for Easy Insertion

It has also been shown [16, 17] that insertion costs, whether done manually or automatically, are reduced if parts are designed so that they are *easy to align, easy to insert, and self-locating* with no need to be held in place before insertion of the next part. Manual insertion costs are also reduced if *access* and *view are not obstructed* while attempting to insert one part into the partially completed assembly. In addition, extra operations (reorienting of a partially completed assembly, for example) should also be avoided.

A part is *easy to align* if insertion is facilitated by well-designed chamfers or other features, a recess for example, such that the required accuracy of alignment and positioning is obtained. Figure 3.39 shows an example of a part that was redesigned to facilitate alignment.

Insertion against a large spring force, the resistance encountered with self-tapping screws, and the resistance encountered when using an interference fit (press fit) are all examples of resistance to insertion. The use of small clearances can also result in resistance to insertion. Jamming and wedging which result during insertion can also be considered

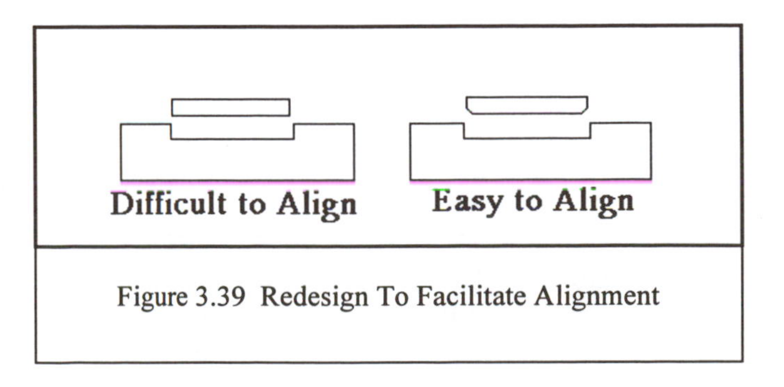

Figure 3.39 Redesign To Facilitate Alignment

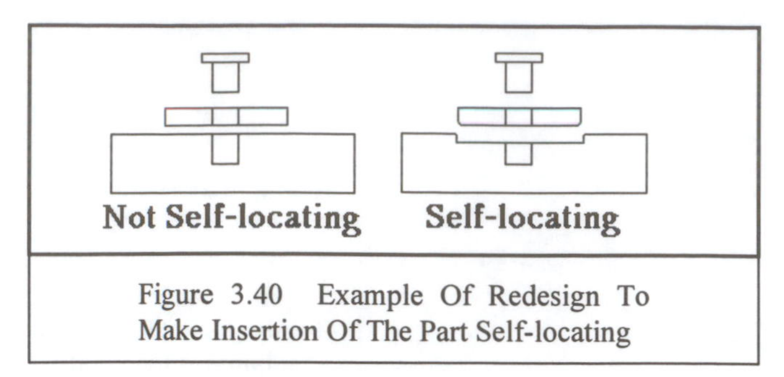

Figure 3.40 Example Of Redesign To Make Insertion Of The Part Self-locating

resistance to insertion. Methods for avoiding jamming in a design are discussed in detail in [17].

Anyone who has tried to do maintenance or repair work on a car has experienced the situation where parts were designed such that both access and vision were obstructed. Such designs should be subjected to 'design for service' considerations and evaluation.

Figure 3.40 shows two designs. In both designs, the assembly operator is to place a block on a plate with the holes aligned. In the original design, shown on the left, the plate is not provided with a recess. Thus, the block is not self-locating and it must be held such that it maintains its position and orientation relative to the plate prior to insertion of a peg (screw) that will secure the block to the plate. In the design shown on the right, the plate is provided with a recess making the placement of the block on the plate self-locating with no need to hold the block before insertion of the peg.

3.9.3 Total Assembly Cost - A Rough Rule of Thumb

Precise quantitative methods are available for use by designers and manufacturing engineers to estimate the time and/or cost to handle and insert parts and assemblies that have been designed [16, 17]. During the early stages of design, however, much more approximate methods can be helpful in guiding design decisions. One approximate method is based on the *Assembly Advisor* shown in Figure 3.41.

In the Assembly Advisor designs are rated as either good, poor, costly, or most costly from an ease of assembly point of view (only). This rough qualitative evaluation depends upon whether or not the part is easy to grasp and manipulate, and on the ease or difficulty of alignment and insertion. The numerical values shown in Figure 3.41 are simply rough approximations of the amount of time in seconds required to handle and insert a part which exhibits the characteristics indicated.

Another rough approximation based simply on part count can also assist in guiding early design. The reasoning is as follows: A "perfectly" designed part from an assembly viewpoint, which is not used as a fastener, will take on average about 3 seconds to handle and insert [17]. On the other hand, a difficult to assemble part, based on the criteria described above, will take about 11-13 seconds to handle and insert. Based on the experience of one of the authors, after analyzing many assemblies and sub-assemblies, it has been observed that it is possible to obtain a rough estimate of the time required to manually assemble a product by simply assuming that, on average, the handling and insertion time per part is from 7 - 9 seconds. Fasteners will take longer to handle and insert than non-fasteners, and some parts will take longer to handle and insert than others, but if at least 10 parts are contained in the assembly, *on average* the total assembly time (including fasteners) will be 7-9 seconds per part or task. The assumption here is that parts are "small" (see footnote on page 3-33) and within easy reach.

The greater the number of parts in the assembly, the better the above estimate is likely to be. Thus, a reasonably well designed product that contains about 20 small parts will likely take about 180 seconds to assemble manually. A comparable product that has only about 10 parts will likely take about 90 seconds to assemble. Thus it is obvious that the

HANDLING	INSERTION: Easily Aligned Easily Inserted*	INSERTION: Not Easily Aligned *or* Not Easily Inserted*	INSERTION: Not Easily Aligned *and* Not Easily Inserted*
Easy to Grasp and Manipulate	Good 4.0	POOR 8.0	Costly 13.0
Not Easy to Grasp and Manipulate	Poor 7.5	Costly 11.5	Most Costly 16.5

*Not easily inserted includes difficulties due to obstructed view and/or access, parts jamming and/or difficulties due to lack of chamfers or part geometry, etc. Resistance to insertion due to press and snap fits is not to be included here.

(Boxed numbers: approximate manual assembly time in seconds)

Figure 3.41 Design for Assembly Advisor. Information in this Advisor is based in part on information in Reference [16]: "Product Design for Assembly" by G. Boothroyd and P. Dewhurst.

product which can have its part count reduced in half will likely have about a 50% reduction in assembly costs. (Note: This general rule of thumb will not necessarily be true for products with very large parts.]

3.9.4 Theoretical Minimum Number of Parts

To reduce not only assembly costs but also manufacturing, inventory, and other associated costs, the minimum number of parts needed for proper functioning of a product should be used. It is therefore very helpful to be able compute or estimate the minimum number required. The following criteria were established in Reference [16], and are included here by permission of G. Boothroyd:

A part is essential only if:

- During the operation of the product, the part moves relative to all other parts already included. Only gross motion should be considered; small motions that can be accommodated by integral elastic elements, for example, are not essential; or
- If it must be made of a material that is different from those of all the other parts already included, or if it must be isolated from these. Only fundamental reasons relating to materials properties are acceptable;
- The part must be separate from all the other parts already included since necessary assembly or disassembly of other separate parts would otherwise be impossible.

These criteria can be used to guide designers as they endeavor to keep the part count to a minimum.

3.10 Summary of DFA Guidelines

In summary, in order to reduce assembly costs and to facilitate both handling and insertion, a designer should make every effort to design parts and products such that:

1. *The minimum number of parts needed for proper functioning of the product is used.* This will usually require the elimination of as many screwing operations as possible and the incorporation of more press and snap fits as a means of fastening. (However, screws are sometimes needed to provide accurate locations and more secure joints. It has also been argued that screws used in efficient robotic assembly can improve the yield (percentage of acceptable products) of an assembly line [24];
2. *Parts are designed so that they are easy to grasp and manipulate with one hand using no grasping tools (i.e. parts do not nest, tangle, are not sticky, sharp, fragile, slippery, flexible, etc.)*;
3. *Parts are end-to-end and rotationally symmetric as much as possible, or else obviously asymmetric;*
4. *Parts are designed so that they are easy to align and to insert (i.e., contain chamfers and/or recesses)*;
5. *For manual assembly, both access and vision are not restricted*;
6. *For automatic assembly, insertion is in a straight line from above.*

3.11 Reducing Part Count By Combining Parts

3.11.1 Results of a Cost Analysis

As discussed in the previous section, one of the best methods for reducing total manufacturing costs is to reduce assembly costs. The best way to reduce assembly costs (as well as other manufacturing costs) is to reduce the number of parts. This can be done either by the outright elimination of parts (for example, by replacing screws, nuts and washers by press/snap fits), or by combining two or more individual parts into a single part. In the latter case, the parts involved are most often injection molded, die cast or stamped. When part reduction by combination of parts occurs, the resulting part is usually more complex than the individual parts, and so the question arises as to whether or not the new complex part is in fact less expensive (considering tooling and processing) than the total cost of the individual parts being replaced, including their assembly.

In [25], a detailed analysis is presented that derives relations needed to compare the cost of producing a single, more complex injection molded or die cast part with the cost of producing and assembling multiple parts for the same purpose. In this section we present only the results of that analysis together with two examples of its application.

To understand the results, the meaning of *relative die construction cost* (C_{dc}) must be understood. This term refers to the ratio of the die construction cost for a part to die construction cost for a simple flat washer. Thus C_{dc} is 1.0 for a simple flat washer, and increases as the size and complexity of the part increases. A rough approximate value for C_{dc} can be found in Figures 3.14 and 3.15; it ranges from 1.0 for simple parts up to about 3.6 for complex parts. The values of C_{dc} found in Figures 3.14 and 3.15 represent average values for the typical kinds of small[1] parts found in consumer products. In Chapter 11 we will present a more detailed methodology for determining the value of C_{dc} and we will find that in extreme cases the value of C_{dc} can reach values even greater than 10.

The following symbols are also needed to present the results:

n = number of parts to be replaced with a single part;

C_{dcx} = the relative die construction cost for the single replacement part;

C_{dci} = the relative die construction cost for the i th original part.

It is shown in Appendix A, that if

$$C_{dcx} < n \tag{3.1}$$

1. "Small" parts have a largest dimension less than about 10 inches (250 mm).

then this is a sufficient condition to conclude that the replacement part will be less expensive than producing and assembling the n multiple parts. For example, if three parts are to be replaced with a single part, and the relative die construction cost for the new replacement part is less than 3.0, then the single replacement part is less expensive than producing and assembling the three original parts. Though Inequality (3.1) appears to involve only die cost, the analysis used to arrive at Equation 3.1 includes processing and material costs as well.

Even if Equation 3.1 is not satisfied (that is, even if C_{dcx} is greater than n) it may still be more economical to combine the parts. To check, we determine the relative die construction costs for each of the original individual parts using either Figure 3.14, or the method described in Chapter 11. That is, we determine the value of each C_{dci}. Then if

$$\Sigma C_{dci} > C_{dcx} \qquad (3.2)$$

we can again conclude that the single replacement part is more economical.

Furthermore, even if the condition in Equation 3.2 is not met, it may even yet be more economical to combine the parts. Whether it is or not depends on the production volume. If the volume to be produced is large enough, then the single replacement part (if it can be made at all and is functionally equivalent) is generally more economical. The method of computing the required production volume is presented in [25] and illustrated in an example below 1.2.

3.11.2 Example 1

Figure 3.42 shows the back cover subassembly for an electric razor. The original subassembly (Design 1) consists of eight parts: the back cover, two side plates which slide into place on the body and are then held in place by the back cover, four screws to secure the back cover to the body, and a label.

The redesigned back cover subassembly (as done by students) is shown as Design 2. It consists of two parts: a redesigned back cover, and a screw. The body is redesigned so that the cover can snap into place; a single screw is provided to assist in securing the cover to the body. While not indicated on the drawing of Design 2, the label has been replaced by lettering molded into the back cover.

There are two variations of Design 2. One is shown in Sectional View-1 and requires an internal undercut to accommodate the snap fit. The other is shown in Sectional View-2 in which the function is accomplished without need for an internal undercut.

We want to determine whether either version of the new more complex back cover design is less costly to produce than the original back cover subassembly which consisted of the four molded parts: the back cover itself, two side plates and a label. We ignore the fact for the moment that four screws are needed with the original design versus one for the redesigns.

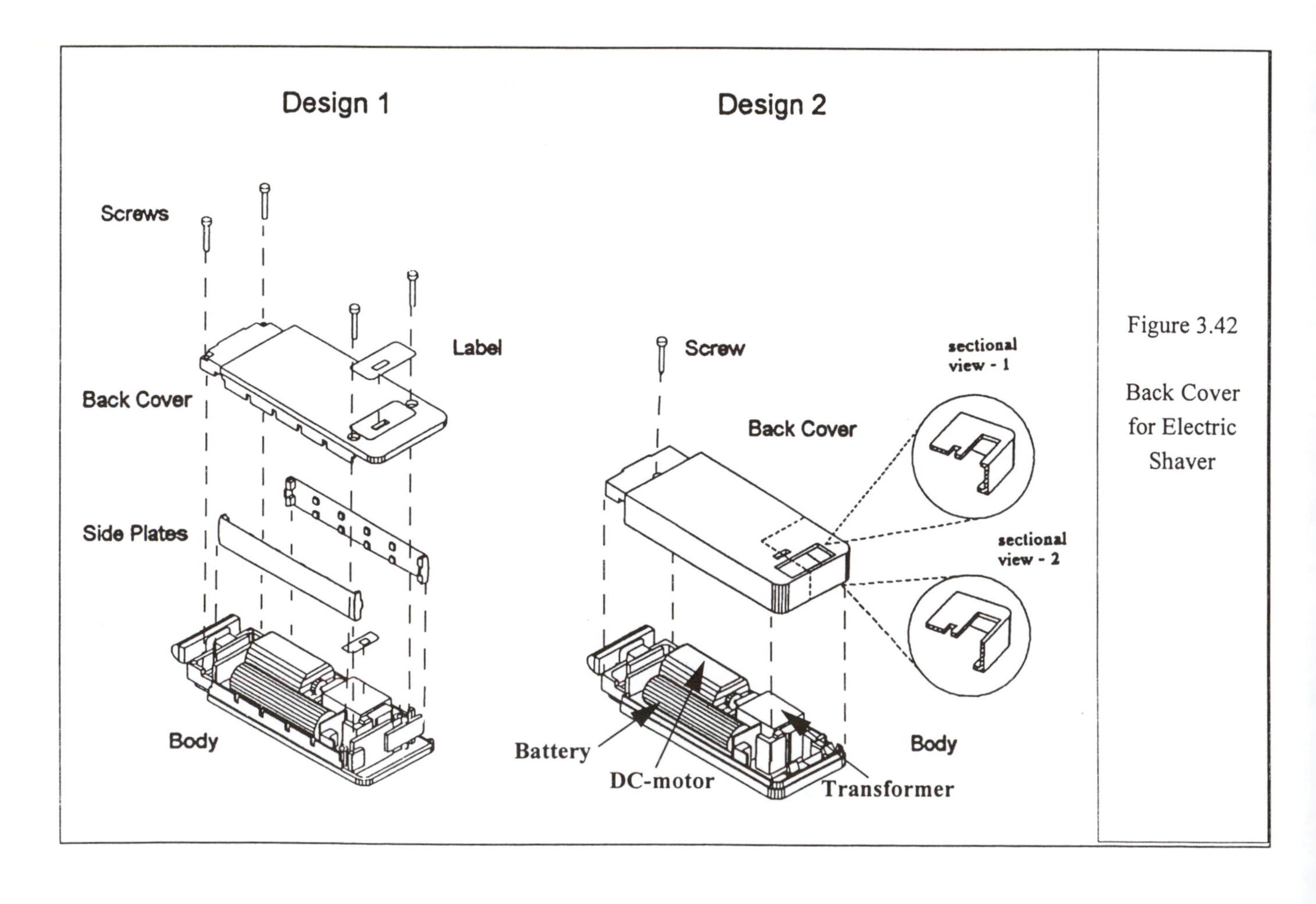

Figure 3.42

Back Cover for Electric Shaver

To compare the costs, we estimate the relative die construction cost for the two designs by using the Injection Molding DFM Advisor, Figure 3.14. From the Advisor, we can see that for the design depicted in sectional view-1, the part is box-shaped with one internal undercut present. Thus, the relative die construction cost, C_{dc}, is approximately 3.2.

For the design shown in sectional view-2, no undercuts are present, thus, the value of C_{dc} is approximately 1.6.

Thus, since n = 4, and in both cases the value of C_{dc} is less than n, then it makes economic sense to replace the four simple parts with this one more complex part. The screws add to the cost savings.

The methodology described in Chapter 11 allows us to determine more precise values of C_{dc}, but the conclusion in this example (that it is more economical to use a single complex part in place of the four simpler parts) does not change.

3.11.3 Example 2

Let us assume that in using the more precise methods discussed in Chapter 11 it is determined that the die construction cost, C_{dcx}, for a replacement part in the above example is 6. Then C_{dcx} would be greater than n, the number of parts it is replacing (4 in this case). Hence, we can not immediately conclude that it is economically viable to replace the four simple parts by this single more complex part.

Let us further assume that by the methods of Chapter 11 the value for C_{dc} for each of the four original parts is no greater than 1.4 (in fact they turn out to be equal to 1.425). Then by use of Eq. 3.2 we see that

$$C_{dci} = 4\,(1.4) = 5.6 < C_{dcx}$$

which implies that we still can not conclude that it is more economical to combine these four parts. (Remember, we still haven't considered the fact that we've eliminate three screws and have fewer parts to assemble.)

To determine the production volume N required to make the single part more economical despite these results, we compute N from:

$$N > 0.8K_{do}(C_{dcx} - n) / K_{eo}\, t_r\, C_{hr} \qquad (3.3)$$

where

- K_{do} = the sum of the tool construction cost K_{dco}) and tool material cost (K_{dmo}) for the reference washer (values for these are given Tables 15.1 and 15.9 of Chapter 15.
- K_{eo} = processing cost for the reference washer (values for these are given Tables 15.1 and 15.9 of Chapter 15.
- t_r = the cycle time for the part relative to the cycle time for the reference washer (A method to determine this is discussed in Chapter 15.)
- C_{hr} = the machine hourly rate for the replacement part relative to the machine hourly rate for the reference washer (A method to determine this value is discussed in Chapter 15).

Readers will not at this point know the meaning of all these terms, most of which are defined in Chapters 11 and 15, but that is not necessary for the point to be made here. Please accept for now that the following values can be obtained readily from the methods presented later:

K_{do} = \$7,000
K_{eo} = \$0.13
t_r = 2.4
C_{hr} = 1.21

Substituting these values into Eq. (3.3) gives the following,

$$N > 0.8(\$7000)(6.0 - 4) / \$0.13(2.4)(1.21) = 29{,}667$$

That is, as long as the production volume is to be greater than about 30,000, then it is more economical to produce the single more complex part than to produce the four simple parts which must then be assembled.

Note that the required production is not especially large for a modern mass produced product, especially one marketed worldwide. Thus one is tempted to conclude that it is almost always better to combine parts when the resulting single part is indeed functionally equivalent and can actually be produced. We know of one case, however, where twenty parts were combined into one awfully complex single part which, it turned out sadly, could not be produced by anyone. That part had to be made in two parts. However, the reduction from twenty to two parts resulted ultimately in great savings in assembly costs. Valuable time was lost, however, in finding out that the single part could not be made, at least not by the vendors with whom they were working.

3.12 Summary and Preview

Summary - Chapter 3

In this chapter on manufacturing for designers (MFD), we introduced and surveyed a number of the most common manufacturing processes. Our purpose is to give designers a qualitative understanding of these processes to help support their design decisions and their communications with manufacturing engineers.

With each process description, we included a discussion of design for manufacturing (DFM) issues, and presented a set of qualitative DFM guidelines especially intended for use early in the design process. In addition, for the cases of injection molding, die casting and stamping, the conditions when part reduction by combining parts is economically advantageous were presented.

In later chapters, more detailed DFM information and methods will be presented to enable designers to perform more quantitative and detailed evaluations of manufacturability of special purpose parts.

Summary - Part I

This concludes the Introduction to the Engineering Design story. You have now been introduced (in Chapter 1) to the main character — Engineering Design — and to her major problems types: conceptual, configuration, and parametric design. You know that she will use guided iteration as a method for solving all these problems.

You were also introduced (in Chapter 2) to the land in which Engineering Design lives -- the Product Realization Process. In addition you were introduced to some of the best practices the people of this land employ to fulfill their desire for high quality, low cost products brought to market quickly to meet customer desires with flexibility.

In Chapter 3, you met DFM and his gang of manufacturing processes. You heard from him some of the guidelines that Engineering Design can use to help shape and evaluate the design of parts. These guidelines are also helpful, at least indirectly, in helping to guide the selection and evaluation of prospective product concepts, which is the subject of Part II (Chapters 4 through 9).

Preview - Part II

We focus our attention now more directly on engineering design as we enter the main parts of the book. There are three Parts organized as follows:

- Part II Chapters 4 - 9 - Engineering Conceptual Design
- Part III Chapters 10-13 - Configuration Design of Parts
- Part IV Chapters 14 -21 - Parametric Design

Chapter 4 describes the engineering conceptual design process and shows how marketing and functional requirements are transformed into a physical concept. The plot, as they say, now begins to thicken.

References

[1] Wick, C. (Editor), "Tool and Manufacturing Engineers Handbook, Volume 2 Forming, 9th edition" Society of Manufacturing Engineers, Dearborn, MI, 1984.

[2] Kalpakjian, S., "Manufacturing Engineering and Technology," Addison-Wesley Publishing Co., Inc. 1989.

[3] Bralla, J. G. (Editor), "Handbook of Product Design for Manufacturing," McGraw-Hill Book Co., NY, 1986.

[4] Schwarz, S. S. and Goodman, S. H., "Plastic Materials and Processes," Van Nostrand Reinhold Co., New York, 1982.

[5] "Modern Plastic Encyclopedia," McGraw-Hill Modern Plastics, Hightstown, NJ, 1991.

[6] "ASM Metals Handbook, Vol. 15, Casting," ASM International, Metals Park, OH, 1988.

[7] Poli, C., Dastidar, P., and Graves, R. A., "Design Knowledge Acquisition for DFM Methodologies," Research in Engineering Design, Volume 4, Number 3, 1992.

[8] Ashby, M. F., "Materials Selection in Mechanical Design," Pergamon Press, New York, 1992.

[9] "ASM Metals Handbook, Vol. 14, Forming and Forging," ASM International, Metals Park, OH, 1988.

[10] Knight, W. A. and Poli, C., "Design for Forging Handbook," Mechanical Engineering Department, University of Massachusetts at Amherst, Amherst, MA, 1984.

[11] Knight, W. A. and Poli, C., "Design for Economical Use of Forging: Indication of General Relative Forging Costs," Annals of the CIRP, Vol 31, 1982.

[12] Gokler, M. I., Knight, W. A., and Poli, C., "Classification for systematic Component and Process Design for Forging Operations," Proceeding of the Ninth North American Manufacturing Research Conference, Pennsylvania State University, State Park, PA, May 1981.

[13] Knight, W. A., and Poli, C. "A Systematic Approach to Forging Design," Machine Design, January 24, 1985.

[14] *A Guide to Aluminum Extrusions*, The Aluminum Association, 818 Connecticut Avenue NW, Washington DC 20006.

[15] Boothroyd, G., "Fundamentals of Metal Machining and Machine Tools," McGraw-Hill Book Company, New York, 1975.

[16] Boothroyd, G., and Dewhurst, P., "Product Design for Assembly," Boothroyd Dewhurst Inc., Wakefield, RI, 1989.

[17] Boothroyd, G., "Assembly Automation and Product Design," Marcel Dekker, New York, 1992.

[18] Yoosufani, Z., "Design of Parts for Ease of Handling," M.S. Project Report, Mechanical Engineering Department, University of Massachusetts at Amherst, Amherst, 1978.

[19] Yoosufani, Z. and Boothroyd, G, "Design for Manufacturability - Design of Parts for Ease of Handling," Report No. 2, NSF Grant Apr 77-10197, Sept. 1978.

[20] Seth, B., "Design for Manual Handling," M.S. Project Report, Mechanical Engineering Department, University of Massachusetts at Amherst, Amherst, MA, 1979.

[21] Seth, B. and Boothroyd, G., "Design for Manufacturability - Design for Ease of Handling," Report No. 9, NSF Grant Apr 77-10197, Jan. 1979.

[22] Boothroyd, G., Poli, C. and Murch, L.E., Automatic Assembly, Marcel Dekker, Inc., New York, NY, 1982.

[23] Boothroyd, G., Poli, C. and Murch, L. E., "Handbook of Feeding and Orienting Techniques for Small Parts," Mechanical Engineering Department, University of Massachusetts at Amherst, Amherst, MA 01003, 1978.

[24] Ulrich, Karl T. et al, "Including the Value of Time in Design for Manufacturing Decision Making," MIT Sloan School of Management Working Paper #3243-91-MSA, Dec. 1991.

[25] Poli, C. and Fenoglio, F., "The Feasibility of Part Reduction in an Assembly," Concurrent Engineering, Vol.1, No. 1, Jan/Feb 1991.

[26] Slade, B. N. *Compressing the Product Development Cycle: From Research to Marketplace*, American Management Association, 135 W. 50th St., New York. Quote by permission of the publisher.

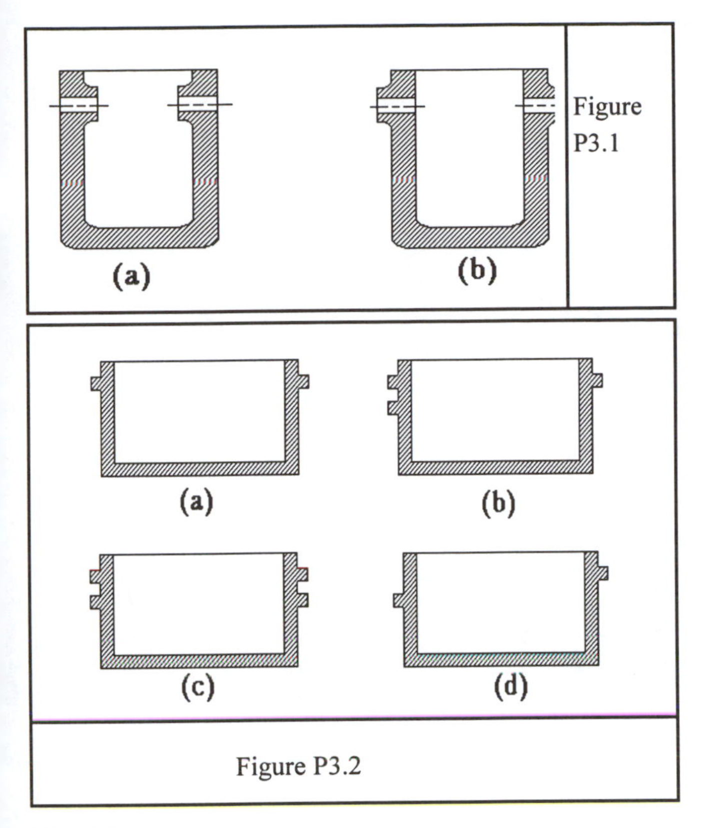

Figure P3.1

Figure P3.2

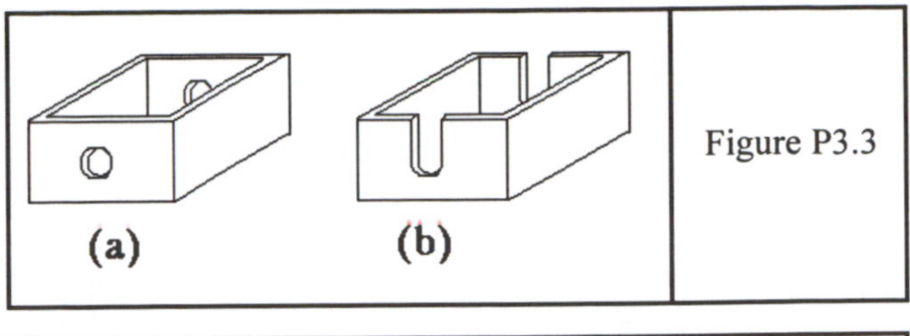

Figure P3.3

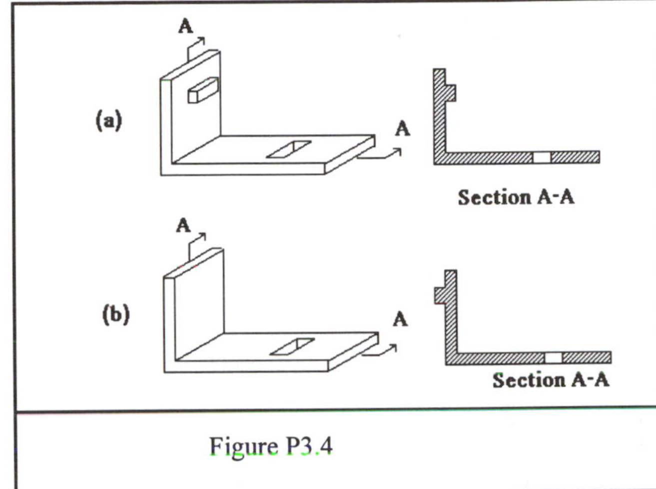

Figure P3.4

Problems

3.1 Figure P3.1 shows the sectional view of two proposed alternative designs for a box shaped part which is enclosed on four sides. Based on the injection molding advisor shown in Figure 3.14, which of the two designs is less costly? What is the approximate savings in tooling cost that one can achieve by using the least costly design? Assume that the wall thickness is the same in both designs.

3.2 Figure P3.2 shows the sectional view of four proposed alternative designs for a box shaped part which is enclosed on four sides. Based on the injection molding advisor shown in Figure 3.14, which of the designs is the least costly? Which of the four designs is the most costly? What is the approximate savings in tooling cost that one can achieve by using the least costly design as compared to the most costly design? Assume that the wall thickness is the same in all designs.

3.3 Figure P3.3 shows the preliminary sketch of two proposed designs. Based on the injection molding advisor shown in Figure 3.14, which of the two designs is less costly? What is the approximate difference in tooling cost between these two designs? Assume that the wall thickness is the same in both design.

3.4 Figure P3.4 shows the preliminary sketch of two proposed designs. Based on the injection molding advisor shown in Figure 3.14, which of the two designs is less costly? What is the approximate difference in tooling cost between these two designs? Assume that the wall thickness is the same ins both designs.

3.5 In an effort to become more competitive, a large automotive company has decided to expand its design-for-manufacturing group. Assume that you have applied for a position with that group. As part of the interview process you have been shown the proposed design of a die casting similar to the part shown in Figure P3.5 What suggestions would you make in order to reduce the cost to die cast the part? What suggestions would you make if the part were to be injection molded?

3.6 Exercise 3.5 for the part shown in Figure P3.6

3.7 Figure P3.7 shows the preliminary sketch of three proposed stamping designs. Based on the stamping advisor shown in Figure 3.26, which of the designs is the least costly to stamp? What is the approximate difference in tooling cost between these two designs? Assume that the wall thickness is the same in all designs

3.8 The proposed design of a stamping is shown in Figure P3.8. The location of hole A relative to the plate on which hole B is located is considered critical. Based on the stamping advisor shown in Figure 3.26, how would you rate this design?

3.9 Repeat exercise 8 for the stamping shown in Figure P3.9. Are there are redesign suggestions that you can make to help reduce the cost to stamp the part.

3.10 Figure P3.10 shows the current design of a stainless steel forging. Assume that you have been assigned the task of redesigning the forging so as to reduce manufacturing costs. What suggestions would you make?

3.11 Figure P3.11 shows the current design of a low carbon steel forging. Assume that you have been assigned the task of redesigning the forging so as to reduce manufacturing costs. What suggestions would you make?

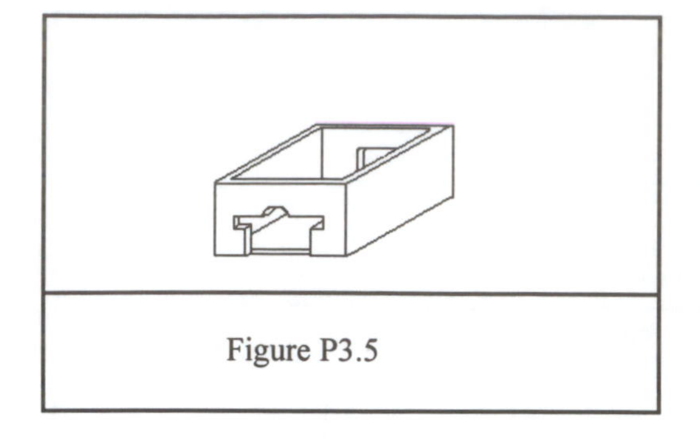

Figure P3.5

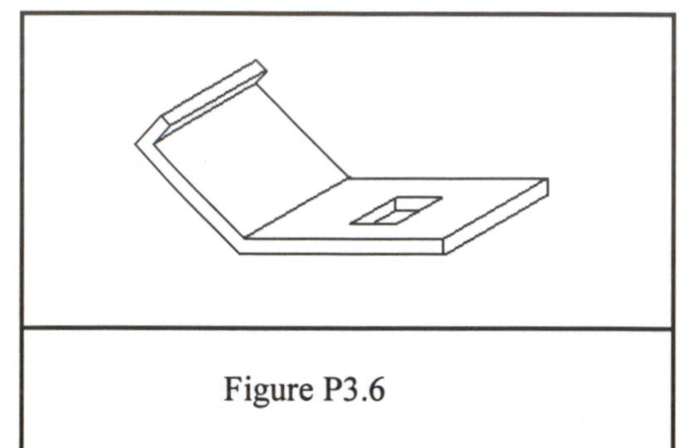

Figure P3.6

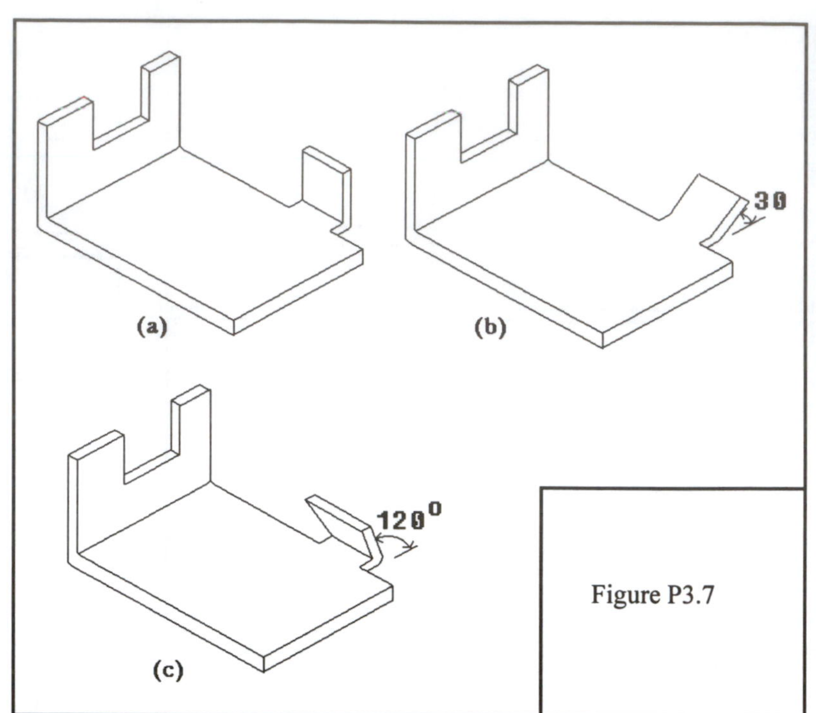

Figure P3.7

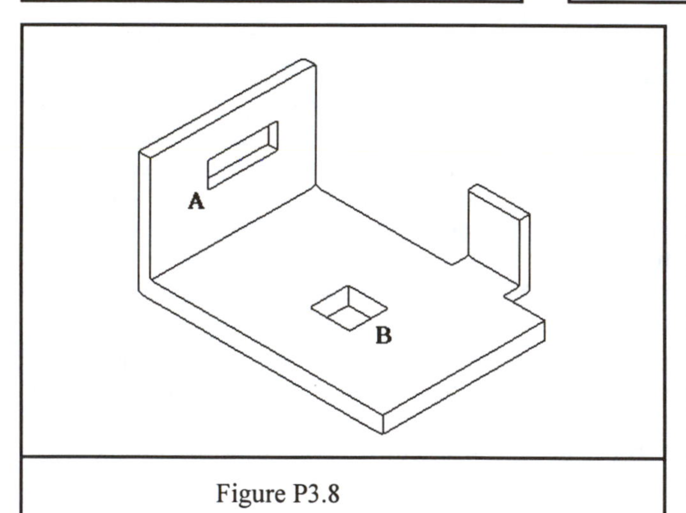

Figure P3.8

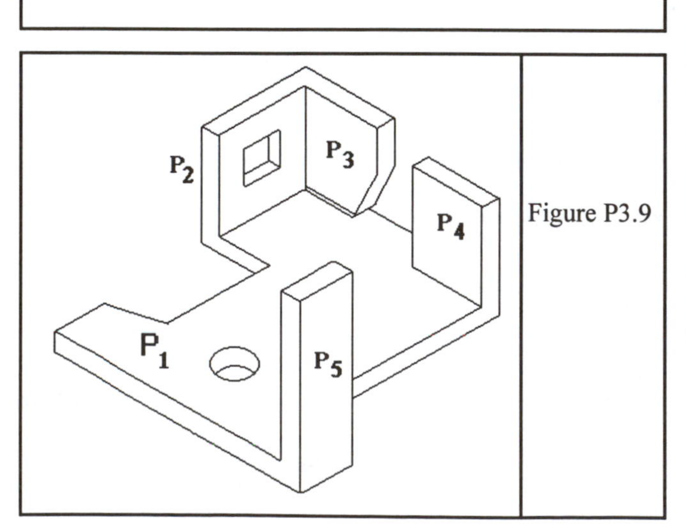

Figure P3.9

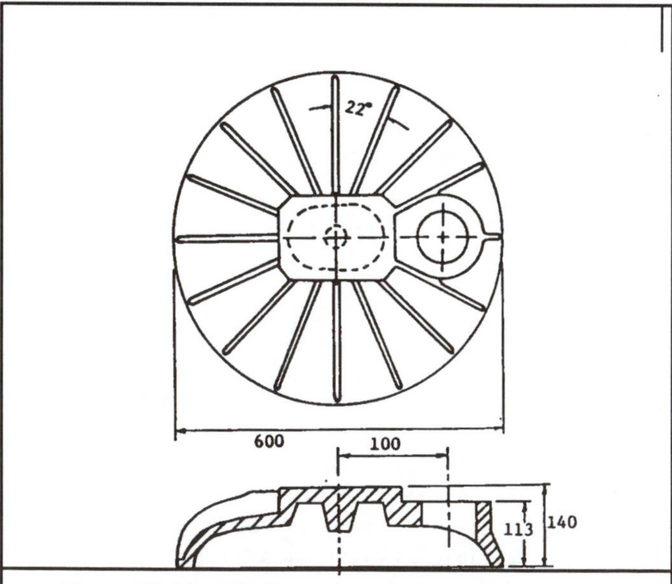

Figure P3.10 All dimensions in mm. Ref: Design for Forging Handbook by W. A. Knight and C. Poli, University of Massachusetts, Amherst, MA., 1981

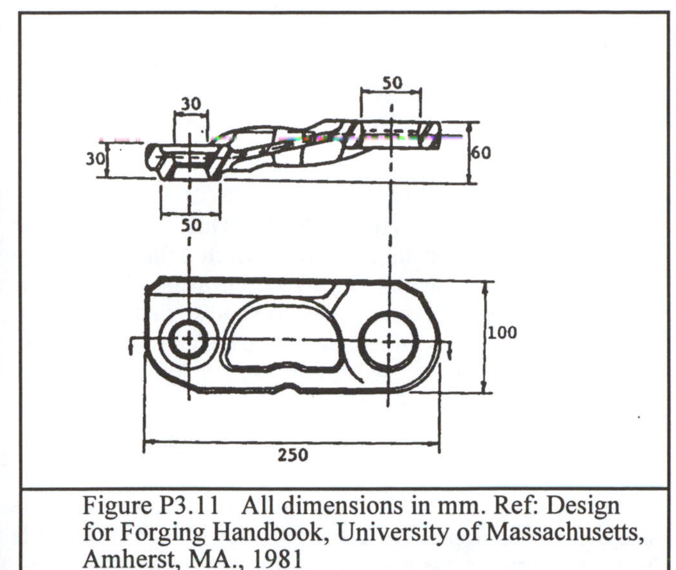

Figure P3.11 All dimensions in mm. Ref: Design for Forging Handbook, University of Massachusetts, Amherst, MA., 1981

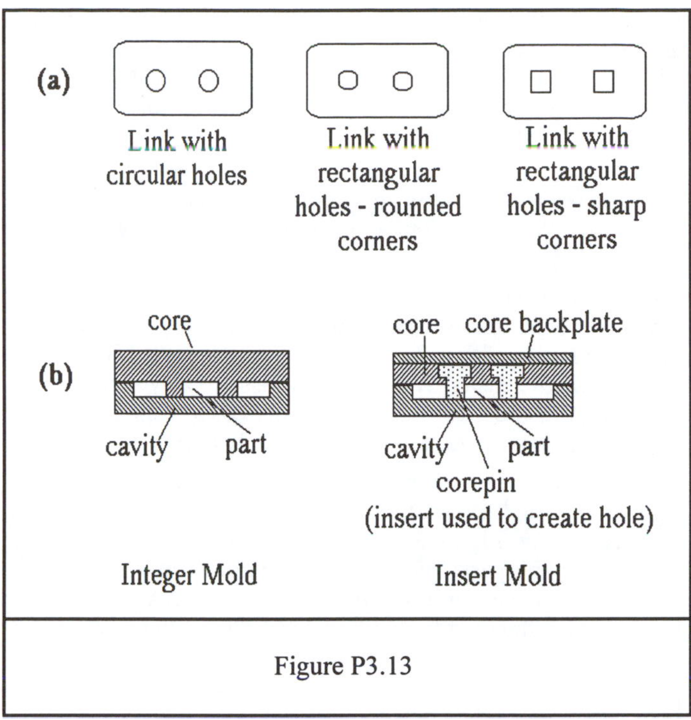

Figure P3.13

3.12 Figure P3.12 shows the current design of an aluminum forging. Assume that you have been assigned the task of re-designing the forging so as to reduce manufacturing costs. What suggestions would you make.

3.13 Figure P3.13 (a) shows three versions of an injection molded link with two holes. Version 1 has circular holes, version 2 has rectangular holes with rounded corners while version 3 has rectangular holes with sharp corners.

Figures P3.13 shows two alternative mold designs which can be used to produce these links. The first version is that of an integer mold in which the projections required to create the two holes are machined directly into the core. Also shown in Figure P3.13 (b) is a second version which is called an insert mold. In this version the projections shown in the first version of the mold are replaced by two core pins which are inserted into the core.

a) Assuming that the link is to be molded using an integer mold, which of the three hole designs would be least difficult to provide in the mold? Which would be the most difficult to provide) Repeat part (a) under the assumption that an insert mold is to be used to produce the link.

3.14 Shown in Figure P3.14 are two links with circular holes. In one design the link is provided with sharp external corners while in the other design the link is provided with rounded external corners. From the point of view of machining the mold required to injection mold the link, which version is easier (less costly) to produce?

3.15 Figure P3.15 shows an injection molded part with two grooves in the side walls. These grooves

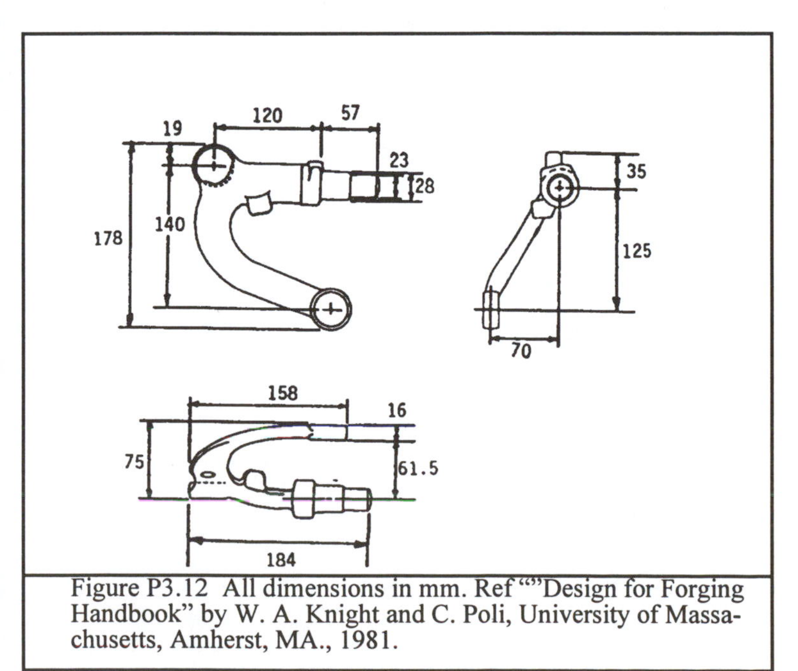

Figure P3.12 All dimensions in mm. Ref “”Design for Forging Handbook” by W. A. Knight and C. Poli, University of Massachusetts, Amherst, MA., 1981.

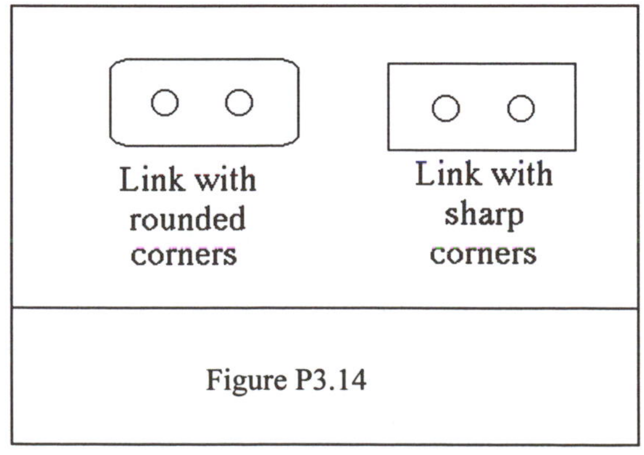

Figure P3.14

are called side shut-offs. Figures P3.15 (c) and (d) show two alternative mold designs for producing the part. From the point of view of machining, which mold is less costly to produce? Are there any difficulties, other than machining difficulties, with either of these molds?

3.16 Figure P3.16 shows two alternative molds which can be used to create a deep box shaped part with a through hole in the base of the box. From a machining point of view, which mold is easier to produce?

3.17 Figure P3.17 shows a drawing of a type of caster assembly which is commonly found on heavy boxes. Using the assembly advisor, Figure 3.42, evaluate the ease or difficulty of assembling the caster. What suggestions would you make in the design in order to reduce assembly costs? Can you estimate the approximate savings in assembly costs?

3.18 Figure P3.18 shows a drawing of a portion of a floppy disk drive which is commonly found in PCs. Using the assembly advisor, Figure 3.42, evaluate the ease or difficulty of assembling the parts shown. What suggestions would you make in the design in order to reduce assembly costs? Can you estimate the approximate savings in assembly costs?

3.19 Figure P3.19 shows the assembly drawing for a common stapler. The entire stapler itself is considered as part number 1 and is shown in Figure (a). The remaining components [shown in Figure P3.19(b)] consist of the following (the dimensions of each component are indicated in mm and are shown in parentheses:

2 - Base (injected molded)

3 - Staple remover (injected molded). This is a component not normally found on staplers. It is inserted through the top of the base and is fastened to the base via a snap fit. The remover can be easily pivoted about one end so that after insertion it rests in a recess contained on the underside of the base (not shown). It can be used to remove staplers from papers which have been stapled together.

4 - Base plate (stamped). Used to fold the staplers as they are used.

5 - Spring plate (stamped). This spring is compressed by the staple holder as staples are used. It is used to return the staple holder and cover to its normal position after staples are used.

6 - Staple holder subassembly (shown in greater detail in Figure P3.19(c)) and P3.19 (d).

7 - Top cover subassembly (shown in greater detail in Figure P3.19(e))

Using the assembly advisor, estimate the difficulty (or ease) of assembling the stapler. The handling and insertion characteristics of each part and subassembly is indicated in the drawings.

3.20 In Section 3.10.2 it was shown, for the case of both injection molding and die casting, that if the relative die construction cost for a replacement part is less than the number of parts being replaced, then it is less costly to replace these individual parts by a single more complex part. Following a procedure similar to the one used in section 3.10.2 show that this same result can be applied to stamped parts.

3.21 A possible method for reducing assembly costs shown in Figure P3.17 is to die cast the plate and two brackets as a single part. Based solely on the potential reduction in overall manufacturing costs, would you recommend that the three parts be combined into one?

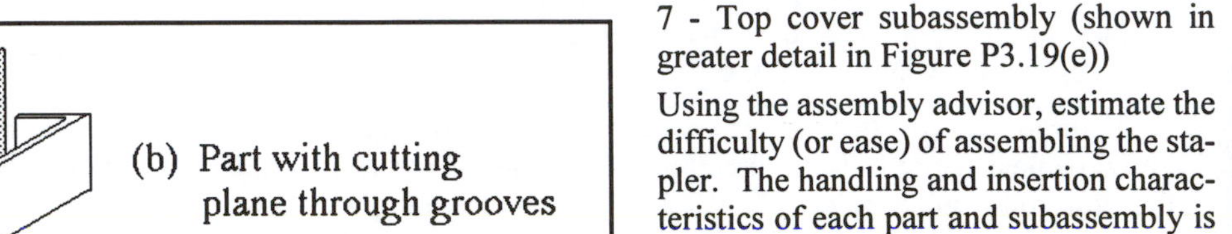

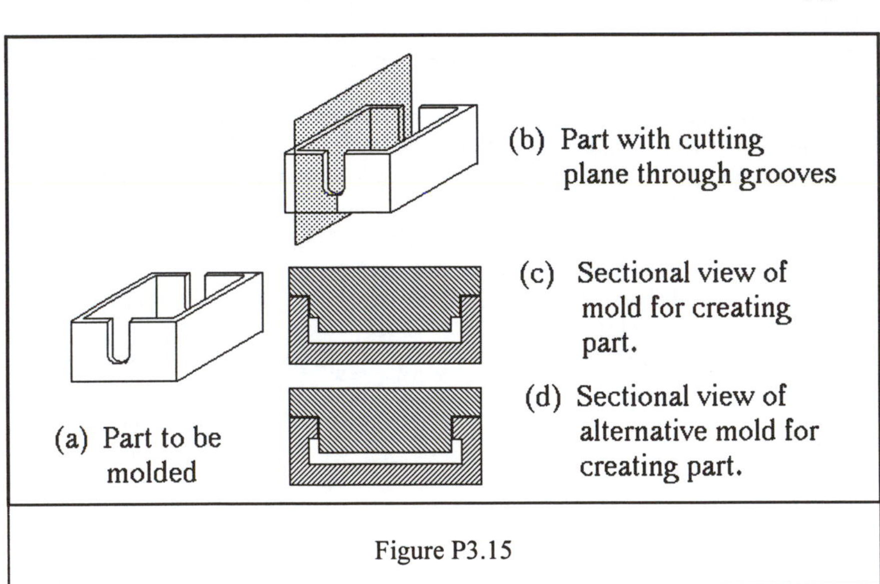

Figure P3.15

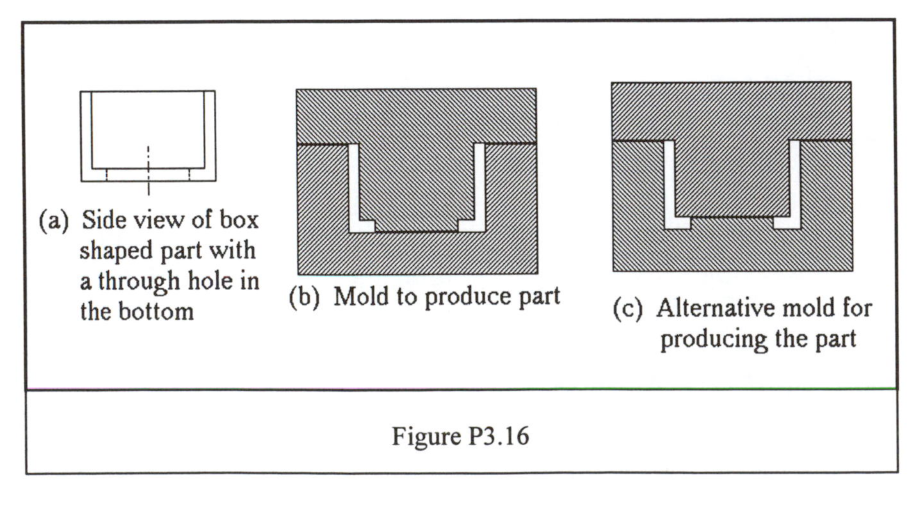

Figure P3.16

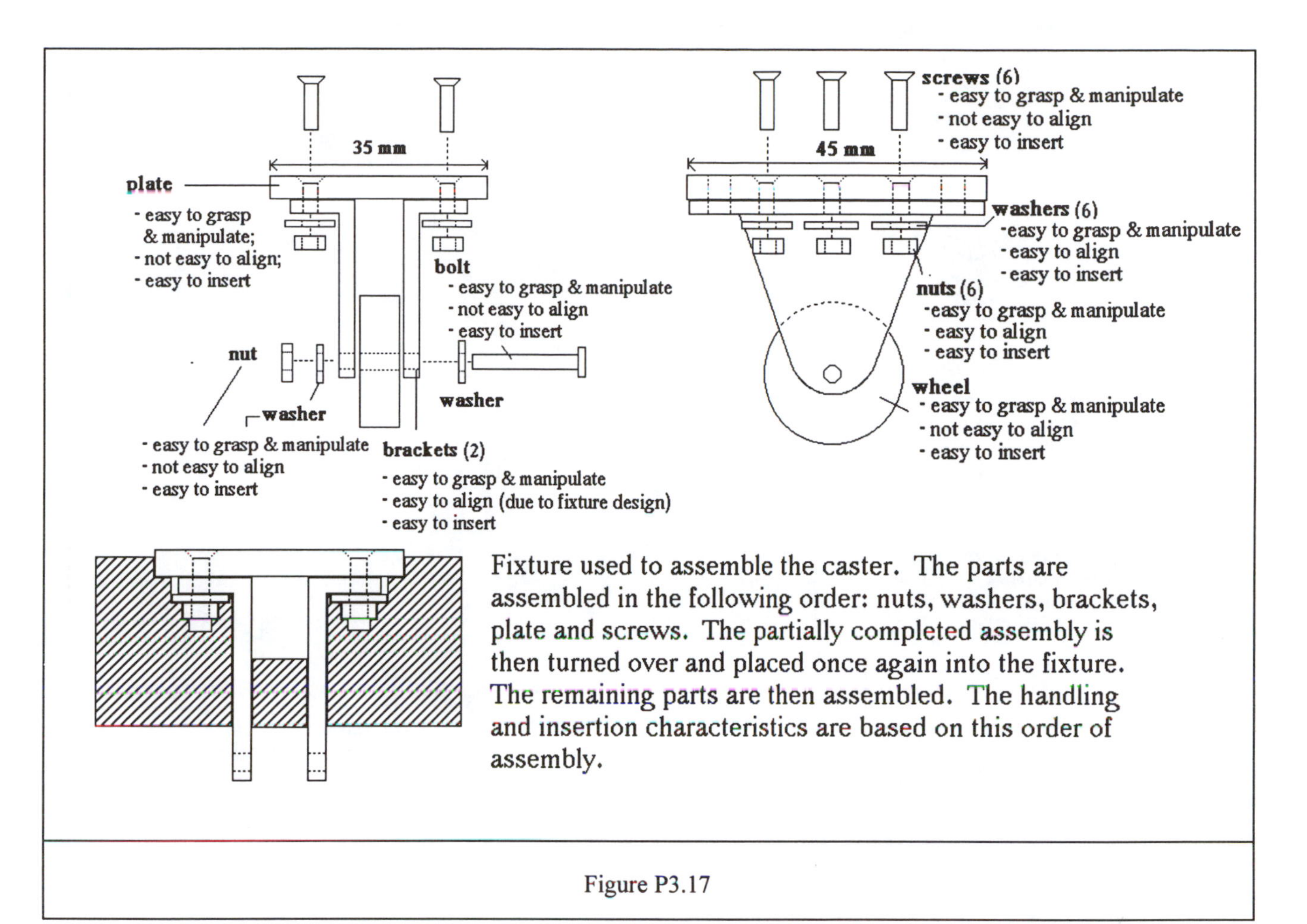

Fixture used to assemble the caster. The parts are assembled in the following order: nuts, washers, brackets, plate and screws. The partially completed assembly is then turned over and placed once again into the fixture. The remaining parts are then assembled. The handling and insertion characteristics are based on this order of assembly.

Figure P3.17

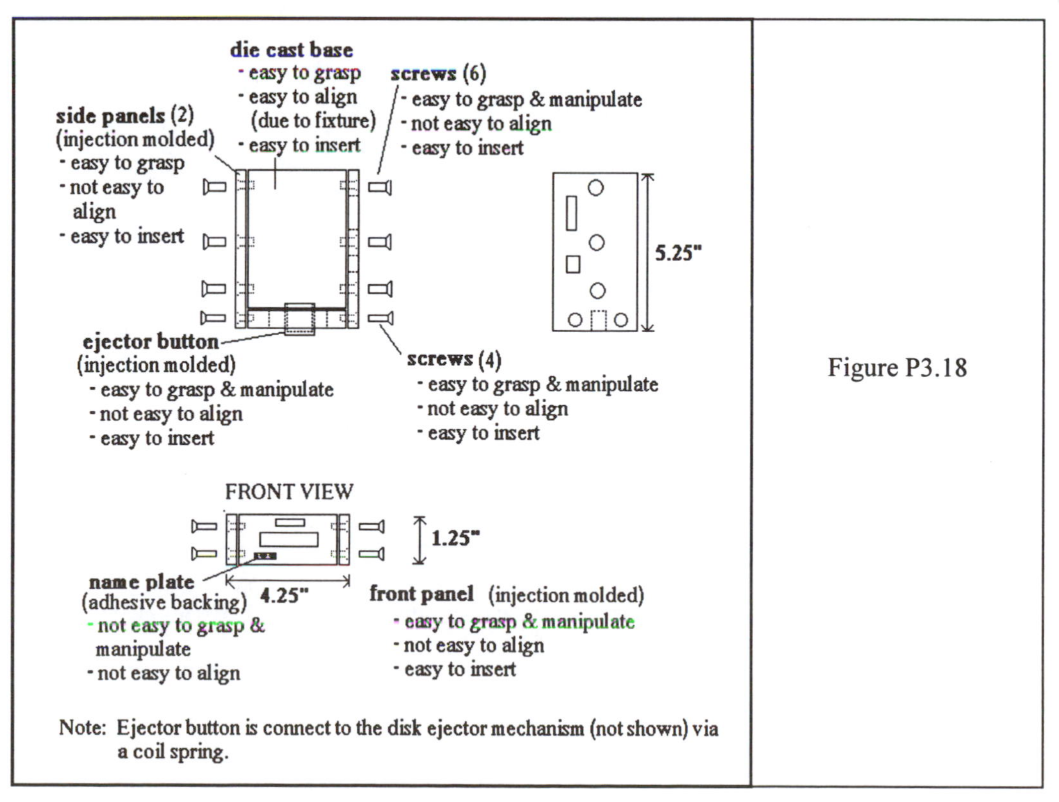

Figure P3.18

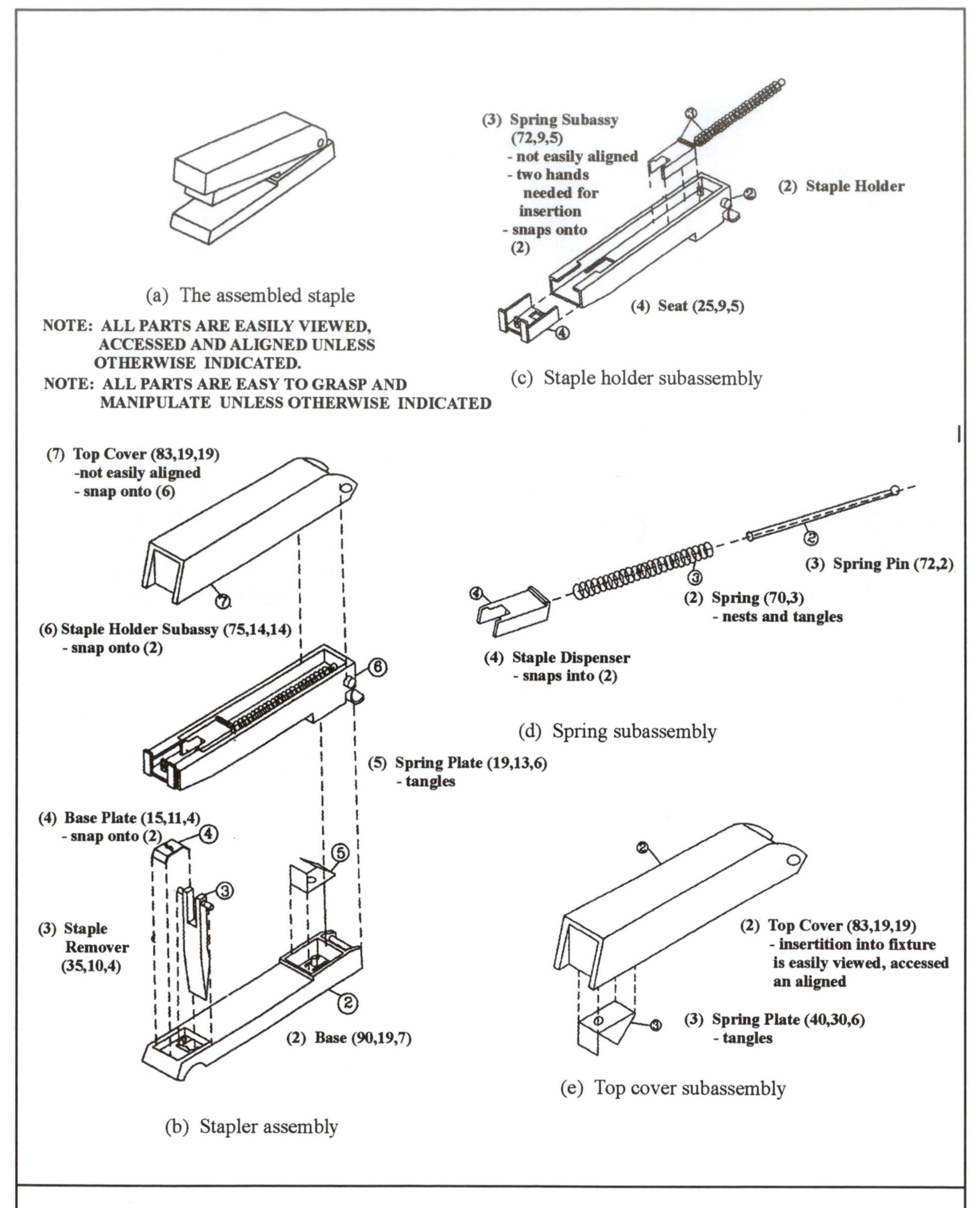

Figure P3.19 (Ref. Assembly Analysis and Comparison of Two Staplers by N. Renganeth and R. Radhakrishnan, Mechanical Engineering Department, University of Massachusetts Amherst, Amherst, MA, May 1991)

Part II Engineering Conceptual Design

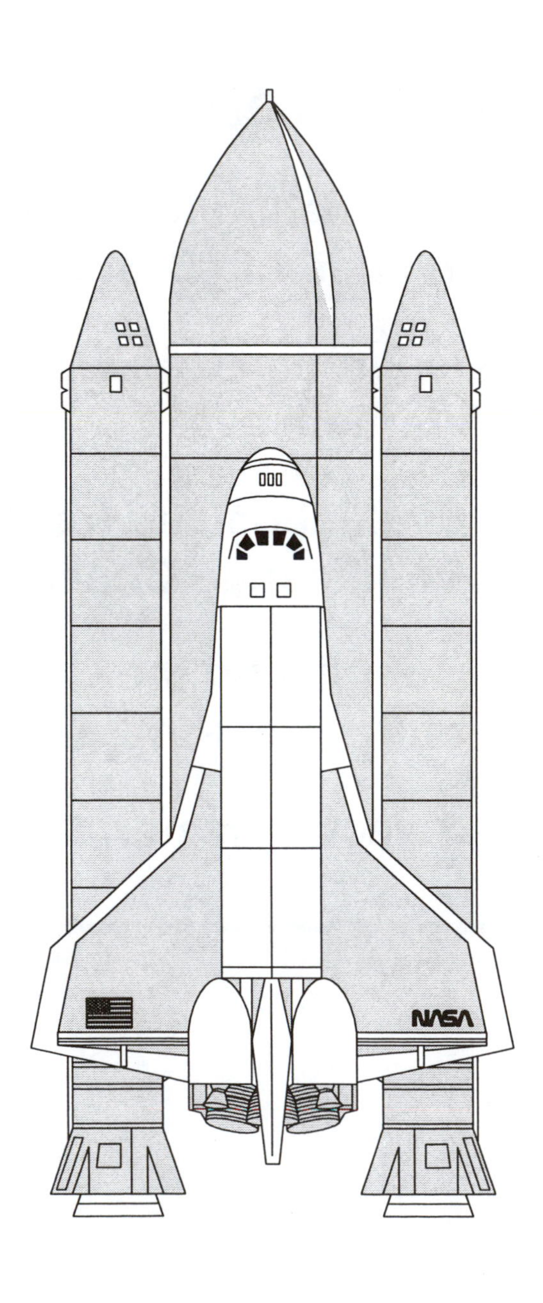
NASA

Chapter Four

From Marketing Concept to Physical Concept

"This phase...takes the statement of the problem and generates broad solutions.....It makes the greatest demands on the designer, and [provides] the most scope for striking improvements.....It is the phase where...the most important decisions are taken."

Michael J. French*[1]
Conceptual Design for Engineers (1985)

4.1 The Nature of the Engineering Conceptual Design Problem

4.1.1 The Starting Place

The starting place for engineering conceptual design varies — slightly — depending on whether the designed object is a product, a sub-assembly of a product, or a component within a product or sub-assembly. We will say more about the nature of these slight variations later in this chapter, and also in subsequent chapters. For now, however, the small differences need not distract us from understanding the essence of engineering conceptual design.

In all cases, the *function* of the designed object (whether product, sub-assembly, or component) is known at the start of engineering conceptual design. In addition to function, of course, there are also other requirements. As a way of formulating engineering conceptual design problems, we will soon (in Chapter 5) gather the functional requirement,s and all the additional requirements, together into an Engineering Design Specification. Then the task of engineering conceptual design is to produce a physical description of the object that will fulfill the required function(s), and meet the other requirements as well.

* This quotation is by permission of its author, and is reprinted from *Conceptual Design for Engineers*, The Design Council, London, 1985.

In engineering conceptual design, we begin with knowledge of *what* the required function is; we add the information about *how* the function will be performed. This is a huge step. It requires that we enter the real world of physical principles, materials, how things work, and how things work together. The amount of information about the designed object that is added during engineering conceptual design is tremendous, and there is not again in the design process an opportunity for such significant choices.

4.1.2 The Goal

The goal of engineering conceptual design is to determine the *physical concept* of the designed object. By the physical concept we mean: (1) information about the *physical principles* by which the object will achieve its principal function(s), and (2) an abstract physical description of the object called its *embodiment*. See Figure 4.1.

For example, suppose the required function is to support a load over an open space. One physical effect by which this might be accomplished is that longitudinal tensile and compressive stresses within a material can support a perpendicular load. This is, of course, a result from beam theory.

One physical embodiment that uses this effect is a long slender member of uniform cross section; we call it a "beam". Note in this example how the physical effect is an integral part of the embodiment; had we elected to make use of only purely tension or compression stresses to support the load, an embodiment called a truss might have resulted.

An embodiment is an *abstract* physical description; few details are provided. We know only, for example, that a "beam" is a long slender member of uniform cross section, but we do not know the cross sectional shape (i.e., its configuration), nor do we know anything about the dimensions. Similarly, we know that a "truss" is an arrangement of connected long slender members in tension or compression, but we do not know the actual arrangement of the members, or their lengths, shapes, or dimensions.

These simple examples show in a general way what is meant by the *physical concept* of a designed object. Simply put, a physical concept includes the physical principles employed and an abstract physical embodiment. We shall see, however, that when the object is a complex assembly of sub-assemblies and components, the physical concept is not so simple as in the examples we have used here to introduce the basic ideas.

4.1.3 About Part II and This Chapter

This chapter is an introduction to Part II of this book, which deals with the application of guided iteration to engineering conceptual design. The other chapters in Part II are:

- Chapter 5 Formulating the Problem - The Engineering Design Specification
- Chapter 6 Generating Conceptual Design Alternatives
- Chapter 7 Evaluating and Redesigning
- Chapter 8 Completing the Conceptual Design of Parts
- Chapter 9 Selecting Materials and Processes

Here in Chapter 4, we describe an aspect of the engineering conceptual design process called *conceptual decomposition.* The entire engineering conceptual design process is affected by the nature of conceptual decomposition, so this introduction helps make all the subsequent chapters in Part II more meaningful.

As noted above, slight variations of the basic engineering conceptual design problem are encountered in the course of engineering design. These variations depend on whether the designed object is:

(1) a new product,

(2) a sub-assembly within a product, or

(3) a component within a product or sub-assembly.

In this chapter we will discuss how these variations arise out of the process of conceptual decomposition.

4.2 The Process of Conceptual Decomposition

4.2.1 Two Approaches to Conceptual Decomposition: Direct and Function-First

The process of conceptual decomposition for either a product (except a single part product) or sub-assembly results in the creation of a set of embodiments of new sub-assemblies and components. For convenience, we'll call them *subsidiary* components and sub-assemblies. They are conceived (or selected) and connected so that they work together to fulfill the functions of the product. One author calls this a *scheme* [2], which is a very good term. Note that the subsidiary components and sub-assemblies do not exist until the designer(s) create them; that is, there exists no prior information that they exist or about what they may be. The

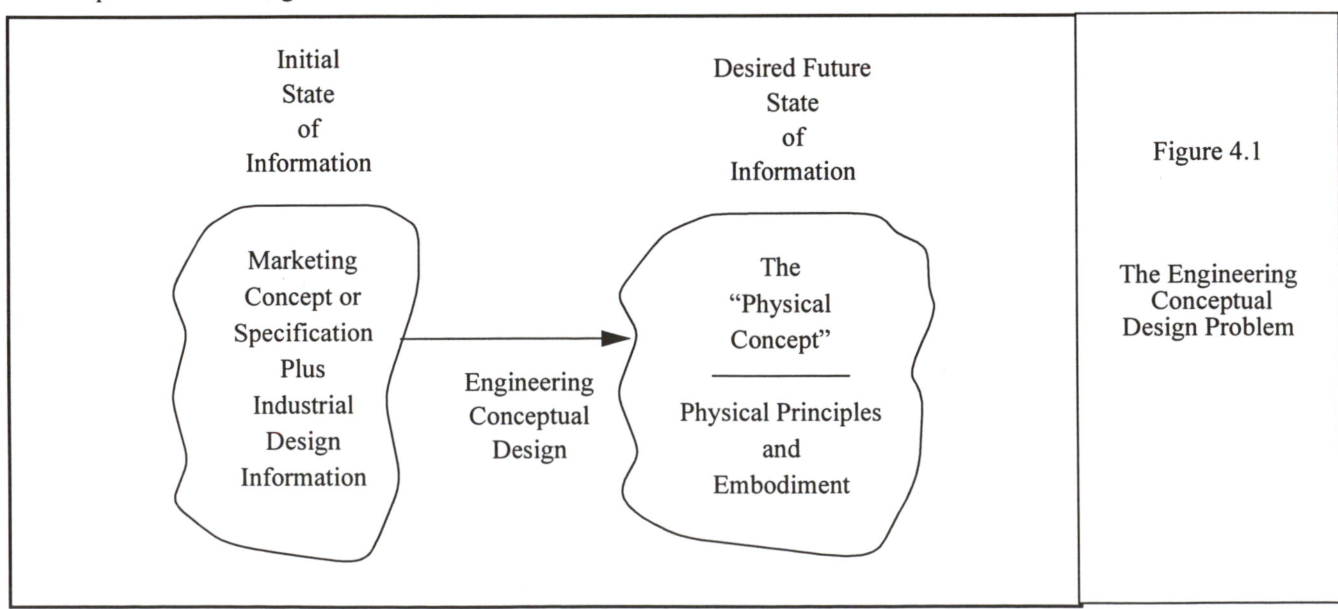

Figure 4.1

The Engineering Conceptual Design Problem

> "The beginning is the most important part of the work."
>
> Plato
> *The Republic (329)*

process is illustrated schematically for a single stage of decomposition in Figure 4.2.

Instead of saying that a product is "decomposed" into subsidiary components and sub-assemblies, it might be more accurate to say that a product is "composed" of the subsidiary components and sub-assemblies. However, it is common practice to use 'decomposition' so we will stick to that.

There are (at least) two basic approaches to conceptual decomposition. One — we'll call it the *direct* decomposition method — is to decompose the product or sub-assembly directly into its subsidiary sub-assembly and component embodiments. For example, we decompose an automobile into, say, embodiments of its engine, drive train, body, suspension system, and (say) steering system.

Another approach is to perform the decomposition process in two distinct steps: (1) first a *functional* decomposition is done in which required sub-functions — purely as functions with no associated embodiments — are identified; (2) then embodiments are identified to fulfill each of the sub-functions. We call this *function-first* decomposition.

Note that the results of the first step in a function-first decomposition say or imply nothing at all about the physical embodiments to be employed to fulfill the sub-functions. This first step identifies *only* the sub-functions required to fulfill the overall function.

In general, there are a number of possible decompositions into sub-functions that could work for most products. Usually these different decompositions will be based on the use of different physical principles to accomplish the overall task. In addition, each functional decomposition may also be achievable by different embodiments. Thus, as discussed in more detail in Chapter 6, a wide variety of alternative solutions can be generated by this approach. Naturally, some of the alternatives will ultimately be judged better than others when they are evaluated in terms of all the various requirements, including functionality.

It is important to note that the sub-functions in a functional decomposition generally are not simply parcelled-out pieces of the product's function. That is, this process is not like cutting up some functional pie where the whole is the sum of the parts. Generally the sub-functions, like the subsidiary components and sub-assemblies, do not exist among the product's functional requirements, though they may in some cases be implied to some extent. Examples of this are given below.

The two step function-first conceptual decomposition process described above (first decomposition into sub-functions followed by determination of embodiments to fulfill them) is a sound idea, and theoretically it may well be the best way to proceed. It is the method touted by the German design research community. See [3, 4, 5, 6, 7, 8]. However, the process is not now often followed consciously or explicitly in the practice of mechanical design, at least in the United States. It is, however, a more common practice in the

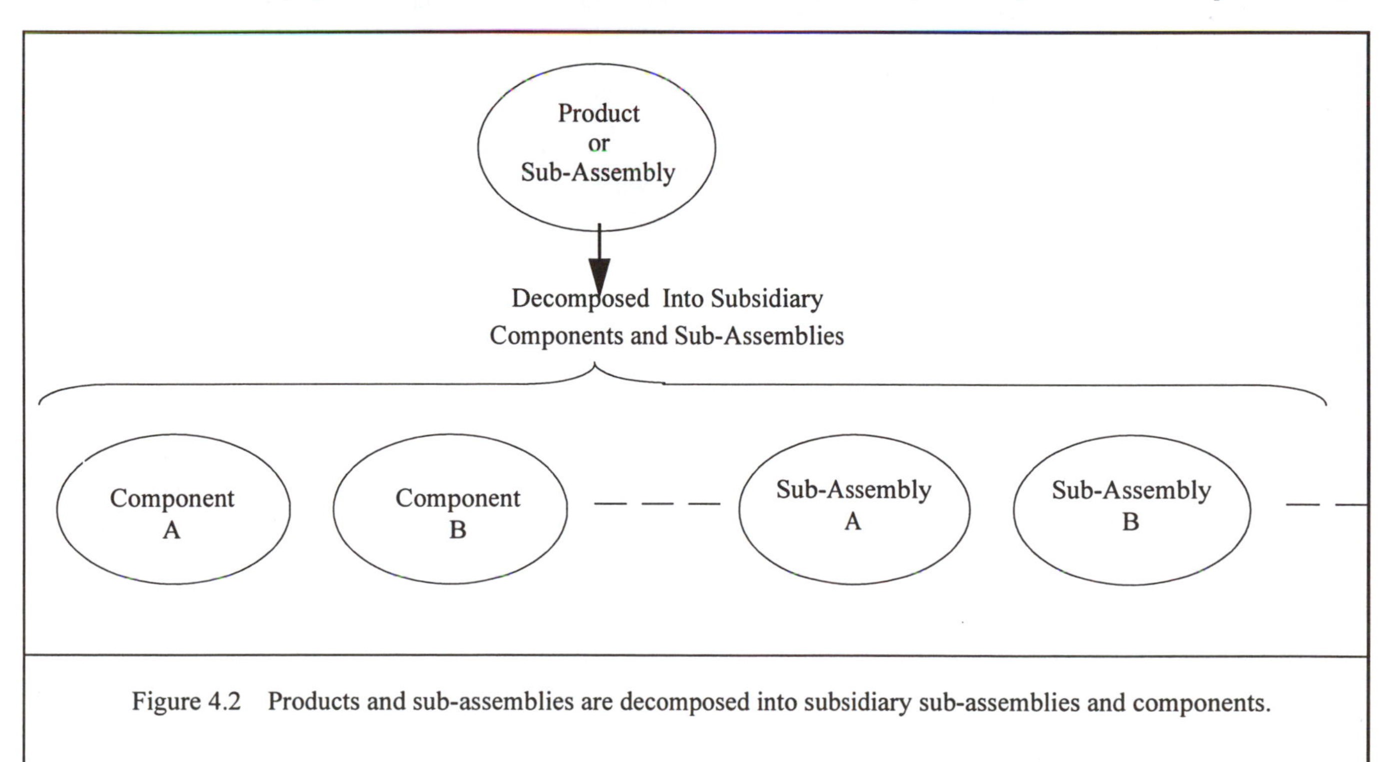

Figure 4.2 Products and sub-assemblies are decomposed into subsidiary sub-assemblies and components.

design of electrical circuits. Most mechanical designers generally follow the direct approach and identify the embodiments directly, often in a sketch. For example, in a bicycle, we directly identify a "handlebar", which is an embodiment (for what function?). And we decompose the other parts and sub-systems without consciously or *explicitly* thinking first about the function each is being created to perform. Note in the bicycle, if we did think first about the function required, we might then be more likely to consider embodiments other than traditional steering systems using handlebars to fulfill the human guidance system.

Example. Consider the design of an automobile braking system. The overall function is to stop the wheel from turning. One direct decomposition into embodiments to fulfill this function would immediately give us a foot pedal, a hydraulic system, and brake shoes and drums.

Alternatively, a function-first approach would not identify any embodiments until after a set of sub-functions has been developed to perform the overall function. In this case, the first function needed is a signal from the operator that the brake is to be applied. In the usual case, this is provided by the driver's brain, leg, and foot. But there are many other ways to provide the signal function. It could be a hand signal, a voice signal, a head signal, and even a signal from a computer with a crash avoidance vision system. Of course, some of these are not very practical (we think *now*) but note how thinking about the signal function as a pure function led us to consider new embodiment possibilities.

Another function to be performed in the brake system is the application of a retarding torque to the wheel. We could achieve this function with the usual shoe, or with a generator, or perhaps an electric brake. The latter is described by Crossley [4] as "the pedal slides on a resistor, and the shoe is replaced by an eddy-current brake, thus eliminating all wear. This (idea) might be thrown out today because it depends on the battery, but wait until electric cars become more common." Note again that by thinking first in purely functional terms, we are led to consider ideas for embodiments that might not otherwise occur to us.

When designers use the direct decomposition method, there is no doubt that they have the functions of the subsidiary sub-assemblies and components in mind, even if unconsciously. That is, what we call the direct method is most likely one of *simultaneous* decomposition into embodiments and their associated functions. Function is not ignored in this process; however, by not considering function separately and explicitly, designers may fail to consider alternative ways to achieve the required functionality.

Which of the two methods described above — direct or function-first — is the best way to perform the decomposition process? As a guiding principle, we recommend that function-first decomposition into sub-functions always be *attempted* before considerations of embodiments, but only insofar as is practical. That is, designers should try decomposition function-first, but not carry the effort compulsively to an extreme. The reason is that thinking about pure function without embodiments is highly abstract, and not (yet) a familiar way for most of us to think. It is therefore difficult to do effectively and efficiently. Nevertheless, we think it should always be attempted before resorting to the direct approach because function-first decomposition can often lead to very good innovative new ideas.

In the next two sub-sections, we discuss conceptual decomposition of products and sub-assemblies primarily using the direct approach. Later in this chapter, and again in Chapter 6, we discuss the function-first approach in some more detail.

4.2.2 Conceptual Decomposition of Products

We begin with the decomposition of products because the other two cases (sub-assemblies and components) actually originate during the conceptual decomposition of products.

For a product, the engineering conceptual design problem (Figure 4.1) has as its initial information state the marketing concept or specification. It also includes the information developed in industrial design. This may or may include information about the product's approximate size, shape, special features, materials, etc.

When developing the physical concept of a product by conceptual decomposition, we create only embodiments of the *principal functional* subsidiary components and sub-assemblies. That is, we create only those components and sub-assemblies that are essential to the required overall functionality of the product. Usually, therefore, we perform only one stage of conceptual decomposition for each product as shown in Figure 4.2. For example, we go no farther than to decompose an automobile into, say, its engine, drive train, body, suspension system, and (say) steering system. (Subsequently, each of these subsidiary sub-assemblies will be conceptually decomposed as a part of their own engineering conceptual design phase.)

In order to explain the physical principles by which a product will work, we describe in qualitative physical terms how the subsidiary components and sub-assemblies will work together to accomplish the required functions of the product. This requires that we describe the *principal functions* of each subsidiary component and sub-assembly.

To understand how the functions of the components and sub-assemblies work together to achieve the overall function, we must also describe the principal relationships or connections that each has with the others. These relationships and connections are called *couplings*. Sometimes couplings are shown in a sketch rather than explained in words.

The most common kind of coupling among subsidiary components and sub-assemblies is physical, as when they must be physically connected or in some special spatial relationship to one another. But there are also energy couplings where power is transmitted from one component or sub-assembly to another. There may also be couplings of force, as when one must provide support for another. And the subsidiary components and sub-assemblies may be required to share (or compete for) scarce or limited resources like space,

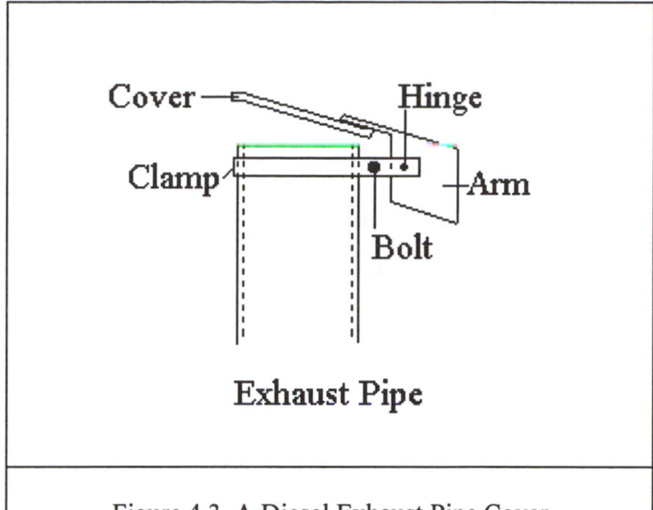

Figure 4.3 A Diesel Exhaust Pipe Cover

weight, or cost. To show and express the functions and couplings of the subsidiary components and sub-assemblies, a sketch is often worth a thousand or more words.

Of course, there is usually more than just a single stage of conceptual decomposition design of most products. As noted above, each of the sub-assemblies created in the first stage of decomposition will ultimately also be conceptually decomposed into its respective sub-assemblies and components. And each of those sub-assemblies will be decomposed into its respective sub-assemblies and components, and so on and on until only components remain. For example, an automobile engine may be decomposed into, among other things, an engine block and a carburetor. Then in turn, the carburetor may be decomposed into, among other things, a float and a cover. Thus the process of conceptual decomposition repeats (or recurs) until no new sub-assemblies are created.

Example. As a very simple example of this process, and the kind of information included, consider the diesel exhaust cover shown in Figure 4.3.

The physical concept of the product as shown has four principal embodiments: the clamp, the cover, the arm, and the hinge bolt. The physical principle that holds the clamp in place on the exhaust pipe is friction resulting from the normal force developed by the nut and bolt passing through the clamp's flange. The physical effect that enables the cover to achieve its function (to cover the opening) is gravity. The cover is lifted up by the pressure of the exhaust gases to allow them to escape. The arm acts as a support for the cover and as a counterweight to reduce the exhaust pressure required to lift the cover. The clamp is physically coupled with the exhaust pipe, and so is the cover when it is resting on the top of the exhaust pipe. The arm and clamp are physically coupled with each other via the hinge. The arm and cover are physically coupled by the spot welds. The basic materials class for all parts is steel, and the arm and cover are to be stamped.

This physical concept is illustrated schematically in Figure 4.4. The figure shows the *decomposition diagram* identifying the principal sub-assemblies and components together with the main couplings. Other alternative physical concepts to fulfill the functional and marketing requirements of this product are no doubt possible.

For a product as simple as the exhaust pipe cover, the physical concept includes information about individual parts. This is not generally true, however, for larger and more complex products. The initial decomposition diagram for a window air conditioner, for example, might include only the mechanical refrigeration unit, the motor, a fan for indoor circulation, a fan for outdoor air circulation, the enclosure, and a control system. Then the decomposition of the refrigeration unit would include the compressor, the evaporator, the condenser, an expansion valve, and the piping. We would thus not get the level of separate parts until after several stages of decomposition.

An example of a coupling in a window air conditioner is the requirement that the outdoor air circulating fan to be able to supply the required amount of air to and through the condenser coils. And of course, all of the subsidiary sub-assemblies and components are coupled in this product in terms of limited space requirements.

Another Example. Consider how the recursive conceptual decomposition process might unfold in the case of a flat plate solar collector for heating water. These collectors consist of an outer frame, one or two transparent covers, a blackened plate with tubes attached that carry the water to be heated, insulation around and behind the plate, and a back cover. A first decomposition could be into two special purpose sub-assemblies, the enclosure and the collector plate, as shown in Figure 4.5. Note that each of these sub-assemblies

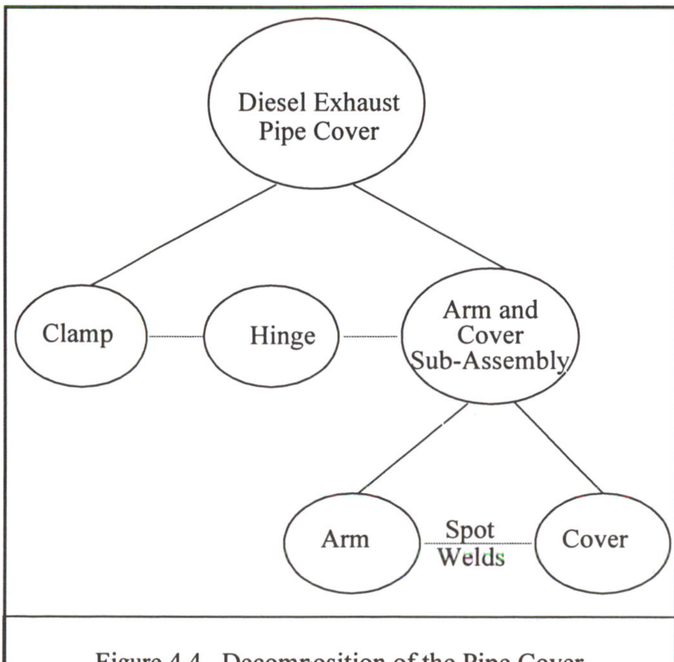

Figure 4.4 Decomposition of the Pipe Cover

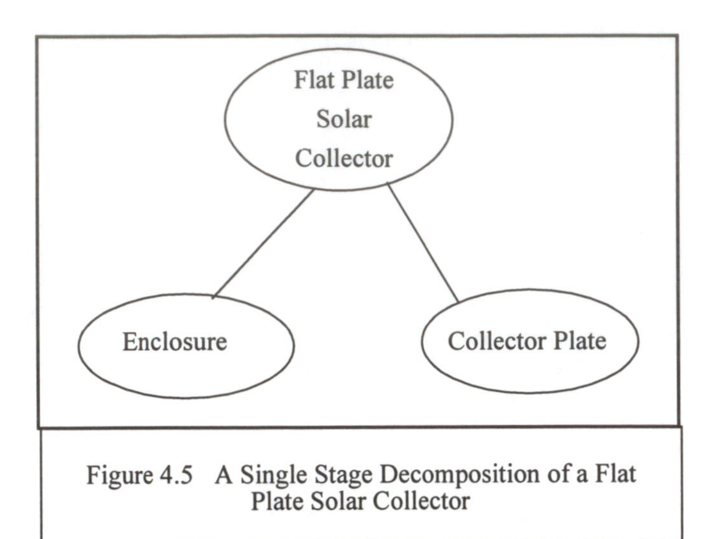

Figure 4.5 A Single Stage Decomposition of a Flat Plate Solar Collector

will have functional requirements unique to its particular role in the overall product, requirements derived from but not the same as those of the whole product. For example, the Engineering Design Specification will have an efficiency requirement, but will say nothing about the required emissivity of the collector plate.

Note that the two special purpose subsidiary sub-assemblies created by the initial conceptual decomposition of the collector are not independent of each other; they are coupled. For example, assuming there is a total cost limit, a more expensive enclosure means there must be a less expensive collector plate, or vice-versa. Parts that must fit together are coupled physically.

After the first stage of decomposition, each newly created subsidiary sub-assembly of the solar collector is subjected to its own engineering conceptual design process. For the enclosure, this results in another stage of conceptual decomposition as shown in Figure 4.6. In this decomposition, the glass covers are created as standard parts (because there are standard glass sheet sizes). The frame is created as another special purpose subsidiary sub-assembly, which again is treated to its own engineering conceptual design resulting in yet another round of conceptual decomposition. And so on — until no special purpose sub-assemblies are created during conceptual decomposition

The components — the parts and standard modules — that are created by the conceptual decomposition of product and their sub-assemblies do not, by their nature, require additional conceptual decomposition. Once created, and their principal functions and couplings identified, they are ready for their own version of engineering conceptual design as discussed briefly below and in more detail in Chapter 8.

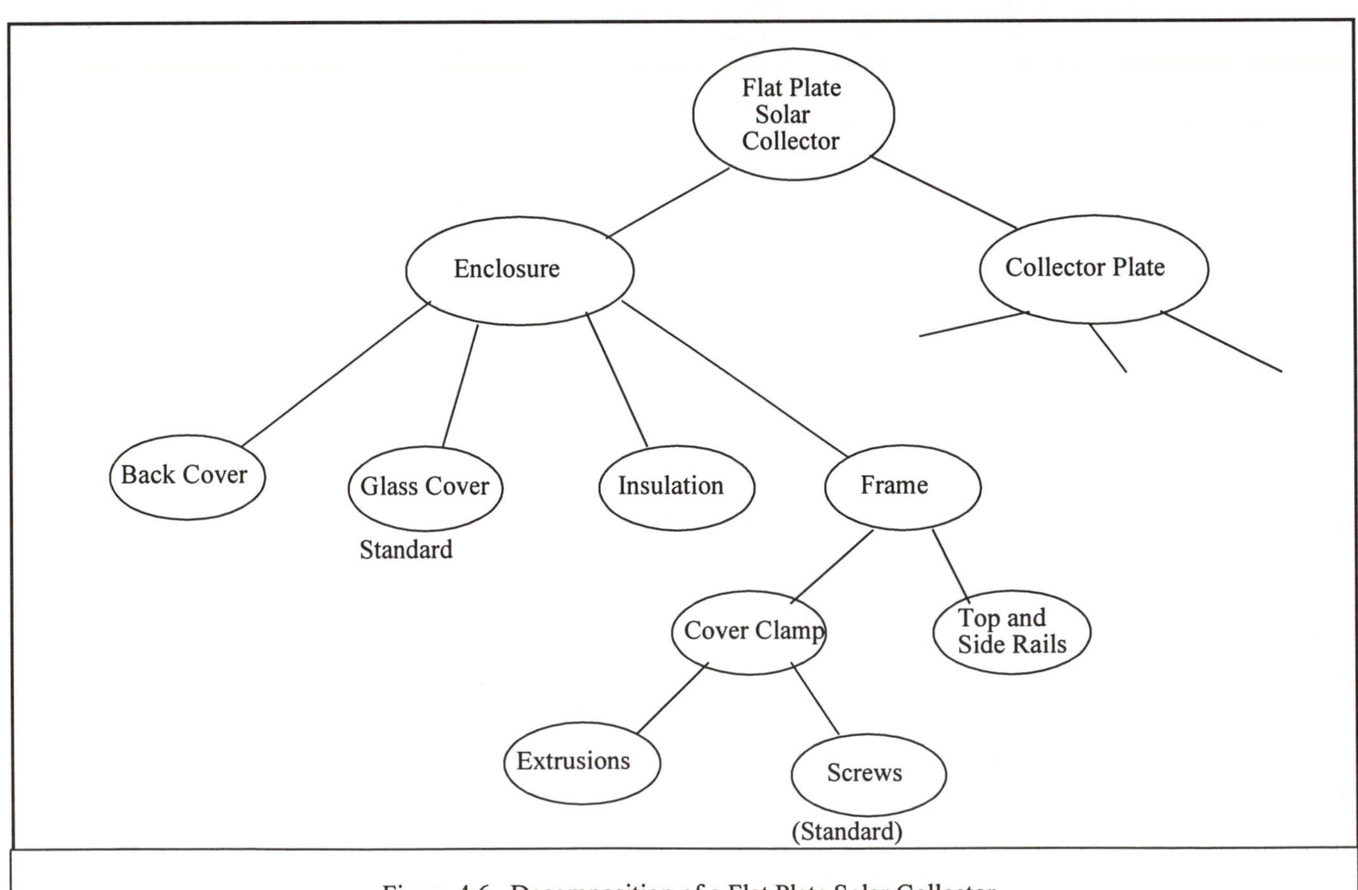

Figure 4.6 Decomposition of a Flat Plate Solar Collector

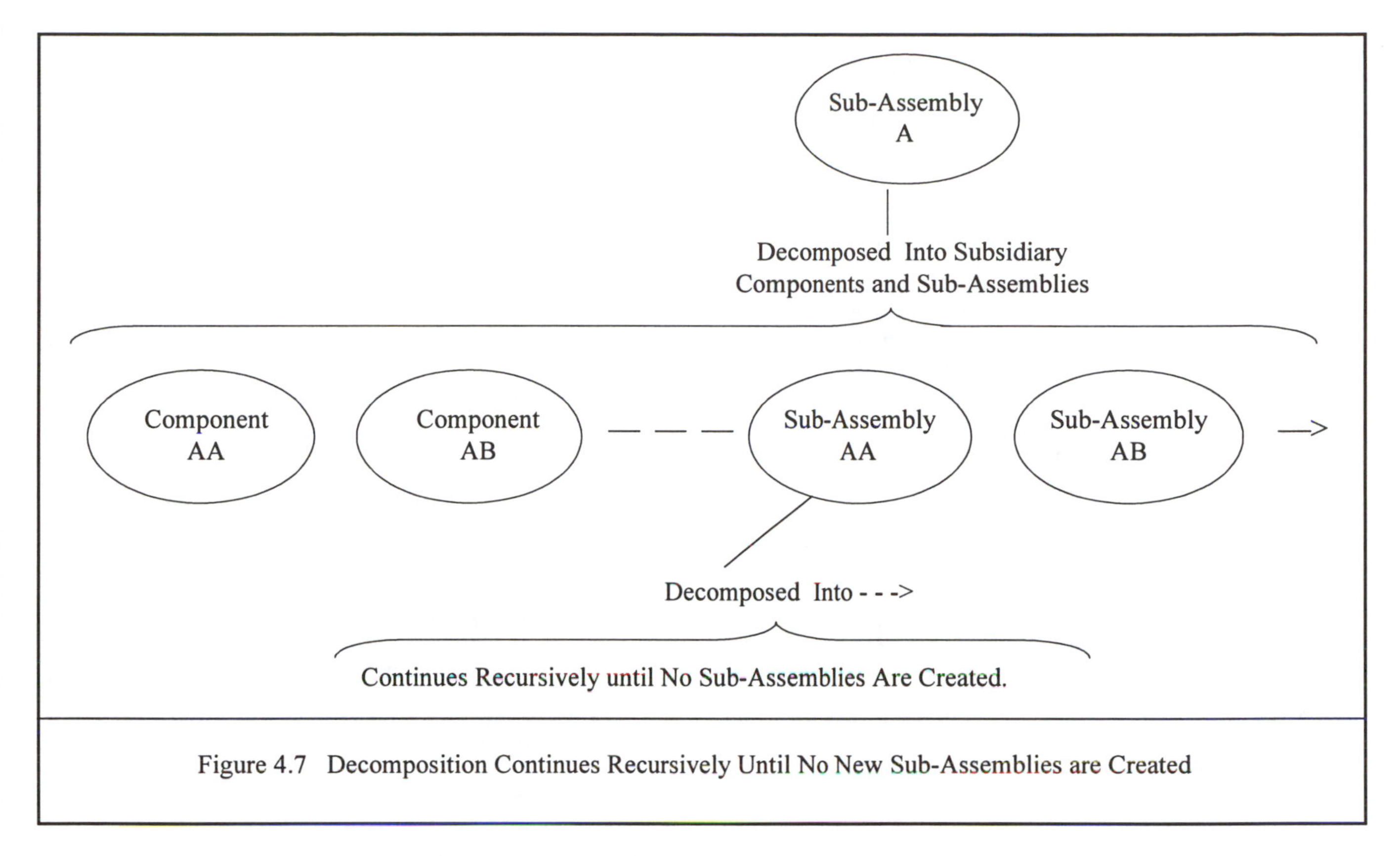

Figure 4.7 Decomposition Continues Recursively Until No New Sub-Assemblies are Created

It is important to note again that the sub-assemblies and components that appear during conceptual decomposition are *new* designed objects. They did not exist prior to their creation during conceptual decomposition, but once created they *are* the designed objects of which the product or sub-assembly is composed. That is, they literally *become* the product. To complete the conceptual design of the product, therefore, we must complete the conceptual design of these new components and sub-assemblies. How we do this, and some of the special problems involved, is discussed in the next two sub-sections. Section 4.2.3 deals with newly created sub-assemblies; Section 4.2.4 deals with new components.

4.2.3 Conceptual Decomposition of Sub-Assemblies

For a sub-assembly that has been created during the conceptual decomposition of a product, the initial state of information is slightly different from the initial state of information for the product as a whole. For sub-assemblies, we not only have the initial information for the whole product, we also have additional, specific information about the special functions and couplings of the sub-assembly within the product.

The functions of newly created sub-assemblies are generally very different from the function(s) of the product. For example, the function of an automobile has to do with providing self-propelled ground transportation for two to six people. The function of the automobile's suspension system, which is certainly an important sub-assembly, is not mentioned explicitly as a function of the automobile. The suspension system has its own special functionality. The same is true of the carburetor; it is a principal sub-assembly of the engine, but its specific functions are not mentioned among the functions of an automobile, or even of an engine.

In general, the functions of subsidiary components and sub-assemblies are narrower and more technical (i.e., having to do with forces, energy or material flow, and so on) than are the functions for a product, which are more closely identified with the in-use purposes of the customers.

Except for this slightly different starting place, the process of engineering conceptual design for sub-assemblies is the same as the process for products. That is, for each sub-assembly we perform a stage of conceptual decomposition just as we did for the product. And we again identify the principal functions and couplings of new subsidiary components and sub-assemblies that are created. And so on.

So long as this process creates new sub-assemblies that require conceptual decomposition, it continues on and on recursively until no new sub-assemblies are created. See Figure 4.7.

4.2.4 Engineering Conceptual Design for Components

The components (special purpose parts, standard parts, and standard modules) that are created by conceptual decomposition do not, of course, require further conceptual decomposition. Their engineering conceptual design is still not complete, however, because the physical principles by which they will fulfill their functions have not yet been considered. We leave the components for now, however, and take them up again in Chapters 5 and 8.

4.2.5 A Principle to Guide Decomposition

It has been argued persuasively [9, 10] that conceptual decomposition is best performed so that the resulting subsidiary sub-assemblies and components are functionally independent, if possible. Failing complete independence, the required couplings should be as few and as simple as possible. This principle provides a possible way to guide the generation of more effective decompositions, and a possible criteria by which competing alternative decompositions may be compared and evaluated. The basic idea is that the decomposition is probably best in which the couplings are as few and as simple as possible.

4.2.6 Real Problems Are Extremely Complex

The way products or sub-assemblies are conceptually decomposed into subsidiary components and sub-assemblies can seem fairly obvious when describing known and relatively simple products with the 20-20 vision of hindsight. When the products or sub-assemblies are complex and/or new, however, and there is the pressure of time, force of habit, and limitations of people and organizations, the best way to perform conceptual decomposition can be far from obvious. There are many options available, and the choices are not always clear — though the consequences may be great.

Consider what might happen, for example, if a sub-assembly created during, say, a third or fourth stage of decomposition turns out to be impractical. Then that entire level of decomposition, with all of its couplings, would be subject to change. A huge number of coupled and subsidiary components and other sub-assemblies designs could be affected. It might even turn out that the whole concept would have to be discarded.

For large or complex products, recursive conceptual decomposition results in a very large number of coupled sub-assemblies and components. Even keeping track of them all, together with all their interactions, is a major problem. Thus the management of information in product realization processes is an active area of research into engineering design, the results of which are sorely needed.

We have described the recursive decomposition process above in a fairly structured, orderly and idealized way. However, it hardly ever, perhaps never, actually takes place so neatly or so sequentially. Some critical sub-assemblies are usually designed in great detail before the design of others is even begun. The couplings of the *huge* number of components and sub-assemblies are generally too numerous to keep track of in an explicit way. One change on one component or sub-assembly can ripple changes throughout the product. Dozens, hundreds, or even thousands of people may be involved. Thus, what can be written about in an orderly way can be, or appear to be, quite chaotic in practice.

An underlying structure does exist, however, even if it isn't as obvious or as neat and clean as we can make it sound in a book using simple examples. We trust that the complexity of participating in the design of products with multitudes of subsidiary components and sub-assemblies is apparent to and appreciated by readers. But we also hope that the basic structure we have presented will help readers to more clearly comprehend and evaluate the decomposition processes encountered in practice.

4.3 Overview of Guided Iteration Applied to Engineering Conceptual Design

4.3.1 Introduction

> "The significance of the Wright Brothers' first flight is not how far or how high or how long the Kitty Hawk flew, but that it flew by the right principles".
>
> John R. Dixon

It is the thesis of this book that the right principles for solving engineering design problems of all types and sizes are those of guided iteration. In this section, we present an overview of how guided iteration is implemented for engineering conceptual design. Then in the next four chapters, we apply this general methodology to the engineering conceptual design phase of engineering design.

4.3.2 Formulating Engineering Conceptual Design Problems: The Engineering Design Specification

Formulating a problem for engineering conceptual design means preparing an *Engineering Design Specification.* Information developed for the Engineering Design Specification is used throughout the rest of the design process; that is, it is used not only for engineering conceptual design but also for configuration and parametric design as well. Thus the Engineering Design "Spec" is extremely important to the ultimate success of the entire project. Normally, therefore, it should be a written document.

The purpose of preparing the Engineering Design Specification is to convert the vague, qualitative, and incomplete information that is generally available at the beginning of engineering conceptual design into a set of specific, quantitative, complete performance requirements for the designed object. As a help in doing this, we make use of a technique called Quality Function Deployment (QFD) [11]. The next chapter (Chapter 5) introduces QFD and discusses the Engineering Design Specification in detail.

4.3.3 Generating Alternatives in Engineering Conceptual Design Problems

If possible, a large number of alternative physical concepts should be generated for possible evaluation as solutions to the engineering conceptual design problems. The case for this is stated very well in [13]. Basically the reason is that the selection of the best possible conceptual alternative is a crucial step in obtaining the best possible final solution, and mistakes at this stage are extremely costly in time if they have to corrected later by backtracking from the configuration or parametric stages. Unfortunately there is a human tendency, strong in some designers and design organizations, to pass quickly through the engineering conceptual stage by considering only one or two of the possible conceptual solutions that are most familiar to the people involved [12]. This procedure very often ignores other possible solutions that may be superior — ones that may be found by competitors who are more thorough.

Specific methods for generating creative alternative physical concepts are discussed in Chapter 5. The methods include: a set of applied creativity methods; performing a systematic search for innovative combinations; and explicit consideration of a wide variety of physical principles, laws, and effects.

4.3.4 Evaluating Alternatives in Engineering Conceptual Design Problems

Several methods useful in comparing and evaluating prospective physical concepts are presented in Chapter 6. These include: Pugh's method [13], a cost-benefit method from Pahl and Beitz [3], and a method called Dominic's method. All of these methods make use of the content of the Engineering Design Specification to develop a list of criteria on which the alternatives are rated and compared. An additional criterion that should be evaluated is the potential need for special tolerances during the manufacturing phase; this issue is discussed briefly later in this chapter.

4.3.5 Guided Redesign in Engineering Conceptual Design Problems

All of the methods available for comparison and evaluation of physical concepts indicate in general and qualitative terms which alternatives are best. In addition, and at least as important, the methods also illuminate the specific characteristics of proposed alternatives that are weak or strong. Thus the attention of designers is guided naturally and directly to changes or refinements that are needed to improve alternatives. With such improvements made, the alternatives can be re-evaluated, and re-designed again if necessary

Ultimately, of course, additional improvements in alternatives may become impossible. Then it must be decided if one or more of the alternative physical concepts is acceptable as is; that is, is one "good enough"? The overall ratings resulting from the methods will give a general indication of how good the alternatives are (and why), but the acceptability decision is ultimately a subjective decision that must be made by the designer, design team, and/or management.

If no alternative is deemed acceptable, then the search for new alternatives must be resumed. But this, too, is now guided by the results of the previous evaluations. After evaluation, we know why new technologies must be found, new materials specified, new manufacturing processes employed, or new physical principles tried. Thus we are guided in our search for new alternative physical concepts that will have a better chance of fulfilling the requirements of the Engineering Design Specification.

4.3.6 Summary of Guided Iteration Methods for Engineering Conceptual Design

A number of specific methods are employed within the guided iteration methodology to solve engineering conceptual design problems. These methods are listed in the table below, and described and illustrated in the next four chapters.

Steps in the Guided Iteration Process	Methods
Problem Formulation	+Quality Function Deployment and House of Quality +Engineering Design Specification
Generation of Alternatives	+Search for Alternative Physical Laws and Effects +Techniques of Creativity +Memory Search +Analogous Problems +Literature searches
Evaluation of Alternatives	+Pugh's Method +Dominic's Methods +Pahl and Beitz Method
Redesign, *Guided by* the Evaluations and Qualitative Physical Reasoning	+Modify Existing Alternative +New Alternatives

4.4 Tolerances at the Conceptual Stage

We have already pointed out above that consideration of any special tolerances that may be required is an issue that should be evaluated, if at all possible, at the conceptual stage. The reason is that very stringent (i.e., *tight*) tolerances are more expensive to produce — sometimes a great deal more expensive. They usually require extra inspection or extra fine control of the process during production. They are also very likely to reduce the yield (percent of acceptable products) in a production process.

Unfortunately, at the conceptual stage, it may not be possible in all cases to anticipate what the tolerance requirements will be on the parts that must ultimately be made. Nevertheless, the effort should be made. If it appears that tighter than standard tolerances *may* be needed to make a concept work well, then that is an important issue to be investigated in more detail during evaluation of the concept. To be surprised later by the need for many and/or special tolerances can be very unpleasant.

It is helpful to be clear about the distinction between design *of* assemblies and design *for* assembly. Design *of* assemblies is what we have been discussing in this chapter. Design *for* assembly (DFA) involves mainly the design of parts so that they can be easily handled and inserted properly into place during the assembly process. Design *for* assembly does involve some design *of* assembly issues such as designing for the minimum number of parts, and so that there is easy access for all parts during the assembly process. (All parts should be insertable from a single direction, for example, if the assembly is to be done automatically.)

There are, however, issues in design *of* assemblies that have little to do with design *for* assembly. One of these is called "stack-up", meaning the way tolerances can add-up in an assembly. Note in Figure 4.8 that, as the part is dimensioned, the tolerance on dimension A in Figure 4.8 (a) is +/- 0.04 inches. It is possible to place a different tolerance on A as shown in Figure 4.8 (b). This tolerancing allows any of the individual 1 inch dimensions still to be +/- 0.01 inches, but limits the accumulated overall variation in length to +/- 0.01 inches. Figure 4.8 (c) shows another variation which designers might use depending on the functional requirements. Designers must obviously be aware of such issues: *tolerancing requires attention to both functionality and manufacturability.*

Research and methods for stack-up analysis are discussed in references [14, 15].

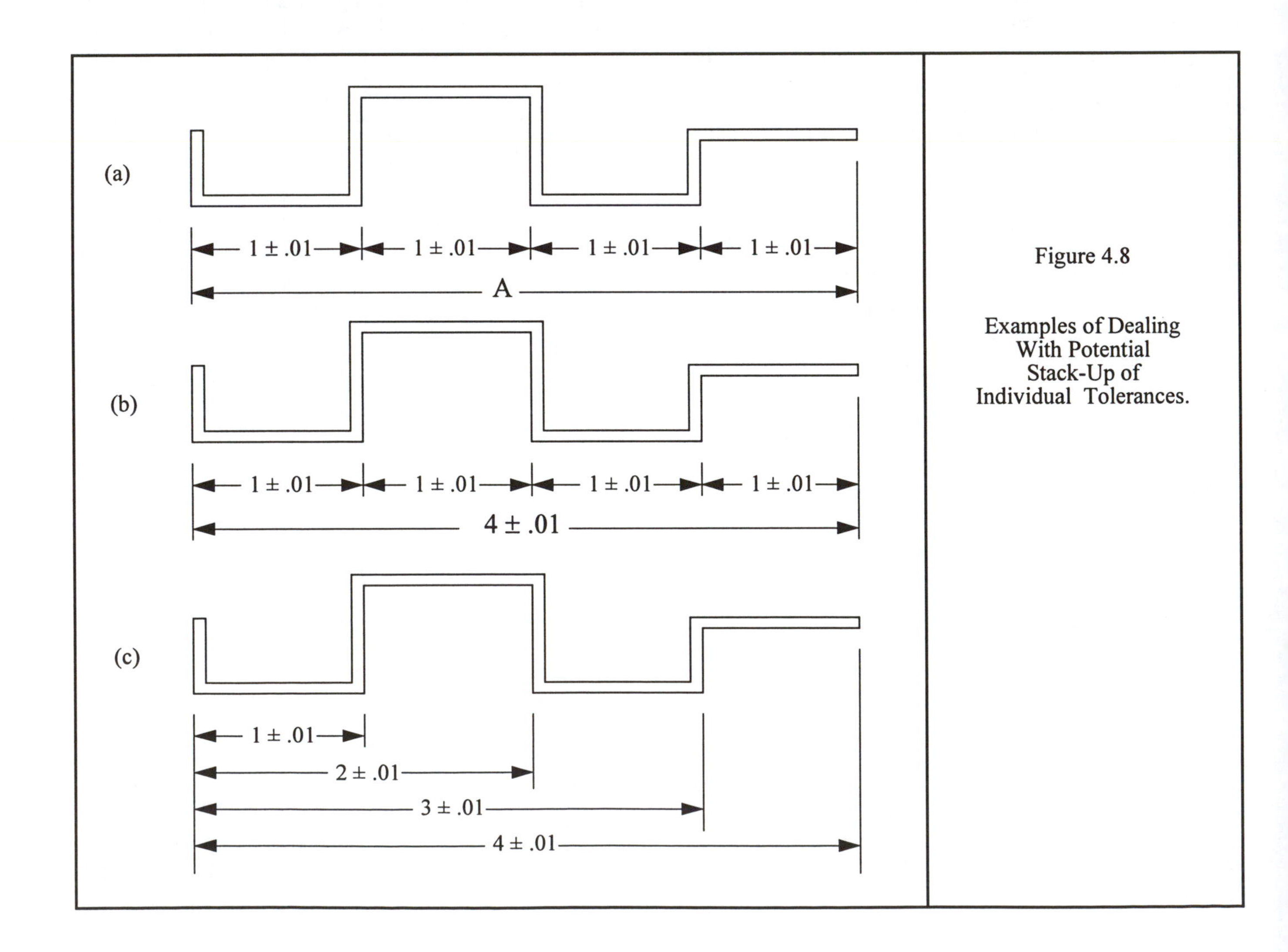

Figure 4.8

Examples of Dealing With Potential Stack-Up of Individual Tolerances.

4.5 Summary and Preview

Summary. This chapter has two main purposes: (1) to introduce and describe the process of conceptual decomposition (both direct and function-first) as it occurs during the engineering conceptual design of products and their subassemblies; and (2) to provide a brief overview of the guided iteration methodology as it is applied to engineering conceptual design.

Preview. In Chapter 5, the Engineering Design Specification is described in detail, and the concept of Quality Function Deployment is introduced as an aid to developing an Engineering Design Specification.

Recommended Reading

Conceptual Design for Engineers, Michael J. French, The Design Council, London, 1985.

Management of Engineering Design, D. J. Leech, Wiley, 1972.

Crossley, E. "A Shorthand Route to Design Creativity" *Machine Design*, April 10, 1980.

Crossley, E. "Defining the Job: First Step in a Successful Design," *Machine Design*, May 22, 1980.

Crossley, E. "A Systematic Approach to Creative Design," *Machine Design*, March 6, 1980.

Crossley, E. "Make Science a Partner," *Machine Design*, April 24, 1980.

Marples, D. L., "The Decisions of Engineering Design", *IRE Transactions on Engineering Management,* Volume EM-8, Number 2, June, 1961.

Problems

1. Develop two ways the first stage of a direct conceptual decomposition for a bicycle (or toaster or drill or other product of your choice) might be done.

2. Develop a functional decomposition for a device whose purpose it is to enable humans to tighten or loosen hexhead bolts. Then show how the crescent wrench, straight wrench, and socket wrench each fulfill the desired sub-functions.

3. Develop a functional decomposition for a bicycle braking system (or other small system of your choice). Develop at least two embodiments for each major sub-function.

References

[1] French, Michael J. *Conceptual Design for Engineers,* The Design Council, London, 1985. Quotation by permission of the author.

[2] Leech, D.J. *The Management of Engineering Design*, Wiley, New York, 1972.

[3] Pahl, G. and Beitz, W. *Engineering Design*, Edited by Ken Wallace, The Design Council, London, England, 1984.

[4] Crossley, E. "A Shorthand Route to Design Creativity" *Machine Design*, April 10, 1980.

[5] Crossley, E. "Defining the Job: First Step in a Successful Design," *Machine Design*, May 22, 1980.

[6] Crossley, E. "A Systematic Approach to Creative Design," *Machine Design*, March 6, 1980.

[7] Crossley, E. "Make Science a Partner," *Machine Design*, April 24, 1980.

[8] Welch, R.V. "Conceptual Design of Mechanical Systems: A Representation and Computational Model" Doctoral Dissertation, University of Massachusetts, Amherst, MA., 1992.

[9] Suh, Nam P. , *The Principles of Design*, Oxford University Press, New York, 1990.

[10] Suh, N. P. and Rinderle, J. R. "Qualitative and Quantitative Use of Design and Manufacturing Axioms", CIRP Annals of Manufacturing Technology, 31(1), pp 333-338, 1982.

[11] Hauser, D.R. and Clausing, D. "The House of Quality." *Harvard Business Review,* May-June, 1988.

[12] Marples, D. L. "The Decisions of Engineering Design", IRE Transactions on Engineering Management, Volume EM-8, Number 2, June, 1961.

[13] Pugh, S. *Total Design: Integrating Methods for Successful Product Engineering*, Addison-Wesley, Reading, MA, 1991.

[14] Chase, K. W. and Greenwood, W.H. "Design Issues in Mechanical Tolerance Analysis", Advanced Topics in Manufacturing Analysis, ASME Winter Annual Meeting, Boston, MA. December, 1987.

[15] Spotts, M. F. *Dimensioning and Tolerancing for Quantity Production*, Prentice-Hall, Englewood Cliffs, N.J., 1983.

Chapter Five

Formulating the Problem: The Engineering Design Specification

"We have defined, briefly, the function of the designer. The first task is to decide what the customer wants....the second is to convert the customer's objective into a numerically detailed design requirement."

D. J. Leech [1]*
Management of Engineering Design (1972)

5.1 Introduction

In this chapter, we describe the first step of the guided iteration process for solving engineering conceptual design problems: the formulation of the problem in engineering terms. The principal method for doing this is preparation of an Engineering Design Specification (or Engineering Spec, or Specification, or just Spec). In this Chapter, we concentrate on preparing Engineering Design Specifications for whole products. In Chapter 8, we discuss preparing Engineering Specs for sub-assemblies and components created during the decomposition of products.

An Engineering Design Specification, whether for product or subsidiary sub-assembly or component, contains statements of the information about a designed object in two major categories:

1. In-use Purposes, and
2. Functional Requirements.

For products, some information is usually available in these categories from the marketing specification and industrial design. In the case of sub-assemblies and components, the principal functions and couplings are also known, at least qualitatively. Indeed, much of the information normally available to us at the beginning of engineering conceptual design is qualitative, incomplete, and/or approximate. That's a problem because, for engineering design, we need quantitative, complete, and accurate information. The purpose, then, of preparing an Engineering Spec is to make the information we have as quantitative, and as complete, as feasible at this stage of the design process.

In preparing an Engineering Design Specification, we are not only formulating the engineering conceptual design problem, we are also preparing a document that will serve configuration and parametric design as well (though aspects of the Spec will no doubt get modified along the way).

In the case of a product, *in-use purposes* include three sub-categories, all verbal statements having to do with the product's anticipated users and mis-users (i.e., customers):

1. The primary purpose(s) to which users will put the product,
2. Any unintended purposes to which the designed object may predictably be put (given that human beings are who they are), and
3. Any special purpose features required or desired.

The sub-categories of functional requirements for a product are qualitative or quantitative *goals and limits* placed on:

1. Product performance
   a. *Functional Performance Requirements* Include But Are Not Limited To:
      i. Capacity (e.g., energy or material flows, forces, etc.),

---

* Quotation from *Management of Engineering Design* by D. J. Leech, Copyright 1972. Reprinted by permission of John Wiley and Sons, Inc.

ii. Input and Output Conditions (e.g., temperature, energy, pressures, flows, power, deflections, forces, etc.),

iii. Efficiency, and

iv. Accuracy and Sensitivity.

b. *Complementary Performance Requirements* include but are not limited to:

Useful Life,
Reliability,
Robustness,
Safety,
Noise,
Side Effects,
Legal Requirements and Codes,
Maintenance Requirements,
Requirements on Users (e.g., knowledge, skill, speed, etc.).

2. The environmental and other conditions under which the product is to perform,
3. Economic issues,
4. Physical attributes,
5. Process technologies,
6. Aesthetics, and
7. Product development time and cost.

We said above that the Engineering Design Specification for a product should be developed as completely and as accurately *"as feasible"* because the Specification is hardly ever fully completed. Indeed, it is normal for a Specification to be subject to change throughout the entire engineering design process. Of course, changes later in the process — if they cause significant redesign — can often be very expensive of time, money, and final product quality. Thus we want the Engineering Design Specification to be as complete and accurate *as feasible* as early as possible. However, it is simply unrealistic to expect that it can be made perfect and complete at this early stage.

The specific information to be supplied in the above categories of an Engineering Design Spec is the main subject of this chapter beginning in Section 5.3. First, however, we introduce a technique called Quality Function Deployment (QFD) [2] since it can help us preparing the Spec.

5.2 Quality Function Deployment (QFD) and the House of Quality: Phase I

5.2.1 Customer Attributes and Engineering Characteristics

Quality Function Deployment (QFD) [2] is a term used to describe a strategy for focusing engineering design attention on quality issues as perceived by customers. The *House of Quality* is a technique for structuring information commonly used to implement QFD. The House of Quality [2] is essentially a matrix (with some associated embellishments) that relates what customers want (expressed as *Customer Attributes*) to technical product characteristics that design engineers can provide (expressed as Engineering Characteristics).

Customer Attributes are usually expressed qualitatively. For example, customers want something to be "easy", or "fast", or "smooth", etc. Engineering Characteristics, on the other hand, are usually quantitative, expressing what has to be provided in the objective physical world physical world in order to satisfy the customers' desires. For example, customers may want a vacuum cleaner that is "easy to carry upstairs". One associated Engineering Characteristic will therefore be its weight (in pounds).

Figure 5.1 shows a skeleton House of Quality matrix. Engineering Characteristics are listed across the top in groups or categories. For example, one group might be materials, another the physical characteristics, another manufacturing, and so on. Customer Attributes are listed along the left side, also in groups or categories. For example, one group might be functions, another features, and so on.

To illustrate again the meanings of the terms *Customer Attribute* and *Engineering Characteristic*, consider this example: In a coffee maker, one attribute customers want is "good coffee" [3]. It isn't the only thing they want; they also want the coffee maker to look good in the kitchen, to be easy to clean, etc. All such customer wishes are called Customer Attributes. They are usually best expressed in terms the customers themselves use.

To identify the Engineering Characteristics that will provide the Customer Attributes, engineers must first determine the functional requirements of the coffee maker that influence the "goodness" of coffee as perceived by customers. These involve (at least) the temperature and flow rate of the water delivered to the coffee grounds, the time the water takes to flow through the grounds, and the temperature of the coffee when it is poured into the cup. Note that these functional requirements are all observable or measurable physical quantities related to the behavior of the coffee maker.

Once the functional requirements for making good coffee are established, every proposed design will have parameters that influence the functional requirements, and thus the quality of the coffee. For the types of coffee makers generally sold, Engineering Characteristics will therefore include the electrical resistance of the heater, the size of the tube supplying hot water to the coffee grounds, and other factors having to do with geometry and materials. Engineering Characteristics are thus design variables that designers can control (at least to some extent) in order to meet the functional requirements -- which in turn will give customers what they want.

At the early stage in the design process that we are considering, before the Engineering Design Specification has been developed, there is not yet a proposed physical product design. Thus, there aren't yet any Engineering Characteristics to discuss. At this stage, therefore, the Engineering Characteristic dimension of the House of Quality matrix cannot yet be addressed. However, determining and stating the desires and needs of customers in terms of explicit Customer Attributes can and definitely should be done at this stage. Doing so focuses engineering design attention on what customers value and expect in a product, and therefore provides important information for developing the various functional requirements in the Engineering Design Specification.

As the design process proceeds, more parts of the House of Quality can be added to keep attention focused on quality; that is, on what the customers want. Thus in later chapters, more will be said about using the House of Quality in other phases of engineering design.

In some firms, it will be the case that the Customer Attributes needed for the House of Quality will have been fully and explicitly identified as a part of the marketing product concept or industrial design phase. Indeed, most should have been. But if not, this must be done as a part of formulating the engineering conceptual design problem. In any case, the desires of customers should be continuously reviewed for possible changes or omissions.

5.2.2 An Example of Customer Attributes

A small protective portable carrying case for 3 1/2 inch computer disks might have this list of Customer Attributes as expressed by potential customers:

- Protects the disks
- Attractive
- Inexpensive
- Can identify the contents
- Stays closed
- Easy to open
- Can't be used as a coaster for coffee cups
- Holds at least six disks
- Doesn't pop open on its own
- Won't break if dropped
- Compact

This is a short list because it is a simple product. However, for larger and more complex products, the list may be much longer and more difficult to obtain and organize.

			ENGINEERING CHARACTERISTICS						
			Material			Physical		Mfg.	– – –
CUSTOMER ATTRIBUTES		Importance	EC-1	EC-2	EC-3	EC-4	EC-5	EC-6	EC - - -
Category 1	CA – 1.1								
	CA – 1.2								
	CA – 1.3								
Category 2	CA – 2.1								
	CA – 2.2								
	CA – 2.3								
CA – 3.1									
CA – 4.1									
CA – – –									
		100							

Figure 5.1 Skeleton House of Quality

Once a complete list of customer attributes is obtained, redundancies should be eliminated as much as possible. In the list above, for example, "stays closed" and "doesn't pop open on its own" are redundant.

With redundancies gone, the next step is to organize the list hierarchically. For example, in the above list, "protects the disks" is general; "stays closed", "can't be used as a coffee cup coaster", and "won't break if dropped" are specific attributes in that general category.

It may be that a general category will have to be created in order to develop a well organized hierarchy of CA's. For example, in the above list, "easy to open", "can identify the contents", and "compact" are all facets of a more general requirement that might be called "convenient to use".

Thus, we organize the list of CAs for this example hierarchically as follows:

- Protects the disks
    - Stays closed
    - Can't be used as coaster for coffee cups
    - Won't break if dropped
- Inexpensive
- Holds at least six disks
- Convenient to use
    - Easy to open
    - An identify the contents
    - Compact
- Attractive

Often there will be a third layer to the hierarchy of CA's. The set of the most specific attributes (lowest level in the hierarchy) in each grouping should all share the characteristic that they could be tested in a single test by a customer who is imagined to have been given a trial product. The test can be subjective, of course. We call such attributes *single dimensional.* For example, "easy to open" meets this test, but "convenient to use" does not since to determine whether or not the product is convenient to use would require a multi-faceted test by the customer.

Notice that the most specific attributes in the above list meet the requirement of single dimensionality. These are:

- Stays closed
- Can't be used as coaster for coffee cup
- Won't break if dropped
- Inexpensive
- Holds at least six disks
- Easy to open
- Can identify the contents
- Compact
- Attractive

Often in developing the House of Quality it is helpful to assign priorities to the primitive customer attributes. This is done numerically in such a way that the total assignments add up to 100. For example, a designer might assign values to the customer attributes of the disk box as follows:

• Stays closed	20
• No coaster use	5
• Won't break if dropped	14
• Inexpensive	10
• Holds at least six disks	7
• Easy to open	20
• External identification	10
• Compact	7
• Attractive	5
TOTAL	100

The House of Quality shown in Figure 5.2 includes these results. Of course, Engineering Characteristics are not included since none has yet been determined at this stage of design. This example is continued in Chapter 7 where the discussion of QFD and House of Quality is also continued.

5.2.3 Competitive Benchmarking.

There is another use for the House of Quality: it can serve as a useful tool in *competitive benchmarking.* As stated in Chapter 2, competitive benchmarking is the process of comparing a firm's existing product or proposed designs to the products of the best competitors in order to establish performance metrics for the firm. The House of Quality can be used as a framework to determine how well the competitors' and the firm's products satisfy what customers want in comparison to proposed new designs.

5.3 Content of the Engineering Design Specification

5.3.1 Introduction

Once customer needs and wants are identified in the form of Customer Attributes, and the House of Quality initiated, the next task in engineering conceptual design is development of the Engineering Design Specification. Some of the information to be included in this Specification will already be available from the marketing product specification and decision processes that have preceded this step. However, the Engineering Design Specification requires additional information, and requires that the information be made more explicit and written down.

As listed above, the Engineering Design Specification includes information in two main categories: *in-use purposes* and the (more objective) *functional requirements.* We now elaborate on the sub-areas of these categories.

5.3.2 Categories of In-Use Purposes

There are three categories of in-use purposes needed in the Engineering Design Specification for products; each of them is discussed briefly here.

Primary Intended Uses By Customers

At the conceptual stage, we use the term *primary intended purpose* to express verbal information about how customers intend to use a designed object. For example, one intended purpose might be "to make coffee." Another might be "to convert electrical to rotating mechanical energy." A third might be "to aid in the catching of a baseball and to protect the hand when doing so". A fourth could be "to remove unwanted hair from the face."

Information about the primary intended purpose of a proposed designed object is a fundamental to the formulation of an engineering conceptual design problem. The intended purpose is the *sine-qua-non* for a functional designed object; it expresses the service that the designed object will provide to its users.

Sometimes a product will have multiple intended purposes, like the baseball glove. For example, a hammer is used to both drive and pull nails; a hotel key is used to open a door and to provide identification in the hotel restaurant. Multiple purposes are often desired even though it is easier to design things that have only a single purpose. But in any case, *any and all* intended purposes that a product must perform to serve its user are an essential category of information in the formulation of conceptual design problems.

Predictable Unintended Uses

Products get used in ways that have little or nothing to do with their primary intended in-use purpose(s). Wrenches get used as hammers, screw drivers as levers or chisels, shelves and chairs as step ladders, car tires as swings for children, chain saws as brush cutters, and so on. People will actually stand on the tops of top-loading washing machines or driers, so the tops of these appliances must be designed to support the weight of a person, maybe even one who is jumping up and down!. When such mis-uses and alternative uses are reasonably predictable — considering the well known ingenuity and occasional folly and fallibility of human beings — then designers *must* design for their *safe mis*-uses. This is essentially a legal as well as a common sense requirement. The logic is: since we can expect that some fool will no doubt sometimes use a wrench as a hammer, we must design the wrench to be a reasonably safe hammer.

Warnings on products and product labels help but do not suffice totally to relieve designers of the legal responsibility to design parts and products that are safe even when mis-used in reasonable (and sometimes even unreasonable) predictable ways. Issues of legal liability are discussed in this book in Chapter 24.

CUSTOMER ATTRIBUTES		Importance	ENGINEERING CHARACTERISTICS										
Protects Disks	Stays Closed	20											
	No Coaster	5											
	Doesn't Break	14											
	Inexpensive	10											
	Six Disks	8											
Convenient to Use	Easy to Open	20											
	External ID	10											
	Compact	8											
	Attractive	5											
		100											

Figure 5.2 House of Quality for Disk Carrying Case

Special Purpose Features

Especially in the case of consumer products, marketing considerations may require, or at least suggest, that the main purpose(s) be embellished or supplemented with certain special features that enhance the value of the product to users. For example, a feature desired in a rechargeable electric shaver may allow it to be directly plugged into a receptacle for recharging — thus eliminating the inconvenience of the cord usually needed for this task. The plug-in feature is not directly related to the in-use purpose — removing facial hairs — but the feature is important for marketing. Another example of a special purpose feature is the re-dial button on touchtone telephones. Another: the no-drip tops on laundry detergent containers.

Special purpose features may be given a priority rating at this stage as *essential, important*, or *desirable*. Often the priorities used in practice are just two: *must* and *want*. However, having only two categories is not as helpful in subsequent design decisions as the three categories suggested.

5.3.3 Categories of Functional Requirements

Functional requirements express goals and limits of a product's performance, economics, environmental effects, physical attributes, aesthetics, manufacturing technologies, and product development time and cost. The requirements should be expressed as quantitatively as possible. The specific categories included in the functional requirements are at least those discussed in this Section, though specific cases may require additional categories depending on the situation.

Performance Requirements

Performance requirements have to do with what a product does, how well it does it, for how long, and so on. Performance requirements come in two major, somewhat fuzzy, categories: requirements that relate directly to function, and requirements that relate only indirectly to function. We call the latter *complementary* performance requirements.

(a) *Functional Performance Requirements* Include But Are Not Limited To:

- Capacity (e.g., energy or material flow rates, forces, etc.)
- Input and Output Conditions (e.g., temperature, energy, pressures, flows, power, deflections, forces, etc.)
- Efficiency
- Accuracy and Sensitivity

(b) *Complementary Performance Requirements* Include But Are Not Limited To:

- Useful Life
- Reliability
- Robustness
- Safety
- Noise
- Side Effects
- Legal Requirements and Codes
- Maintenance Requirements
- Requirements on Users (e.g., knowledge, skill, speed, etc.)

Some performance requirements involve a number of issues. Safety, for example, is likely to depend on a number of factors. Others, however, involve only a single parameter such as cost, weight, efficiency, or the like. We call these parameters *performance parameters*. Many performance parameters can be expressed numerically, but it is only necessary that they can be evaluated, whether that evaluation is numeric or subjective (as, say, excellent, good, fair, etc.).

A given performance requirement may involve one, two, or more performance parameters. For example, performance parameters related to reliability might be average annual repair costs, time lost for repairs, or (in a lawn mower, for instance) the number of pulls for starting.

Environmental Requirements

Environmental requirements refer to such conditions as the temperature, humidity, corrosive elements, noise, dirt, vibration, electric or magnetic fields, and the like that may be present in the environment when the designed object is to be used or stored. Extremes or variations in these conditions, especially cyclic variations, may be especially important.

How products will be disposed of, and how the disposal will influence the environment (e.g., air, water, etc.) is also an increasingly important requirement. Indeed, design for *dis*-assembly and disposal or recycling is becoming an essential concern and field of study.

If the product causes any possible pollution of air, water, etc., during its lifetime of use, then this must also be addressed in the Engineering Spec.

Physical Requirements

Most engineering conceptual design specifications include requirements on the physical attributes that describe the designed object physically. These may include, for example, the weight, size, shape, surface finish, and the like.

Economic Requirements

All design problems include requirements of an economic nature. These may include tooling costs, initial product cost, maintenance costs, return on investment, cash flow, break even time, and others. See Chapter 23.

Aesthetic Requirements

In many products, there are aesthetic issues involved in marketing. These may express how a product's design is to conform to a firm's image with its customers; e.g., for quality, style, uniqueness, or high performance. Alternatively, they may describe the way in which a product line's identity is differentiated from competitive lines. Issues of trademark and distinctive trade dress are discussed in Chapter 24.

Aesthetic goals are, of course, difficult to express and evaluate, but it is important that the effort be made. Aesthetic concerns can and often do have profound effects on material selection, manufacturing processes, and even the manner in which functions are fulfilled. Because they are often described in highly subjective terms by marketing people and industrial designers, engineers are often uncomfortable and impatient with these issues. However, subjective and aesthetic issues are an important part of competitive product realization, and working closely with marketing people and product concept designers is as important for engineering designers as is working with the manufacturing group. It is the task of engineers to convert the more subjective aesthetic terms into objective engineering requirements as needed.

Manufacturing Technologies

If there are processing restrictions — perhaps, for example, certain manufacturing processes cannot be used, or others must be used — then these must be stated. There is no point spending time designing parts to be made on machines that are not available.

Product Development Time and Cost

Though for mass produced product, product development cost is not often an important long term financial issue, there can nevertheless be short term issues of capital availability that restrict the funds for product development. Designers may sometimes think such restrictions are overly severe, and that the restrictions are "penny wise and pound foolish" (and they will sometimes be right!). Nevertheless, if the restrictions are there, they must be respected and are therefore best stated up front as part of the Engineering Design Specification. The reason is that the project must not develop solutions that require more capital funds for development than are actually going to be made available.

More than the cost, however, the *time* required for product development is likely to cause designers serious difficulties. Since speed to market is often critical, any information related to product development time must also be stated in the Engineering Design Specification. Time requirements can therefore often influence the design choices that are made, since choices that result in long product development times may be impossible, or fruitless, to implement.

5.3.4 Expressing Functional Requirements

Functional requirements — which prescribe the goals and limits on performance and complementary performance issues — can be expressed in several ways in an Engineering Design Specification:

1. Qualitatively,
2. As extremum goals, with or without limits,
3. As target values with a tolerance, and
4. As ranges.

The following discussion defines and provides examples to illustrate each of these types.

Qualitative Requirements

Qualitative functional requirements, though they do not provide the quantitative information that may ultimately be needed, nevertheless indicate important information early about how relatively severe a particular requirement may be. Qualitative requirements are verbal statements, often expressed with words like high, low, moderate, fast, slow, etc.

An example of a qualitatively stated requirement is: The material should be consistent with the quality image of the firm's product line. Another example: the "feel" of the product should give customers confidence in its ability to perform as expected. Another: Sintering is to be used to make the part.

Since qualitative requirements are stated in words, often without any numbers, this is sometimes a source for concern among engineers who must ultimately design the artifact with numbers, for specific processes, and with specific materials. However, at the engineering conceptual stage, it is often not necessary that every functional requirement be stated quantitatively. It may be enough at this stage to say, for example, that the product must operate "at extremely high temperatures", or "at temperatures in excess of 500 F", or "under severe vibrations", etc. Ultimately, of course, such approximate requirements will have to be made concrete, but to get started on the right track it may only be necessary to state the requirement qualitatively.

When quantitative information is readily available, however, it should be stated. But at the conceptual stage, there is sometimes no need to go to great time-consuming lengths to gather the detailed information needed to define everything as quantitatively as will be necessary later. It *is,* however, most important to include at least the mention of *all* the issues that will be relevant to the design and manufacture of the product, even if quantitative data are not yet available or well established.

There is an important point in the previous paragraph: whether or not quantitative, or even qualitative, information can be provided about some factor at this stage, it is crucial that all the possibly relevant performance and complementary factors be listed. It is far more likely that forgotten or ignored issues will lead to subsequent failure (or disaster) than it is that an error in some remembered issue will lead to similar troubles. What designers forget to include almost always comes back to haunt them — and their customers.

Extremum Goals, With or Without Limits

An example of an extremum goal is: The cost should be as low as possible. Another example: The efficiency should be as high as possible.

Extremum goals have no limits on *how* high or low their value should be; it should just be *as* high or low *as possible*. However, sometimes an extremum goal does have a limit. For example: The weight should be as small as possible, but can not exceed about ten pounds maximum.

Limits on extremum goals can be exact or fuzzy. The previous example illustrated a fuzzy limit ("about" ten pounds). An exact limit would be: The weight should be as small as possible, but not more than (exactly) ten pounds.

Target Goals, With a Tolerance

An approximate target goal might be that the power output of an engine should be five horsepower. With a tolerance, this might be "five horsepower plus or minus 0.25 horsepower." Or "five horsepower plus 0.25 and minus zero horsepower."

Ranges, With Exact or Fuzzy Limits

An example of a range goal with an exact physical limit is: the length must be between 3.2 and 6.7 inches. Another example: the weight may not exceed 3.5 pounds. Note that the latter is not quite the same as a limited extremum goal; it does not say the weight should be as large as possible up to 3.5 pounds. It says, in effect, that the weight can be anywhere in the range from zero to 3.5 pounds.

An example of range requirements, but with fuzzy limits is: The power consumed *should* be between 5.2 and 5.6 horsepower, but *must* not be less than 5.0 horsepower or exceed 6.0 horsepower. Another: The volume *should* be between 5.2 and 5.3 cubic feet, but *must* be more than 3.9 and less than 5.6 cubic feet.

Figure 5.3 illustrates graphically the nature of some of these types of required goals and limits. In the sketches there — which we call "satisfaction plots" — we plot degree of satisfaction (from unacceptable to fully satisfied) versus the value for a performance parameter achieved by a design. For example, in Figure 5.4, we have illustrated that for some designed object, weight cannot exceed 15 pounds and that any weight less than 10 pounds is fully satisfactory.

5.3.5 "Constraints"

Sometimes the word *constraint* is used to refer to one or more, or even generically to all, of the above types of functional requirements. However, because the word "constraints" is used in so many different ways (e.g., to include both goals and limits) by so many different authors, we try to avoid its use in this book in connection with functional requirements.

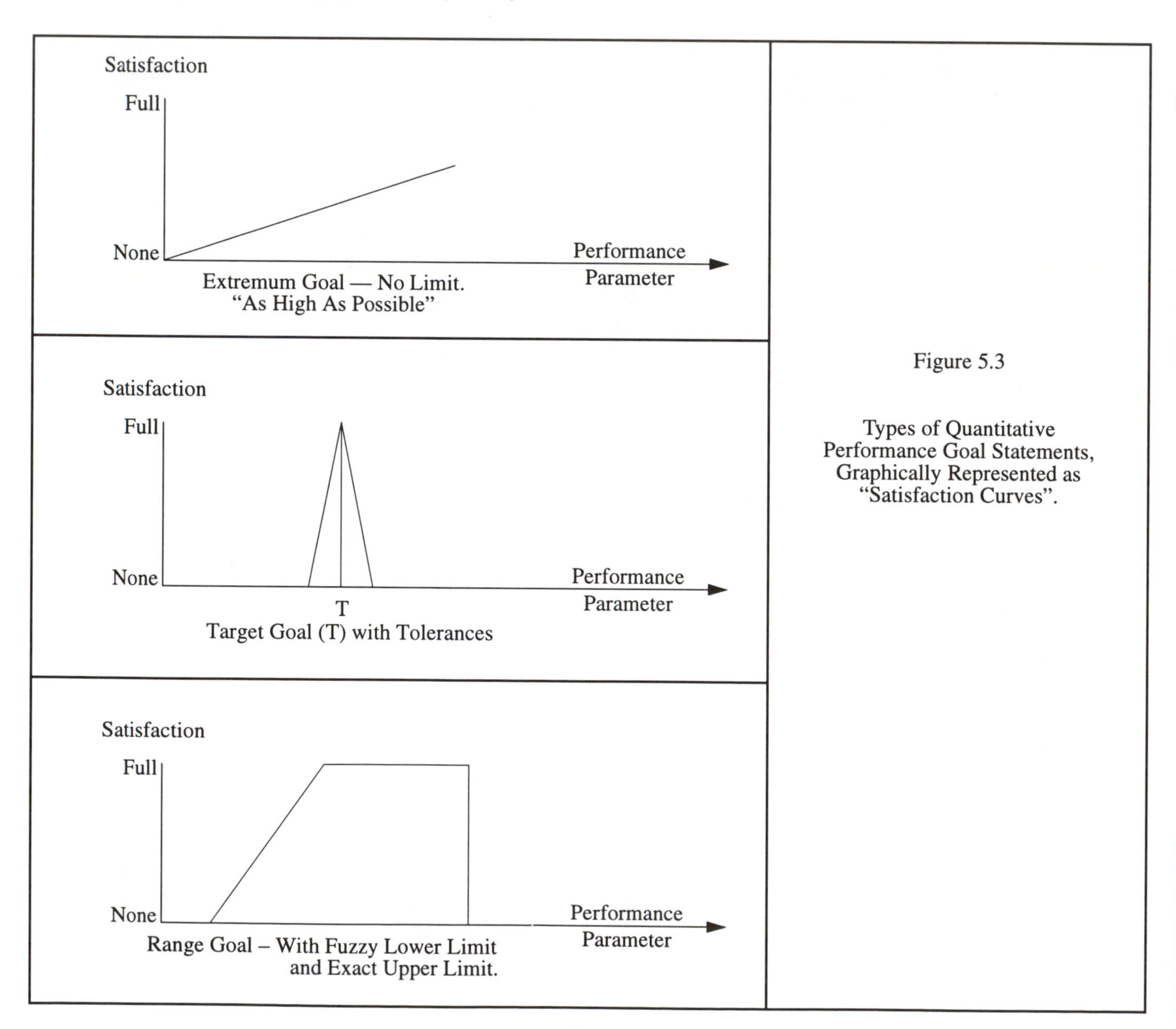

Figure 5.3

Types of Quantitative Performance Goal Statements, Graphically Represented as "Satisfaction Curves".

5.3.6 Completeness and Invariability of the Engineering Design Specification

As noted above, an Engineering Design Specification should never be thought of as complete or invariable at any stage in the design process. It is, in fact, never complete and invariable until the design is complete, and it may not be complete and invariant even then. Some information desired simply isn't available at the outset of most projects. Some information that is available may turn out to be wrong. It is not uncommon that issues which later must be included are not apparent at early stages. It can also happen that when the design is complete, the Spec has to be changed to conform to the actual design.

Some specification writers often try to specify *everything* about a product before design begins in absolutely precise terms, down to each and every smallest detail. This is hardly a least commitment policy, and sets the stage almost certainly for multitudes of engineering change orders (ECOs) as well as huge time and cost overruns. Though Specifications should be as complete as possible on *what* the designed object is to do, under what circumstances, and any other truly essential information about requirements, Specs should say as little as possible about *how* the requirements are to be met. In other words, when the Engineering Design Specification writers try also, in effect, to do too much of the design, the result often will not be a happy one. The policy of least commitment is a good one.

On the other hand, there must be an Engineering Spec. Unfortunately, sometimes there isn't. Moreover, the Spec should be as complete and as quantitative as reasonably possible at the outset. And then it must be kept as up to date as possible as the design proceeds. Unforeseen changes or revelations can be extremely expensive of time and money, and can seriously damage the quality of the results.

And remember: Include mention of every possible relevant issue, even if little or no information is yet available to establish a specific, quantitative requirement. Any issues you forget to mention at all are likely to come back to haunt you in very nasty ways!

5.4 Example: A Portable Wind Chill Meter

1. In-use purpose

 a. Primary intended purpose:

 For skiers and winter hikers: to be able to determine the so-called “wind chill” factor quickly and conveniently.

 b. Unintended uses:

 None anticipated, though winter campers are likely to find some!

 c. Special Features:

 Could indicate both temperature and wind velocity separately as well as the combined wind chill,

 Readings could be independent of the meter's orientation with the wind.

2. Functional requirements

 a. Performance requirements

 (i) Functional performance requirements:

 Indicated reading should be accurate to within one degree using U.S. Weather Bureau formula for wind chill.

 The meter should provide a reading after no more than twenty seconds for reaching equilibrium with the outdoor environment.

 The meter should perform with the above accuracy from 32 F to 50 F, and from wind velocities from 10 mph to 60 mph. Accuracy should not be affected by mis-orientation with the wind direction up to 20 degrees angle.

 (ii) Complementary performance requirements:

 The expected life in normal use should be at least ten years (except that batteries, if used, can have a life of six months.)

 The reliability of the meter should be such that no more than one meter per thousand sold will be returned for repairs or replacement during the first year of use.

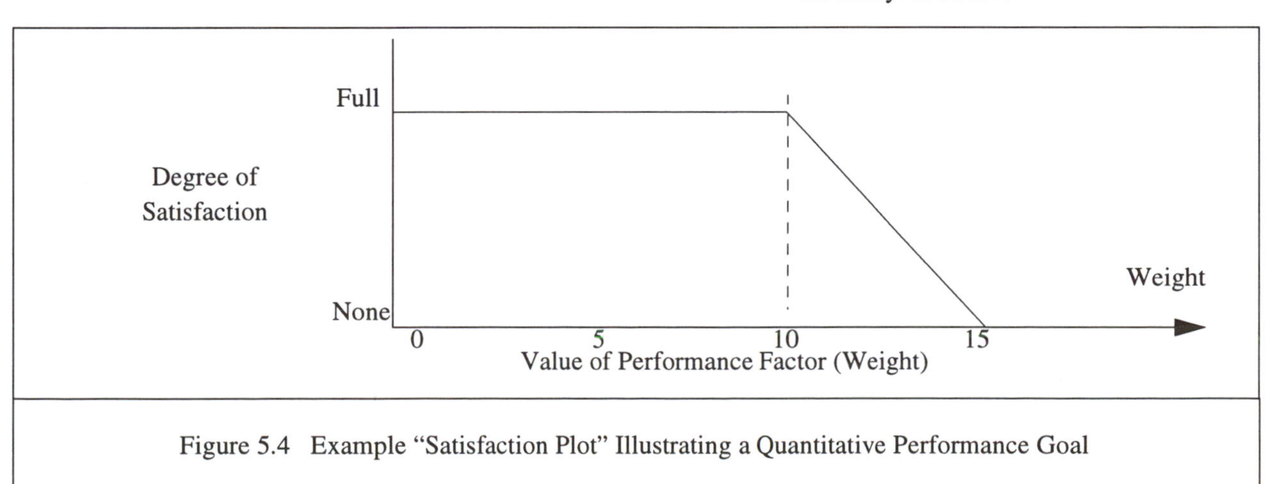

Figure 5.4 Example “Satisfaction Plot” Illustrating a Quantitative Performance Goal

The meter must be able to sustain shocks and pressures without damage when being carried in jacket pockets or backpacks, or dropped into snow or onto hard ice from five feet. It must be rustproof and waterproof to immersion for one minute in cold water with no manual drying afterwards.

There should be no way that a hiker or skier who falls could be cut by sharp edges or breaking glass from the meter.

Except for battery replacement (which should be doable without tools in less than three minutes by the user), no maintenance should be required.

The meter should be conveniently readable by people with normal vision.

b. Environmental conditions

See above for in-use environmental conditions. The meter must also survive shipping summer storage up to 120 F and 80% relative humidity without damage.

c. Economic issues

The manufactured cost in volumes of 5000 must not exceed $10.00.

d. Physical attributes

To be carried in a ski or hiking jacket pocket, the meter must not weigh more than 16 ounces or be larger than 3 inches by 5 inches by 3/4 inches. Smaller and lighter would be better.

e. Process technologies

No restrictions.

f. Aesthetics

The meter should look like a rugged, reliable, accurate device compatible with the quality equipment needed by skiers and hikers.

g. Product development time and cost

The product should be designed and prototypes tested by <specified date>. Design and development cost, including engineering, model shop, and laboratory should not exceed <specified amount>.

Please note in this example how quantitatively the requirements are expressed. It doesn't say that the meter should be accurate, it says it should be accurate to one degree. It doesn't say the meter should be reliable, it says how many meters can be returned for repairs during the first year.

5.5 Engineering Design Spec for Supplementary Design Project A

Following is an Engineering Design Spec prepared by a student team for Supplementary Design Project A:

In-use purposes

(a) The primary purpose(s) to which users will put the product:

To construct on-site the structural framework for customized factory work benches, tables for automation equipment (e.g., parts feeders), and work stations (especially for manual assembly);

To modify the above re-using the materials.

(b) Any unintended purposes to which the designed object may predictably be put:

Support climbing, standing and sitting by workmen;

Support the dropping of tools and equipment.

(c) Any special purpose features required or desired:

Enable routing of electrical power, air lines, and the like;

Convenient attachment of light fixtures, hose clamps, hangers, shelves, etc.

Functional Requirements

(a) Product Performance

(i) Functional

A reference table 1.80 m long, 1.05 m high, and 0.90 m wide must:

Support equipment weighing 3500 N dropped from a height of 75 mm;

Withstand horizontal forces on the top with horizontal deflection no greater than 2 mm;

Withstand an average hammer blow to the midpoint of any member with no loss of structural integrity.

Shelves must support 3500 N with deformations less than 1 mm;

Vibrations??

Others??

The reference table should be able to be modified on-site to a size of 1.5 m long, 0.90 m high, and 0.75 m wide without new materials with two men in one hour using readily available tools and equipment.

(ii) Complementary

The useful life with normal use should exceed ten years;

The structures should require no maintenance;

The materials must be non-combustible;

There must be no sharp edges or corners that could cut hands or catch loose clothing.

(b) The environmental and other conditions under which the product is to perform:

The materials used must not be structurally degraded by oil, paint, solvents, water, coffee, and the like;

The materials used must be suitable for use at temperatures to 120 F and humidities to 100%.

(c) Economic issues:

The reference table, fully assembled on site, should cost no more than $100.

(d) Physical attributes.

(e) Process technologies:

The product line should be manufacturable at existing company facilities (i.e., injection molding, die casting, plastic and aluminum extrusion, or sheet metal stamping);

Assembly should require no welding, gluing or the like.

(f) Aesthetics:

The appearance must be attractive and consistent with the atmosphere of quality, modern automated assembly, robotics, computers, and the like.

(g) Product development time and cost:

The conceptual and configuration design must be completed in ten weeks (i.e., this semester!).

In the above Spec, note that in several categories there are open questions. For example, this team did not know how to specify a requirement regarding vibration. However, by just including vibration among the requirements, they have reminded themselves of this issue and that they must consider it in more detail later. Perhaps one team member was assigned to find out more about what is required and how to express it.

Note, too, how the team tried to be quantitative. Actual temperatures, costs, weights, and the like are specified. The use of a "reference table" is an interesting way to get a metric on some rather difficult issues. It also has some dangers: perhaps the reference table itself will be satisfactory, but some variation may not. Ultimately the strength requirements implicit in the table should be converted to more fundamental parameters like tensile strength, bending moments, critical buckling strengths, and so on.

5.6 Summary and Preview

Summary. The first step in the guided iteration process for all the stages of engineering design — conceptual, configuration, and parametric — is to formulate the problem. For engineering conceptual design, that formulation takes the form of preparing a detailed Engineering Design Specification starting with whatever information is available from the marketing studies and from the Industrial Design phase. As noted, this information will vary greatly from firm to firm, and from product to product. However, regardless of what is initially available, the Engineering Design Specification must contain the specific and quantitative information needed to guide the rest of the design process in a way that results in a product that meets the needs of customers, the manufacturing process, and the company.

The Specification will not be absolutely complete after its initial preparation — and it may not yet have completely detailed, quantitative information — but once the major requirements for the designed object are understood and defined, then (but only then), it is time to think about alternative conceptual solutions. Leaping to solutions before the Specification is as complete as feasible, no matter how obvious they may seem, either closes the door on other possibly better solutions, or prejudices the preparation of the Engineering Design Specification so that only the preconceived solution works. We should always practice least commitment.

The Engineering Design Specification therefore is (or should be) neutral about what the conceptual solution might be. It should dictate or imply a preference for no physical form, and it should not require that particular physical principles be employed by the solution — unless these are dictated by customer or marketing requirements. The Spec should not influence any more than absolutely necessary how the designed object will meet the functional and other requirements. The Spec states what the requirements are, not the means by which they will be achieved. It is then the designers' task during the rest of the conceptual design process to find the physical concept or embodiment that best meets the Specification.

The Engineering Design Specification is never cast in stone. It is a dynamic document; it can be changed and added to. It usually is. Of course, changes that cause delays or cost money can be very troublesome. But excessive rigidity can also be a problem. Reason and good sense will find the happy medium.

Designers must beware especially of issues that may be forgotten or ignored. They are the most dangerous ones. Design engineers should talk to the marketing people and to customers, and get all those customers attributes needed for the House of Quality. Manufacturing engineers must also be consulted, along with everyone else who may have information relevant to the design requirements.

Students preparing an Engineering Design Specification for the first time sometimes have difficulty stating requirements *without* also including how the requirements will be met. Note in the above example that almost nothing is said about how the requirements will be satisfied. Materials are not specified, means of sensing temperature or velocity are not specified, manufacturing process are not specified, and so on. This allows designers as much freedom as possible to develop the best possible product as the design process proceeds.

Preview. After problem formulation, the next step in the guided iteration design process is the generation of alternatives. This is therefore the subject of the next chapter.

Other Recommended Reading

Total Design, Stuart Pugh, Addison-Wesley, 1991.

Management of Engineering Design, D. J. Leech, Wiley, 1972.

Engineering Design Methods, Nigel Cross, Wiley, 1989.

Conceptual Design for Engineers, M. J. French, The Design Council, Springer-Verlag, 1985.

References

[1] Leech, D. J. *The Management of Engineering Design*, John Wiley and Sons, Inc., New York, 1972. Quote by permission of the publisher.

[2] Hauser, D.R. and Clausing, D. "The House of Quality", *Harvard Business Review,* May-June, 1988.

[3] Freeze, K. "Braun AG: The KF 40 Coffee Machine", Design Management Institute Case Study, Design Management Institute, Boston, MA 02111, 1990.

Problems

5.1 Describe the incandescent light bulb conceptually. Don't use the term "light bulb".

5.2 "Post-It'™ notes from 3M company are a huge marketing success. Write a conceptual design specification, and describe the notes conceptually. Your specification should not forecast or "telegraph" the *Post-It* solution. Can you think of any other conceptual solutions to your specification?

5.3 Write an Engineering Design Specification for which the following would be possible conceptual solutions: pencil, ball-point pen, quill pen, crayon, magic marker, bookcase, v-belt drive system, you name it.

5.4 Write a physical concept description of the lead pencil. (Note: For a very interesting history of the pencil, see *The Pencil* by Henry Petroski, a Civil Engineer.)

5.5 Write an Engineering Design Specification for which an automatic overhead garage door *opener* is a solution.

5.6 Write another Engineering Design Spec for the garage door opener but stress the marketing need for a solution that is very easy to install.

5.7 List the Customer Attributes for starting the House of Quality for which of the following designed objects are example solutions: carrying case for CDs, ball point pens, bicycle, toaster.....

Chapter Six

Generating Conceptual Design Alternatives

"Disciplined thinking focusses inspiration rather than blinkers it."

Gordon L. Glegg [1]*
The Design of Design (1971)

6.1 Approaches to Generating Conceptual Alternatives

In this chapter, we assume that an Engineering Design Specification has been prepared, and that now the task is to generate a set of alternatives for the physical concept (or embodiment) that will meet the stated requirements. Then in the next chapter, we consider how to compare and evaluate the alternatives that are generated. We consider here primarily the generation of alternatives for products and their sub-assemblies. Parts are discussed more specifically in Chapter 8. However, a number of the creative methods described in this chapter are general and apply as well to parts as to products and sub-assemblies. Indeed, in some cases the methods apply also to non-technical areas of life as well.

In Chapter 4, we described the process of decomposition of a product or sub-assembly into subsidiary sub-assemblies and components, and showed how decomposition can be performed directly in a single step resulting in a hardware description, or in two steps, the first of which is a purely function-first decomposition. We noted that one reason for doing this is that it not only leads to more alternatives, it also may lead to novel ideas that might not otherwise be considered.

Function-first decomposition is one method for generating conceptual alternatives. There are also other methods:

(1) examining the potential use of alternative physical laws and effects,

(2) use of one or more well known applied techniques for creative idea generation, and

(3) conducting an information search.

In this chapter, we describe and illustrate each of these methods as well as function-first decomposition. We also discuss briefly some ways that engineers can improve their personal creativity in general over the long term.

6.2 Function-First Decomposition

6.2.1 Functional Decomposition of an Existing Product

Note: Readers interested in this subject are urged to supplement the discussion here with study of the more detailed presentation in *Engineering Design* by Pahl and Beitz [2]. See also the articles in *Machine Design* by Crossley [3, 4, 5, 6]. These source and others referenced in Chapter 4, contain references to other valuable work on the subject of function in conceptual design.

Before considering the more difficult task of using function-first decomposition to create an original design, it is possible to learn much from the functional decomposition of an existing product. As an example, consider a particular electrical powered, free standing auxiliary room heater. The

---

* Quotation from *The Design of Design* by Gordon Glegg, Copyright 1971, Cambridge University Press. Reprinted by permission of the Cambridge University Press.

> "Form ever follows function."
>
> Louis Henri Sullivan (1896)

unit sits on the floor, has an electrical cord and plug to an electrical wall outlet, an electrical resistive heating element, and a fan to circulate air from the room across the element and back into the room. Figure 6.1 shows the first stages of one way this product might be decomposed into hardware; that is, into sub-systems, components and parts. Figure 6.2 shows a schematic diagram of the major hardware sub-systems and components. In this kind of hardware decomposition, functionality is only implicit.

In contrast, Figure 6.3 shows the first stages of a function-first decomposition for a heater of this type. When we take this approach, the manner in which the various sub-functions shown might be accomplished is still an open question; that is, the hardware sub-systems, components, and parts that might be selected to perform the various sub-functions are not yet identified. We have thus left our options open for possible new and different solutions.

For example, we could choose to provide for supporting the unit, not with legs to the floor, but by hanging it from the ceiling or by a cantilever beam from a wall. We might fulfill the energy distribution function by radiation or by natural convection rather than by a forced flow of air. Or we might convert the electrical energy to thermal energy with a motor and brake rather than with a resistive type element. Not all of these alternatives are practical, perhaps, but the point is that the function-first decomposition encourages us to consider the possibilities. Thus this function-first decomposition process sometimes leads to innovations that might otherwise be overlooked.

It should be noted how alternative ways to meet a given sub-function generally employ, not just different hardware, but also different physical principles and effects. For example, regarding energy distribution from the space heater, thermal radiation is a fundamentally different mechanism of heat transfer from convection. And a brake uses the effect of mechanical friction rather than electrical resistance to generate heat. We will have more to say later in this chapter (Section 6.3) about the role of exploring alternative physical principles and effects in generating alternatives.

In general, when a product exists or is to be substantially adapted from another similar product, a functional decomposition can be fairly readily developed from the existing product. Then one can use the resulting functional diagram as a starting place to look for better solutions. This can be done in several ways:

1. By searching for new ways to meet the functions using different physical principles and hardware. Using radiation instead of convection to distribute

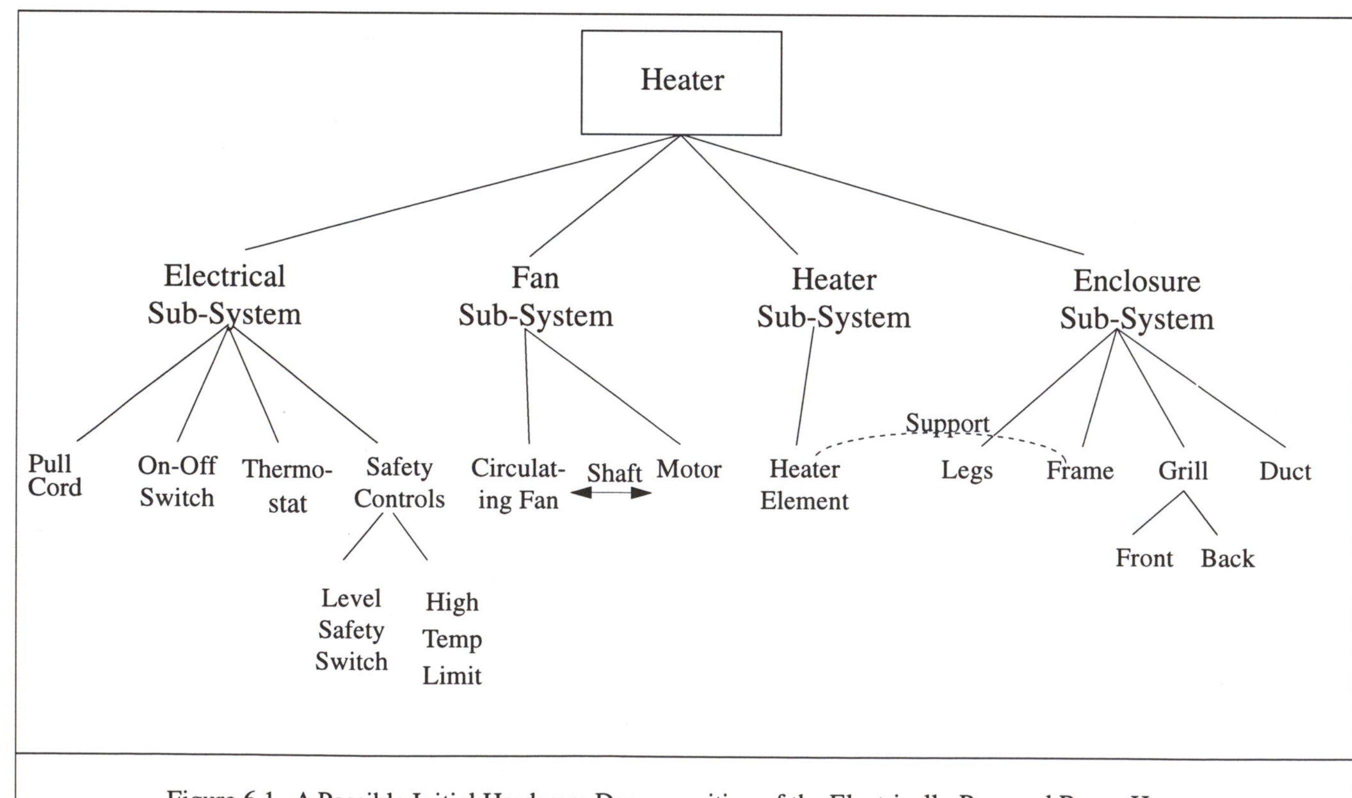

Figure 6.1 A Possible Initial Hardware Decomposition of the Electrically Powered Room Heater

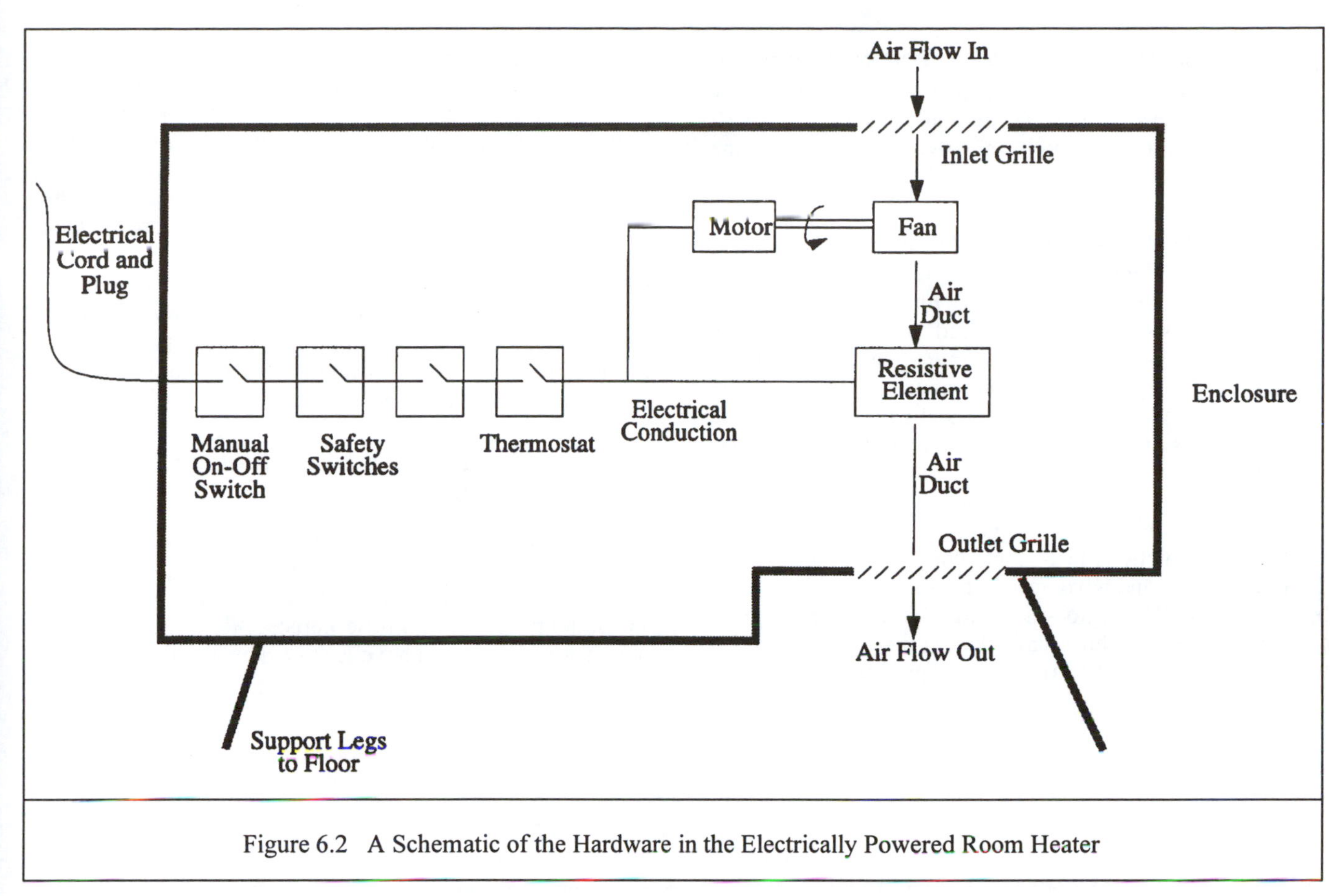

Figure 6.2 A Schematic of the Hardware in the Electrically Powered Room Heater

Figure 6.3 An Initial Functional Decomposition of the Space Heater

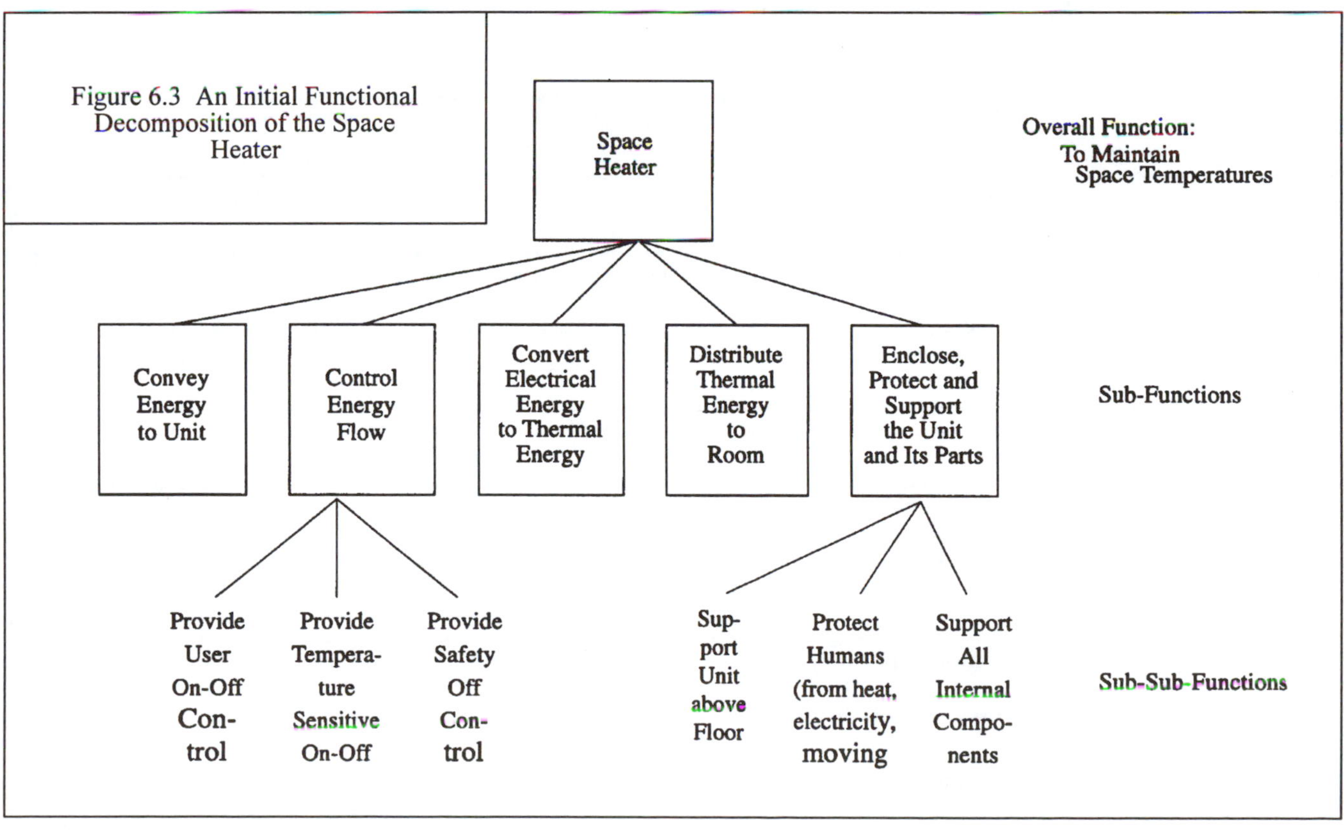

energy from the space heater is an example of this;

2 By searching for ways to modify the functional decomposition by (i) combining functions or (ii) separating functions into two or more sub-functions; or

3. By starting from scratch to develop a substantially different functional decomposition as if the product were completely new.

As an example of combining functions, suppose that equipment exists for making bread by the conventional sequential process of mixing the ingredients, kneading the mixture, allowing time for rising, punching out entrapped air, forming the dough into loaves, and baking. A functional diagram of this process would show all the various functions (i.e., mixing, kneading, etc.) as separate functions, and then the temptation would surely be there to meet them separately. Recently, however, designers have found a way to fulfill all these functions in a single container (at least on the domestic scale), and have thus found a new market for home bread makers. All one does is pour in the (unmixed) ingredients, turn on the machine, and take out the bread after a few hours. By combining functions, a new product was born

Separating previously combined functions can also sometimes lead to good new ideas, or to the re-discovery of good old ones. For example, consider the door knob mechanism that combines latching and locking into a single product. The trouble is, the latch bolt in these devices is rather easy to slide back in order to gain illegal entry. Clearly better security is provided by separating the latching from the locking functions, the latter being performed by a separate bolt that is more difficult to foil.

Before we go on to discuss the process of function-first decomposition for new products, it will be helpful to organize the categories and refine the terminology that we use for discussing function.

6.2.2 Categories and Terminology for Functionality

When developing a functional decomposition diagram, the top or beginning level generally expresses functionality in user-oriented terms, perhaps better referred to as "in-use purposes". The in-use purpose, for example, of the room heater discussed above is "to maintain the temperature

Device	Input(s)	Function	Other Effects	Output(s)
Nozzle	Fluid flow	Increase velocity of fluid	Decrease pressure of fluid	Fluid flow
Motor	Electrical Energy	Convert electrical energy to rotating mechanical energy	Thermal energy generated	Rotating mechanical energy
Pump	Fluid flow and mechanical energy	Increase pressure of fluid	Change direction of flow? Thermal energy	Fluid flow
Gear	Rotating mechanical energy	Change speed of rotation	Change direction of rotation	Rotating mechanical energy
Electric resistance element	Electrical energy	Convert to thermal energy	Element increases in temperature	Thermal energy
Pencil	Mechanical energy (Force in motion)	Transfer lead from pencil to paper		Lead deposit on paper
Expansion valve in refrigeration system	Liquid refrigerant	Reduce temperature of refrigerant	Reduce pressure of refrigerant	Refrigerant (two phase mixture)
Lever or wrench	Energy (Force or torque in motion)	Increase magnitude of force or torque		Energy (Force or torque in motion)
Switch	Mechanical energy (force in motion)	Separate or join contacts	Flow of electrical enabled or stopped	Position of contacts moved
Room thermostat	Flow of room air	Separate or join contacts	Flow of electricity enabled or stopped	Position of contacts moved

Figure 6.4 Functionality of Some Common Devices

Input	Functional Device	Output
Material Flow --------->		--------> Material Flow
and/or	Behavior	and/or
Energy Flow --------->	or	--------> Energy Flow
and /or	Action	and/or
Information Flow		--------> Information Flow
or Signal --------->		or Signal

Figure 6.5 Generalized Inputs, Outputs, and Behaviors of Functional Devices

of the room". When we decompose such top level functions into sub-functions and sub-sub-functions, the terminology used to describe the functions becomes increasingly technical — and the functions themselves become simpler to express. Ultimately, as the decomposition proceeds, we get to the most basic of technical functions, like "convert electrical to thermal energy"' or "increase the magnitude of a force".

It is helpful in thinking about function to distinguish between the *purpose* of a device and its *physical behavior* or *action*. For example, the purpose of a switch is to control the flow of electricity in on-off fashion. The physical behavior of the switch is to move — we say 'open' or 'close' — the contacts mechanically. Obviously, the behavior or action enables achievement of the purpose. We will generally use the term function to designate physical behavior, though it is usually obvious which is being referred to.

To help introduce the terminology of function, we will consider the functionality of a number of common devices as shown in Figure 6.4

The possible list of devices that might be listed in Figure 6.4 — together with their specific inputs, functions, and outputs — is, of course, infinite. However, if we generalize a bit to *classes* of inputs, then the numbers are not very large at all. Indeed, inspection of Figure 6.4 suggests that all inputs and outputs to functional devices could be generalized to either material or energy flows.

In most of the devices listed in Figure 6.4, the behavior not only achieves the purpose but also involves the same energy and material flows expressed in the purpose. In a motor, the purpose is energy conversion, and the behavior involves flows of energy being converted from one form to another. The purpose of a nozzle is to increase the velocity of a fluid, and the behavior involves the flow of fluid.

There is, however, something different about the switch and thermostat in this respect. Strictly speaking, the inputs in these devices are energy or material flow, and the outputs are material flow (the contacts are material that moves). However, the purpose does not have to do with conversion of energy or the flow of material. The purpose is to *control* the flow of electrical energy, whereas the behavior is the physical motion of the parts. Indeed, the input energy involved in moving the parts is tiny and unrelated to the amount of electrical energy flow. In other words, the behaviors and the purposes of the switch and thermostat are disjoint in a way that is not true with the other kinds of devices.

The inputs and outputs to devices like switches and thermostats (and other types of sensors and control devices) is therefore best thought of as *information* — or as a kind of *signal*. (See [3].) That is, our fingers that flip the switch provides a signal (or are the response to a signal from the brain) along with the energy that moves the contacts. In any case, it is the signal (or the information in the signal), not the energy required, that is important to functionality in this case. The same is true of the thermostat: the flow of room air through the thermostat provides information or a signal that the room has (say) cooled. Then the action or behavior of the thermostat moves the contacts.

Therefore, in general, it can be argued that the inputs and outputs to all functional devices or systems are one or more of three general types:

1. Material flow,
2. Energy flow, or
3. Signals (or Information flow).

Schematically, this can be represented as shown in Figure 6.4.

The behaviors or actions of functional devices can also be generalized into a relatively small number of types. Note, for example, that in Figure 6.5, the behaviors might be generalized to converting (the form) of energy, increasing (the velocity or pressure) of material, signalling to enable or block the flow of electricity, and so on. Of course, Figure 6.4 is a small list and many other types of behaviors exist. Examination of many such devices by researchers and engineers interested in a functional approach to conceptual design has lead to identification of five basic types as described in Pahl and Beitz [2] which may be summarizes as follows:

1. Change, convert or transform,
2. Increase (amplify) or decrease magnitude,

3. Connect (join) or disconnect (separate),
4. Conduct or channel (or block), or move location, and
5. Store or take from storage

Readers should note that these generalized behaviors or actions (i.e., functions) can be taken on any form of input (material, energy, or signal) and produce any form of output (material, energy, or signal) as shown in Figure 6.6.

The first of these could be accomplished by an electric motor or a brake. The second could be accomplished by a steam boiler or condenser, or by a pump. The third could be accomplished by a valve.

The above generalized structure and terminology creates, when all the combinations are considered, a number of basic functional building blocks. With these building blocks, we can decompose a product's overall function into sub-functions, sub-sub-functions, and so on. Then, we can explore how functions can be combined or separated to obtain different functional decompositions.

Since one is committed only to functions, not to the hardware that achieves the functions, it should be apparent that this process implements the policy of least commitment very nicely. It also leads to exploration of many alternative conceptual design possibilities.

6.2.3 Application to a Product

When an existing product is to be modified for improvement, or a new product adapted from a relatively similar existing product, then an initial functional decomposition can be developed based on the existing product as described above. However, when a new product is being conceived, then the functional decomposition must be done from scratch beginning with only the top level function or in-use purpose of the product. This is more difficult, and requires careful, incisive thought about what sub-functions (at each level during decomposition) are required to achieve the functions at the level above.

As an example, we consider the conceptual design of a braking system for an automobile (see [5]). The overall function is to take a signal input and decrease the velocity of the automobile. The signal will somehow originate with the driver, at least usually. To perform the deceleration, a force must be applied to the automobile. No matter how this is done, it will require energy. Thus we can represent the function as shown in Figure 6.7(a).

There are any number of ways to use energy to apply a retarding force to a moving automobile. We want to keep our options open on how this might be done by thinking of it, not as some hardware system, but only as a function to be performed by some as yet unselected equipment using as yet unspecified physical principle(s). Thus we can think of the function as involving conducting the energy and then converting it to a form for decelerating the vehicle. See Figure 6.7(b).

There are a variety of ways the energy can be converted to a force to retard the automobile. Here, for example, are some of the ways (there are no doubt others) that the function of applying a retarding force to the automobile might be accomplished:

1. By air friction (e.g., by deploying a parachute),
2. By ground friction (e.g., by lowering a brake shoe to the pavement),
3. By a forward thrust jet,
4. By transferring the auto's kinetic energy to storage [See Fig.6.7(c).],
5. To a flywheel,
6. By a generator to a battery
7. By friction force on the wheels (dissipates kinetic energy to thermal),
8. By mmechanical friction (the conventional solution),
9. By fluid friction (in the manner of a fluid transmission),
10. By an eddy current brake (this idea is adapted from unpublished notes by Professor Erskine Crossley),
11. By reversing the engine or its thrust.

Input	Functional Device	Output
Material Flow ----------> and/or Energy Flow ---------> and /or Information Flow or Signal --------->	• Change or Transform • Increase (amplify or Decrease Magnitude • Connect (join) or Disconnect (separate) • Conduct, channel or block, or move location • Store or Take from Storage	--------> Material Flow and/or --------> Energy Flow and/or --------> Information Flow or Signal

Figure 6.6 Generalized Functions of Functional Devices

We must also determine a source for the energy input, and depending on its form, a way to convert the energy input into the form required by the deceleration function. Sources of energy could include:

1. The driver (motion of leg or arm)
2. Hydraulics (liquid or air)
3. Electrical
4. From storage in flywheel or battery or other.

The signal may also be provided in more than one way:

1. The driver (motion of leg, arm, or hand; voice),
2. From an automatic crash avoidance system.

The course we follow now is to look for the best ways (according to the goals of the project as outlined in the Engineering Design Spec) to perform and combine the various functions. As decisions are made about one or more of the functions, it is often the case that additional functions appear through the separation of functions. For example, suppose we choose to use the force and displacement of the driver's leg and foot as the initial source of energy, converting it to hydraulic energy. Then we will have a "conducts energy" function to conduct the energy from the pedal to the wheel, and another energy conversion function in order to convert the hydraulic energy back to the mechanical energy needed to apply a frictional force (or torque) to the wheel. This is essentially the functionality shown in Figure 6.7 (b). But if we want to store the energy of the vehicle as we slow it down, we would need another function ("store energy") as shown in Figure 6.7 (c).

6.2.4 Comments on the Function-First Approach

The function-first approach to conceptual design is essentially this: (1) First decide what has to be done purely in terms of functions, decomposing the functions into sub-functions until simple functions are obtained; then (2) look for physical principles and effects that can be employed to achieve those functions; then (3) look for specific hardware or types of hardware to employ the principles and achieve the functions. In effect, you design the object in terms of its required functionality and the physical principles to be used. Only lastly do you think about hardware to fulfill the functions.

This approach is not easy to implement in all situations. One has to learn to think in abstract terms about pure function and about physical principles and effects. But this is a way of thinking that can be cultivated and learned. The advantage is that the method facilitates the examination of options that may not have been considered if one leaps quickly to the selection of specific principles or (worse) to

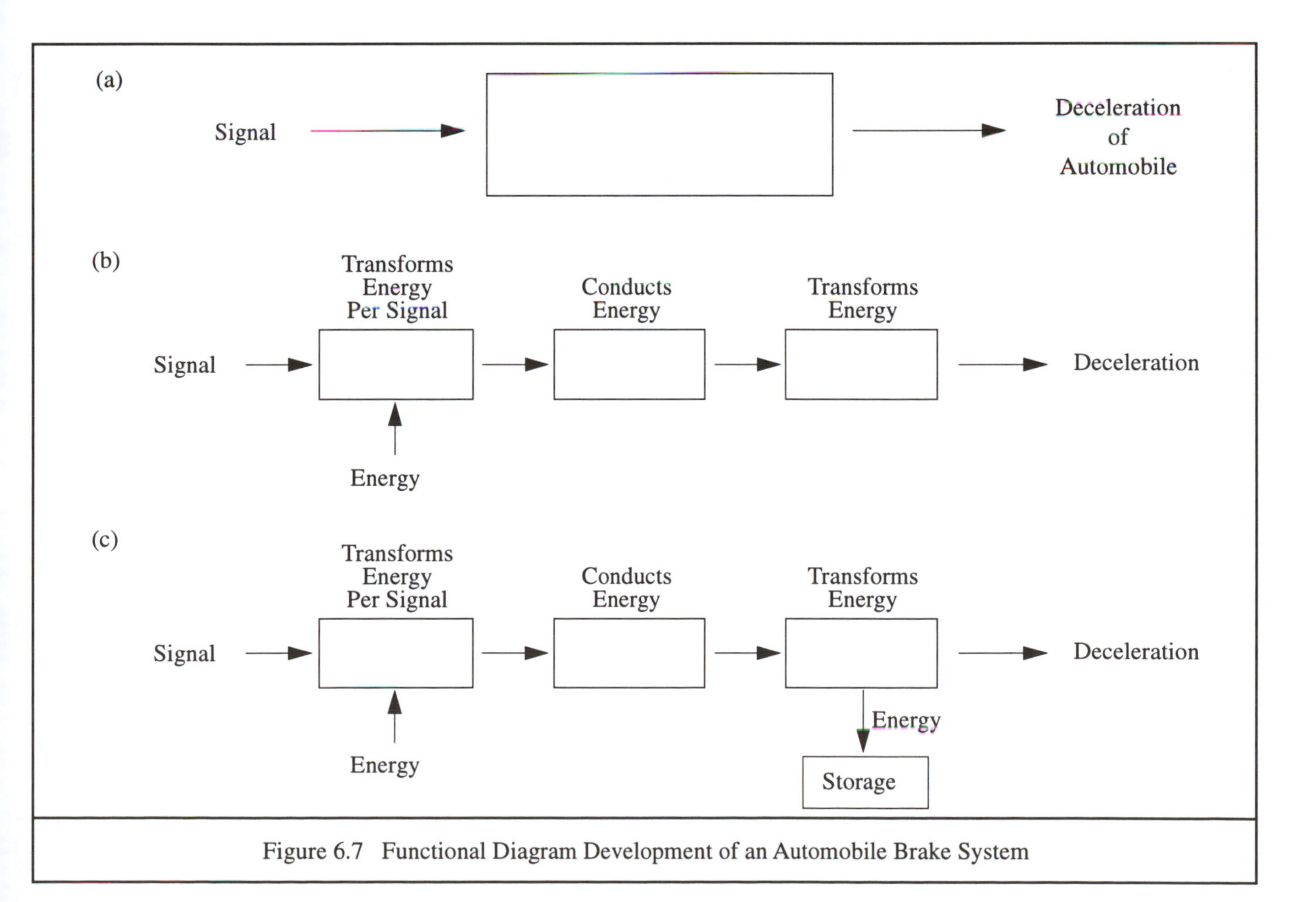

Figure 6.7 Functional Diagram Development of an Automobile Brake System

the selection of specific hardware. Thus we encourage designers always to try function-first decomposition as a part of the conceptual design solution process. The results will sometimes be very gratifying.

A variation on the function-first approach is to begin with a hardware decomposition, and then identify the functions that each sub-assembly and component fulfill. Then one can look for alternative ways to fulfill the identified functions, and for ways to separate or combine functions. This approach may not be quite as general as a purely function-first approach, and hence may miss some alternatives, but it is somewhat less abstract and hence a bit easier to implement.

6.3 Searching Via Alternative Physical Laws and Effects

Note that an important part of the description of an embodiment is the physical principle or principles on which it operates. Therefore, another — and *very* powerful -- method for generating creative ideas for physical concepts or embodiments to meet a given Engineering Design Specification is based on the fact that new conceptual solutions often employ different (i.e., previously unused for this purpose) physical laws and effects. By considering the use of alternative physical principles and phenomena, a designer can sometimes become aware of the possibility of a new conceptual solution.

In this section, we present a long but nevertheless incomplete list of physical laws and effects. For a more complete list, and a great deal more discussion, interested readers are encouraged to consult the original sources in Hix [7] and Collins [8].

The list of physical laws and effects is not only helpful in finding alternative solution concepts, it is also useful as a checklist to insure that no forgotten physical behavior or side effect is going to happen to interfere with other solution possibilities.

Selected Physical Laws.

Listed here are some established physical laws which can be employed to accomplish design goals. Of course, if readers or designers are not familiar with the laws — when they apply, what physical quantities they govern, and so on — then the list will be of little practical value except as an initial guide to further study. Readers are encouraged to refer to [7] not only for a more complete list, but also for good brief descriptions of the laws. Courses, texts, and references in physics, chemistry, and engineering science can also be consulted for more complete understanding. This is a major reason that engineering students must get a solid grounding in basic science.

Note: The following lists of selected physical laws and effects is adapted from Hix [7] and reproduced with permission of the General Electric Company.

Conservation Laws:

Conservation of Energy (in many forms).)

Conservation of Mass

Conservation of Mass and Energy ($E = mc^2$)

Conservation of Momentum (Linear, angular)

Newton's Laws:

First Law

Second Law ($F = ma$)

Third Law

Ohm's Law ($E = IR$).

Thermodynamic laws:

First Law (Conservation of Energy: heat, work, and internal energy)

Second Law (Entropy increases, and work potential decreases in isolated systems; efficiency of heat engines is limited)

Third Law (Can't reach absolute zero temperature)

Archimedes' Principle.

Heat Transfer:

Conduction Law

Convection

Radiation (Stephan-Boltzman Law)

Electro Magnetic Laws.

Periodic Table.

Selected Physical Effects.

A physical *effect* is a reaction to some other physical action, whether or not the reasons are understood and explained in laws. As with the laws listed above, the list below is incomplete; readers are referred to the references and basic science books for more details and better explanations. The point here is that designers who would be innovative and creative in conceptual design should have as broad a knowledge and understanding of as many of the known physical laws and effects as possible.

- Absorption
- Adsorption
- Anisotropy
- Buoyancy Effects
- Barkhausen Effect
- Doppler-Fizeau Effect
- Thermoelectric Effects
- Diffraction
- Cavitation
- Volta Effect
- Friction (Dry, rolling, fluid viscosity, Stokes law)
- Critical Pressure and Temperature
- Diffusion
- Diffraction
- Photoelectric Effects

- Electro-kinetic Phenomena
- Piezoelectric Effect
- Thermal Expansion
- Gauss Effect
- Hall Effect
- Hertz Effect
- Joule Thomson Effect
- Incandescence
- Photovoltaic Effects
- Pinch Effect
- Photoelastic Effect
- Raoult's Law
- Resistance-Temperature Effect
- Surface Tension and Capillarity
- Thermoelastic Effect
- Ultrasonics
- Wien Effect
- Stress-Strain Effects (Hooke's law, elastic limit, Poisson's ratio,
- Creep, fatigue, work hardening, photoelasticity, stress corrosion)

Example

A beam is one way to support a load over an open space. Suppose for a moment that the other ways also familiar to us were not yet known, and consider how searching for different physical laws and effects might lead us to them. Beams make use of the principle that a load perpendicular to a long slender structural member can be carried by longitudinal bending stresses in the member (accompanied by a little bending).

By looking for ways to carry the load using different laws and effects, we can conceive different structural concepts. For example, if we consider using the effects of tension and compression only (no bending), we are led to both the truss and suspension concepts. If we explore buoyancy as an effect, we are led to pontoons or helium balloons.

Another Example

Most fluid flowmeters are based on fluid mechanics phenomena. However, mass and heat flow phenomena can also be used to measure flow rate in devices called *thermal flowmeters*. Thermal flowmeters are extremely accurate and can be used in certain applications where traditional flowmeters are inappropriate. The concept of a thermal flowmeter might be generated by looking for other physical phenomena possibly useful in flow measurement — in this case, convection heat transfer.

Here are some products or processes that might have been conceived by considering different physical laws or effects to accomplish some required function: jet planes, ball-point pens, transistors, laser printers, fluorescent lights, microwave ovens, assembly lines, air powered tools, and thermo-electric refrigerators.

6.4 Applied Techniques of Creative Idea Generation

6.4.1 Introduction

Creativity (or, if you wish, *inventiveness*) is a rather nebulous term that refers generally to the ability of an individual, group, or organization to generate and implement ideas that are simultaneously *new, unique, and useful*. We include *useful* because we are designing functional parts and products, not art objects. Ideas do not have to be *immediately* useful to be creative in an engineering design sense, but they do have to be potentially useful at some point in time.

A number of rather specific methods or techniques have been developed and proven effective in generating new, unique, and useful ideas. These include the following that are to be discussed:

- Brainstorming;
- Systematic Search for New Combinations;
- Techniques of *Synectics* (Gordon [9]), including:
- Inversion
- Fantasy
- Analogy
- Empathy;

6.4.2 Brainstorming

Brainstorming [10] is a well known technique of applied creative idea generation, but it is (unfortunately) more talked about than used, at least in its most effective, pure form. The reason for its non-use and mis-use is that brainstorming is difficult for many people to do. Also, it is less useful for highly technical or specialized engineering problems than it is for more general, less technical problems. However, there are times when it can be used to great advantage in engineering design, especially at the engineering conceptual stage, and design engineers should understand how to use it at those times.

Brainstorming is generally considered a method to be used by a group. The basic brainstorming methodology is as follows. The group assembles in a comfortable, private place with a tape recorder. The problem is explained. Then the group members begin suggesting as many solutions as possible in as free and open a way as possible. Sessions may last an hour or more. After the session, the tape is transcribed, and then analyzed for possibly useful suggestions by the project team.

It is not necessary that the brainstorming team and the design project team consist of all the same people; in fact, it

> "Nothing is more dangerous than an idea, when its the only one we have."
>
> Emile August Chartier
> *Systeme des Beaux-Arts (1920)*

is best to include some individuals in the brainstorming session who are not familiar with all the details of the specific problem to be solved, though they should generally familiar with the technical field or marketing issues involved.

There are specific rules to brainstorming that must be followed if brainstorming is to be expected to work well:

Rule 1. There must be *no criticism, evaluation, or negative comments* about ideas expressed. In this statement, *no* means *NO!* Such comments will inhibit team members from advancing other ideas, and limit the number of new ideas that will be obtained.

Rule 2. *As many ideas as possible* must be collected. In this statement, *many* means *MANY!* Group members must speak out freely with the ideas that occur to them without inhibitions caused by concern for whether the ideas initially appear *useful or not*. The project group will evaluate them later. The session itself is not the time for evaluation; it is the time to collect *lots* of ideas. Ideas should thus be allowed to flow in a kind of group "stream of consciousness" manner, allowing one idea to lead to another. The more ideas the better.

Rule 3. *Wild, "off the wall", unusual ideas are wanted*. In this statement, *unusual* means *UNUSUAL!* The reason is that such ideas may very well *lead to* some other more practical ideas even if they do not turn out to be practical in themselves.

These rules are not easy for many people, especially some engineers, to follow. We are very used to evaluating immediately every idea we hear for its practicality. We also tend to evaluate our own ideas even before we speak them. Brainstorming, however, requires that such evaluation, which is usually negative or critical, be temporarily set aside -- something that is hard to do. Fortunately, it only takes a little practice with a competent leader for most people to learn to suspend evaluative judgement long enough for the purposes of a brainstorming session. Then the process becomes fun as well as a very effective way to generate creative ideas.

Though brainstorming was designed for groups, and works best in groups that follow the rules, it can also be done by individuals working alone. Just get a tape recorder and follow the rules yourself: think wild, don't criticize your own ideas, and generate as many ideas as possible. Analyze and sort them out the next day. It works very well — *if* you *really* follow the rules!

6.4.3 An Example Brainstorming Session

Below is an example brainstorming session carried out by engineering students. The session is adapted by permission of McGraw-Hill Book Company, New York, from *Design Engineering: Inventiveness, Analysis, and Decision Making* by J. Dixon [10]. The problem to be solved is to separate green (unripe) tomatoes from ripe tomatoes quickly and automatically in large numbers. This is a real problem. Tomato farmers in some places are finding it more economical to pick all their plants of all their tomatoes at one time.

> "What is the use of a book, Alice thought, without pictures or conversations."
>
> Lewis Carroll
> *Alice's Adventures in Wonderland (1865)*

This can be done by machine. But then they have all the ripe and unripe tomatoes mixed up in a hopper. How do they separate them?

TOM: We separate them by color. A color meter ought to be practical.

ED: Emissivity or reflectivity - green ones ought have a higher reflectivity.

MARY: Hardness. We squeeze them -- easy -- or poke them.....

DICK: Electrical conductivity....

JOAN: Electrical resistance....

ED: Magnetism....

TOM: Mysticism. A Swami of some kind gets the ripe ones to float up out of the hopper.

JOAN: Float them on water....

DICK: Tomatoes must be mostly water -- and have the specific volume of water?

TOM: Do they float or sink?

DICK: Ravioli. When you first put it in the pot, when its frozen, it sinks. Then as it thaws it comes to the top...

MARY: Maybe some tomatoes float and some sink?

ED: Maybe the ripe ones float and green ones sink -- or vice-versa...

JOAN: So we separate them by how they float or sink in water...

DICK: Wouldn't have to be water....

TOM: Could be urine....

GROUP: Ugh!

TOM: Well, they used to measure the specific gravity of urine by whether a little ball would float or sink in it....Maybe tomatoes......

MARY: We could use some kind of oil -- non-toxic....

JOAN: Or salt water.

MARY: Size; won't the green ones be smaller?

TOM: Size and weight ought to correlate.

JOAN: That's specific volume....

DICK: X-rays -- size of the seeds or something like that...

ED: Smell, odor...

DICK: Sound -- Can you hear a tomato?

MARY: Can a tomato hear?

TOM: Potatoes!

GROUP:Huh?

TOM: Rhymes with tomatoes.....

MARY: Potatoes have eyes....

JOAN: If tomatoes had eyes they could see and separate themselves.

ED: Unless they were color blind...

TOM: The potatoes could see the tomatoes and be in charge of the separating...

JOAN: People can see. Let some guy look at them and put them in differentbins.

DICK: Heat -- infrared radiation....

ED: Thermal conductivity....

TOM: Get a juggler -- the ones he drops are the ripe ones....

JOAN: Use statistics -- somehow....

MARY: Just shake the hopper -- won't the lighter ones rise?

ED: Blow air through as you shake....

TOM: Use random numbers -- let 3s and 7s be ripe....

This session went on for a while, but we stop here. Note how ideas are put forward, picked up, and followed in a free-wheeling way. The lack of criticism makes this possible. Afterwards a list of possible solutions can be compiled for evaluation and investigation. A tape recorder or stenographer is a must for effective use of brainstorming.

6.4.3 Systematic Search for New Combinations

Many creative ideas for conceptual solutions are the result of forming new combinations of other concepts or embodiments that had not been previously related in the same way. Knowing this — that is, knowing that *new combinations* of things, ideas, and processes often lead to new solutions — one can explicitly look for new combinations.

To carry out a search for new combinations, usually we have first to identify the major functional dimensions of the search. For example, In looking for a new solar heating system concept, the three major functions involved are (a) energy collection, (b) energy storage, and (c) the energy distribution.

As another example of functional dimensions, consider the design of a devise to measure and display the speed, orientation, and direction of a golf club at the moment of impact with the ball. In this case, the functional dimensions might be: (a) the speed detection method; (b) the orientation detection method; (c) the direction detection method; (d) the computation technology; and (e) the display technology.

Next, for each of the identified dimensions, as many alternative methods as possible should be generated for performing each function. A brainstorming session for each function might help generate a good list of these alternatives.

Now each combination of alternatives (one for each function) becomes a possible way for the product to be designed. This can be a huge number equal to:

$$\text{No. of Combinations} = M_1 \times M_2 \times M_3 \times \ldots\ldots\ldots\ldots \times M_n$$

where n is the number of functions and M_i is the number of alternatives for the ith function. However, in part because the number is large and some of the combinations unlikely, there may well be combinations among them that are new and creative. Thus we search systematically and exhaustively for these combinations.

When there are only two functional dimensions, a search matrix can be constructed on paper that illustrates the process. See Figure 6.8. When N is greater than two, a two-dimensional view is not feasible so the various alternatives must be listed, even though it might be a long list.

6.4.4 Synectics

Synectics is a term coined in [9] to describe a set of four techniques for idea creation that includes: inversion, fantasy, analogy, and empathy. These techniques are especially powerful in generating creative conceptual alternatives, and readers are encouraged to begin trying them out whenever possible. A designer's ability to use them to advantage will grow with repeated effort and practice.

The underlying principle in using all four Synectic techniques is that new ideas come from taking a different view or perspective on a problem. Therefore, each of the four methods provides a different way to approach and view the problem and its possible solution.

Inversion. Inversion provides a new viewpoint by 'inverting' (in the most general sense) the problem or proposed solutions in one or more of several possible ways -- turn things inside-out or upside-down, stop moving parts and move stationary parts, and so on. Inversion is especially useful in finding new decomposition possibilities for concepts, and for finding new configurations. For example, parts or features of parts can be imagined turned upside down or inside out, or sideways, or at some fresh angle. Things inside can be placed outside, and vice-versa. Things on the left, or behind, can be put on the right, or in front. Parts that are stationary can be made to move while those moving can be made stationary. And so on. Even materials can be "inverted"; that is, changed or exchanged for other materials (e.g., plastics for metals).

Rather obviously, inversion is just a mechanism for stimulating designers to consider a whole range of new ways of possibly accomplishing a required function; that is, a different perspective. And it works very well to do just that.

Here are some examples of products whose conceptual or configuration design might have been developed by inversion: Rear engine autos and airplanes; heat pumps (inverted refrigerators); the computer mouse (invert the joy stick and what do you get?).

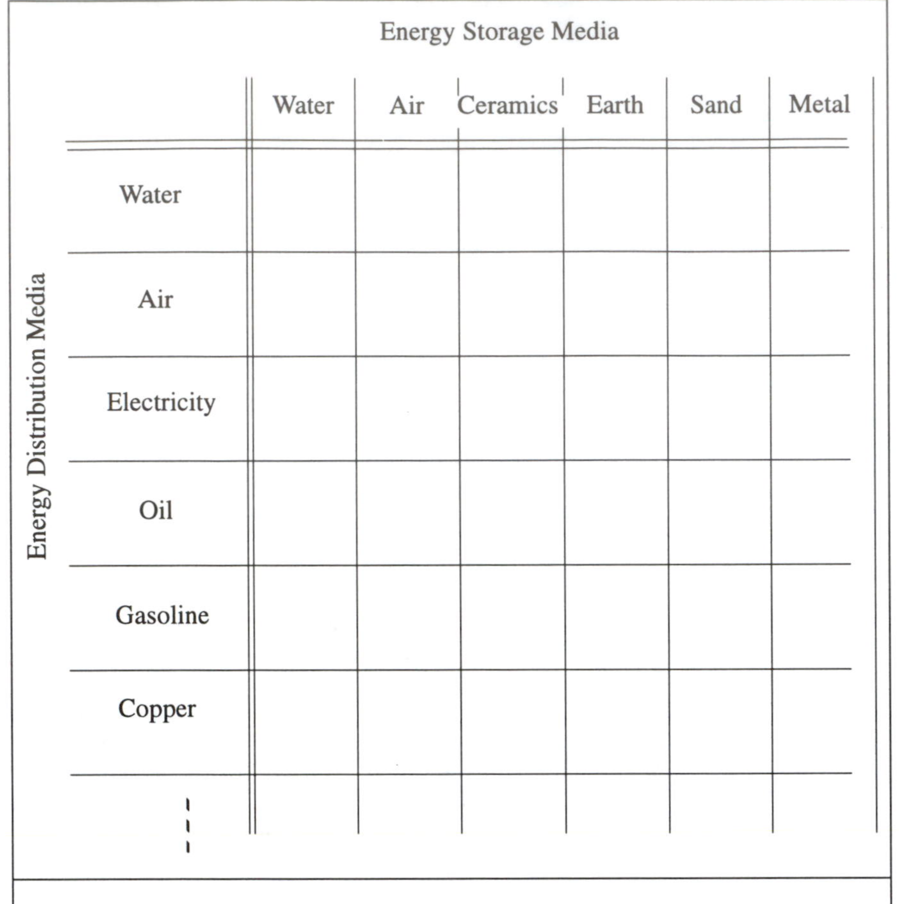

Figure 6.8 Example Matrix for Systematic Search for Combinations for a New Solar Heating System Concept

Fantasy. Using fantasy as an idea generation method means *imagining* or *wishing* that something exists or is possible which appears in reality to be impractical or impossible. The fantasy or wish may, for example, violate some physical law, or be a material with extraordinary properties. That is okay; wish for it anyway. It may lead to some unforeseen practical solution.

Using fantasy is harder to do than inversion because it requires that we suspend our sense of reality temporarily in order to give ourselves a new view of the problem and its solution. This is not so easy to do. Fantasy works, however, because sometimes the solutions that we fantasize turn out not to be impossible, or the fantasies stimulate us to consider other new ideas that are practical.

> "There's no use in trying," she said: "one can't believe impossible things."
>
> "I daresay you haven't had much practice," said the Queen."When I was your age, I always did it for half-an-hour a day. Why, sometimes I've believed as many as six impossible things before breakfast."
>
> Lewis Carroll
> *Through the Looking Glass (1872)*

In a sense, every new product or process idea or improvement is the result of someone's fantasy. Consider the first Polaroid camera, first Xerox machine, first automobile, first airplane, first ball point pen, and so on. And there will continue to be new firsts.

Some other examples of conceptual ideas that might have been created by fantasy are: silly putty; Post-it notes [Trademark 3M Company]; helicopters; spreadsheet software; telephone; "smart" products of various kinds.

Analogy. New ideas often are developed by analogy to other fields, other solutions, or other physical phenomena. All that is required is to think about the problem at hand in general enough terms so that the characteristics it has in common with other situations become apparent. Though engineers will most often employ analogy with other engineered products or processes, probably the largest store of likely ideas is in the natural world. In the millions of types of plants and animals, nature has developed millions of structures, pumps, mechanisms, control systems, and so on. For this reason, a good course in general biology is a valuable elective for would-be creative design engineers.

Here are some conceptual solutions that might have been developed by using analogy: peristaltic pump, ball joint, reservoir-dam-water power plant, rope swing.

Empathy. Using empathy is another Synectics technique that, like fantasy, requires engineers to be very open to their imaginations. Empathy means putting one's self in another's place. Thus, in the case of conceptual design, empathy involves identifying physically and personally (i.e., empathizing) with the part, product or process that is to be created. It thus requires imagining that our body and/or mind must actually perform the function(s) required of the (not yet designed) designed object. Then we must state how that feels and what we would need or do if we were to do the task. This may often require some fantasizing as well as empathy.

Using empathy as a technique is difficult for people who are inhibited in a group of colleagues, and who isn't? We have to trust our colleagues to be kind, or we have to be very thick-skinned, or we just have to try it alone. When empathy can be done, it can be a very powerful method for gaining a new perspective on a problem, and hence in generating new creative conceptual solutions.

Examples of products or processes that might have been developed by use of empathy are: socket wrenches; anti-lock brakes; sewing machine heads.

Here is an example of a session that uses some of the above mechanisms. It is adapted here by permission of McGraw-Hill Book Company, New York, from *Design Engineering: Inventiveness, Analysis, and Decision Making* by J. Dixon [11]. The time frame of the problem is back before audio cassettes and compact disks were available. High quality sound systems at the time used 33 rpm records and had delicately balanced tone arms to get the contact between the needle and the record just right. This equipment was sensitive to vibrations. A student in a class was having trouble with this problem in his high-rise dorm room, which vibrated from truck traffic (among other causes) enough to disturb the tone arm and interfere with proper needle tracking. What could be done?

BOB: Here's my problem: The table I have my turntable on is kind of wobbly. The needle skips and jumps whenever anything touches the table, or sometimes if I move around the room, you know, kind of vigorously. It even jumps around sometimes when a truck goes by. I need a good way to fix it.

AL: Why don't you get a sturdy table?

BOB: Well, that's not the problem, really. The whole building shakes! I have the same trouble when I put the turntable on the floor.

SUE: Aren't there any vibration isolators built into it?

BOB: Yes, there are four springs - one at each corner of the table.

JOHN: Are they the right springs? Maybe they need to be stiffer, or springier?

ANN: Is it moving vertically or horizontally?

Note: This group is getting the problem defined better.

BOB: I think the problem is mostly from horizontal movement, but I'm not sure. The springs probably take care of vertical vibrations?

SUE: So is the problem that the springs are designed to deal with vertical vibration only, but you also have horizontal vibrations?....So we need a concept to provide some horizontal isolation as well?

PETE: How about horizontal springs?

ANN: Springs don't have to be either horizontal or vertical. We could set them in at an angle and still just have four springs....

Note: *There was a kind of inversion in this suggestion: changing the direction of the springs. Empathy is used next.*

JOHN: Problem solved; next problem.

SUE: Hold on, now. We've just begun. Let's look for some other ideas by using some of the mechanisms. Al, you're good at the empathy thing. Stand in there and hold that turntable steady while I shake the floor.

AL: (Stands, puts his hands over his head to hold up the imaginary turntable; he imagines his feet are on the shaking floor.) Well, I need swivel hips or something, or swivel joints in my shoulders, to keep the turntable still while my feet get pushed around. Like a waiter with a tray on a ship in rough water.

PETE: A ball joint?

AL: (Still empathizing, holding up the imaginary turntable.) I really need to disconnect my arms from my feet, except for the vertical force needed to hold the tray -- I mean turntable.

JOHN: A ball joint ought to be good for something here. Or just some balls. What about having a bunch of balls -- like in "lazy Susans"....

Note: *Analogy has been used, and is used again below.*

ANN: I think they have the ball bearings in a circular track, but the balls could be spread out in uniform pattern....The turntable could be free to roll around on them; its own inertia would keep it pretty still.

PETE: We could make it heavier....

SUE: We need a kind of "tablecloth effect" -- like when you pull a tablecloth out from under some dishes....

AL: (Back to his seat.) The tablecloth works because the dynamic coefficient of friction is low. If we had a low coefficient of friction.....

JOHN: The ball bearings give us that....

BOB: There are other ways to get low friction. How about just lubricating it? Float the turntable in an oil bath.

AL: What you really need is a magic carpet! A nice calm magic carpet that just floats on the air. Not connected to the building.

Note: *Fantasy.*

SUE: Well, you could use air lubrication.

ANN: That would be springy in the vertical direction too. An air cushion one way, and an air bearing the other way. Like one of those floating vacuum cleaners. But where would the air come from?

PETE: Never mind that now.

JOHN: Some kind of balloon, or air mattress?

SUE: What about just a pillow?

AL: I still think the ball idea has something in it we haven't got out of it yet.

BOB: We lose the vertical isolation.

ANN: What about bouncy balls? Soft. That would act like a vertical spring, and a horizontal lubricant.

SUE: We need the balls to be soft vertically, but hard horizontally.

Note: *Fantasy was just employed.*

PETE: Good idea! Some kind of "silly putty" balls that act one way or another depending on what you do to them.

JOHN: Let's try empathy on a different part. The real problem is that the needle -- which is attached to the tone arm -- won't stay in the groove when the turntable box gets bumped.

BOB: So give the arm more inertia, make it heavier. Oh, no, that would ruin the record...

JOHN: (Gets down on the floor and stretches out one arm, pointing his fingers down to simulate the needle. Let's say I'm the tone arm. My fingers are the needle.

AL: That's not quite it, John. The arm is balanced from a pivot.....

JOHN: Okay, so am I. I can imagine that. Now shake the floor, I mean the table. I've got to keep up with the motion to keep the needle in the groove. I need the arm to be lighter, not heavier, so I can move easier.

SUE: Lighten the tone arm?

ANN: You've got to keep the pressure of the needle on the record very steady.

BOB: I can do that by adjusting the pivot and counterbalance. I think reducing the weight of the arm will help.

PETE: Have we given up on springs? What about hanging it from springs?

Note: *Inversion was used.*

JOHN: Springs at an angle....

SUE: What's the advantage?

PETE: I don't know. We could have longer springs? Hang the whole thing from the ceiling?

ANN: Add a damper of some kind. You'd get a slow sway, but no quick movements to move the tone arm.

This example is not brainstorming; the problem solvers were deliberating trying to use analogy, fantasy, empathy, and inversion to stimulate new ideas. Like in brainstorming, however, they did well to keep criticism and negative evaluations fairly well out of the discussion.

6.5 Searching for Information

6.5.1 Literature Searches

Another extremely effective way to get creative ideas for solving a conceptual engineering design problem is to search the literature for reports of products, ideas, and research that has already been done somewhere in the world. Fortunately there are many resources out there to help us with these searches. Unfortunately, these resources are not used nearly as widely or as effectively by engineers as they should. Here are some of the available sources of information and ideas:

Trade Magazines

Examples of general design oriented trade magazines are *Mechanical Engineering, Design News, Machine Design, IEEE Journal of Design.*

Each issue of these magazines contain literally hundreds of advertisements for products, each of which may one day be useful as part of a solution to some problem you will encounter. There are also articles; if you read the interesting ones (to you), and look through the entire magazine, you may not remember any particular idea, but you will remember that it exists. That will be enough to get you started finding it.

There are also trade magazines in special fields (e.g., packaging, solar energy, electronics); in fact, just about every imaginable technical field has a trade journal.

Most magazines publish an annual index to their advertisers and to their technical articles that can be searched when the topic of the search is known. However, when searching for new ideas, we often don't know what name or title to look for. There is thus no substitute for regular reading or at least surveying these trade magazines. And to do this in our busy lives, it is just *essential that we have our own subscriptions.* It simply does not work to resolve to read the one at the library, or to plan to borrow a colleagues.

Trade magazines are a tremendous source of ideas. Spending even an hour with each issue will stock our memories with all kinds of products, concepts, and specific techniques which we can draw on when a new idea is needed.

Handbooks

There are handbooks of data, procedures, and other technical information that should be readily available. Examples are *Mark's Mechanical Engineering Handbook*, *Eshbach's Handbook on Engineering Fundamentals*, *Kent's Mechanical Engineering Handbook*, *Metal's Handbook* (from the American Society for Metals), and a number more. The information in such references is extensive. We all need at least one readily available. Look in your library.

Industry Manuals

Just about every manufacturing industry has an association that publishes books of valuable data, and sometimes very useful design procedures. There are just too many of these to list, but they are invaluable sources of information about specific product lines like gears, springs, motors, belt and chain drives, and most everything else.

Research Journals

Examples of general coverage engineering design research journals are:

Research in Engineering Design,

ASME Journal of Engineering Design,

Manufacturing Review,

Journal of Engineering Design,

Artificial Intelligence in Engineering, Design, and Manufacturing,

Systems Automation: Research and Applications

Concurrent Engineer: Research and Applications

Computer-Aided Design

Research oriented journals generally do not carry advertising, and the articles are not as likely as those in trade magazines to have ideas that will solve immediate problems. However, every firm has core technologies (e.g., combustion, classes of materials, mechanisms, etc.) whose fundamentals are central to their product lines. Just as there are trade journals in rather specific topics, so too there are research journals for rather specific fields that correspond roughly to such core technologies. There is, for example, an *ASME Journal of Heat Transfer*, an *ASME Journal of Fluid Mechanics*, and so on.

As with trade journals, there is no substitute for having a personal subscription to those that relate specifically or generally to your work. The value of just one good idea from any of these sources will pay the cost of the subscriptions many, many times over.

Published Data Bases

Examples of published data bases of literature are The Engineering Index and The National Technical Information Service (NTIS). These are references to the technical literature. Through them you can find articles on thousands of subjects published in hundreds of journals. It can be tedious looking for what you want because you have to look year by year. But it is often very rewarding; it is amazing how much work useful in design has been accomplished and published. And the information in these data bases is now included in computerized data bases that are a bit easier to access.

Computerized Data Bases

Most company, University, and public libraries will have access, perhaps even on-line access, to computerized data bases of technical (and other) information. These are, more or less, computerized versions of published data bases like the ones described above.

The computerized data bases are extremely useful in searching for whatever has been published about a particular subject. The services are convenient, but they do generally require a modest payment. And they require the assistance of a knowledgable reference librarian who is familiar with the system operation and with the way information is organized in the system.

It would be difficult to over-emphasize the importance to engineering designers of being able and willing to make use of these computer-based information resources. Try it.

6.5.2 People as Sources of Information and Ideas

People designers know and work with are also very important sources of information and ideas. Colleagues can be consulted directly, or the subject of interest can be brought up at lunch or coffee breaks. Don't be afraid to ask: "Say, Mary, do you know anything about <whatever> — or know anyone who does?" She may know, or remember seeing something in a magazine, or know that Joe Doaks somewhere is an expert in that subject.

Vendors are another great source of information and ideas. Of course, vendors may tend to push their own products, but it is also true that they want to be generally helpful. (It's a natural human tendency, and besides, they want your good will in the future.) Vendors get around to a lot of other companies, see a lot of different ways to do many things, and they usually want to help a potential customer even if it has nothing directly to do with an immediate or future sale.

Engineering designers who have worked even a short time will begin to develop a vendor network. In addition, every engineering designer can use several resources to access other vendors. These are, for example, the *Yellow Pages* of the telephone book and the *Thomas Register*. The latter, available in virtually every library, is a set of huge books providing information about companies that manufacture products of every imaginable kind. Access is by the name of the product, so you have to know at least something about what you are looking for. But there are cross references and lots of key words to assist finding something, or someone, who may be helpful.

Other people may also be able to help: customers, friends in other companies or industries, old friends from college. Ask them! Even if they don't know, they may know someone who does.

6.6 Creativity Inhibitors

So far in this chapter we have discussed ways to stimulate the production of creative ideas for solving conceptual engineering design problems. It is also important to identify the factors that interfere with creativity, so that these can be reduced or eliminated. The most important inhibitors are: psychological set, fear, and pressure of time.

Set. Set as used here means, essentially, the force of habit, perhaps hardened by vested interest or just plain reluctance to change. Habits develop because they have resulted in success in the past, or have been satisfying in some way. In design, there is definitely a strong tendency to use old ideas that have worked reasonably well in the past. Up to a point, this is fine. It saves time, the technology is well known so it reduces risk, and people are comfortable with it.

However, it must also be remembered that competitors are out there looking hard for new and better ways. Thus we can never really give up on the search for new products, new improvements in old products, new materials, new processes, new features, and new technology of all kinds. When people in a firm get so comfortable with old solutions that they fail to evaluate them objectively, and fail to search for new and better ways, there is great danger to the firm.

Set is not just a cognitive (i.e., intellectual) problem. Set has a psychological component that actually prevents people from seeing their own unquestioning, habitual behavior objectively. The arguments they use to support the old ideas, and the reasons they give to show that new ideas are either bad or unnecessary will sound reasonable. But they may not be; they may simply be the rationalization for set.

Fortunately, set is reasonably easy to avoid for anyone who really wants to avoid it. All one must do is honestly be aware of its possible existence. Ask yourself: "Am I being set here?", and you will often find that is enough for you to look for some new ideas. In other words, just to be conscious about set to avoid it.

Fear. The techniques of brainstorming and synectics obviously require people who are confident in themselves, and not afraid to make a mistake, to suffer minor embarrassments, or to say something that might be considered stupid or impractical. Thus people who are afraid of these feelings are not very good at using the techniques.

Fear works against creativity in other ways, too. Fear of failure is the big one: What if the new idea does not work out? What if we become personally identified with the new idea, and then it fails? Wouldn't we then be seen as failing, too?

Such fears are not irrational, but the risks can be minimized. Though one person may very often originate an idea, seldom is that person solely responsible for the idea's adoption. Usually responsibility gets spread to the team as a whole and also managers are involved. Anyway, new ideas, if they are risky, should not be incorporated into active product development projects until they have been well tested in the laboratory and the technology understood.

And don't forget the opposite fear: What if the competitors get the new idea, and it works for them?

Pressure of Time. Pressure of time, quite naturally, causes designers and design teams to adopt old solutions rather than to take the time to explore new ideas that might work better, but might not. It is most likely that intense pressure of time is here to stay in the product realization process. Of course, we have argued several times in this book that an active product development project is not the time to explore new and untested ideas. But they must be explored constantly in every firm that expects to get and stay on top.

What pressure of time can do, however, is prevent even the testing of new ideas off to the side in the laboratory. Then the new ideas and the technologies on which they are based never become ready and available for a product. There must be people available who have the time to study and develop new ideas so that they are ready for quick, safe, and reliable incorporation into a product.

Lack of Reward. If people who generate new ideas are not rewarded for them, then of course they will tend to keep the ideas to themselves. Strange as it may seem, some firms are not organized to reward innovation. This is, of course, a management issue. However, design engineers who feel themselves to be on the more creative side should investigate how a prospective employer rewards new ideas as they come along.

Failure to Patent. One way to reward a design engineer or team for a new idea is to patent it. This is not only a source of pride for the people involved, but it may also prove to be financially rewarding. Again, a surprising number of companies do not pursue patents, tending instead to try to keep the ideas secret. More often than not, secrecy fails. And even when it doesn't, there is a strong tendency for new ideas to be developed in different places at nearly the same time.

Patents are discussed in more detail in Chapter 24.

6.7 Longer Term Enhancement of Personal Creativity

Personal creativity is often based on a great memory together with the ability to generalize technical situations and to see in them possible new applications and new combinations of old ideas, an intellectual combination sometimes referred to as either "genius" or "experience" or both.

In the short run, when faced with a conceptual design problem that must be solved right now, there is little one can do to add to the personal memory store. Either something useful is in there or it isn't; either it can be recalled by some technique, or it can't. But in the longer term, there are ways we can enhance our useful memories of potential solutions to conceptual design problems. What we must do is consciously observe a lot of design solutions, and consciously attempt to generalize the solutions we have observed. In other words, we must consciously — constantly, over the years — purposefully build up our "experience"; that is, what we know and can remember.

The reason for this is that, for most people, it is not enough just to have the experience. We must learn from the experiences we have, and to do that most of us must make a conscious effort to be explicit about what we are learning. We also need to think about how ideas and methods can be made general so they can be applied again in the future in a somewhat different situation.

Here are some ways to observe a lot of design solutions and build your longer term personal experience:

Subscribe to trade magazines and research journals. And look through every one that comes.

Keep a memory support notebook. After every completed design task with which you are involved or which you have observed (whether conceptual, configuration, or parametric) write down the key ideas of the solution and how they were arrived at. Look especially for the general ideas. The writing takes only a few minutes but fixes the ideas in your memory.

Practice curiosity. Look, *really* look, at the designed objects in your environment, and the new ones you come across. There are, for example, chairs, bicycles, soda cans, lamps, and a million other things. Don't just take them for granted; *see* them in terms of design and manufacturing. *Ask questions.* Ask yourself explicitly about (materials, function, physical principles, manufacturing processes, etc. Write down the key ideas in your memory notebook.

Just do these things regularly, over time, and it is guaranteed that you will become a more creative engineering designer.

6.8 Standard Versus Special Purpose Parts and Assemblies

A physical concept or embodiment includes information about whether the designed object is to be a special purpose part or assembly as distinguished from a standard one. For example, it might be decided that the physical concept be or include a "screw" or a "motor." Alternatively, we might decide — for one reason or another — to design a special purpose part or assembly to perform the functions of these, or other, standard parts or assemblies. We might believe, for example, that a special purpose part or assembly could be better tailored to the specific needs of the designed object.

This is occasionally true, but only *very* occasionally. The reason standard parts are standard is that they have been found generally useful in a wide variety of situations. Moreover, the firms that make them are experienced specialists in the design and manufacture of the standard parts and assemblies they sell.

Thus if the selected physical laws and effects for an alternative can be embodied in a standard part or assembly, or if the physical concept is one that can be configured as a standard part or module, then using an available standard part or module is very likely to be the best solution. A very good design rule is:

Use standard parts and modules wherever possible.

Usually standard parts and modules will be purchased from vendors outside a company; in this case, we say they are *outsourced.*

As noted, though it is most likely, it is not *absolutely* certain that a standard part or module will always be the best solution. Here are some reasons it may not be: More parts, even if they are standard parts, are usually more costly overall than fewer parts, standard or not (though time issues must also be considered); potential outside sources (i.e., vendors) may be less reliable suppliers than needed for this product; the quality of the standard designed objects from possible outside sources may be uncertain or too variable; or perhaps your firm does not want others to know what is being done. It could be, too, that using a standard part or module causes other design changes that result in poorer overall performance or higher costs. All of these points will generally be revealed when the alternatives are evaluated as described in the next chapter.

However, at the conceptual design stage, the rule above cannot be ignored without a very good explanation. Unless there is a compelling reason, the best practice is to include standard parts and modules among the alternatives to be evaluated if at all possible.

It may be that outside sources of standard parts or modules appear at this stage to be more expensive than making the part or module in-house, but this is most unlikely if the true costs of inventory, overhead, and the like are known.

One manufacturing firm, by long tradition, makes for itself nearly every one of the parts, including many standard parts, used in its large and varied line of products. The firm therefore has an inventory of literally thousands of parts. The plant manager asserts that "Because another company making these parts would make a profit, we are bound to pay more buying from outside than it costs us to make them." This conclusion is very doubtful, though the antiquated accounting procedures in the firm make it impossible to determine true costs of not only production, but also overhead, space, management, and inventory associated with all these parts.

6.9 Continuation of Conceptual Design for Supplementary Design Project A

Every engineering design project is different. Some offer a wide variety of options at the conceptual stage, but few configuration options. Some involve intense work at the parametric level while for others the dimensions are more or less obvious. This project provides relatively few options at the level of conceptual development of an assembly to meet the requirements of the Engineering Spec (See end of Chapter 5) There are many more options when the cross sectional configuration of the members is considered.

As a part of their conceptual design process, one student group developed a simple decomposition diagram as shown in Figure 6.9, and the associated hardware schematic shown in Figure 6.10. Note that in these diagrams, little is implied about the shape of the members or brackets, except that the members are long and thin. Also, nothing is said about materials.

In terms of function, a possible first level decomposition is shown in Figure 6.11. Whereas the hardware decomposition in Figure 6.9 has implicitly assumed that the members are going to be long and thin, the function diagram makes no such assumptions. In terms of functionality, the vertical members could be sheets or trusses as easily as long and thin (implying beams and columns). The connecting function, assumed in Figure 6.9 to be done by "brackets", could as well be done several other ways; for example, with clamps, pins, or by some clever geometric fit that require no special parts.

Here are some of the concepts developed by another student team:

1. An assembly of long, thin members and intersecting brackets, like a "Tinker Toy" or "Erector Set". (This is essentially the concept shown in Figure 6.9.)

2. A hung system: the bench tops are hung from chains or rods from a grid of overhead supporting beams.

3. A cantilever system: bench tops are supported from cantilever beams supported from structure built to the walls.

4. A combination of (2) and (3).

5. A "bar stool" design: each top is supported by a single round (or other shape) column that can be moved as desired to locate the tables as desired. Long tables could use

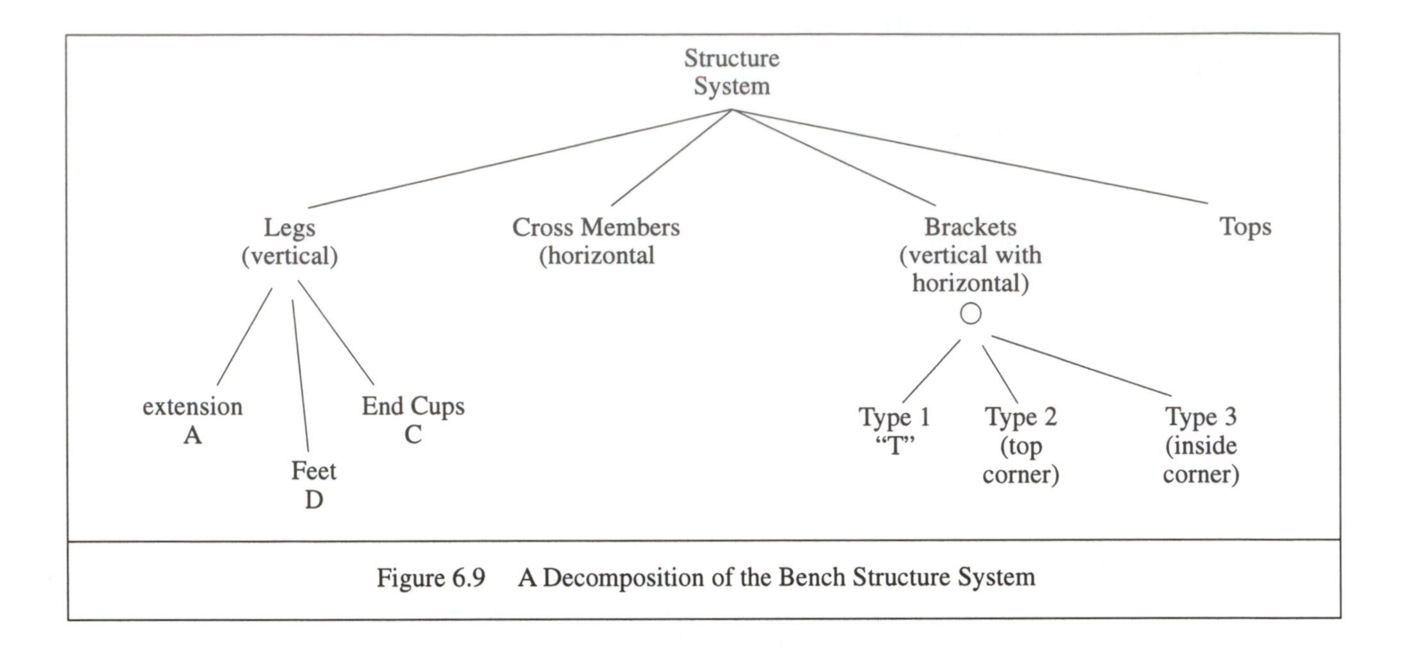

Figure 6.9 A Decomposition of the Bench Structure System

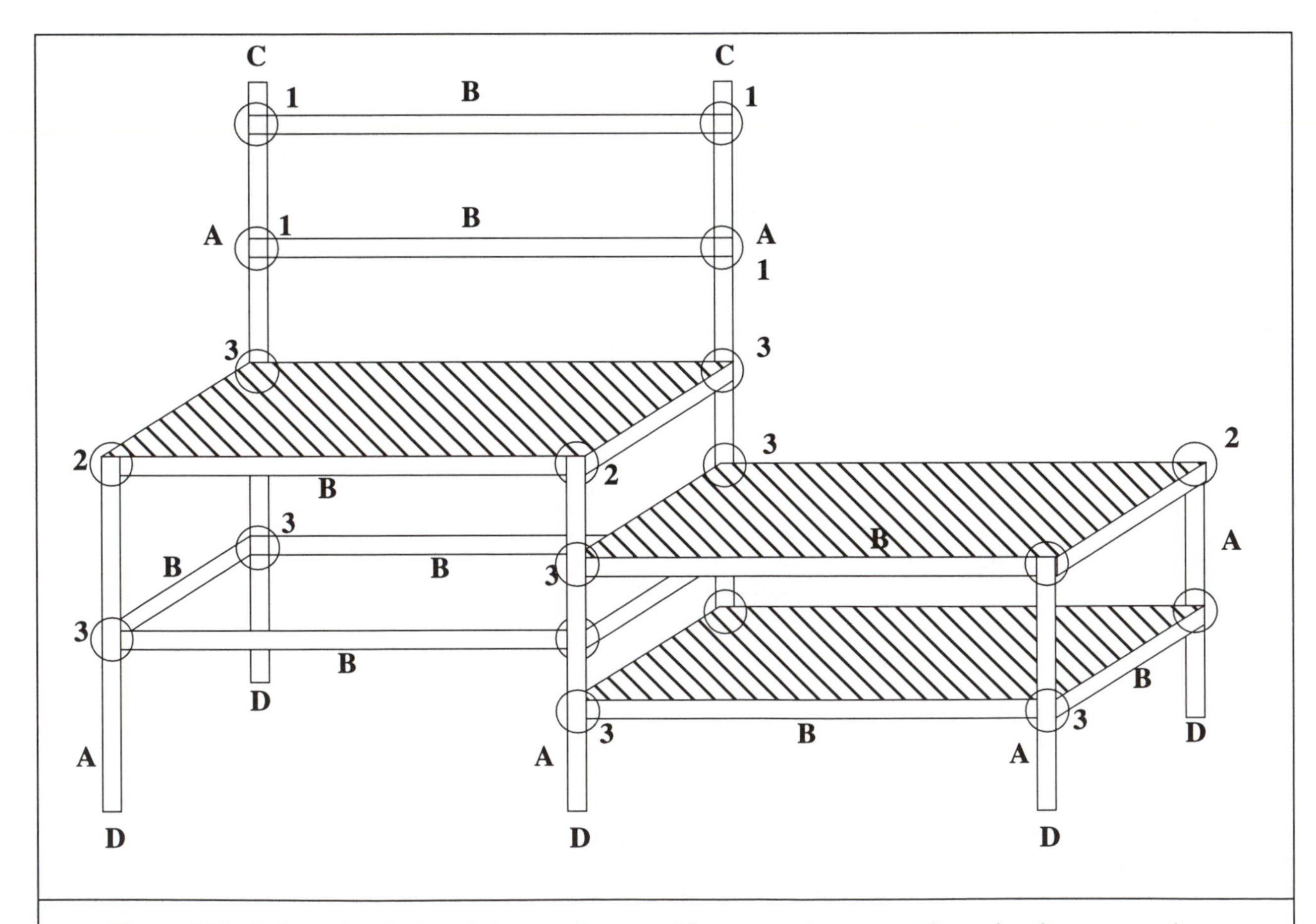

Figure 6.10 Schematic of a Bench System Concept. There are only two members plus three connecting brackets; these may yet be combined as the design proceeds.

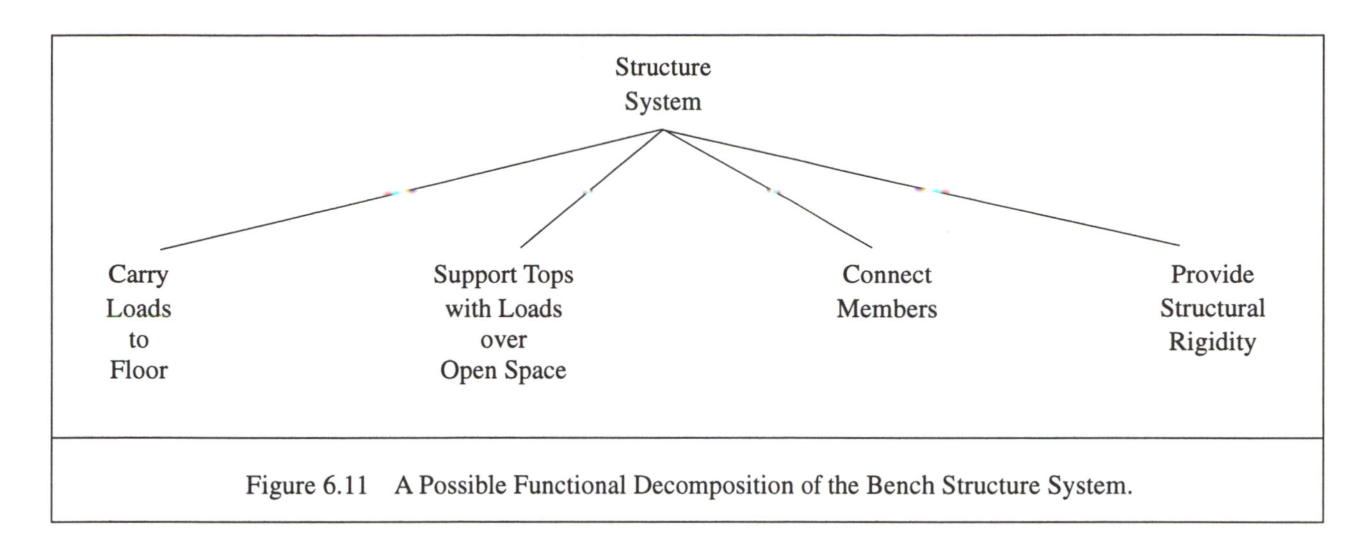

Figure 6.11 A Possible Functional Decomposition of the Bench Structure System.

two or more supporting columns.

6. A "milk crate" design: standard sized 'boxes' are made to be stacked and interlocked in various ways to create tables of desired sizes and heights.

7. A "saw horse" system: standard saw horses are made to support the table tops.

8. An "L-shaped support" design in which standard L-shaped legs are made large enough to be stable standing on their ends. Four of these can then be located at the corners to support any sized table top.

Readers can no doubt add to this list.

A major issue at the conceptual level in this problem is the choice of basic materials and associated manufacturing processes. It is relatively straight forward to generate a list of likely possibilities: steel, aluminum, wood, or plastic. Others (e.g., ceramics) can be excluded for obvious reasons based on the Spec.

6.10 Summary and Preview

> "The greatest invention of the nineteenth century was the invention of the method of invention."
>
> Alfred North Whitehead
> *An Introduction to Mathematics (1911)*

Summary. Generating alternatives is especially important in the conceptual design stage; it is important but less so in configuration design, and not very important at all in parametric design. At the conceptual stage, if good ideas are missed, there will not likely be time to return to them once the process moves on. Moreover, the quality of the final product is very intimately involved with the nature of the conceptual solution selected. Of course, concepts not generated as alternatives will not be available for evaluation and consideration. Thus, this is the time in the process for creativity and inventiveness.

Being creative in engineering design is not just an accident of birth. As with all human activities, some people will be better at it than others, but there are many methods available to support the generation of alternative concepts. They include:

- Function-First Decomposition
- Searching for Alternative Physical Laws and Effects
- Brainstorming;
- Systematic Search for New Combinations;
- Techniques of Synectics including:
- Inversion
- Fantasy
- Analogy
- Empathy
- Searching for New Information From the Literature, and From
- Other People

Each of these has been discussed in this chapter.

There are also ways to improve your individual creativity over the longer run. Mostly this involves getting lots of exposure to ideas and different ways of doing things, and making sure that you learn from this exposure.

Finally in this chapter, we pointed out that not everything should be invented anew. When standard parts and modules are available that will do the job, be sure they are in the set of alternatives to be evaluated.

Preview. Once a set of alternatives has been generated for meeting the requirements of the Engineering Design Specification, the next step in the guided iteration process is to evaluate those alternatives so that the best can be determined and selected. Thus, methods of evaluating engineering conceptual design alternatives are the subject of the following chapter.

References

[1] Glegg, Gordon L., *The Design of Design*, Cambridge University Press, 1969. Quotation by permission of the publisher.

[2] Pahl, G. and Beitz, W. *Engineering Design*, Edited by Ken Wallace, The Design Council, London, England, 1984.

[3] Crossley, E. "A Shorthand Route to Design Creativity" *Machine Design*, April 10, 1980.

[4] Crossley, E. "Defining the Job: First Step in a Successful Design," *Machine Design*, May 22, 1980.

[5] Crossley, E. "A Systematic Approach to Creative Design," *Machine Design*, March 6, 1980.

[6] Crossley, E. "Make Science a Partner," *Machine Design*, April 24, 1980.

[7] Hix, C. F., and Alley, R.P. *Physical Laws and Effects*, Wiley, New York, 1958. Material adapted with permission of General Electric Co.

[8] Collins, J.A., Hagan, B.T., and Bratt, H.M. "The Failure-Experience Matrix", *ASME Journal of Engineering for Industry*, Volume 98, 1976.

[9] Gordon, W.J.J. *Synectics: The Development of Creative Capacity,* Harper and Row, New York, 1961.

[10] Osborne, A. *Applied Imagination*, Scribner, New York, 1953.

[11] Dixon, J.R. *Design Engineering: Inventiveness, Analysis, and Decision Making*, McGraw-Hill, 1966.

Other Recommended Reading

Professional Creativity, E. Von Fange (1959) Prentice Hall, Englewood Cliffs, NJ.

Creative Engineering Design, Harold R. Buhl (1960) The Iowa State University Press, Ames, Iowa.

Disciplined Creativity, R. L. Bailey, Ann Arbor Science Publishers, Ann Arbor MI, 1979.

Synectics: The Development of Creative Capacity, Gordon, W.J.J., Harper and Row, New York, 1961.

Problems

6.1 Write an embodiment description for the standard component we call a spring. (Or paper clip, gear, motor, you name it).

6.2 Describe one or more of these objects as an embodiment: wheelbarrow, chain saw, bookcase, etc.

6.3 Go to your library and arrange for a computer data base search on the subject of energy storage -- or temperature control, heat exchangers, adhesives, or something that interests you. (Note: This may cost some money!)

6.4 List one example of a product or process that might have been developed by application of each of the Synectic techniques.

6.5 By looking for combinations, find as many conceptual solutions as you can for a system that will cut brush -- or measure wind chill, or provide *interesting* exercise, or something of your own choice.

6.6 Develop an Engineering Design Specification for an easy to install automatic overhead garage door opener. Develop a set of conceptual solution alternatives.

6.7 By exploring the possible use of different physical laws and effects, develop some conceptual solution alternatives for non-lethal self defense or defense of home from intruders. (Or for separating green from red tomatoes, or for thermometers, or for a functional device of your own choice.)

6.8 By personal brainstorming, list as many conceptual solutions as you can for:

Fastening sets of papers together,

Providing natural light in homes and offices,

Walking on water,

Preventing drunk driving,

Providing for personal non-lethal self-defense,

Some problem of your own choosing.

Chapter Seven

Evaluation and Redesign of Engineering Concepts

"What is a cynic? A [person] who knows the price of everything, and the value of nothing."

Oscar Wilde
Lady Windermere's Fan (1892)

7.1 Introduction

At every design stage, the quality of a solution is determined to a great extent by the nature and competence of the evaluations performed on the alternatives that have been generated. In this chapter, several methods for evaluating and comparing proposed conceptual design solutions are described and illustrated.

None of these evaluation methods for conceptual designs is without its shortcomings. The analysis methods are not nearly as rigorous or as formal as the engineering analysis procedures available for evaluation of designs at the parametric design stage (such as, for example, beam theory or finite element analysis). In essence, the methods available for evaluating engineering concepts are little more than ways to garner and structure subjective opinions. However, the methods presented here are the best available, and they work quite well when employed by conscientious, knowledgable people. Certainly to make use of them is a far better policy than to select an alternative without careful evaluation.

The criteria by which proposed conceptual designs are evaluated vary somewhat with the application, and with the content of the Engineering Design Specification. One important evaluation criterion, however, is *always* present even though it may not be mentioned explicitly in the Specification. That criterion is *ease of manufacturability.* Though manufacturability is implicit in concern for cost, manufacturability is so important to competitiveness that it should be considered explicitly.

Evaluating the manufacturability of a concept or embodiment is not easy to do. One reason is that the information available about the designed object at this stage is quite abstract; the details of configuration and size that often strongly influence manufacturability are not yet established. Nevertheless, an evaluation can and should be made to the extent possible in order to minimize future problems.

The conceptual design evaluation methods described here are:

- Pugh's method,
- Dominic method,
- Quality-Function Deployment, and
- Pahl and Beitz method.

We turn attention now to the first and most widely used of these approaches: Pugh's method.

7.2 Pugh's Method

A method for evaluating concepts has been pioneered and used successfully by Professor Stuart Pugh [1]. It is implemented by establishing an evaluation team and setting up a matrix as shown in Figure 7.1.

The various physical concepts or embodiments to be compared and evaluated are identified across the top of the matrix. Each of these embodiments is to be described at the same level of abstraction, and each in a rough sketch. The major functional elements and their connections are generally included in the sketches; otherwise the sketches need not be detailed.

As shown in Figure 7.1, the criteria by which the

		Alternative Embodiments				
		1	2	3	4	→
Evaluation Criteria	Criterion A		S			
	Criterion B		S			
	Criterion C		S			
	Criterion D		S			
	↓		↓			

Figure 7.1 Skeleton Matrix for Pugh's Method of Concept Selection

physical concepts are to be evaluated are identified along the left side of the matrix. It is important that these be the correct criteria, though there is no formal method for selecting them. The Engineering Design Specification will suggest many possibilities, and manufacturability should definitely be one. Discussions among the members of the evaluation team will add others. From all the possible criteria that might be included, evaluators must select those that are most relevant to physical concept evaluation. Using too many criteria makes the evaluation task tedious; but if any important criteria are omitted, the best concept may not be selected. Thus a strong effort should be made to make the evaluation criteria complete, clear and reasonably independent of each other.

One of the physical concept alternatives (preferably one that is well understood) is chosen as the *datum* or *reference* case. Evaluations of the other physical concepts will be compared to this datum physical concept. As shown in Figure 7.1, the letter S (or number 0) is placed in every box in the column applying to the datum physical concept.

Usually the project team does the evaluation, though others can be called in to assist. To perform the evaluations, all the alternative physical concepts are discussed by the evaluation team. As a result of discussion — and even argument — the evaluators decide as a group how each alternative is to be rated on each criterion. Depending on their conclusions, a plus (for better than the datum), minus (for poorer than the datum), or S or zero (for about the same as the datum) is placed in the corresponding boxes in the matrix.

As a simple example, consider the concepts available for transmitting power from the pedals of a bicycle to the rear axle. We might do it with gears, v-belts, chain drives (the usual solution today), an electric generator-motor combination, by fluid power, by compressing air, by linkages, and possibly others. See Figure 7.2.

Evaluation criteria in this example might include cost, weight, size, reliability, efficiency, force required, braking ability, customer appeal, and perhaps others.

Figure 7.2 shows some likely assignments of the pluses and minuses for this example. The results perhaps indicate the reason the chain drive is such a popular solution for this problem. However, we can be sure that eventually something better will come along, probably involving a computer (i.e., a "smart" drive).

When the evaluation of each physical concept is complete, a score for each can be computed. This is done by simply counting and recording the pluses, minuses, and zeros for each alternative. Don't try to convert these into a single average; just record and examine how many pluses, minuses, and S's (or zeros) there are for each concept being considered.

When this is done, the evaluation team will usually see certain patterns. Some alternatives will be obviously inferior. There may be several that look good, but for different reasons. Further discussion can often sort out the best concept or two that should receive additional attention.

On the other hand, after some analysis and revisions, it may turn out to be desirable or necessary to repeat the evaluation procedure several times. The team can cull out the poorer options, expand and refine the criteria, and/or refine the remaining better design alternatives before beginning again. A different physical concept can be used as the datum. Each time, the team better understands each alternative, and also gets a better understanding of the advantages and disadvantages of each.

Evaluation Criteria		Alternative Embodiments								
		Gears	V-Belt	Chain Drives	Motor Generator	Fluid Power	Compressed Air	Linkages		
Cost		–	+	s	–	–	–	s		
Weight		–	+	s	–	–	–	s		
Size		–	s	s	–	–	–	–		
Reliability		+	–	s	s	–	–	s		
Efficiency		s	–	s	–	–	–	s		
Force		s	s	s	–	–	–	s		
Incorporate Braking		s	s	s	+	+	+	s		
Customer Appeal		–	–	s	–	-	–	–		
Results	Pluses	1	2	–	1	1	1	1		
	Same As	3	3	8	1	0	0	5		
	Minuses	4	3	–	6	7	7	2		

Figure 7.2 Example of Pugh's Method Applied to Bicycle Drives

Pugh's method has an excellent record of success in actual industrial projects. The discussions and arguments are important, and compensate for the obviously subjective criteria that are possible at this stage. A possible shortcoming is that all criteria are treated as if they were of equal importance when, of course, they are not in most situations. The method is nevertheless a superb practical tool for focussing on the important issues, and for generating the right kinds of discussions among evaluators. Readers are urged to learn more about it by consulting the original reference by Pugh [1].

7.3 The Dominic Method

The Dominic method is named for a computer program called Dominic that performs parametric design See Howe [2] The method has its roots implicitly in decision theory. See, for example, [3]. It employs a matrix of alternatives and evaluation criteria the same as in Pugh's method, but the Dominic method does the rating and evaluation differently. It is more complex than Pugh's method; on the other hand, the Dominic method can also provide somewhat more detailed evaluation information.

Instead of pluses, minuses, and zeros (or S's), the Dominic method requires that each alternative be rated on each criterion by the evaluator or evaluation team as Excellent, Good, Fair, Poor, or Unacceptable. It also requires that each evaluation criterion be assigned a priority rating as High, Moderate, or Low. (Sometimes it is more convenient to think of the priority levels as Very Important (or Crucial), Important, and Desirable.) When these ratings and priority assignments have been made, a table of all the evaluation ratings can be prepared as in the following (fictitious) example.

The example assumes there are three conceptual alternatives (A, B, and C) and eight evaluation criteria: three with high priority, two with moderate priority, and three with low priority). The table below shows that for alternative A, the three high priority criteria are evaluated as Good, Good, and Fair; the two Moderate priority criteria are rated as Excellent and Good; and the three Low Priority criteria are rated as Fair, Fair, and Poor.

Alternative	High Priority Criteria	Moderate Priority Criteria	Low Priority Criteria
A	G, G, F	E, G	F, F, P
B	E, E, P	G, G	E, G, F
C	E, G, G	E, G	G, G, F

Now for each alternative (A, B, and C), an overall evaluation is obtained by using Figure 7.3. In the Figure, we have plotted the individual evaluations for each of three alternatives from the information in the table above. Instructions for obtaining an overall evaluation from the individual evaluations are included in the figure. Note that the overall evaluation for alternative B is Poor; for alternative A is Fair; and for alternative C is Good.

We can see readily from Figure 7.3 that if the single Poor rating (circled) for alternative B could somehow be improved, the overall rating for alternative B would be improved considerably. Note that all the other criteria for B are above the Excellent line. Thus, the results help guide redesign efforts, in this example pointing to the issue that most needs design attention to improve Alternative B.

The location of the dividing lines shown in Figure 7.3 is arbitrary; evaluation groups are free to draw them any way they wish in order to help them discriminate among alternatives and get overall ratings that have the most meaning. The ones shown, however, are a reasonable place to start.

An advantage of the Dominic approach is that it (like Pugh's Method) does not use numerical weightings and numerical priority ratings. Though engineers feel comfortable with numerically based evaluations, numerical approaches are not the best ways to make qualitative comparisons and decisions. The categories recommended in the Dominic approach have proved to work very well in evaluating designs.

Sometimes the Dominic approach will lead to ties; that is, to cases where two or more alternatives have the same rating. Usually, evaluation groups will be able to make the choice in these cases, though tie-breaking schemes can be developed. One such scheme, for example, is to break ties on the basis of which alternative has the best individual ratings on the highest priority criteria.

7.4 Using QFD to Evaluate Concepts

In Chapter 5, we introduced the idea of quality-function deployment (QFD) and the House of Quality as an aid to formulating conceptual design problems. The process of QFD begun there can now be extended in order to serve engineering concept or embodiment evaluation.

As described in Chapter 5, the first step in using QFD and the House of Quality to evaluate conceptual alternatives is to develop a complete set of customer attributes (CA's) for the product. The next step is to prioritize these attributes. This is done numerically by the evaluation team in such a way that the numerical assignments add up to 100. Thus for five attributes, the prioritizing might look like this:

Attribute 1	10 points
Attribute 2	25 points
Attribute 3	20 points
Attribute 4	5 points
Attribute 5	40 points
Total	100 points

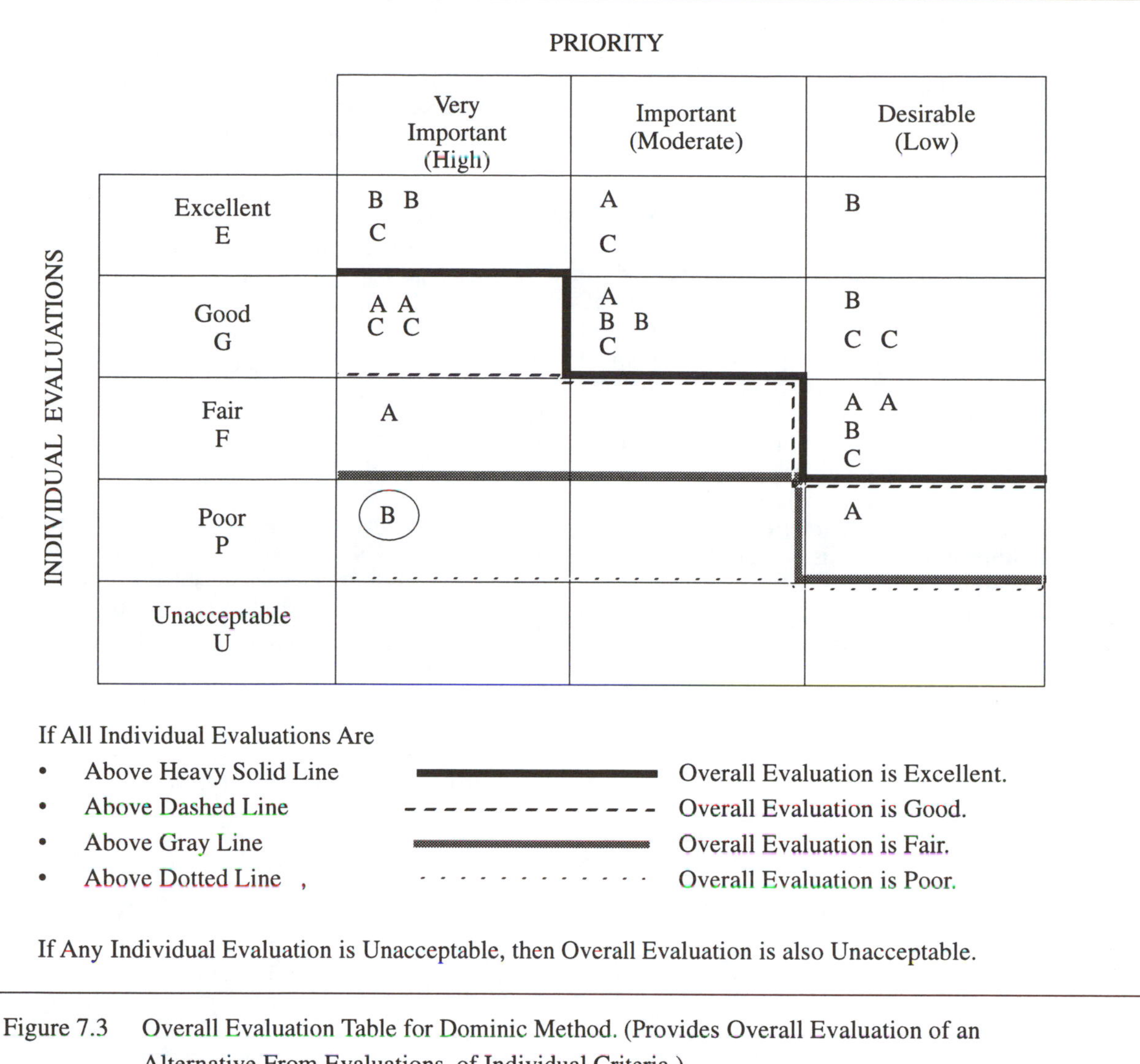

Figure 7.3 Overall Evaluation Table for Dominic Method. (Provides Overall Evaluation of an Alternative From Evaluations of Individual Criteria.)

With the prioritizing of the customer attributes done, the next step is to list and organize the various *engineering characteristics* (EC's) for each alternative concept or embodiment that is to be evaluated that influence the customer attributes. We need a list of EC's that are (a) measurable or clearly observable, and (b) directly affect one or more CA's.

As an example in Chapter 5, we listed the Customer Attributes for a small disk carrying case. See Section 5.2.2 and Figure 5.2. Continuing that example, a list of possible engineering characteristics for a particular disk carrier design that is to be made of a single injection molded part might be as follows:

- force to open
- material impact resistance
- results of impact drop test
- size of the case
- material color
- material surface finish
- shape of the case
- number of disks held
- method of identification of contents
- cost to manufacture

These fundamental engineering characteristics can be grouped into logical categories for convenience as follows:

- *Material*
  - Impact Resistance
  - Color
  - Cost
- *Physical*
  - Size
  - Shape
  - Surface Finish
- *Manufacturing Cost*
  - Tooling Cost

Processing Cost

- *Mechanics*

  Impact drop test

  Force to open

- *Method of Identification*

Now the basic House of Quality matrix can be extended for this product as shown in Figure 7.4.

To perform an evaluation, entries are made in the matrix by the evaluators that indicate the strength of relationship between each Customer Attribute and each Engineering Characteristic for the physical concept being evaluated. It may also be desirable to indicate, with a plus or minus sign, whether the relationships are positive or negative. Another bit of data included might be the source of the information on which the relationship is based; that is, is it subjective, based on statistical analysis, or the result of engineering analysis? Readers are referred to the original literature for additional refinements to the House of Quality methodology [4].

In the disk carrying case example being developed here, we have indicated only the strength or sensitivity of the relationship and the sign (where applicable). Strength or sensitivity is indicated as High, Moderate, or Low. Where no relationship exists, no entry is made.

We can now use the resulting information in the matrix as an aid to evaluating a proposed concept, or for comparing several alternative concepts. There are several ways to do this; one way is to simply use the new information as an aid when using the Pugh method described above. However, we favor the following approach, though it is more complex.

With all this done, the basic evaluation is an answer to the question: *How well can the Engineering Characteristics of the proposed physical concept be developed so as to meet the desired Customer Attributes?*

That is, from the House of Quality matrix, we can easily construct a series of questions of the following form:

How well does the <*concept name*> enable us to provide for a suitable <*Engineering Characteristic*> that will result in a quality product in terms of <*Customer Attribute*>*?* To illustrate, in our disk case example, one such question is: How well does the *injection molded single part concept* enable us to provide for a suitable *force to open* that will result in a quality product in terms of *staying closed?*

The questions raised in this way can be ranked in importance by consideration of the Relative Importance assigned to the Customer Attribute and by the strength of the

			ENGINEERING CHARACTERISTICS										
			Material			Physical			Mfg.		Mechanics		ID
CUSTOMER ATTRIBUTES		Importance	Impact Resistance	Color	Cost	Size	Shape	Finish	Tool Cost	Process Cost	Drop Test	Force to Open	Method
Protects Disks	Stays Closed	20					L				M	H+	
	No Coaster	5					M	L					
	Doesn't Break	14	H+				M				H		
	Inexpensive	10			H-	M			M	M			
	Six Disks	8				H							
Convenient to Use	Easy to Open	20					L					H-	
	External ID	10											H
	Compact	8				H-	M						
	Attractive	5		H		M		H					M
		100											

Figure 7.4 House of Quality for Disk Carrying Case

relationship shown in the matrix. In the sample question above, this is a very important question since "stays closed" is one of the most important Customer Attributes, and since the "force to open" is strongly related to that CA.

It must be remembered here that the answers to the questions can hardly be precise or certain at the conceptual stage of design. The product has not been designed yet; it exists only as information about the abstract embodiment of a physical concept. However, by directing the evaluation team's attention to the critical Engineering Characteristics and their relationship to the Customer Attributes, an evaluation of a concept can be focussed on the right issues.

7.5 Evaluating Manufacturability of Concepts

One important evaluation issue does not appear in the list of Customer Attributes: manufacturability. Customers don't really care whether or not something is easy or difficult to manufacture, though they do care, of course, how much the product costs when it reaches the marketplace. For this reason, we recommend that the costs of materials, tooling, and processing also be evaluated at the conceptual stage.

The issue of manufacturability includes more than just direct costs. Time-to-market is also crucial. And other issues of manufacturability can, and hence should, be evaluated qualitatively at the conceptual stage. It matters, for example, whether a product's performance will be impaired during its life due to one or more predictable manufacturing imperfections. And will tighter than normal tolerances be required?

To evaluate the manufacturability of a conceptual alternative, a second matrix can be developed — we call it the Conceptual Stage Manufacturability Advisor. A skeleton of a typical matrix is shown in Figure 7.5.

The basic matrix lists Product Characteristics across the top and includes generally at least those shown in Figure 7.5: Production Quantity, Material Properties, Operation and Maintenance Issues, Tolerances, Geometric Issues, and Other. (Readers should understand that there may be changes, deletions, or additions required for particular applications.)

		Product Characteristics									
		Quantity	Materials Required	Operation and Maintenance	Tolerance			Geometry			Other
Manufacturing Attribute					Number	Non-Std.		Size	Complexity		
Tooling	Cost										
	Lead Time										
	Life										
	Maintenance										
Processing	Cycle Time										
	Quality Control										
	Process Control										
Materials	Cost										
	Availability										
	Processability										
Assembly	Handleability										
	Insertion										
Other											

Figure 7.5 Skeleton Conceptual Design for Manufacturability Advisor

Manufacturing Attribute		Product Characteristics												
		Production Quantity	Material Properties			Operation Issues		Tolerance			Geometry Issues			Other
			Impact Resistance	Colorable	Rigid	Molded-In Hinge		Number	Non-Standard		Size	Complexity	Fits	
Tooling	Cost					H								
	Lead Time					M								
	Life													
	Maintenance													
Processing	Cycle Time													
	Quality Control					H								
	Process Control					H								
Materials	Cost		M											
	Availability													
	Processability					M								
Assembly	Handleability													
Other														

Figure 7.6 Conceptual Design for Manufacturability Matrix for Plastic Disk Case

Manufacturing Attributes for the proposed concept are listed vertically along the left side. These include issues of tooling, processing, materials, assembly, and others that may apply in specific cases.

In the various cells of the Conceptual Stage Manufacturability Advisor, evaluators can record the level of difficulty or risk (e.g., as High, Moderate, or Low) that each Product Characteristic creates for each Manufacturing Attribute. If there are no special problems, then no entry need be made. The level of the difficulty or risk is expressed as H for high, M for moderate, and L for low. Figure 7.6 shows how the matrix might be completed for the injection molded single piece (hinge molded-in) disk carrying case. (Note: The molded-in hinge is called a "living hinge", and it has become a fairly common and relatively easy-to-do feature, though this was not the case with the first attempts.)

We can certainly see readily from the matrix that the molded-in (living) hinge is (or was at the time) a potential risk. Thus steps taken immediately to deal with and reduce this risk are important. For example, can we find designers or molders who have done this before and so can tell us about the pitfalls? Or can a mold flow simulation be performed to study the problem in more detail?

As with Pugh's method, using a House of Quality or Conceptual Stage Manufacturability Advisor matrix to perform concept evaluation really requires that the evaluation team interpret the results by *thorough* discussion. These methods are aids to evaluation; they help call attention to issues that may need require attention. They are not substitutes for knowledge and exchange of information and ideas.

7.6 Competitive Benchmarking

More information can be attached to the basic House of Quality shown in Figure 7.4 in order to make it useful as a tool for comparing a company's existing or proposed products with one or more competitors; that is, for competitive benchmarking. At the bottom of the matrix, objective measures of the engineering characteristics are listed for the company's product as well as any competitors. See Figure 7.7.

The values for the Engineering Characteristics of the disk carrying case shown in Figure 7.7 are only illustrative guesstimates here, but in a real study would be the actual measured or computed values. Obtaining these actual, quantitative values for Engineering Characteristics enables engineers to determine how their existing or proposed products function to deliver the Customer Attributes compared to how competitors achieve the same goals. Of course, this kind of benchmarking can only be done after parametric design has been done.

7.7 The Pahl and Beitz Method for Concept Evaluation

Two German engineers, Gerhard Pahl and Wolfgang Beitz, in their pioneering 1976 book entitled *Engineering Design* [5] have developed a more technical approach to concept evaluation than either Pugh's method, Dominic's method, or the House of Quality. Space here allows only a summary of the methodology here.

(Note: Serious students and practitioners of engineering design are strongly urged to obtain a personal copy of the Pahl and Beitz book and to read it thoroughly. It presents an

			Engineering Characteristics										
			Material			Physical			Mfg.		Mechanics		ID
Customer Attributes		Importance	Impact Resistance	Color	Cost	Size	Shape	Finish	Tool Cost	Process Cost	Drop Test	Force to Open	Method
Protects Disks	Stays Closed	20					L				M	H+	
	No Coaster	5					M	L					
	Doesn't Break	14	H+				M				H		
	Inexpensive	10			H-	M			M	M			
	Six Disks	8				H							
Convenient to Use	Easy to Open	20					L					H-	
	External ID	10											H
	Compact	8				H-	M						
	Attractive	5		H		M		H					M
		100											
					$/ lb	$inch^3$	–	–	$	$/unit		lb F	
OUR PRODUCT				Tan	.59	16		Matte	8600	.06		1–3	
COMPETITOR A				Black	.86	15		Smooth	8000	.03		1–2	
COMPETITOR B				Blue	.42	18		Matte	7000	.06		0.5–1	
NEW PROTOTYPE				Grey	.48	16		Matte	9000	.05		0.5–1	

Figure 7.7 House of Quality for Disk Carrying Case

extremely useful, highly systematic approach to all phases of engineering design. The practice of engineering design certainly owes much to their important work.)

The Pahl and Beitz method of concept evaluation is based on a *use-value* analysis. It involves a series of steps as outlined below. In our abbreviated presentation here, much is omitted that is important to full appreciation and actual application of the method. Again: students of engineering design are urged to consult the original text.

Step 1 - List the Evaluation Criteria

This is, as the name implies, a list of the factors on which the concept is to be evaluated. The Engineering Design Specification is a source of information for creating this list. Additional information about the customer attributes used in QFD should also be consulted. The list cannot omit any essential factors, but at the same time it cannot be so long that the evaluation is overly detailed and cumbersome.

In the Pahl and Beitz method, the criteria must be expressed in such a way that they can subsequently be assigned numerical evaluations from 0 to 1. That is, the factor itself is not listed per se (e.g., life), but it is expressed in terms of a desirable quality (e.g., long life, low weight, reliable) so that an evaluation near zero will indicate Very Poor and a value near unity indicates Excellent.

Step 2 - Assign Numerical Weighting Factors to the Evaluation Criteria

The weighting factors are also numerical from 0 to 1, and they express the relative importance of the criteria to the overall evaluation. The sum of all the factors must be 1.

Pahl and Beitz recommend the construction of an hierarchical objective tree as an aid to assigning the weighting factors. To illustrate, in the example of the disk carrying case, the objective tree might look as shown in Figure 7.8 (a) and (b). Though there are only three levels in this example, more complex problems may well have four or even five levels.

Weights of the categories *at each level for each objective* must add to 1.0. Thus the weights assigned to the three objectives at Level 1 (protects disks, convenience, and cost) must add to 1.0. Similarly, the weights of the two sub-objectives under Protects Disk must also to 1.0, and so on. See Figure 7.8 (c) for illustrations of how this might be done for the disk case example.

Note that to get the overall importance of a lower level factor (e.g., Easy to Open), we need only multiply the weights given in its hierarchy. For example, for Easy to Open, the overall weight is: 0.5 x 0.4 = 0.20.

Thus, for the disk carrying case, we have the following weighting factors for the various evaluation criteria (after simplifying the problem a little):

Protects Disk = 0.4
Stays Closed = 0.2
Does Not Break if Dropped = 0.2
Convenient to Use = 0.4
Easy to Open = 0.2
Compact Size = 0.06
Attractive = 0.04
Carries 6 Disks = 0.1
Cost = 0.2

Step 3 - Assign Operational Measures to Each Evaluation Criteria

Pahl and Beitz call these "parameters". They are simply expressions (quantitative if possible, but verbal if necessary) of what one would measure, or what information one would gather, in order to determine a measure of the evaluation factor. In the case of "long life", for example, the operational measure is "life: hours" or "life: years".

Now, parametric values are estimated or computed for each measure or parameter. Since at the conceptual design stage there is as yet little detailed information available about the design, this may require rough computations or order of magnitude analyses. As a last resort, the estimates may have even to be verbal such as excellent, good, fair, or whatever is appropriate to the particular criteria. However, this then becomes, in effect, a part of the next step.

Step 4 - Assign Numerical Evaluation Values to the Individual Criteria

In this step, an evaluation is made, necessarily subjective in most cases, of how well the concept is expected to perform on each of the identified evaluation criteria. Pahl and Beitz present two possible scales for doing this, one providing five categories, and one providing ten categories. In both methods, the evaluations are made verbally but then converted to numbers for use in the overall evaluation.

The five category system is based on the German Guideline VDI 2225 [6], The categories and associated number values (NVs) are:

Unsatisfactory	Number Value = 0
Just Tolerable	Number Value = 1
Adequate	Number Value = 2
Good	Number Value = 3
Very Good (Ideal)	Number Value = 4

In the ten category system, the number values (NVs) are:

Absolutely Useless Solution	Number Value = 0
Very Inadequate Solution	Number Value = 1
Weak Solution	Number Value = 2
Tolerable Solution	Number Value = 3
Adequate Solution	Number Value = 4
Satisfactory Solution	Number Value = 5
Good Solution with a Few Drawbacks	Number Value = 6
Good Solution	Number Value = 7
Very Good Solution	Number Value = 8
Solution Exceeding the Requirement	Number Value = 9
Ideal Solution	Number Value = 10

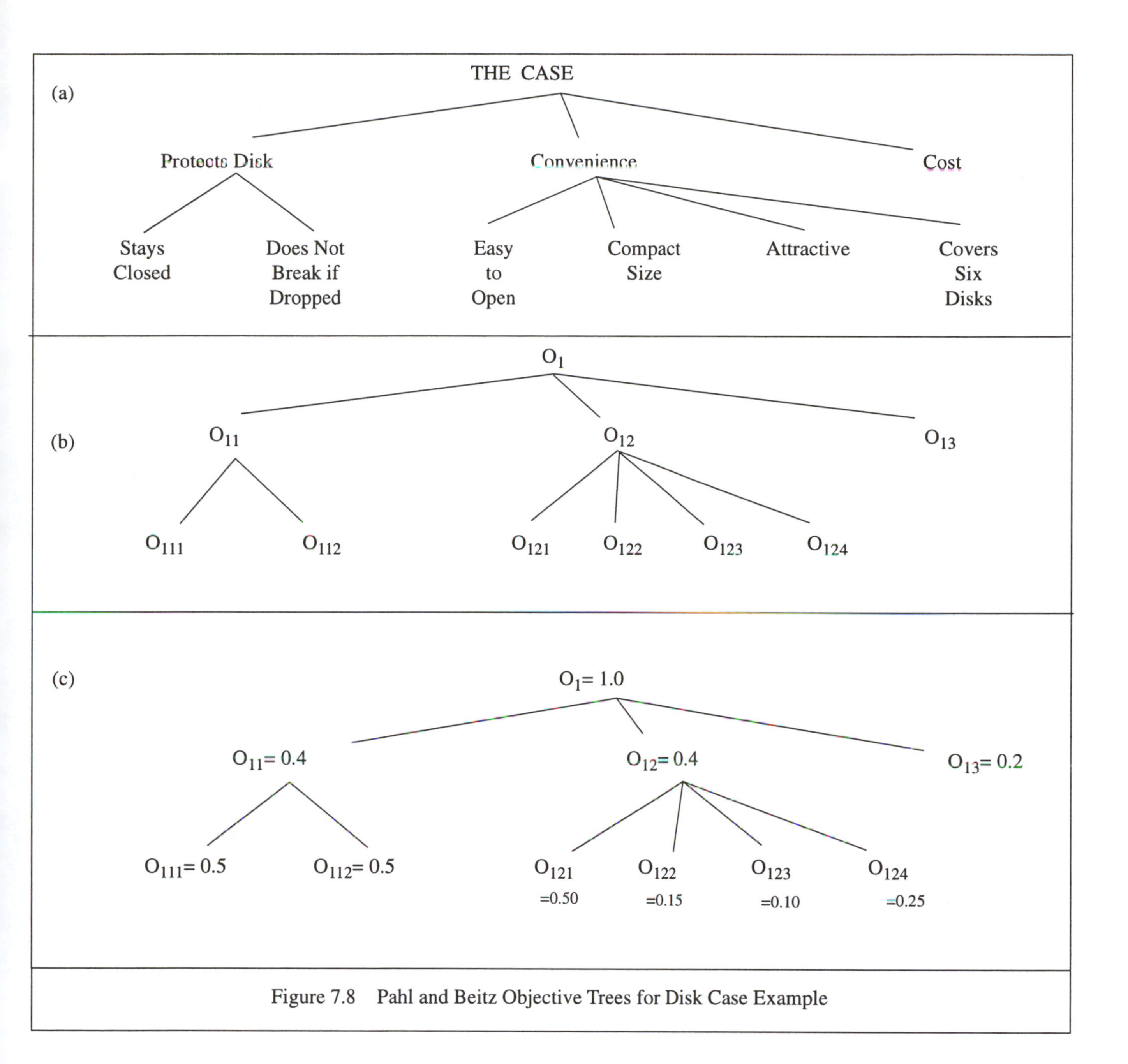

Figure 7.8 Pahl and Beitz Objective Trees for Disk Case Example

To illustrate in the disk carrying case example, let us assume that the evaluations (NV's) assigned to the plastic injection molded concept are (using the five point scale above):

Stays Closed	Adequate	NV = 2
Does Not Break	Very Good	NV = 4
Easy to Open	Good	NV = 3
Compact Size	Very Good	NV = 4
Attractive	Very Good	NV = 4
Carries Six Disks	Very Good	NV = 4
Cost	Very Good	NV = 4

Step 5 - Obtain an Overall Evaluation

Now with weights (Ws), from Step 4 and number values (NVs) obtained earlier in Step 2 established for the individual evaluation criteria, an overall evaluation of the concept is obtained by summing the products of those weights and values. That is:

Overall Weighted Value = OWV

$$OWV = \sum_{i} (W \cdot NV)_i$$

where i refers to the i th evaluation factor For our example, the overall evaluation is:

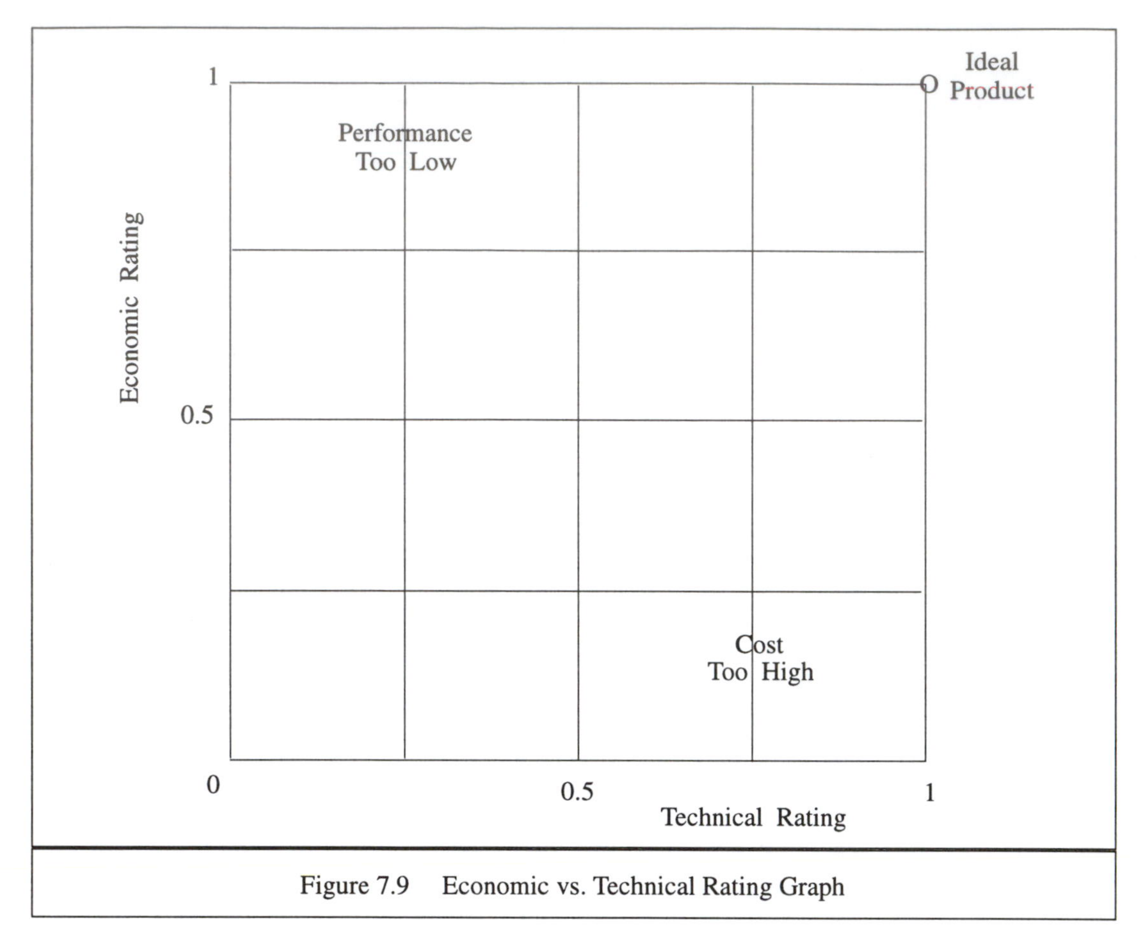

Figure 7.9 Economic vs. Technical Rating Graph

$$
\begin{aligned}
OWV &= 2 \times .2 + 4 \times .2 + 3 \times .2 + 4 \times .06 \\
&\quad + 4 \times .04 + 4 \times .1 + 4 \times .2 \\
&= 3.04
\end{aligned}
$$

Step 6 - Compare the Alternatives

Obviously, the larger the OWV, the better the proposed alternative is expected to be. However, Pahl and Beitz discuss several supplementary forms of overall rating that can be obtained from the preceding analysis and that can be used to compare various concept alternatives.

Of most interest here is a supplementary method that establishes a Technical Rating (0 to 1) using the above analysis, and an Economic Rating (also 0 to 1) as described below. The Technical Rating is found by computing the ratio of the OWV of the alternative to the OWV of a 'perfect' concept, which will usually have an OWV of 4.0. In our example, therefore, the Technical Rating is 3.04/4.0 = 0.76.

To obtain an Economic Rating, at least rough cost estimates must be possible -- not always the case at the conceptual level of design. However, when they are available, the Economic Rating is the ratio of the estimated cost of the alternative to the cost of an 'ideal' concept. The Economic Rating is thus a number between 0 and 1.

Next the two ratings — Technical and Economic — are plotted as shown in Figure 7.9. The location of different alternatives on this graph provides insight into their advantages and shortcomings. Pahl and Beitz also point out that the graph provides a useful way to see the effects of design changes (hopefully improvements!) since "before and after" rating points can be compared readily.

Step 7 - Consider Uncertainties

Pahl and Beitz note that errors are possible in the above procedure. There can be errors of judgement, and errors caused by lack of knowledge. There can also be errors associated with the psychology of the people doing the estimates. They therefore caution against putting more confidence in the final numerical ratings than is justified by the procedures from which they were developed. The numerical ratings are guides, not absolute evaluations or predictors.

Step 8 - Search for Weaknesses

A concept may receive a high or relatively high overall rating, but at the same time it may be weak on one or more individual criteria. It is relatively easy to search for these criteria, and it must be done. These weaknesses may be fatal to the concept, and they will certainly be issues that will need special attention during development if the concept is selected.

Concepts / Evaluation Criteria	A "Sticks and Brackets" Steel – Heavy	B "Sticks and Brackets" Alum. Ext.	C "Sticks and Brackets" Steel – Light (Rolled)	D "Milk Crate" Plates	E "Bar Stool" Steel Alum.
Material Life in Industrial Environment	O	S	S	–	S
Modular Potential Number of Different Parts	O	S	S	+	+
Ease of Assembly on Site	O	+	+	+	+
Ease of Modification For Size and Shape of Tables	O	+	+	+	–
Strength and Rigidity Potential	O	–	–	–	–
Flexibility in Creating Custom Sizes and Shapes	O	S	S	–	–
Expected Cost of Assembled Tables	O	+	+	+	+
Manufacturability	O	+	+	+	+
Appearance in Automation Setting	O	+	S	–	–
Maintenance	O	S	S	–	S
+'s	O	5	4	5	3
–'s	O	1	1	3	4

Figure 7.10 A Pugh's Method Evaluation Matrix for Design Project A

7.8 Redesigning Conceptual Alternatives

Any of the evaluation method described in this chapter will reveal and point clearly to the specific weaknesses of the alternative engineering concepts being evaluated. If none of the alternatives is acceptable, then the evaluation results can be used to guide re-design. Or it may be that more alternatives will have to be generated by returning to the methods of Chapter 6.

7.9 Continuation of Conceptual Design for Design Project A

At the conclusion of Chapter 6, we listed a number of concepts for the table support system, and noted that several material-process combinations could also be considered. Here we will reduce those lists to the most promising, and then show the results a Pugh's method evaluation of them. The results shown here are essentially those of a student team, though there has been some combining and editing.

Figure 7.10 shows the Pugh's method matrix, but of course does not show all the discussion that went into selecting the criteria, the concepts to evaluate, and the evaluations themselves. Readers will surely have ideas of their own about these choices.

A few comments for explanation may be helpful:

> The "bar stool" alternative was assumed to require some special stiffening of the table tops which made them difficult to modify in size. The stability problems with this concept are also obvious. With enough of the supporting cyl-

inders provided for stability, the concept approaches the "sticks and brackets" concept.

> The "heavy" steel concept assumed that steel pipe or angle iron or channel could be used. While this can be cut on site, it is much easier to cut and handle aluminum extrusions (easily cut with a table saw and special blade) or the lighter weight rolled steel forms.

> Wood was not even included as a material choice because it is obviously inappropriate and impractical in an industrial automation environment.

> "Rigidity" included both the rigidity potential of the individual components, and the rigidity potential of the completed structure. Thus, the milk crates might be made strong enough to be okay individually, but assembled they could become like a house of cards.

The choice here is clearly the "sticks and brackets" concept using either aluminum extrusions or rolled steel.

7.10 Summary and Preview

Summary. In this chapter, we have presented several methods for evaluating and comparing proposed alternative engineering concepts or embodiments. They are not perfect methods, they must be implemented by people with knowledge of marketing, design, manufacturing, and more. They are not 'plug-and-chug; they require judgement and discussion. But they work well when applied seriously.

Pugh's method and the Dominic method are no doubt the most readily usable, and provide good information in return for a modest effort. There are other methods described in [7, 8].

Preview. As explained in Chapters 4 and 6, the generation of physical concepts or embodiments, except when the stage of single components is reached, involves repeated decomposition of a product or special-purpose assembly into several sub-assemblies or components. As each level of decomposition is completed, a set of new physical concepts or embodiments is generated at the next level.

This process is repeated until there are no more special purpose sub-assemblies requiring decomposition. During this process, there generally will be a number of standard parts, standard modules, and special purpose parts created.

When these newly "born" parts and standard modules are created as portions of physical concepts at the various levels of decomposition, their in-use purposes and essential functional couplings are defined as part of the concept of which they are a part. However, beyond this information, little has been established to support the configuration and parametric design of the parts and standard modules that are created by decomposition. For example, there is as yet no Engineering Design Specification for them, and thus detailed functional requirements are not yet established. Moreover, in the case of parts, the basic materials and manufacturing classes have not been determined even though this information is normally required as a part of their conceptual design.

Therefore we have a little more work to do to complete the conceptual design of parts and standard modules. We do this is the next two chapters. In Chapter 8, we discuss the development of the Engineering Design Specification for parts and standard modules created during the repeated decomposition process. Then in Chapter 9 we discuss the process of selecting basic materials and manufacturing processes for parts. That will complete Part II of the book — Engineering Conceptual Design — and we will then (in Chapter 10) begin the configuration design of parts.

References

[1] Pugh, S. *Total Design: Integrating Methods for Successful Product Engineering*, Addison-Wesley, Reading, MA., 1991.

[2] Howe, A. E., Cohen, P. R., Dixon, J. R., and Simmons, M. K., "Dominic: A Domain-Independent Program for Mechanical Design", The International Journal of Artificial Intelligence in Engineering, Volume 1, Number 1, 1986.

[3] Keeney, R. L. and Raiffa, H., Decisions With Multiple Objectives, Wiley, New York, 1976.

[4] Hauser, D.R. and Clausing, D. "The House of Quality." *Harvard Business Review,* May-June, 1988.

[5] Pahl, G. and Beitz, W. *Engineering Design*, Edited by Ken Wallace, The Design Council, London, England, 1984.

[6] VDI 2225. VDI Society for Product Development and Marketing, Committee for Systematic Design, VDI Design Handbook, Beuth Verlag, D-1000, Berlin, Germany, 1986.

[7] Howlett, E., Ulrich, K., and Eppinger, S. "Teaching Note on Concept Selection for Product Design", Sloan School of Management, M.I.T., Cambridge, MA., 1991.

[8] Ulrich, K. T. and Eppinger, S. D. *Product Design and Development*, McGraw-Hill Book Co., New York, 1995.

Recommended Reading

Leech, D.J. *The Management of Engineering Design*, Wiley, New York, 1972.

Problems

7.1 To illustrate our evaluation method, consider three possible alternatives for the disk carrying case: a one piece injection molded case with molded-in hinge and snap latch; a padded vinyl "soft pack" with a zipper opening; and a one-piece cardboard case with a flip-top lid more or less like a Crayola™ box. Use Pugh's method to compare the alternatives.

7.2 Repeat 7.1 using the House of Quality and/or the Pahl and Beitz method, and/or the Dominic method.

7.3 Use the methods for evaluation in this chapter to evaluate alternative concepts do-it-yourself variable size bookcases, or adjustable garden hose nozzles, or some other product that interests you.

Chapter Eight

Completing the Conceptual Design of Special Purpose Parts

"The Park (Central Park in New York City) throughout is a single work of art and as such is subject to the primary law of every work of art, namely, that it shall be framed upon a single, noble motive, to which the design of all its parts, in some more or less subtle way, shall be confluent and helpful."

F. L. Olmsted and C. Vaux
Report submitted with their first prize design (1858)

8.1 An Engineering Design Specification for Parts

Not only in works of arts, but also in engineering design, parts must always contribute effectively and integrally to the overall concept and function of the whole design. Each part is important to the whole, but each part also has a life of its own.

For one thing, parts are manufactured *as parts*. Also, parts have their own special functionalities. Up to this point, however, we have considered primarily products and their sub-assemblies. As we have pointed out, products and sub-assemblies are subjected to a process of recursive decomposition that continues until there are no more sub-assemblies to be decomposed; that is, until there are only parts and standard modules to be designed or selected. In this chapter, we begin to consider more specifically the design of these individual parts and standard modules.

To begin, we must first develop a complete Engineering Design Specification for these entities. Though some of the information in the specifications of the sub-assemblies out of which parts and standard modules are created will apply, much of that information is simply not directly relevant to individual parts. A great deal of the information needed to design parts and standard modules is not in the Specifications of their parent assemblies.

As an example, consider a bracket that is to be part of the hinge on the door of a domestic clothes dryer. The door sub-assembly will have been part of the initial decomposition of the drier. The hinge sub-assembly will have been part of the decomposition of the door sub-assembly. The bracket will appear as part of the decomposition of the hinge sub-assembly. As a part of the hinge's conceptual design, we will know the in-use function of the bracket, and how it is coupled to the other parts of the hinge and/or dryer cabinet or frame. However, though we perhaps have some additional knowledge about the bracket, we will not yet have recorded it in a Specification for the bracket itself. For example, what forces must it support? What environmental conditions will it experience? What are the basic size and shape requirements and limits? Such information is critical to the design of the bracket (a special purpose part) but is not part of the Specification for the drier itself, or even of the hinge sub-assembly.

In addition, though knowledge of the basic material class (e.g., steel, aluminum, plastic, etc.) and manufacturing process (e.g., stamping, extrusion, casting, injection molding, etc.) is usually needed for configuration design of a part, we will not yet have made these selections.

In this chapter and the next, therefore, we discuss how to complete the conceptual design of the parts and standard modules created during the Engineering Conceptual Design of assemblies and sub-assemblies. The main focus in these two chapters is on special purpose parts, though the process is essentially the same for standard parts and modules. There are two main tasks to be done: (1) we must prepare an Engineering Design Specification for the parts; and (2) we must select their basic material class and manufacturing process. This chapter deals with the Engineering Spec, the next chapter deals with materials and process selection.

8.2 The In-Use Purposes of Parts

Though products (and to some extent their sub-assemblies) have a wide variety of unique in-use purposes, there are a only a relatively limited number of in-use purposes for parts to perform. Though they do occasionally have other in-use purposes, or combined in-use purposes, most parts transmit forces or heat, or provide some sort of barrier. Below is a comprehensive list — not necessarily absolutely complete — of in-use purposes served by typical parts together with a few examples of parts or features of parts that perform them.

In-Use Purposes of Parts or Features of Parts:

- *Transmit or Support Force(s) or Torque(s):*

  Brackets; beams; struts; columns; bolts; springs; knobs...

  levers; wheels, rollers, handles, parts that fasten, hold, or

  clamp such as bolts, screws, and nails; bosses...

- *Transmit or Convert Energy:*

  Heat — heat fin, electric resistance heating element...

  Mechanical Power — shaft, connecting rod, gears...

  Electricity — wire, light bulb element...

- *Provide a Barrier (For Example: Reflect, Cover, Enclose, or Protect):*

  Light — wall.

  Heat — thermal insulator, thermal reflecting surface...

  Electricity — electrical insulator, magnetic shield...

  Sound — wall, sound absorbing wall surface...

  Material — wall, cover, enclosure...

- *Allow Passage (For Example, of light, rods, shafts, wires, pipes, etc.):*

  Holes, windows, grooves, etc.

- *Control or Regulate the Passage of:*

  Fluids — nozzles, orifices, pipes, ducts...

  Light — shutter, wheel...

- *Indicate:*

  Clock hands, instrument needle, color, embossing

- *Locate or Guide:*

  Grooves, holes, bosses, tabs, slots...

- *Aid Manufacturing:*

  Fillets, gussets, ribs, slots, holes...

- *Add Strength or Rigidity (i.e., Stiffen):*

  Ribs, fillets, gussets, rods...

- *Reduce Material Use:*

  Windows or holes in walls, ribs...

- *Provide Connection or Contiguity (so a part can be a single part):*

  Walls, rods, ribs, gussets, rods, tubes...

When the configuration and parametric stages of design for a part are reached, it is of course essential to know the in-use purpose(s) that the part is to serve. Since reducing the number of parts is always an important goal, it is therefore helpful to combine as many such purposes as possible into a single part.

The in-use purposes of parts are an essential part of the Engineering Design Specification for parts, but still more information is required to complete the Spec.

In particular, we need to develop more quantitative functional requirements for the part just as we did for products and their sub-assemblies as described in Chapter 5.

8.3 Completing the Engineering Specification for Parts

As outlined in Chapter 5, the Engineering Design Specification contains information about the designed object in two major categories: (1) *in-use purposes*; and (2) *functional requirements*. For parts, the in-use purposes fall into two sub-categories:

1. The primary in-use purpose(s) of the part;
2. Any special purpose features required or desired by the designer to aid functionality or manufacturability.

The sub-categories of information about functional requirements are qualitative or quantitative *goals and limits* placed on:

(a) Performance;

(b) The environmental conditions under which the part is to perform;

(c) Economic issues;

(d) Physical attributes;

(e) Process technologies;

(f) Aesthetics;

(g) Part development time and cost; and

(h) Anticipated special tolerance requirements.

For parts, the primary in-use purposes are usually among those listed in the preceding section.

Usually, the meaning and interpretation of the categories of functional requirements for parts is the same as described in Chapter 5 for assemblies and sub-assemblies. In the case of parts, of course, the idea of "performance" has a more narrow interpretation. The functional requirement related to a part's performance is generally a quantitative statement of the in-use purposes; that is, How much heat must be transferred?, How large are the forces to be supported?, and so on.

There are, however, certain factors that are especially important to the functional requirements of many parts because the factors relate directly to many common mechanical failures such as those resulting from fatigue, stress concentrations, impact loading and the like. Thus it is important to emphasize in the Engineering Design Specification for the part all the information relevant to these potential failure modes. Remember, what a designer forgets to consider is most likely to be the cause of future trouble.

Among the factors to be certain are included are the following:

- The ranges of loads (hence stresses) to be encountered,
- Manner of loading: static or fluctuating,
- The number of load applications expected (lifetime),
- The velocity of the loading (i.e., impact loading including torsional impacts),
- Unforeseen loads due to such conditions as bearing,
- Misalignment, higher than expected belt or bolt,
- Tensions.............and so on.

There are also operating conditions related to the part in service that should be emphasized in the Engineering Design Specification. Designers *must* know the *true* operating conditions that parts will encounter. These include:

- Vibrations,
- Corrosion, including (say) possible decarburization of steels,
- Temperature of the part under stress;
- Allowed elastic deformations for proper function,
- Misalignment, or other assembly or operating conditions that could cause unexpectedly high stresses.

8.4 An Example

In-Use Purpose

Primary Purpose:

To provide a convenient means for the human operator of a garden roto-tiller to pull and push the spring loaded wire that engages and disengages the clutch.

Couplings:

The part must attach to the cylindrical roto-tiller handle for its support. It will also, of course, have a force or torque applied by the operators hand during its operation, and it must attach to and move the clutch control wire.

Special Purpose Features:

The part should be comfortable to the hand, and should enable the operator to move the control wire without letting go of the roto-tiller handle. (*Note:* Though not discussed in this book, the field of ergonomics studies the capabilities and needs of humans as they relate to machine interactions.)

Functional Requirements

Performance:

The force required to push and pull the clutch control wire is approximately three pounds-force.

The control wire must move through a distance of approximately 1.5 inches.

The number of operating cycles (i.e., engaging or disengaging the clutch) is approximately 30,000 in the life of the product.

The operation of the part to work the clutch will not be sudden. However, the part may be subjected to impact and other loads during handling and moving (e.g., when being moved into and out of storage, or into and out of, say, a pick-up truck).

Environmental Conditions:

The part will be subjected to outdoor temperate zone conditions of rain, humidity, and temperatures. During storage, temperatures from -10 F to 100 F are very likely.

There is likely contact with soil, fertilizer, lime, etc. used in gardens.

Economic Issues:

Cost is not a major issue with this part, but should be kept as low as possible consistent with quality, long term performance.

Physical Attributes:

The envelop of this part will be perhaps about six to eight inches long by about two to three inches by about one inch.

Process Technologies:

No restrictions.

Aesthetics

The part should appear like an integral part of the rest of the roto-tiller, most of which will be steel, either unpainted or painted a single color. It is important that the part appear sturdy, consistent with rugged outdoor work equipment.

Development Time and Cost

The design of the part should be completed (including prototype testing and accurate cost estimate) within three months. Production should be able to begin within one year.

Tolerances

No special tolerances are anticipated.

8.5 The Physical Concept or Embodiment of a Part

It will be remembered that once an Engineering Design Specification has been formulated, the goal of the engineering conceptual design process is to generate, evaluate, and select a *physical concept or embodiment* of a designed object to meet the Specification. As described in Chapter 5, the physical concept or embodiment to be developed contains the following information about the designed object:

- Whether the designed object or artifact is a special purpose part, standard part, standard module, or special purpose assembly or sub-assembly.
- The physical principles, laws, or effects that govern the designed object's operation or behavior;
- Its *abstract* form (i.e., its approximate envelop size and shape);
- *If a standard or special purpose part*, the basic material class (e.g., steel, aluminum, plastic, wood, etc.) and basic manufacturing process (e.g., injection molding, die casting, stamping, etc.) to be used.
- *If a standard part or standard assembly or module*, the name of its basic type (e.g., electric motor, gasoline engine, spring, screw, etc.)
- *If a special purpose assembly,* a list and sketch of the major *functional* elements including their (i) functions within the assembly, and (ii) their qualitative coupling requirements with other components and sub-assemblies.

Identifying the physical laws and the abstract shape for a part is no different from doing so for an assembly of sub-assembly. We have not yet, however, said anything about the process of selecting the basic material class and basic manufacturing process for parts. This, therefore, is the subject of the next chapter.

8.6 Summary and Preview

Summary. This short Chapter is a kind of watershed in this book. In earlier chapters, we dealt primarily with products and assemblies through the Engineering Conceptual Design stage. In this chapter, we have begun to consider specifically the design of the individual parts of which the products and sub-assemblies are composed. The first step is to complete an Engineering Design Specification for each of these parts.

Preview. In the next chapter, we continue the focus on conceptual design of parts. Specifically, we discuss how to select basic materials and manufacturing processes for parts.

Problems

8.1 Develop and evaluate several conceptual solutions for the roto-tiller handle described in Section 8.4.

8.2 Develop an Engineering Design Specification for one or more of the following parts: eye glass tine, broom handle, head of a hammer, the main structure of a carpenter's level. Or select a part of some product that interests you.

Chapter Nine

Selecting Materials and Processes

"Selecting a material for a part involves more than selecting a material that has the properties to provide the necessary service performance; it is also intimately connected with the processing of the material into a finished part."

George Dieter [10]*
Engineering Design (1983)

9.1 Introduction

When it has been determined that a designed object is to be a special purpose part, the task of Engineering Conceptual Design includes determining the basic material class (e.g., steel, thermoplastic, aluminum, etc.) and the basic manufacturing process (e.g., injection molding, stamping, extrusion, etc.) to be used. In this chapter we describe a methodology for making these important choices at the conceptual engineering design stage. In later chapters, we will discuss how these general choices are made more specific during configuration and parametric design.

Selecting a material class and manufacturing process for a part is a bit like the old chicken-egg argument: Which comes first? Should we select a material first, and then a process, or vice-versa? Either way, ultimately it is the combination of material and process that must work together during the design, production, and use of a part to meet the requirements of the Engineering Design Specification.

The task of selecting materials and processes is not an easy one. One reason is that there are so many materials and processes to choose from. For each material, there are dozens of material properties that have to be considered. And for each process, there are a variety of process capabilities and limitations to be considered in relation to the design and production requirements of the part. Moreover, the choice is also influenced by issues having to do with concerns such as safety, cost, availability, codes, disposal, and so on.

To make the problem even harder, the choice of material and process is influenced by a part's size, shape and geometry, but at the conceptual stage there is little information available about either the configuration or the dimensions. For example, stronger materials can lead to thinner walled parts and, as we learned in Chapter 3, certain processes can better cope with geometrically complex parts. We must nevertheless make basic material-process selections at the conceptual stage before it is feasible to go on to the configuration and parametric stages. We do not always make only a single selection, but the field of possibilities is usually reduced to no more than two or three.

Despite the fact that a part's configuration and parametric design have not yet been determined, there is still information available at the conceptual stage that can be used to guide material and process selection. A great deal of this information is recorded in the Engineering Design Specification. For example, the environmental conditions under which the part must perform are specified, and whether the part will be subjected to relatively small or large forces is also known.

In addition to information stated explicitly in the Specification, more can be inferred with reasonable correctness. For example, the approximate size, shape, and degree of geometric complexity are generally understood qualitatively even though the details of configuration and parametric design have not been addressed specifically.

In this chapter, we present a general methodology for

---

* Quotation from *Engineering Design: A Materials and Processing Approach* by George Dieter, Copyright 1983, by permission of McGraw-Hill Book Co.

selecting one or more material-process combinations for special purpose parts at the conceptual stage. Though this method and some of the data used to support it will also be useful in refining the material-process selection at later stages of the design process, the discussion in this chapter is essentially limited to the conceptual stage. A more detailed approach to material selection can be found in [1].

Though some data are given, data on all materials and processes are not included in this book. There are simply too many. Thus, some of the most commonly used materials and processes are included, and it is assumed that readers will have taken courses or have access to information which will provide more complete coverage. The major goal here is to provide a *methodology* for materials and process selection. As always, designers are strongly urged to consult with materials and manufacturing experts about material-process selections before considering the selection "final".

9.2 Two Approaches

9.2.1 Overview

There are two approaches to determining candidate material-process combinations for a part. Designers can use either approach depending on which is most natural to the part being designed. Both approaches end up at the same point. The two approaches are (a) material-first, and (b) process-first.

In the *material-first approach*, designers begin by selecting a material class — guided by the requirements of the application. Then processes consistent with the selected material are considered and evaluated — guided by production volume and information about the size, shape, and complexity of the part to be made.

In the *process-first approach*, designers begin by selecting the manufacturing process — guided by production volume and information about the size, shape, and complexity of the part to be made. Then materials consistent with the selected process are considered and evaluated — guided by the part's application.

In the next two sub-sections, an overview of each of these approaches is described. Then in the remainder of this chapter these approaches are described in greater detail.

9.2.2 Material-First Overview

In the material-first approach, application-related criteria derived from such issues as the environment in which the part will be used, the relative strength required, safety or code requirements, and so on, are used to select a set of candidate material classes. These criteria are discussed more completely in Section 9.4. In addition, to assist with material selection, Appendix 9-A contains three tables (Tables 9A.1, 9A.2, and 9A.3) that list a number of material properties and property ranges for broad classes and sub-classes of materials. The values are approximate and intended only to assist designers with trial material choices. More exact and complete tables of properties can be found in the References.

The organization of the Tables is as follows:

- Table 9A.1 - Properties of Selected Cast Metals (Alloys of aluminum, magnesium, copper, zinc, and steel)
- Table 9A.2 - Properties of Selected Wrought Metals (Alloys of aluminum, magnesium, zinc, and steel)
- Tables 9A.3 - Properties of Selected Plastics (Thermoplastics and thermosets)

Table 9.1 illustrates that with each material class, there is an associated set of feasible processes as shown in the column labelled Level III for metals and Level II for plastics. For example, if the material class selected is thermoplastics, then Table 9.1 (c) indicates that the feasible processes are injection molding, extrusion, extrusion blow molding, rotational molding, and thermoforming.

Now refer to Table 9.2 (a, b, c). Assuming the part is to be made of a material in a selected class, information about the part size, general shape, and complexity can be used to rule out some processes and point favorably to others. For example, Table 9.2(c) indicates that injection molding is usually not a good choice if production volumes are less than about 10,000.

In summary, the materials-first approach to materials-process selection maps first from application information to a class of materials by using material data and properties similar to those given in Tables 9A.1, 9A.2, and 9A.3. (More will be said about this in Sections 9.3 and 9.4.) By using Table 9.1, we are then able to select processes consistent with the materials selected based on application information. Then information about the part is added, and the result mapped into a process type using the guidelines presented in Table 9.2 (a, b, c).

The following illustrates the approach in rough schematic form:

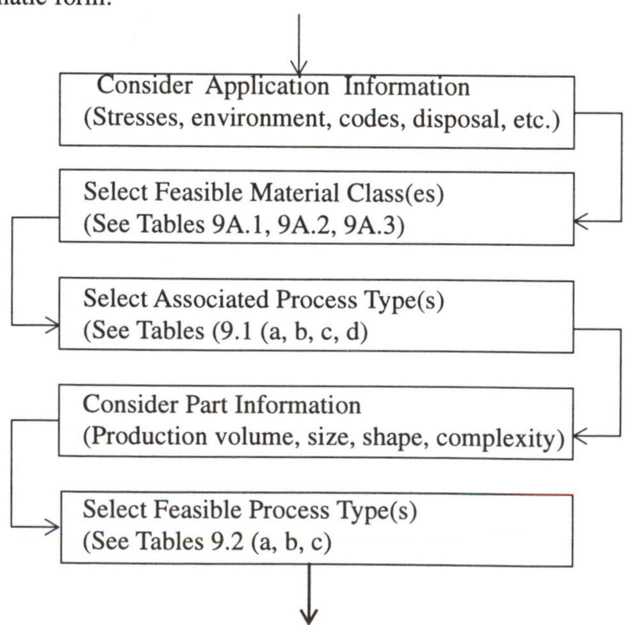

Level I Material	Level II Material	Level III Materials *Processes*	Level IV Material
Metals	Cast	Steels (carbon, alloy, stainless) *Sand Casting* *Investment Casting* *Centrifugal Casting* *Ceramic Mold Casting* *Forging*	ASTM A27-81 ASTM A352-80 ASTM A148-80 ASTM A297
		Aluminum and Magnesium Alloys *Sand Casting* *Investment Casting* *Centrifugal Casting* *Die Casting* *Permanent Mold Casting* *Plaster Mold Casting* *Shell Mold Casting* *Ceramic Mold Casting* *Forging*	Aluminum A380.0 Aluminum A413.0 Aluminum 201.0 Magnesium AZ91D Magnesium AZ63A
		Copper Alloys *Sand Casting* *Investment Casting* *Centrifugal Casting* *Plaster Mold Casting* *Shell Mold Casting* *Ceramic Mold Casting* *Forging*	C94800 C84400 C80100 C81400
		Zinc Alloys *Sand Casting* *Centrifugal Casting* *Die Casting* *Permanent Mold Casting* *Plaster Mold Casting* *Ceramic Mold Casting* *Forging*	SAE 903 SAE 925 Alloy 7 ILZRO 16

Table 9.1(a) Preliminary Material and Process Selection: Metals: Cast

Level I Material	Level II Material	Level III Materials *Processes*	Level IV Material Examples
Metals	Wrought	Steels (carbon, alloy, stainless) *Stamping* *Forging* *Extrusion* *Rolling* *Drawing*	SAE 1008 - Hot rolled SAE 1008 - Cold rolled SAE 2330 - Cold drawn 301 Stainless 410 Stainless
		Aluminum and Magnesium Alloys *Stamping* *Forging* *Extrusion* *Rolling* *Drawing*	Aluminum 2024 Aluminum 2124 Aluminum 1100 Aluminum 6061, 6063 Aluminum 7075 Magnesium AZ61A-F Magnesium AZ80A-75
		Copper Alloys *Stamping* *Forging* *Extrusion* *Rolling* *Drawing*	C23000 C37700 C11000
		Zinc Alloys *Stamping* *Forging* *Extrusion* *Rolling* *Drawing*	ZN-0.08PB ZN-1Cu Z300

Table 9.1(b) Preliminary Material and Process Selection: Metals: Wrought

Level I Material	Level II Material *Processes*	Level III Material Examples	Level IV Material Examples
Plastics	Thermoplastics *Injection Molding* *Extrusion* *Extrusion Blow Molding* *Rotational Molding* *Thermoforming*	ABS	Magnum 213 (Dow) Cycolac T (GE)
		ABS Glass Reinforced	AbsafiL G 1200/30 (Akzo Eng.)
		ABS Carbon Reinforced	J-1200CF/20 (Akzo Eng.)
		Polystyrene	Polysar 410 (polysar) PS 318 (Huntsman)
		Polycarbonate	Lexan 101 (GE) Makrolan 2800 (Mobay)
		Polycarbonate Glass Reinforced	R-40FG (Thermofil)
		Polycarbonate Carbon Rein-forced	R-40F-5100 (Thermofil)
		Acetal	Delrin 900 (DuPont) Celcon M25 (Celanese)
		Acetal Glass Reinforced	Thermofil G-40FG (Thermofil)
		Polyamide Nylon 6/6 Nylon 6 Nylon 6 Glass Reinforced Nylon 6/6 C Reinforced	 Adell AS-10 (Adell) Zytel 408 (DuPont) Ashlene 830 (Ashley) CR1401 (Custom Resins) NyLafil G 3/0(Akzo) Ultramid A3WXH (BASF)
		Polyethylene	Chevron PE1008.5 (chevron)
		Polypropylene	Excorene pp 122f (Exxon)

Table 9.1(c) Preliminary Material and Process Select: Plastics: Thermoplastics

Level I Material	Level II Material *Processes*	Level III Material Examples	Level IV Material Examples
Plastics	Thermosets *Compression Molding* *Transfer Molding*	Alkyd	Durez 24668 (Occidental)
		Alkyd Glass Reinforced	Glaskyd 2051B (Am. Cyan-imid)
		Epoxy Glass Reinforced	Rogers 2004 (Rogers) Rogers 1961 (Rogers)
		Polyester Glass Reinforced	Durez 30003 (Occidental)

Table 9.1(d) Preliminary and Process Selection: Plastics: Thermosets

Wrought Process	Materials	Production Volume	Part Size	Shape Capability
Stamping	Wrought Aluminum Steel Copper Brass	Minimum quantity 10,000 to 20,000. For smaller volumes and simple geometries, low cost steel rule dies can be used in place of progressive dies.	Generally less than 450 mm (18 inches). Larger sizes are done using die lines in lieu of progressive dies.	Moderate complexity is possible; however, molding and casting are capable of producing more complex shapes.
Forging	Aluminum Magnesium Copper Steel Titanium Superalloys	For medium and large sized forgings a production volume of 1000 to 10,000. For small forgings (under 1/2 lb.) a production volume of 100,000 may be required.	Maximum size is generally less than 800 mm (32 inches).	Moderate shape complexity. No internal undercuts possible. External undercuts limited to simple depressions.
Extrusion	Wrought Aluminum Steel Copper Magnesium Zinc	1000 (larger parts) to 100,000 (small parts)	From 1/4 to 10 or 12 inches (6 to 300 mm) in diameter . Lengths can be very long.	Constant cross section

Table 9.2(a) Preliminary Process and Material Selection: Wrought Processes

Casting Process	Materials	Production Volume	Part Size	Shape Capability
Die Casting	Generally aluminum and zinc. Brass and magnesium are also die cast.	Generally greater than 10,000.	Limited by size of available die casting machine. Maximum part size usually less than 600 mm (24 inches).	Almost any shape is possible. Internal undercuts should be avoided for practical reasons.
Investment Casting	Steel Stainless Steel Aluminum Magnesium Brass, Bronze Ductile Iron	Generally less than 10,000.	Maximum size generally less than 250 mm (10 inches).	Same as die casting
Sand Casting	All common metals	Minimum quantity between 1 and 100.	No maximum size. Size limited by carrying capacity of crane.	Almost all shapes are possible. External undercuts are limited due to need to extract pattern from sand.
Centrifugal Casting	Most metals		Large - usually over 100 pounds.	Generally rotationally symmetrical, but non-rotational parts are possible.
Permanent Mold Casting	Aluminum Zinc Magnesium Brass	Minimum quantity about 1000.	About the same as die casting.	About the same as die casting.
Plaster Mold Casting	Mainly aluminum and copper	Minimum quantity about 10	Generally limited to parts weighing less than 100 lb.	Undercuts are difficult to provide.
Ceramic Mold Casting	All common metals	Minimum quantity about 10.	Generally limited to parts weighing less than 100 lb.	Undercuts are difficult to provide.

Table 9.2(b) Preliminary Process and Material Selection: Casting Processes

Plastic Process	Materials	Production Volume	Part Size	Shape Capability
Injection Molding	Thermoplastics (Unfilled, Reinforced)	Generally greater than 10,000. However, less expensive tooling can be used for smaller production volumes. Seldom used for volumes less than 1000.	Limited by size of available injection molding machine. Maximum part size usually less than 600 mm (24 inches).	Almost any shape is possible, including internal and external undercuts.
Compression Molding; Transfer Molding	Thermosets (Unfilled, Reinforced)	Same as injection molding.	Same as injection molding.	Same as injection molding.
Extrusion	Thermoplastics (Unfilled, Reinforced)		Limited by size of extruder. The most common extruders have maximum diameters of about 200 mm (8 inches). Some are available up to 12 inches, or more.	Constant cross section.
Extrusion Blow Molding	Thermoplastics		Generally between 1 oz. and 1 gallon. Maximum size about 55 gallons.	Hollow thin walled parts. Minor undercuts okay.
Rotational Molding	Thermoplastics Thermosets (some)	Low (compared to injection molding)	Limited by size of molding machine. Usually must fit within 5 foot diameter sphere.	Hollow thin walled parts.
Thermoforming	Thermoplastic sheets and films		Usually 1 foot to 6 feet. However, some as large as 10 feet by 30 feet.	Simple flat and boxed shaped parts. Holes and openings cannot be formed; secondary operations required for these.

Table 9.2(c) Preliminary Process and Material Selection: Plastic Processes

9.2.3 Process-First Overview

In the process-first approach, the first step is to select a candidate process type (or types) using information available about production volume and about the part's approximate size, shape, and complexity. Information to support this selection is presented in Table 9.2. Intelligent use of Table 9.2 requires a basic understanding and knowledge of the information presented in Chapter 3. For example, if the part is a long part which has a constant cross-sectional area, then extrusion is a possibility. If the part has a complex geometric shape (a telephone housing for example), then molding or casting is required.

Once a process has been selected, the next step is to use application information to rule out or point favorably to a material class associated with the selected process. Note that in column 2 of Table 9.2 (a, b, c, d), there is a set of material classes associated with each process type. Thus, if we decide to use molding to produce a telephone housing then we can select from among a thermoplastic and a thermoset.

Schematically, the process-first approach looks like:

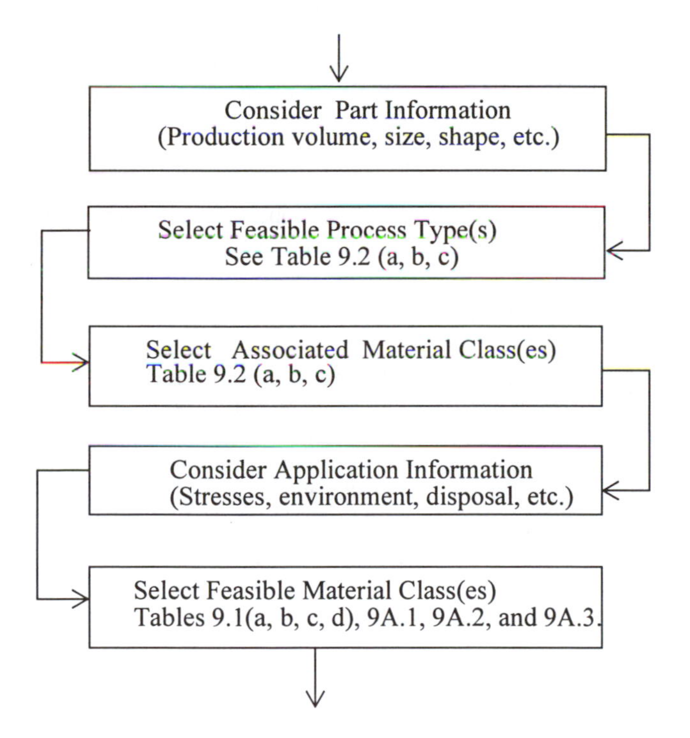

9.3 A Hierarchical Organization of Material Alternatives - Table 9.1

9.3.1 Introduction

At the conceptual stage, though it may occasionally happen, it is unlikely to be necessary to go so far as to propose specific materials such as 1030 hot rolled steel or 6063-T5 aluminum or Zytel 408 (a nylon 6/6 polyamide thermoplastic made by duPont). The reason we don't generally need such specificity yet is that the evaluation of the physical concept seldom depends on it. The finer discriminations can therefore generally be postponed (consistent with least commitment) until the configuration or parametric stage.

With this in mind, Table 9.1 has been prepared to support the selection process by organizing materials into a hierarchy of classes. The Table lists the most commonly used materials in four levels from very broad (e.g., metals or plastics) to very specific (e.g., Lexan 101). Here are two examples that illustrates the meaning of the four levels:

	Example A	Example B
Level I	Metal	Plastic
Level II	Wrought	Thermoplastic
Level III	Aluminum	ABS
Level IV	Aluminum Alloy 6061	ABS (Dow) Magnum 213

9.3.2 Level I of the Hierarchy

At the highest level (I) of the class hierarchy in Table 9.1, the material classes are

- metals,
- plastics,
- ceramics,
- wood, and
- concrete.

Only metals and plastics are considered below this top level in this book.

Metals Vis-a-Vis Plastics

Tables 9A.1, 9A.2 and 9A.3 list some of the most commonly used metals and plastics along with their more important mechanical and physical properties. It is evident from these data that, in general, plastics tend to be less dense (hence, lighter in weight than a comparable metal part), less costly, better able to resist corrosion, and better insulators than metals. In addition, as we learned in Chapter 3, plastics can be processed into almost any conceivable shape, with any desired surface finish (from mirror to textured) — and often requires no finishing.

Metals, however, tend to have better mechanical properties. The elastic modulus of steel, for instance, is almost 60 times higher than that of an unreinforced rigid plastic, and the moduli of aluminum, zinc, and copper alloys are about 20 times higher than unreinforced plastic. (The moduli for reinforced plastics are about three times that of unreinforced plastics.) Also, steels have higher temperature capabilities, better thermal and electrical conductivity, and are capable of being processed by a very fast processing method: stamping. On the other hand, carbon steels usually need costly protective procedures such as pickling, galvanizing, priming and painting to prevent rust.

Metals and plastics, especially reinforced plastics, often compete head to head as the material choice. In recent years, plastic has been used increasingly to replace or substitute for metal. Some examples are: automobile bumpers, automobile fuel tanks, lawn tractor hoods, wheel covers, instrument panels, and electric staple guns. At present a plastic (long fiber nylon) automobile jack is being evaluated by auto makers in both the U.S. and Japan. The reasons for this trend have to do with a variety of factors, including weight (to help auto makers meet the U. S. Government's average fuel economy requirements), corrosion resistance, insulating effects, consolidating parts (to reduce assembly costs), and improving ergonomics (i.e. user comfort).

Because of the wide variety of possible applications and evaluation criteria, it is not possible to state any very specific rules about when plastic or metal will be the preferred choice. When the mechanical properties (strength at operating temperatures, creep, impact resistance, fatigue resistance, etc.) of plastics are adequate for the task, however, their other desirable properties will usually make them the choice. However, the *total* cost (material cost, tooling cost and processing cost) of a plastic part is not necessarily less than that of a functionally equivalent metal part.

9.3.3 Level II of the Hierarchy

At the second level in the hierarchy (column 2 of Table 9.1) metals are classified as cast or wrought, and plastics as thermoplastics and thermosets.

Metals: Cast vis-a-vis Wrought

Two of the most important manufacturing considerations in selecting a material for a design application are: (1) how easily can the material be formed into a finished part; and (2) how might the material's mechanical and physical properties change during the forming of the part.

In forging, for example, as we learned in Chapter 3, a 'thin' cup shaped part is formed by squeezing (or hammering) a metal billet between two die halves which are attached to a hammer or press. Actually, the metal in a forged part will first go through a couple of intermediate stages: an ingot is cast, then worked (i.e., wrought) by rolling or extruding the ingot into a billet, and finally it is forged.

In casting, recall from Chapter 3, a cup-like part is produced by pouring or injecting molten metal into a die. The solidification process that follows results in parts whose metal structure is usually composed of large grains, small voids and small pockets of impurities called inclusions.

Working the metal alloy from an as-cast ingot to a wrought state — as in forging — is accompanied by a marked improvement in its mechanical properties. The associated plastic deformation refines the cast structure of the metal, closes the shrink voids which occur, and breaks up the inclusions. Since the porosity and inclusions which occur in castings are often the source of fractures, their removal in wrought metal is responsible for considerable increased ductility and strength in alloys *of the same composition.* For non-ferrous metals this increase in ductility can be seen by comparing the Elongation Percent for cast and wrought metals in Tables 9A.1 and 9A.2.

The increase in tensile and yield strengths of wrought metals over those of a cast metal *of the same composition* is not easily seen by comparing the values shown in Tables 9A.1 and 9A.2. The difficulty comes in finding cast and wrought alloys of the same composition. It is certainly possible, however, to find a cast alloy which exhibits greater ductility and strength than a wrought alloy. Tables 9A.1 and 9A.2 show, for example, that a high carbon steel casting can have superior mechanical properties to those of a low carbon steel sheet. However, for the same composition, wrought metals do have superior mechanical properties.

It is interesting that cast and wrought carbon steel can be ordered from suppliers simply by specifying, not the composition, but only the desired mechanical properties. Producers can adjust the composition of the steel to meet specified mechanical properties. Thus, given that it is possible to select a cast metal that has mechanical properties equal to a wrought metal, the question becomes — what should guide designers in the choice between cast and wrought?

Of course, we should always consider a casting process (hence cast metals) when it will be more economical. In addition, however, we should consider a cast metal when (a) parts have complex shapes and/or hollow sections, (b) large parts (sand castings can produce parts of almost any size), or when (c) overall manufacturing costs can be reduced by combining several simple parts into a single more complex part and thereby reduce assembly costs.

Conversely we should select a wrought metal (and consequently a wrought process) when the geometry of the part is simple, and relatively fast and/or inexpensive processes such as stamping or extrusion can be used to produce the geometry.

Plastics: Thermoplastics and Thermosets

Table 9A.3 contains a representative list of some of the more commonly used thermoplastics and thermosets. It is evident from the data contained in this table that, generally, thermosets have a much higher flexure modulus. Thus, thermosets have greater rigidity than thermoplastics. In addition, some thermosets, alkyd for example, have good electrical properties (including arc resistance), low water absorption, and retention of electrical properties at elevated temperatures. Thus, alkyds are most often used for automotive distributor caps, coil caps, and circuit breakers. Other thermosets are used for electrical connectors and switch gears, and circuit boards.

Thermoplastics, however, are faster and easier to process. The cycle time is strongly dependent on wall thickness, but most parts made of thermoplastic materials have cycle times between 15 and 60 seconds. The cycle time for thermosets parts, however, ranges between 30 seconds and 5 minutes because the material cures relatively slowly during the molding process. As implied in Chapter 3, the tooling costs for injection molding, compression molding and transfer molding are comparable; thus the lower processing costs for injection molded parts generally makes thermoplastics the plastic of choice when it is capable of satisfying functionality requirements.

9.3.4 Level III of the Hierarchy

At Level III, metals are subdivided into the various broad categories such as steel (carbon, alloy and stainless), aluminum alloys, and copper alloys. Plastics at Level III are subdivided into the various classes of thermoplastics and thermosets such as polycarbonates and polyesters. Usually Level III is about the right one for specifying and evaluating conceptual design alternatives for most special purpose parts.

Metals

Tables 9A.1 and 9A.2 contain a representative sample of some of the most commonly used ferrous and non-ferrous metals and metal alloys. A study of these tables shows that, in general, the non-ferrous alloys (aluminum, magnesium, copper, and zinc) are much more resistant to corrosion, lighter in weight (less dense), and are better thermal and electrical conductors than steel. In addition, the non-ferrous alloys are usually also easier (and less costly) to fabricate since they have lower tensile and yield strengths than steel. They are also usually easier (and less costly) to cast since they have a lower melting temperature than steel. Indeed, because of steel's high melting temperature, some casting processes (e.g., die casting) are not suitable for steel.

As a class, however, Tables 9A.1 and 9A.2 show that ferrous metals have higher tensile and yield strengths, and a higher modulus, than non-ferrous alloys. Table 9A.2 shows, for example, that:

	1015 Hot Rolled Steel	1100-H18 Aluminum
Modulus, E (psi)	30×10^6	10×10^6
Tensile Strength (psi)	50×10^3	24×10^3
Yield Strength (psi)	7.5×10^3	22×10^3

Thus, similar parts made of steel are stiffer than those made of aluminum or any of the other non-ferrous alloys. Consequently, for similar parts made of aluminum and steel and subjected to the same loading, the steel component ($E = 30 \times 10^6$ psi) will deflect only about one-third that of the aluminum component ($E = 10 \times 10^6$ psi). Thus, aluminum components must be designed in configurations that provide additional stiffness. Alternatively, aluminum, magnesium, copper and zinc alloys can be strengthened and (strain) hardened by heat treatment, cold rolling or drawing.

Low carbon steels (less than 0.20% carbon) are used primarily in the production of bars, sheets, strips and drawn tubes. Low carbon steels possess good formability (the ductility, as seen by its percent elongation, is high) and are adaptable to low-cost production techniques such as stamping. For this reason, low carbon steels are primarily used in the production of parts for many consumer products.

Increasing the carbon content of steel increases its strength and hardness but decreases its ductility and formability, thereby increasing the cost to process it. Mild steel (carbon content between 0.15% and 0.30%) is used in the production of structural sections, forgings, and plates. High carbon steel (greater than.80% carbon) has low ductility, but its hardness and wear resistance are high. It is thus used in the production of hammers, dies, tool bits, etc.

Alloy steels are steels which contain small amounts of elements like nickel, chromium, manganese, etc., in percentages that are greater than normal (in carbon steel). The alloying elements are added to improve strength or hardness or resistance to corrosion.

Plastics

At Level III, plastics are divided into various broad family classes such as ABS, acetal, nylon, etc. Table 9A.3 contains a representative sampling of these plastics in both their neat (i.e. unmodified by the addition of fillers or powders) state, as well as in a composite material state. In the *composite* state, fine fibers (0.01 mm diameter) of additives (e.g., glass fibers) are chopped to very short lengths and mixed with the resin in order to improve its mechanical properties. Table 9A.3 shows that in every case the strength and modulus of the resin is increased by the addition of glass fibers. See, for example, how the strength and modulus of nylon 6 with 30% glass compares with that of nylon 6 without glass.

Metal powders and fibers are also mixed with plastics to improve their electrical and thermal conductivities, and to provide EMI (electromagnetic interference) shielding.

In general, the term *composite* includes plastics which come with additives, fillers and reinforcing agents like glass fibers, carbon, or graphite. A composite can also have additives of metal, wood, foam, or other material layers, in addition to the resin/fiber combination. For the purposes of this chapter, composites are restricted to resins which contain reinforcing additives to improve the mechanical and thermal properties of the resins.

A study of Table 9A.3 shows that polycarbonate has the highest impact resistance listed. In fact, *polycarbonate* has the highest impact resistance of any rigid, transparent plastic. For this reason, and its relative ease of producibility, it is used in the production of many consumer products that might get dropped or struck. Examples are: computer parts and peripherals, business machine housings, power tool housings, vacuum cleaner parts, sports helmets, windshields, lights, boat propellers, etc.

Table 9A.3 shows that Nylon™ (DuPont) is resistant to oils and greases and most common solvents. It has high strength (for a plastic) and a high modulus, and it has a relatively high maximum service temperature. The coefficient of friction of Nylon is also low, and for this reason it is frequently used in the production of bearings, bushings, gears, and cams.

Acetals are strong, stiff and highly resistant to abrasions and chemicals. Like nylon they have a low coefficient of friction when placed in contact with metals. They are used in the production of brackets, gears, bearings, cams, housings, and plumbing.

9.3.5 Level IV of the Hierarchy

At Level IV of the hierarchy in Table 9.1, we have listed only a small number of specific examples of the material classes in Level III. Examples of metals are listed according to specific alloy grade (e.g., ASTM A27 6535). The mechanical and physical properties of a representative sampling of metals are given in Tables 9A.1 and 9A.2. While Level III is about the right level for evaluating conceptual design alternatives, the selection of a specific material based on specific mechanical and physical properties must be done at Level IV. The selection at Level IV can usually be delayed until the parametric stage of design, though in some cases it must be made sooner.

For plastics, the physical and mechanical properties given in Table 9A.3 are based on the generic version of a given plastic. For this reason, a range of values is generally given for each class of resins (such as ABS or Nylon 6). The existence of a range of values is partially due to the difference in properties between commercial brands.

As noted above, Tables 9A.1, 9A.2 and 9A.3 provide only a sampling of materials and material properties. References [2, 3, 4, 5, 6, 7] must be consulted for a more complete listing, for more complete and accurate material properties, and for additional assistance in the selection of material and process combinations for various applications.

9.4 Application Issues in Selecting Materials

How and where a part is to be used — that is, the conditions of stress and environment it will encounter when in use — obviously have a great influence on the choice of a suitable material for the part. In this section, we list and discuss briefly a number of the most common application issues. There may well be others that will be relevant in certain applications so designers must be thorough in considering other possible issues in addition to those listed. (Remember, what you forget to consider is the biggest danger!)

9.4.1 Forces and Loads Applied

Magnitude

To a certain extent, of course, the internal stresses experienced by the materials of parts can be reduced by competent configuration design. However, it is still generally true that the larger the external forces, the greater the need for stronger materials. Thus the magnitude of external forces is a factor relevant to the required tensile and compressive yield strength of materials. Though steels are generally the strongest, other metal alloys and plastic composites can also be made very strong.

Creep

If steady loads are applied over a long period of time (i.e., months or years), and if the temperature of the part is elevated, then creep (the slow continuous deformation of material with time) must be considered a possibility.

At room temperature, the deformation or strain for most metals (excluding lead) is a function only of the stress. As the temperature increases, materials subjected to loads which cause no plastic (permanent) deformation at room temperature, may now deform plastically (creep) depending upon the combination of stress, temperature, and time. Since most metals (and ceramics) have high melting temperatures (over 2700 F for steel, about 1100 F for aluminum), they start to creep only at temperatures well above room temperatures. A general rule of thumb is that metals begin to creep when the operating temperature is greater than 0.3-0.4 times the melting temperature [8].

Creep is often a more important concern with plastics. In fact, many plastics creep at room temperature. Plastics have no well defined melting point; thus, for plastics the important temperature to consider is known as the *glass transition temperature,* T_g Roughly speaking, below the glass transition temperature a plastic is in a glassy, brittle state; above T_g it is in a rubbery state. Well below T_g, plastics do not creep. Examples: The glass transition temperature for Nylon 6 is about 50 C (122 F), while that for polycarbonates is about 150 C (302 F). For epoxies, T is usually greater than 100 C (212 F).

When designing with plastics that support loads, therefore, temperature is an especially important consideration. Unfortunately, creep data for specific plastics is still difficult to find in the open literature, and so must be sought from the resin manufacturer.

Impact Loads

Under certain conditions, even ductile materials (as defined by the uniaxial tensile test) will have a tendency to behave in a brittle manner. This is particularly true when parts are subjected to impact or sudden dynamic loads. Thus, if a part in use will be subjected to impact loads, then the impact strength of the material to be used is a relevant design evaluation issue.

Impact strengths are measured by either a Charpy test or an IZOD test [9]. While it is difficult to relate the results of these tests directly to design requirements, in general, materials that have high impact resistance also have high strength, high ductility and high toughness. These impact strengths are perhaps more useful in comparing various types of grades of metals or plastics within the same material family, and not as useful in comparing one metal with another metal or one plastic with another plastic.

Cyclic Loads and Fatigue

Many products and components (the door handle on a car, the latch mechanism on a brief case, a tennis racquet, a paper clip, etc.) are subjected to repeated fluctuating loads. Despite the fact that the loads on these parts and products may fall well below the tensile strength of the material, and even below the yield strength of the material, the door handle falls off, the latch comes apart, the tennis racquet breaks, and the paper clip fails. Such failures are usually due to fatigue. In fact, fatigue failures are responsible for many failures in mechanical components.

If stresses in a material fluctuate, then the concept of *endurance limit* becomes relevant. When the *average* internal stress is zero, it has been found that there will be no fatigue failure if the fluctuating load amplitude is kept below a critical value called the *endurance limit* or *endurance stress*.

Tables 9A.1 and 9A.2 show values for the endurance limit for some metals and metal alloys. It is clear that in general steel is better than the non-ferrous metals. In fact, aluminum alloys do not have an endurance limit.

While plastics also exhibit fatigue, for many plastics there is no well-defined fatigue strength (the stress level at which the test specimen will sustain N cycles prior to failure) or endurance limit. Thus, as in the case of metals, large numbers of cyclic stress reversals can be expected to cause failure in plastic parts even if the applied stress is low. Also, in the case of plastics, the mode of deformation and the strain rate has a much more profound effect on the results than they do for metals. In addition, high frequency cyclic loading can cause the plastic to warm up and soften, thus further reducing the load-carrying of the part.

9.4.2 Deformation Requirements

If the degree of a part's deformations as a result of the applied loads are crucial, then the resistance of material to deformation will be a factor. The relevant material property is the Modulus of Elasticity, E. Low values of E will result in a part which, when loaded, will have large deflections (as compared to the same part/load configuration of a part made of a material with a large value of E). At times this may be desirable — for example in the design of springs and beams which are to act as springs. The value of E also affects the natural frequency of a part or assembly. For example, a beam or spring with a low E (i.e. low stiffness) has a lower natural frequency than one with a high modulus.

As seen from the values for the modulus of a representative sampling of metals, metal alloys and plastics in Tables 9A.1-9A.3, the values of E can be summarized as follows:

Engineering Thermoplastics (no additives or fillers):

$$E = 0.3 - 0.5 \times 10^6 \text{ psi } (2000\text{-}3500 \text{ MPa})$$

Engineering Thermoplastics (glass reinforced):

$$E = 1.3 - 2.0 \times 10^6 \text{ psi } (9000 - 14{,}000 \text{ MPa})$$

Non-Ferrous Metal Alloys:

$$E = 10\text{-} 20 \times 10^6 \text{ psi } (68950 - 13{,}800 \text{ MPa})$$

Ferrous Metals and Metal Alloys:

$$E = 30 \times 10^6 \text{ psi } (206{,}850 \text{ MPa})$$

9.4.3 Other Application Factors

In addition to the force and deformation requirements discussed above, there are many other factors related to application that may have to be considered. Among them are:

In-Service Temperature

Is it low? Is it elevated? Will the material be capable of functioning at these temperatures? Metals can be used at higher temperatures than plastics. Thermosets can be used at higher service temperatures than thermoplastics. References [2, 3, 4, 5, 6, 7], as well as other references, can assist us in determining the effect of temperatures on materials.

Exposure to Ultra-Violet (Sunlight)

Will the optical properties of the material, especially plastic materials, change after exposure to sunlight. Once again References [2, 3, 4, 5, 6, 7] as well as others can assist us in determining the effect of sunlight and other elements on materials.

Exposure to Moisture (Fresh water, sea water)

Will the exposure be intermittent or continuous? Will the part or product be immersed in water? Again References [2, 3, 4, 5, 6, 7] contain information that can assist us in determining which metals and metal alloys can best resist corrosion and moisture, and the effect of moisture on materials. For example, according to [3], "Aluminum alloys of the 1xxx, 3xxx, 5xxx, and 6xxx series are resistant to corrosion by many natural waters." Reference [3] goes on to state that service experience with these same wrought aluminum alloys in marine applications demonstrates their good resistance and long life to sea water.

Chemicals (Acids, alkalies, etc.)

Will the part or product be subjected to chemicals? Will the materials we select be able to resist corrosion by these chemicals?

Weather

Some materials are better able to deal with atmospheric conditions than others. Ferrous alloys need protection since they corrode when exposed to air. Most aluminums have excellent resistance to atmospheric corrosion and in many outdoor applications they do not require a protective coating. Stainless steels are suitable for exposure to rural and industrial atmospheres. Plastics, of course, do not typically corrode when exposed to atmospheric conditions. Again, references [2, 3, 4, 5, 6, 7] are excellent sources of information concerning the effect of atmospheric conditions on materials.

Insulating and Conducting Requirements

These can include both electrical and thermal transfers. In thermal cases, note that heat may be transferred by conduction, convection, and radiation.

Transparency and Color Requirements

Safety and Legal Requirements

The issues may included flammability, Food and Drug Administration regulations, Underwriters Laboratory standards, National Sanitation Foundation requirements, etc.)

The list above is just a brief list of things that one should consider. In specific, real cases, there generally are others. Remember, don't forget to consider *everything*! The issue that you forget to consider will often come back to haunt you!

9.5 Cost

Once one or more materials have been selected then, as explained above, an indirect selection of alternative processes also occurs. For example, if we decide that stainless steel must be used to produce a part of moderate geometric complexity (thin walls, external undercuts, several ribs and/or bosses, etc.) then we have ruled out, as discussed in Chapter 3, the possibility of using die casting. If external undercuts are present and if these undercuts are due to projections, then forging is also eliminated as a possible means of forming the part. (If the projections are due to circular holes, then forging is still a possibility.) Thus, from Table 9.1 it is seen that we are essentially left with choosing from among sand casting, investment casting and ceramic mold casting. If on the other hand, an aluminum alloy could have been selected, then several other casting processes would have been available to choose from, including die casting.

While the selection of one material or alloy may be based in part on material cost, it is the total cost of a part that should be considered when making the final selection. And the total cost of the part is a function not only of material cost but also of tooling cost and processing cost. While one may not always be free to select that material process combination which results in the lowest overall part cost, elimination of that material process combination should only be done with good reason.

References [2, 3, 4, 5, 6] contain a comprehensive listing of materials along with description of designations, mechanical properties, thermal properties, composition, and typical uses as well as the processes best suited for the various metal alloys and grades of plastics. Those references should be used in conjunction with Tables 9.1 - 9.2 and Tables 9A.1 - 9A.3.

The material prices listed in Tables 9A.1 - 9A.3 are incomplete. The prices of plastics, both thermoplastics and thermosets, are easy to obtain. A pricing update appears each month in the trade journal, *Plastics Technology*. Metal prices are more difficult to obtain. A partial and incomplete list of base metal prices can be found in each issue of *American Metal Market*. The actual cost of a metal depends, however, on metallurgical requirements, dimensions and shape (bar, plate, channel, etc.), tolerances, surface treatment, thermal treatment, and quantity. The actual price of a metal can only be obtained by getting a quotation from a vendor.

9.6 Part Information in Selecting Processes - Table 9.2

The number of possible manufacturing process alternatives, though not as huge as the number of material class alternatives, it is nevertheless substantial. Table 9.2 contains a list of the more commonly used processes together with the associated material classes and data about practical production volumes, part sizes, shapes, and approximate complexity.

The first group of processes in Table 9.2(c) are melting and molding processes which can be used with both thermoplastics and thermosets. This group of processes offers probably the greatest flexibility in geometric shape complexity. Because of the faster processing times achievable when using thermoplastics, injection molding is generally the process of choice for "engineering type" plastic parts.

The group of processes listed in Table 9.2(b) are casting processes. Casting also offers a great deal of flexibility in shape complexity but, because of the difficulty of producing internal undercuts, they offer somewhat less shape capabilities than molding. However, as we have learned in Chapter 3, undercuts, especially internal undercuts, should be avoided as much as possible. Some casting processes can be used to cast almost any metal (sand casting, for example) while others, such as die casting, can be used only to cast only a few metals. Because die casting has a faster cycle time than other casting processes and because it is more automated than other processes, it is the casting process of choice when aluminum, zinc or magnesium will satisfy the necessary material requirements.

The group of processes listed in Table 9.2(a) are the wrought processes of stamping, forging and extrusion. These processes offer less flexibility in our choice of shapes; however, among all manufacturing processes, stamping is the fastest. Based on material cost and the mechanical properties of metals, low carbon cold rolled steel is generally the material choice for stampings.

For each process listed in Table 9.2 a representative list of the materials which can be formed by the process, along with minimum production volumes, part sizes and a rough guideline as to the geometric shape capabilities of each process is given. Thus, based on considerations of size, production volume, and shape, a preliminary process selection (and indirectly a preliminary material selection) can be made from the list contained in Table 9.2.

In using Table 9.2, designers involved in the design of parts for consumer products should keep in mind that from a strictly economic point of view, a product is not likely to be successful unless it has a minimum production volume of 10,000 - 20,000. The most economic processes to use at

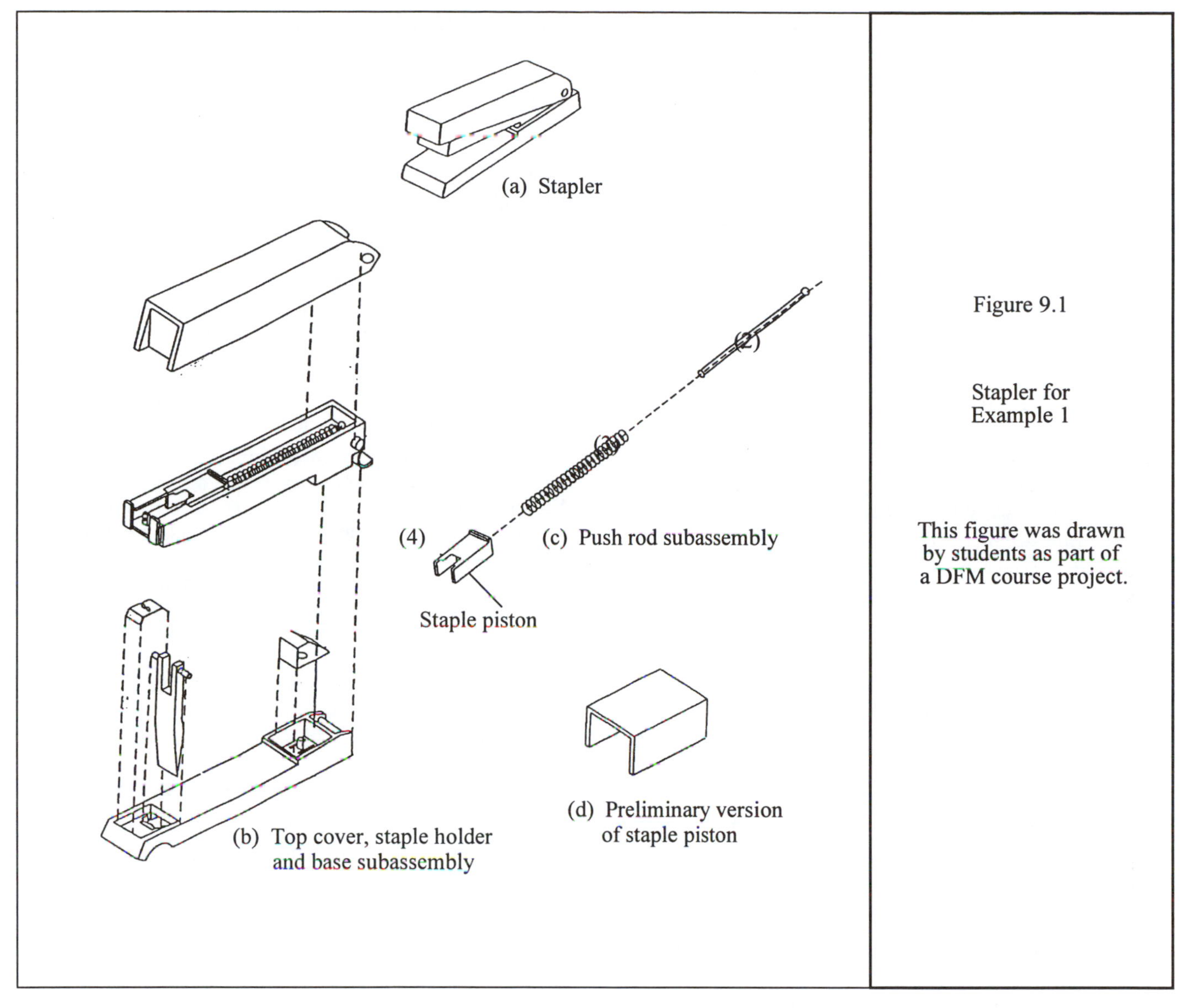

Figure 9.1

Stapler for Example 1

This figure was drawn by students as part of a DFM course project.

these production volumes are injection molding, die casting and stamping.

While forging can be used at production volumes as low as 5,000, it is usually more costly. Forgings are typically used only when a part is to be subjected to high shock and fatigue, often as parts of machines.

Thus, from a practical point of view, we will in a large majority of designs be faced with selecting from among a cold rolled steel stamping, an aluminum die casting and an engineering thermoplastic molding.

Part size can also eliminate processes from consideration. For example, due to machine size limitations, most die castings, plastic moldings and stampings are less than 450 mm to 600 mm in length. Most forgings are less than 800 mm in length. Parts larger than this are often fabricated by joining or fastening together standard shapes (sheets, tubes, beams, etc.) to form the necessary geometry.

9.7 Examples

9.7.1 Example 1 - Process First Approach

Figure 9.1a shows a drawing of a generic type of small hand held stapler of the type we might typically carry around in a brief case or purse. The maximum dimension of the stapler is less than 4 inches (100 mm). Figure 9.1b shows the top cover, base, and staple holder subassemblies of one brand of stapler on the market. For the purposes of this example, we focus on a portion of the staple holder subassembly, namely, the push rod subassembly shown in Figure 9.1(c). Various versions of this subassembly are found in all staplers since its function is to push the staples forward as they are used.

Let us assume that we are involved in the design (or redesign) of a small stapler which is to compete with the one discussed above. Let us also assume that we are at the early stage of the design, and that the materials and processes for the various components have not yet been selected. The exact geometrical shapes of the components have not been determined. For the purposes of this example we will concentrate on the preliminary selection of a process and materials combination to produce an alternative design for the staple piston shown in Figure 9.1c.

A study of various staplers will show that a wide variety of staple piston geometries exists. In general, however, at the early design stages they are all basically a simple U-shaped part similar to the one shown in Figure 9.1d. The exact geometrical configuration of this part, as well as the other components in the stapler, will depend in part on the processes and materials chosen to produce the various components. Thus, let's determine the alternative material/process combinations that could be used to produce a U-shaped part for this application.

From a study of Table 9.2, as well as our knowledge of Chapter 3, we know that the following plastic forming processes are all capable of producing a U-shaped part:

injection molding, compression molding, transfer molding, extrusion.

Also, the following metal forming processes can produce the part:

die casting, investment casting, sand casting, permanent mold casting, plaster mold casting, ceramic mold, stamping, forging and extrusion.

Let's assume that we have estimated that we will produce about 20,000 of these staplers. Let us further assume that we will need some method of connecting the push rod (part 2), and possibly the spring (part 3), to the piston (part 4). Thus we may need the facility for producing an undercut in the part. Therefore, based on production volume and the possible need to produce one or more undercuts in the part, Table 9.2 indicates that we should eliminate forging, extrusion, investment casting, sand casting, plaster mold casting and ceramic mold casting from further consideration.

We are now left with injection molding, compression molding, transfer molding, die casting, permanent mold casting, and stamping as possible processes for producing our U-shaped part. From Table 9.2 we see that using injection molding implies the use of a thermoplastic material while using compression molding or transfer molding implies that thermosets could be used. Die casting and permanent mold casting permit the use of aluminum or zinc. Stamping allows us to consider the use of wrought aluminum, steel, copper or brass.

Staplers of this type are typically used in school or office environments. Thus, they are not subjected to corrosive environments and are not subjected to high temperatures. Therefore, from a service environment point of view, there does not appear to be a need for a thermoset polymer. In addition, since we are aware (from Chapter 3) that the use of thermosets requires considerably longer processing times than thermoplastics, we eliminate the use of thermosets from further consideration. From Table 9.1 (or Table 9.2) we see that this in turn eliminates from further consideration the use of compression molding and transfer molding.

Since we are not concerned about corrosion, Tables 9A.1 and 9A.2 indicate that we can also eliminate from consideration the need to use more expensive materials such as stainless steel, brass, and copper. We are basically left, therefore, with selecting either a thermoplastic (and hence using injection molding), or aluminum or steel. The use of aluminum implies the use of either die casting or stamping, while the use of steel implies using stamping.

In summary, then, by use of Table 9.2 and our knowledge of Chapter 3, we arrive at the following possible process/material combinations for producing the staple piston:

- •Injection molding/thermoplastic,
- •Die casting or permanent mold casting/cast aluminum,
- •Stamping/wrought aluminum or wrought steel.

To reduce the above list of material/process combinations still further, other factors must be taken into consideration. Among these factors is, of course, cost. Later, in section 9.8, we will learn that for parts whose wall thickness is less than 2 mm (0.08 inches), injection molding is generally less costly than die casting. This then allows us to reduce the process selection to one of either injection molding or stamping. At this early stage in the design process it would be difficult to decide which of these two processes would result in a less costly part. However, if cost is the overriding issue, then the technique discussed in section 9.9 could be used to help decide between injection molding and stamping.

9.7.2 Example 2 - Material First Approach

Let us approach the materials/process selection for this same stapler part using a materials first approach.

Again we assume that there is no need to consider the use of materials which are particularly good at high temperatures and/or in a corrosive environment. Thus, as explained in the previous example, Tables 9A.1, 9A.2, and 9A.3 allow us to eliminate from consideration the need to use materials such as copper, brass, stainless steel and thermosets. This leaves us with thermoplastics, aluminum (cast and wrought) and steel (cast and wrought).

From Table 9.1, and our knowledge of Chapter 3, we see that using thermoplastics implies that injection molding and extrusion are capable of providing the necessary U-shape to the part. We also see that the use of aluminum implies that sand casting, investment casting, ceramic mold casting, forging, extrusion and stamping can be used. Using aluminum also allows us to add for consideration the use of die casting and permanent mold casting to our list of possible processes.

Once again, we assume a production volume in the vicinity of 20,000 and the possible need to produce one or more undercuts in order to assemble the push rod and spring to the piston. Hence, based on the information provided in Table 9.2, we eliminate from further consideration the processes of extrusion, forging, plaster mold casting and ceramic mold casting.

Thus, as in the previous example, we arrive at the same conclusion: that at this stage of the design the following material/process combinations are possible for producing a U-shaped part:

- Thermoplastics/injection molding,
- Cast aluminum/die casting or permanent mold casting,
- Wrought aluminum/stamping,
- Wrought steel/stamping.

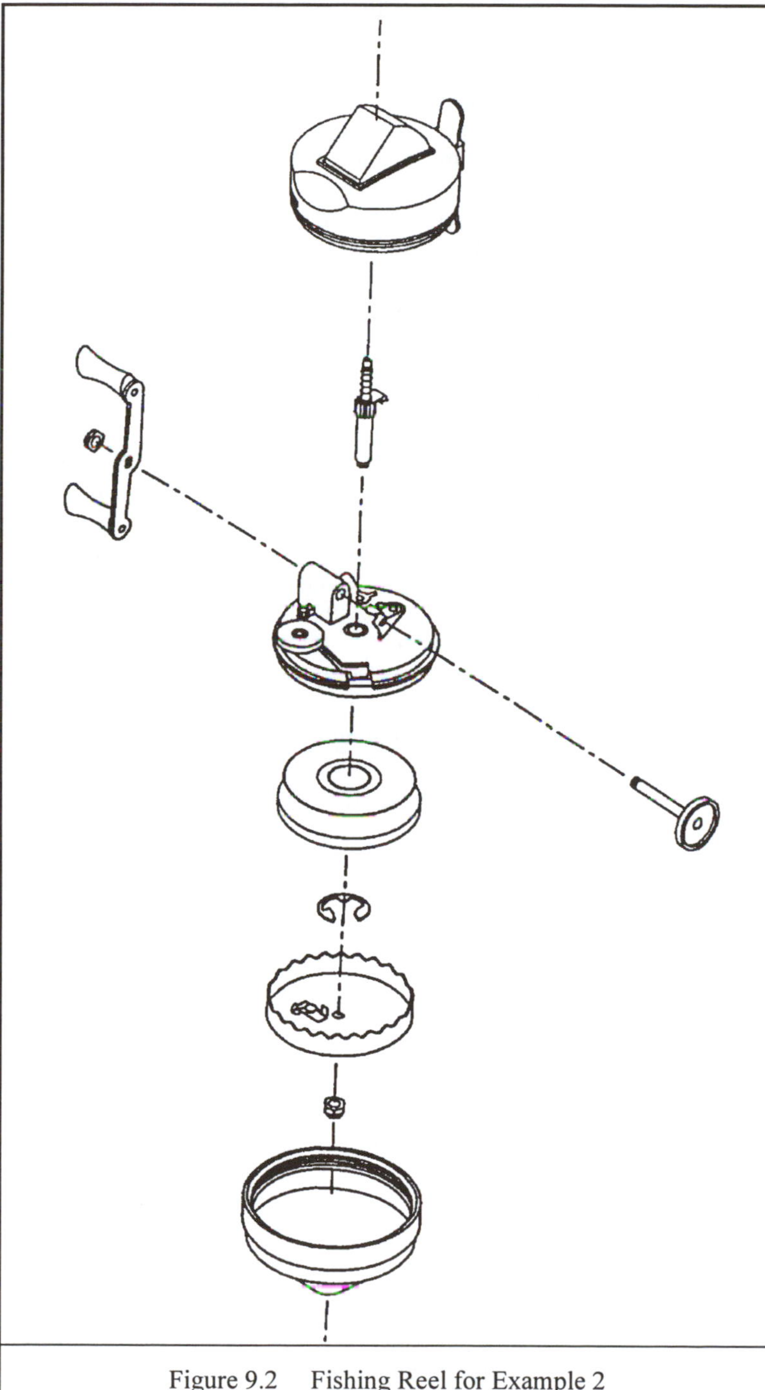

Figure 9.2 Fishing Reel for Example 2
Drawn by students for DFM course project.

9.7.3 Example 3

Figure 9.2 shows an assembly drawing of a fairly common type of fishing reel. One of the subassemblies in this reel is the crank handle which consists of a crank, two knobs, a nut, and two rivets for connecting the knobs to the crank. For the purposes of this example we will concentrate on the crank. It will be assumed that 25,000 of these reels are to be produced, and that the reel is intended for use in streams and lakes where fish are relatively small.

The forces to which the crank will be subjected are not large (certainly under 20 pounds). The material should be one that will not rust. For these reasons, it appears that the part could be made from either a plastic or a metal which does not easily rust.

In spite of the fact that injection molding may be the least costly method for producing the crank, let us assume that for marketing reasons we reject its use because of the perception by many fishermen that metal is better.

While the reel is not primarily intended for use in salt water fishing, we need to consider the possibility that it may at times be used there. Thus, the metal to be used should be one that would resist salt water corrosion. From Tables 9A.1 and 9A.2 we conclude that aluminum or stainless steel, cast or wrought, could be used. From Table 9.2, and our knowledge of Chapter 3, we see that any one of a number of casting processes could be used to produce the part from either aluminum or stainless steel. In addition, we know that stamping could also be used to produce the same geometry.

From the discussions in Chapter 3 of stamping and the various casting processes, it is easy to reason that for a production volume of 25,000, reasonably high production rates (with their lower processing costs) may be desirable. Thus, from the point of view of higher production rates, die casting and stamping are the processes of choice. Thus, it is concluded that the following material/process combinations could be used to produce the crank, namely

- Aluminum/die casting,
- Aluminum/stamping,
- Stainless steel/stamping.

As discussed earlier, other factors must be considered in order to reduce this list still further. Among these factors would be the selling cost of the reel and the material to be used for the remaining components. For example, all component parts should be capable of resisting corrosion to about the same level. Thus, to make some components of aluminum and some of stainless steel would probably not make sense. From Table 9A.2 we can also see that stainless steel is considerably more costly than aluminum. Thus, if cost is a major factor (and it usually is) then aluminum and stamping are probably the material and process of choice.

9.7.4 Example 4

As our final example, let's consider we are involved in the design of a small storage and cleaning case for an individual's contact lenses. These cases have two separate cylindrical compartments, one for each of two lenses. There is a cap for each of the compartments which screws onto the main case and is tightened by hand to prevent leakage of the neutral saline solution used for soaking the lenses during sterilization. The sterilization temperature of the case is controlled at 220 F.

Let's begin the initial materials/processes selection using a processes first approach. We know that the case must be designed as a box-shaped enclosure with two separate compartments. Thus, the case must either be fabricated by assembling individual components, or it must be produced as a one-part assembly.

As we learned in Chapter 3, to minimize assembly costs, we should design the case with a minimum number of parts. Again, from our knowledge of Chapter 3, we know that the only processes capable of producing the main body of a lens case with two cup shaped enclosures (minus the two covers) as a single part are casting, injection molding, transfer molding and compressions molding.

From Table 9.2 we see that compression molding and transfer molding, processes which are slower than injection molding, imply the use of a plastic thermoset. The temperature and corrosion environments in this case are not severe enough to warrant the need for a thermoset; thus, we reject the use of compression molding and transfer molding. We are left with deciding between injection molding a thermoplastic material or casting a metal.

Because of the need to resist corrosion by saline solution, Tables 9A.1 and 9A.2 indicate that aluminum or stainless steel would be the materials of choice. If this case is to sell for something in the order of a couple of dollars, then with a comparison of the material costs (shown in Tables 9A.1-9A.3) for aluminum, stainless steel and engineering thermoplastics, we can eliminate stainless steel from further consideration. Thus, we once again arrive at the choice between an aluminum die casting or an injection molded thermoplastic.

Once again, based on information provided in Section 9.8, it is easy to conclude that, based on price, injection molding is the process of choice. Consequently, a thermoplastic is the material of choice.

The choice of which particular thermoplastic to use must be based on additional considerations such as:

- The impact strength required. In this case, normal impact strength is needed. That is, normal handling, occasional dropping, but no sharp or heavy blows.
- The material will come into contact with moisture/ steam at 220 F.
- With the caps screwed on for eight to ten hours at 220 F, creep must be considered a factor.
- The flexural modulus is somewhat important to keep the threads from deforming, but again there are no large external loads to consider.
- Must have FDA approval the same as food additives.

The information and data provided in Table 9A.3 of this chapter is insufficient to make a selection of a specific thermoplastic to satisfy all of these requirements. Table 9A.3 is simply a sampling of a few thermoplastics with only some of their properties. To select the specific material to satisfy all of the functional requirements, we must refer to a reference such as [7].

9.8 Injection Molding vs. Die Casting

9.8.1 Introduction

The examples discussed in the previous section indicated that we are often faced with choosing between (i) a thermoplastic injection molded part, or (ii) a geometrically similar die cast aluminum part. The purpose of this section is provide additional information, based solely on cost, to assist in the selection between two such parts. It is assumed here that the size, shape, wall thickness, and dimensions of subsidiary features (such as holes) are identical.

In the analysis that follows, the subscript 'a' will refer to an aluminum die casting while the subscript 'p' will refer to a plastic injection molding.

9.8.2 Relative Cost Model

We define ΔK_{ap} as follows:

ΔK_{ap} = [Cost to produce a part as an aluminum die casting]

— [Cost to produce the same part as a thermoplastic injection molding] (9.1)

or, symbolically,

$$\Delta K_{ap} = K_{ta} - K_{tp}$$
$$= (K_{ma} + K_{da}/N + K_{ea}) - (K_{mp} + K_{dp}/N + K_{ep}) \quad (9.2)$$

where:

K_{ta} = the total cost to produce the part as an aluminum die casting;

K_{tp} = the total cost to produce the part as a thermoplastic injection molding;

K_{ma}, K_{mp} = material cost of the part;

K_{da}, K_{dp} = tool or die cost required;

N = production volume; and

K_{ea}, K_{ep} = processing cost for the part.

If ΔK_{ap} is greater than zero, then it is more economical to produce the part as an injection molded part.

From Chapter 3, we can see from a study of the IM Advisor (Figure 3.14) and the Die Casting Advisor (Figure 3.15) that the tooling costs for moldings and die castings are essentially the same. Thus, Equation (9.2) reduces to:

$$\Delta K_{ap} = (K_{ma} + K_{ea}) - (K_{mp} + K_{ep}) \tag{9.3}$$

Now define

$$\Delta K_{m} = K_{ma} - K_{mp} \tag{9.4}$$

Hence, Equation (9.3) becomes

$$\Delta K_{ap} = \Delta K_{m} + (K_{ea} - K_{ep}) \tag{9.5}$$

Unfortunately, metal prices published in the open literature are sparse, and not as reliable as plastic prices. In general, because metal prices depend on a multitude of factors, true metal prices can only be obtained by direct quotation from vendors when the metals are needed. In spite of this, however, from the data provided in Tables 9A.1 - 9A.3 it is seen that the cost per unit volume for an unreinforced engineering thermoplastic is less than that of aluminum. It is also seen that the cost of reinforced plastics are in the same price range as aluminum. Thus, we can at least say that ΔK_{m} is generally a positive quantity.

Now Equation 9.5 shows that if the processing cost for injection molding is less than the processing cost for die casting; that is:

If

$$K_{ep} < K_{ea} \tag{9.6}$$

then ΔK_{ap} is certainly positive and it is less costly to injection mold than to die cast.

We now introduce two reference parts, one for injection molding and one for die casting. We will use these reference parts to compare processing times (see Figures 3.14 and 3.15) and hence processing costs for alternative competing designs. Let K_{epo} and K_{eao} represent the processing cost for the aluminum die cast reference part and the injection molded reference part, respectively.

If

$$C_{ep} = K_{ep}/K_{epo} \tag{9.7}$$

and

$$C_{ea} = K_{ea}/K_{eao} \tag{9.8}$$

where C_{ep} and C_{ea} represent the cost of a part relative to the reference part, then Equation 9.6 gives:

$$C_{ep}K_{epo}/C_{ea}K_{eao} < 1 \tag{9.9}$$

Hence, when Equation (9.9) is satisfied, ΔK_{ap} is positive and it is more economical to injection mold than to die cast.

Based on data provided in Chapter 15 (Tables 15.1 and 15.9) which are reproduced below as Tables 9.6 and 9.7, it can be shown that:

$$K_{epo}/K_{eao} = (16)(27.53)/(17.23)(62.57) = 1/2.5$$

Thus, Equation (9.9) reduces to the following condition:

When

$$C_{ep}/C_{ea} < 2.5 \tag{9.10}$$

then it becomes more economical to mold than to die cast.

That is, when the relative processing cost for the injection molded version of the design is less that 2.5 times the relative processing cost for the die casting version of the design, then it is more economical to injection mold.

The relative processing cost for a part can be expressed as the product of the cycle time for the part relative to the cycle time for the reference part, t_r, and the machine hourly rate for the machine required to process the part to the machine hourly rate for the machine required to process the reference part, C_{hr}. That is

$$C_{e} = t_{r}C_{hr} \tag{9.11}$$

Substituting Equation (9.11) into Equation (9.10) gives

$$(C_{hrp}/C_{hra})(t_{rp}/t_{ra}) < 2.5 \tag{9.12}$$

In Chapter 15 (section 15.15) we also show that machine hourly rates and, consequently, relative machine hourly rates C_{hrp} and C_{hra} are functions of the machine tonnage required to mold or die cast the part. The machine tonnage, in turn, is a function of the projected area of the part normal to the direction of die closure. In both injection molding and die casting, the tonnage required is approximately 3 tons per square inch. Thus, comparable parts require comparable sized machines. The relative hourly rates for the two machines, however, are not equal. Table 9.8 shows how C_{hrp} and C_{hra} vary with machine tonnage.

The IM Advisor and the Die Casting Advisor of Chapter 3 (Figure 3.14 and 3.15) can be used to estimate the relative cycle times for injection molded parts, t_{rp}, and for die castings, t_{ra}. A more accurate method for estimating these relative cycle times can be found in Chapter 15.

Table 9.6 Relevant Data for the Injection Molded Reference Part

Part Material	Polystyrene
Material Cost, K_{po}	1.46×10^{-4} c/mm^{3}(1)
Part Vol, V_o	1244 mm^3
Cycle time, t_o	16 s (2)
Mold Machine Hourly Rate, C_{ho}	\$27.53(3)

(1) *Plastic Technology,* June 1989, (2) Data from collaborating companies, (3) *Plastic Technology,* July 1989

Table 9.7 Relevant Data for the Die Cast Reference Part

Part Material	Aluminum
Material Cost, K_{po}	0.0006 cents/mm^3
Part Vol, V_o	1885mm^3
Cycle time, t_o	17.23 s
Mold Machine Hourly Rate, C_{ho}	\$62.57 /h

9.8.3 Results

In Chapter 15 we will be introduced to the concept of partitionable and non-partitionable parts. Very briefly, partitionable parts are those parts that can be easily and completely divided (except for add-ons like bosses, ribs, etc.) into a series of simple plates (see Figure 15.8 for example). Non-partitionable parts are those parts whose complex geometries make it difficult to partition into a series of simple plates.

Figure 15.6 of Chapter 15 shows an example of a non-partitionable part. Partitionable parts have exterior plates and interior plates. Simply stated, the exterior plates of a part are its peripheral side walls whereas interior plates are non-peripheral side walls and the base of the part. Methods for partitioning parts and classifying plates are discussed in detail in Chapter 15. For now we concentrate only on the results relevant to comparing costs.

Using the data contained in Chapters 3 and 15, along with the data provided in Table 9.8, the plots shown in Figures 9.4 through 9.8 were obtained. In obtaining these results, it is assumed that the plate controlling the cycle time is always an external plate, and that tolerances are easy to hold. Since internal walls do not play a major role in determining the cycle time for injection molded parts, but do often control the cycle time for die castings, the results depicted in Figures 9.4 - 9.8 are, from the point of view of injection molding, conservative. In addition, the effect of part surface quality or surface finish is ignored. This is done since it is difficult to compare moldings and die castings based on part surface finish requirements. Also, since engineering thermoplastics can not generally be used for parts whose wall thickness is greater than 5 mm, the results shown in these figures are restricted to wall thickness less than 5 mm. In general, when the wall thickness is greater than 5mm, one must use either foamed materials or metals.

Table 9.8 Machine Tonnage and Relative Hourly Rate

	Injection Molding (1)	Die Casting (2)
Machine Tonnage	Relative Hourly Rate, C_{hrp}	Relative Hourly Rate, C_{hra}
< 100	1.00	1.00
100-299	1.19	1.11
300-499	1.44	1.25
500-699	1.83	1.47
700-999	2.87	2.02
> 999	2.93	2.11

(1) Data published in *Plastic Technology*, June, 1989,
(2) Data obtained from die casting vendors, June, 1989

We see from an examination of Figures 9.4 - 9.8, that for parts whose cycle time is controlled by relatively thin plates (wall thickness less than 2 mm), it is always less costly to mold the part. As the size of the part increases (i.e., as C_{hrp}/C_{hra} increases) and the wall thickness increases, then the certainty that injection molding is less costly becomes less.

For values of $(C_{hrp}/C_{hra})(t_{rp}/t_{ra}) > 2.5$

it is still not necessarily true that an injection molded part is more costly. In this situation, Equation (9.5) must be used in place of Equation (9.12) in order to determine which of the two processes is more economical. In this case, the exact cost of the material must be known and, for the die cast version, the nature (internal or external) of the plate controlling the cycle time must be determined

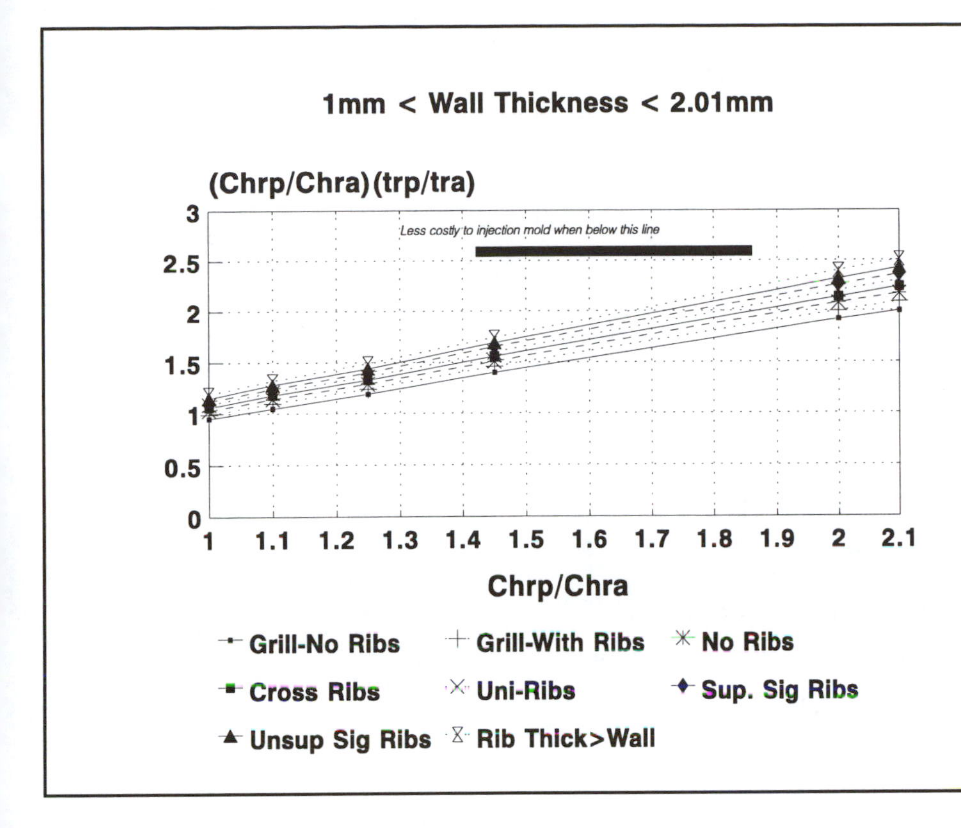

Figure 9.3

Partitionable Parts

1 mm < Wall Thickness < 2.01 mm

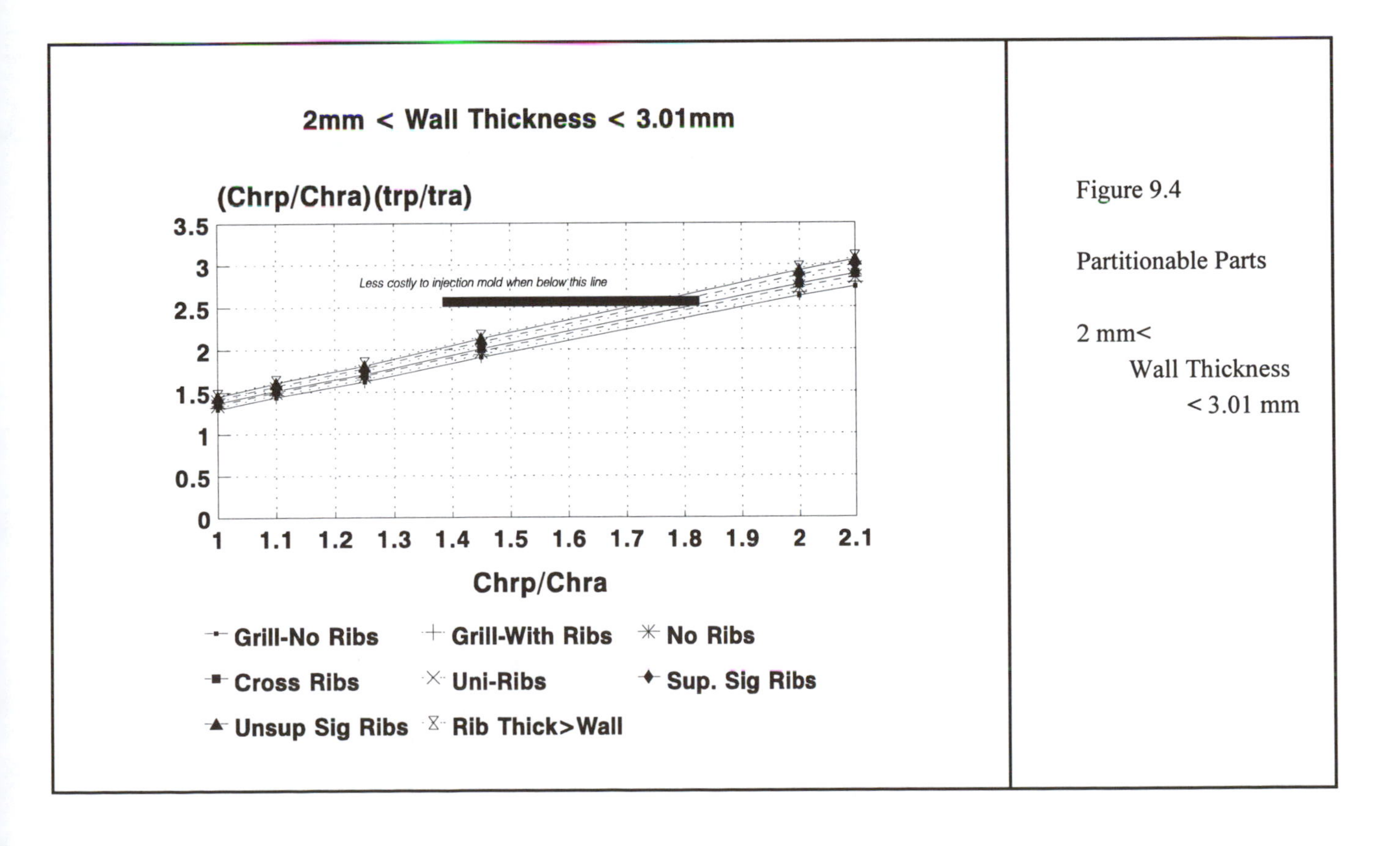

Figure 9.4

Partitionable Parts

2 mm< Wall Thickness < 3.01 mm

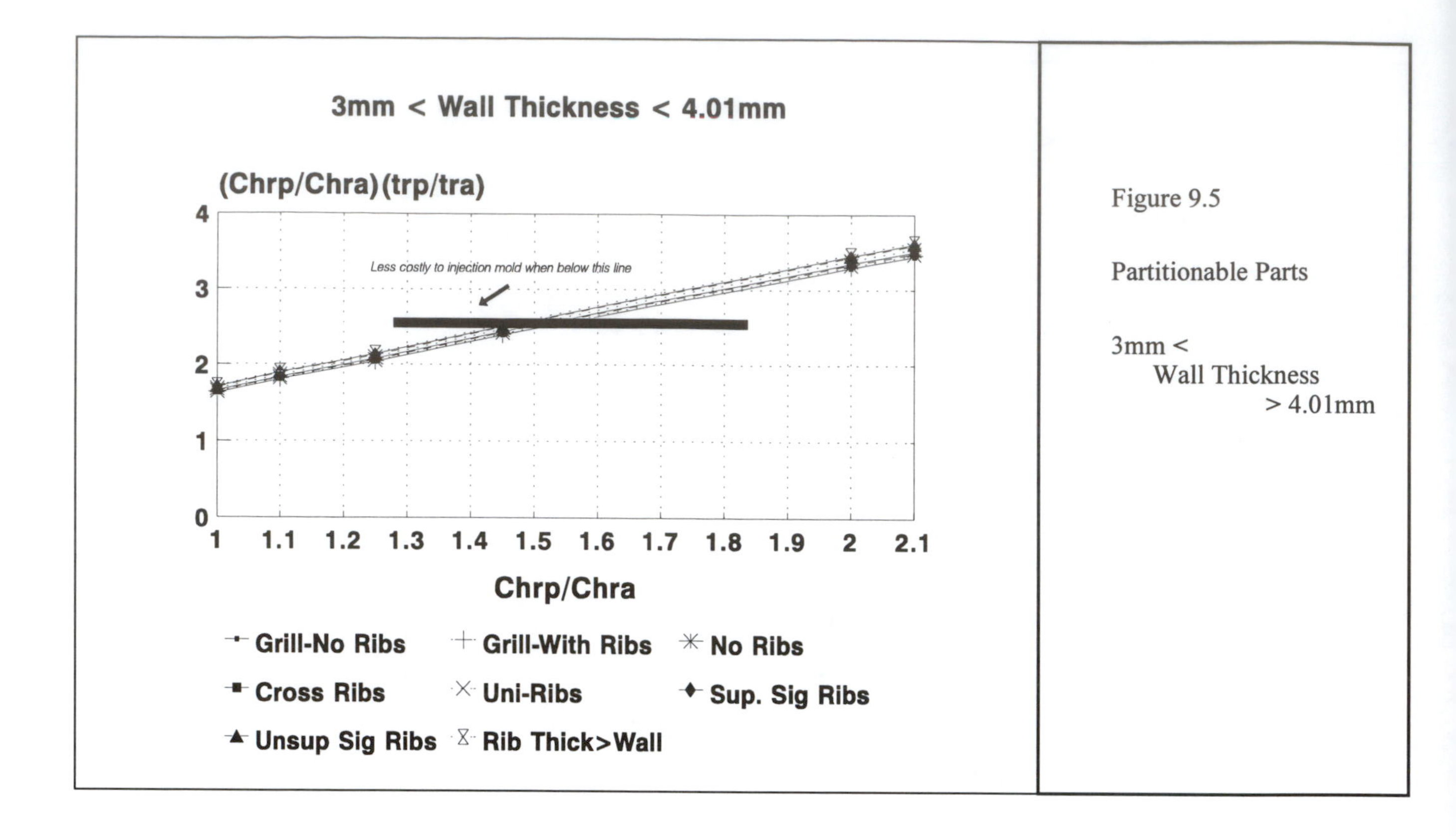

Figure 9.5

Partitionable Parts

3mm < Wall Thickness > 4.01mm

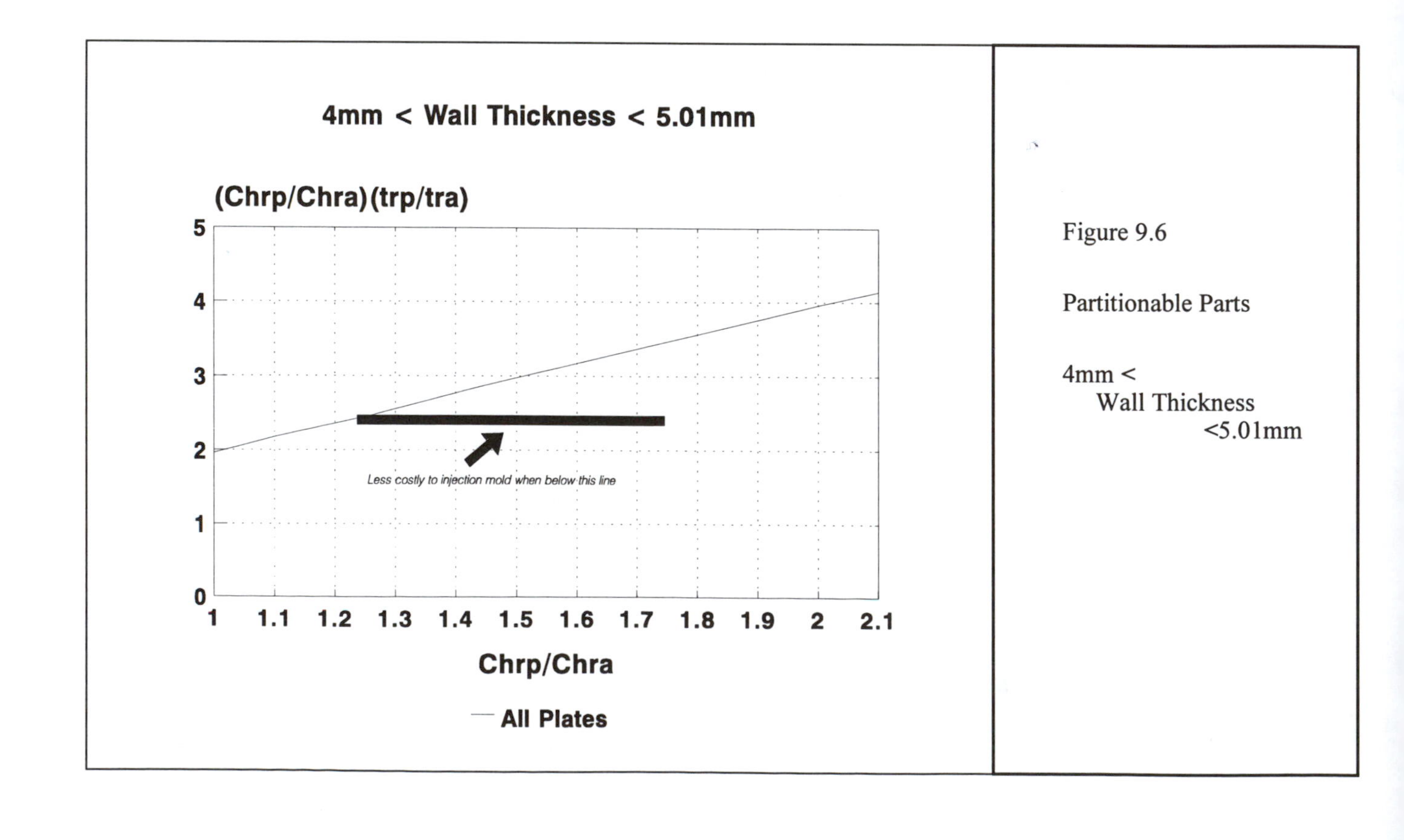

Figure 9.6

Partitionable Parts

4mm < Wall Thickness <5.01mm

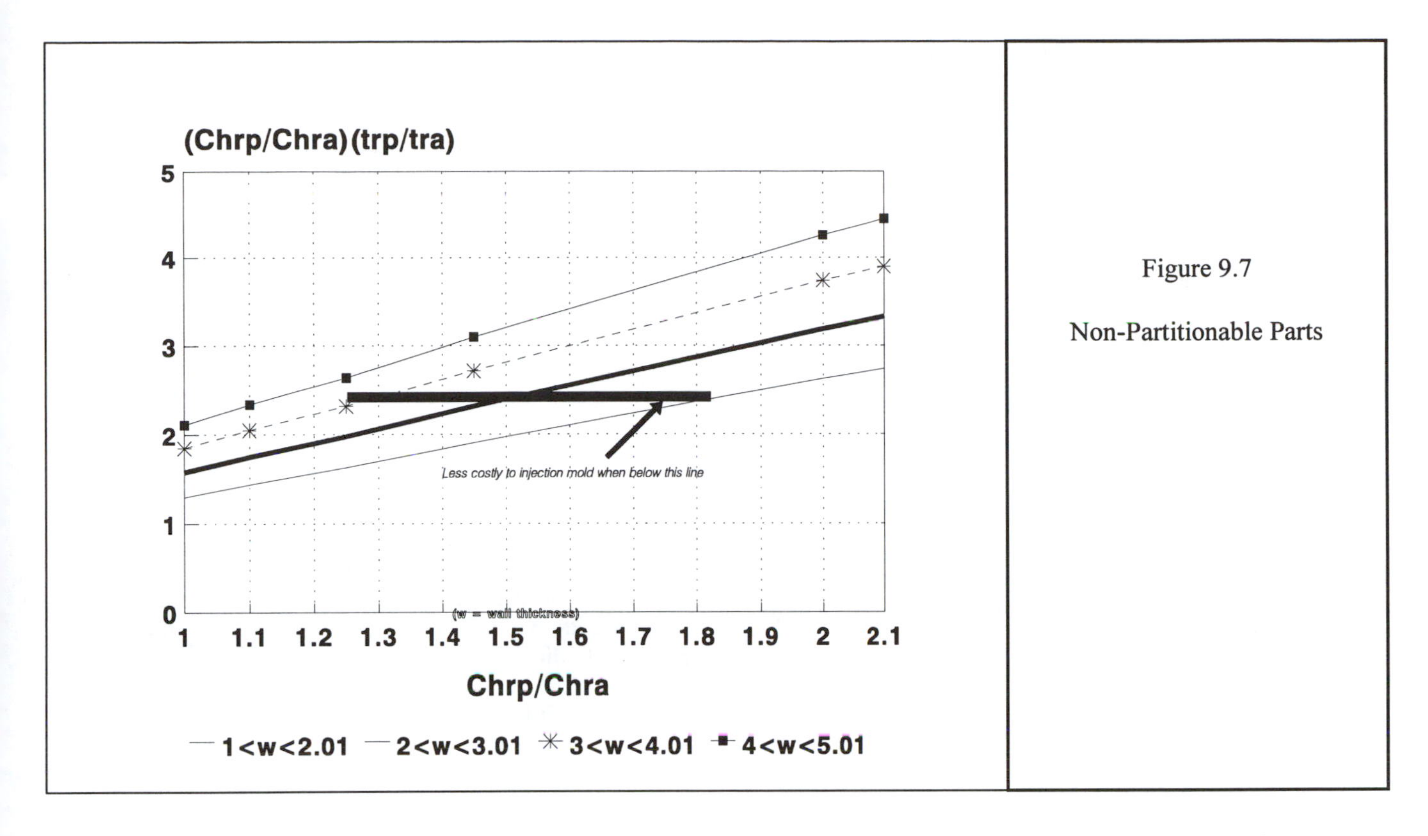

Figure 9.7

Non-Partitionable Parts

9.8.4 Example

Consider the crank shown in Figure 9.2. In section 9.7.3 the possibility of injection molding this crank was rejected for marketing reasons. We assume here we wish to re-examine this issue, but strictly from the point of view of manufacturing cost. It is assumed that functionally a crank with a wall thickness of 3.5 mm will suffice.

The crank is a flat part with a flat parting surface between the two mold halves. Thus, from Figures 3.14 and 3.15 we find that for a wall thickness of 3.5 mm, $t_{rp} = 2.4$ and $t_{ra} = 1.9$.

A “ball park” figure for the projected area of the crank normal to the direction of mold closure is 1875 mm^2 or 3 in^2. Thus, the tonnage required is about (recall that about 3 tons per square inch is required to mold or cast a part)

$$F = 3A_p = 3(3) = 9 \text{ tons}$$

Hence, from Table 9.8 we see that the relative machine hourly rate for both versions of the design are 1.0. Hence,

$$C_{hrp}/C_{hra} = 1; \quad (C_{hrp}/C_{hra})(t_{rp}/t_{ra}) = 1.3$$

and from Figure 9.6 we conclude that it is more economical to mold this part from plastic than to die cast it from aluminum.

9.9 Injection Molding vs. Stamping

9.9.1 Introduction

In the previous section we determined, for the case of two geometrically identical parts, when it would be more economical to mold than to die cast. The purpose of this section is to discuss, for two 'functionally' equivalent geometries, when it is more economical to injection mold a part and when it is more economical to stamp a part that has a functionally equivalent geometry. We use the phrase *functionally equivalent* here because it is seldom possible to stamp a part that has exactly the same geometry as an injection molded part. The differences in the physics of the two processes make certain geometries more difficult to produce in stamping than in injection molding. For example, it is simple to form an injection molded box-shaped part with each of the walls connected to the adjoining wall. In stamping, while a box shaped part could be drawn to the same geometry, it would be easier to *wipe form* the walls as discussed in Chapter 3. In this situation the adjoining walls would not be connected to each other; they would, however, be connected to the base. Thus, when a box shaped part whose walls are not connected will satisfy the functional requirements, then we can think of these two parts as being 'functionally' equivalent.

9.9.2 Relative Cost Model

In this section, the subscript 's' will be used to indicate stamping and the subscript 'p' will indicate injection molding. A stamping will be less costly to produce if ΔK_{ps}, the difference in cost between an injection molded a part and a functionally equivalent stamped part, is greater than zero; that is, if

$$\Delta K_{ps} = (K_{mp} + K_{dp}/N + K_{ep}) - (K_{ms} + K_{ds}/N + K_{es}) \quad (9.13)$$

is a positive quantity.

9.9.3 Case 1 - Stamping More Economical

Processing Costs

We argue here that processing costs for stampings are always less than processing costs for injection molding, assuming of course that the parts are more or less functionally equivalent. The logic is as follows.

As pointed out in Chapter 3, the slowest stamping press is capable of delivering about 24 strokes per minute. Since one part is produced with each stroke, a part is produced about every 2.5 seconds. One way to look at this, therefore, is to say that the cycle time for a stamping on a slow press is 2.5 seconds. (More typically, however, stamping presses can deliver 90 strokes per minute which results in a cycle time of 0.67 seconds.)

Also from Chapter 3, the fastest cycle time for an injection molded part is about 10-15 seconds. More typically, however, the cycle time is about 30-50 seconds. Thus, the fastest cycle time for an injection molded part is at least 4 to 5 times greater than that of the slowest stamping press.

The machine hourly rate for a large stamping press which operates at about 24 strokes per minutes is about $65 per hour. (The hourly rate for a more typical 100 ton press which operates at about 90 strokes per minute is about $30 per hour.) Multiplying the hourly rate by the cycle time shows that the cost per stamping produced on the 24 stroke per minute machine is about 4.5 cents/part.

The hourly rate for the smallest injection molding machine is about $27 per hour. Thus, if we assume a fast cycle time of 10 seconds, then the processing cost per part is 7.5 cents/part. Thus, processing costs for stampings are always less than processing costs for injection molding. That is,

$$\Delta K_e = K_{ep} - K_{es} > 0 \quad (9.14)$$

and $$\Delta K_{ps} = (K_{mp} + K_{dp}/N) + \Delta K_e - (K_{ms} + K_{ds}/N) \quad (9.15)$$

hence, if

$$K_{ms} + K_{ds}/N < K_{mp} + K_{dp}/N \quad (9.16)$$

then ΔK_{ps} is a positive quantity, and it is less costly to stamp.

Material Cost

As indicated earlier, the prices of metals are somewhat difficult to obtain. However, the end user price (i.e., the cost to a stamper who buys in quantities of 10,000 lb or more) for drawing quality carbon steel sheet, C_{cs}, is about 8.8 cents/in^3 to 12 cents/in^3. The end user prices for 6061 aluminum, C_{al}, and 304 stainless steel, C_{ss}, are about 20 cents/in^3 and 48 cents/in^3, respectively.

Comparing these prices with those for engineering thermoplastics (C_p) given in Table 9A.3 it is seen that (particularly for fiber reinforced thermoplastics)

$$C_p = C_{cs} \quad (9.17)$$

$$2C_p = C_{al} \quad (9.18)$$

$$4C_p = C_{ss} \quad (9.19)$$

In general, the sheet thickness used in stampings is about one-half the wall thickness size of a comparable molding. Thus, the volume of material required for a stamping is about half the volume of plastic material required for a comparable molding. Thus,

$$K_{mp} = C_p V_p = C_p(2V_s) \quad (9.20)$$

and $$K_{ms} = C_m V_s \quad (9.21)$$

where C_m is the price per unit volume for metal.

Carbon Steel vs. Plastic. The material cost for a part molded from an engineering thermoplastic is given by Equation (9.20), while the material part cost for a part stamped from carbon steel is given by the expression

$$K_{ms} = C_{cs} V_s \quad (9.22)$$

Since $C_p = C_{cs}$, then comparing equations (9.20) and (9.22) gives $$K_{mp} = 2K_{ms} \quad (9.21)$$

Thus, if $$K_{ds} < K_{dp} \quad (9.24)$$

then ΔK_{ps} is positive and it is less costly to produce the part as a stamping. If Equation (9.24) is not satisfied, this does not imply that a stamping is necessarily more costly; it simply means that a more detailed comparison of costs must be carried in order to determine if ΔK_{ps} is positive.

Aluminum vs. Plastic. Once again it is assumed that the amount of aluminum required to produce a stamping is about half the plastic material required to produce an equivalent molding. Thus, in general,

$$2C_p = C_a \quad (9.18)$$

Hence, $$K_{mp} = V_p C_p = 2V_s(C_{al}/2) = K_{ms}$$

Thus, if $K_{ds} < K_{dp}$ (9.25)

then ΔK_{ps} is positive and an aluminum stamped part is less costly to produce than a molded part.

Stainless Steel vs. Plastic. Similarly,

$$K_{mp} = V_p C_p = 2V_s(C_{ss}/4) = K_{ms}/2$$

Thus if $$K_{ds} < K_{dp} - K_{ms}/2N \qquad (9.26)$$

then ΔK_{ps} is a positive value and, once again, a stainless steel stamped part is less costly to produce than a molded part.

9.9.4 Case 2 - Injection Molding More Economical

Processing Cost

As indicated above, the processing costs for a stamped part are considerably less than those for an injection molded part. Thus, we will ignore it here. We define ΔK_{sp} as the difference in cost to stamp a part and to injection mold a functionally equivalent part. Then if:

$$\Delta K_{sp} = (K_{ms} + K_{ds}/N) - (K_{mp} + K_{dp}/N + K_{ep}) \qquad (9.27)$$

is positive, a molded part is less costly to produce than a comparable stamped part.

Material Cost

Carbon Steel vs. Plastic. From Equation (9.23) the part material cost for a molded part is about twice the part material cost for a carbon steel stamped part. Thus, Equation (9.18) indicates that if

$$K_{dp} + N(K_{ep} + K_{mp}/2) < K_{ds} \qquad (9.28)$$

then ΔK_{sp} is positive and molding a part is less costly.

Aluminum vs. Plastic. In this case, as shown above:

$$K_{mp} = K_{ms}$$

Thus, If $$K_{dp} + NK_{ep} < K_{ds} \qquad (9.29)$$

Then ΔK_{sp} is greater than zero, and once again molding a part is less costly.

Stainless Steel vs. Plastic. In this case

$$K_{mp} = K_{ms}/2$$

Consequently,

If $$K_{dp} + N(K_{ep} - K_{mp}) < K_{ds} \qquad (9.30)$$

So ΔK_{sp} is positive and molding a part is less costly.

9.9.5 Tooling Cost

Use of Equations. (9.24) to (9.30) requires an estimate of the absolute (not relative) tooling cost required to produce the part and an estimate of the material cost. Unfortunately, while the approximate data provided in Chapter 3 and the more precise data base provided in Chapters 11 and 12 are sufficiently precise to allow a cost comparison between alternative designs *within* a given manufacturing process, they are not sufficiently precise to permit a comparison of alternative designs *between* two different processes. The reason is that, in using the group technology approach of Chapters 11 and 12, each cell of the matrices contains groups of parts which are similar but not identical. The relative tooling cost data provided are for a representative part which would fall within that cell. The absolute tooling costs for all parts which would fall within that cell, however, are slightly different.

A more precise technique for determining whether or not a part is cheaper to mold or to stamp is currently under development at the University of Massachusetts Amherst. Preliminary results indicate that as the number of features (add-ons) on a part increases and as the number of bend stages required to form the stamped part increases, the more likely it is that the part will be less costly to mold.

For example, consider the parts (a link and a bent link) shown in Figure 9.8. The injection molding tooling costs for these two parts is not drastically different (Figure 9.9). However, for stamping, as the part increases in complexity (Figure 9.10a being the least complex and Figure 9.10b the most complex), the number of stations required to stamp the part increases and consequently the size and complexity of the tooling also increases (Figure 9.11). The result is that tooling costs increase significantly. Preliminary findings indicate that while it may be less costly to stamp the design shown in Figure 9.9a, as the part increases in complexity (Figure 9.9b) it is more likely that these designs are more economical to mold.

(a) Straight link

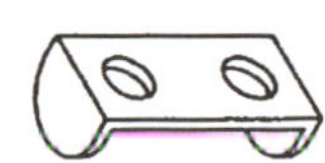

(b) Bent link

Figure 9.8

Straight and Bent Links

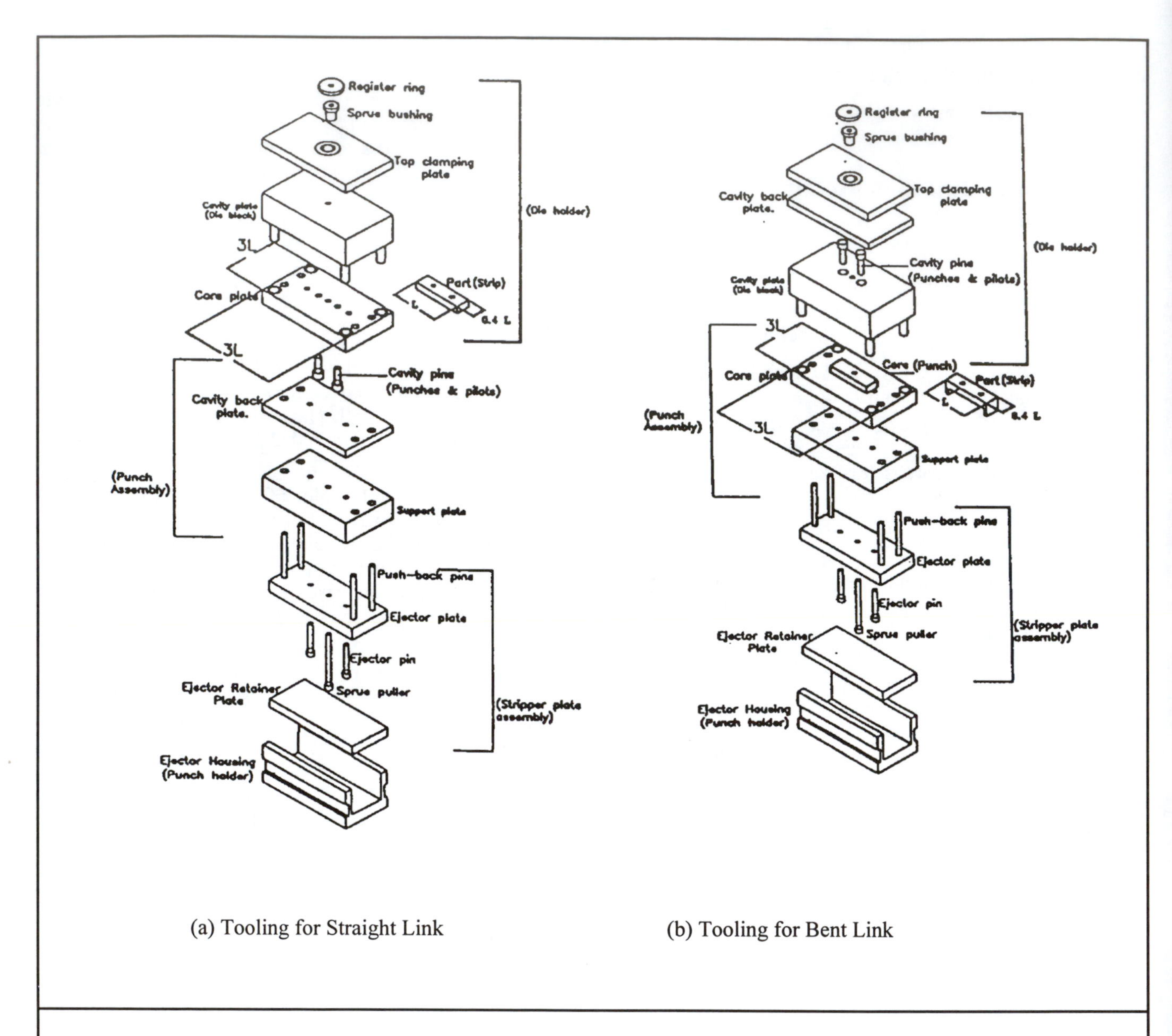

(a) Tooling for Straight Link

(b) Tooling for Bent Link

Figure 9.9 Tooling for Injection Molded Link

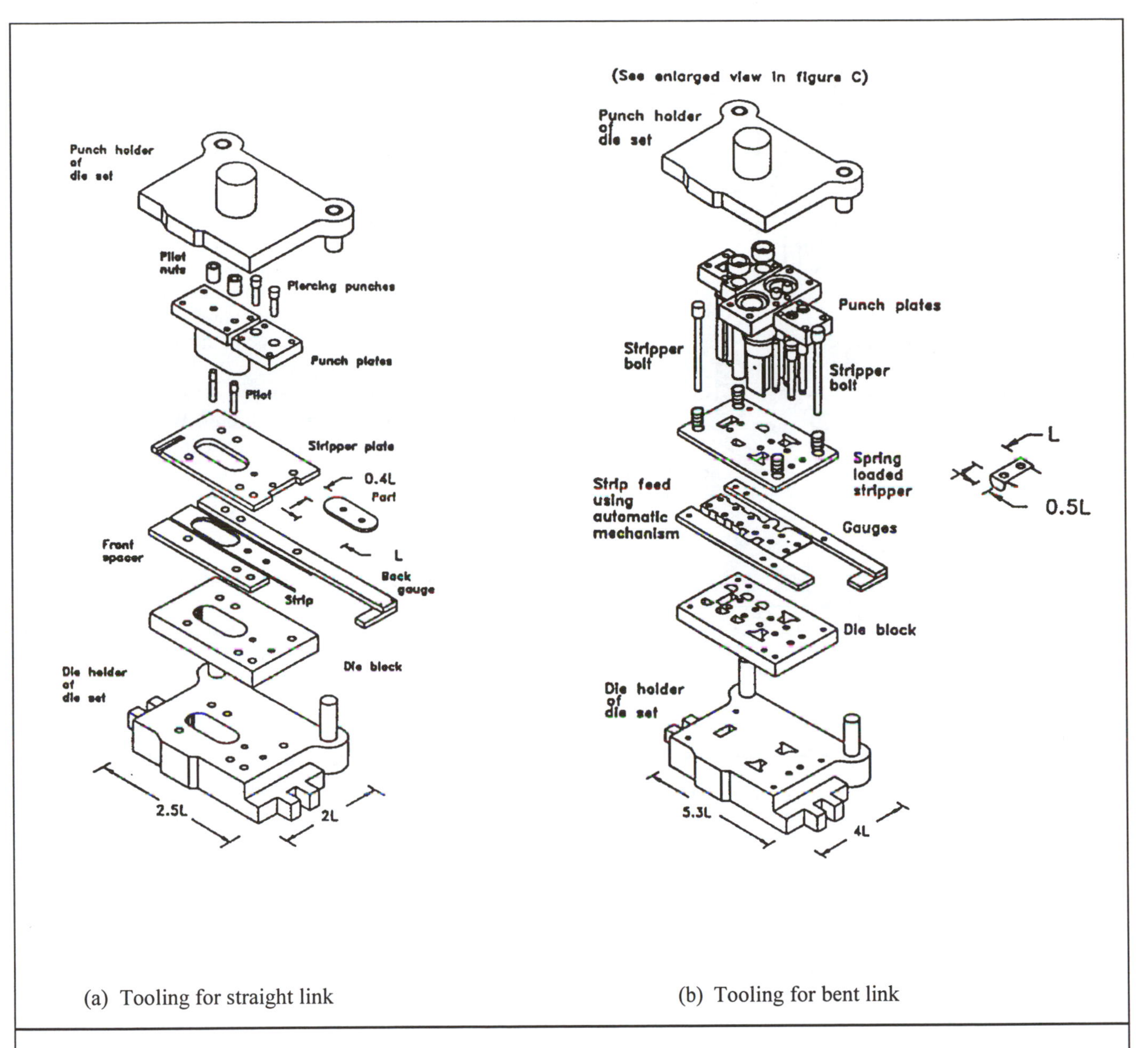

(a) Tooling for straight link

(b) Tooling for bent link

Figure 9.10 (a) and (b) Tooling for Stamped Link

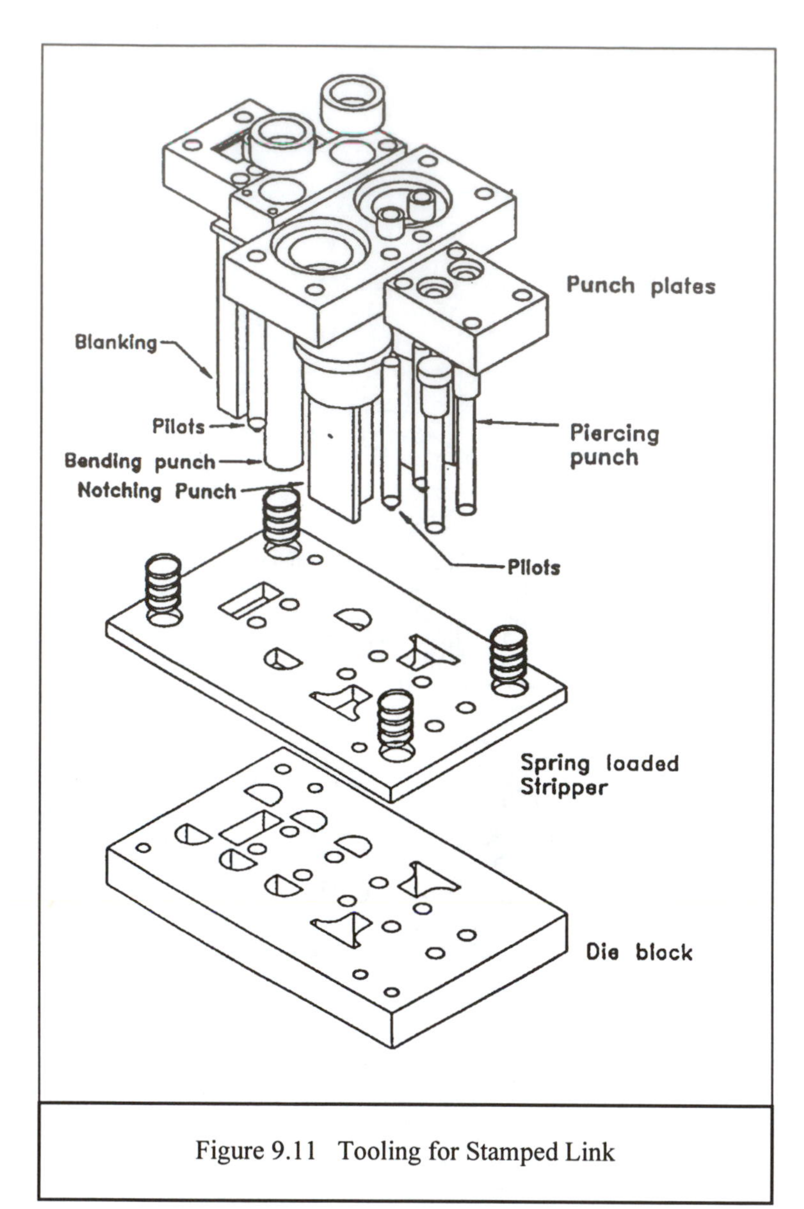

Figure 9.11 Tooling for Stamped Link

to be made. *Then* materials consistent with the process and the part's application are considered.

Designers can use either approach depending on which is most convenient or natural to the specific part being designed. Both approaches end up at the same point. As always, designers are strongly urged to consult with materials and manufacturing experts about material-process selections before considering the selection 'final'.

It is not always possible to reduce the choice of material/process combination to a single alternative. It is thus often necessary to move on to configuration design with two or three possibilities. This will usually necessitate designing configurations for each of the material-process combinations since different materials and processes may well lead to different optimal configurations. Of course, there will be common features, but configuration design must be done with a material-process combination (at least to Level III) in mind.

Preview. This completes Part Two, the discussion of conceptual engineering design. We turn now to the use of guided iteration for the configuration design of parts.

9.10 Summary and Preview

Summary. In this chapter, we presented a general methodology for selecting one or more material-process combinations for special purpose parts at the conceptual stage. Two approaches were presented for determining candidate material-process combinations for a part, namely, (a) material-first; or (b) process-first.

In the material-first approach, designers begin by selecting a material class guided by the requirements of the application. *Then* processes consistent with the selected material are considered and evaluated guided by production volume and information about the size, shape, and complexity of the part to be made.

In the process-first approach, designers begin by selecting the process guided by production volume and information about the size, shape, and complexity of the part

References

[1] Ashby, M. F., *Materials Selection in Mechanical Design,* Pergamon Press, New York, 1992.

[2] *Metals Handbook, 9th Edition - Volume 1, Properties and Selection: Iron and Steels*, American Society of Metals, Metals Park, OH, 1978.

[3] *Metals Handbook, 9th Edition - Volume 2, Properties and Selection: Nonferrous Alloys and Pure Metals*, American Society of Metals, Metals Park, OH, 1979.

[4] *Metals Handbook, 9th Edition - Volume 3, Properties and Selection: Stainless Steels, Tool Materials and Special-Purpose Metals*, American Society of Metals, Metals Park, OH, 1980.

[5] *Modern Plastic Encyclopedia '91*, Modern Plastics, Hightstown, NJ, October 1990.

[6] *Manufacturing Handbook and Buyers' Guide, 1991-1992*, Plastics Technology, New York, NY, 1991.

[7] "Engineered Materials Handbook, Vol. 2, Engineering Plastics," American Society of Metals, Metals Park, OH, 1988.

[8] Ashby, M. F. and Jones, D. R. H., *Engineering Materials 1, An Introduction to their Properties and Application,* Pergamon Press, Oxford, England, 1980.

[9] Kalpakjian, Serope, *Manufacturing Engineering and Technology*, 2nd ed. Addison-Wesley Publishing Co., 1992.

[10] Dieter, George. *Engineering Design, A Materials and Processing Approach*, McGraw-Hill Book Company, New

Problems

9.1 Assume that the caster shown in Figure P9.1 is at the early configurational design stage and that neither the part material nor the processes to be used to produce the component parts has been decided upon. Assume that the caster is to be used on dollies utilized in moving "heavy" boxes being unloaded from ships docked in various harbors around the world. For an estimated production volume of 50,000:

a) Which material/process combinations would you consider as possibilities for producing the brackets?

b) Which material/process combinations would you consider as possibilities for producing the plate?

Note: In deciding on alternative material/process combinations for the parts, you should allow the geometry of the parts to vary slightly from that shown in Figure P9.1 in order to take advantage of alternative processes.

9.2 In an effort to reduce assembly costs, a suggestion has been made to combine the two brackets and the plate shown in Figure P9.1 into a single part. Which material/process combinations would you consider a possibilities for producing the redesigned part?

Note: In deciding on alternative material/process combinations, you should allow the geometry of the part to vary slightly from that shown in Figure P9.1 in order to take advantage of alternative processes.

9.3 Figure P9.3 shows the proposed design for the spinner head being utilized as part of the spinning reel shown in Figure 9.2. 20,000 reels will be produced. What material/process combinations would you consider as possibilities for the part? *Note*: The diameter of the head is about 60 mm and the length is about 15 mm.

9.4 Assume that you are designing a small kitchen scale (maximum weight to be measured is about 2 pounds). Assume that you are at the stage of considering alternative material/process combinations for the 'food carrier' and the "food carrier support" as shown in Figure P9.4.If 50,000 of these scales will be built what material/process combinations would you consider as possibilities for the two parts? The dimensions of the food carrier are about 4 inches by 2 inches by 0.5 inches.

9.5 Repeat problem 4 for a food carrier and support produced as a single part.

9.6 Tables 9A.1, 9A.2 and 9A.3 give a representative sample of materials with some of their mechanical and physical properties.

a) In spite of the fact that a product would be subjected to a corrosive environment, can you list applications where you might select a high density corrosive material over a low density "non-corrosive" material?

b) Can you suggest applications when a low density material would be a better choice than a high density material?

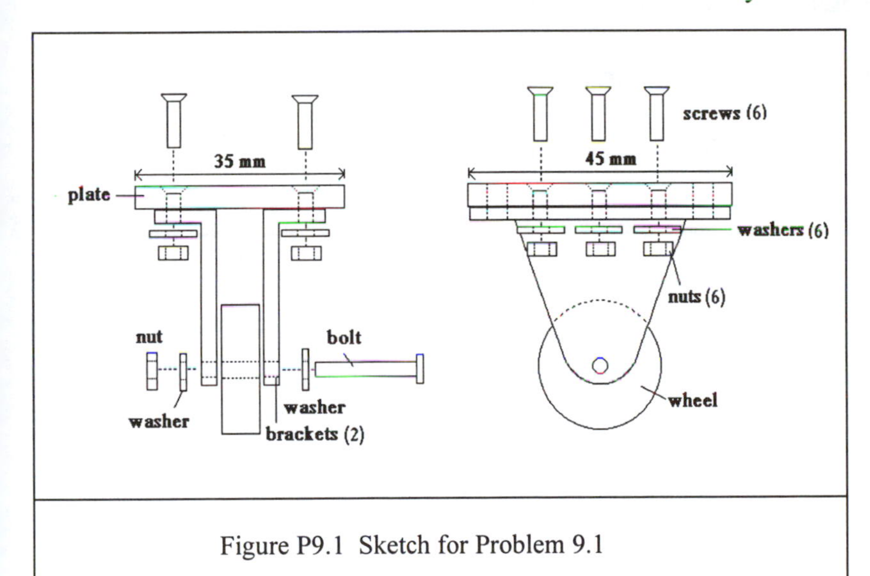

Figure P9.1 Sketch for Problem 9.1

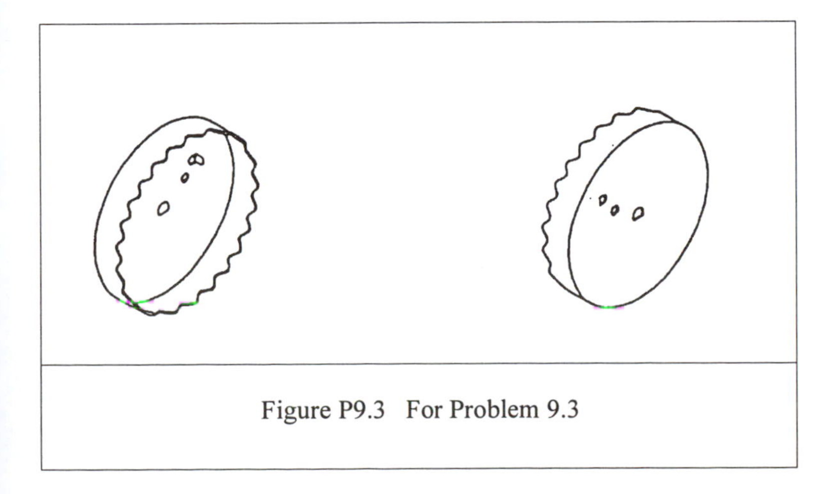

Figure P9.3 For Problem 9.3

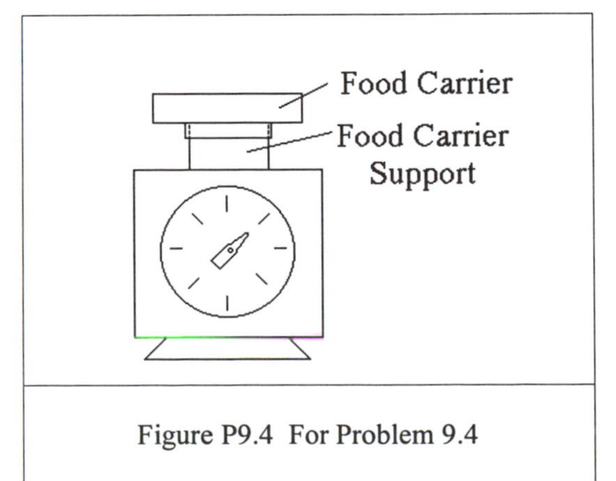

Figure P9.4 For Problem 9.4

Appendix 9.A

Some Properties of Selected Materials

Material	Casting Processses Used In	Corrosion Resistance	Cost cents per cubic in.	Tensile Strength kpsi	Yield Strength kpsi	Sheer Strength kpsi	Modulus Mpsi	Elongation Percent	Hardness Brinell	Endurance Limit kpsi	Density lb/in^3	Thermal Conductivity, 68 deg. F (BTU-ft/hrft^{2}F)	Coef. of Thermal Expansion 68-212 deg. F (10^{-6} in/in/F)	Electrical Conductivity percent IACS	Typical Uses
Aluminum Alloys															
A380.0	Die casting	Good	7.8	47	23	28	10.3	4	80		0.097	58	11.7	25	Auto parts, motor frames, housings, vac. cleaners. Most widely used al die casting alloy.
A413.0				35	16	25	10.3	3.5	80		0.096	70	11.4	31	Misc. parts. Used where resistance to corrosion needed.
201	Sand, permanent, and investment.	T7 temper recommended for opt resistance to stress corrosion cracking.		65	55		10.3	9	110		0.101		10.7	30	Structural members, aerospace housings, truck and trailer castings.
Magnesium Alloys															
AZ91D	Die casting	Good resistance to atmosphere, attacked by salt water unless finished.		34		20	6.5	3	63	14	0.065	31	14	13	
AZ63A	Sand, permanent mold.		10.8	29	14	18		6	50	12	0.065	39	14	15	
Copper Alloys															
C80100	Centrifugal, continuous, investment, plaster, and sand.	Corrosion and oxidation resistant.		25	9		17	40	44	9	0.323	226	9.4	100	
C81400				52	40		16	17		15	0.319	182	10	82	Electrical parts that meet RWMA Class II standards.
C84400				38	14		12	14	35	55	0,314	49	11	16.4	Low press, valves and fittings. General hardware supplies.
Zinc Alloys															
SAE903	Die casting	Excellent		41		31	12.4	10	82	6.9	0.238	65.3	15.2	27.5	Auto parts, household utencies, office equip., toys.
Steel															
Low carbon (<.2%)	Sand, investment, continuous, ceramic, and centrifugal.	Rust easily. Corrosion rate increased if salt is present.		45 - 65	25 - 45		30	23 - 30	115 - 140	19 - 22	0.283	8 - 8.3			
Medium carbon				65 - 95	35 - 50										
High carbon (.5 - 1.0%)				100 - 125	50 - 60										
Alloy				68 - 125	38 - 170			32	137 - 401	32 - 88					
Stainless	Sand, ceramic & centrifugal.	High resistance.		69 - 120	36 - 100		25 - 29	2 - 60	140 - 390		0.272 - 0.29	6.4 - 10.3			

Table A9.1 Properties of Selected Cast Alloys
(1 psi = 6895 Pa; 1 kpsi = 6.895; Example 10 kpsi (6.895) = 69 MPa)

Material	Processses Used In	Corrosion Resistance	Cost cents per cubic in.	Tensile Strength kpsi	Yield Strength kpsi	Sheer Strength kpsi	Modulus Mpsi	Elongation Percent	Hardness Brinell	Endurance kpsi	Density lb/in^3	Thermal Conductivity, 68 deg. F (BTU-ft/hrft^{2}F)	Coef. of Thermal Expansion 68-212 deg. F (10^{-6} in/in/F)	Electrical Conductiv-ity percent IACS	Typical Uses
Aluminum Alloys															
1100 (Soft) Annealed	Stamping, extrusion and forging.	High to rural, industrial, and marine atmosphere, and good to others.		13	5	9	10	35 - 45	23	5	0.098	128	13	59	Sheet metal parts, drawn or spun parts.
1100-H14 (Half-Hard)				18	17	11	10	9 - 20	32	7			13.1		
1100-H18 (Hard)				24	22	13	10	5-15	44	9			13.1	57	
2014 Annealed		High to rural, poor to sea water, and others.		27	14	18	10.6		45	13	0.101	111	12.8	50	Heavy duty forgings, plate and extrusions for aircraft fittings, wheel and major structural components; truck frame and truck components.
2014-T4				62	42	38	10.6	20	105	20		111	12.8	34	
2014-T6				70	60	42	10.6		135	18		111	18.8	40	
6061 Annealed		High to good.	6.9	18	8	12	10	25 - 340	30	9	0.098	104	13	47	Trucks, places where high stress and corrosion res. needed.
7075 Annealed		Good		33	15	22	10.4	17	60		0.101		13.1		Structural components where high stress and corrosion resistance needed.
Magnesium Alloys															
AZ61A-F	Extrusion, forging.	Good to atmo-sphere. Attacked by sea water		38 - 46	21 - 33	20	6.5	7 - 17	55	20	0.065	34	14	12.5	General purpose forgings and extrusions.
Copper Alloys															
C23000 Annealed	Stamping, extrusion	Fair	50	39	10	31	17	48	F56		0.316	92	10.4	20	Electrical conduits, sockets; fastners, pipes, costume jewelry.
C23000 Hard				70	57	42	17	5	B77						
C37700-Annealed	Forging	Good-excellent		52	20	35	15	45	F78		0.305	69	11.5	27	Forgings of all kinds.
Zinc Alloys															

Carbon Steel, Alloy Steel and Stainless Steel – Continued on Next Page

Table A9.2(a) Properties of Selected Wrought Alloys
(1 psi = 6895 Pa; 1 kpsi = 6.895; Example 10 kpsi (6.895) = 69 MPa)

Material	Processses Used In	Corrosion Resistance	Cost cents per cubic in.	Tensile Strength kpsi	Yield Strength kpsi	Sheer Strength kpsi	Modulus Mpsi	Elongation Percent	Hardness Brinell	Endurance kpsi	Density lb/in^3	Thermal Conductivity, 68 deg. F, (BTU-ft/hrft^{2}F)	Coef. of Thermal Expansion 68-212 deg. F (10^{-6} in/in/F)	Electrical Conductiv-ity percent IACS	Typical Uses
Carbon Steel															
Rods, Bars, Forgings															
1015 (low carbon) Hot Rolled				50	27.5			27.5	101						
1015 (low carbon) Cold Drawn				56	47			47	111						
1045 (med carbon) Hot Rolled	Forging, extrusion	Poor		82	45		30	16	163		0.283	27	8.1	18	
1045 (med carbon) Cold Drawn				91	77			12	179						
1095 (high carbon) Hot Rolled				120	66			10	248						
1095 (high carbon) Cold Drawn				99	76			10	197						
Sheets															
Commercial	Stamping	Poor		52	38			30	55		0.283	27	8.1	18	
Drawing Quality				44	27			42	42						
Alloy Steel															
Rods, Bars, Forgings															
1340	Forging, stamping	Poor		100 - 282	76 - 235			9 - 25	235 - 578						
Sheets															
ASTM A606 Cold Rolled	Stamping	Improved resistance		65	45			22							
Stainless Steel															
302 Annealed	Stamping	Excellent, including food products		90	40			50	Rb85				9.6	73	General purpose.
304 Cold Rolled	Stamping, forging		42.6	110	75		28	60	240		0.29	9.4			
316 Annealed	Stamping	Best of all standard stainless.	57.7	84	42			50	Rb79				8.9	74	Parts exposed to severe corrosive media and stressed parts subject to high temperatures.

Table A9.2(b) Properties of Selected Wrought Alloys – *Continued From Previous Page*

(1 psi = 6895 Pa; 1 kpsi = 6.895 Mpa; Example: 10 kpsi (6.895) = 69 MPa)

Material	Process Used In	Chemical Resistance	Cost cents per cubic in.	Tensile Strength kpsi	Flexural Strength kpsi	Tensile Modulus kpsi	Flexural Modulus kpsi	Hard-ness-Rockwell	Impact Izod-Notch ft-lb/in	Density lb/in^3	Thermal Conductivity, 68 deg. F (BTU-ft/hrft^2-F)	Coef. of Thermal Expansion 68-212 deg. F (10^{-6} in/in/F)	Typical Uses
Thermoplastics													
ABS Medium Impact	Injection molding and extrusion.	High to acqueous acids, alkilis and salt.	3.4	6.3 - 8.0	9.9 - 11.8	340 - 400	350 - 400	80	24 - 4.0	0.038	0.96 - 2.16	3.2 - 4.8	Appliance parts; office, lawn and garden equipment; toys.
Acetal	Injection molding, extrusion, blow molding, and rotational molding.	Excellent to most. Poor for strong acids and alkalis.	6.4	10	14.1	520	410 - 450	80	1.5	0.052	1.56	4.5	Appliance parts, gears, bushings, auto and plumbing parts.
Acetal - 20 % glass			8.4	8.5	16.5	1300	800	110	0.8	0.056		2.0 - 4.5	
Nylon 6		Resists weak acids, alcohol, and common solvents.	5.9	5.5 - 13	10.0 -11.6	200 - 500			0.8 - 3.0	0.039	1.2	1.6 - 8.3	Bearings, gears, bushings, rod, tubing.
Nylon 6 - 30 % glass			7.7	22 - 26	26 - 34	1000 - 1450			2.3 - 3.0	0.05	1.2 - 1.7	1.2 - 3.0	General purpose parts requiring stiffness.
Nylon 6/6		Attacked by strong concentrations of mineral acids.	6.5	11.8		385 - 475	410		1.0	0.041	1.7	1.7	Bearings, gears, bushings, rod tubing.
Nylon 6/6 - 30% glass			9.8		26 - 35	1400 - 2000	1300		2.2	0.05	1.5	1.5	Bearings, gears, bushings, rod tubing.
Polycar-bonate	Injection molding and extrusion.	Resists weak acids and alkilis, oils and grease.	6.7	8.5 - 9.0	12 - 14.2	325 - 340	310 - 350	63	12 - 18	0.04	1.35 - 1.41	3.75	Electrical parts, portable tool housings, lenses, sporting goods, impellers, and auto parts.
Polycar-bonate - 40% glass	Blow molding and thermo-forming.		10.4	23	27	1680	1400	50	2.5	0.055	1.53	0.93	
Thermosets													
Alkyd	Compression and transfer molding.	Resistant to weak acids. Unattacked by organic liquids (alcohol, fatty acids, and hydrocarbons).	5.5	7 - 8	19 - 20	1950	2500	70 - 75	2.2	0.079	4.2 - 7.2	1 - 3	Encapsulation of resistors, coils and small electronic parts, switches, relays, connectors, sockets, circuit breakers, parts for transformers, motor controllers, and auto ignition systems.
Alkyd and glass				5 - 9	12 - 17	2250	2500	70 - 80	8 - 12	0.073	2.4 - 3.6	1 - 3	
Epoxy and glass		Highly resis-tant to water and bases.		8 - 11	19 - 22		1500 - 2500	75 - 80	0.4 - 0.5	0.069	1.2 - 6	1 - 2	Electrical molding such as condensers, resistors, coils, etc.

Table A9.3 Properties of Selected Plastics

(1 psi = 6895 Pa; 1 kpsi + 6.895 MPa; Example: 10 kpsi (6.895) = 69 MPa)

Part III Configuration Design of Parts

Chapter Ten Formulating the Problem and Generating Alternatives

Chapter Eleven Evaluating Part Configurations for Manufacturability: Injection Molding and Die Casting

Chapter Twelve Evaluating Part Configurations for Manufacturability: Stamping

Chapter Thirteen Overall Evaluation and Redesign of Part Configurations

This page is intentionally blank

Chapter Ten

Formulating the Problem and Generating Alternatives

"By a small sample we may judge the whole piece."

Miguel de Cervantes
Don Quixote de la Mancha (1605)

10.1 Introduction to Part III - Configuration Design of Parts

As Cervantes notes, you can't make a quality whole from inferior parts. Getting the right configuration is a critical step in getting a quality part.

In general, to *configure* something means to arrange the relative positions of its elements. In engineering design, we expand this general definition to include not only arranging the elements but also to determining what the elements are to be. In an assembly or sub-assembly, the "elements" are the "major functional components and sub-assemblies". In the case of parts, the "elements" are called *features*, a concept we will come back to shortly in this chapter.

In the case of assemblies and sub-assemblies, configuration of the major functional sub-assemblies and components is done as an integral part of conceptual design. When an assembly or sub-assembly is decomposed into its components and sub-assemblies, and the associated couplings are specified, we are in effect determining and arranging its elements; that is, we are configuring the assembly. We discussed this process in Part II. For parts, however, we have not yet discussed the process of configuration.

In Part II, a *physical* concept for a product or part was developed. Before the conceptual design stage, we knew only the *requirements* for the product, but there was no information about it as a physical entity. By applying guided iteration, we created a physical concept. This constituted a great deal of progress towards a solution to the engineering design problem, but there is still much to do. There is still a great deal of information needed to complete the design.

In engineering conceptual design, we first expanded our options by generating a number of alternatives. Then we narrowed the options by performing evaluations and making choices. We decided to use certain functional or hardware decompositions instead of others; we decided to employ certain physical principles or effects instead of others. In our recursive decomposition of the product and its sub-assemblies, we created abstract parts to meet certain purposes in conjunction (i.e., coupled) with other parts. For these parts, we chose certain broad classes of materials over others. We also decided, at least tentatively, on the manufacturing process to use. But we have determined very little about their size and shape.

The decisions about parts made during engineering conceptual design imply an approximate size for each part, and perhaps even an approximate or general shape. But to manufacture a part, we need to do more than *imply* the size and shape: we need complete and precise information. Here in Part III, therefore, we will consider how to design the basic shape of parts. We will not yet determine exact dimensions; that is the subject of Part IV of the book.

In both Part III and Part IV we will use the basic guided iteration methodology to achieve our design goals. The basic steps are, as always: problem formulation, generation of alternatives, evaluation, and redesign guided by the results of the evaluations. For the design of the shape or configuration of parts, the methods used to implement these steps (and the chapters where they are presented) are summarized as follows:

- Problem Formulation (Chapter 10):

The Part Configuration Requirements Sketch.

- Generation of Alternatives (Chapter 10):
  Qualitative Reasoning From Physical Principles,
  Qualitative Reasoning About Manufacturing, Especially Tooling, and
  Qualitative Reasoning About Tolerances.
- Evaluation of Manufacturability (Chapters 11 and 12):
  Quantitative Estimation of Relative Tooling Cost for Injection Molded, Die Cast, and Stamped Parts.
- Evaluation (Chapter 13):
  Tolerance Plan Evaluation,
  Pugh's Method and Dominic's Method, and
  Qualitative Reasoning About Risks.
- Redesign (Chapter 13):
  Guided By Evaluation Results.

It should be noted that Chapters 11 and 12 are about the *quantitative* estimation of the relative *tooling costs.* It is remarkable that quite good estimates of these costs can be made from information about only the configuration of parts. That is, even though at the configuration stage we do not yet have exact dimensions, we can nevertheless estimate its relative tooling cost, and relate these costs to aspects of the part's configuration. Later, once dimensions are known, we can then also estimate relative *processing* costs. This is the subject of Chapters 15 and 16.

This Chapter. In this chapter, we deal with the first two steps of the guided iteration process for configuration design of parts: how part configuration problems are formulated and how alternative solutions are generated.

Though new engineering graduates are seldom given sole responsibility for the design of a whole product or assembly, they are often assigned the task of designing parts. One prominent manufacturer assumes that an engineer can design, on average, about four parts per year. Thus a typical part design can be expected to involve about three person-months. This may seem a lot of time and effort for just one part, but remember, design is almost always a team effort so there are meetings to attend, memos to read and write, searches for (and waiting periods) for information, and the inevitable changes in the Specifications, couplings, and special requirements that occur along the way. Design is fun, but it not without its frustrations.

The initial information available for beginning configuration design is, of course, the final information available from the conceptual design phase. Therefore, at least in theory, we begin part configuration design with the following known information:

1. (a) The *Product* Marketing Concept or Specification, and/or
   (b) Its In-Use Purpose(s) and Couplings;
2. The Engineering Design Specification for the *Part*;
3. The Embodiment of the Part, Including:
   (a) The physical principles, laws, or effects by which it will meet its functional requirements; and
   (b) The abstract form; and
   (c) The basic material classes and manufacturing processes to be used.

Though all of the above information theoretically will be available to begin a part design, it is not realistic to expect that it always will be. If there is no Spec, for example, then of course the first task is to develop one.

The goal of configuration design is, as the term implies, to determine a "configuration" for the part. We thus turn in the next section to a discussion of what is meant by "configuration".

10.2 What is a *Part Configuration*?

10.2.1 Features

The task of part configuration design is to select or determine what *features* the part will possess and how those features are to be arranged and connected. Features include:

- *Walls* of various kinds;
- *Add-ons* to walls such as holes, bosses, notches, grooves;
- Ribs, etc.;
- *Solids* elements such as rods, cubes, tubes or spheres; and
- *Intersections* among the walls, add-ons, and solid elements.

When alternative configurations are generated, they must be evaluated *as configurations* before we move on to the parametric design stage where exact dimensions are assigned. That is, each alternative configuration needs a separate evaluation of its configuration *per se* for its performance and manufacturability. We therefore need to have an idea of what constitutes a configuration, and when two proposed configurations are essentially the same, or different.

As noted above, we consider a special purpose part to be made up of *features* that include: *walls*; *add-ons* to walls such as holes, bosses, notches, grooves, ribs, etc.; *solid elements* such as rods, blocks, cubes, tubes or spheres; and *intersections* among the walls, add-ons, and solid elements. In order for two parts to have the *same configuration*, we temporarily ignore add-ons and the intersection types, and then require three conditions:

(a) the same feature types must be present in the configuration, and

(b) the features present must have the same spatial arrangement and connectivity. In other words, the intersections, whatever their type, connect the same corresponding features, and

(c the *relative* dimensions of or between the features, or of the whole part, must be within specified limits or ranges.

Parts that meet only the first two of these requirements will be said here to have the same topology[1], but not the same configuration.

A few simple examples will make the above meanings clear. In the case of a beam, configuration has to do with the cross-sectional shape. An I-Beam, for example, is one of many possible configurations for a beam. An I-Beam has three walls (two flanges and a web) arranged as in the letter I; moreover, the relative dimensions among the features of real I-Beams are all reasonably consistent. The beams with cross-sections shown in Figure 10.1 are thus all I-Beam configurations because they all have flanges and webs arranged in exactly the same manner, and because their relative dimensions (e.g., height to width and thickness to width) all meet an acceptable criteria for an "I-Beamness".

The cross sections shown in Figure 10.2, however, are not the same configuration because, though they also have three walls as sub-parts, the arrangement or connectivity of those walls is different.

Figure 10.3 are also not classified as the same I-Beam configurations even though their three sub-parts are arranged the same as in an I-Beam. In this case, their relative dimensions do not meet the required ranges expected for an I-Beam. For an I-Beam, we would expect that the web and the flanges would have thicknesses of comparable order of magnitude, surely no more than a factor of two or three different. We would also expect that the web and flange thicknesses would be small compared with the overall height and width of the cross section. We might say, a bit informally, that the above beams have the same topology but not the same configuration.

The definition of "configuration" here will leave some readers with an uncomfortable feeling because it is somewhat fuzzy; that is, exactly where one configuration class ends and another begins is not precisely or formally specified. To illustrate this fuzziness, which of the beam cross-sections shown in Figure 10.4 are in the I-Beam class and which (if any) are not?

Though such fuzziness is possibly a theoretical issue, it is not a practical problem. In practice, it is usually quite obvious when two configurations with the same topology are different enough to require individual evaluation as solutions to the design problem at this stage. We can, however, if need be, develop a more precise definition — even though it is arbitrary. Here, for example, is a rule one might use:

1. This is not a mathematically rigorous use of the word topology. Strictly speaking, forms have the same topology if they can be produced by certain transformations such as stretching, bending, etc. In our use here, only wall length and thickness changes are allowed.

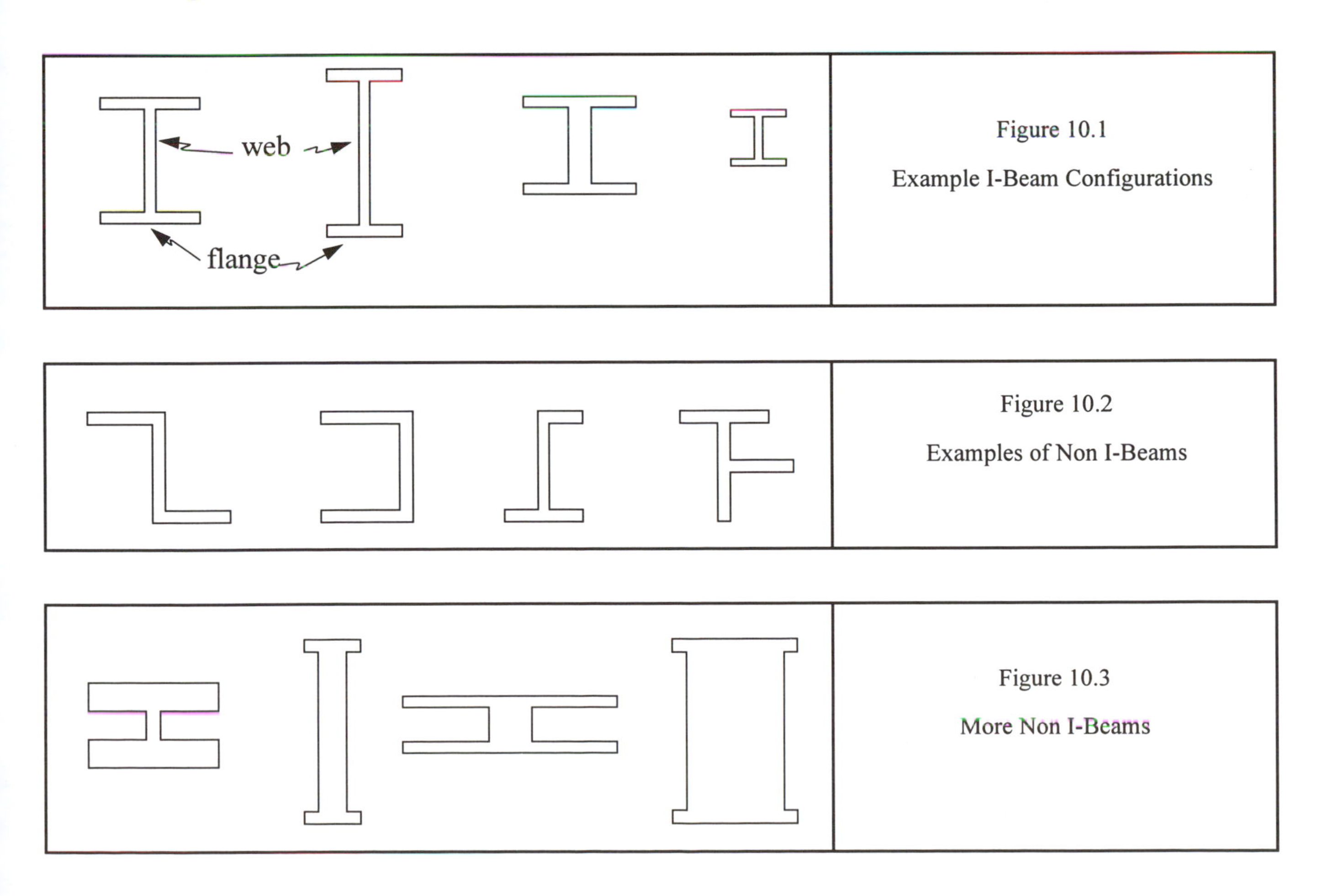

Figure 10.1

Example I-Beam Configurations

Figure 10.2

Examples of Non I-Beams

Figure 10.3

More Non I-Beams

Two part configurations with the same topology are the same configuration if the ratios of any two dimensions on one part are no more than X% different from the same ratios on the other part. If more than two parts are being considered, then one of the parts must be used consistently as a reference to compute ratios. X might be, say, 50%.

For example, if we use the beam cross section in Figure 10.4(a) as the reference, then cross sections (b) and (c) have the same configuration, but cross section (d) does not. Its ratio of flange length to thickness is more than 50% greater than the same ratio in the reference configuration. In general, however, such preciseness is not needed.

10.2.2 Configurations of Standard Parts and Modules

When a physical concept or embodiment is described as a standard part (e.g., spring or screw), the task of part configuration design is to select a particular sub-class or type to be used. For example, if the conceptual solution is a "spring", there are several types — configurations — of springs available: coil or helical springs, leaf springs, beams, Belleville springs, and others. There are also many types or configurations of other standard parts, such as screws, beams, and the like. Configuration design for such standard parts involves making the selection from among these types. Then in parametric design, the numerical parameters of the selected type of standard part or are determined. Parametric design of a helical spring, for example, involves determining the number of coils, the wire diameter, etc.

In the case of standard modules, there are also decisions about types or classes to be made. There are many types of motors, for example, and many types of pumps and other standard modules. Before the exact parameters for these modules can be made, the basic types — that is, their configurations — must be selected.

10.3 The Goal of Configuration Design: Establishing *Attributes* for Parametric Design

10.3.1 Special Purpose Parts

The final desired information state in a part configuration problem for a special purpose part is (a) identification of the features (including add-ons) that make up the part, (b) specification of the arrangement or connectivity of those feature, (c) specification of any approximate required relationships of the dimensions of the features or between the features, and (d) a listing of the *attributes* of the part that must be given values in the next stage of design (i.e., parametric design). Attributes may be numeric (as in sizes or distances) or non-numeric (as in material choices).

Another way to view the idea of attributes in relation to configurations is to note that if two designs have the same attributes, then they have the same "topology". If the attributes have values that are reasonably similar (say within 50% or so), then they also are the same configuration.

Part configurations in mechanical design are usually described by a sketch or by well understood terms. Saying "I-Beam", for example, is sufficient to describe that particular beam configuration. When "I-Beam" is stated, we get an

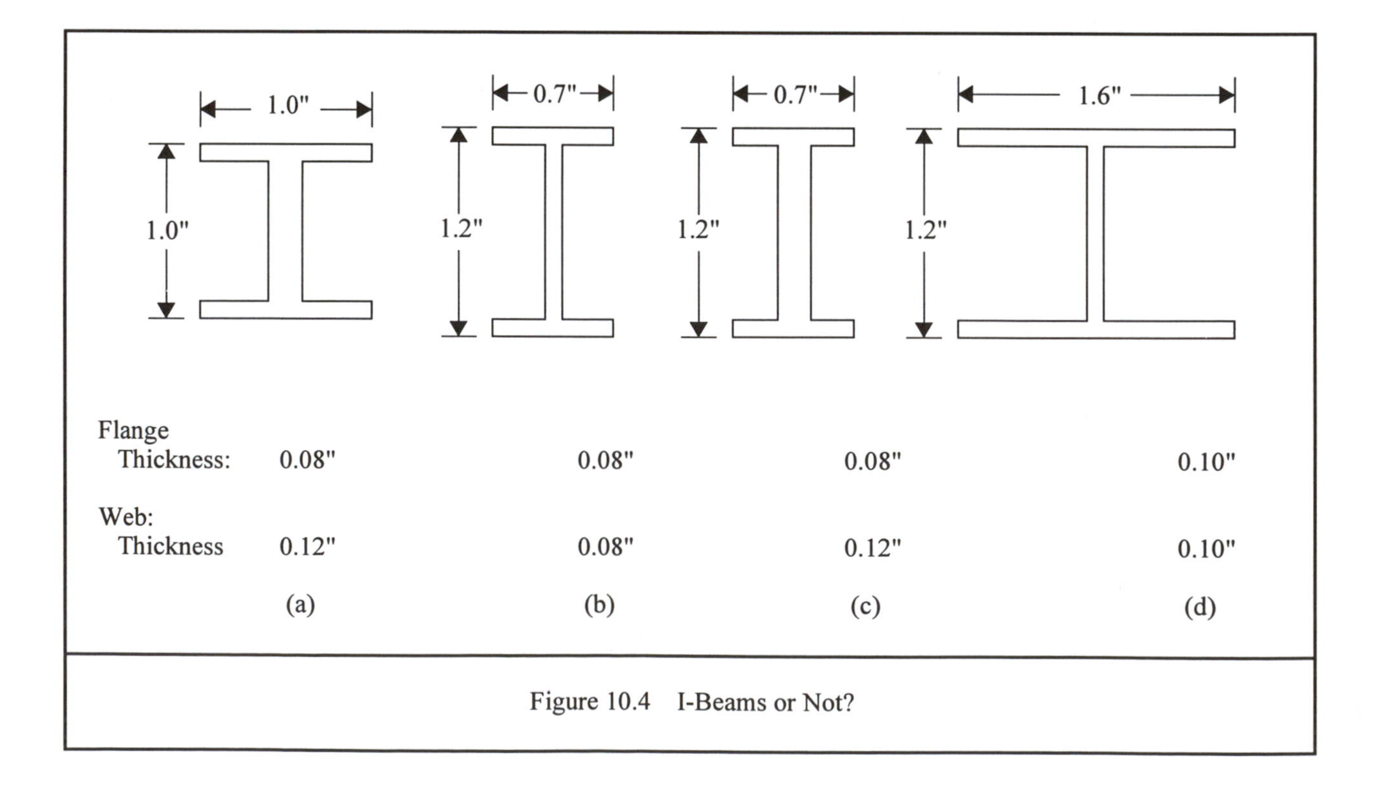

Figure 10.4 I-Beams or Not?

idea of the ranges of relative dimensions involved. Not all names, however, work so well. For example, if we say "L-Bracket", it could mean any of the L-brackets shown in Figure 10.5, and more, but some of these would not be considered in the same configuration class. Thus a sketch is helpful, if not necessary, in defining the configuration of such a generic part.

For standard part types, there are generic names, but for most special purpose parts, we don't have a generally accepted term for the configuration. Thus, we *must* provide a sketch to describe it.

Attributes

Once the features and their connectivity of a part are specified, whether by name or sketch, there will then exist a set of *attributes* which can be explicitly identified and which make up (along with knowledge of connectiveness) the part's configuration. In the case of an I-Beam, for example, the attributes are:

Beam height
Beam width
Web thickness
Flange thickness
Fillet radii
Material

The sets of *values* assigned to these attributes must enable the configuration to meet the functional requirements for the beam, and must be selected also with manufacturability explicitly in mind. It is the task of parametric design, the subject of Chapters 14 through 21 to assign values to attributes created at the configuration stage.

In the second of the L-brackets shown in Figure 10.5, the attributes are the width, the length of wall1, the length of wall2, thickness (if the same for both walls), and the wall intersection angle (assuming it doesn't have to be a right angle to meet a requirement in the Engineering Design Specification).

Sometimes it is desirable to include certain *global attributes* of parts. For example, a global attribute of a box is its volume. The dimensions of the rectangular envelope of a part is another global attribute. The volume of the material used may be another. Usually global attributes are computable from the feature dimensions, and so their inclusion among the attributes is possibly redundant. Including them may, however, be convenient when they are directly involved in meeting some aspect of the part's functionality. A container, for example, may have some required volume. And any part may have limits on its size or weight.

10.3.2 Attributes of Standard Parts and Modules

When the basic type or class of a standard part or module is selected, the attributes are automatically determined. That is, the attributes simply come along with the choice of configuration. For example, when a spring is determined to be a helical spring, the attributes are its coil diameter, the wire diameter, the wire material, the number of turns, and the end conditions. If the spring configuration had been selected to be, say, a leaf spring, then quite a different set of attributes would result — but what those attributes are is known.

10.4 Formulating Part Configuration Problems

With the background provided above, we can now begin to discuss the process of guided iteration for special purpose part design. We begin, of course, with the formulation of the problem which includes the following steps:

1. Review the Relevant Engineering Design Specifications.

If the part is a portion of a product, review also the Engineering Design Specification for all parent sub-assemblies and

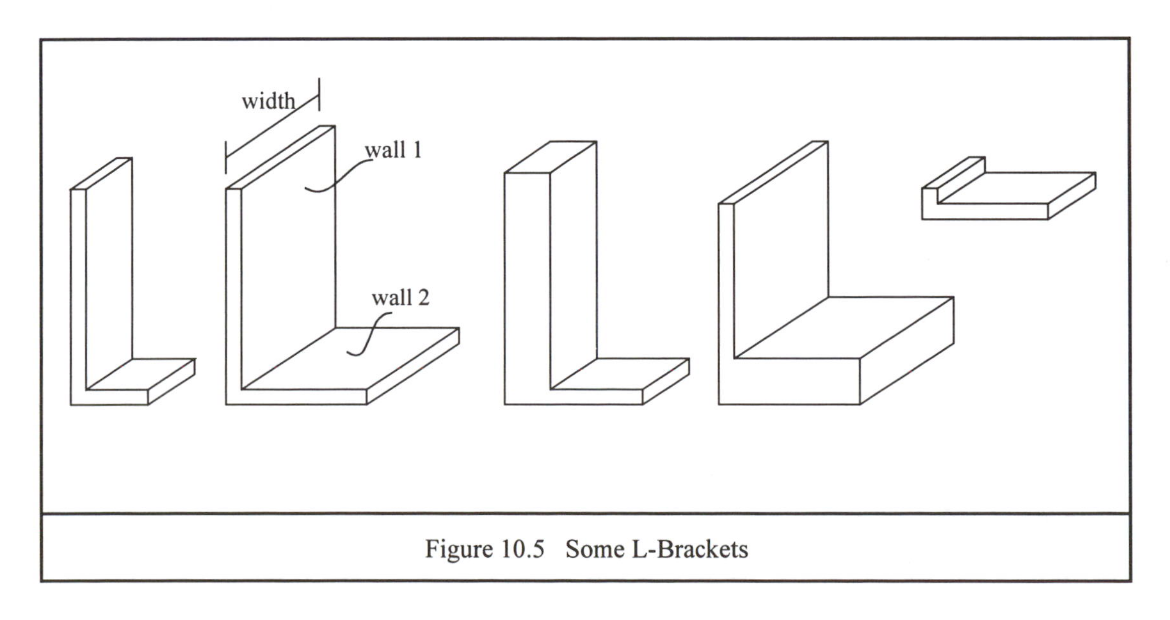

Figure 10.5 Some L-Brackets

assemblies. It must never be forgotten that parts must be compatible with the whole, and serve the purposes and meet the other requirements of the whole.

2. Identify and Describe the Couplings

Identify and describe the nature of any couplings that the part has with other parts or sub-assemblies, and with human users or service people (if any). We must know about physical couplings so that the mating features can be made compatible. We also must know about any space or weight limits that might have to be shared or compromised with other parts. And of course we also need to know about relevant forces, energy flows, material flows, and the like.

3. Can the Part Be Eliminated or Combined?

Reducing the number of parts is almost always the very best way to reduce assembly, inventory, and overhead costs. The trade-off between the number of parts and individual part complexity is discussed in Chapter 3. It is almost always less costly to make and assemble fewer, more complex parts than it is to have a high part count. However, this general rule has exceptions and must be examined. The added complexity, for example, may delay production while the more complex tooling is being made.

4. Can a Standard Part or Module be Used?

As we have pointed out in Section 6.7, if the Engineering Design Specification can be met with a standard part, then using a standard part is very likely to be the most cost effective solution. We repeat: a good design rule is:

Use standard parts and modules wherever possible.

However, this does not override the more important rule to reduce the part count. A standard part is generally less costly overall than a special purpose one, but two standard parts may not be less costly (all costs considered) than two standard ones.

With the above preliminary steps completed, we are ready to continue the formulation by developing the part configuration requirements' sketch.

10.5 The Part Configuration Requirements Sketch

Just about the only way to configure a new special purpose part is to begin — by sketching *something*. Then you continue by sketching it again and again and again in guided iteration fashion. That is, you generate an initial configuration of the part in a sketch (on paper or CAD system), then evaluate it for performance and manufacturability, and then redesign (by re-drawing) as needed as many times as necessary until an acceptable solution is obtained.

> "In creating, the only hard thing's to begin."
>
> James Russel Lowell
> *A Fable for Critics (1848)*

This sounds easy enough, and actually it is. In fact, the hardest thing about this process for new designers seems to be the beginning. Thus we concentrate a bit here on how to do that. There are three basic ways to begin: (1) sketch an initial configuration of some kind — any kind — even if it doesn't meet all the requirements; (2) start with a previous design that is close to what is needed; and (3) develop a part configuration requirements' sketch.

The first of these approaches is difficult for many students and new designers, and for good reason: as a method it provides no guidelines for where or how to start, thus creating an obstacle to starting at all. By far, the second approach is the most common in industry where many similar parts will have been designed. (Whether the previous designs are readily accessible when they are needed, either as drawings or CAD representations, is another question.) For new parts, and especially for new engineers designing new parts, we recommend the third approach: developing a *part configuration requirements' sketch.*

A *part configuration requirements' sketch* is made roughly (but only roughly) to scale showing the approximate external forces, flows, mating features of coupled parts, and the like, that the part must accommodate. The sketch therefore shows the essential surroundings of the part, and locates loads, support points or areas, heat or other energy flows, adjacent parts, forbidden areas, and so on. At this stage, the part configuration requirements' sketch does not show any features of the part to be designed.

The ability to sketch — that is, to express ideas for parts graphically in approximate form — is essential to part configuration design. Students who do not feel confident making sketches (it should be fun) should take the trouble to learn how to do it. It *can* be learned, even if you feel you have no 'artistic' ability.

As a simple example of the development and use of a part configuration requirements' sketch, suppose a bracket is to be designed to support a weight at some distance from an available vertical support wall. We assume also that there are, say, several pipes between the weight and the wall, and that the bracket is to be some nominal, uniform width (the dimension into the paper). Then the problem formulation sketch in this case is essentially two-dimensional as shown in Figure 10.6(a).

It is not always the case that the nature and locations of the physical couplings with surrounding parts are known at the time a part must be designed. In such cases, designers can design the coupling intersections more or less any reasonable way they want — with the following proviso. One should try to be as accommodating as possible to the designers of the adjacent parts. Therefore, always keep the couplings as simple as possible. For one reason, *you* may be the designer of the coupled part or parts. For another, if easy, functional, manufacturable couplings are not provided, you may have to redesign *your* part. Design involves teamwork.

But design also requires commitment. Even if the to-be-coupled parts are not yet established, designers will often have to proceed as best they can. Thus it may be necessary to fill in imaginary boundaries or couplings with adjoining parts in order to complete the requirements sketch. Just remember to communicate these assumptions (or wishes) to those who will be responsible for designing the mating parts, and know that these boundaries may get changed later on. Thus, pay special attention to the nature of couplings created; try to keep them not only simple, but also easily modifiable without major changes to the rest of the part.

10.6 Generating Alternative Configuration Solutions

Often, instead of generating a number of alternative configurations before evaluating any that seem promising, designers will generate only one alternative, evaluate it subjectively, and then redesign it by erasing (on paper) or revising (in a CAD system). When this is done, the design process evolves as a number of small iterative redesign steps.

There are, however, three difficulties with the process described in the above paragraph: one is that some creative or less obvious configurations may be missed; another is that the trial configurations generated along the way are lost through erasure or deletion on the CAD system; and a third is that the evaluations may not be as thorough and technically sound as they could be. In simple or familiar cases, these difficulties usually pose no practical problems. Designers can visualize new starting places, they can remember intermediate steps that might be promising alternatives, and evaluation issues are simple and straight forward.

Even in a simple problem, however, there may be dozens of alternatives that can be generated. In more complicated situations, there can easily be hundreds. We therefore recommend the following procedure which will make the generation of many alternatives easier to do — and the various alternatives easier to keep track of. The procedure is described below primarily as a paper and pencil operation, but can certainly be carried out on a CAD system. Either way, designers must have a system for keeping track of, storing, and retrieving the many trial configurations that may be generated during the process.

1. Make a number of copies of the part configuration requirements' sketch described above. (On a computer, make, name, and store the requirements' sketch in a readily accessible file.)

In the bracket example, that means making a number of copies of Figure 10.6(a)

2. *Ignoring contiguity* (that means ignoring the fact that the part ultimately must be one part with all its features somehow connected together), locate on one of the configuration requirements sketches the *minimum essential* material "chunks" (or sub-parts or partial features, usually partial walls, though there could also be a boss or the surroundings of a hole) needed to meet the part's requirements. If there are choices for where to place material, arbitrarily pick one that seems most likely to result in the *simplest* possible final part when contiguity is finally accommodated. We will call the result of this step a *non-contiguous part configuration.*

In the example of Figure 10.6(a), this means placing a wall section under the load (to support it), and another wall section along the wall for attachment. There are different possibilities for the location along the wall so we just pick one that seems will result in a simple part. One possibility is shown in Figure 10.6(b).

3. In our example, and in most problems, obviously there are several possible non-contiguous part configurations. Therefore, using other copies of the requirements sketch, generate additional feasible arrangements of non-contiguous part configurations while staying within the realm of engineering sense. That is, not every imaginable non-contiguous configuration is actually generated, though

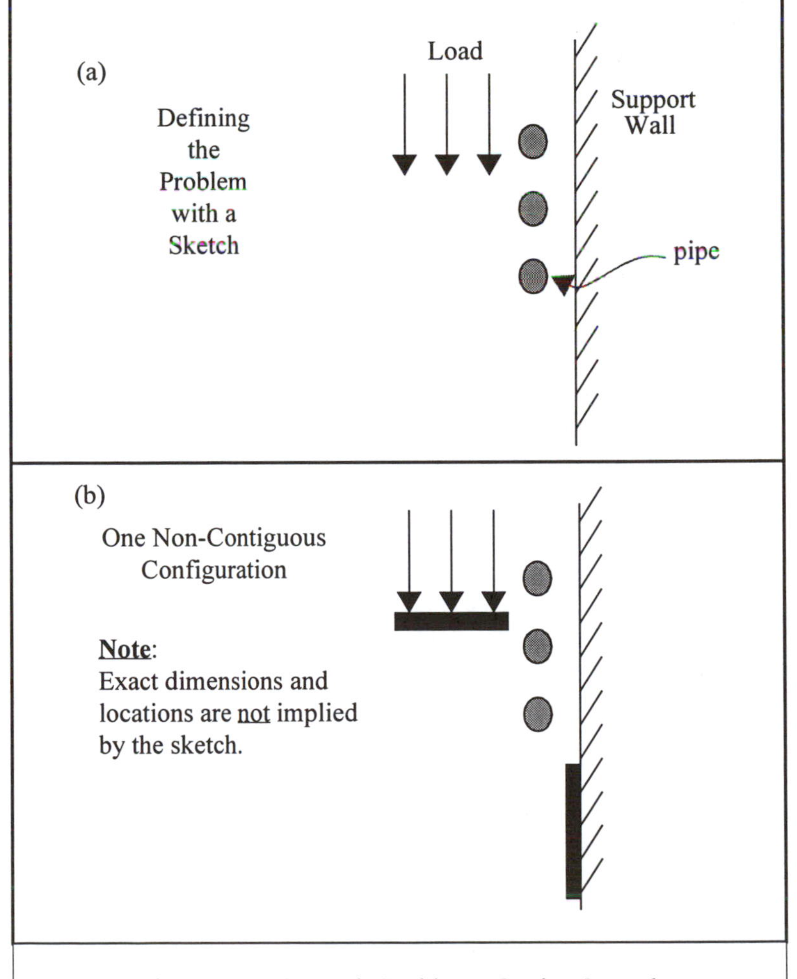

Figure 10.6 Example Problem: Getting Started

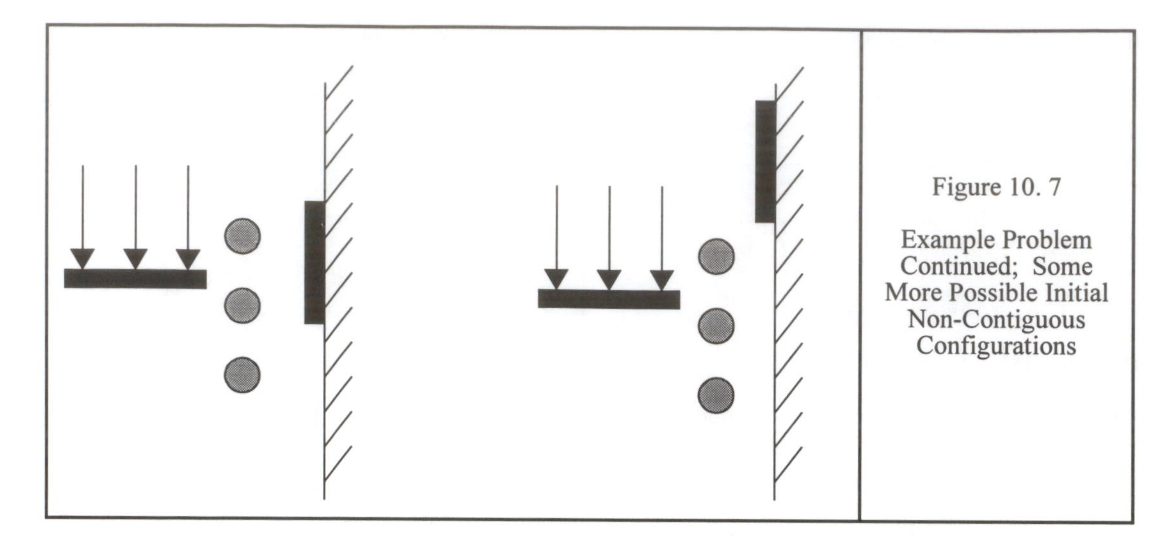

Figure 10. 7

Example Problem Continued; Some More Possible Initial Non-Contiguous Configurations

as many as practicable should be considered. Limiting the alternatives is guided by qualitative physical reasoning, by goals of simplicity and minimum use of materials, and by knowledge of the manufacturing process to be used.

For our simple example, the attachment to the wall can be re-located. Two more possibilities are shown in Figure 10.7.

4. Make a number of copies of each of these non-contiguous alternatives (or store each in the computer). In complex situations, this may very well involve some additional qualitative reasoning to exclude, at least temporarily, some of the alternatives that do not seem to present good opportunities for leading to a final good solution. Such culling out is appropriate, else an unmanageable number of remote possibilities will be generated. But do not permanently throw away any alternative sketches of configurations that are not being pursued; one of them may be needed later.

5. Now, in the selected non-contiguous solution alternative sketches, add the minimal material needed to achieve contiguity. Don't be concerned with exact dimensions — establish only a contiguous configuration. That is, be concerned only with what features are there, with how they are connected or related, and just their approximate dimensions. Also, do not be concerned much with physical practicality — just get the features identified and connected up so the part becomes a single, contiguous part that will perform the required functions assuming the required materials and manufacturing processes are available. Make several (at least two or three) of these contiguous part designs. These are, of course, possible final configurations, but they have not yet been subjected to careful and thorough evaluation and redesign.

A few possible contiguous solutions for one of the non-contiguous solutions are shown in Figure 10.8.

6. At this point, if it is necessary, but *only* if it necessary, refine the material selection to Level IV in Table 9.1. Usually it is not necessary, so "least commitment" prevails.

For our example, we have decided that the material specified in the physical concept is aluminum, and that no additional material information is needed at this time. We can select the exact aluminum alloy later. The process selected is die casting.

7. Next generate additional feasible contiguous configurations *guided simultaneously by*:

a. Qualitative physical reasoning regarding the functionality of the part; and

b. Qualitative physical reasoning regarding the manufacturability of the part.

These two subjects — qualitative physical reasoning about function and qualitative physical reasoning about manufacturability — as they relate to part configurations are discussed in the next two Sections of this chapter. More detailed and quantitative evaluation of configurations for their manufacturability is discussed in Chapters 11 and 12. Methods for complete configuration evaluation, including both function and manufacturability, are presented in Chapter 13.

8. From the full set of feasible configurations that has been generated, select several — those that most clearly satisfy the qualitative reasoning principles — for more detailed evaluation and redesign as described in Chapters 11, 12 and 13. If none of these configurations is found acceptable upon detailed evaluation and redesign, then it may be necessary to return to step 7, or step 6, or even (in very difficult cases) to consider asking for changes in the configuration problem formulation sketch or in the Engineering Design Specification. Such changes, of course, cannot be made unilaterally. Everyone on the design team affected by a proposed change in a part must be involved in such a decision.

In complex cases, if the above process is followed rigorously, a lot of paper — too much — can be generated. Some common engineering sense must be used to avoid getting the process bogged down in paper and bookkeeping. Ridiculous possibilities need not be pursued, though perhaps the reason they are 'ridiculous' should be identified. Alternatives with attractive characteristics (efficient material use, easy manufacture, and elegant functionality) should obviously be given priority. At the same time, however, it is good practice to explore enough different non-contiguous possibilities far enough to insure that even better options are not being ignored or dismissed.

10.7 Qualitative Physical Reasoning to Guide Generation of Alternatives

As just noted, the procedure outlined above has the potential for generating a very large number of alternative configurations. There must be some way to guide and limit the generation process in a way that provides us with a smaller set of only the most likely to be successful alternative configurations. Then we can subject this smaller set to more detailed evaluation and redesign. Though at the configuration stage of part design, we do not yet have actual dimensions — the size and spatial relationships among the features are still only approximate — we can still get useful guidance using *qualitative physical reasoning.*

Qualitative in this context means, essentially, without numbers (though orders of magnitude of numbers are certainly involved). Thus, qualitative reasoning fits configuration design evaluation well because configurations are themselves largely without numbers. Nevertheless, even without numbers, the basis of qualitative reasoning is still rooted in fundamental physical principles. It is much, much more objective than guesses or feelings.

To illustrate how qualitative reasoning works in a familiar example, consider again an I-Beam as an example. We already believe that an I-Beam is a good configuration for a beam. But why?

The answer requires application of some qualitative knowledge from strength of materials. We know that for a member carrying a transverse load, a bending moment is created, and that the outer fibers (relative to the neutral axis) are the more stressed. Thus, beams with material concentrated in the outer portions make more efficient use of material than beams with more of their material located nearer the neutral axis. In an I-Beam, the flanges carry the bulk of the load, and the web is included primarily to provide contiguity, though it also carries transverse shear loads which are occasionally also important. And the intersections between flange and web help support torsion. (If torsional loading is a major issue, one might consider a box configuration since its closed section makes it stiffer in torsion.)

Thus by understanding the physical principles of beam theory, we can qualitatively guide the generation of feasible configurations for a beam: the more an alternative

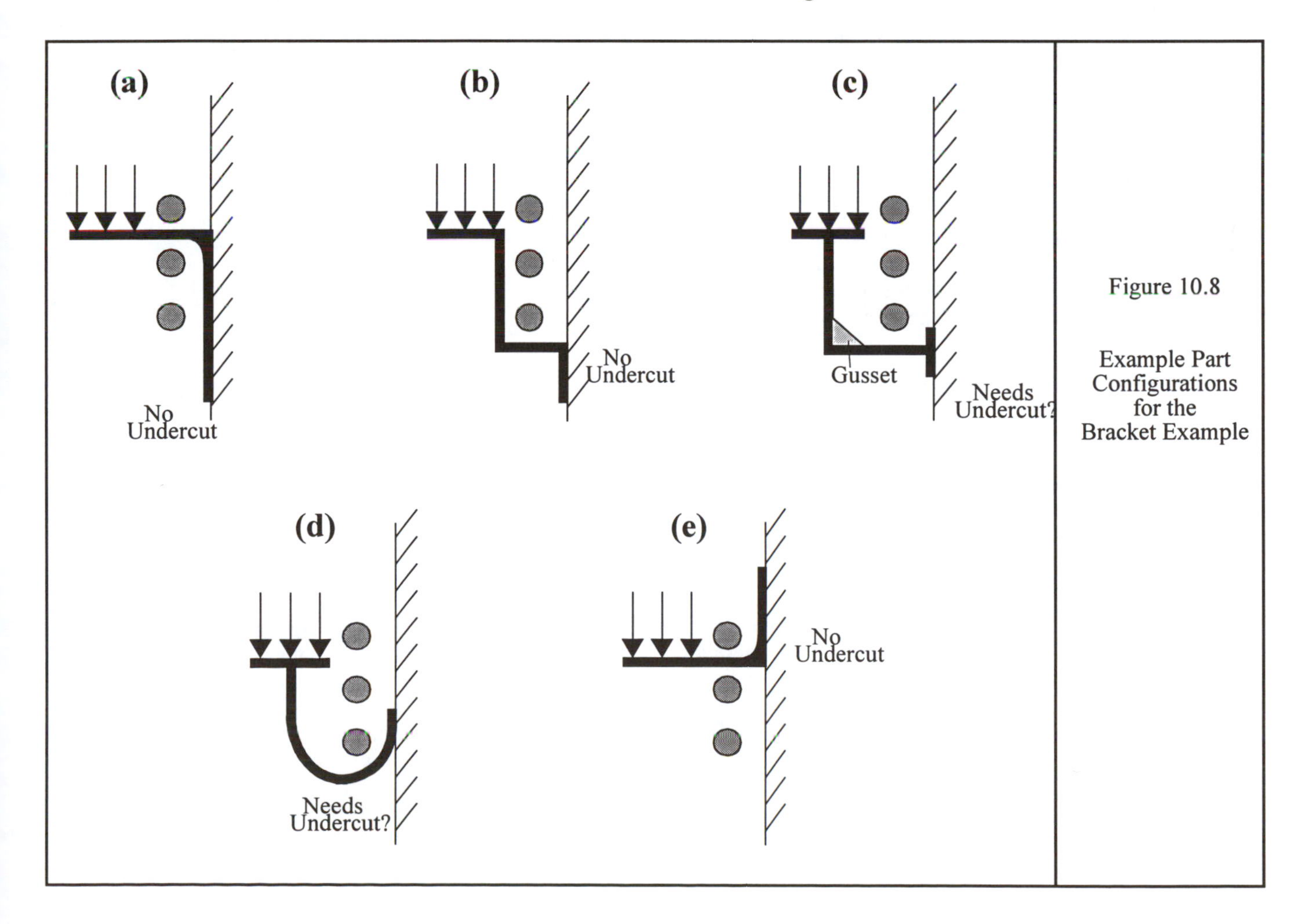

Figure 10.8

Example Part Configurations for the Bracket Example

can place material away from the neutral axis, the better. In other words, the more like an I-Beam, the better.

Using a similar approach — that is, reasoning qualitatively about the physical principles involved — we can also evaluate other types of configurations for functionality. Following are some more examples.

Material Use

Good configurations will enable the use of as little material as possible; that is, they use material efficiently to achieve functionality. The I-Beam is, of course, a good example of this. In the case of a part whose function is to transmit a force, a *very* common part function, this means that the stresses in the part should be uniform and as high as is safe. Having relatively lightly stressed material in some sections of such a part is a waste of material, at least from a functional point of view in supporting forces. Thus, we try to select configurations that have the potential for uniformly high stresses throughout. This generally means that tension and compression are better than bending for supporting and transmitting forces.

There are exceptions to this general guideline, however, and manufacturing must also be considered. In a forging, for example, it may be that reducing the amount of material used results in the need for more dies and higher loads (to move more material greater distances, and to produce the thin sections). The added dies and processing costs may be greater than the cost of material saved.

The issues of efficient material use in heat transfer are discussed below.

Columns

Columns *can* fail in compression, but mostly they fail by either lateral or torsional buckling. The latter is a special concern because it is often neglected, allowing columns to be designed that are perfectly safe in compression and even lateral buckling, but which fail (sometimes surprisingly) by twisting.

We know from strength of materials again that the cross-sectional moment of inertia is a key to buckling resistance. For lateral buckling of a column, the minimum moment, considering 360 degrees' worth of possibilities, is the only one that matters. Thus a configuration that has a more or less constant moment of inertia regardless of orientation will be the most efficient. Clearly, round (or at least symmetrical), is the most efficient configuration.

For torsional buckling, it is the torsional moment of inertia that matters. Intersections of walls increase this moment somewhat, but the most effective configuration against torsion is to create sections that are hollow. This has the effect of moving material away from the axis of twist, somewhat analogous to the logic of an I-Beam. Thus if torsional buckling is a concern, a hollow symmetrical section may be helpful.

Brackets and Trusses

We use the term *bracket* to refer to any part whose function is to support a load or loads, transferring the force to some supporting structure, usually a wall of some kind. Brackets often must be designed with spatial constraints in mind; that is, there are regions of space that may not be trespassed by the bracket material — like the pipes in Figure 10.6(a). But whatever the constraints, some qualitative rules can help guide the generation of effective alternatives. A few examples of such rules follow.

Tension and compression are more efficient ways to transmit forces from one place to another than is bending. When bending is required, make the "beam" as short as possible and make the cross section as much like an I-Beam as possible. Avoid sharp corners so as to reduce stress concentrations; in fact, round the corners and intersections of walls whenever possible. If this isn't possible, providing add-ons like fillets and gussets can help.

In most trusses, loads are carried by compression and tension; thus trusses tend naturally to be efficient configurations from the standpoint of material use.

Sometimes, a truss-like structure must be designed to connect loads at oddly distributed points to supports at other locations. Configurations that have the potential for using as much tension and compression as possible (as distinguished from bending) are desirable. Thus placing members in the same directions as the loads often helps [1].

Inside Corners

As noted above, to reduce stress concentration, select configurations that allow for rounded corners or that can add-on fillets. As we shall see, this rule almost always supports ease of manufacturability also.

Heat Transfer

The configuration of parts whose function involves the transfer of heat can also be evaluated, in part, by qualitative physical reasoning. We know that heat is transferred by conduction, convection, and/or radiation. To the extent that the mechanism employed is conduction, then configurations that result in the lowest possible temperature gradients are the best. This is accomplished by providing the potential for using a high thermal conductivity material and by providing for a suitably large, direct path. Configurations that provide opportunities for reducing contact resistances to adjacent parts involved in the heat flow are also helpful.

When convection is the primary mechanism, to increase heat transfer we need configurations that provide large surface areas adjacent to the surrounding medium. (This is the reason fin configurations are often used.) If the configuration can help induce turbulence in the surrounding medium, that is also helpful. In the case of fin-like parts, the surface area is naturally large, but we must also consider the need for conduction of heat to the extremities of the fin. Keeping a uniformly high temperature would require a tapered fin, but this raises manufacturing difficulties. Fins that are too long and thin cannot be manufactured easily, but this is an issue that can get sorted out during the parametric design stage.

Radiation is an important heat transfer mechanism requiring that the heat exchanging surfaces be as large as possible, and be 'visible' to each other. The materials should have the potential for being made with surfaces of high emissivity and absorptivity.

Whenever possible, of course, configurations that can make use of multiple heat transfer mechanisms may be advantageous, but trade-offs here are not always obvious until computations can be made at the parametric design stage.

Insulating Against Heat Transfer

To insulate against heat transfer, we must try to prevent or reduce all three of the heat transfer mechanisms: conduction, convection, and radiation. To reduce conduction, this means configurations that can reduce or eliminate direct paths for heat conduction, especially when the material has a high thermal conductivity.

To block convection, we must prevent the circulation of the surrounding gas, usually air. Most insulating materials thus create a large number of very small pockets for gas, which (if it can't circulate to produce convection heat transfer) is a poor conductor of heat.

An interesting configuration to discuss here in the context of qualitative reasoning about configurations is the so-called 'Thermos' bottle. It virtually eliminates both conduction and convection through the sides and bottom by employing an evacuated chamber (no air to circulate). And by using very low emissivity surfaces, it reduces radiation greatly. Indeed, almost all of the heat loss from such a bottle is through the corked opening, which should therefore be kept as small as possible.

Residential windows make another interesting example of configurations that must resist heat transfer. Configurations of metal windows that allow a direct metal (or even wood) path from outside to inside are poor, of course. The concept of providing an air space between two panes in such windows is based on qualitative reasoning about heat transfer: the air space is an insulator, but it does not interfere with light flow.

10.8 Qualitative Reasoning About Manufacturability Issues

In addition to qualitative physical reasoning about functionality, configurations of parts are also strongly influenced by manufacturing issues. An overview of a number of manufacturing processes, and design for manufacturability (DFM) guidelines for them has been presented in Chapter 3. Many of the issues discussed in that chapter are rather directly relevant to part configuration design, and hence can be used to guide the generation of readily manufacturable alternatives. It is recommended that students take time now to again review the various DFM guidelines in Chapter 3.

Tolerancing of part designs so that the parts will both function well and be manufacturable also has important implications at the configuration stage. The more tolerances specified that must be met, and the tighter those tolerances are, the more difficult and costly manufacturing becomes. Therefore, in Section 10.10 we discuss tolerance issues as they relate to various manufacturing issues. Before we can have that discussion, however, we must introduce information about the types of tolerances that are employed. The best way to do this is within the context of the American National Standard on Dimensioning and Tolerancing (ANSI) [2]. Thus we introduce the ANSI Standard first.

10.9 Tolerances: The ANSI Standard

10.9.1 Introduction

The American National Standard on Dimensioning and Tolerancing (ANSI Y14.5M-1982) presents a set of standards for tolerance representation that should be followed by designers. The Standard comprises definitions and rules to help designers express their design intent adequately and unambiguously. In this book, we will only introduce the most basic elements of the Standard, which in its totality is rather long. All new designers are urged to obtain, read, and keep a copy of the Standard close by.

10.9.2 Linear and Angular Dimensions

The simplest, easiest, and most common tolerances are applied to linear and angular dimensions, represented as either limit dimensions or as plus and minus tolerances. Review Figures 1.4 and 1.5.

The effects of tolerancing can be subtle and unforeseen. Note, for example, that the tolerance applied to the diameter of the hole in Figure 10.9(a) allows the variations shown in Figure 10.9(b). Also, the issues created by tolerance stack-up (See Figure 1.4) are often not at all obvious without careful study and analysis.

10.9.3 Definition of Terms

Size Feature. A *size feature* in the ANSI Standard is either a cylindrical or spherical surface, or a set of two parallel surfaces, associated with a dimension that defines the size of the feature (e.g., a diameter or distance between the planes).

Basic Dimensions. A *basic dimension* describes a *theoretically exact* value of a geometric characteristic such as size, position, or orientation. Basic dimensions are used as locations for establishing tolerances on other dimensions. Basic dimensions are placed in small boxes as shown by the box containing the letter B in Figure 10.10.

Datum Features. Dimensions and their tolerances are often expressed in reference to what are called *datum features*. A datum is a *theoretically exact* point, axis, or plane which is used as origin from which geometric characteristics of a part are established. As a very simple example, the

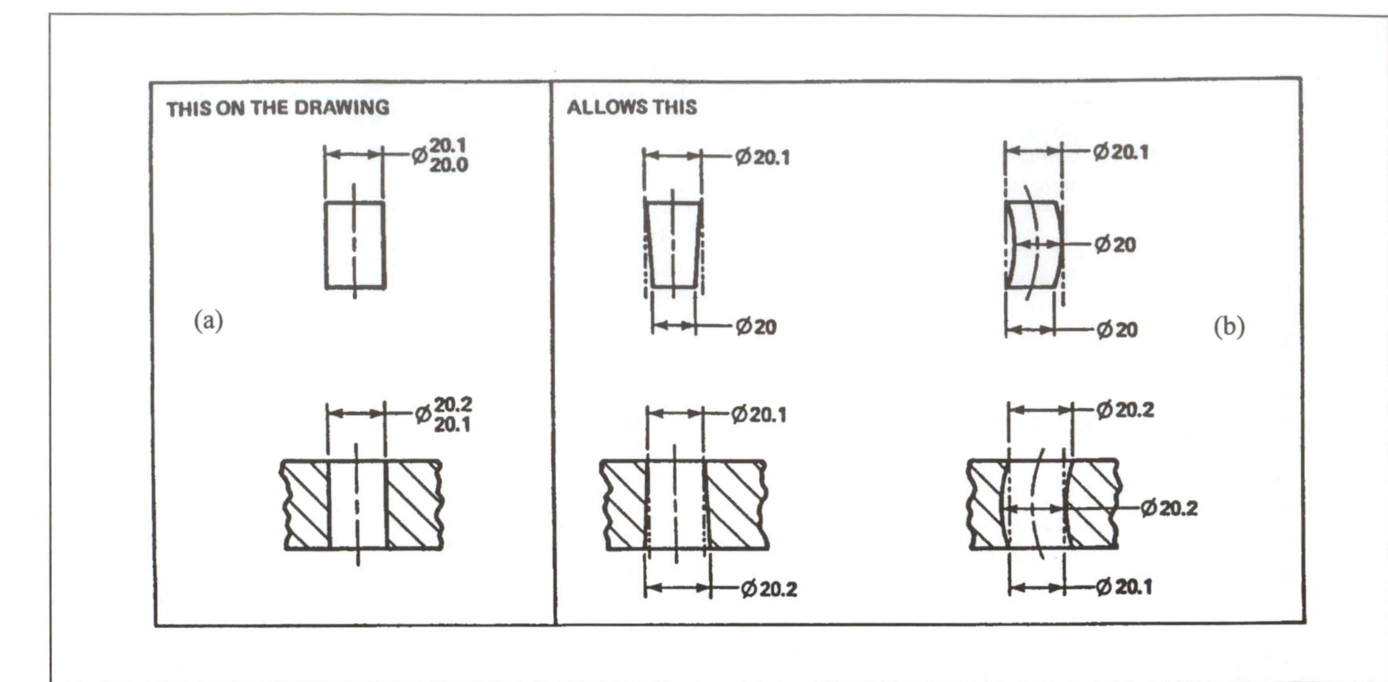

Figure 10.9 Extreme Variations of Form Allowed by Size Tolerance. Reprinted from ANSI Y14.5M - 1982, "Dimensioning and Tolerancing", by permission of American Society of Mechanical Engineers, New York.

datum for the basic dimension 2.0 shown in Figure 10.10 is the theoretical plane designated by the symbol B.

10.9.4 Classes of Tolerances and Symbols

The methods for representing tolerances discussed above are limited to linear and angular dimensions. A more comprehensive system of tolerance representation has been developed for other classes of tolerances. There are thirteen types of geometric characteristics to which tolerances may be applied, and a symbolic convention is described in the Standard for use in representing each of the various types.

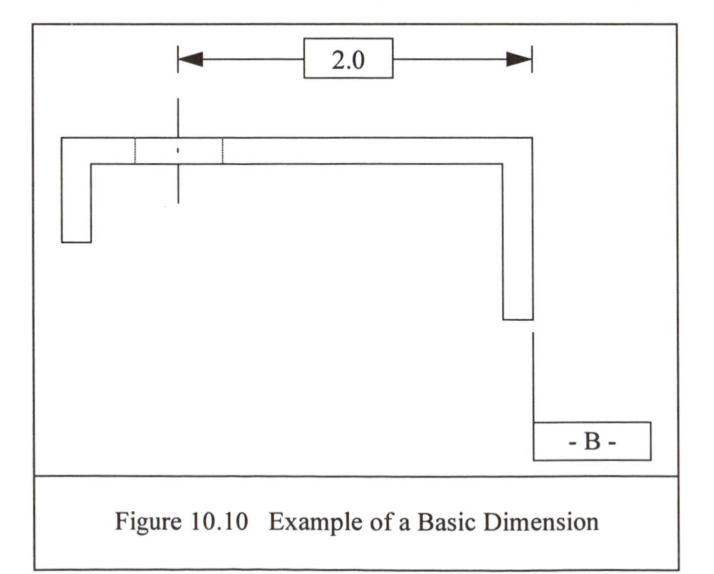

Figure 10.10 Example of a Basic Dimension

See Figure 10.11.

Location Tolerances. A positional tolerance defines a zone within which the center, axis, or center plane of a size feature is permitted to lie. Position tolerances must be referred to a datum.

It is to be noted here that in a system of tolerancing in which the axis of a hole is given a +/- tolerance in two coordinate directions, the allowed zone is rectangular (or square if the tolerance is the same value in both directions). Thus the axis of the hole can be off its specified location more in the diagonal direction than in the coordinate directions. In the ANSI system, the allowed zone can be specified (more logically) to be a circle so that the axis cannot deviate more in one direction more than another.

Orientation Tolerances. Orientation tolerances include the geometric characteristics of angularity, perpendicularity, and parallelism. Orientation tolerances can be applied to features of size (i.e., there need not necessarily be a datum involved).

Form Tolerances. Form tolerances include the geometric characteristics of straightness, flatness, circularity, and cylindricity. Reference to datum features is not necessary.

Runout Tolerances. Runout has to do with variations of surfaces when parts are rotated about a specified reference axis (e.g., the flat end of a shaft). Readers are referred to the ANSI Standard for more information on runout tolerances.

Profile Tolerances. A profile is an outline (e.g., straight lines, arcs etc.) Readers are referred to the ANSI

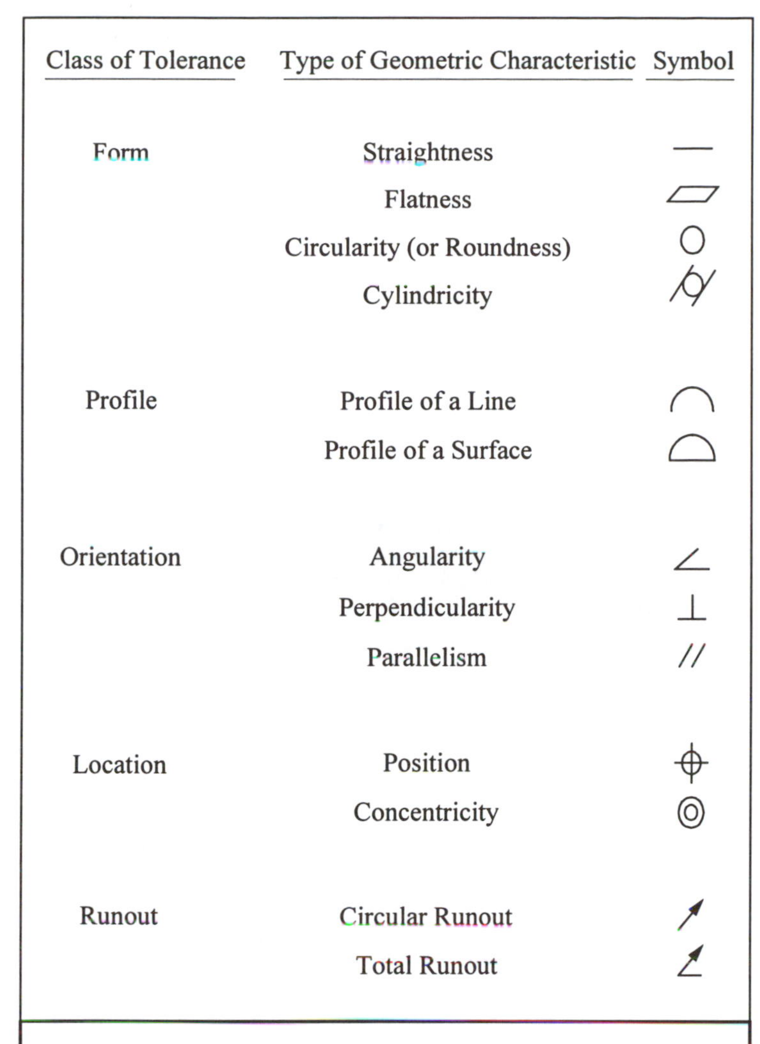

Class of Tolerance	Type of Geometric Characteristic	Symbol
Form	Straightness	—
	Flatness	⏥
	Circularity (or Roundness)	○
	Cylindricity	⌭
Profile	Profile of a Line	⌒
	Profile of a Surface	⌓
Orientation	Angularity	∠
	Perpendicularity	⊥
	Parallelism	//
Location	Position	⌖
	Concentricity	◎
Runout	Circular Runout	↗
	Total Runout	⌰

Figure 10.11 Classes of Tolerances and Symbols

Standard for a description of profile tolerancing of an object in a specified plane.

In addition to the symbols representing geometric characteristics, there are also symbols used to provide additional explanation of the tolerance. In Figure 10.12 we show only a sample of these added symbols.

Origins. Sometimes it is necessary for functional reasons to denote the origin of a toleranced dimension. Specifying an origin is illustrated in Figure 10.13(a). The origin symbol is a small circle drawn at the end of the dimension line. With the origin drawn as in Figure 10.13 (a), then the result shown in Figure 10.13(b) is allowed because the dimension and tolerance always have to be applied from the origin.

10.9.5 Feature Control Frames.

Following the ANSI Standard, designers use a structure called a *Feature Control Frame* to express and represent tolerances. Examples of feature control frames for several types of tolerances are shown in Figures 10.14 through 10.20. These figures also include notes that help to illustrate the meaning of the various tolerance types.

The generic structure of a control frame is shown in Figure 10.21.

10.10 Tolerances and Manufacturing Processes

10.10.1 Tolerance Issues in Injection Molding and Die Casting: Configuration Stage

The following features of parts in the configuration design of an injection molded or die cast part influence the ability of the process to attain tolerance specifications [3]. We refer to them as *tolerance influence features*.

- Undercuts,
- Unsupported walls and unsupported significant projections,
- Supported walls and supported significant projections,
- Primitive connections or intersections, and
- Parting plane.

Some of these features (e.g., undercuts) are clearly identifiable at the configuration stage. Others may only be known for sure to exist at the parametric stage. However, even in these cases, it is often apparent at the configuration stage, even without exact dimensions, that the *potential* for adverse influence on tolerancing exists.

The presence of significant variations in wall thicknesses will also influence tolerances, but we consider it more a parametric issue. See Sections 15.9.2 and 15.22 for additional discussion of tolerances in injection molding and die casting.

Trade associations — (for example, the Society of the Plastics Industry (SPI) — also describe the features of parts that effect the ability of processes to hold tolerances easily. SPI considers two types of tolerances: dimensional tolerances like size and location, and geometric tolerances that refer to orientation.

Figure 10.22 [3] shows a table that summarizes, in general, the relative qualitative strength of the effect that the presence of feature types have on the ease of achieving tolerances in the various tolerance classes for both injection molding and die casting. The relative strength of influence of the feature is stated as strong, moderate, or small. A blank in the table indicates very little to no influence.

By listing the above tolerance influence features and by presenting Figure 10.22, we do not imply that the pres-

Symbol	Interpretation
Ⓜ	Apply the tolerance at the maximum material condition
Ⓛ	Apply the tolerance at the least material condition
⌀	Apply to the diameter
R	Apply to the radius
↧	Apply to the depth dimension
□	Applies to the square shape

Figure 10.12 Additional Explanatory Symbols. Note: The "maximum material condition" is the condition in which the feature contains the maximum amount of material allowed by stated dimensions and tolerances; for example, a minimum hole diameter contains the maximum material allowed. The "least material condition" is the condition in which the feature contains the least amount of material allowed; for example, a minimum shaft diameter.

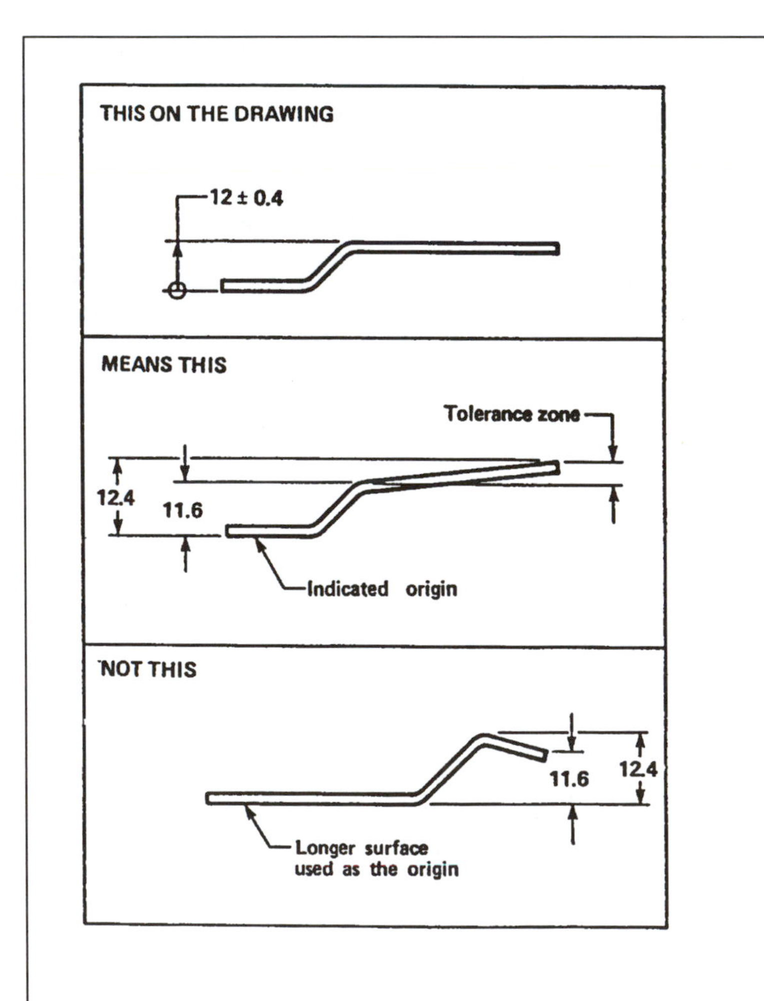

Figure 10.13

Relating Dimensional Limits to an Origin.

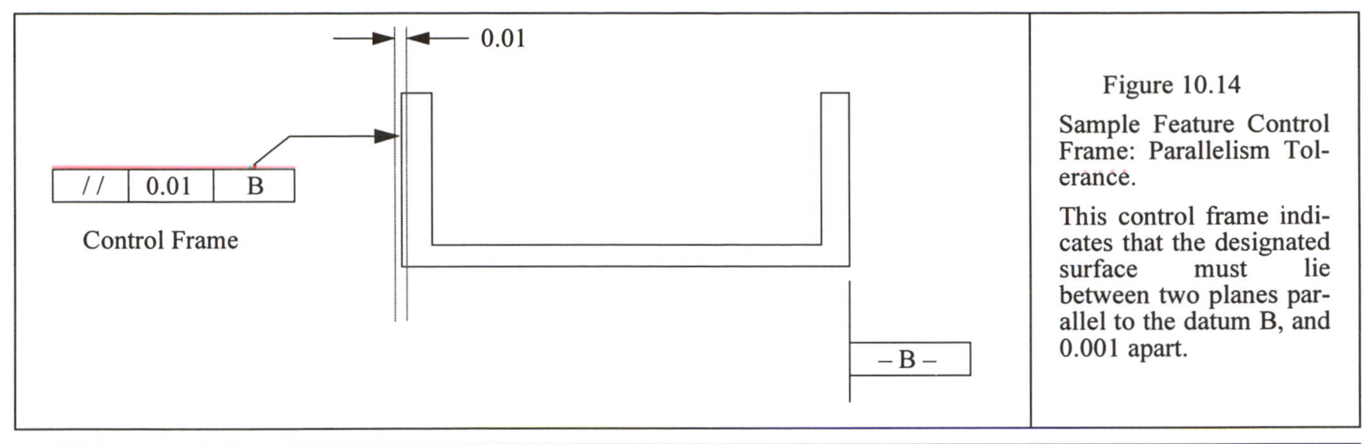

Figure 10.14

Sample Feature Control Frame: Parallelism Tolerance.

This control frame indicates that the designated surface must lie between two planes parallel to the datum B, and 0.001 apart.

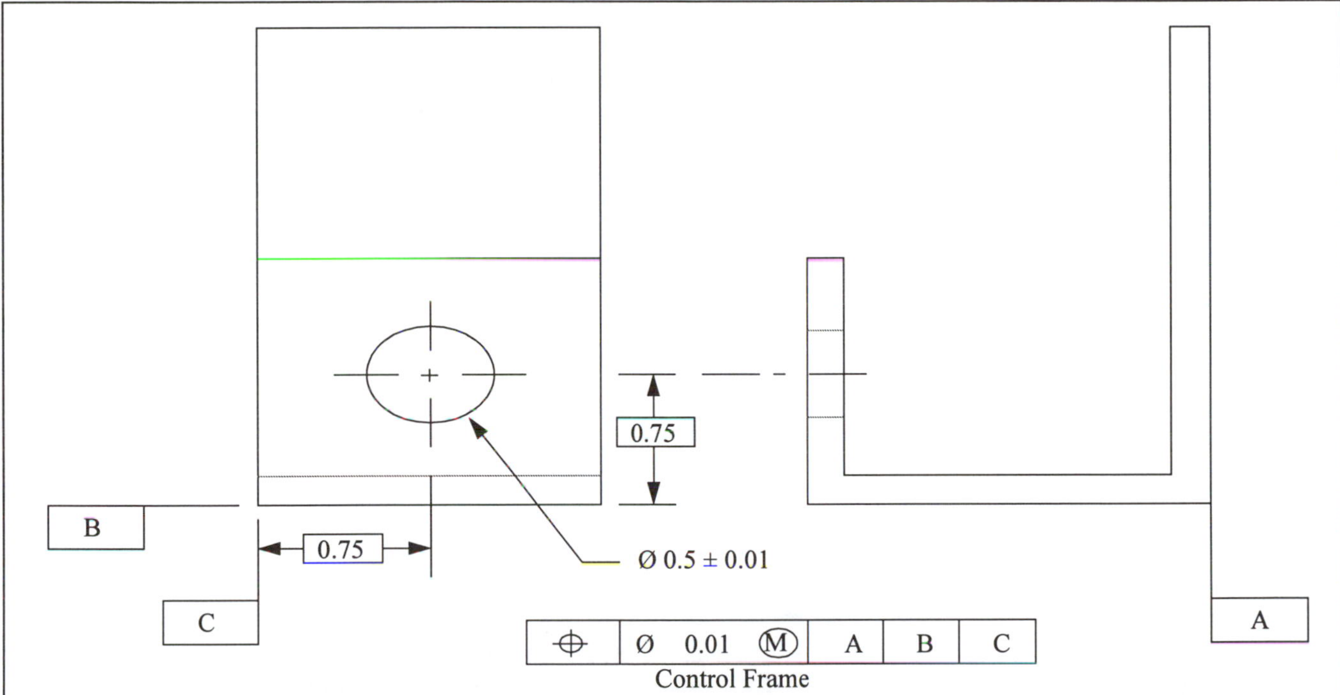

Figure 10.15 Example Control Frame: Positional Tolerance. The hole size tolerance is +/- 0.01. The control frame indicates that the location of the hole axis is to be positioned first perpendicular to datum A, then in relation to datum B, as shown, and finally in relation to datum C. The tolerance zone for the axis location is a circle of diameter 0.01 at the maximum material condition (i.e., the smallest hole diameter).

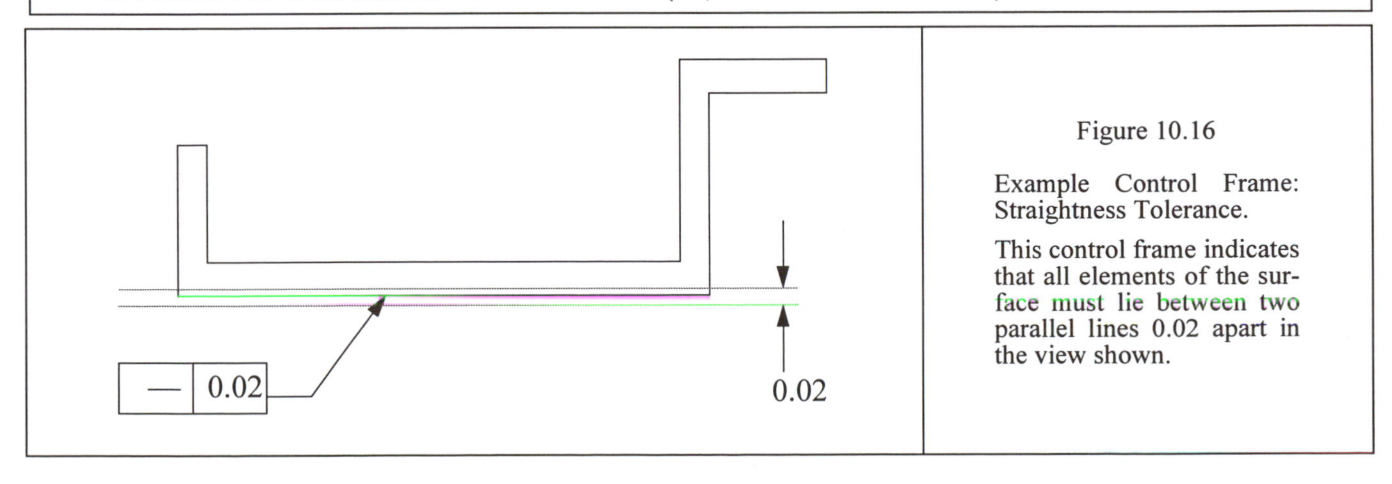

Figure 10.16

Example Control Frame: Straightness Tolerance.

This control frame indicates that all elements of the surface must lie between two parallel lines 0.02 apart in the view shown.

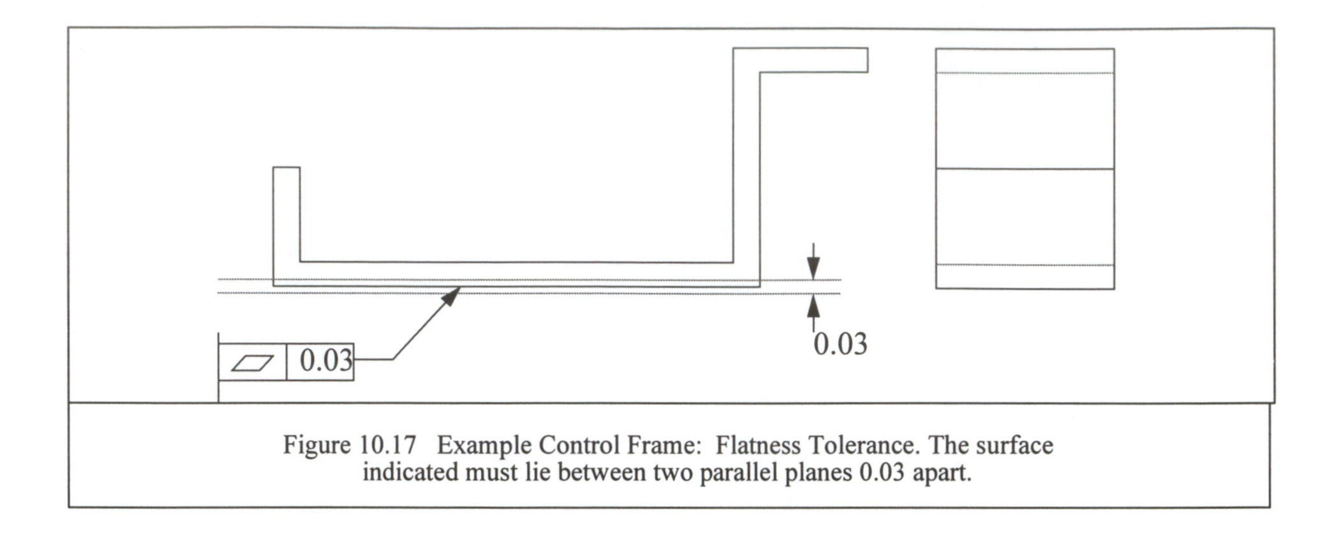

Figure 10.17 Example Control Frame: Flatness Tolerance. The surface indicated must lie between two parallel planes 0.03 apart.

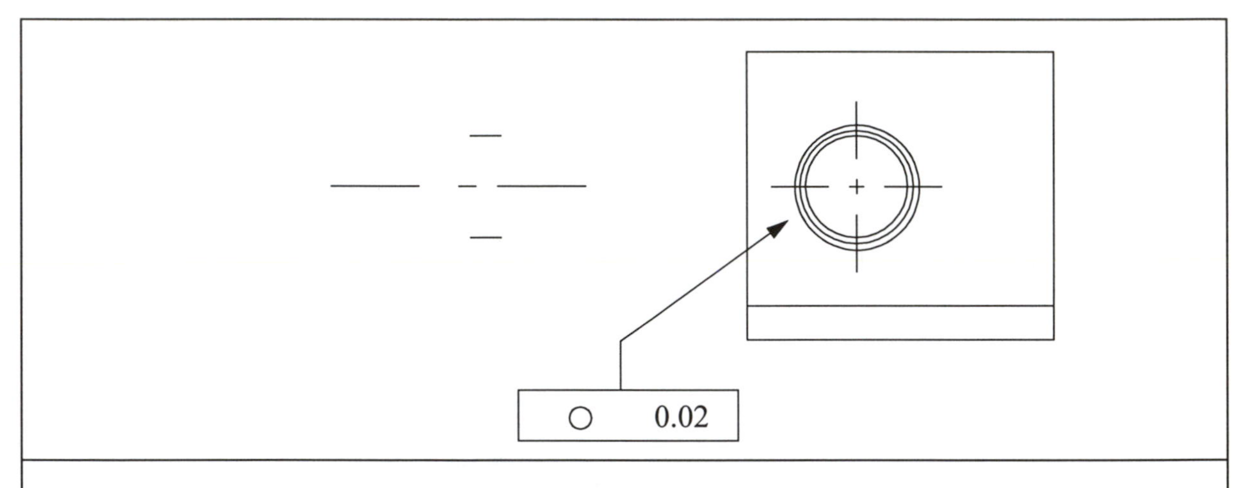

Figure 10.18 Example Control Frame: Circularity. The control frame indicates that all points on the perimeter of the hole must lie between two concentric circles, one having a radius 0.02 larger

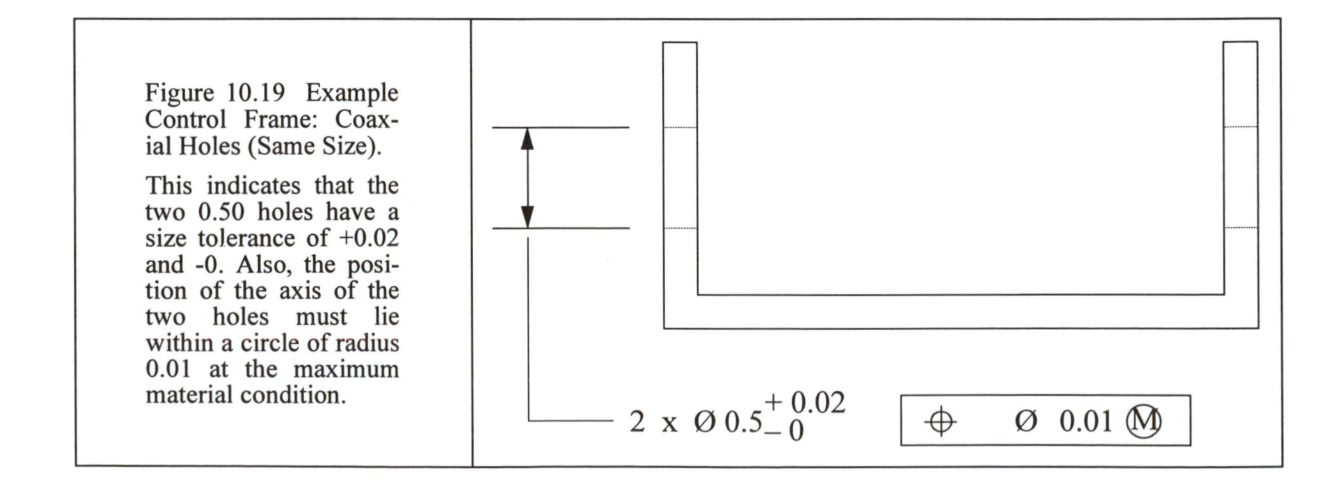

Figure 10.19 Example Control Frame: Coaxial Holes (Same Size).

This indicates that the two 0.50 holes have a size tolerance of +0.02 and -0. Also, the position of the axis of the two holes must lie within a circle of radius 0.01 at the maximum material condition.

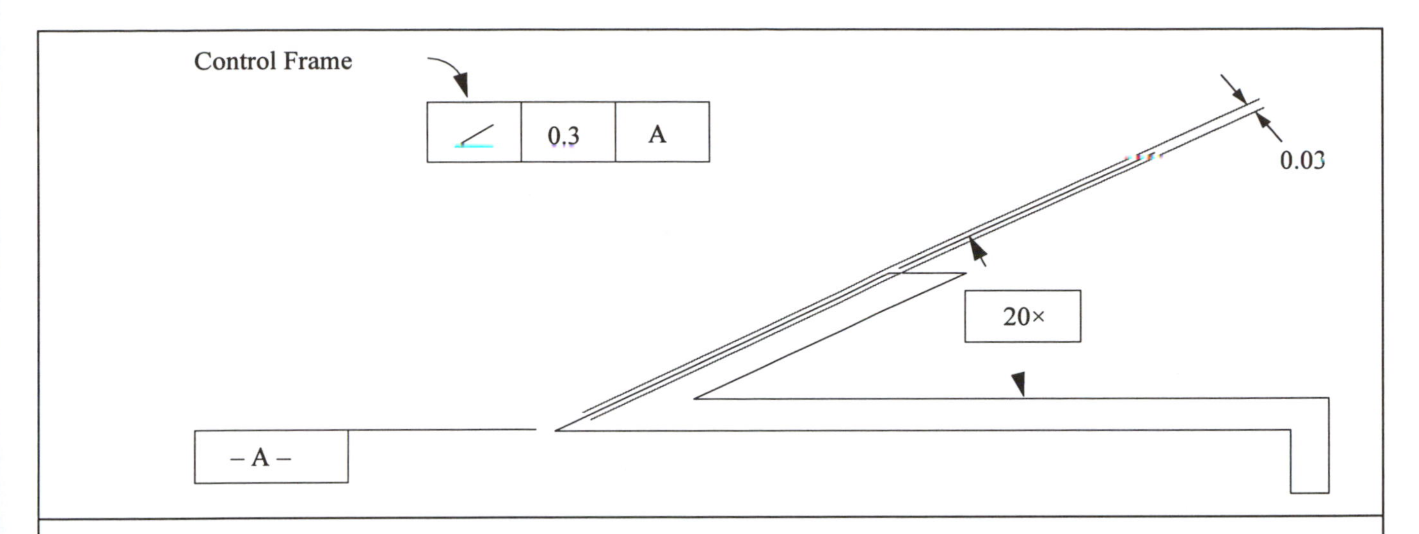

Figure 10.20 Example Feature Control Frame: Angularity Tolerance. This indicates the designer's intention that the sloping surface must lie between two parallel planes 0.03 apart inclined at 20 degrees to the datum plane A. If there were additional size dimensions and tolerances applied, those would also have to be honored.

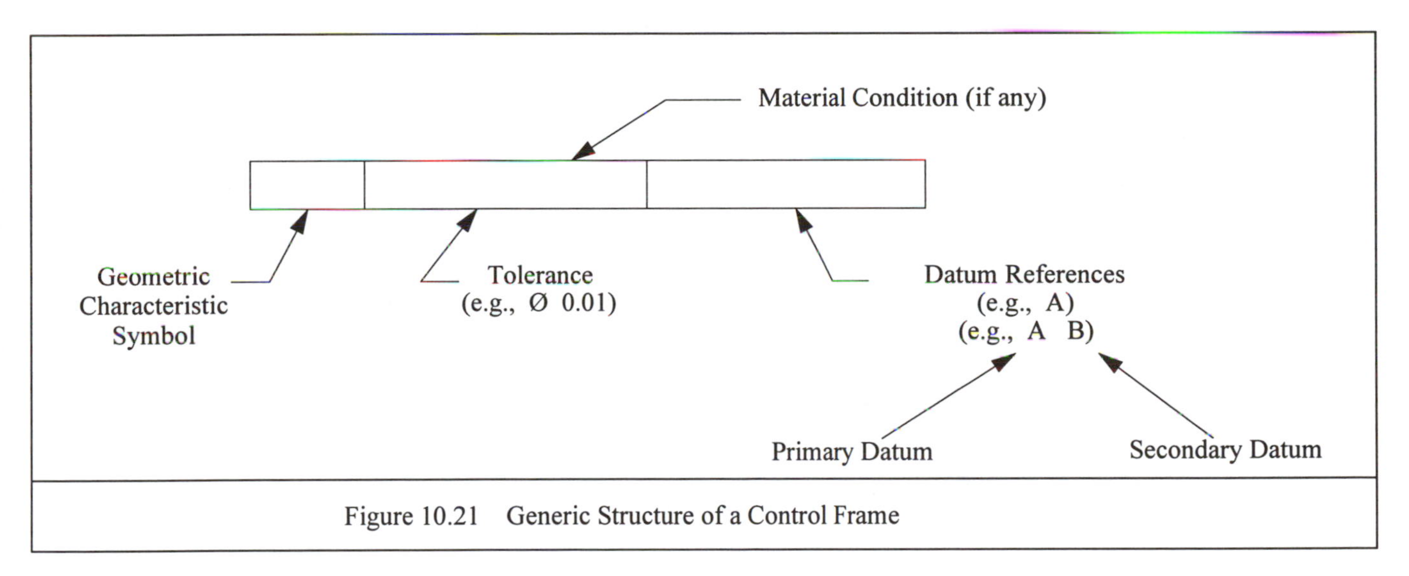

Figure 10.21 Generic Structure of a Control Frame

ence of a feature in a part *necessarily* leads to the tolerancing difficulties or added costs. The variable capability of molders and the quality of equipment make exact prediction impossible. We do imply, however, that the presence of the listed features is a signal that designers should investigate the possible impact on tolerancing. Moreover, the *number* of features with a Strong (and/or Moderate) influence on a tolerance class should be kept as small as possible.

Undercuts. Undercuts are defined and described in Chapters 3 and 11. As shown in Figure 10.22, their presence has a strong adverse effect on the ability to produce fine tolerances in almost every tolerance class. The main reason is that the side action units used to manufacture parts with undercuts must themselves have clearances and tolerances to enable them to move.

Parting Plane. Size dimensions that extend from one side of the parting plane to the other are relatively difficult to control closely. The reason is that there is an added tolerance due to the parting plane itself. Moreover, the mold will generally expand and contract during the molding process, adding to dimensional uncertainty and consistency.

Unsupported Walls and Unsupported Significant Projections. Unsupported walls and unsupported significant projections are defined in Chapter 11. The angularity of a projection that is unsupported is difficult to control closely. The straightness is also hard to control, especially if the projection is long. Therefore, dimensions involving the exact location of the extremities of a projection are also a problem to control.

As an example, the overall length of the part shown in Figure 1.5 is difficult to control because of all the unsupported walls included within the dimension.

Supported Walls and Supported Significant Projections. Tolerances associated with walls and significant projections (e.g., a tall boss) that are supported (say, by a gusset) are much easier to control.

Non-Significant Projections. Non-Significant projections are defined in Chapter 11. These projections, because they are usually shorter, are less difficult to control than unsupported projections, but still have some influence on the ability to control angularity tolerances.

10.10.2 Tolerances Issues in Stamping: Configuration Stage

The tolerance influence features in stamping are:

•Side-action Features (Undercuts)

•Closely spaced features

•Multiple Bends

Side-action features and closely spaced features are defined the same as in injection molding and die casting. Tolerances involving multiple bends (e.g., on dimensions across one or more bends) are difficult to control in stamped parts because the bends themselves are difficult to control exactly, in part due to springback of the sheet after bending.

Figure 10.23 [3] shows a table that summarizes, in general, the relative qualitative effect that the presence of each feature type has on ease of achieving tolerances in the various tolerance classes for both injection molding and die casting. The relative influence of the feature is stated as strong, moderate, or small. A blank in the table indicates little to no influence.

10.10.3 Tolerance Issues in Aluminum Extrusions: Configuration Stage

ANSI has developed clear information about standard tolerances for aluminum extrusions as shown in Figure 14.8 [4]. This information is presented as parametric rather than configuration information, but the data imply some guidelines about configuration design of extrusions, and there are also guidelines published in the same reference that include some configuration stage information. Some of these guidelines have to do with ease or cost of manufacture, but many also are related to tolerance control. Indeed, many DFM guidelines, whether they explicitly say so or not, are related to tolerance control.

Configuration guidelines for ease of tolerance control in aluminum extrusions include:

- Avoid tight straightness tolerances on long, thin wall sections. Add small ribs to help achieve straightness.
- Angles of unsupported wall sections are difficult to control precisely. Hence, tolerances on size dimensions involving long legs are difficult to control. If necessary for tolerance control, support long legs with webs connecting to other parts of the extrusion. (This may create a more costly hollow shape, but if tolerances are critical, it may be necessary.)
- Using uniform wall thickness throughout the cross section makes all dimensions easier to control.
- Rounding inside corners makes control of angles easier to control than when corners are sharp.
- The dimensions of symmetrical shapes are generally easier to control than non-symmetrical shapes.

Note how some of the issues here are similar to those in injection molding; the reason is that both are internal flow processes.

10.10.4 Tolerance Issues in Forging: Configuration Stage

Shrinkage and warpage contribute largely to the difficulties of controlling tolerances in forging. By designing forged parts so that they have simple shapes and common proportions, designers can help minimize these effects. It is also important to provide for well rounded corners, fillets, and edges. Thin sections, such as webs, should be kept to a minimum in number, and should be made as thick as possible.

As in injection molded and die cast parts, dimensions that extend across the forging die closure parting plane are more difficult to control than those on one side only of the parting plane.

10.10.5 Design for Tolerances (DFT): General Guidelines for Configuration Stage

In addition to the process-specific guidelines outlined above, there are some DFM design guidelines relevant to tolerances -- we could call them *DFT* guidelines -- that apply to all processes:

- Designers should consider tolerances with not only the function in mind, but also with knowledge of the standard tolerance capabilities of the proposed manufacturing process.
- Designers should minimize the need for and use of tolerances; not every dimension requires a special tolerance. The fewer tolerances that are specified, the easier the part will be to manufacture.
- Where tolerances are required, designers should attempt to avoid the need for tolerances that are tighter than standard for the proposed process.
- If non-standard tolerances cannot be avoided, then the number of them should be minimized.
- void redundant dimensions and consider stack-up effects in evaluating tolerances.

A method for evaluating proposed tolerance plans for trial part designs is described and illustrated in Chapter 13.

Tolerance Class	Tolerance Influence Features				
	Unsupported Walls and Unsupported Significant Projections	Supported Walls and Unsupported Significant Projections	Non-Significant Projections	Parting Plane	Undercuts
Size	—	—	—	Moderate	—
Position	Strong	Moderate	Weak	Strong	Strong
Concentricity	Strong	Moderate	Weak	Strong	Moderate
Angularity	Strong	Moderate	Weak	Strong	Moderate
Perpendicularity	Strong	Moderate	Weak	Strong	Moderate
Parallelism	Strong	Moderate	Weak	Strong	Moderate
Flatness	Strong	Moderate	Weak	Strong	Moderate
Circularity	Moderate	Weak	—	Strong	Moderate
Cylindricity	Moderate	Weak	—	Strong	Moderate
Straightness	Moderate	Weak	—	Strong	Moderate

Figure 10.22 Relative Strength of Tolerance Features for Injection Molding and Die Casting

10.11 Example Configuration Design: Extension Cord Carrier

10.11.1 Information Available From Industrial and Engineering Conceptual Design

The product in this example is a real one, manufactured and distributed by Bayco Products Company, Richardson, Texas. However, the figures on marketing, costs, production volume used in the example are entirely imaginary.

From the Product Marketing Spec, we imagine that the following has been learned:

> Surveys have indicated that a large number of tradespeople (e.g., home builders, electricians, masons, etc.) and do-it-yourself homeowners would purchase an inexpensive carrying and storage device for electrical extension cords. The in-use purpose, therefore, is convenient handling and storage of a long (say, up to 150 feet) electrical cord.

> If the product can be sold, say, at a retail price of $1.50, the expected first year sales are 1,000,000 units. Distribution costs and wholesale and retail mark-ups require that the product be manufactured for a cost of no more than 50 cents per unit.

From an Engineering Design Specification, we learn:

> Special purpose features include provision of a convenient method(s) for hanging the storage device and cord on, say, a garage wall or in a panel truck. A handle should be comfortable for carrying. There should be an easy way to grip the plug and socket ends of an extension cord so they don't flop around and cause the cord to unwrap.

> Unintended uses include storing or hanging things from the device other than extension cords (e.g., clothes, 'heavy' tools, rope, chain, etc.) Of course, the more of these that can be accommodated without a negative effect on the primary purpose, the better.

> The service environment includes normal atmospheric (summer and winter) temperatures, home garages, panel trucks, and the like. The device should not rust or corrode if left out of doors.

> A 150 foot extension cord weighs about 10 pounds. The outside diameter is about 5/16 inches.

From the selected physical concept/embodiment, we learn that the product is to be a single thermoplastic injection molded truss-like part.

10.11.2 Formulating the Configuration Design Problem

1. Review the Relevant Specifications.

   This has just been done above.

2. Identify the Couplings.

   This device has couplings with: the human hand for carrying; a nail or hook, and the associated wall, for hanging storage; the wrapped cord; and the cord ends (electrical plugs and sockets).

3. Can the Part Be Eliminated or Combined?

   No. (It is a single part product.)

4. Can a Standard Part Be Used?

   No. (Nothing exists to meet the special purpose requirements.)

5. Prepare a Part Configuration Requirements Sketch.

   A possible requirements sketch is shown in Figure 10.24(a) and (b).

In (a), which is roughly pictorial, shows two possible hand locations, the wrapped cord, the cord ends, and sample hooks for wall storage. Of course, the locations of these elements are not at all fixed by this sketch, and the wrapped cord would not have to be in the relatively long, narrow shape shown; it could have a more rectangular shaped wrapping. Figure 10.24(b) shows a two-dimensional sketch of the forces that the part must support (with the hand in one position), but again there is no intent to establish exact magnitudes or locations with the sketch.

Note that these configuration requirements sketches

	Tolerance Influence Features		
Tolerance Class	Side-Action Features (Undercuts)	Closely Spaced	Across Multiple Bends
Size	—	—	—
Position	Strong	Weak	Strong
Concentricity	Strong	Weak	Strong
Angularity	Weak	Strong	Weak
Perpendicularity	Weak	Strong	Weak
Parallelism	Weak	Strong	Weak
Flatness	—	—	—
Circularity	Strong	Strong	—
Cylindricity	Strong	Strong	—
Straightness	—	—	—

Figure 10.23 Relative Strength of Tolerance Influence Features for Stamping

are a bit like 'negatives' of the design; they show the future designed object's surroundings, but not the object itself. And they are not to any exact scale; there may be different ways to construct them in a configuration sense also. For example, in the example being considered, the hands can be in different orientations to the wrapped cord, and the location of the plus and socket holders is quite optional. But the sketch indicates what the part must do, what it must interact with in some way, and is thus a good way to start the configuration design process.

10.11.3 Generating Alternative Configurations

Non-Contiguous Alternatives. Though there could be others, for purposes of this example, we have selected only the requirements configuration shown in Figure 10.24 (b) as the basis for generating the non-contiguous part design shown in Figure 10.25. Notice how material has been placed to support the loads from the hand and cord, and to accommodate the hooks or nails. We have not yet included provisions for the plug and socket holders.

The part here is essentially two-dimensional. This problem presents us with two different configuration design problems: the overall configuration of the truss, and the cross sectional configuration of the elements of the truss. First we will consider the overall configuration.

Contiguous Alternatives. Now we join up the disconnected chunks of the non-contiguous design in order to make a set of contiguous alternatives. Figure 10.26 (a) shows probably the simplest contiguous design based on Figure 10.25. Figure 10.26 (b) results from a bit of qualitative reasoning about supporting the forces created by the wrapped cord. And Figure 10.26(c) also is based on additional reasoning about deformations in the box-like truss of Figure 10.26 (b). The triangles in (c) are more rigid.

There are, of course, many more configurations that can evolve from this beginning. One that was used in a successful product is shown in Figure 10.27.

We leave it as an exercise for students to design the configurations of the cross section of the truss' features.

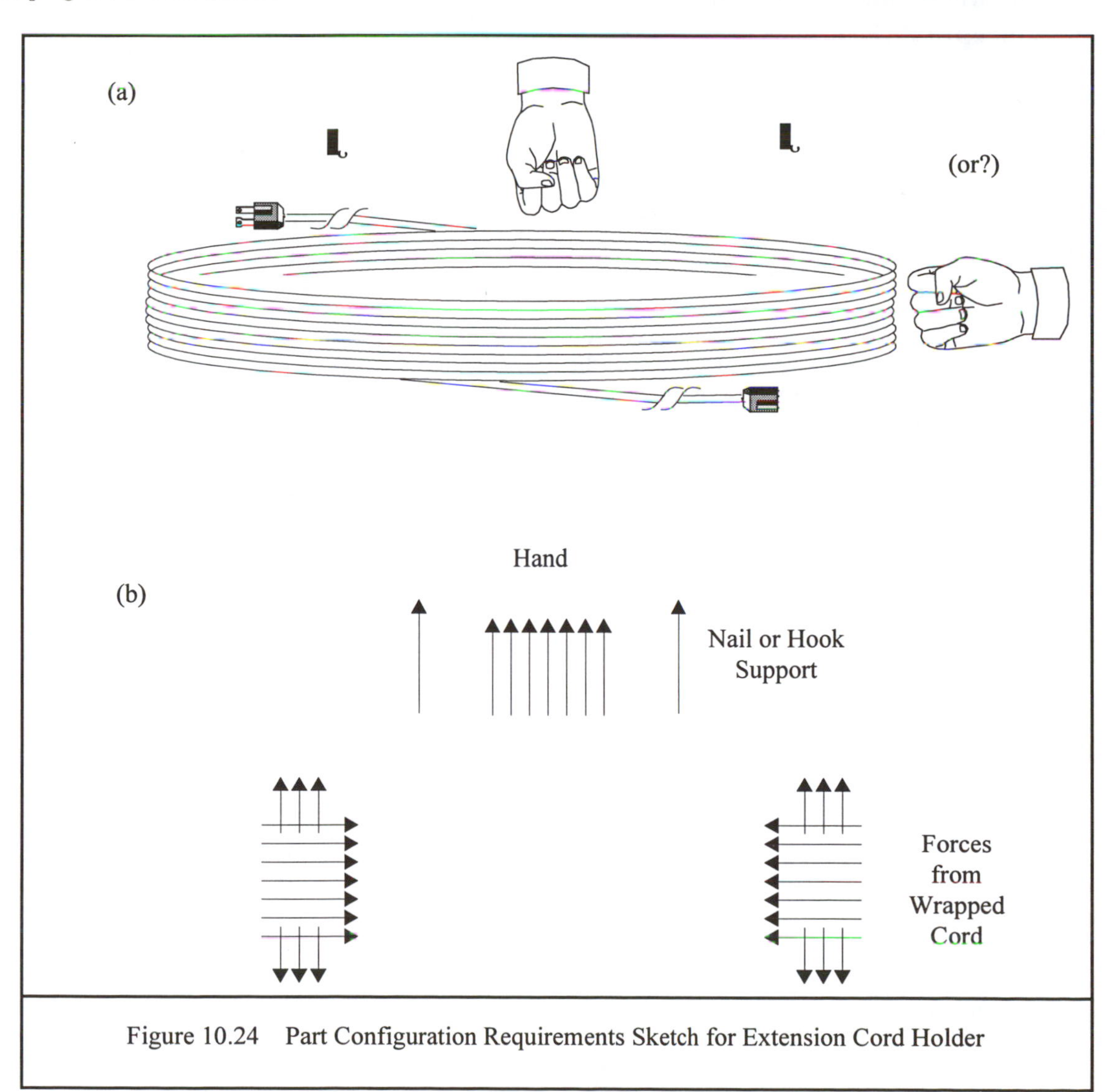

Figure 10.24 Part Configuration Requirements Sketch for Extension Cord Holder

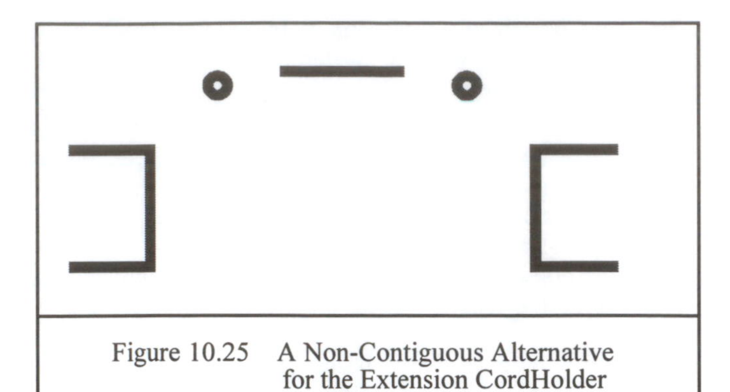

Figure 10.25 A Non-Contiguous Alternative for the Extension CordHolder

10.12 Example: Section of Enclosure for Laboratory Instrument

10.12.1 Information Available From Industrial and Engineering Conceptual Design

In this example, the main section of the enclosure for a laboratory instrument has been designed, but a significant gap in the enclosure remains. The material selected for the enclosure as a whole, and for the section still to be designed, is a thermoplastic to be made by injection molding. The section must match the rest of the enclosure in style and appearance. It must complete the enclosure and also provide support for three electronic components to be mounted inside the section.

This part, then, is one that has been "born" as a result of decomposing the enclosure into two parts. In this case, the material and process selection are obviously inherited from the choices made for the enclosure as a whole. Other factors, such as the color and surface texture of the part will also be taken directly from the Spec for the enclosure as a whole. From the Engineering Design Spec for the part, we learn that the weight of the electronic components is approximately three ounces, and we get the nearly exact dimensions of brackets proposed for their mounts. The concept is that these components will be mounted on bosses molded on the inside of the enclosure section to be designed.

10.12.2 Formulating the Problem

1. Review the Relevant Specifications.

   This has just been done above.

2. Identify the Couplings.

   This device has couplings with the main section of enclosure, and with the three electronic components to be mounted on it.

3. Can the Part Be Eliminated or Combined?

   Well, apparently not, though it might be worth investigating whether a single enclosure section could be designed instead of the sectional one proposed.

4. Can a Standard Part Be Used?

   No.

5. Prepare a Part Configuration Requirements Sketch.

   We will assume here that a similar part is not available as a starting place, and that we want to begin by developing a part configuration requirements' sketch.

A possible requirements sketch is shown in Figure 10.28. The sketch shows the main section of the enclosure as it has so far been designed. The sketch also shows proposed locations for the brackets proposed to hold the three electronic components. (Note: The sketch shown does not attempt to show the details of the exposed thin wall sections which, in the real part, may have a small tongue and groove or lap joint configuration for mating with adjoining wall sections.)

It should be noted that the amount of knowledge a part designer has about the kind of couplings shown in Figure 10.28 will vary greatly. Consider, for example, the process that has apparently gone before resulting in the design

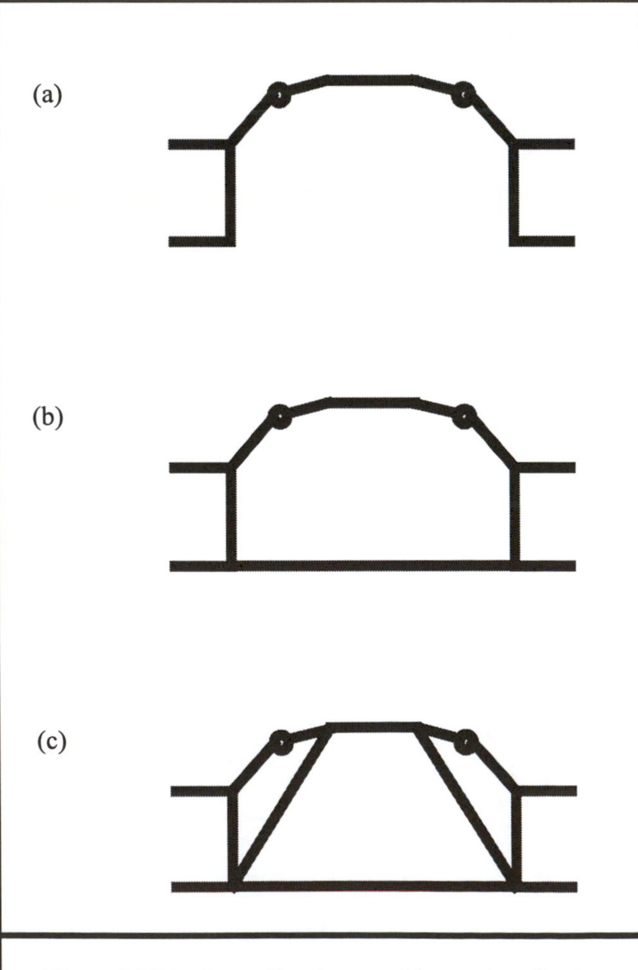

Figure 10.26 Some Contiguous Alternatives for the Extension Cord Holder

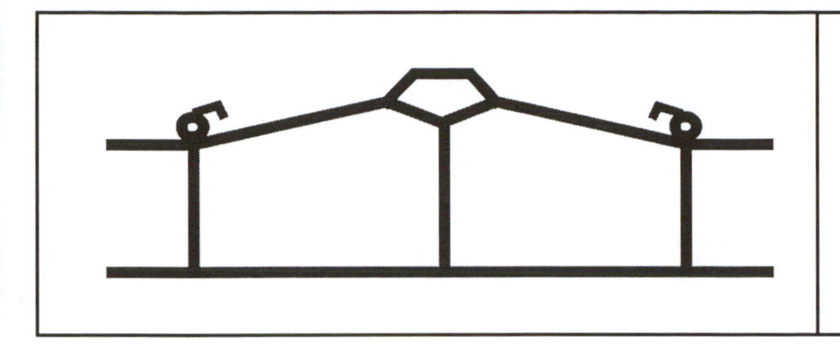

Figure 10.27

This is close to the configuration of the commercially successful product patented and marketed by Bayco Products Co., Tyler, Texas.

of the main enclosure section shown. Since the fill-in section had obviously not yet been designed, the specific nature and location of the coupling between the two sections could not have been sketched to support that design process. Something, therefore, had to be assumed. We have then a kind of "Which comes first, the chicken or the egg" situation. In this case, "Which comes first, the main section or the smaller piece"? If the smaller piece isn't designed, then it cannot be used as a part of requirements' sketch for the main part, and vice-versa. In this case, the designer of the first part to be tackled has, in effect, to design the intersection between the two in order to get started. As we have noted earlier, he or she should keep this intersection (i.e., coupling) as flexible and as simple as possible.

Note how any assumptions made about the intersection will strongly influence the shape of the adjoining part. Also note how important it is to us in this example that the main section designer leaves us with a simple coupling situation, and with as much flexibility as possible to make changes that affect only the local region of the coupling rather than the whole other part. As we have said: couplings should be as simple as possible; and couplings should be as independent as possible of each other and the functions of the part and other parts.

Since the forces in this case are very small, they really have no influence on the configuration. Any enclosure section rigid enough to support itself will support the light weight components.

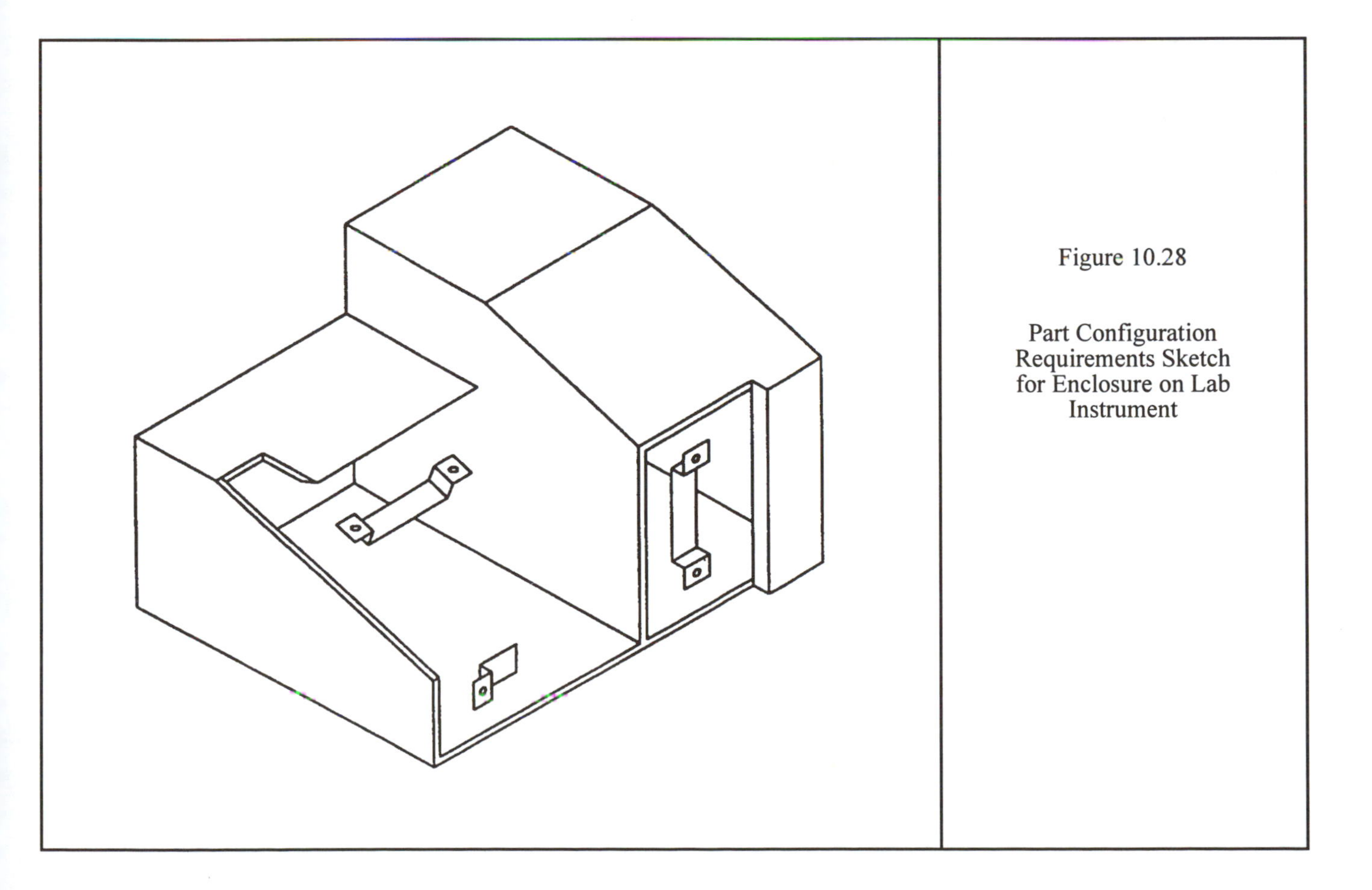

Figure 10.28

Part Configuration Requirements Sketch for Enclosure on Lab Instrument

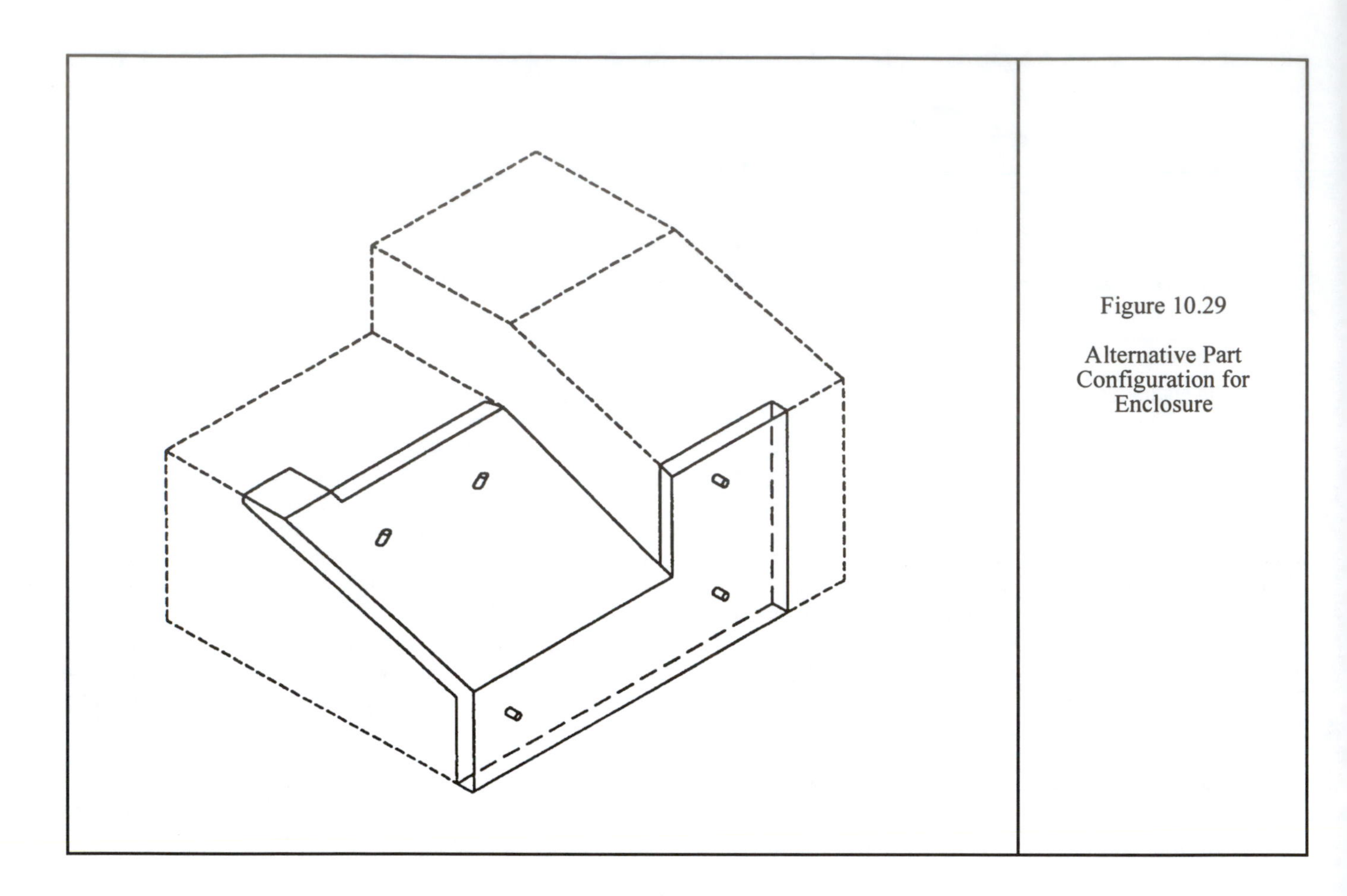

Figure 10.29

Alternative Part Configuration for Enclosure

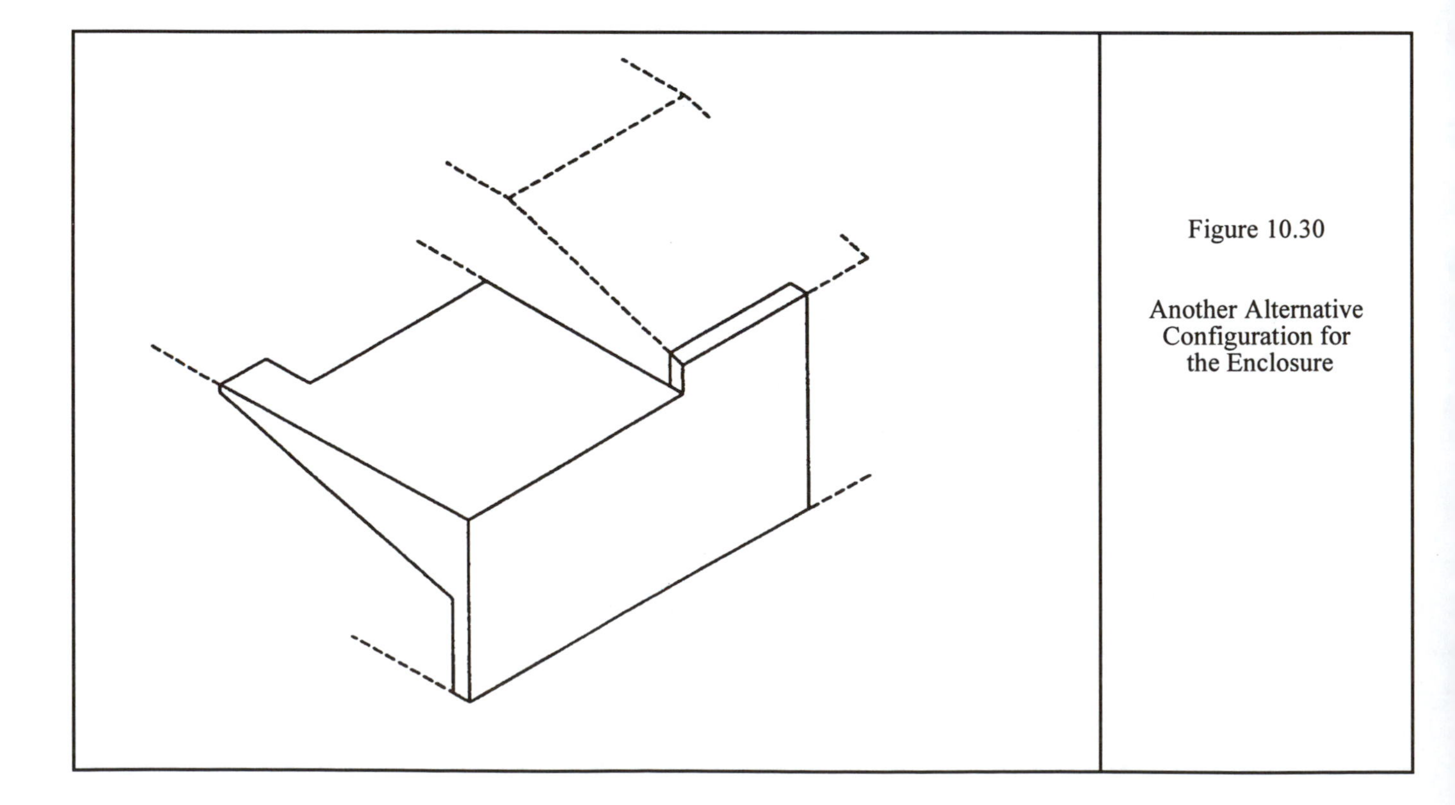

Figure 10.30

Another Alternative Configuration for the Enclosure

10.12.3 Generating Alternatives

In this example, in which options have been restricted by the prior design of the main enclosure section, there appear to be few reasonable alternatives possible. Two possibilities are shown in Figures 10.29 and 10.30. Note that Figure 10.30 requires that the location of one of the electronic components be moved slightly. Before using this configuration, therefore, we would have to make sure that this small change does not confound the design of some other parts or sub-assembly of the whole product. Another problem with Figure 10.30, perhaps, is that aesthetically it appears rather "clunky" with its horizontal top wall instead of the nicely inclined one in Figure 10.29 which matches the incline on the main enclosure section.

10.13 Continuation of Design Project A

Like the extension cord carrier, the factory table system presents two kinds of configuration problems. One is to configure the cross section of the aluminum extrusion(s) to be used as legs, cross pieces, and the like. The other is to configure the connections among these parts; that is, to determine the manner in which they will be connected. Figure 10.31 shows a number of possible cross sections generated by several student teams working on this project. Figures 10.32 and 10.33 show two approaches to the assembly configuration.

In this case, the extrusions and the connectors (the connectors will likely be cast aluminum) must really be designed concurrently. The reason is that the manner of physical coupling between the extrusions and the connectors is a major design issue. Of course, the extrusions must be sufficiently stiff in bending and twisting to support the expected loads, and this must be provided for in the extrusion's cross section. Though parametric design can produce thicker or thinner walls in the extrusion, if the cross sectional configuration is not an effective one in efficiently carrying the loads, no assignment of parametric values will be able to make the design a good one.

Students doing this project should try to find an opportunity to handle several aluminum extrusions that are not hollow. Many hardware stores sell some standard aluminum extruded shapes. Note the strength of the non-hollow cross sections in torsion. If you can find cross sections that are hollow, note how that configuration adds considerable torsional stiffness, and how it is also effective in carrying bending loads. (A square hollow beam is like an I-Beam with two webs).

The more hollow sections within a cross section, the better the torsional strength. On the other hand, the cost of the extrusion per pound also increases because hollow sections, especially with multiple hollows, and are more difficult to manufacture. Dies for complex hollows are also more expensive, but this is not a major concern in this case because of the expected volume.

10.14 Summary and Preview

Summary. Configuring parts correctly so that they meet their functions efficiently *and* can be easily manufactured and assembled is crucial to effective product design. Of course, if parts can be eliminated or combined with other parts, that is most desirable. And if parts, especially standard parts, can be outsourced, that is often the best course. But there will still be many parts that will have to be specially designed and produced.

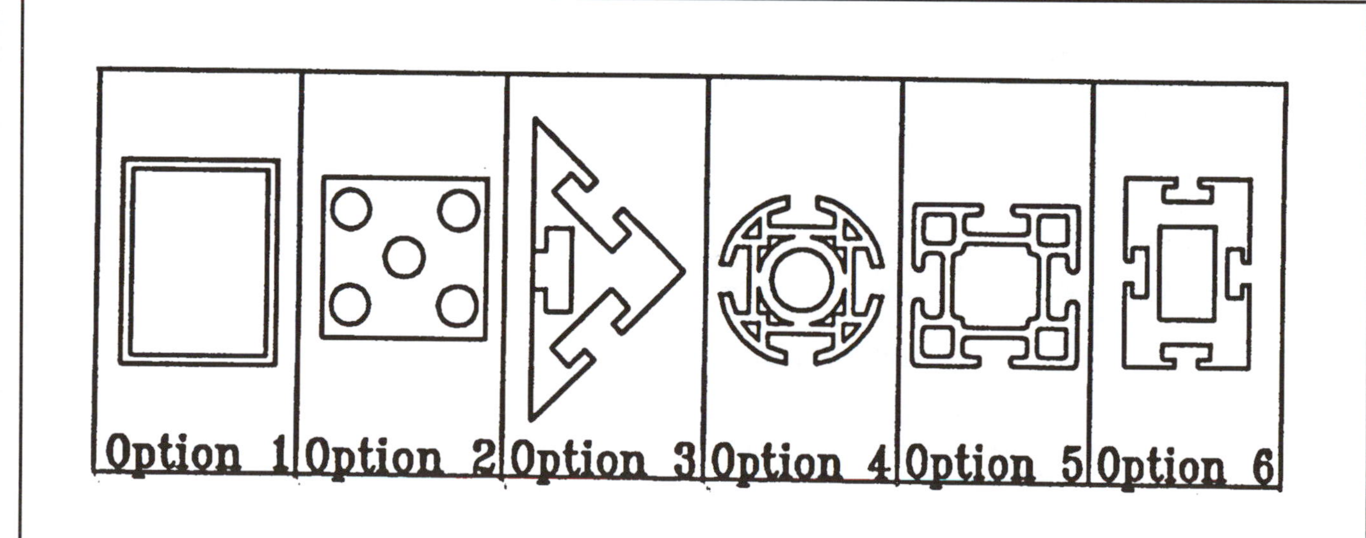

Figure 10.31 A Student Team's Alternative Configurations for Extruded Members

Figure 10.32

A Student Team's Alternative Configurations for Corner Parts

In some cases, the requirements of new parts can be closely matched with previously produced parts. In these cases, it is efficient and helpful to learn as much as possible from the previous design and manufacturing experience. Sometimes the existing drawings or CAD representations can be used. The opportunities for time and money savings offered by starting a design with a previous drawing should not be missed. At the same time, designers should not short circuit the process of considering thoroughly the requirements and specifications for the new part. As always, issues overlooked are likely sources of future trouble.

Designing configurations involves balancing the needs of functionality with the needs of manufacturing. Since there are only approximate numerical values for attributes available at the configuration stage, the reasoning to produce an optimum balance is necessarily qualitative. This is one reason it is so important for designers to understand both physical principles and manufacturing processes in fundamental ways.

In this chapter, we have presented a rather structured method for formulating and generating alternatives for configurations of special purpose parts. The intent is to help students and new designers to get started in part design without mental blocks (or even panic!), and to avoid major mistakes of either omission or commission. But experienced or inexperienced, designers should always keep *all* the requirements of the relevant Engineering Design Specifications in mind, *and* always keep the nature and DFM guidelines of the manufacturing process in mind.

Preview. In formulating and generating alternatives for part configurations, qualitative reasoning about manufacturability is required as discussed in this chapter. But much more detailed manufacturability evaluation is needed before a part configuration is finally selected for parametric design. In the next two chapters, therefore, we present more rigorous methods for evaluating the manufacturability of part configuration designs for injection molded, die cast, and stamped parts.

References

[1] Nevill, G. E., Jr. and Paul, G. H., Jr. "Knowledge-Based Spatial Reasoning for Designing Structural Configurations", Proceedings of the ASME Computers in Engineering Conference, Vol 1, 1987.

[2] ANSI Y14.5M - 1982, "Dimensioning and Tolerancing", American Society of Mechanical Engineers, New York.

[3] Fathaillal, A. K "Feature-Based Representations to Support the Evaluation of Component Tolerances for Manufacturability", M. S. Thesis, University of Massachusetts, 1992.

[4] "American National Standard Dimensional Tolerances for Aluminum Mill Products", ANSI-H35.2-1993, The Aluminum Association, Inc., Washington, D.C., 1993.

Problems: See Page 10-30

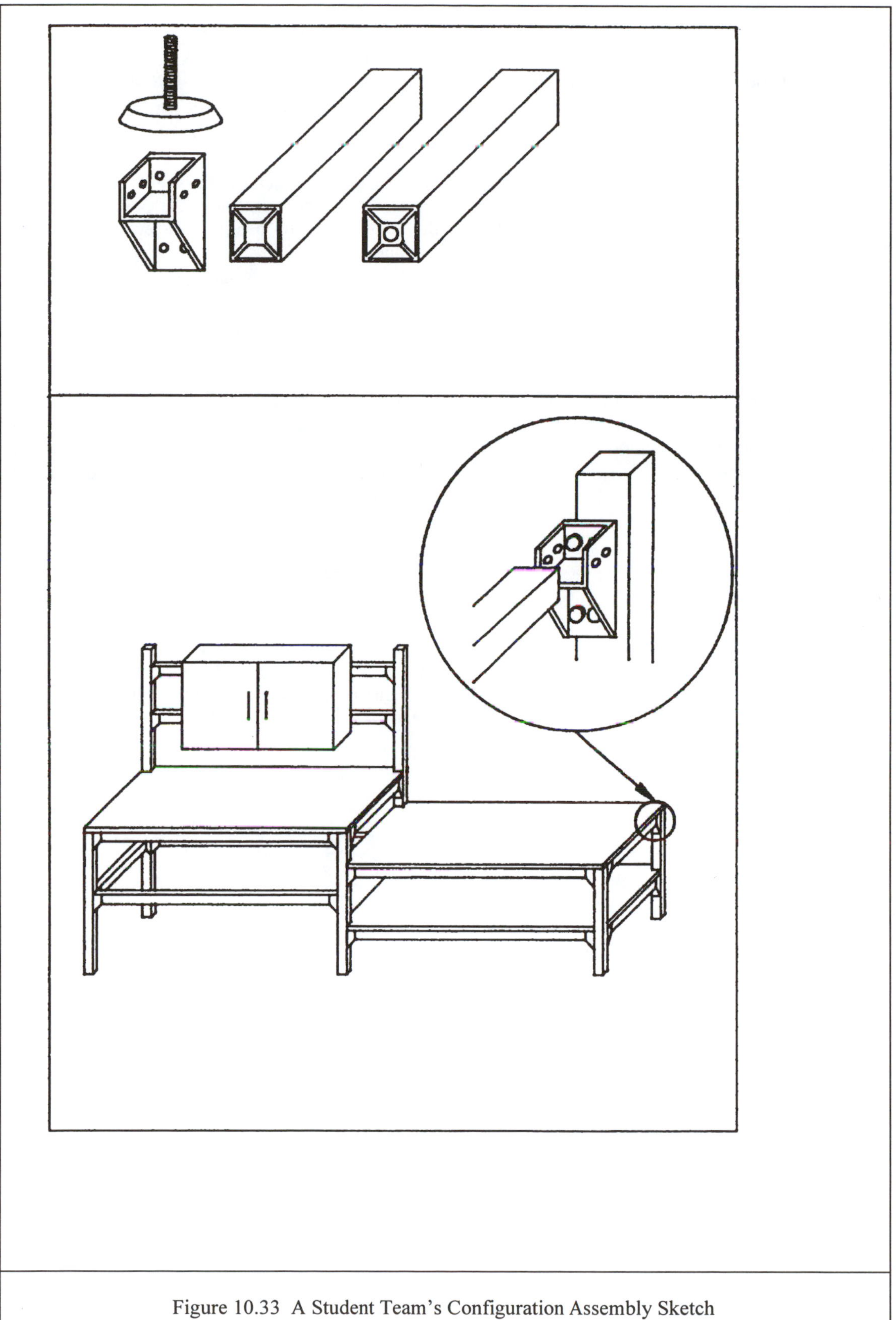

Figure 10.33 A Student Team's Configuration Assembly Sketch

Problems

10.1 Define a set of attribute value ranges for a configuration class of: coffee cups, or screw drivers, or lead pencils, or bolts, or shafts, or bookcases, or some product line of your choice.

10.2 A box beam has four walls arranged to form a rectangular hollow. Draw a set of three box beams of similar configuration. Draw three more that, though they have the same topology, have a different configurations.

10.3 What are the attributes of a box beam, of a computer disk carrying case, of a bicycle wheel, of a lead pencil, of a paper clip, or of some product line of your choice.

10.4 In Figure P10.4, the two part features with their small loads F and P are to be supported by an injection molded part that is to be fastened by bolts to the wall. Load F is about one inch out from the wall, and the left edge of part B is about two inches from the wall. Develop a configuration for the part to perform this function.

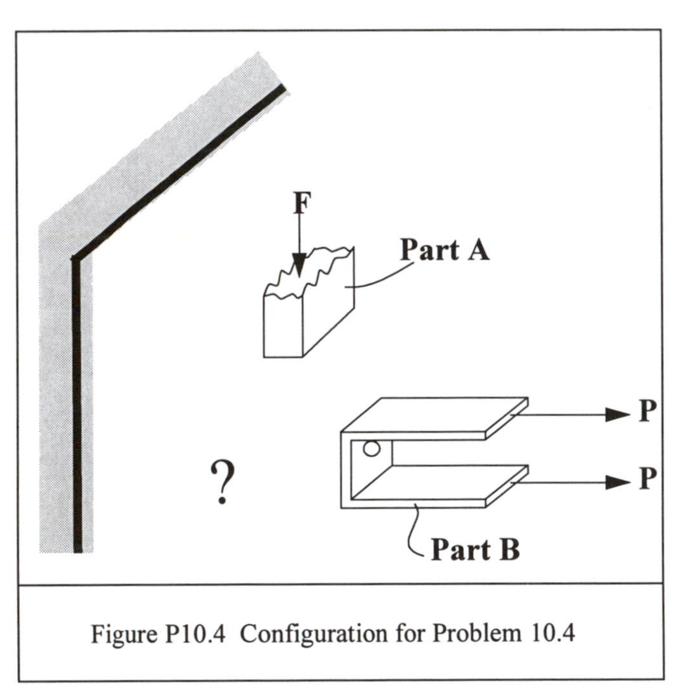

Figure P10.4 Configuration for Problem 10.4

10.5 The physical concept requires provision of a force proportional to the displacement of a rod (i.e., of a 'spring'). Sketch as many different spring configurations as you can.

10.6 List the attributes for a leaf spring, a bolt, a baseball bat, a U-bracket, a paper airplane, or some object of your choice.

10.7 A sheet of metal is to be folded into the shape of an open rectangular box. Show the possible configurations of the sheet prior to bending. Select any one, and list its attributes.

10.8 Prepare at least two part configuration requirement sketches for each of the following: an extruded picture frame, a one piece molded towel rack, the parts of the enclosure for a stamped electric baseboard heater, a clothes' hanger, a gutter (for collecting and directing rain from a roof), or some other part of your choosing.

10.9 Select one of the parts from Problem 9 and develop several trial configurations. Discuss how you were guided and limited by qualitative physical and manufacturability reasoning.

10.10 Develop the configuration stage for an indoor Christmas tree stand. This product require several parts, and you will first need to develop a concept.

10.11 Two aluminum extrusions are to be designed to be the frame of a solar collector. The desired locations of the enclosing glass and other components of the collector are shown in Figure P10.11. Prepare at least three possible part configuration sketches. Assume that the glass must be put in place after the four sides of the frame have been assembled.

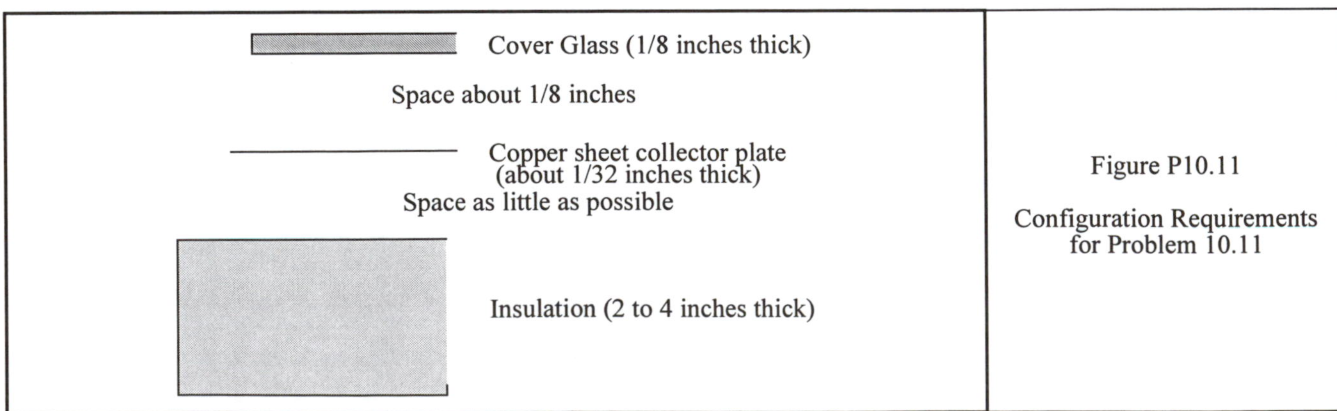

Figure P10.11

Configuration Requirements for Problem 10.11

Chapter Eleven

Evaluating Part Configurations for Manufacturability: Injection Molding and Die Casting

"A little neglect may breed great mischief.......for want of a nail the shoe was lost; for want of a shoe the horse was lost; for want of a horse the rider was lost....."

Benjamin Franklin
Preface: Courteous Reader (1758)

11.1 Introduction

A little neglect of DFM can also breed great mischief. The devil can often be in the details.

The procedures described in the preceding chapter will result in several possible part configurations whose generation has been guided by qualitative reasoning related to both function and their manufacturability. In this chapter and the next, we describe methods by which designers can perform more formal *quantitative* manufacturability evaluations of parts at the configuration stage. Later, in Chapter 13, we will discuss ways to consider both functional and manufacturing evaluations simultaneously.

The ability to evaluate part manufacturability at the configuration stage — before exact dimensions have been determined — is important because, in practice, design decisions made at this stage often become essentially irrevocable.= Thus, before moving on to the task of assigning values to attributes (which can sometimes be time consuming), it is important for designers to be as certain as possible that the configuration selected is the best possible one, considering both function and manufacturability.

In this book, we present detailed, quantitative manufacturability methods for three manufacturing domains: injection molding, die casting, and stamping. Design for manufacturability issues in other processes (such as assembly, forging, extrusion, and others) are also covered, but not in as much detail as these three. As noted earlier, the essential reason for concentrating on these particular manufacturability domains is that they account for more than 70% of the special purpose parts found in consumer products.

It should always be remembered that the best method for reducing assembly costs is to reduce the number of parts in an assembly. This is often accomplished by combining several individual parts into one (sometimes more complex) part using either injection molding, die casting or stamping. To do this by taking full advantage of the capabilities of the process, but without exceeding those capabilities, requires that designers be able to perform detailed DFM analyses.

This chapter is devoted to DFM methods for injection molding and die casting; Chapter 12 deals with DFM for stamping.

11.2 Estimating Relative Tooling Costs for Injection Molding

The cost of an injection molded part consists of three sub-costs: tooling (or mold) cost, K_d/N; processing cost (or equipment operating cost), K_e; and part material cost, K_m.

$$\text{Total Cost of a Part} = K_d/N + K_e + K_m \quad (11.1)$$

where K_d is the total tooling cost for a part and N is the number of parts to be produced with the mold -- that is, the production volume.

At low production volumes (N less than, say,

20,000), the proportion of the part cost due to tooling is often relatively high compared with processing costs since the total cost of tooling, K_d, is divided by a small value of N. As the production volume increases, however, the total tooling cost (K_d) does not change; consequently the tooling cost per part decreases, while the material and processing cost per part remains essentially the same. Thus, at high production volumes, the proportion of total cost due to tooling is relatively low; the major costs are due to processing and materials.

With only configuration information available, we can do little to estimate processing costs (K_e) or part material costs (K_m). However, we can make a reasonably accurate estimate of tooling costs (K_d). In fact, if we restrict ourselves to relative tooling cost then the analysis is accurate enough to permit a comparison between competing designs. As pointed out above, tooling costs are often important, and the fact that we can perform a DFM analysis of them at the configuration stage (before parametric design) is useful. The analysis helps designers identify the features of proposed configurations that contribute most significantly to tooling costs so that the features can be eliminated, or at least their negative impact on manufacturing cost reduced.

Relative Tooling Cost (C_d). The DFM methodology to be presented here determines tooling cost of injection molded parts as a *ratio* of expected tooling costs to the tooling costs for a reference part. This ratio is called *relative cost.* Thus, relative to a reference part, total die costs are:

$$C_d = \frac{\text{Cost of Tooling for Designed Part}}{\text{Cost of Tooling for Reference Part}}$$

That is: $C_d = (K_{dm} + K_{dc}) / (K_{dmo} + K_{dco})$ (11.2)

where K_{dmo} and K_{dco} refer to die material cost and die construction cost for the reference part. (In this case, the reference part is a flat washer with OD = 72 mm, ID = 60 mm and t = 1 mm. The approximate tooling cost for this reference par — in the 1991-92 time frame — is about $7,000 including about $1,000 in die material costs.)

Equation (11.2) can be written as

$$C_d = K_{dm} / (K_{dmo} + K_{dco}) + K_{dc} / (K_{dmo} + K_{dco})$$

$$= A\,(K_{dm} / K_{dmo}) + B(K_{dc} / K_{dco}) \quad (11.3)$$

where

$$A = (K_{dmo} / (K_{dmo} + K_{dco})$$

$$B = (K_{dco} / (K_{dmo} + K_{dco})$$

Based on data collected from mold makers, a reasonable value for A is between 0.15 - 0.20 and a reasonable value for B is between 0.80 - 0.85. For our purposes we will take A and B to be 0.2 and 0.8, respectively. Hence, Equation (11.3) becomes:

$$C_d = 0.8\,C_{dc} + 0.2\,C_{dm} \quad (11.4)$$

where C_d is the total die cost of a part relative to the die cost of the reference part, C_{dc} is the die construction cost for the part relative to the die construction cost of the reference part, and C_{dm} is the die material cost for the part relative to the die material cost of the reference part.

In this Section, we show how to determine the relative tooling construction costs (C_{dc}). Then the following section deals with the relative tool material costs (C_{dm}).

Relative Tooling Construction Costs (C_{dc}). To estimate relative tool construction costs for a part, designers must understand in some detail the complex relationships between the part and its mold. Certain features and combinations of features result in more complex molds and hence higher tooling costs. It may be that, in order to meet a part's function, such features or their combinations cannot be changed or eliminated, but in many cases they can be — saving time and money. In any case, designers should *know* the tooling costs their designs are causing, and make every attempt to reduce them.

The time required for tooling to be designed, manufactured, and tested is also a factor, though there is as yet no analyses to estimate it. In general, however, the higher the cost of tooling, the longer the time required for making the tool.

Relative tooling construction cost is computed here as the product of three factors:

$$C_{dc} = C_b\,C_s\,C_t \quad (11.5)$$

where

C_b = The approximate relative tooling cost due to size and basic complexity;

C_s = A multiplier accounting for other complexity factors called *subsidiary* factors;

C_t = A multiplier accounting for tolerance and surface finish issues.

We will now discuss how to compute these factors, and present examples of their use. Readers will be rewarded with an easy-to-use method of design for manufacturing principles and practices for injection molded and die cast parts. And much of what is learned will be useful in understanding other manufacturing domains as well. In order to use the methodology, however, a reader must be familiar with a number of concepts related specifically to the manufacture of injection molded parts. Though there appear to be a large number of them (they are explained in the next subsections), the concepts are individually relatively easy to understand. All are explained as they are introduced.

(1 in = 25.4 mm; 100 mm = 3.94 in)

Key: Flat Parts value above the slanted line, Box-Shaped Parts value below (example: first digit 6, 5.37 / 6.28)

BASIC COMPLEXITY (each cell: Flat Parts / Box-Shaped Parts)

FIRST DIGIT				SECOND DIGIT →	L ≤ 250 mm (4)				250mm < L ≤ 480mm				L > 480 mm		
				Number of External Undercuts (5)	zero	one	two	More than two	zero	one	two	More than two	zero	one	More than one
					0	1	2	3	4	5	6	7	8	9	10
Parts without Internal Undercuts (1)		Parts whose peripheral height from a planar dividing surface is constant (2)	Part in one half (3)	0	1.00 / 1.64	1.23 / 1.87	1.38 / 2.02	1.52 / 2.16	1.42 / 2.89	1.65 / 3.12	1.79 / 3.27	1.94 / 3.41	1.83 / 4.28	2.07 / 4.51	2.33 / 4.77
Parts without Internal Undercuts (1)		Parts whose peripheral height from a planar dividing surface is constant (2)	Part not in one half (3)	1	1.14 / 1.86	1.37 / 2.09	1.52 / 2.24	1.66 / 2.38	1.61 / 2.99	1.84 / 3.22	1.99 / 3.37	2.13 / 3.51	2.09 / 4.42	2.32 / 4.66	2.58 / 4.92
Parts without Internal Undercuts (1)		Parts whose peripheral height from a planar Dividing Surface is not constant - or - parts with a non-planar Dividing Surface (2)		2	1.28 / 1.92	1.51 / 2.15	1.66 / 2.29	1.80 / 2.44	1.81 / 3.38	2.04 / 3.61	2.19 / 3.76	2.33 / 3.90	2.34 / 5.01	2.58 / 5.24	2.84 / 5.50
Parts with Internal Undercuts (1)	On Only One Face of the Part	Parts whose ONLY Dividing Surface (2) is planar, or parts whose peripheral height from a planar dividing surface is constant		3	2.33 / 3.19	2.57 / 3.43	2.71 / 3.57	2.86 / 3.72	2.75 / 4.44	2.98 / 4.68	3.13 / 4.82	3.27 / 4.97	3.17 / 5.83	3.40 / 6.07	3.66 / 6.33
Parts with Internal Undercuts (1)	On Only One Face of the Part	Parts whose peripheral height from a planar Dividing Surface is not constant - or - parts with a non-planar Dividing Surface (2)		4	2.98 / 3.73	3.21 / 3.97	3.36 / 4.11	3.50 / 4.26	3.52 / 5.20	3.75 / 5.43	3.89 / 5.58	4.04 / 5.72	4.04 / 6.82	4.28 / 7.06	4.54 / 7.32
Parts with Internal Undercuts (1)	On More Than One Face of the Part	Parts whose ONLY Dividing Surface (2) is planar, or parts whose peripheral height from a planar dividing surface is constant		5	4.20 / 5.37	4.43 / 5.61	4.58 / 5.75	4.72 / 5.89	4.62 / 6.62	4.85 / 6.86	4.99 / 7.00	5.14 / 7.14	5.03 / 8.01	5.27 / 8.24	5.53 / 8.51
Parts with Internal Undercuts (1)	On More Than One Face of the Part	Parts whose peripheral height from a planar Dividing Surface is not constant - or - parts with a non-planar Dividing Surface (2)		6	5.37 / 6.28	5.60 / 6.52	5.74 / 6.66	5.89 / 6.81	5.90 / 7.74	6.13 / 7.98	6.28 / 8.12	6.42 / 8.27	6.43 / 9.37	6.67 / 9.60	6.93 / 9.86

Figure 11.1 Classification System for Basic Tool Complexity, C_b. Notes (1)-(5) are found in Appendix 11A.

11.3 Determining Relative Tooling Construction Costs Due to Basic Part Complexity (C_b)

11.3.1 Overview

Values for C_b — the relative tooling cost factor due to basic part complexity — are found in the interior boxes of the matrix in Figure 11.1. The numbers above the slanted lines in the boxes in Figure 11.1 apply to flat parts; those below the slanted line apply to box-shaped parts. Note that the value for C_b in the upper left corner of the matrix for a flat part is 1.00; thus this box corresponds to the cost of the reference part.

Readers should note that, in general, values for C_b decrease significantly as one moves up and to the left in the matrix. This fact will help guide designers to redesigns that reduce tooling costs.

11.3.2 The Basic Envelope

Figure 11.1 requires that the part to be evaluated be classified as either *flat* or *box shaped.* (This is done because in general, box shaped parts require more mold machining time and hence result in higher tool construction costs than flat parts.) In order to determine whether a part is flat or box shaped, we determine the ratio of the sides of the *basic envelope* for the part. The basic envelope is the smallest rectangular prism which completely encloses the part.

The lengths of the sides of the basic envelope are denoted by L, B, and H, where $L \geq B \geq H$. See Figure 11.2. A part is considered flat if L/H is greater than about 4; otherwise it is considered box shaped.

In order not to overestimate the amount of mold machining time required, in determining the basic envelope, small, isolated projections are ignored if their greatest dimension parallel to the surface from which they project is less than about one-third times the envelope dimension in the same direction (see Figure 11.3). This is done so that a part which is basically flat when the projection is ignored, is not classified as a box shaped part. If more than one projection exists, each should be examined separately.

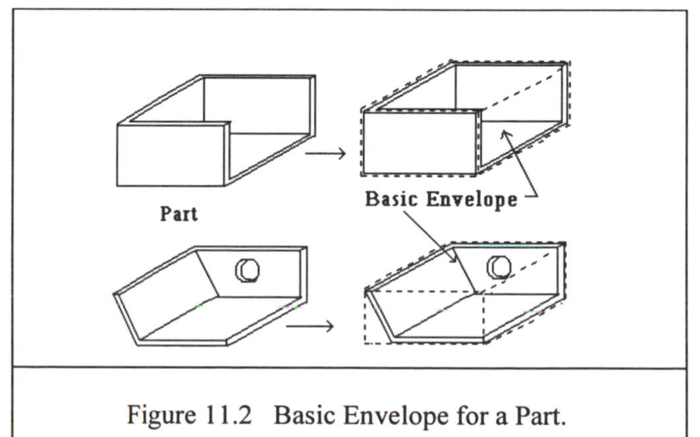

Figure 11.2 Basic Envelope for a Part.

11.3.3 The Mold Closure Direction

As noted briefly in Chapter 4, designers of injection molded parts *must* consider the direction of mold closure in order to be able to design for manufacturability. The reason is that the orientation of certain part features and configurations in relation to the mold closure direction can have an important influence on tool construction costs. This is also reflected in the fact that knowledge of the mold closure direction is essential for designers to use Figure 11.1, and hence to estimate tooling construction costs.

Knowledge of the mold closure direction is also essential in order to identify and possibly redesign the features that may be causing high tooling costs.

Recessed Features. In order to determine the best or most likely direction of mold closure, it is necessary to understand the meaning of a *recessed feature.* A recessed feature is any depression or hole in a part, including also depressed features that come about due to closely spaced projecting walls. Figure 11.4 shows some examples of recessed features. Also shown in Figure 11.4 are sectional views of the molds that can be used to produce these features. The part shown in Figure 11.4(d) is not considered to have a recess because the direction of mold closure does not affect basic tool construction difficulty.

Holes and Depressions. Depressions are pockets, recesses, or indentations of regular or irregular contour that are molded into a portion of an injection molded part. Holes are prolongation of depressions that completely penetrate some portion of the molding.

Circular holes and depressions can be formed either by an *integer mold* in which the projections required to create the two holes are machined directly into the core half of the mold as shown in Figure 11.4(f), or by an insert mold. In the case of an insert mold [Figure 11.4(g)], the projections shown in the first version of the mold are replaced by a core pin which is inserted into the core.

Rectangular holes and irregularly shaped holes can also be formed by the use of either integer molds or insert molds, but the cost to create the tooling to form these holes is more costly (See Problem 3.13).

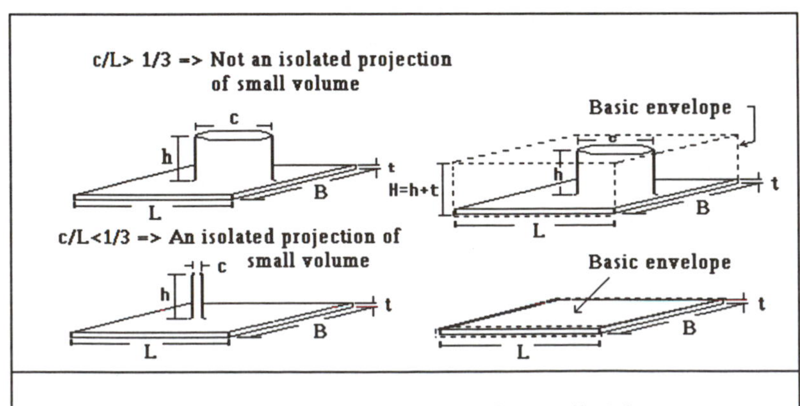

Figure 11.3 Isolated Projections of a Small Volume.

Projections. A feature that protrudes from the surface of a part is considered a projection. The most common examples are ribs and bosses.

A *rib* is a narrow elongated projection with a length generally greater than about three times its width (thickness), both measured parallel to the surface from which the feature projects [see Figure 11.5(a)] and a height less than six times its width. Ribs may be located at the periphery or on the interior of a part or plate. A rib may run parallel to the longest dimension of the part (a longitudinal rib) or it may run perpendicular to this dimension (a lateral rib). Radial ribs and concentric ribs are also common. A rib may be continuous or discontinuous, or it may be part of a network of other ribs and projecting elements. The direction of a rib corresponds to the direction of movement necessary to extract the rib from an imaginary cavity perpendicular to the length that conforms to the rib. If the height of a narrow elongated projection is greater than six times it width then the projection is considered a wall.

Non-peripheral ribs and non-peripheral walls are generally created by milling or EDMing, a cavity in either the core half or cavity half of the mold [see Figure 11.5(a)]. If the minimum rib thickness is greater than or equal to about 3 mm (0.125 inches) then the cavity is machined by milling, otherwise it is machined by the EDM process.

Two closely spaced longitudinal or lateral ribs, that is two ribs whose spacing is less than three times the rib width, are usually formed by first milling a cavity in the core half of the plate and then using an insert to form the two closely spaced ribs [see Figure 11.5(b)]. The cost to create this *cluster* of two closely spaced ribs is about equal to the cost to create a single rib.

A *boss* is an isolated projection with a length of projection which is generally less than about three times its overall width, the latter measured parallel to the surface from which it projects. A boss is usually circular in shape but it can take a variety of other forms called knobs, hubs, lugs, buttons, pads, or "prolongs."

Bosses can be solid or hollow. In the case of a solid circular boss, the length of the boss and its width are both equal to the boss diameter. The boss is created by simply milling a hole in the core half of the mold [Figure 11.5 (c)]. In the case of a circular hollow boss, a pin is used to create the hole [Figure 11.5 (d)]. While the width of the boss is still equal to the outside diameter, the boss thickness is equal to the difference between the outside and inside radii.

Bosses, and sometimes ribs, are supported by ribs of variable height called gusset plates. In the case of bosses, these support are radially located as shown in Figure 11.5(e). These ribs are machined in simultaneously using the EDM process. For this reason this *cluster* of ribs costs about the same to create as a single rib.

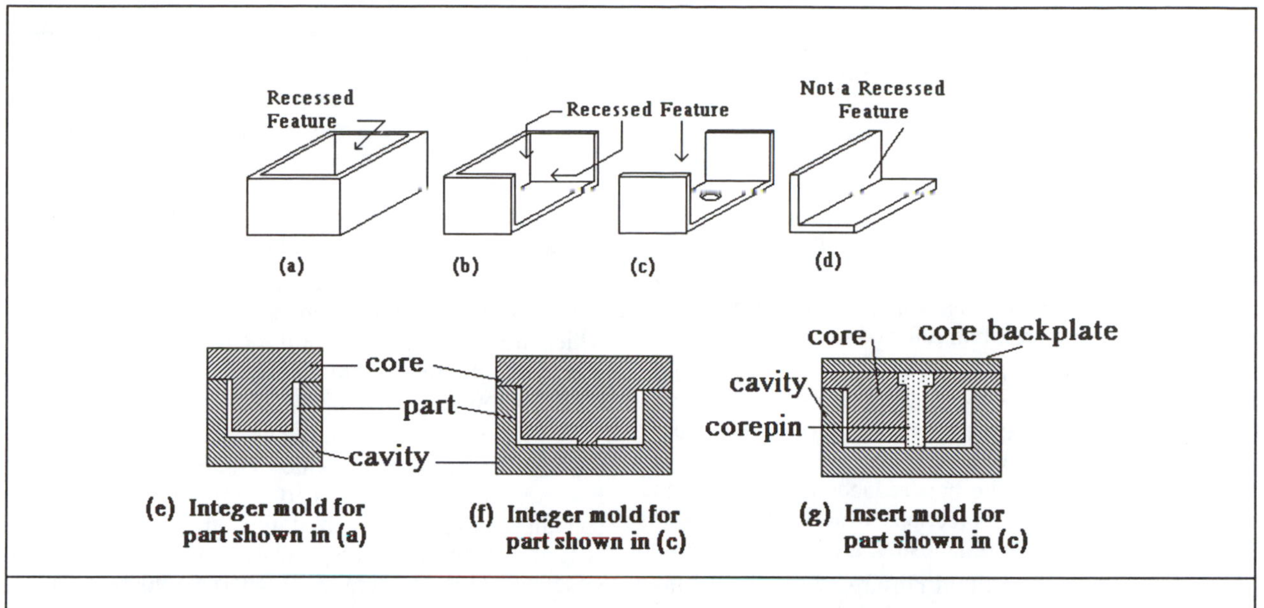

Figure 11.4 Examples of Parts with Recessed Features and Sectional Views of their Molds.

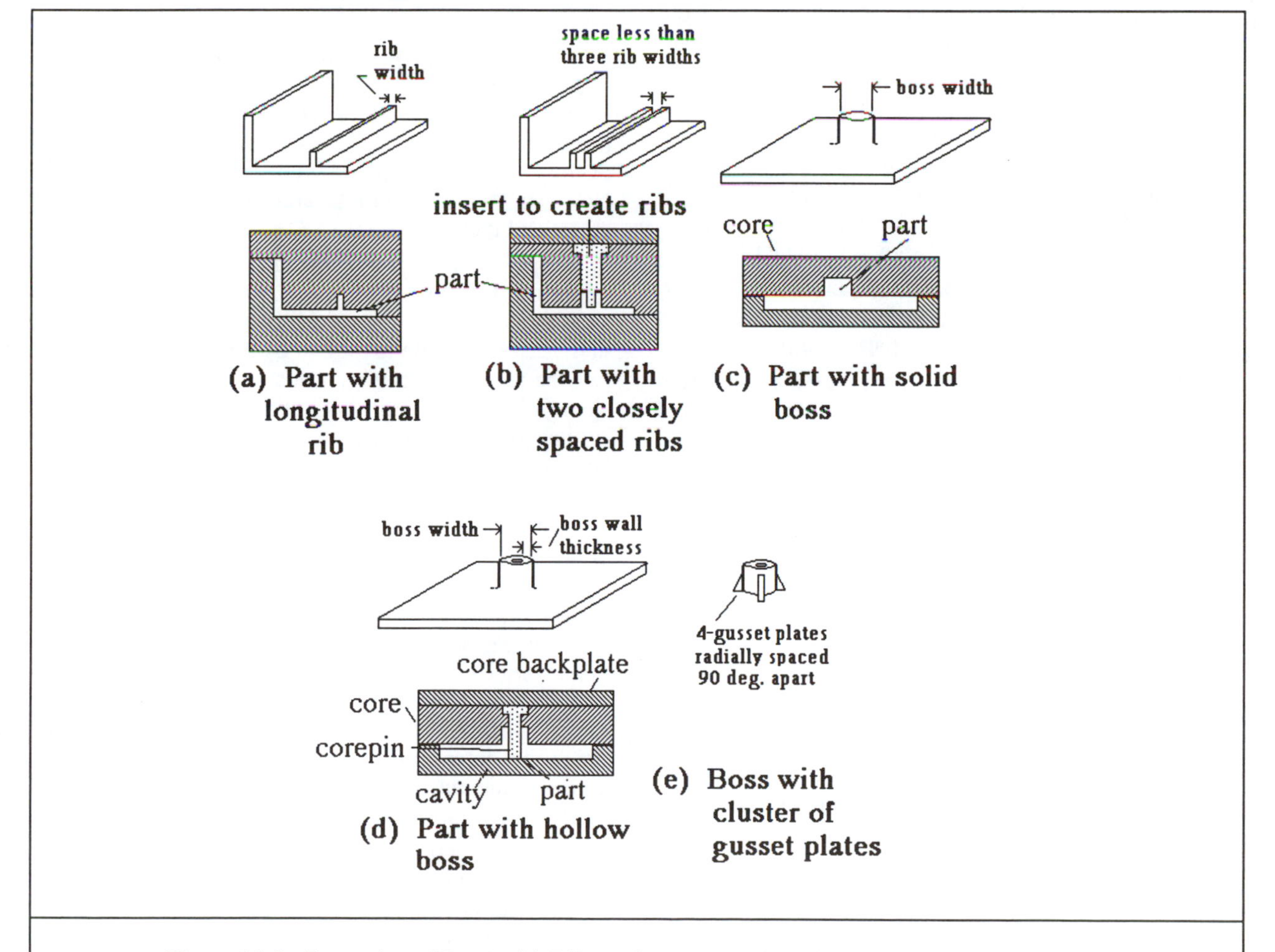

Figure 11.5 Examples of Parts with Ribs and Bosses, and Sectional Views of their Molds.

Dividing and Parting Surfaces.

One reason determining the direction of mold closure is so crucial to tooling cost evaluation is that it affects the location of the *parting surface* between the mold halves. The mold closure direction and the parting surface location together establish which recessed features can be molded in the direction of mold closure, and which (because they are not parallel to the mold closure direction) will require special tooling in the form of side action units or lifters in order to permit ejection of the part. These subsidiary features (i.e., holes, projections, etc.) are often referred to as *add-ons*.

In order to determine where a parting surface should be located, we will introduce the concept of *dividing surface*. First, however, we must define the meaning of the *principal shape*. The principal shape of a part (Figure 11.6) is the solid geometric shape to which the part reduces if all holes and depressions, as well as all projections which are normal or parallel to the direction of mold closure, are neglected.

Given a direction of mold closure, the dividing surface (Figures 11.7) is defined as an imaginary surface, in one or more planes, through the principal shape of the part for which the portion of the part on either side of the surface can be extracted from a cavity conforming to the form of the outer shape of the portion in a direction parallel to the direction of mold closure.

If the dividing surface is in one plane only, it is called a *planar dividing surface.* In general, the dividing surface which results in the least costly tooling is the one that should be used as the parting surface in the construction of the tooling. Figure 11.8 shows the parting surface that was used in the tooling for the box shaped part shown in Figure 11.7.

A dividing surface is a potential parting surface. A part may have several dividing surfaces, but of course a mold when constructed has only one parting surface.

Designers who understand the process of injection molding can usually plan for a convenient mold closure direction quite readily with just a little thought and study of the part. For new designers, or designers with little knowledge of injection molding, an algorithm useful for determine the best or most likely (i.e., most economical) direction of mold closure is presented in Figure 11.9.

11.3.4 Undercuts

In general, *undercuts* are combinations of part features created by recesses or by projections whose directions are *not parallel to the mold closure direction.* Undercuts are classified as either *internal* or *external*.

By reference to Figure 11.1, readers should note how the number of external undercuts increases the part's tooling cost by moving the part's location to the right in the Figure. Similarly, note how the number of internal undercuts moves a part's place in the matrix downward, thus also increasing tooling cost.

Internal Undercuts. Internal undercuts are recesses or projections on the inner surface of a part which, without special provisions, would prevent the mold cores from being withdrawn in the line of closure (often called the line of the *draw*). See Figure 11.10.

To permit withdrawal of the core when there is an internal undercut, hardened steel pins (called form pins) are built into the cores. Alternatively, cores called split cores must be constructed in two or more parts. (See Figure 11.11 for an illustration of a form pin.) Both split cores and form pins add to the complexity and hence cost of the tooling.

External Undercuts. In general, external undercuts are holes or depressions on the external surface of a part which are not parallel to the direction of mold closure (Figure 11.12). Some exceptions to this generalization are discussed below.

The number of external undercuts in a part is equal to the number of surfaces which contain external undercuts. For example, (a) shows a part with only one surface which contains an external undercut; thus, the number of undercuts is 1. In (b), however, the part has two surfaces with external undercuts; thus, the number of external undercuts is 2.

In addition, projections located on the external surface of a part such that a single mold dividing surface (planar or nonplanar) can not pass through them all, are also considered external undercuts.

As with internal undercuts, the presence of external undercuts requires special provisions to allow for ejection of parts from the mold cavity. To permit ejection, a steel member called a side cavity or side core must be mounted and operated at right angles to the direction of mold closure. However, this solution also adds to tooling complexity and cost. See Figure 11.13.

Side Shutoffs. In some situations, a hole or a groove in the side wall of a part can be molded without the need for side action cores. Figure 11.14 shows two examples. In these cases, the hole is formed by a portion of the core abutting the face of the cavity. Such holes are called *simple side shutoffs* because contact between mold halves occurs on one surface only. *Complex side shutoffs* occur when contact between mold surfaces occurs on more than one plane. A tab (Figure 11.14) is an example of a complex side shutoff. To determine whether a hole or depression is a side shutoff or an undercut, the following test can be applied:

With a solid plug conforming to the exact shape of the inner surface of the part already inserted, imagine the part inserted into a plug conforming to the exact shape of the outer surface of the part. If the outer plug can now be removed, by the use of straight line motion parallel to the direction of mold closure, the hole is considered a side shutoff. If the outer plug can not be removed, the hole is considered an undercut.

A part with *isolated* grooves and cutouts on the external surface of a part can also sometimes be considered a part with side shutoffs and constant peripheral height, rather than a part whose peripheral height is not constant. A groove or cutout is considered isolated if its dimension normal to the direction of mold closure is less than 0.33 times the envelope dimension in the same direction (see Figure 11.14).

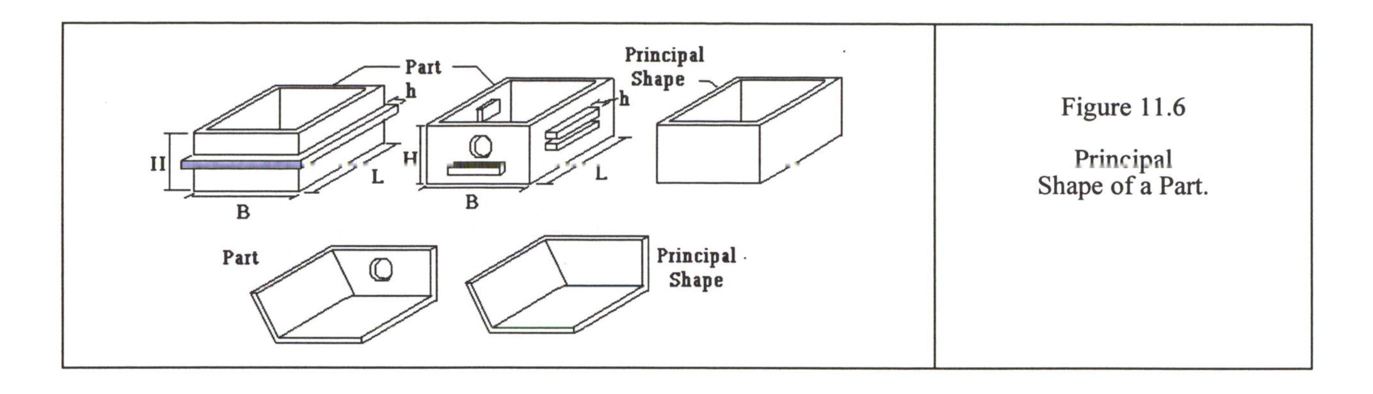

Figure 11.6

Principal Shape of a Part.

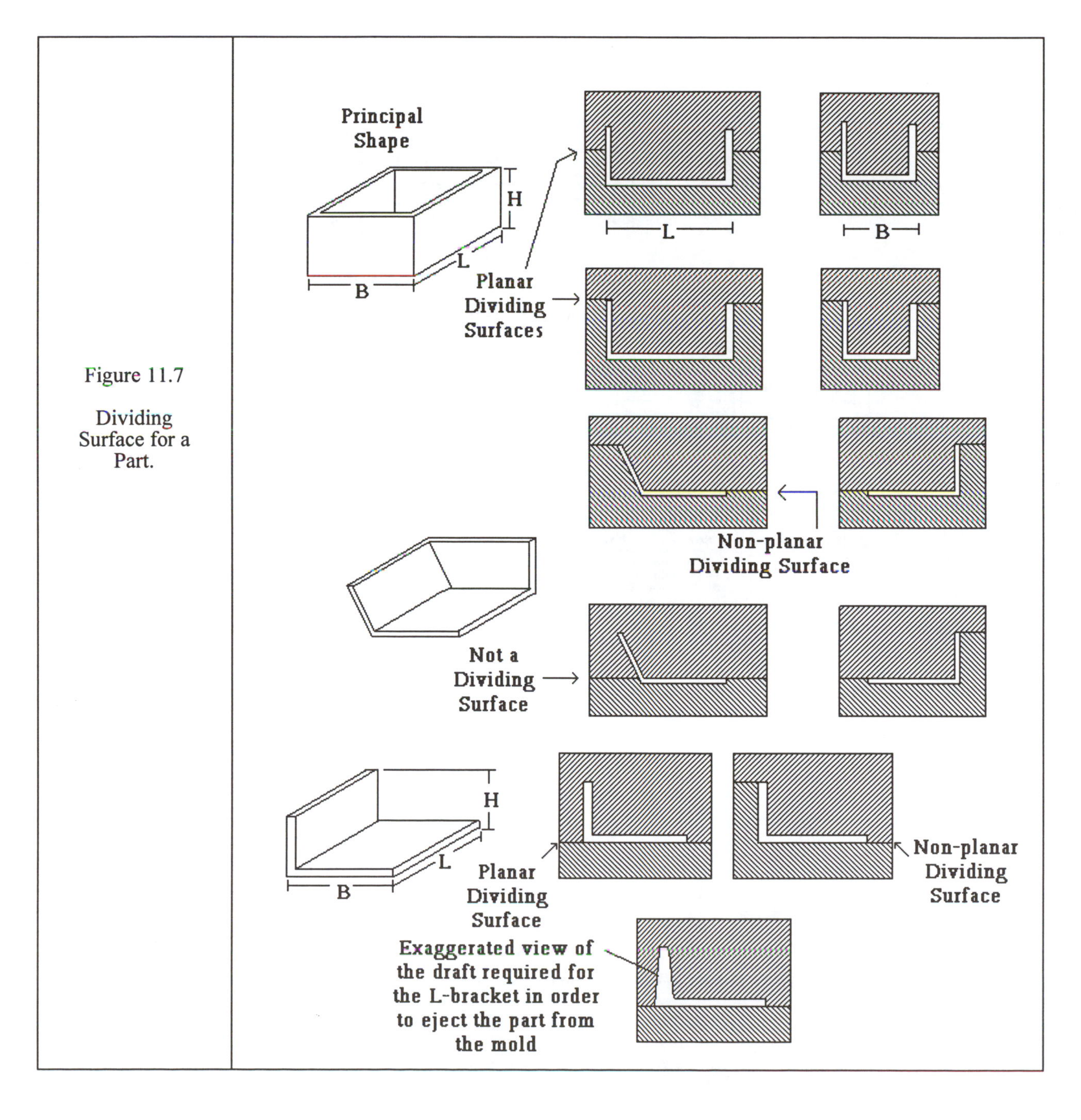

Figure 11.7

Dividing Surface for a Part.

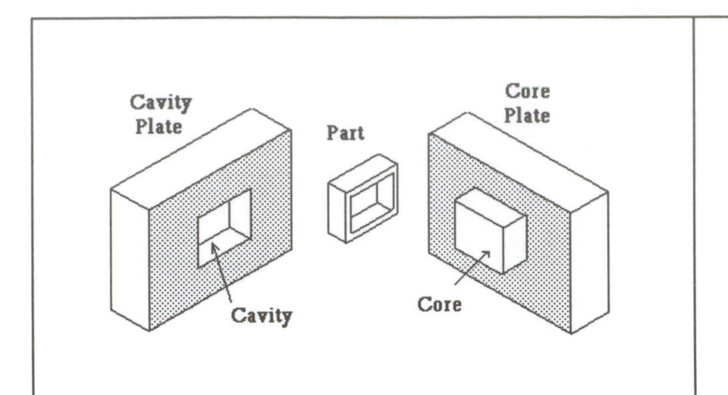

Figure 11.8
Tooling for Box Shaped Part.

Illustrates parting surface between the core and cavity halves of the mold.

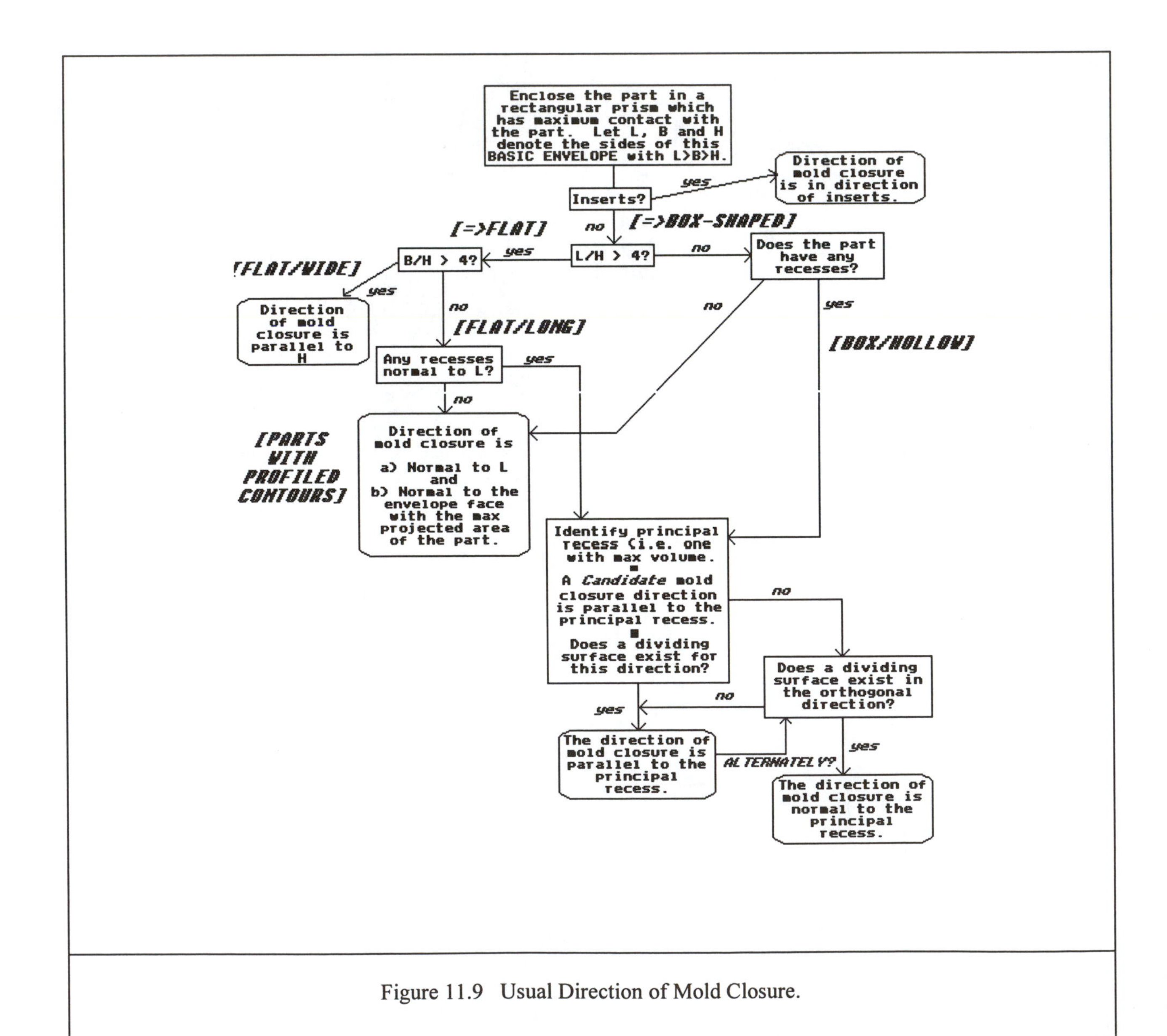

Figure 11.9 Usual Direction of Mold Closure.

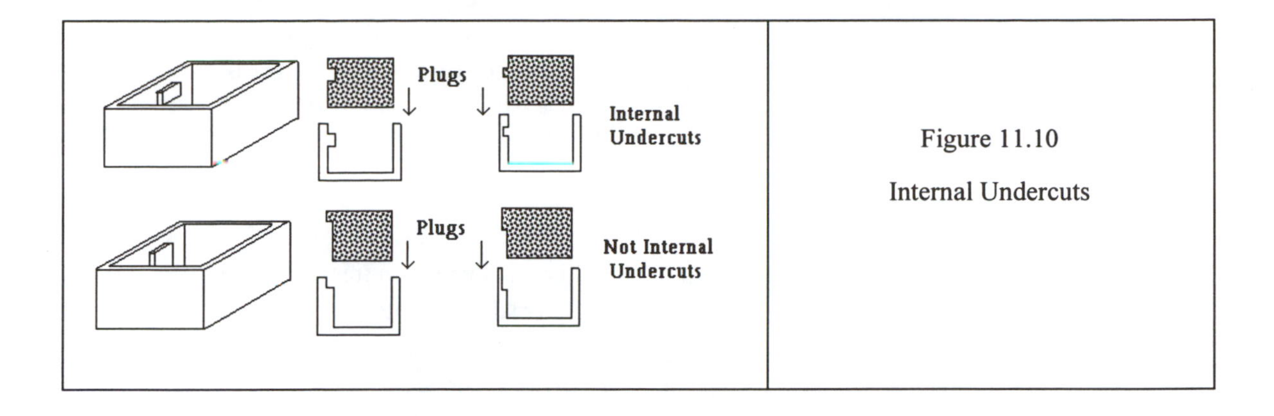

Figure 11.10

Internal Undercuts

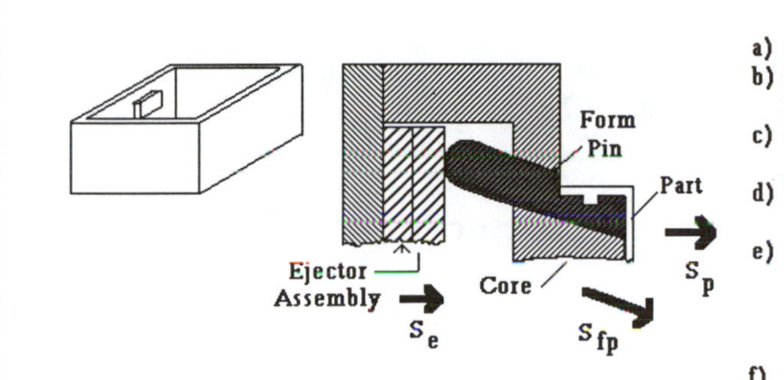

a) Plastic solidifies and shrinks onto core.
b) Mold is opened and part is withdrawn from the cavity.
c) Ejector plate is moved relative to core as shown by vector S_e
d) Form pin slides in the direction shown by vector S_{fp}
e) The part moves in a direction parallel to motion of the ejector plate, is lifted from the form pin so that the projection is withdrawn from the form pin cavity.
f) The part is removed or blown off the form pin

Figure 11.11 Form Pin Used to Produce Internal Undercut

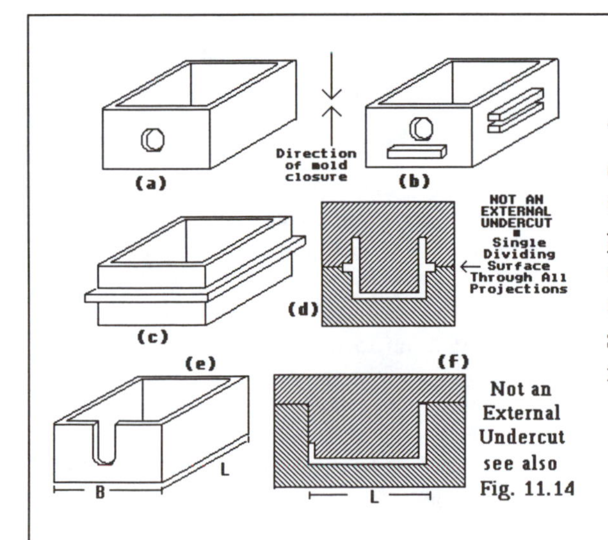

The number of external undercuts in a part is equal to the number of surfaces which contain external undercuts. For example, (a) shows a part with only one surface which contains an external undercut; thus, the number of undercuts is1. In (b), however, the part has two surfaces with external undercuts; thus, the number of external undercuts is 2.

Figure 11.12 External Undercuts

11.3.5 Other Factors Influencing C_b

Parts Molded in One Half the Mold. Mold costs are influenced to some extent by the amount of machining that must be done to create the core and cavity hollow sections. If the part cavity can be molded entirely in one half of the mold, then the other part of the mold needs no special machining. A flat part is said to be *in one half the mold* when the entire part is on one side of a planar dividing surface. Thus a box shaped part is said to be in one half the mold (Figure 11.15) when the principal shape of the part lies entirely on one side of a planar parting surface.

Peripheral Height. While the L-shaped part shown in Figure 11.7 does have a planar dividing surface and could be molded using a planar parting surface, in general this would not be done. To use a planar parting surface would require a taper (draft) on the vertical wall, the wall parallel to the direction of mold closure, so that the part can be easily removed from the mold (See Figure 11.7).

To avoid the need for a taper, a non-planar parting surface is generally used. To indicate situations where a non-planar parting surface would probably be used, even if a planar dividing surface exists, the concept of a constant peripheral height is introduced. For the box shaped part shown in Figure 11.7, a planar dividing surface exists and the peripheral height as measured from a planar dividing surface is constant (i.e. the wall height is constant), thus, a planar parting surface is used to produce the part (Figure 11.8). However, for the L-shaped part, the peripheral height from the planar dividing surface is not constant (i.e. the wall height is zero on three of the four peripheral surfaces) and, thus, a non-planar parting surface would be used to construct the tooling.

11.3.6 Entering and Using Figure 11.1

The value of C_b can readily be determined from Figure 11.1 given the following information, all of which can be found easily by methods explained above:

1. The longest dimension of the basic envelop, L;
2. The number of external undercuts;
3. The number and location (on one or more faces of the part) of internal undercuts;
4. Whether or not the part will be made in one half of the mold;

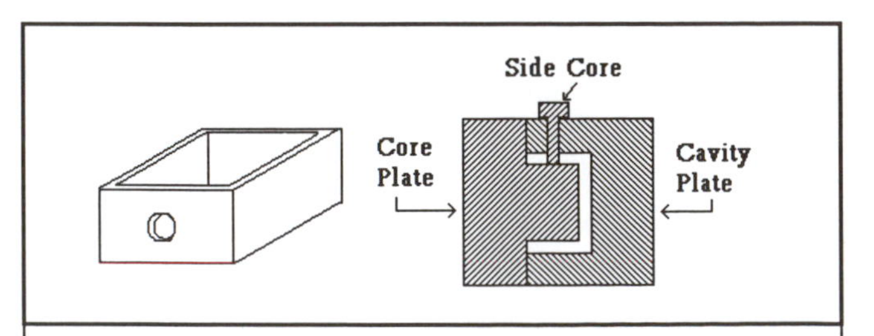

Figure 11.13 Side Core Used to Create an External Undercut Necessitated by a Circular Hole.

5. Whether the dividing surface will be planar or not;
6. Whether or not the part's peripheral height from a planar dividing surface is constant or not; and
7. Whether the part is flat or box shaped.

For example, refer to Figure 11.1 and consider two parts with the following characteristics:

	PART A	PART B
1. Longest dimension (mm)	400	200
2. External Undercuts	0	3
3. Internal undercuts (faces)	0	1
4. Dividing surface	Planar	Planar
5. Peripheral height	Constant	Constant
6. Part in One Half?	Yes	-
7. Flat or Box Shaped	Flat	Box

Readers should verify that the relative cost is 1.42 for Part A and 3.72 for Part B.

11.4 Determining C_s

As noted previously, the relative tooling construction cost for a part is found from

$$C_{dc} = C_b\, C_s\, C_t \tag{11.5}$$

where C_b = The approximate relative tooling cost due to size and basic complexity;

C_s = A multiplier accounting for other complexities called subsidiary factors;

C_t = A multiplier accounting for tolerance and surface finish issues.

In the preceding section, we showed how C_b is determined. In this section, we show how to obtain a value for C_s.

Features like ribs, bosses, holes, lettering, etc., aligned with the mold closure direction contribute to mold complexity. We refer to the number and complexity of such features as *cavity detail.* Figure 11.16 shows the method for rating the cavity detail as low, moderate, high, or very high.

In addition to the level of cavity detail, C_s is influenced by the complexity and number of external undercuts. Table 11.1 requires only that a judgment be made about whether *extensive* undercut complexity exists or does not exist. External undercuts other than unidirectional holes or depressions are considered extensive since the creation of such tooling is more costly. Figure 11.17 provides an example of a part that clearly has extensive external undercut complexity.

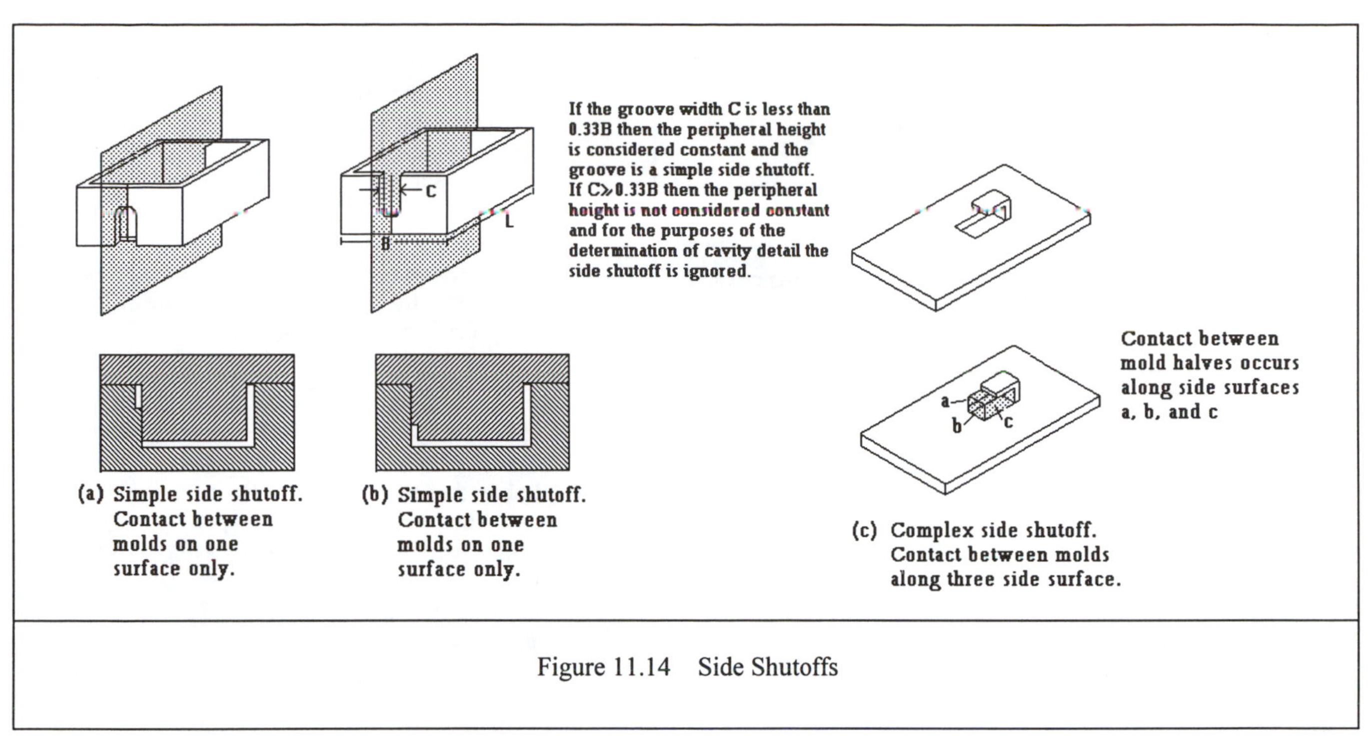

Figure 11.14 Side Shutoffs

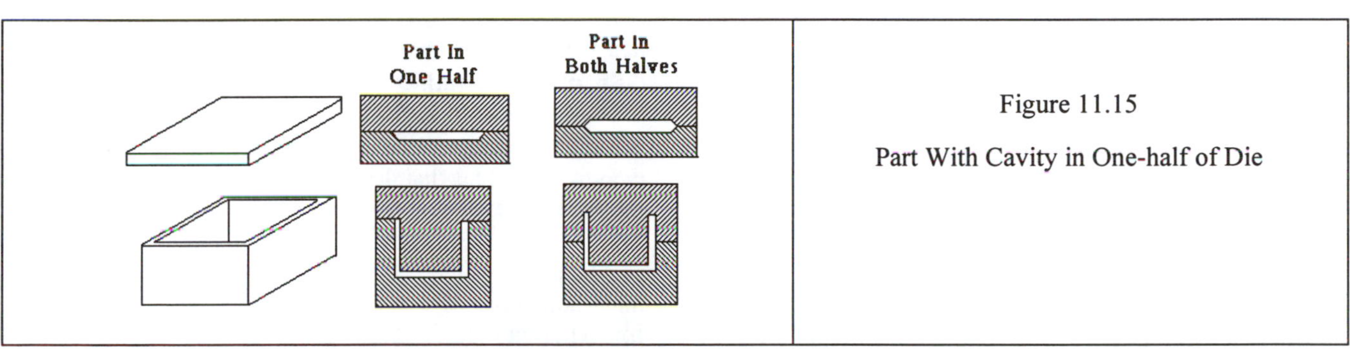

Figure 11.15

Part With Cavity in One-half of Die

Feature		Number of Features (n)	Penalty per Features	Penalty
Holes or Depressions	Circular		2n	
	Rectangular		4n	
	Irregular		7n	
Bosses	Solid (8)		n	
	Hollow (8)		3n	
Non-peripheral ribs and/or walls and/or rib clusters (8)			3n	
Side Shutoffs	Simple (9)		2.5n	
	Complex (9)		4.5n	
Lettering (10)			n	
			Total Penalty	

SMALL PARTS (L ≤ 250 mm)

Total Penalty ≤10 => Low cavity detail
10 < Total Penalty ≤20 => Moderate cavity detail
20 < Total Penalty ≤40 => High cavity detail
Total Penalty >40 => Very high cavity detail

MEDIUM PARTS (250 < L ≤ 480 mm)

Total Penalty ≤15 => Low cavity detail
15 < Total Penalty ≤30 => Moderate cavity detail
30 < Total Penalty ≤60 => High cavity detail
Total Penalty >60 => Very high cavity detail

LARGE PARTS (L > 480 mm)

Total Penalty ≤ 20 => Low cavity detail
20 < Total Penalty ≤ 40 => Moderate cavity detail
40 < Total Penalty ≤ 80 => High cavity detail
Total Penalty > 80 => Very high cavity detail

1 in = 25.4 mm; 100 mm = 3.94 in

Figure 11.16 Determination of Cavity Detail. Notes (8)-(10) can be found in Appendix 11A.

THIRD DIGIT			FOURTH DIGIT: EXTERNAL UNDERCUT COMPLEXITY — Without Extensive (7) External Undercuts (5)	With Extensive (7) External Undercuts (5)
SUBSIDIARY COMPLEXITY			0	1
CAVITY DETAIL (6)	Low	0	1.00	1.25
	Moderate	1	1.25	1.45
	High	2	1.60	1.75
	Very High	3	2.05	2.15

Table 11.1 Subsidiary Complexity Rating, C_s, for Injection Molding. Notes (5) - (7) can be found in Appendix 11A on page 11-26.

11.5 Determining C_t

The effects of surface finish requirements and the strictness of required tolerances on relative tool construction costs are accounted for by the factor, C_t, which is obtained from Table 11 2.

11.6 Using the Part Coding System to Determine C_b, C_s, and C_t

When analyzing a part for entry into Figure 11.1 and Tables 11.1 and 11.2, it is convenient to make use of the part coding system that has been developed for this purpose. The coding system involves six digits that, in effect, describe the part in the fashion of group technology. Here are the descriptions of the meaning of the digits in the coding system and their interpretation.

For Figure 11.1:

First Digit (0 - 6): The first digit in the coding system identifies the row in Figure 11.1 (for C_b) that describes the part. It is fixed by (1) the number of faces with internal undercuts, (2) by whether the part is in one-half the mold, (3) by whether the dividing surface is planar or non-planar, and (4) by whether the peripheral height is constant from the dividing surface.

Second Digit (0 - 9): The second digit identifies the column in Figure 11.1 that describes the part. It is thus fixed by the part size (L) and by the number of external undercuts.

Together, the first and second digits locate the place in Figure 11.1 where the value of C_b is found. (Remember: the values above the slanted line in that Figure refer to flat parts; values below the line refer to box shaped parts.).Readers should verify that the first two digits of the code for parts A and B described just above are, respectively, 0-4 and 3-3.

FIFTH DIGIT			SIXTH DIGIT: TOLERANCES, T_a — Commercial	Tight
			0	1
SURFACE FINISH R_a	SPI 5-6	0	-	-
	SPI 3-4	1	1.00	1.05
	TEXTURE	2	1.05	1.10
	SPI 1-2	3	1.10	1.15

Table 11.2 Tolerance and Surface Finish Rating, C_t, for Injection Molding.

For Table 11.1:

C_s is determined from Table 11.1 by the third and fourth digits of the coding system as follows:

Third Digit (0 - 2): The third digit in the coding system identifies the row in Table 11.1 (for C_s) that describes the part. It is determined by the level of cavity detail as determined from Figure 11.16

Fourth digit (0 - 1): The fourth digit identifies the column in Table 11.1 that describes the part. It is determined by the extent of external undercut complexity.

As an example, readers should verify from Figure 11.16 that the penalty factor for a part with five radial ribs, three hollow bosses, three simple side shut-offs, and localized lettering is 32.5, resulting in a level of cavity detail for a large part (L > 480 mm) of Moderate. Also verify from Table 11.1 that C_s for such a part with extensive external undercuts and moderate cavity detail is 1.45. (The third and fourth coding digits for this part are 1 and 1, respectively.)

For Table 11.2:

Fifth digit (0 - 3): The fifth digit identifies the row in Table 11.2 (for C_t) that describes the part. It is fixed by the nature of the required surface finish

Sixth digit (0 - 1): The sixth digit identifies the column in Table 11.2 that describes the part. It is fixed by whether the tolerances required are commercial or tight.

11.7 Total Relative Tooling Construction Cost

As defined earlier, the total relative mold construction cost is:

$$C_{dc} = C_b \; C_s \; C_t \tag{11.5}$$

and it is determined as shown above at the configuration

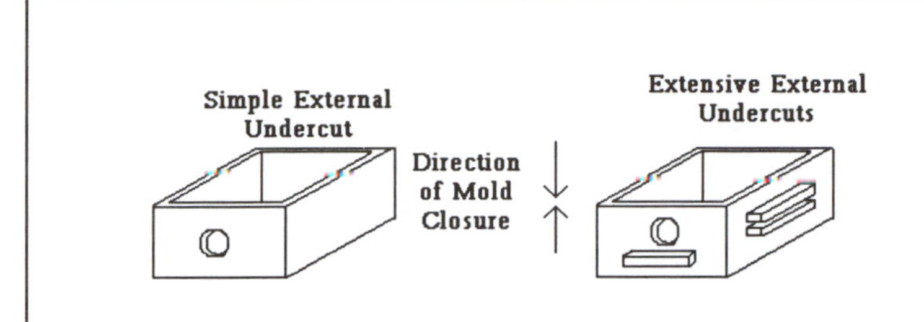

Figure 11.17

External undercuts caused by features other than circular, unidirectional holes are considered extensive external undercuts because the tooling is more costly to create.

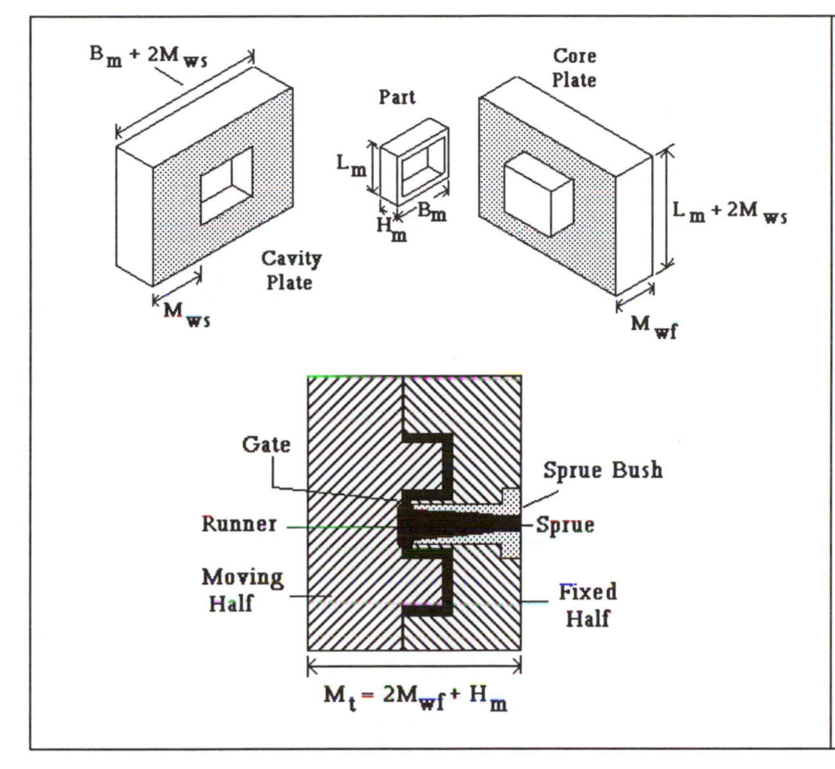

Figure 11.18

Mold Dimensions for Two Plate Mold.

stage of part design, that is prior to any detailed knowledge concerning part dimensions, ribs sizes, wall thickness, etc. The results show very clearly and simply the aspects of a part's design that contribute most heavily to tooling construction costs. In Figure 11.1, for example, designers can see clearly that parts should be redesigned if possible to move the rating up and to the left in the matrix, and then compute approximately how much can be saved. Removing undercuts accomplishes this goal, as do other simplifying changes that reduce detail or eliminate the need for special finishes or tight tolerances.

11.8 Relative Mold Material Cost

In order to compute total relative tooling costs, we must be able to estimate the mold material cost as well as its construction costs. This is relatively easy to do from a knowledge of the approximate size of a part -- which in turn dictates the required size of the mold. Referring to Figure 11.18, we define the following mold dimensions:

M_{ws} = Thickness of the mold's side walls,

M_{wf} = Thickness of the core plate,

L_m and B_m = The length and width of the part in a direction normal to the mold closure direction,

H_m = The height of the part in the direction of mold closure,

M_t = The required thickness of the mold base.

With these definitions, the following equations can be used sequentially to determine the projected area of the mold base, M_a, and the required thickness of the mold base, M_t, which in turn are used to obtain the relative mold material cost (C_{dm}) from Figure 11.20.

C = Value obtained from Figure 11.19

$$M_{ws} = [0.006\, C\, H_m^{4}]^{1/3} \tag{11.6}$$

$$M_{wf} = 0.04\, L_m^{4/3} \tag{11.7}$$

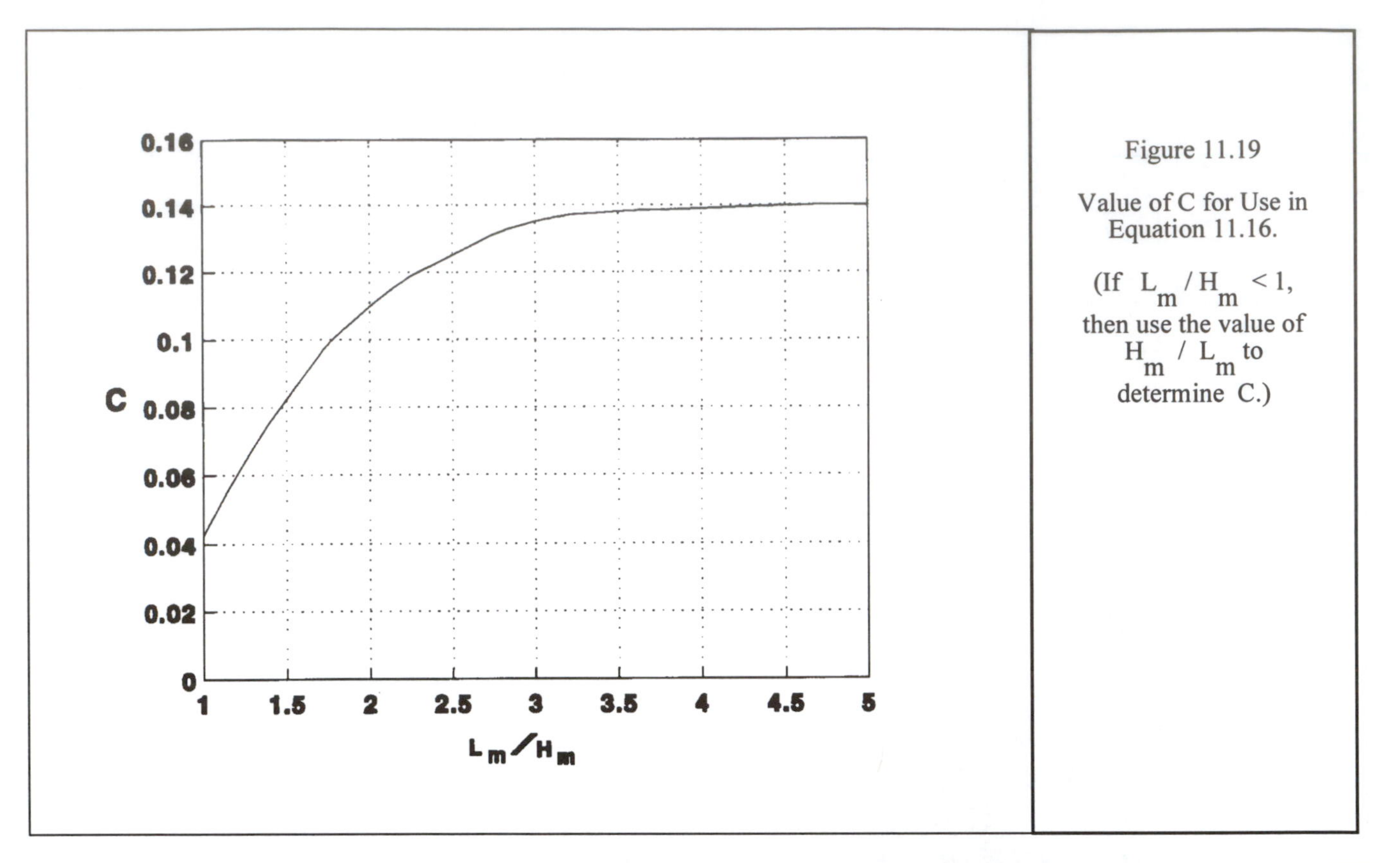

Figure 11.19

Value of C for Use in Equation 11.16.

(If $L_m / H_m < 1$, then use the value of H_m / L_m to determine C.)

$$M_a = (2\,M_{ws} + L_m)\,(2\,M_{ws} + B_m) \tag{11.8}$$

$$M_t = (H_m + 2\,M_{wf}) \tag{11.9}$$

C_{dm} = Value obtained from Figure 11.20

Total Relative Mold Cost. The total relative mold cost is determined from Equation (11.4):

$$C_d = 0.8\,C_{dc} + 0.2\,C_{dm} \tag{11.4}$$

where C_d is the total mold cost of a part relative to the mold cost of the standard part, C_{dc} is the mold construction cost relative to the standard and C_{dm} is the mold material cost relative to the mold material cost of the standard part.

11.9 Multiple Cavity Molds

The above discussion and equations apply to single cavity molds only. For the case of multiple cavity molds, the mold construction costs for a mold consisting of n_c cavities, $C_{dc}(n_c)$, is approximately given by the following expression:

$$C_{dc}(n_c) = C_{dc}\,(0.73 n_c + 0.27) \tag{11.10}$$

While the projected area of the mold base depends on the actual layout of a multiple cavity mold, it is assumed here that projected area is roughly given by the product of the projected area for a single cavity mold, M_a, times the number of cavities n_c, i.e.,

$$M_a(n_c) = M_a\,n_c \tag{11.11}$$

11.10 Example 11.1

11.10.1 The Part

As a first example we consider the part shown in Figure 11.21. The only dimensions shown are those which indicate the general overall size of the part as well as rough dimensions and location of the ribs which appear on the side walls of the part. The principal shape of the part is shown in Figure 11.22.

Commercial tolerances will be satisfactory, and the required surface finish is (Society of the Plastics Industry) SPI-3, which coincides with the low gloss finish found on most industrial products.

11.10.2 Relative Tooling (Mold) Construction Cost

Basic Complexity. The dimensions of the basic envelope of this part are L = 180 mm, B = 50 mm, H = 50 mm. Since L/H is less than 4, the part is box shaped.

If the direction of mold closure is assumed to be in the direction of the recess, then a planar dividing surface exists for the principal shape of the part. Planes AA and BB are just two of the many planar dividing surfaces that exist for the principal shape of this part (i.e., just two of the many surfaces that could be used to separate or part the two halves of the mold). Initially, dividing surface AA is taken as the parting surface. In this case there are no internal undercuts, the peripheral height from the planar dividing surface (AA) is constant, and the part is in one half of the mold. Thus, the first digit of the coding system is 0

With dividing surface AA, there are two external undercut. Since L is less than 250 mm, the second digit is 2. With the first two digits being 0 and 2, Figure 11.1 indicates a value for C_b of 2.02.

Since one of the major methods available for reducing mold manufacturability costs is to reduce the number of external and internal undercuts, the tooling cost for the part is re-examined using BB as the planar dividing surface.

In this case, the peripheral height from BB is still constant; however, the part is no longer in one half the mold. Thus, the first digit is 1.

Since BB passes through both external projections, they are no longer considered undercuts. Thus, the second digit is 0.

Thus with BB as the dividing plane, from Figure 11.1 we get a value for C_b of 1.86. This lower value of C_b indicates that the use of BB as the parting plane will result in a lower basic tool construction cost

Subsidiary Complexity:

Since there are no ribs, bosses, holes, depressions, etc., in the direction of mold closure, cavity detail is low and the third digit of the coding system is 0.

The fourth digit is also 0 since with BB as the parting plane there are no external undercuts. Thus from Table 11.1 we find the multiplying factor, C_s, due to subsidiary complexity is 1.00.

Surface Finish/Tolerance: The part has a surface finish of SPI-3 and commercial tolerances are used. Thus, the fifth and sixth digits are 1 and 0, respectively giving a value for C_t from Table 11.2 of 1.00.

Total Relative Mold Construction Cost C_{dc}:

$$C_{dc} = C_b\, C_s\, C_t = 1.86\,(1)\,(1) = 1.86$$

11.10.3 Relative Mold Material Cost

From Figure 11.19, for L_m/H_m of 3.6, C is 0.138. Thus, the thickness of the mold wall is given by:

$$M_{ws} = [0.006CH_m^{\,4}]^{1/3} = [0.006(0.138)(50)^4]^{1/3} = 17.3$$

mm and the thickness of the base is:

$$M_{wf} = 0.04L_m^{\,4/3} = 0.04(180)^{4/3} = 40.7 \text{ mm.}$$

Consequently, the projected area of the mold base is

$$M_a = [2(17.3)+180][2(17.3)+50] = 18155 \text{ mm}^2$$

and the required plate height is

$$M_t = [50 + 2(40.7)] = 131.4 \text{ mm.}$$

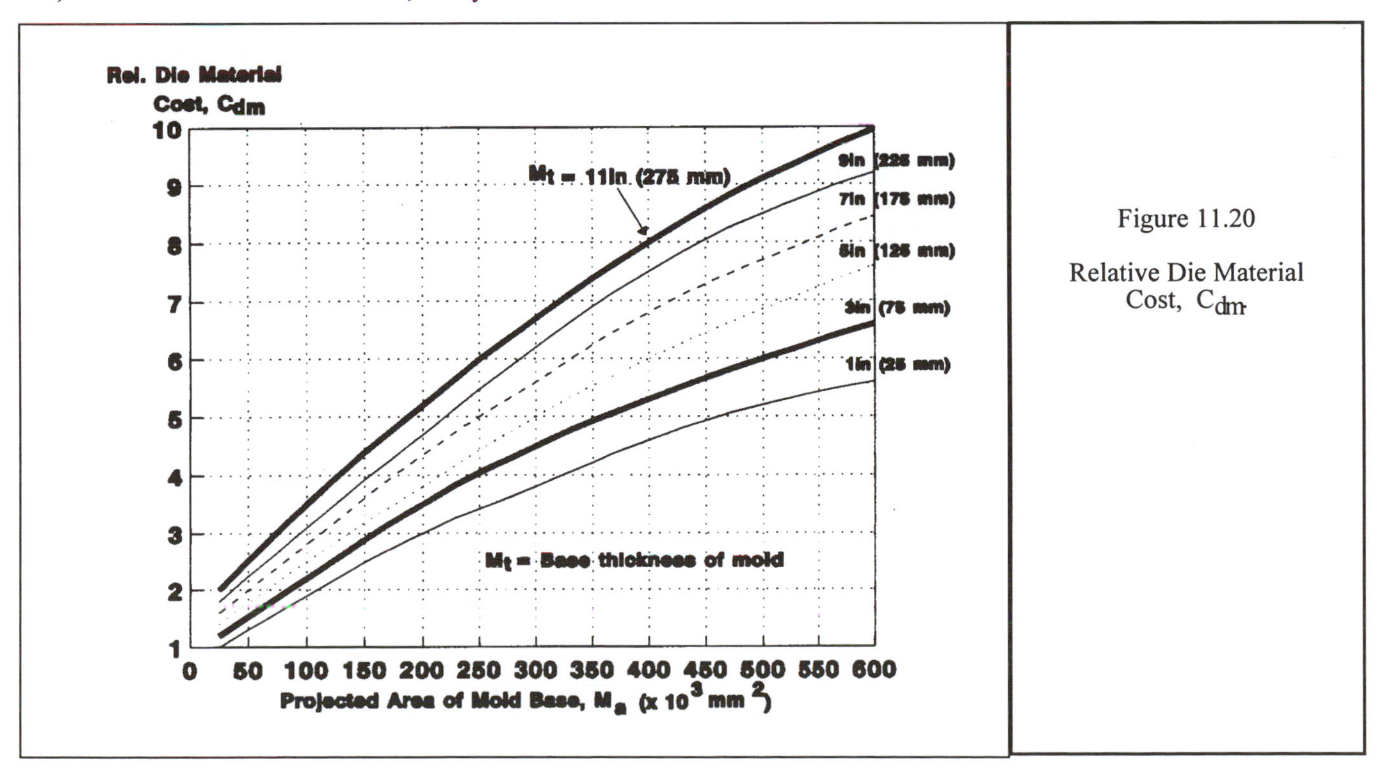

Figure 11.20

Relative Die Material Cost, C_{dm}

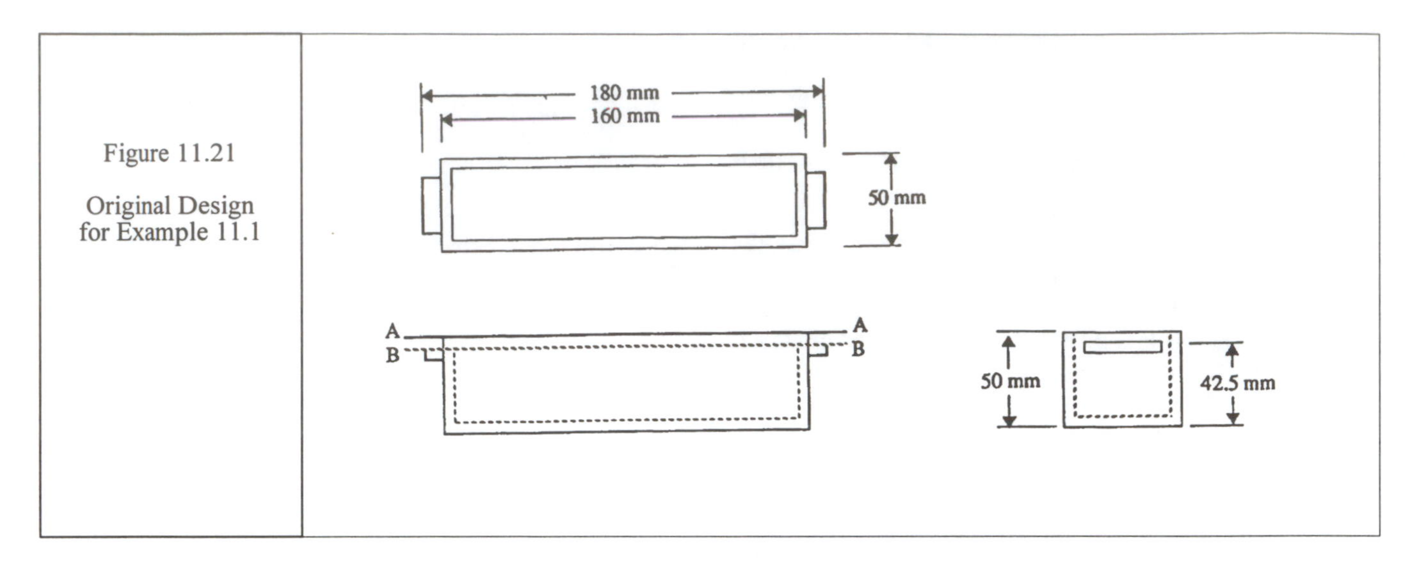

Figure 11.21

Original Design for Example 11.1

From Figure 11.20, the relative mold material cost, C_{dm}, for this part is approximately 1.5, and the total relative mold cost is

$$C_d = 0.8\, C_{dc} + 0.2\, C_{dm} = 0.8\,(1.86) + 0.2\,(1.5) = 1.79$$

11.10.4 Redesign Suggestion

The mold manufacturability costs for this part can be reduced slightly if the part is in one half the mold. This can be done by moving the two side projections to the top; that is, so that the tops of the projections are tangent to plane AA. In addition, relocating the side projections in this manner avoids the need for "reverse" taper or draft.

11.11 Example 11.2

11.11.1 The Part

As a second example we consider the part shown in Figure 11.23. Once again, the only dimensions shown are those that indicate the overall size and shape of the part. Detailed dimensions concerning wall thickness, holes sizes, etc., are not available at the configuration stage. The principal shape of the part is shown in Figure 11.24.

11.11.2 Relative Mold Construction Cost

Basic Complexity. Initially, the dimensions of the basic envelope of this part are:

$$L = 55 \text{ mm}, \quad B = 40 \text{ mm}, \quad H = 20 \text{ mm}.$$

It is possible that the projections indicated in Figure 11.23 are such that the largest dimensions parallel to the surface from which they project are less than 0.33 times the envelope dimension in the same direction. If so they are isolated projections of small volume, and would consequently be ignored in determining the basic envelope. Hence, the dimensions of the basic envelope of the part in this case would be:

$$L = 55 \text{ mm}, \quad B = 40 \text{ mm}, \quad H = 16 \text{ mm}.$$

In either case, L/H is less than 4, the part is box shaped.

If the direction of mold closure is assumed to be in the direction of the major recess (i.e., normal to the LB plane of the Basic Envelope), then a non-planar dividing surface (AA) exists for the principal shape of the part.

Since there are no internal undercuts, the first digit of the coding system is 2.

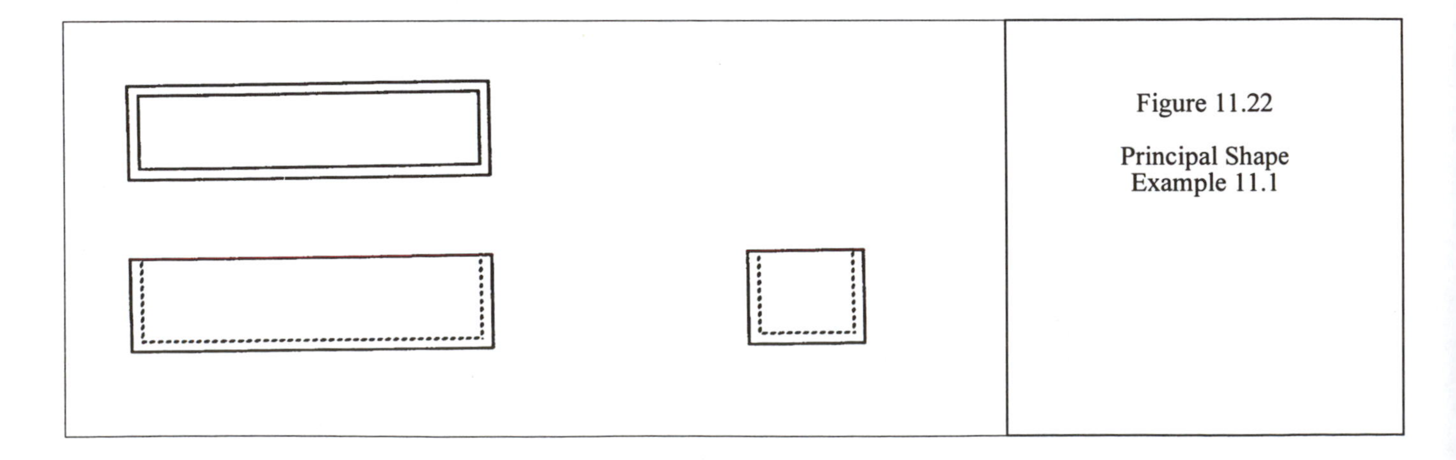

Figure 11.22

Principal Shape Example 11.1

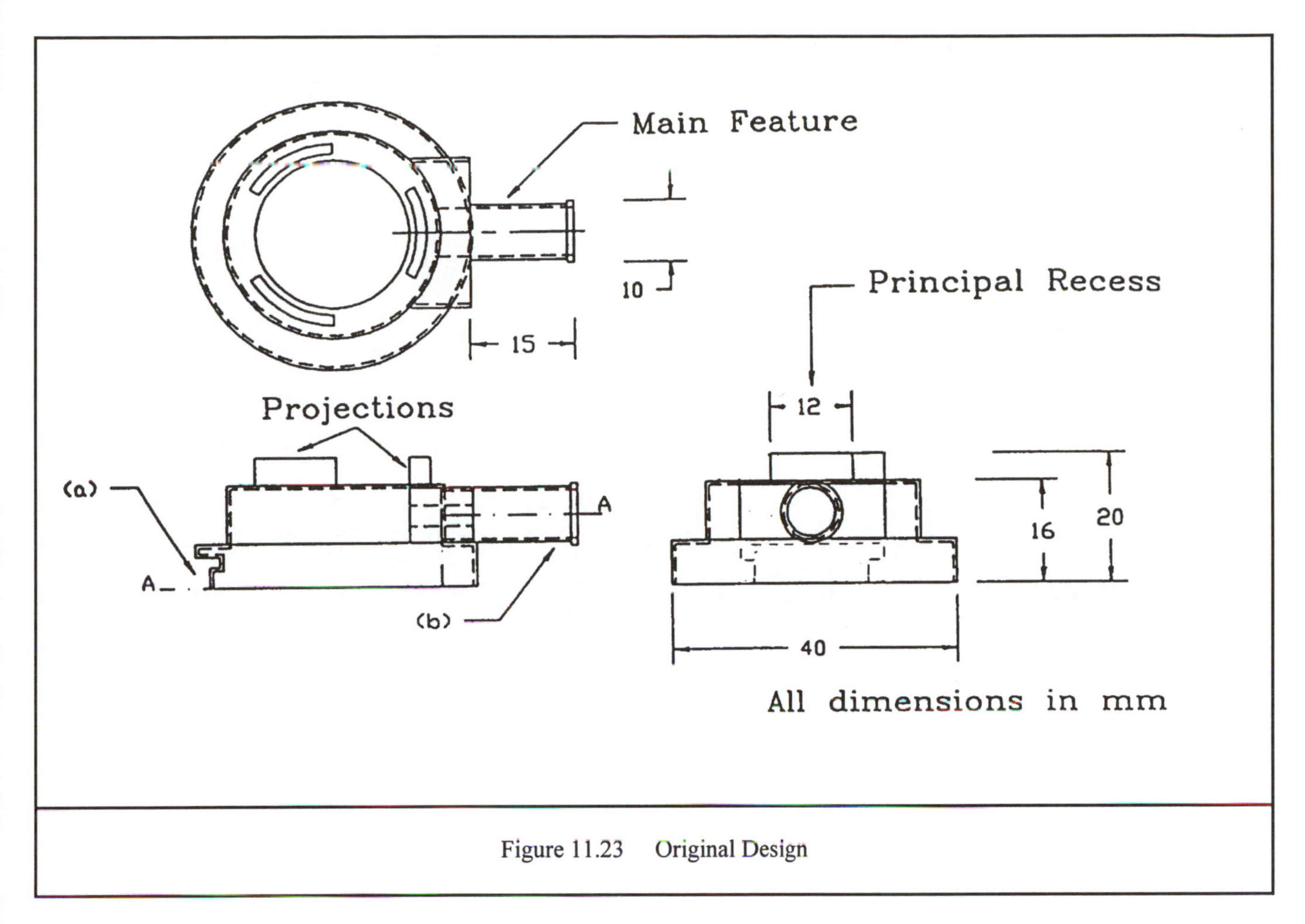

Figure 11.23 Original Design

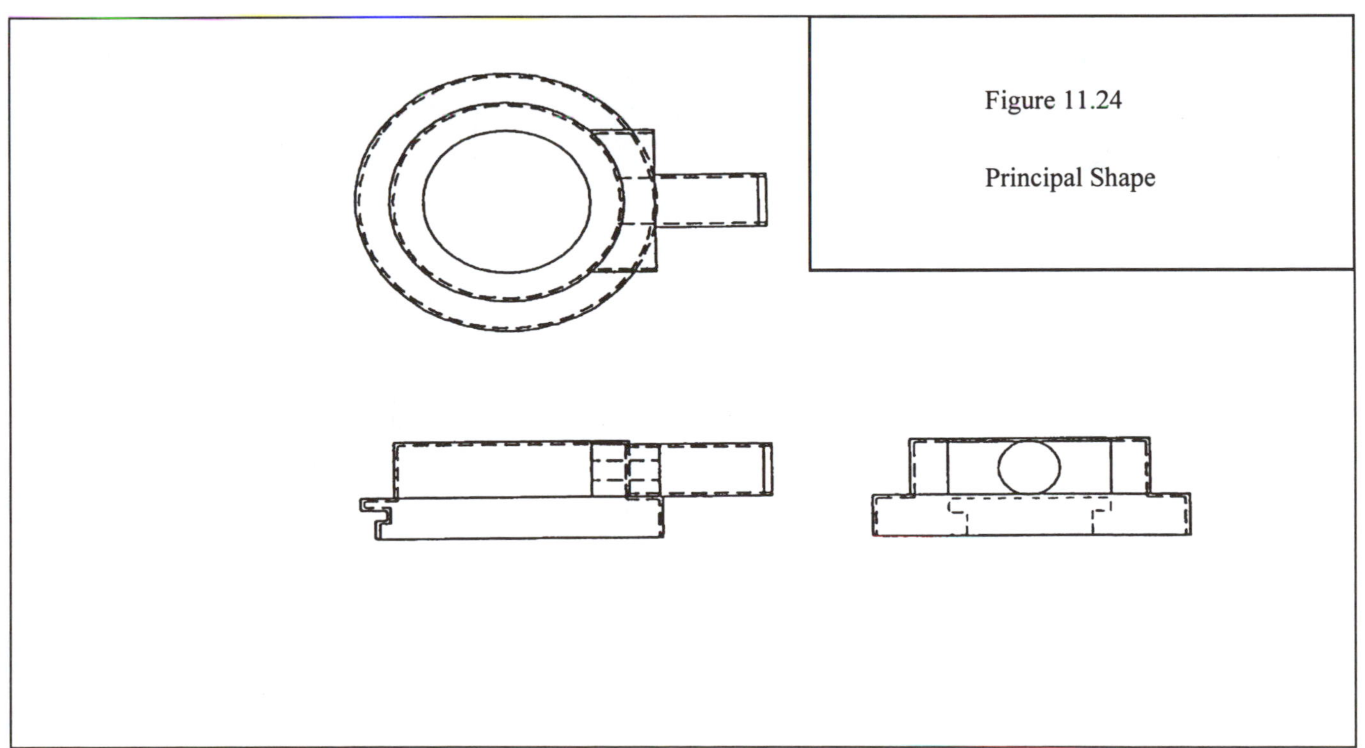

Figure 11.24

Principal Shape

There are two external undercuts present, and L is less than 250 mm. Thus, the second digit is 2.

With the first digits of 2 and 2, the value of C_b from Figure 11.1 is 2.29.

Subsidiary Complexity. There is one set of concentric ribs, and one circular hole in the direction of mold closure. Thus the total penalty for this small part is 5, cavity detail is low and the third digit is 0.

The fourth digit is 1 because the external undercuts are extensive. Thus from Table 11.1, the factor C_s, due to subsidiary complexity, is 1.25.

Surface Finish/Tolerance. The part has a surface finish of SPI-3 and commercial tolerances are used. Thus the fifth and sixth digits are 1 and 0, respectively. Thus from Table 11.2, C_t is 1.00.

Total Relative Mold Construction Cost C_{dc}

$$C_{dc} = C_b C_s C_t = 2.29(1.25)(1) = 2.86$$

11.11.3 Relative Mold Material Cost

Since L_m = 55 mm and H_m = 20 mm, then

$$L_m/H_m = 2.75$$

and from Figure 11.19, C is 0.13. Thus, the thickness of the mold wall is

$$M_{ws} = [0.006CH_m^4]^{1/3} = [0.006(0.13)(20)^4]^{1/3} = 5.0 \text{ mm}$$

and the thickness of the base is

$$M_{wf} = 0.04L_m^{4/3} = 0.04(55)^{4/3} = 8.4 \text{ mm.}$$

Consequently, the projected area of the mold base is

$$M_a = [2(5.0)+55][2(5.0)+40] = 3250 \text{ mm}^2$$

and the required plate height is

$$M_t = [20 + 2(8.4)] = 36.8 \text{ mm}$$

Hence, from Figure 11.20, the relative mold material cost, C_{dm}, for this part is approximately 1.0.

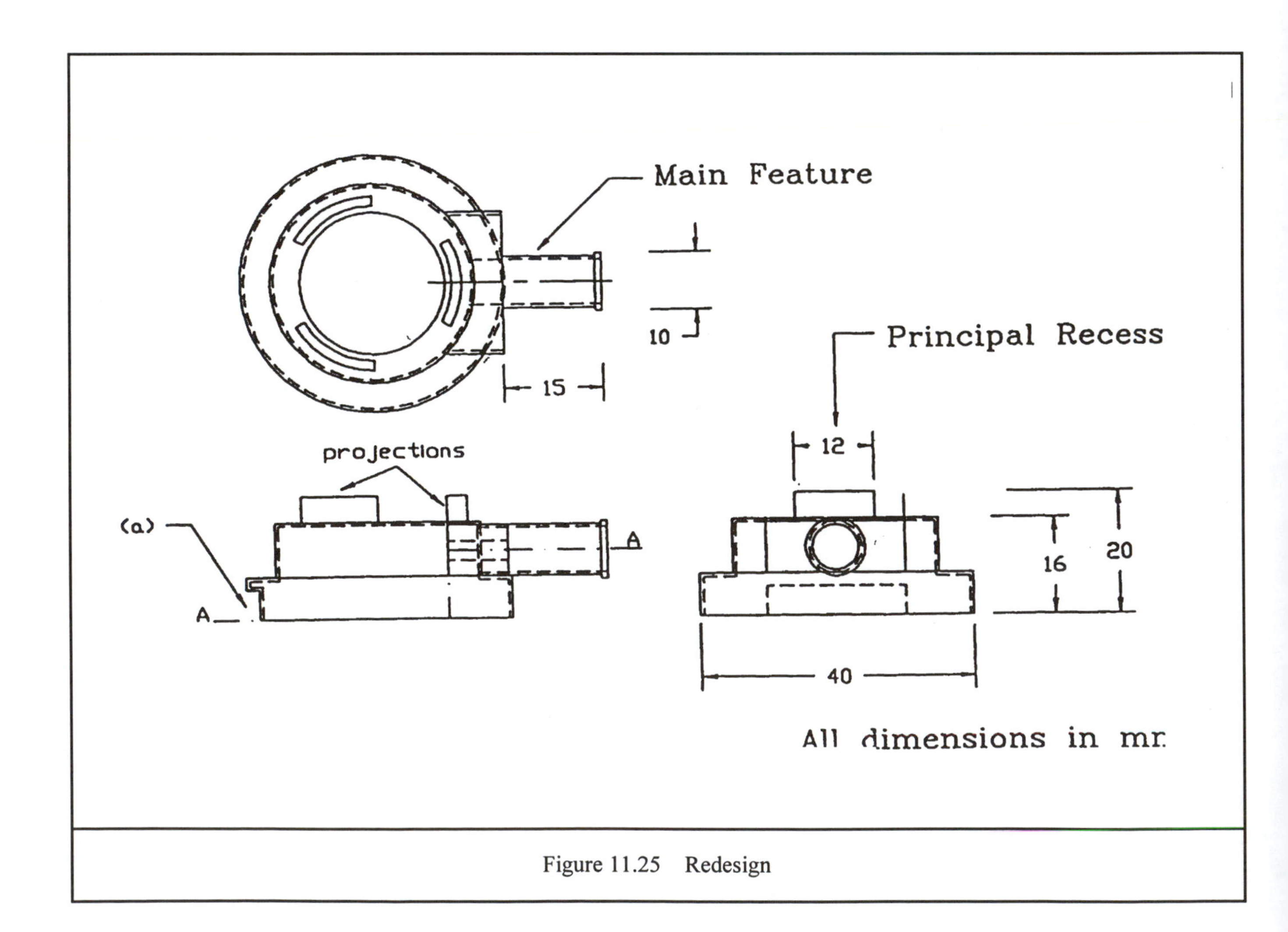

Figure 11.25 Redesign

11.11.4 Total Relative Mold Cost

$$C_d = 0.8\,C_{dc} + 0.2\,C_{dm} = 2.49$$

11.11.5 Redesign Suggestions

The two features causing tooling complexity cost in this case are the two external undercuts (a) and (b). If external undercut (a) can be eliminated as shown in Figure 11.25, then the new basic complexity code becomes B21, and C_b becomes 2.15.

With this redesign, the cavity detail remains low, and the third digit is still 0. The remaining external undercut does not constitute an extensive undercut. Hence, the fourth digit is 0, and C_s is 1.00.

With these values, the new total mold construction cost C_{dc} becomes

$$C_{dc} = (2.15)\,(1)\,(1) = 2.15$$

which is a 25% reduction in mold construction costs, and about a 23% reduction in total mold cost.

11.12 Die Cast Part Cost

11.12.1 Overview

Because the processes of injection molding and die casting are so similar, much of what has been said concerning the influence of part geometry and tolerance specifications on injection molding tooling costs applies equally well to die casting.

In die casting, as in injection molding, the three major cost components of a part are material cost, tooling cost and processing cost. These three cost components are influenced, to varying degrees, by the geometry, size, and material of the part as well as by subsidiary factors such as part quality requirements. The same coding system for tooling costs applies, with only minor modifications, to die casting.

The purpose of the sections that follow is to discuss the application of the previous coding system and cost model to die casting. The definition of terms introduced earlier will not be repeated here.

As in injection molding, the total relative mold construction cost is given by

$$C_{dc} = C_b\,C_s\,C_t \qquad (11.5)$$

11.12.2 Relative Tooling Construction Cost

Figures 11.26 and 11.27, as well as Tables 11.3 and 11.4, apply to die casting and are used in precisely the same way that Figures 11.1 and 11.16, and Tables 11.1 and 11.2, are used for injection molding. As indicated in Figure 11.26, internal undercuts and internal threads generally can not be die cast. This is due primarily to the combination of high temperatures, high pressure, and high injection velocities required for die casting. For such features, these result in rapid wear of the form pins used to create internal undercuts which, in turn, creates excessive flash that is difficult and costly to remove. In addition to the problem of flash removal, in the case of internal threads, high shrinkage makes extraction of the thread forming unit very difficult.

All of the tooling used in die castings is produced with essentially the same surface finish. The effect of molten metal on the surface of the dies is such that deterioration of the surface occurs so quickly that various grades of surface finish are unwarranted. Thus, while the part itself may have different levels of surface quality (as will be explained in Chap 15), all of the tooling is produced with the same surface finish. This fact is reflected in Table 11.4.

As in the case of injection molding, most of the tooling used to produce mold castings is composed of pre-engineered, standardized mold base assemblies and components. Thus, the curve developed earlier for determination of the relative mold material cost, C_{dm}, is equally applicable here. (Figure 11.20)

11.12.4 Total Relative Mold Cost

The total relative mold cost of a part, C_d, is given by the expression

$$C_d = 0.8\,C_{dc} + 0.2\,C_{dm} \qquad (11.4)$$

FOURTH DIGIT

	SUBSIDIARY COMPLEXITY		EXTERNAL UNDERCUT COMPLEXITY	
			Without Extensive (7) External Undercuts (5)	With Extensive (7) External Undercuts (5)
			0	1
THIRD DIGIT	CAVITY DETAIL (6)	Low 0	1.00	1.25
		Moderate 1	1.25	1.45
		High 2	1.60	1.75
		Very High 3	2.05	2.15

Table 11.3 Subsidiary Complexity Rating, Cs, for Die Casting. Notes (5)-(7) are in Appendix 11A.

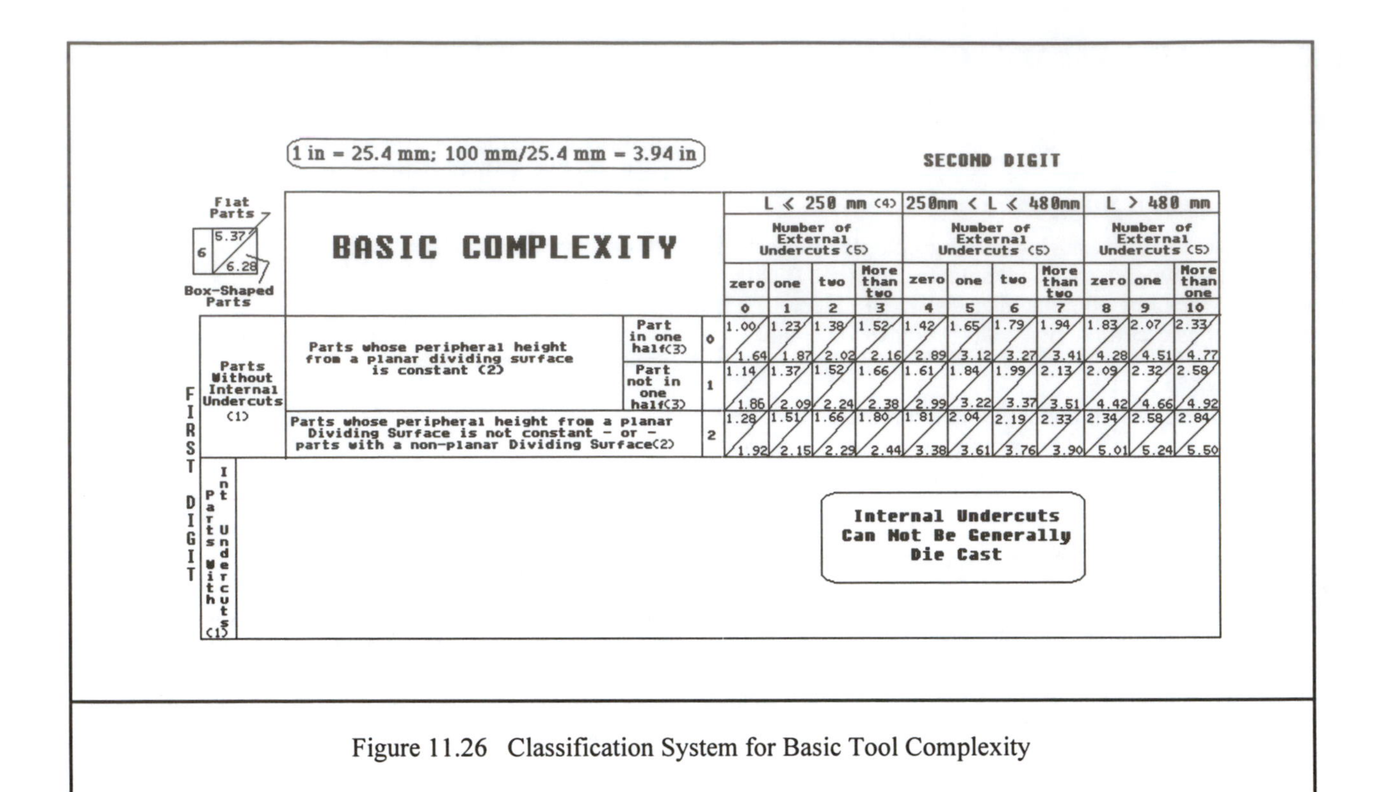

1 in = 25.4 mm; 100 mm/25.4 mm = 3.94 in

SECOND DIGIT

Flat Parts: 6, 5.37 / Box-Shaped Parts: 6.28

BASIC COMPLEXITY

FIRST DIGIT				L ≤ 250 mm (4)				250mm < L ≤ 480mm				L > 480 mm		
				Number of External Undercuts (5)				Number of External Undercuts (5)				Number of External Undercuts (5)		
				zero	one	two	More than two	zero	one	two	More than two	zero	one	More than one
				0	1	2	3	4	5	6	7	8	9	10
Parts Without Internal Undercuts (1)	Parts whose peripheral height from a planar dividing surface is constant (2)	Part in one half(3)	0	1.00 / 1.64	1.23 / 1.87	1.38 / 2.02	1.52 / 2.16	1.42 / 2.89	1.65 / 3.12	1.79 / 3.27	1.94 / 3.41	1.83 / 4.28	2.07 / 4.51	2.33 / 4.77
		Part not in one half(3)	1	1.14 / 1.86	1.37 / 2.09	1.52 / 2.24	1.66 / 2.38	1.61 / 2.99	1.84 / 3.22	1.99 / 3.37	2.13 / 3.51	2.09 / 4.42	2.32 / 4.66	2.58 / 4.92
	Parts whose peripheral height from a planar Dividing Surface is not constant - or - parts with a non-planar Dividing Surface(2)		2	1.28 / 1.92	1.51 / 2.15	1.66 / 2.29	1.80 / 2.44	1.81 / 3.38	2.04 / 3.61	2.19 / 3.76	2.33 / 3.90	2.34 / 5.01	2.58 / 5.24	2.84 / 5.50
Parts With Internal Undercuts (1)	Internal Undercuts Can Not Be Generally Die Cast													

Figure 11.26 Classification System for Basic Tool Complexity

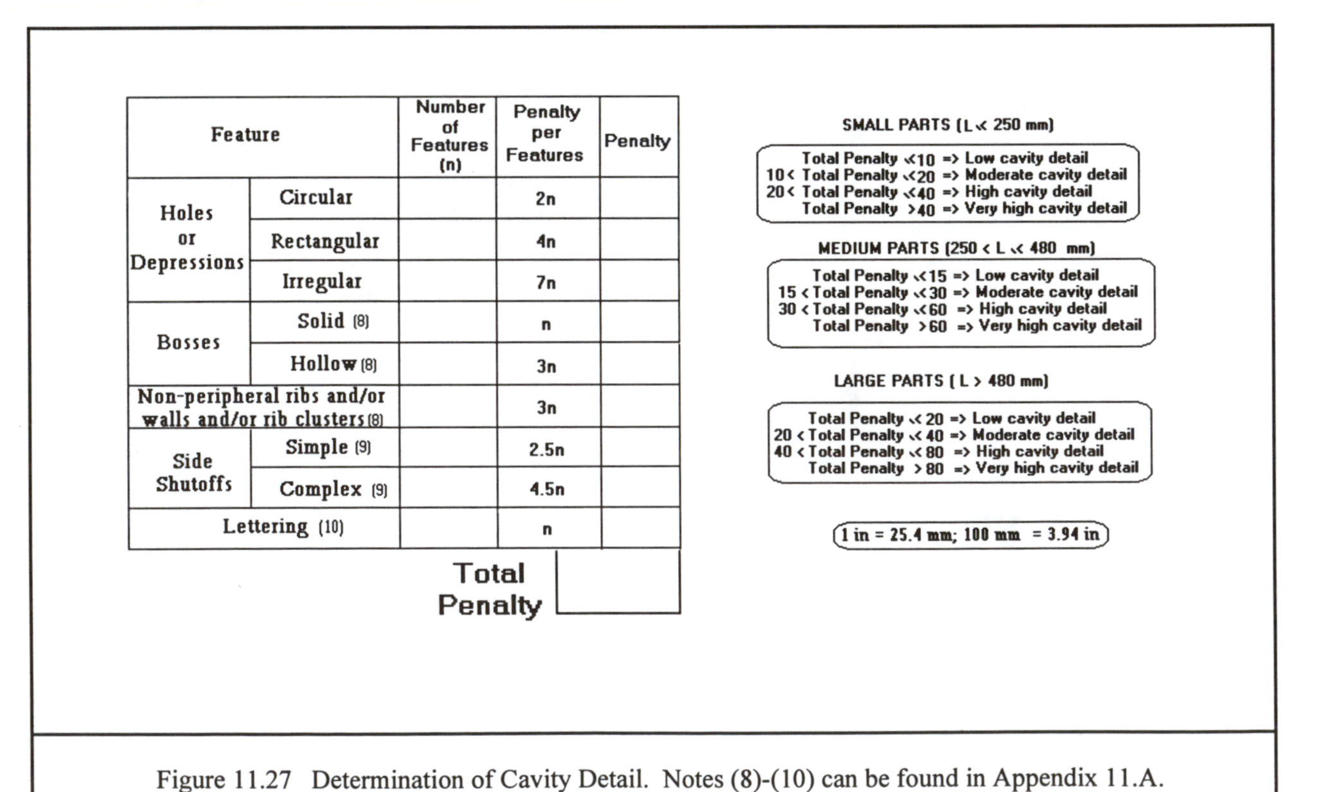

Feature		Number of Features (n)	Penalty per Features	Penalty
Holes or Depressions	Circular		2n	
	Rectangular		4n	
	Irregular		7n	
Bosses	Solid (8)		n	
	Hollow (8)		3n	
Non-peripheral ribs and/or walls and/or rib clusters (8)			3n	
Side Shutoffs	Simple (9)		2.5n	
	Complex (9)		4.5n	
Lettering (10)			n	
			Total Penalty	

SMALL PARTS (L ≤ 250 mm)

Total Penalty ≤ 10 => Low cavity detail
10 < Total Penalty ≤ 20 => Moderate cavity detail
20 < Total Penalty ≤ 40 => High cavity detail
Total Penalty > 40 => Very high cavity detail

MEDIUM PARTS (250 < L ≤ 480 mm)

Total Penalty ≤ 15 => Low cavity detail
15 < Total Penalty ≤ 30 => Moderate cavity detail
30 < Total Penalty ≤ 60 => High cavity detail
Total Penalty > 60 => Very high cavity detail

LARGE PARTS (L > 480 mm)

Total Penalty ≤ 20 => Low cavity detail
20 < Total Penalty ≤ 40 => Moderate cavity detail
40 < Total Penalty ≤ 80 => High cavity detail
Total Penalty > 80 => Very high cavity detail

1 in = 25.4 mm; 100 mm = 3.94 in

Figure 11.27 Determination of Cavity Detail. Notes (8)-(10) can be found in Appendix 11.A.

11.13 Example Relative Tooling Cost

11.13.1 The Part

Consider the part shown in Figure 11.28. While the wall thickness of the part is specified along with the boss height and boss thickness, for an evaluation of the relative tooling cost of the part, these dimensions are not necessary.

11.13.2 Relative Tool Construction Cost.

Basic Complexity: The length of the projections parallel to the surface of the part, c, is 10 mm. The ratio of c to the envelope dimension parallel to c is less than 1/3; so the projections are isolated projections of small volume. Consequently, the dimensions of the basic envelope are:

L = 160 mm, B = 130 mm, H = 10 mm.

Since L/H is greater than 4, the part is flat.

The direction of mold closure is normal to the LB plane; thus a planar dividing surface exists for the principal shape of the part.

No internal undercuts are present, the part has a planar dividing surface, the peripheral height from that dividing surface is constant, and the part is in one half. Thus, the first digit is 0.

FIFTH DIGIT	SURFACE FINISH R_a	SIXTH DIGIT: TOLERANCES, T_a — Commercial 0	Tight 1
	0	–	–
	1	1.00	1.05
	2		
	3		

Table 11.4 Tolerance and Surface Finish Rating, C_t, for Die Casting

There are no external undercuts and L is less than 250 mm. Hence, the second digit is also 0.

Thus, from Figure 11.26, the basic mold manufacturability cost, C_b, is 1.00.

Subsidiary Complexity: There are 8 radial ribs (24 penalty points), 1 concentric rib (3 penalty points), 3 hollow bosses (9 penalty points), and 1 circular hole (2 penalty points). Although it is not indicated in the above drawing, there is lettering on the underside of the part (1 penalty point). Consequently, cavity detail is high, and the third digit is 2.

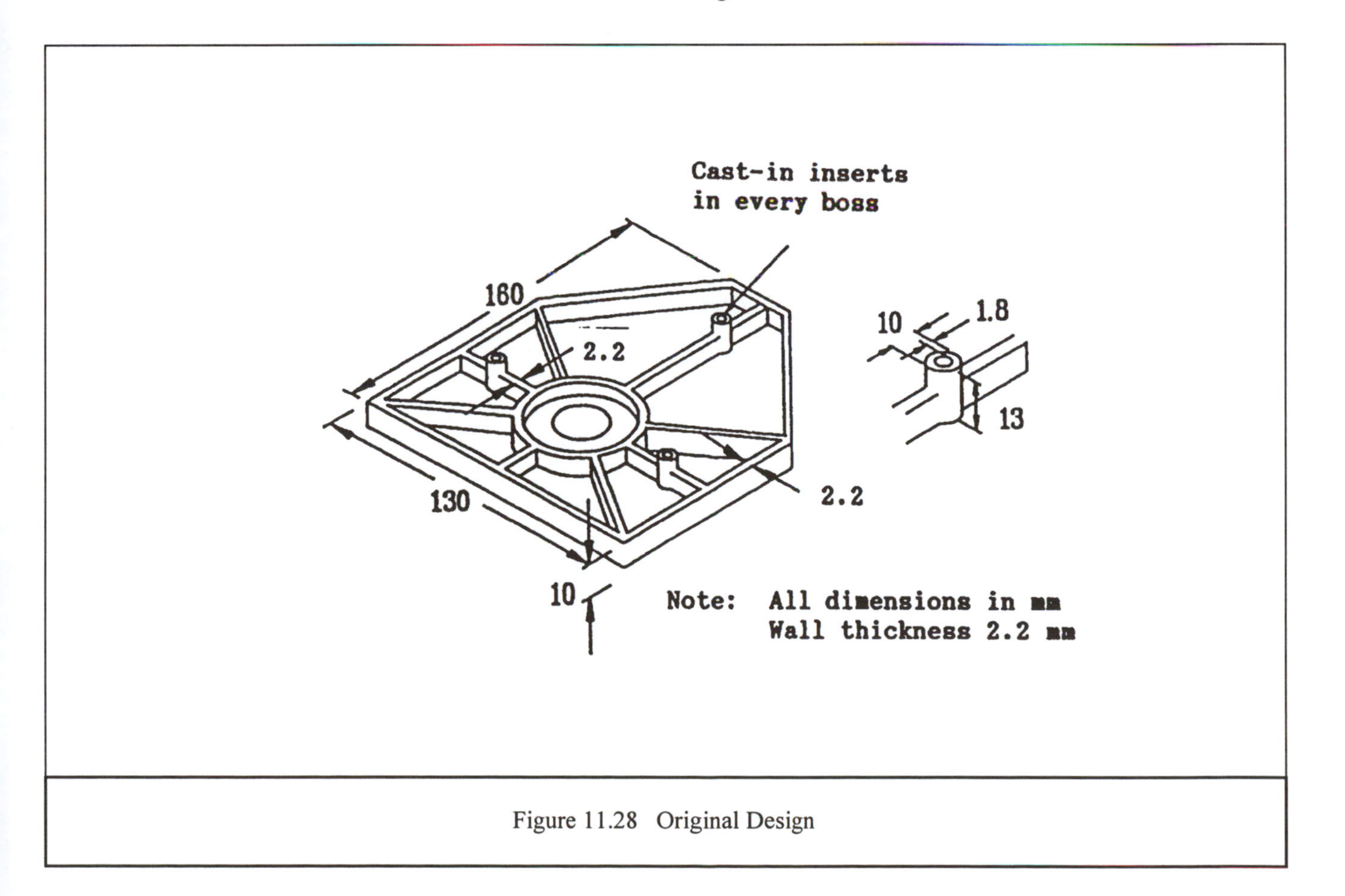

Figure 11.28 Original Design

Since no external undercuts are present, the fourth digit is 0. Thus the multiplying factor, C_s, due to subsidiary complexity, is 1.60.

Tolerance: Commercial tolerances are used; thus, C_t is 1.00.

Total relative mold construction cost, C_{dc}.

$$C_{dc} = C_b\,C_s\,C_t = 1.00\,(1.60)\,(1) = 1.60$$

11.13.3 Relative Mold Material Cost

From Figure 11.19, for L_m/H_m (160/23) of 6.9, C is 0.14. Thus, the thickness of the mold wall is

$$M_{ws} = [0.006\,C\,H_m^{\,4}]^{1/3} = [0.006(0.14)(23)^4]^{1/3} = 6.2 \text{ mm}$$

and the thickness of the base is

$$M_{wf} = 0.04\,L_m^{\,4/3} = 0.04\,(160)^{4/3} = 34.7 \text{ mm}$$

Consequently, the projected area of the mold base is

$$M_a = [2(6.2)+160][2(6.2)+132] = 24{,}894 \text{ mm}^2$$

and the required plate height is

$$M_t = [23 + 2(34.7)] = 92.5 \text{ mm}$$

Hence, from Figure 11.20, the relative mold material cost, C_{dm}, for this part is approximately 1.5.

11.13.4 Total Relative Mold Cost

$$C_d = 0.8\,C_{dc} + 0.2\,C_{dm} = 0.8\,(1.60) + 0.2\,(1.5) = 1.58$$

11.13.5 Redesign Suggestions

Mold manufacturability costs can be reduced if the cavity detail can be reduced. For example, if the radial ribs are removed the cavity detail would be reduced from high to moderate, and C_s would become 1.25. The total relative mold construction cost, C_{dc}, then becomes 1.25. This is a 22% reduction in mold construction costs.

11.14 Worksheet for Relative Tooling Cost

The determination of the relative die construction costs, the relative die material costs and the overall relative die costs is a straight forward, though some times cumbersome, procedure. A Worksheet that is helpful is shown on page 11-27. To illustrate the use of the Worksheet, it has been filled out on page 11- 23 for the part shown in Figure 11.23 (Example 11.2).

The Worksheet shown on page 11-27 may be copied.

11.15 Summary and Preview

Summary. This chapter has described a systematic approach for calling designers' attention to those features of injection molded and die cast parts which tend to increase the tooling cost to manufacture parts — and for estimating the relative costs of tooling. The system employs a six digit coding system for determining total relative tooling cost, which groups parts according to similarity in tool construction difficulty. The system highlights those features which significantly increase cost so that designers can minimize difficult to produce features.

Using the methodology presented, designers can perform a tooling cost evaluation of a proposed part using only the information available at the configuration stage of part design. That is, the evaluation can be performed from only the knowledge of whether certain features are present or absent and, if present, their approximate location and orientation. Dimensions are not needed. The methodology points out what features or arrangements of features contribute to the cost so that the direction of improved redesign is made apparent.

Preview. In the next Chapter, a similar methodology is developed for use with stamped parts. Then in Chapter 13, methods for evaluation of configuration designs that include both functional and manufacturability issues are presented.

References

[1] Poli, C., and Fernandez, R., "Coding System for Mold Manufacturability," Mechanical Engineering Department, University of Massachusetts at Amherst, Amherst, MA, 1989.

[2] Poli, C., Escudero, J. and Fernandez, R., "How Part Design Affects Injection Molding Tool Costs," Machine Design, Nov. 24, 1988.

[3] Poli, C., Kuo, S. M. and Sunderland, J. E., "Design for Injection Molding Coding System for Relative Cycle Time," Mechanical Engineering Department, University of Massachusetts at Amherst, Amherst, MA, 1989.

[4] Poli, C., Kuo, Sheng-Ming, and Sunderland, J. E., "Keeping a Lid on Mold Processing Costs," Machine Design, Oct. 26, 1989.

[5] Kuo, Sheng-Ming, "A Knowledge-Based System for Economical Injection Molding," Ph.D. Dissertation, University of Massachusetts at Amherst, Amherst, MA, Feb. 1990.

[6] Escudero, Juan R., "Two Methods to Assess the Effect of Part Design on Tooling Costs in Injection Molding," Mechanical Engineering Department, University of Massachusetts at Amherst, Amherst, MA, 1988

[7] Dym, J. B., "Product Design with Plastics," Industrial Press, New York, 1983

[8] Shanmugasundaram, S. K., "An Integrated Economic Model for the Analysis of Mold Cast and Injection Molded Parts," M.S. Final Project Report, Mechanical Engineering Department, University of Massachusetts at Amherst,

Worksheet For Relative Tooling Costs

Injection Molding and Die Casting

Original Design

(a) Relative Die Construction Cost

Basic Shape: L = ***55***; B = ***40***; H = ***16***; => Box/Flat ***Box***

Basic Complexity: 1st Digit = ***2***; 2nd Digit = ***2*** => C_b = ***2.29***

Subsidiary Comp: 3rd Digit = ***0***; 4th Digit = ***1*** => C_s = ***1.25***

T_a/R_a: 5th Digit = ***1***; 6th Digit = ***0*** => C_t = ***1.00***

Total Relative Die Construction Cost, $C_{dc} = C_b C_s C_t$ = ***2.86***

(b) Relative Die Material Cost:

L_m = ***55***; B_m = ***40***; H_m = ***20***; Die Closure Parallel To ***H***

L_m/H_m = ***2.75***; thus, C = ***0.13***

$M_{ws} = [0.006CH_m^4]^{1/3} = [0.006(0.13)(20)^4]^{1/3} = 5\ mm$

$M_{wf} = 0.04L_m^{4/3} = 0.04(55)^{4/3} = 8.4\ mm$

$M_a = (2M_{ws} + L_m)(2M_{ws} + B_m) = [2(5) + 55][2(5) + 40] = 3250\ mm^2$

$M_t = (H_m + 2M_{wf}) = [20 + 2(8.4)] = 36.8\ mm$

Thus, C_{dm} = ***1.00*** and $C_d = 0.8\,C_{dc} + 0.2C_{dc}$ $= 0.8(2.86) + 0.2(1) = 2.49$

Redesign Suggestions:

Eliminate external undercut (a) as shown in Figure 11.25

Basic Complexity: 1st Digit = ***2***; 2nd Digit = ***1*** => C_b = ***2.15***

Subsidiary Comp: 3rd Digit = ***0***; 4th Digit = ***1*** => C_s = ***1.0***

T_a/R_a: 5th Digit = ***1***; 6th Digit = ***0*** => C_t = ***1.0***

$C_{dc} = C_b C_s C_t$ = ***2.15***; $C_d = 0.8\,C_{dc} + 0.2\,C_{dm}$ $= 0.8(2.15) + 0.2(1) = 1.92$

% Savings = (2.49 – 1.92) / 2.49 = 0.23 = 23 %

Amherst, MA 0903, August 1990.

[9] Poli, C, Fredette, L. and Sunderland, J. E., "Trimming the Cost of Mold Castings," Machine Design, March 8, 1990.

[10] Fredette, Lee, "A Design Aid For Increasing The Producibility of Mold Cast Parts," M.S. Final Project Report, Mechanical Engineering Department, University of Massachusetts at Amherst, Amherst, MA, 1989.

[11] Shanmugasundaram, S. K., "An Integrated Economic Model for the Analysis of Mold Cast and Injection Molded Parts," M.S. Final Project Report, Mechanical Engineering Department, University of Massachusetts at Amherst, Amherst, MA, 1990.

Problems

11.1 Determine the relative die cost for the part shown in Figure P11.1. Assume that the part has a surface finish of SPI-3 and that commercial tolerances are to be used.

11.2 Redesign the part shown in Figure P11.1 so that tooling costs can be reduced. What are the savings in tool costs that you achieved by redesigning the part?

11.3 Determine the relative die cost for the part shown in Figure P11.3. Assume that the part has a surface finish of SPI-3 and that commercial tolerances are to be used.Problems, Continued

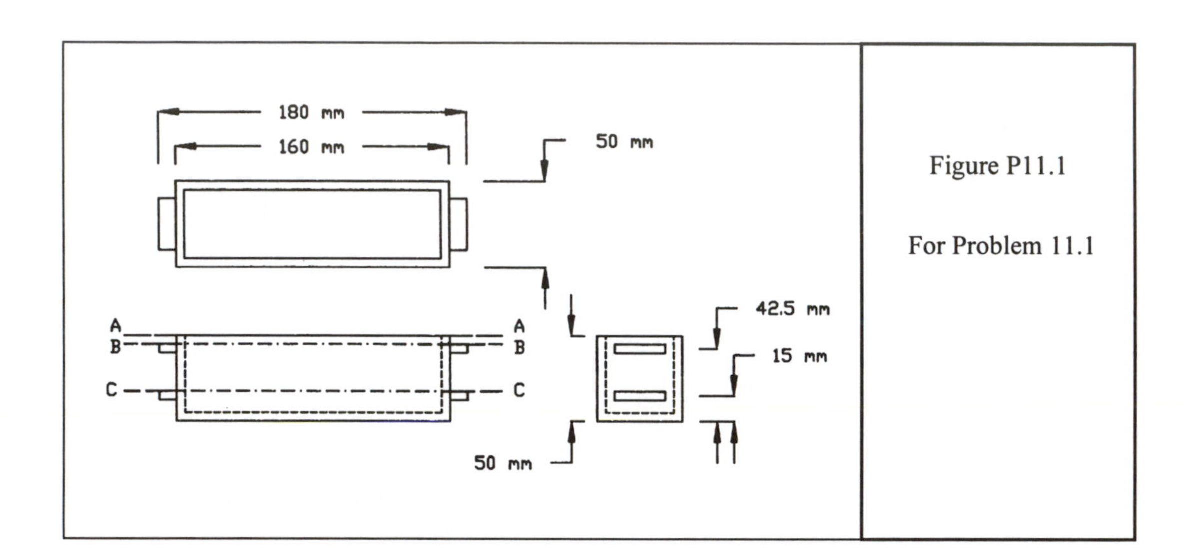

Figure P11.1

For Problem 11.1

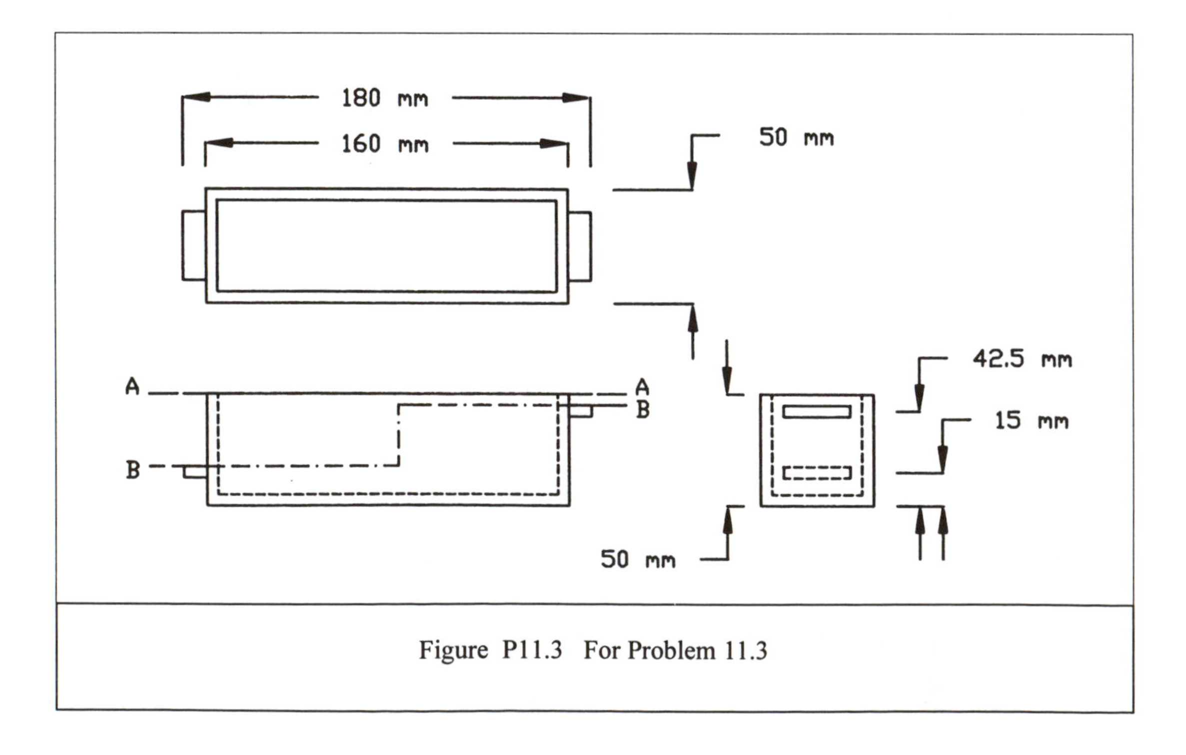

Figure P11.3 For Problem 11.3

11.4 Redesign the part shown in Figure P11.3 so that tooling costs can be reduced. What are the savings in tool costs that you achieved by redesigning the part?

11.5 Determine the relative die cost for the part shown in Figure P11.5. Assume that the part has a surface finish of SPI-3 and that commercial tolerances are to be used.

11.6 Redesign the part shown in Figure P11.5 so that tooling costs can be reduced. What are the savings in tool costs that you achieved by redesigning the part?

11.7 Determine the relative die cost for the part shown in Figure P11.7. Assume that the part has a textured surface finish and that tight tolerances are to be used.

11.8 Redesign the part shown in Figure P11.7 so that tooling costs can be reduced. What are the savings in tool costs that you achieved by redesigning the part?

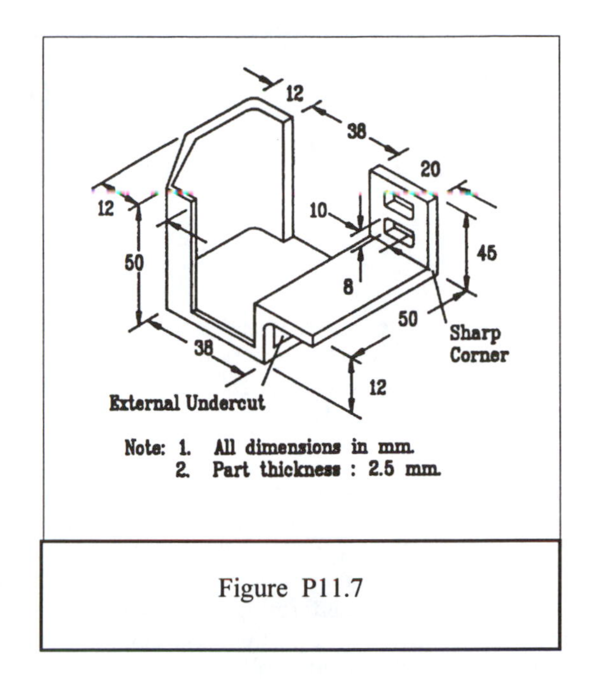

Figure P11.7

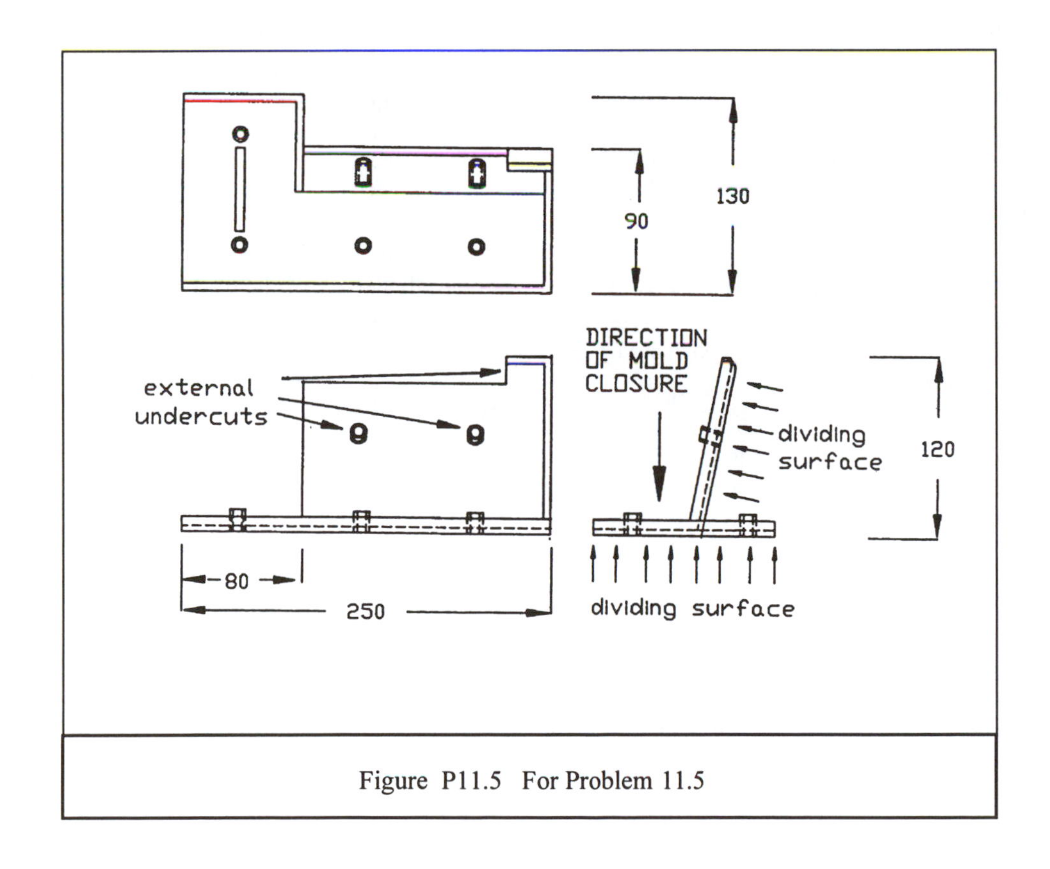

Figure P11.5 For Problem 11.5

Appendix 11.A

Notes for Figures 11.11, 11.16, 11.26, 11.27, and Tables 11.1 and 11.3.

(1) ***Internal undercuts*** are recesses or projections on the inner surface of a part which prevents solid plugs, conforming to the exact shape of the inner surface of the part, from being inserted. See Figure 11.10.

(2) ***Dividing surface.*** Given a direction of mold closure, a dividing surface is defined as an imaginary surface, in one or more planes, through the principal shape of the part, for which the portion on either side of the surface can be extracted from a cavity, conforming to the complementary form of the outer shape of the portion, in a direction parallel to the direction of mold closure. If the dividing surface is in one plane only it is regarded as a planar dividing surface.

The peripheral height from a planar dividing surface is considered constant if the height does not vary by more than 3 times the wall thickness. See Figure 11.7.

(3) For ***box-shaped*** parts, the part is in ***one half*** when the principal shape of the part lies entirely on one side of a planar dividing surface. For ***flat parts***, the part is in ***one half*** when the entire part is on one side of a planar dividing surfac. See Figure 11.15.

(4) L is the longest dimension of the basic envelope of the part. If H is the smallest dimension of the basic envelope, then when L/H is greater than 4, the part is considered ***flat***, otherwise it is considered ***box-shaped***

(5) ***external undercuts*** are holes or depressions on the external surface of a part which are not parallel to the direction of mold closure. Projections which are on the external surface of a part and are such that a single dividing surface, *planar or nonplanar,* can not pass through all of them are also considered external undercuts.

The ***number of external undercuts*** is equal to the number of surfaces bearing unidirectional holes and/or depressions not in the direction of mold closure and/or projections that prevent a single dividing surface from passing through all of the projections. See Figure 11.12.

(6) ***Cavity detail*** is a measure of the concentration of features parallel to the direction of mold closure. Typical features that increase cavity detail are ribs, bosses and holes. See Figure 11.16.

(7) ***External undercuts*** other than unidirectional circular holes or depressions are considered extensive external undercuts. See Figure 11.17.

(8) A ***rib*** is a narrow elongated projection with a length generally greater than about three times its width (thickness), both measured parallel to the surface from which the feature projects, and a height less than six times its width. Ribs may be located at the periphery or on the interior of a part or plate. Peripheral ribs are not included in the rib count.

A narrow elongated projection with a height greater than six times its width is considered a ***wall***.

A cluster of two closely spaced longitudinal ribs or lateral ribs (that is, two ribs whose spacing is less than three times the rib width) are counted as one rib.

A ***boss*** is an isolated projection with a length of projection which is generally less than about three times its overall width, the latter measured parallel to the surface from which it projects. A boss is usually circular in shape but it can take a variety of other forms called knobs, hubs, lugs, buttons, pads, or "prolongs."

Bosses can be solid or hollow. In the case of a solid circular boss, the length of the boss and its width are both equal to the boss diameter.

Bosses, and sometimes ribs, are supported by a cluster of ribs of variable height called gusset plates. This cluster of ribs is treated as one rib or one cluster of ribs. See Figure 11.5.

(9) Holes in a component that do not need to be classified as undercuts are considered ***side shutoffs.*** Side shutoffs can be simple or complex. Isolated grooves or cutouts on the external surface of a part are also considered side shutoffs. A groove or cutout is considered isolated if the dimension of the cutout normal to the direction of mold closure is less than 0.33 times the envelope dimension in the same direction. Penalties due to simple side shutoffs are not considered for parts whose first digit is 2, 4, or 6. See Figure 11.14.

(10) All words and symbols at one location on the part are classified as a single ***lettering entity*** since the entire lettering pattern on the tooling will be made using one electrode.

Worksheet For Relative Tooling Costs

Injection Molding and Die Casting

Original Design

(a) *Relative Die Construction Cost*

Basic Shape: L = ___________; B = _________; H = ________; => Box/Flat _________

Basic Complexity: 1st Digit = ________; 2nd Digit = _________ => C_b = ________

Subsidiary Comp: 3rd Digit = ________; 4th Digit = _________ => C_s = ________

T_a/R_a: 5th Digit = ________; 6th Digit = _________ => C_t = ________

Total Relative Die Construction Cost, $C_{dc} = C_b C_s C_t$ = ___________

(b) Relative Die Material Cost:

L_m = _______; B_m = _______; H_m = _______; Die Closure parallel to _______

L_m/H_m = ________; thus, C = _________

$M_{ws} = [0.006CH_m^4]^{1/3}$ = ____

$M_{wf} = 0.04L_m^{4/3}$ = ____

$M_a = (2M_{ws} + L_m)(2M_{ws} + B_m)$ = ____

$M_t = (H_m + 2M_{wf})$ = ____

Thus, C_{dm} = _____________ and $C_d = 0.8C_{dc} + 0.2C_{dm}$ = ______________________

Redesign Suggestions:

Basic Complexity: 1st Digit = ________; 2nd Digit = ________ => C_b = _________

Subsidiary Comp: 3rd Digit = ________; 4th Digit = ________ => C_s = _________

T_a/R_a 5th Digit = ________; 6th Digit = ________ => C_t = _________

$C_{dc} = C_b C_s C_t$ = ____________; $C_d = 0.8C_{dc} + 0.2C_{dm}$ = ______________________

% Savings = ____

This page available for notes, sketches and calculations.

Chapter Twelve

Evaluating Part Configurations for Manufacturability: Stamping

"You can't think and hit at the same time."

Yogi Berra

12.1 Introduction

Yogi is no doubt right in baseball, but engineering designers must do at least two things at once: consider both function and DFM. In this chapter, we consider the specifics of doing DFM at the configuration stage for stamped parts.

As indicated in Chapter 3, stamping is a process by which thin-walled metal parts are formed from sheets of metal. Parts are formed by a series of cutting and forming operations performed in combination punch and die sets mounted in mechanical or hydraulic presses. While other sheet metal forming processes exist (stretching, spinning, deep drawing, etc.) [1], more sheet metal parts are made by stamping than by any other sheet metal process.

As in the case of injection molded and die cast parts, stampings are generally designed by product designers working for large consumer products corporations. However, the parts are usually produced by small (typically less than 100 employees) custom stampers. Unlike injection molding and die casting, most stampers also make the tooling necessary to produce the parts.

One purpose of this chapter is to present a systematic approach for identifying, at the configuration design stage, those features of parts which significantly affect the cost of stampings. The goal is to learn how to design so as to minimize difficult-to-stamp features.

We also present a methodology for estimating the relative tooling cost of proposed stamped parts based on configuration information. Unlike injection molding and die casting, however, where the total relative tooling cost can be determined at the configuration stage of design, in stamping only the maximum and minimum relative tooling cost can be determined. The reason is that some of the details required to finalize the relative tooling cost can only be determined after parametric design. For example, in injection molding and die casting, any hole which is not parallel to the direction of mold closure is an undercut. In stamping, on the other hand, whether or not such a hole is an undercut (called *side action features* in stamping) depends on its dimensions and tolerances. Nevertheless, a great deal of information concerning tooling costs for stampings can be determined at the configuration stage, and this information can and should be used to help guide the generation and selection of part configurations before parametric design is done.

12.2 Estimating the Relative Cost of Stamped Parts

12.2.1 Total Cost

The total cost of a stamped part consists of: tooling cost, K_d/N; processing cost (or equipment operating cost), K_e; and material cost, K_m. That is:

$$\text{Total Cost} = K_e + K_d/N + K_m \qquad (12.1)$$

Because of the high speed of the presses used to produce stamped part (24 strokes/min to 700 strokes/min) cycle times of less than 1 second are generally achievable. For this reason, even at high production volumes, the proportion of cost due to processing is low.

When only the configuration of the part is known, we can only estimate K_d. Later (in Chapter 16) when more exact dimensions have been established through parametric design, we will show how to compute K_e and K_m.

12.2.2 Total Relative Tooling Cost

As in the case of injection molding and die casting, the total tooling costs are made up of die material cost, K_{dm}, and die construction cost, K_{dc}. Thus, relative to a reference part, total die costs are:

$$C_d = \frac{\text{Cost of Tooling for Designed Part}}{\text{Cost of Tooling for Reference Part}}$$

That is: $$C_d = (K_{dm} + K_{dc})/(K_{dmo} + K_{dco}) \qquad (12.2)$$

where K_{dmo} and K_{dco} refer to die material cost and die construction cost for the reference part. As shown in Chapter 11, this equation can be written as:

$$C_d = K_{dm}/(K_{dmo} + K_{dco}) + K_{dc}/(K_{dmo} + K_{dco})$$

$$= A\,(K_{dm}/K_{dmo}) + B(K_{dc}/K_{dco}) \qquad (12.3)$$

where

$$A = (K_{dmo}/(K_{dmo}+K_{dco})$$
$$B = (K_{dco}/\,(K_{dmo} + K_{dco})$$

Based on data collected from stampers, a reasonable value for A is between 0.15 - 0.20, and a reasonable value for B is between 0.80 - 0.85. We will take A and B to be 0.2 and 0.8, respectively. Hence, Equation (12.3) becomes

$$C_d = 0.8\,C_{dc} + 0.2\,C_{dm} \qquad (12.4)$$

where C_d is the total die cost of a part relative to the die cost of the reference part, C_{dc} is the die construction cost for the part relative to the die construction cost of the reference part, and C_{dm} is the die material cost for the part relative to the die material cost of the reference part.

If the reference part is a stamped washer (OD = 50 mm; ID = 10 mm; t = 1.5 mm) made of low carbon cold rolled steel (CRS), then the tool construction cost is about $5,500, and the die material cost is about $1,000.

12.3 The Part Coding System for Stamping: An Overview

12.3.1 The Matrices

Figures 12.1, 12.2, and 12.3 are coding matrices used for estimating the relative cost for stamped parts. Each Figure applies to a different type of stamping operation.

- Figure 12.1 applies to basic operations of shearing and the forming of local features like holes and bosses.
- Figure 12.2 applies to straight bending (wiping) operations.
- Figure 12.3 applies to contour forming; that is, to operations involving curved or other non-straight bends.

Two values are found in each cell of each of the coding matrices: the upper number is called the *basic relative tool construction cost*, C_b; the lower number is the *number of active stations required*, N_a. When C_b is multiplied by factors to account for part material type and sheet thickness, the result is the relative tool construction cost, C_{dc}. The values of C_b include both design and build hours for the tooling.

Since stamped parts are made by a sequence of operations, values obtained from the three figures (i.e., C_{b1} and N_{a1} from Figure 12.1, C_{b2} and N_{a2} from Figure 12.2, and C_{b3} and N_{a3} from Figure 12.3) are all used as explained below to estimate the total relative tooling cost and the relative die material cost for a part.

12.3.2 The Coding Digits

Entering the coding matrices of Figures 12.1, 12.2, and 12.3 to obtain the three C_b and three N_a values is done by first determining the values of six digits that classify the part as follows:

The *first digit* classifies parts based on the amount of die detail required. We describe how to determine the first digit in Section 12.5.2 below.

The *second digit* takes into account the part size and the complexity of the periphery of the part. Determining the second digit is explained in Section 12.5.3.

The third and fourth digits are related to the complexity and size of bends in the stamping. The *third digit* assesses bend complexity. The *fourth digit* classifies the part according to the length of the longest bend and the presence of critical features near the bend. Obtaining the third and fourth digits is described in Sections 12.5.4 and 12.5.5.

The fifth and sixth digits are related to the complexity of any non-straight (i.e., contour) bends in the stamping. The *fifth digit* classifies the part according to whether it has such bend lines and their nature. The *sixth digit* accounts for the length of these bends and the presence of critical features in or near them. (See Sections 12.5.6 and 12.5.7

In addition to these six coding digits required for Figures 12.1, 12.2, and 12.3, two digits are included in the system to account for increases in tooling costs due to the type of material used and the material's sheet thickness. These digits — the seventh and eighth — are discussed in the following sub-section.

1 in = 25.4 mm
100 mm/25.4 mm = 3.94 in

C_{b1} — 1.00
N_{a1} — 2

				SECOND DIGIT						
				L_{ul} <100 mm (9)		100 < L_{ul} < 200 (9)		200 < L_{ul} < 450 (9)		L_{ul} > 450 (9)
				L_s < 1.4 (10)	L_s > 1.4 (10)	L_s < 1.4 (10)	L_s > 1.4 (10)	L_s < 1.4 (10)	L_s > 1.4 (10)	
				0	**1**	**2**	**3**	**4**	**5**	**6**
Unfolded parts (1) with uniform sheet thickness less than or equal to 6.5 mm, with or without non-peripheral features (2) whose direction is parallel to the sheet thickness (3) with or without holes whose diameter is greater than the sheet thickness, without significant projections (4).	Parts with LOW die detail (5).	Parts without closely spaced features (6), narrow projections or narrow cutouts (7).	**0**	1.00 2	1.20 2	1.30 2	1.50 2	1.56 2	2.45 2	*Use Sheet Metal Fabrication*
		Parts with only one type of closely spaced protruding feature (8). - OR - Parts with closely spaced non-protruding features and/or narrow cutouts and/or narrow projections (7), without closely spaced protruding features	**1**	1.29 4	1.49 4	1.59 4	1.79 4	1.88 4	2.77 4	
		Parts with more than one type of closely spaced protruding feature (8), with or without closely spaced non-protruding features, narrow cutouts or narrow projections - OR - Parts with only one type of closely spaced protruding feature with closely spaced non-protruding features and/or narrow cutouts and/or narrow projections.	**2**	1.58 5	1.78 5	1.88 5	2.08 5	2.20 5	3.09 5	
	Parts with MEDIUM die detail (5)	Parts without closely spaced features (6), narrow projections or narrow cutouts (7).	**3**	2.33 5	2.53 5	2.63 5	2.83 5	3.32 5	4.21 5	
		Parts with only one type of closely spaced protruding feature (8). - OR - Parts with closely spaced non-protruding features and/or narrow cutouts and/or narrow projections (7), without closely spaced protruding features	**4**	2.62 6	2.82 6	2.92 6	3.12 6	3.64 6	4.53 6	
		Parts with more than one type of closely spaced protruding feature (8), with or without closely spaced non-protruding features, narrow cutouts or narrow projections - OR - Parts with only one type of closely spaced protruding feature with closely spaced non-protruding features and/or narrow cutouts and/or narrow projections.	**5**	2.91 7	3.11 7	3.21 7	3.41 7	3.96 7	4.85 7	
	Parts with HIGH die detail (5).	Parts without closely spaced features (6), narrow projections or narrow cutouts (7).	**6**	2.96 6	3.16 6	3.26 6	3.46 6	3.60 6	4.49 6	
		Parts with only one type of closely spaced protruding feature (8). - OR - Parts with closely spaced non-protruding features and/or narrow cutouts and/or narrow projections (7), without closely spaced protruding features	**7**	3.25 7	3.45 7	3.55 7	3.75 7	3.92 7	4.81 7	
		Parts with more than one type of closely spaced protruding feature (8), with or without closely spaced non-protruding features, narrow cutouts or narrow projections - OR - Parts with only one type of closely spaced protruding feature with closely spaced non-protruding features and/or narrow cutouts and/or narrow projections.	**8**	3.54 8	3.74 8	3.84 8	4.04 8	4.24 8	5.13 8	
Unfolded parts (1) without uniform sheet thickness, or unfolded parts with uniform sheet thickness greater than 6.5 mm, or unfolded parts with non-peripheral features (2) whose direction is not parallel to the sheet thickness (3) or unfolded parts with holes whose diameter is less than the sheet thickness, or unfolded parts with significant projections (4), or unfoldable parts (1)			**9**	*Parts like this are not usually stamped and/or can not generally be stamped*						

Figure 12.1 Classification System for Shearing and Local Features

1 in = 25.4 mm
100 mm/25.4 mm = 3.94 in

C_{b2} — 0.79
N_{a2} — 2

FOURTH DIGIT

THIRD DIGIT				$L_b \leq 100$ mm (17)				100 mm $< L_b \leq 200$ mm (17)				200 mm $< L_b \leq 450$ mm (17)			
				Parts without side-action features (18)		Parts with side-action features (18).		Parts without side-action features (18)		Parts with side-action features (18).		Parts without side-action features (18)		Parts with side-action features (18).	
				Parts without features near the bend line (19).	Parts with features near the bend line (19)	Parts without features near the bend line (19)	Parts with features near the bend line (19)	Parts without features near the bend line (19).	Parts with features near the bend line (19).	Parts without features near the bend line (19).	Parts with features near the bend line (19).	Parts without features near the bend line (19).	Parts with features near the bend line (19)	Parts without features near the bend line (19).	Parts with features near the bend line (19).
				1	2	3	4	5	6	7	8	9	10	11	12
Parts with straight bends (11) without multiple plate junctions (12) without a primary plate connected on all sides by overbent plates (13) — Parts with one bend stage	All bends in the same direction (14)	Parts with all bend angles ≤ 90° (16).	1	0.79 2	0.99 3	1.79 4	1.99 5	0.94 2	1.23 3	1.94 4	2.23 5	1.20 2	1.49 3	2.20 4	2.49 5
		Parts with 90° < bend angles ≤ 105°	2	1.40 3	1.60 4	2.40 5	2.60 6	1.79 3	2.08 4	2.79 5	3.08 6	2.05 3	2.34 4	3.05 5	3.34 6
		Parts with bend angles > 105° (16).	3	1.94 3	2.14 4	2.94 5	3.14 6	2.22 3	2.51 4	3.22 5	3.51 6	2.48 3	2.77 4	3.48 5	3.77 6
	All bends not in the same direction (14)	Parts with all bend angles ≤ 90° (16).	4	1.07 3	1.27 4	2.07 5	2.27 6	1.34 3	1.63 4	2.34 5	2.63 6	1.60 3	1.89 4	2.60 5	2.89 6
		Parts with 90° < bend angles ≤ 105°	5	1.68 5	1.88 6	2.68 7	2.88 8	2.19 5	2.48 6	3.19 7	3.48 8	2.45 5	2.74 6	3.45 7	3.74 8
		Parts with bend angles > 105° (16).	6	2.22 5	2.42 6	3.22 7	3.42 8	2.62 5	2.91 6	3.62 7	3.91 8	2.88 5	3.17 6	3.88 7	4.17 8
	Parts with two bend stages (15)	Parts with all bend angles ≤ 90° (16).	7	1.26 3	1.46 4	2.26 5	2.46 6	2.05 3	2.34 4	3.05 5	3.34 6	2.31 3	2.60 4	3.31 5	3.60 6
		Parts with 90° < bend angles ≤ 105°	8	1.87 5	2.07 6	2.87 7	3.07 8	2.90 5	3.19 6	3.90 7	4.19 8	3.16 5	3.45 6	4.16 7	4.45 8
		Parts with bend angles > 105° (16).	9	2.41 5	2.61 6	3.41 7	3.61 8	3.33 5	3.62 6	4.33 7	4.62 8	3.59 5	3.88 6	4.59 7	4.88 8
	Parts with more than two bend stages (15)	Parts with all bend angles ≤ 90° (16).	10	1.86 5	2.06 6	2.86 7	3.06 8	3.16 5	3.45 6	4.16 7	4.45 8	3.42 5	3.71 6	4.42 7	4.71 8
		Parts with 90° < bend angles ≤ 105°	11	2.47 7	2.67 8	3.47 9	3.67 10	4.01 7	4.30 8	5.01 9	5.30 10	4.27 7	4.56 8	5.27 9	5.56 10
		Parts with bend angles > 105° (16).	12	3.01 7	3.21 8	4.01 9	4.21 10	4.44 7	4.73 8	5.44 9	5.73 10	4.70 7	4.99 8	5.70 9	6.00 10
Parts without straight bends (11), or parts with multiple plate junctions (12), or parts with a primary plate connected on all side by overbent plates (13).			13	*Parts like this can not be wipe formed*											

Figure 12.2 Classification System for Wipe Forming (Bending). Notes (14) - (19) can be found in Appendix 12.A

SIXTH DIGIT

1 in = 25.4 mm
100 mm/25.4 mm = 3.94 in

C_{b3} — 1.05
N_{a3} — 2

FIFTH DIGIT					$L_f \le 100$ mm (26)				100 mm $< L_f$ (26)			
					Parts without side-action features(18)		Parts with side-action features(18).		Parts without side-action features(18)		Parts with side-action features(18).	
					Parts without features on/near the form (27)	Parts with features on/near the form (27)	Parts without features on/near the form (27)	Parts with features on/near the form (27)	Parts without features on/near the form (27)	Parts with features on/near the form (27)	Parts without features on/near the form (27)	Parts with features on/near the form (27)
					1	2	3	4	5	6	7	8
Parts with non-straight bends(11), or radius forms (20), or shallow draws(21), without deep draws or multiple plate connections(12), or non-circular, stepped or tapered recesses(22), or curved channels, or doubly curved sections(23).	Parts with non-straight bends(11), without radius forms(20), without shallow draws(21).			1	1.05 2	1.48 3	2.05 3	2.48 4	Sufficient data not available for such parts			
	Parts with radius forms(20), without non-straight bends (11), without shallow draws(21).	Radius form angle<180° (24)	R/t<15 (25)	2	1.12 3	1.60 4	2.17 4	2.65 5				
			R/t>15 (25)	3	1.37 2	1.85 3	2.42 3	2.90 4				
		Radius form angle >180° (24)		4	Four slide form tools used							
	Parts with shallow draws(21), without radius forms(20), without non-straight bends(11).			5	1.28 4	1.71 5	1.71 5	2.55 6				
Parts with multiple plate connections(11), or non-circular, or stepped or tapered recesses(22), or curved channels or doubly curved sections(23), or deep draw, or any combination of non-straight bends (11), or radius forms(20), or shallow draws(21).				6	Parts like this are not considered here							

Figure 12.3 Classification System for Contour Forming. Notes (11) - (27) can be found in Appendix 12.A

Table 12.1 Classification System for Part Material

Material	Digit 7	F_{mc}	F_{mb}	F_{mf}	Comments
Soft CRS, Low C	0	1.00	1.0	1.0	1008-1012 tempered rolled or annealed last
Soft Red Metal	1	0.98	1.0	1.0	C23000 (red brass) annealed; C26000 (cartridge brass) annealed
Hard Red Metal	2	0.99	1.1	1.2	C23000 (red brass) cold rolled; C26000 (cartridge brass) cold rolled
Soft Al Alloy	3	1.00	1.0	1.0	1100-0; 5052-0
Hard Al Alloy	4	1.00	1.1	1.2	1100-H18; 5052-H18
Hard CRS, Low C	5	1.01	1.1	1.2	1008-1012 half-hard (Rockwell 70-85; tensile strength 380-520 Mpa
Medium Stainless	6	1.02	1.1	1.2	302 annealed
Hard Stainless	7	1.04	1.1	1.2	304 cold rolled

12.3.3 Effect of Part Material Type

Using the first six digits and Figures 12.1, 12.2, and 12.3, designers can obtain values for:

1. C_{b1}-- the relative die construction cost due to shearing and local forming operations (Figure 12.1);
2. C_{b2}-- the relative die construction cost due to bending (wipe forming) operations (Figure 12.2);
3. C_{b3}-- the relative die construction cost for contour forming operations (Figure 12.3)

With these values, the total basic relative tooling construction cost, C_b, can be obtained from the following equation:

$$C_b = C_{b1} F_{mc} + C_{b2} F_{mb} + C_{b3} F_{mf} \quad (12.5)$$

where F_{mc}, F_{mb} and F_{mf} account for the effect of part material on the basic relative tool construction cost. Harder workpiece materials require more tool maintenance (especially tool regrind). In addition, some materials exhibit more springback and hence more experimentation (called tryout) and learning is required to obtain satisfactory results. Finally, there is a 'nervous' factor (i.e., uncertainty) for using materials other than soft cold roll steel or soft aluminum. In all cases, maintenance charges, learning, and nervous factors are charged up front as part of tool construction cost.

Values for F_{mc}, F_{mb} and F_{mf} are determined from Table 12.1. Note that the various material types correspond to a different value for the seventh digit of the coding system.

12.3.4 Effect of Sheet Thickness

Die construction costs are also affected by the sheet thickness of the part. Sheet thickness is the thickness of the stock metal material used. For shearing operations, the clearance between the punch and die must be less than the sheet thickness. For sheet thickness less than 0.005" (0.125 mm), adequate clearances between the punch and die are difficult to achieve, raising tool fabrication costs. In addition, thin material can buckle when being fed in a progressive die. Hence, such a stock thickness is considered *critically thin*.

While a stock thickness greater than 0.125" (3.00 mm) has only a negligible effect on tool fabrication costs, it does significantly increase tool material costs. In addition, thick stock which is coiled for use on progressive dies has undesirable wrinkles. Hence, such a stock thickness is often considered *critically thick*.

The factors F_t and F_{dm} account for the effects of sheet thickness on the die construction costs and the die material cost, respectively. Thus, the relative die construction cost, C_{dc}, is given by:

$$C_{dc} = C_b F_t \quad (12.6)$$

Values for F_t are obtained from Table 12.2. Note that in the Table different sheet thicknesses correspond to different values for the eighth digit in the coding system.

The method for estimating die material cost, C_{dm}, and the role that F_{dm} plays in the determination of C_{dm} is discussed in Section 12.7.

Table 12.2 - Classification System for Sheet Thickness

t (mm)	Digit 8	F_t	F_{dm}
0.125 - 3	0	1.0	1.0
< 0.125	1	1.7	1.0
> 3	2	1.0	1.4

12.3.5 Example

To insure reader understanding of the basic method for using the coding system for estimating the relative tooling cost of stamped parts, we present a simple example of its use.

Problem:

Suppose that the first six digits for a part (we explain below in this chapter how values for these digits are obtained) are 3, 2, 6, 10, 2, and 3 respectively. Also suppose that the part material is to be a hard aluminum alloy (such as 1100-H18) and that the sheet thickness is to be 0.156 mm.

(a) What are the seventh and eighth digits?

(b) What is the relative tool construction cost?

Solution:

(a) From Tables 12.1 and 12.2, the seventh and eighth digits are 4 and 0, respectively.

(b) From Figure 12.1, using the first and second digits, we get C_{b1} = 2.63.

(c) From Figure 12.2, using the third and fourth digits, we get C_{b2} = 3.17.

(d) From Figure 12.3, using the fifth and sixth digits, we get C_{b3} = 2.17.

(e) From Table 12.1, we get F_{mc}, F_{mb} and F_{mf} to be 1.0, 1.1, and 1.2 respectively.

Thus $C_b = C_{b1} F_{mc} + C_{b2} F_{mb} + C_{b3} F_{mf}$

$= (2.63)(1.0) + (3.17)(1.1) + (2.17)(1.2) = 8.72$

From Table 12.2, we get F_t = 1.0.

Thus $C_{dc} = C_b F_t = (8.66)(1.0) = 8.66.$

12.4 Terminology and Features Important in Stamping

12.4.1 Introduction

In order to better understand how to design a stamping, and to determine the first six digits of the part coding system, it is necessary for designers to know and recognize the types of features that can be stamped and how they are created by the process. This requires knowing the meaning of a number of terms and concepts associated specifically with stamped parts. This section, therefore, is devoted to describing the kinds of features that are commonly stamped, and to introducing the terms used to describe these features and the punch and die sets used to produce them.

Students and other readers are encouraged also to review the material on stamping processes in Chapter 3, Section 3.5.

12.4.2 Die Block, Die Set, and Closure Direction

As indicated in Chapter 3 (Section 3.5.3) a die is a collection of parts which, when mounted on a press, can produce a desired stamping. While each die is custom-made to produce a given stamping, numerous die components are identical or similar in design. Figures 3.21 and 3.22 show a typical die set and the die assembly to produce a simple link.

As shown in Figure 3.21, the die set (which can be commercially purchased as a unit) consists of punch and die holders, guide pins, guide pin bushings, and heel blocks (not shown in Figure 3.21). The die block (shown in Figure 3.22), usually made of heat treated steel, contains the desired machined recesses that give the stamping its shape. The die block is mounted onto the die holder. The punches (or male part of the die) are usually the upper members of the die set and are attached to the punch holder. The sheet metal strip from which the stamped part is made is fed in a direction normal to the direction of die closure. The sheet metal strip consists of both the part (at various stages of forming) and a carrier strip, which carries the partially completed stamping from one station of the die to another. The carrier strip does not form a portion of the final part, but merely conveys the partially formed part from station to station.

12.4.3 Basic Envelope

The basic envelope of a part is the smallest rectangular prism which completely encloses the part, neglecting any small protruding features. (Section 12.4.7 defines 'protruding features' precisely.)

The lengths of the sides of the basic envelope are denoted by L, B and H (L≥B≥H). Figure 12.4 shows the basic envelope of a stamped part, with dimensions marked.

12.4.4 Part Size

If we imagine that a bent part is flattened (remember,

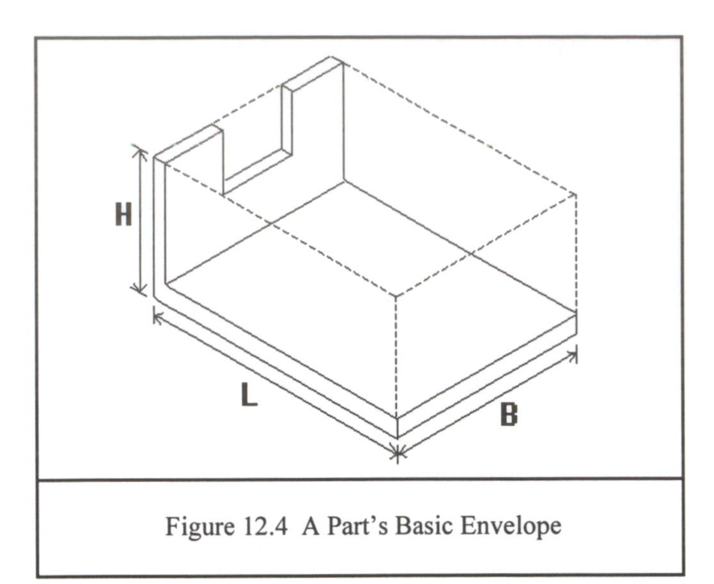

Figure 12.4 A Part's Basic Envelope

prior to bending the sheet is in fact flat) then the largest dimension of the flat envelope (Figure 12.5), L_{ul}, is a measure of the part size. Usually the carrier strip is designed in such a way that the part is laid out on the strip with L_{ul} perpendicular to the direction of the strip feed. This is done to keep the length of the die as small as possible. Thus, the width of the die is an important measure of the part size. An example of such a layout is shown in Figure 9.11(b) where the strip layout and tooling required to produce a U-shaped part, whose bend lines are parallel to L_{uw}, are shown.

On occasion there are geometric constraints that prevent laying out of the part with L_{ul} perpendicular to the direction of strip feed. For example, if the bends lines for the U-shaped part shown in Figure 9.11(b) were instead parallel to L_{ul}, then it would be difficult to form the bends and simultaneously to provide the carrier strip without laying the part out such that L_{uw} is perpendicular to the direction of strip feed.

12.4.5 Flat, Bent, and Contour Formed Parts

If the dimension H of the basic envelope is equal to the thickness of the stamping, then the part is *flat*. Any part that is not flat is considered bent and/or contour formed.

A part is considered *bent* if the part contains only bends which are produced by localized plastic deformation in a narrow strip along a straight bend line. In general the radius of curvature of such bends is less than four times the sheet thickness. Figure 12.4 shows a bent part. We will say more about bent parts in the next section.

Parts which contain either non-straight bend lines (Figure 12.6), radius forms (Figure 12.7), or shallow draws (Figure 12.8) are referred to as *contour formed* parts. Unlike bending, where the plastic deformation is confined to a narrow strip along the bend line, in contour forming extensive plastic deformation occurs over the entire folded area.

Radius Forms. For a radius form, the length of the longest side of the smallest rectangular prismatic envelope that encloses the radius formed portion of a part is denoted as the *form length* L_f (Figure 12.9). For non-straight bend lines, the form length is the approximation of the longest non-straight bend line of the part (Figure 12.6). For shallow draws, L_f is the diameter of the drawn feature (Figure 12.8).

Also for shallow draws, the form length, L_f, is an indicator of the size of the form punch and die block required to form the part. Parts within the high L_f category require more time for fabrication of their die components, and also have increased die material cost.

The *radius form angle* is the angle formed by the radius form (Figure 12.9). If the angle is greater than 180 degrees, the part is said to have an *overform* (Figure 12.10).

If a part is overformed, then additional stations or dies are needed to achieve the desired geometry. Often, the part is radius formed to an angle of 180 degrees in one stage, and then overformed by either a concave punch that encloses the form from outside or by cam-actuated form tools.

12.4.6 Bends, Bend Sizes and Bend Angles

Bends. Bending involves localized plastic deformation in a narrow strip along the line of the bend. A part with straight bends is shown in Figure 12.11.

Most straight bends with bend angles of less than 90° are achieved by using a spring loaded pressure pad to clamp the workpiece to the die and then using the descending punch to wipe the workpiece over the edge of the die (Figure 12.11). The die radius is used to impart the bend radius to the workpiece. This process is referred to as *wipe forming*.

Bend Size. The length of the longest bend in the part is denoted as the bend size, L_b. The bend may be one continuous bend as shown in Figure 12.11 or it may be made up of several individual collinear bend lines as shown in Figure 12.12.

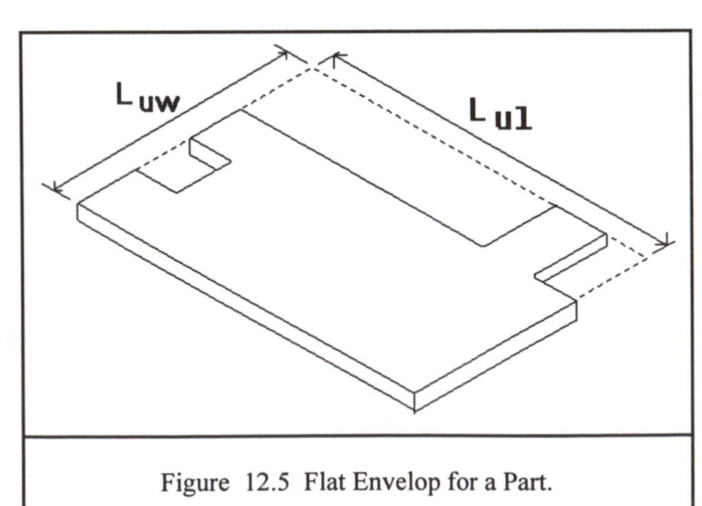

Figure 12.5 Flat Envelop for a Part.

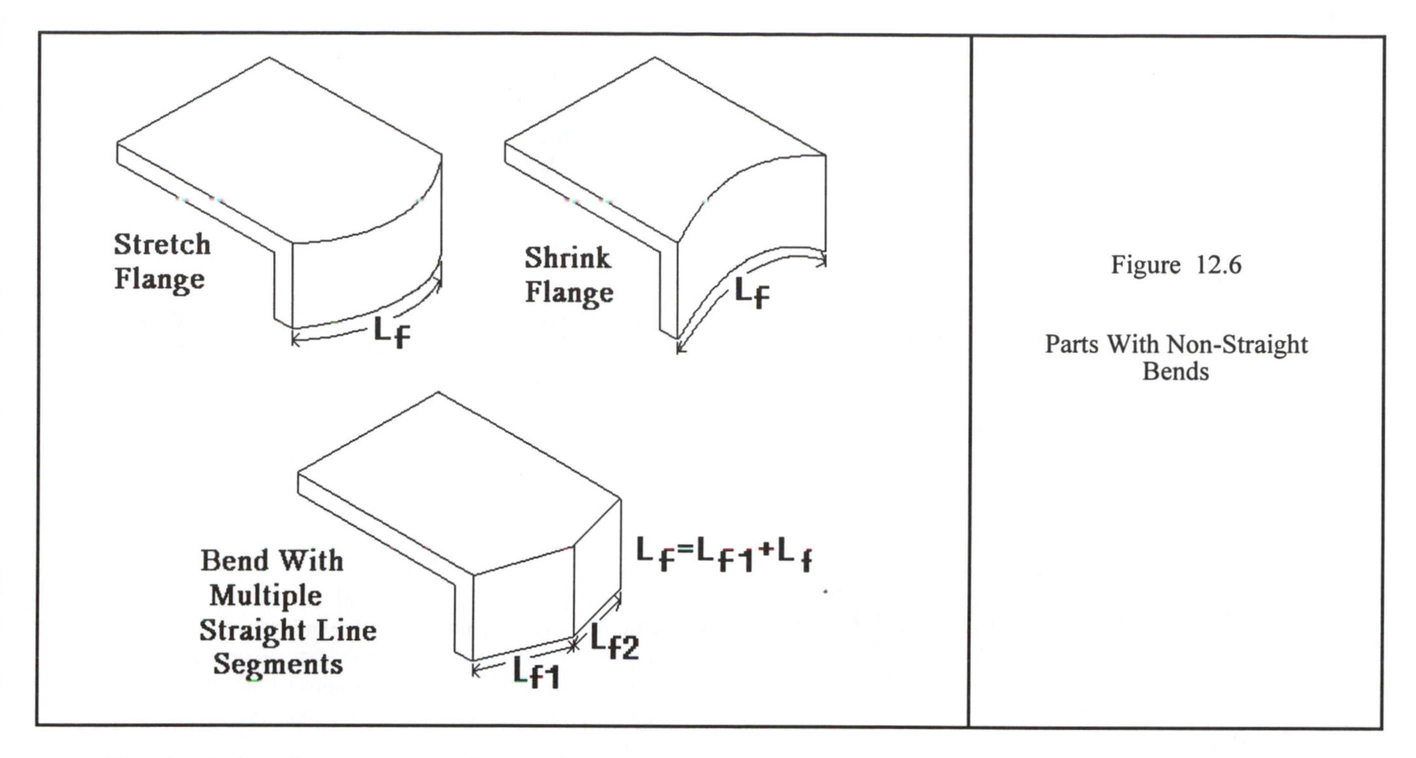

Figure 12.6

Parts With Non-Straight Bends

The bend size, L_b, represents the maximum bend length and indicates the size of the form tool required to wipe the bend or bends at a single station (for a progressive die), or the size of the secondary form die (in compound and secondary type tooling). (Readers may review the difference between a progressive die and a compound die in Section 3.5.3 of Chapter 3). As the length of a bend increases, the size of the wiping punch and die block increase proportionately. This increases the number of hours needed to fabricate these die components — and raises the die material cost. Long bends also require heel blocks to counter the increased horizontal component of the bending force. Parts with long bends will frequently lead to the need for sectioned die blocks, both for ease of maintenance and to avoid distortion under heat treatment. Such construction raises fabrication time owing to increased assembly time. Bends that are longer than 18" (455 mm) in length are placed in a separate size category, and considered to be cost drivers, for the reasons discussed.

Bend Angle. The angle through which a part is folded about a bend line during wipe forming is called the bend angle (Figure 12.13).

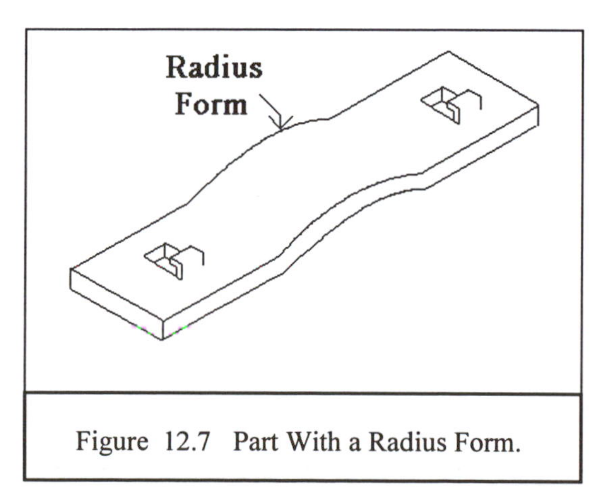

Figure 12.7 Part With a Radius Form.

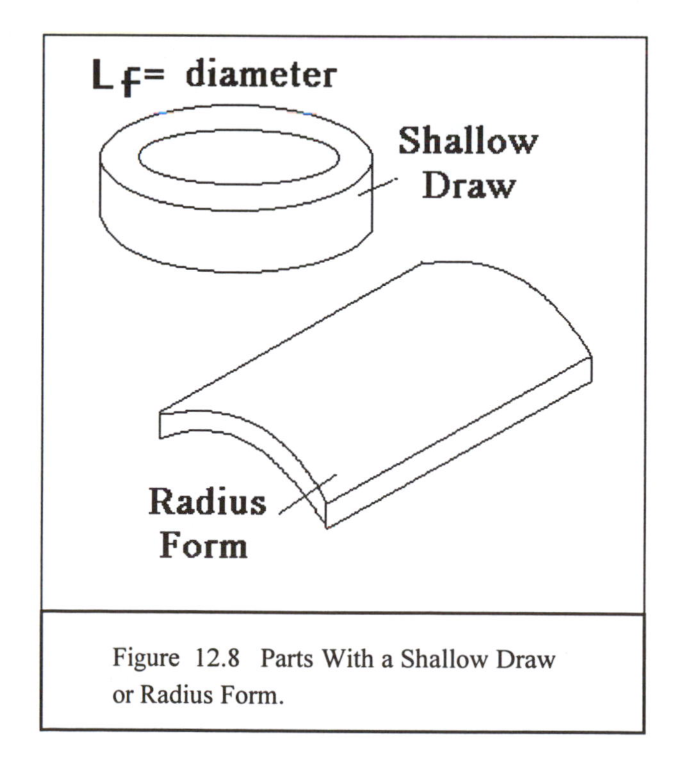

Figure 12.8 Parts With a Shallow Draw or Radius Form.

12.4.7 Protruding and Non-Protruding Features

In general, features in stamping include those that appear on the surface and those that appear on the periphery. Surface features include holes, ribs, tabs, embosses, etc. Periphery features include notches, steps, small projections, hems, curls, etc.

There are two basic types of surface features: (a) protruding, and (b) non-protruding. A *non-protruding feature* is one which remains flush with the plane of the sheet. A hole is an example of a non-protruding feature.

Protruding Features. A *protruding feature* is one which extends out of the plane of the portion of the stamping on which it is located. Examples of protruding features include embossings, dimples, tabs, extruded holes, lance and form type tabs, and stamped lettering (Figure 12.14).

To create a protruding feature, a deformation process other than shearing must be involved, though shearing may occur simultaneously with or precede the local deformation. Thus, an extruded hole is created by localized extrusion after the hole is pierced, leading to a small boss that protrudes out of the plane. Embossing is a forming operation that results in an impression that protrudes out of the plane of the part, but has no central hole.

An *extruded hole* is one that is generated at one station using a specially stepped punch that first shears a smaller hole and then follows through to deform the local area around the hole into a projection (Figure 12.15). Alternately, the hole could be pre-pierced in a separate station, and then the edges extruded at a second station.

This procedure creates, in effect, extruded bosses which are often used for joining by tapping or thread-forming the projecting hole for use as a nut. Such bosses can also be used for alignment or reinforcement. Extruded holes are also referred to as flanged, collared, embossed or drifted holes.

A *semi-perf* is a small — often about 3 mm diameter — button-like projection (Figure 12.16). It is frequently used as a locator during the assembly phase, or in subsequent fastening operations (such as when resistance spot welding). Semi-perfs are also known as partial extrusions, rivet lug forms, dimples, bumps or partial slugs.

Embossings, Figure 12.17, are small, shallow formed projections on the surface of stamped parts. Raised letters, beads, and ribs* are examples of embossings.

Embossings are produced by localized forming with the amount of deformation depending on the design and depth of the projection. Sheet thickness remains unchanged, and embossings have smooth contours. Embossings call for complementary punch and die shapes, and these components

* A rib is a narrow elongated projection whose length, l, is generally greater than three times its width, b. Ribs are usually added to stiffen or reinforce a stamped part.

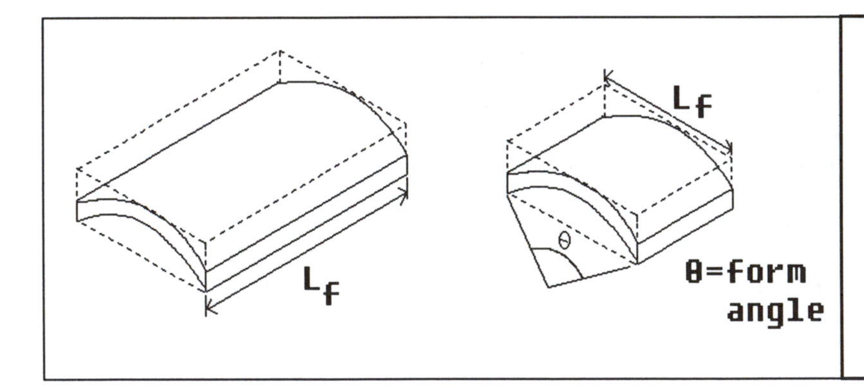

Figure 12.9

Form Length for Radius Formed Parts.

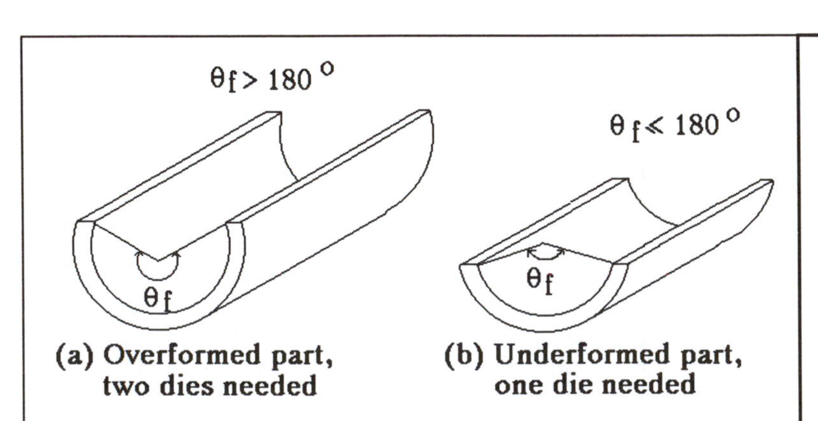

Figure 12.10

Two Radius Formed Parts

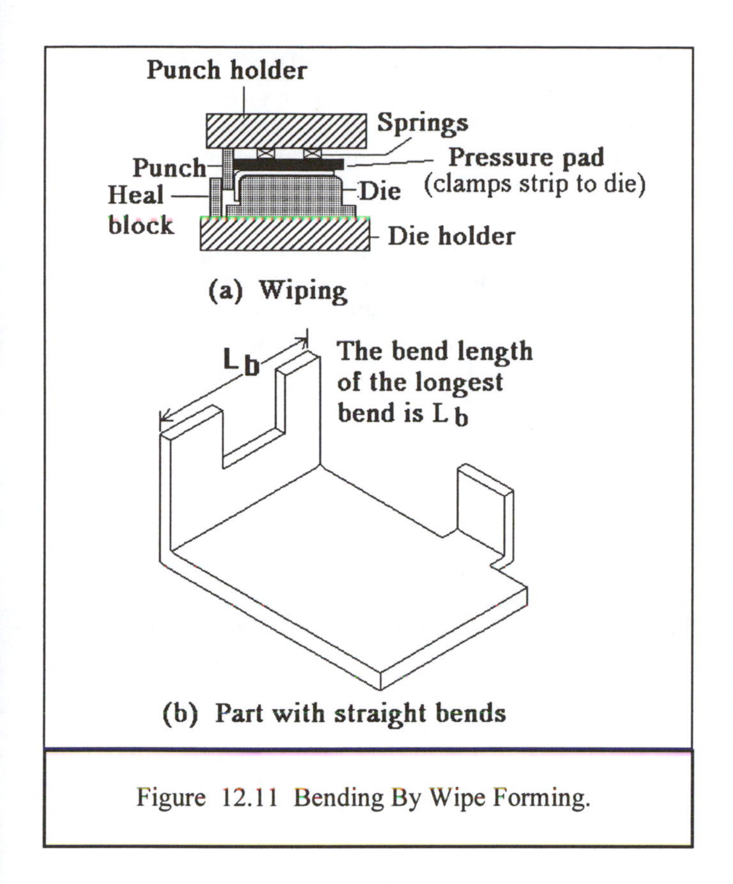

Figure 12.11 Bending By Wipe Forming.

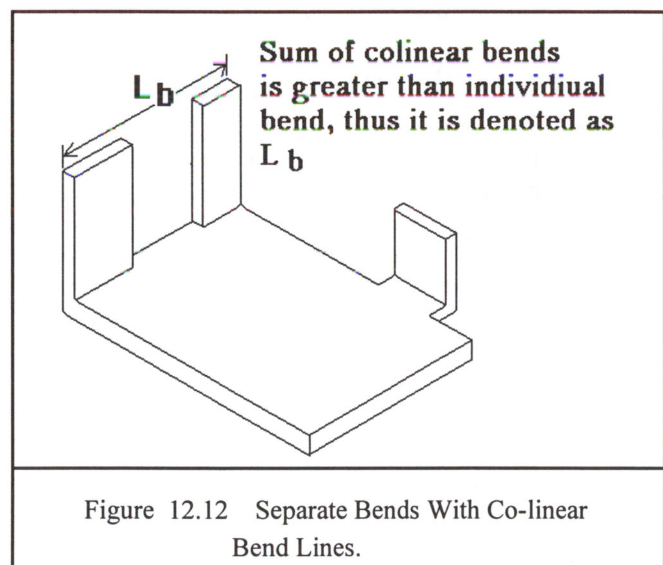

Figure 12.12 Separate Bends With Co-linear Bend Lines.

may be incorporated into a die as inserts. The operation is not generally done simultaneously with any shearing operation, and requires a separate station or secondary die.

Circular, rectangular, and oval holes and embosses can be produced using standard off-the-shelf punches and die buttons. Non-standard shapes, Figure 12.18, call for specially machined punches and dies, and they increase tool fabrication hours. Some irregular shapes may be generated by a sequence of regular punch hits, but they add to the number of

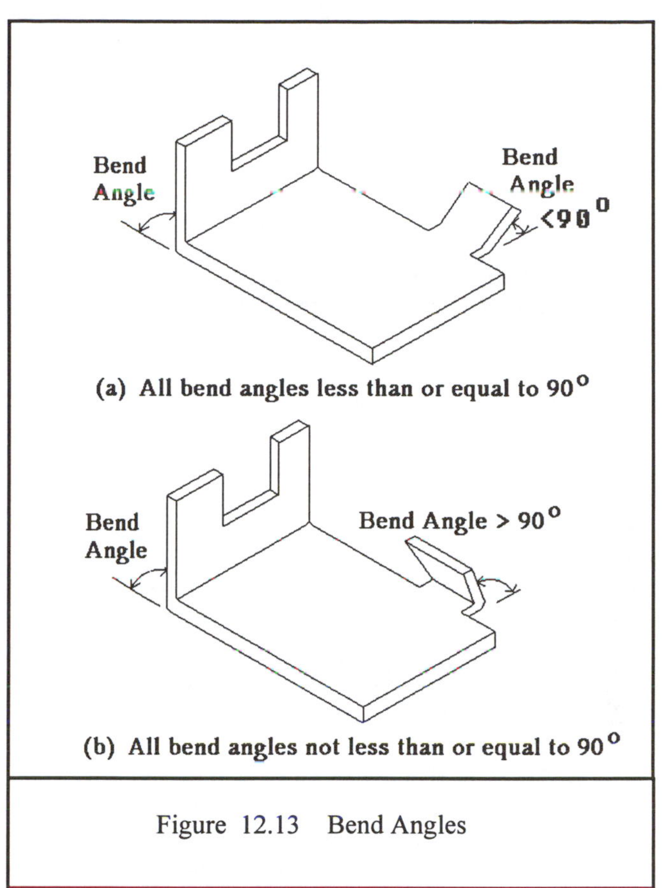

Figure 12.13 Bend Angles

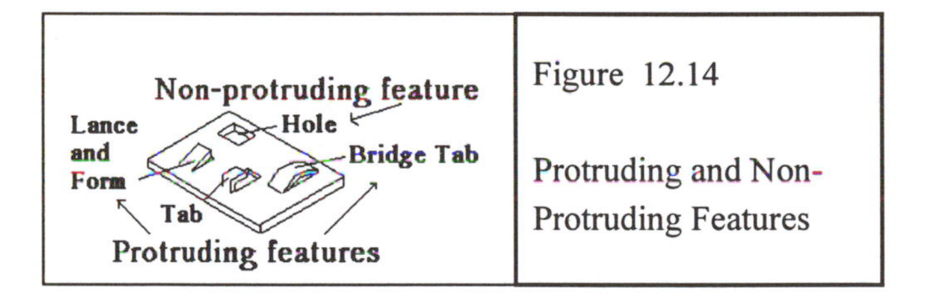

Figure 12.14

Protruding and Non-Protruding Features

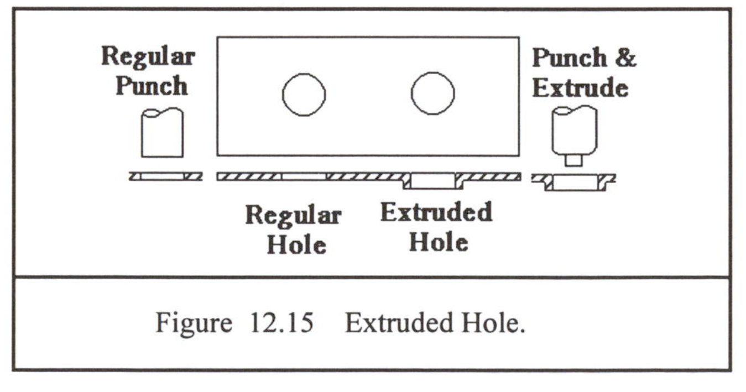

Figure 12.15 Extruded Hole.

stations or secondary dies needed.

Lance forms (Figure 12.19) are partial cuts in the stamping that are formed into projections.

Lance forming involves a combination of shearing (or lance cutting) followed by local forming to create the feature (which resembles a pocket-shaped opening). Lance forming is commonly used to create louvers, or to raise metal from

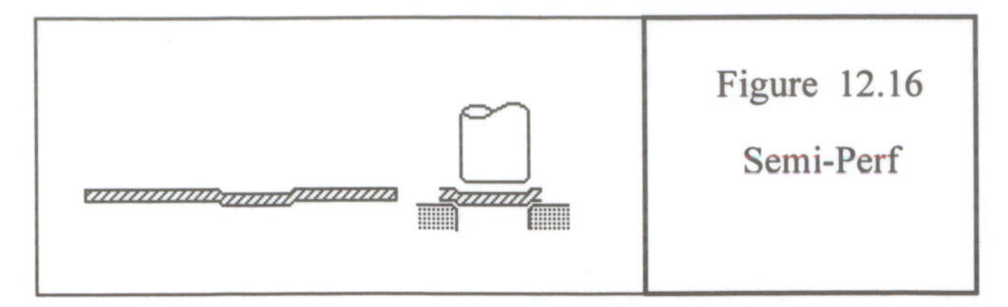

Figure 12.16 Semi-Perf

the stamping surface to provide a channel for wire. Lance forming punches have a cutting edge for shearing the lance cut, as well as a forming surface to form up the edge of the cut.

Curls (Figure 12.20) are features that result from rolling the edge of a stamped part.

The curl diameter should be 10 to 20 times stock thickness. Curling is usually preceded by wiping (i.e., bending) the edge. A curling punch (with a groove cut into it corresponding to the curl) rolls the bent edge into the desired shape, while the part is gripped with a pressure pad. A curl requires at least two stations; hence it is a complex feature from the tooling standpoint.

Sizing is the procedure used to bring some dimension to a desired size by squeezing the metal to the desired dimension. Sizing is also used to flatten or smooth the areas around holes and projections.

Sizing is carried out using an open die, consequently, there is little, if any, restriction to metal flow. In addition, the amount of metal moved is relatively small compared to the volume of the workpiece. In progressive dies, sizing is accomplished in a separate station with mating punch and die. Figure 12.21 shows a sized feature.

While a sized feature is not really a protruding feature, like most protruding features, sizing is a localized forming operation and is therefore included here.

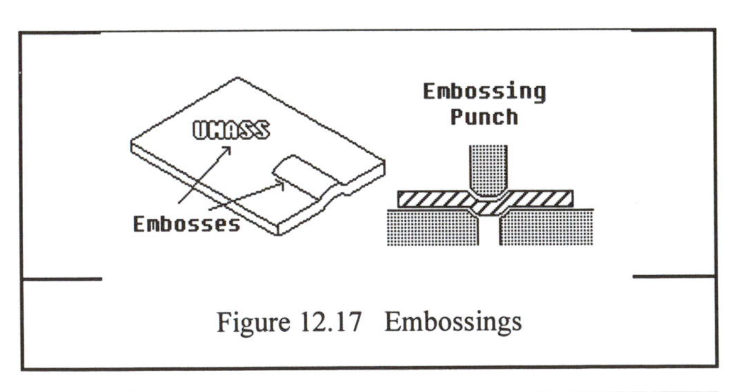

Figure 12.17 Embossings

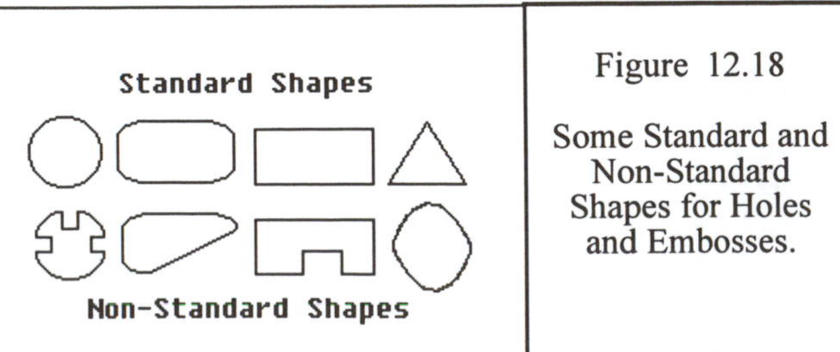

Figure 12.18 Some Standard and Non-Standard Shapes for Holes and Embosses.

12.4.8 The Form Zone and Springback

Form Zone. Features which lie on a portion of the part perpendicular to the direction of die closure are considered to be in or near a *form zone* if any of the following conditions apply:

(1) A feature (hole, ribs, tab, etc.) whose shape or location must be preserved is either on a straight or non-straight bend line or within three sheet thicknesses of a straight or non-straight bend line;

(2) A feature (hole, ribs, tab, etc.) lies within the radius formed portion of the part and its shape or location must be preserved.

(3) The radius of a radius form is tightly toleranced.

If (1) or (2) applies, then the feature must be created in a separate secondary die after the part is bent or radius formed, or in a separate station in a progressive die. If (3) applies, fabrication complexity will increase owing to increased tryout. Figure 12.22(a) shows a part with features in the form zone, Figure 12.22(b) shows a feature near a straight bend line.

Springback. The ratio R/t -- where R is the radius of curvature of a radius formed elemental plate, and t is the stock thickness -- is a measure of the amount of springback that the radius form generates. Parts with significantly different values for this ratio are shown in Figure 12.23. A part with a radius form of large curvature and made of thin stock will be more "springy" than a part that is made of thick stock and with a "tight" radius. Springback increases fabrication complexity by significantly raising tryout.

12.5 Estimating Relative Tooling Cost

12.5.1 Factors Influencing Relative Tooling Costs

From Equation (12.2) we see that tooling costs (K_d) depend on the cost of the die material (K_{dm}) and the cost to construct the die (K_{dc}). Based on data collected from custom stampers [2]-[5], the cost to construct a stamping die is a function of:

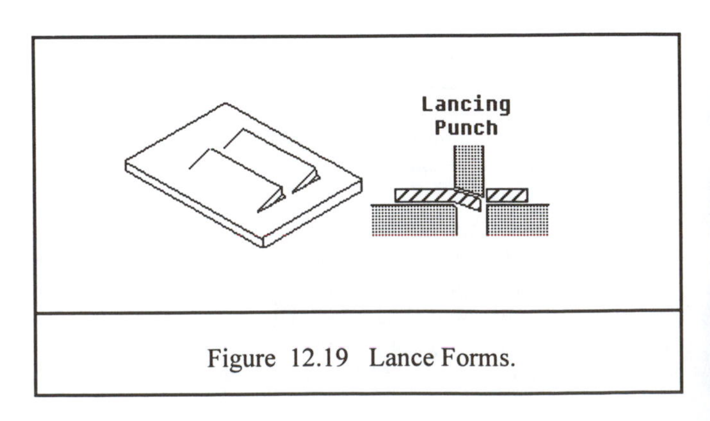

Figure 12.19 Lance Forms.

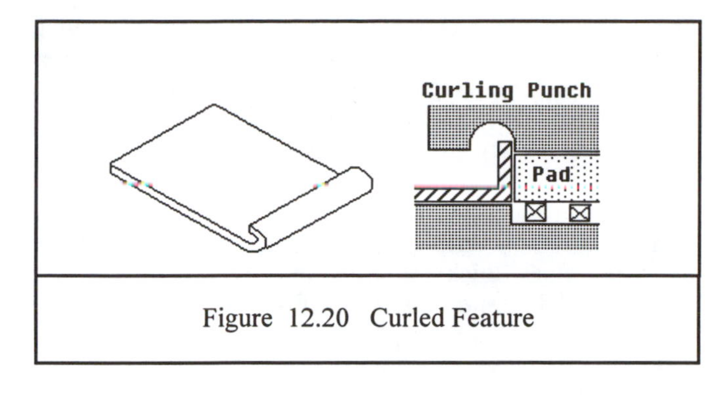

Figure 12.20 Curled Feature

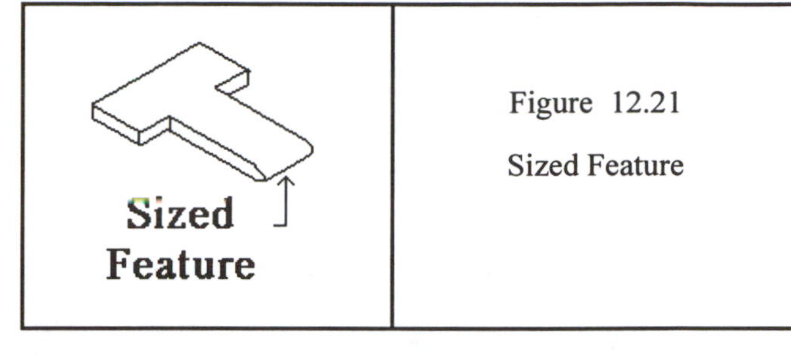

Figure 12.21

Sized Feature

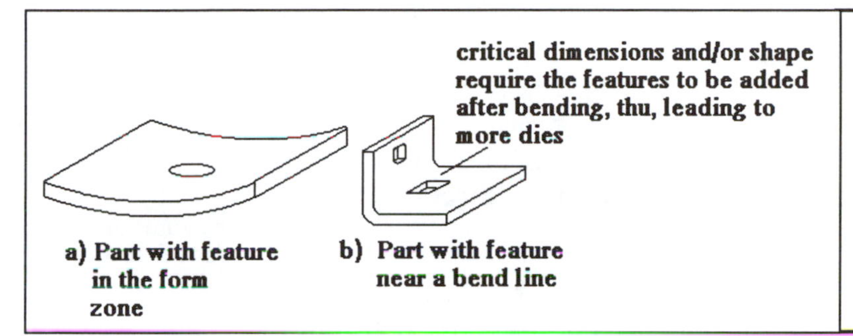

Figure 12.22

Part With Feature in the Form Zone Near or At a Bend Line.

- part size, L_{ul}
- bend size, L_b
- peripheral complexity (to be defined section 12.5.3),
- the number of active stations, N_a, required,
- part material, and
- sheet thickness used to produce the part.

The effects of these factors are incorporated into the data found in Figures 12.1, 12.2, and 12.3. As noted above, the first six digits of the coding system guide entry into these Figures. We have explained in Sections 12.2 and 12.3 above how to use the values for C_b obtained from the Figures. Values for N_a are used to estimate the relative die material cost as explained in Section 12.6 below. We can now explain in detail how each digit is obtained.

12.5.2 Determining the First Digit

(1) To obtain the first digit (for Figure 12.1), the first step is to determine whether or not the part can feasibly be stamped at all. Parts, when unfolded, can generally be stamped when:

(a) they have a uniform sheet thickness less than 6.5 mm, and

(b) the direction of all non-peripheral features is parallel to the sheet thickness (i.e., when there are no extruded holes not perpendicular to the sheet.)

Parts with these characteristics can generally be stamped, and their first digit will be between 0 and 8 inclusive. Other, essentially nonstampable, parts have a first digit of 9. Parts with holes having diameters less than the sheet thickness (these holes must be machined in later), or parts which contain extruded holes or ribs whose height is greater than four times the sheet thickness are examples of parts which can not be stamped.

(2) The next step in determining the first coding digit is to determine the level of die detail (number of holes, embossings, lance forms, tabs, etc.) required to produce the part. Figure 12.24 provides information for determining the amount of die detail present. All parts are placed into one of three die detail categories (Low, Medium, or High) according to the number of features. As the number of features increases (i.e., as the die detail increases) the number of active stations required also increases. Consequently, tooling

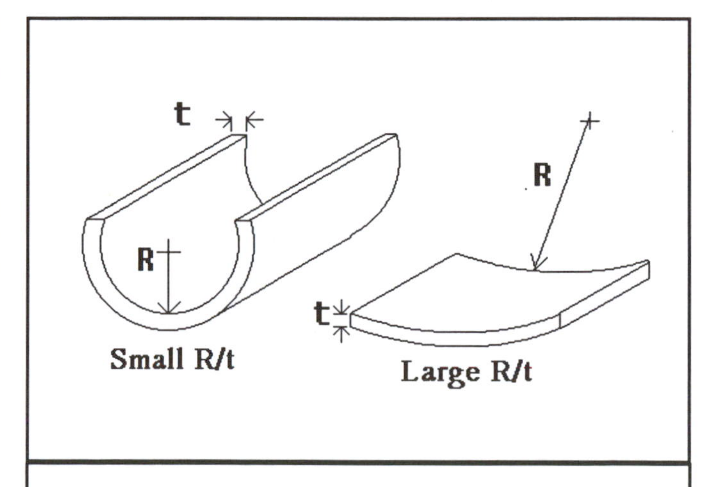

Figure 12.23 Two Radius Formed Parts With Differing Springback Properties.

cost goes up.

Each feature type is assigned a certain penalty per feature, and this is multiplied by the number of features of its type to obtain the total penalty associated with each feature type. The penalties are then summed to obtain a gross penalty associated with the part. The gross penalty is used to help the user assign the part to an appropriate die detail category. (Note that the die detail categories change based on the part size.)

To make use of Figure 12.24, the following terms must be understood:

(a) *Number of Distinct Features and the Direction of Features.* For ease of maintenance and construction, it is common practice among stampers not to produce more than one feature type at a given die station. Thus, holes and tabs, for example, are produced at separate die stations. Hence, as the number of distinct features increases, the number of die stations also increases. This increases tooling costs directly by requiring that more dies be produced.

In the case of protruding features, if two or more features are on the same surface of the stamping (and project out in opposite directions), then the features are said to be in opposite directions (Figure 12.25). Such combinations of features cause a torque on the die set at that station, and may need to be formed in separate stations or with separate secondary dies. Alternatively, by using pressure pads of varying stiffness together with specially sized tooling, these features can be formed in a single station. However, this leads to a die that needs more fabrication hours than a die for a part with the features are all on the same face.

(b) *Closely Spaced Features.* Two non-peripheral features which are less than three times the stock thickness apart, or any non-peripheral feature that is less than three times the stock thickness from the sheet periphery, are defined as *closely spaced* features (Figure 12.26).

In the case of compound and secondary tooling, if features are closely spaced with respect to each other or to the edge, the part of the die block that supports the stamping

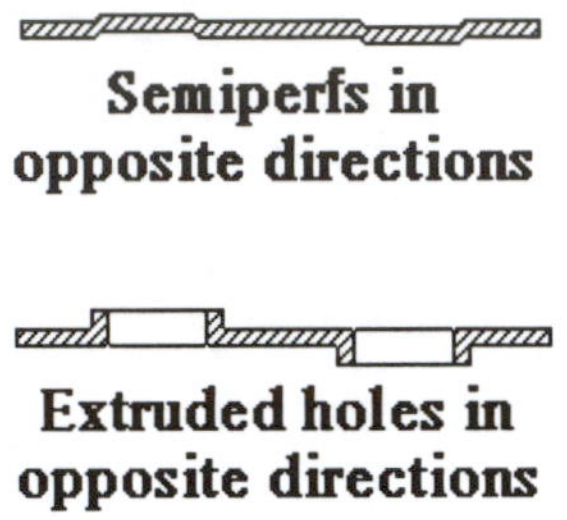

Figure 12.25

Features in Opposite Directions

Feature	Number of Features (n)	Opposite Directions? (Y/N)	Penalty
Standard Hole		n / n	
Non-standard Hole		2n / 2n	
Coin		3n / 3n	
Standard Emboss		(n+1) / n	
Non-standard Emboss		2(n+1) / 2n	
Extruded Hole		2(n+1) / 2n	
Lance Form		3(n+1) / 3n	
Curl		3(n+1) / 3n	
Curl		3(n+1) / 3n	
Hem		4(n+1) / 4n	
Semi-Perf		(n+1) / n	
Tab		3(n+1) / 3n	
Lettering		2(n+1) / 2n	
		Total Penalty	

Small Parts (L ≤ 100 mm)

Total Penalty ≤ 4 => Low Die Detail
4 < Total Penalty ≤ 8 => Medium Die Detail
Total Penalty > 8 => High Die Detail

Medium Parts (100 mm < L ≤ 200 mm)

Total Penalty ≤ 6 => Low Die Detail
6 < Total Penalty ≤ 12 => Medium Die Detail
Total Penalty > 12 => High Die Detail

Large Parts (L > 200 mm)

Total Penalty ≤ 10 => Low Die Detail
10 < Total Penalty ≤ 17 => Medium Die Detail
Total Penalty > 17 => High Die Detail

Figure 12.24

Determination of Die Detail

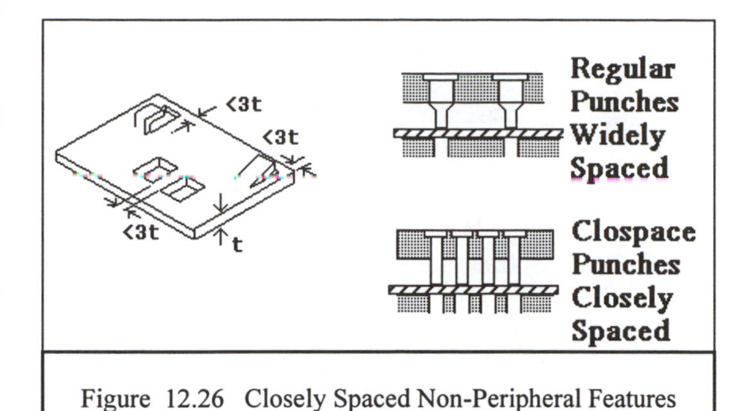

Figure 12.26 Closely Spaced Non-Peripheral Features

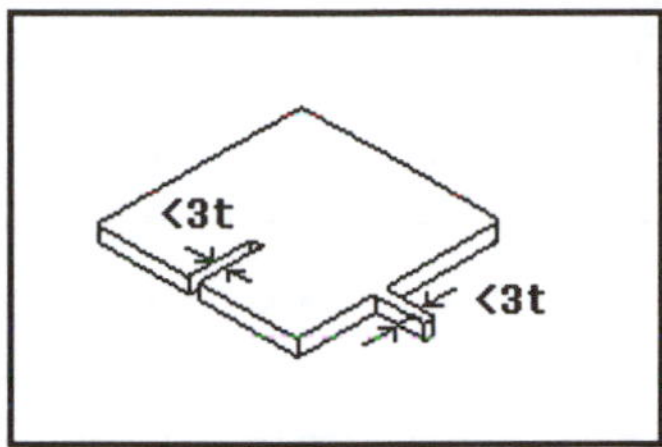

Figure 12.27

Part With Narrow Cutout and Narrow Projection

between these features will have a thin section that tends to crack or break in use. Moreover, the punches used to generate these features may have to be crowded together in too close a manner. In addition, if the space between features is limited, there may be distortion from insufficient space to clamp the part (with spring loaded strippers or pads) in between the features. As a result of these constraints, closely spaced features necessitate the use of secondary dies, and they add to tooling complexity.

In the case of progressive dies, closely spaced features lead to the need for additional stations in order to keep the tool strong and to allow room for the punches.

When closely spaced features are present, half of the features (every other one) are produced at one station and the remaining features are produced at a separate station. Hence, for each distinct feature type which is closely spaced, an additional active station is required.

(c) *Narrow Cutouts and Narrow Projections.* Any projections on the peripheral edge of the part that have a minimum width less than three times the stock thickness are considered narrow (Figure 12.27). Similarly, any cutout or notch in a cut edge whose width is less than three times the stock thickness is also considered narrow.

In general, a peripheral cutout (notch) or projection is imparted to the part when the part is blanked or separated from the strip at the final station. In the case of a narrow cutout or a narrow projection, these features would require a narrow die wall or a narrow portion of the blanking punch. Both of these situations can result in damage to the blanking punch. Thus, the narrow cutout is imparted in a separate station using a narrow punch, while the narrow projection is imparted in two stages, the first putting a notch on one side of the projection and the second removing the remaining material.

(3) The next step in determining the first digit is to consider whether or not any non-peripheral features are closely spaced with respect to each other, or are near the edge of the part. The presence of narrow projections or cutouts in any of the cut edges is also accounted for in the determination of the first digit. See Figure 12.1.

Because many of the design details (exact dimensions and tolerances, for example) for a stamping are not known until parametric design, it may not be possible at the configuration stage to determine whether features are closely spaced, or whether peripheral cut-outs and projections are narrow, or whether holes at bent faces are side action features. Thus, at this stage, the first digit for a stampable part may sometimes be restricted to values of 0, 3 and 6; and the fourth digit to values of 1, 5 and 9; and the sixth digit to values of 1 and 5. This in turn implies that only the minimum number of active stations, N_a-min, and the minimum relative tool construction cost, C_b, have been determined. That is, this implies that the part will be designed so that tooling costs will be minimized.

12.5.3 Determining the Second Digit

Determining the second digit for Figure 12.1 requires knowledge of the part size, L_{ul}, and the peripheral complexity. Part size is defined above in Section 12.4.4. The remainder of this sub-section describe how to determine the level peripheral complexity.

Peripheral Complexity. When possible, the final

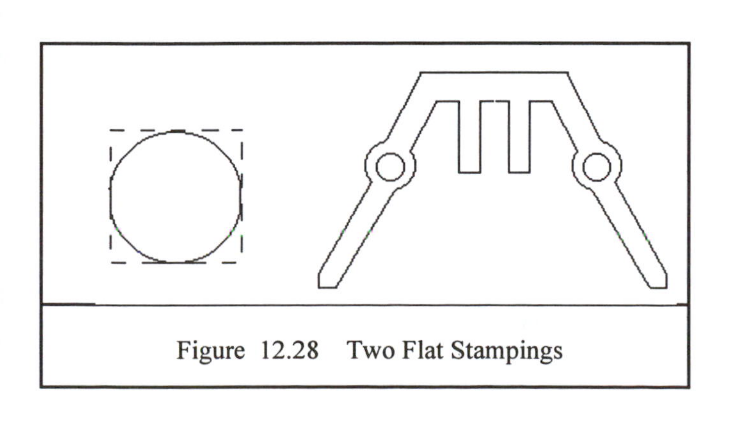

Figure 12.28 Two Flat Stampings

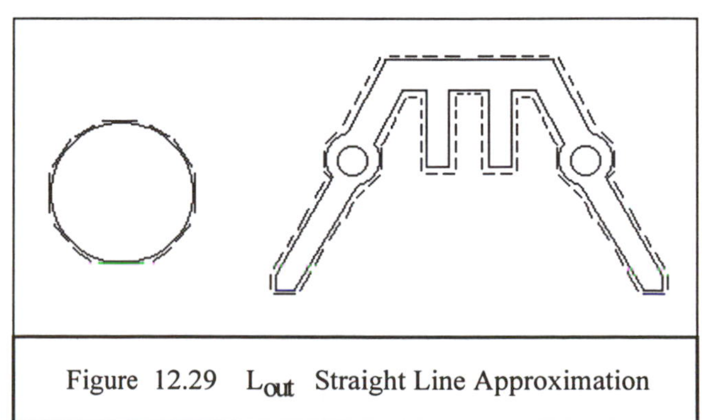

Figure 12.29 L_{out} Straight Line Approximation

peripheral shape of a flat stamping (Figure 12.28) is imparted at the last station of a progressive die using a blanking punch. For a bent part, the peripheral shape of the unbent portion is also imparted by use of a blanking punch at the final die station. The blanking punches and the corresponding dies are usually formed by use of wire electro discharge machining (EDM). The cost of producing the punches and dies is directly related to the peripheral complexity of the part being produced.

To get an estimate of the EDM costs of producing the die cavity, therefore, we compute a peripheral complexity ratio (L_s) defined as:

The peripheral length of a part prior to bending or contour forming is:

$$L_s = \frac{L_{out}}{\text{Perimeter of smallest flat envelop enclosing the part}}$$

where L_{out} is the peripheral length of the part prior to bending or contour forming. Usually, the value of L_{out} can be determined by approximating the part periphery by a series of piecewise connected straight line segments as shown in Figures 12.28 and 12.29. For non-flat parts with complex formed and bent profiles, the process of flattening and measuring L_{out} may prove difficult.

The perimeter of the flat envelop

$$2(L_{ul} + L_{uw})$$

is computed as shown in Figure 12.5. In general the values of L_{ul} and L_{uw} can be easily determined by simply visualizing the flattened shape of the part. For a large proportion of formed and bent stampings, visualization of the flattened part is not difficult.

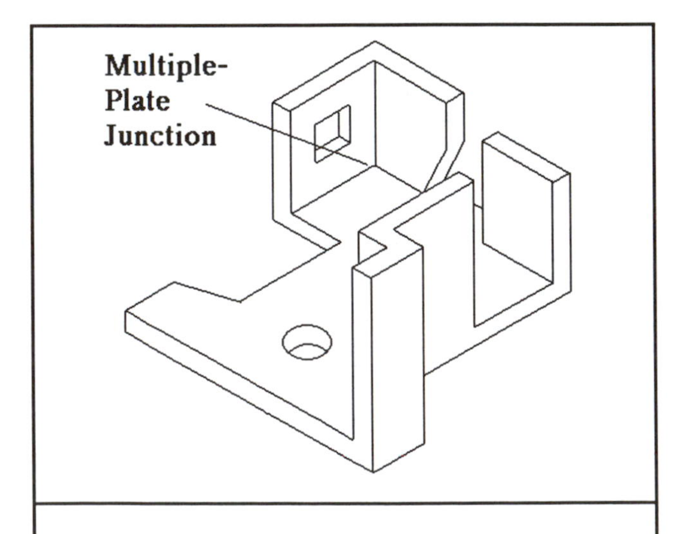

Figure 12.30 Part With a Multiple Plate Junction

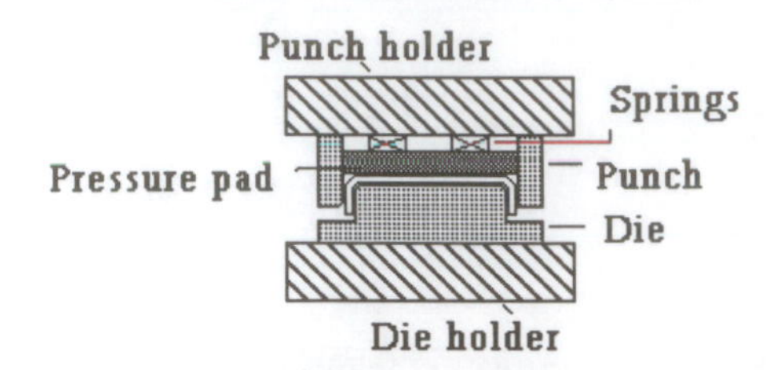

(a) Tooling for bends in the same direction

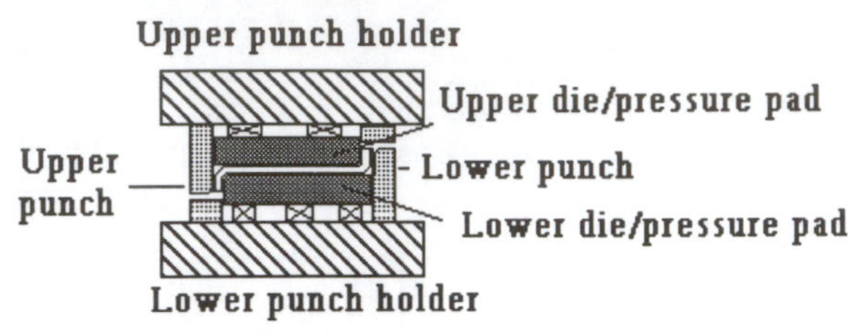

(b) Tooling for bends not in the same direction

Figure 12.31 Tooling for Bend Angles Less Than or Equal to 90 Degrees.

Thus:

$$L_s = L_{out} \,/\, 2(L_{ul} + L_{uw}) \quad (12.7)$$

In other words, the greater the deviation in L_{out} from the perimeter of the smallest rectangular flat envelope enclosing the part, the greater the number of wire EDM hours required to produce the tooling.

12.5.4 Determining the Third Digit

The third and fourth digits are related to the complexity of any bends imparted to the stamping. The third digit assesses bend complexity based on the number of stages needed to achieve the bent shape, the presence of bends in opposing directions, and the presence of any bends greater than 90 degrees.

(1) The first step in determining the third digit to be used in Figure 12.2 is to determine whether the bends can be feasibly wipe-formed. Parts without straight bends (Figure 12.6), parts containing bends which can not be formed by wipe forming (Figure 12.7), and parts with a multiple plate junction (Figure 12.30), are examples of bends which can not be formed by wipe forming. These parts have a third digit of 13. Parts that can feasibly be formed have a third digit of from 1-12 inclusive.

(2) Assuming the part can be wipe formed, the second step in determining the third digit is to determine the number of bend stages and, if there is only one bend stage, whether all the bends are or are not in the same direction.

The number of active stations required to produce the

bends for a given part depends upon the arrangement or configuration of the bends. Since some bends can be wiped simultaneously at the same die station, the number of required bend stages is not necessarily equal to the number of bends. A method for determining the number of bend stages required is discussed in Section 12.6 below.

For parts with two bends which can be formed at a single station, those bends in the same direction require only one set of pressure pads. Parts with bends in opposite directions require two sets of pressure pads with different spring constants and specially sized dies, all of which raise the die construction cost (Figure 12.31). Two stations are sometimes used to impart bends in opposite directions.

(3) The final step in determining the third digit is to estimate the bend angle(s). The angle through which a part is folded about a bend line during wipe forming is called the bend angle (Figure 12.13). Parts with bend angles less than 90° can be bent at one station using inexpensive tooling similar to that shown in Figure 12.31.

For parts with bend angles between 90° and 105°, two bend stations are generally required. At the first station the part is bent through 90° as shown in Figure 12.31; at the following station it is bent again (overbent) as shown in Figure 12.32. Because of the limited overbend, relatively inexpensive tooling can be used to carry out the overbend.

If the bend angle is greater than 105°, two operations are again required — though costs are higher. At the first station the part is bent through 90°, while at the second station more expensive cam actuated slides are required (Figure 12.32). Figure 12.32(b) also shows a rotary bender being used at one station to achieve the bend in one operation. Rotary benders, however, are not typically used by stampers.

12.5.5 Determining the Fourth Digit

The fourth digit, also for use in Figure 12.2, depends upon (a) bend size, (b) whether or not there are side action features, and (c) whether or not there are features near the bend lines. We have already defined part size. Side-action features and features near bend lines are discussed below.

Side Action Features. Features of a bent part, usually holes or slots, that do not lie in a plane perpendicular to the direction of die closure, and which must be accurately located or aligned (± 0.2 mm), must be created after bending by use of cam actuated tools (Figure 12.33). If these features are created before bending, then it may not be held to true position after bending. Such features are called *side-action features* and are equivalent to the undercuts which occur in injection molding and die casting.

Features Near the Bend Line. See Figure 12.33. Any feature (hole, rib, tab, etc.) that lies in a plane perpendicular to the direction of die closure, and is within three sheet thicknesses of a bend line, and whose shape or location must be preserved, may have to be imparted in a separate secondary die after the part is bent, or in a separate station in a progressive die. In either case, tooling costs increase.

12.5.6 Determining the Fifth Digit

The fifth and sixth digits are related to the complexity of any contour forms imparted to the stamping. The fifth digit classifies the part according to whether the part has contour forms; that is, non-straight bend lines, radius forms, or shallow draws. The complexity of any radius form is also accounted for by the presence of overforms, and the determination of the ratio of form radius to sheet thickness.

The fifth digit is used in Figure 12.3 to find C_{b3}. However, only certain parts are considered in the coding system. These parts have non-straight bends (Figure 12.6), or radius forms (Figure 12.7), or shallow draws (Figure 12.8). They do not have non-circular, or stepped, or tapered recesses, or features (such as multiple plate connections as shown in Figure 12.30) that must be deep drawn. These parts have a fifth coding digit of 1 through 5 inclusive. Other parts, not included in the system, have a fifth digit of 6. See Figure 12.3.

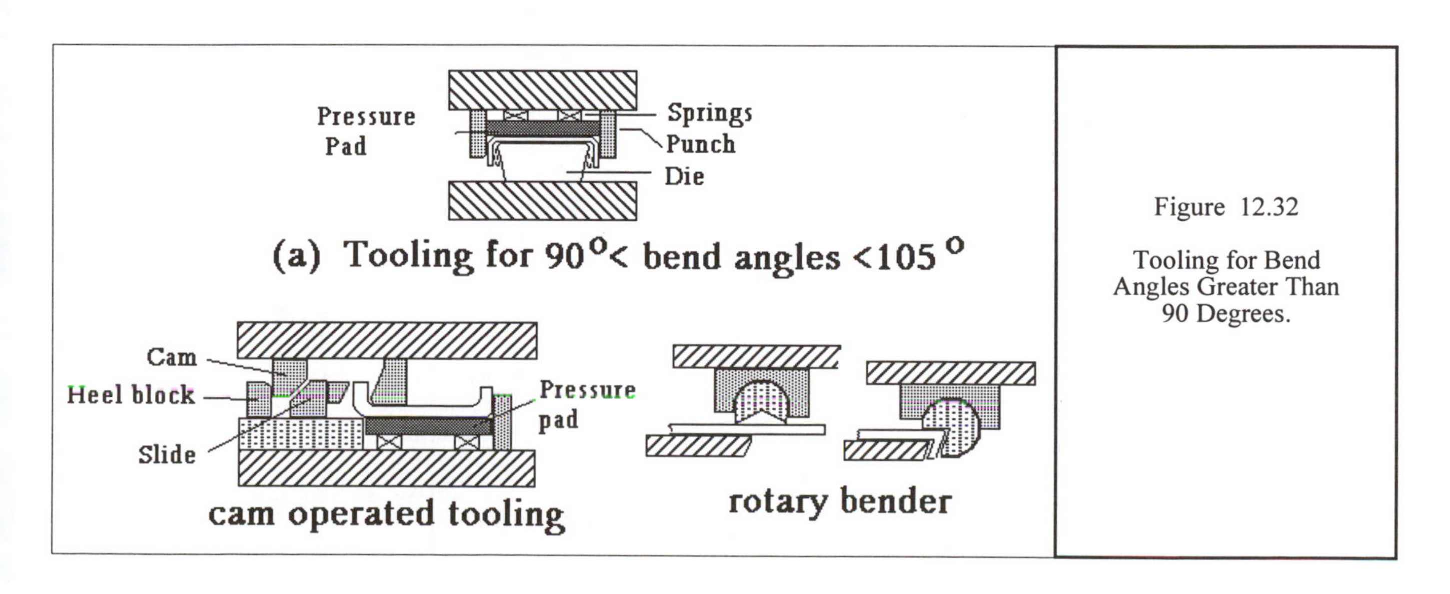

Figure 12.32

Tooling for Bend Angles Greater Than 90 Degrees.

The next step in determining the fifth coding digit is to consider the nature of the bends (straight or non-straight, radius forms and their angles), and the presence of shallow draws. These considerations will fix the fifth digit at a value from 1 to 5 inclusive. The definitions of terms needed are shown in Figures 12.6, 12.7, 12.8, and 12.9.

As explained before, if a radius form subtends an angle less than 180°, the operation can be performed at one station. If the angle is greater than 180°, a second die is needed after first forming through 180°. In general, as the radius increases, springback becomes more important. This results in increased planning and tryout of forming dies and more engineering hours and tool build hours.

12.5.7 Determining the Sixth Digit

The sixth digit accounts for the length of the contour form and the presence of critical features in or near the form area. The information needed for the sixth digit is the same as for the fourth digit.

12.6 The Number of Bend Stages

12.6.1 Introduction

One of the difficulties encountered in using the coding system -- especially in determining the third digit -- is estimating the number of bend stations (or stages) required to produce the part. In this section, we present a method for estimating this number. The method requires first partitioning the part into *elemental plates.*

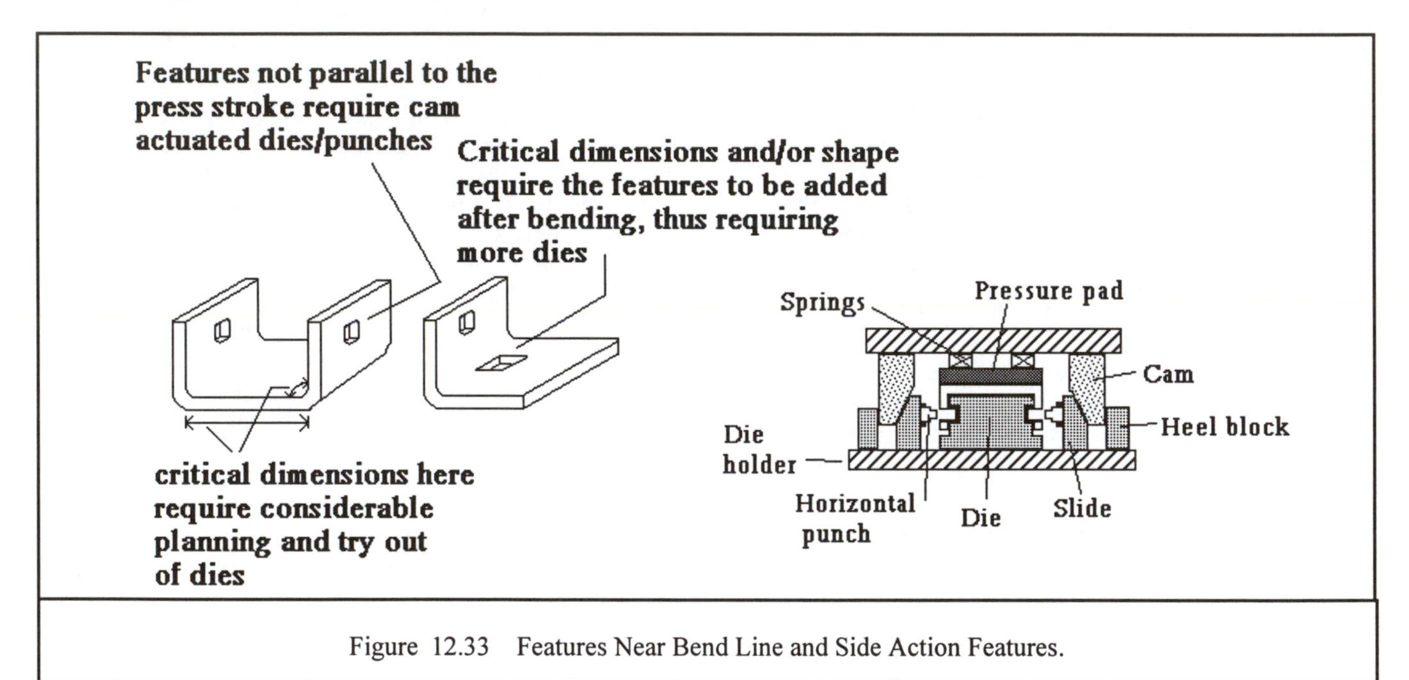

Figure 12.33 Features Near Bend Line and Side Action Features.

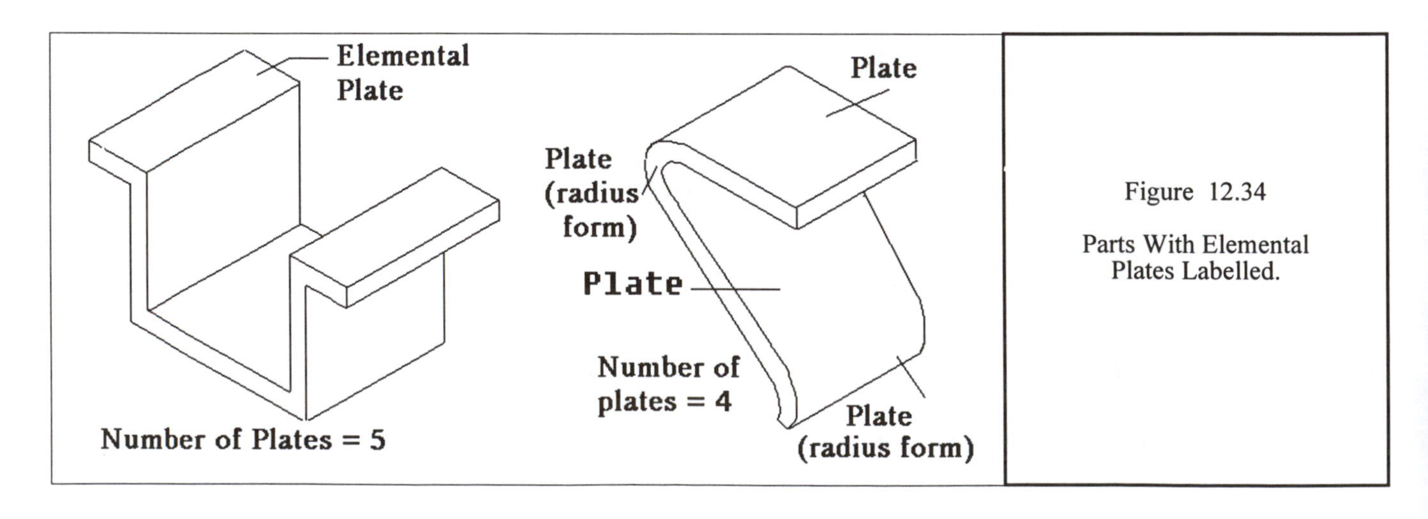

Figure 12.34

Parts With Elemental Plates Labelled.

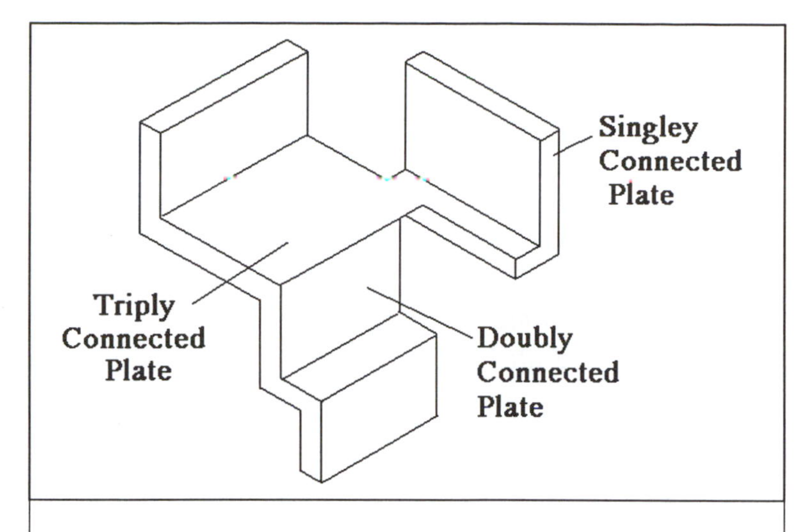

Figure 12.35 Parts With Singly and Multiply Connected Plates

12.6.2 Part Partitioning - Elemental Plates

A stamped part that is not flat can be decomposed into a set of *elemental plates*. Each plate is a part of the stamping bounded either by the part periphery, bend lines, radius forms, or a combination of these.

An elemental plate has a constant spatial orientation (all vectors normal to the surface point in the same direction), except in the case of a radius form. A radius form is an elemental plate that has a constant radius of curvature and all surface normals point to the center of curvature of the form. Figure 12.34 shows two parts with elemental plates labelled.

Movement along the part surface from one plate to another implies the crossing of a bend line or the boundary of a radius form. A flat part is a plate in itself.

An elemental plate that is connected (by a bend line or the boundary of a radius form to only one other plate is called a *singly connected elemental plate.* Similarly, a plate that is connected to two other plates is called a *doubly connected plate*, and so on. Figure 12.35 shows a part with singly, doubly and triply connected elemental plates.

The *primary plate* of a part is the plate that is most likely to be placed parallel to the plane of the die block (either in a progressive or secondary die). As noted earlier, the direction of die closure will always be normal to the plane of the die block.

Finding the Primary Plate. The primary plate of a part is determined by searching for an elemental plate with the following properties in this approximate order of significance:

1) The elemental plate with the largest area as shown in Figure 12.36;

2) If the part has a complex profile with several bends, then the primary elemental plate is the elemental plate that is surrounded on its boundary by the largest number of bend lines. Figure 12.37 shows a part with such an elemental plate;

3) If no single elemental plate has a significantly larger area than the others, and the part does not have a complex profile with several bends, then the primary elemental plate is the elemental plate with the maximum number of internal features (holes, slots or protruding features). Figure 12.38 shows a part with such an elemental plate.

12.6.3 Approximate Number of Stages Required to Form the Bends

In order to code a bent stamping it is necessary to know the number of stages needed to fold the part to the desired final geometry. This establishes the number of dies or stations needed to form the bends. To do this for any general bent stamping the following procedure [3] has been found useful.

The part shown in Figure 12.39 will be used to illustrate the procedure. The part has all its elemental plates numbered as specified by the procedure.

STEP 1 - Remove All Singly Connected Plates.

Excluding the primary elemental plate,

Figure 12.36

Primary Plate Based on Largest Area

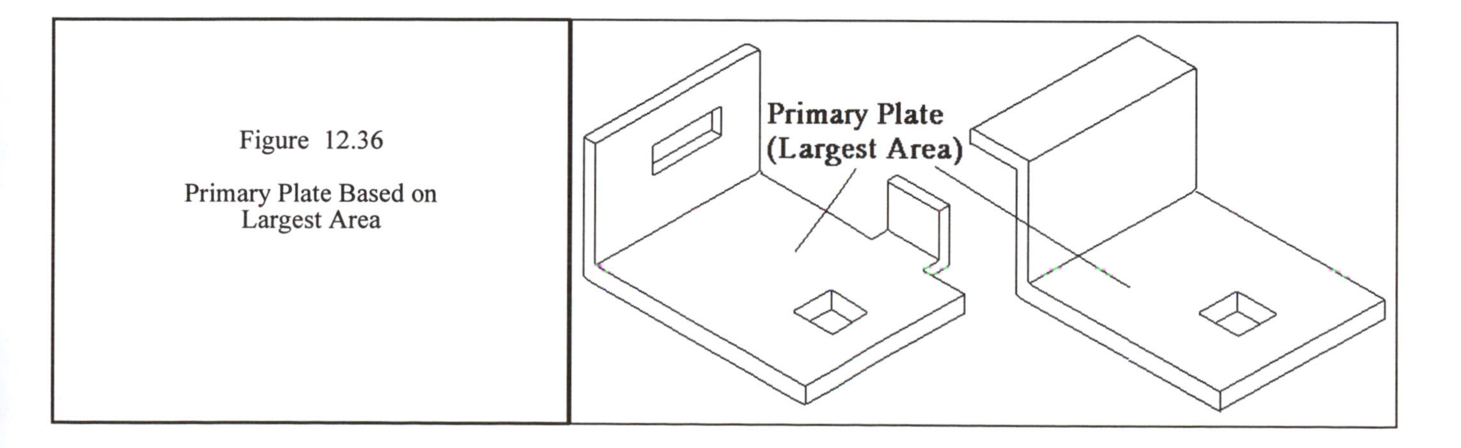

determine all the singly connected elemental plates of the stamping and number them as P_{ij}, where the subscript i is the current step number, and j is the number of the currently identified singly connected elemental plate.

Remove all the singly connected elemental plates identified above, namely, P_{11}, P_{12}, P_{13} and P_{14}. The part after removing these elemental plates is shown in Figure 12.40.

STEP 2 - Remove the Singly Connected Plates for the Modified Part.

Identify all the singly connected elemental plates in the modified part as shown in Figure 12.40. Once again, the primary elemental plate is not to be considered. The only candidate elemental plate is the one numbered P_{21}. Remove this elemental plate from the part, thus modifying it a second time. The truncated part is shown in Figure 12.41.

STEP 3 - Repeat Step 2.

The only singly connected elemental plate in the modified part shown in Figure 12.41 is P_{31}. Remove this elemental plate. The truncated part is shown in Figure 12.42.

After the third step, the truncated part is the primary elemental plate. The procedure terminates when the primary elemental plate is the only remaining elemental plate. The number of steps required to reach this state is the number of bend stages. In this example, the number of bend stages is 3.

This process of identifying the singly connected elemental plates, removing them from the part, and repeating the process until the primary elemental plate is the only elemental plate left can be applied to any bent stamping. The result is a simple approximation to the number of required bend stages without considering tooling details. Figure 12.43 shows two parts with multiple bend stages.

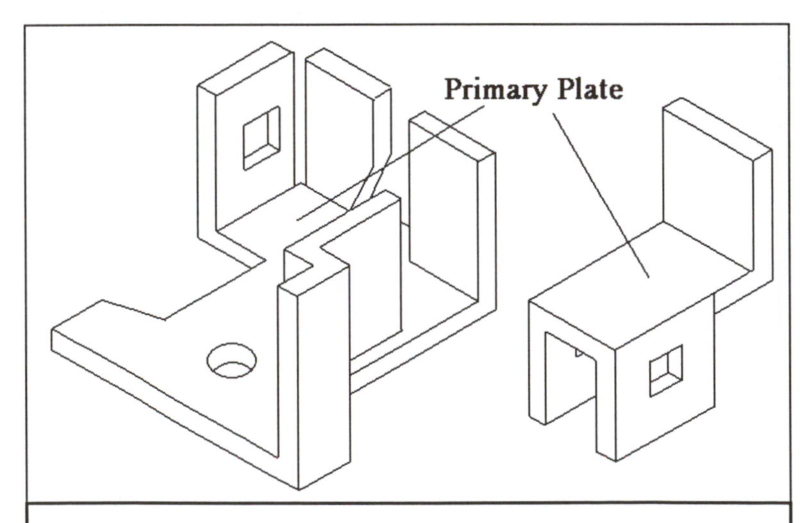

Figure 12.37 Primary Plate Based on the Maximum Number of Bend Lines Surrounding the Plate.

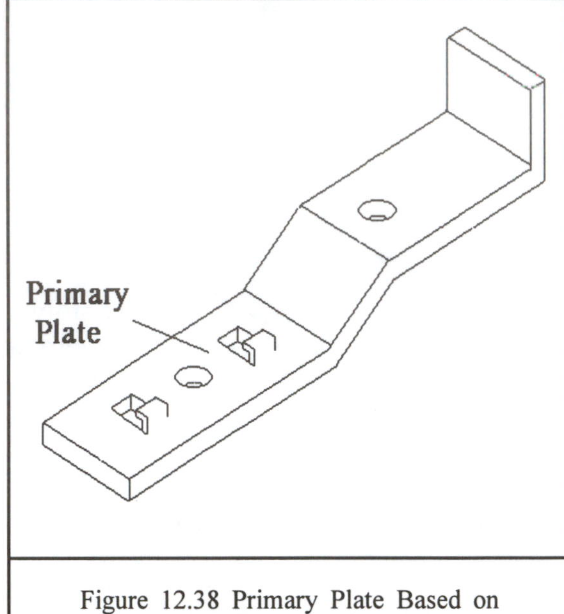

Figure 12.38 Primary Plate Based on the Maximum Number of Features

12.6.4 Bend Directions

For a part with a single bend stage, if all the bend lines are in the same plane, and all the singly connected elemental plates are on the same side of the primary plate, then the bends are in the *same direction.*

Parts with all the bends in the same direction can be formed by wiping down all the singly connected elemental plates using a relatively simple forming tool which has wiping punches, die block, and a pressure pad. The pressure pad is placed below the multiply-connected elemental plate when wiping the singly-connected elemental plates in a direction opposite to that of the press stroke (called "wiping up"). The pad is placed above the multiply-connected elemental plate when wiping the singly-connected elemental plates in the same direction as the press stroke (called "wiping down"). Figure 12.44 shows parts that have all their bends in the same direction.

Parts with bends not in the same direction will have singly connected elemental plates on both sides of the primary plate. These parts require more complex form tooling to form them at a single progressive station or in a single secondary form die. Typically such tooling includes specially sized punch and die steels, and two pressure pads of differing spring constants.

Another processing strategy is to form the

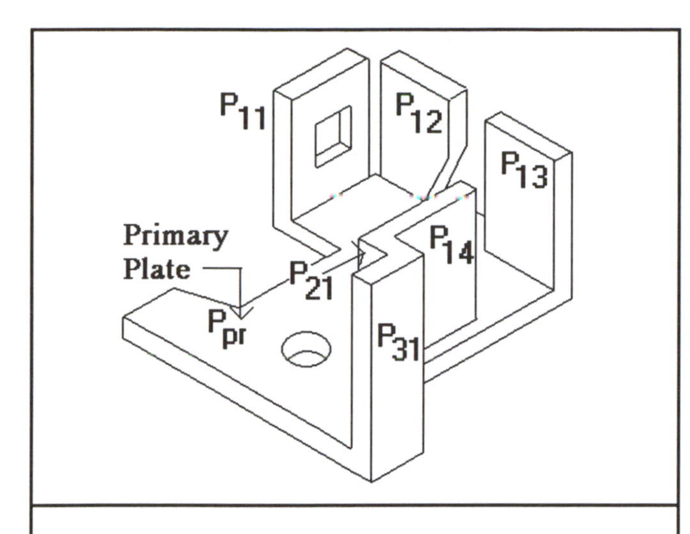

Figure 12.39 Part With Elemental Plates Numbered

part at more than one station (wiping down at one station, and wiping up at another), or using two secondary form dies (wiping down in both dies with re-oriention of the part in between dies). Either way, such a bend configuration significantly raises fabrication hours for tooling, and the part in Figure 12.45 is an example.

If parts do not have all bend lines in the same plane, they can not be bent to the final shape in a single step. The part is now evaluated using the scheme described below to determine the approximate number of stages required. The output of this scheme is the minimum, not necessarily the total, number of stages required to form the part.

12.6.5 Determination of Bend Stages from Part Sketch

In general, during the initial stages of design we will not know the points of separation between the various elemental plates which form a bent part. For example, in place of the part shown in Figure 12.39, one is more likely to encounter the part shown in Figure 12.46. The part in Figure 12.46 can not be stamped. Instead, to produce the part with sheet metal as indicated in the drawing would require the use

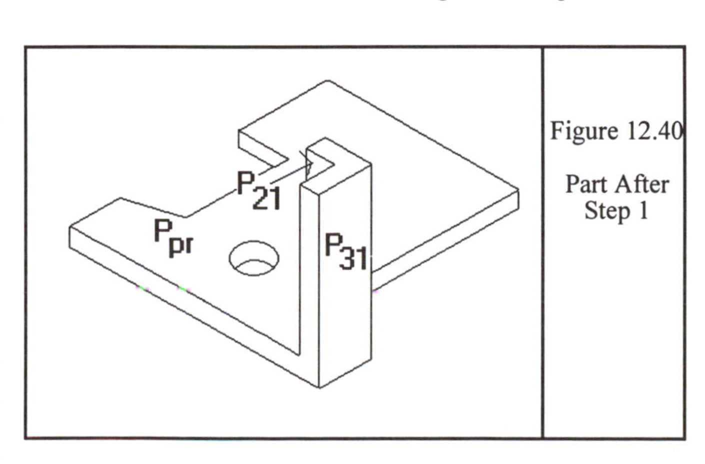

Figure 12.40 Part After Step 1

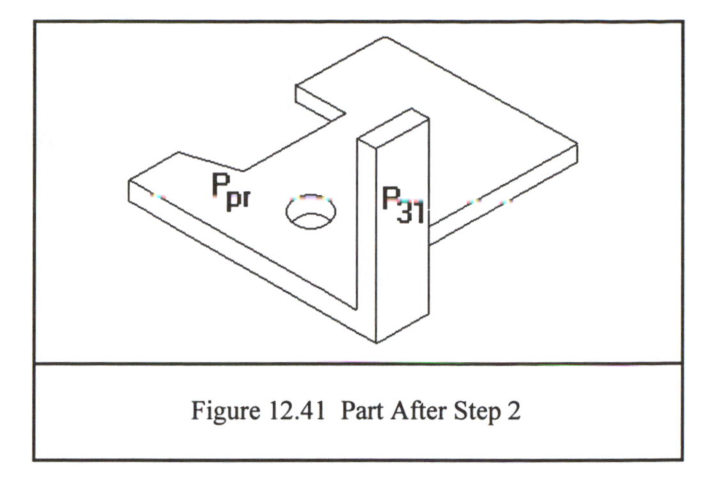

Figure 12.41 Part After Step 2

of deep drawing or wipe forming followed by one or more welding operations. Deep drawing is a process which involves high tooling costs due to extensive tryout, high processing costs, and slow cycle time of hydraulic presses. To produce the part by stamping, then, some of the junctions of the plates which constitute the part must be separated from each other.

Figure 12.47 shows two alternatives. In the first, plates P_{11} and P_{12} are separated from each other but connected to the primary plate, P_{pr}. Plates P_{14}, P_{21} and P_{31} are connected to each other. However, P_{14} and P_{21} are separated from the primary plate.

In the second alternative, P_{12} is connected to P_{11} but separated from plate P_{pr}. In both alternatives, three bend stages are required. Thus, from the standpoint of tooling, the two designs are approximately the same.

In some situations, the number of bend stages required for alternative designs will not be the same. For example, if the part did not contain plates P_{21} and P_{14}, then the design shown in Figure 12.48 (a) would require one more bend stage than the one shown in Figure 12.48(b). Except for

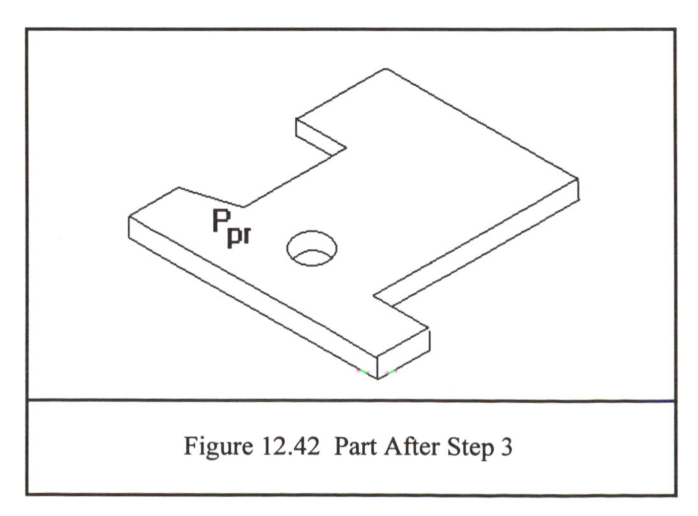

Figure 12.42 Part After Step 3

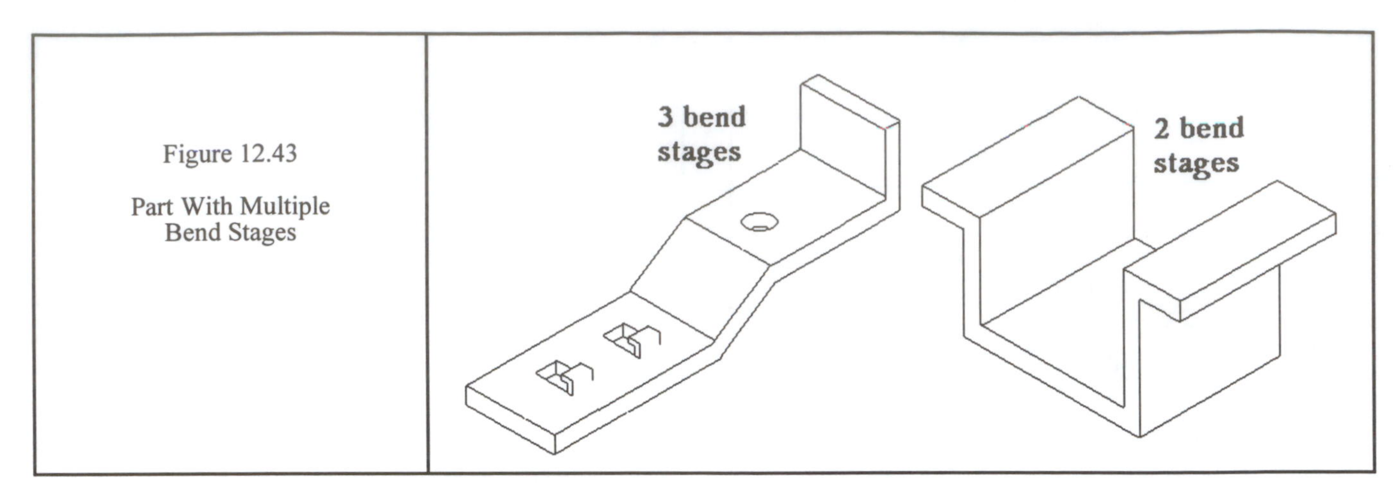

Figure 12.43

Part With Multiple Bend Stages

situations where functionality does not permit, parts should be designed using the minimum number of bend stages. Thus, the design shown in Figure 12.48(b) should be selected.

Non-stampable Parts. In designing stamped parts, one must take care not to design parts which simply can not be made. (At best, it is embarrassing!)(Figure 12.49 shows plates P_{14}, P_{21}, and P_{31} separated from each other but connected to P_{pr}. This part can not be produced as shown because of the interference between plates P_{21} and P_{14} when they are unfolded.

12.7 Relative Die Material Cost

12.7.1 Progressive Dies

Die material costs consist of the cost of the die set (punch holder, die holder, guide posts, bushings — Figure 12.50), tool steel (die block, punches, die inserts, cams, back gage, finger stop, etc.), and soft or machine steel (punch plate for holding individual punches, stripper plates and non-critical parts). As mentioned earlier, some items are standard and can be catalog purchased. These include the die set itself, punches, die buttons, pilots, springs, etc. In general, 25% of the total cost of the die material is due to the die set. The remaining 75% is due to tool steel and soft material. Hence, a good estimate of the total die material cost can be taken as 4 times the cost of the die set required.

Assuming that a 'one-up' tool is being used (i.e., that only one part at a stroke is being made), then the die material cost is a function of the size of the die set. The size of the die set is, in turn, a function of the number of stations required to produce the part. The number of active stations required is obtained from Figures 12.1-12.3.

In addition to active stations, most progressive dies also make use of idle stations. This is done in order to strengthen the die, to avoid closely spaced punches, and to distribute the cutting load over a larger area. The number of idle stations required depends on the unbent width of the part, L_{uw}. When L_{uw} is small, more idle stations are added in order to avoid closely spaced punches. Table 12.3 can be used to determine the total number of stations required.

Table 12.4 can be used to determine the size of the die set, S_{ds}. Substituting the value of S_{ds} in to the following equation [2], gives the die material cost for the part relative to the die material cost of the reference washer.

$$C_{dm} = F_{dm}[2.7\,S_{ds} / (25.4)^2 + 136] / 260.2 \qquad (12.8)$$

The value of F_{dm} accounts for the effect of sheet thickness on

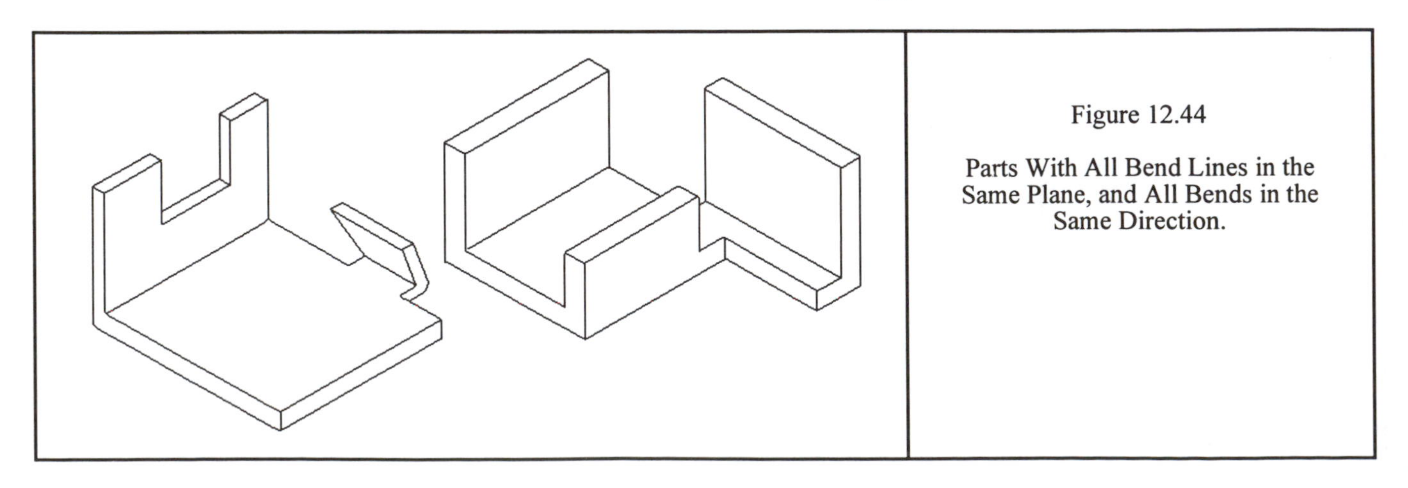

Figure 12.44

Parts With All Bend Lines in the Same Plane, and All Bends in the Same Direction.

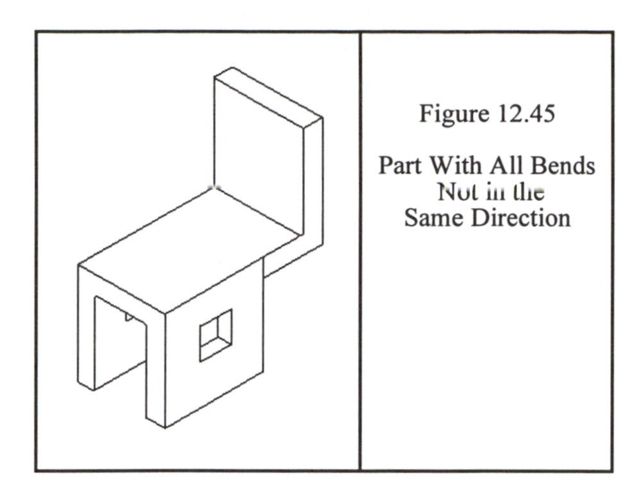

Figure 12.45

Part With All Bends Not in the Same Direction

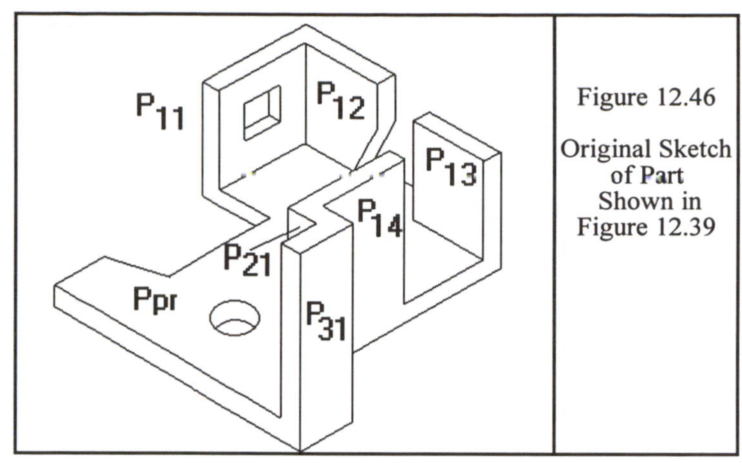

Figure 12.46

Original Sketch of Part Shown in Figure 12.39

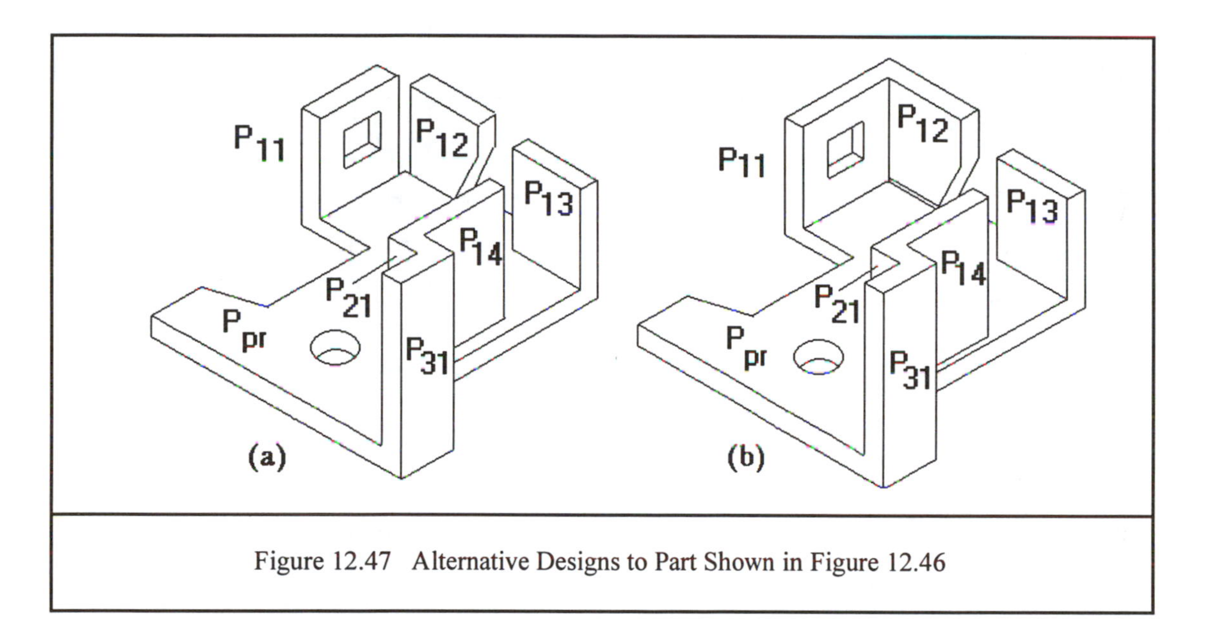

Figure 12.47 Alternative Designs to Part Shown in Figure 12.46

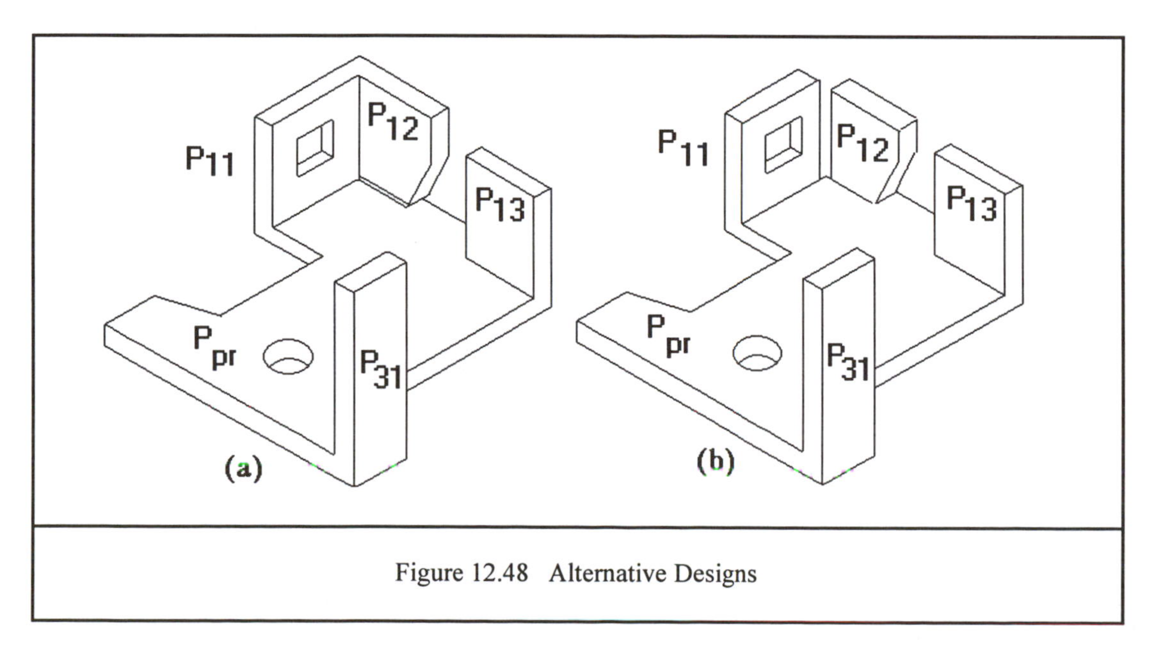

Figure 12.48 Alternative Designs

Table 12.3 Determination of Number of Stations Required [2]
Number of Active Stations, $N_A = N_{a1} + N_{a2} + N_{a3} =$
Number of Idle Stations, $N_I = N_{I1} + N_{I2} + N_{I3} =$ (a) For $L_{uw} \leq 25$ mm $N_{Ij} = 1.5 (N_{aj} - 1), j = 1, 2, 3$ (b) For 25 mm < L_{uw} < 125 mm $N_{I1} = 0$ (part without curls) $N_{I1} = 2$ (part with curls) $N_{I2} = (N_{a2} - 1)$ $N_{I3} = (N_{a3} - 1)$ (c) For $L_{uw} \geq 125$ mm, $N_{I1} = N_{I2} = N_{I3} = 0$
Total Number of Stations, $N_S = N_A + N_I$

Table 12.4 Die Block and Die Set Size [2]

Bottom Plate of Die Set

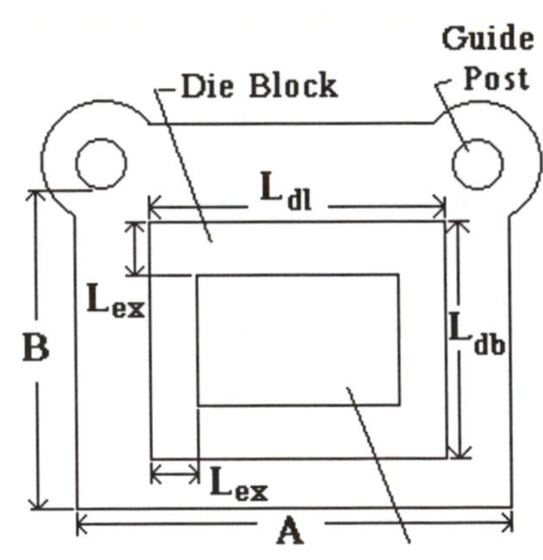

Die Block Size (mm)

$L_{dl} = N_S L_{uw} + 2 L_{ex} =$

$L_{db} = L_{ul} + 2 L_{ex} =$

where $L_{ex} = 25$ if $N_S L_{uw} \leq 75$

$L_{ex} = 37$ if $75 < N_S L_{uw} < 250$

$L_{ex} = 50$ if $N_S L_{uw} \geq 250$

Die Set Size:

$A = L_{dl} + 25 =$

$B = L_{db} + 25 =$

$S_{ds} = AB =$

die material cost and is obtained from Table 12.2.

12.7.2 Example

Problem: For a part whose six digit stamping code is 3, 2, 6, 10, 2 and 3, determine the number of active and idle stations required in a progressive die. In addition, assuming that the flat envelope dimensions, L_{ul} and L_{uw}, are 90 mm and 65 mm, respectively, determine the die set size, S_{ds}.

Solution:

(a) From Figure 12.1, for first and second digits of 3, 2 we get $N_{a1} = 5$.

(b) From Figure 12.2, for third and fourth digits of 6, 10 we get $N_{a2} = 6$.

(c) From Figure 12.3, for fifth and sixth digits of 2, 3 we get $N_{a3} = 4$.

Thus, the total number of active stations, $N_a = 15$.

(d) Since $L_{uw} = 65$ mm, then from Table 3 we see that $N_{i1} = 0$, $N_{i2} = 6\text{-}1 = 5$, and $N_{i3} = 4\text{-}1 = 3$. Thus, the total number of idle stations required is 8.

(e) The total number of stations, N_s, required for a progressive die is the sum of the active and idle stations or 15 and 8 which is 23.

(f) From Table 12.4 we see that since $N_s L_{uw}$ is greater than 250 mm, then L_{ex} is 50 and

$$L_{dl} = 23(65) + 2(50) = 1595 \text{ mm}$$

$$L_{db} = 90 + 2(50) = 190 \text{ mm}$$

$$A = L_{dl} + 25 = 1620 \text{ mm}$$

$$B = L_{db} + 25 = 215 \text{ mm}$$

$$S = AB = 348{,}300 \text{ mm}^2$$

12.8 Example 1 - Relative Tooling Cost

As a first example, we consider the part shown in Figure 12.51. Assume that we've sketched the part as shown so that the separation between the various elemental plates is as indicated in Figure 12.51. Since we are at the configuration design stage, there are several alternative approaches we could use to analyze the part.

In one approach, we could assume the worst possible conditions. In this case, we treat holes A, B, and C as side-action features, and dimension the cutout so it is narrow. The advantage to this approach is that we obtain a comparison between the most costly design and one which is less costly.

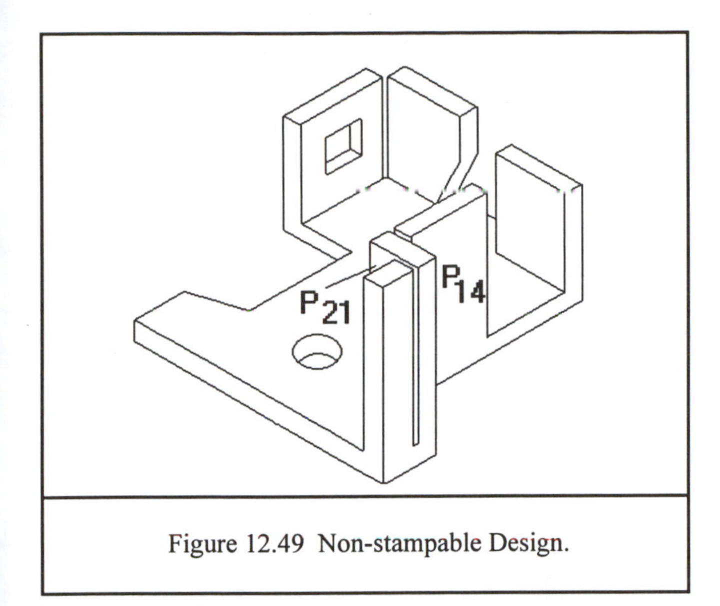

Figure 12.49 Non-stampable Design.

Alternatively, we could assume the best possible conditions. In this case, we assume that we do not know yet whether holes A, B, and C are to be side-action features. We also assume that the dimensions of the cutout are not yet known, and consequently that it is not to be treated as narrow. With these assumptions, we obtain a lower bound with regard to tooling cost, and can investigate the possibility of redesigns that will lower the cost to produce the part still further.

Finally, we could take a middle approach. In this case, we assume that we know that the location of holes A and C are not critical, but we're uncertain about hole B. We also assume that, since we are not certain about the precise dimensions of the cutout, we will treat it as narrow. Thus, we will obtain a comparison between a design which treats the cutout as narrow and hole B as critical, with one that does not treat then as narrow and critical, respectively.

In this example, we will assume the best possible conditions and obtain the minimum relative tooling cost. Thus we are assuming that all dimensions are approximate and that none of the features are closely spaced, near a bend, or result in side-action features. In addition, the cutout present will not be assumed narrow.

Relative Die Construction Cost

Shearing and Local Forming - The First and Second Digits

The part has a uniform sheet thickness less than 6.5 mm, and hole diameters greater than the 1.6 mm thickness of the sheet. In addition, the non-peripheral features are in a direction normal to the sheet thickness. Hence, the part is stampable, and the first digit is between 0 and 8.

Using Figure 12.24, it is seen that the part contains 4 standard holes (4 penalty points), and 2 semi-perfs in the same direction (2 penalty points). Thus, the total penalty in this case is 6. Since the largest dimension of the flat envelope, L_{ul}, is about 92 mm, the part contains medium die detail. This indicates that the first digit is 3, 4, or 5.

The features are considered widely spaced and the cutout is considered wide. Thus, since there are no narrow projections, the first digit is 3.

Since L_{ul} is about 92 mm, the second digit it 0 or 1.

The peripheral length of the part before bending, L_{out}, is approximately 350 mm while the width of the flat envelope of the part is about 65 mm. Hence, the peripheral complexity of the part, L_s, is

$$L_s = L_{out}/2(L_{ul} + L_{uw}) = 350/2(92 + 65) = 1.13 < 1.4$$

Thus, the second digit is 0 and, from Figure 12.1, the relative tool construction cost due to shearing and local forming is C_{b1} = 2.33. The number of active stations required is N_{a1} = 5.

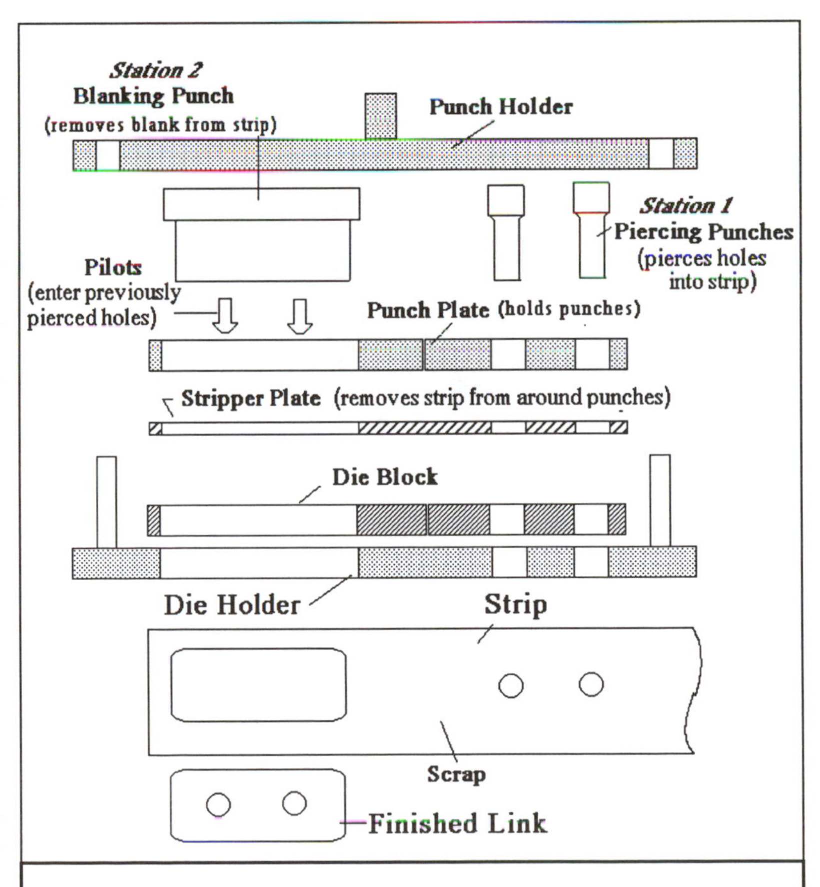

Figure 12.50 Some Components Included in Die Material Cost.

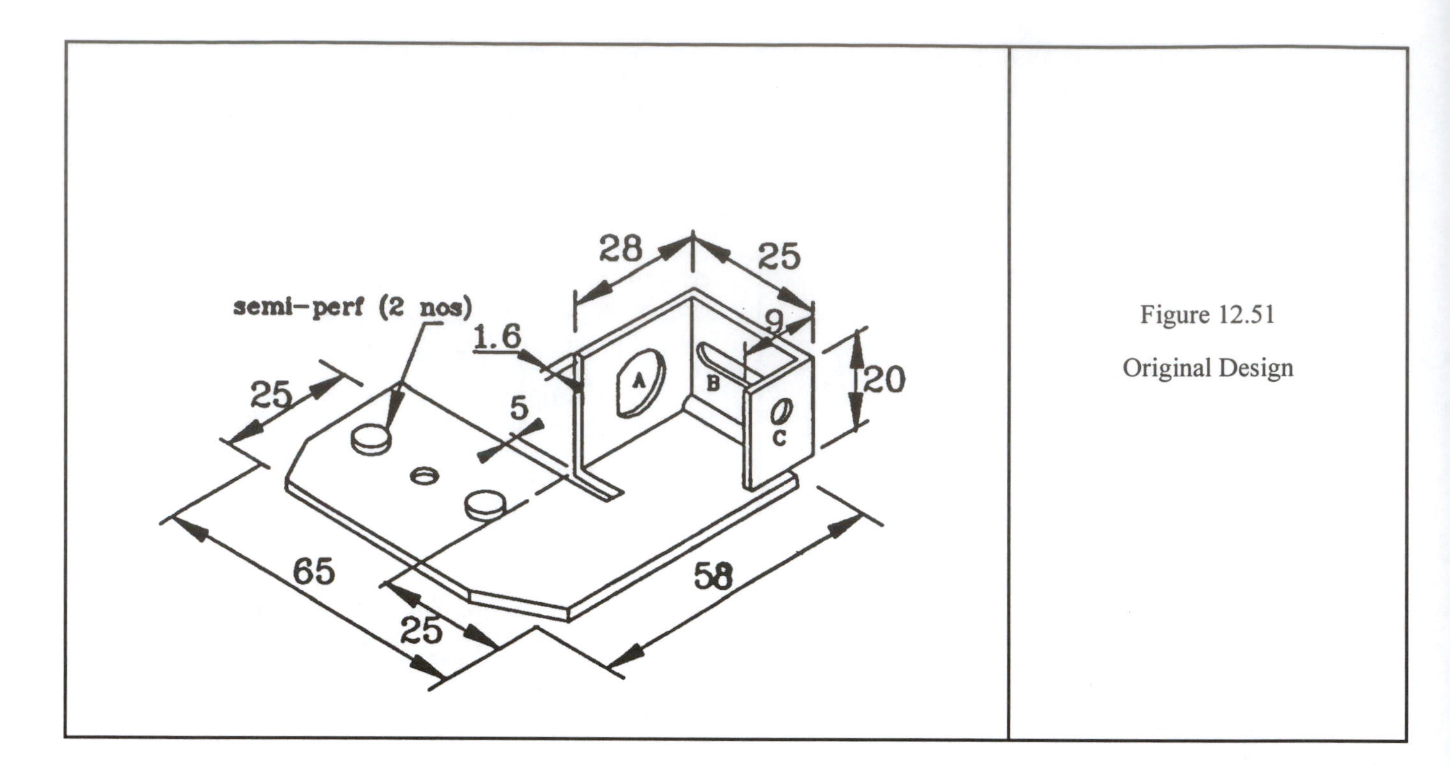

Figure 12.51

Original Design

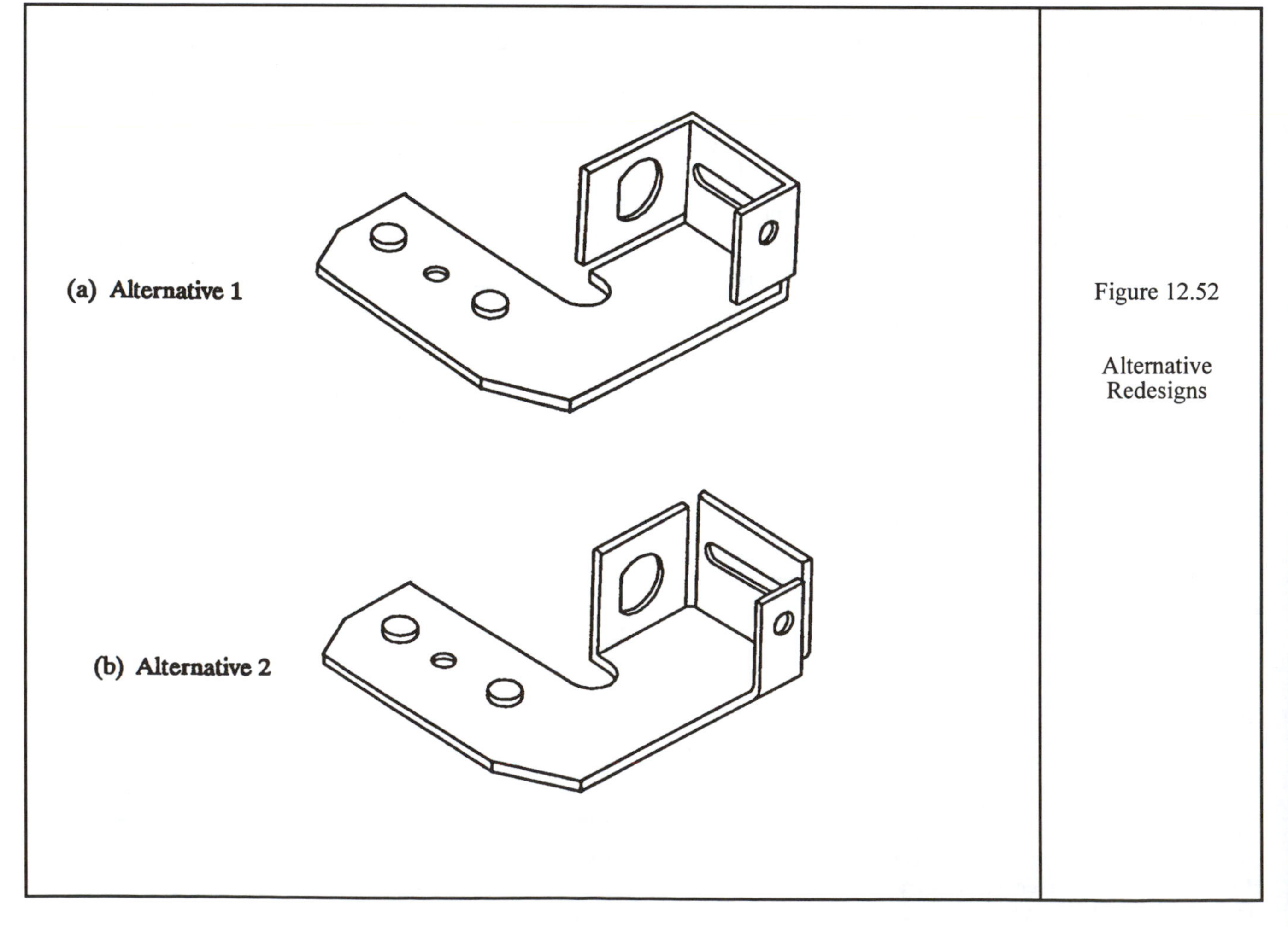

Figure 12.52

Alternative Redesigns

Wipe Forming (Bending) - The Third and Fourth Digits

According to the criteria specified in Figure 12.2, the part is bendable. Hence, the third digit lies between 1 - 12.

The part has three bend stages, thus, the third digit is 10, 11 or 12. The maximum bend angle is 90°; thus, there are no overbends and the third digit is 10.

The maximum bend length is 25 mm, no side action features are present and no features are near bend lines; thus the fourth digit is 1, $C_{b2} = 1.86$, and $N_{a2} = 5$.

Contour Forming - The Fifth and Sixth Digits

There are no contour or radius forms present; thus C_{b3} and N_{a3} are both 0.

Material and Sheet Thickness

The part material is cold rolled steel 1.6 mm thick. Therefore, from Tables 12.1 and 12.2, the seventh and eight digits are both 0 and the factors which account for workpiece material and thickness, F_{mc}, F_{mb}, F_{mf}, F_{dm} and F_t are all equal to 1.0.

Basic Relative Die Construction Cost

Substituting into Eq. (12.4) gives

$$C_b = 2.33(1.0) + 1.86(1.0) = 4.19$$

for the basic relative die construction cost.

Total Relative Die Construction Cost

Substituting into Eq. (12.6) gives

$$C_{dc} = C_b F_t = 4.19$$

for the relative die construction cost.

Relative Die Material Cost

The total number of active stations required for this part is

$$N_a = N_{a1} + N_{a2} + N_{a3} = 5 + 5 + 0 = 10$$

Using Table 3, since $L_{uw} = 65$ mm, and since no curls are present, then no idle stations, N_{i1}, are required for shearing and forming.

Since $N_{a2} = 5$, the number of idle stations required during the wipe forming stages of the part are

$$N_{i2} = (N_{a2} - 1) = 4$$

Thus the total number of stations required, N_s, is 14.

From Table 12.4, the die block size required is

$$L_{dl} = N_s L_{uw} + 2L_{ex} = 14(65) + 2(50) = 1010 \text{ mm}$$

$$L_{db} = L_{ul} + 2L_{ex} = 92 + 2(50) = 192 \text{ mm}$$

Hence, the die set size, S_{ds}, is

$$S_{ds} = (L_{dl} + 25)(L_{db} + 25) = 1035(217) = 224{,}595 \text{ mm}^2$$

Thus, since $F_{dm} = 1$, the relative die material cost is, from Eq. (8):

$$C_{dm} = [2.7S_{ds}/(25.4)^2 + 136]/260.2 = 4.14.$$

Total Relative Die Cost

The total relative die cost is obtained from Eq. (12.4):

$$C_d = 0.8C_{dc} + 0.2C_{dm} = 0.8(4.19) + 0.2(4.14) = 4.18$$

Redesign Suggestions

The second digit in this case is 0. Thus, from Figure 12.1 we see that little can be done to reduce the tooling cost for this part by reducing the peripheral complexity, L_s. Similarly, since it has been assumed at this design stage that none of the features result in side action features, and that none of the features are near bend lines, then the fourth digit can not be further reduced. Hence, no reduction in tooling cost can be achieved in this manner.

A study of Figures 12.1 and 12.2 shows, however, that with a third digit of 10, the major cost driver is the number of bend stages required. This number can be reduced by reducing the number of multiply-connected plates. The number of multiply-connected plates can be reduced by redesigning the part as shown in Figure 12.52.

The alternative redesign shown in Figure 12.52 has no multiply connected plates, and hence requires only one bend stage. In addition, all of the bends are in the same direction; hence, the third digit is now 1. For this redesigned part, $C_{b2} = 0.79$ and $N_{a2} = 2$. In this case, $C_b = 3.12$, $C_{dc} = 3.12$ and $N_a = 7$. This results in a reduction of die construction cost of approximately 25%.

The reduction in the number of active stations from 10 to 7 will also result in a reduction in the die material cost. In this case, N_{i1} is still 0 while N_{i2} is reduced to 1. Hence the total number of stations, N_s, required is reduced from 14 to 8. The new die block set is then approximately 684 mm (L_{dl}) by 185 (L_{db}), and the die set size, S_{ds}, becomes 148,890 mm^2. The die material cost, C_{dm}, for Alternative 2 is, therefore, 3.11. This is a 23% savings in die material cost. The overall die cost for the redesigned part is

$$C_d = 0.8(3.12) + 0.2(3.11) = 3.12$$

This is an overall savings in tooling cost of 25%.

12.9 Example 2

As a second example, consider the part shown in Figure 12.53. To analyze this part, we will take a middle of the road approach. That is, except for holes A (see Figure 12.54), we will assume that the other holes are not critical and they will therefore, not be treated as side-action features. While the details of holes A are not as yet known, we will assume that we know enough about this design that even at this configurational stage of design, we know that their location is critical and thus they will be treated initially as side-action features.

Relative Die Construction Cost

Shearing and Local Forming - The First and Second Digits

The original design (sketch) of the part is shown in Figure 12.53. Figure 12.54 shows the details of the hidden features.

The part has a uniform sheet thickness less than 6.5 mm, and hole diameters greater than the 2.0 mm thickness of the sheet. In addition, the non-peripheral features are in a direction normal to the sheet thickness. However, upon unfolding the part, in order to determine the size of the flat envelope, it is seen that elemental plates M and N overlap (Figure 12.55). Hence, the part is not stampable and the first digit is 9.

Redesign Suggestions

If the length of the plates M and N are reduced as shown in Figure 12.56, the plate is now stampable. This redesigned part will now be analyzed.

Shearing and Local Forming

To determine the amount of die detail, it is necessary to check for the presence of side-action features. In this case the location of holes A were specified to have a tight location tolerance from the bend line. Hence, hole A must be formed after bending and is a side-action feature. Thus, hole A is not included in the determination of die detail.

Using Figure 12.24, it is seen that the part contains three standard holes (3 penalty points), and 2 semi-perfs in the same direction (2 penalty points). Thus, the total penalty in this case is 5. Since the largest dimension of the flat envelope, L_{ul}, is about 186 mm, the part contains low die detail. This indicates that the first digit is 0, 1, or 2.

Since there are no closely spaced features, no narrow projections, and no narrow cutouts, the first digit is 0.

Since L_{ul} is 186 mm, the second digit is 2 or 3.

The peripheral length of the part before bending, L_{out}, is 934 mm, while the width of the flat envelope of the part is 143 mm. Hence, the peripheral complexity of the part, L_s, is

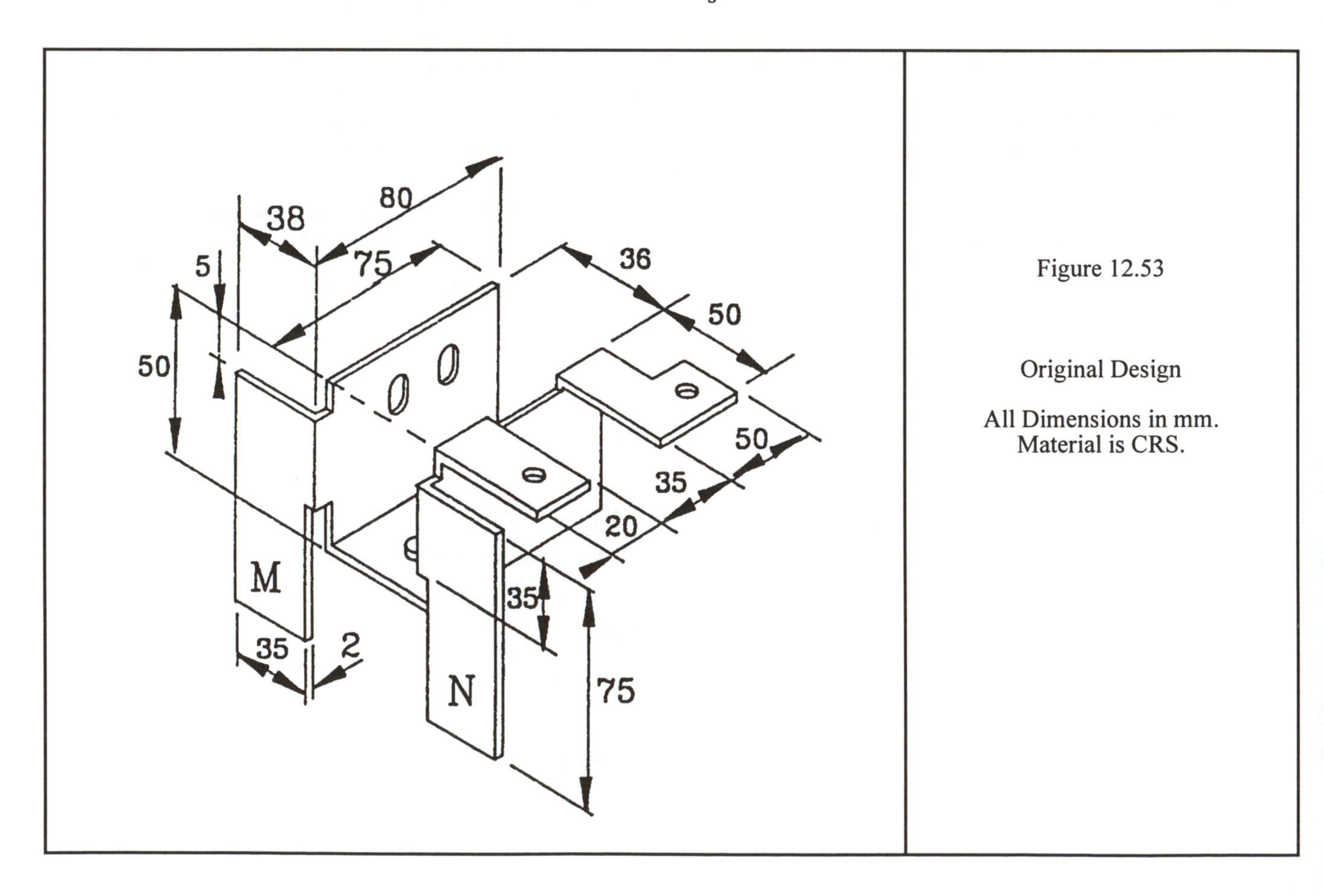

Figure 12.53

Original Design

All Dimensions in mm.
Material is CRS.

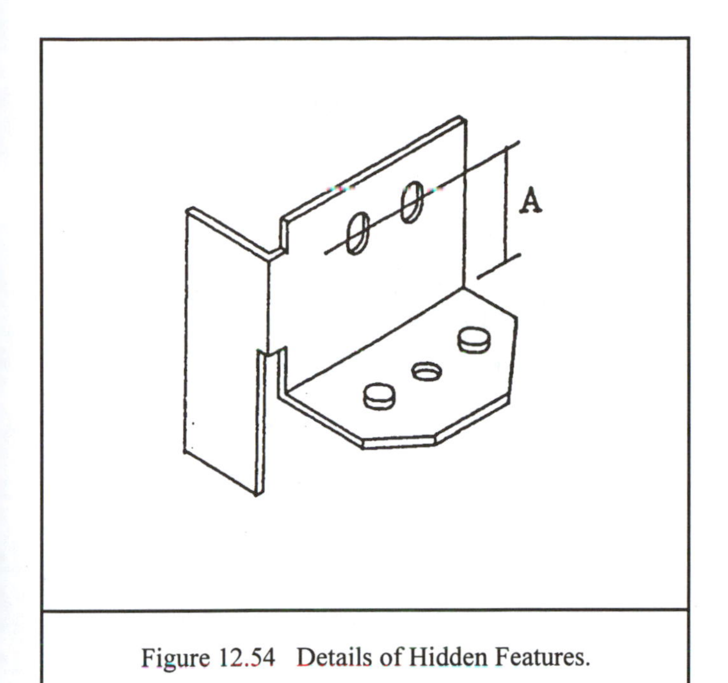

Figure 12.54 Details of Hidden Features.

$L_s = L_{out}/2(L_{ul} + L_{uw}) = 934/2(186 + 143) = 1.42 > 1.4$

Thus, the second digit is 3 and, from Figure 12.1, the relative tool construction cost due to shearing and local forming is C_{b1} = 1.50 and the number of active stations required, N_{a1} = 2.

Wipe Forming (Bending) - The Third and Fourth Digits

According to the criteria specified in Figure 12.2, the part is bendable. Hence, the third digit lies between 1 - 12.

The part has two bend stages, thus, the third digit is 7, 8, or 9. The maximum bend angle is 90°, thus, there are no overbends and the third digit is 7.

The maximum bend length is 80 mm, and as indicated before one side action feature is present. There are no features near the primary plate bend lines, thus, the fourth digit is 3, C_{b2} = 2.26 and N_{a2} = 5.

Contour Forming

There are no contour forms or radius forms present, thus, the fifth and sixth digits are both 0. Hence, C_{b3} and N_{a3} are both 0.

Material and Sheet Thickness

The part material is cold rolled steel 2.0 mm thick. Therefore, the seventh and eight digits are both 0 and the factors which account for workpiece material and thickness, F_{mc}, F_{mb}, F_{mf}, F_{dm} and F_t are all 1.0.

Basic Relative Die Construction Cost

Substituting into Eq. (12.4) gives

$$C_b = 1.50(1.0) + 2.26(1.0) = 3.76.$$

The Total Relative Die Construction Cost

Substituting into Eq. (12.6) gives

$$C_{dc} = C_b F_t = 3.76$$

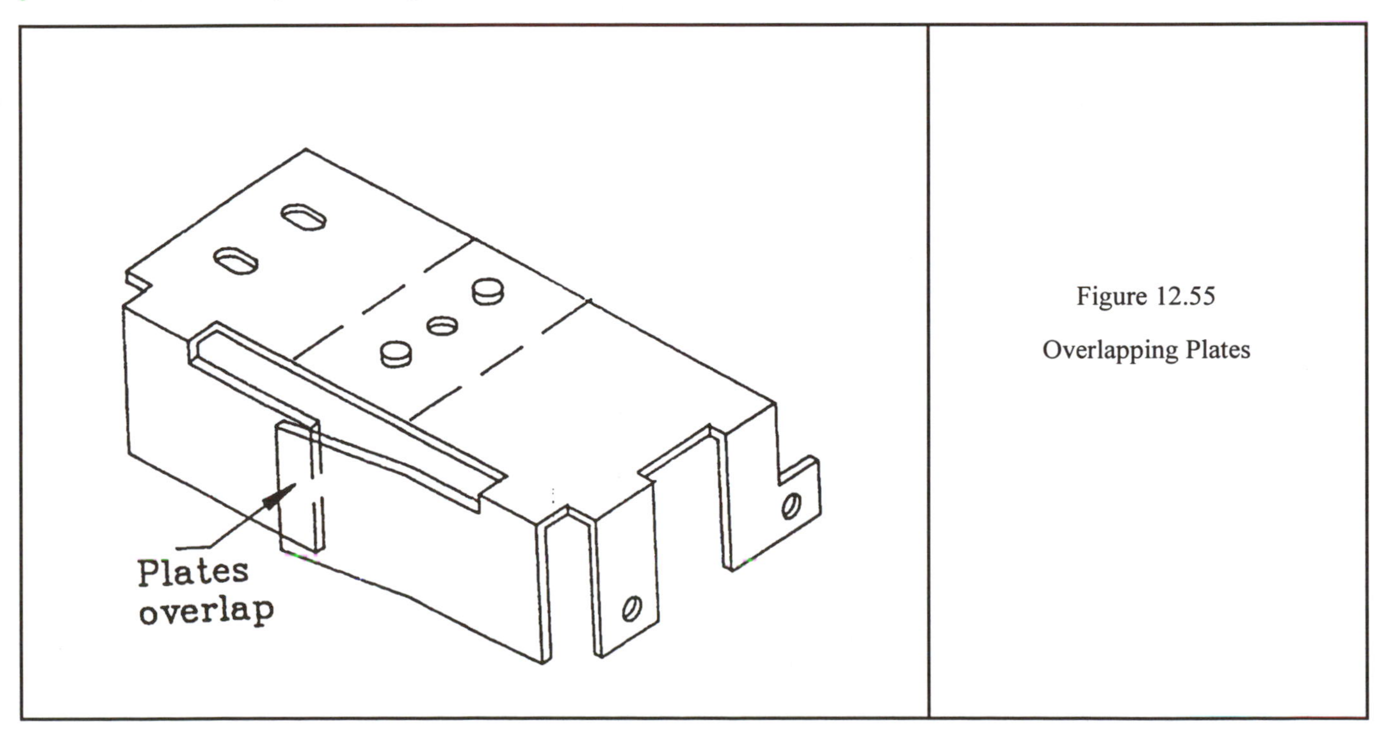

Figure 12.55

Overlapping Plates

Relative Die Material Cost

The total number of active stations required for this part is

$$N_a = N_{a1} + N_{a2} = 2 + 5 = 7$$

Using Table 12.3 it is seen that since L_{uw} = 143 mm and since no curls are present, then no idle stations are required. Thus, the total number of stations required, N_s, is 7.

From Table 12.4, the die block size required is

$$L_{dl} = N_s L_{uw} + 2L_{ex} = 7(143) + 2(50) = 1101 \text{ mm}$$

$$L_{db} = L_{ul} + 2L_{ex} = 186 + 2(50) = 286 \text{ mm}$$

Hence, the die set size, S_{ds}, is

$$S_{ds} = (L_{dl} + 25)(L_{db} + 25) = 1126(311) = 350{,}186 \text{ mm}^2$$

Thus, the relative die material cost is, from Eq. (12.8), (recall $F_{dm} = 1$)

$$C_{dm} = [2.7S_{ds}/(25.4)^2 + 136]/260.2 = 6.16$$

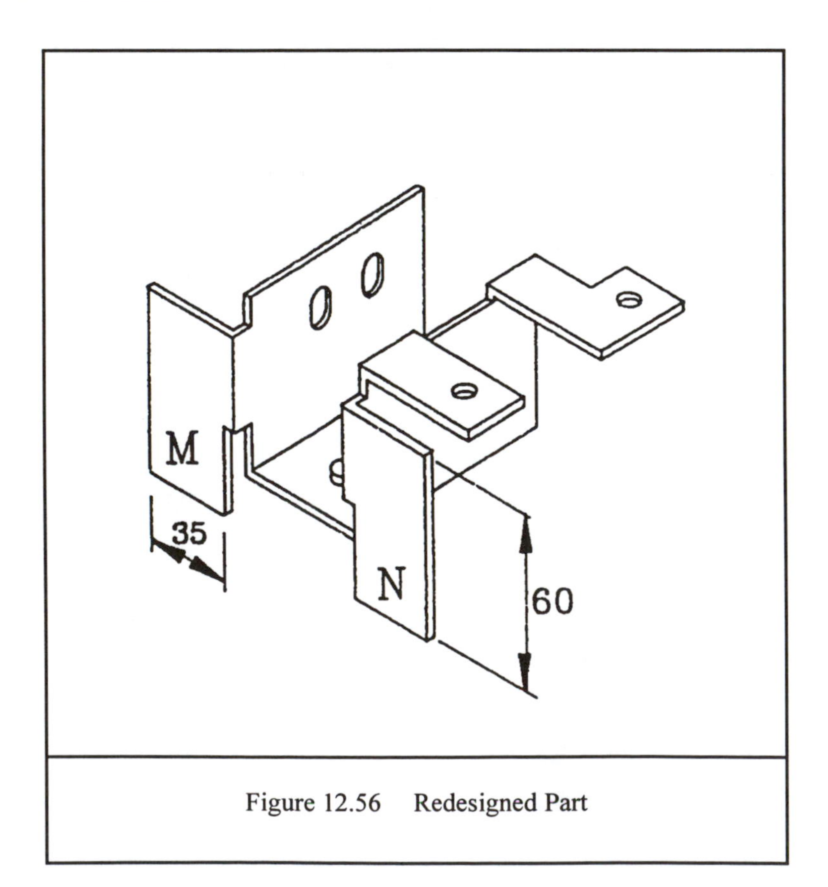

Figure 12.56 Redesigned Part

Total Relative Die Cost

The total relative die cost is obtained from Eq. (12.3), thus,

$$C_d = 0.8C_{dc} + 0.2C_{dm} = 0.8(3.76) + 0.2(6.16) = 4.24$$

12.10 Worksheet for Relative Tooling Cost (Stamping)

To facilitate the calculation of the relative tool costs for stamping, the worksheet shown below can be used. The copy shown below has been completed for Example 1 (Figure 12.51). A blank copy of the worksheet is available in Appendix 12.B and may be reproduced for use with this book.

WORKSHEET FOR RELATIVE TOOLING COSTS — STAMPING

Original Design:

Example 1

(a) Relative Die Construction Cost

Flat Envelope: L_{ul} = ***92***; L_{uw} = ***65***; Die Complexity: ***Medium (4 holes,2 semi-perfs)***

Peripheral Complexity: L_{out} = ***350***; $L_S = L_{out} / 2 (L_{ul}+L_{uw})$ = ***350 / 2(92+65) = 1.13 < 1.4***

Shear Complexity:

1st Digit = ***3***; 2nd Digit = ***0*** => C_{b1} = ***2.33***; N_{a1} = ***5***

Wipe Form Complexity: ***3 bend stages; angles < 90°; L_b < 100***

3rd Digit = ***10***; 4th Digit = ***1*** => C_{21} = ***1.86***; N_{a2} = ***5***

Contour Form: 5th Digit = ***0***; 6th Digit = ***0*** => C_{b3} = ***0***; N_{a3} = ***0***

Effect of material: ***CRS, Table 12.1***

7th Digit = ***0***; F_{mc} = ***1***; F_{mb} = ***1***; F_{mf} = ***1***

Effect of sheet thickness: ***t = 1.6 mm, Table 12.1***

8th Digit = ***0***; F_t = ***1***; F_{dm} = ***1***

Basic Complexity, $C_b = C_{b1}F_{mc} + C_{b2}F_{mb} + C_{b3}F_{mf}$ = ***2.33(1.0) + (1.86)(1.00) = 4.19***

Total Relative Die Construction Cost, $C_{dc} = C_bF_t$ = ***4.19 (1.00) = 4.19***

(b) Relative Die Material Cost:

Active Stations, $N_a = N_{a1} + N_{a2} + N_{a3}$ = ***5 + 5 = 10***

Idle Stations, N_{i1} = ***0***; N_{i2} = ***5 – 1 = 4***; N_{i3} = ***0***;

Total Stations, $N_S = N_a + N_i$ = ***14***

$L_{dl} = N_s L_{uw} + 2L_{ex}$ = ***14(65) + 2(50) = 1010***; $L_{db} = L_{ul} + 2L_{ex}$ = ***92 + 2(50) = 192***

$A = L_{dl} + 25$ = ***1035***; $B = L_{db} + 25$ = ***217***; $S_{ds} = AB$ = ***224, 595***

$C_{dm} = F_{dm} [2.7S_{ds}/(25.4)^2 + 136]/260.2$ = ***4.14***

$C_d = 0.2C_{dm} + 0.8C_{dc}$ = ***0.2(4.14) + 0.8(4.19) = 4.18***

Redesign Suggestions:

Shearing: 2nd digit = 0 so little to be gained by reducing peripheral complexity

Bending: No side action features, so can't reduce fiourth digit

3rd digit= 0 —> Savings can be achieved by reducing bend stages

WORKSHEET FOR RELATIVE TOOLING COSTS — STAMPING

Redesign:

(a) Relative Die Construction Cost

Flat Envelope: L_{ul} =______; L_{uw} =______; Die Complexity: ______________

Peripheral Complexity: L_{out}=______; $L_S = L_{out} / 2 (L_{ul}+L_{uw})$ = ______________

Shear Complexity:

1st Digit =______; 2nd Digit =______ => C_{b1}=______ ; N_{a1} =______

Wipe Form Complexity:

3rd Digit =______; 4th Digit =______ => C_{21}=______; N_{a2}=__________

Contour Form: 5th Digit=______; 6th Digit =______ => C_{b3}=______; N_{a3}=__________

Effect of material:

7th Digit=______; F_{mc}=______; F_{mb}=______; F_{mf}=______

Effect of sheet thickness:

8th Digit = ______; F_t=______; F_{dm} =______

Basic Complexity, $C_b = C_{b1}F_{mc} + C_{b2}F_{mb} + C_{b3}F_{mf}$ =______________

Total Relative Die Construction Cost, $C_{dc} = C_bF_t$ = ______________

(b) Relative Die Material Cost:

Active Stations, $N_a = N_{a1} + N_{a2} + N_{a3}$ = ______________

Idle Stations, N_{i1} =______; N_{i2} =__________; N_{i3} =______;

Total Stations, $N_S = N_a + N_i$ =______

$L_{dl} = N_s L_{uw} + 2L_{ex}$ = ______________; $L_{db} = L_{ul} + 2L_{ex}$ = __________

$A = L_{dl} + 25$ =______; $B = L_{db} + 25$ =______; S_{ds} = AB =__________

$C_{dm} = F_{dm} [2.7S_{ds}/(25.4)^2 + 136]/260.2$ = ______________

$C_d = 0.2C_{dm} + 0.8C_{dc}$ =______________

% Savings:

12.1 Summary and Preview

Summary. The purpose of this chapter is to present a systematic approach for identifying, at the configuration design stage, those features of parts which significantly affect the tooling cost of stampings. In addition, the goal is to learn how to design so as to minimize difficult-to-stamp features. Therefore, a methodology for estimating the relative tooling cost of proposed stamped parts based on configuration information was presented. Unlike injection molding and die casting, however, where the *total* relative tooling cost can be determined at the configuration stage of design, in stamping only a range of relative tooling costs can be determined. Nevertheless, a great deal of information concerning tooling cost can be determined at the configuration stage, and this information can be used to help guide the generation and selection of part configurations before parametric design is done.

Preview. In the next chapter, methods for evaluation of configuration designs that include both functional and manufacturability issues are presented.

References

[1] Wick, C., (editor) "Tool and Manufacturing Engineers Handbook, Vol 2, Forming," Society of Manufacturing Engineering, Dearborn, MI, 1984.

[2] Mahajan, P. V., "Design for Stamping - Estimation of Relative Die Cost for Stamped Parts," MS Final Project Report, Mechanical Engineering Department, University of Massachusetts at Amherst, Amherst, MA 1991.

[3] Dastidar, P. G., "A Knowledge-Based Manufacturing Advisory System for the Economical Design of Metal Stampings," Phd Dissertation, Mechanical Engineering Department, University of Massachusetts at Amherst, Amherst, MA 1991.

[4] Poli, C., Dastidar, P. G., and Mahajan, P. V., "Design for Stamping, Part I - Analysis of Part Attributes that Impact Die Construction Costs for Metal Stampings," Proceeding of the ASME Design Automation Conference, Miami, Sept. 1991.

[5] Poli, C., Mahajan, P. V., and Dastidar, P. G., "Design for Stamping, Part II - Quantification of on the Tooling Cost for Small Parts," Mechanical Engineering Department, University of Massachusetts at Amherst, Sept. 1991.

Problems

12.1 In determining the relative die material cost for the part shown in Figure 12.51 (Example 1 - Original Design), it was assumed that the strip layout was such that the longest flat dimension of the part, L_{ul}, was perpendicular to the direction of strip feed. Determine the relative die material cost for this same part if L_{ul} is laid out parallel to the direction of strip feed. What is the total relative die cost in this case?

12.2 In determining the relative die material cost for the Alternative 2 design [Figure 12.52 (b)] of the part considered in Problem 1 above, it was assumed that the strip layout was such that the longest flat dimension of the part, L_{ul}, was perpendicular to the direction of strip feed. Determine the relative die material cost for this same part if L_{ul} is laid out parallel to the direction of strip feed. What is the total relative die cost in this case? Which of the two layouts (i.e. L_{ul} parallel to the direction of strip feed, or L_{ul} perpendicular to the direction of strip feed) seems more feasible?

12.3 The part shown in Figure P12.3 is made of a soft cold rolled steel. The location of holes A and C are not critical. Distortion of hole B is not permitted. Determine:

a) The relative die construction cost for the part.

b) The relative die material cost for the part.

c) The overall die cost for the part.

12.4 The part shown in Figure P4 is made of a soft cold rolled steel. The dimension A is 9 mm. It is an important dimension that must be held to within +/-0.02 mm. Determine:

a) The relative die construction cost for the part.

b) The relative die material cost for the part.

c) The overall die cost for the part.

12.5 For the part shown in Figure P12.3, suggest an alternative design for the part and estimate the savings in tooling cost as a result of the redesign.

12.6 For the part shown in Figure P12.4, suggest an alternative design for the part and estimate the savings in tooling cost as a result of the redesign

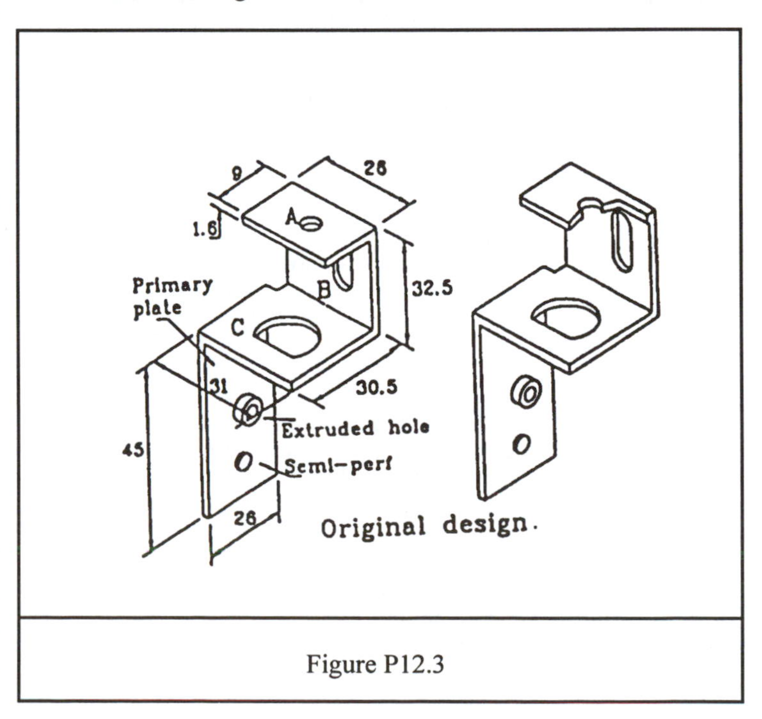

Figure P12.3

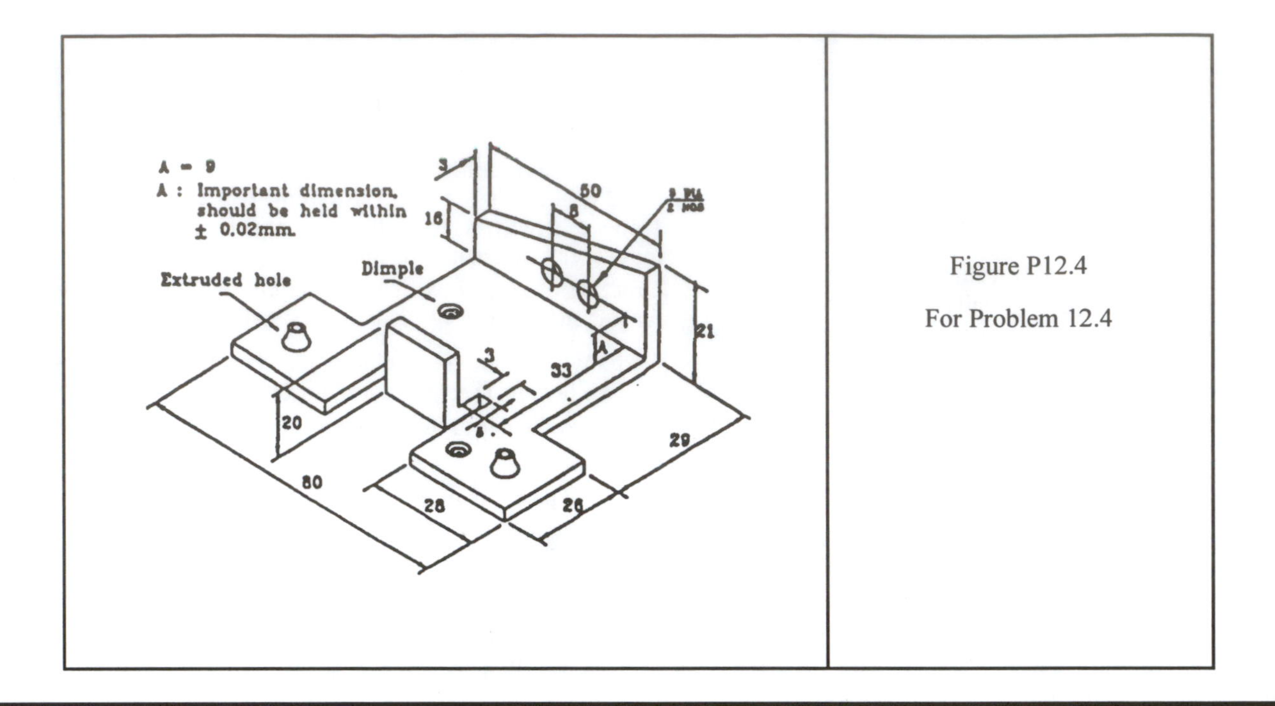

Figure P12.4

For Problem 12.4

Appendix 12.A

Notes For Use With Figures 12.1 - 12.3

(1) An unfolded part is the flat part that is obtained after all the plates that comprise the part have been made coplanar. Parts which, when unfolded, contain one or more plates which overlap are considered unfoldable. See Figures 12.53 and 12.55.

(2) Features such as holes, embosses, tabs, etc. which do not lie on the periphery of a part are considered isolated projections and/or recesses. These features are some times called non-peripheral features. See Figure 12.14.

(3) The axis of a non-peripheral feature is the straight line along which a solid plug which conforms to the internal shape of the feature will have to move in order to be inserted into the feature. For tabs and lance forms, the internal shape is the sheared part of the feature. See Figures 12.17 and 12.19.

(4) A significant projection is a non-peripheral feature (excluding tabs and lance forms) that protrudes from the surrounding plate by a distance greater than four times the sheet thickness. Such features generally exceed the height that can be achieved by a local forming operation.

(5) Die detail is a measure of the concentration of non-peripheral protruding and non-protruding features on a stamped part. As the number of features increases, the number of stations required to produce the part increases.

(6) Two non-peripheral features which are less than three times the stock thickness apart from each other, or a non-peripheral feature that is less than three times the stock thickness from the part periphery are considered closely spaced features. See Figure 12.26.

(7) Any projection or cutout on the peripheral edge of a part that has a minimum width less than three times the stock thickness are considered narrow.

A part with closely spaced non-narrow notches can be considered as a part with narrow projections. Similarly a part with closely space non-narrow projections can be considered as a part with narrow cutouts.

(8) A protruding feature is one which protrudes out of the plane of the particular portion of the stamping on which it is located. Some examples of protruding features include embossings, dimples, tabs, extruded holes, lance forms, and stamped lettering.

(9) L_{ul} is the longest dimension of the flat envelope of the part.

(10) L_s is the ratio of the total sheared length of the part periphery, L_{out}, over the perimeter of the flat envelope, $2(L_{ul} + L_{uw})$. If the part is frame-like, with large internal hole(s), then the perimeter of the holes is added to the length of the part periphery while determining L_{out}.

(11) A part with a straight bend is a part that is not flat and has been folded about a straight line with the radius of curvature of this fold less than four times the sheet thickness. Such bends are normally formed by wiping.

A part with a non-straight bend is a part that is not flat and which has been folded about a curved line, or a line that is a series of straight line segments, with the radius of curvature of the fold less than four times the sheet thickness. See Figure 12.6.

(12) A multiple plate junction is a junction on a part where three or more non-separated plates meet. Such parts are usually drawn. See Figure 12.30.

(13) If a part has a primary plate which is connected on all sides to overbent plates, then strip feeding in a progressive die is impossible. In addition, part removal from a secondary die will cause distortion. Such configurations are not suitable for stamping operations.

(14) If all the bend lines are in the same plane and all the singly connected elemental plates are on the same side of the multiply connected elemental plates, then the bends are in the same direction. See Figures 12.44 and 12.45.

(15) Parts that require two bend stages are those parts for which the bend stage algorithm described in the instructions yields an output of two. Such parts will usually require about 2 secondary form dies or 2 progressive form stations to produce. Similarly, parts that require more than two bend stages are those parts for which the bend stage algorithm yields an output of three or more. These parts will usually require 3 or more secondary form dies, or progressive form stations to produce. See Figure 12.43.

(16) The rotation between portions of the part on either side of a bend line in the plane containing both portions is the bend angle. It represents the angle through which a part is folded about a bend line while wipe forming.

(17) L_b is the length of the longest bend line on the part. The bend may be one continuous bend or it may be made up of several individual collinear bend lines. See Figure 12.12.

(18) A side-action feature is a feature, usually a hole or slot, that does not lie perpendicular to the plane of die closure (i.e. in or parallel to the primary plate) and which must be accurately located (+/- 0.2 mm) or aligned. Such features must be imparted to the part after bending in order to hold such accurate dimensional and/or geometric specifications. Since the tools that produce these features must move at an angle to the direction of die closure, expensive cam-actuated slides must be used. See Figure 12.33.

(19) Any feature (holes, ribs, tabs, etc) in the primary plate that is within three stock thicknesses of a bend line, and whose shape or location must be preserved is said to be near the bend line.

If such a feature is imparted to the part while it is flat, then shape or locational accuracy cannot be guaranteed owing to plastic deformation near the bend line while the bending operation occurs. Hence, these features necessitate the use of an additional post-form secondary die or additional post-form progressive station. See Figure 12.33.

(20) A part which is curved about a single axis, and the radius of curvature is greater than four times the sheet thickness is considered a radius formed part. See Figure 12.7.

(21) A part with an extensive cup-shaped depression is considered a drawn part. If the depth of the cup is less than its radius, the draw is considered a shallow draw. See Figure 12.8.

(22) If a part has an extensive non-circular depression, or if the depression is stepped or tapered by an angle larger than 3 degrees, then the part can not be produced by conventional shallow cup drawing.

(23) Curved channels are parts with channel-like cross-sections (duct or open tube for example) that are curved or bent about one of its transverse axes. Such parts are formed in multiple operations using special equipment like wiper and roll forming machines. Doubly curved parts (car hoods and fenders for example) are parts which are curved about two axes. Such parts are formed with special equipment like stretch forming and stretch drawing presses.

(24) If the radius formed portion of the part is considered to be part of an imaginary cylinder whose radius is the same as the radius of curvature of the radius form, then the radius form angle is the angle subtended by this part at the center of the cylinder. See Figure 12.9.

(25) If R denotes the radius of curvature of the radius formed portion of a part and t denotes the plate thickness, then if R/t is greater than 15, the form will exhibit considerable springback.

(26) Parts without features on/near the form are parts that have features whose shape and/or location must be accurately maintained, and which lie on plates that are connected to a contour form and within 3t of the adjacent contour formed plate, or that lie on a contour formed plate with a direction that is parallel to the direction of die closure.

WORKSHEET FOR RELATIVE TOOLING COSTS — STAMPING

Design:

(a) Relative Die Construction Cost

Flat Envelope: L_{ul} =______; L_{uw} =______; Die Complexity: ____________________

Peripheral Complexity: L_{out}=______; $L_S = L_{out} / 2 (L_{ul}+L_{uw})$ = ____________________

Shear Complexity:

1st Digit =______; 2nd Digit =______ => C_{b1}=______ ; N_{a1} =______

Wipe Form Complexity:

3rd Digit =______; 4th Digit =______ => C_{21}=______; N_{a2} =__________

Contour Form: 5th Digit=______; 6th Digit =______ => C_{b3}=______; N_{a3}=__________

Effect of material:

7th Digit=______; F_{mc}=______; F_{mb}=______; F_{mf}=______

Effect of sheet thickness:

8th Digit = ______; F_t=______; F_{dm} =______

Basic Complexity, $C_b = C_{b1}F_{mc} + C_{b2}F_{mb} + C_{b3}F_{mf}$ =____________________

Total Relative Die Construction Cost, $C_{dc} = C_bF_t$ = ____________________

(b) Relative Die Material Cost:

Active Stations, $N_a = N_{a1} + N_{a2} + N_{a3}$ = ____________________

Idle Stations, N_{i1} =______; N_{i2} =____________; N_{i3} =______;

Total Stations, $N_S = N_a + N_i$ =______

$L_{dl} = N_s L_{uw} + 2L_{ex}$ = ____________________; $L_{db} = L_{ul} + 2L_{ex}$ = __________

$A=L_{dl}+25$=______; $B = L_{db} + 25$ =______; S_{ds} = AB =__________

$C_{dm} = F_{dm} [2.7S_{ds}/(25.4)^2 + 136]/260.2$ = ____________________

$C_d = 0.2C_{dm} + 0.8C_{dc}$ =____________________

If a Redesign: % Saving =

Redesign Siggestions:

Chapter Thirteen

Overall Evaluation and Redesign of Part Configurations

"A theme that appears over and over in this book is that quality must be built in at every design stage. It may be too late, once the plans are on their way."

W. Edwards Deming* [1]
Out of the Crisis (1982)

13.1 Introduction

"Quality at every design stage" requires evaluation at every stage. This chapter is about evaluating designs for quality at the configuration stage. In Chapter 10, we generated configurations and performed some preliminary evaluations for functionality. In Chapters 11 and 12, quantitative DFM methods for evaluating and comparing the expected relative tooling costs of proposed injection molded, die cast, and stamped part configurations were described. These DFM methods expand on those described in Chapter 3 by providing greater detail about the features of parts that require costly tooling. When a designer would like also to have a qualitative evaluation (in terms of, say, Good, Fair, Poor, and Most Costly), it can be obtained by mapping from the design's location in Figures 11.1, 11.26, 12.1, and 12.2 to the equivalent location in Figures 3.14, 3.15, 3.26(a), and 3.26(b), respectively.

Thus, using the information in this book and/or other references, we can begin this chapter with the understanding that designers have evaluated proposed part configurations fairly rigorously on issues related to manufacturability. Remember, however, that since we are still at the configuration stage, these evaluations can be for tooling only. We can not yet make evaluations related to processing because that evaluation requires parametric information. The quantitative estimation of relative processing costs is the subject of Chapters 15 and 16.

In evaluating and selecting the best configuration for a part design, manufacturability issues are not the only ones that must be taken into account. There are also important functional issues as discussed in Chapter 10 including efficient use of materials, and minimizing risk of mechanical failure. And an issue that relates to both function and manufacturing is the ability of the proposed process to hold the tolerances specified in order for the part to fulfill its function properly.

In this chapter, we describe two methods for taking both manufacturing and functional issues into account in making overall evaluations of part configurations. The evaluation methods described here are essentially adaptations of Pugh's method and the Dominic method described in Chapter 7. These methods are not the only ones that might be used. Different methods will appeal to different firms or groups of designers, and different methods will be more or less suited to different types of situations. Students, however, should learn to use at least one method and then improve or extend their knowledge as needed for their own purposes.

13.2 Evaluation Criteria for Part Configurations

For both Pugh's method and Dominic's method, the evaluation of configurations begins by establishing the evaluation criteria. While no general list of criteria can be presented that will include, for certain, all the issues that might be important in every application, many of the following criteria are often significant in evaluating part configurations:

1. Functionality: Potential for Meeting the Functional Requirements,
2. Potential for Efficient Use of Materials,

* From *Out of the Crisis*, by W. Edwards Deming, 1986. Quotation by permission of MIT Center for Advanced Engineering Study, Cambridge, MA.

3. Potential for Reducing Mechanical Failure Risks,
4. Analyzability at the Parametric Level,
5. Manufacturability, Including:
   (a) Ability of the Process to Hold the Specified Tolerances,
   (b) Ease of Handling and Insertion for Assembly,
   (c) Time to Obtain and Test Tooling, and
   (d) Relative Tooling Cost.

Readers should note that evaluations of items 5(b) and 5(d), if desired, can be obtained from the methods outlined in Chapters 3, 11, and 12. Items 1, 2, and 3 are discussed in Chapter 10. Tolerances are also discussed in Chapter 10 and in the next section of this chapter. Other issues must be evaluated in the context of the evaluation methods as described below.

We will henceforth assume in this chapter that the evaluation criteria for the specific application have been established, and that separate qualitative evaluations for each criterion can be obtained, say, in terms of Good, Fair, Poor, Most Costly. With these separate evaluations available for each evaluation criterion, we then use either Pugh's method or the Dominic method as illustrated below to perform an overall evaluation — and, of course, to guide redesign.

First, however, we must consider how to evaluate designs from the special point of view of tolerancing.

13.3 A Method For Evaluating the Manufacturability of Proposed Tolerancing Plans

13.3.1 Overview

A method for evaluating proposed tolerancing plans on trial part designs for injection molded, die cast, and stamped parts has been developed in Fathaillal, 1992 [2]. We present it in this Section in slightly modified form, using injection molding as a reference.

Figure 13.1 illustrates a worksheet useful in implementing a tolerance evaluation. We will describe the procedure in general, and then apply it to example parts. Referring to Figure 13.1:

Column 1. In the first column, each tolerance specified on the part is designated as I, II, III, etc. We will refer to the total number of tolerances specified and listed as N_t, which is entered in the space provided at the left bottom of the sheet.

Column 2. For each specified tolerance, the *type of tolerance* is listed in Column 2. This evaluation method recognizes the following tolerance types: position, concentricity, angularity, perpendicularity, parallelism, flatness, circularity, cylindricity, straightness, and size.

Column 3. For each specified tolerance, the qualitative *tightness* required is listed in Column 3. The possibilities are: tight (or fine), commercial (or standard), and coarse. *Tight* means that the tolerance required is tighter than standard; *coarse* means that the specified tolerance is looser than standard.

Column 4. For each of the specified tolerances, the *tolerance influence features* on the part are listed in Column 4. These are the features of the part that have an influence on the ability of the manufacturing process to hold a required tolerance level on a specified tolerance of the indicated type. There may be one or more influence features for each specified tolerance, though there are seldom more than two or three.

For injection molded and die cast parts, the possible types of tolerance influence features are: unsupported walls and unsupported significant projections, supported walls and supported significant projections, non-significant projections, the parting plane, and undercuts. For stamped parts, the possible tolerance influence features are side action features (undercuts), closely spaced features, and multiple bends.

Column 5. For each specified tolerance, the *strength of the influence* of each individual tolerance influence feature is determined from Figure 10.22 for injection molded parts, and Figure 10.23 for stamped parts. Results are listed in Column 5 as Strong, Moderate or Weak.

For example, Figure 10.22 shows that the effect of an unsupported wall feature on a angularity type tolerance related to that wall in an injection molded part is Strong.

Column 6. For each specified tolerance, the *level of difficulty* of achieving each specified tolerance is determined from Figure 13.2 and recorded in Column 6 of Figure 13.1 as Very High Difficulty, High Difficulty, Medium Difficulty, Low Difficulty, Not Difficult. The estimates in Figure 13.2 apply approximately to all tolerance types in injection molded, die cast, and stamped parts. If there are more than two tolerance influence features, then the two with the strongest influence can be used in Figure 13.2 to obtain the level of difficulty for Column 6.

For example, suppose a specified standard tolerance that has two influence features, one of which is Strong and one of which is Weak, the level of difficulty in achieving the tolerance is estimated from Figure 13.2 to be of Medium Difficulty.

The numbers in parentheses in Figure 13.2 are called Manufacturing Tolerance Difficulty Points; they are used as explained below.

Column 7. Column 7 records an empirical index called Manufacturing Tolerance Difficult Points. The Points assigned to each specified tolerance are:

- Not Difficult = 0
- Low Difficulty = 1
- Medium Difficulty = 2
- High Difficulty = 3
- Very High Difficulty = 4

For the part as a whole, the sum of the Manufacturing Tolerance Difficulty Points for the proposed tolerancing plan is entered at the bottom in Figure 13.1.

Column 1	Column 2	Column 3	Column 4	Column 5	Column 6	Column 7
Tolerance Number	Tolerance Type	Relative Tightness	Tolerance Influencing Features	Strength of Influences of Individual Tolerance Influencing Features	Level of Difficulty in Achieving the Specified Tolerance	Manufacturing Tolerance Difficulty Points
I						
II						
III						

Total Number of Tolerances (N_t)____________

Total Manufacturing Difficulty Points____________ Overall Evaluation ____________

Figure 13.1 Worksheet for Tolerance Evaluation

Overall Tolerance Plan Evaluation

An overall evaluation for the proposed tolerancing plan is obtained from Figure 13.3. In this Figure, the overall evaluation is based on (1) the number of tolerances and (2) the total Manufacturing Tolerance Difficulty Points, which in term are based on the number and strength of influence of the *tolerance influence features*.

This evaluation method for tolerancing plans should not be considered precise, but only as roughly indicative of the approximate difficulty that proposed tolerances will present during manufacturing. Process capabilities for holding tolerances vary from plant to plant, and designers *must* consult with the manufacturing experts. The method here is intended as an early guide to assist designers in calling attention to designs that may present potentially difficult tolerancing plans.

When implementing the method, designers should keep track of the information that will help guide redesign. For example, the tolerance influence features that increase the difficulty of achieving each specified tolerance should be noted. These will usually be features with influences that are Strong (or Moderate) in Figures 10.22 and 10.23. Perhaps the configuration can be modified to reduce their influence. Also, in general, note that both the tightness of specified tolerances (Figure 13.2) and the number of specified tolerances (Figure 13.3) have a great effect on the evaluation of tolerancing plans. Redesign to eliminate the need for a tight tolerance, or even to eliminate the need for any assigned tolerance at all, will improve the design's manufacturability — at least from a tolerancing point of view. An example of the method is presented in the next section.

Combinations of Influence Features on the Specified Tolerances	Tightness of Specified Tolerance		
	Coarse	Standard (Commercial)	Fine
No Influence Feature	Not Difficult (0)	Not Difficult (0)	Low Difficulty (1)
One or Two of Weak Influence	Low Difficulty (1)	Low Difficulty (1)	Medium Difficulty (2)
One of Moderate Influence or One of Moderate Influence and One of Weak Influence	Low Difficulty (1)	Low Difficulty (1)	Medium Difficulty (2)
Two of Moderate Influence or One of Strong and One of Weak or One of Strong Influence	Low Difficulty (1)	Medium Difficulty (2)	High Difficulty (3)
Two of Strong Influence or One of Strong Influence and One of Moderate Influence	Medium Difficulty (2)	High Difficulty (3)	Very High Difficulty (4)

Figure 13.2 Estimating the Level of Difficulty of Producing a Tolerance Based on Tightness and Types of Influence Features

13.3.2 An Example Tolerancing Evaluation

As an example of applying the evaluation method outlined above, we evaluate the tolerancing plan shown on the injection molded part configuration shown in Figure 13.4 [2]. The holes in the vertical walls of this U-channel are required to be coaxial in order, say, to support a shaft. Since we are at the configuration stage, the feature control frame shown can not show a numerical tolerance value. It can, however, indicate that the location of the holes from datum A will have a commercial tolerance level. The "tight" tolerance applies to the coaxiality. We imagine for this example that these tolerances, plus a tolerance on the size of the holes, are the only ones required for proper functionality.

We begin the evaluation by listing the individual tol-

Total Manufacturing Difficulty Points	Number of Tolerances, N_t					
	1	2	3	4	5	6
0	Excellent	Excellent	Excellent	Good	Fair	Poor
1	Excellent	Excellent	Excellent	Good	Fair	Poor
2	Excellent	Excellent	Excellent	Good	Fair	Poor
3	Good	Good	Excellent	Good	Fair	Poor
4	Fair	Good	Good	Good	Fair	Poor
5		Fair	Good	Good	Fair	Poor
6		Fair	Fair	Fair	Fair	Poor
7		Poor	Fair	Fair	Fair	Poor
8		$	Poor	Fair	Poor	$
9			Poor	Poor	Poor	$
10			$	Poor	$	$
11			$	Poor	$	$
Š12			$	$	$	$

NOTE:
"$" in the table indicates a potentially very difficult to manufacture tolerancing plan. Designers should consult with manufacturing experts, and try to redesign to reduce the number and/or difficulty of tolerance control.

Figure 13.3 Rating for Tolerance Plans Based on Number of Tolerances and Manufacturing Difficulty Points

erances that the proposed design requires. See Figure 13.5. Tolerances I and II are taken to be the size tolerances on the two hole diameters. Tolerances III and IV are the position tolerances on the locations of the holes relative to datum plane A. Tolerance V is the coaxiality tolerance for the holes. The information on these tolerances is shown in Columns 1, 2, and 3 of the worksheet in Figure 13.5.

In Column 4 of the worksheet, we list the tolerance influence features of the part relevant to each of the five listed tolerances. All the tolerances are influenced by the fact that the holes are undercuts and lie on an unsupported wall. For the coaxial tolerance, we might list two unsupported walls and two undercuts as tolerance influence features, but we do not in this case because the effects of all these features are hardly going to be independent of one another.

The entries in Column 5 are obtained from Figure 10.22. They represent the relative strength of influence of the individual tolerance influence features on the specified tolerances.

Column 6 reports the level of difficulty of achieving each specified tolerance considering the number and strength of influence of the combined tolerance features. The values are obtained from Figure 13.2.

Column 7 lists the Manufacturing Tolerance Difficulty Points corresponding to the difficulty levels in Column 6. The sum of the points is recorded at the bottom right.

The overall evaluation of this tolerancing plan is found from Figure 13.3 to be "$". This indicates that the proposed plan is likely to be extremely difficult or expensive to achieve in practice, and that designers should therefore make a serious attempt to reduce the difficulty through redesign of the configuration.

One possible redesign in this case is to eliminate the undercuts by using side shut-offs, thus creating slots open to the top rather than holes. We could also support the wall projections with gussets plates. Then the tolerance evaluation worksheet would be as shown in Figure 13.6. The overall tolerance plan evaluation is now Fair, a considerable improvement. Moreover, so long as five tolerances are specified, an evaluation of Fair is the best possible evaluation.

It should be noted that the overall evaluation of manufacturability with respect to tolerances depends upon a combination of the number of tolerances specified and their individual difficulty. Again, the moral for designers is: minimize the number of tolerance specifications required, avoid tighter than standard tolerances, and minimize the effects of tolerance influence features on the tolerances that must be specified.

13.3.3 Evaluating the Manufacturability of Proposed Tolerancing Plans For Stamped Parts

The method described above can be used as well for stamped parts, except that Figure 10.23 is used instead of Figure 10.22.

13.3.4 Evaluating Tolerance Plans for Other Processes

Relatively simple qualitative methods for evaluating tolerance plans like the one presented above for injection molded, die cast, and stamped parts have not (to our knowledge) been developed for other processes. For other processes, the general configuration design evaluation methods

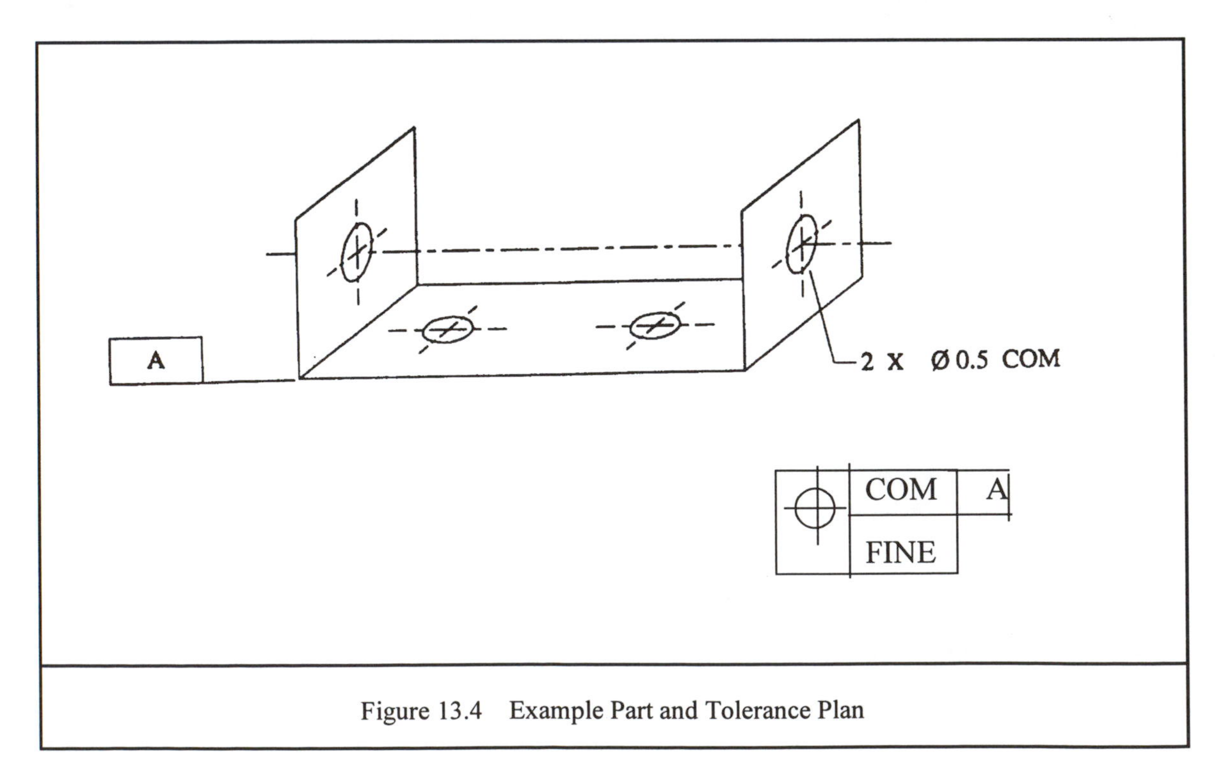

Figure 13.4 Example Part and Tolerance Plan

Column 1	Column 2	Column 3	Column 4	Column 5	Column 6	Column 7
Tolerance Number	Tolerance Type	Relative Tightness	Tolerance Influencing Features	Strength of Influences of Individual Tolerance Influencing Features	Level of Difficulty in Achieving the Specified Tolerance	Manufacturing Tolerance Difficulty Points
I	Size	Com	—	—	Not Difficult	0
			—	—		
II	Size	Com	—	—	Not Difficult	0
			—	—		
III	Position	Com	Undercut	Strong	High Difficulty	3
			Unsupp. Wall	Strong		
IV	Position	Com	Undercut	Strong	High Difficulty	3
			Unsupp. Wall	Strong		
V	Position (Co-Axial)	Fine	Undercut	Strong	Very High Difficulty	4
			Unsupp. Wall	Strong		

Figure 13.5 Total Number of Tolerances = 5; Total Manufacturing Difficulty Points = 10; Overall Evaluation = "$"

Column 1	Column 2	Column 3	Column 4	Column 5	Column 6	Column 7
Tolerance Number	Tolerance Type	Relative Tightness	Tolerance Influencing Features	Strength of Influences of Individual Tolerance Influencing Features	Level of Difficulty in Achieving the Specified Tolerance	Manufacturing Tolerance Difficulty Points
I	Size	Com	—	—	Not Difficult	0
			—	—		
II	Size	Com	—	—	Not Difficult	0
			—	—		
III	Position	Com	—	—	Not Difficult	0
			—	—		
IV	Position	Com	—	—	Not Difficult	0
			—	—		
V	Position Co-Axial	Fine	—	—	Low Difficulty	1
			—	—		

Figure 13.6 Number of Tolerances = 5; Manufacturing Difficulty Points = 1; Overall Evaluation = Fair

described in this chapter have to include tolerancing as one of the evaluation criteria. Evaluations of the tolerancing plans in these cases will have to more subjective than for injection molding, die cast, or stamped parts, but will generally include consideration of the number of tolerances required, the presence of features that influence the ability of the process to hold the specified tolerances, and the tightness of the tolerances specified.

13.4 Part Configuration Evaluation Using Pugh's Method: Example

So far in this chapter we have been discussing the evaluation of tolerancing only. In the remainder of the chapter, tolerancing is only one of the evaluation criteria to be considered. We will now consider evaluation of the entire design of proposed part configurations.

Pugh's Method, which was described in some detail in Chapter 7 as a way of evaluating conceptual designs, is equally applicable for evaluating part configurations. Indeed, the methodology is used in the same way. We illustrate its use with an example.

The example part is an electrical box; that is, an enclosure to contain and protect electrical switches, relays, circuit breakers, and the like. These boxes are typically about 2 to 3 inches deep, and vary in height and width from, say, 3 by 3 inches to, say, 16 by 36 inches. We assume that the proposal is to make a series of boxes using an injection molded engineering thermoplastic instead of the traditional stamped steel sheet metal design.

The configuration issue to be investigated in this example is not the whole product but just the method of attaching a cover to the box. We therefore will assume that all the requirements of electrical, fire, and building codes regarding the use of plastic materials for this application have been investigated and satisfied. It can be argued that this example is actually a concept evaluation as much as a configuration evaluation since the different alternatives employ different methods for hinging. Our purpose is to describe the method, however, so it doesn't matter either way.

We begin the evaluation process at the point where the method and configuration of the attachment method for a cover must be evaluated, and assume that the following alternatives have been generated:

A. A "living" hinge. The hinge is a very thin section molded in to connect both the bottom and top of the box; the thin section flexes as required when the top is raised or lowered. A familiar example of a living hinge is a 3-ring notebook, which has two living hinges, one for the front cover and one for the back;

B. A three piece hinge (two plates and a drop in hinge pin) analogous to those used on wooden doors;

C. Same as B except a bolt, washers, and nut used instead of a pin;

D A "piano" type hinge (See description below);

E. No hinge. The cover is screwed on to the box with four screws;

F. A "cup and post" hinge as shown in Figure 13.7. The cups can be molded into the box, the posts molded into the cover;

G. A "snap-on hinge" as shown in Figure 13.8.

Alternative G makes use of properties of plastics to create "snap fits", the basic idea of which is shown in Figure 13.9. A protruding part feature (labelled A in Figure 13.9) is attached to a member that deforms — usually in bending — while being slid over or into another part (labelled B in Figure 13.9) until A snaps into place. Snap fits are easy to assemble, but sometimes difficult to disassemble if access is not provided for the finger or a tool to bend the first part (A) up again so it can pass back over the retaining ridge (on B).

The proposed snap-on hinge (Alternative G) for the electrical box uses a variation of the basic concept of snap fits. The flat section on the otherwise round post (C) to be molded into the box allows the circular finger-like feature (D) on the cover to be pushed over as the finger deforms slightly causing the opening to increase enough for the finger

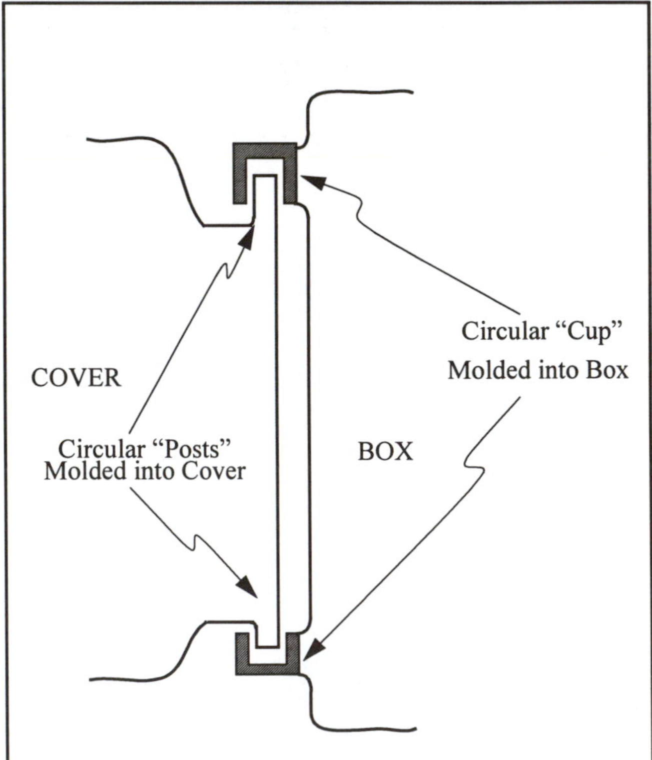

The cover is easily put in place by first inserting the top post all the way up into the top cup. Then the bottom post can drop into the bottom cup.

Figure 13.7 Configuration of a Cup and Post Hinge.

to pass over the post. After the finger snaps into place over the post, it is difficult (though possible) to force it back off. Figure 13.10 shows an assembled cover and box using this idea, though not to scale. This is still configuration design so dimensions are only approximate. The post and fingers must be ultimately sized during parametric design to accommodate the snap fit assembly and to act as a smoothly functioning hinge that provides for the cover to seat with the box appropriately. Figure 13.10 shows one hinge; there must always be at least two.

A matrix to implement a first cut Pugh's method evaluation of these configurations is shown in Figure 13.11. In the matrix, we have separated manufacturing from functionality issues, though this would not necessarily have to be done. The pinned door-type hinge has been selected as the reference. Here are some of the considerations that have led to the evaluations shown in the evaluation matrix:

A) Living Hinge

Advantages: The one piece assembly means no assembly costs. Also, it is possible to have *strippable* internal undercuts for the indentations that function to hold the box closed if performance requirement isn't too high. (A 'strippable' internal undercut is one which can be formed without the need for movable form pins, slides, etc.; in a sense, from a manufacturing point of view, it really isn't an internal undercut.)

Disadvantage: A non-uniform wall thickness is required to produce the living hinge. This can lead to non-uniform shrinkage which in turn lead in turn to warping. The effects of non-uniform wall thickness are discussed in greater detail in Chapter 15. A very long living hinge is likely to present special molding problems not encountered with shorter hinges.

B) Three-piece hinge

Description: The top and bottom of the box are provided with hollow cylindrical projections (external undercuts) with their locations staggered so that when the lid is put on top, the openings in the projections line up so it is possible to slide a pin through them (like a door hinge).

Advantages: It is possible to make the entire box using a uniform wall thickness so that warping should not be a difficult problem.

Disadvantages: Now two molds are needed in order to produce the two individual parts. Also, we now have at least a three piece assembly (lid, base and pin), and both the lid and the base require complex tooling to produce the external undercuts required for the hinge.

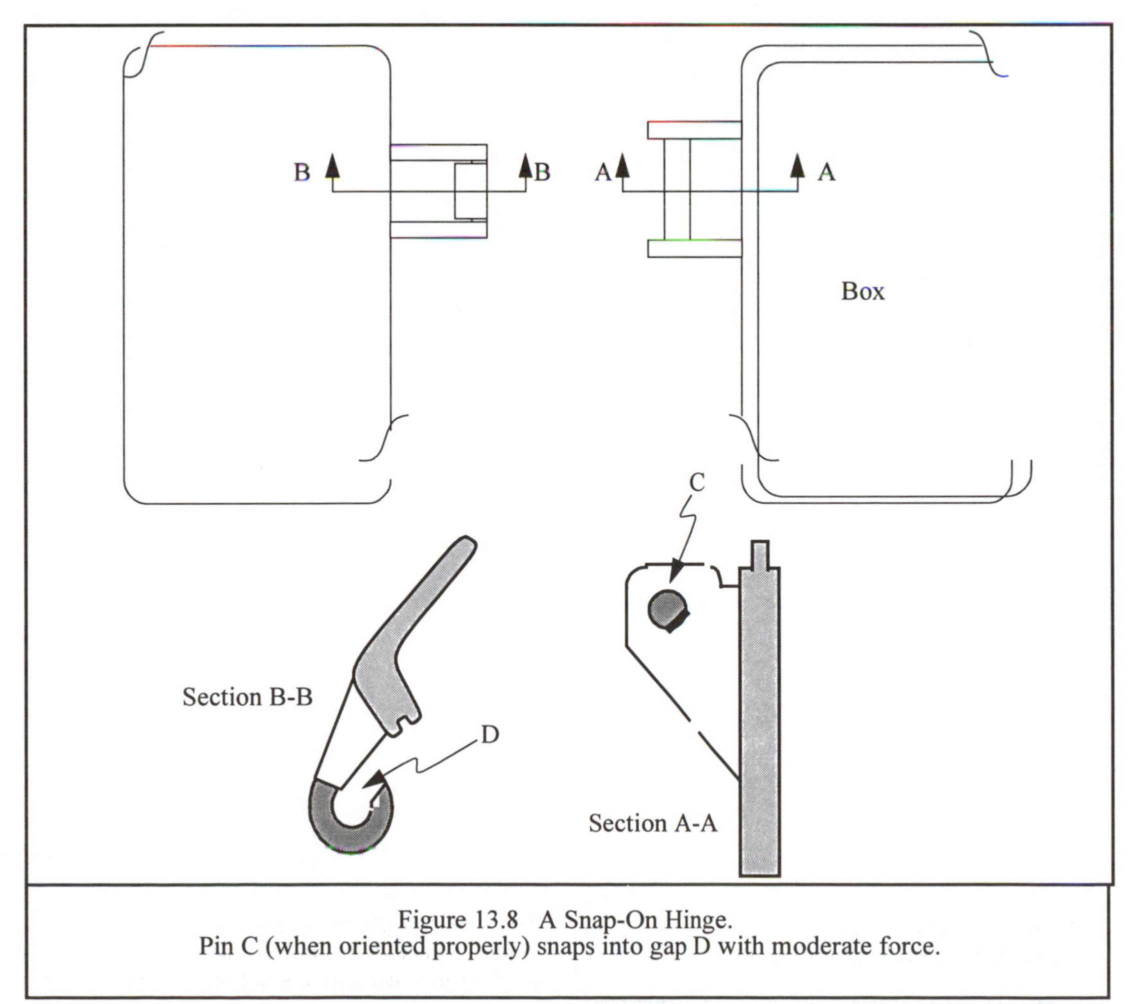

Figure 13.8 A Snap-On Hinge.
Pin C (when oriented properly) snaps into gap D with moderate force.

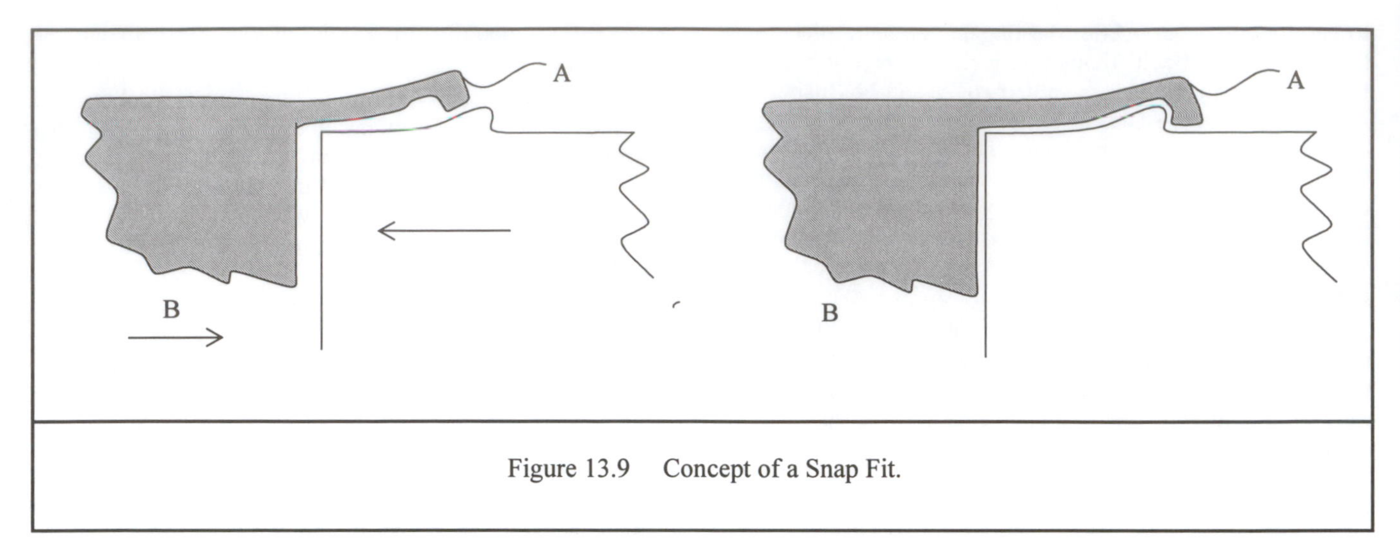

Figure 13.9 Concept of a Snap Fit.

C) Same as (b) but using bolt, washers and nut instead of pin

Advantages: Same as (b).

Disadvantages: We now have at least a 5-part assembly. The tooling requirements are no better than (b).

D) Piano hinge

Description. A 'piano hinge' is a very long, hinge with two very long hinge plates and a very long pin. They are used on pianos, and on parts where a strong, full length hinge is required. They are manufactured as standard assemblies and can be purchased in various sizes and any reasonable desired length.

Advantages: This would make a very strong hinge. Since it would be purchased and assembled as a single unit, it makes a fairly simple solution to the hinge problem.

Disadvantages: Assembly costs are encountered since a number of screws would be required for fastening the piano to the box and cover. It would be expensive. There would have to be hinges purchased (or cut) in various lenghts for each sized box.

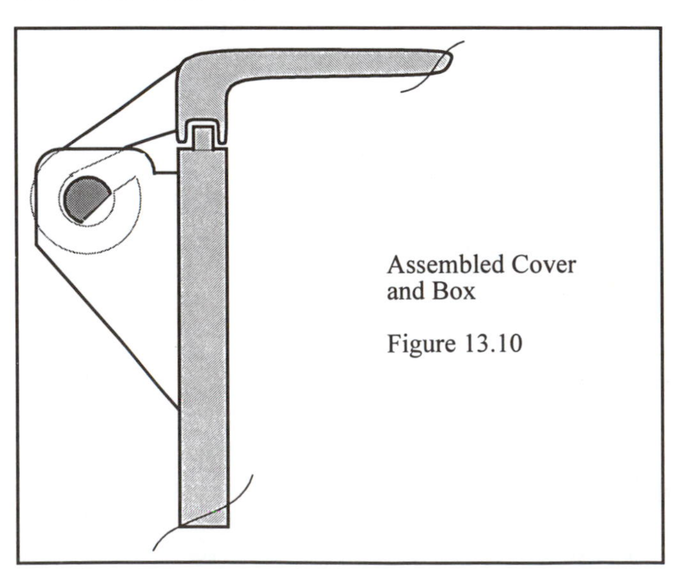

Assembled Cover and Box

Figure 13.10

E) No Hinge (i.e., screw the lid to the base)

Advantages: The parts can be molded with uniform wall thickness leading to no processing difficulties. No external undercuts are required - leading to simplified tooling. (Two molds are required to produce the individual lid and base.)

Disadvantages: Six parts which must be assembled. (Note: In Section 3.11, it was shown that it is almost always less expensive to utilize more complex tooling in order to reduce the number of parts.)

F) Cup and Post Hinge

Advantages: Uniform wall thickness.

Disadvantages: Box requires expensive tooling for the internal undercuts. The two parts must be assembled. Boxes might be difficult to make secure when closed.

G) Snap-on Hinge

Advantages: There can be a relatively uniform wall thickness. Also, this can have one less undercut than alternative (b) and (c) and, therefore, relative to them, it is better.

Disadvantages: Though this is a two piece assembly, and one-piece would be better, two pieces that snap together is the next best thing.

Examination of the evaluation matrix shows clearly that alternatives E and F can be eliminated from further consideration, though for different reasons. The living hinge (Alternative A) can also be eliminated because of its poor evaluation in terms of functionality. (Note, however, that in the computer disk case product, where performance is not so critical, the living hinge rates high on functionality.) Alternative C is clearly inferior to B, so we can eliminate C also.

This leaves Alternatives B, D, and G for further consideration. It is probably the case that the advantages of the piano type hinge are overkill in terms in terms of functionality, and hence the added cost provides little useful benefit. The snap-on hinge then has the considerable advantage of requiring no extra hardware parts. Its one disadvantage, an external undercut, is not especially important in this case since the expected production volume is very high. We

Evaluation Criteria	A Living Hinge	B Standard 3-Piece	C 3-Piece With Bolt	D Picture	E Screwed On	F Cup and Post	G Snap-On
Functionality							
Strength	–	S	0	+	0	–	0
Security (Tamper Proof)	–	S	–	–	–	–	0
Long Life	0	S	0	0	0	–	0
Smooth Operating	0	S	0	0	–	0	0
East of Operation	0	S	0	0	–	0	0
Cost of Hardware	+	S	–	–	+	+	+
Manufacturing							
Tooling Cost	–	S	0	0	0	+	+
Processing Cost	–	S	0	0	0	0	0
Assembly Cost	+	S	–	–	–	0	+
Pluses	2	-	0	1	1	2	3
Minuses	4	-	3	2	4	3	0

Figure 13.11 Pugh's Method Matrix for Evaluation of Hinge Alternatives

would therefore tend to select this alternative, though a more detailed evaluation should be performed to insure that there are no lurking unforeseen problems.

13.5 Example Using the Dominic Method

As in applying Pugh's method to configuration evaluation, the Dominic method is used in essentially the same manner for configuration designs as it is for conceptual designs. Remember, however, that in contrast to Pugh's method, Dominic's method requires that a priority (e.g., High, Moderate, or Low) be assigned to each evaluation criteria. Again we illustrate with an example, this one taken from [3].

The proposed design configuration is a hub whose function is to support a central through shaft inside a cylinder as shown in Figure 13.12. Though final dimensions are not yet determined, the exterior diameter of the hub as shown will be about ten inches; the shaft diameter will be about two inches.

Two concepts are under consideration, both made of aluminum: one made as a forging; the other as a die casting. In either case, several configurations can be proposed. These are:

Hub Shape #1: Has radial stiffening ribs and transverse mounting holes through lugs as shown in Figure 13.13. This hub (as well as Hubs 2 and 3 in Figures 13.14 and 13.15)) is mounted to the cylinder with bolts as shown in Figure 13.17. The bolt holes can be produced as part of a casting but will have to be drilled after forging.

Hub Shape #2: Has a full circular peripheral rib (or 'skirt') instead of stiffening ribs. See Figure 13.14. Again the transverse bolt holes will have to be drilled after forging.

Hub Shape #3: Has neither stiffening ribs nor a peripheral rib or skirt. The transverse holes are mounted in lugs as in Hub #1. See Figure 13.15.

Hub Shape #4: This hub has stiffening ribs and longitudinal mounting holes, and hence requires a redesign of the cylinder to provide a mounting surface. (Note: Mounting the hub to the cylinder is not shown.)

To use Dominic's method for evaluating these four alternatives, we first establish the evaluation criteria and their priorities. For example, in this case they might be:

	Criterion	Priority
A.	Tooling Cost	Low
B	Processing Cost	Moderate
C.	Material Cost	High
D.	Secondary Operations	High
E.	Stiffness	Moderate
F.	Dimensional Accuracy	Moderate
G.	Resists Corrosion	High
H.	Ease of Mounting	Low
J.	Redesign of Cylinder	Moderate

Next we rate each of the alternatives as Excellent, Good, Fair, Poor, or Unacceptable on each of these criteria. For example, for the forged hubs (designated as 1F, 2F, 3F, and 4F):

	Criterion	Hub 1F	2F	3F	4F
A.	Tooling Cost	Fair	Good	Good	Fair
B.	Processing Cost	Fair	Good	Good	Fair
C.	Material Cost	Good	Poor	Poor	Good
D.	Secondary Operations	Poor	Poor	Poor	Good
E.	Stiffness	Good	Good	Good	Good
F	Dimen. Accuracy	Fair	Fair	Fair	Fair
G.	Resists Corrosion	Good	Good	Good	Good
H.	Ease of Mounting	Fair	Fair	Fair	Good
J.	Redesign Cylinder	Exc	Exc	Exc	Poor

For the die cast hubs, designated as 1DC, 2DC, 3DC, and 4DC:

	Criterion	1DC	2DC	3DC	4DC
A.	Tooling Cost	Fair	Fair	Fair	Exc
B.	Processing Cost	Good	Unacc	Unac	Good
C.	Material Cost	Exc	Fair	Fair	Exc
D.	Secondary Operations	Fair	Fair	Fair	Fair
E.	Stiffness	Good	Good	Good	Good
F	Dimen Accuracy	Fair	Fair	Fair	Fair
G.	Resists Corrosion	Good	Good	Good	Good
H.	Ease of Mounting	Fair	Fair	Fair	Good
J.	Redesign Cylinder	Exc	Exc	Exc	Poor

Here are some of the considerations that have resulted in the above evaluations:

Hub 1F

Disadvantages: Because of the relatively thin ribs (for the forging process), forging this part will probably require about 4 die sets. In any event it will require at least one more die set than either Hub 2F or Hub 3F. This additional die set will require an extra forging operation (as compared to Hub 2F and Hub 3F) thereby resulting in increased processing costs relative to these two alternate designs.

In addition, because of the thin ribs, the dies will wear faster.

The ribs will also require more force in order to push the workpiece material into the die cavities. This will result in the need for a large press or hammer (as compared to Hub 2F and Hub 3F) with an added increase in processing costs.

Because the side holes can not be forged, a secondary machining operation is required.

Advantages: This version of the design, like alternative Hub 4F, uses less material and is lighter in weight than the other two alternatives. However, the savings in material costs are more than compensated for by increased tooling and processing costs.

Hub 2F and Hub 3F

Disadvantages: These two versions use more material and are heavier in weight than the other two versions. However, the increase in material costs is more than compensated for by savings in tooling and processing costs.

Because the side holes can not be forged, a secondary machining operation is required.

Advantages: The main advantage here is the decrease in tooling and processing costs due to the elimination of the ribs. At least one less die will be required to form these parts, the die wear will be less severe, and a smaller hammer or press will be required to forge the part. All of this leads to lower tooling and processing costs.

Hub 4F

Disadvantages: Almost all of the disadvantages discussed for Hub 1F apply here. The one exception is the need to utilize a secondary machining operation here to provide the holes. Since the holes are no longer external undercuts, the holes can be initially forged. Some finish machining may be required, but finish machining may be required on all of the forged versions.

Advantages: Same as Hub 1F.

Hub 1D

Disadvantages: The main disadvantage of this design is the number of external undercuts required. According to the Die Casting Advisor (Chapter 3, Figure 3.15) this results in a manufacturing evaluation of Poor. Compared to forging, however, a secondary operation is not required to produce these external undercuts; thus overall we rate tooling as Fair.

If the hub is to be used in a location where appearance

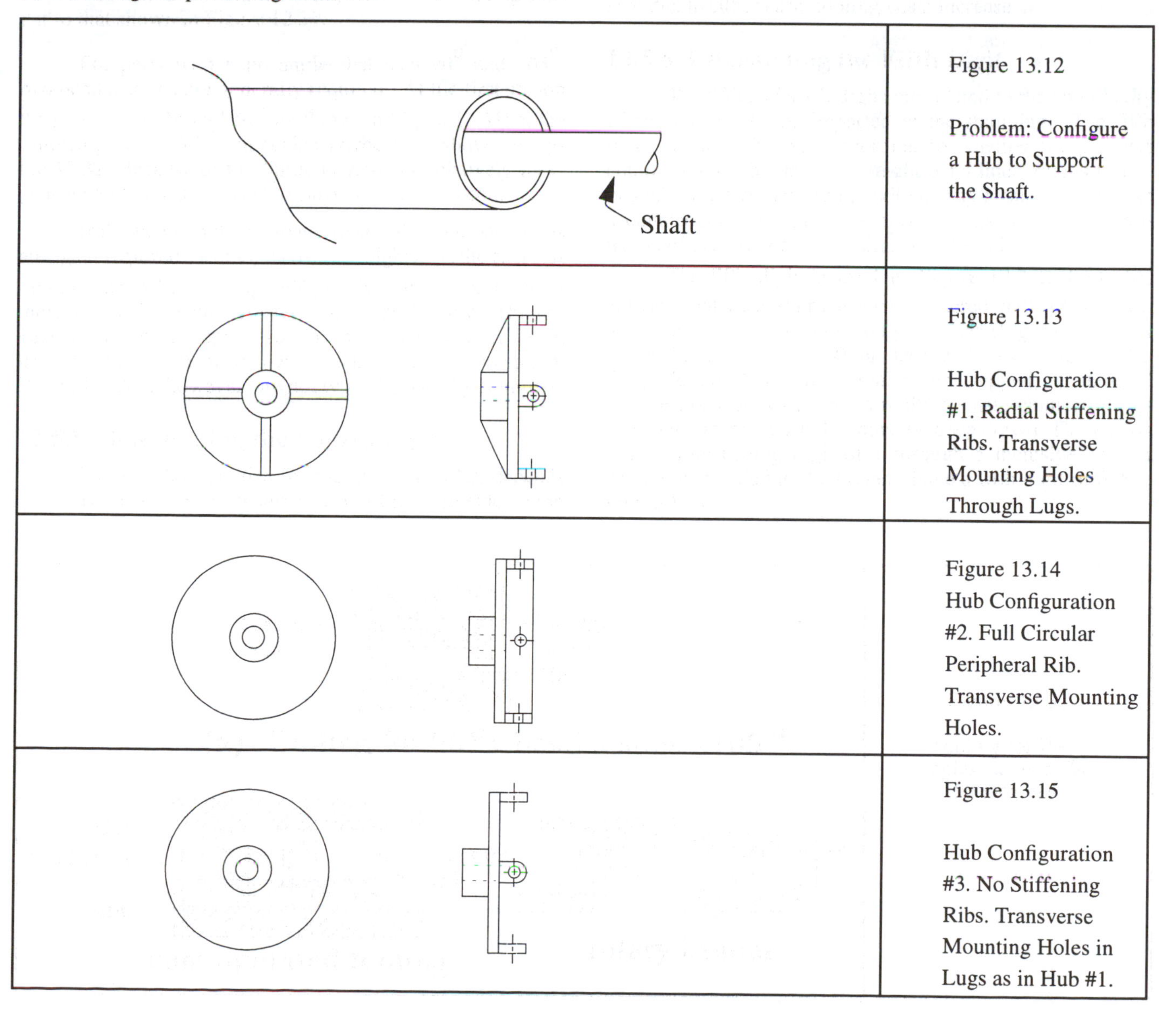

Figure 13.12

Problem: Configure a Hub to Support the Shaft.

Figure 13.13

Hub Configuration #1. Radial Stiffening Ribs. Transverse Mounting Holes Through Lugs.

Figure 13.14
Hub Configuration #2. Full Circular Peripheral Rib. Transverse Mounting Holes.

Figure 13.15

Hub Configuration #3. No Stiffening Ribs. Transverse Mounting Holes in Lugs as in Hub #1.

If the hub is to be used in a location where appearance is not important, then secondary painting or plating operations will not be required — then secondary operations can, perhaps, be limited to burr removal.

Advantages: This design has low material cost, is light in weight and can be produced with a uniform wall thickness. Compared to forging, the secondary operations that will be required are less.

Hub 2DC and Hub 3DC

Disadvantages: These two designs are not really acceptable for die casting. The elimination of the radial ribs and the use of a relatively thick 'truncated cone' leads to wall thickness which exceed the normal capabilities of the die casting process.

Hub 4DC

Advantages: This design will have a low tooling cost since the undercuts have been eliminated. Processing costs are no worse than Hub 1DC.

Disadvantages: Based on Fig. 3.15 (Die Casting Advisor) and the information provided in Chapter 11, at this configurational design stage, there are no particular disadvantages to this design from the point of view of manufacturing.

The Dominic evaluation matrix and a summary of the results are shown in Figure 13.18.

The evaluation indicates that designs 4F and 4DC are most desirable if a way can be found to redesign the cylinder to accommodate the mounting bolts without too much added cost. (Note how the issues of coupling are influencing the evaluation of alternative designs for this part; this is not uncommon.) Design 1DC is also desirable, and there are no secondary operations required. The final choice, then, may be determined by whether the secondary operations on 1DC or the cylinder redesign for 4F and 4DC can be accomplished most easily. Ultimately, a second evaluation may have to be done to compare these two alternatives again after these issues are addressed.

13.6 Looking for the *Unknown* Risks

All evaluation methods are in some sense balancing potential benefits against risks. Therefore, the notion of *explicitly* looking for technical and manufacturing risks in a design is useful. The idea is to look for those previously unquestioned assumptions that have been made while generating the configuration and while doing the evaluations outlined above. That is, we want especially to look for issues that may so far have been overlooked.

In the design of early flat plate solar collectors, ther-

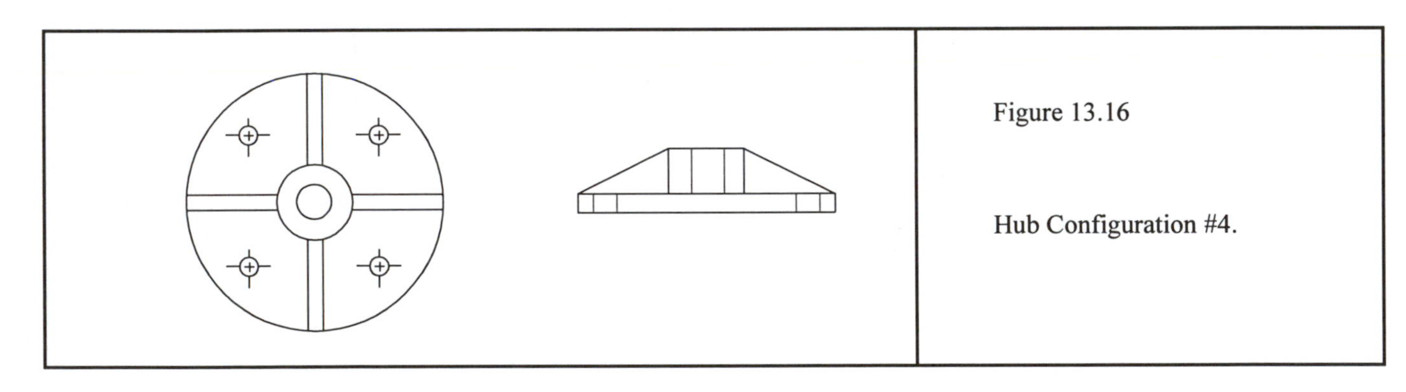

Figure 13.16

Hub Configuration #4.

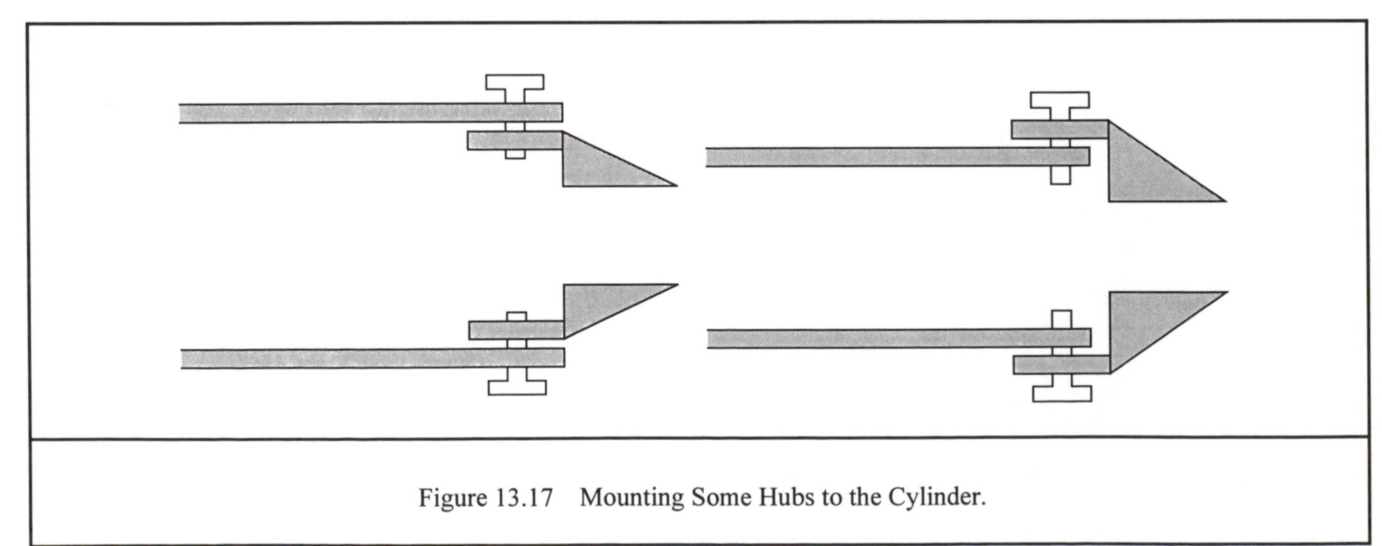

Figure 13.17 Mounting Some Hubs to the Cylinder.

Individual SEP Evaluations	Importance Level of Individual Evaluations		
	High	Moderate	Low
Excellent	1D-C	1F-J 2F-J 3F-J 1D-J 2D-J 3D-J	4D-A
Good	1F-C 2F-G 3F-G 4F-C 1F-G 4F-D 4F-G 1D-G 2D-G 3D-G 4F-C	2F-B 3F-B 4F-E 2F-E 1D-E 1D-G 2D-G 4D-B	2F-A 3F-A 4F-A 4D-H
Fair	2D-D 3D-D 1D-D 2D-C 3D-C 4D-D	1F-B 2F-F 3F-E 4F-B 1F-F 4F-F 3D-F 4D-F 1D-F 2D-F 3D-E	1F-A 2F-H 3F-H 4F-A 1F-H 1D-H 2D-A 3D-H 1D-A 2D-H 3D-A
Poor	1F-D 2F-C 3F-C 2F-D 3F-D	4F-J 4D-J	
Unacceptable		2D-B 3D-B	

Configuration Alternative	Overall Rating	What to Redesign to Improve Overall Rating
1F	Poor	D - Eliminate secondary operations?
2F	Poor	C - Reduce material costs? D - Eliminate secondary operations?
3F	Poor	C - Eliminate secondary operations? D - Eliminate secondary operations?
4F	Fair	J - Redesign cylinder?
1D	Fair	D - Eliminate secondary operations?
2D	Unacceptable	B - Reduce processing costs?
3D	Unacceptable	B - Reduce processing costs?
4D	Fair	J - Redesign cylinder?

Figure 13.18 Dominic Matrix (Above) and Summary of Evaluations of Alternative Hub Configurations

mal efficiency required that there be two transparent cover sheets, usually glass. The first collectors built using two glass sheets often failed because their inner sheet broke as a result of thermal stresses; the inner glass easily got above 300 F. To correct this, one firm elected to replace the inner glass sheet with a special flexible, transparent, ultra-violet resistant plastic sheet that was produced conveniently in rolls. In feel, the material was much like common polyethylene sheet, though a bit stiffer. Computations showed that the thermal efficiency would be even better than using a glass sheet, and collectors would be lighter and less expensive, too. Being flexible, the sheets would not break under thermal stress.

An aluminum extruded frame was designed with provision for attaching the sheet into the frame. The edges of the plastic sheet, stretched taut, were pressed into a slot and held in place with a snap-fit thin rod. Assembly went well, and a number of collectors were built and sold.

The plastic did not shatter in service., and the efficiency was improved as expected. However, to the great surprise of the designer (and his boss!), the sheets wrinkled permanently and noticeably when they cooled after getting hot the very first time. Though subsequent tests showed that the wrinkling did not affect the performance, the collectors could not be sold because of the appearance of the wrinkles. Those that were sold had to be repaired. Fortunately a feasible (though costly) way was found to replace the wrinkled sheets with a glass sheet installed to allow properly for its thermal expansion so it did not break. But the company, which had the highest efficiency collector and solar hot water system on the market at the time, subsequently failed. -- in part because of the expense of repairing these collectors.

The designer in this case took a huge and unnecessary risk, and he did it essentially subconsciously. He did not stop to ask what might go wrong, and to examine honestly where he was operating beyond his knowledge. Thus, he did not discuss the application with the manufacturer of the plastic sheet until after the trouble was encountered -- too late. It turned out that the wrinkling was easily predictable by the scientist who had developed the material. A single phone call just to ask the manufacturer if there might be problems in the application would have saved a great deal of difficulty.

The risk was present in the ignorance of the designer about the material's response to alternate heat and cold. The risk was present in what the designer *failed to consider.* Remember, what designers fail to consider is likely to be the source of future troubles. So are risks taken that are not faced openly and squarely.

In assessing risks at the configuration stage, it must be remembered that we have only information about the configuration; we do not yet have exact dimensions. Values for wall thickness, corner radii, and so on, are not established until the next stage (parametric design). Thus, what we have to look for now is what the possibilities are, assuming that the part will be dimensioned as effectively as it can be.

Here is a partial list of possible risk-revealing questions. Students and readers are encouraged to add to the list from their own knowledge as it grows with experience. Hopefully, the addition knowledge will not be derived from *unexpected* failures of parts!

What are the most likely ways the part might fail in service?

Excessive Stress? Can the part be dimensioned to keep stresses below yield levels?

Fatigue? If there will be cyclic loads, can the configuration be dimensioned so as to keep the internal stresses below the fatigue limit?

Stress Concentrations? Can the part be dimensioned to keep local stress concentrations low?

Buckling? If a possibility, can the configuration be dimensioned to prevent it?

Unexpected Shocks or Loads? What unexpected dynamic loads might be encountered in service or in assembly? Can these be handled?

What are the most likely ways the part might not meet its expected functionality?

Tolerances? Is the configuration such that functionality will be especially sensitive to the actual tolerances that can be expected in a production situation? Are too many special (tight) tolerances required to make the part work well?

Creep? If creep is a possibility, will it result in loss of functionality?

Strain and Deformation? If performance is sensitive to retention of size and shape, can the part be dimensioned to preserve the required integrity?

Thermal Deformations? Might thermal expansion or contraction cause the configuration to perform inadequately?

Might there be unforeseen difficulties with handling and assembly?

Dimensions? Might the part end up being dimensioned such that previous sassumptions about assembleability become invalid?

Tangling? Might the parts tangle if dimensioned in some way? What about possible processing problems?

Will the available machines be able to make the part?

Production runs? Are the desired production runs consistent with the machines and expected costs?

Tooling wear? Is tooling wear or maintenance a possible problem?

Weld lines? If the process is a flow process, can weld lines be located appropriately? (A weld line is formed when a material flow must divide -- say around a hole -- and then rejoin. The weld lines tend to be less strong, and more subject to fatigue failures.)

Geometric compatibility?

Is the part geometrically compatible with its adjoining parts? What could go wrong is this regard? What if there is a small change in this part, or an adjoining part; can the configuration accommodate the change without major redesign?

What about the effects of tolerances of the *adjoining* parts?

Materials?

Is the material selected compatible with the configuration?

Will standard raw material supplies be of adequate quality?

Has the material been thoroughly investigated for its use in *this* particular application? Are there previous uses in similar applications? Have experts in the material's properties and processing been consulted?

Is the material compatible with the rest of the product?

Designer and Design Team Knowledge?

What if....? Has every possible, unfortunate, unlikely, unlucky, even stupid *"What if....."* situation been considered?

Where are the places the designer or design team is working without adequate knowledge? Where is there ignorance of materials, forces, flows, temperatures, environment, etc.? Where are there guesses, hopes, fears, and assumptions instead of knowledge: Materials? Stresses? Fastening methods? Manufacturing process? Tolerances? Costs? Adjoining parts? Environmental conditions? *Where **Are** the Risks?*

Risk cannot be avoided. No matter how conservative a configuration design is, there will be some risk, some places where absolutely complete information does not exist. Indeed, good designs will have some reasonable risk or they will be *too* conservative, too costly, too heavy, and so on. We cannot avoid *all* risk; we try to avoid *unknown* risk.

13.7 Redesign Guided By Evaluations

At the conceptual design stage, evaluations by Pugh's method and Dominic's method point clearly to the aspects of prospective designs that need redesign -- or to the need for whole new alternatives. This is also the case at the configuration stage: the evaluations point almost directly to the strengths and deficiencies of the alternatives evaluated. Thus redesign is easily guided by the evaluations.

13.8 Continuation of Supplementary Design Project A

All the student teams used Pugh's method to evaluate their proposed alternative configurations. Figure 13.19 shows one of the evaluation matrices.

It is most likely in a project of this type that two, or even three, configurations would be studied in more detail at the parametric level. If one of the simpler, lighter extrusions can be dimensioned to support the required loads, then it could be used. On the other hand, quality is an extremely important issue for this product, and so small savings in cost will not compensate for very much risk of poor performance (e.g., wobbly or hard to assemble and disassemble tables.)

13.9 Summary and Preview

Summary. The methods available for evaluating part configurations are essentially the same as the methods available for evaluating conceptual alternatives. We have described the use of Pugh's method and the Dominic method. The issues involved, however, are now less abstract. When evaluating part configurations, we can make use of a great deal of qualitative physical reasoning about function, failure modes, tolerances, and manufacturability. Fortunately, the methods accommodate the inclusion of such issues very well. Thus, parts can be designed by the guided iteration method that perform well and are readily manufacturable.

Preview. This concludes Part III of this book. In our story of Engineering Design, concepts have been designed (Part II), and now part and standard component configurations have been designed or selected (Part III). Much of the reasoning so far has involved only a few numbers and very little computation. Now that we have established part and component configurations, however, it is time to begin to assign numbers. There is also other specific information (e.g., exact materials) that must be determined. We call the process of determining the exact numeric and other values by the term *parametric design.*

Parametric design, like conceptual and configuration design, is done by the basic methodology of guided iteration. In addition, there are two special methods of parametric design that can sometimes be used: optimization and design for robustness. Basically, these methods are variations on guided iteration, but the methods integrate and merge the steps so that they appear as a single (and different) process rather than as the usual set of sequential guided iteration steps. We thus treat these special methods separately in Chapters 18 and 19. Chapters 14, 17, 20, and 21 deal with the application of the general method of guided iteration, using Dominic's method, to perform parametric design.

Design for manufacturability continues, as always, to be an important issue. Manufacturability issues can be included readily into the guided iteration methodology when doing parametric design of special purpose parts. Sometimes manufacturability can also be included in optimization methods. And manufacturing issues are incorporated naturally into design for robustness methods.

For special purpose parts, separate design for manufacturing (DFM) evaluations can be made for relative *processing* costs when parametric values have been proposed. (Recall that DFM evaluations for relative *tooling* costs could be made based only on configuration information.) Chapters 15 and 16 describe how to make processing cost estimates for injection molded, die cast, and stamped parts.

The parametric design methods presented in Part IV are most useful for special purpose and standard parts, standard modules, and small relatively simple assemblies (where all the attributes can be taken as a single set of attributes).

	Slot	Plates	Brackets
Cost	–	S	+
Weight	+	S	–
Weight Capacity	–	S	S
Temperature Range	–	S	+
Vibration	–	S	+
Maintenance	–	S	+
Safety	–	S	+
Modularity	+	S	S
Manufacturability	–	S	+
Aesthetic	S	S	+
Strength	–	S	S
Minimal Parts	S	S	S
Standard Parts	–	S	S
Results	2 +		7 +
	9 –		1 –
Best Configuration	Brackets		
Key	+ Good S Standard – Bad		

Figure 13.19 A Student Team's Evaluation Matrix for Design Project A.

For larger and more complex assemblies, practical generalized formal parametric design methods unfortunately do not yet exist for performing parametric design as a whole. Companies get the job done by their own ad hoc methods that have worked, more or less well, for them. Selected parts, components, and small assemblies get parameters assigned somehow, and then these get coupled as reasonably as possible to the others. Doing parametric design on entire large assemblies is an area of engineering design that requires more research before a general methodology can be proposed and tested.

Though we cannot deal as a whole with parametric design of huge and very complex products, we can nevertheless do a great deal with the parametric design of parts, components, and small assemblies. This is important. The larger products are made up of parts, components, and small assemblies that must be properly designed in their own right. And we have excellent methods for doing this that are the subject of Part IV.

References

[1] Deming, W. Edwards, *Out of the Crisis*, 1986. Quotation by permission of MIT Center for Advanced Engineering Study, Cambridge, MA.

[2] Fathaillal, A. "Feature-Based Representations to Support the Evaluation of Component Tolerances for Manufacturability" Master of Science Thesis, University of Massachusetts, Amherst, May 1992.

[3] Knight, W. A.and Poli, C. "A Systematic Approach to Forging Design", *Machine Design*, January 24, 1985.

Problems

13.1. The injection molded part shown as a configuration only in Figure P13.1 has three unsupported bosses as shown. As designed, for some functional reason the relative location of the tops of the bosses must be finely controlled with tolerances on D_1, D_2, D_3, and angle alpha. Moreover, it is also necessary that surface A be flat to standard tolerance.

Evaluate this tolerancing plan with the bosses unsupported, and again with the bosses supported. What redesign possibilities would you suggest be investigated?

13.2. The stamped part shown in Figure P13.2 as a configuration only is designed so that the two angles will require standard tolerance control. Moreover, the design requires that A and B, and surfaces C and D, be tightly controlled for parallelism.

Evaluate this tolerancing plan.

How might "springback" cause a manufacturing problem?

13.3. Figure P13.3 shows a configuration requirements sketch for a door guide for a pair of sliding wood doors. The guide must be able to be fastened to the floor and allow the doors to slide easily (into and out of the paper). The doors are hung so they do not need vertical support. These sliding doors are sold in several standard thicknesses from 3/4 inches to 1-3/8 inches, so it may be necessary to produce several guides -- or else make a guide assembly that can accommodate different thicknesses.

Develop several configurations for the door guide, and perform an overall evaluation of them using Pugh's or Dominic's method.

13.4. An inexpensive (selling price less than $2.00) stamped metal garden trowel as shown in Figure P-13.4 requires a handle. Develop and evaluate several configurations assuming the concept calls for an injection molded handle. Use Pugh's or Dominic's method.

13.5. A drywall inside corner trowel is a a thin, smooth stainless steel part folded to approximately a 90° angle. See Figure P13.5. A handle assembly is available that consists of a partly hollow wooden grip with a tapered metal cone fastened by friction to its smaller, open end. The task is to configure the design of the bracket(s) to connect the handle assembly to the trowel. The concept is a stamped metal bracket that will be spot welded to the trowel. The bracket is to be driven into the open end of the handle assembly. Develop several configurations for the connecting bracket(s) and perform an overall evaluation of them.

13.6 The concept is to have an injection molded spool large enough that a long extension cord can be wound up on it. The spool is to be mountable on a workbench. When the cord is wanted, it is simply pulled off as the spool turns. A crank arm is to be used to wind up the cord onto the spool. See Figure P13.6. Develop several configurations for the part or assembly that will support the spool above the bench. Evaluate your alternatives.

13.7. See Figure P13.7. Configure and evaluate several alternatives for connecting road signs to posts as shown.

13.8.Consider the design of tops and handles for plastic trash barrels. Configure a top and handle combination so that the handle can serve both for carrying the barrel and for clamping on the top for storage.

13.9. See Figure P13.8. Small commercial food mixers require a convenient way to add ingredients during mixing. Pouring them in directly is difficult because of the size and location of the mixer head interfere. Moreover, ingredients splashing out of the bowl during mixing is also a problem. A concept has been proposed as follows: There will be a two piece injection molded assembly of half-rings that fit around the top of the mixer bowl and interlock with each other to form a full ring. The ring thus formed can be put in place (or removed) without removing the bowl or the paddles, and it will serve as a splash guard. On the front of the forward ring, a chute can be molded in that serves as a kind of ramp down which ingredients can be added to the bowl (even during mixing) safely and easily. Configure and evaluate several alternatives for the two parts of the ring.

13.10. Explain how the "configuration" of an assembly and the "configuration" of a part are similar and how they are different. Use examples.

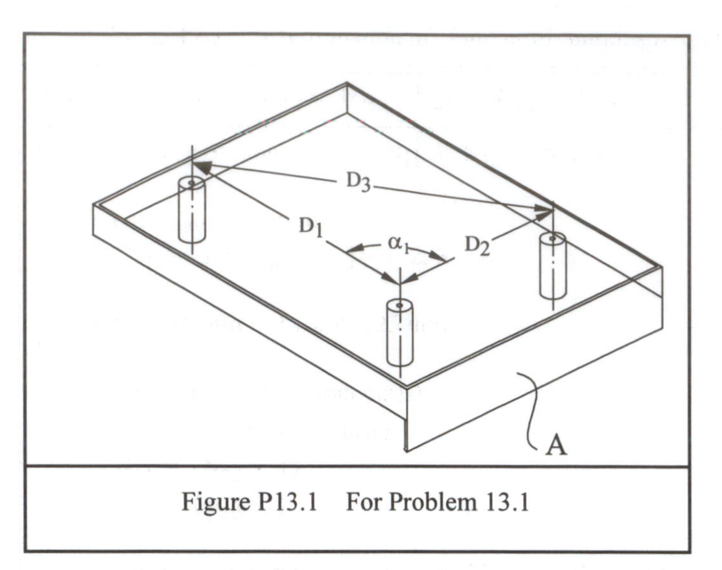

Figure P13.1 For Problem 13.1

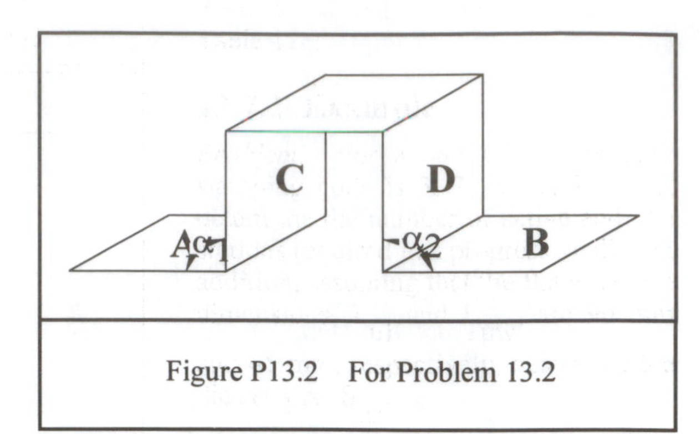

Figure P13.2 For Problem 13.2

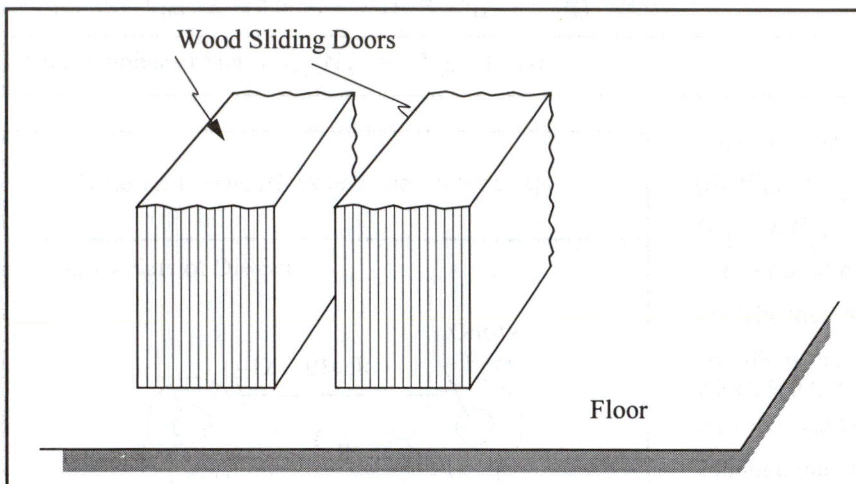

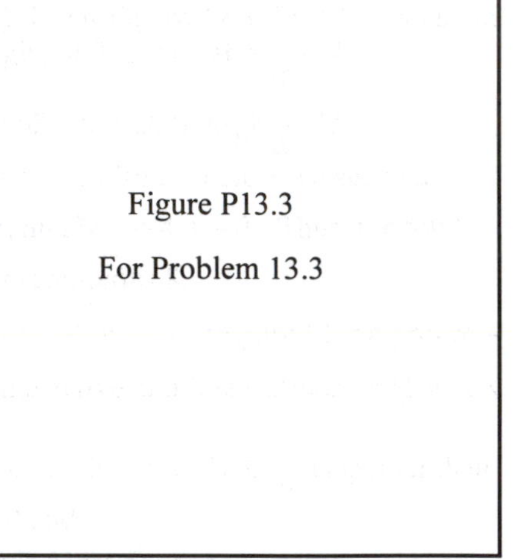

Figure P13.3

For Problem 13.3

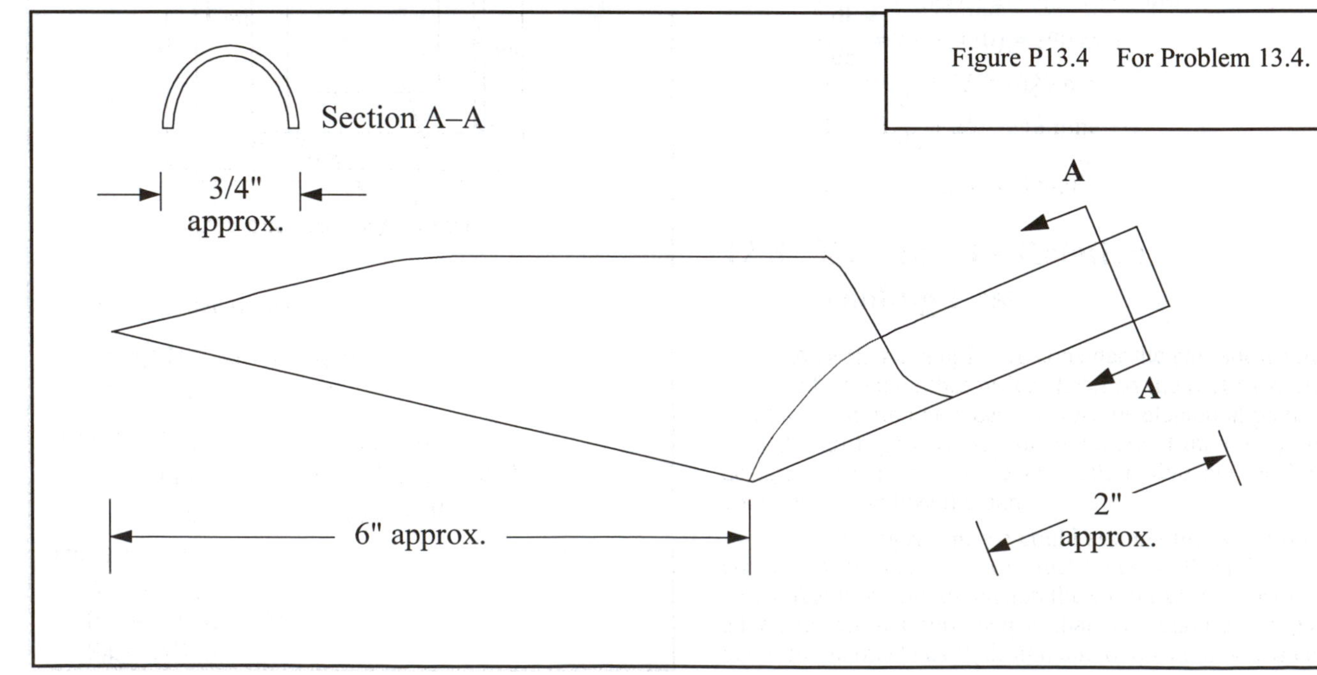

Figure P13.4 For Problem 13.4.

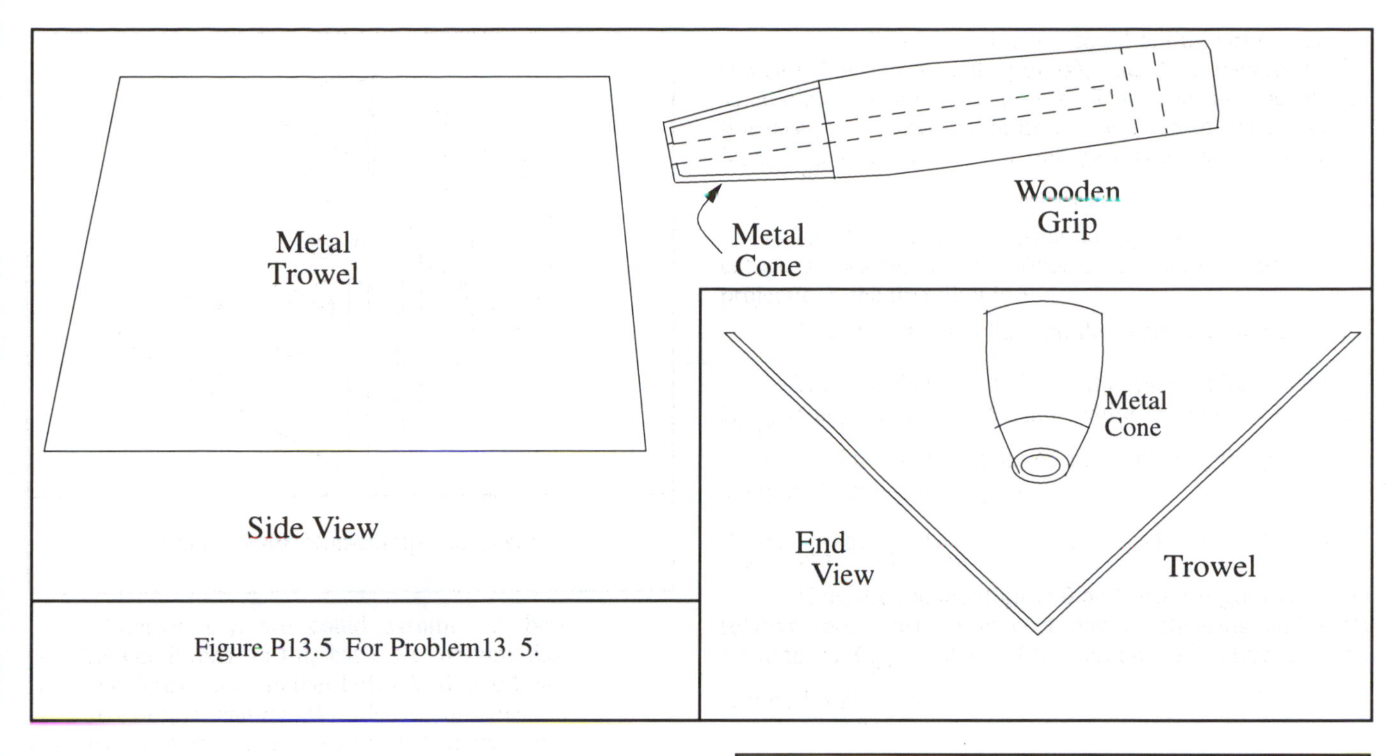

Figure P13.5 For Problem13. 5.

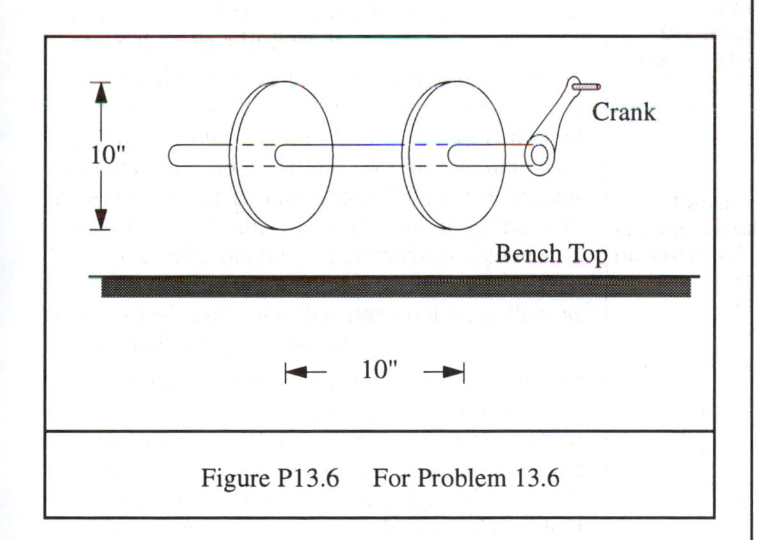

Figure P13.6 For Problem 13.6

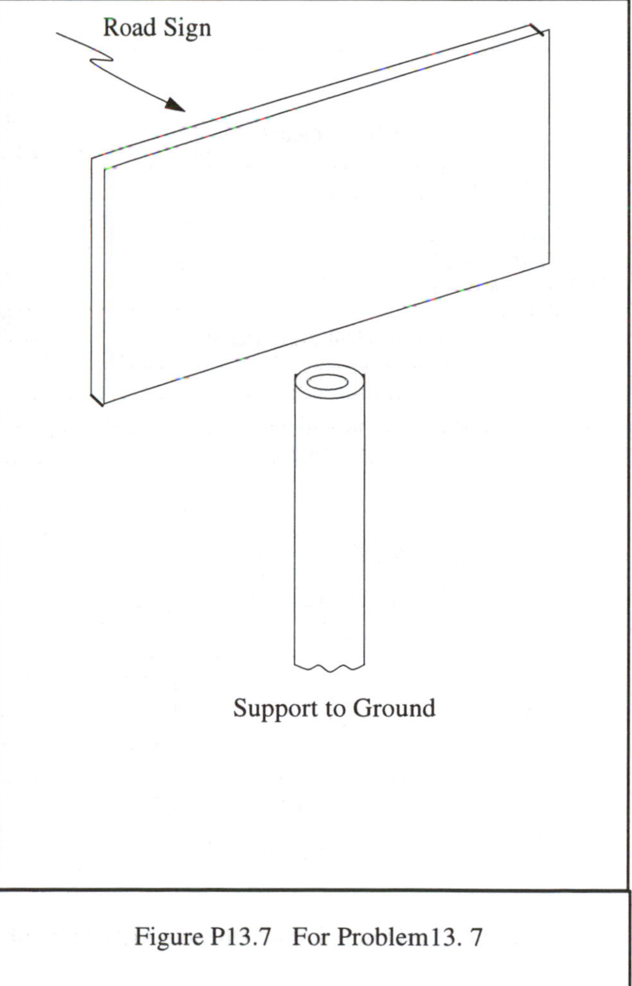

Figure P13.7 For Problem13. 7

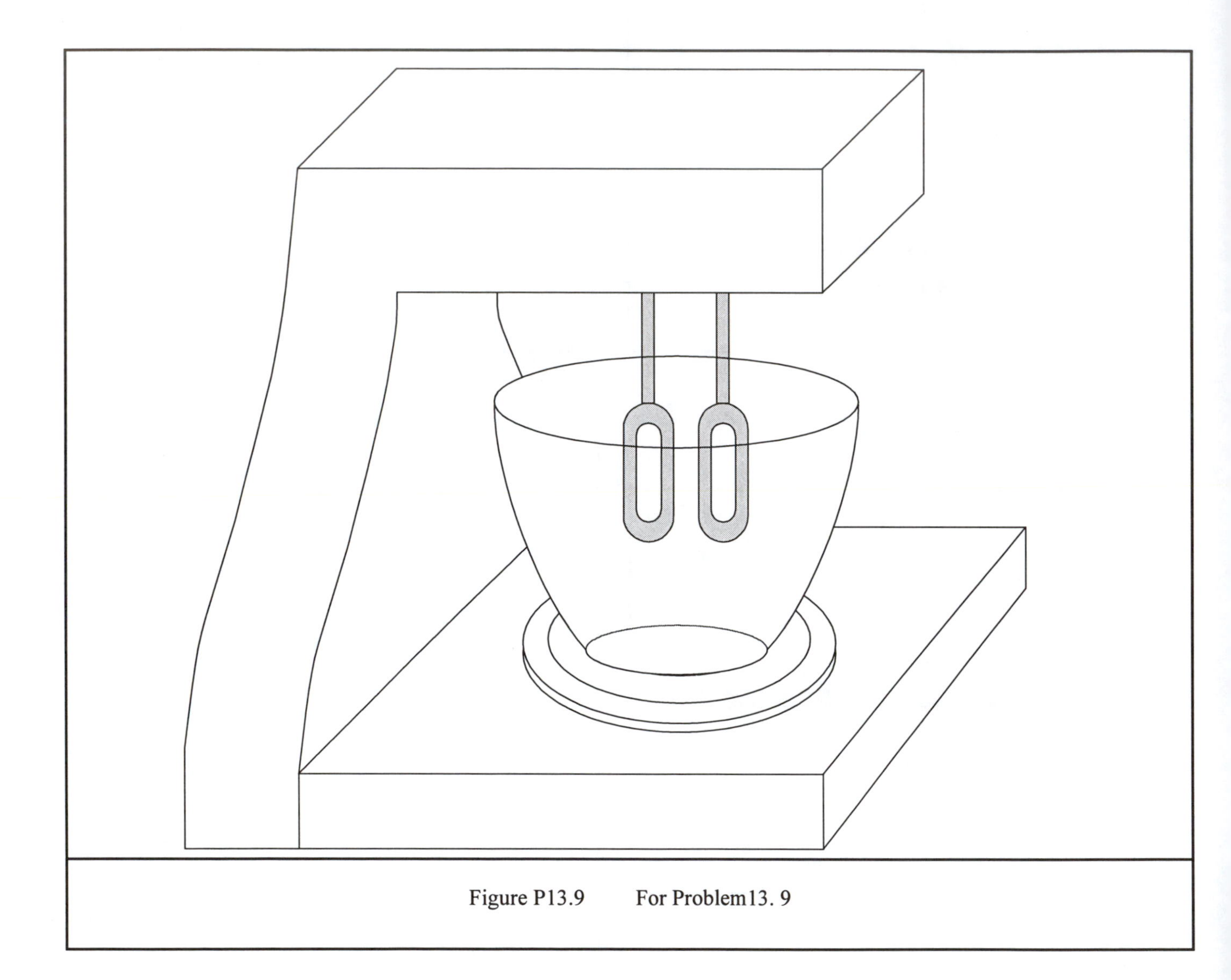

Figure P13.9 For Problem13. 9

Part IV Parametric Design of Components

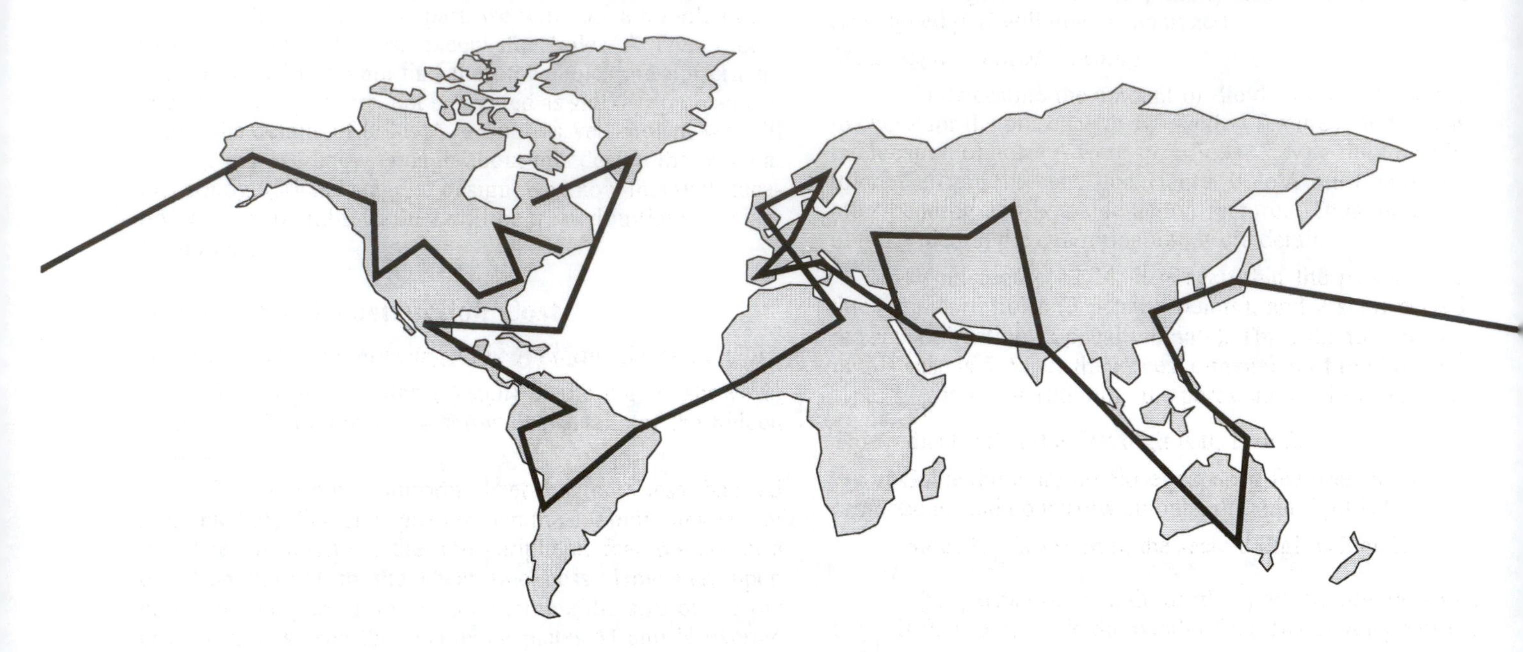

Chapter Fourteen

Introduction to Parametric Design

"The best safeguard is to have a map laid out in advance of the various routes you can take. This book is mine. You may very well have your own and a better one. The fatal thing is to have none."

Gordon L. Glegg
*The Development of Design(1981)**

14.1 Overview of Part IV

In Chapter 2, we outlined the *guided iteration* methodology for solving engineering design problems, and described three major types of engineering design problems: conceptual, configuration (of parts), and parametric. Chapters 4 through 9 deal with application of guided iteration to engineering conceptual design. Chapters 10 through 13 concern the configuration design of parts. Now in Part IV — Chapters 14 through 21 — we turn our attention to parametric design.

In solving conceptual and configuration design problems, there is little need for numerical computation. The reason is that descriptions of physical concepts (or embodiments) and configurations do not involve much numerical information. Moreover, evaluations at these levels do not require much numerical or quantitative analysis and computation. Except for the evaluation of relative tooling costs for special purpose parts at the configuration stage, evaluations of concepts and configurations are based on qualitative reasoning about physical principles and manufacturing processes.

In parametric design, however, numerical computations become much more important. The attributes identified at the configuration stage of parts become the design variables for parametric design, and their values must now be determined. These values are mostly, though not exclusively, numerical. In accomplishing this task, we encounter many more numbers, equations, and analyses. We also find that relative manufacturing processing costs (as distinguished from tooling costs) are sensitive to the values assigned, so that relative processing costs must now be considered along with functionality as a part of parametric design.

Application To All Components. The discussion of engineering conceptual design in Part II included all kinds of functional designed objects, including special purpose assemblies and all components. The discussion of configuration design in Part III, however, was focussed primarily on special purpose parts. (Remember, the decomposition of special purpose assemblies into functional sub-assemblies (i.e., their "configuration") is included within their conceptual design phase.) In considering parametric design, we can again expand the discussion to include not only special purpose parts, but also standard parts and standard assemblies as well. In other words, we deal here with the parametric design of all components. Most of the methods presented here can also apply to small special purpose assemblies; thus, only large special purpose assemblies are excluded.

The parametric design of large special purpose assemblies is excluded simply because there is (at this writing) no articulated methodology for carrying it out. Determining the attributes and their values of the many components of a complex special purpose assembly (with all its sub-assemblies and parts), has as yet no theoretical foundation sufficient to support development of a methodology. The process is accomplished in practice essentially by trial and error, somewhat like guided iteration, but without much guidance and little prescribed methodology. The people, teams, and teams

* Quotation from *The Development of Design* by Gordon L. Glegg, Copyright 1981, with permission of Cambridge University Press.

of teams that do it, do the best they can. They muddle through and they get the job done, often very well, but they operate in very poorly charted waters. Continuing research will eventually provide the methodology needed, but this will take more time.

In contrast, we have a number of powerful methods available for the parametric design of components and small assemblies. In part IV, we describe three of these methods:

- Guided Iteration (Chapters 17, 20, and 21),
- Optimization (Chapter 18), and
- Taguchi methods™* (Chapter 19).

We also describe DFM methods for evaluating the manufacturing processing costs for special purpose parts. Since these costs are quite sensitive to the values assigned to attributes, they must be considered a part of parametric design. The topics covered are:

- DFM methods for relative processing costs for injection molded and die cast parts (Chapter 15), and
- DFM methods for relative processing costs for stamped parts Chapter 16).

About this Chapter. In this chapter, we introduce and compare guided iteration, optimization, and Taguchi's method™. We also discuss some issues that are relevant to parametric design regardless of which method is used, including the important supportive role of engineering science and analysis. We argue that parametric design should be based, not merely on informal "experience" or "the seat of the pants", but on a design methodology, proper analyses, and specific quantitative knowledge of DFM.

* "Taguchi method" is a trademark of the American Supplier Institute, Detroit, Michigan.

14.2 The Goal of Parametric Design

14.2.1 Values for the Design Variables

The goal of parametric design is to find the values for the *design variables* that will produce the best possible design considering both function and manufacturing. A *design variable* is an attribute established at the configuration stage whose value is under control of the designer, though feasible or possible values may fall within some limits. The design variables for which values must be found may include dimensions, tolerances, materials, and others (for example, surface finish or color). Most attributes of concern will, however, have numeric values; that is, dimensions and tolerances. Thus, the primary concern of parametric design is with finding the values of numeric design variables.

We use the symbol x to represent the value of a design variable. In formulating parametric design problems, design variables are identified including:

- names, symbols, and units;
- range of allowed values; and
- whether they are continuous or discrete. If discrete, the

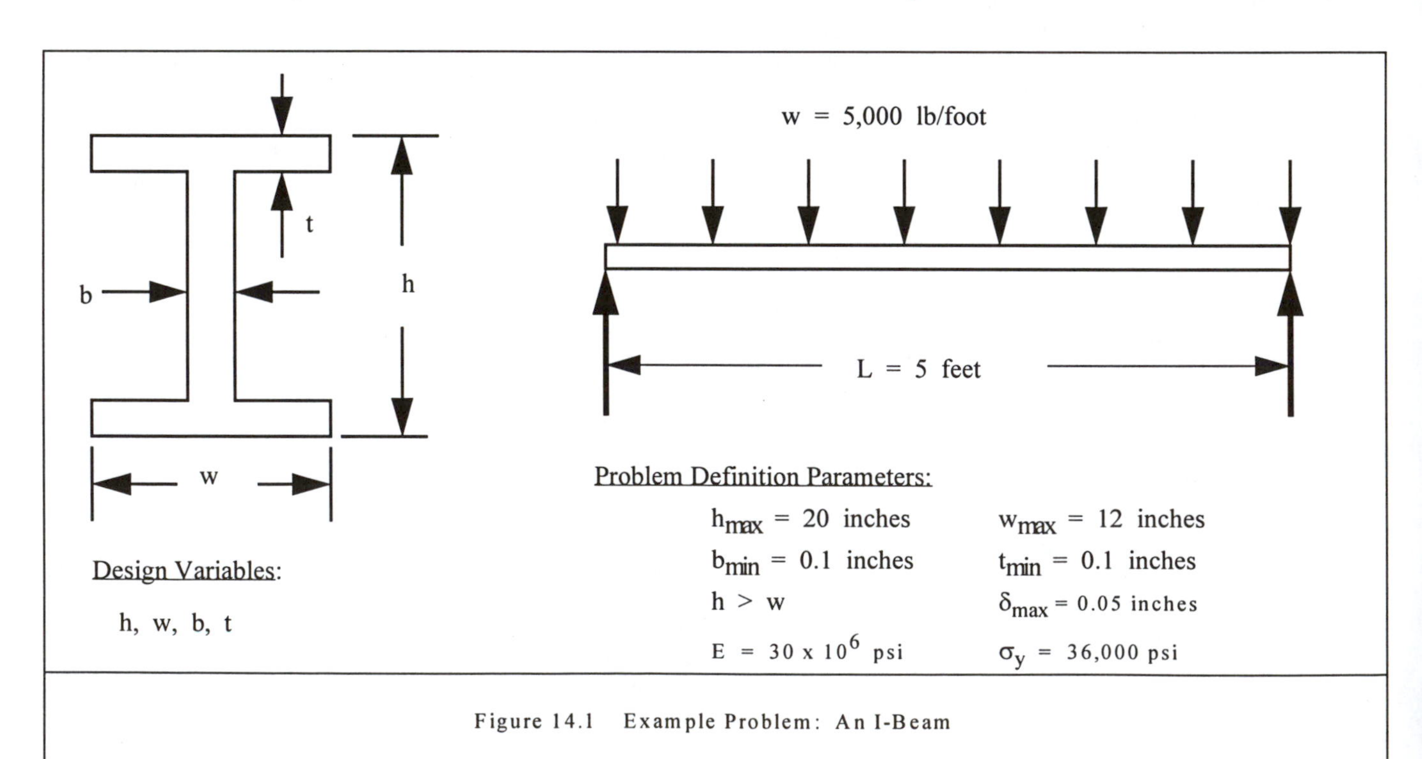

Figure 14.1 Example Problem: An I-Beam

permissible values are also included.

Information about limits and other conditions on the design variables is often available in the Engineering Design Specification.

I I I I I I I I I I I I*

14.2.2 Example: An I-Beam

As an example of identifying design variables, consider the case of an I-Beam. The attributes and hence the parametric design variables are (See Figure 14.1):

- beam height, h,
- beam width, w,
- web thickness, b,
- flange thickness, t,
- fillet radii, r (not shown in Figure 14.1), and
- material (a specific type of steel).

To keep this illustrative example short and simple, we will henceforth omit the radii of the fillets as design variables, and we will also assume that a specific material has been selected. The design variables in this example are thus all numeric. After illustrating how to solve such all-numeric problems, we will discuss the solution of problems with non-numeric design variables.

For each of the design variables, we also indicate the ranges of allowed values and whether the variables are continuous or discrete. For this example, we assume the following:

Design Variable	Symbol	Allowed Range	Type
beam height	h	< 20 inches	continuous
beam width	w	< 12 inches	continuous
web thickness	b	> 0.1 inches	continuous
flange thickness	t	> 0.1 inches	continuous

In most real situations, a steel I-Beam would not be designed as a special purpose part as we are doing in this example. Instead, a beam would be selected from the set of standard I-Beam sizes manufactured and carried in stock by suppliers. However, so that we can introduce and illustrate the guided iteration process for parametric design in a very familiar problem area, we will assume that this I-Beam is to be specially made.

This example will be continued as we consider methods of solution for parametric design problems. First, however, there are some other preliminary issues to discuss.

*I I I I I I I I I I I I

* Strings of symbols will be used at times in the book to indicate the beginning and end of an example, especially when the example is continued in non-contiguous fashion.

14.3 The Role of Analysis in Parametric Design

The solution to parametric design problems, whatever method is used, is generally heavily dependent on analyses; that is, on solutions to equations or sets of equations that describe or simulate the expected physical behavior of proposed trial designs. In guided iteration, for example, analysis equations can provide some of the essential information needed to support the evaluation step. Optimization methods require analyses. And as we shall see, Taguchi™ methods make use of analyses, simulations, or experimentation to provide evaluation information.

Earlier in this book, we stressed that analysis *serves* design. This is especially true at the parametric stage. Without analytical procedures to provide estimates of the performance of trial designs, designers would have no way to get the critical quantitative information needed to evaluate those designs. Without such information, designing at the parametric stage would be, in effect, a guessing game. This is called "designing blind", and it is *not* fun. It forces designers either to be too conservative or to take unknown risks, or both. It follows also that sloppy analysis leads to sloppy design. An incorrect analytical result can lead directly to failed parts or products. (You certainly hope, don't you, that bridge or aircraft designers are basing their parametric designs on correct stress and vibration analyses?)

It should be noted, however, that an analysis of a design cannot be performed until there is a design to analyze. That is, analysis cannot be used to create a design. Designers employing a design process create trial designs, which are then analyzed. Then the analysis results are evaluated in relation to the design goals, and the design redesigned and re-analyzed as needed.

Engineering students learn to do analysis as they solve textbook problems. The problems help students learn about the physical principles in engineering science courses like statics, strength of materials, dynamics, thermodynamics, fluid mechanics, etc. The problems in these textbooks almost always start with the design of something as a given. Many students also learn finite element methods of analysis. Unfortunately, it is not especially critical in an academic course that problems get solved correctly; only homework or exam grades are affected. But in a real design situation, if there are analytical errors, the design may fail causing possibly serious economic or human injury. This is the reason it is so important for engineering students to learn the basic physical principles, to learn analytical methods, and to learn to be accurate and reliable in using them.

Another problem with textbook analysis problems from a design viewpoint is that usually all the engineering assumptions are given; that is, the models to be analyzed are provided by the problem statements. It is given, for example, that materials are "perfectly" elastic, walls are adiabatic, fluids are frictionless, processes are isothermal, and that, say, beam theory applies. This is not the case in a design situations. Designers, and their analysis expert helpers, must make the right assumptions and develop proper models so that the results from the idealized analytical models are relevant to the real design.

At the parametric design level, there are some wonderful and powerful analysis procedures available to serve design. Modern design engineers need to know what these methods can do, and how to make use of them either themselves or with the aid of analysis specialists. However, and this is an important point, *the evaluation and decision making that constitutes the design process is independent of the nature of the analytical procedures used in their support.* A trial structure design may be analyzed by beam theory or by finite element procedures; the *process* of evaluation and design decision making is the same. A trial heat exchange component may be analyzed by computer simulation or by conventional methods; the *process* of evaluation and design decision making is the same.

This book is primarily about that common design process involving evaluation and decision making. Nevertheless, in examples, we include the analytical procedures needed to support the evaluations. But the particular analytical method selected for these examples is not important to the design process itself. If a finite element analysis is more accurate, we might get a better design, but the design process would be the same. For this reason, and to keep the presentation of the process itself as clear as possible, we have not included finite element methods in the parametric design examples. Readers should note, however, that such methods are important analytical procedures that can be used to great advantage in support of the parametric design process.

To illustrate this point more concretely, we refer back to Figure P11.5. Consider the case where there are significant normal forces on the two bosses on the vertically sloping plate on the part shown. Given a trial design (plate thicknesses, corner radii, material, etc.), we might analyze the resulting stresses and deflection by using beam theory (modelling the plate as a cantilever beam, say), by plate theory, or by a finite element method. The results of any of these analyses would give us quantitative information about stresses and deflections. That information, together with DFM information, would then be the basis for evaluation and redesign. No doubt, if properly done, the finite element results would be more accurate in this case (and in most cases where the geometry is complex), but the design process is the same regardless of which analysis method is used.

Another point to be made about analysis procedures in design is this: they can sometimes be extremely expensive of time and money to perform. Therefore, designers must know how accurate and/or reliable they really need the analyses to be in order for them to support properly the design decisions that must be made.

In the design of one extruded component, an analytically oriented engineer spent a great deal of time and money doing finite element heat transfer analysis on a part design. The result of designing based on the heat transfer analysis was a design that could in no way be actually extruded. The design was limited by manufacturing concerns, not by thermal analysis. If the manufacturing limitations had been recognized at the outset, a simple and inexpensive heat transfer analysis would have been more than sufficient to insure an optimum design.

In another case, however, the designer of an extruded frame element forgot to evaluate the performance of a design for twisting, and so no torsional analysis was performed. The result was a frame that was inadequate due to twisting when loaded. In this case, the detailed finite element analysis was needed.

The moral: Analysis serves design. Since several kinds of analyses are generally available, designers have to know what kinds of analyses to perform, when to perform them, how to perform them properly and accurately, and how to use the results to support the design process tasks of evaluation, decision making, and iterative redesign.

14.4 The Role of DFM in Parametric Design

As we learned in Chapters 11 and 12, the relative tooling cost of an injection molded, die cast or stamped part can be determined during the configuration stage of a design. We need know only the approximate dimensions of the part and whether or not certain features, such as bosses, ribs, holes, etc. are present, and if present, their approximate location and orientation. For example, although it is unlikely one would injection mold or die cast an I-Beam, let us assume that the I-Beam example discussed in section 14.2 is such that it can be die cast of aluminum. From the configuration of an I-Beam, we know at the configuration stage of design that regardless of the specific dimensions of the web and flanges, the part would have a planar dividing surface, low cavity detail, and no undercuts. With this information (we know from the information presented in Chapter 11) that the relative tooling cost for this part would depend primarily on the length of the beam, and for that length there is little that need be, or can be, done to the part geometry to reduce tooling cost.

Once we arrive at the parametric design stage, we begin to assign exact dimensions to the I-Beam. Initially, the dimensions may be determined via a functionality analysis. For example, we may use the fourth order beam equation to determine the maximum deflection of the beam, and based on the maximum deflection, we may alter the beam dimensions so that functionality requirements can be met with minimum material. In the case of injection molding and die casting, we know from the preliminary DFM Advisors presented in Chapter 3 that once we assign wall thicknesses to the web and flanges, we will affect the machine cycle time of the part. In the next chapter, we will also see that as the cycle time increases, the processing cost also increases. Thus, while functionality issues may point to the desirability of a large flange thickness, manufacturability issues may dictate the desirability of a small flange thickness.

In the next two chapters we will introduce expanded versions of the DFM Advisors presented in Chapter 3. We will show, for example, that the flange thickness and the fillet radius both effect the cycle time for an injection molded I-Beam. The flange thickness has a direct effect on the machine cycle time while the fillet radius, r, affects the ability to maintain tolerances. Consequently both influence part yield and the effective cycle time. Large fillet radii will make tolerances easier to maintain, and thereby increase part yield. Smaller radii have the opposite effect. In fact, the use of small fillet radii may require the use of gusset plates in order to maintain a reasonable part yield. In this particular case, the use of large fillet radii may be desirable from both a functionality point of view (stress concentrations will be reduced) and from a manufacturing point of view.

The point here is that, at the parametric design stage, the part and feature dimensions will effect *both* the functionality of the part and its manufacturability. Consequently, design for functionality and DFM must be carried out concurrently.

For ease of illustration in this section, we have used a die cast I-Beam as an example of a part to be designed to satisfy both functionality and DFM. In general, for special purpose parts, the geometry of the part will be more complex than that of the I-Beam, and the process used will more likely be either injection molding or stamping. The role of DFM in parametric design, however, remains the same. Of course, for more complex parts the functionality analysis and the DFM analysis may be more difficult to carry out; a finite element analysis may be required. The geometry of the part may be such that ribs and bosses may be present, and the layout and dimensions of these projections can have a significant influence on processing costs (and even on tooling costs in the case of stamping). Functionality may dictate the need for tall, thick bosses unsupported by gusset plates, while DFM may dictate the desirability of shorter thinner bosses supported by gusset plates. Designers must have the knowledge and design methodology to make the proper trade-offs, considering both functionality and DFM.

14.5 Tolerances at the Parametric Stage of Design

14.5.1 Introduction

In Chapters 10 and 13, we discussed tolerances related to part design at the configuration stage. In this section, we extend that discussion to the parametric design of parts. There is also additional discussion of tolerances at the parametric stage in Chapters 15 and 16.

As we noted in Chapter 1, there is as yet little research of a formal or systematic nature to help designers assign tolerances. Research in tolerance synthesis is on-going, however, and designers should watch the research and trade literature carefully for new results as they are published.

In Chapter 19, we will discuss the application of Taguchi methods™ to tolerance design. However, that discussion is essentially process-independent except that knowledge of commercial or standard tolerances for the process involved is needed.

In this section, we present tolerance information related to manufacture of parts in several processes. Readers should understand that processes are being continuously improved so that the approximate data presented may well get outdated with time. Moreover, process capability varies from machine to machine, and from firm to firm. Thus the figures presented here should be considered as approximate only, and designers should certainly check with manufacturers for more accurate and timely values.

We do not discuss the relationship between part tolerances and the stack up issues that occur when parts are assembled into sub-assemblies and products. Here we discuss only those issues related to tolerances for individual parts.

The following discussion largely assumes that the design of a part is given. Remember, however, that the point of having and using such information is not only to support estimation of what tolerances can be held economically, but also to support redesign of parts so that specified tolerances are more easily held.

14.5.2 Tolerances in Injection Molding

Tolerances of injection molded parts can be fairly closely controlled, but certainly not as closely as in machined parts. Tolerances depend on the ability of the molding machines to maintain constant molding conditions (e.g., temperatures, pressures, etc.) and on the sensitivity of the material to any process variations. That is, given a design, tolerances depend in part on the equipment and in part on the material being molded. Variations in dimensions can also be created by non-uniform shrinkage which may vary with the direction of resin flow in the mold.

Unless demanded by functionality, specifying tolerances can be unnecessarily costly. Tolerances increase inspection costs, and may require secondary finishing. Tight tolerances are especially to be avoided unless absolutely necessary. Because of shrinkage and warpage that is difficult to predict precisely, mold dimensions are hard to determine. Experimentation with molds may be required, as well as possible adjustment of mold sizes using metal sprays.

The Society of the Plastic Industry publishes industry standards for tolerances in injection molded parts [1]. Figures 14.2 and 14.3 show recommended "production tolerances at the most economical level" for two different materials, one a thermoplastic and one a thermoset. The reference(which is updated regularly) contains similar data for many other materials. There is additional discussion of tolerances in injection molding in Section 15.9.2.

14.5.3 Tolerances in Die Casting

As in other processes, achieving a large number of tolerances, especially if some are tighter than standard, can be both costly and time consuming. Specify a tolerance only on those dimensions that require one for functionality. If possible, expect and allow some other dimensions to be even out of standard range.

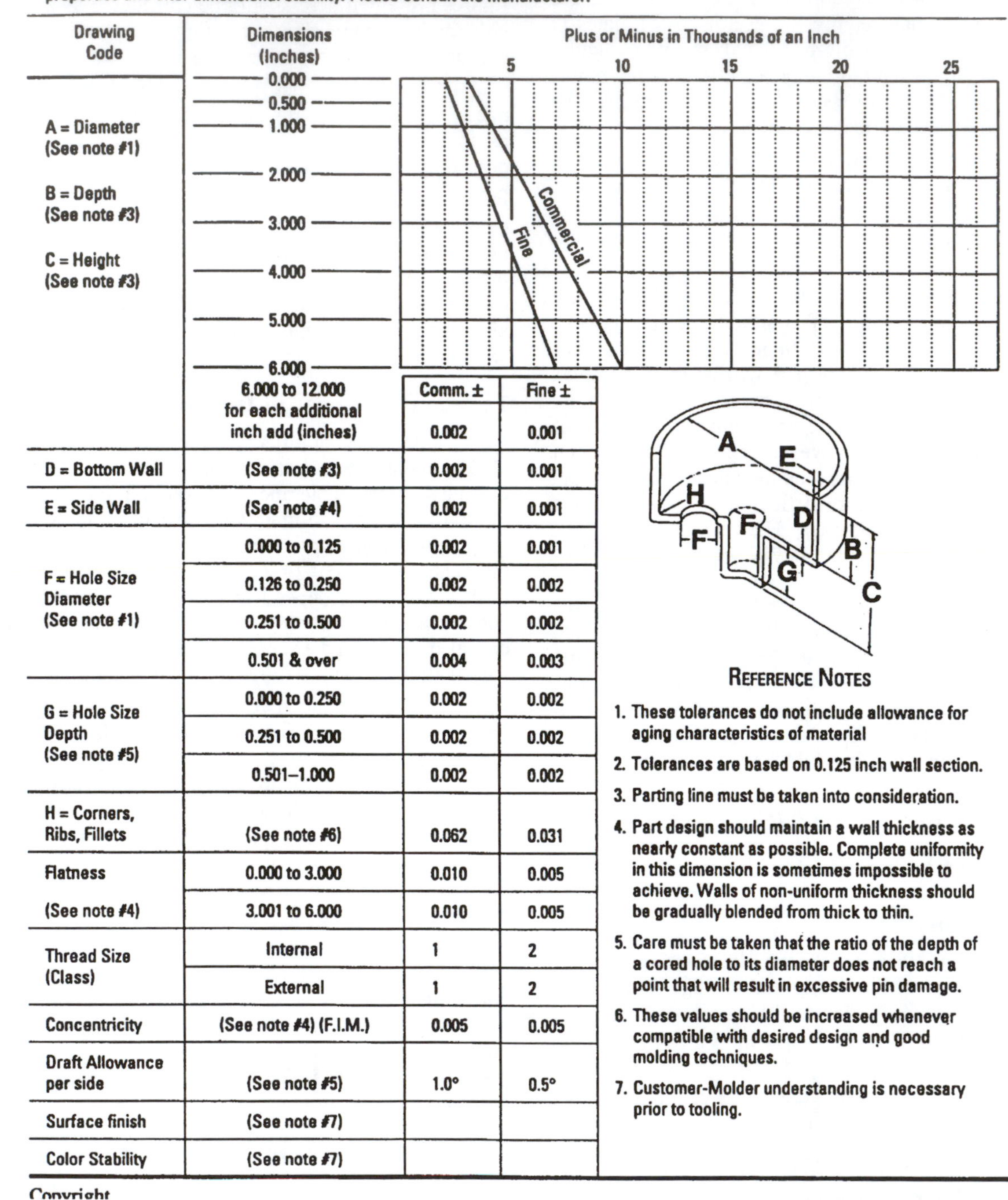

Standards & Practices of Plastics Molders

Material Alkyd (Thermoset)

Note: The *Commercial* values shown below represent common production tolerances at the most economical level. The *Fine* values represent closer tolerances that can be held but at a greater cost. Any addition of fillers will compromise physical properties and alter dimensional stability. Please consult the manufacturer.

Drawing Code	Dimensions (Inches)	Comm. ±	Fine ±
	6.000 to 12.000 for each additional inch add (inches)	0.002	0.001
D = Bottom Wall	(See note #3)	0.002	0.001
E = Side Wall	(See note #4)	0.002	0.001
F = Hole Size Diameter (See note #1)	0.000 to 0.125	0.002	0.001
	0.126 to 0.250	0.002	0.002
	0.251 to 0.500	0.002	0.002
	0.501 & over	0.004	0.003
G = Hole Size Depth (See note #5)	0.000 to 0.250	0.002	0.002
	0.251 to 0.500	0.002	0.002
	0.501–1.000	0.002	0.002
H = Corners, Ribs, Fillets	(See note #6)	0.062	0.031
Flatness	0.000 to 3.000	0.010	0.005
(See note #4)	3.001 to 6.000	0.010	0.005
Thread Size (Class)	Internal	1	2
	External	1	2
Concentricity	(See note #4) (F.I.M.)	0.005	0.005
Draft Allowance per side	(See note #5)	1.0°	0.5°
Surface finish	(See note #7)		
Color Stability	(See note #7)		

Reference Notes

1. These tolerances do not include allowance for aging characteristics of material
2. Tolerances are based on 0.125 inch wall section.
3. Parting line must be taken into consideration.
4. Part design should maintain a wall thickness as nearly constant as possible. Complete uniformity in this dimension is sometimes impossible to achieve. Walls of non-uniform thickness should be gradually blended from thick to thin.
5. Care must be taken that the ratio of the depth of a cored hole to its diameter does not reach a point that will result in excessive pin damage.
6. These values should be increased whenever compatible with desired design and good molding techniques.
7. Customer-Molder understanding is necessary prior to tooling.

Figure 14.2 Tolerance Standards for Alkyd, a Thermoset. Reprinted from *Standards and Practices of Plastic Molders*, 1993, with permission of the Society of the Plastics Industry, Inc., Washington, D.C.

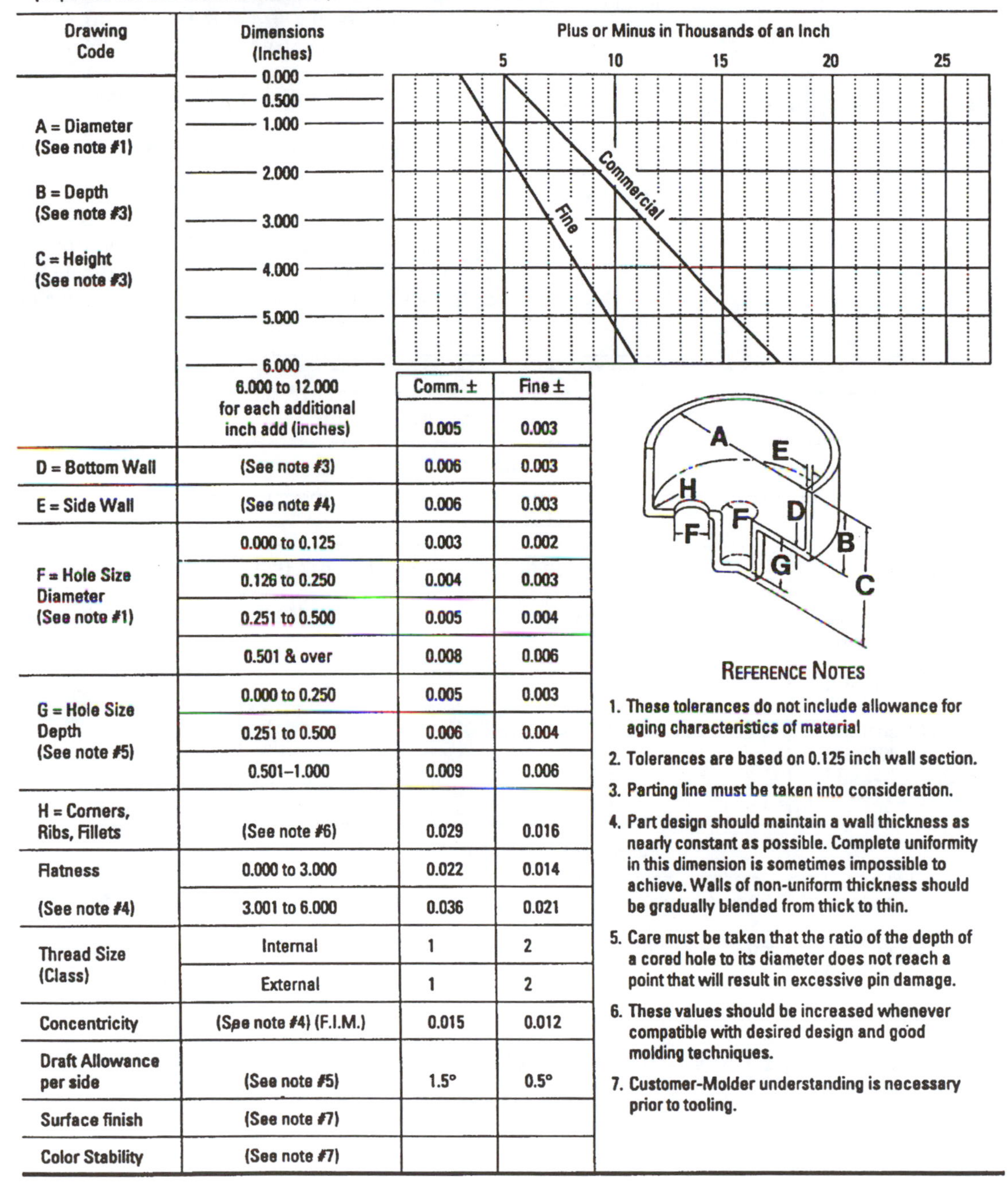

Standards & Practices of Plastics Molders

Material
Polypropylene
(PP)

Note: The *Commercial* values shown below represent common production tolerances at the most economical level. The *Fine* values represent closer tolerances that can be held but at a greater cost. Any addition of fillers will compromise physical properties and alter dimensional stability. Please consult the manufacturer.

Drawing Code	Dimensions (Inches)	Comm. ±	Fine ±
	6.000 to 12.000 for each additional inch add (inches)	0.005	0.003
D = Bottom Wall	(See note #3)	0.006	0.003
E = Side Wall	(See note #4)	0.006	0.003
F = Hole Size Diameter (See note #1)	0.000 to 0.125	0.003	0.002
	0.126 to 0.250	0.004	0.003
	0.251 to 0.500	0.005	0.004
	0.501 & over	0.008	0.006
G = Hole Size Depth (See note #5)	0.000 to 0.250	0.005	0.003
	0.251 to 0.500	0.006	0.004
	0.501–1.000	0.009	0.006
H = Corners, Ribs, Fillets	(See note #6)	0.029	0.016
Flatness	0.000 to 3.000	0.022	0.014
(See note #4)	3.001 to 6.000	0.036	0.021
Thread Size (Class)	Internal	1	2
	External	1	2
Concentricity	(See note #4) (F.I.M.)	0.015	0.012
Draft Allowance per side	(See note #5)	1.5°	0.5°
Surface finish	(See note #7)		
Color Stability	(See note #7)		

Reference Notes

1. These tolerances do not include allowance for aging characteristics of material
2. Tolerances are based on 0.125 inch wall section.
3. Parting line must be taken into consideration.
4. Part design should maintain a wall thickness as nearly constant as possible. Complete uniformity in this dimension is sometimes impossible to achieve. Walls of non-uniform thickness should be gradually blended from thick to thin.
5. Care must be taken that the ratio of the depth of a cored hole to its diameter does not reach a point that will result in excessive pin damage.
6. These values should be increased whenever compatible with desired design and good molding techniques.
7. Customer-Molder understanding is necessary prior to tooling.

Figure 14.3 Tolerance Standards for Polypropylene. Reprinted from *Standards and Practices of Plastic Molders*, 1993, with permission of the Society of the Plastics Industry, Inc., Washington, D.C.

The manufacturing difficulties in die casting are due to shrinkage that is difficult to predict exactly, to thermal expansion and contraction of the mold, to variations in processing conditions, and so on. Tighter than standard tolerances are sometimes achievable after experimentation by molders, but of course are more costly.

Figures 14.4 through 14.6 list standard tolerances for die cast parts in different materials. There is additional discussion of tolerances in die casting in Chapter 15, Section 15.22.

14.5.4 Tolerances in Stampings

The same admonitions to minimize the number of required tolerances and to avoid tight tolerances as much as possible apply to stamping as well as to the other processes described above. Figure 14.7 lists recommended standard tolerances for wipe formed parts.

14.5.5 Tolerances in Aluminum Extrusions

A good discussion of design issues in aluminum extrusions is found in [2]. A table of standard tolerances is reproduced here in Figure 14.8.

14.5.6 Tolerances in Forgings

Tolerances in forgings are influenced by shrinkage and warpage, and on the size of the forging as well. Dimensions across the parting plane are more difficult to control. A good discussion of forging tolerances can be found in and complete and detailed information is available in [4, 5, 6].

14.5.7 Comparing Tolerances in Various Processes, Including Machining

Tolerances in various machining processes are found in many references such as [7]. For comparison purposes, Figures 14.9 [8] and 14.10 [9] show approximate tolerances ranges for a number of machining processes as well as some

	Die-casting alloy, mm (in)			
	Zinc	Aluminum	Magnesium	Copper
For critical dimensions				
Dimensions to 25 mm (1 in)	±0.08 (±0.003)	±0.10 (±0.004)	±0.10 (±0.004)	±0.18 (±0.007)
Each additional 25 mm over 25 to 300 mm (each additional in over 1 in to 12 in)	±0.025 (±0.001)	±0.038 (±0.0015)	±0.038 (±0.0015)	±0.05 (±0.002)
Each additional 25 mm over 300 mm (each additional in over 12 in)	±0.025 (±0.001)	±0.025 (±0.001)	±0.025 (±0.001)	
For noncritical dimensions				
Dimensions to 25 mm (1 in)	±0.25 (±0.010)	±0.25 (±0.010)	±0.25 (±0.010)	±0.35 (±0.014)
Each additional 25 mm over 25 to 300 mm (each additional in over 1 in to 12 in)	±0.038 (±0.0015)	±0.05 (±0.002)	±0.05 (±0.002)	±0.08 (±0.003)
Each additional 25 mm over 300 mm (each additional in over 12 in)	±0.025 (±0.001)	±0.025 (±0.001)	±0.025 (±0.001)	

Figure 14.4 Recommended Tolerances for Die-Casting Dimensions Determined By Cavity Dimensions in Either Half of the Die. From *Handbook of Product Design for Manufacturing* by James G. Bralla, Copyright 1986, reprinted by permission McGraw-Hill Book Company [3].

Projected area of die casting, cm² (in²)	Additional tolerances, die-casting alloy, mm (in)			
	Zinc	Aluminum	Magnesium	Copper
Up to 300 (50)	±0.10 (±0.004)	±0.13 (±0.005)	±0.13 (±0.005)	±0.13 (±0.005)
300–600 (50–100)	±0.15 (±0.006)	±0.20 (±0.008)	±0.20 (±0.008)	
600–1200 (100–200)	±0.20 (±0.008)	±0.30 (±0.012)	±0.30 (±0.012)	
1200–1800 (200–300)	±0.30 (±0.012)	±0.40 (±0.015)	±0.40 (±0.015)	

***Parting-line tolerances, in addition to linear-dimension tolerances, must be provided when the parting line affects a linear dimension. The above tolerances are to be added to linear tolerances worked out for a dimension as provided in Table 5.4-7.**

Figure 14.5 Parting Line Tolerances in Addition to Linear Dimension Tolerances, Based on Single Cavity Die. From *Handbook of Product Design and Manufacturing* by James G. Bralla, Copyright 1986, by permission of McGraw-Hill Book Company [3].

other processes (e.g., injection molding). The tolerances ranges shown in Figure14.9 apply to a 1 inch dimension.

The data shown in Figure 14.10 illustrate an important point that we have not yet made in our discussions of tolerances. It is that achieving tighter tolerances and achieving smoother surfaces both require greater attention to manufacturing processing, and hence tend to go hand in hand. That is, producing tighter tolerances generally also produces smoother surfaces, and vice-versa. It must be noted, however, that manufacturing and equipment capabilities vary, so designers must check on the actual capabilities available to produce the parts being designed.

14.5.8 Fits

When parts must fit together (a common example is when a shaft must fit into or through a hole) the tolerances on both parts must be considered simultaneously. If the largest allowed shaft will not readily fit into or move appropriately in the smallest allowed hole, assembly may not be possible or the required function may be impaired. On the other hand if the smallest allowed shaft fits too sloppily into the largest allowed hole, function may be impaired in some of the shaft-hole combinations that will be assembled.

Projected area of die-casting portion, cm² (in²)	Additional tolerances, die-casting alloy, mm (in)			
	Zinc	Aluminum	Magnesium	Copper
Up to 60 (10)	±0.10 (±0.004)	±0.13 (±0.005)	±0.13 (±0.005)	±0.25 (±0.010)
60–120 (10–20)	±0.15 (±0.006)	±0.20 (±0.008)	±0.20 (±0.008)	
120–300 (20–50)	±0.20 (±0.008)	±0.30 (±0.012)	±0.30 (±0.012)	
300–600 (50–100)	±0.30 (±0.012)	±0.40 (±0.015)	±0.40 (±0.015)	

***Moving-die-part tolerances, in addition to linear-dimension tolerances and parting-line tolerances, must be provided when a moving die part affects a linear dimension.**

Figure 14.6 Moving-Die-Part Tolerances, in Addition to Linear Dimension Tolerances. From *Handbook of Product Design and Manufacturing* by James G. Bralla, Copyright 1986, by permission of McGraw-Hill Book Company [3].

Standard practices for designing fits have been established. A few are presented here to introduce readers to the issues and types of numbers involved. For complete details, see references [10]. In general, fits will have one or more of three basic functions: allowing constrained motion between the mating parts; providing for location of one of the parts; or providing for holding the parts together under stress. The main types of fits used for these functions are, respectively, running (or sliding) fits, locational fits, and force fits. Within each of these types of fits, there are a number of classes depending on the clearances or interferences allowed by the tolerances assigned.

Running fits allow for sliding or turning between the parts. In running fits, the tightest class fit is called RC1. For a 1 inch nominal diameter in this class, the hole diameters produced may vary from 1.0000 to 1.0004 inches. The shaft sizes produced may vary from 0.99945 to 0.9997 inches. In this case, therefore, note that the smallest allowed clearance between the shaft and the hole in an assembly is 0.0003 inches, and the largest is 0.00095 inches. These fits will result in no noticeable "play" between the parts, and will provide for quite accurate location. The shaft will slide or turn, but not easily. Other classes of running fits do allow for easy relative motion. See also Figure 14.11.

Locational fits provide for less clearance than running fits, and even allow for some overlap; that is, they allow the possibility that some shafts produced may be slightly larger than some of the holes produced.

Force fits ensure tolerances such that all the shafts produced are larger than the holes produced so that assembly requires that shafts be forced into the holes. For example, in a Class FN1 force fit, for a one inch nominal hole, the hole diameter may vary from 1.000 inches to 1.0005 inches, and the shaft may vary from 1.0008 inches to 1.0012 inches. Note that the smallest amount of interference (the amount by which the shaft is larger than the hole) is thus 0.0003 inches, and the largest interference is 0.0012 inches. These fits will require a small amount of force to assemble, and will provide for light duty permanent assemblies.

Much more interference can be specified in force fits. In Class FN5, for example, the smallest amount of interference for a one inch shaft is 0.0013 inches and the largest allowed is 0.0033 inches. These amounts of interference will require heavy forces to assemble, or else assembly by shrink fitting. Parts with this much interference can support heavy stresses without relative movement. See also Figure 14.12.

	Closest tolerance	Normal production tolerance
Angle of bend of formed legs		
Special dies	**±½°**	**±1°**
Press brake or other universal tooling	**±1°**	**±2°**
Bending brake	**±2°**	**±3°**
Height or length of bent leg (dimension *H*)	**±0.2 mm (0.008 in)**	**±0.4 mm (0.016 in)**
Height of extruded holes	**±0.2 mm (0.008 in)**	**±0.3 mm (0.012 in)**
Deviations from flatness	**±0.3%**	**±0.5%**

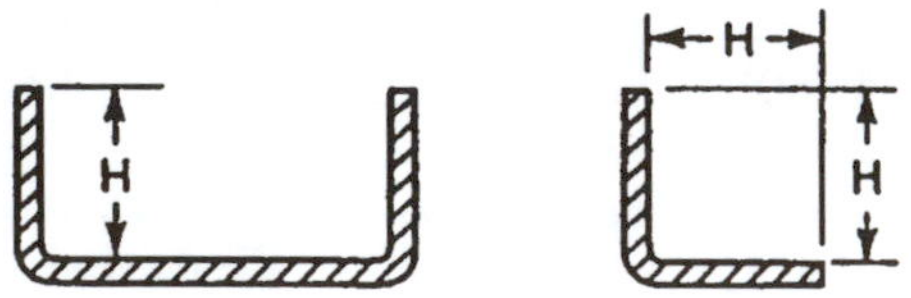

Figure 14.7 Recommended Tolerances for Dimensions of Wipe Formed Stampings.
From *Handbook of Product Design and Manufacturing* by James G. Bralla,
Copyright 1986, by permission of McGraw-Hill Book Co.

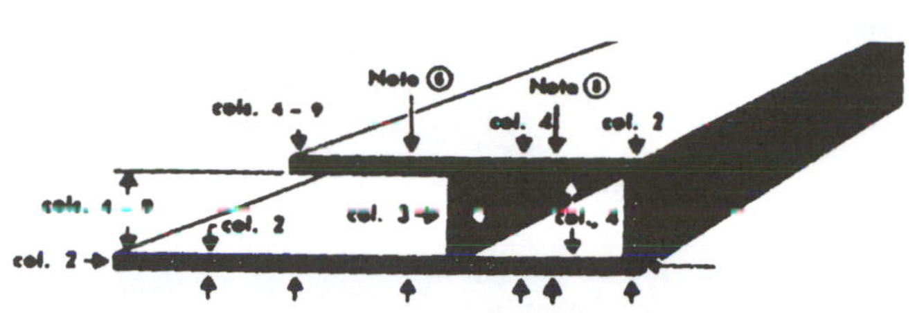

TOLERANCE ②③ —In. Plus and Minus

SPECIFIED DIMENSION in.	METAL DIMENSIONS: ALLOWABLE DEVIATION FROM SPECIFIED DIMENSION WHERE 75 PER CENT OR MORE OF THE DIMENSION IS METAL ⑨⑩				SPACE DIMENSIONS: ALLOWABLE DEVIATION FROM SPECIFIED DIMENSION WHERE MORE THAN 25 PER CENT OF THE DIMENSION IS SPACE ⑥⑧											
	All Except Those Covered By Column 3		Wall Thickness ④ Completely ⑤ Enclosing Space 0.11 Sq In. and Over (Eccentricity)		At Dimensioned Points 0.250-0.624 Inches from Base of Leg		At Dimensioned Points 0.625-1.249 Inches from Base of Leg		At Dimensioned Points 1.250-2.499 Inches from Base of Leg		At Dimensioned Points 2.500-3.999 Inches from Base of Leg		At Dimensioned Points 4.000-5.999 Inches from Base of Leg		At Dimensioned Points 6.000-8.000 Inches from Base of Leg	
	Col. 2		Col. 3		Col. 4		Col. 5		Col. 6		Col. 7		Col. 8		Col. 9	
Col. 1	Alloys 5083, 5086, 5456	Other Alloys	Alloys 5083, 5086, 5456	Other Alloys	Alloys 5083, 5086, 5456	Other Alloys	Alloys 5083, 5086, 5456	Other Alloys	Alloys 5083, 5086, 5456	Other Alloys	Alloys 5083, 5086, 5456	Other Alloys	Alloys 5083, 5086, 5456	Other Alloys	Alloys 5083, 5086, 5456	Other Alloys
	CIRCUMSCRIBING CIRCLE SIZES LESS THAN 10 INCHES IN DIAMETER															
Up thru 0.124	.009	.006	±15% of specified dimension; ±.090 max ±.015 min	±10% of specified dimension; ±.060 max ±.010 min	.013	.010	.015	.012	....	....	....	....	....	....	....	....
0.125-0.249	.011	.007			.016	.012	.018	.014	.020	.016	....	....	....	....	....	....
0.250-0.499	.012	.008			.018	.014	.020	.016	.022	.018	.024	.020	....	....	....	....
0.500-0.749	.014	.009			.021	.016	.023	.018	.025	.020	.027	.022	....	....	....	....
0.750-0.999	.015	.010			.023	.018	.025	.020	.027	.022	.030	.025	.035	.030	....	....
1.000-1.499	.018	.012			.027	.021	.029	.023	.032	.026	.036	.030	.041	.035	....	....
1.500-1.999	.021	.014			.031	.024	.033	.026	.038	.031	.043	.036	.049	.042	.057	.050
2.000-3.999	.036	.024			.046	.034	.050	.038	.060	.048	.069	.057	.080	.068	.092	.080
4.000-5.999	.051	.034			.061	.044	.067	.050	.081	.064	.095	.078	.111	.094	.127	.110
6.000-7.999	.066	.044			.076	.054	.084	.062	.104	.082	.121	.099	.142	.120	.162	.140
8.000-9.999	.081	.054			.091	.064	.101	.074	.127	.100	.147	.120	.182	.145	.197	.170
	CIRCUMSCRIBING CIRCLE SIZES 10 INCHES IN DIAMETER AND OVER															
Up thru 0.124	.021	.014	±15% of specified dimension; ±.090 max ±.025 min	±15% of specified dimension; ±.090 max ±.015 min	.025	.018	.027	.020	....	....	....	....	....	....	....	....
0.125-0.249	.022	.015			.026	.019	.029	.022	.035	.028	....	....	....	....	....	....
0.250-0.499	.024	.016			.028	.020	.032	.024	.038	.030	.058	.050	....	....	....	....
0.500-0.749	.025	.017			.030	.022	.035	.027	.049	.040	.068	.060	....	....	....	....
0.750-0.999	.027	.018			.031	.023	.039	.030	.057	.050	.079	.070	.099	.090	....	....
1.000-1.499	.028	.019			.033	.024	.043	.034	.069	.060	.089	.080	.109	.100	....	....
1.500-1.999	.036	.024			.046	.034	.056	.044	.082	.070	.102	.090	.122	.110	.182	.170
2.000-3.999	.051	.034			.061	.044	.071	.054	.097	.080	.117	.100	.137	.120	.197	.180
4.000-5.999	.066	.044			.076	.054	.086	.064	.112	.090	.132	.110	.152	.130	.212	.190
6.000-7.999	.081	.054			.091	.064	.101	.074	.127	.100	.147	.120	.167	.140	.227	.200
8.000-9.999	.096	.064			.106	.074	.116	.084	.142	.110	.162	.130	.182	.150	.242	.210
10.000-11.999	.111	.074			.121	.084	.131	.094	.157	.120	.177	.140	.197	.160	.257	.220
12.000-13.999	.126	.084			.136	.094	.146	.104	.172	.130	.192	.150	.212	.170	.272	.230
14.000-15.999	.141	.094			.151	.104	.161	.114	.187	.140	.207	.160	.227	.180	.287	.240
16.000-17.999	.156	.104			.166	.114	.176	.124	.202	.150	.222	.170	.242	.190	.302	.250
18.000-19.999	.171	.114			.181	.124	.191	.134	.217	.160	.237	.180	.257	.200	.317	.260
20.000-21.999	.186	.124			.196	.134	.206	.144	.232	.170	.252	.190	.272	.210	.332	.270
22.000-24.000	.201	.134			.211	.144	.221	.154	.247	.180	.267	.200	.287	.220	.347	.280

Figure 14.8(a) Standard Tolerances for Extruded Rod, Bar, Shapes, and Tube, Except for Shapes in TS510, T4510, T6510, and T8510 Temper 7. From *ANSI-H35.2-1993, American National Standard Dimensional Tolerances for Aluminum Mill Products,* by permission of The Aluminum Association, Inc. Washington, D.C.
See Figure 14.8(b) on the next page for explanatory notes.

14.6 Why We *Need Methods* for Parametric Design

At the parametric design stage, there is a tendency in practice to avoid the use of formal methods. Experienced designers tend to rely on "experience" and what has worked before. Students, on the other hand, often attempt to perform parametric design by manipulating (i.e., "juggling") the set of analysis equations that predict or describe the behavior and performance of the designed object. That is, the attempt is made to solve these equations simultaneously or sequentially for values of the design variables, and hence to solve the parametric design problem by analysis. No doubt this approach — we'll call it *equation juggling* — is popular because students have learned to depend on it in engineering science courses where it works so well on textbook analysis problems.

There is little we can say to experienced designers to encourage the use of more formal methods of parametric design. If their experience has resulted in generalized knowledge and understanding that can be applied in new situations, and it works, then that is fine. It might be noted, however, that the slow adaptation of Taguchi's method™ (or *some* method) of robust design by U.S. industry was an important factor enabling competitors to design and produce reliable products that captured a number of important markets. The lesson is: Experience that "works" does not necessarily *work well enough* to beat competitors who are continually learning new and better methods.

For design problems, equation juggling may at first *appear* easier or quicker than the guided iteration, optimization, or Taguchi methods™ to be presented. Equation juggling seductively proposes a way to avoid the difficult decision making and messy value judgments involved in for-

①These Standard Tolerances are applicable to the average shape; wider tolerances may be required for some shapes and closer tolerances may be possible for others.

②The tolerances applicable to a dimension composed of two or more component dimensions is the sum of the tolerances of the component dimensions if all of the component dimensions are indicated.

③When a dimension tolerance is specified other than as an equal bilateral tolerance, the value of the Standard Tolerance is that which would apply to the mean of the maximum and minimum dimensions permissible under the tolerance.

④Where dimensions specified are outside and inside, rather than wall thickness itself, the allowable deviation (eccentricity) given in Column 3 applies to mean wall thickness. (Mean wall thickness is the average of two wall thickness measurements taken at opposite sides of the void.)

⑤In the case of Class 1 Hollow Shapes the standard wall thickness tolerance for extruded round tube is applicable. (A Class 1 Hollow Shape is one whose void is round and one inch or more in diameter and whose weight is equally distributed on opposite side of two or more equally spaced axes.)

⑥At points less than 0.250 inch from base of leg the tolerances in Col. 2 are applicable.

⑦Tolerances for extruded shapes in T3510, T4510, T6510, T73510, T76510 and T8510 tempers shall be as agreed upon between purchaser and vendor at the time the contract or order is entered.

⑧The following tolerances apply where the space is completely enclosed (hollow shapes):
For the width (A) the tolerance is the value shown in Col. 4 for the depth (D).
For the depth (D) the tolerance is the value shown in Col. 4 for the width (A).
In no case is the tolerance for either width or depth less than at the corners (Col. 2, metal dimensions).

D Depth
A Width

Example—Alloy 6061 hollow shape having 1 x 3 inch rectangular outside dimensions: width tolerance is ±0.021 inch and depth tolerance ±0.034 inch. (Tolerances at corners, Col. 2, metal dimensions, are ±0.024 inch for the width and ±0.012 inch for the depth.) Note that the Col. 4 tolerance of 0.021 inch must be adjusted to 0.024 inch so that it is not less than the Col. 2 tolerance.

⑨These tolerances do not apply to space dimensions such as dimensions "X" and "Z" of the example (right) even when "Y" is 75 percent or more of "X." For the tolerance applicable to dimensions "X" and "Z," use Col. 4, 5, 6, 7, 8 or 9, dependent on distance "A."

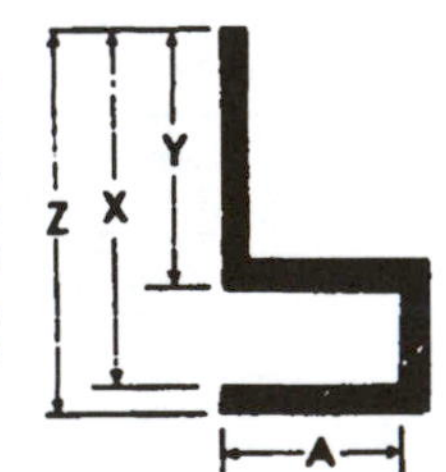

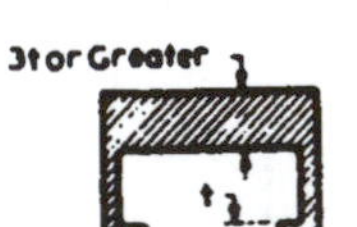

⑩The wall thickness tolerance for hollow or semihollow shapes shall be as agreed upon between purchaser and vendor at the time the contract or order is entered when the nominal thickness of one wall is three times or greater than that of the opposite wall.

Figure 14.8(b) Explanatory Notes for Figure 14.8(a).
From *ANSI H35.2-1993* by permission of The Aluminum Association, Washington, D.C.

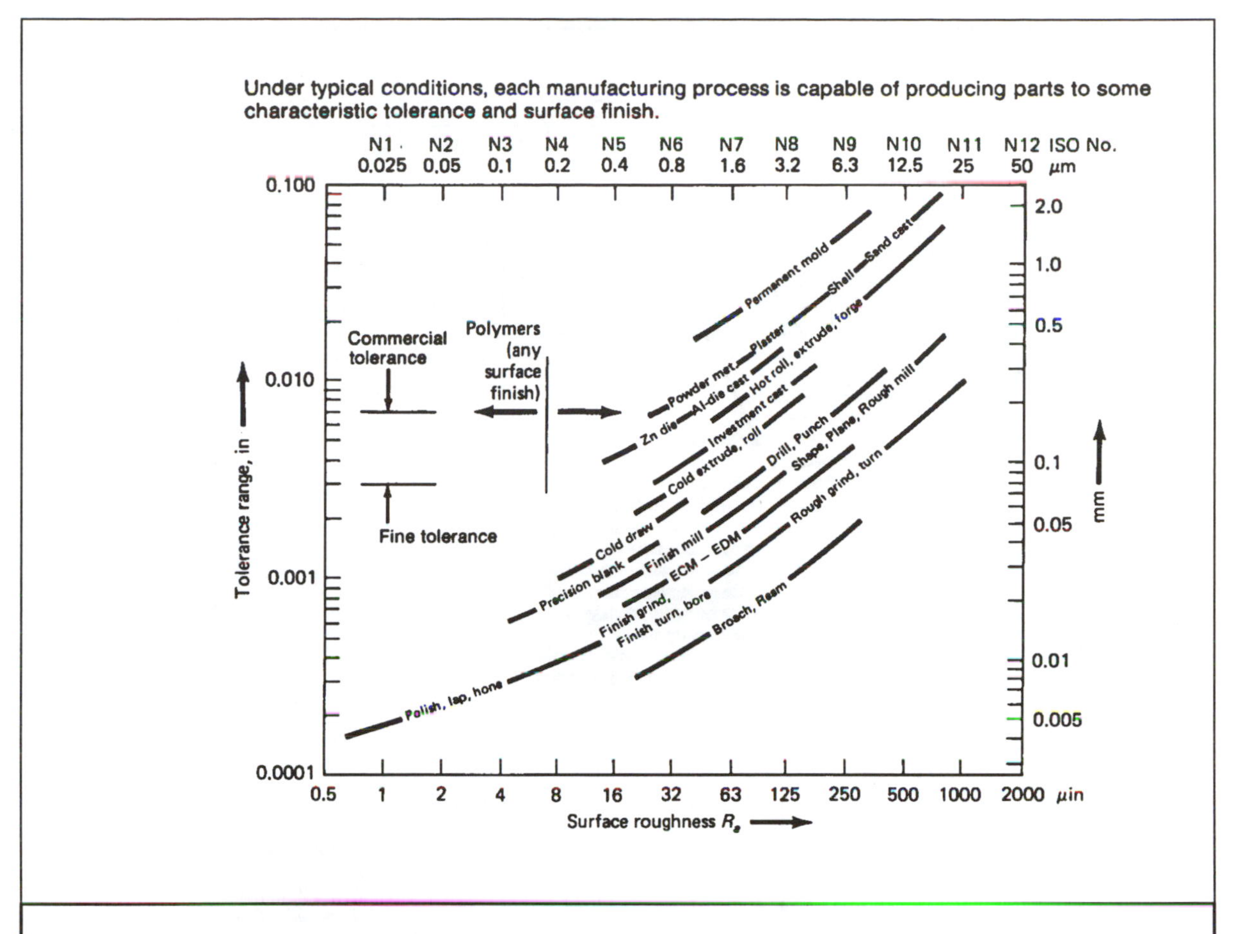

Figure 14.9 Tolerance Correlations with Surface Roughness for Manufacturing Processes. Reprinted from *Introduction to Manufacturing Processes*, Second Edition, by J. Schey, Copyright 1987 by permission of McGraw-Hill Book Company, New York.

mulating and solving problems by other methods. However, equation juggling really offers only the false hope that all one has to is "plug and chug" on the equations to get a design. Real design problems, however, require much more thought and knowledge supported by valid problem solving methods.

Equation juggling as a parametric design methodology is little more than random, unguided hunting around for a lucky solution. Once in a while, it will work, but mostly it does not; it just wastes valuable time. Using equation juggling is like being lost in the woods with no plan for finding a way out. One wanders around, often in circles, hoping. In parametric design, methods such as guided iteration, optimization, or Taguchi's method *must* be employed for effective problem solving. And issues like tolerances, robustness, cost, and DFM must be considered as well as the equations of engineering science.

As an example of the ineffectiveness of equation juggling as a method, suppose we try it on a simple I-Beam design. Refer to Figure 14.1 for the problem definition. Assume that the material has a maximum allowed material stress of 36,000 psi and a modulus of elasticity is $E = 30 \times 10^6$ psi. The maximum allowed deflection is specified as 0.050 inches. What we have to do is determine the numerical values for the beam's attributes, which in this case are:

beam height, h,

beam width, w,

web thickness, b, and

flange thickness, t.

The equations that describe a proposed beam's behavior or performance are:

$$\text{Maximum Stress} = Mh/2I = WL^2h/16\,I \qquad (14.1)$$

$$\text{Maximum Deflection} = 5WL^4/384\,E\,I \qquad (14.2)$$

where $$I = wh^3/12 - [(w - b)(h - 2t)^3]/12 \qquad (14.3)$$

Naturally, we would like to minimize the cost of the beam, tantamount to making the cross sectional area (A) as small as possible.

$$\text{Area} = 2tw + b(h - 2t) \qquad (14.4)$$

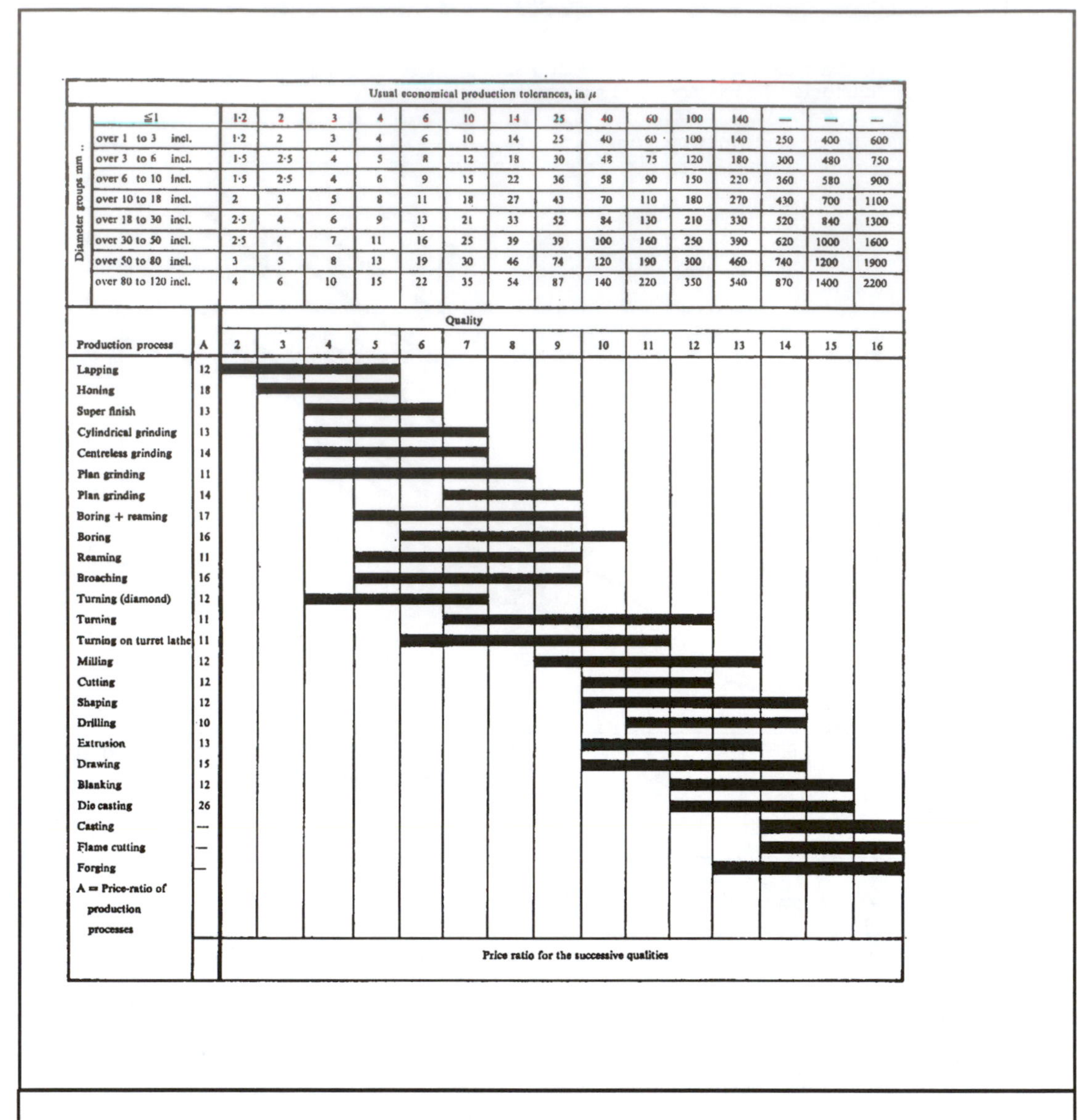

Diameter groups mm :	Usual economical production tolerances, in μ														
≦1	1·2	2	3	4	6	10	14	25	40	60	100	140	—	—	—
over 1 to 3 incl.	1·2	2	3	4	6	10	14	25	40	60	100	140	250	400	600
over 3 to 6 incl.	1·5	2·5	4	5	8	12	18	30	48	75	120	180	300	480	750
over 6 to 10 incl.	1·5	2·5	4	6	9	15	22	36	58	90	150	220	360	580	900
over 10 to 18 incl.	2	3	5	8	11	18	27	43	70	110	180	270	430	700	1100
over 18 to 30 incl.	2·5	4	6	9	13	21	33	52	84	130	210	330	520	840	1300
over 30 to 50 incl.	2·5	4	7	11	16	25	39	39	100	160	250	390	620	1000	1600
over 50 to 80 incl.	3	5	8	13	19	30	46	74	120	190	300	460	740	1200	1900
over 80 to 120 incl.	4	6	10	15	22	35	54	87	140	220	350	540	870	1400	2200
Quality	2	3	4	5	6	7	8	9	10	11	12	13	14	15	16

Production process	A
Lapping	12
Honing	18
Super finish	13
Cylindrical grinding	13
Centreless grinding	14
Plan grinding	11
Plan grinding	14
Boring + reaming	17
Boring	16
Reaming	11
Broaching	16
Turning (diamond)	12
Turning	11
Turning on turret lathe	11
Milling	12
Cutting	12
Shaping	12
Drilling	10
Extrusion	13
Drawing	15
Blanking	12
Die casting	26
Casting	—
Flame cutting	—
Forging	—

A = Price-ratio of production processes

Price ratio for the successive qualities

Figure 14.10 Grades of Tolerances Economically Feasible [8]
From *Handbook of Precision Machinery* by A. Davidson, Copyright 1970,
reprinted by permission of McGraw-Hill Book Company, New York.

We begin by trying an equation juggling approach suggested by several engineering students. They note correctly that a good design will occur when the maximum stress in the beam is at the yield stress, thus using the material as fully as possible, and when the deflection is at the maximum allowed so that the beam is not over-designed. This leads to the following two equations where both MaxStress (the maximum actual stress in the beam) and MaxDeflection (the maximum actual deflection) have the values established by the limits in the material and problem specifications:

$$\text{MaxStress} = Mh/2I = WL^2h/16\,I = 36{,}000 \text{ psi} \quad (14.5)$$

$$\text{MaxDeflection} = 5WL^4/384\,E\,I = 0.05 \text{ inches} \quad (14.6)$$

Since W and L and E are known (See Figure 14.1) the second of these can be solved directly for the moment of inertia, an apparently encouraging development! The result is

$$I = 46.8 \text{ inches}^4.$$

With I known, the first equation above can then be solved for the beam height, h, giving:

$$h = 18 \text{ inches}.$$

This seems a fairly tall I-Beam, but that's what the equations give so it must be right, right? Wrong, as we shall see.

	Smallest Allowed Clearance (inches)	Largest Allowed Clearance (inches)
Class RC–1 (Tight - No "plug")	0.00030	0.00095
Class RC–3 (Close Running)	0.00080	0.00210
Class RC–6 (Free Running)	0.00160	0.00480
Class RC–9 (Loose Running)	0.00700	0.01550

Figure 14.11 Sample Clearances for Running and Sliding Fits (Nominal size 0.71 to 1.19 inches. "Clearance" is the difference between the diameter of a (larger) hole and (smaller) shaft.

	Smallest Allowed Interference (inches)	Largest Allowed Interference (inches)
Class FN 1 (Light drive fits)	0.00030	0.00120
Class FN 3 (Heavy drive fits)	0.00080	0.00210
Class FN 5 (Shrink fits)	0.00130	0.00330

Figure 14.12 Sample Interferences Allowed for Force and Shrink Fits (nominal size 0.95 to 1.95 inches). "Interference" is the difference between the diameter of a (larger) shaft and a (smaller) hole.

There are still three design variable values to be determined (w, b, and t), and two equations (14.4 and 14.5) are still available. Thus, we either need another equation or some other way to establish one of the design variables. In this problem, we can make use of qualitative physical reasoning to provide a value for the web thickness, b. Since its contribution to resisting the applied bending moment is small, b should surely be as small as possible to keep the area as small as desired. Thus we set b to its smallest allowed value. In this case:

$$b = 0.1 \text{ inches.}$$

Now we have two equations and two unknowns, but the equations cannot be solved directly. We can, however, easily set up a simple trial and error process to find the values of b and w that minimize the area. Some equation jugglers will resist to the death using a trial and error process to the death, and instead will hunt in vain for another equation. The associated frustration sometime leads to the *invention* of another equation that has no engineering validity or relevance to the problem! However, we want to give the equation juggling method a fair test, and since a little trial and error will lead to a solution, we proceed.

The plan for the trial and error process will be:

1. Assume a value for w;
2. Compute a value for t from equation (14.3)
3. Compute a value for Area from (14.4)
4. On a graph, plot w as abscissa, Area as ordinate, and return to step 1.

With this procedure, it should soon become apparent what assumed value for w will lead to a minimum value for area, and the problem will be solved. Using a computer spreadsheet would save time. The results follow.

Since (with h = 18), w can vary from quite small up to 18 inches, we start with a assumed value of w = 12 inches, which gives a reasonable ratio of h to w. Then from equation (14.3), we get

$$t = -0.00093 \text{ inches.}$$

The negative value for t is not very encouraging, but we noted that in solving equation (14.3), a smaller value for w might lead to a positive value for t. So we try w = 6 inches. This gives a value for t of essentially zero. Still discouraging, but moving in the right direction.

Therefore, we try an even smaller value for w. For w = 2, the resulting value for t is *still* essentially zero. Based on this result (we are being directed to values for t as small as possible) we could logically at this point decide to set t to its minimum allowed value of 0.1 inches, and then solve equation 14.4 directly for w. The result is:

$$w = -0.012 \text{ inches.}$$

Clearly we are being forced to conclude that there is no need for any flange at all on this 18 inch tall I-Beam. Since this is absurd (and such a beam would have poor torsional properties) we know we need a shorter beam, one with a height less than 18 inches. But how much less? We can only guess since there is really nothing to guide our choice. Using equation juggling as a method, there are no other equations to juggle. It appears we are stuck. But let us continue to hunt around (as equation jugglers usually do) trying to use the analysis equations to find a solution.

We will learn in Chapter 18 that optimum solutions

typically have only one of the constraints fully met (we call that one an *active* constraint), so we can use that knowledge to reason that probably we will not find a solution that has both beam MaxStress equal to the yield stress *and* MaxDeflection equal to the maximum allowed by the problem (0.05 inches).

To explore this approach, we can put the known values of W, L, and E into equations 14.1, 14.2 and 14.3, juggle a bit, and get the following results:

$$I = 2.34 / \text{MaxDeflection} \quad (14.7)$$

$$h = (1 / 3338)(\text{MaxStress} / \text{MaxDeflection}) \quad (14.8)$$

where MaxStress and MaxDeflection refer as before to the actual stress and deflection in a proposed beam. Now we can set the value for MaxDeflection (the single active constraint) to its limit 0.050, again giving $I = 46.8$. However, we do not set any value for MaxStress. Perhaps we can compute it. From the first equation, we have:

$$h = \text{MaxStress} / 166.9. \quad (14.9)$$

This says that the beam height we want depends on the stress we want. We could, in this case, try again to perform a trial and error process for minimum area as a function of assumed values for h. The plan for this trial error is a little more complex than before, but it can still relatively easily be done:

1. Guess a value for h and compute MaxStress from Equation 14.9 above;
2. Guess a value for w (or guess b, take your pick);
3. From equation 14.3, compute b (or w);
4. From equation 14.4, compute area - and plot it on a graph of w vs. area;
5. Return to step 2 and repeat as many times as necessary to find the minimum area for the assumed value of h;
6. Return to step 1 and guess another value for h, and repeat as many times as necessary to find the combination of h, w, and b that gives a minimum area for the assumed value of deflection = Maximum Deflection.

Doing the above procedure is essentially performing optimization the hard way. It is probably *possible* to do this in a simple problem like the I-Beam (support from a computer spreadsheet would help speed up the guessing game), but it is not a practical approach in more complex problems. Moreover, it should be remembered that the assumption that the solution is limited by deflection (rather than by stress) is not always correct. In the specific example being used here, with an allowed deflection of 0.05 inches, deflection is in fact the limiting factor. However, for some (at this point unknown) specified larger value of allowed deflection, it would be the stress that limits the solution. In this case, then, the above process will have been totally incorrect and a waste of time. It would have to be repeated starting with the limiting stress. And in most real problems, there are even more constraints to be tried as the active one.

In addition to the kinds of difficulties with the equation juggling approach just illustrated, it has another serious shortcoming. It does not support trade-off decisions by the designer on performance issues. That is, there is no way, other than blind guessing, for designers using equation juggling to control or influence the problem solving process in order (say) to accept a little higher cost (area) in order to get a little better deflection (or vice-versa). In many design problems, there are such trade-offs to be made among not just two, but a number of performance factors. Equation juggling provides no way for doing this important design task.

Perhaps even more importantly, equation juggling does not allow trade-offs with manufacturing issues except *after* the equations have been juggled and solved, if they can be solved. That is, equation juggling does not support the trade-offs needed for *simultaneous* design for function and manufacturability. As we shall see in the next few chapters, both Taguchi methods and the guided iteration method can incorporate these important trade-offs easily, and optimization can also sometimes do so.

In summary, juggling analysis equations is *not* a way to do parametric design. We need more effective problem solving methods, and equation juggling is hardly even a method. We must use methods that support design decisions and knowledge much more effectively — methods like guided iteration, optimization, and Taguchi Methods™.

14.7 Guided Iteration for Parametric Design of Components: Formulating the Problem

As with engineering conceptual and configuration design, parametric design problems for components can be solved by the general method of guided iteration. The steps in the guided iteration process are, as always:

- Formulation of the problem,
- Generation of alternatives,
- Evaluation, and
- Redesign *guided by* physical reasoning and the results of the evaluations.

The specific methods used to implement these steps in parametric design are different from the methods used in conceptual and configuration design. The process is described and illustrated in Chapter 17. Additional examples are presented in Chapters 20 and 21.

14.8 Optimization Methods for Parametric Design

Optimization is a well developed field of study that is the subject of whole courses and books. There are excellent texts and reference books on optimization available for use by designers; examples are [11, 12, 13, 14]. The more technically advanced manufacturing firms will likely have optimization experts with which designers and design teams can consult. There are also computer programs available; see, for example, [15].

In this book, only one chapter — Chapter 18 — is devoted to optimization. Thus, we are able present there only a few of the more basic optimization methods. Though exceptions are noted in Chapter 18, in general, optimization methods are useful when the following conditions are met:

1. The design variables (defined the same as for guided iteration) are all *numeric and continuous*. In this case, optimization methods are not only possible but likely to be effective and efficient. If not, optimization can still possibly be used, but some adaptations will be required;

2. A *single* function (called the "Criterion Function" or "Objective Function") can be written in terms of the design variables which expresses the overall quality or goodness of a trial design. Often this single function is cost, though in some cases it can be weight, efficiency, robustness, or some other performance factor;

In the case of I-Beam example begun above, the criterion function would probably be cost, which would be proportional to weight in most cases. Thus the cross sectional area of the beam could be used; that is, the criterion function (U) would be:

$$U = A = 2\,t\,w + (h - 2t)\,b.$$

3. Analytical procedures are available and practical for computing the criterion function given values for the design variables;

4. All limits on the design variables and all required relationships involving the design variables can be expressed as equations or inequalities. These equations are typically called "constraints" in optimization literature.

In the I-Beam example, the constraints might be expressed as:

h(max) = 20 inches

h(min) > 0

w(max) = 12 inches

w(min) > 0

b(min) = 0.1 inches

t(min) = 0.1 inches

MaxDeflection = 0.05 inches

MaxStress = 36,000 psi

$h \geq w$

Note here that the maximum allowed stress and deflection are included as constraints.

We elaborate on the formulation of optimization problems in Chapter 18.

Optimization Vis-a Vis Guided Iteration. When there are two or more design variables, guided iteration is itself a kind of manual ad hoc optimization process. It prescribes how to select a design variable (or variables) to change, and by how much to change it (or them) in order to improve the quality of the design. Guided iteration does not purport to optimize exactly, however, only to reach a satisfactory solution near the true optimum. Such a solution is sometimes called a *satisficing* solution. [16]. Though the solution is not precisely optimum, it satisfies the designer.

Guided iteration keeps the designer in close touch with the physics of the problem. Knowledge of the domain and use of qualitative physical reasoning aids the process. Evaluations refer directly to the various goals of the designer. On the other hand, optimization methods are primarily mathematical processes. Once the criterion function and the constraints are defined, the solution process can be done by prescribed procedures and computer programs.

Though there are certainly times in the implementation of optimization methods that physical insight and interpretations are needed, it is more or less the case that guided iteration is basically a physically based process aided by mathematics, whereas optimization is basically a mathematical process aided by physical reasoning. The difference is thus more a matter of emphasis and style than of the absolute nature of the two processes.

Guided iteration has an advantage in some problems since it allows the designer more conscious and explicit control over trade-offs among two or more conflicting goals. Guided iteration can also include goals whose evaluations are not quantifiable numerically. However, when optimization methods can be used, and there are many problems where they can, then they are very powerful.

Sub-Optimization. As shown above, the criterion function in an I-Beam design problem is rather easily developed. However, in more complex, realistic parametric design problems, an appropriate criterion function often cannot be so readily formulated to meet the conditions required by optimization techniques. Nevertheless, sometimes certain sub-parts of problems can be solved by optimization. This is called *sub-optimization,* and it can often be used to considerable advantage.

As an example of sub-optimization, consider the design of a post and beam structure. The design of the whole structure includes non-numeric variables (materials, arrangements of posts) and a complex set of performance factors impossible to reduce to a single criterion function (cost, deflection, maximum stress, assembly costs, time required, etc.). It may very well be possible, however, to find the optimum beam sizes to use for a given configuration of posts by applying optimization to beam design as a separate, isolated problem. Also, the posts could perhaps be optimized separately. In such a case, we would say that the beam design and post design have been sub-optimized.

Sub-optimization can be effective and helpful, but there is a caveat. It is important for designers to recognize that one cannot optimize a whole problem solution by dividing it up into sub-problems, each of which is sub-optimized separately. *Sub-optimization of all the sub-parts of a system does not in general lead to optimization of the whole system.* Still, sub-optimization can be used to advantage in situations where any adverse effect from a sub-optimized section on the whole system is negligible or acceptable.

For example, in the post and beam problem mentioned above, sub-optimizing the beam and post designs for a given post placement is useful. On the other hand, a different post placement might lead to a better overall design.

14.9 Statistically Based and Taguchi Methods™ for Parametric Design

Methods from the field of statistics can also be used to perform parametric design in some cases. We will discuss only the so-called Taguchi method™ [17, 18] in this book because it is fairly easily applied, and because its use is now fairly common. Moreover, it has a good record of successful application. However, students should know that the Taguchi method™ is not the only way that statistics can be used to support parametric design [19, 20].

Genichi Taguchi is a Japanese engineer-statistician whose incisive, creative thinking about parametric design has had a profound effect on the quality of parts and products [21, 22, 23]. His methods have been widely used by Japanese manufacturers in the 1970s but were not introduced into the United States until the 1980s. Taguchi developed methods for assigning values to attributes, and then (in a second, separate step) for assigning manufacturing tolerances to the values. We discuss these techniques in some detail in Chapter 19. In this section, we provide only an introduction and overview sufficient for designers to make an initial determination about whether to employ the techniques or to proceed with optimization or guided iteration.

Taguchi's methods are an application of the more fundamental field of statistics, including especially the principles of design of experiments. Engineering design students should study these fields in order to understand the underlying principles. Moreover, there are ways other than Taguchi's methods for applying statistics to design for robustness, and no doubt new and improved techniques will be developed in the future. Learning the fundamentals will enable students to keep abreast of the new developments easily.

Robustness, Noise Factors, and Control Factors. The overall evaluation criterion in Taguchi's techniques is called *robustness*. Robustness refers to how consistently a component or product performs under variable conditions in its environment and as it wears during its lifetime. The variable conditions under which a product must function may include, for example, a range of temperatures, humidity, or input conditions (e.g., voltages, flow rates).

Robustness also refers to the degree a product's performance is immune to normal variations in manufacture; that is, to variations in materials and processing. Even the most closely controlled manufacturing processes vary enough so that no two products are precisely the same.

It is common to use the terms *noise* or *noise factors* for the uncontrollable variable conditions of environment, wear, and manufacture. Thus another way to describe robustness is to say that it is the degree to which the performance of product is insensitive to noise factors.

Noise factors, which the designer cannot control, are not to be confused with the design variables, whose values the designer *can* control. Design variables are called *control factors* in Taguchi methods. Though designers have no control over the noise factors, the ranges over which noise factors vary are usually reasonably predictable. Noise factors and their ranges are, in effect, types of problem definition parameters -- though they were essentially neglected before Taguchi and other statistical methods were introduced.

It should be noted that robustness can be thought of as being expressed by one or a set of expressions for the *rate of change of a Solution Evaluation Parameter with respect to a noise variable*. For example, in the beam design problem, suppose that in some application, the quality of the beam's performance depends (among the other things) on the robustness of the beam's deflection in relation to environmental temperature as a noise factor. Then we would want to add "the rate of change of deflection with respect to temperature" to the list of Solution Evaluation Parameters (SEP's). With this additional SEP, we could then proceed with guided iteration as usual. (Note that solution by optimization would be impossible in this case unless a way could be found either to include both cost and robustness in the single criterion function — or unless either robustness or the cost could somehow be expressed as a constraint leaving the other to be the single criterion function.

When robustness criteria, or some expressions thereof, are included among the evaluation parameters in a problem's formulation, then designers will often wish to use Taguchi (or other specialized statistical) techniques to assign parametric values. Sometimes optimization can be used as an alternative (as explained above), and guided iteration can *always* be used. However, the statistically based approaches were developed expressly for this purpose, and are very effective.

Strategies to Achieve Robustness. To achieve robustness in the face of the environmental and other noise factors, there are two different strategies that may be followed. One strategy is to design the product so that the performance of sensitive parts is insulated from the noise conditions. For example, if the performance of some part or sub-assembly is sensitive to temperature, then the part or sub-assembly might literally be shrouded in thermal insulation or even provided with a means of temperature control. Or, in a similar way, if some part or sub-assembly is sensitive to vibration, it might be provided with vibration isolation.

Alternatively, steps might be taken to remove or reduce the source of such noise (e.g., eliminate the cause of the temperature variations or the source of the vibration).

Both insulating the part or product from the noise, and eliminating the source of the noise, are called *reduce the noise* strategies.

An alternative design strategy is to *reduce the consequences* of the noise. In this approach, the noises are accepted as given, but the product is designed so that its lifetime performance is as insensitive to them as possible. For example, instead of thermally insulating the part or parts whose performance is sensitive to temperature, those parts can be designed so their performance is not significantly impaired by the expected temperature variations.

Often, of course, both "reduce the noise" and "reduce the consequences" strategies may be used simultaneously, but reducing the noise is usually a considerably more expensive solution. Thus it only makes good sense to do as much as possible to design products and components to reduce the consequences of noise before going to the trouble and expense of protecting them from or removing the noises. This is especially true since reducing the consequences of the noise can often be done effectively enough simply by making a judicious selection of values for the design variables using Taguchi or other statistically based methods.

Robustness Vis-a-Vis Performance. Robustness is the inverse of the *variability* of performance. That is, when performance variability (considering the effects of the noise factors) is low, a design is robust. However, robustness is not the only design goal. Performance itself is also important. We would not usually want, for example, a very robust design that performed very poorly. On the other hand, we also do not want a design that performs in a superior way, but only in very narrow ranges of environmental and manufacturing noise. That is, we generally want parts and products to have *both* good performance *and* high robustness. Thus, our the goal is a suitable compromise or trade-off between performance and robustness.

Taguchi techniques have a built-in trade-off methodology for determining the "best" combination of performance and robustness. It is called the *signal to noise ratio.* Though it is a reasonable criterion, designers using Taguchi methods have no control over it. Nevertheless, the Taguchi techniques have a very good track record for producing excellent overall results.

A disadvantage of Taguchi's method is that only a few values of the design variables over a limited range can be considered. Another disadvantage is that in many cases, experimentation is required to obtain the performance results. When the cost and time required for experimentation is large, the disadvantage is obvious. Where analysis and/or simulation can be used instead of experimentation, that will usually be both quicker and less expensive. The use of statistical methods and proper design of experiments can, in most cases, make experimentation more efficient.

Using Taguchi's methods is not the only way to achieve robust designs that also perform well. Robustness (that is, variability of performance) can often be included among the other SEPs when using guided iteration [24]. And it is also sometimes possible to construct a criterion function that includes robustness as well as performance, thus making optimization an available method, too.

14.10 Summary and Preview

Summary. This chapter has introduced three methods for performing parametric design of components and small assemblies: guided iteration, optimization, and Taguchi's method™. Each of the methods is discussed in greater detail and with examples in subsequent chapters: guided iteration in Chapters 17, 20, and 21; optimization in Chapter 18; and Taguchi methods in Chapter 19.

Preview. Before we go into more detail in how to implement these parametric design methods, however, some important issues of design for manufacturing (DFM) at the parametric stage must be discussed. For special purpose parts to be injection molded, die cast, or stamped, we have already (in Chapters 11 and 12) presented quantitative methods for estimating relative tooling costs. This was done with part configuration information only. Once parameters are assigned, relative processing costs can be estimated. Methods for doing this, again for injection molded, die cast, and stamped parts, are presented in Chapters 15 and 16. Thus we turn attention again to DFM so that issues of manufacturability can be well in the designers mind as parametric design proceeds.

References

[1] *Standards and Practices of Plastics Molders,* The Society of The Plastics Industry, Inc., Washington, DC 20005, 1993.

[2] *American National Standard Dimensional Tolerances for Aluminum Mill Products, ANSI-H35.2-1993*, The Aluminum Association, Inc. 818 Connecticut Avenue, N.W., Washington, D.C., 1979.

[3] Bralla, J.G. *Handbook of Product Design for Manufacturing*, McGraw-Hill Book Company, New York, 1986.

[4] *Forging Design Handbook*, American Society of Metals, Metals Park, Ohio, 1972.

[5] Sheridan, S.A. *Forging Design Handbook*, American Society of Metals, Metals Park, Ohio, 1972.

[6] Wilson, F. W. *Die Design Handbook*, McGraw-Hill Book Company, New York, 1965.

[7] Trucks H.E. *Designing for Economical Production*, Society of Manufacturing Engineers, Dearborn, Michigan, 1975.

[8] Schey, J. Introduction to Manufacturing Processes, 2nd Edition, McGraw-Hill Book Company, 1987.

[9] Davidson, A. *Handbook of Precision Engineering,* McGraw-Hill Book Company, New York, 1970.

[10] ANSI B4.1-1967 "Preferred Limits and Fits for Cylindrical Parts" Published by ASME, New York, 1967.

[11] Wilde, D. J. *Globally Optimal Design*, John Wiley and Sons, New York, 1978.

[12] Reklaitis, A. G., Ravindran, A., and Ragsdell, K. M. *Engineering Optimization*, John Wiley and Sons, New York, 1983.

[13] Papalambros, P. Y. and Wilde, D. J. *Principles of Optimal Design*, Cambridge University Press, Cambridge, England, 1988.

[14] Arora, J. S. *Introduction to Optimum Design*, John Wiley and Sons, New York, 1989.

[15] Optdes and OptdesX are trademarks of Design Synthesis, Inc., 3883 N. 100, East Provo, UT 84604, 1990.

[16] Simon, H. *Sciences of the Artificial*, MIT Press, Cambridge, MA, 1969

[17] Taguchi, G. "The Development of Quality Engineer-

ing", *The American Supplier Institute Journal*, Vol. 1, No. 1, pp 5-29.

[18] Taguchi, G, and Clausing, D., "Robust Quality", *Harvard Business Review,* January-February,1990, pp 65-75.

[19] Box, G.E.P., Bisgaard, S., and Fung, C. "An Explanation and Critique of Taguchi's Contributions to Quality Engineering", *Quality and Reliability International*, Vol. 4, pp121-131, 1988.

[20] Box, G.E.P., "Signal-to-Noise Ratios, Performance Criteria, and Transformations" (with Discussion), *Technometrics*, Vol. 30, No. 1, February, 1988.

[21] Peace, G. S.*Taguchi Methods*, Addison-Wesley, Reading, MA, 1993.

[22] Ross, P. J. *Taguchi Techniques for Quality Engineering*, McGraw-Hill, New York, 1988.

[23] Phadke, M. S. *Quality Engineering Using Robust Design*, Prentice Hall, Englewood Cliffs, N.J., 1989.

[24] Orelup, Mark. "Dominic III: Incorporating Taguchi's Philosophy, Material Choice, and Memory in the Design of Mechanical Components", Ph D Dissertation, University of Massachusetts, May 1992.

Other Recommended Reading

Wilde, D. J. *Globally Optimal Design*, John Wiley and Sons, New York, 1978.

Taguchi, G, and Clausing, D. "Robust Quality", *Harvard Business Review,* January-February, 1990, pp 65-75.

Problems

14.1 Golf club shafts and heads are manufactured separately by different processes. The shaft is then epoxied into a circular opening called a "hossle" about an inch deep that is cast into the neck of the head. The nominal outer diameter of most steel golf shafts where they are inserted into the hossle is 0.370 inches. What do you think the tolerance might be on the outer diameter of the shaft? What do you think might be the nominal diameter of the opening in the head and its tolerance?

14.2 Suppose that it has been decided that the configuration of a post is to be a hollow circular pipe. (a) For guided iteration, what would you take to be the design variables? (b) For optimization, what would you choose for the criterion function? (c) For optimization, what are the constraints?

14.3 A long steam pipe of given dimensions is to be insulated. What are the design variables for the insulation? What would be the criterion function for optimization?

Chapter Fifteen

Evaluating Parametric Designs for Manufacturability: Injection Molding and Die Casting

"Economic manufacturing does not just happen. It starts with design and considers practical limits of machine tools, processes, tolerances and finishes."

H. E. Trucks*[11]
Designing for Economical production (1987)

15.1 Injection Molded Part Costs

15.1.1 Introduction

As we learned in Chapters 11 and 12, the first stage of a manufacturability evaluation for injection molded, die cast, or stamped parts is an evaluation of tooling costs. This can be done at the configuration design stage where only approximate dimensions, locations and orientations of features are known. At the parametric stage, making use of the near final dimensions, locations and orientation of features, a manufacturing evaluation of the relative cost to *process* a part can be made. Then the total cost of a part can be computed as the sum of the per part tooling costs, processing costs, and material cost.

15.1.2 Processing Costs

Processing costs (sometimes called operating costs) are the charges for use of the injection molding machine. They depend on the machine hourly rate, C_h ($/hr), and the effective cycle time of the process, t_{eff}. The effective cycle time is the machine cycle time, t, divided by the production yield, Y. Production yield (or just "yield") is the fraction of the total parts produced that are satisfactory, and hence usable. Thus:

$$\text{Processing cost per part} = C_h\, t_{eff} = C_h\, t / Y \qquad (15.1)$$

where Y = Production Yield (usable parts/total parts produced).

Part surface "quality" requirements and tolerances are the main causes for variations in production yield. A low yield reduces the number of acceptable parts that are produced in a given time, and thus increases the effective cycle time to a value higher than the actual machine cycle time, t. Increases of 10% to 30% in the effective cycle time for a given part are typical. The reasons for this increase are discussed in greater detail in Section 15.9

The *relative processing cost* is the cost of producing a part *relative to* the cost of producing a reference part. Relative processing cost, C_e, can be expressed as:

$$C_e = t\,C_h / t_o\,C_{ho} = t_r\,C_{hr} \qquad (15.2)$$

where t_o and C_{ho} represent the cycle time and the machine hourly rate for the reference part. C_{hr} represents the ratio C_h/ C_{ho}, and t_r is the *total relative cycle time for the part compared with the reference part*. That is,

* Quotation reprinted from *Designing for Economical Production* by H.E. Trucks, Second Edition, 1987, with permission of the Society of Manufacturing Engineers, Dearborn, MI.

$$t_r = t / t_o \qquad (15.3)$$

The reference part in this case is the same flat washer used as a in Chapter 11: a 1 mm thick, flat washer whose outer and inner diameters are 72 mm and 60 mm, respectively. Additional data (part material, material cost, tooling cost, etc.) for the reference part are given in Table 15.1

Table 15.1 Data for the Reference Part

Material	Polystyrene
Material Cost (K_p)	1.46×10^{-4} cents/ mm^3 (1)
Part Volume (K_p)	1244 mm^3
Die Mat Cost (K_{dmo})	$980 (2)
Die Construction Time (Includes design and build hours)	200 hours
Labor Rate (Die Construction)	$30/hr (2)
Cycle Time (t_o)	16 seconds (2)
Mold Machine Hourly Rate (C_{ho})	$27.53 (3)

(1) *Plastic Technology*, June 1989.
(2) Data from collaborating companies, 1989.
(3) *Plastic Technology*, July 1989.

15.1.3 Material Costs

The material cost for a part, K_m, is given by

$$K_m = V\ K_p \qquad (15.4)$$

where V is the part volume and K_p is the material cost per unit volume. Thus, if the subscript "o" is used to indicate the reference part, then the relative material cost can be expressed as

$$C_m = K_m / K_{mo} = (V/V_o)(K_p/K_{po}) = (V/V_o)C_{mr} \qquad (15.5)$$

Table 15.2 contains the relative material prices for the most often used engineering thermoplastics. The prices are all relative to polystyrene.

15.1.4 Total Cost

The total production cost of a part, K_t, can be expressed as the sum of the material cost of the part, K_m, the tooling cost, K_d/N, and processing cost, K_e. Thus,

$$K_t = K_m + K_d / N + K_e \qquad (15.6)$$

where K_d represents the total cost of the tool, and N represents the production volume.

If the manufacturing cost of a reference part is denoted by K_o, then the relative total cost of the part, C_r, can be expressed as:

$$C_r = (K_m + K_d/N + K_e) / K_o \qquad (15.7)$$

In the remainder of this chapter, we present methods for computing these relative costs — that is, relative processing cost, relative material cost, and total relative part cost — for injection molded and die cast parts. Relative processing, relative material, and total relative costs for stamped parts will be discussed in the following chapter.

A prerequisite step to determining total relative processing cost is the determination of the total relative cycle time. Thus we begin in the next section with relative cycle time.

Table 15.2 Relative Material Prices, C_{mr}, for Thermoplastics. Based on material prices in *Plastics Technology*, June 1990.

Material	C_{mr}
ABS	1.71
Acetal	2.92
Acrylic	1.54
Nylon 6	2.79
Polycarbonate	2.96
Polyethylene	0.71
Polypropylene	0.62
Polystyrene	1.00 (Reference)
PPO	2.33
PVC	0.62

15.2 Determining Total Relative Cycle Time (t_r) for Injection Molded Parts - Overview

Statistical studies of processing costs as a function of part geometries have led to a part coding system for determining injection molding processing costs similar to the coding system used for tooling costs. The mechanics of using the system are explained in the next several sub-sections. The overall result is that the total relative cycle time relative to the reference part, t_r, is obtained as a function of three parameters:

- The *basic relative cycle time*, t_b,
- An *additional relative cycle time*, t_e, due to the presence of inserts and internal threads, and
- A *multiplying penalty factor*, t_p, to account for the effects of part surface quality and tolerances.

In terms of these parameters, the *total relative cycle time relative to the reference part*, t_r, is given by

$$t_r = (t_b + t_e)\, t_p \tag{15.8}$$

The basic relative cycle time, t_b, is given by values found in one of the three matrices shown in Figure 15.1, which define the first and second digits of the coding system. Note that to use this Figure, the meaning of a number of basic terms must be understood including: partitionable and non-partitionable parts; slender (S), non-slender (N), and frame-like parts; elemental plates; part thickness (w); grilles and slots; ribs, and types of ribs; gussets; *significant* ribs and bosses; and "easy" versus "difficult" to cool parts. We also must recall the definition of the basic envelope of a part which was introduced in Chapter 11.

The additional relative time due to inserts and internal threads, t_e, is found from Table 15.3. Table 15.3 defines the third digit of the coding system.

The time penalty factor, t_p, to account for surface requirements and tolerances, is found in Table 15.4, which defines the fourth and fifth digits in the coding system. To use Table 15.4, we must be able to distinguish between tolerances that are "easy" to hold and those that are more "difficult" to hold. We must also be able to distinguish surface finish requirements that are "low" from those that are

1 in = 25.4 mm; 100 mm = 3.94 in

SECOND DIGIT

Wall Thickness

(a) Slender Partitionable Parts, S

				w < 1 mm	1mm ≤ w ≤ 2mm	2mm < w ≤ 3mm	3mm < w ≤ 4mm	4mm < w ≤ 5mm	w > 5mm
				0	1	2	3	4	5
Parts with $L_u/B_u > 10$ (1) or Frames (2)	Plates with $L_u/2w < 100$ (3) without lateral projections (4)	Without ribs	0	Difficult to fill or eject	1.00	1.35	1.70	2.55	Use foamed materials (9)
		With ribs	1		1.10	1.45	1.85	2.70	
	Plates with $L_u/2w > 100$ and/or plates with lateral projections (4)	Without ribs	2		1.15	1.55	2.00	2.85	
		With ribs	3		1.25	1.82	2.20	3.85	

(b) Non-Slender Partitonable Parts, N

FIRST DIGIT					w < 1 mm	1mm ≤ w ≤ 2mm	2mm < w ≤ 3mm	3mm < w ≤ 4mm	4mm < w ≤ 5mm	w > 5mm
					0	1	2	3	4	5
Parts with $L_u/B_u \le 10$ (1)	Plates without significant rib (5) or significant bosses (6) with or without non-peripheral ribs or bosses	Plates which are grilled or slotted (7)	Without non-peripheral ribs	0	Difficult to fill or eject	1.68	2.39	3.11	3.82	Use foamed materials (9)
			With non-peripheral ribs	1		1.82	2.53	3.25	3.96	
		Plates which are not grilled or slotted (7)	Without non-peripheral ribs	2		1.96	2.67	3.39	4.10	
			With concentric or cross ribbing	3		2.10	2.81	3.53	4.24	
			With radial or unidirectional ribbing	4		2.24	2.96	3.67	4.39	
	Plates with significant ribs (5) and/or significant bosses (6)	Plates with rib and/or boss thickness less than the wall thickness (8)	Ribs/bosses supported by gusset plates	5		2.38	3.10	3.81	4.53	
			Ribs/bosses not supported by gusset plates	6		2.52	3.24	3.95	4.67	
		Plates with rib and/or boss thicknes (8) greater than or equal to the wall thickness		7		2.66	3.38	4.09	4.81	

(c) Non-Partitionable Parts

			w < 1 mm	1mm ≤ w ≤ 2mm	2mm < w ≤ 3mm	3mm < w ≤ 4mm	4mm < w ≤ 5mm	w > 5mm
			0	1	2	3	4	5
Parts which are not partitionable	Easy to cool (10)	0	Difficult to fill or eject	2.66	3.38	4.09	4.81	Use Foamed Materials
	Difficult to cool (10)	1		3.56	4.50	5.47	6.40	

Figure 15.1 Classification System for Basic Relative Cycle Time, t_b (Injection Molded Parts)

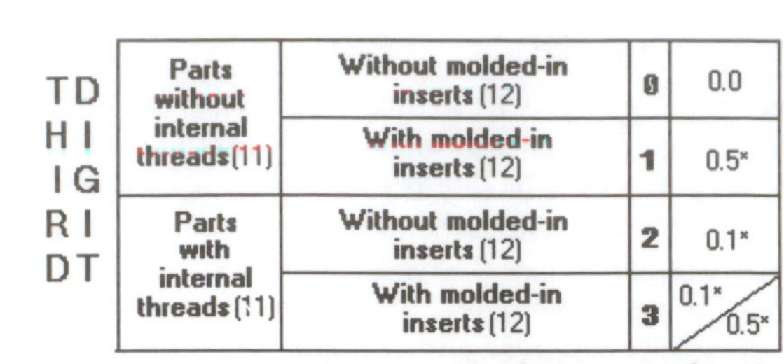

THIRD DIGIT			
Parts without internal threads (11)	Without molded-in inserts (12)	0	0.0
	With molded-in inserts (12)	1	0.5*
Parts with internal threads (11)	Without molded-in inserts (12)	2	0.1*
	With molded-in inserts (12)	3	0.1* / 0.5*

* per insert or per thread

Table 15.3 Additional Relative Time, t_e, Due to Inserts and Internal Threads (Injection Molding)

"high". These issues are discussed and explained below.

As in the tooling evaluation system, the value of the basic relative cycle time, t_b, for the reference part is 1.00; it can be found in the upper left hand corner for slender or frame-like parts. (The washer chosen as the reference part is a frame like part.) Note that the values for t_b increase significantly as one moves down and to the right of the matrix. This information, as with tooling, helps guide designers to redesigns that can reduce relative cycle time, and, hence, processing costs.

In the next few sub-sections, we explain the meaning of all the terms needed, and then illustrate the use of the coding system in the evaluation of several example parts.

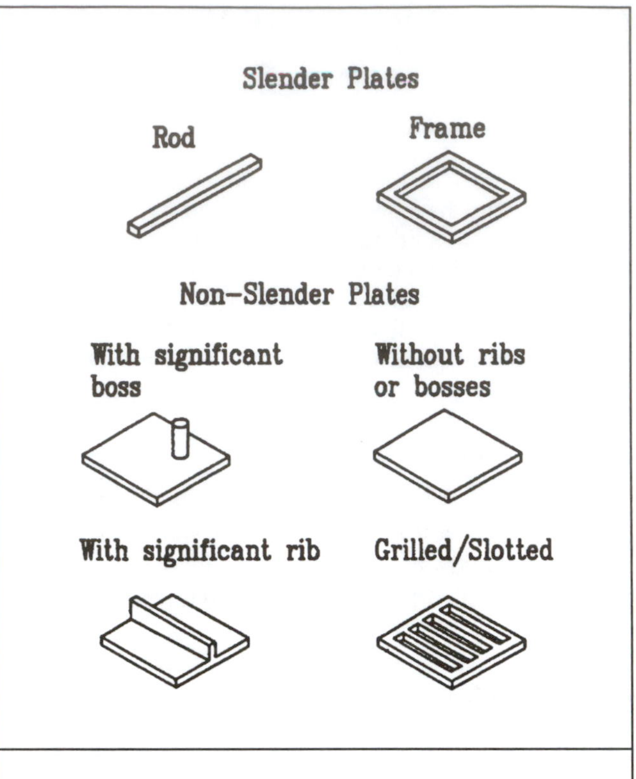

Figure 15.2 Examples of Elemental Plates

15.3 Determining the Basic Part Type: The First Digit

Figure 15.1, which is used to find the basic relative cycle time, t_b, is in three sections: (a) slender partitionable parts, (b) non-slender partitionable parts, and (c) non-partitionable parts. This section presents the definitions of these basic part types.

Partitionable parts are parts that can be easily and completely divided (except for add-ons like bosses, ribs, etc.) into a series of *elemental plates*. An elemental plate is a contiguous thin flat wall section whose edges are either not connected to other plates, or are connected via distinct intersections (e.g., corners). An elemental plate may have add-on features like holes, bosses, or ribs.

Examples of several types of elemental plates are shown in Figure 15.2. (Note: The meaning of the word "significant" in Figure 15.2 will be explained shortly.) A method for partitioning a part into its elemental plates is described in the next section.

For *each* elemental plate in a part, we will determine a cooling or solidification time. The plate with the longest cooling time controls the machine cycle time of the entire part. Thus, every elemental plate should be carefully designed so that its individual solidification time is minimized.

Partitionable parts are further classified as either *slender or frame-like* (S), or *non-slender* (N).

To distinguish quantitatively between slender and non-slender parts (see Figure 15.3 for a qualitative distinction), we consider the basic envelope of the part as shown in Figure 15.4. Given a basic envelope with dimensions L, H, and B (see Section 11.3.2), a *slender* partitionable part is one for which

$$L_u / B_u \geq 10,$$

where

$$L_u = L + B,$$

FOURTH DIGIT			FIFTH DIGIT: Tolerances not difficult to hold (14)	Tolerances difficult to hold (14)
			0	1
Plate surface requirements (13)	Low	0	1.00	1.20
	High: 1mm < w ≤ 2mm	1	1.30	1.43
	High: 2mm < w ≤ 3mm	2	1.22	1.41
	High: 3mm < w ≤ 4mm	3	1.16	1.37
	High: 4mm < w ≤ 5mm	4	1.10	1.32

Table 15.4 Time Penalty Due to Surface Requirements and Tolerances (Injection Molding)

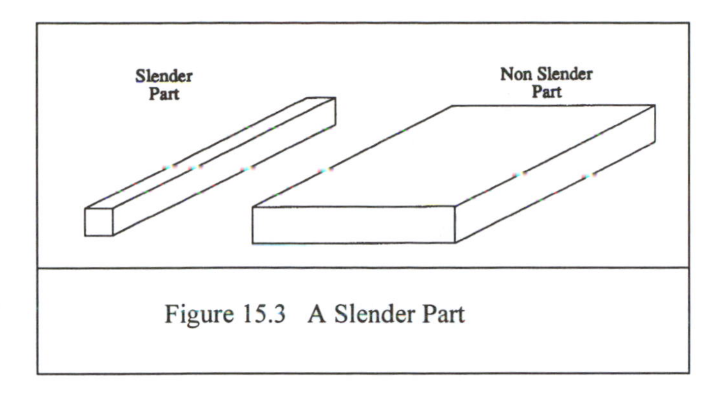

Figure 15.3 A Slender Part

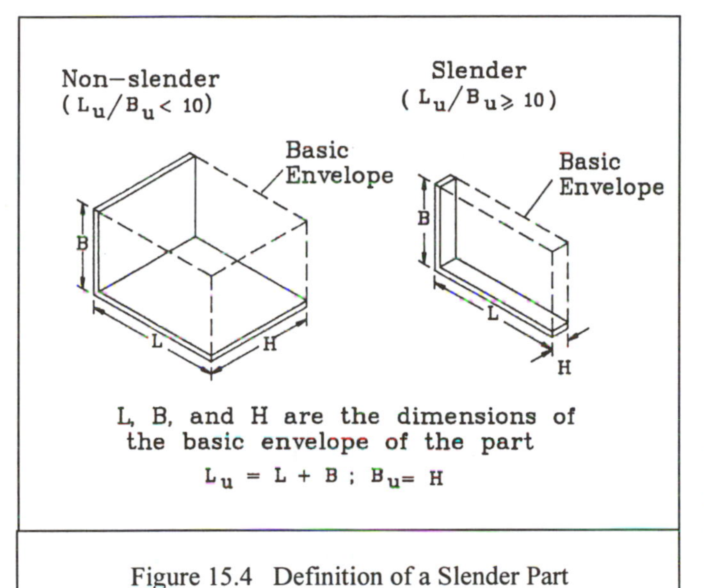

Figure 15.4 Definition of a Slender Part

Figure 15.5 Frames

and

$$B_u = H.$$

For parts with a bent or curved longitudinal axis, the *unbent length*, L_u, is the maximum length of the part with the axis straight (Figure 15.4). The width of this unbent part is referred to as B_u.

Frames are parts or elemental plates which have a through hole greater than 0.7 times the projected area of the part/plate envelope, and whose height is equal to the wall thickness (Figure 15.5).

Molds for slender parts and frames are generally more difficult to fill than molds for non-slender parts; however, slender parts are generally easier to cool.

Non-partitionable parts include parts with complex geometries, or parts with extensive subsidiary features such that they cannot be easily partitioned into elemental plates. We will also consider as non-partitionable parts that have simple geometric shapes but contain certain difficult to cool features. Examples of non-partitionable parts are shown in Figures 15.6 and 15.7. A more complete discussion of non-partitionable parts is given in Section 15.5.

15.4 Partitioning Partitionable Parts

Part partitioning is a procedure for dividing a part into a series of elemental plates similar to the ones shown in Figure 15.2. The procedure works best on parts with an uncomplicated geometrical shape, and features which can be cooled using cooling channels, bubblers and/or baffles as described below.

To partition or divide a part into a series of plates (see Figure 15.2) we proceeds as follows:

a) Determine the principal shape of the part and

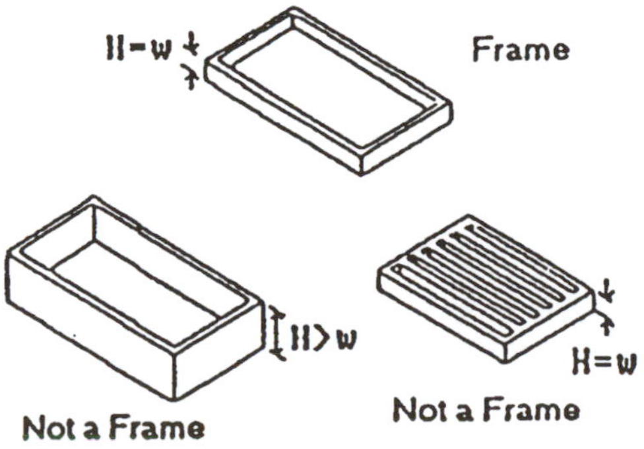

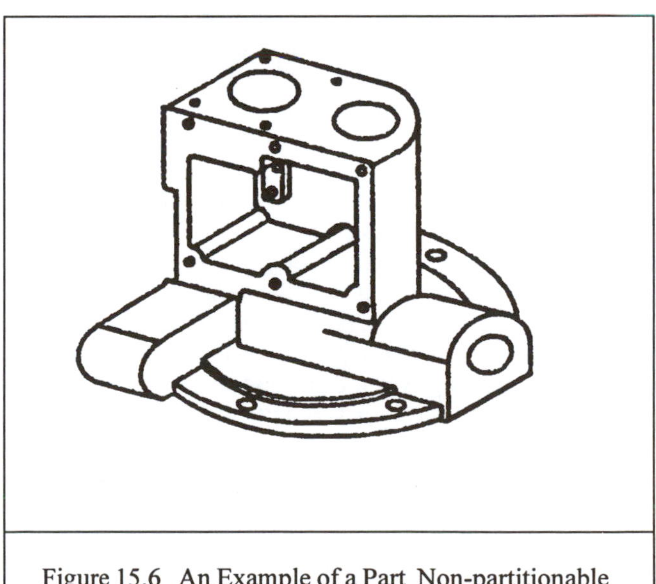

Figure 15.6 An Example of a Part Non-partitionable Due to Geometric Complexity

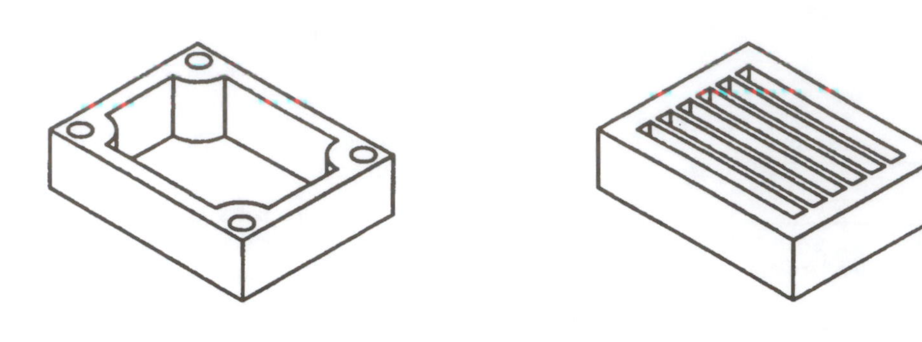

Figure 15.7 Examples of Parts Non-partitionable Due to Difficult to Cool Features. For the part on the left, the size and depth of the cored holes make them difficult to cool. For the part on the right, the depth and spacing of the plates shown make them difficult to cool.

whether the part is slender or non-slender. (Recall from Section 11.3.3 that the principal shape of a part is the solid geometric shape to which the part reduces if all holes, depression, and projections normal or parallel to the direction of mold closure are neglected.)

b) Divide the principal shape into a series of plates.

c) Assign the corresponding add-on features (bosses, ribs, etc.) to the appropriate elemental plates.

Figure 15.8 shows a hollow rectangular prismatic part comprised of four side walls and a base. The four walls and the base have equal and constant wall thicknesses. Also shown in Figure 15.8 is the principal shape of the part along with two alternative divisions of the part into elemental plates. To understand that the divisions are equivalent, it is necessary to understand how parts are cooled.

Figure 15.9 shows a schematic of a cooling system for such a part. While the side walls (*external plates*) are efficiently cooled by fluid channels running parallel to the walls, the base (*internal plate*) needs to be cooled by more sophisticated but less efficient units such as baffles and bubblers. Thus, while plates (1), (2), (3) and (5) in Figure 15.8 are geometrically similar easier to cool external plates, plate (5) is a more difficult to cool internal plate, while the others are easier to cool external plates. Although plate (4) is also an external plate, it is a grilled plate.

While the above explanation indicates that a difference in cooling time should exist between internal and external plates, in practice the increase in cycle time for injection molded parts was not found to be statistically significant. This may be due in part to the fact that most engineering type partitionable parts have an L/H ratio small enough so that reasonable cooling occurs in any case.

Figure 15.10 shows the partitioning of some additional parts whose *part envelope* is rectangular. (The part envelope is the smallest cylinder or rectangular prism that can completely enclose the part. When the choice is not clear, the one that has the maximum contact with the part should be chosen.) Figure 15.11 shows the partitioning of

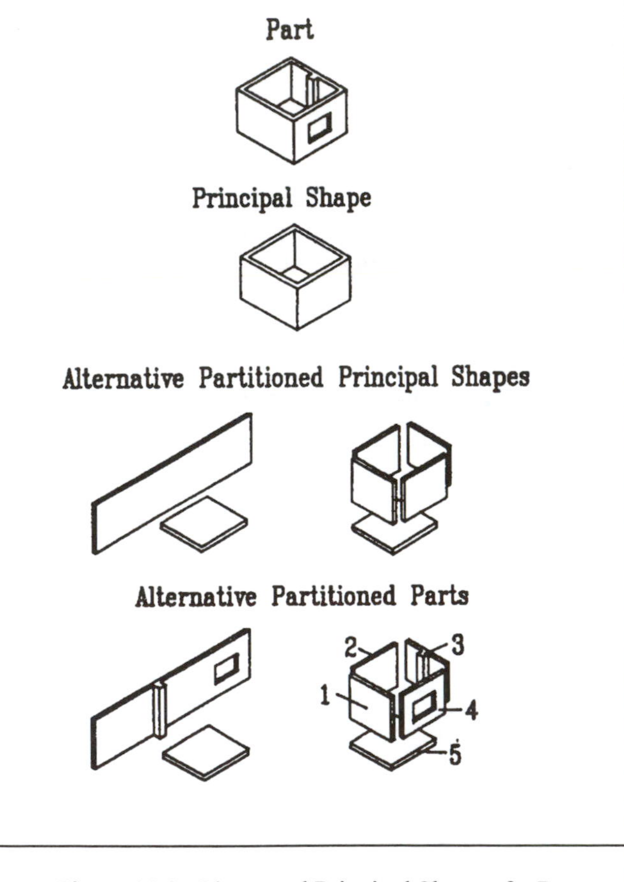

Figure 15.8 Plates and Principal Shape of a Part

parts whose part envelope is cylindrical. The partitionings illustrated in Figure 15.11 assume that the cylinders have a constant wall thickness, and similar sets of subsidiary features (holes, projections, etc.). It is also assumed that the diameter of the cylinder is greater than 12.5 mm. A smaller diameter would result in a part that would be extremely diffi-

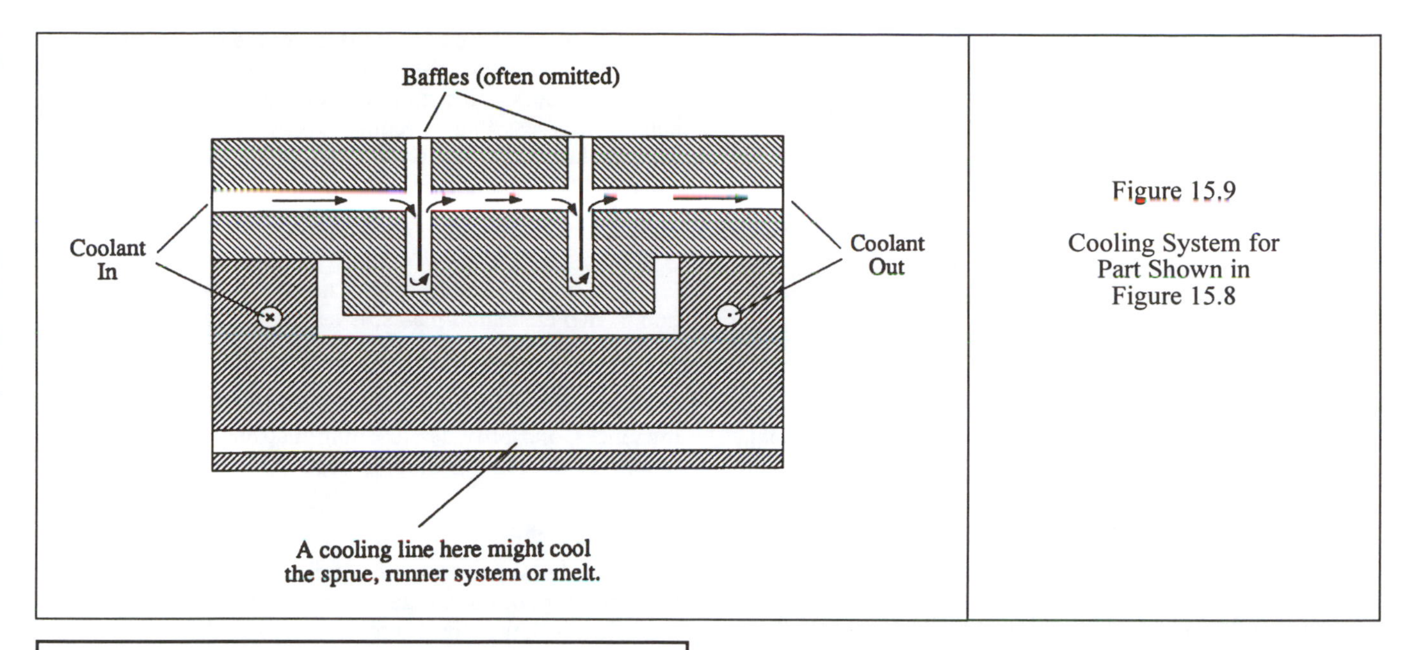

Figure 15.9

Cooling System for Part Shown in Figure 15.8

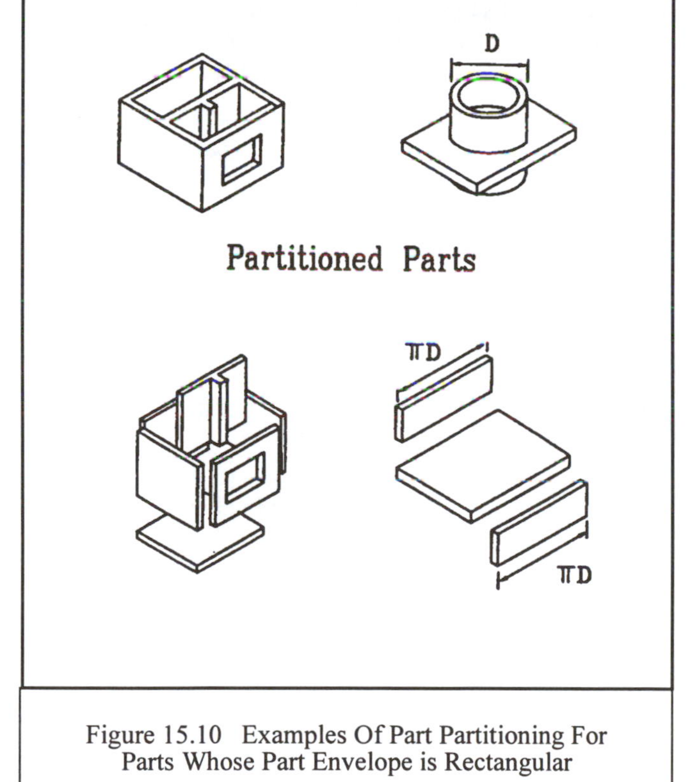

Figure 15.10 Examples Of Part Partitioning For Parts Whose Part Envelope is Rectangular

cult to cool. If the wall thickness of the cylinder were not constant, the same partitioning could be used; however, one would use the maximum wall thickness in determining the relative cycle time of the elemental plate.

15.5 Non-Partitionable Parts

Not all parts are partitionable. Some contain geometrically complex shapes (Figure 15.6) while others contain some extremely difficult to cool features (Figure 15.7) where even baffles and bubblers cannot be used. (In general, if the internal diameter of a cavity or hole is less than 12.5 mm, bubblers cannot be used.) These parts are generally harder to cool and may present difficulties in meeting other requirements related to warping, surface finish, tolerances, etc. These types of parts have relative cycle times greater than those predicted by use of the coding system applied to slender and non-slender partitionable parts. The coding system can, however, be used to determine the lower bound for the relative cycle time of non-partitionable parts. This is done by determining:

a) The relative cycle time for that portion of the part which is "easy-to-cool", and

b) The relative cycle time for the most difficult to cool feature.

The lower bound will be the larger of the values determined in (a) and (b). The actual value could be 25% to 50% higher than the values obtained via the coding system.

15.6 Other Features Needed To Determine the First Digit

15.6.1 Introduction

The ability to identify (and partition) slender and non-slender partitionable parts enables us to determine whether the basic relative cycle time, t_b, will be found in section (a), (b), or (c) of Figure 15.1. However, to completely identify the first digit and get the actual value of t_b, we must also be able to determine the presence or absence of such features as ribs (and their types), lateral projections, grilles and slots, and gussets. Therefore, in this section, we define the meaning of these terms as used in the coding system.

15.6.2 Ribs

Ribs and bosses are also discussed in Section 11.3.3, including descriptions of the molds used to create them.

Types of Ribs. When ribs are present, cycle time has been found to depend upon the types of ribs present. See Figure 15.12. *Multidirectional ribs* and *concentric ribs* provide greater rigidity than *unidirectional ribs* and *radial ribs*. For this reason multidirectional ribbing is often preferred as a means of stiffening parts with thin walls. (A properly designed thin wall with ribs can be lighter than a non-ribbed thicker wall with the same stiffness. In addition, plates with multidirectional ribbing have less tendency to warp than plates with unidirectional ribbing, thereby making shorter cycle times possible.) From the point of view of machine cycle time, *peripheral ribs* can be treated as walls because they do not increase the localized wall thickness of the part.

Significant Ribs. We designate ribs designed such that $[3w < h < 6w]$ or $[b > w]$ as *significant ribs* (Figure 15.13). Significant ribs tend to increase the machine cycle time, because of increased local wall thickness at their base, and, in the case of unsupported unidirectional ribs, make tolerances difficult to hold. In addition, such ribs can be difficult to fill, and may result in shallow depressions (called *sink marks*) on the surface of the part under the rib due to the collapsing of the surface following local internal shrinkage. To avoid sink marks, ribs should be designed so that (a) the rib height, h, is less than or equal to three times the localized wall thickness, w, and (b) the rib width, b, is less than or equal to the localized wall thickness.

15.6.3 Gussets

A rib of variable height, usually present at the junction of two elemental plates, is called a gusset plate (Figure 15.14). A gusset plate can also be present at the junction between a projection (boss, rib, etc.) and the wall to which it is attached. Gusset plates facilitate mold filling, help hold tolerances, and provide stiffening that may permit a reduction in the thickness of bosses and walls.

15.6.4 Bosses

For bosses, cycle time is influenced by whether or not they are supported by gusset plates (See Figure 15.14.) Plates with large (tall) projections are difficult to cool because these features significantly increase localized wall thickness, are difficult to fill, and have a tendency to warp. Significant projections should be supported by gusset plates

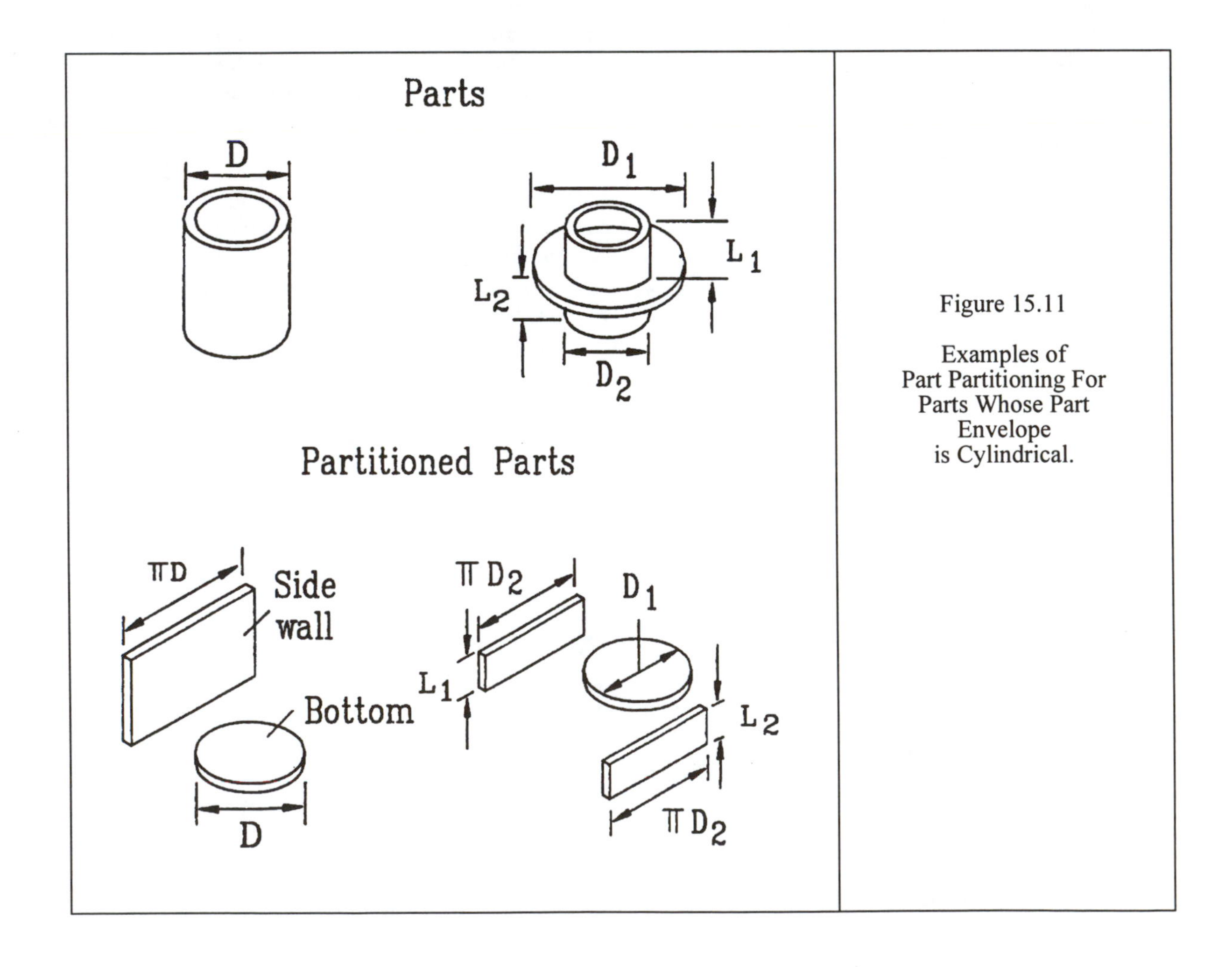

Figure 15.11

Examples of Part Partitioning For Parts Whose Part Envelope is Cylindrical.

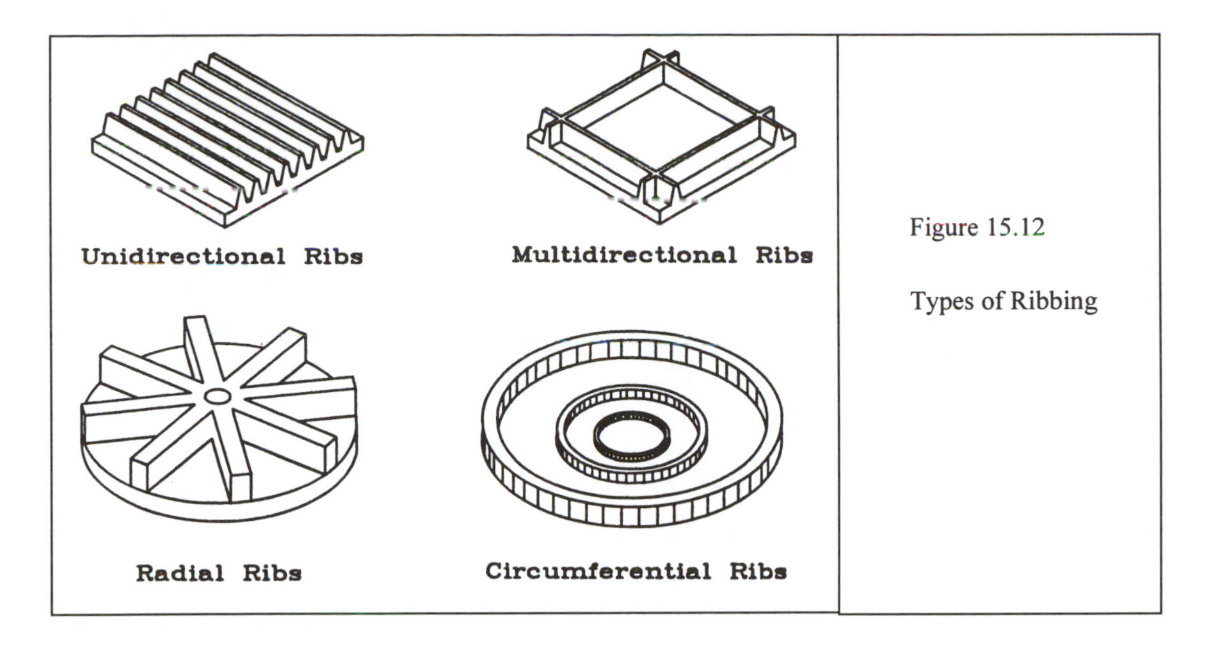

Figure 15.12

Types of Ribbing

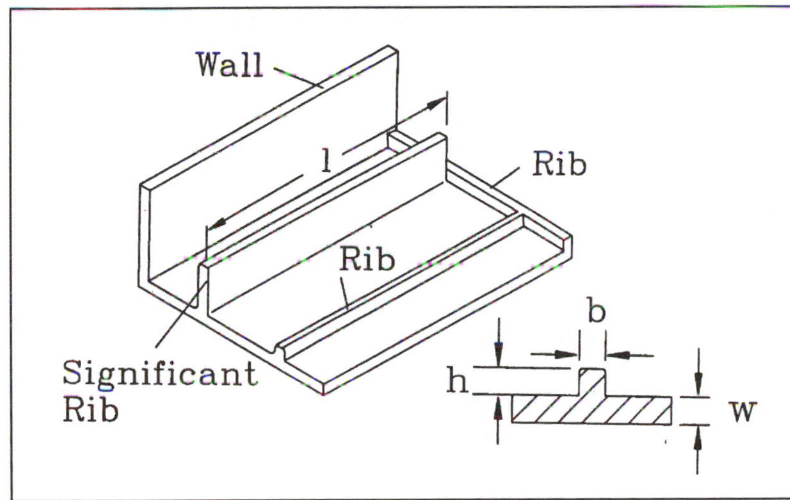

Figure 15.13 Ribs

The length of the rib is determined by length l, width b, and height h. The wall thickness of the plate is given by w. A significant rib is one where $3w < h < 6w$ or $b > w$.

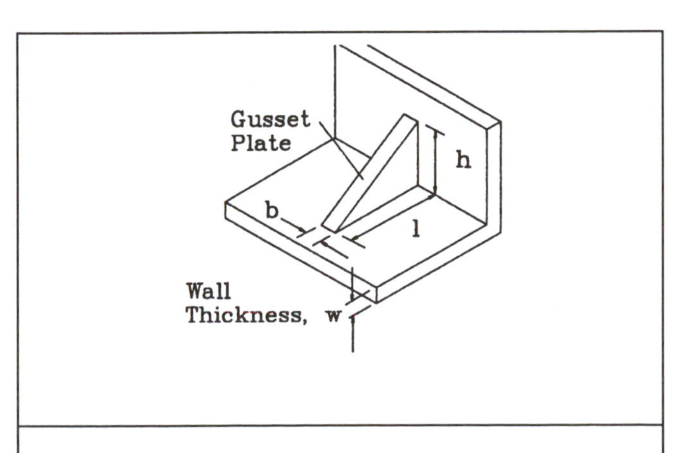

Figure 15.14 Gusset Plate. The thickness of the gusset plate is denoted by b, the height by

to reduce their tendency to deflect due to residual stresses.

Significant Bosses. We designate bosses designed such that [$3w < h$] or [$b > w$] as *significant bosses* (Figure 15.15). Significant bosses tend to increase the machine cycle time, make tolerances difficult to hold, and they can be difficult to fill. In addition, significant bosses often cause sink marks.

15.6.5 Grilles and Slots

Elemental plates containing: a) multiple through holes, b) no continuous solid section with a projected area greater than 20 percent of the projected area of the plate envelope, and c) whose height is equal to its wall thickness are considered grilled or slotted (Figure 15.16). Such plates are low in strength and usually have low surface gloss.

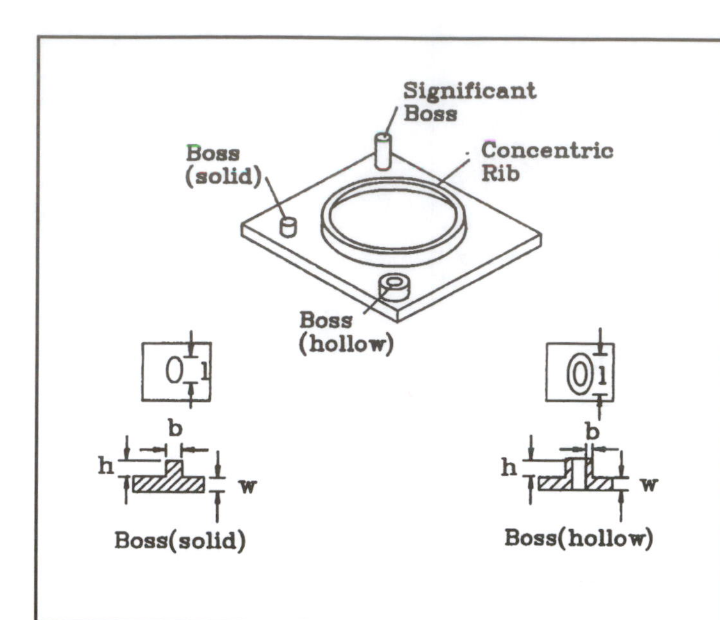

Figure 15.15 Bosses

The boss length is denoted by l, the boss width by b, the boss height by h, and the wall thickness of the plate by w.

15.6.6 Lateral Projections

Lateral projections are add-on features which protrude from the surface of a slender plate in a direction normal to the longitudinal axis (Figure 15.17). For long slender parts, such projections are difficult to fill.

15.7 Wall Thickness - The Second Digit

The above definitions will enable designers to determine the basic part type, and to establish the first digit of the coding system in Figure 15.1. Determining the second digit is essentially based on the largest wall thickness (w) of the elemental plates.

In the case of non-partitionable parts, two wall thicknesses need to be used. We need to determine the cycle time based on the easy to cool features and the cycle time based on the difficult to cool features. Thus, one wall thickness is the largest wall thickness of all the easy to cool portions of the part. The other is the largest wall thickness of all the difficult to cool features.

15.8 Inserts and Internal Threads - The Third Digit

Refer again to Table 15.3. Note that to obtain the value of t_e we need only determine whether or not the part has internal threads and inserts. The meaning of internal thread is obvious. *Inserts* are metal components added to the part prior to molding the part. Inserts are added for decorative purposes, to provide additional localized strength, to transmit electrical current, or to aid in assembly or subassembly work. The use of inserts increases the machine cycle time.

15.9 Surface Requirements and Tolerances - The Fourth and Fifth Digits

15.9.1 Surface Requirements

Hot molds produce glossier surfaces and result in longer cycle times. Cool molds yield duller finishes and result in shorter cycle times. In addition, and more importantly, high surface gloss requirements may greatly reduce production efficiency and yield because of a higher rejection rate resulting from visible sink marks, jet lines and other surface flaws.

The preferred surface for a part is usually one that is produced from a mold having a Society of Plastics Industry (SPI) finish of 3 or 4. Such parts tend to be used on industrial products where high gloss is not required. Parts requiring a high gloss and good transparent clarity are produced from molds having an SPI of 1 or 2. For parts requiring a textured surface and low surface gloss, a mold having a SPI of 5 or 6 is used.

For the purposes of the coding system, the part surface requirement (See Table 15.4) is considered *High* when

>Parts are produced from a mold having a SPI surface finish of 1 or 2.

>Sink marks and weld lines are not allowed on an untextured surface.

In the coding system, parts without high surface requirements are considered to have *low* surface requirements.

A statistical analysis of piece parts collected from several molders shows that the adverse effect of a high gloss requirement is more significant on thin parts than on thick parts. A possible explanation is as follows:

The colder the mold or the melt, the more viscous the

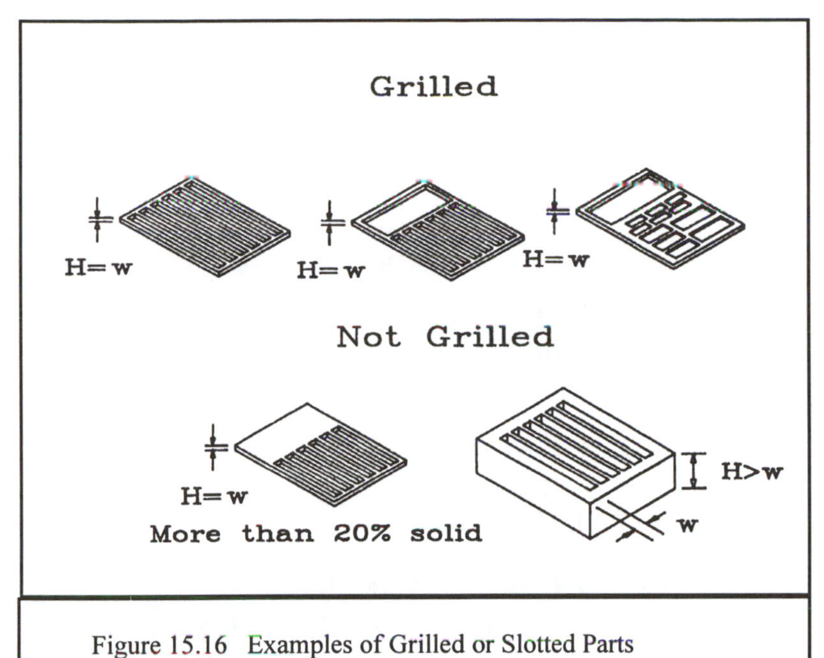

Figure 15.16 Examples of Grilled or Slotted Parts

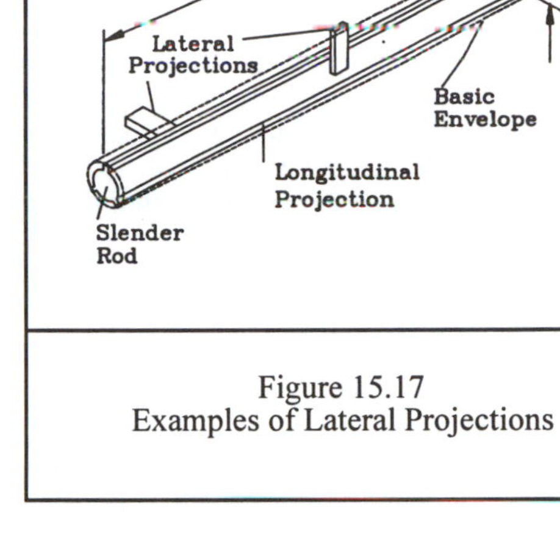

Figure 15.17
Examples of Lateral Projections

flow. The more viscous the flow, the greater the tendency to leave visible "flow marks" on the surface of the part. These flow marks make it difficult to obtain a good surface gloss. Since thin parts cool faster than thick parts, poor surface gloss is more likely to occur on thin parts.

15.9.2 Tolerances

There are two types of tolerance requirements: (a) *dimensional* and (b) *geometric*. Dimensional tolerances refer to tolerances on the length, width, and height of a part, as well as on the distance between features. Geometric tolerances refer to tolerances on flatness, straightness, perpendicularity, cylindricity, etc.

For dimensional tolerances, a standard exists prepared by the Society of Plastic Industry, and each material supplier converts their data to suit a specific material. Thus, to determine whether the tolerances specified are tight or commercial, we must refer to either the data published by the Society of the Plastic Industry for the material in question, or to data supplied by the resin manufacturer. No industry wide standard exists as yet for geometric tolerances.

Part yield is influenced by part tolerance requirements. Part yield decreases when part tolerances are relatively difficult to hold. Part tolerances are considered difficult to hold under the following conditions.

a) External undercuts are present.

b) A tolerance is specified across the parting surface of the dies. (The high pressures (10,000 psi) cause slides and molds to move slightly making tolerances across the moving surfaces difficult to hold.)

c) The wall thickness is not uniform. (Thick sections connected to thin sections tend to shrink more than the thin sections. This is because the thick sections continue to cool down and shrink after the thin sections have solidified This variation in shrinkage can result in part warpage. When warping is a problem, the cycle time of a part is increased to allow the part to be more rigid when it is ejected.)

d) Unsupported projections (ribs, bosses) and walls are used. (Unsupported projections can bend making tolerances between them difficult to hold.)

e) More than three tight tolerances, or more than five commercial tolerances are required. (Tolerances should be specified *only* where *absolutely* necessary. As the number of tolerances to be held increases, the proportion of defective parts produced increases.)

15.10 Using the Coding System: Overview

To determine the relative cycle time for a part using the coding system shown in Figure 15.1 and Tables 15.3 and 15.4, we proceed as follows.

a) Determine whether or not the part is partitionable.

b) If the part is partitionable, partition it into elemental plates, and assess the relative cycle time for each plate. *The plate with the largest relative cycle time controls the cycle time for the part.*

c) If the part is not partitionable, we first code the part using the maximum wall thickness of the part. Then we code the part a second time using the most difficult to cool feature. *The feature with the largest relative cycle time controls the cycle time for the part.*

Several examples are presented in Sections 15.12-15.14 below.

15.11 Effect of Materials on Relative Cycle Time

In theory, the machine cycle time depends on the material used. In practice, however, there is little significant difference in cycle time between geometrically similar engineering parts made of different materials. Although it is true that some materials will solidify faster than others, there are other factors that tend to cause the actual cycle time to be very similar for the parts with the same geometry but made of different materials. Some of these factors are:

a) The most important objective for a precision molder is the production of parts with satisfactory dimensional stability and high gloss. Fast molding cycles cause greater variations in mold and melt temperatures than slow cycles. The large temperature variations often result in poor dimensional stabilities, rejected parts, and hence lower yield. Furthermore, fast cycles require lower mold and melt temperatures that are likely to cause difficulties in producing parts with high gloss.

Therefore, in order to produce parts with high gloss and dimensional stability, precision molders are likely to manufacture parts at nearly the same rate for rapidly solidifying materials as they would for materials that solidify at a slower rate.

b) Many engineering parts are produced in annual production volumes of 5,000 to 10,000, and in batch sizes of 500 to 1,000. These batch sizes can be produced in 2 to 4 shifts. For such small batches, it is not practical to try to optimize the cycle time for each part/material combination.

c) Each machine has its own peculiar characteristics and idiosyncrasies. Thus, a part made of the same material but molded on two different machines under the same set of operating conditions, will require different cycle times for the same high gloss and dimensional stability. Consequently, optimal operating conditions will vary from machine to machine. The likelihood that a given part would, for each batch, be run on the same machine is very low.

15.12 Example 15.1: Determination of Relative Cycle Time for a Partitionable Part

15.12.1 The Part

As the first example, consider the part shown in Figure 15.18. In addition to the general overall size of the part, the wall thickness and the size of the ribs and bosses are given. These dimensions are not needed in order to estimate the relative tooling cost for this part, but they must be known in order to determine the relative cycle time.

15.12.2 The Basic Part Type

In this example, the length of the projections parallel to the surface of the part, c, are 10 mm. Since c/L is less than 1/2 then, noting Figure 11.3, these projections are considered *isolated projections of small volume*. Consequently, the dimensions of the basic envelop are:

$$L = 160, \quad B = 130, \quad H = 10.$$

Since the part is straight, the unbent dimensions L_u and B_u are equal to L and B, respectively. Thus, $L_u/B_u < 10$ and the part is *non-slender*. We indicate the fact that the part is non-slender by using the letter N.

15.12.3 Part Partitioning

This part consists of a single elemental plate; thus, no partitioning is needed.

15.12.4 Relative Cycle Time

Basic Relative Cycle Time: The length, l, the width, b, and the height, h, of each projection (boss) is 10 mm, 1.8 mm, and 13 mm, respectively. The plate thickness, w, is 2.0 mm and uniform. Thus, $h/w > 3$, and the projections are considered *significant bosses*. Since the boss thickness (b) is less than the wall thickness (w), and they are not supported by gusset plates, the first digit is 6.

The maximum plate thickness is 2.0 mm; hence, the second digit is 1 and the basic code is N61. Thus, the *basic relative machine cycle time*, t_b, is 2.52. (Remember, relative cycle time is the cycle time relative to the reference part, and it is therefore dimensionless.)

Internal Threads/Inserts: There are no internal threads, but there are three molded-in inserts (one in each boss). Hence, the third digit is 1, and the *additional relative time*, t_e, is 0.5 per insert, or 1.5 for the part.

Surface Gloss/Tolerances: Let's assume that the part requires a low surface gloss (which corresponds to a SPI surface finish of 3 on the mold). Then the fourth digit is 0. Because the bosses are unsupported, it is difficult to hold tolerances between them. Hence, the fifth digit is 1, and the *penalty factor*, t_p, is 1.20.

Relative Effective Cycle Time: The relative effective cycle time is given by substituting into Equation (15.8):

$$t_r = (t_b + t_e)t_p = (2.52 + 1.5)1.2 = 4.82$$

15.12.5 Redesign Suggestions

The relative cycle time can be reduced in several ways. One method is to provide gusset plates to support the bosses. In this case, the basic code becomes N51, and the fifth digit becomes 0. Thus, the relative machine cycle time is reduced to 2.38, t_p becomes 1, and the relative effective cycle time becomes 3.88, a 19% reduction.

A second method for reducing the relative cycle time

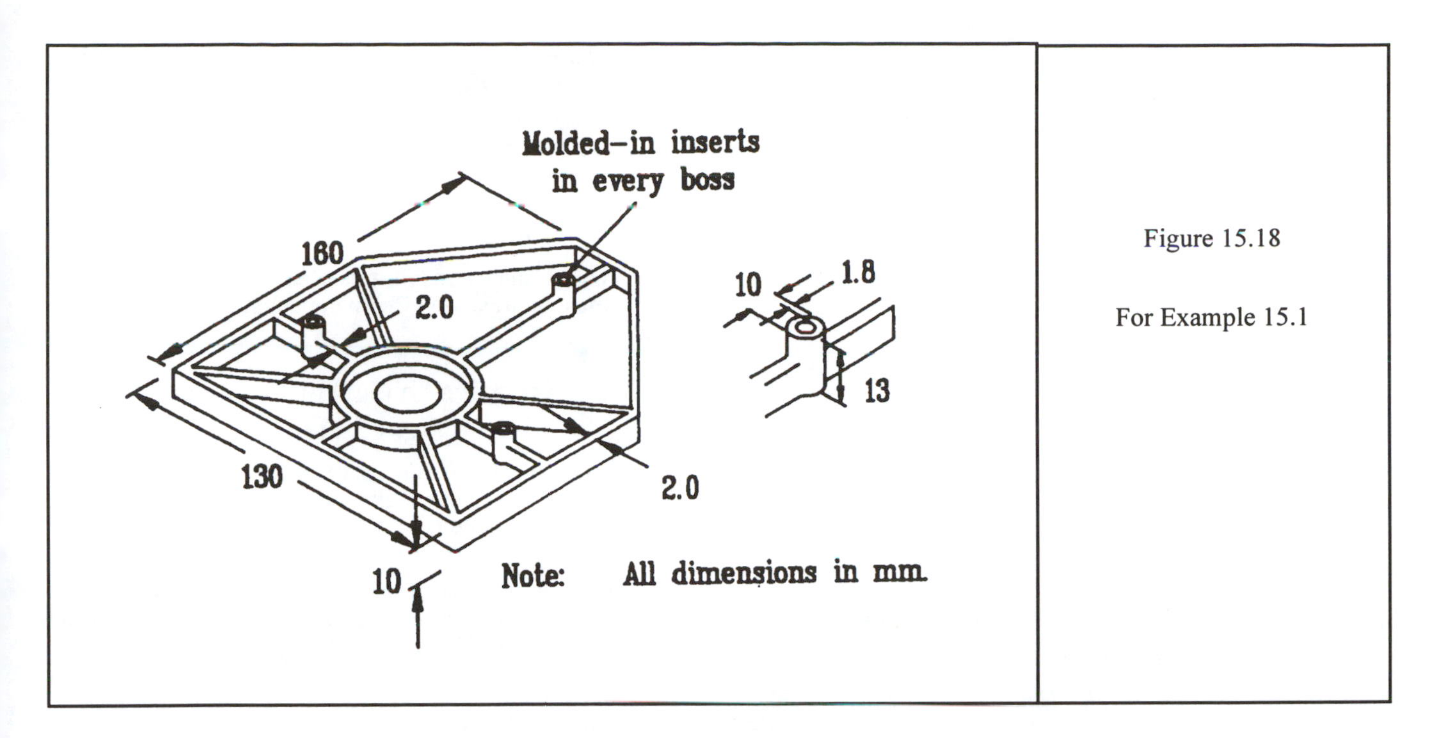

Figure 15.18

For Example 15.1

would be to reduce the height of the bosses so they become non-significant bosses. In this case, the new basic complexity code becomes N41, the fifth digit becomes 0, and the new relative machine cycle time is reduced to 3.74. This is a 22% reduction.

If the bosses are left as originally designed, but the molded-in inserts are removed, t_e is reduced to 0. Then the relative effective cycle time becomes 3.02, a 37% savings.

Readers should note that the magnitude of these savings, created by relatively small design changes, can be extremely important for high volume parts.

15.13 Example 15.2: Determination of Relative Cycle Time for a Partitionable Part

15.13.1 The Part

As a second example, consider the part in Figure 15.19. In addition to the overall size of the part, the wall thickness is given. Since the height of the "peripheral projection" in this case is greater than 6w, it will be treated as a wall rather than a rib. (From the point of view of the coding system, all peripheral projections, including ribs, will be treated as peripheral walls.) The part geometry is rather simple and is, thus, partitionable.

15.13.2 Basic Part Type

Since $L_u = L + H = 130$ mm and $B_u = B = 50$ mm, then $L_u/B_u = 2.6$, and the part is non-slender (N).

15.13.3 Part Partitioning

This part can be partitioned in two ways. It can be partitioned as shown in Figure 15.20 into two separate elemental plates. Alternatively, since the wall thickness of the part is constant and both plates are external plates, the part can be treated as a single plate as shown in Figure 15.21.

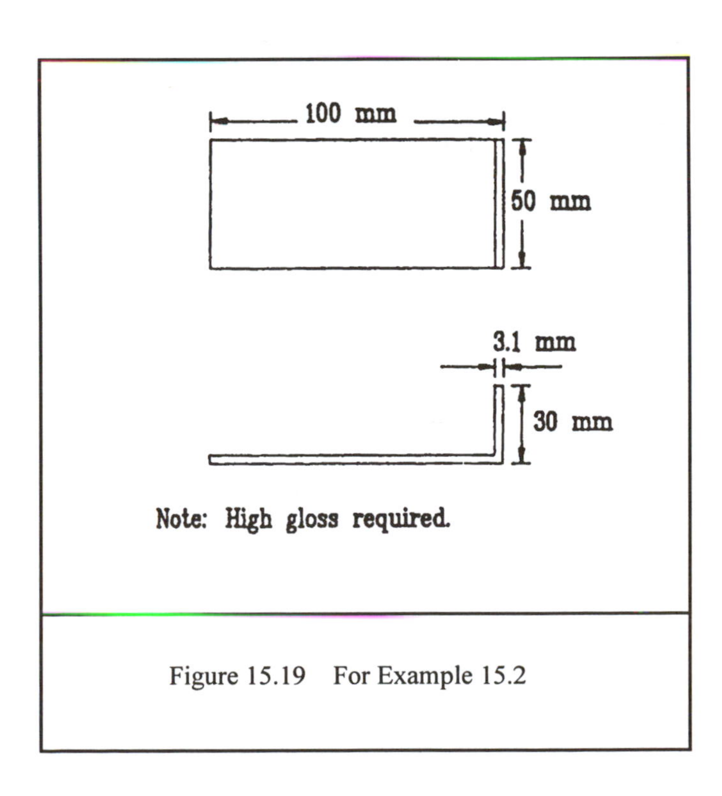

Figure 15.19 For Example 15.2

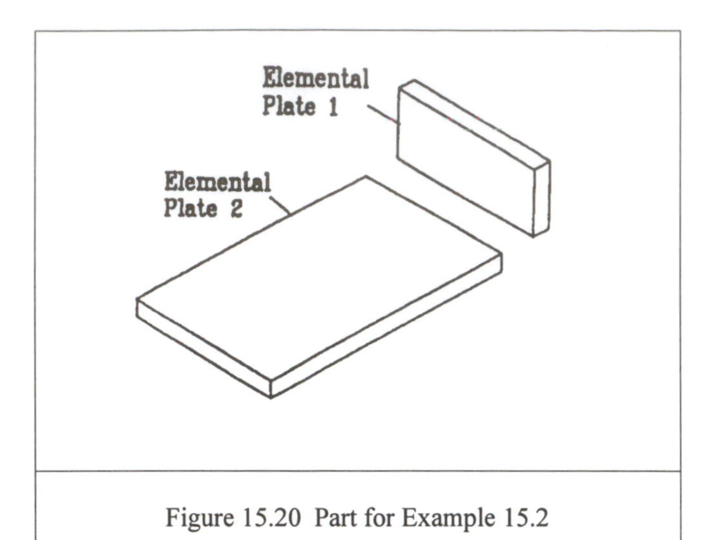

Figure 15.20 Part for Example 15.2

15.13.4 Relative Cycle Time - Elemental Plate 1 (Fig. 15.20)

Basic Relative Cycle Time: The plate is neither grilled nor slotted, and no projections are present. Thus, the first digit is 2. The wall thickness is 3.1 mm, hence, the second digit is 3. The basic code is N23. Thus, the basic relative machine cycle time, t_b, is 3.39.

Internal Threads/Inserts: There are no internal threads nor molded-in inserts. Hence, the third digit is 0 and the relative extra mold opening time, t_e, is 0.

Surface Gloss/Tolerances: The part is assumed to require a high surface gloss (SPI surface finish of 1 or 2 on the mold) thus, the fourth digit is 3. The 90° angle between the side wall and the bottom is difficult to hold because of the unradiused (sharp) corner and the lack of supporting gusset plates. Thus, the part tolerance is difficult to hold, the fifth digit is 1 and the penalty, t_p, is 1.37.

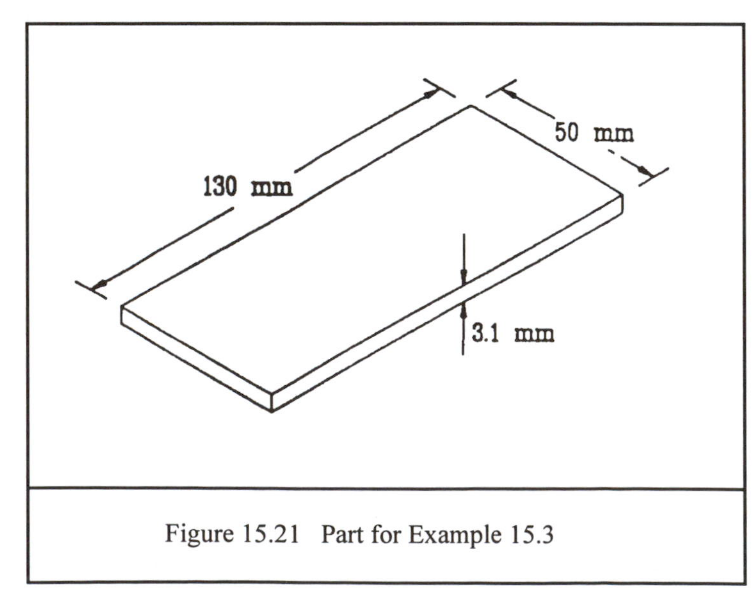

Figure 15.21 Part for Example 15.3

Relative Effective Cycle Time: The relative effective cycle time for the original design is:

$$t_r = (t_b + t_e)t_p = (3.39)1.37 = 4.64$$

15.13.5 Relative Cycle Time - Elemental Plate 2 (Fig. 15.20)

The complete code for this plate is identical to that for plate 1, thus, the relative effective machine cycle time is the same.

15.13.6 Relative Cycle Time - Elemental Plate (Fig. 15.21)

Basic Relative Cycle Time: Once again, the plate is neither grilled nor slotted and no projections are present. The code for this plate is also N23 and the relative machine cycle time is 3.39.

Internal Threads/Inserts: There are no internal threads nor molded-in inserts. Hence, the third digit is 0 and the relative extra mold opening time, t_e, is 0.

Surface Gloss/Tolerances: The part is assumed to require high surface gloss, thus, the fourth digit is 3. The 90° angle between the side wall and the bottom is difficult to hold because of the unradiused (sharp) corner and the lack of supporting gusset plates. Thus, the part tolerance is difficult to hold, the fifth digit is 1 and the penalty is 1.37.

Relative Effective Cycle Time: The relative effective cycle time for the design is:

$$t_r = (t_b + t_e)t_p = 4.64$$

15.13.7 Redesign Suggestions

The two cost drivers in this case are the high surface gloss requirement and the difficult to maintain 90° angle between the side wall and the bottom surface. If the part surface can be replaced by a textured surface (SPI-3 on the mold) the fourth digit becomes 0. If in addition the corner between the two plates can be radiused or if supporting gusset plates can be used, the fifth digit becomes 0 and the new penalty is raised to 1.0. Thus, the new relative effective cycle time is 3.39; a 27% reduction.

15.14 Example 15.3: Relative Cycle Time for a Non-Partitionable Part

15.14.1 The Part

The part and all of the dimensions necessary for a determination of the relative cycle time are shown in Figure 15.22.

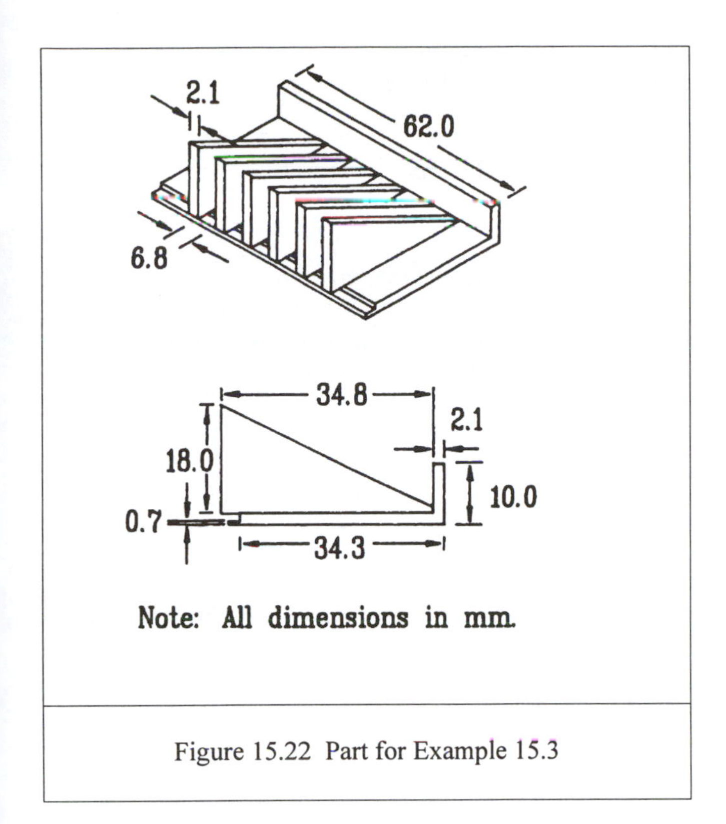

Figure 15.22 Part for Example 15.3

15.14.2 Part Partitioning

The closely spaced "fins" are difficult to cool, because the spacing between fins is less than 12.5 mm. Thus, in spite of its rather simple geometry, the part is considered *non-partitionable.*

15.14.3 Relative Cycle Time

Basic Machine Cycle Time: Since this part is non-partitionable, the basic code must be found for the thickest easy to cool feature as well as the thickest difficult to cool feature. However, since the thickness of the part is uniform, only the difficult to cool feature needs to be coded. Thus, since the wall thickness is 2.1 mm, the first and second digits are 1 and 2, respectively. Hence, the basic code for the part is NP12, and the basic relative time, t_b, is 4.50.

Internal Thread/Inserts: There are no internal threads or molded-in inserts; thus, third digit is 0, and $t_e = 0$.

Surface Gloss/Tolerances: The presence of the fins makes sink marks unavoidable on the plate to which the fins are attached. Thus, the fourth digit is 2. The lack of gusset plates supporting the closely spaced fins makes the tolerance on the distance between fins difficult to hold; hence, the fifth digit is 1. Therefore, the penalty is 1.41

Relative Effective Cycle Time: The relative effective cycle time is:

$$t_r = (t_b + t_e)t_p = (4.50 + 0)(1.41) = 6.34.$$

15.14.4 Redesign Suggestions

For non-partitionable parts, design improvements are often difficult to make without major changes in part geometry. One possible improvement in the current case, in the sense of reducing processing costs, is to increase the spacing between fins so that the cooling difficulty can be reduced.

It is worth pointing out that the part was coded as though the spacing between fins is critical, and that sink marks are not acceptable. However, practical concerns may be such that stringent tolerance and surface requirements are not necessary. This would change both the fourth and fifth digits to 0, and the relative effective cycle time would also change accordingly.

15.15 Relative Processing Cost

The preceding discussion and examples have shown how to compute the relative part cycle time, t_r. This is a critical requirement in estimating relative processing costs. As noted in the first section of this chapter, the relative processing cost is given by:

$$C_e = t_r C_{hr} \qquad (15.2)$$

where C_{hr} represents the ratio C_h/C_{ho}. C_{ho} represents the machine hourly rate for the reference part, and C_h is the machine hourly rate ($/h). The relative machine hourly rate, C_{hr}, can be determined from Table 15.5, but it is first necessary to determine the injection molding machine size (tonnage) required to mold the part.

The machine force required to mold a part is approximately two to five tons per square inch, depending on the material to be molded and the projected area of the part normal to the direction of mold closure. In general, it is assumed that a machine whose tonnage, F_p, exceeds three times the projected area of the part expressed in in^2 will suffice. Thus,

Table 15.5 Machine Tonnage and Relative Hourly Rate (Data published in *Plastic Technology*, June, 1989.) Injection Molding.

Machine Tonnage	Relative Hourly Rate
< 100	1.00
100 - 299	1.19
300 - 499	1.44
500 - 699	1.83
700 - 999	2.87
> 1000	2.93

the required machine tonnage is approximately

$$F_p = 3\ A_p \qquad (15.9)$$

where the projected area, A_p, is in square inches. Or:

$$F_p = 0.005\ A_p \qquad (15.10)$$

where the projected area is in mm^2.

15.16 Relative Material Cost

As noted in the first section of the chapter, the material cost for a part, K_m, is given by

$$K_m = V\,K_p \qquad (15.4)$$

where V is the part volume and K_p is the material cost per unit volume, and the relative material cost is:

$$C_m = (V / V_o)\ C_{mr} \qquad (15.5)$$

Values for C_{mr} are given above in Table 15.2.

15.17 Total Relative Part Cost

As stated earlier in this chapter (Section 15.1.4), the total production cost of a part, K_t, is now computed as the sum of the material cost of the part, K_m, the tooling cost, K_d/N, and the processing cost, K_e:

$$K_t = K_m + (K_d / N) + K_e \qquad (15.6)$$

where K_d represents the total cost of the tool, and N represents the production volume.

If the manufacturing cost of a reference part is denoted by K_o, then the total cost of the part relative to the cost of the reference part, C_r, can be expressed as:

$$C_r = (K_m + K_d/N + K_e) / K_o. \qquad (15.7)$$

If K_{mo}, K_{do}, and K_{eo} represent the material cost, tooling cost, and equipment operating cost for the reference part, then C_r can be written as follows:

$$C_r = (K_m/K_{mo})(K_{mo}/K_o) + (N_o/N)(K_d/K_{do})(K_{do}/K_o)(1/N_o) + (K_e/K_{eo})(K_{eo}/K_o).$$

If $f_m = K_{mo}/K_o$,

$f_d = K_{do}/N_oK_o$, and

$f_e = K_{eo}/K_o$,

which represent the ratio of the material cost, tooling cost and processing cost of the reference part to the total manufacturing cost of the reference part, and if

$C_m = K_m/K_{mo}$,

$C_d = K_d/K_{do}$, and

$C_e = K_e/K_{eo}$

are, respectively, the material cost, tooling cost and equipment operating cost of a part relative to the material cost, tooling cost and equipment operating cost of the reference part, then the above equation becomes

$$C_r = C_m f_m + C_d f_d + C_e f_e \qquad (15.11)$$

where it is assumed that $N = N_o$. The value for C_d is obtained from Equation (11.4) as described in Chapter 11, while the values for C_e and C_m are obtained from Equations 15.2 and 15.5, respectively.

If it assumed that only single cavity molds will be used, then the values for f_m, f_d, and f_e can be obtained from the following equations::

$$f_m = \frac{V_oK_{po}}{V_oK_{po} + (K_{dco} + K_{dmo})/N + t_oC_{ho}} \qquad (15.12)$$

$$f_d = \frac{(K_{dco} + K_{dmo})/N}{V_oK_{po} + (K_{dco} + K_{dmo})/N + t_oC_{ho}} \qquad (15.13)$$

$$f_e = \frac{t_o \cdot C_{ho}}{V_oK_{po} + (K_{dco} + K_{dmo})/N + t_oC_{ho}} \qquad (15.14)$$

The values of f_m, f_d, and f_e represent the proportion of the total relative part cost due to part material, tooling, and processing, respectively. For the data shown in Table 15.1 for the reference part, the above expressions reduce to the following:

$$f_m = N / (3.84x10^6 + 68N) \qquad (15.15)$$

$$f_d = (3.84 \times 10^6) / (3.84 \times 10^6 + 68N) \qquad (15.16)$$

$$f_e = 67\,N / (3.84x10^6 + 68N) \qquad (15.17)$$

Plots of Equations (15.15) - (15.17) are shown in Figure 15.23. Note that at low production volumes, less than 10,000, the proportion of part cost due to tooling is more than 80% of the total cost of the part, while the proportion of

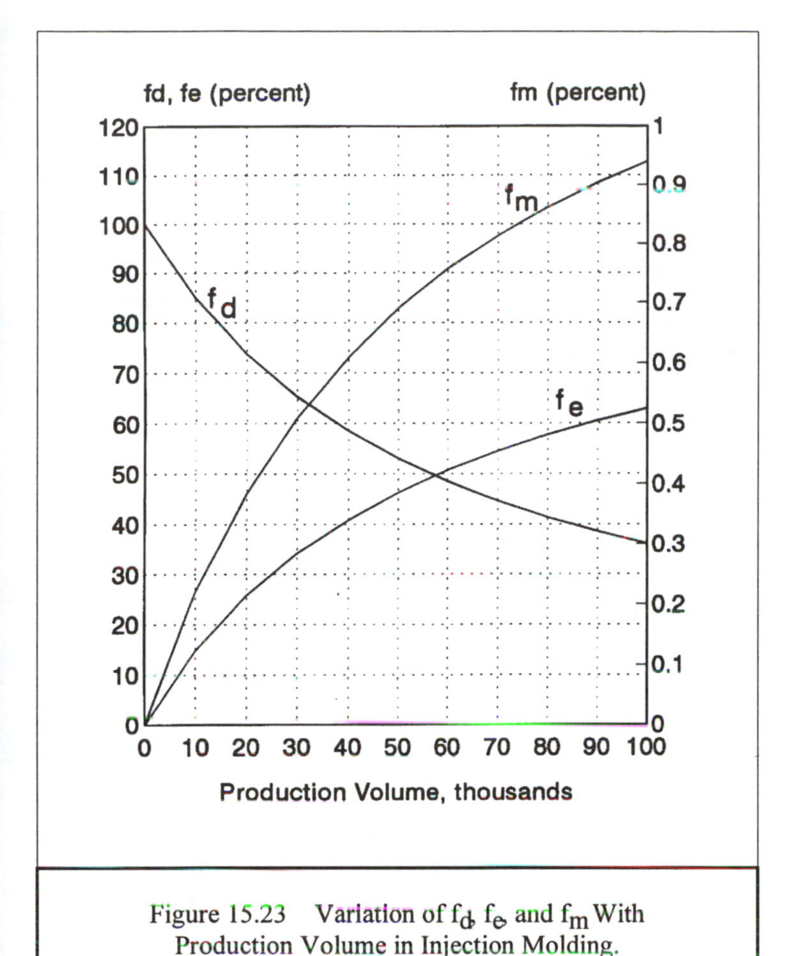

Figure 15.23 Variation of f_d, f_e, and f_m With Production Volume in Injection Molding.

the cost due to processing is less than 15% of the total part cost. At large production volumes, say 100,000 parts, the proportion of part cost due to tooling drops to less than 40%, while the proportion of cost due to processing rises to more than 60%.

15.18 Example 15.4: Determination of the Total Relative Part Cost

The part shown in Figure 15.24 is a made of polycarbonate. The wall thickness is a uniform 3.5 mm, and the ribs are, according to the classification system, not significant.

As originally designed, the total relative die construction cost, C_d, is found to be 2.32. (See Exercise 1 of Chapter 11.)

The relative cycle time, t_r, for this part can be shown to be:

$$t_r = 3.67(1.20) = 4.40.$$

The projected area of the part is 9,000 mm^2; hence, from Equation (15.10), the required clamping force is:

$$F_p = 0.005A_p = 45 \text{ tons}.$$

Consequently, from Table 15.5, the relative machine hourly rate, C_{hr}, is 1.00. Thus,

$$C_e = t_r C_{hr} = (4.40)(1) = 4.40.$$

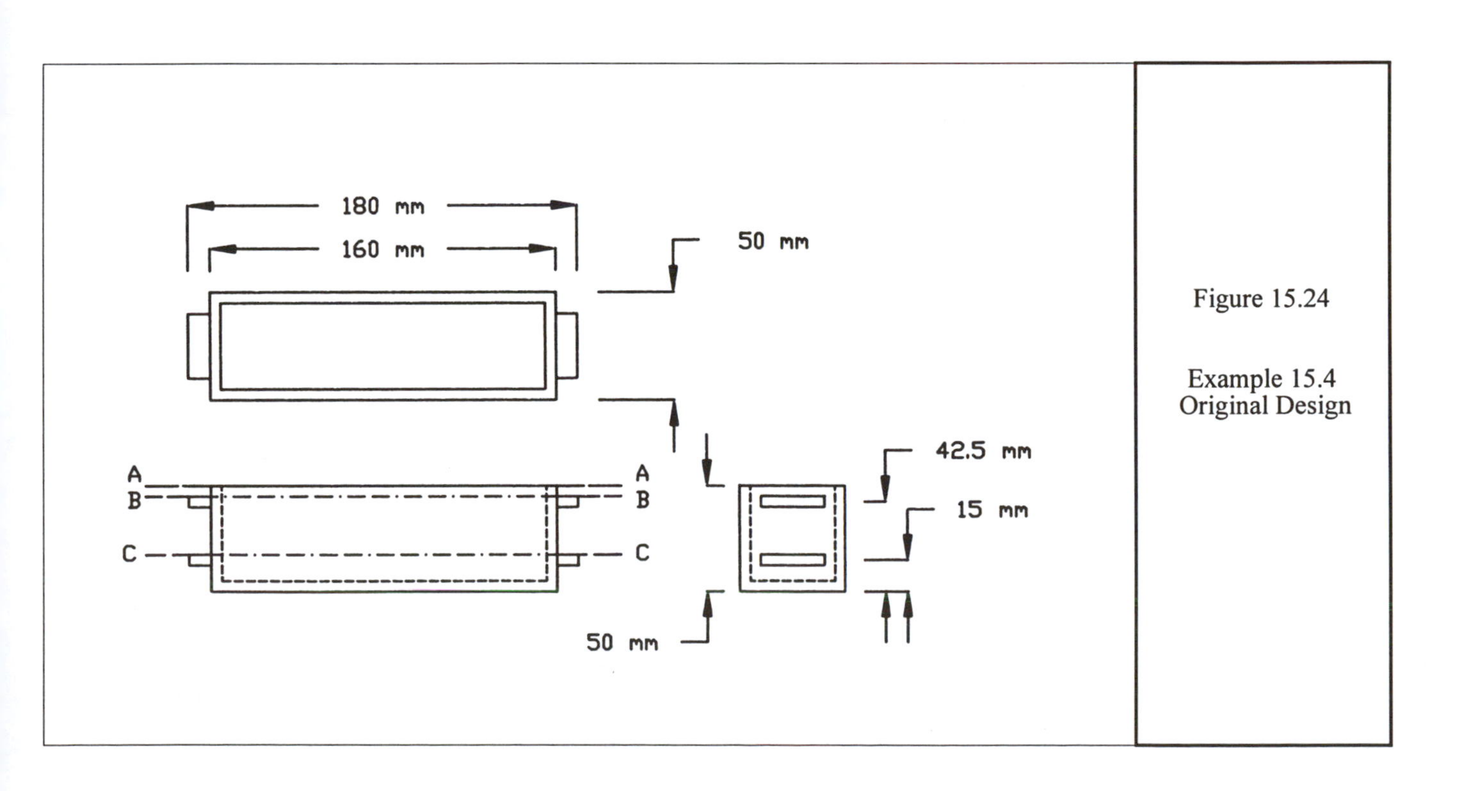

Figure 15.24

Example 15.4 Original Design

The part volume is approximately 105,000 mm^3. Since the material is polycarbonate, from Table 15.2 the material price, C_{mr}, is 2.96. From Equation (15.5) the relative material cost for the part is

$$C_m = (V/V_o)C_{mr} = (105{,}000/1244)(2.96) = 250$$

where $V_o = 1244\ mm^3$ is the volume of the reference part.

Production Volume = 10,000

If the production volume of the part is 10,000 pieces, then from Figure 15.23,

$$f_m = 0.002,\ f_d = 0.85,\ f_e = 0.15,$$

and the total relative cost of the original part, C_r, is, from Equation 15.11:

$$C_r = (0.002)250+(2.32)0.85+(4.40)0.15 = 3.13.$$

If the part is redesigned as shown in Figure 15.25, then C_d is reduced to 1.65 (See Exercise 2 of Chapter 11). The values for t_r, C_{hr}, C_e, and C_m remain the same. Thus,

$$C_r = (0.002)250+(1.65)0.85+(4.40)0.15 = 2.56.$$

This is an overall savings of some 18% in piece part cost.

Production Volume = 100,000

If the production volume of the part is 100,000 pieces, then from Figure 15.23:

$$f_m = 0.009,\ f_d = 0.36,\ f_e = 0.64$$

and the total relative cost of the original design, C_r, is, from Equation (15.11):

$$C_r = (0.009)250+(2.32)(0.36)+(4.40)(0.64) = 5.90.$$

For the redesigned part,

$$C_r = (0.009)250+(1.65)(0.36)+(4.40)(0.64) = 5.66.$$

Thus, for a production volume of 100,000 parts, the savings is only 4.3%. This reduction in savings is due to the fact that, in this particular case, all of the savings were due entirely to a reduction in tooling costs.

You may have been surprised that the relative cost of the part in this example increased when the production volume increased. There is a tendency to expect that, since the absolute part cost will decrease when the production volume increases, then the relative part cost will also decrease. Recall, however, that it has been assumed here that the production volume of both the part and the reference part are the same. Thus, as the production volume increases, the absolute cost of both the the reference part and the designed part are decreasing, but the cost of the reference part relative to the reference part is increasing. This is due to the fact that the absolute cost of the simple washer is decreasing faster than the absolute cost of the designed part.

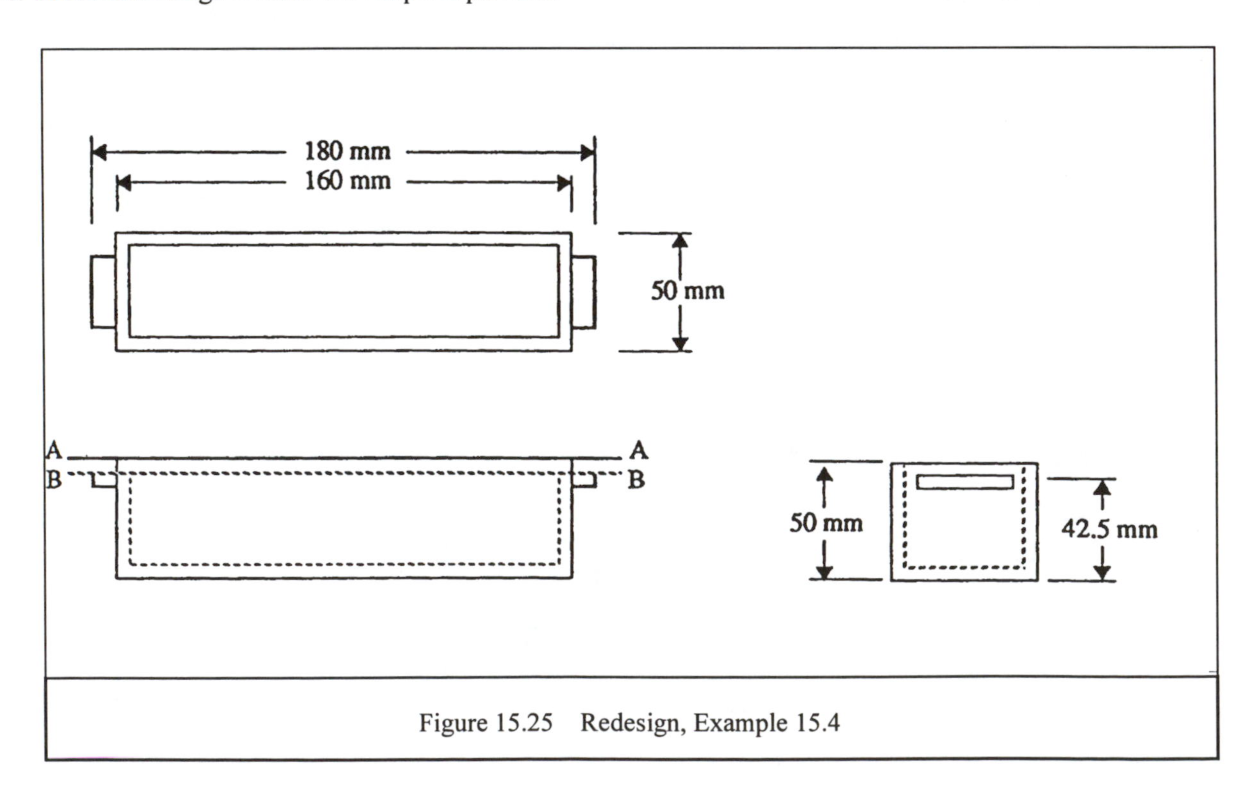

Figure 15.25 Redesign, Example 15.4

15.19 Die Cast Part Costs - Overview

As has already been stated, the processes of injection molding and die casting are somewhat similar. In fact, they are sufficiently similar that the classification system for the determination of relative cycle time developed for injection molding, with only minor modifications, is applicable to die casting.

As in the case of injection molding, die cast parts are classified as partitionable or non-partitionable. In addition, as in the case of injection molding, for those parts which can be partitioned into elemental plates, machine cycle time is affected by:

a) Wall thickness,

b) The type of elemental plate (slender, framed, grilled, etc), and

c) The type of subsidiary features (ribs, bosses, etc.),

Unlike the case of injection molding, however, the relative cycle time for an elemental plate depends upon whether or not the elemental plate is an internal (difficult to cool) plate or external (easy to cool) plate. In addition, unlike injection molding, wall thicknesses up to 14 mm are commonly die cast.

Figure 15.26 shows the part coding system and data base used to account for the above factors on machine cycle time for die cast parts.

15.20 Production Yield and Effective Cycle Time

As in all manufacturing situations, production yield is defined as the percentage of acceptable parts produced in a given period of time. As in injection molding, surface 'quality' requirements of a part are the main cause for variations in part yield. A low production yield reduces the number of acceptable parts that are produced in a given time, and thus increases the *effective cycle time* to some value which is higher than the actual machine cycle time.

In die casting, as in injection molding, mold temperatures affect the surface quality of a part. Hot molds yield a smoother more flaw free surface and result in a longer cycle time. Cool molds result in parts with more visible flaws such as flow lines, blisters, etc., and result in a shorter cycle time. In addition, and more importantly, high surface quality requirements may greatly reduce production yield because of a higher rejection rate due to visible sink marks, flow (jet) lines, and other surface flaws.

Part yield is also influenced by part tolerance requirements. Part yield decreases — and effective cycle time increases — when part tolerances are *difficult-to-hold*. Tolerances are discussed in more detail below.

15.21 Surface Finish

Although a standard surface smoothness is produced on tooling when it is constructed, the surface of the parts produced from the tooling can vary, as explained above, due to diverse operating conditions. In addition, die age and die temperature can have serious effects on the as-cast surface finish of the resulting part.

As-cast surface finish requirements can be divided into three categories: *mechanical grade, paint grade* and *high grade*.

A *mechanical grade* surface is one where all surface defects are acceptable. Such parts are generally used in a location where appearance is not important. A mechanical grade surface is also acceptable if the part is to be subjected to a secondary operation, such as barrel tumbling, which modifies the original surface and completely obscures any initial defects.

A *paint grade* surface requirement refers to a part on which minor surface defects are allowable. Parts with paint grade surface finish are generally painted to make the surface defects less noticeable. Parts whose strength would be reduced to an unacceptable level if produced with a mechanical grade finish are sometimes upgraded to paint grade and left unpainted. Once again such parts are generally used in a location where appearance is not important.

A *high grade* surface finish is one in which no surface defects are acceptable. These parts are generally subjected to a secondary plating operation which would make surface flaws highly noticeable.

15.22 Part Tolerances

For dimensional tolerances an industry standard exists. This standard was prepared by the American Die Casting Institute. Thus, to determine whether the tolerances specified are tight or commercial, one must refer to the data published by this Institute. No industry wide standard exists for geometrical tolerances.

Part yield is influenced by part tolerance requirements. Part yield decreases when part tolerances are difficult-to-hold. Part tolerances which are difficult to hold in injection moldings are also difficult to hold in die casting. Thus, the presence of external undercuts, non-uniform wall thickness, unsupported projections and walls make part tolerances difficult to hold. In addition, the specification of tolerances across the parting surface of the dies and/or the specification of more than three tight or five commercial tolerances makes tolerances difficult to hold.

As is always the case, tolerances should be specified only where absolutely necessary. As the number of tolerances to be held increases, molders and die casters have difficulty in maintaining these tolerances and the proportion of defective parts produced increases. In general, molders and die casters can routinely hold 2 or 3 tolerances and have difficulty in maintaining 5 or 6 tolerances.

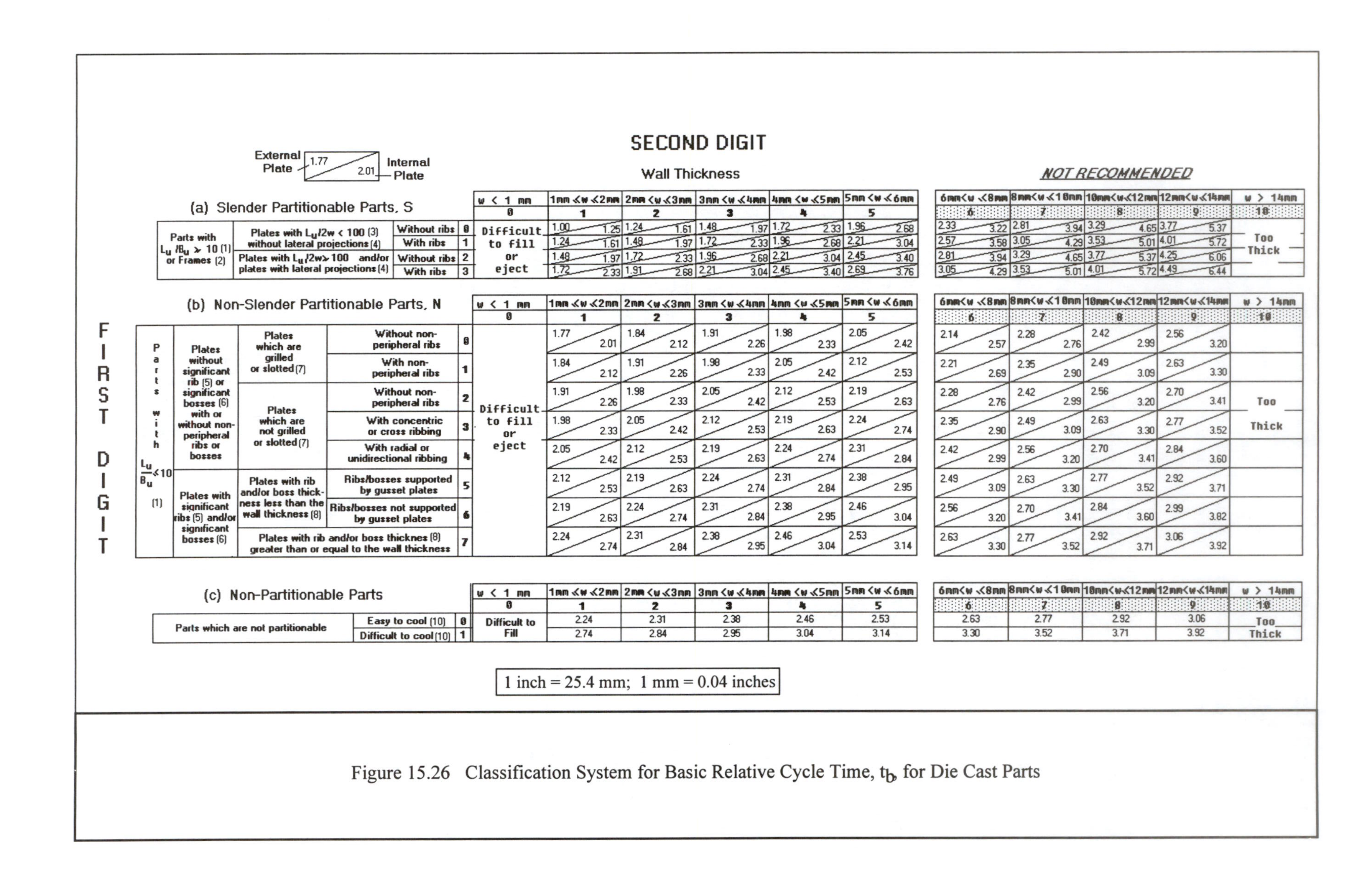

SECOND DIGIT

Wall Thickness

NOT RECOMMENDED (columns 6–10)

Each cell gives: External Plate / Internal Plate (e.g. 1.77 / 2.01)

(a) Slender Partitionable Parts, S

FIRST DIGIT				0: w < 1 mm	1: 1mm ≤ w ≤ 2mm	2: 2mm < w ≤ 3mm	3: 3mm < w ≤ 4mm	4: 4mm < w ≤ 5mm	5: 5mm < w ≤ 6mm	6: 6mm < w ≤ 8mm	7: 8mm < w ≤ 10mm	8: 10mm < w ≤ 12mm	9: 12mm < w ≤ 14mm	10: w > 14mm
Parts with $L_u/B_u > 10$ (1) or Frames (2)	Plates with $L_u/2w < 100$ (3) without lateral projections (4)	Without ribs	0	Difficult to fill or eject	1.00 / 1.25	1.24 / 1.61	1.48 / 1.97	1.72 / 2.33	1.96 / 2.68	2.33 / 3.22	2.81 / 3.94	3.29 / 4.65	3.77 / 5.37	Too Thick
		With ribs	1		1.24 / 1.61	1.48 / 1.97	1.72 / 2.33	1.96 / 2.68	2.21 / 3.04	2.57 / 3.58	3.05 / 4.29	3.53 / 5.01	4.01 / 5.72	
	Plates with $L_u/2w > 100$ and/or plates with lateral projections (4)	Without ribs	2		1.48 / 1.97	1.72 / 2.33	1.96 / 2.68	2.21 / 3.04	2.45 / 3.40	2.81 / 3.94	3.29 / 4.65	3.77 / 5.37	4.25 / 6.06	
		With ribs	3		1.72 / 2.33	1.91 / 2.68	2.21 / 3.04	2.45 / 3.40	2.69 / 3.76	3.05 / 4.29	3.53 / 5.01	4.01 / 5.72	4.49 / 6.44	

(b) Non-Slender Partitionable Parts, N

FIRST DIGIT					0: w < 1 mm	1: 1mm ≤ w ≤ 2mm	2: 2mm < w ≤ 3mm	3: 3mm < w ≤ 4mm	4: 4mm < w ≤ 5mm	5: 5mm < w ≤ 6mm	6: 6mm < w ≤ 8mm	7: 8mm < w ≤ 10mm	8: 10mm < w ≤ 12mm	9: 12mm < w ≤ 14mm	10: w > 14mm
Parts with $\frac{L_u}{B_u} \le 10$ (1)	Plates without significant rib (5) or significant bosses (6) with or without non-peripheral ribs or bosses	Plates which are grilled or slotted (7)	Without non-peripheral ribs	0	Difficult to fill or eject	1.77 / 2.01	1.84 / 2.12	1.91 / 2.26	1.98 / 2.33	2.05 / 2.42	2.14 / 2.57	2.28 / 2.76	2.42 / 2.99	2.56 / 3.20	Too Thick
			With non-peripheral ribs	1		1.84 / 2.12	1.91 / 2.26	1.98 / 2.33	2.05 / 2.42	2.12 / 2.53	2.21 / 2.69	2.35 / 2.90	2.49 / 3.09	2.63 / 3.30	
		Plates which are not grilled or slotted (7)	Without non-peripheral ribs	2		1.91 / 2.26	1.98 / 2.33	2.05 / 2.42	2.12 / 2.53	2.19 / 2.63	2.28 / 2.76	2.42 / 2.99	2.56 / 3.20	2.70 / 3.41	
			With concentric or cross ribbing	3		1.98 / 2.33	2.05 / 2.42	2.12 / 2.53	2.19 / 2.63	2.24 / 2.74	2.35 / 2.90	2.49 / 3.09	2.63 / 3.30	2.77 / 3.52	
			With radial or unidirectional ribbing	4		2.05 / 2.42	2.12 / 2.53	2.19 / 2.63	2.24 / 2.74	2.31 / 2.84	2.42 / 2.99	2.56 / 3.20	2.70 / 3.41	2.84 / 3.60	
	Plates with significant ribs (5) and/or significant bosses (6)	Plates with rib and/or boss thickness less than the wall thickness (8)	Ribs/bosses supported by gusset plates	5		2.12 / 2.53	2.19 / 2.63	2.24 / 2.74	2.31 / 2.84	2.38 / 2.95	2.49 / 3.09	2.63 / 3.30	2.77 / 3.52	2.92 / 3.71	
			Ribs/bosses not supported by gusset plates	6		2.19 / 2.63	2.24 / 2.74	2.31 / 2.84	2.38 / 2.95	2.46 / 3.04	2.56 / 3.20	2.70 / 3.41	2.84 / 3.60	2.99 / 3.82	
		Plates with rib and/or boss thicknes (8) greater than or equal to the wall thickness		7		2.24 / 2.74	2.31 / 2.84	2.38 / 2.95	2.46 / 3.04	2.53 / 3.14	2.63 / 3.30	2.77 / 3.52	2.92 / 3.71	3.06 / 3.92	

(c) Non-Partitionable Parts

FIRST DIGIT			0: w < 1 mm	1: 1mm ≤ w ≤ 2mm	2: 2mm < w ≤ 3mm	3: 3mm < w ≤ 4mm	4: 4mm < w ≤ 5mm	5: 5mm < w ≤ 6mm	6: 6mm < w ≤ 8mm	7: 8mm < w ≤ 10mm	8: 10mm < w ≤ 12mm	9: 12mm < w ≤ 14mm	10: w > 14mm
Parts which are not partitionable	Easy to cool (10)	0	Difficult to Fill	2.24	2.31	2.38	2.46	2.53	2.63	2.77	2.92	3.06	Too Thick
	Difficult to cool (10)	1		2.74	2.84	2.95	3.04	3.14	3.30	3.52	3.71	3.92	

1 inch = 25.4 mm; 1 mm = 0.04 inches

Figure 15.26 Classification System for Basic Relative Cycle Time, t_b, for Die Cast Parts

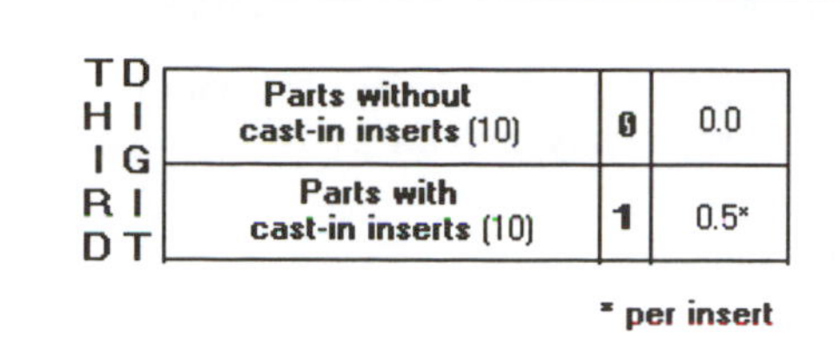

THIRD DIGIT			
	Parts without cast-in inserts (10)	0	0.0
	Parts with cast-in inserts (10)	1	0.5*

* per insert

Table 15.6 Additional Relative Time, t_e, Due to Inserts.

			FIFTH DIGIT	
			Tolerances not difficult to hold (14)	Tolerances difficult to hold (14)
FOURTH DIGIT			0	1
Cast surface finish (11)	Mechanical Grade	0	1.00	1.05
	Paint Grade	1	1.11	1.16
	High Quality	2	1.16	1.22

Table 15.7 Time Penalty, t_p, Due to Surface Requirements and Tolerances for Die Cast Parts.

Tables 15.6 and 15.7 provide the part coding system and data base used to account for the effect of part surface finish and tolerance on the relative cycle time of aluminum die castings.

15.23 Example 15.5: Determination of Relative Cycle Time for a Partitionable Part

15.23.1 The Part

The part is shown in Figure 15.27. All of the necessary part dimensions, wall thickness, rib and boss sizes, etc. are specified. In addition, the part requires a high grade surface finish since it is to be subjected to a secondary plating operation by the customer. The part also contains three cast-in inserts.

15.23.2 Part Partitioning

The part consists of a single elemental plate and, thus, no partitioning is needed.

15.23.3 Relative Cycle Time

Basic Machine Cycle Time: As in Example 15.1, neglecting the three isolated projections of small volume, the dimensions of the *basic envelope* of this part are:

$L = 160, B = 130, H = 10.$

Since the part is straight, then the unbent dimensions, L_u and B_u, are equal to L and B, respectively. Thus, $L_u/B_u < 10$ and the plate is non-slender (N).

The length, l, the width, b, and the height, h, of each projection (boss) is 10 mm, 1.8 mm, and 13 mm, respectively. The plate thickness, w, is 2 mm. Thus, $h/w > 3$, and the projections are considered significant bosses, Since the

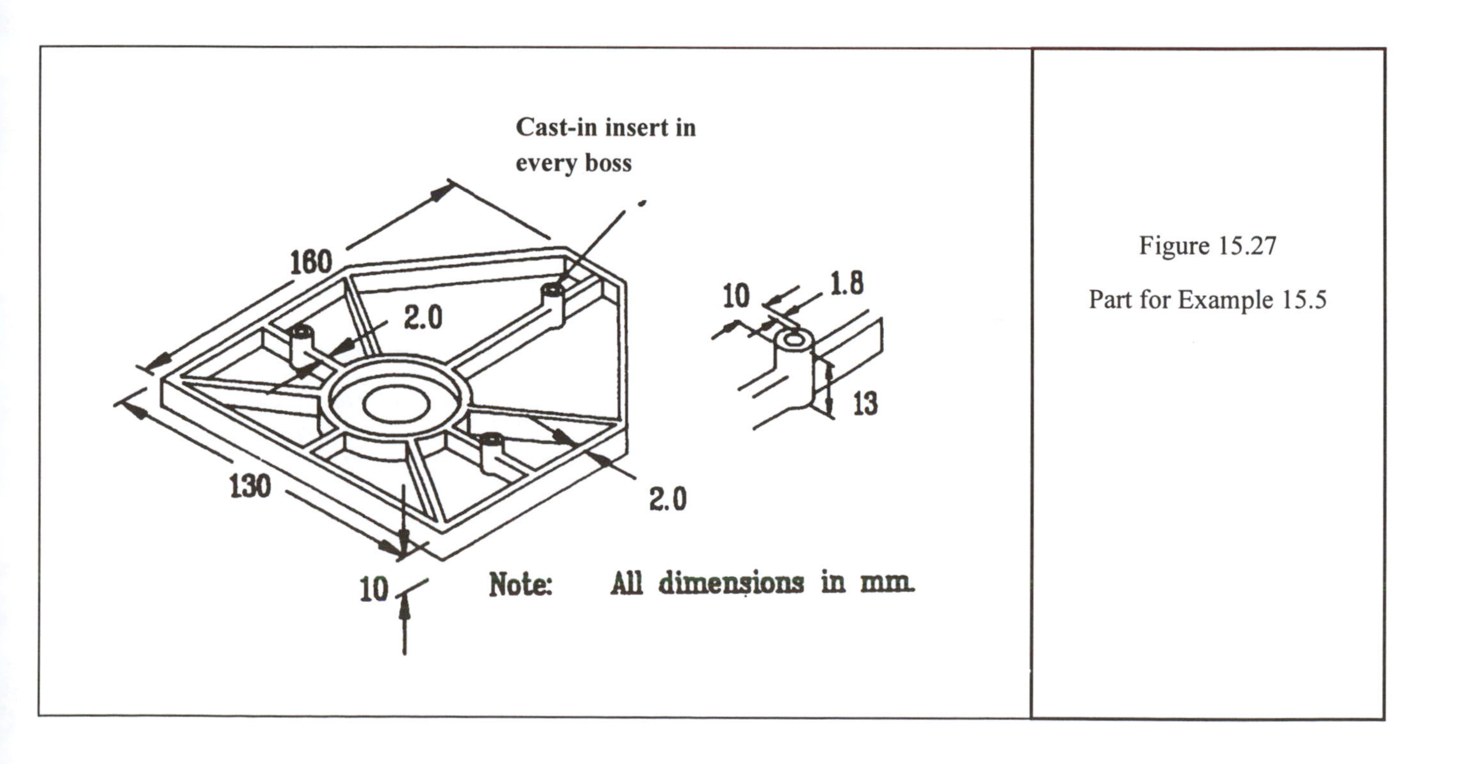

Figure 15.27

Part for Example 15.5

boss thickness is less than the wall thickness and they are not supported by gusset plates, the first digit is 6.

The nominal plate thickness is 2 mm; hence, the second digit is 1. The plate, as discussed earlier in Section 15.4, is an external plate; thus, the basic code is N61E and the basic relative machine cycle time, t_b, is 2.19.

Inserts: There are three cast-in inserts. Hence, the third digit is 1, and the relative extra mold opening time, t_e, is 0.5 per insert, or 1.5 for the part.

Surface Finish/Tolerances: Since a high grade surface finish is required, the fourth digit is 2. Because the bosses are unsupported, it is difficult to hold tolerances between them. Hence, the fifth digit is 2, and the production penalty, t_p, is 1.22.

Relative Effective Cycle Time: The relative effective cycle time for the original design is:

$$t_r = (t_b + t_e)t_p = (2.19 + 1.5)(1.22) = 4.50.$$

15.23.4 Redesign Suggestions

The relative cycle time can be reduced in several ways. One method is to provide gusset plates to support the bosses. In this case, the basic code becomes N51E, and the fifth digit becomes 0. Better still, the bosses could be redesigned so that they become nonsignificant. The basic code then becomes N41E, and t_b becomes 2.05. In this case, the fifth digit is still 0, and t_r becomes 4.1202, about an 8% reduction.

A more significant method for reducing the relative cycle time would be to remove the cast-in inserts. If this is done, t_e is reduced to 0, and the relative effective cycle time becomes 2.5, about a 44 % reduction.

15.24 Total Relative Part Cost

As in the case of injection molding, the total production cost of a die casting is:

$$K_t = K_m + K_d/N + K_e. \qquad (15.6)$$

As before, K_d represents the total cost of the tool, and N represents the production volume.

The total cost of the part relative to the cost of the reference part, C_r, is given by Equation (15.11), or

$$C_r = C_m f_m + C_d f_d + C_e f_e \qquad (15.11)$$

where C_m, C_d, and C_e represent the material cost, tooling cost and processing cost of a part relative to the reference part. The reference part in this case is an aluminum die casting, a washer whose OD = 65 mm, ID = 55 mm, and whose thickness is 2 mm.

The values of f_m, f_d, and f_e represent the proportion of the total relative part cost due to part material, tooling, and processing, respectively, and are given by Equations (15.12) - (15.14).

The value for C_d, C_e, and C_m are obtained from the following expressions:

$$C_d = 0.8C_{dc} + 0.2C_{dm} \qquad (15.18)$$

$$C_e = t_r C_{hr} \qquad (15.19)$$

$$C_m = (V/V_o)C_{mr} \qquad (15.20)$$

The values for C_{dc}, C_{dm}, are obtained by the methods discussed in Chapter 11, while the value of t_r is obtained by use of Figure 15.26, the classification system for die casting and illustrated in the example above. The value for C_{hr} for die casting machines is obtained from Table 15.8.

Since, in general, the vast majority of die castings used for non-decorative purposes are made of aluminum, the only material considered here and the only material for

Table 15.8 Machine Tonnage and Relative Hourly Rate (Data obtained from the die casting industry, June, 1989

Die Casting	
Machine Tonnage	Relative Hourly Rate C_{hr}
< 100	1.00
100-199	1.05
200-299	1.08
300-399	1.12
400-499	1.17
500-599	1.21
600-699	1.26
700-799	1.29
800-899	1.33
900-999	1.38
1000-1199	1.45
1200-1499	1.59
>1500	1.73

Table 15.9 Relevant Data for the Reference Part

Part Material	Aluminum
Material Cost	0,0006 Cents/mm^3
Part Volume, V_o	1885 mm^3
Die Material Cost	$980.
Die Construction Time	200 hours
Labor Rate, Die Construction	$30. / hour
Cycle Time	17.23 Sec
Die Cast Machine Hourly Rate	$62.57 / hr

which the data base contained in Figure 15.26 is applicable to is aluminum. Thus, $C_{mr} = 1$ and

$$C_m = V / V_o \qquad (15.21)$$

Table 15.9 contains some relevant data for the reference part.

Using that data the expressions for f_m, f_d, and f_m reduce to the following expressions for single cavity molds

$$f_m = \frac{N}{27.5N + 6.2x10^5} \qquad (15.22)$$

$$f_d = \frac{6.2x10^5}{27.5N + 6.2x10^5} \qquad (15.23)$$

$$f_e = \frac{26.5N}{27.5N + 6.2x10^5} \qquad 15.24)$$

Plots of Equations (15.22) - (15.24) are shown in Figure 15.28. As in the case of injection molding, at low production volumes, less than 10,000, the proportion of part cost due to tooling is more than 70% of the total cost of the part, while the proportion of the cost due to processing is less than 25% of the total part cost. At large production volumes, say 100,000 parts, the proportion of part cost due to tooling drops to less than 20%, while the proportion of cost due to processing rises to almost 80%. Thus, at low production the emphasis on redesign should be on lowering tooling costs. At high production volumes, the emphasis on redesign should be to lower the processing costs.

15.25 Example 15.6: Determination of the Total Relative Part Cost

For the part considered earlier in Figure 15.27, it was determined in Section 11.13 that the relative tool cost for the part is $C_d = 1.58$. In addition, it was determined above that the relative cycle time for the part is $t_r = 4.50$.

The volume of the part is about 52,000 mm^3; hence,

$$C_m = V/V_o = 52{,}000/1855 = 27.6$$

As in injection molding, the press tonnage required is given by the expression

$$F_p = 0.005\ A_p$$

where it is assumed that A_p is in mm^2. In this case A_p = 20,000 mm^2 so:

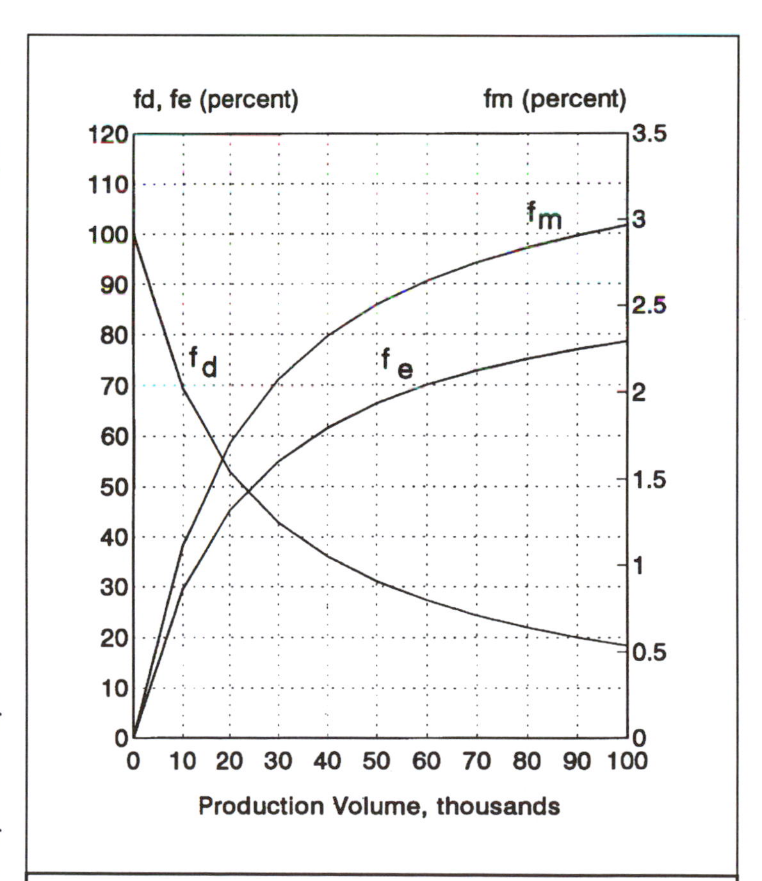

Figure 15.28 Variation of f_d, f_e and f_m With Production Volume (Aluminum Die Casting)

$F_p = 100$ tons,

and from Table 15.8, $C_{hr} = 1.05$. Thus,

$$C_e = t_r C_{hr} = 4.73.$$

Production Volume = 10,000. For a production volume of 10,000, from Figure 15.28 it is seen that

$$f_m = 0.011, f_d = 0.70\ f_e = 0.30.$$

Thus,

$$C_r = 27.6(0.011) + 1.58(0.70) + 4.73(0.30) = 2.83$$

If the part is redesigned as discussed in Section 11.13, and 15.23, then $C_d = 1.30$, $t_r = 2.6$ and $C_e = 2.6$. Thus,

$$C_r = 27.6(0.011) + 1.30(0.70) + 2.6(0.30) = 1.99$$

This is an overall savings of some 30% in piece part cost.

Production Volume = 100,000. If the production volume of the part is 100,000 pieces, then from Figure 15.28,

$$f_m = 0.03, f_d = 0.20, f_e = 0.79,$$

and the total relative cost of the original design, C_r, is:

$$C_r = 27.6(0.03) + 1.58(0.20) + 4.50(0.79) = 4.70.$$

For the redesigned part,

$$C_r = 27.6(0.03) + 1.30(0.20) + 2.6(0.79) = 3.14.$$

Thus, for a production volume of 100,000 parts, the overall savings in part cost is almost 30%. This increase in savings is due to the fact that the redesign of the part reduced the relative cycle time. Thus, as the production volume increased, the overall savings also increased.

15.26 Worksheet for Relative Processing Cost (Injection Molding and Die Casting)

To facilitate the calculation of the relative processing cost and the overall relative part cost, a worksheet has been prepared. The copies shown on the next two pages have been completed for Example 15.1 (Figure 15.18). A blank copy of the worksheet is available in Appendix 15.C and may be reproduced for use with this book.

15.27 Summary and Preview

Summary. The purpose of this chapter was to present a systematic approach for identifying, at the parametric design stage, those features of the part which significantly affect the processing cost of injection molded and die cast parts. The goal was to learn how to design so as to minimize difficult-to-process features.

A methodology for estimating the relative processing cost of proposed injection molded or aluminum die cast parts based on parametric information was presented. In addition, a method for estimating the overall relative part cost for these same parts was introduced.

Preview. In the next Chapter a method for evaluation of the relative processing cost and overall relative costs of stamped parts is presented.

References

[1] Poli, C., Kuo, S. M. and Sunderland, J. E., "Design for Injection Molding - Coding System for Relative Cycle Time," Mechanical Engineering Department, University of Massachusetts at Amherst, Amherst, MA, 1989

[2] Poli, C., Kuo, S. M., and Sunderland, J. E., "Keeping a Lid on Mold Processing Costs," Machine Design, Oct. 26, 1989.

[3] Kuo, Sheng-Ming, "A Knowledge-Based System for Economical Injection Molding," Phd Dissertation, University of Massachusetts at Amherst, Amherst, MA 01003, Feb. 1990.

[5] Poli, C. and Fredette, L., "Product Design for Economical Die Casting," Mechanical Engineering Department, University of Massachusetts at Amherst, Amherst, MA, 1989

[6] Dym, J. B., "Product Design with Plastics," Industrial Press, New York, 1983

[7] Shanmugasundaram, S. K., "An Integrated Economic Model for the Analysis of Die Cast and Injection Molded Parts," M.S. Final Project Report, Mechanical Engineering Department, University of Massachusetts at Amherst, Amherst, MA 01003, August 1990.

[8] Poli, C, Fredette, L. and Sunderland, J. E.,"Trimming the Cost of Die Castings," Machine Design, March 8, 1990

[9] Fredette, Lee, "A Design Aid For Increasing The Producibility of Die Cast Parts," M.S. Final Project Report, Mechanical Engineering Department, University of Massachusetts at Amherst, Amherst, MA, 1989

[10] Shanmugasundaram, S. K., "An Integrated Economic Model for the Analysis of Die Cast and Injection Molded Parts," M.S. Final Project Report, Mechanical Engineering Department, University of Massachusetts at Amherst, Amherst, MA, 1990

[11] Trucks, H. E., *Designing for Economical Production*, Society of Manufacturing Enginers, Dearborn, MI, 1987.

Problems (See page 15-27)

Worksheet - Relative Processing Cost
(Injection Molding and Die Casting

Original or Redesign:	***Example 15.1***				
L_u = 160	***B_u = 130***	***L_u/B_u = 1.23 < 10***		Slender / **<u>Non-Slender?</u>**	
Digit	Plate 1	Plate 2	Plate 3	Plate 4	Plate 5
Ext/Int					
First	***6***				
Second	***2***				
t_b	***3.24***				
Third	***1***				
t_e	***3(0.5)=1.5***				
Fourth	***0***				
Fifth	***1***				
t_p	***1.2***				
t_r	***(3.24+1.5) 1.2= 5.69***				
Relative Cycle Time	***5.69***				
Redesign Suggestions	***Support bosses to reduce first digit and fifth digit OR reduce boss height to make them non-significant. Could eliminate inserts - but these would need to be inserted after molding - thus little to be gained.***				
Relative Material Cost:	V =	V_o =	C_{mr} =	C_m =	
Relative Tooling Cost:	C_d = ______________ (From Previous Calculations)				
Total Relative Cost	N =	A_p =	F_p =	C_{hr} =	
	$C_e = t_r C_{hr}$ = _____	f_d =	f_e =	f_m =	
	$C_r = C_d f_d + C_e f_e + C_m f_m$ = ______________				
% Savings in Processing Costs:					
% Savings in Overall Costs:					

Worksheet - Relative Processing Cost
(Injection Molding and Die Casting

Original or Redesign: ***Version 1***

L_u=	B_u=	L_u / B_u=	Slender / Non-Slender?

Digit	Plate 1	Plate 2	Plate 3	Plate 4	Plate 5
Ext/Int					
First	***5***				
Second	***2***				
t_b	***3.10***				
Third	***1***				
t_e	***1.5***				
Fourth	***0***				
Fifth	***0***				
t_p	***1***				
t_r	***3.10(1.5)= 4.6***				
Relative Cycle Time	***4.6***				
Redesign Suggestions					

Relative Material Cost:	V =	V_o=	C_{mr}=	C_m=
Relative Tooling Cost:	C_d= ______________________ (From Previous Calculations)			
Total Relative Cost	N =	A_p=	F_p=	C_{hr}=
	$C_e = t_r Chr$ = ______	f_d=	f_e=	f_m=
	$C_r = C_d f_d + C_e f_e + C_m f_m$ = ______________________			

% Savings in Processing Costs: ***(5.69 - 4.60) / 5.69 = 19 %***

% Savings in Overall Costs:

Problems

15.1 The hollow cylindrical part shown in Figure P15.1 is made of nylon 6 in a mold having a SPI finish of 3. Using the group technology based methodology discussed in this chapter, estimate the relative cycle time for the part. What redesign suggestions would you make in order to reduce the relative cycle time of the part? Assume commercial tolerances.

15.2 Using the group technology based methodology discussed in this chapter, estimate the relative cycle time for the part shown in Figure P15.2 if it is produced with commercial tolerances. Assume that the part is made of nylon 6 in a mold with a SPI finish of 3. The maximum wall thickness of the part is 2.5 mm. The minimum wall thickness of the part is 1.5 mm

15.3 Redesign the part shown in Fig P15.2 so that the cycle time of the part is reduced. What is the percent reduction in cycle time due to this redesign?

15.4 Using the group technology based methodology discussed in this chapter, estimate the relative cycle time for the transparent part shown in Figure P15.4. Assume commercial tolerances. Can you suggest at least one way to reduce the cycle time?

15.5 The mold used to produce the part shown in Figure P15.5 is assumed to have a surface finish of SPI-3. Estimate the relative cycle time for the part under the assumption that the part is made of polycarbonate. Can you make any redesign suggestions that would reduce the relative cycle time required to produce the part and, hence, the processing cost for the part? The wall thickness of the difficult to cool feature is 2.5 mm.

15.6 For the part shown in Figure P15.2 determine the total relative part cost for production volumes of 25,000 and 100,000.

15.7 Assume that the part shown in Figure P15.2 is redesigned so as to reduce both the tooling cost (Problem 8 of Chapter 11) and processing cost (Problem 3 above). Determine the percent savings in cost achieved by this redesign for production volumes of 25,000 and 100,000.

15.8 Assume that the part shown in Figure P15.2 is to be an aluminum die casting. Determine the relative cycle time for the part under the assumption that it is to be produced with a mechanical grade finish.

15.9 Assume that the part shown in Figure P15.5 is to be an aluminum die casting. Determine the relative cycle time for the part under the assumption that it is to be produced with a paint grade finish.

15.10 A study of Figures 15.1 and 15.26 of this chapter shows that the group technology based coding system for injection molded part and die castings is essentially the same. These figures also show that the relative cycle time data base for injection moldings is higher than those for a geometrically equivalent die casting. This difference is particularly true as the parts become more difficult to cool. Does this seem reasonable? Explain.

15.11 For Example 15.4 it was assumed that the cycle time for the reference part, t_o, is 16 s (see Table 18.1). Assume that the cycle time t_o is 10 s and determine for the part analyzed in Example 15.4 the reduction in costs between the original design and the redesign for production volumes of 10,000 and 100,000.

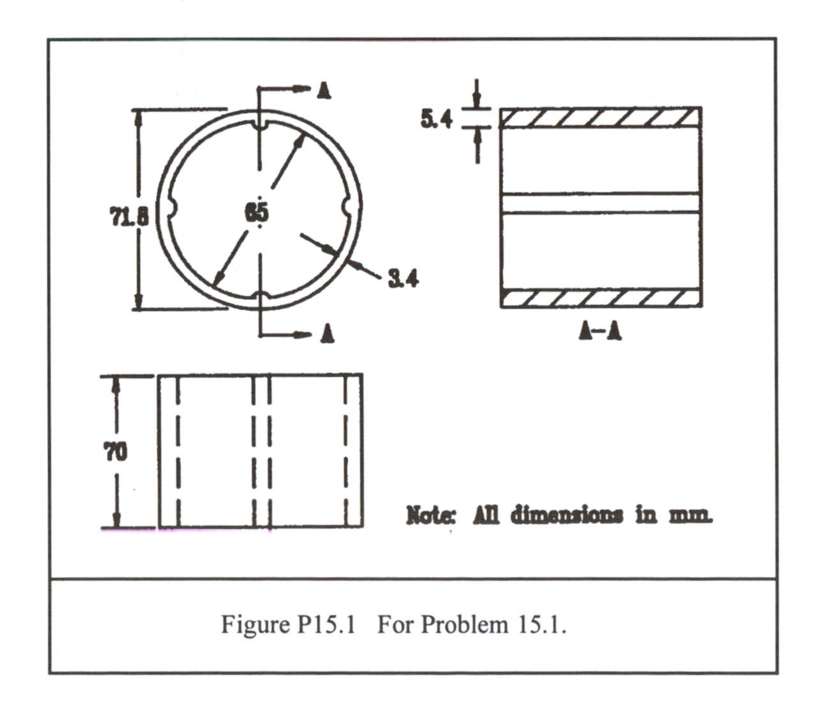

Figure P15.1 For Problem 15.1.

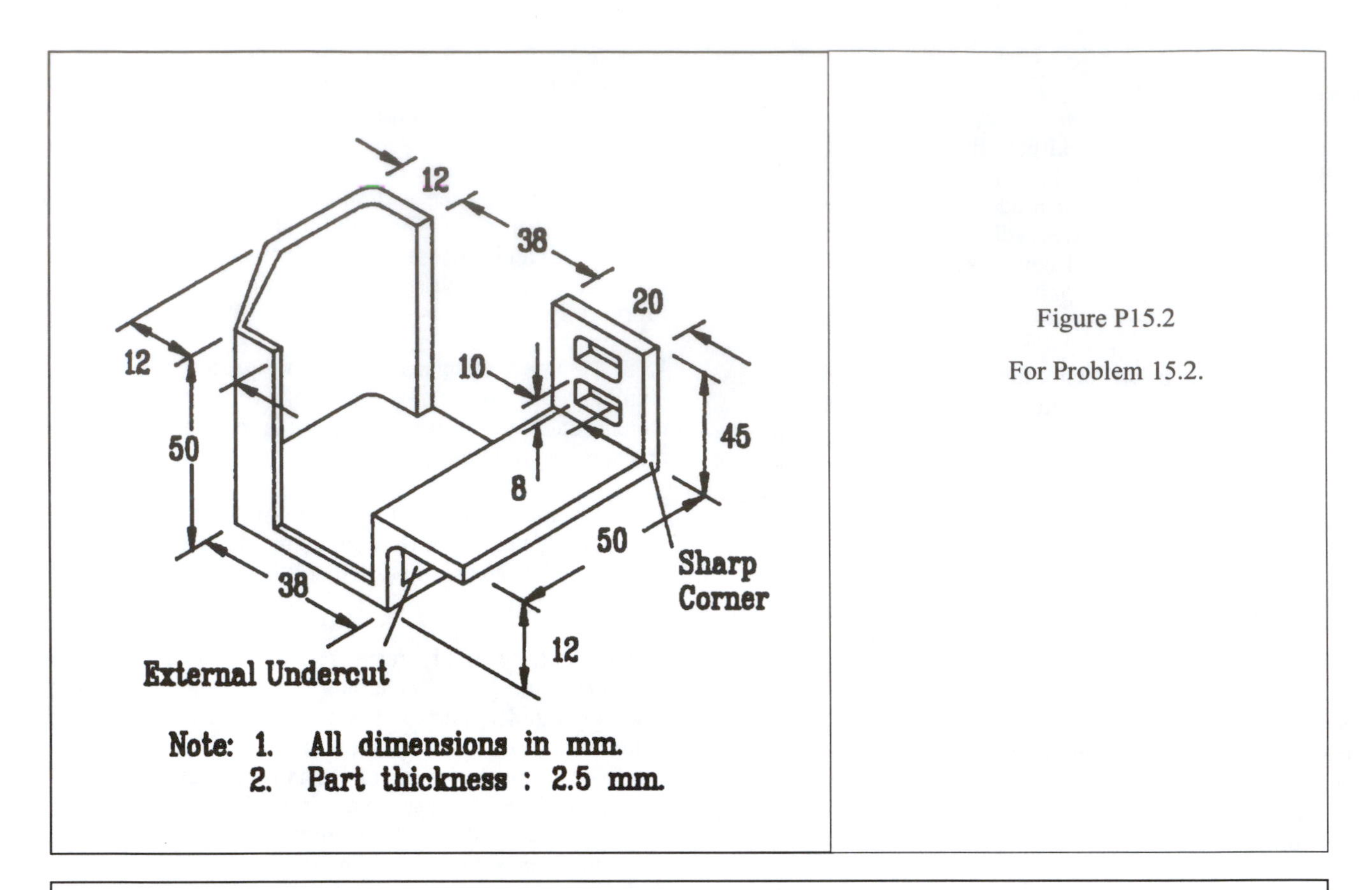

Figure P15.2

For Problem 15.2.

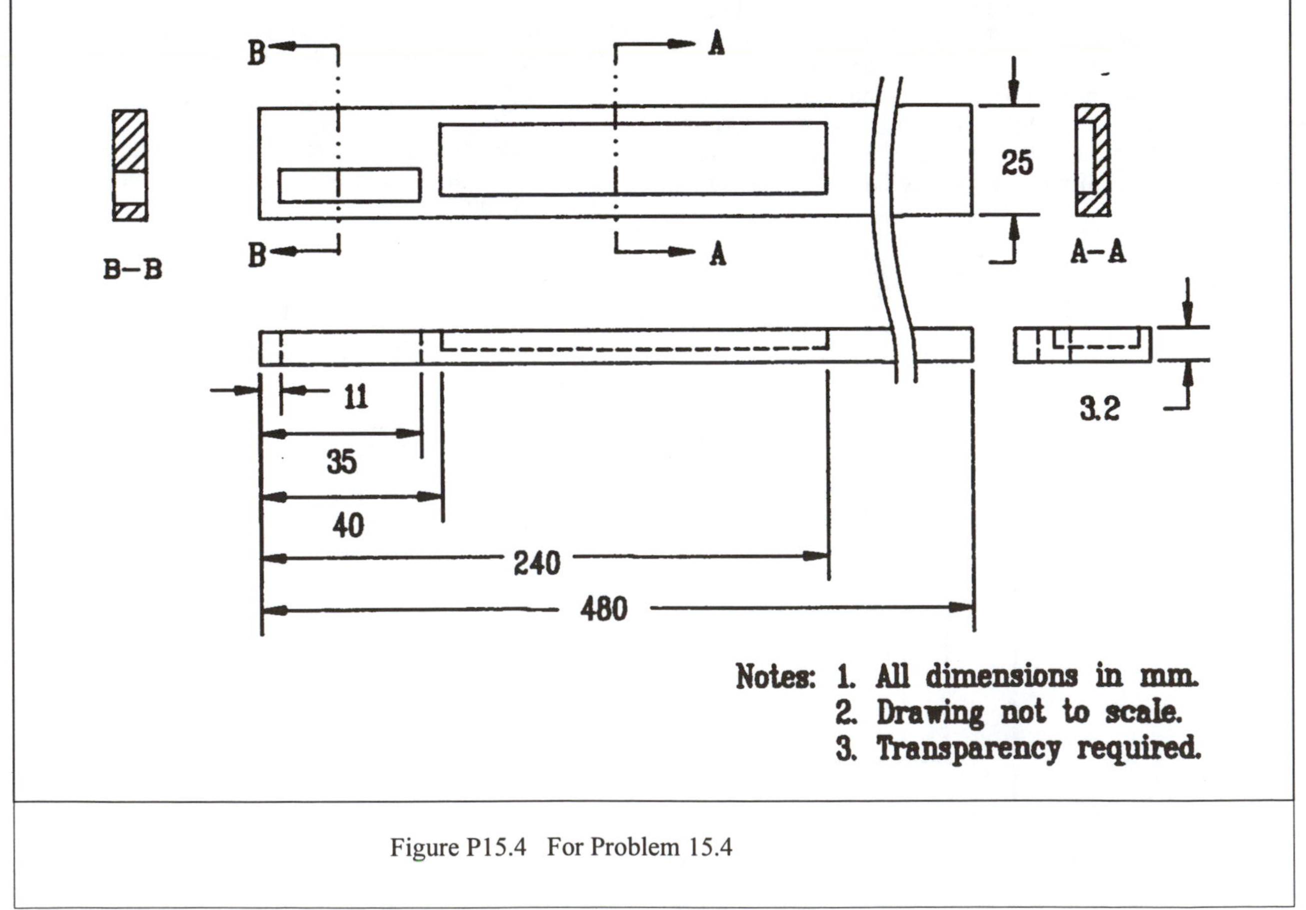

Figure P15.4 For Problem 15.4

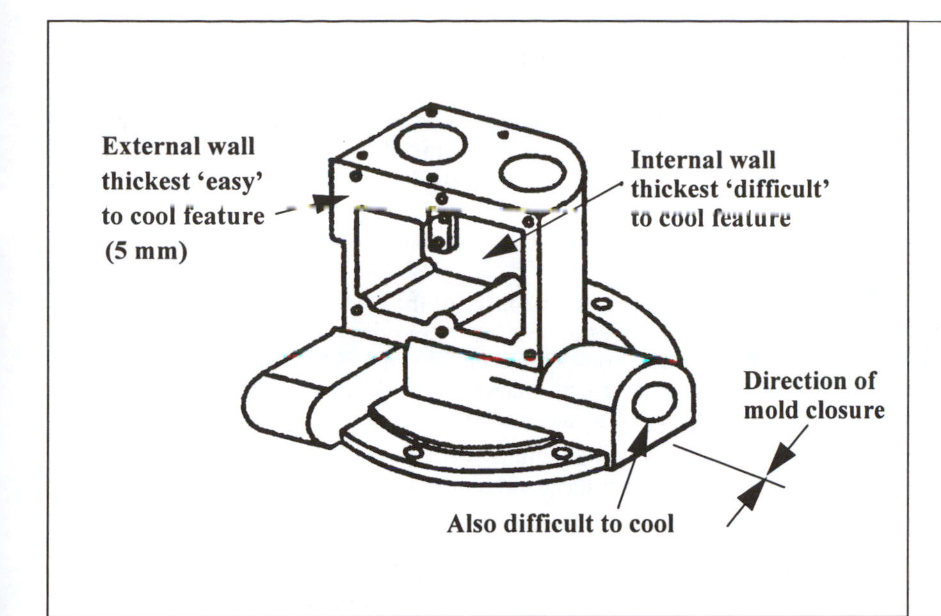

Figure P15.5

For Problem 15.5

Appendix 15.A

Notes for Figure 15.1, Tables 15.3 and 15.4.

1) For parts or elemental plates with a bent or curved longitudinal axis, the ***unbent length***, L_u, is the maximum length of the part with the axis straight. The width of this unbent part is referred to as B_u. See Figure 15.4.

2) ***Frames*** are parts or elemental plates which have a through hole greater than 0.7 times the projected area of the part/plate envelope and whose height is equal to its wall thickness. See Figure 15.5

3) The ***thickness*** of the elemental plate is denoted by w. For parts/plates where the thickness is not constant, w is the maximum thickness of the plate or part.

4) ***Lateral projections*** are shape features which protrude from the surface of a slender plate (rod) in a direction normal to the longitudinal axis. For long slender parts such projections are difficult to fill. See Figure 15.17.

5) A ***rib*** is a narrow elongated wall like projection whose length, l, is greater than three times its width, b. Ribs may be located either at the periphery or on the interior of a part/plate. A rib may be continuous or discontinuous or part of a network of other ribs and projecting elements. To avoid sink marks ribs should be designed so that (a) the rib height, h, is less than or equal to three times the localized wall thickness, w, and (b) the rib width, b, is less than or equal to the localized wall thickness. Such ribs are considered as ***non-significant ribs***. Ribs designed such that [$3w < h < 6w$] or $b > w$ are called ***significant ribs***. Significant ribs tend to increase the machine cycle time and make tolerances difficult to hold. For the purposes of this system, peripheral ribs are treated as walls.

6) A ***boss***, like a rib, is a projecting element; however, its length, l, is less than three times its width, b. It takes a variety of forms such as a knob, hub, lug, button, pad, or 'prolong'. A boss should be designed such that (a) the boss height, h, is less than or equal to three times the localized wall thickness and (b) the boss width, b, should be less than or equal to the localized wall thickness. These types of bosses are considered ***non-significant bosses***. Bosses designed such that $3w < h$ or $b > w$ are called ***significant bosses***. Significant bosses tend to increase the machine cycle time and make tolerances difficult to hold. See Figure 15.15.

(7) An elemental plate with

a) multiple through holes,

b) no continuous solid section with a projected area greater than 20 percent of the projected area of the plate envelope, and

c) whose height is equal to its wall thickness is called ***grilled/slotted***. See Figure 15.16.

(8) The wall thickness referred to here is the localized wall thickness (See also Notes 6 and 7).

(9) The use of ***foamed materials*** results in a surface gloss which is generally less acceptable than the one which results from the use of thermoplastic materials. To improve the surface gloss of parts made with foamed materials, parts are usually subjected to secondary finishing operations (painting, etc.). In addition, the minimum thickness achievable with foamed materials is greater than that obtainable with thermoplastics. For these reasons foamed materials are generally not used for parts whose wall thickness is less than or equal to 5 mm.

In addition, ***engineering thermoplastics*** are not generally used if the wall thickness is greater than 5 mm because the shrinkage of these plastics becomes difficult to control when w > 5 mm.

(10) Features such as ***holes or depressions*** which have an internal diameter smaller than 12.5 mm are considered difficult to cool.

(11) Holes or depressions with internal grooves or undercuts such that a solid plug that conforms to the shape of the hole or depression cannot be inserted are called internal undercuts. Such restrictions prevent molding from being extracted from the core in the line of draw. When these internal undercuts take the form of internal threads, a special unscrewing mechanism is used. When the number of threads becomes large, the time required to unscrew the mechanism can significantly increase the machine cycle.

(12) ***Inserts*** are metal components added to the part prior to molding the part. These metal components are added for decorative purposes, to provide additional localized strength, to transmit electrical current, and to aid in assembly or subassembly work. The use of inserts increases the machine cycle time.

(13) For the purposes of the present coding system, the ***part surface requirement*** is considered high when parts are produced from a mold having an SPI/SPE surface finish of 1 or 2, and/or sink marks and weld lines are not allowed on an untextured surface.

Parts without high surface requirements are considered to have low surface requirements.

(14) Part ***tolerances*** are considered difficult-to-hold if:

a) External undercuts are present.

b) The wall thickness is not uniform.

c) A tolerance is required across the parting surface of the dies.

d) Unsupported projections (ribs, bosses) and walls are used.

e) More than three tight tolerances or more five commercial tolerances are required.

Appendix 15.B

Notes for Figure 15.26, Tables 15.6 and 15.7

(1) to (8) are identical to notes (1) to (8) for injection molding (classification system for relative cycle time) and are contained in Appendix A.

(9) Features such as holes or depressions which have an internal diameter smaller than 12.5 mm are considered ***difficult to cool.***

(10) ***Inserts*** are metal components added to the part prior to casting the part. These metal components are added for decorative purposes, to provide additional localized strength, and to aid in assembly or subassembly work. The use of inserts increases the machine cycle time.

(11) As-cast surface finish requirements can be divided into three categories, namely, *mechanical grade, paint grade* and *high grade.*

A ***mechanical grade*** surface is one where all surface defects are acceptable. Such parts are generally used in a location where appearance is not important. A mechanical grade surface is also acceptable if the part is to be subjected to a secondary operation, such as barrel tumbling, which modifies the original surface and completely obscures any initial defects.

A ***paint grade*** surface requirement refers to a part on which minor surface defects are allowable. Parts with paint grade surface finish are generally painted to make the surface defects less noticeable. Parts whose strength would be reduced to an unacceptable level if produced with a mechanical grade finish are sometimes upgraded to paint grade and left unpainted. Once again such parts are generally used in a location where appearance is not important.

A ***high grade*** surface finish is one in which no surface defects are acceptable. These parts are generally subjected to a secondary plating operation which would make surface flaws highly noticeable.

(12) Part ***tolerances*** are considered difficult-to-hold if:

a) External undercuts are present.

b) The wall thickness is not uniform.

c) A tolerance is required across the parting surface of the dies.

d) Unsupported projections (ribs, bosses) and walls are used.

e) More than three tight tolerances or more than five commercial tolerances.

Worksheet - Relative Processing Cost
(Injection Molding and Die Casting

Original or Redesign:

L_u =	B_u =	L_u / B_u =	Slender / Non-Slender?

Digit	Plate 1	Plate 2	Plate 3	Plate 4	Plate 5
Ext/Int					
First					
Second					
t_b					
Third					
t_e					
Fourth					
Fifth					
t_p					
t_r					
Relative Cycle Time					

Redesign Suggestions				
Relative Material Cost:	V =	V_o =	C_{mr} =	C_m =
Relative Tooling Cost:	C_d = ____________ (From Previous Calculations)			
Total Relative Cost	N =	A_p =	F_p =	C_{hr} =
	$C_e = t_r C_{hr}$ = ______	f_d =	f_e =	f_m =
	$C_r = C_d f_d + C_e f_e + C_m f_m$ = ____________			

% Savings in Processing Costs:

% Savings in Overall Costs:

This page available for notes, sketches, and calculations

Chapter Sixteen

Evaluating Parametric Designs For Manufacturability: Stamping

"The product designers also need to get some hands-on experience with manufacturing processes.....so they develop an appreciation for how their design will be manufactured."

Smith, P.G. and Reinertsen, D. G.*[5]
Developing Products in Half the Time (1991)

16.1 Introduction

In Chapter 12, we expressed the total cost of a stamped part in terms of: (1) tooling cost per part, K_d/N; plus (2) processing cost, K_e; plus (3) material cost, K_m. At the configuration stage, since the final dimensions and tolerances of the part are not yet established, only tooling cost can be estimated . (In the case of stamping, it may be possible to obtain only a lower bound on the tooling cost.) At the parametric design stage, however, the final dimensions and tolerances are established. Thus, the total relative tooling cost can now be more accurately obtained by the methods presented in Chapter 12. In addition, at the parametric stage, the relative processing cost, the relative material cost, and thus the overall relative part cost can be obtained.

In this chapter, we concentrate on methods for evaluating the relative processing cost and the overall relative part cost for stamped parts made on progressive dies. In general, progressive dies are used for parts whose unfolded length, L_{ul}, is less than 100 mm. For values of L_{ul} between 100 mm and 200 mm, both progressive dies and die lines of compound and single dies are commonly used. For parts larger than 200 mm, die lines are more common than progressive dies. This is due primarily to press capacity constraints that arise as the size of progressive dies increase.

* Quotation from *Developing Products in Half the Time*, by P. G. Smith and D. G. Reinertson, 1991, published by Van Nostrand Reinhold, New York, by permission of P. G. Smith.

The analysis that follows is restricted to the use of progressive dies. It has been shown that for production volumes in excess of about 100,000, parts are most economically produced using progressive dies [1]. In many cases, even for production volumes down to about 10,000, the use of progressive dies can result in a lower production cost.

16.2 Relative Processing Cost, C_e

If t_{cy} represents the effective cycle time of the press, and C_h represents the machine hourly rate, then the relative processing cost, C_e, is:

$$C_e = t_{cy} C_h / t_o C_{ho} = t_r C_{hr} \qquad (16.1)$$

where t_o and C_{ho} represent the cycle time and the machine hourly rate for the press used to produce the reference washer introduced in Section 12.2.2.

As Equation (16.1) indicates, to estimate the processing cost, we must first estimate the relative machine hourly rate, C_{hr}, and the relative cycle time, t_r. Since both these parameters require an estimate of the force required of the press to perform the stamping operations, we present next a procedure for estimating the required press tonnage.

16.3 Determining Press Tonnage, F_p

The press size required depends on the length of the press bed, A (which must be large enough to accommodate the die set), and the force required to stamp the part. The force required is a function of the number and type of features being formed, the size of the part, and the workpiece material.

The required (or rated) press tonnage[1], F_p, is generally taken as approximately 50% larger than the computed force, F, in tons, required to form the part. The added 50% is an approximate factor to account for uncertainties in the material properties and other estimates. That is:

$$F_p = 1.5\ F \qquad (16.2)$$

where F is the estimated total force in tons required to form the part. The required force, F, is the sum of several factors:

$$F = F_s + F_b + F_{fl} + F_{fo} + F_{dr} \qquad (16.3)$$

where F_s = Forces for Shearing

F_b = Forces for Bending

F_{fl} = Forces for Flanging

F_{fo} = Forces for Radius Forming

F_{dr} = Forces for Drawing

The shearing force required, F_s, is the sum of stripper forces, F_{st}, required to remove the strip from around the shearing and blanking punches, and the separation force, F_{out}. The separation force F_{out} is the force required to separate the part from the strip at the last station of the progressive die, and the force required to shear and form the individual non-peripheral features. The force required to create the individual non-peripheral features can be approximated by multiplying the sum of F_{st} and F_{out} by a factor, X_{dd}, which depends on die detail (Section 12.5.2). Values for X_{dd} can be obtained from Table 16.1. Thus, the required shearing force is given by

$$F_s = (F_{out} + F_{st})(1 + X_{dd}) \qquad (16.4)$$

where

$$F_{out} = 1.5\ L_{out}\ \sigma_s t \qquad (16.5)$$

$$F_{st} = 21 L_{out} t \qquad (16.6)$$

where F_{st} = the force in Newtons

σ_s = the shear strength of the material in MPa

L_{out} = the length of the periphery of the part in mm

t = sheet thickness in mm.

The 1.5 factor in Equation (16.5) is used to account for any peripheral features (cutouts, notches, etc.) which occur on the periphery of the part and not accounted for by L_{out}. If no notches are present, then Equation (16.5) will overestimate the force required. However, given the approximate nature of the calculations performed here, this is not considered a serious limitation. The other forces in Equation (16.3) above are computed as follows:

$$F_b = \sigma_t L_{bt} t/18 \qquad (16.7)$$

$$F_{fl} = 0.25 L_{fl}\ \sigma_t t \qquad (16.8)$$

$$F_{fo} = L_{fo}\ \sigma_t t \qquad (16.9)$$

$$F_{dr} = D_p t\, \sigma_t (d_o/D_p - 0.7) \qquad (16.10)$$

where

σ_t is the tensile strength of the material in MPa,

L_{bt} is the sum of all the straight bend lengths,

L_{fl} is the total length of all the non-straight bend lines,

L_{fo} is the perimeter of the radius forms,

d_o is the diameter of a drawn feature before drawing,

and D_p is the diameter of the feature after drawing.

Values for the shear strength and tensile strength of some common materials can be determined with the help of Table A9.2 of Chapter 9. When shear strength is unknown but tensile strength is known, an approximate value for the shear strength is obtained by multiplying the tensile strength by a value between 0.65 and 0.85. For high values of tensile strength, one should use about 0.85 as the multiplying factor. For low values of tensile strength we should use 0.65. When in doubt use 0.7.

Table 16.1 Values of X_{dd} for Use in Equation 16.4.

Die Detail	X_{dd}
Low	0
Medium	0.22
High	0.28

1. Since most presses in the U.S. are rated in tons, the English system of units is used here.

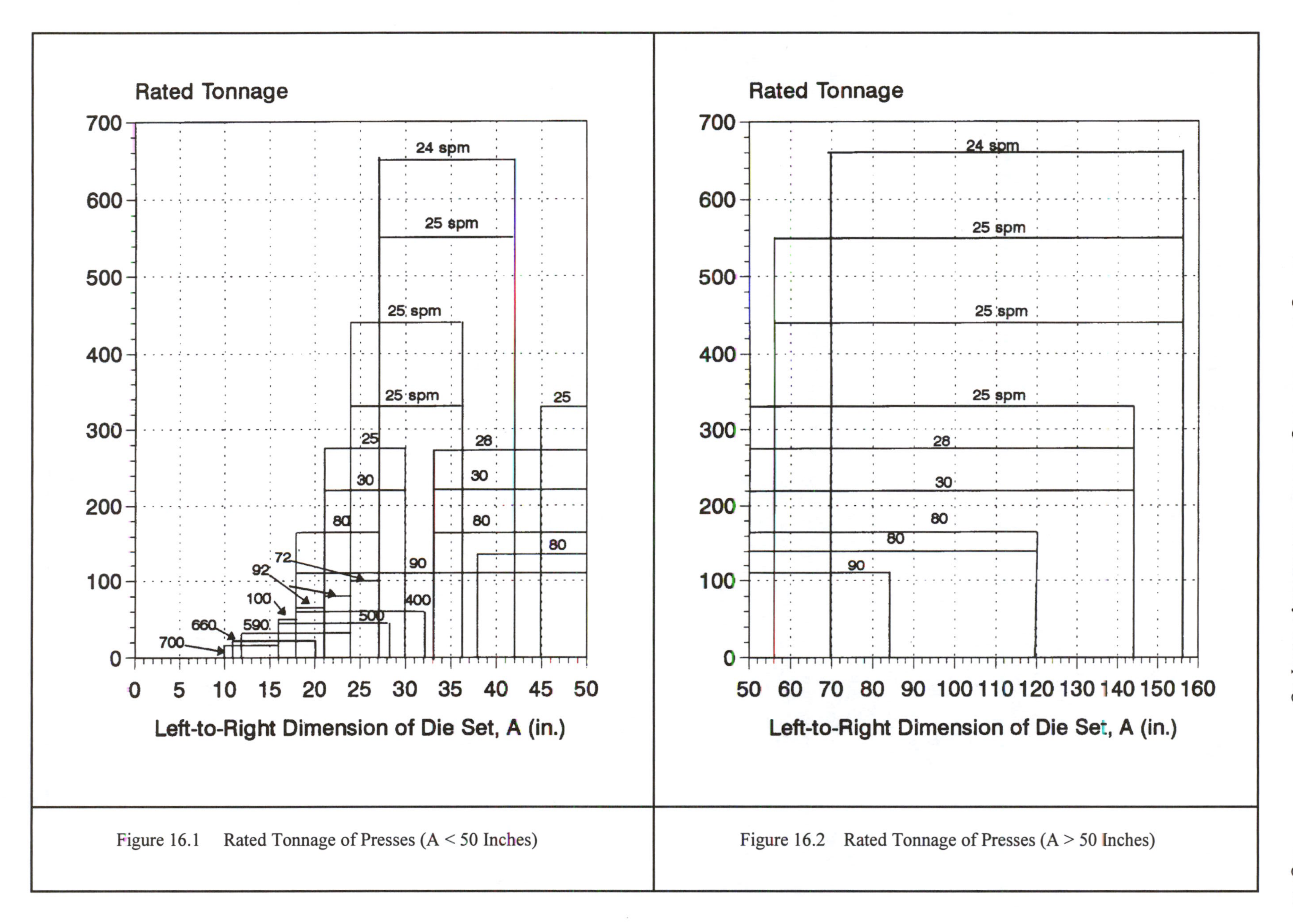

Figure 16.1 Rated Tonnage of Presses (A < 50 Inches)

Figure 16.2 Rated Tonnage of Presses (A > 50 Inches)

16.4 Press Selection

Given the length of a die set, A, and the total force F required to stamp a part, a press can be selected from Figures 16.1 or 16.2. Then Figure 16.3 can be used to determine the relative hourly rate, C_{hr}, for the selected press.

Figures 16.1 and 16.2 are used as follows. Enter the Figure at the value of A along the abscissa. Then move vertically upward at this value until the first horizontal line is reached that has a rated tonnage equal to or greater than the required forceF_p. This horizontal line will correspond to a press with a speed expressed in strokes per minute (SPM) shown along the horizontal line. The rated tonnage of the press can be read along the ordinate at the level of the horizontal line. The rated tonnage, SPM value, and length A determine a press whose hourly rate is given in Figure 16.3. The vast majority of stampers using progressive dies use coiled stock which is automatically fed to the die. Occasionally, the stock is manually fed.

By continuing to move upward in Figures 16.1 or 16.2 at the value of A, other presses (larger, usually slower, and more expensive) can also be selected. Generally speaking, however, the smallest press capable of producing the forces required will result in the lowest overall part cost.

16.5 Determining the Relative Cycle Time

The section above describes how a press can be selected, and its hourly rate determined. To establish part cost, however, we must also know the rate at which useful parts can be produced. The press has a speed (SPM), but this is not the same as the rate at which useful parts are produced, even though it is assumed in what follows that one part is completed with each stroke of the press. (This is done with what are called a "one-up" coil fed progressive dies.) The reasons are: (1) not all parts produced are actually useful, and (2) it takes time to set-up the progressive dies for actual production, during which time no parts are produced at all.

We define the cycle time, t_{cy}, as the average time required to produce a useful part (or the *effective* cycle time). The relative cycle time, t_r, is the average cycle time of a part compared with the average time to produce the standard part. Thus, relative cycle time is given by:

$$t_r = t_{cy} / t_o$$

where t_o is the cycle time for the reference part and t_{cy} is the effective cycle time for the part to be produced. In the

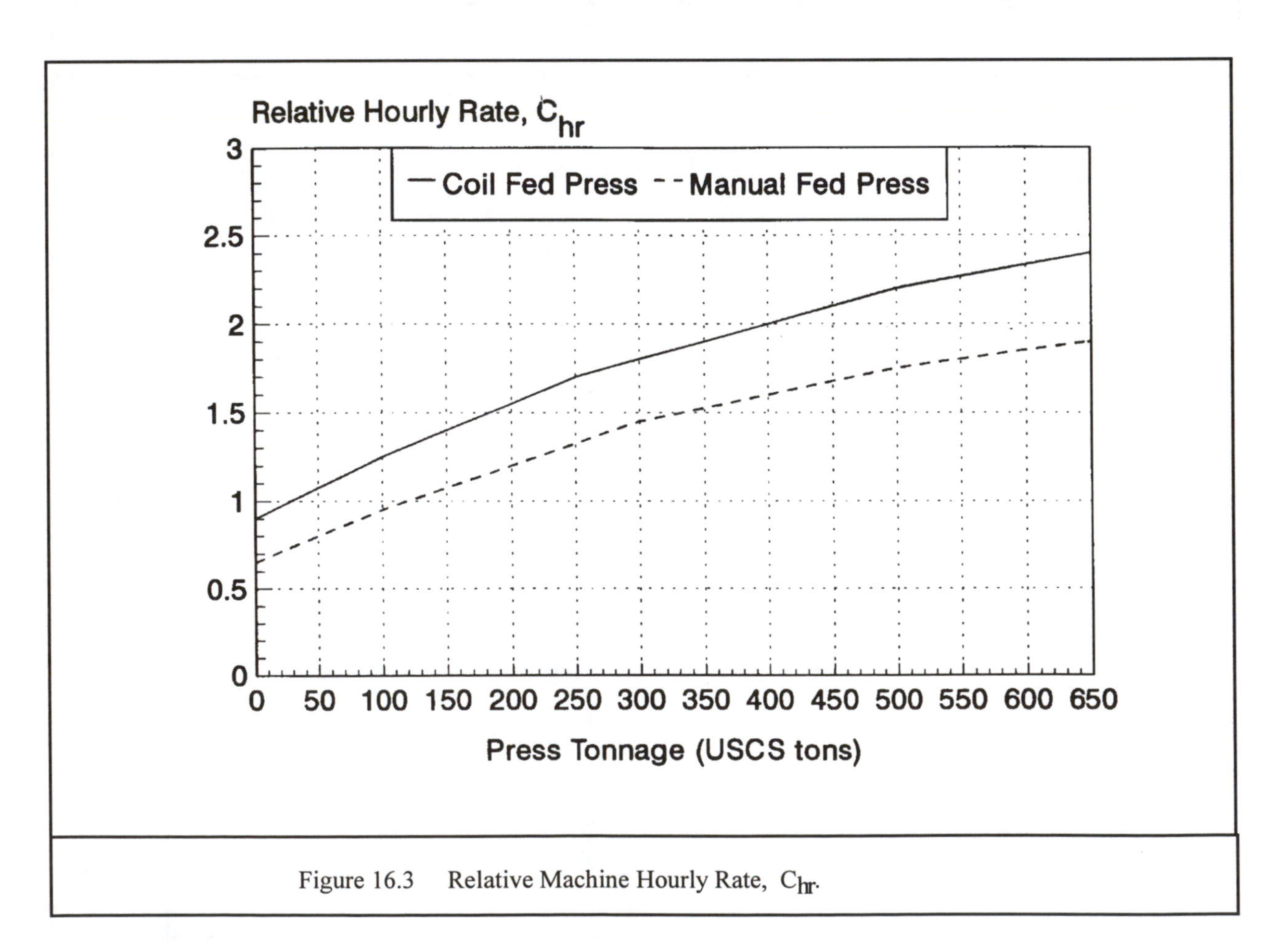

Figure 16.3 Relative Machine Hourly Rate, C_{hr}.

case of stamping, the cycle time required to produce the reference washer is 0.234 sec. (See Table 16.3.)

The effective cycle time, t_{cy}, in seconds, for a given press is given by [1]:

$$t_{cy} = 3600\,[1 / (F_{eff}\,60\ \text{SPM}) + t_{setup}/N] \qquad (16.11)$$

where F_{eff} is the press efficiency (i.e. the proportion of time the press is up) and accounts for the downtime of the press due both routine maintenance of the press and dies. SPM is the number of strokes per minute achievable by the press, t_{setup} is the approximate setup time in hours (approximately 1 hour in many cases), and N is the production volume.

As noted in Section 16.2 above, the relative part cost is given by:

$$C_e = t_r\ C_{hr}.$$

With the methods described above, both t_r and C_{hr} can be computed.

16.6 Example 16.1: Relative Processing Cost for a Part

The part shown in Figure 16.4 is made of a commercial quality, cold rolled steel. The relative tool construction cost for this part was determined earlier in Section 12.8. It was assumed there that holes A, B and C were not side action features, and that the cutout shown was not narrow.

Let us assume that the dimensions and tolerances of the part have been finalized such that holes A, B and C remain as non side-action features and that the cutout remains wide. Let us now determine the relative processing cost for the part.

The length of the perimeter of the unfolded part, L_{out}, was found in Chapter 12 to be about 350 mm. From Table A9.2 of Chapter 9, the approximate shear strength of a commercial quality carbon steel sheet is about 306 MPa (i.e., 0.85 times tensile strength). Since the sheet thickness is 1.6 mm, then from Equation (16.5), F_{out} is:

$$F_{out} = 1.5(350)(306)(1.6) = 257 \text{ kN}.$$

From Equation (16.6), the stripping force, F_{st}, is:

$$F_{st} = 21(350)(1.6) = 11.8 \text{ kN}.$$

Previously (Chapter 12) the part was found to have medium die detail. Hence, from Table 1, $X_{dd} = 0.22$. Thus, the total force required for shearing, F_s, is:

$$F_s = (F_{out} + F_{st})(1 + X_{dd}) = (257 + 11.8)(1.22)$$
$$= 328 \text{ kN}.$$

The total length of all the bends, L_{bt}, is 65 mm. Since the tensile strength of a commercial quality carbon steel sheet (Table A9.1 of Chapter 9) is about 360 MPa, then from Equation (16.7):

$$F_b = (65)(1.6)(360)/18 = 2.1 \text{ kN}.$$

Thus, the total force requirement is:

$$F = F_s + F_b = 330.1 \text{ kN}.$$

The minimum press force required, F_p, is obtained from

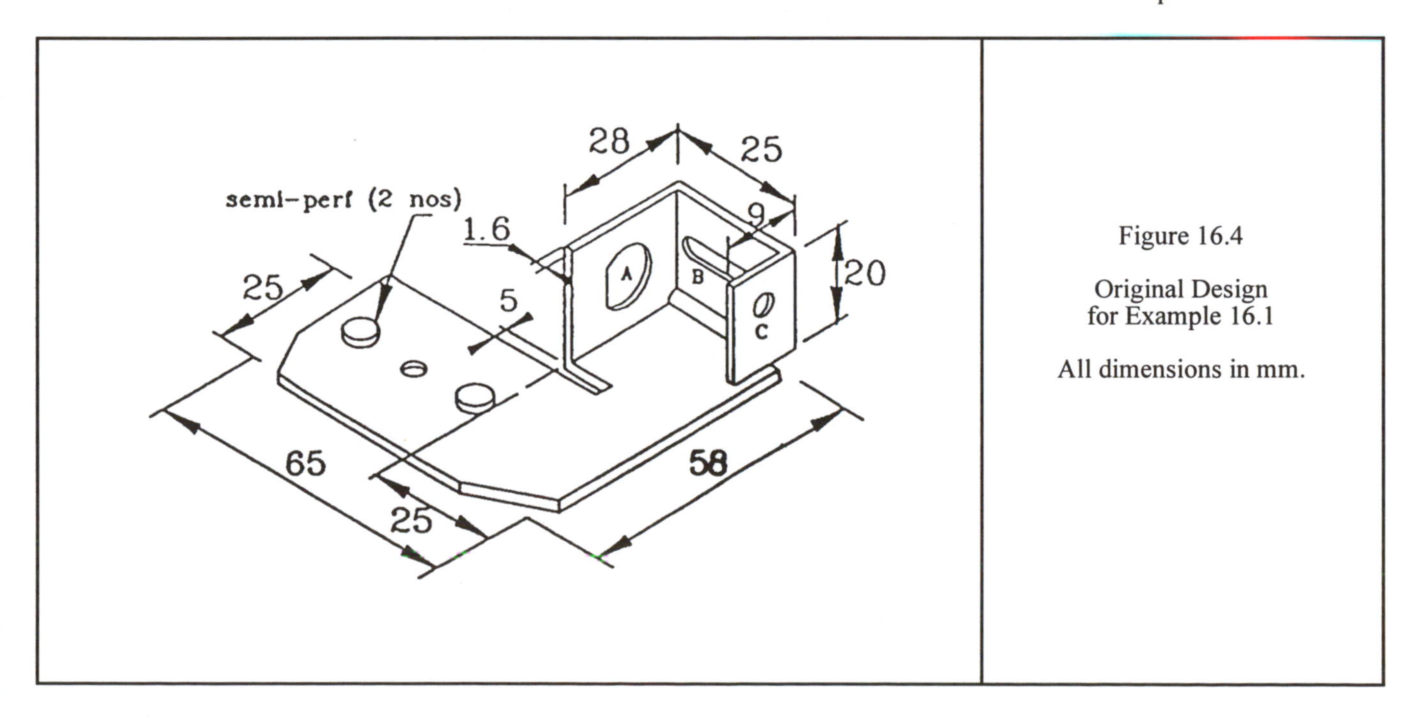

Figure 16.4

Original Design for Example 16.1

All dimensions in mm.

Equation (16.2): $F_p = 1.5F = 495$ kN = 56 tons.

The die set length A was determined earlier (Chapter 12) to be given by:

$$A = L_{dl} + 25 \text{ mm} = 1035 \text{ mm} = 40.8 \text{ in.}$$

Hence, from Figure 16.1, the smallest press which can satisfy both the required tonnage and the required die set length, is a 110 ton press with a speed of 90 strokes per minute.

From Figure 16.3, the relative hourly rate for this press, C_{hr}, is about 1.3.

Production Volume = 10,000

A reasonable value for the press efficiency, F_{eff}, is 0.75. Thus, from Equation (16.11), the effective cycle time for the press, at a production volume of 10,000, is:

$$t_{cy} = 3600[1/0.75(60)(90) + 1/10{,}000] = 1.25 \text{ sec.}$$

Therefore, since the cycle time required to produce the reference washer is 0.234 seconds (see Table 16.3):

$$t_r = t_{cy}/t_o = 1.25/0.234 = 5.34.$$

The relative processing cost for the part is, thus:

$$C_e = t_r C_{hr} = 5.34(1.3) = 6.94.$$

Production Volume = 100,000

For a production volume of 100,000:

$$t_{cy} = 0.924 \text{ sec}$$

$$t_r = 3.95$$

$$C_e = 3.95(1.3) = 5.13$$

(As an exercise, students should confirm the values shown for a production volume of 100,000).

16.7 Example 16.2: Relative Processing Cost for a Part

Figure 16.5 shows the redesigned version of the part shown in Figure 16.4. Because L_{out} and L_{bt} do not significantly change, and since the amount of die detail remains at medium, the minimum required press force is still about 56 tons.

Because the number of stations required to produce this redesigned part has been reduced (see Section 12.8), the value of A is about 28 in. In this case, Figure 16.1 indicates that a 60 ton press with a speed of 400 spm can be used. The relative hourly rate for this machine is about 1.2.

Production Volume = 10,000

From Equation (16.11) the effective cycle time for this press, at a production volume of 10,000, is

$$t_{cy} = 3600[1/0.75(60)(400) + 1/10{,}000] = 0.5647 \text{ s}$$

Therefore,

$$t_r = t_{cy}/t_o = 0.56/0.234 = 2.39$$

The relative processing cost for the part is, thus,

$$C_e = t_r C_{hr} = 2.39(1.2) = 2.87$$

This is a 65% reduction is processing cost.

Production Volume = 100,000

For a production volume of 100,000, it can be shown that

$$t_{cy} = 0.236 \text{ sec}$$

$$t_r = 1.01$$

$$C_e = 1.01(1.2) = 1.21$$

This is about a 75% reduction in processing cost.

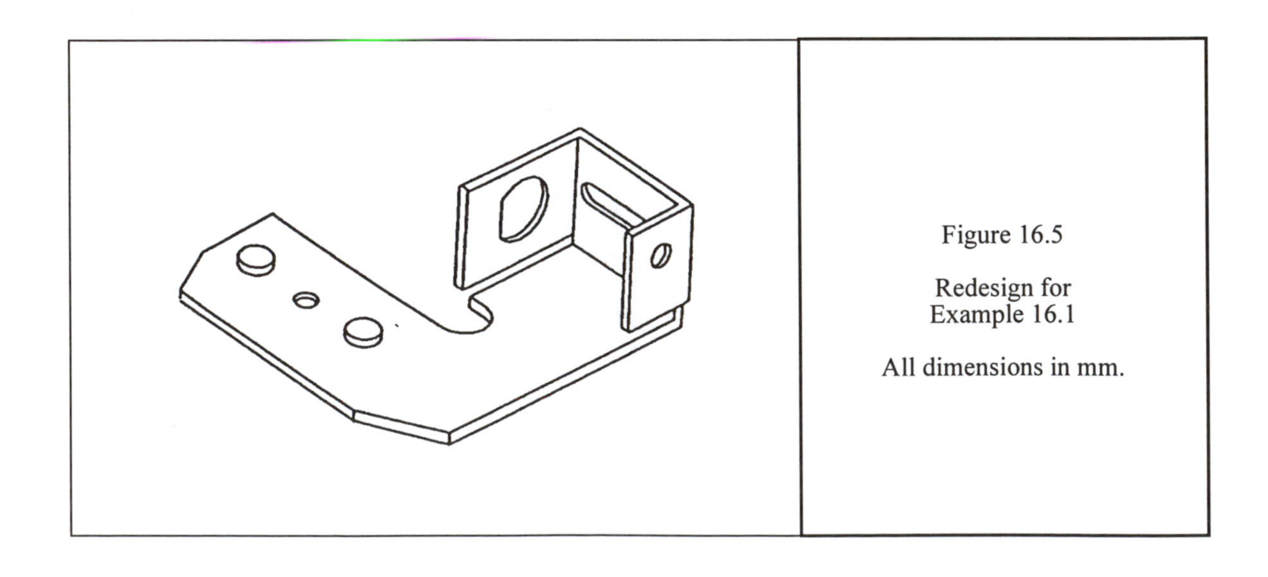

Figure 16.5

Redesign for Example 16.1

All dimensions in mm.

Table 16.2 Some Relative Material Prices For Sheet Metal, C_{mr}

Material	C_{mr}
Hot rolled, oiled steel	0.84
Low carbon CRS, killed	0.97
High carbonCRS, spring	0.98
Low carbon CRS, half hard	0.98
Low carbon CRS	1.00
Low carbon CRS, galvanized	1.39
Low carbon CRS, zinc coated	1.41
Al alloy	1.79
Annealed CRS, spring	1.86
Stainless steel	5.09
Cu sheet, electrolytic tough	6.04
Phorphor bronze, spring	8.41
Cu-Be alloy	34.08

16.8 Relative Material Cost

The expression for the relative material cost for a stamped part, C_m, is identical to Equation (15.5):

$$C_m = K_m / K_{mo} = (V / V_o) C_{mr} \qquad (16.12)$$

where V and V_o are the volume of the part and the reference part, respectively. C_{mr} is the price of the material relative to low carbon cold rolled steel (CRS). Table 16.2 contains the relative material prices for some of the most commonly stamped materials.

16.9 Total Relative Part Cost

As in the case of injection molding, the total production cost of a stamped part, K_t, can be expressed as the sum of the material cost of the part, K_m, the tooling cost, K_d/N, and equipment operating cost (processing cost), K_e.Thus,

$$K_t = K_m + K_d/N + K_e \qquad (16.13)$$

where K_d represents the total cost of the tool, and N represents the production volume.

If K_{mo}, K_{do}, and K_{eo} represent the material cost, tooling cost, and equipment operating cost for the reference part, then as shown in Section 15.17, C_r can be written as:

$$C_r = C_m f_m + C_d f_d + C_e f_e \qquad (16.14)$$

where f_m, f_d, and f_e represent the ratio of the material cost, tooling cost and processing cost of the reference part to the total manufacturing cost of the reference part. Values for f_m, f_d, and f_e are given by the following three equations:

$$f_m = \frac{V_o \cdot K_{po}}{V_o \cdot K_{po} + (K_{dco} + K_{dmo}) / N + t_o C_{ho}} \qquad (16.15)$$

$$f_d = \frac{(K_{dco} + K_{dmo}) / N}{V_o \cdot K_{po} + (K_{dco} + K_{dmo}) / N + t_o C_{ho}} \qquad (16.16)$$

$$f_e = \frac{t_o \cdot C_{ho}}{V_o \cdot K_{po} + (K_{dco} + K_{dmo}) / N + t_o C_{ho}} \qquad (16.17)$$

C_d, C_m, and C_e are, respectively, the tooling cost, material cost and equipment operating cost of a part relative to the material cost, tooling cost and equipment operating cost of the reference part. Values for C_d can be obtained from Equation (12.4), which is repeated below for convenience. Values for C_m and C_e can be obtained from Equations (16.12) and (16.1), respectively.

$$C_d = 0.8C_{dc} + 0.2C_{dm} \qquad (12.4)$$

For the data shown in Table 16.4 for the reference part, the above expressions reduce to the following:

$$f_m = N / (3.06x10^5 + 1.08N), \qquad (16.18)$$

$$f_d = 6561 / (6561 + 0.023N), \qquad (16.19)$$

$$f_e = 0.0018 N / (6561 + 0.023N). \qquad (16.20)$$

Plots of Equations (16.18) - (16.20) are shown in Figure 16.6.

Note that at low production volumes, less than about 30,000, the proportion of part cost due to tooling is more than 90% of the total cost of the part, while the proportion of the cost due to processing is less than 1% of the total part cost. At larger production volumes, 100,000 parts for example, the proportion of part cost due to tooling still remains relatively high, about 75%, while processing costs still remains at a relatively low 1.7%.

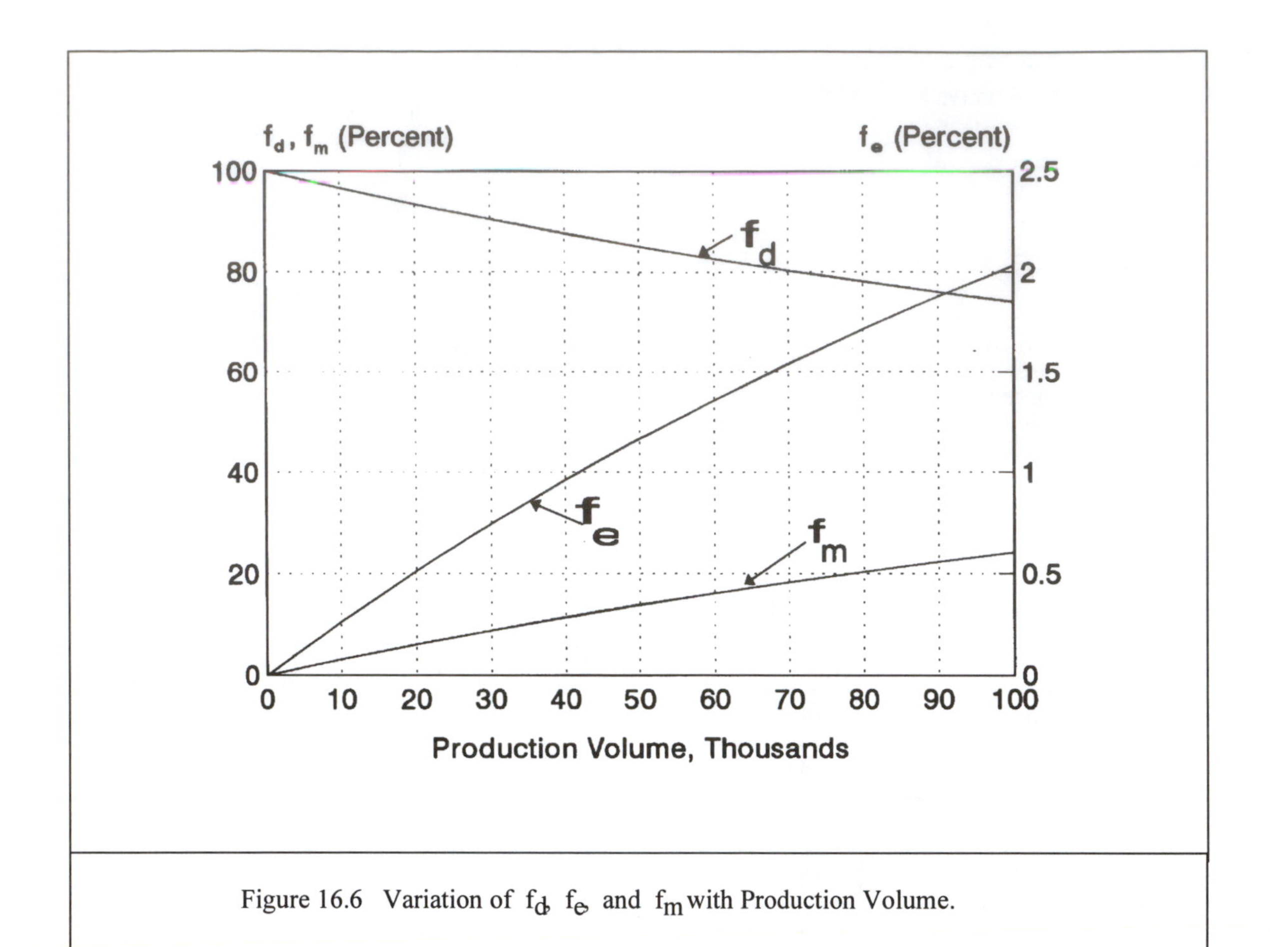

Figure 16.6 Variation of f_d, f_e, and f_m with Production Volume.

Use of Equations (16.18) - (16.20) shows that at a production volume of 2 million, f_m constitutes 82 percent of the part cost, and part material is the dominant concern. The proportion attributable to tooling, f_d, drops to 13%. The proportion of part cost due to processing still remains at a low 5% of the total part cost.

Thus, unlike injection molding and die casting, processing costs in stamping do not make up a significant proportion of part costs. This is primarily due to the high speed (short cycle time) of the presses used.

16.10 Example 16.3: Total Relative Part Cost for a Part

As originally designed, the total relative die construction cost, C_d, for the part shown in Figure 16.4 (Example 12.1, Chapter 12) was found to be 4.18.

For a production volume of 10,000, the relative processing cost, C_e, for the part was found (Example 16.1) to be 6.94, while for a production volume of 100,000, C_e was found to be 5.13.

The volume of the part material needed to make this part is:

$$V = L_{ul}L_{uw}t = 90(65)(1.6) = 9360 \text{ mm}^3.$$

Thus, using the values of V_o and C_{mr} found in Table 16.3:

$$C_m = (V/V_o)C_{mr} = (9360/3750)(1.0) = 2.50.$$

Production Volume = 10,000

At a production volume of 10,000, the cost fractions are (from Figure 16.6):

$$f_m = 0.032; \quad f_d = 0.97; \quad f_e = 0.002.$$

Thus, from Equation (16.14):

$$C_r = 0.032(2.50) + 0.97(4.18) + 0.002(6.94) = 4.14.$$

For the redesigned version of this part (Figure 16.5), C_d = 3.08 and, from Example 16.2, C_e = 2.87. Thus:

$$C_r = 0.032(2.50) + 0.97(3.08) + 0.002(2.87) = 3.07$$

This is a 26% savings is overall part cost. This is identical to the savings in tooling cost found in Example 12.1 of Chapter 12. The savings are due to the fact that, at this production volume, tooling cost is the dominant factor.

Production Volume = 100,000

At a production volume of 100,000, the cost fractions are (from Figure 16.6):

$f_m = 0.241; \quad f_d = 0.739; \quad f_e = 0.020$

Thus, from Equation (16.15), for the original design:

$C_r = 0.241(2.50) + 0.739(4.17) + 0.020(5.13) = 3.79.$

For the redesigned version of this part (Figure 2), $C_d = 3.08$ and, from Example 16.2, $C_e = 1.21$. Thus:

$C_r = 0.241(2.50) + 0.739(3.08) + 0.020(1.21) = 2.90.$

This is a 23% savings is overall part cost. Once again the overall savings in part cost is due almost entirely to the savings in tooling cost.

Table 16.3 Relevant Data for the Reference Part

Part Material	CRS
Material Cost , K_{po}	5.72×10^{-4} cents/mm^3(1)
Part Volume, V_o (2)	3750 mm^3
Die Material Cost, K_{dmo}	\$1041 (3)
Die Construction Time (Includes design and build)	138 hrs (3)
Labor Rate (Die Construction)	\$40/hr (3)
Cycle Time (t_o)	0.234 sec
Press Hourly Rate (C_{ho})	\$27.2 (3)

(1) *Material Price Update*, Xerox Corp., Jan. 1991; (2) Based on total material required to make the part; i.e., ($L_{ul}L_{uw}t$); (3) From collaborating companies.

16.11 Worksheet for Relative Processing Cost (Stamping)

To facilitate the calculation of the relative processing cost and the overall relative part cost, a Worksheet has been developed. The versions shown on the next two pages have been completed for Example 16.1 (Figure 16.4). A blank copy of the Worksheet is printed on Page 16-13, and may be reproduced for use with this book.

16.12 Summary

Summary. In this chapter, we have concentrated on methods for evaluating the relative processing cost and the overall relative part cost for stamped parts made on progressive dies. It was shown that the processing cost of a stamped part represents a very small percentage of the overall part cost. This is due to the high speed of presses used which results in very short cycle times. For this reason, for production volumes less than 100,000 most of the savings achievable in the design of stampings occurs due to the reduction in tooling costs. At very high production volumes (2 million, say), material costs predominate, and part design and strip layout design to reduce material consumption become essential.

Preview. In the next chapter, and in the four chapters that follow, we turn our attention to methods of designing at the parametric stage. These include guided iteration, optimization, and Taguchi Methods™. We begin in Chapter 17 with guided iteration.

References

[1] Dastidar, P. G., "A Knowledge-Based Manufacturing Advisory System for the Economical Design of Metal Stampings," Phd Dissertation, Mechanical Engineering Department, University of Massachusetts at Amherst, Amherst, MA, 1991.

[2] Dallas, Daniel B.,"Metricating the Pressworking Equations, Part 1," Pressworking: Stampings and Dies, Society of Manufacturing Engineers, Dearborn, MI, 1980.

[3] Poli, C., Dastidar, P. G., and Mahajan, P. V., "Design for Stamping, Part I - Analysis of Part Attributes that Impact Die Construction Costs for Metal Stampings," Proceedings of the ASME Design Automation Conference, Miami, Sept. 1991.

[4] Poli, C., Mahajan, P. V., and Dastidar, P. G., "Design for Stamping, Part II - Quantification of on the Tooling Cost for Small Parts," Proceedings of the ASME Design Theory and Methodology Conference, Phoenix, Arizona, Sept. 1992.

[5] Smith, P. G. and Reinertsen, D. G., Developing Products in Half the Time, Van Nostrand Reinhold, New York, 1991. Quotation by permission of the authors.

Problems: See Page 16-12

Worksheet
Relative Processing Cost and Relative Overall Cost — Stamping

Design: ***Original***

Relative Processing Cost (linear dimensions in mm):

Material = ***CRS***; σ_s = ***306*** MPa; σ_t = ***360*** MPa; L_{out} = ***350***

t = ***1.6***; $F_{out} = 1.5\, L_{out}\, \sigma_s t$ = ***257*** kN; $F_{st} = 21\, L_{out}\, t$ = ***11.8*** kN

Detail = ***Medium***; X_{dd} = ***0.22***

$F_s = (F_{out} + F_{st})(1 + X_{dd})$ = ***(257 + 11.8) (1.22) = 328 kN***

L_{bt} = ***65***; Fb = $\sigma_t L_{bt}\, t / 18$ = ***2.1*** kN; $F_{fl} = 0.25\, L_{fl}\, \sigma_t$ = ______ N

$F_{dr} = Dp\, \sigma_t\, (d_o/D_p - 0.7)$ = ______ N; $F_{fo} = L_{fo}\, \sigma_t\, t$ = ______ N

$F_p = 1.5\, F = 1.5\, (F_s + F_b + F_{fl} + F_{fp} + F_{dr})$ = ***1.5 (328 + 2.1) = 495 kN = 56 tons***

(Note: 1 ton = 8.89 kN)

A = ***40.8*** (From die material calculation, in inches)

Press = ***110*** tons; ***90*** SPM; C_{hr} = ***1.3***

N = ***10,000***; $t_{cy} = 3600[\, 1 / F_{eff}\, (60\ \text{SPM}) + 1 / N]$ = ***1.25 s***

$t_r = t_{cy} / 0.234$ = ***5.34***; $C_e = t_r\, C_{hr}$ = ***5.34 (1.3) = 6.94***

N = ***100,000***; $t_{cy} = 3600[\, 1 / F_{eff}\, (60\ \text{SPM}) + 1 / N]$ = ***0.924 s***

$t_r = t_{cy} / 0.234$ = ***3.95***; $C_e = t_r\, C_{hr}$ = ***3.95 (1.3) = 5.13***

Relative Material Cost:

C_{mr} = ***1***; $V = L_{ul}\, L_{uw}\, t$ = ***90 (65) (1.6) = 9360*** mm^3

$C_m = V / V_o\, C_{mr}/3750$ = ***(9360 / 3750)(1.0) = 2.50***

Total Relative Part Cost:

N = ***10,000***; f_m = ***0.032***; f_d = ***0.47*** f_e = ***0.002***

Cr = fm Cm + fd Cd + fe Ce = ***0.32 (2.50) + 0.97(4.17) + 0.002(0.94) = 4.14***

N = ***100,000***; f_m = ***0.241***; f_d = ***0.739*** f_e = ***0.020***

$C_r = f_m\, C_m + f_d\, C_d + f_e\, C_e$ = ***.241(2.50)+ .738(4.17) + 0.020(5.13) = 3.79***

% Savings:

Worksheet

Relative Processing Cost and Relative Overall Cost — Stamping

Design: ***Redesign***

Relative Processing Cost (linear dimensions in mm):

Material = ________; σ_s = ________ MPa; σ_t = ________ MPa; L_{out} = ________

t = ________; $F_{out} = 1.5\, L_{out}\, \sigma_{st}$ = ________ kN; $F_{st} = 21\, L_{out}\, t$ = ________ kN

Detail = ________; X_{dd} = ________

$F_s = (F_{out} + F_{st})(1 + X_{dd})$ = ________

L_{bt} = ________; $F_b = \sigma_t L_{bt}\, t / 18$ = ________ kN; $F_{fl} = 0.25\, L_{fl}\, \sigma_t$ = ________ N

$F_{dr} = D_p\, \sigma_t (d_o/D_p - 0.7)$ = ________ N; $F_{fo} = L_{fo}\, \sigma_t\, t$ = ________ N

$Fp = 1.5\, F = 1.5\, (F_s + F_b + F_{fl} + F_{fp} + F_{dr})$ = ***56 tons (No significant change from original)***

(Note: 1 ton = 8.89 kN)

A = ***28*** (From die material calculation, in inches)

Press = ***60*** tons; ***400*** SPM; C_{hr} = ***1.2***

N = ***10,000***; $t_{cy} = 3600[\,1 / F_{eff}\,(60\ \text{SPM}) + 1 / N]$ = ***0.565 sec***

$t_r = t_{cy} / 0.234$ = ***2.39***; $C_e = t_r\, C_{hr}$ = ***2.87***

N = ***100,000***; $t_{cy} = 3600[\,1 / F_{eff}\,(60\ \text{SPM}) + 1 / N]$ = ***0.2365***

$t_r = t_{cy} / 0.234$ = ***1.01***; $C_e = t_r\, C_{hr}$ = ***1.21***

Relative Material Cost:

C_{mr} = ***1***; $V = L_{ul}\, L_{uw}\, t$ = ***78(74)(1.6)= 9235 (approx)*** mm^3

$C_m = V / V_o\, C_{mr}/3750$ = ***2.46 (= 2.50 approx)***

Total Relative Part Cost:

N = ***10,000***; f_m = ***0.032***; f_d = ***0.97*** f_e = ***0.002***

$C_r = f_m\, C_m + f_d\, C_d + f_e\, C_e$ = ***0.032(2.5) + 0.97(3.08) + 0.002(2.87) = 3.07***

N = ***100,000***; fm = ***0.241***; fd = ***0.739*** fe = ***0.020***

$C_r = f_m\, C_m + f_d\, C_d + f_e\, C_e$ = ***0.241(2.5) + 0.739(3.08) + 0.20(1.21) = 2.90***

% Savings:

26% for N = 10,000; 23% for N = 100,000

Problems

16.1 For a production volume of 50,000 parts, determine the relative processing costs for the part shown in Figure 16.P1. Assume that the part is made of soft cold rolled steel. (Note: The relative tooling costs for this part were determined in Problem 3 of Chapter 12)

16.2 Repeat the calculations of Problem 16.1 for a production volume of 1,000,000.

16.3 For a production volume of 50,000 parts, determine the relative processing costs for the part shown in Figure 16.P3. Assume that the part is made of soft cold rolled steel. Note: The relative tooling costs for this part were determined in Problem 4 of Chapter 12.

16.4 Repeat the calculations of Problem 16.3 for a production volume of 1,000,000.

16.5 For the redesigned version of the part shown in Figure 16.P1 (Problem 5 of Chapter 12) what are the relative processing costs for the part at production volumes of 50,000 and 1,000,000?

16.6 Determine the overall relative part cost for the part considered in Problems 1 and 5.

16.7 Determine the savings in overall part costs for the part considered in Problem 1 if it is redesigned as discussed in Problem 5 of Chapter 12.

17.8 Determine the savings in overall part costs for the part considered in Problem 16.3 if it is redesigned as discussed in Problem 6 of Chapter 12.

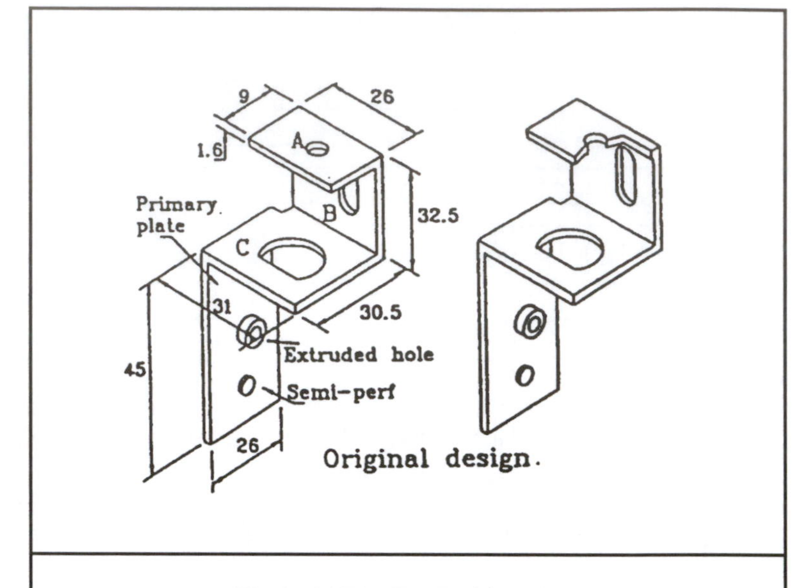

Figure 16.P1 For Problem 16.1

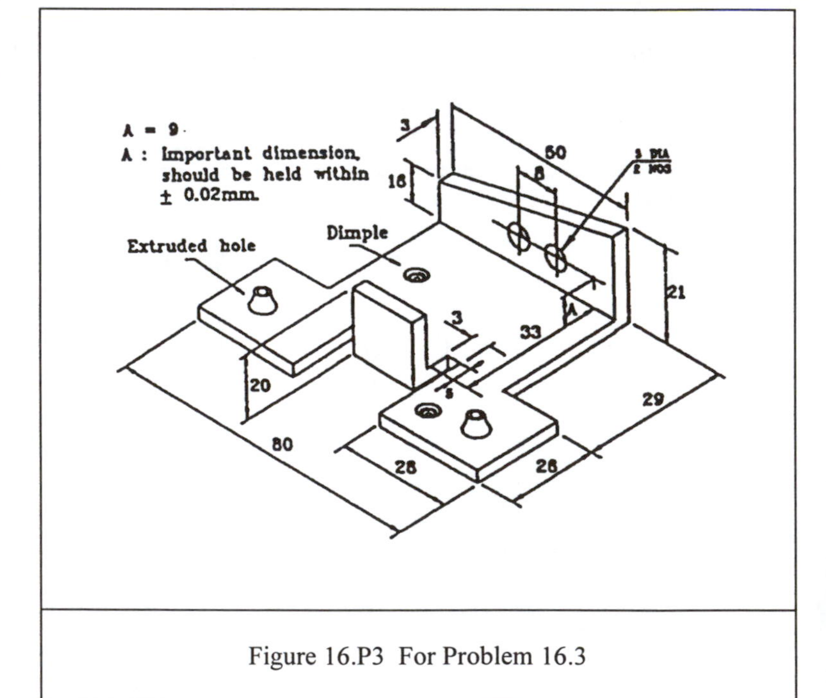

Figure 16.P3 For Problem 16.3

Worksheet
Relative Processing Cost and Relative Overall Cost — Stamping

Design:

Relative Processing Cost (linear dimensions in mm):

Material = ______________; s_s = __________ MPa; s_t = _____________ MPa; L_{out} = _________

t = __________; $F_{out} = 1.5\ L_{out}$ sst = __________________kN; $F_{st} = 21\ L_{out}\ t$ = ____________kN

Detail = __________________; X_{dd} = ___________________

$F_s = (F_{out} + F_{st})(1 + X_{dd})$ = ____

L_{bt} = ____________ ; $F_b = s_t L_{bt}\ t / 18$ = __________kN; $F_{fl} = 0.25\ L_{fl}$ tt = ________________N

$F_p = 1.5\ F = 1.5\ (F_s + F_b + F_{fl} + F_{fp} + F_{dr})$ = _______________________________________
(Note: 1 ton = 8.89 kN)

A = ____________________ (From die material calculation, in inches)

Press = ___________________ tons; _________________ SPM; Chr = ______________________

N = ________________; $t_{cy} = 3600[\ 1 / F_{eff}(60\ SPM) + 1 / N]$ = ___________________________

$t_r = t_{cy} / 0.234$ = __________________; $C_e = t_r C_{hr}$ = ___________________________

N = ________________; $t_{cy} = 3600[\ 1 / F_{eff}(60\ SPM) + 1 / N]$ = ___________________________

$t_r = t_{cy} / 0.234$ = __________________; $C_e = t_r C_{hr}$ = ___________________________

Relative Material Cost:

C_{mr} = __________; $V = L_{ul} L_{uw} t$ = _____________________________ mm3

$C_m = V / V_o C_{mr}/3750$ = ____

Total Relative Part Cost:

N = ______________; f_m = ___________; f_d = ______________ f_e = __________________

$C_r = f_m C_m + f_d C_d + f_e C_e$ = ____

N = ______________; f_m = ___________; f_d = ______________ f_e = __________________

$C_r = f_m C_m + f_d C_d + f_e C_e$ = ____

% Savings:

This page is intentionally blank.

Chapter Seventeen

Parametric Design By Guided Iteration

"Experience keeps a dear [expensive] school,
but fools will learn in no other."

Benjamin Franklin
Poor Richard's Almanac (1743)

17.1 Introduction

As Benjamin Franklin implies, it isn't experience per se that matters. The important thing is what we learn from experience — especially what knowledge we gain that we can generalize and apply in new and different situations.

In Chapter 14, we introduced parametric design and showed how important it is for designers to have methods for assigning values to attributes; a "seat of the pants" approach is just not good enough for engineering design in today's competitive world. In this book, we present three methods for parametric design: guided iteration, optimization, and Taguchi Methods™. This chapter is devoted to guided iteration.

The steps in guided iteration process are, as always:

1. Formulation of the problem,
2. Generation of alternatives,
3. Evaluation, and
4. Redesign *guided by* physical reasoning and the results of the evaluations.

How these steps can be implemented to perform parametric design by guided iteration is now described.

17.2 Formulating the Problem for Guided Iteration

17.2.1 Information Required in the Formulation.

The formulation of a parametric design problem for solution by guided iteration requires specifying four categories of information:

- Problem definition parameters (PDPs),
- Design variables,
- Solution evaluation parameters (SEPs), and
- Dependencies.

Each of these is discussed in the sub-sections that follow.

17.2.2 Specifying the Design Variables

As defined in Section 14.2, parametric design variables correspond to the attributes of a design that require values. Examples are dimensions, tolerances, materials, and perhaps others. In formulating parametric design problems, design variables are identified including:

- Names, symbols, and units;
- Range of allowed values; and
- Whether they are continuous or discrete. If discrete, the permissible values are also included.

17.2.3 Specifying the Problem Definition Parameters (PDPs)

A problem definition parameter (PDP) is a parameter whose value imposes a specific condition the component will encounter in use or in manufacture, or a specific requirement of its functional operation. PDPs thus define the *specific* parametric design problem to be considered. In the formulation, PDPs are listed, including their names, symbols, units, and any limits on feasible assigned values. We use the symbol z to refer to a PDP value.

The PDPs and their values in a parametric design problem are almost always found in the Engineering Design Specification among the various requirements. They can include both functional and manufacturing requirements (e.g., limits on tooling costs).

IIIIIIIIIIIIII

17.2.4 Continuing the I-Beam Example

Recall that in the simple I-Beam example started in Chapter 14, the numerical design variables (see Figure 14.1) are:

Design Variable	Symbol	Allowed Range	Type
beam height	h	< 20 inches	Continuous
width	w	< 12 in	Continuous
web thickness	b	> 0.1 in	Continuous
flange thickness	t	> 0.1 in	Continuous

For this I-Beam, assuming static loading and normal atmospheric conditions, the PDPs and their values for this example are:

Beam Length	L = 5 feet
Load type(s) (i.e., concentrated, distributed) and locations	Uniformly distributed
Load value(s) and locations	13,000 pounds/foot; Full length of beam
Support type(s) (i.e., built in, simple, etc.) and locations	Simply supported at each end
Restrictions on maximum deflection	Max deflection = 0.05 inches
Limits or goals on cost	As low as possible
Any special environmental conditions	None
Material is 1020 steel	E = 30,000,000 psi Yield Stress: 36,000psi

This example will be re-visited from time to time throughout this chapter.

IIIIIIIIIIIIIIIII

17.2.5 Specifying the Solution Evaluation Parameters (SEPs).

What is a Solution Evaluation Parameter? In the guided iteration process, proposed solutions must be evaluated and compared in terms of their quality. To accomplish this step in the parametric stage, we define a set of solution evaluation parameters (SEPs) whose values are related to the quality of a design. They are, in effect, the evaluation criteria. In most real design problems, there are a number of SEPs to be considered (e.g., cost, different aspects of functional performance, weight, life, etc.) We use the symbol y to designate SEP values.

In formulating a parametric design problem for guided iteration, there are three types of information associated with solution evaluation parameters (SEPs) that are also included in addition to the name, units, and symbols for the SEP itself:

1. The importance level: Very Important (or Critical), Important, or Desirable:
2. The methods — often analysis methods — by which the value of the SEP can be obtained given values for the PDPs and design variables; and
3. The relationship between the *value* of the SEP and its *evaluation* in terms of quality. We usually express the evaluation as Excellent, Good, Fair, Poor, or Unacceptable.

Every part or product has a number of individual, more or less identifiable, factors which contribute to its overall quality. In the case of parts, the list of quality-related factors is generally technical rather than consumer or in-use purpose related. They will be factors like cost, weight, size, ability to hold tolerances, corrosion resistance, and many more. Any such factor is called a Solution Evaluation Parameter (SEP).

In some cases, the evaluation of an SEP is done without computation of its numerical value. For example, the manufacturability of a trial part design may simply be evaluated directly by some qualitative means as Excellent, Good, Fair, or Poor (perhaps using information like the DFM Advisors in Chapter 3). Guided iteration works perfectly well in these cases since it uses the evaluations of the SEPs, not the values, as the basis for its redesign decisions.

IIIIIIIIIIIIII

I-Beam Example, Continued. For the I-Beam example, the SEPs are:

- Maximum actual stress in the beam, MaxStress, psi,
- Maximum actual deflection of the beam, MaxDeflection, inches,
- Cross Sectional Area (A) of the beam (in this case, a surrogate for material cost) (inches^2).

To make this example a bit more interesting, we will also assume there is a required relationship that the beam height must be greater than or equal to the beam width. Thus there is an SEP:

- Height-Width Ratio, h/w.

IIIIIIIIIIIIIIII

Obtaining Values for SEPs. For each solution evaluation parameter (SEP), there must be a procedure specified for computing the value of the SEP from the values of PDPs and design variables. As discussed above, we call these procedures by the general term *analysis methods* because they often include equations from the realm of engineering science (e.g., methods for calculating stresses and deflections in beams, heat flow, fluid flows and pressure drops, etc.) However, the term "analysis" here is not intended necessarily to be limited to engineering science procedures. Cost estimates, for example, are also needed to obtain the value of the cost SEP; thus, a cost analysis procedure may be included. There may also be manufacturability analyses involved providing evaluations of how well the proposed design satisfies ease of manufacturability criteria. There may also be aesthetic issues, and while there are not mathematical procedures for estimating the value of aesthetic characteristics, procedures can still be defined. For example, a panel of critics or consumers might be surveyed.

Even for the technical criteria, the values of all SEPs need not be numerical. A specific material will be a value in cases where the material is an attribute derived from the configuration stage. Another kind of non-numeric value occurs when a specific member of a general type or class must be determined. For example, exact motor types might be required.

Sometimes the evaluation of an SEP is determined directly. That is, the analysis procedures for some SEPs will result directly in an evaluation such as Excellent, Good, Fair, Poor, or Unacceptable. For example, if an SEP werethe delivery time for parts to be purchased from vendors, the delivery prospects might be put into categories of Excellent, Good, Fair, etc. Sometimes, aesthetic or marketing factors must be evaluated using methods that might be called "analysis", but are more subjective than the quantitative methods of engineering science.

IIIIIIIIIIIIII

I-Beam Example, Continued. In the case of our I-Beam example, the analytical procedures needed can be from beam theory in strength of materials:

For the maximum stress SEP:

$$\text{Stress} = \sigma_y = Mh/2I = WL^2h/16\,I \qquad (17.1)$$

For the maximum deflection SEP:

$$\text{Deflection} = \delta = 5WL^4/384\,E\,I \qquad (17.2)$$

where:

$$I = wh^3/12 - [(w - b)(h - 2t)^3]\,/12 \qquad (17.3)$$

For the area SEP:

$$A = 2\,t\,w + b\,(h - 2t) \qquad (17.4)$$

For the height to width ratio SEP: h/w.

Note that the value for each of these can be computed from values for the PDPs and design variables.

IIIIIIIIIIIIIIII

Obtaining an Evaluation From a Value. To obtain an evaluation once the value of an SEP has been computed, it is helpful to distinguish between two types of SEPs which we will refer to as (a) soft and (b) hard.

Soft SEPs. Suppose that along a horizontal axis we plot the values of an SEP, y_n. Also assume that we can associate the values with a quality level — that is, an evaluation — from Unacceptable to Excellent. This is shown schematically in Figure 17.1(a).

We can show the same information in more graphic form by using the vertical axis to indicate quality levels as shown in Figure 17.1(b).

However, since the locations of the transitions between quality levels (Poor, Good, etc.) are only subjective estimates, it is convenient to smooth out the steps into a straight line as shown in Figure 17.1(c). The region of the steps or sloping line can be called a *fuzzy* region. This region is important because it is the region where *trade-offs* can be made between or among the quality levels of the various SEPs.

The graphs shown in Figure 17.1 — and there are more examples in Figure 17.2 — are called *satisfaction graphs* because they relate the value of an SEP to the degree of satisfaction designers associate with that value. The Good, Fair, and Poor ranges in the fuzzy regions in Figure 17.2 are omitted to keep the graphs from being cluttered.

Hard SEPs. Hard SEPs do not have fuzzy (or soft) regions in their satisfaction graphs. That is, they do not have degrees of satisfaction; they are either satisfied or not satisfied. An example of a satisfaction graph for a hard type SEP at the minimum acceptable level is shown in Figure 17.2(d).

Hard SEPs often result from the existence of required relationships among design variables and/or the PDPs, and the SEPs. Sometimes these required relationships, when expressed as equations or inequalities, are called *constraints.* Required relationships may derive from physical principles or geometry, or they may be empirically based. Required relationships restrict the freedom of the designer to set values of the design variables in some way.

Required relationships can also originate in the need for a part to fit with some other part in an assembly. For example, if a shaft must fit into a hole with a running fit, then the shaft must be smaller than the hole, whatever their sizes.

Functional requirements may also create required relationships. A very simple example illustrates this point: If a container is to be made to have a specified volume, then once the configuration is determined to be, say, a rectangular box, there will be a required relationship among the attributes: the height times width times depth must be equal to the required volume. Whatever, height, width, and depth we select, they must satisfy the limits set by this required relationship.

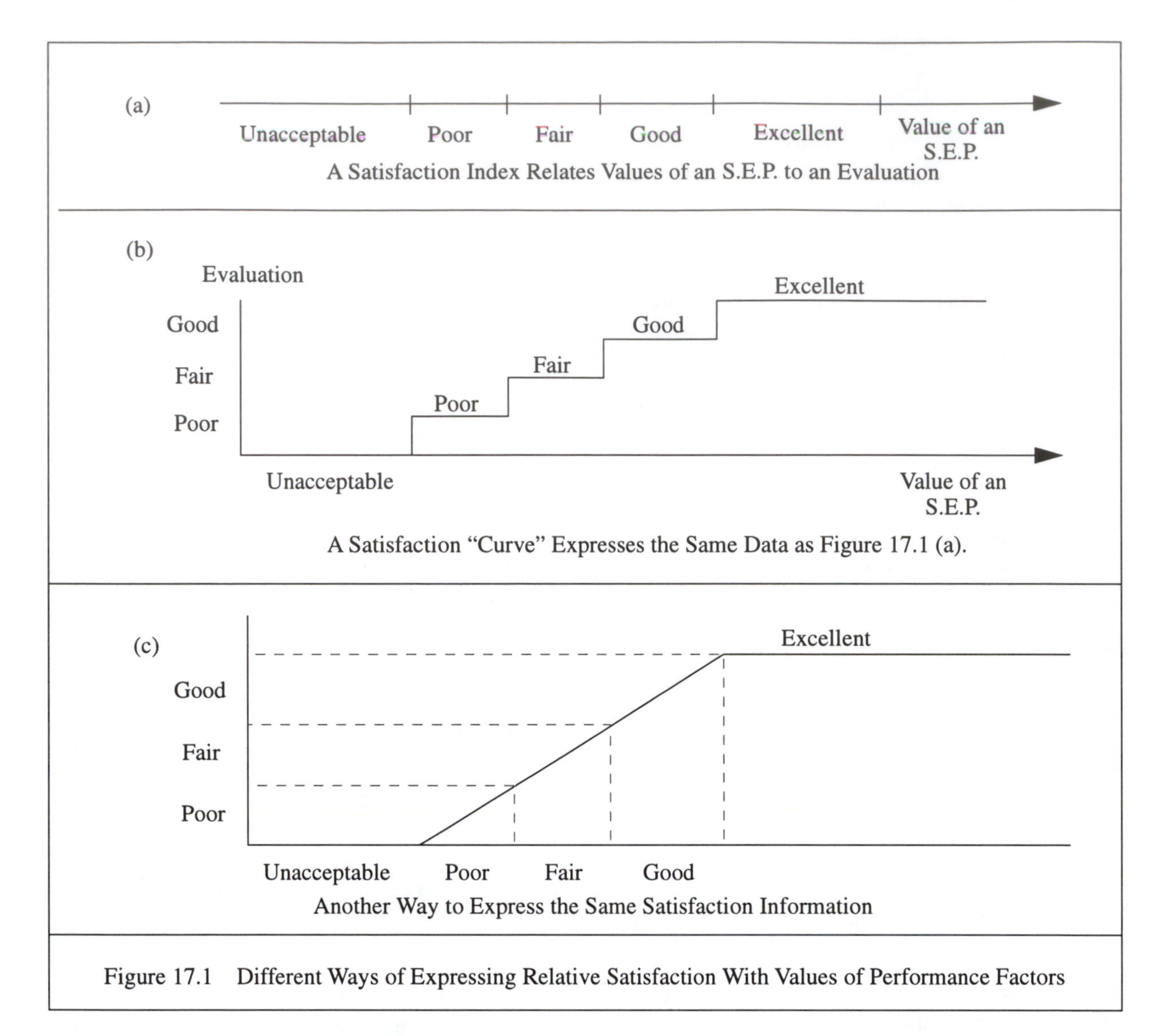

Figure 17.1 Different Ways of Expressing Relative Satisfaction With Values of Performance Factors

Sometimes required relationships result from manufacturing considerations. In injected molded parts, for example, the dimensions of easily manufactured bosses (in the sense of lower processing costs) are limited by the thickness of the walls to which they are attached. In general, the boss height should be no more than three times the local wall thickness, and the boss width should be no more than the local thickness. In the next chapter, bosses that meet these requirements are referred to *non-significant* bosses.

IIIIIIIIIIIIIII

I-Beam Example Continued. Possible Satisfaction Curves for the I-Beam example are shown in Figure 17.3. Note that some important decisions have been made in constructing these curves. For example, In Figure 17.4 (a), it has been decided not to accept designs that result in a stress greater than 35,000 psi even though the nominal yield stress is 36,000 psi. This is a small safety factor to allow for variations in materials and manufacture of the beam. Also, end conditions and loadings may not be exactly as specified.

In Figure 17.3(a), it has also been decided that designs which result in a maximum stress less than about 14,000 psi should not be accepted. This is to discourage over-design. We would surely prefer designs that use the material more effectively. This decision must be considered tentative, however. The 14,000 level may need to be adjusted later during the iterative design process in order to get a meaningful effect.

In Figure 17.3(b), though a deflection of up to 0.05 inches is allowed, note that deflections above 0.045 begin to considered less desirable. This, too, is a kind of safety factor.

In Figure 17.3(c), the shape of the Satisfaction Curve suggests that the smaller the area the better, and requires that the area be no greater than 7.0 inches^2. In most cases, designers will be sufficiently familiar with the problem domain that such limiting values can be assigned reasonably well. If they aren't, the problem can still be solved as we shall see later. In any case, such limits may need to be adjusted during the guided iteration process in order to sup-

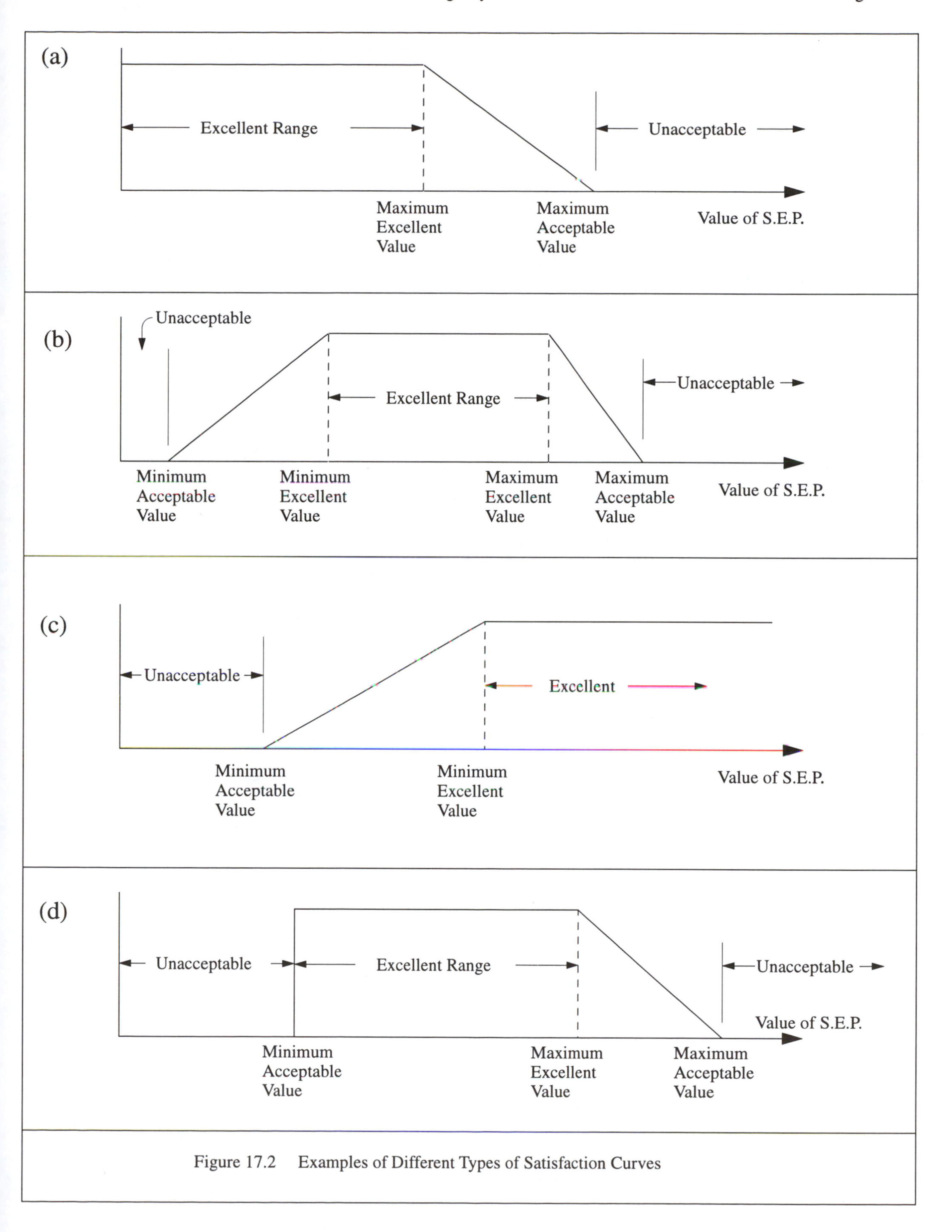

Figure 17.2 Examples of Different Types of Satisfaction Curves

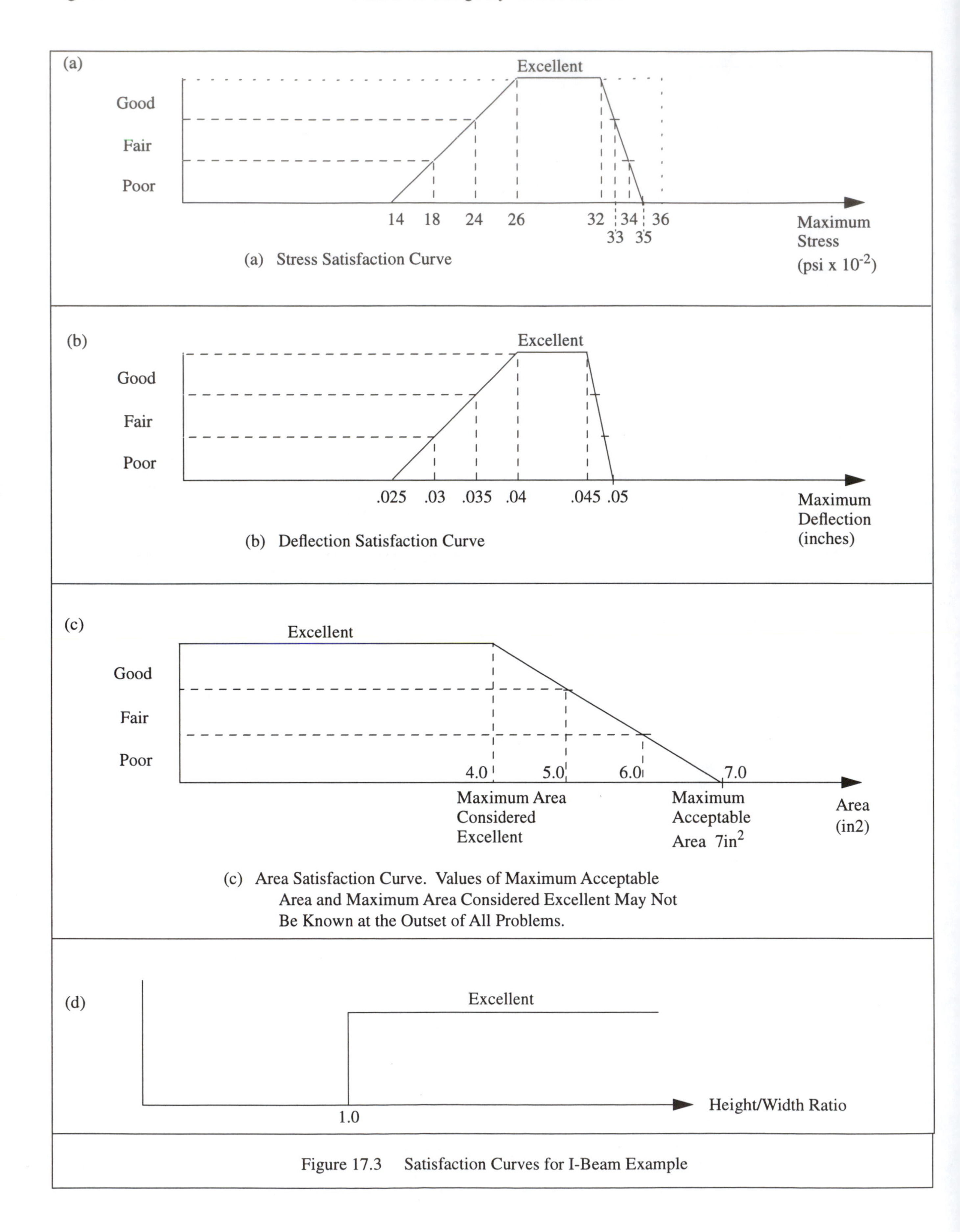

Figure 17.3 Satisfaction Curves for I-Beam Example

port the desired trade-off decisions properly. This point is discussed and illustrated in more detail below.

Note how the Satisfaction Curve in Figure 17.3(d) represents the hard SEP, h / w.

IIIIIIIIIIIIIIII

17.2.6 Specifying Initial Dependencies

The Concept of Dependency. In performing parametric design by guided iteration, designers must iteratively adjust the values of the design variables until a set of values is obtained that results in at least a satisfactory design. To guide this process, there must be available some information about how the various design variables influence the SEPs. For example, suppose that during the parametric design of an I-Beam, the deflection (an SEP) of the current trial design beam is too great. To reduce the deflection, should the designer change the web thickness, the flange thickness, the beam height, or the beam width? And by how much? Or perhaps a different material should be selected?

To provide the information needed to answer such questions, the concept of *dependency* has been developed. A dependency (d_{mn}) is defined as the following relationship between the value of a solution evaluation parameter (y_n) and the value of a design variable (x_m):

$$y_n = K x_m^{d_{mn}} \tag{17.5}$$

where K is a constant of proportionality, n denotes the nth SEP, and m denotes the mth design variable. Note that the first subscript in d refers to the particular design variable involved, and the second subscript refers to the particular SEP involved.

The purpose of creating this concept of dependency is to provide a convenient way for designers to be able to perceive, and especially to be guided by, available knowledge of how individual SEPs — which as a group determine the quality of a design — are related to individual design variables, over which designers have some direct control. For example, in the case of beam deflection, knowledge of dependencies between the SEP deflection and the individual design variables can be a great help to designers in deciding what change in what design variable values to select for the next trial design.

Another way to state the concept of dependency is this: A dependency (d_{mn}) is a numerical factor that indicates approximately how a particular SEP (y_n) is influenced by a particular design variable (x_m).

The relationship between design variables and SEPs expressed in Equation 17.5 is arbitrary; we could define any relationship we wish to work with. None will be perfect. Dependency as defined in Equation 17.5 is related to and behaves somewhat like (but not exactly like) a partial derivative of an SEP with respect to a design variable. Both the dependency and the partial derivative indicate the direction and magnitude of the influence of the particular design variable on the particular SEP. However, dependencies as defined in Equation 17.5 have some advantages as explained in the next paragraph.

A dependency expressed as d_{mn} is dimensionless. A value of zero for a dependency indicates no relationship between a given design variable x and solution evaluation parameter y. That is, if (say) d_{25} = zero, changing the design variable x_2 has no effect on the SEP y_5.

The way dependency is defined in Equation 17.5 also means that a dependency value of $d = +1$ indicates a positive linear relationship between the specified x and y. That is, if d_{25} is +1, then a 50% increase in design variable x_2 will result in a 50% increase in SEP y_5. If a dependency is -1, then the value of the SEP is inversely proportional to the value of the design variable. In either case, the larger values of d indicate a stronger influence while smaller magnitudes indicate a weaker influence.

To illustrate the concept of dependency with a simple example, consider a case in which there are two design variables (x_1 and x_2) and two SEPs (y_1 and y_2). Assume that we are at a point in the parametric problem solving process where we estimate the four dependencies in the problem to be as follows:

$$d_{11} = 4$$

$$d_{12} = 0.005$$

$$d_{21} = 0.3$$

$$d_{22} = -1.0$$

(Note that in Equation 17.5, the first subscript refers to the design variable, the second to the SEP.)

Now further assume that what is needed to improve this design is to increase the value of the SEP y_2 by about 10%. (Perhaps this change would, for example, improve the design from a Fair to a Good overall evaluation.) To accomplish this change in y_2, we can change either of the design variables, x_1 or x_2. By inspecting the relevant dependencies, d_{12} and d_{22}, we see that d_{12} is very, very small. Thus it would require a huge change in x_1 to effect the change we want in y_2. Though this does not absolutely rule out using design variable x_1 to effect the desired change, the very small dependency value often means that some other design variable would be more effective.

If we consider changing x_2 as a way to change y_2, we note that the dependency d_{22} is negative, so to increase y_2 we will have to decrease x_2. The amount of the required decrease in x_2 can be estimated by converting Equation 17.5 to an equivalent form for y_2 and x_2:

$$y_2(i) / y_2(i+1) = [(x_2(i) / x_2(i+1)]^{d_{21}} \tag{17.6}$$

		Design Variables					
		X_1	X_2	X_3		X_i	
Solution Evaluation Parameters	Y_1	d_{11}	d_{21}	d_{31}		d_{i1}	
	Y_2	d_{12}	d_{22}	d_{32}		d_{i2}	
	Y_3	d_{13}	d_{23}	d_{33}		d_{i3}	
	Y_j	d_{1j}	d_{2j}	d_{3j}		d_{ij}	

Figure 17.4 (a) General Dependency Table

	X_1	X_2
Y_1	$d_{11} = 4$	$d_{21} = 0.3$
Y_2	$d_{12} = 0.005$	$d_{22} = -1.0$

Figure 17.4 (b) Numerical Dependencies

Later we sometimes express this relationship also as:

$$y_2\,(\text{old}) / y_2\,(\text{new})) = [(x_2\,(\text{old}) / x_2\,\text{new})]^{d_{21}} \qquad (17.7)$$

In this example, suppose that the current (old) value of x_1 is 0.56. Since we want to increase y_2 by 10%, we can compute the change required in x_2 by solving Equation 17.7:

$$1 / 1.1 = (0.56 / x_2(\text{new}))^{-1.0}$$

The new value to try for x_2 is $x_2(\text{new}) = 0.509$.

In summary: Dependencies are dimensionless numbers that indicate approximately how individual design variables relate to individual SEPs. (See Equations 17.5, 17.6 and 17.7.) A positive dependency indicates a positive influence; a negative dependency indicates a negative influence. A zero dependency indicates no influence. A dependency of +1 indicates a linear influence. Larger or smaller values indicate stronger or weaker influences.

Dependency Tables. In complex problems, there are many dependencies. (If there are m design variables and n SEPs, then there are mn dependencies.) To help keep track of the dependencies, and to make them convenient to use by inspection, they can be recorded in a tables, which we call a *dependency table*. A skeleton dependency table is shown in Figure 17.4(a). A dependency table for the four dependencies listed above is shown in Figure 17.4(b). A blank table for the I-Beam example is shown in Figure 17.5.

Initial Values for Dependencies. To establish an initial dependency table for a problem, values for the dependencies associated with each x-y pair are estimated and recorded in the matrix. Since we may not know exact values at this initial stage, we make educated initial guesses (i.e., we use qualitative physical reasoning) for as many d values as possible. As we shall see, it does not matter that these initial values may be incorrect, though the closer they are to the correct values, the less computation will be required to arrive at a final acceptable solution for the x's.

Even though designers may not know exact values for the dependencies, it is usually possible to enter estimated initial values in the dependency table. Remember, this is a problem that has already been taken through the conceptual and configuration stages. Usually designers have an approximate idea of how strongly the various SEPs are influenced by the various design variables. Fortunately, even if some (or all) the values are off the mark, a good solution will still be found — though probably not as quickly.

IIIIIIIIIIIIIII

I-Beam Example, Continued. Let's consider how to establish initial dependency values for the I-Beam example begun above. As a first step, consider the dependencies involving the SEP area (A) From Figure 17.5 we see that these dependencies are d_{13}, d_{23}, d_{33}, and d_{43}. First consider d_{13}; How will changes in the design variable beam height h affect or influence the SEP beam area, A?

Is the dependency positive or negative? Since increasing h will increase the area (all other things constant), the dependency d_{13} will be positive. It certainly won't be zero, since a change in h will cause a change in A. Will the dependency be greater or less than +1? Since doubling h will not result in a corresponding doubling of A, we know the value is between zero and 1.

This is about all we can reason out for sure about the value of d_{13}. Nevertheless, we can still reason further about its approximate value between 0 and 1. Is the influence of h relatively small or relatively large compared with a linear relationship? Well, as one thinks of an I-Beam, changing its height adds to the area only by extending the length of the web. Nothing else changes, not the thicknesses or the flange width. Thus changing h is not a major factor in the total area. We might therefore assign an initial value of about (say) 0.2 to 0.3 or so. Remember, if we are off the mark with this estimate, it will not affect the final solution -- though poor initial guesses may cause some extra time and effort along the way.

Consider now the dependency between beam width w and beam area, d_{23}. Clearly this dependency is also positive and non-zero. Increasing w causes an increase in A. Will the dependency be larger or smaller than +1. For the same reasons as stated in connection with beam height, it will be less than +1. Will it be larger or smaller than d_{13}? Well, increasing the beam width causes both flanges to increase in length, an effect that seems stronger than just increasing the web length. Thus, d_{23} is probably larger than d_{13}. By how much, we can really only estimate. Perhaps an initial value for d_{23} about 0.5 or 0.6 could be used.

By similar qualitative physical reasoning, reasonable initial values for dependencies related to the area, A, can be determined. Readers should take a moment to make estimates of their own for d_{33}, and d_{43}.

Let us now consider another of the dependencies in the I-Beam problem: d_{42}. Note that this is the dependency between the SEP deflection and the design variable, t (the flange thickness). It is certainly not zero; a change in t will have an effect on the deflection. Will it be plus or minus? Since increasing t will certainly stiffen the beam (giving it more material away from the neutral axis), the deflection will decrease. Thus the dependency will be negative. Will it be greater or less than -1? We really can't answer this without some knowledge of strength of materials and/or without remembering some example beam problems that were solved in that domain. If we happen to remember, then we know that beam deflection is highly sensitive to beam height, more than just linear. We might also reason this out qualitatively from beam theory by remembering how important it is to get as much material as possible away from the neutral axis. Since t is the flange thickness, it could well be expected to have a very strong effect on deflection. We might therefore expect the dependency to be, say, in the range from about -1.5 to - 3.

Indeed, if we go back to the equation for deflection, we see that t^3 appears in the denominator of the deflection equation. For the deflection in our example:

$$\text{Deflection} = 5\,W\,L^4 / 384\,E\,I$$

where $I = w\,h^3 / 12 - [(w - b)(h - 2t)^3] / 12$

	x_1 h	x_2 w	x_3 b	x_4 t
y_1 σ	d_{11}	d_{21}	d_{31}	d_{41}
y_2 δ	d_{12}	d_{22}	d_{32}	d_{42}
y_3 A	d_{13}	d_{23}	d_{33}	d_{43}

Figure 17.5 Blank Dependency Table for the I-Beam Example.

	x_1 h	x_2 w	x_3 b web	x_4 t flange
y_1 σ	d_{11} -1	d_{21} -0.5	d_{31} -0.1	d_{41} -1
y_2 δ	d_{12} -2	d_{22} -0.5	d_{32} -0.1	d_{42} -2
y_3 A	d_{13} +0.2	d_{23} +0.6	d_{33} +0.2	d_{43} +0.9

Figure 17.6 Initial Dependency Table for the I-Beam Example

However, the t^3 term is not alone in the denominator so its influence will not be as strong as a -3 dependency suggests. Thus, as a first estimate for d_{42}, we might take its value to be -2.

Again remember that if our estimates of initial dependency values are off the mark, it may slow down the subsequent solution process, but it will not prevent us from getting a good final solution.

Figure 17.6 shows an initial dependency table for the I-Beam example with values obtained from a junior engineering class. Readers should compare the values presented with their own estimates.

IIIIIIIIIIIIIIII

17.3 Next Steps in the Guided Iteration Process for Parametric Design

With a problem formulated, we can proceed with the

guided iteration process. The procedure used to solve parametric design problems by guided iteration is:

1. *Generation of Alternatives*: An initial trial design is selected.

2. *Evaluation:* The trial design is evaluated. A method for performing the evaluation is described below. Normally, unless a designer is awfully lucky[1], the initial evaluation is found to be Unacceptable.

(3) *Iterative Guided Redesign:* The initial dependency table is employed to guide selection of a new trial design. (We show how this is done below.)

(a) The new trial design is evaluated. If it is acceptable, the process is complete. If not:

(b) The dependency table is updated. (We show how to do this updating below.)

(c) The updated dependency table is studied to guide selection of a new trial design.

(d) Return to Step 2.

Each of these steps is elaborated on below, and illustrated by example. First, however, an issue that sometimes affects the solution process (i.e., coupling) is discussed.

17.4 Coupling

When a solution evaluation parameter (SEP) is influenced to approximately the same degree (order of magnitude) by two or more design variables, then those design variables are said to be *coupled* with respect to that SEP.

The presence, and to some extent the degree, of coupling can be ascertained by inspection of the dependency table. Refer to Figure 17.6. Note that for the stress SEP, design variables h and t have dependencies of -1, and that this value is considerably larger than the other dependencies. We say therefore that the design variables h and t are coupled with respect to the stress SEP.

Actually, since none of the dependencies with respect to stress are zero, then strictly speaking the design variables are all coupled to some extent. However, the two larger ones in this case are most important, and will serve the purposes of this discussion.

(Readers should check their understanding of how to identify coupling by finding another obviously coupled pair of design variables in Figure 17.6.)

Coupling complicates the solution of parametric design problems, sometimes seriously. Ideally, we would like to have the design variables completely *uncoupled.* That is, we would like each solution evaluation parameter to be influenced by only one design variable. If this were the case, the dependency table (if the x's and y's were arranged appropriately in the table) would look like the one shown in Figure 17.7. In such a nice problem, the designer could set x_1 to the value needed to get a desired y_1, x_2 to the value needed to get a desired y_2, and so on. Of course, arranging for such a situation has to be accomplished, if it can be, at the conceptual and configuration stages.

	x_1	x_2	x_3
y_1	d_{11}	0	0
y_2	0	d_{22}	0
y_3	0	0	d_{33}

Figure 17.7 Uncoupled Dependencies

	x_1	x_2	x_3	x_4
y_1	d_{11}	0	0	0
y_2	d_{12}	d_{22}	0	0
y_3	d_{13}	d_{23}	d_{33}	0
y_2	d_{14}	d_{24}	d_{34}	d_{44}

Figure 17.8 Decoupled Dependencies

Readers interested in the issues of coupling and their useful role in design may consult [1].

Almost as wonderful for designers at the parametric stage is the case shown in Figure 17.8. There, though the design variables are not completely uncoupled, designers can solve the parametric problem sequentially. That is, x_1 is set to obtain a desirable value for y_1. Then, given this value for x_1, x_2 is set to give a desirable value for y_2. And so on. In this case, the design variables are said to be *decoupled.* Again, arranging for such a design has to be accomplished at the conceptual and configuration stages.

Though most desirable, it is not necessary for the zero dependencies shown in Figures 17.7 and 17.8 to be actually zero in order to facilitate the solution process. If they are only considerably smaller than the others, solution will be generally easier and quicker than otherwise.

When an SEP with highly coupled design variables also has a relatively narrow range of acceptability in its satisfaction curve, then solving a parametric design problem can be especially difficult.

Sometimes it is possible to reformulate a problem by redefining the design variables and/or solution evaluation

1. Designers are seldom lucky; there are many more ways to design something wrong than there are to design it right.

parameters in such a way that coupling is eliminated or reduced; that is, so that the dependency matrix is more like Figure 17.6 or 17.7. We will illustrate this process in the v-belt drive design example later in this chapter.

17.5 Generating Alternatives for Guided Iteration Solution of Parametric Design Problems

In both engineering conceptual design and configuration design, an emphasis was placed on generating and comparing a relatively large number of alternative solutions, including possibly creative ones. We wanted to be certain that some potentially better concept or configuration alternative was not overlooked. Parametric design, however, is different; we must generate only one initial trial solution to get the process started. After that, the process proceeds iteratively until an at least satisfactory (often near-optimum) solution is reached.

The trial design alternative needed to get started with guided iteration can literally be *any* set of values for the design variables within the limits allowed by the problem formulation. The closer this initial design is to an acceptable near-optimal solution, however, the easier and quicker the iterative solution process will be. But only rarely will the process fail due to a poor choice of starting points.

It may be, in unusual situations, that a different initial trial design will lead to a different final solution. Though this is theoretically possible, and does occasionally happen, in the vast majority of real problems, designers have a good idea of reasonable design variable values. Starting with these reasonable values seldom leads to solutions off the desired track.

IIIIIIIIIIIIII

For example, in the beam problem, these starting initial design values selected by a student in a junior mechanical engineering design class were quite practical despite their lack of first hand knowledge of the domain:

h = 12 inches,
w = 8 inches,
b = 0.5 inches,
t = 0.5 inches.

IIIIIIIIIIIIIIII

17.6 Evaluation

17.6.1 Solution Evaluation Parameters (SEPs)

A key step in the guided iteration process is *always* evaluation. In the formulation of a parametric design problem, the Solution Evaluation Parameters (SEPs) have been identified, and corresponding Satisfaction Curves developed. The Satisfaction Curves provide the means by which each individual SEP can be evaluated — usually as Excellent, Good, Fair, Poor, or Unacceptable. We still need, however, a method for obtaining an evaluation of each trial design *as a whole*. To do this, we will adopt the Dominic method for evaluation described in Chapter 7. The first step therefore is to assign importance levels to the various SEPs.

17.6.2 Importance Levels for the SEPs

Usually a range of three importance levels is sufficient, though two or four may be used. Generally, we use:

(1) Very important (or Critical),
(2) Important, and
(3) Desirable.

Assigning importance levels to SEPs is a matter for the designers' best judgment. For example, in a space application, weight may be *very important* while cost is only *desirable*. In an automobile, both cost and miles per gallon may be *very important* with *acceleration* important or desirable (though in a racing car, these would be different).

There is no requirement in the methodology for how the SEPs are to be distributed among the importance levels. For example, all (or none) of the SEPs can be *very important* if that is the way the design team feels, though it generally works out that there are some SEPs in at least two importance categories.

Though the assignment of importance levels is a judgment call by designers, so long as reasonable judgment is used, the final results are usually not very sensitive to reasonable differences in the choices made.

IIIIIIIIIIIIII

I-Beam Example, Continued. For the I-Beam example, the SEPs and the importance levels [in brackets] selected are:

Area (Surrogate for Cost)* [Very Important],
Maximum stress when loaded [Desirable],
Maximum deflection when loaded [Important].

IIIIIIIIIIIIIIII

17.6.3 Applying the Dominic Method to Get an Overall Evaluation

With importance levels (very important, important, desirable) established, and a satisfaction level (excellent, good, fair, poor, unacceptable) determined from analysis for each individual SEP, the stage is set for obtaining an overall evaluation of the trial design using the Dominic method.

Refer to the matrix, called the *Overall Evaluation Matrix*, in Figure 17.9. To determine an overall evaluation of

---

* Area can be used as a direct measure of cost in this case. Area is proportional to weight, and weight proportional to cost. Cost is almost always one of the SEPs.

a design, the individual SEP evaluations are located in the matrix as shown in Figure 17.10. Then the overall evaluation is determined by the zones indicated along the right side. For example, for a trial design to receive a Good rating overall, *all* its individual ratings must lie *above* the line defining the Good zone. This means that all Very Important SEPs must have a Good or better rating, all Important SEPs must have a Good or better rating, and all Desirable SEPs must have a Fair or better rating.

In Figure 17.10, the overall rating for the design whose individual SEP evaluation shown is thus Fair.

It should be noted how the overall evaluation matrix in Figure 17.10 not only provides an evaluation, but also helps guide redesign. From the location of the individual SEP evaluations, it is easy to see which individual SEPs should be improved to improve the overall design. In the case shown in Figure 17.10, for example, if y_1 can be improved from Fair to Good, then the overall evaluation will increase to Good also. The same cannot be said of y_2 and y_5, however; improving them causes little overall improvement. Note, too, that y_3 may be a candidate for some trade-off in order to accomplish an improvement in y_1; y_3's individual rating could fall a notch without adversely effecting on the overall rating.

Obviously, with such a rating system, there will be different designs that result in the same rating. These can be sorted out by the designer subjectively, or some more formal tie-breaking scheme can be used. One way, for example, to break ties is: "Break ties by giving the highest rating to the design with the most Very Important SEPs that are Excellent. If there is still a tie, then compare the Important SEPs that are Good, and so on."

Another way to break ties might be to devise a method that places the highest ranking on those with the fewest Poor ratings in selected categories of importance.

The arrangement of overall evaluation zones given in Figures 17.9 and 17.10 is by no means the only possible or logical arrangement. Figure 17.11 shows another that is just as feasible; it allows lower standards for the individual SEP ratings in order to get a higher overall rating.

IIIIIIIIIIIIII

I-Beam Example, Continued. In our I-Beam example, the values for the individual SEPs that result from the initial trial design are:

$I = wh^3/12 - [(w - b)(h - 2t)^3]/12 = 831.9 \text{ inches}^4$

$\text{Maximum Stress} = Mh/2I = WL^2h/16I = 3510 \text{ psi}$

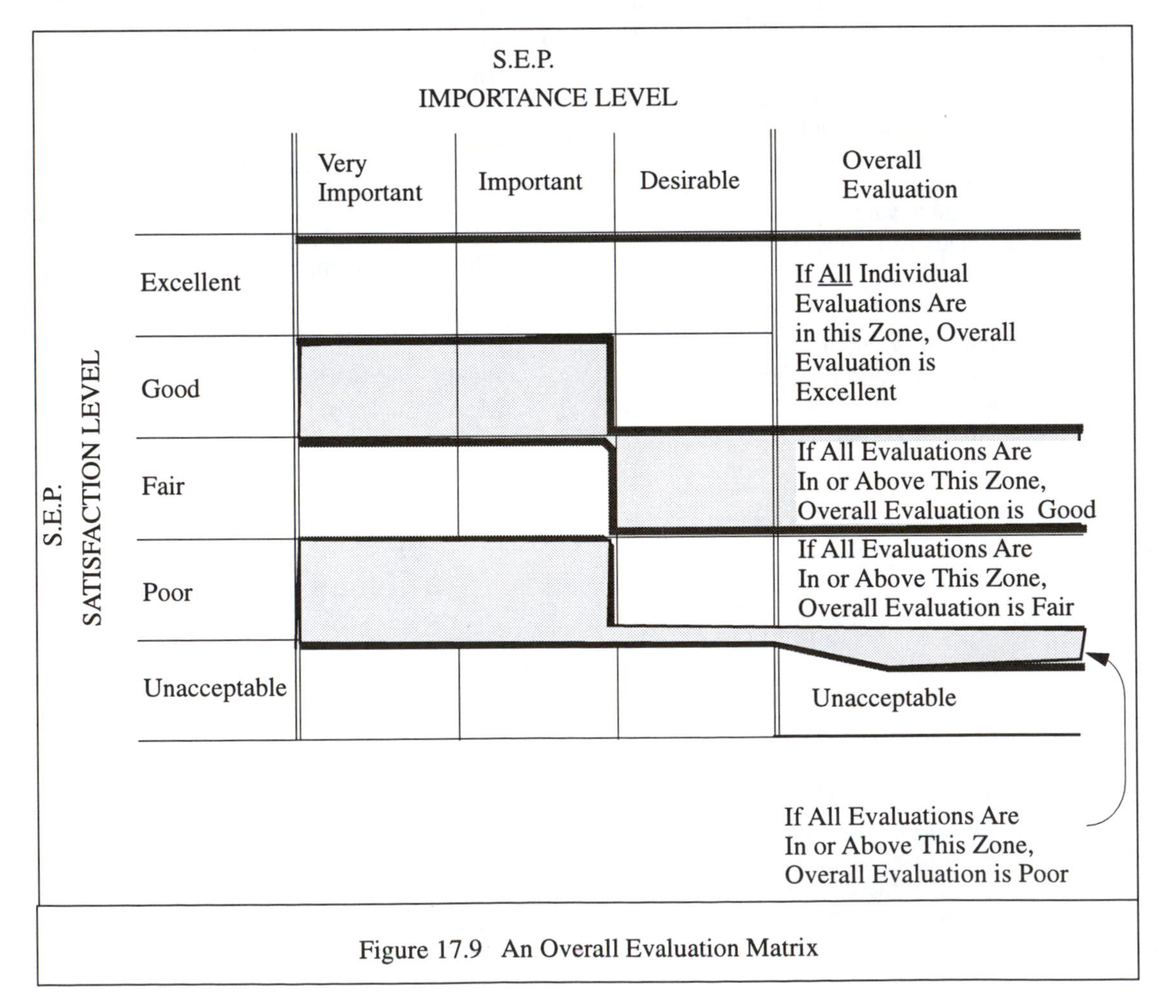

Figure 17.9 An Overall Evaluation Matrix

Maximum Deflection = $5WL^4/384\,E\,I$ = 0.0073 inches

Area = 2tw + b(h - 2t) = 13.5 sq. inches

Referring to the satisfaction curves for the I-Beam (Figure 17.3), we find that these evaluate to satisfactions of:

Max Stress: Unacceptable

Deflection: Unacceptable

Area: Unacceptable.

Obviously, therefore, this initial trial gets an overall rating of Unacceptable, and hence the beam will have to be redesigned.

IIIIIIIIIIIIIIII

It should be noted above that both the maximum stress and the maximum deflection are much less than the values allowed by the specifications. We say they are *overdesigned* because the beam needs to be made smaller — not larger — in order to move them into the satisfied region. These SEPs are not satisfied because, in effect, too much material is being used too ineffectively.

There is sometimes confusion about the difference between the *value* of an SEP and the *evaluation* of an SEP. Note that, for example, the *value* of the SEP stress here is 3510 psi. Its *evaluation* at this value is Unacceptable. The value, which is determined by analysis, is a fundamentally different concept from the evaluation, which is a design issue.

17.7 Guided Redesign

17.7.1 The General Case

Once an initial trial design is evaluated, the next step is redesign *guided by* the evaluation results and dependencies. The evaluation matrix and the dependency table can be used to guide the selection of new values for one or more of the design variables. Then the evaluation is repeated, and the cycle continues until an acceptable design is obtained.

We will assume here that only one design variable is to be changed at a time, though this will not always be the case in practice. Sometimes, it is fairly clear from studying the dependency table how to change two or more design variables to improve the design faster than with a sequence of single changes. To start, however, it is easier to learn the method working with one design variable change at a time. And anyway, this approach will often lead quickly to an acceptable solution, though perhaps not as fast as when making effective multiple design variable changes.

In the first cycle, the initial dependency table and the satisfaction ratings of the SEP's are used to guide the next trial design. In subsequent iterations, the dependency table is first updated, and then used with the SEP satisfactions to guide decisions. The procedure is to examine the evaluation matrix to select one or more candidate SEPs whose improved individual evaluation would improve the overall evaluation. In Figure 17.10, y_1 is one such SEP.

Then the dependency table provides good insight into

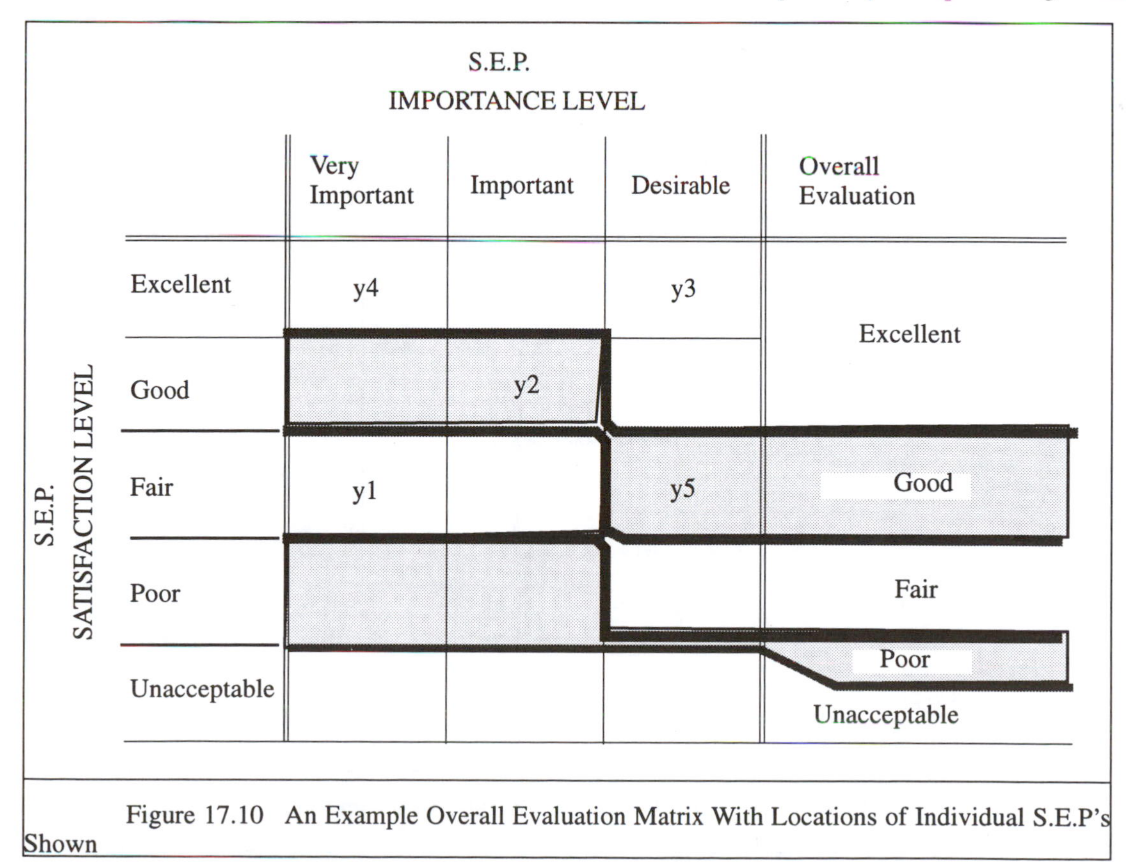

Figure 17.10 An Example Overall Evaluation Matrix With Locations of Individual S.E.P's Shown

which design variables to change in order to cause the desired effect on the selected evaluation parameter(s). The larger a dependency value, the more effect a small change in the design variable will have on the SEP. The signs also indicate the direction of the effect on an SEP of the various design variables.

Note that the dependencies also guide determination of the amount by which the design variables should be changed.

Updating the Dependencies. Equation 17.7 is repeated here as Equation 17.8. It is used both to update the dependency values and, with that done, to determine by how much to change a selected design variable value:

$$y_n(old) / y_n(new) = [(x_m(old) / x_m(new)]^{d_{mn}} \quad (17.8)$$

At the conclusion of every trial design change, for the design variable that was changed (say x_m), we know both the old and new values: $x_m(old)$ and $x_m(new)$. From the results of the previous trial, we have values for all the SEPs (i.e., all the y(old)'s. And now we can compute new values for all the SEPs; i.e., the y(new)'s. Thus we can solve Equation 17.8 for a new set of dependencies involving x_m.

Suppose, for example, that in the design of a widget, there are two SEP's (i.e., y_1 and y_2). Assume they are cost and weight, respectively, and that at the conclusion of an iteration their values are:

$$y_1(old) = \$300 \quad \text{and} \quad y_2(old) = 150 \text{ pounds.}$$

Also assume that one of the design variables [say, $x_4(old)$] is the part's wall thickness and that its value is $x_4(old) = 0.15$ inches. Then suppose we change x_4 to $x_4(new) = 0.12$ inches, and the new values that result for the y's are:

$$y_1(new) = \$250 \quad \text{and} \quad y_2(new) = 80 \text{ pounds.}$$

Then we can solve the above equation for d_{41} and d_{42} For d_{41}:

$$y_1(old) / y_1(new) = [(x_4(old) / x_4(new)]^{d_{ij}}$$

$$300 / 250 = [0.15 / 0.12]^{d_{41}}$$

$$d_{41} = 0.817$$

Readers should verify their understanding of this procedure by computing d_{42}. (The correct result is $d_{42} = 2.817$.)

Computing How Much to Change a Selected Design Variable. Equation 17.8 can also be solved in a different way to compute a magnitude for a proposed change in the selected design variable. Suppose that we wish to reduce the cost (y_1) of the widget above to \$225 by changing wall thickness (x_4). We know that the dependency (d_{41}) is now 0.817. So equation 17.8 is:

$$250 / 225 = [0.12 / x_4(new)]^{0.817}$$

Solving for $x_4(new)$: $x_4(new) = 0.1055$.

By changing x_4 to 0.1055, we should realize approximately

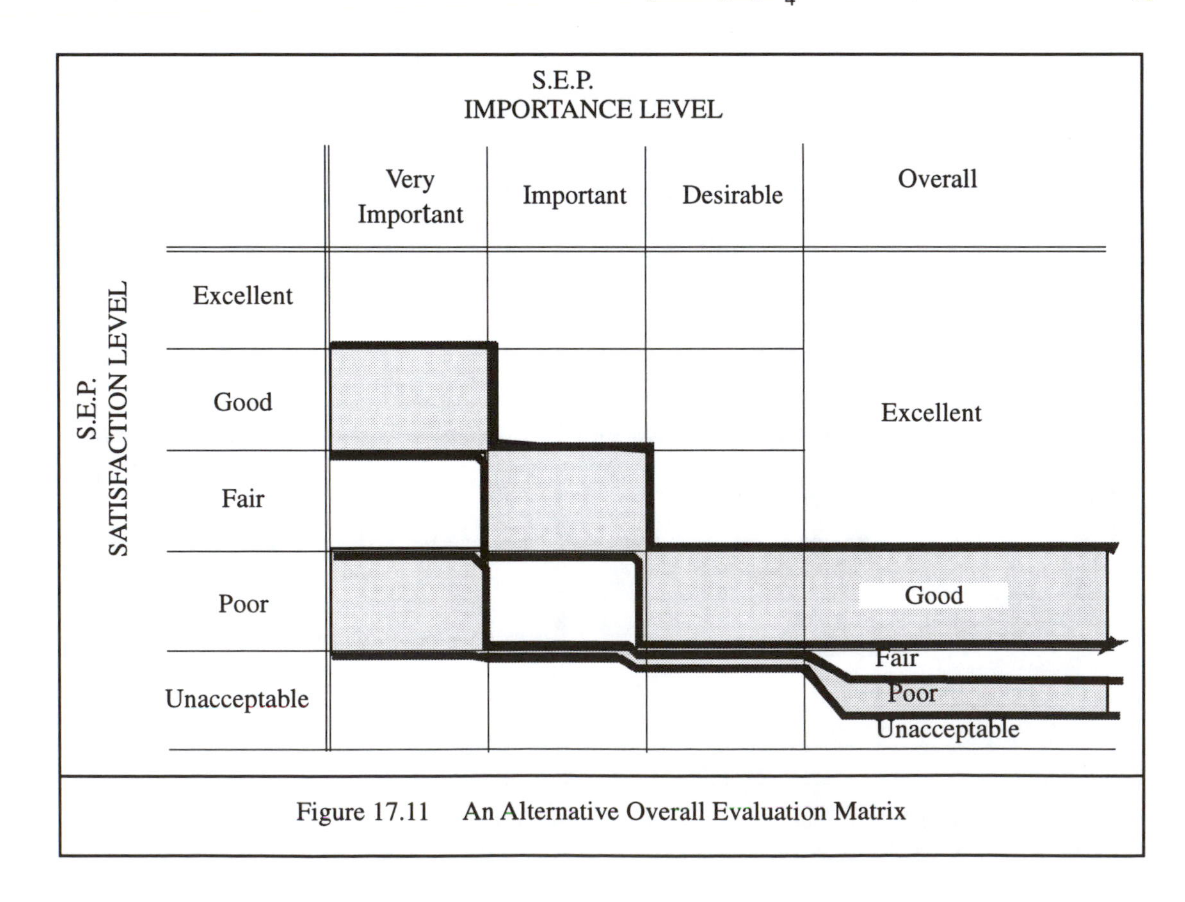

Figure 17.11 An Alternative Overall Evaluation Matrix

the improvement in y_1 that is wanted. In real problems, the results will be only approximately as anticipated because the actual relationships between design variables (x's) and the SEPs (y's) are not exactly as the dependencies assume.

Estimating the Effect of a Design Variable Change on Other SEPs. Once a design variable has been selected for change in order to improve the performance of a particular SEP, and the amount of the change computed, then using the applicable dependencies it is possible to predict the effect that the proposed change will have on one or more other SEPs. It may be, for example, that a change in one design variable will improve one SEP, but make others much worse.

Of course, making such predictions is not always necessary. The effects of the proposed change on all the SEPs will be computed in the next step of the process. If something undesirable happens, the designer can simply back up a step, and change the selected design variable or the amount of the change.

However, there are times when such predictions (even if they are only rough approximations) can save some computation time and effort. One of these times is when two design variables are closely coupled. In this case, it will often be desirable to change both of the involved design variables at once in order to avoid destroying the overall evaluation of the design.

IIIIIIIIIIIIIIII

17.7.2 I-Beam Example, Continued

Returning again to the I-Beam example, we saw above that the results of the first (i.e., the initial) trial design indicate that all three evaluation parameters are Unacceptable, and hence in need of change. In particular, stress should be larger, deflection greater, and area smaller. Looking only at the algebraic signs of the initial dependencies (Figure 17.6) indicates that any of the design variables, if made smaller, will have the desired *direction* of effects on stress, deflection, and area. Variables h and t have the most effect (largest dependency value) on stress and deflection, so they are preferred.

Looking at the values of the dependencies for t and h indicates that decreasing either has approximately the same effect on stress and deflection, but that decreasing t has a greater effect on reducing the area. Since we want ultimately to decrease area as much as possible, it appears that t is the better choice.

At this early stage of the iterative redesign process, it is arbitrary whether we choose to improve stress or deflection, so we elect deflection. To get deflection up to an Excellent rating, it needs to be 0.04 inches. We can estimate the change in h needed to accomplish such a change by using the dependency relationship Equation 17.8 as follows:

$$y(\text{old}) / y(\text{new}) = [\, (x(\text{old}) / x(\text{new})\,]^{d}$$

$$\text{deflection(old)} / \text{deflection(new)} = [\, t(\text{old}) / t(\text{new})\,]^{-2}$$

$$0.00731 / 0.04 = [\, 0.5 / t(\text{new})]^{-2}$$

$$t(\text{new}) = 0.215 \text{ inches.}$$

This approximate calculation suggests changing the value of beam flange thickness t to 0.215 inches. Since we are at an early stage of the redesign process where dependencies are still only rough estimates, we decide to round off the value to 0.20 inches. Thus our new trial design is

h = 12 inches
w = 8 inches
b = 0.5 inches
t = 0.2 inches

With these values, from Equations 17.1 through 17.4 we get the following values for the evaluation parameters:

Max Stress = 6396 psi
Deflection = 0.013 inches
Area = 9.0 inches

Referring again to the Satisfaction curves for this example (Figure 17.3), we can see that, stress and deflection are improved — not overdesigned as much as before — and area is also improved. However, all three SEPs are still Unacceptable. (Note that their value changed, but their evaluation did not.) Thus, more iterations are required, and so we begin by updating the dependency table based on the results of the previous trial.

The results of each iteration enable us to update the column of the dependency table associated with the design variable that was changed. In this case, the design variable changed was the flange thickness, t. The new dependencies for the t column are computed as follows:

For d_{41}: $\quad \text{stress(old)} / \text{stress(new)} = [\, t(\text{old}) / t(\text{new})\,]^{d_{41}}$

$$3510 / 6396 = [\,(0.5 / 0.2)\,]^{d_{41}}$$

$$d_{41} = -0.65$$

For d_{42}:

$$\text{deflection(old)} / \text{deflection(new)} = [\,(0.5 / 0.2)\,]^{d_{42}}$$

$$0.00731 / 0,013 = [\,(0.5 / 0.2)\,]^{d_{42}}$$

$$d_{42} = -0.628$$

In a similar way, we get: $\quad d_{43} = 0.44$

The updated dependency table is shown in Figure 17.12.

I-Beam Example - Third Trial The task again now is to select a design variable to change using the results of the previous evaluations and the new dependency table for guidance. The effect of changing t in the previous trial was very good, but it did not go far enough. The new dependencies for t are not as large as our original guesses, and now not as favorable as those for h in having the desired effect on stress and deflection.

However, we again select t just for the purpose of seeing how far we can go with it. We know that it is limited by manufacturing to 0.1 inches. We have already changed it (with beneficial effects) from 0.5 to 0.2, so we decide to change it all the way to its limit of 0.1. The results of this change are:

Max Stress = 8992 psi [Unacceptable - Overdesign]
Deflection = 0.019 inches [Unacceptable - Overdesign]
Area = 7.5 inches [Unacceptable]

Again referring to the Satisfaction Curves in Figure 17.3, we see that all of these SEPs are still Unacceptable, though all are moving in the right direction.

The updated dependency table is shown in Figure 17.13.

I-Beam Example - Fourth Trial. We still have to increase stress, increase the deflection, and decrease the area in order to begin to get acceptable designs. The updated dependency table clearly points now to h as the design variable to change in order to have the desired effect. Moreover, by using the values in the table, we can estimate by how much to change h in order to have a particular effect.

We elect to use a proposed change in deflection from its current value of 0.019 to 0.04 because that would give us an excellent evaluation for deflection. The computation is:

$$\text{deflection(old)} / \text{deflection(new)} = [\, h(\text{old} / h(\text{new})\,]^{-2}$$

$$0.019 / 0.04 = [\, 12 / h(\text{new})]^{-2}$$

$$h(\text{new}) = 8.27 \text{ inches}$$

We round this off so that the new trial design is:

h = 8 inches
w = 8 inches
b = 0.5 inches
t = 0.10 inches

Computing the new values for the evaluation parameters, and getting their satisfaction levels gives:

Max Stress = 16760 psi [Poor]
Deflection = 0.052 inches [Unacceptable]
Area = 5.5 inches [Fair]

It will be noted that the deflection reached 0.052 inches, though we had expected something closer to 0.040 inches. One reason is that the relationship between deflection and height is not exactly as assumed by equation 17.8. In addition, the value of the dependency used was only our original guess; it had never yet been updated to a more accurate value. Thus we should expect only approximate results. As we get closer to a solution, our predictions will get better.

Now referring to Figure 17.3, we begin to see some progress in the evaluation levels, which are:

Max Stress -> [Poor]
Deflection -> [Unacceptable]
Area -> [Fair]

Though stress and area have been moved into the acceptable range, deflection is still unacceptable. Thus the overall evaluation is still Unacceptable, and we must proceed to another iteration.

First we update the dependency table as shown in Figure 17.14. Readers should check one or more of the values to insure their understanding of how the new dependencies are computed. (The new value for d_{12} is –1.836.)

I-Beam Example - Fifth Trial. We now clearly want to make a change that moves the value of deflection into the acceptable range. The previous change — from h =12 to h = 8 — did the right things, but went too far with deflection. The dependency table also points again to h as the design variable to change, though we'll need to increase it now to get a smaller value of deflection.

		x_1 h	x_2 w	x_3 b	NEW x_4 t
y_1	σ	d_{11} -1	d_{21} -0.5	d_{31} -0.1	d_{41} -.58
y_2	d	d_{12} -2	d_{22} -0.5	d_{32} -0.1	d_{42} -.62
y_3	A	d_{13} +0.1	d_{23} +0.9	d_{33} +0.2	d_{42} +.44

Figure 17.12

Updated Dependency Table - I-Beam After Second Trial Design

		x_1 h	x_2 w	x_3 b	NEW x_4 t
y_1	σ	-1	-0.5	-0.1	-.50
y_2	d	-2	-0.5	-0.1	-.55
y_3	A	+0.1	+0.9	+0.2	+.26

Figure 17.13 Updated Dependency Table After Third Trial in I-Beam Example

To determine by how much to change h we use the dependency value between deflection and h, and again try for a deflection of 0.04 inches:

$$\text{deflection(old)} / \text{deflection(new)} = [\, h(\text{old} / h(\text{new})\,]^{-.2.5}$$

$$0.052 / 0.04 = [\, 8 / h(\text{new})]^{-.2.5}$$

$$h(\text{new}) = 8.89 \text{ inches.}$$

We round this off to 9 inches. The new design is thus:

h = 9 inches

w = 8 inches

b = 0.5 inches

t = 0.1 inches

Computation gives:

Max Stress = 14060 psi

		NEW x_1 h	x_2 w	x_3 b	x_4 t
y_1	σ	-1.5	-0.5	-0.1	-.50
y_2	d	-2.5	-0.5	-0.1	-.55
y_3	A	+.77	+0.9	+0.2	+.26

Figure 17.14 Updated Dependency Table After Fourth Trial in I-Beam Example

Deflection = 0.039 inches

Area = 6.0 inches

Evaluation using Figure 17.3 gives:

Max Stress -> [Poor]]

Deflection -> [Good]

Area -> [Fair]

Entering these values into an overall evaluation matrix as shown in Figure 17.15 results in an overall evaluation of Fair for this design. The updated dependency table after the fifth trial is shown in Figure 17.16.

Is this "Fair" design acceptable? Certainly not at this early stage. Though we may ultimately be unable to improve on it, we probably will. We should certainly try. We have hardly done much exploration; note that two of the design variables have not even been modified from their original values, which were in this example little more than educated guesses.

We leave the continuation of this example as an exercise for students. One student's results are summarized in Figure 17.17; this final design (h = 11.7; w = 6; b = t = 0.1) resulted in an overall evaluation of Good.

IIIIIIIIIIIIIIIII

Readers should note that during the example above, the design process consisted of making decisions about (a) which design variables to change and (b) by how much. The analyses served these choices by providing numerical information about the values of the SEPs. Then the evaluations and dependencies guide the design choices.

The first time the process described here seems cumbersome, but after doing a couple of problems, the use of Equation 17.7 becomes clear, and students find the process straight forward. Spread sheets can often be used to great advantage to reduce the computation required, including updating the dependencies. We will show this in examples below.

17.8 Recapitulation of the Guided Iteration Method: Numeric Design Variables

In outline form, here is a summary of the steps for solving parametric design problems by guided iteration:

1. Problem Formulation

a. Identify the problem definition parameters (PDPs) (z's) and specify their symbols and units.

b. Identify the design variables (x's) and specify their limits, symbols and units.

c. Identify the solution evaluation parameters (SEP's) (y's) and their symbols and units. This will include identifying any required relationships (constraints):

i. Importance Levels (Very Important, Important, or Desirable)

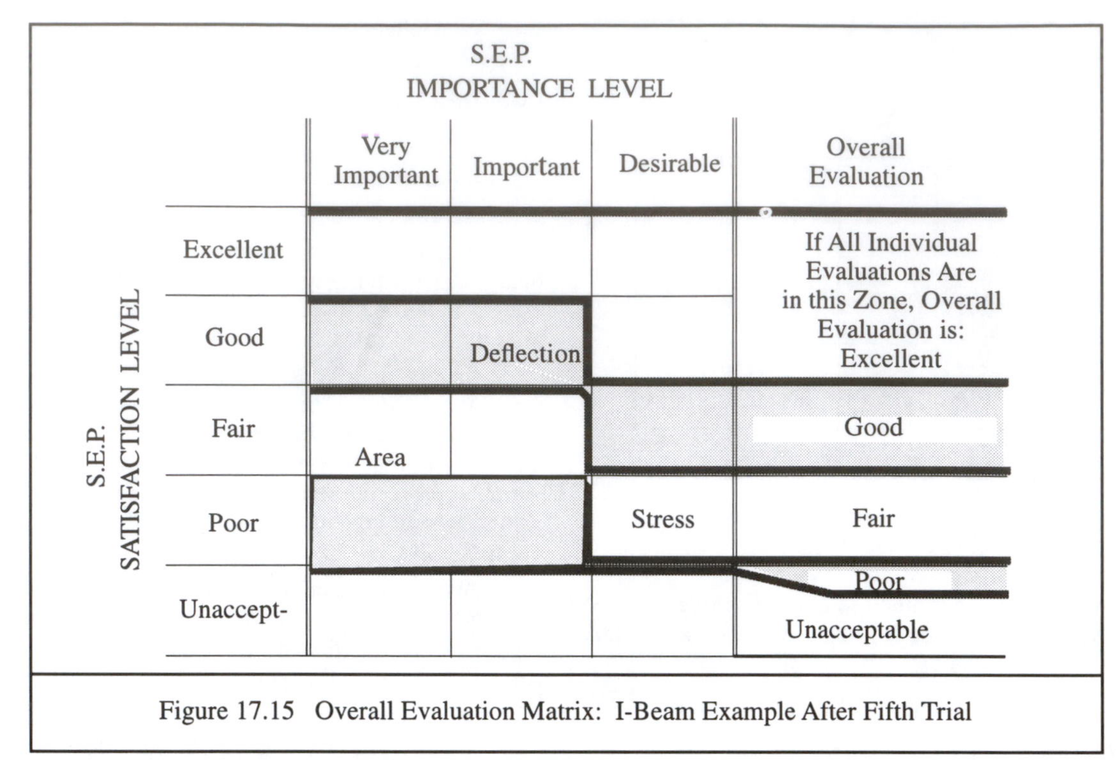

Figure 17.15 Overall Evaluation Matrix: I-Beam Example After Fifth Trial

ii. Satisfaction Curves

iii. Analysis Procedures for Computing SEP values.

d. Establish an initial dependency table, and estimate initial values for the dependencies.

2. *Generate an Initial Trial Design.*

3. *Evaluate the initial trial design.* This includes computing values for the SEP's, finding satisfactions for the SEP's, and determining an overall evaluation using Figure 17.10 or some alternative. Note that a method (e.g., Dominic's method must be established by which an overall evaluation of the trial design can be obtained from the individual SEP evaluations.)

4. *Perform Guided Redesign:*

a. Using the SEP satisfactions, the overall evaluation, and the dependency table, select a design variable (or variables) to change. Use Equation 17.7 to estimate by how much to change the selected design variable(s) in order to improve the design as desired.

b. Evaluate the new design. If acceptable, the problem is solved and ready for communication of the results.

c. If not acceptable, then update the dependency table and return to step 5.a.

17.9 Non-Numeric Design Variables

The above problem solving methodology assumes that all the design variables are numeric. However, often there are non-numeric attributes that also must have values assigned during parametric design. Material choices are a typical example. Another example is the determination of the available types of v-belts are to be used in a v-belt drive.

When non-numeric design variables are present, then trial values must be chosen for them as a part of the initial trial design. Then the problem is worked as an all numeric problem (with the non-numeric variable(s) unchanged) until the best possible design for the given non-numeric value(s) is achieved. If a suitable result is obtained, it may not be necessary to repeat the procedure with a different value for the non-numeric design variable value.

		(NEW) x_1 h	x_2 w	x_3 b	x_4 t
y_1	σ	-1.5	-0.5	-0.1	-.50
y_2	d	-2.4	-0.5	-0.1	-.55
y_3	A	+.74	+0.9	+0.2	+.26

Figure 17.16 Updated Dependency Table After Fifth Trial in I-Beam Example

Trial Number	Design Variables					Evaluation Parameters			Overall Evalua-tion
	h	w	b	t	I	σ	δ	A	
1	12	8	0.5	0.5		3510(U)	.0074(U)	13.5(U)	U
2	12	8	0.5	0.2	177.0	6350(U)	.013(U)	9.0(U)	U
3	12	8	0.5	0.1	125.0	9000(U)	.019(U)	7.5(U)	U
4	8	8	0.5	0.1	44.7	16800(P)	.052(U)	5.5(F)	U
5	9	8	0.5	0.1	60.0	14060(P)	.039(U)	6.0(F)	F
6	9	6	0.5	0.1	52.1	16200(P)	.045(E)	5.6(F)	F
7	9	6	0.1	0.1	29.4	28700(F)	.080(U)	2.1(E)	U
8	11	6	0.1	0.1	46.1	22400(F)	.051(U)	2.3(E)	U
9	11.7	6	0.1	0.1	53.0	20700(F)	.044(E)	2.35(E)	G

Figure 17.17 Summary of Results for the I-Beam Example

Since each selection of a new value for a non-numeric design variable value can result in a great deal of work solving the corresponding numeric problem, designers must be careful to make good choices and to limit the need to explore many non-numeric alternatives. Of course, if the numeric guided iteration solution process is supported by a computer for the computations, then this is less difficulty.

In summary, the solution process is modified as follows if there are non-numeric design variables:

1. Select values for the non-numeric design variables,
2. Solve the problem as a purely numeric problem,
3. If needed, select a new value or values for the non-numeric design variables. Return to Step 2.

17.10 Another Example: A V-Belt Drive System

We illustrate the guided iteration parametric design process now in another domain: v-belt drives. There are several more examples in Chapters 20 and 21.

Refer to Figure 17.18. Belt drives (of several types) and chain drives are examples of an embodiment that employs a pulley-like wheel to transfer torque from a shaft to a moving member in tension (the belt or chain). At the other pulley or sprocket, the tensile force creates a torque which is transferred to a second shaft. We think of the different types of belts and chains as configurations in this domain.

The belts and pulleys that are used to create v-belt drives systems are standard components. To design a system requires that designers select a set of standard pulleys (sheaves) and belts. Pulleys come in standard diameters and there are standard belt sectional shapes that we will refer to as belt types. (In this book, we will refer to three types: 3v, 5v, and 8v, the latter being the larger.) There can be any number of pulleys from one to a dozen or more.

As shown in Figure 17.18, the design variables are:

Belt type, bt,

Number of belts, M,

Drive pulley diameter, D_1,

Load pulley diameter, D_2, and

Length of belt(s), l.

Manufacturer's catalogs not only list all the sizes and types of belts and pulleys available, they also provide a design procedure which will result in satisfactory designs. However, the expected life of a design is not taken into account explicitly in these procedures. Presumably the designs that result give lives of from one to two years in typical use.

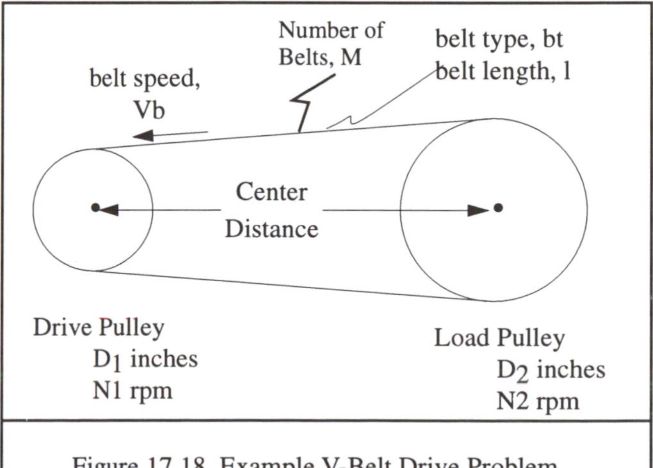

Figure 17.18 Example V-Belt Drive Problem

There are also commercial computer programs available that generate v-belt solutions, and books and research papers also provide design methodologies. The analysis procedure in [4] has been used for the analysis procedures (especially for belt life) used in the following example.

1. Formulate the Problem

a. Identify the design variables:

Drive pulley diameter, D_1, inches,

Load pulley diameter, D_2, inches,

Belt length, l, inches,

Number of belts, M,

Center distance, CD, inches,

Belt type, bt (3v, 5v, or 8v).

The last of these, belt type, is a non-numeric variable obtainable directly from manufacturers catalogs from the problem definition parameters. Thus, henceforth here we will assume that belt type is known, and work the problem as an all numeric problem.

The pulley diameters and belt length are not continuous variables, but are available only in discrete sizes. For example, one manufacturer lists about forty pulley diameters from 2.2 inches to 60 inches. Approximately fifty belt lengths are available. As we design a v-belt system, we will often do analyses assuming that the variables are continuous, and then select the closest values that correspond to commercially available pulleys and belts.

b. Identify the problem definition parameters (PDPs):

Drive Speed (i.e., rotational speed of pulley delivering the power, usually from a motor or engine) RPM

Desired Load Speed (i.e., rotational speed of the pulley being driven, such as one connected to a fan or pump), RPM

Horsepower to be transmitted, HP

Desired Belt Life, Hours (of service)

Center Distance maximum (CDmax) and minimum (CDmin)

Belt Velocity maximum (Vmax) and minimum (Vmin)

Force on Shafts (Maximum)

Center distance limits are important, and should be made no more restrictive than absolutely necessary to accommodate required machine placement. If these limits are too restrictive, suitable designs may not even be possible.

Belt velocity is important to belt-pulley adhesion, but only if it is too small or too large within quite a wide range.

In this section, we will be working an example in which the Problem Definition Parameters are:

Drive speed: $N_1 = 1500$ rpm

Load speed: $N_2(\text{desired}) = 900$ rpm

Horsepower: HP = 400

Center distance: CD(min) = 12 inches

CD(max) = 32 inches

Desired Life: Life(desired) = 8000 hours

Desired belt velocity: V(min) = 1500 fpm

V(max) = 7000 fpm

c. Identify the solution evaluation parameters (SEP's) and their importance levels:

(i) Soft SEPs

Actual belt life - Life(act) (Not the same as the *desired* belt life) [Very Important]

Actual load speed - $N_2(\text{act})$ (Not the same as *desired* load speed.) [Important]

Cost [Very Important]

Belt Velocity - V [Desirable]

Center Distance - CD [Desirable]

Force - F (This is the lateral force exerted on the pulley shafts due to the tension in the belts.) [Important]

(ii) Hard SEPs

There is also a required geometric relationship among CD, l, D_1, and D_2. Thus one of these variables is a hard SEP defined by the relationship:

$$CD = 1/4 \{ B + [B^2 - 2 (D_2 - D_1)^2]^{1/2} \}$$

where $B = l - (\pi / 2) (D_1 + D_2)$

d. Establish an Initial Dependency Table:

See Figure 17.19. The values shown are estimates based on very rough qualitative reasoning. Remember, we need approximate values to get started, but as the design process proceeds, we will compute new values as we go along.

Here are some examples of the kind of qualitative physical reasoning that can help establish the initial dependencies shown in Figure 17.19. Consider first d_{31}: To what extent will the belt length (design variable 3) influence the actual load speed (SEP 1)? None, of course; the belt length has nothing to do with the speeds, so this dependency is zero.

Now consider d_{11}: How strongly does the diameter of the drive pulley (design variable 1) influence the rotational speed of the load pulley? We know that as we change the drive pulley diameter (but keep the RPM the same), the belt must travel faster directly in proportion to the diameter. If we double the drive pulley diameter, the load rotational speed will also double. Thus the dependency is +1.

Reasoning to make estimates of the above two dependencies was relatively easy; the values of the dependencies are almost obvious. So let's try some less obvious ones: d_{16},

	inches D_1	inches D_2	inches L	M
N2 (actual) rpm	d11 1	d21 -1	d31 0	d41 0
Belt Velocity fpm	d12 1	d22 0	d32 0	d42 0
Life hrs.	d13 2	d23 1	d33 1	d43 4
Force lbs.	d14 0.5	d24 0.5	d34 0	d44 0.5
CD in.	d15 -0.3	d25 -0.2	d35 0.5	d45 0
Cost $	d16 .2	d26 .3	d36 .4	d46 1

Figure 17.19 Initial Dependencies V-Belt Example

d_{26}, and d_{46}. For d_{16}, how strongly does the drive pulley diameter influence the cost? The cost consists of drive pulley, load pulley, and belts. Generally the drive pulley is smaller than the load pulley. Since the drive pulley is only one of three factors in the cost, we would expect this dependency to be less than 1.0. How much less, we really do not know unless we have previous knowledge of the domain. A value anywhere in the range 0.2 to perhaps 0.5 would be a reasonable place to start.

We have selected 0.2 as shown in Figure 17.19 because we expect the larger load pulley to have a somewhat greater influence. We give it a value of 0.3 for a start.

Dependency d_{46} has been assumed larger because the number of belts influences not only the cost of the belts but also the cost of the pulleys which must accommodate more belts. The value of exactly 1.0, however, is just assumed to be in a reasonable range.

Note once again that these initial dependencies need only be somewhat reasonable; they do not need to be especially accurate. The guided iteration method will almost always recover from poor initial assumptions, though it may require more computation. Also, note again that though students are often confronted with problem domains with which they are unfamiliar, in practice this is seldom the case. Having designed the component at hand through conceptual and configuration stages, designers usually have a fairly good idea of how the design variables will more or less influence the SEPs. When they don't, then a best guess is all that is needed to get the solution process going.

As described earlier, at this point it is a good idea to inspect the dependency table for coupling. Perhaps a less coupled formulation can be found. There is a very good example of this in this problem.

Note in Figure 17.19 how D_1 and D_2 are closely coupled with respect to load speed N_2(act). Here is what can happen along the way in the problem solving process: Since only D_1 and D_2 can be used to bring the load speed N_2 into its acceptable range, suppose we use them and get N_2 satisfactory. Now suppose later that we want to use either D_1 or D_2 for some other purpose; otherwise we have only belt length and number of belts to work with. Looking at the dependency values, we might want to use D_1 to improve Life, or D_2 to improve Force. Say we do use D_1, then Load Speed will most likely go well out of range since its dependency value with D_1 is +1. In order to bring Load Speed (N_2) back, we would have immediately to change D_2 by an amount directly in proportion to the change we just made in D_1. In other words, to keep N_2, with its narrow range of acceptability, more or less on track, we can't change D_1

	D_1/D_2	D_1	L	M
N2 (actual)	d11 1	d21 0	d31 0	d41 0
Belt Velocity V	d12 0	d22 1	d32 0	d42 0
Life	d13 -0.2	d23 2	d33 1	d34 4
Force	d14 +0.1	d24 0.5	d34 0	d44 0.5
CD	d15 +0.1	d25 -0.3	d35 0.5	d45 0
Cost	d16 -0.1	d26 0.2	d36 0.4	d46 1

Figure 17.20 Revised Initial Dependency Table

without also changing D_2. Thus we say they are coupled, and as a result we will be able to change either D_1 or D_2 by only very tiny amounts without causing the load speed to move to Unacceptable. The only alternative is to change both D_1 and D_2 at the same time in order to keep N_2 satisfactory.

One way to make this kind of difficulty more tractable is to redefine the design variables. In the v-belt problem, instead of using D_1 and D_2, we might redefine the design variables as D_1 and the ratio D_1/D_2. Then the initial dependency table will be as shown in Figure 17.20. In this reformulation, D_2 is no longer a design variable. If the designer changes D_1, there is no affect on N_2(act) since D_1/D_2 does not change. Of course, changing D_1 while D_1/D_2 remains constant results in a change in D_2, but this does not affect the design process since D_2 is no longer a design variable. To influence N_2(act), the designer works with D_1/D_2.

In the remainder of this example, we will not use the reformulated design variables, but will keep on with the formulation as started above.

2. For Each SEP

a. Importance levels as indicated above;

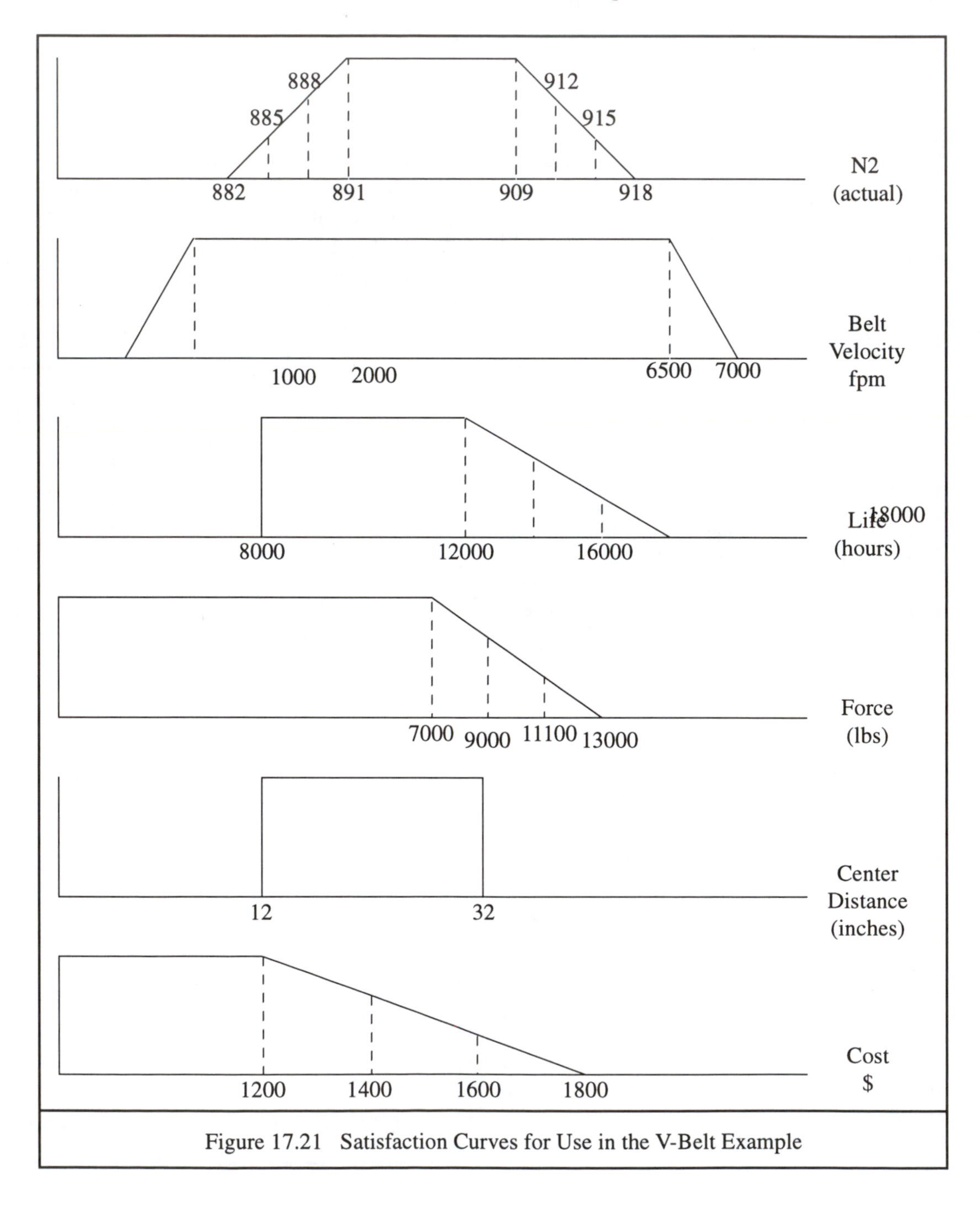

Figure 17.21 Satisfaction Curves for Use in the V-Belt Example

b. Satisfaction curves as shown in Figure 17.21;

c. Analysis procedures:

For Load Speed (act):

$$N2(act) = (N_1)(D_1)/(D_2) \text{ rpm}$$

For Belt Velocity:

$$V = (\pi / 12)(N_1)(D_1) \text{ feet per minute (fpm)}$$

For Shaft Forces:

Research and experimentation has established the following values for type 8v belts. For the values for other belt types, consult [2].

$k_c = 3.288$ $k_b = 4830$ $Q = 3638$ $x = 12.629$

Then the belt tensions are:

$$F1 = t1 + tb1 + tc \quad \text{and} \quad F2 = t1 + tb2 + tc. \quad \text{pounds}$$

where: $tb1 = kb / D_1$ and $tb2 = kb / D_2$. pounds

and $tc = kc\,(V / 1000)^2$ pounds

and $t1 = tc + \{ [(33000)(C1)(HP)] / [(M)(V)(C1 - 1)]$
}

where $\ln C1 = (2\ A\ )(0.512)$

and $\cos A = (D_1 - D_2) / (2\ CD)$

For Life:

$$Life = L / \{ (720)(V) [(1 / Y1) + (1 / Y2)] \} \text{ hours}$$

where $Y1 = (Q / F1)^x$ and $Y2 = (Q / F2)^x$

For Cost: The data presented here were derived from a manufacturer's price sheet in 1983. It is likely that the form of the cost functions has not changed, but it is absolutely certain that the constants have increased. Thus these costs are not be used except for example purposes within this book. The data given are for 8v belt types only.

Cost of 8v belts = (0.47 L) $ per belt;

Cost of 8v pulleys ($ per belt) =

$$(19 + 11.3\,D) + (M - 4)(0.1)(19 + 11.3\,D)$$

where M is the number of belts and D is the pulley diameter.

For Center Distance:

$$CD = 1/4 \{ B + [B^2 - 2 (D_2 - D_1)^2]^{1/2} \}$$

where $B = L - (\pi / 2)(D_1 + D_2)$

Note 1: The analysis procedures required for v-belt problems are considerably more extensive than for the I-beam, and it is nearly impossible to proceed without computer support to do them quickly. In the rest of this example, calculations of the values of evaluation parameters have been done by computer.

Note 2: Readers should also note that doing equation shuffling to a solution is indeed impossible in this case.

d. Method for obtaining overall evaluation from evaluations of the individual designs: We will use Dominic's method as described earlier.

3. Generate an Initial Trial Design.

For our example, we select these values for the initial design:

$D_1 = 18$ inches,

$D_2 = 21$ inches,

$L = 100$ inches,

$M = 10$.

4. Evaluate the Current Trial Design

Using a computer program to compute the values of the evaluation parameters that result from the initial trial design gives the results shown below. The letters in the brackets are the SEP evaluations obtained from Figure 17.20:

N2(act)	= 1285 rpm	[U]
V	= 7070 fpm	[U]
Life	= 3580 hours	[U]
Force	= 9450 lbs	[F]
CD	= 19.31 inches	[E]
Cost	= $ 1240	[G]
Overall:		Unacceptable

	NEW inches D_1	inches D_2	inches L	M
rpm N2 (actual)	1	-1	0	0
fpm Belt Velocity	1	0	0	0
Life hrs.	+.66	1	1	4
Force lbs.	.65	0.5	0	0.5
CD in.	-.54	-0.2	0.5	0
Cost $	.23	.30	.4	1

Figure 17.22 V-Belt Dependencies After Second Trial

5. Perform Guided Iteration

Second Trial Design. Since load speed [N2(act)] from the first trial is unacceptable, we elect to try to fix it. (We'd like it to be close to 900 rpm.) It is obvious that the only way to effect a change in load speed is with either D_1 or D_2. Since D_2 seems large already, we choose to use D_1. The dependency of unity means that:

$$N2(act)(old) / N2(act)(new) = D_1(old) / D_2(new)$$

$$1285 / 900 = 18 / D_1(new)$$

$$D1(new) = 12.6 \text{ inches.}$$

The closest commercial pulley is 12.5 inches, so the second trial design is:

D_1 = 12.5 inches

D_2 = 21 inches

L = 100 inches

M = 10

Evaluating this design gives:

N_2(act)	= 893 rpm	[E]
V	= 4910 fpm	[E]
Life	= 2030 hours	[U]
Force	= 7460 lbs	[G]
CD	= 26.6 inches	[E]
Cost	= $ 1123	[E]

Overall: Unacceptable (But much improved)

The updated dependency table is shown in Figure 17.22. We show here only one of the computations as an example. For the D_1-Life dependency:

$$Life(old) / Life(new) = [\ D_1(old) / D_1(new)\ ]^{d_{13}}$$

$$2580 / 2030 = [\ 18 / 12.5\ ]^{d_{13}}$$

$$d_{13} = 0.66$$

Third Trial Design. Since belt life is now the only evaluation parameter that is not at least acceptable, we decide to improve it. Examining the dependency table indicates that the way to have a fairly large beneficial effect on life is to increase the number of belts. (Readers should do the qualitative physical reasoning that explains this.) Since we need a fairly large improvement in life (to at least 8000 hours), we elect to work with M as the design variable to be changed.

From the dependencies in Figure 17.22:

$$Life(old) / Life(new) = [\ M(old) / M(new)\ ]^4$$

$$2030 / 8000 = [\ 10 / M(new)]^4$$

$$M(new) = 17.4$$

Thus the new trial design is:

D_1 = 12.5 inches

D_2 = 21 inches

L = 100 inches

M = 14

Evaluating this design gives:

N_2(act) = 893 rpm [E]

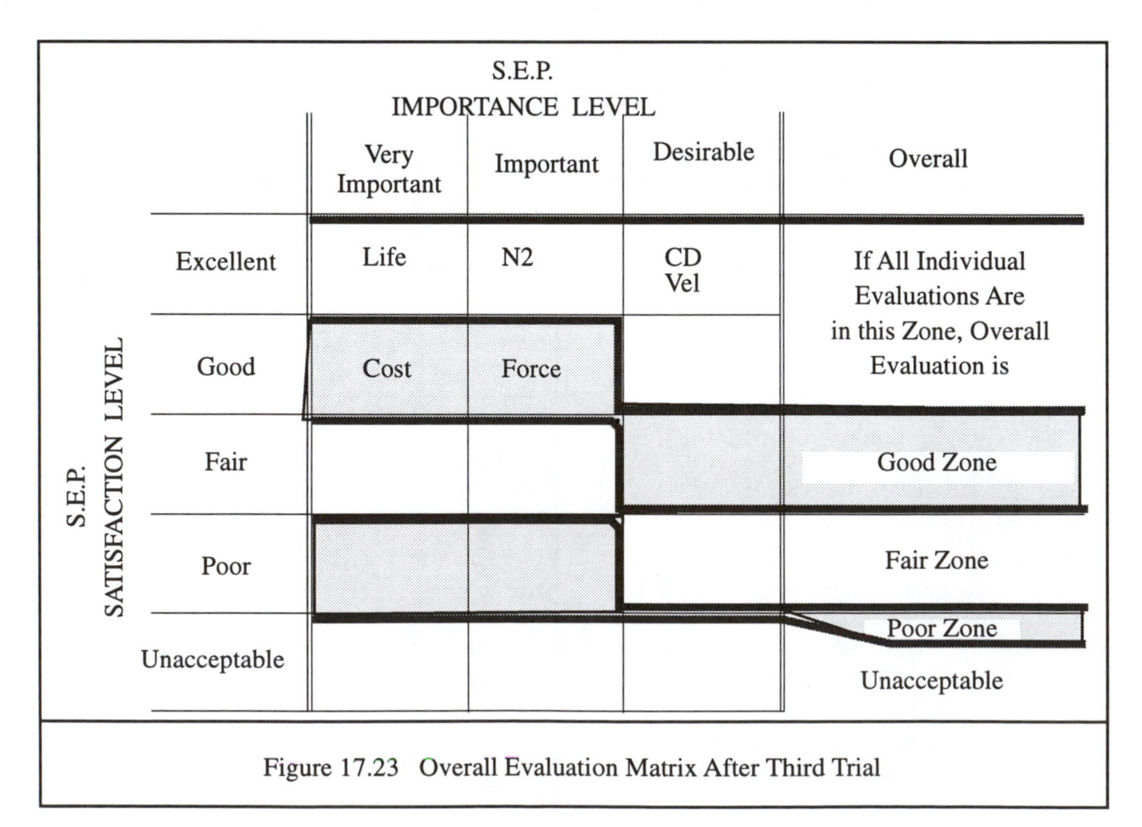

Figure 17.23 Overall Evaluation Matrix After Third Trial

	inches D_1	inches D_2	inches L	NEW M
rpm N_2 (actual)	1	-1	0	0
fpm Belt Velocity	1	0	0	0
Life hrs.	+.66	1	1	4.4
Force lbs.	.65	0.5	0	0.5
CD in.	-.54	-0.2	0.5	0
Cost $	.23	.30	0.4	.82

Figure 17.24 Dependencies After Third Trial

V = 4910 fpm [E]
Life = 9800 hours [E]
Force = 8700 lbs [G]
CD = 26.6 inches [E]
Cost = $ 1290 [G]

Now using the overall evaluation matrix shown in Figure 17.23, we get an overall evaluation of Good. The new dependency table is shown in Figure 17.24.

Fourth Trial Design. We can try again to improve on the cost. The number of belts has the largest dependency value relative to cost, so we decide to try M = 13.

Thus the new trial design is:

D_1 = 12.5 inches
D_2 = 21 inches
L = 100 inches
M = 13

Evaluating this design gives:

N_2(act) = 893 rpm [E]
V = 4910 fpm [E]
Life = 7130 hours [U]
Force = 8390 lbs [G]
CD = 23.6 inches [E]
Cost = $ 1200 [G]

This design did improve the cost, but now the belt life has fallen below the acceptable range. The overall evaluation therefore is again Unacceptable. There is little else that can be done, we will have to accept the third trial design with its overall evaluation of Good.

Examination of the last dependency table indicates that we might improve cost by decreasing the pulley diameters, D_1 and D_2. However, 12.5 inches is the smallest diameter pulley made for type 8v belts. For the same reason, we cannot reduce belt length to help the overall evaluation because 100 inches is the shortest 8v belt produced.

17.11 Changing Multiple Design Variables

In the preceding discussion and examples of the guided iteration method for parametric design, we have changed only one design variable in each new trial design. Sometimes, it is obvious from the signs and magnitudes of the dependencies how to change two (or even three) design variables simultaneously in order to arrive at an acceptable design quicker. Sometimes, also, two design variables are tightly coupled so that changing one essentially necessitates a corresponding change in the other. This is the case, for example, with D_1 and D_2 in the v-belt problem above.

Changing multiple design variables can improve the efficiency of the process, but there is a disadvantage, too. After a multiple change, there is no way to update the dependency table with the data obtained from the new SEPs. Each SEP will have been influenced by all of the design variables changed. The only way to compute new dependency values is to perform "experiments" by changing one design at a time by a minimal amount, compute a new set of performance values, and then get the dependency in the usual way. If there is computer support for the computations, this is not difficult or time consuming. But without such support, the computations can become time consuming.

17.12 Selecting Standard Components By Guided Iteration

The design of a product, special purpose assembly, or sub-assembly usually involves both the design of special purpose parts and the selection of standard parts and components like bolts, motors, shafts, springs, and the like. There is both a configuration stage and a parametric stage to the selection process. There are, for example, different types (i.e., configurations) of bolts, motors, springs, etc. These are generally selected by a combination of qualitative reasoning about the geometry, functionality, and cost of the required component. Guided iteration is used; that is, you must select a trial type, evaluate it for the multiple purposes involved, and then revise your selection based on the evaluation.

Pugh's method can usually be used nicely for the evaluation. Inputs from vendors can also be helpful.

Once a type has been selected, then the particular size or model within that type must be selected. This is a parametric design task, and it can be done by guided iteration. Often, of course, the design variables can take on only the discrete values manufactured as standard. That is, only certain bolt diameters are available, only certain v-belt lengths and pulley sizes are available, and so on. Sometimes it is easiest to solve the problem assuming that the variables are continuous, and then select a standard size nearest the result. This is the approach taken in the examples above. Other times, using only the available standard values as trial designs presents no difficulties. Indeed, usually it does not as in the bolt design examples in Chapter 20.

In either case, the guided iteration methodology can be employed as outlined in this chapter.

17.13 Non-Quantitative Solution Evaluation Criteria

In some situations there are important non-quantitative evaluation criteria — such as aesthetic issues — which numerical or objective satisfaction curves can not be constructed. Yet these factors may need to be evaluated for each trial design and traded off against quantitative SEPs like cost, efficiency, weight or performance. In these cases, guided iteration can still be employed though not with the usual precision in using numerical dependencies to guide redesign.

Even though an SEP is subjective, an evaluation can still be determined in terms of Excellent, Good, Fair, Poor, or Unacceptable. For aesthetic evaluations, for example, an artist, industrial designer, or panel of consumers may be consulted. Once the subjective issue has been evaluated separately, then an overall evaluation of a trial design can be performed in the usual manner using the Dominic method. However, working back from the evaluations to a redesign suggestion that will improve the subjective factor (without harming the others too much) is not readily supported by a dependency table. However, some guidance can be obtained by keeping a separate record of the subjective SEP evalua-

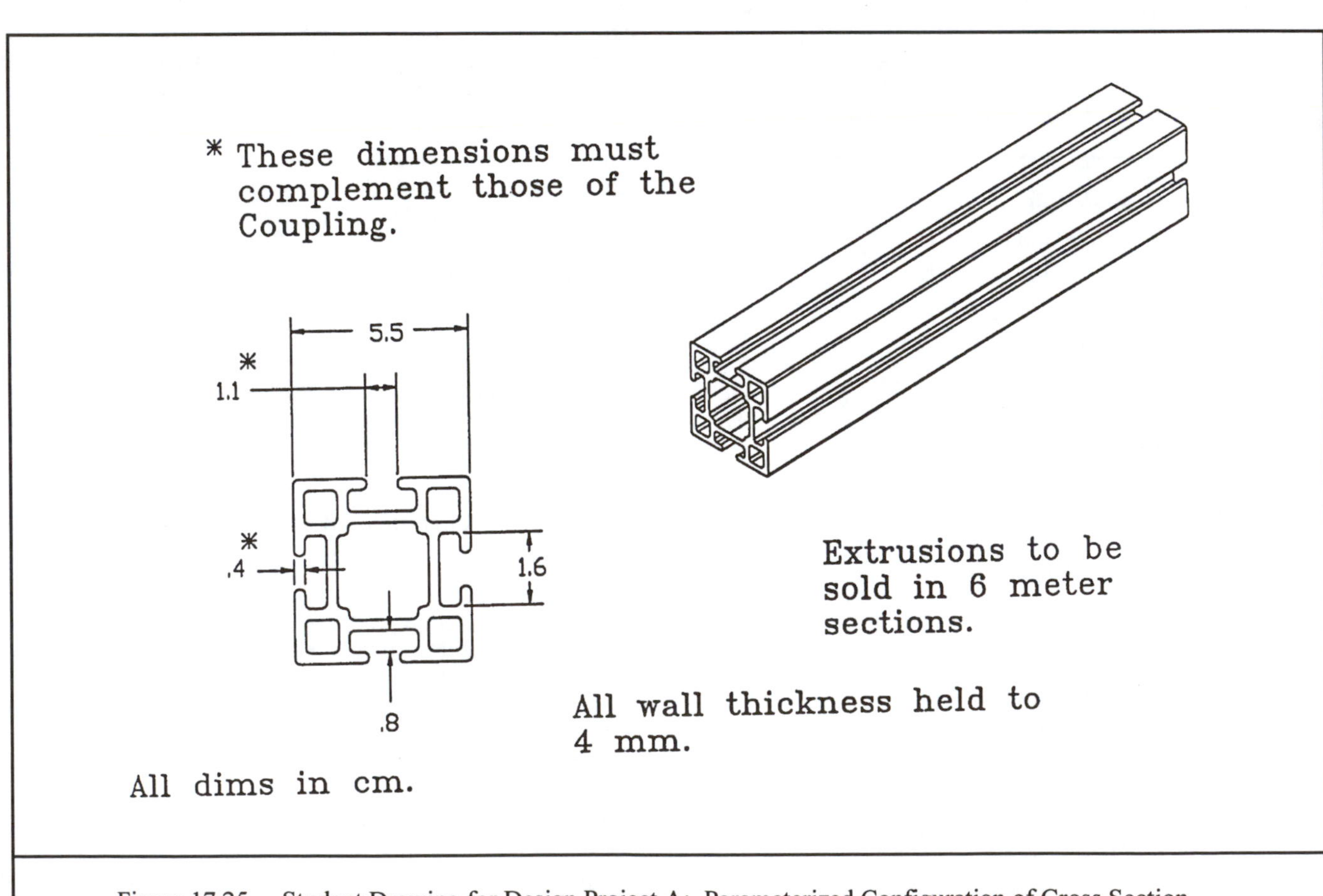

Figure 17.25 Student Drawing for Design Project A: Parameterized Configuration of Cross Section

tions in a table as follows:

SEP Eval	Design Var A	Design Var B	Design Var C	Etc

Usually a designer will be able, after several trials, to see a pattern in the relationship between the design variable values and the subjective evaluations. If shape or form is the issue, the ability to sketch the designed object after each trial, perhaps using a CAD or solid modeler, will be of great benefit. Though guidance won't be precise, once a rough pattern is recognized, at least some guidance is possible.

When the subjective SEP has to do the aesthetics of form, changing the design variables to selected ratios may be helpful. The reason is that the visual attractiveness of form is often a function of the proportions of the various dimensions.

17.14 Supplementary Design Project A, Continued

At the parametric stage, dimensions and tolerances must be assigned to the attributes of the parts whose configurations were determined earlier. In a one semester project, coupled with a course, there is insufficient time for student groups to do complete structural analyses of the various parts. In some cases, relatively simple beam theory can be employed to analyze simplified sections giving approximate ball-park results.

We present here selected results from two student groups. The dimensions shown are incomplete, and some are impractical. They should be taken only as initial trials, not as final recommended values.

Figures 17.25, 17.26, 17.27, and 17.28 show a number of parts generated by one configuration design which have been given some initial dimensions. There are a number of problems with the dimensions and the drawings. However, the Figures illustrate the general idea of what parametric design must do: make the dimensions as complete and as precise as needed for functionality -- and also make them economical to manufacturable. (As an exercise, readers should criticize these drawings.)

Figure 17.29 shows results from another group. This

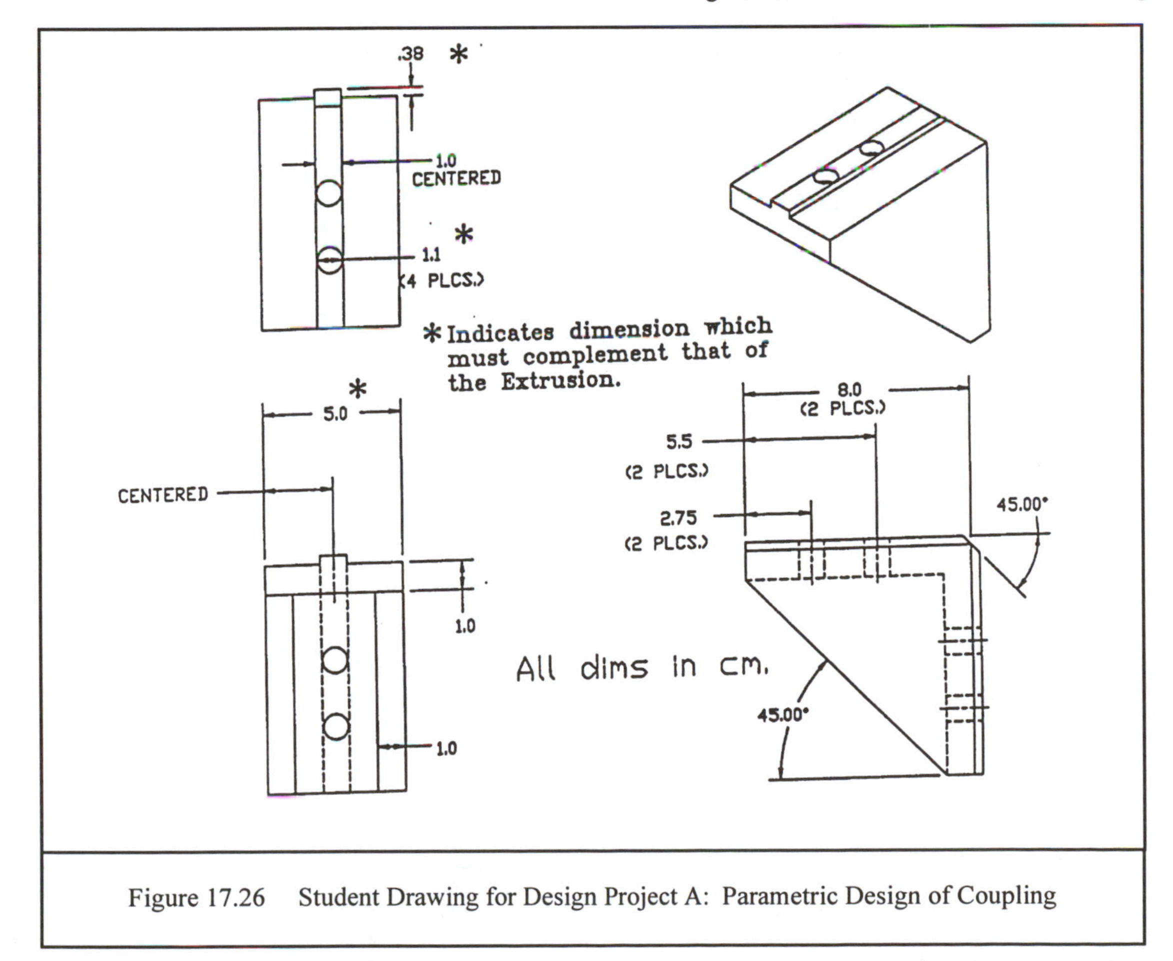

Figure 17.26 Student Drawing for Design Project A: Parametric Design of Coupling

group used a computer optimization program to find the minimum wall thickness of a square hollow beam required to support a static load (worker plus machine) on an eight foot span. Since their beam configuration is stronger than a simple square hollow, the thickness determined supported by that analysis provides only an order of magnitude for the wall thickness — though no doubt a safe one assuming dynamic loads can be ignored. (As an exercise, readers should criticize these drawings.)

17.15 Summary and Preview

Summary. At the parametric design stage, guided iteration is a designer controlled and operated (often computer supported), multiple objective satisficing method that can cope with most kinds of design variables and a variety of evaluation criteria to achieve a satisfactory solution. Robustness issues can usually be included among the Solution Evaluation Parameters if desired [3]. Thus the method is very generally applicable to many types of parametric design problems.

A number of examples of applying the guided iteration method to standard mechanical parts and components are described in Chapter 20. More examples, but from the thermal-fluids area, are described in Chapter 21.

Preview. When a designer encounters parametric design problems, guided iteration is not the only methodology Engineering Design available. The other most powerful methods are (1) optimization and (2) statistically based methods, especially Taguchi methods. [TM] The next chapter is an introduction to optimization; Chapter 19 deals with Taguchi methods.

References

[1] Rinderle, J. R. and Krishnan, V., "Constraint Reasoning in Concurrent Design", Design Theory and Methodology - DTM '90, ASME DE-Vol 27, Proceedings of the 2nd International Conference on Design Theory and Methodology, Chicago, Illinois, September, 1990.

[2] Spotts, M. F., *Design of Machine Elements*, Prentice-Hall, Englewood Cliffs, New Jersey, 1985.

[3] Orelup, M. F., "Dominic III: Incorporating Taguchi's Philosophy, Material Choice, and Memory in the Design of Mechanical Components", Ph. D. Dissertation, University of Massachusetts, Amherst, 1992.

[4] Motherway, J.E., "V-Belt Drive Design With a Micro-Computer", *Computers in Mechanical Engineering*, Volume 2, Number 1, July 1983.

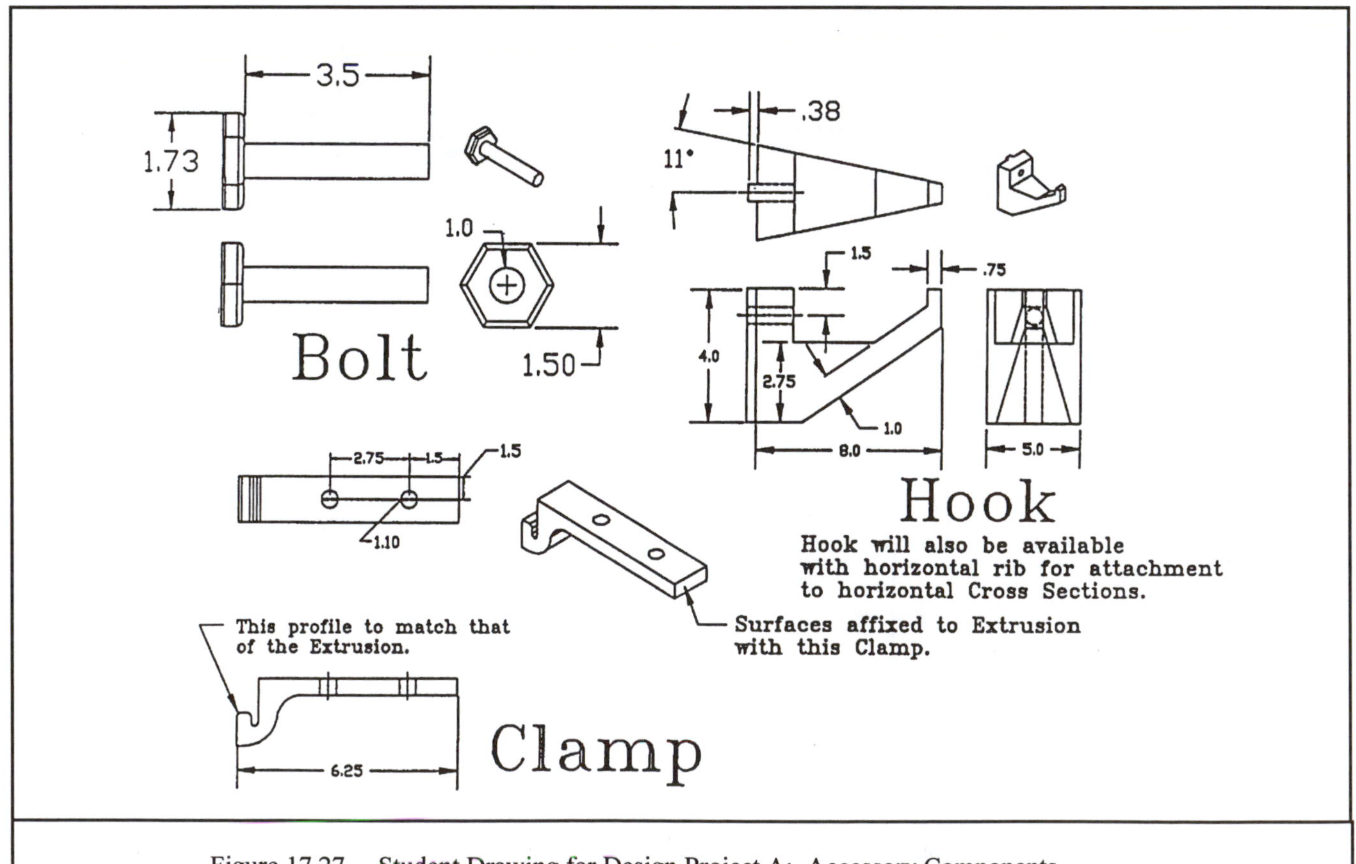

Figure 17.27 Student Drawing for Design Project A: Accessory Components

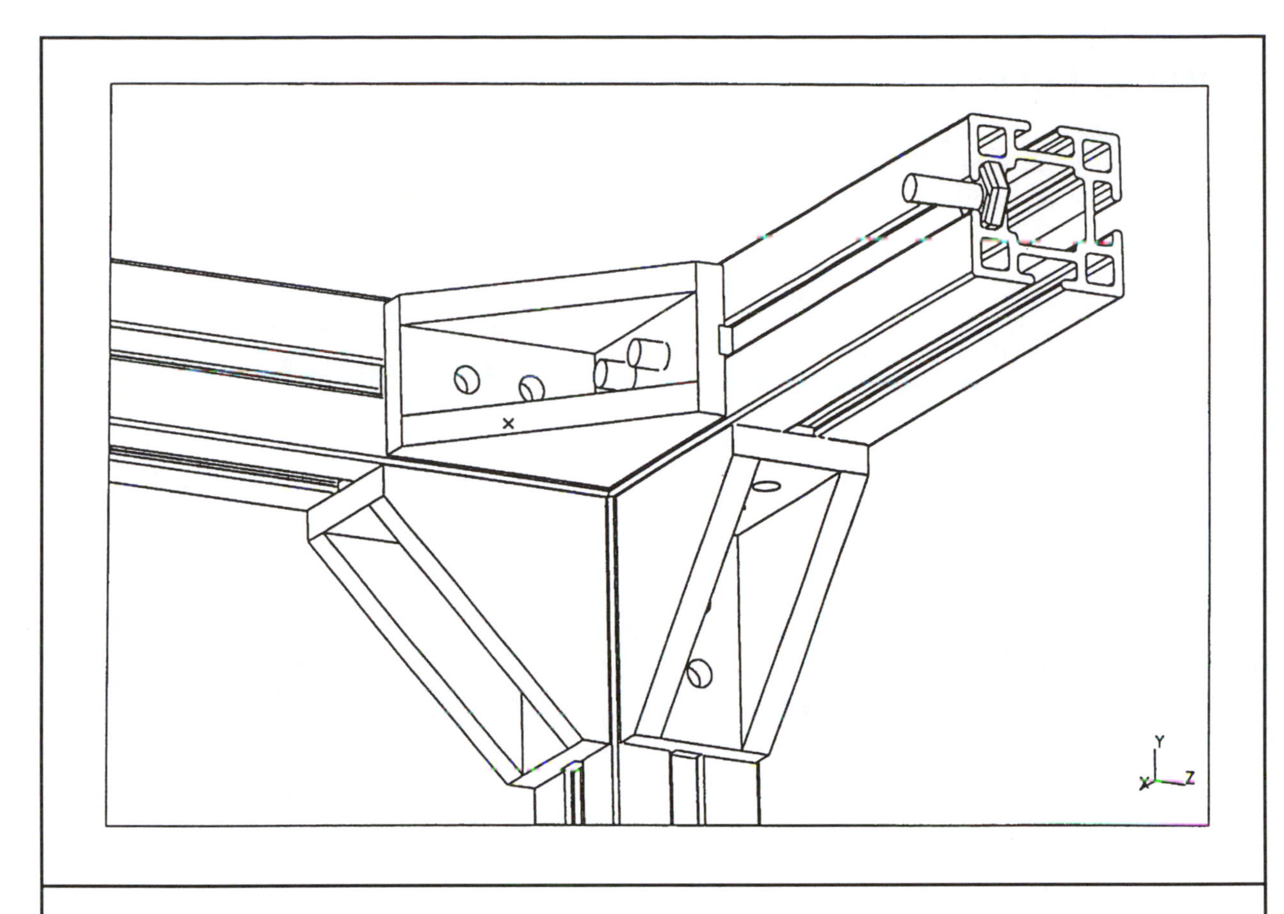

Figure 17.28 Student Drawing for Design Project A: A Solid Model of Corner Assembly

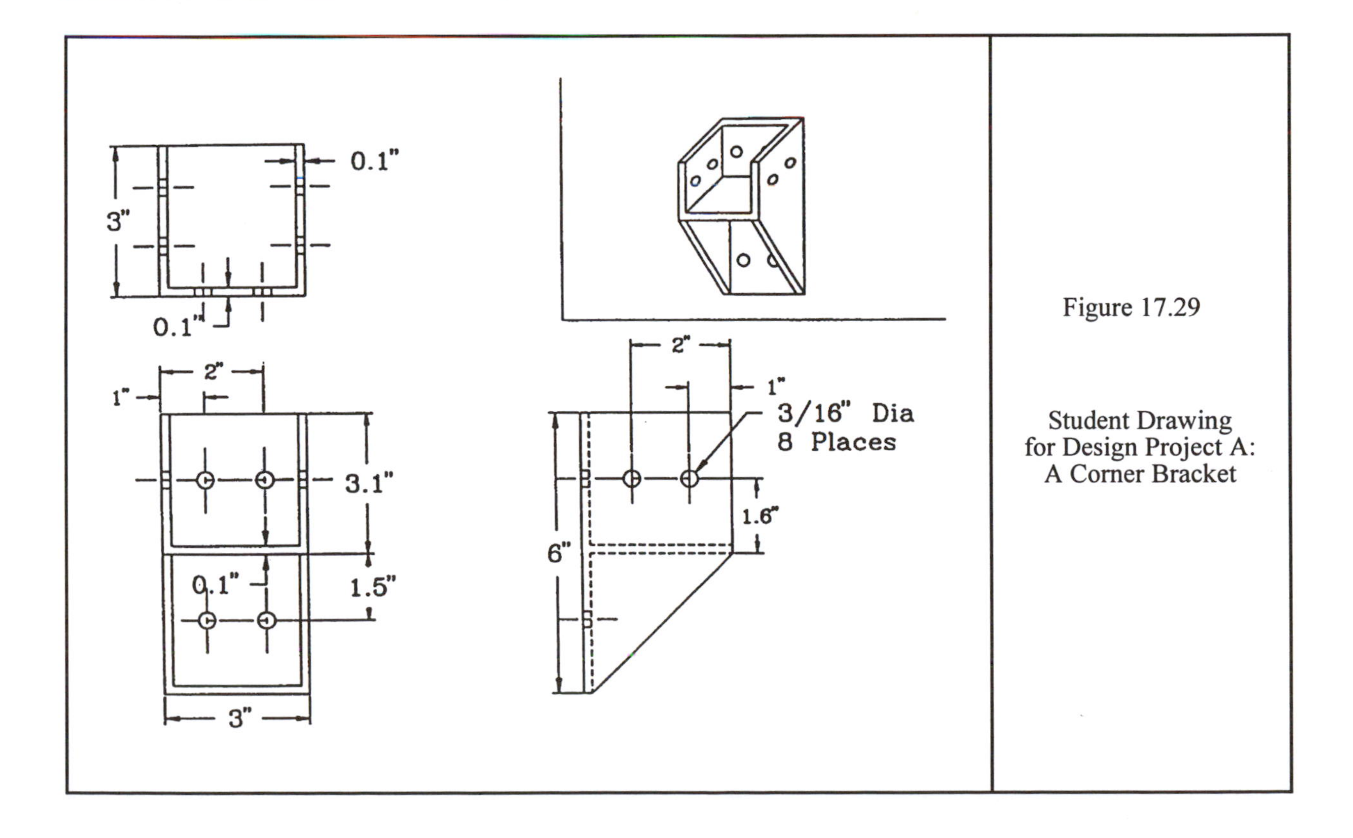

Figure 17.29

Student Drawing for Design Project A: A Corner Bracket

Supplementary Reading

The Structure of Human Decisions, D. W. Miller and Martin K. Starr, Prentice-Hall, Englewood Cliffs, NJ, 1967.

Decisions With Multiple Objectives, R. W. Keeney and H. Raiffa, John Wiley and Sons, New York, 1976.

Problems

17.1 Devise an over all evaluation matrix similar to Figure 17.9 but make it one that sets a higher standard for the quality rating of trial designs.

17.2 Use guided iteration to design an I-beam if the problem definition parameters are as follows:

Length of the beam [L = 12 feet]

Load type(s) [uniformly distributed]

Load location(s) and load value(s)
[3000 pounds per foot; full length of beam]

Support type(s) and locations(s)
[simply supported at each end]

Any restrictions on the maximum beam deflection
[maximum allowed deflection is 0.06 inches]

Any limitations or goals on cost
[as low as possible]

Any special environmental conditions [None]

Material is Aluminum 6061-T6.

17.3 Use guided iteration to design an I-beam if the problem definition parameters are as follows:

Length of the beam [L = 8 feet]

Load type(s) [concentrated]

Load location(s) andvalue(s): [1000 pounds at L = 8 feet]

Support type(s)) and locations(s)
[cantilevered, built in at L = 0.]

Any restrictions on the maximum beam deflection
[maximum allowed deflection is 0.02 inches]

Any limitations or goals on cost [as low as possible]

Any special environmental conditions [None]
Material is 1020 steel

E = 30,000,000 psi; Yield Stress = 36,000 psi.]

17.4 For the I-beam problem used an example in this Chapter, sketch satisfaction curves for stress and deflection that do not provide for consideration of over-design.

17.5 Assume that the design variables in the preliminary simplified design of a cast iron flywheel for a shear press are: the mean diameter (D, feet), width of the flywheel rim (a, inches), and the depth of the flywheel rim (b, inches). This approach assumes that the kinetic energy of the flywheel is due entirely to the rim — a fair assumption for a preliminary design. See Figure P17.5(a).

The problem definition parameters are: the rotational speed of the flywheel at the start of a cut (N_1 = 200 rpm) and the energy required per cut, (E_c = 100,000 foot-pounds.) The density of cast iron is d = 450 pounds per cubic foot.

The Solution Evaluation Parameters (SEPs) are: the weight of the flywheel (W, pounds), the maximum rim velocity (V_1, feet per second), and the fractional reduction in

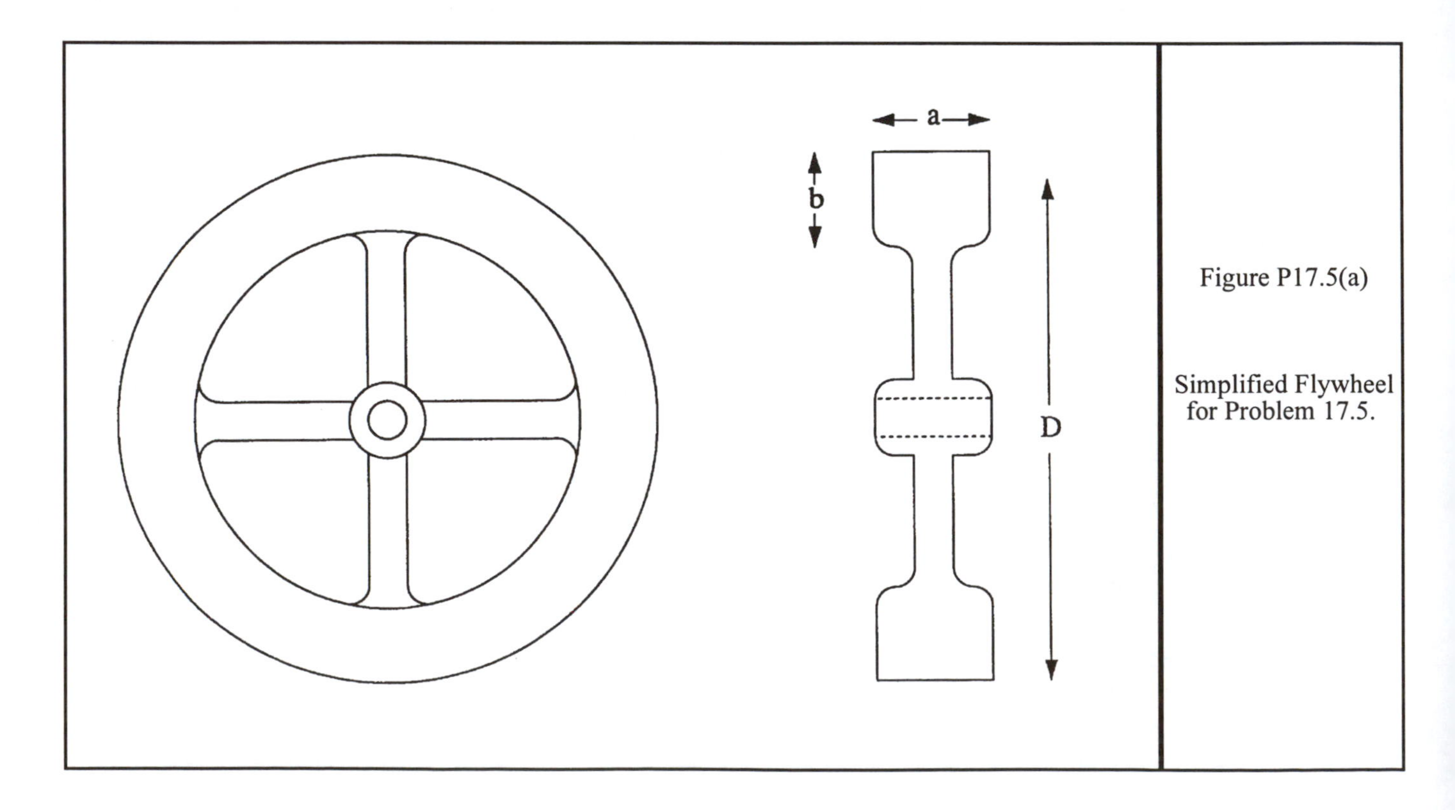

Figure P17.5(a)

Simplified Flywheel for Problem 17.5.

speed during the part of each stroke that the shear is cutting, R. Proposed initial satisfaction curves for these SEPs are shown in Figure P17.5(b).

The limit on the rim velocity shown in its satisfaction curve reflects the outward tensile stresses created as the wheel rotates at high speed. Cast iron has a tensile strength of only about 19,000 psi. It would reach this value at a rim velocity of about 435 fps. (What factor of safety is reflected in the proposed satisfaction curve for rim velocity?)

The equations for computing SEP values from the design variables and PDPs are as follows:

$$W = \pi D a b d / 144$$

$$V_1 = N_1 \pi D / 60$$

$$R = (V_1 - V_2) / V_1$$

where V_2 is the velocity of the rim immediately after a cut. In addition, from dynamics we know that the kinetic energy (foot-pounds) of the wheel at a given rim velocity V is

$$E = W V^2 / 2 g$$

Thus $E_c = W (V_1^2 - V_2^2) / 2 g$

Use guided iteration to find values for the design variables.

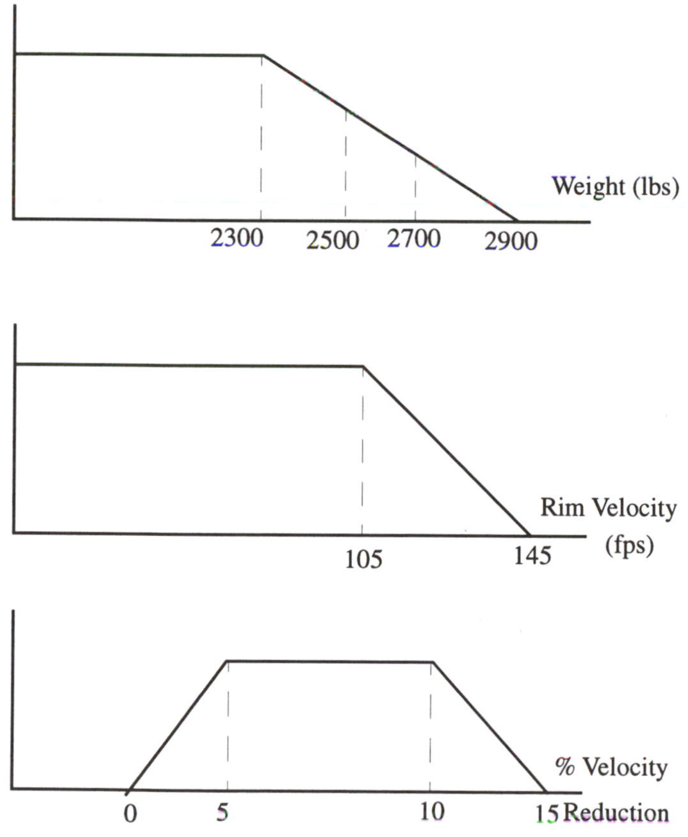

Weight (lbs)
2300 2500 2700 2900

Rim Velocity (fps)
105 145

% Velocity Reduction
0 5 10 15

Figure P17.5(b) Satisfaction Curves for Problem 17.5.

17.7 A tubular steel column sixteen feet high is to carry a vertical load of 1000 pounds. The column is to be built-in at the base, but is free at the top. The minimum feasible wall thickness is 0.10 inches. The maximum allowed outside diameter is 4 inches. Formulate the problem and design the column by guided iteration.

17.8 Use guided iteration to do the parametric design of a truncated cone cup. The volume is 16 cubic inches. The SEPs are (1) surface area which is to be as small as possible, and (2) the appearance, which is to be as pleasant as possible. (Note: The surface area of the side of a truncated cone is:

Area =0.5(Sum of the perimeters of the bases) (Slant height)

(Note: Since the appearance is more directly related by proportions, you might want to change one or more of the design variables to a ratio.)

17.9 We wish to determine the parameters of a small stylish shiny white plastic wastebasket for an executive office. It has been decided to configure the wastebasket as an elliptical cross section with straight vertical sides. The required volume is 1100 cubic inches. What dimensions do you recommend? Do the think the color and texture of the surface might affect your result? How might this product be manufactured if the quantities are to be large?

The area of an ellipse is

$$A = 0.7854 A B$$

where A and B are the major and minor axes respectively. The perimeter of an ellipse is $P = 2 A E$ where E is a function of k:

$$k = 1 - (B^2 / A^2)$$

The relationship between E and k is approximately

k	E
0.999	1.001
0.97	1.07
0.94	1.12
0.90	1.17
0.85	1.23
0.80	1.28
0.75	1.32
0.70	1.36
0.60	1.42
0.40	1.51
0.20	1.55
0	1.57

This page available for notes, sketches, and calculations.

Chapter Eighteen

Introduction to Optimization

"The intersection of engineering design with optimization theory generates the discipline of optimal design.......The *optimal design* of a device is defined as the feasible plan that makes it as good possible according to some quantitative measure of effectiveness."

Douglass J. Wilde *[1]
Globally Optimal Design (1978)

18.1 Introduction

Optimization is an advanced, mathematically sophisticated field of study supported by a large body of theory, and a wide range of application methods. Optimization methods can find the values of design variables that will produce a maximum or minimum of a specified mathematical function subject to given constraints. Thus, when designers can define their objective suitably in terms of a single mathematical function of the design variables, then optimization becomes a feasible method for performing parametric design.

In this book, we provide only a brief introduction to optimization. There are excellent texts and reference books on the subject [1, 2, 3, 4]. There are also computer programs that can be used for performing optimization [5]. And there are optimization methods called *integer programming* that can solve some kinds of configuration and parametric problems simultaneously; see, for example [6].

In this Chapter, we explain and illustrate how to formulate problems for solution by optimization, and how to use some of the basic and relatively simple optimization methods. We do not make an attempt to present optimization theory fully, and many optimization methods — especially the more mathematically advanced ones — are beyond our scope here. For more depth and breadth of understanding about optimization theory and methods of application, interested readers are urged to consult one or more of the references cited above. Also, in most larger design and manufacturing firms, there will be optimization experts and/ or consultants available to support designers who have problems suitable for optimization solutions.

* From *Globally Optimally Design* by Douglass J. Wilde, John Wiley and Sons, Interscience, Copyright 1978, reprinted by permission of John Wiley and Sons.

18.2 Formulating Optimization Problems

18.2.1 Steps in the Formulation

The steps in the formulation of optimization problems are as follows:

1. Specify the design variables. These are the same as in guided iteration;
2. Specify the problem definition parameters (PDPs), also the same as in guided iteration;
3. Specify the criterion function. (Refer to Chapter 14, Section 14.8);
4. Specify the procedures by which the value of the criterion function can be computed when given values for the design variables and PDPs;
5. Specify the constraints.

18.2.2 Design Variables

Most optimization methods have been developed for finding values of continuous numerical design variables. Problems with one or more non-continuous (i.e., discrete) design variables are often solved *as if* the variables were continuous. Then when the optimum is found, methods exist for selecting the best discrete values near the optimum continuous value(s).

When there is one (or more) non-numeric design variable, the only thing that can be done is to select a trial value for the non-numeric variable(s), optimize as in a normal numeric problem, and repeat until an overall optimum is found. This is essentially the same procedure used for non-numeric variables in guided iteration.

Suppose, for example, the choice of materials for a structural component is one of the design variables. The problem can be formulated for a given material, and the optimum design determined for that material. Then a second material can be selected and the process repeated. And so on. This sounds cumbersome, but in most cases there are only a few feasible materials to be tried. Moreover, if an automated optimization process is employed, the process need not be time consuming for the designers involved.

18.2.3 Criterion Function

Guided iteration allows any number of SEPs, but optimization permits only one, a kind of grand SEP called the criterion (or objective) function. One way to develop the criterion function is to first specify the SEPs as is done for guided iteration. Then, one of the soft SEPs formulated as an extremum type SEP is often the logical choice as the criterion function. That is, this SEP is re-formed as needed so that it is "the larger the better" or "the smaller the better." This is then the function that will be optimized (minimized or maximized). Other SEPs are then converted into hard type SEPs, and made constraints for optimization purposes.

The criterion function is a function of the design variables. Its value is a measure of the quality or goodness of a design. Symbolically:

$$\text{Criterion Function: } U = U(x_1, x_2, x_3, \ldots\ldots, x_n) \qquad (18.1)$$

where x_1, x_2, x_3, etc., are the design variables. The task of optimization, therefore, is to find the set of values for the design variables that either minimizes or maximizes the value of the criterion function, U. For example, if U happens to be a cost function, then we would want to find values for the x's that result in a minimum value for U.

The requirement that there must be a *single* criterion function means that problems with multiple, conflicting evaluation criteria involved in overall quality must be modified for solution by optimization methods. Such problems, and most design problems are included, are called multi-objective problems; see [7]. Optimization methods are not able to make trade-offs when several factors — often conflicting — contribute to quality. For example, if quality is determined by a combination of cost, weight, efficiency, and corrosion resistance — each of which has its own range of satisfaction and level of importance — then it is difficult (if not impossible) to construct a single, valid criterion function involving all of these.

Using Weighting Factors to Get a Single Criterion Function From Multiple Criteria. Sometimes attempts are made by designers to construct a single criterion function from multiple criteria by forming a weighted sum of the several individual criteria. That is, a function U_1 is developed that measures design quality based on, say, cost. This function is given a weighting factor of w_1. Then another quality factor is developed as U_2, perhaps based on reliability, and given a weight of w_2. And so on. Then an overall criterion function is formulated as:

$$U = w_1 U_1 + w_2 U_2 + \ldots\ldots\ldots\ldots + w_n U_n.$$

$$U = \sum_i w_i \cdot U_i \qquad (18.2)$$

Though this is a tempting approach to many engineers, such a procedure is *not* generally recommended. The reason is that there is no valid way to determine the values for the weighting factors. Doing so subjectively (e.g., important U's get bigger weights than less important ones) in no way assures that an optimum is obtained that reflects the true relative value judgments of the designer. One gets only the illusion of a solution meeting the desired criteria.

Using Ratios to Get a Single Criterion Function From Multiple Criteria. Another method sometimes used to combine multiple criteria into a single objective function is to combine the individual criteria into ratios. For example, with two criteria, one of which correlates directly with quality, and one of which correlates inversely with quality, we might form a single criterion as:

$$U = U_1 / U_2$$

This approach, like the use of weighting factors, also leads only to the illusion of a desired solution. To see why, consider which of the following — or some other arbitrary ratio — should be the one used in this case?

$$U = U_1 / U_2$$

$$U = (U_1)^2 / U_2$$

$$U = (U_1)^{1.7} / (U_2)^{1/2}$$

18.2.4 Constraints

With the criterion function specified, the next task in the formulation of an optimization problem is the specification of *constraints*. Getting a complete and correct — but not overdone — set of constraints is important. Optimization methods find the values of the design variables that produce an optimum (minimum or maximum) value of the criterion function, *subject to the constraints.* Different constraints lead to different solutions; a wrong, incomplete, or overdone set of constraints can lead to wrong solutions.

Constraints may simply specify limits on the ranges of values that a design variable may take on. Or, a constraint may express some more complex required relationship among the design variables, possibly also involving the problem definition parameters. An example of such a constraint is the geometric relationship among center distance, pulley diameters, and belt length in the v-belt domain discussed in Chapter 17.

Constraints are the same as hard SEPs. Soft SEPs — the ones not used as the criterion function — must be converted to hard SEPs for solution by optimization methods. Thus, the "fuzzy" (sloping) regions in the satisfaction graphs become vertical lines. The limits can be at any value the designer chooses, but optimization methods will find no solutions outside the exact specified allowed ranges. Optimization does not do trade-offs; in return for this loss, it finds a more exact optimum *for the problem it is given.* It is up to the designer to formulate the right problem.

Constraints in an optimization formulation are expressed in relationships of two types: (1) equalities; and (2) inequalities. Mathematically, we express the constraints as:

Equality constraints:

$$C_1 = C_1 (x_1, x_2, x_3..........) = 0$$

$$C_2 = C_2 (x_1, x_2, x_3..........) = 0$$

$$C_n = C_n (x_1, x_2, x_3..........) = 0 \quad (18.3)$$

Inequality constraints:

$$L_1 = L_1 (x_1, x_2, x_3..........) < 0$$

$$L_2 = L_2 (x_1, x_2, x_3..........) < 0$$

$$L_m = L_m (x_1, x_2, x_3..........) < 0 \quad (18.4)$$

Equality constraints, in effect, reduce the number of design variables in a problem. If the equations are simple enough, substitutions can be made to eliminate one or more of the design variables from the criterion function. This can reduce the amount of computation required considerably.

It may be noted that when a variable is to be limited to a range, two inequality constraint equations of the type listed above are needed. For example, if x_5 is limited to values between 10 and 20, then we write constraints as:

$$(10 - x_5) < 0 \quad \text{and} \quad (x_5 - 20) < 0.$$

IIIIIIIIIIIII

18.2.5 I-Beam Example, Continued

In the I-Beam example begun in Chapter 14, the criterion function (U) has been specified as:

$$U = A = 2\,t\,w + (h - 2t)\,b$$

and the other conditions of the problem have been specified as constraints:

h(max) = 20 inches

h(min) > 0

w(max) = 12 inches

w(min) > 0

b(min) = 0.1 inches

t(min) = 0.1 inches

MaxDeflection = 0.05 inches

MaxStress = 36,000 psi

$h \geq w$

Note that the criterion function is easily computed when the design variable values are known. From beam theory, we can express the maximum deflection and stress in terms of the design variables and PDPs:

$$\text{MaxDeflection} = 5WL^4/384\,E\,I$$

$$\text{MaxStress} = Mh/2I = WL^2h/16\,I$$

where $I = wh^3/12 - [(w - b)(h - 2t)^3]/12.$

This problem is now easily solvable by several optimization methods, including some available as software.

IIIIIIIIIIIIIIIII

I_I_I_I_I_I_I_I_I_I_I_I_I_I_I

18.2.6 Another Example - A Rectangular Tank

As another simple example of formulating a problem for optimization, consider this problem which has been adapted from an exercise in a Westinghouse Electric Company R & D brochure.

The Problem. One open top rectangular welded steel tank (L x W x H) is to be constructed for hauling 400 cubic yards of ore across a river. A barge is available to make as many trips as are needed to complete the transfer. The tank will have no value after the ore has been moved. The material for the tank will cost $10.00 per square foot. Welding will cost $5.00 per foot. Each round trip across the river costs $140.00. Space and handling limitations require that the tank be no longer than 9 feet and no higher than 6 feet. To meet delivery requirements, there is time for no more than 50 trips. Formulate the problem for optimization to find the optimum tank dimensions and number of trips.

Formulation:

The design variables are:

Length of tank: L (feet)

Width of Tank: W (feet)

Height of Tank: H (feet)

Number of Trips: N

The problem definition parameters are:

Volume of Ore: V (cubic feet) [400 x 27]

Cost of Sheet: C_m ($ / square foot)

Cost of Welding: C_w ($ per foot)

Cost per Trip: C_t ($)

The criterion function is:

$$U = \text{Total Cost} = \text{Cost of Tank} + \text{Cost of Trips}$$
$$= \text{Cost of Material} + \text{Cost of Welds} + \text{Cost of Trips}$$
$$= [L\ (2H + W)] + C_m\ (2HW)$$
$$+ C_w\ (4H + 2L) + N\ C_t$$

The constraints and required relationships are:

$$N = V / HWL$$
$$H \leq 6$$
$$L \leq 9$$
$$W \leq 9$$
$$N \leq 50$$

With this formulation, a number of optimization methods could be used to find the solution.

18.2.7 Review of Formulation Steps

In review, the steps in formulating a problem for optimization are:

1. Specify the design variables, the same as in guided iteration;
2. Specify the problem definition parameters (PDPs), also the same as in guided iteration;
3. Specify the criterion function (U). Remember, this is a *single* function that expresses quantitatively the overall quality of the design defined by the specified design variables;
4. Specify the procedure(s) by which the value of the criterion function can be computed when given values for the design variables and PDPs; and
5. Specify the constraints. These are restrictions on possible values of the design variables, or required relationships among them.

Before taking up solution methods, there are two related topics to be considered first: boundedness and coupling.

18.3 Boundedness

In formulating optimization problems, it is possible to develop a set of *constraints* that prevent an optimum from being found. Difficulties may appear during the solution process if the constraints are such that there is no feasible solution within the constraints. Also, difficulties arise if solution possibilities are *unbounded*, meaning that the value of the criterion function can be increased in a maximization problem (or decreased in a minimization problem) without limit by allowed (i.e., unconstrained) changes in one or more of the design variables.

There are formal mathematical theories in optimization texts and references that can be used in complex cases to detect these circumstances, but the theories are beyond our scope. Here, we can provide only a conceptual understanding of these issues. For a complete treatment, one or more of the optimization references should be studied.

Over-constrained. Recall that constraints can be either equalities or inequalities. When there are more independent equality constraints than design variables, an optimization problem is said to be *over-constrained.* In such a case, there are no solutions to the optimization problem.

For example, consider the I-Beam problem shown in Figure 14.1. The four design variables are h, w, b, and t. We want to optimize the cross-sectional area, A (in this case, to find the minimum area) where

$$\text{Area} = 2\,t\,w + b\,(h - 2\,t).$$

Suppose that, for some reason, we have the following five equality constraints imposed on the problem:

1. Specified Stress = $M\,h / 2\,I = W\,L^2\,h / 16\,I$
2. Specified Deflection = $5\,W\,L^4 / 384\,E\,I$

   where $I = w\,h^3 / 12 - [\,(w - b)\,(h - 2\,t)^3\,] / 12$
3. $b = t$
4. $h^2 + w^2 = 144$
5. $h - 2\,t = 7.5$ inches

This problem as posed has no solution because it is over-constrained, though unrepentant equation jugglers may like to have a try at solving it.

Under-constrained. On the other hand, in some cases, there may be inadequate constraints (equality and/or inequality) so that the value of the criterion function can be increased (or decreased) indefinitely. In this case, the criterion function is said to be *unbounded*, and the problem is said to be *underconstrained.* No optimum solution can be found in this case, either.

Again consider the I-Beam problem as an example. If b is not given a lower limit, optimization methods will generally try to make it zero. But this is physically impossible, so the methods will not produce a practical result.

Fully Constrained. A third possibility is that the problem may be *fully constrained* so that there is only one answer possible. This occurs when there are as many design variables as independent equality constraints. A solution can be found, essentially by simultaneous solution of the set of constraint equations, but then this is not really an optimization problem.

Suppose in the I-Beam problem that only the first four of the five constraints listed above are involved. Then this is an equation jugglers delight: a solution can be found by solving the four equations simultaneously. But, unfortunately, such nice problems seldom occur in practice.

Active Constraints. A constraint is said to be *active* if it influences the solution. Said another way, if a constraint can be removed without there being any effect on the solution (i.e., on the optimum), then that constraint is said to be inactive. Any constraints that are not active at the optimum point can be removed — not considered at all — thus simplifying the problem formulation.

An active inequality constraint that is at its extreme limiting value at the optimum point (that is, it is effectively in use as an equality at the optimum) is said to be *critical*. In every optimization problem, there must be a least one active constraint that is critical at the optimum point. If this were not the case, then one or more of the design variables could continue to be changed in some way to improve the objective function — until a limit of some constraint is reached.

In the I-Beam problem, the lower limit on b will turn out to be a critical constraint because b will be at its limiting value at the optimum solution.

18.4 Coupling

In Chapter 17, we introduced the concept of coupling, and described how to identify it from inspection of a dependency table. Everything said in Chapter 17 about the desirability of reducing coupling by redesigning the object, or by redefining design variables, applies here as well. If a constraint expresses a narrow range of acceptability, or if two or more design variables are highly coupled with respect to an SEP, then optimization methods are likely to have difficulty, though they can still succeed.

18.5 Monotonicity

Monotonicity is a useful concept that can be employed to help simplify optimization problems [1, 3].

An objective (or criterion) function is said to be monotonic with respect to a design variable if the algebraic sign of the dependency between the objective function and the design variable is invariant within the scope of the problem domain. A design variable can also be monotonic with respect to the criterion function if a positive change in the design variable always has the same effect (whether positive or negative) on the criterion function.

Note that the objective function in the I-Beam example given by

$$\text{Area} = 2\,t\,w + b\,(h - 2\,t)$$

is monotonic with respect to w, b, and h. No matter what the values of the other variables, increasing any of these will result in an increase in A.

The situation with the variable t is different, however. An increase in t will sometimes cause A to increase and sometimes cause it to decrease (if b were somehow to become greater than w). Thus A is not monotonic with respect to t.

It has been shown [1, 3] that every design variable monotonic with respect to the objective function *must* be limited appropriately by some constraint — or else the solution will be unbounded. For example, in a maximization problem, if there is a positive monotonic design variable, then that design variable must be limited to some maximum value by a constraint. If it isn't, then that design variable can be increased without limit, causing a corresponding unlimited increase in the objective function (that is, unboundedness).

In a minimization problem, if there is a negatively monotonic design variable, then it must be limited to some minimum value. If it isn't, the design variable, and hence the criterion function also, can be decreased without limit.

In this book, we have indicated that design variables, when they are defined in the problem formulation, should routinely have their allowed ranges (maximum and minimum values) specified. This practice automatically prevents the kind of unboundedness discussed here.

Monotonicity theory also states that if a design variable does not appear in the objective function, then either (a) that design variable must be bounded above and below at the optimum by an active constraint, or (b) not bounded at all. In the latter case, any constraints which are monotonic with respect to that design variable can be deleted from the problem formulation.

These monotonicity concepts can sometimes be used to simplify, or even solve, certain optimization problems. Thus optimization problem formulations should be studied with them in mind before time consuming computation procedures are employed.

I\ I\ I\ I\ I\ I\ I\ I\ I\

18.6 Example - A Simple Structure

To illustrate the formulation issues discussed above, we consider a very simple example. Figure 18.1 shows a structure ready for parametric design; the configuration issues have been settled. The weight W is to be supported at a distance L_1 from the wall. The configurations of the tensile and compression members (L_2 and L_1,respectively) are solid round, and the members are made of the same material with known material properties which, we assume for convenience, to be those of 6063-T6 aluminum. The initial four steps in the formulation are as follows:

1. The design variables are the angle a and the diameters D_1 and D_2. The range of possible values for a is 20 to 80 degrees. The limits of both D's are from 0.1 inches to 2.0 inches.
2. The problem definition parameters are: the material properties (E and Yield Stress), the length L_1, the given configuration, and the weight W to be supported.
3. (a) Soft type solution evaluation parameters (SEPs) are:

   i.) Cost. As a surrogate for material cost, we will use the volume of the two supporting members which we would like to keep as small as possible.

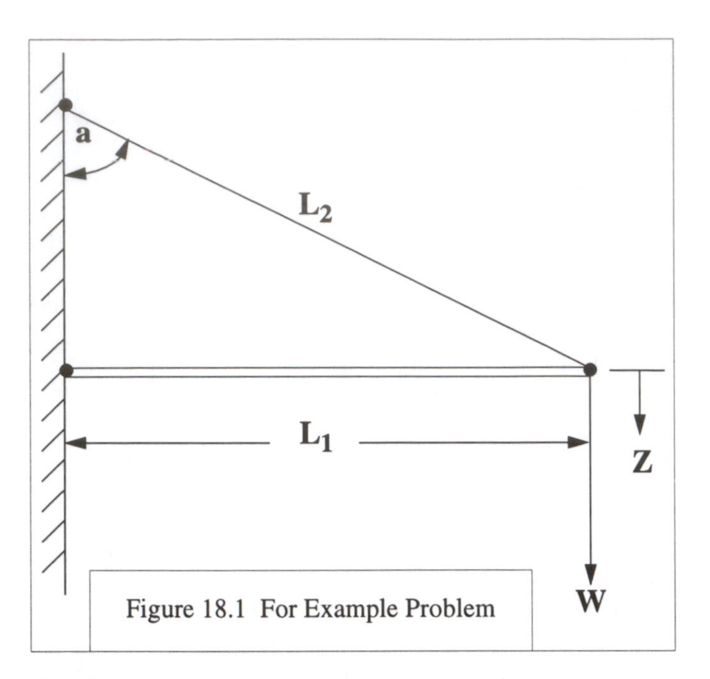

Figure 18.1 For Example Problem

The satisfaction graph is shown in Figure 18.2(a), though we cannot yet specify values for V_3 or V_{max} in the graph. We won't ever need to specify them if we make volume the criterion function and solve the problem by optimization.

ii.) The vertical deflection (Z) of the weighted end of the compression member which, for some imaginary reason in this example, we would like to see as small as 0.005 inches but certainly no more than 0.01 inches. See Figure 18.2(b).

iii.) Also, the maximum stress in the two members (tension in member 2 and compression in member 1) must be less than the yield stress of the material. We have shown this as a soft SEP in Figure 18.2(d).

(b) Hard SEPs:

There is a required relationship expressing the fact that the compression member must not buckle under its compressive load. The critical

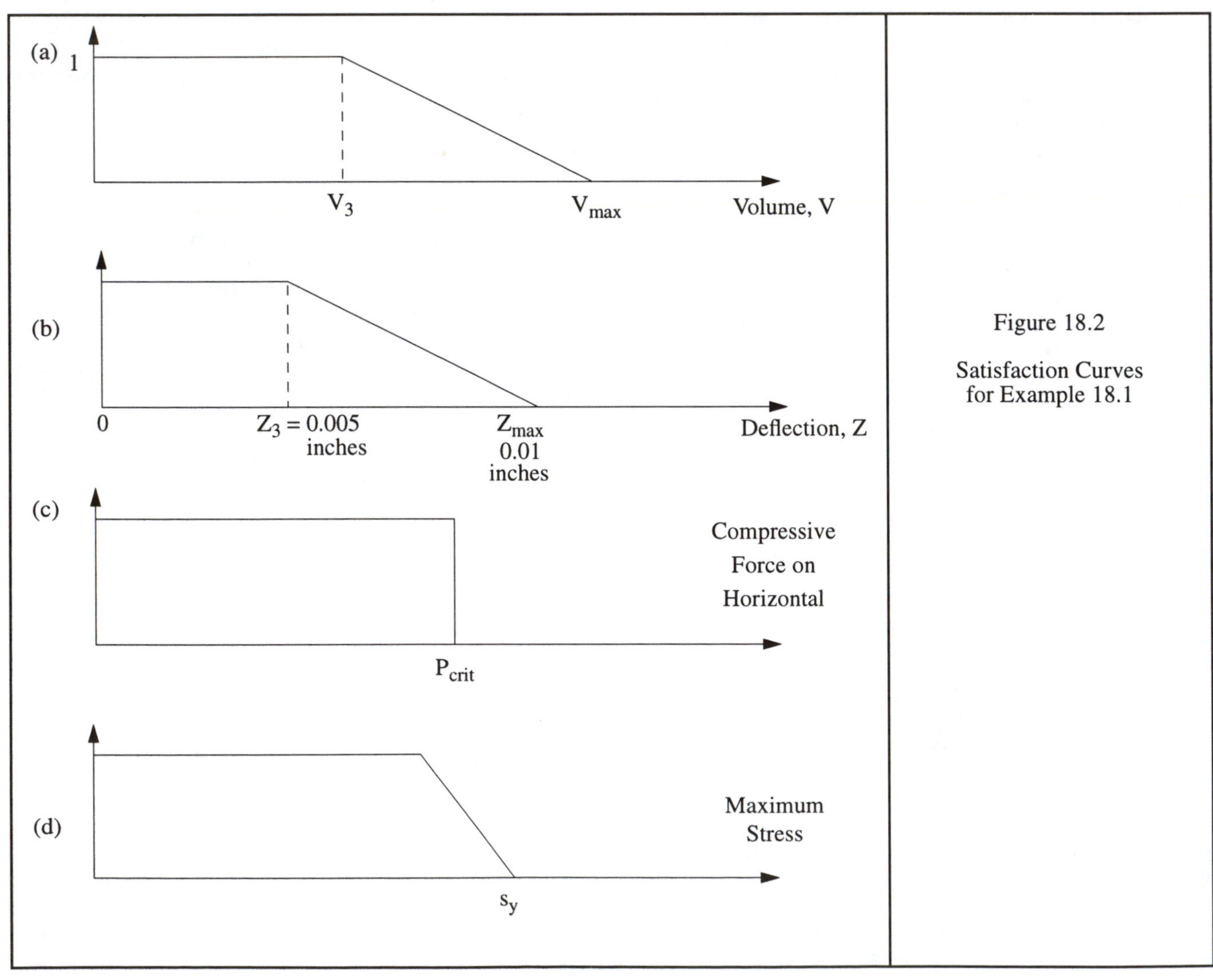

Figure 18.2

Satisfaction Curves for Example 18.1

S.E.P.s	Design Variables		
	a	D_1	D_2
Volume V	+0.1	0.4	0.5
Deflection Z	-0.2	-0.3	-0.4
Force F_1	-0.2	0	0

Figure 18.3

Initial Dependencies for Example 18.1

buckling force for this simply supported column is:

$$Pcrit = (\pi)^2 \; E \; I \, / \; (L_1)^2$$

This can be expressed as a hard SEP as shown in Figure 18.2(c). In words: the compressive force on the horizontal member is an SEP that must be less than the critical force.

4. An initial dependency table might look as shown in Figure 18.3. This is needed if we are going to solve the problem by guided iteration, but is not absolutely needed for optimization. However, the table might be helpful in revealing coupling that could be reduced or eliminated by redefining the design variables.

5. Examine the design variables for possible simplifications.

   The dependency table indicates some coupling between D_1 and D_2 with respect to both volume V and, to a lesser degree, deflection Z. However, redefining them (as, say, D_1 and D_2/D_1) does not have an effect in this case. Thus, we shall leave the design variables as previously defined.

6. In this case, we would select the volume (a surrogate for cost) as the criterion function. That is:

$$U = V = A_1 L_1 + A_2 L_2$$

   where $A_1 = \pi (D_1)^2 / 4$ and $A_2 = \pi (D_2)^2 / 4$.

7. We have to reformulate the deflection and stresses as hard SEPs so they will serve as constraints in the optimization problem. We do this as:

   $0 < Z \leq Zmax$ where $Zmax = 0.01$ inches;

   Max Compressive Stress in 1 = Yield Stress = 31,000 psi;

   Max Tensile Stress in 2 = Yield Stress = 31,000 psi.

8. The analytical equations needed are as follows:

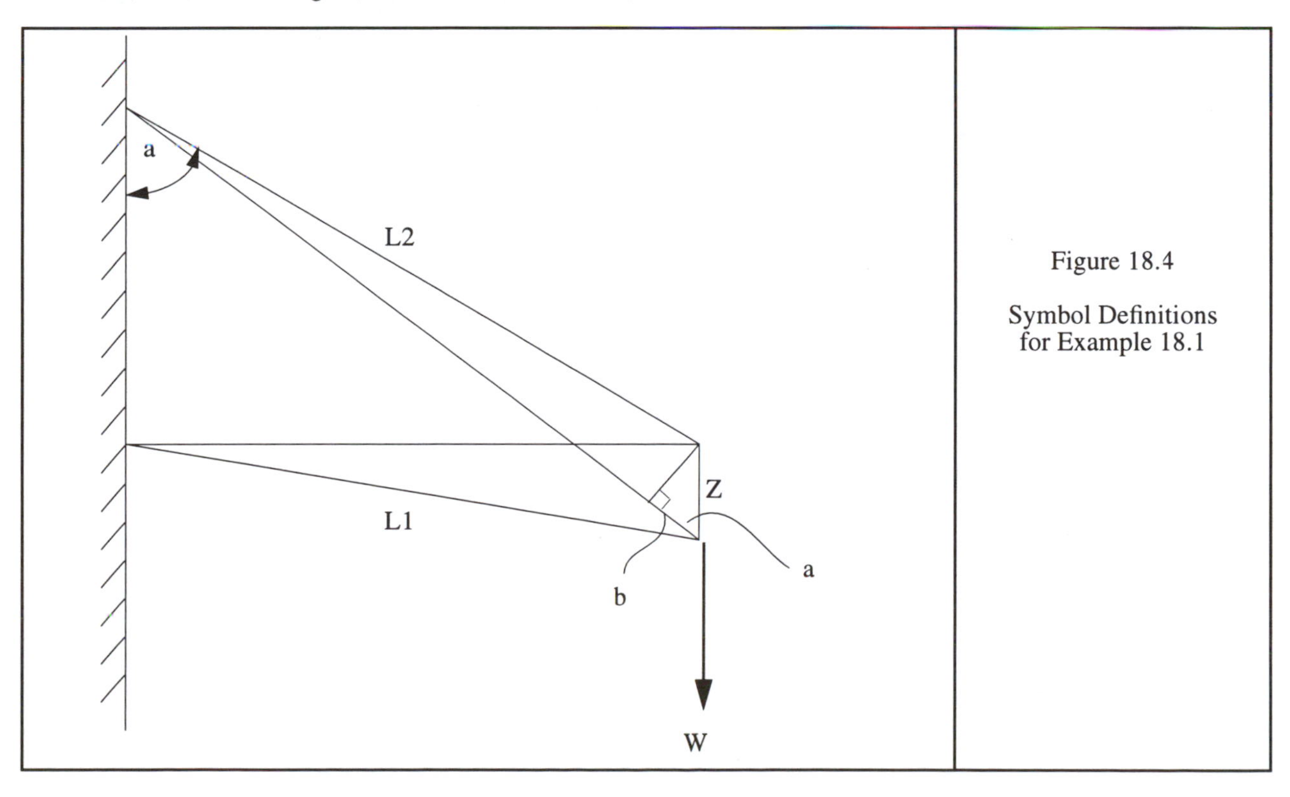

Figure 18.4

Symbol Definitions for Example 18.1

For the criterion function: $U = A_1 L_1 + A_2 L_2$

where

$A_1 = \pi (D_1)^2 / 4$ and

$A_2 = \pi (D_2)^2 / 4.$

For the compressive force, F_1: $F_1 = W \tan a$

For the tensile force, F_2: $F_2 = W / \cos a$

For the critical force: $Pcrit = (\pi)^2 E I_1 / (L_1)^2$

where $I_1 = (\pi)(D_1)^4 / 64.$

For the tensile stress in member 2: $S_2 = F_2 / A_2$

For compressive stress in member 1: $S_1 = F_1 / A_1$

For the vertical drop of the load from the horizontal, refer to Figure 18.4. Note that b is the elongation of the tension member neglecting the higher order effect on Z due to shortening of member L_1 (This is reasonable assuming D_1 turns out to be considerably larger than D_2.) Then:

$Z = b / \cos a$

where b is the elongation of tensile member 2 given by:

$b = (F_2 L_2) / (A_2 E).$

The important material properties for 6063-T6 aluminum are:

E = 10,000,000 psi S(yield) = 31,000 psi.

As an example of the use of these equations, consider a case when the problem is defined by:

W = 300 pounds L_1 = 36 inches

and a trial design has been selected as:

a = 45 degrees D_1 = 1 inch D_2 = 0.5 inches.

Then the analytical equations just above give the following results:

A_1 = .7854 inches^2

A_2 = .1964 inches^2

F_1 = 300 lbs

F_2 = 424.3 lbs

b = 0.011 inches

Z = 0.00778 inches

S_1 = 382 psi

S_2 = 2160 psi

I_1 = 0.0491 inches^4

Pcrit = 3739 lbs

U = 38.27 inches^3

Inspection of the values for the hard SEPs (constraints) shows that all of them are satisfied. However, there is no way of knowing at this point whether or not U is at or even near its optimum value. In fact, we can say for sure that it is not at its optimum since none of the constraints is critical.

The problem has now been formulated as an optimization problem and can thus be solved by one or more optimization methods (some are discussed below) or by computer programs employing these methods. (We could also use the guided iteration method from Chapters 14 and 17 to solve this example problem.)

/\ /\ /\ /\ /\ /\ /\ /\ /\

Optimization problems can be categorized depending on the nature of the design variables, criterion function, and constraints. And there are many solution methods available depending on the type of problem. Experts in optimization can formulate problems in order to match effective solution methods with problem types, and thus implement solutions efficiently. In complex situations, therefore, designers should consult with an optimization specialist.

In many cases, however, designers can formulate and solve parametric problems by optimization. Therefore, next we consider solution methods for a series of optimization problem types, starting with the simplest. For additional, more powerful methods needed for more complex problems, readers are referred to the literature and to optimization specialists.

18.7 Solving Optimization Problems By Differentiation

18.7.1 The General Case

A useful way of thinking about many maximization problems is as an analogy to a hill to be climbed — the optimum is at the top. Some optimization techniques are, in fact, called "hill climbing" methods. (As we have noted, guided iteration is also a kind of hill-climbing method.) If the problem is to minimize (instead of maximize) the criterion function, then we can think of a valley to be descended instead of a hill to be climbed. This hill (or valley) analogy is easy to visualize in problems with one design variable, or even two. When there are n-dimensions, however, the analogy is still helpful as a way of thinking about certain optimization methods.

Suppose U is a function that can be differentiated in closed form with respect to the x's (rare in real problems, but it does happen). Then the optimum will likely (but not certainly, since constraints may interfere) be at the point where the derivative (i.e., the slope) is zero. That will be at the top of the "hill" (or bottom of the valley), providing of course that the "hill" has a top, and only one top.

To find an optimum in these cases, we ignore the constraints temporarily. Then we differentiate the criterion function with respect to each of the design variables, setting the results to equal to zero. The resulting set of simultaneous equations created defines the optimum:

$$\frac{\partial U}{\partial x_1} = 0$$

$$\frac{\partial U}{\partial x_2} = 0$$

|

|

|

$$\frac{\partial U}{\partial x_2} = 0 \qquad (18.5)$$

============

18.7.2 Example - Hot Water Pipe

As an example, we consider the problem of finding the optimum pipe diameter (D) and insulation thickness (x) for a pipe to carry hot water over some distance [8]. We assume the following (simplified) cost functions:

Cost of Pumping $= K_1 / D^5$

Cost of Pipe $= K_2 D$

Cost of Heat Lost $= K_3 / x$

Cost of Insulation $= K_4 D x$.

where the constants (K's) are known constants. Then the criterion function (total cost) in terms of the design variables D and x is:

$$U = K_1 / D^5 + K_2 D + K_3 / x + K_4 D x.$$

Differentiating U with respect to x and setting to zero gives:

$$\frac{\partial U}{\partial x} = - K_3 / x^{*2} + K_4 D^* = 0$$

where x^* is the value of x at the optimum, and D^* is the value of D at the optimum. Differentiating U with respect to D and setting to zero gives:

$$\frac{\partial U}{\partial x} = - 5 K_1 / D^{*6} + K_2 + K_4 x^* = 0.$$

The above two equations can now be solved simultaneously for D^* and x^*. For example, if $K_1 = 10$, $K_2 = 1$, $K_3 = 1$, and $K_4 = 1$, then we find that

$$D^* = 1.75 \quad \text{and} \quad x^* = 0.75$$

This is the optimum assuming that these values for D^* and x^* are within the constraint limits set for them. If there are constraints on the design variables that restrict the feasible region so as to rule out one or more of them taking on the value(s) obtained from solving the above set of equations, then some additional work will have to done to determine the optimum.

============

18.8 Problems With a Single Design Variable

18.8.1 Introduction

Problems with only a single design variable are generally easy to solve by one or more of several optimization methods. We devote space to this type of problem here, not just to explicate specific solution methods, but because the methods and associated issues (e.g., feasible range and boundedness) are easy to understand in this context, and are also relevant to problems with two or more design variables as well.

////////////////////////////////

18.8.2 Example - Optimum Thickness of Insulation

As a first example, consider the case of determining the optimum value of insulation thickness for a wall. Cost is the criterion function, and it consists of the cost of the insulation, $K_1 x^{1/2}$, plus the cost of energy lost through the wall, K_2 / x. K_1 and K_2 here are known constants that determine the costs when x is known.

Here we are looking for a minimum, so the analogy is to seek the lowest point in a valley rather than the highest point on a hill. The criterion function is:

$$U = K_1 x^{1/2} + K_2 / x.$$

where x is the insulation thickness. Assume $K_1 = 1000$ \$/inch$^{1/2}$ and $K_2 = 4000$ \$-inches. The only constraint is that x must be greater than or equal to + 1.0, and may not exceed 3 inches. These constraints are expressed as:

$$L_1 = x - 1 > 0$$

$$L_2 = x - 3 < 0.$$

These constraints establish a feasible range for the solution: x must lie between 1 and 3 inches. Or, in other words, the constraints bound x on both its low and high sides.

Solution by Differentiation. One way to solve this problem is — as outlined in the previous section — to ignore the constraints temporarily and differentiate U with respect to x, setting the result equal to zero. This gives:

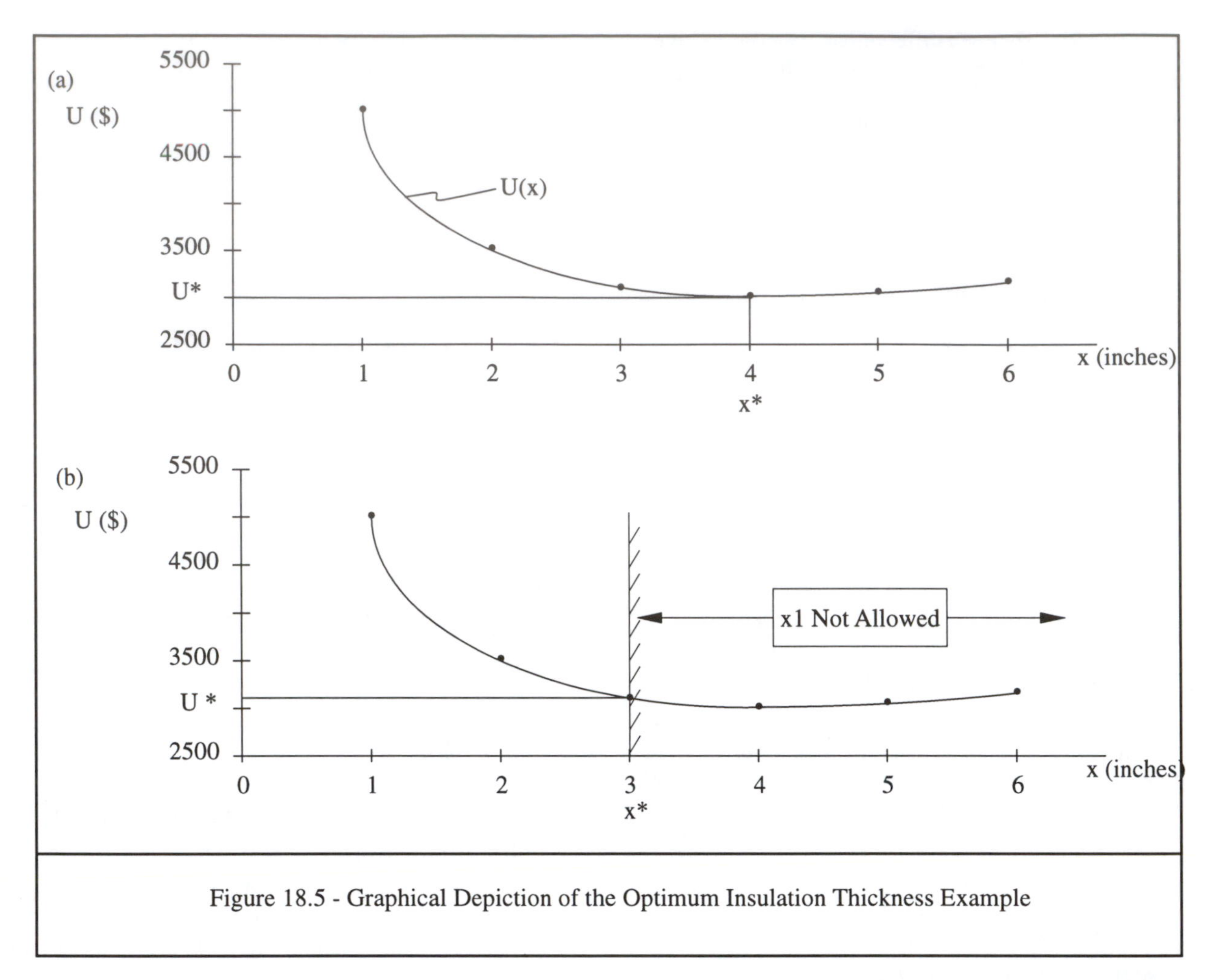

Figure 18.5 - Graphical Depiction of the Optimum Insulation Thickness Example

$$\frac{\partial U}{\partial x} = 1/2\, K_1 \, (x^*)^{-1/2} - K_2 / (x^*)^2 = 0$$

where x* is the value of x at the optimum point. Solving for x*:

$$x^* = (2 \; K_2 / K_1)^{2/3} = 4 \text{ inches.}$$

See Figure 18.5(a).

Now the value for x* obtained above is outside the feasible solution range for x. In this case, it is easy to see from the plot that the optimum value for U (i.e., U*) is at the boundary where x = 3. And it will usually, but not necessarily always, be the case in problems of this type that the best value of a design variable that is constrained in this way will be the value at the closest boundary. See Figure 18.5(b).

Imagine for a moment that the upper limit on x is 6 inches instead of 3 inches. Then the optimum found by differentiation (x* = 4) would be the solution. This would be called an *internal* optimum because the optimum is inside the feasible region rather than on one its boundaries. Most of the time, however, optimum solutions are on a boundary rather than internal.

//////////////////////////////

Graphical-Numerical Solutions. Optimization problems with a single design variable can usually be readily solved simply by plotting values of the design variable against values of the criterion function. We have to use intelligent trial values of x when constructing the plot, however, or a huge number of unnecessary and possibly time consuming computations for U may be required. Usually we have some idea of the order of magnitude of the optimum range, so we outline here an efficient way to converge to a solution starting from a value far below the optimum that is to be found. The method is called the *golden search* method because it is based on the "golden search" ratio — 1.618 — discovered by the ancient Greeks [3].

The golden search method can be used to obtain values for a graphical solution, or it can be incorporated into a computer algorithm for numerical solution. In either case, the method is: first, *bracket* (or bound) the optimum point into a known interval (I) on x; then, search and continuously reduce the size of the interval (I) until an optimum is found to the desired accuracy.

Bracketing the optimum. Suppose we are looking for the value of x that gives a minimum for U, and also suppose our initial guess for x is $x_0 = a$. (If x is bounded on its low side by zero, the value of a can be any small, positive number.) The first thing we must do is bracket the optimum; that is, assuming a is below the optimum, we must determine another value for x that we know is above the optimum.

We begin the process of bracketing by computing and recording or plotting the value of U corresponding to $x_0 =$ a. Call it $U_0 = U(a)$. See Figure 18.6 (a).

For the next point, we use $x_1 = x_0 + 1.618\ x_0$, and again plot the resulting value for $U(x1) = U_1$. Assuming a is below the optimum value for x, this value of U will be less than U_0. Next we use:

$$x_2 = x_1 + 1.618^2\ x_1 \quad (18.6)$$

and plot the corresponding U_2. Thus in general:

$$x(n + 1) = x(n) + 1.618^n\ x(n) \quad (18.7)$$

We continue this process until we find that

$$U(n) > U(n - 1); \quad (18.8)$$

that is, until U begins to increase. At this point we know that the minimum is bracketed by x(n - 2) and x(n). The difference between these values for x is now called the search interval, I. That is:

$$I = x(n) - x(n - 2) \quad (18.9)$$

Searching the Interval. Now we must search the

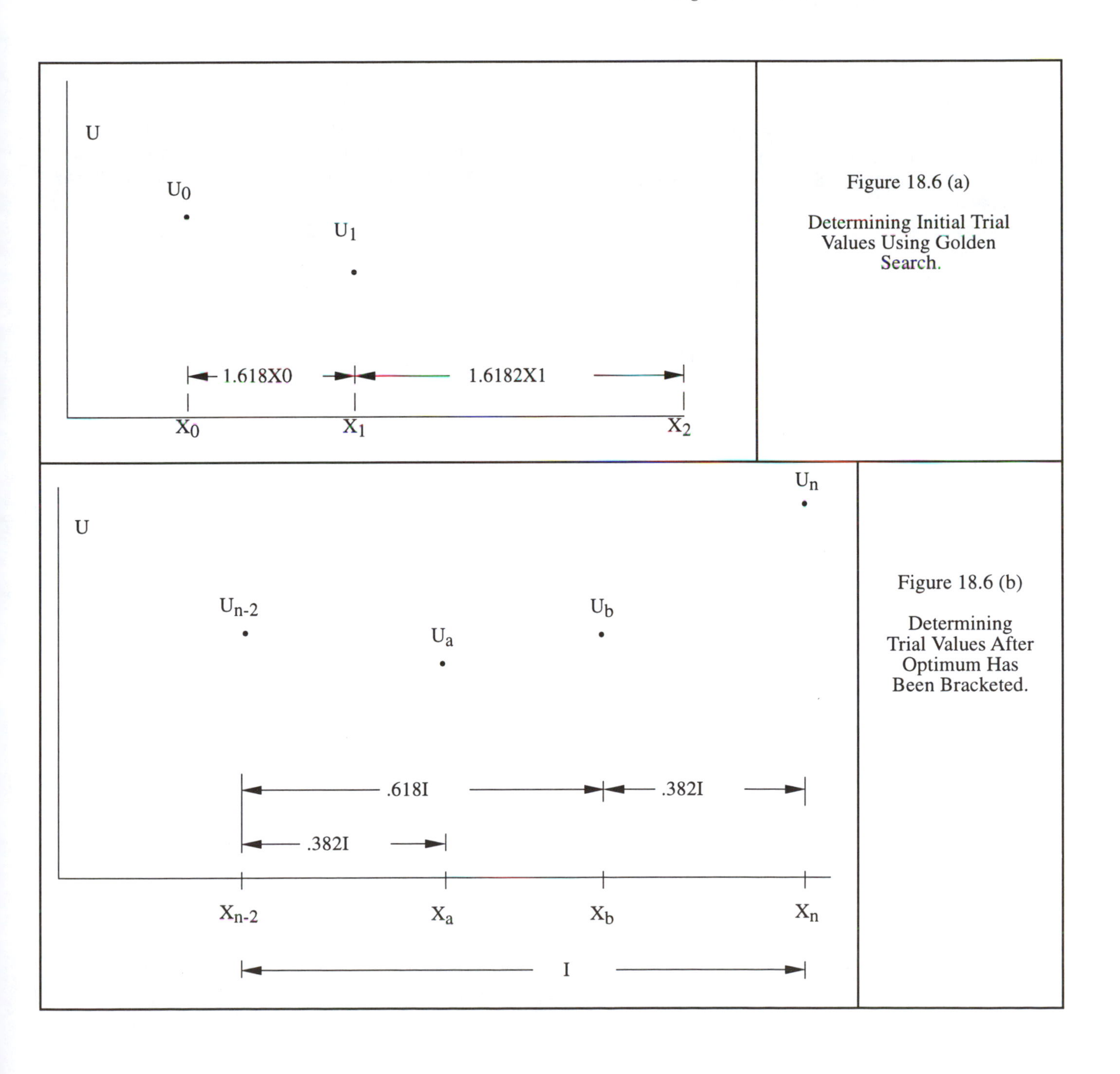

Figure 18.6 (a)

Determining Initial Trial Values Using Golden Search.

Figure 18.6 (b)

Determining Trial Values After Optimum Has Been Bracketed.

interval I for the value of x that produces the minimum value for U. We do this by determining the following two intermediate values for x and U. Refer to Figure 18.6 (b) and note that I = 0.618 I + 0.382 I.

$$x(a) = x(n-2) + 0.382\,I \qquad (18.10)$$

$$x(b) = x(n-2) + 0.618\,I \qquad (18.11)$$

The corresponding values for U are called U(a) and U(b) as shown in Figure 18.6(b).

Now we can determine a new, smaller interval. If

$$U(a) > U(b),$$

then the optimum x is between x(a) and x(n). We can redefine the interval accordingly to be I = x(n) - x(a) and search the new interval as just described.

If $U(b) > U(a)$,

then the optimum is between x(n - 2) and x(b) so we redefine the interval accordingly and search the new interval as just described.

The "golden" process described will converge very quickly to a minimum. It is stopped when the interval I is small enough to satisfy the degree of accuracy required for the optimum.

+++++++++++++++++

18.8.3 Example - A Welded Tank

As an illustration of the above process, consider the following problem: A welded tank is to be constructed in the form of a cylinder with a flat top and bottom. The material costs $0.025 per square inch, and welding costs $0.100 per inch. The overall height of the tank may not exceed 8 feet. The volume of the tank must be 20 cubic feet. Design the tank for minimum cost of material and welding.

Design variables: radius r (inches) and height h (inches)

Problem definition parameters: Given configuration, cost functions

Criterion Function:

$$U = \text{Cost}$$

$$= 0.025\,[\,2\pi r^2 + 2\pi r h\,] + 0.100\,[\,(4\pi r) + h\,]$$

Equality constraint: Volume = $\pi r^2 h$ = 34,560 cubic inches

Inequality constraints: $r > 0$

$h > 0$

$h < 96$ inches

Though there appear at first glance to be two design variables in this problem (r and h), the equality constraint on volume can be used to eliminate one of them. Solving it for h, substituting into the criterion function (and rearranging) gives:

$$U = 0.314\,r^2 + 1728/r + 1.257\,r + 1100/r^2$$

Starting with an initial trial value of r = 1, the golden search method generates the following results:

Trial	r	U
1	1	2830

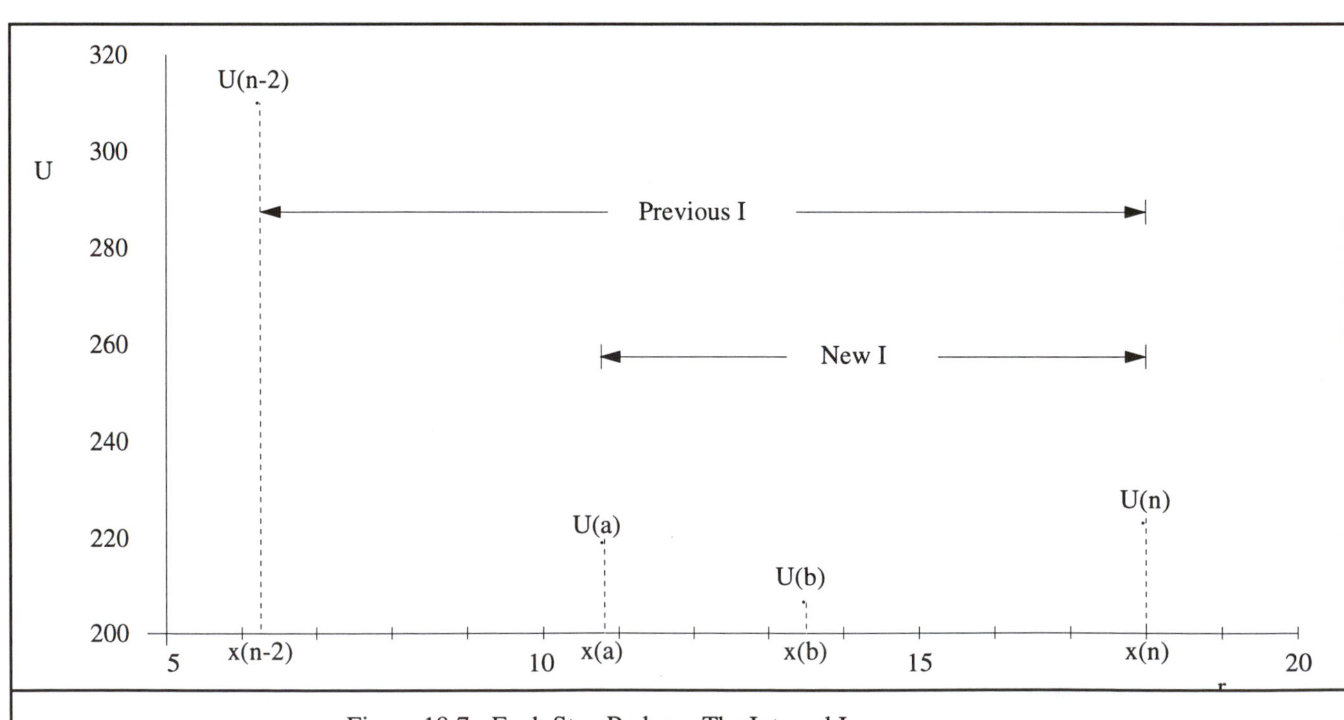

Figure 18.7 - Each Step Reduces The Interval I

Trial	r	U
2	1.618	1491
3	2.618	826
4	4.236	480
5	6.554	311
6	11.09	217
7	17.94	223

With the 7th trial, we have gone past the minimum. The interval I as defined in Figure 18.7 is 17.94 - 6.553 = 11.387. Thus the set of trials corresponding to those in Figure 18.7 are as follows:

Trial	r	U
n - 2	6.553	311
a	10.90	2191
b	13.59	208
n	17.94	223

Note that though we show four trials here, only two of them require new computation for U. The new interval I is now 17.94 - 10.90 = 7.04, and the values for x in the next set of trials are therefore as follows. Note that though we show four trials here, only one of them (x(b) = 15.25) requires new computation for U:

Trial	r	U
n - 2	10.90	311
a	13.59	219
b	15.25	208
n	17.94	223

The new interval I is now 15.25 - 10.90 =4.35, and:

Trial	r	U
n - 2	10.90	219
a	12.56	210
b	13.59	208
n	15.25	210

This process can continue until we get as close to an optimum as we desire. In this case, after only a few trials, we can see that the optimum U is very close to 208 and that the optimum value of r is about 13.5 inches. The height of tank corresponding to this r is about 60 inches, so the constraint on h is satisfied.

+++++++++++++++++

18.9 Problems With Two or More Design Variables

18.9.1 Introduction

There are too many optimization methods for multiple design variable problems available for all to be included here. Moreover, many of them require mathematics beyond the background (though not the ability) of most engineering undergraduates. We will therefore present here only a sampling of methods, our selections being motivated by the level of complexity, and the general usefulness in mechanical engineering design. Readers are referred to the references and to optimization specialists or software for a wider range of more powerful techniques.

When it can be done, of course, the method of differentiation described above can be used for *any* number of variables. The other methods to be discussed below are: alternating single-variable search; the method of steepest ascent (or descent), linear programming, and methods for shape optimization.

18.9.2 Alternating Single-Variable Search

We have shown above how, when there is only a single design variable, to conduct a numerical search for the optimum using the golden search method. This method can be extended to two or more design variables, though it can become rather "brute force" and computation intensive. The support of a computer (especially with a spreadsheet) can be very helpful.

The search procedure is simply to cycle through the design variables one at a time, selecting one for variation while holding the others constant. The optimum value for each variable is found by maximizing U in a single-variable search with that variable; then another variable is selected.

This is repeated, one variable after another, until all variables have been run through. It will almost always be necessary to repeat the cycle several times through all the variables in order to get near the true optimum.

The optimum can be detected when running through a cycle of design variable changes produces very little improvement in the value of U.

In terms of the hill-climbing analogy, this process is like following a selected constant compass direction (corresponding to changing a single design variable) as long as it takes us up hill. When it no longer takes us up, we switch to another direction (i.e., design variable), and so on. Thus we essentially zig-zag up the hill. Of course, we have to hope that the hill is "well behaved"', without false summits, plateaus that could trap us, or precipices that could prevent us from proceeding. In any case, it is a slow process when there are a large number of variables.

18.9.3 The Method of Steepest Ascent (Descent)

Instead of zig-zagging in directions defined by a random sequence of single design variables, the method of steepest ascent enables us to advance more or less always in the direction that is most directly up the hill; that is, in the direction that has the steepest slope.

To find this direction, we must differentiate the criterion function with respect to each of the design variables. Since this differentiation cannot often be done analytically in most real world problems, we will describe the steepest ascent method here as a numerical process. We define the starting point with the following values of the design variables:

Initial design variable values:

$$x_{10}, x_{20}, x_{30} \ldots\ldots\ldots\ldots x_{n0},$$

where the second subscript refers to the starting state.

The finite difference derivatives at the starting point are:

$$\left(\frac{\partial U}{\partial x_1}\right)_0 \approx \left\langle\frac{\Delta U}{\Delta x_1}\right\rangle_0$$

$$\left(\frac{\partial U}{\partial x_1}\right)_0 \approx \left\langle\frac{\Delta U}{\Delta x_1}\right\rangle_0$$

$$\vdots$$

$$\left(\frac{\partial U}{\partial x_1}\right)_0 \approx \left\langle\frac{\Delta U}{\Delta x_1}\right\rangle_0$$

We can compute these values by selecting small delta x's and computing the resulting delta U's from the expression for the criterion function.

Then to get new values for the design variables, we use the following:

$$x_{11} = x_{10} + K \cdot \left\langle\frac{\Delta U}{\Delta x_1}\right\rangle_0$$

$$x_{21} = x_{21} + K \cdot \left\langle\frac{\Delta U}{\Delta x_2}\right\rangle_0$$

$$\vdots$$

$$x_{n1} = x_{n0} + K \cdot \left\langle\frac{\Delta U}{\Delta x_n}\right\rangle_0$$

where K is a constant still to be determined.

Note in the above set of equations how the larger the derivative, the greater will be our movement in that direction.

The value of K will specify how much we will move in the selected direction. Logically, we should move as far as possible — as long as the value of U is increasing. That is, we want to perform a single-variable optimization with K as the single variable. When we get this optimum value for K, we use the equations above to determine the new design variable values, and repeat the process by computing new derivatives, etc.

In terms of the hill climbing analogy, this method enables us to proceed in a direction that has the steepest uphill direction until that direction is no longer increasing our elevation (the value of the criterion function). Then we find the new steepest direction, and so on. When the "hill" is reasonably well behaved, the method of steepest ascent (or descent) will converge more rapidly than alternating single-variable search to an optimum. However, there is a price to be paid in terms of additional computation. Finding K each time is an optimization problem in itself, and though it a single variable problem, there can still be considerable computation involved. The support of a computer to perform the computations is essential in most cases.

18.9.4 Linear Programming

Occasionally a problem is *linear,* meaning that the design variables appear in the criterion function, and in all constraints, only to the first power. Linearity happens fairly regularly in optimization problems in the field called operations management, but it can also happen to lucky mechanical designers on occasion.

When a problem is linear, the criterion function can be written as follows:

$$U = c_1 x_1 + c_2 x_2 + c_3 x_3 + ... + c_n x_n \qquad (18.12)$$

where the c's are known constants. If, in addition, the design variables are limited to positive (or zero) values, and the constraints can be written as:

$$a_{11} x_1 + a_{12} x_2 + \qquad + a_{1n} x_n \leq b_1$$

$$a_{21} x_1 + a_{22} x_2 + \qquad + a_{2n} x_n \leq b_2$$

$$.$$

$$.$$

$$a_{m1} x_1 + a_{m2} x_2 + \qquad + a_{mn} x_n \leq b_m \qquad (18.13)$$

where the a's and b's are constants, then the problem is called a *linear programming* problem.

We discuss in this book only linear programming problems with two design variables, primarily so that students will be able to recognize the problem type. There are a number of solution methods in the references cited above for problems with more than two variables.

When a linear program problem has only two design variables, a graphical solution is possible, and understanding how it is obtained is helpful for visualizing more complex linear programming solutions. A simple example will illustrate the point. Consider a case when the criterion function is

$$U = 2.5 \ x_1 + 3.5 \ x_2$$

and the constraints are:

$$x_1 > 0$$

$$x_2 > 0$$

$$x_1 < 2.0$$

$$x_2 < 6$$

$$x_1 + 2 \ x_2 < 4$$

If we construct a plot of x_1 versus x_2 as in Figure 18.8, then the constraints can be used to rule out areas where solutions are not allowed. For example, the first two constraints rule out regions of negative x_1 and x_2. The third constraint rules out the region above $x_1 = 2$. The fourth rules out the region above $x_2 = 5$. The last constraint rules out the region where $x_1 + 2 \ x_2$ is greater than 4, as shown by the sloping line in Figure 18.9.

We can immediately see in the figure the region where solutions are feasible as permitted by the constraints. Note that the fourth constraint is not an active one.

Linear programming theory can show that the optimum will be at one of the corners, or (by coincidence) along one the lines defining the feasible region. To find which corner, we plot lines of constant values for U. In this example, we did this by letting $x_1 = x_2 = 1$. Then we solve the criterion function equation for U, which is then $U = 2$. Next with $U = 2$, and $x_1 = 0$, we find $x_2 = -1$. This defines the straight line labelled $U = 2$ in Figure 18.8.

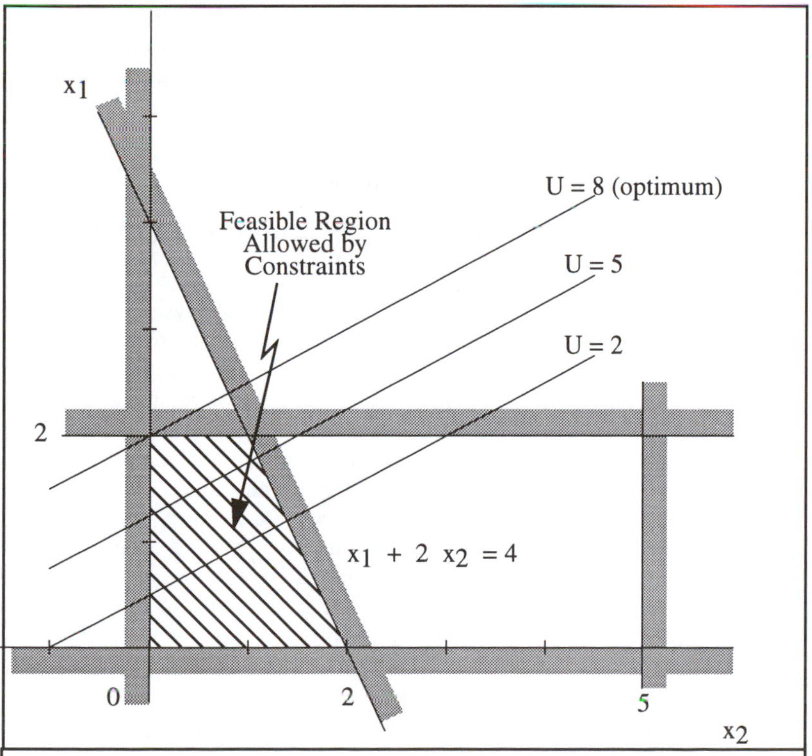

All lines of constant U will be parallel, so we can draw a family of lines for constant U as shown. By inspection we can see that the line with maximum U will pass through the upper left corner of the valid region. This corresponds to $x_2 = 0$ and $x_1 = 2$, so that is the optimum solution in this example.

Note that the first and third constraints listed above are critical; that is, they are active at the optimum point.

When there are more than two design variables, of course this graphical method is not feasible — though the principles are the same. There are several computational methods for finding a solution, but they are complex and described in detail in many textbooks so we will omit them here. There are also computer programs that can be used to solve linear optimization problems.

Figure 18.8 Graphical Depiction of a Linear Programming Problem

18.10 Structural Shape Optimization

18.10.1 General Issues

Structural shape optimization is a sub-field of optimization that has essentially become a field itself. In a sense, any optimization problem whose goal is to find the dimensions of a load carrying member is a structural shape optimization problem. However, when the term "structural shape optimization" is used, it most generally implies the use of finite element methods incorporated with various optimization techniques to determine the minimum weight of some special purpose structural part. In this text, we cannot do more than provide a brief introduction to the subject. Our purpose is to insure that students of design recognize situations when structural shape optimization methods might be useful, and to understand conceptually what those methods can do. For designers who can perform finite element analyses, the design process can use the analysis results either iteratively with guided iteration or with optimization methods to determine the optimum.

Some structural shape optimization problems can be solved more or less automatically by existing finite element programs available from software vendors. These programs may, for example, remove nodes (equivalent to removing material) from parts of the member that are lightly stressed, and thus by iteration evolve a shape that is more uniformly stressed near the maximum allowed.

There are also optimization programs available incorporating finite element capability (including automatic mesh generation) that allow designers to define a criterion function in terms of attributes of the shape. Constraints can also be defined. Then the programs find the values of the attributes that optimize the specified criterion function.

A simple, commonly visible example of shape optimization can be found in the tapered light poles and sign carrying beams now seen along many highways. Not only are these tapered structural elements aesthetically pleasing but they also use less material because they become smaller as the bending moment becomes smaller, thus using material more effectively to carry stress. For a given material and configuration (cross section, often circular hollow, and taper, probably linear) finding the optimum taper is a structural shape optimization problem. Total weight would be the criterion function. The design variables would be the diameter and thickness (as functions of position along the member) and the degree of taper. Constraints would include the maximum stress, the maximum allowed deflection, minimum wall thickness, and no doubt some other limits based on the capability of the manufacturing process. Finite element analysis would be required to determine the values of stress, deflection, and so on, for any trial set of design variables. But the problem is formulated in the manner of other optimization problems, and solution methods exist.

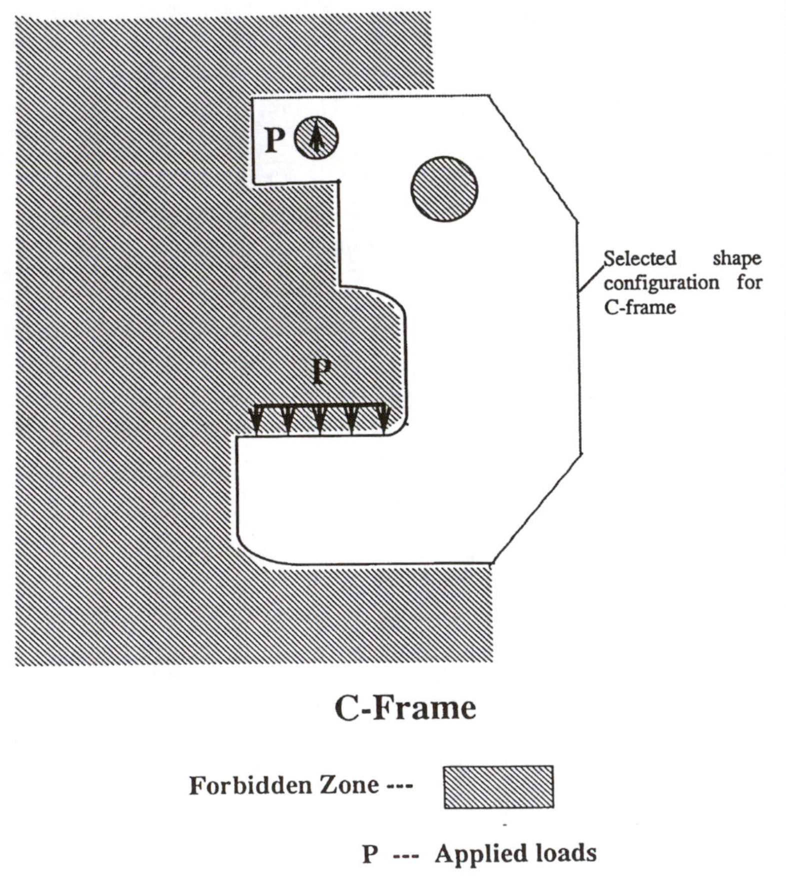

P
P
Selected shape configuration for C-frame
C-Frame
Forbidden Zone ---
P --- Applied loads

Figure 18.9 Configuration for Punch Press Example

GGGGGGGGGG

18.10.2 Example - A Punch Press Frame

An example of a how structural shape optimization can be used to advantage is described below. This example is adapted from Zhu [9]. The problem is to reduce the weight of a large structural frame used in very heavy punch presses. The problem as shown here is not solved in all its possible complexity, but is simplified to enable a quicker, though approximate, solution. The exterior configuration of the part, which is very large, and the load and support points, are shown in Figure 18.9. The size and location of the holes is fixed. The basic shape is not to change, though the dimensions may change.

The seven design variables selected (X_1 through X_7) for the optimization are shown in Figure 18.10, which defines the part's configuration. The regions shown all are to have the constant thickness throughout their region in the final design. However, the thicknesses of the various regions may be different from the others.

Note the simplification that has been done in formulating this problem in this way. There could have been a different problem defined to optimize the dimensions of the part with many more design variables, allowing for a better optimum, but at a price of more computation. In this case, the designer decided to try this simple problem first.

The criterion function in this problem is the total weight of the frame. Constraints include limits on the deflection of the frame, and limits on the allowed stresses in the frame. Both the deflections and stresses can be computed by a finite element analysis when trial values for the design variables are specified.

Figure 18.11 shows a high level schematic of the solution method in which an optimization program interfaces with a finite element program. The results are shown in Figure 18.12, together with the finite element mesh used. A 15% reduction in weight was accomplished.

In the example just shown, the initial design and the final design were close enough together that the finite element mesh did not need to be modified as the solution proceeded. When it does need to be changed to maintain the accuracy of the analysis, this makes the solution more difficult. This difficulty can be handled, however, and very powerful methods are being developed as the research in this field continues.

GGGGGGGGGGGG

18.11 Summary and Preview

Summary. This chapter is an introduction only to the application of optimization to parametric design problems. Optimization is a mathematical field that can find the value

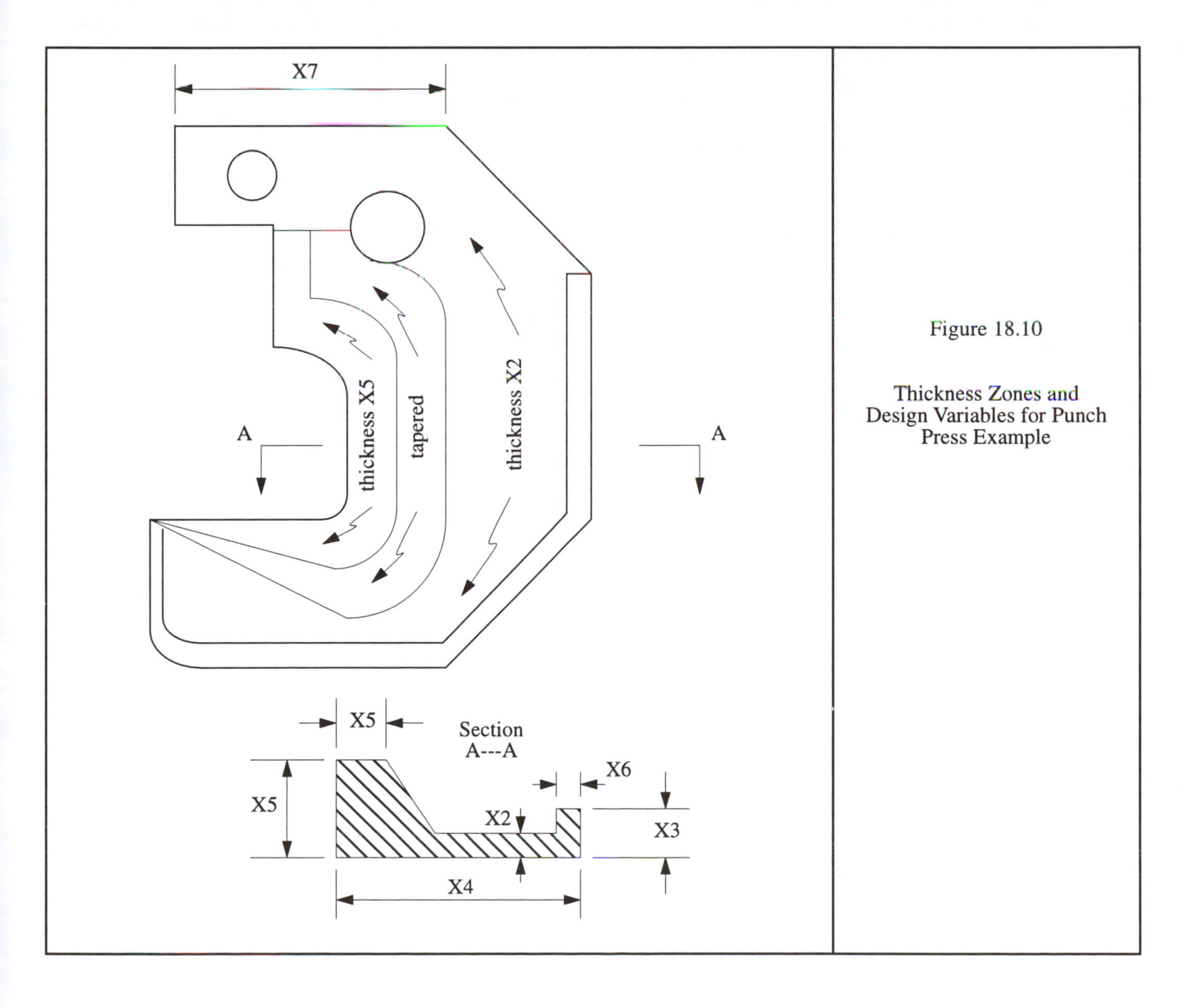

Figure 18.10

Thickness Zones and Design Variables for Punch Press Example of design variables that will make an objective function an optimum — subject to the given constraints. It is up to the designer to formulate the design problem so that the results of the mathematical optimization process are correct and meaningful to the actual design problem. When this can be done, then optimization provides very powerful methods to support parametric design.

Some of the simpler optimization methods have been presented in this chapter. Many others, including more advanced methods, are provided in the references. Also, software programs are available, and there are experts in the field who can help. Our purpose here has been to insure that students and other readers can recognize when optimization may be useful, to be able to make an initial formulation of an optimization problem, to solve relatively simple problems, and to know when to get expert help.

Preview. We present three methods in this book for performing parametric design: guided iteration, optimization, and design for robustness using Taguchi's Methods™. The next chapter deals with the last of these.

There are also ways other to perform design for robustness. In fact, robustness can sometimes be expressed as in an objective function for optimization. And robustness can always be included as one of the evaluation criteria in guided iteration. Moreover, there are statistical methods other than Taguchi's that can be used. However, Taguchi's Method™ is popular, has a good track record, is relatively easy to use (at least sometimes), and designers should definitely have an understanding of what it can do, of its logic, and its limitations.

References

[1] Wilde, Douglass, *Globally Optimal Design*, Wiley-Interscience, John Wiley and Sons, New York, 1978. Quotation by permission of John Wiley and Sons.

[2] Arora, J. S., *Introduction to Optimum Design*, John Wiley and Sons, New York, 1989.

[3] Papalambros, P. Y. and Wilde, D. J., *Principles of Optimal Design*, Cambridge University Press, Cambridge, England, 1988.

[4] Reklaitis, A. G., Ravindran, A., and Ragsdell, K. M., *Engineering Optimization*, John Wiley and Sons, New York, 1983.

[5] Optdes and OptdesX are trademarks of Design Synthesis, Inc., 3883 N. 100, East Provo, UT 84604.

[6] Grossman, I. E., "Mixed Integer Programming Approach for the synthesis of Integrated Process Flowsheets", Computers and Chemical Engineering, Vol. 9, No.1, pp 463-482, 1985.

[7] Sandgren, E. "A Multi-objective Design Tree Approach for Optimization Under Uncertainty", ASME DE-Vol 19-2, Proceedings of the 1989 ASME Design Automation Conference, Montreal, Quebec, Canada, September, 1989.

[8] Dixon, J. R., *Design Engineering: Inventiveness, Analysis, and Decision Making*, McGraw-Hill, 1966.

[9] Zhu, B., "Structural Shape Optimization of a 'C' Frame Punch Press", M.S. Thesis, Xian Jiao Tong University, Xian, P.R.China, 1988.

[10] Doebelin, E. O., *Measurement Systems*, McGraw-Hill, New York, 1966

Recommended Supplementary Reading

Arora, J. S., *Introduction to Optimum Design*, John Wiley and Sons, New York, 1989.

Papalambros, P. Y. and Wilde, D. J., *Principles of Optimal Design*, Cambridge University Press, Cambridge, England, 1988.

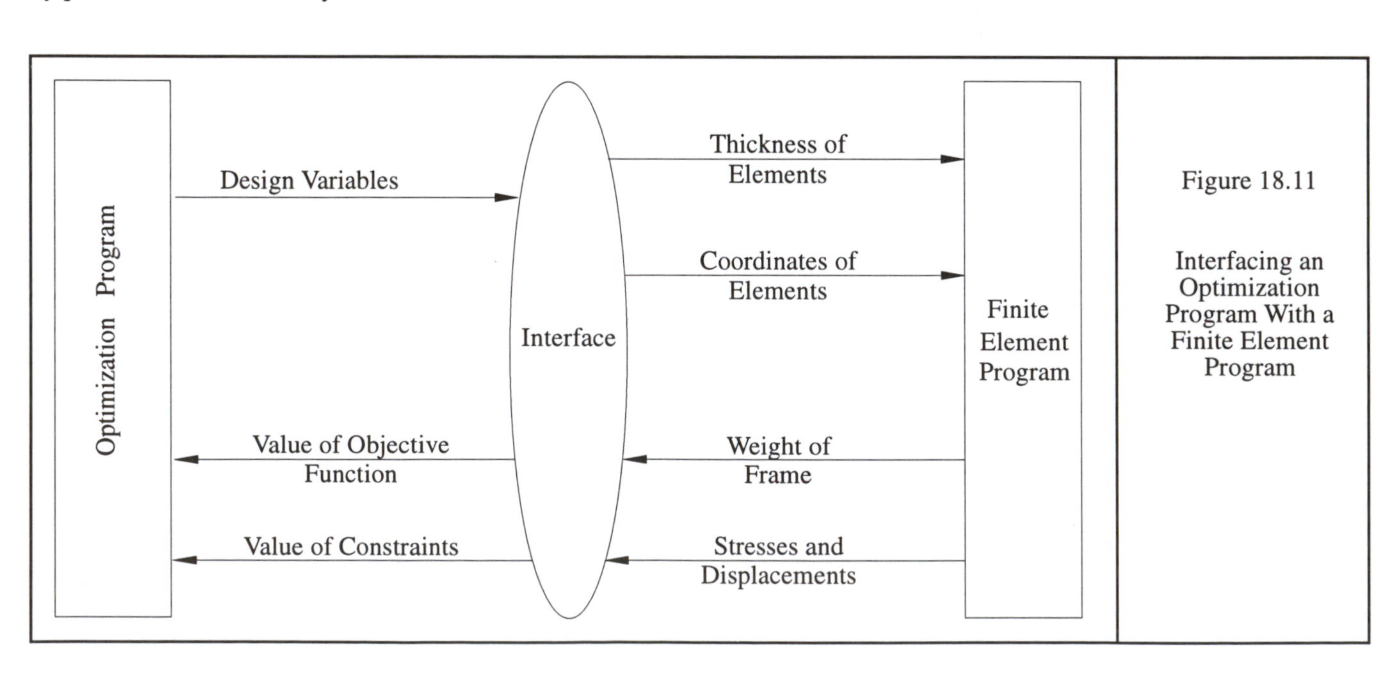

Figure 18.11

Interfacing an Optimization Program With a Finite Element Program

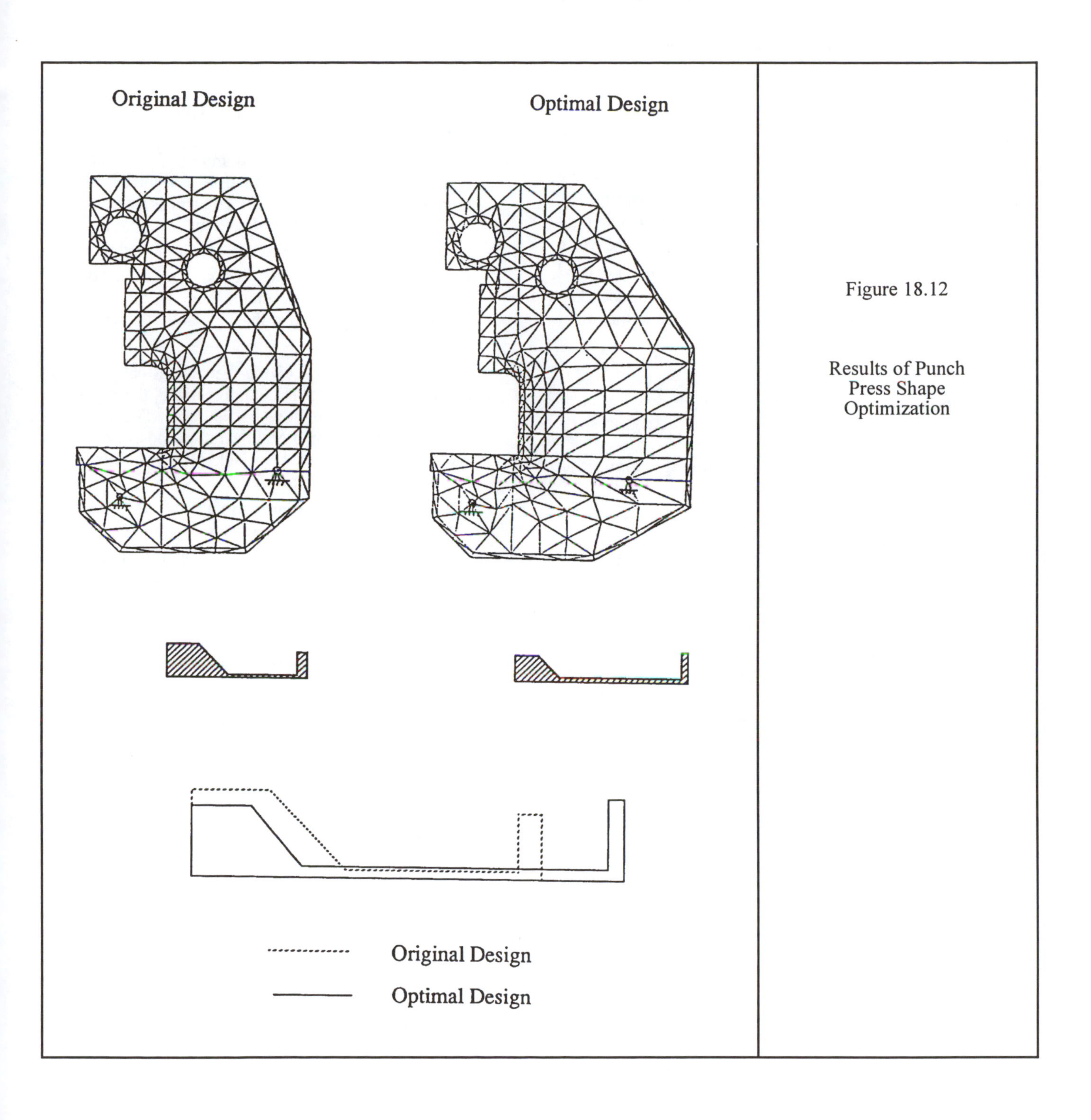

Figure 18.12

Results of Punch Press Shape Optimization

Problems

18.1 (a) Formulate Problem 17.2 or 17.3 for solution by optimization.

(b) Solve the problem as formulated in Part (a).

18.2 (a) Formulate Problem 17.5 for solution by optimization.

(b) Solve the problem as formulated in Part (a).

18.3 Formulate the V-Belt drive problem described in Section 17.10 as an optimization problem.

18.4 A plastic part has two design variables, t_1 and t_2. For reasons of functionality, t1 must not be less than 0.05 inches and t_2 must not be less than 0.10 inches. More over, t_1 cannot exceed 0.10 inch and t_2 cannot exceed 0.20 inches. Tooling costs are essentially independent of the values chosen for t_1 and t_2, but other manufacturing costs in cents per part are:

Processing: $C_p = 100\ t_2$

Material: $C_m = 20\ t_1 + 40\ t_2$

Assembly: $C_a = 10 + 100\ (t_2 - t_1)$

(a) Formulate as a linear programming optimization problem.

(b) Solve the problem as formulated.

18.5 A welded steel tank is to cylindrical with hemi-spherical top and bottom. The volume must be 50 cubic feet, and the internal pressure will be 400 psi. Material costs $0.12 per cubic inch. Welding cost $2.20 per inch of length per inch of thickness. Metal is available in any thickness from 1/8 inch to 1 inch. The maximum available sheet size is 15 feet by 15 feet. The minimum cylinder diameter that can be rolled is 18 inches. The maximum tensile strength of the welded joints is 30,000 psi.

Formulate this problem as an optimization problem.

18.6 Develop a different configuration for the structure shown in Figure 18.10, define the attributes, and describe how you perform parametric design using (a) guided iteration and (b) optimization assuming you had a finite element program available.

18.7 A soda can is to be designed for minimum material cost using a given material and prescribed wall thickness, t. The ends are twice as thick as the side walls. The diameter of the can cannot exceed 3.5 inches, or be less than 2 inches. The can's height must be at least 4 inches, but cannot exceed 8 inches. The volume held must hold at least 12 fluid ounces.

Formulate this problem for solution by optimization. Solve by a numerical search method.

18.8 A large flat-ended cylindrical hot water tank with a given diameter (D) and height (H) is to be insulated. The problem is to find the optimum thickness (t) of insulation. We can assume that the insulation cost is proportional to its thickness times the tank surface area (A). The annual energy lost will be:

$$E = K\ A / t$$

The annual cost of this energy will be

$$C_a = E\ C_e$$

Both K and C_e are known parameters. Assume a 20 year life of the system, and a 5% discount rate (See Chapter 23). Formulate this problem as an optimization problem.

18.9 Figure P18.9 shows a configuration for a thickness measuring instrument to be used for measuring the thickness, t, of thin parts. The concept is that the pressure P_1 can be used to indicate the x; note that if x is increased, P_1 will decrease. It is assumed that h is accurately known, so that if x can be accurately determined by the instrument, then t will be established. It is assumed also that P_s and T_s are known and kept constant.

A simplified analysis of the instrument can be made assuming that the air is incompressible, the pipes and tanks rigid, and the flow frictionless. These are not bad assumptions for initial design purposes, though a more precise analysis is possible [10] and experimental calibration will be needed.

We note that (for incompressible flow) the mass flow through orifice S (with diameter D_s) will be equal to the mass flow through the annular opening at point A where the exit tube opening has a diameter of D_a. That is:

$$G_s = G_a$$

From fluid mechanics principles applied to orifice s:

$$G_s = C_s\ (\pi\ D_s^2 / 4)\ [\ 2\ d\ (P_s - P_1)\ ]^{1/2}$$

where C_s = Orifice S discharge coefficient

d = density of air

P_o = ambient pressure.

There is, in effect, an annular orifice at the base of the instrument. The mass flow can be estimated from:

$$G_a = C_a\ (\pi\ D_a\ x)\ [\ 2\ d\ (\ P_1 - P_0\ )\ ]^{1/2}$$

where C_a = Orifice at A discharge coefficient.

The two discharge coefficients will not be exactly equal in practice, but if the openings are small compared with the pipe size, they will both approach a value of about 0.6. For preliminary design purposes, we can take them to be equal. Then equating the mass flows with $P_0 = 0$ psig gives:

$$P_1 = P_s / [\ 1 + 16(\ D_a^2\ x^2 / D_s^4)\ ]$$

This provides us with a design relationship between x and P_1 as needed. Since we would like the instrument to be as sensitive as possible, we would like to design so that the sensitivity

$$\text{Sensitivity} = \frac{\partial P_1}{\partial x}$$

is at its maximum.

(a) Formulate this problem as an optimization problem.

(b) Using differentiation, find the relationship that exists among the design variables at the point of maximum sensitivity.

(c) For an instrument to measure values of x in the vicinity of 0.0020 inches, find values of the design variables for maximum sensitivity if $P_S = 25$ psig and the constraints include:

Maximum $D_S = 1/16$ inch

Minimum $D_S = 1/48$ inch

Maximum $D_a = 1/16$ inch

Minimum $D_a = 1/64$ inch

(d) Which constraint is active at the optimum? What is the sensitivity at the optimum?

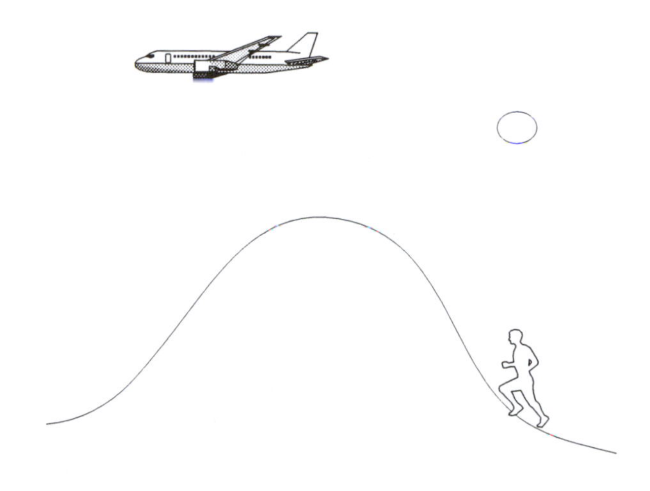

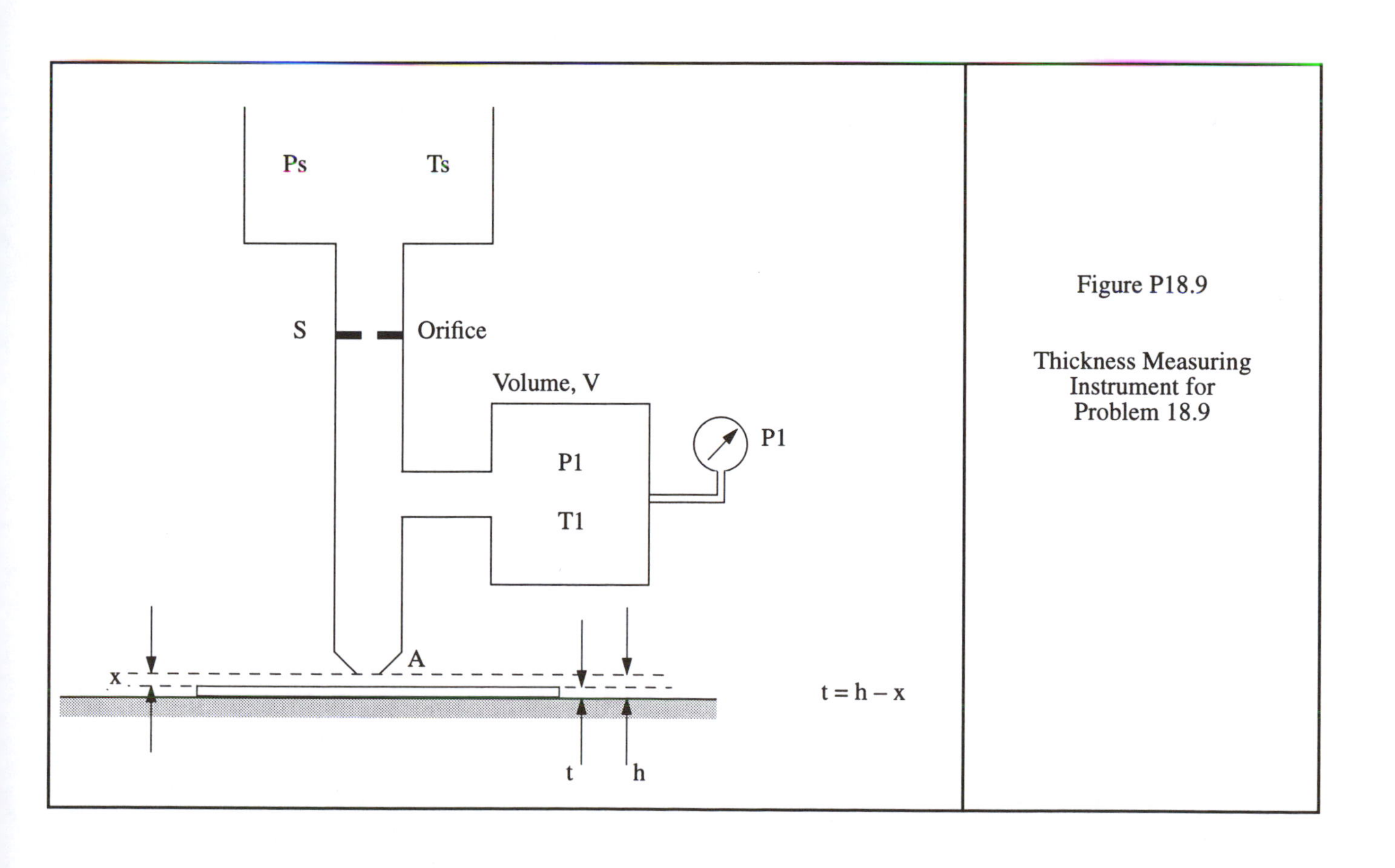

Figure P18.9

Thickness Measuring Instrument for Problem 18.9

This page is intentionally blank.

Chapter Nineteen

Introduction to Taguchi Methods™

"Have you heard of the wonderful one-hoss shay,
That was built in such a logical way
It ran a hundred years to the day,
And then, of a sudden, it ———"

Oliver Wendell Holmes
The One-Hoss Shay (The Deacon's Masterpiece) (1858)

19.1 Introduction to Design for Robustness

The degree to which a product's quality of performance is insensitive to variations in manufacture, in-service wear, and in-service environmental variations is called its *robustness*. Robustness is one of the important criteria on which trial parametric designs may be evaluated. Robustness can normally be included as a Solution Evaluation Parameter (SEP) in guided iteration, and can also sometimes be the objective or criterion function for optimization. Other methods used for parametric design for robustness are statistically based, and the best known is commonly called the *Taguchi Method™*.* It is the main subject of this Chapter.

All statistically based methods of parametric design for robustness are based to a large extent on research from the field called *design of experiments,* a branch of statistics. In a single chapter, we can provide only an introduction to design of experiments in essentially "how to do it" form related specifically to its use in parametric design. Interested students are encouraged to study a book on design of experiments [1], or to take a full course on the subject. For industry readers, there are short courses available through various universities and professional societies.

* *Taguchi Method* is a trademark of the American Suppliers Institute, Dearborn, Michigan.

In our previous discussions of robustness, we considered that it could be one of the several factors to be considered in the total performance of a part, component, or product. The Taguchi Method takes a somewhat different approach. It establishes a composite measure called (unfortunately) the *signal-to-noise* ratio that provides for a fixed built-in trade-off between performance and robustness. Qualitatively, we might write this as:

$$\text{Robustness} \propto \text{Signal-to-Noise Ratio}$$

$$\text{Signal-to-Noise Ratio} \propto \frac{\text{Performance}}{\text{Variability of Performance}}$$

where variability of performance is due to variations in manufacturing, environmental conditions, and wear. Later in this chapter, we will define the signal-to-noise function more quantitatively, and show how it is used to establish design variable values.

There is no theoretical reason to support this particular function — that is, signal to noise ratio — chosen by Taguchi as the way to trade off performance with robustness. Indeed, there is *no* best or "correct" function. The Taguchi signal-to-noise ratio is, however, reasonable — and it has worked well in practice.

Taguchi's approach to parametric design also includes a method for the assignment of tolerances. Before tolerances are assigned, design variable values are established using the assumption that standard manufacturing tolerances and material property variations are in effect. Then the effects of the variations allowed by these standard tolerances and material tolerances are analyzed and predicted, often based on the results of experimental studies. Tolerances and material specifications are then made tighter than standard if, but only if, the standard ones are inadequate to maintain a desired level of performance and robustness. Taguchi provides a method for finding the tolerances required to meet the performance standards. Final decisions about tolerances then must also consider the trade-off between cost and performance.

Before we can discuss the Taguchi Method of parametric design to problems, readers must first learn a little about the basic methodology used in design of experiments. Therefore, in the next section we present an introduction to the basic application methodology of that field.

Note: An excellent but still brief presentation of Taguchi's Method can be found in "Statistical Methods and Applications", by J.G.Elliott, published by Allied Signal Corporation [2]. Students are encouraged to get a copy of this booklet for their own library. More complete treatments are recommended for designers who require a more fundamental understanding. See, for example, references [3, 4, 5]. Papers and publications with helpful explanations and interesting examples include references [6, 7, 8, 9, 10, 11].

19.2 Some Applied Tools for Designing Experiments — and a Simple Introductory Example

19.2.1 Problem Formulation.

Our general goal in this section is to show how to determine the extent to which a desired performance outcome is influenced or affected by such factors as the design variables, processing conditions, material properties, environmental factors, etc. To this end, the problem formulation requires first that a function indicating performance be established. This function, called the *Result*, is the measure of the outcome that we care about; it may be the life of a part or product, or it may be efficiency, power output, deflection, weight, etc. The Result thus plays a role somewhat analogous to the criterion function in optimization.

We must also establish the *variable factors* that we believe will influence the Result. These need not be numerically valued variables; they can, for example, be which vendor supplied the materials, or whether the product was produced on the night or day shift, etc.

As an example, we consider a case of planning the process for heat treating aluminum extrusions made of alloy 6063. Our *Result* in this case will be the tensile strength of the extrusions after heat treatment.

19.2.2 Example - Heat Treating Aluminum Extrusions

In this aluminum extrusion example, we have decided as a start that the two most important variable factors influencing the Result are:

A - Oven temperature, and

B - Time in the oven.

We want now to perform a small set of experiments in order to learn about the relative influence of the variable factors A and B on the Result. Next, therefore, we must decide how many discrete values of the variable factors we will use in the experiments. The usual number is two or three, though more are possible. The more there are, the more experiments are needed, adding significantly to the cost and time involved.

Generally, we have a good idea of the feasible range the values of the variable factors might span. If we use two variables, we would then choose values somewhere near the extremes of this range. If we use three variables, we would add a selection near the center of the range. Using three variables will obviously give better results, but the cost (time and money) must be considered.

For simplicity at this point, let us say we elect to use two values for the variable factors. (Later we will consider the effect of electing to use three values.) We must now establish what these values are to be, but after this we will refer to the values selected as only level 1 or level 2. The values for levels 1 and 2 we choose for the extrusion heat treating example are:

	Levels	
	1	2
Oven Temp, F	300	500
Time in Oven, hrs	3	5

Now with the problem formulated in this way, what we would like to know is: What experiments should we conduct in order to determine the relative effects of oven temperature and time on the Result?

In this example, there are only four experiments we *can* run: (1) 500 F, 3 hrs; (2) 500 F, 5 hrs; (3) 300 F, 5 hrs; and (4) 300 F, 3 hrs. We will have to run them all in this very simple case. However, in more complex situations, with more variable factors and/or more levels, the methodology of design of experiments will enable us to get the information we need with many fewer experiments than would be required if we ran all possible combinations. Using a simple example at this stage, however, enables the method to be explained without a confusing mountain of numbers.

19.3 Orthogonal Arrays and Interactions

19.3.1 Arrays

The basic tool for achieving the goal of deciding which experiments to run is called an *orthogonal array.* An orthogonal array is a matrix that shows, if interpreted correctly, what experiments should be run in certain cases to get the information desired. In Appendix 19A of this chapter, there are a selected number of orthogonal arrays presented taken from "Orthogonal Arrays and Linear Graphs" by G. Taguchi and S. Konishi [8] reproduced by permission of the American Supplier Institute, Dearborn, MI.

Choosing the most appropriate array from the library of arrays is a key step in applying the Taguchi Method, and we will discuss how to make the choice in the next few pages. First, however, we will see how to use the arrays selected by continuing with the extrusion heat treating example.

Assume that the array called L4 in Appendix C has been selected. The array is shown in Figure 19.1 and shows an experimental design with three columns labelled 1, 2, and 3 for variable factors. We generally assign the variable factors to the columns, and in this problem we will subsequently assign variable factors (A and B) to two of them.

	Columns (Variable Factors)			
	1	2	3	Result
Run 1	1	1	1	
Run 2	1	2	2	
Run 3	2	1	2	
Run 4	2	2	1	

Interaction Table

Column 1	2	3
(1)	3	2
	(2)	1
		(3)

Figure 19.1 Array L4 (See Appendix 19A (From "Orthogonal Graphs and Linear Arrays" by G. Taguchi and S. Konishi. Reproduced by permission of the American Supplier Institute, Dearborn, Michigan.

This L4 array prescribes that four experiments (labelled Run 1, Run 2, etc.) should be conducted, and it shows whether the level 1 or level 2 values of the variable factors (when assigned to the columns) should be used in the experiments. For example, in Run 1, the value of all three variable factors should be set to their Level 1; in Run 4, the value of the variable factors in columns 1 and 2 should be set to their Level 2 whereas the variable in column 3 should be set to its Level 1. In our heat treating example, level 1 for temperature is 300 F; level 2 is 500 F.

The information we will be able to get from the results of the prescribed set of experiments will tell us about the relative *strength of the influence* that each of the variable factors assigned to the columns — temperature and time in our example — has on the Result. Making the assignments of variable factors to the columns properly is a critical (and sometimes tricky) task. In the extrusion example, we have two variable factors (A - temperature, and B - time) each which can be assigned to either column 1, 2, or 3.

19.3.2 Interactions

In making the assignment of variable factors to the columns, engineers must consider the possibility (actually, likelihood) that the effects of the variable factors on the Result are not always independent of each other. For example, the influence of heat treating time on the Result (strength) may be different at different oven temperatures. (Actually, this is the case.) When variable factors are interdependent, they are said to *interact.*

To help detect interactions, and determine their effects, the arrays are designed so that some of the columns measure the influence of the interactions between other variable factors assigned to other columns. In array L4, for example, column 3 detects the interaction between the variable factors in columns 1 and 2.

To tell us about how the arrays detect interactions, there is associated with each array in Appendix 19A a table showing its interactions. Note the interaction table below array L4 in Figure 19.1. This interaction table is interpreted as follows: Starting with (1) as it appears in the table — here it refers to column 1 in the array — and reading across to column 2, we find that column 3 measures the interaction of columns 1 and 2.

In a similar way, starting with (1) and reading across to column 3, we find that column 2 measures the interaction of columns 1 and 3. And, starting with (2) and reading across to column 3, we find that column 1 measures the interaction of columns 2 and 3.

19.3.3 Assigning the Variable Factors to Columns.

In assigning variable factors to columns in the extrusion heat treating example, we can assign temperature and time to any of the columns we wish, but then the remaining column will measure the interaction effect between temperature and time. That is, the unassigned column will give us information about the strength of the interaction between temperature and time on the Result.

Though as just noted, we *could* assign variable factors in the example arbitrarily to *any* of the three columns, it is best to go about the assignment of variable factors in a consistent way. This will be most helpful in larger, more complex situations. Here is a way recommended in the excellent manual by J. G. Elliott [2]:

1. Write the variable factors (which we will now designate as A, B, C, etc.) along a horizontal line in an order such that the variable factors whose interactions you want to study are next to one another. In our example, with only two variable factors (temperature A, and time B) that is easy:

```
        (AB)
A -------------------------- B
```

where (AB) designates an interaction between A and B. In more complex cases, getting the variable factors arranged so that those with interactions of interest are adjacent to one another may take a little trial and error. There are examples later in the Chapter.

2. Assign the lowest numbered column that is not yet assigned (that will be column 1 when you begin, of course) to the leftmost unassigned variable factor in the diagram above. We show the column assignment below the variable factor in the diagram. That is, assign column 1 to variable factor A:

```
        (AB)
A --------------------------- B
1
```

3. Assign the next lowest numbered column not yet assigned (that will be column 2 when you begin) to the next variable factor in the diagram. That is, assign column 2 to variable factor B:

```
        (AB)
A ----------------------- B
1                         2
```

4. Now use the Interaction Table to determine the lowest numbered column that represents the interaction of the two variable factors just assigned. In our example that is column 3:

```
        (AB)
A ------------------------- B
1           3               2
```

5. Go back to step 2 and repeat until all column assignments are made.

Later in the chapter, we will illustrate this procedure in more complex situations.

To continue with the extrusion heat treating example, we can make the following assignments for the three factor variables:

Column 1	Factor A	Oven Temp
Column 2	Factor B	Oven Time
Column 3	Interaction of A and B	Interaction of Temp and Time

With this done, we next perform the experiments prescribed by the array. The Results assumed in our extrusion heat treating example are shown in Figure 19.2. The 1's and 2's in column 3 are the levels designated in the L4 array for that column; they are used as described below to determine the strength of the AB interaction.

	Columns 1 (A)	2 (B)	3 (AB)	Result Strength psi
Run 1	300 F	3 hrs	1	6,000
Run 2	300 F	5 hrs	2	7,000
Run 3	500 F	3 hrs	2	12,000
Run 4	500 F	5 hrs	1	10,000

Figure 19.2

Results of Extrusion Heat Treating Experiments

19.3.4 Analyzing the Experimental Results.

There are several methods of analyzing the data obtained from the sets of experiments to get the information we want. Two of these methods are informal methods that give approximate but often useful information. The third — using a method called Analysis of Variations (ANOVA) is more formal and more precise — and more work. In this book we will present only the two informal methods, but ANOVA is not difficult to learn, and design students should definitely study it either in a course on statistics or on their own.

The first informal analysis method involves using a simple, very rough computation to estimate the effect of each of the variable factors, including the interactions. This is done as follows.

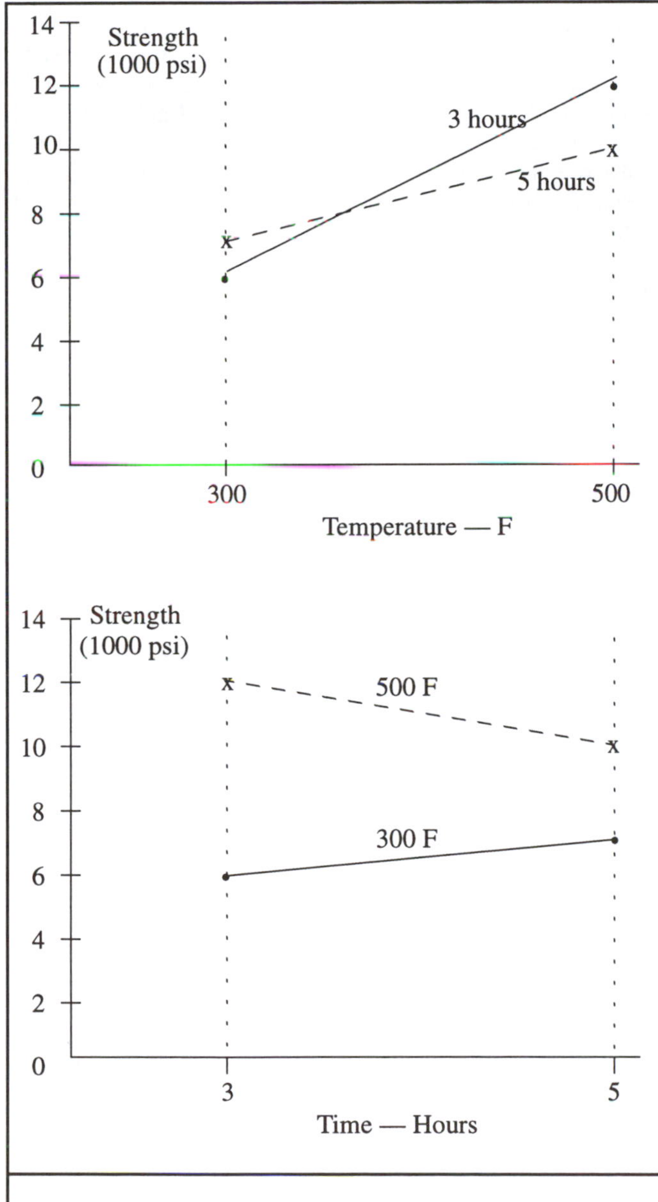

Figure 19.3 Illustrating the Interactions in the Heat Treating Example.

Effect of Factor A = (Average Result For A at Setting 2)
MINUS
(Average Result for A at Setting 1)

For the Results shown in Figure 19.2, the Average Result for Variable A (temperature) at Setting 2 (500F) is:

(12000 + 10000) / 2 = 11000 psi.

Similar expressions can be written for the other factors. In the example, therefore, the effects are:

Effect of Temp = (Average of Level 2 Values)
MINUS
(Average of Level 1 Values)

= (11000) - (65000) = 4500 psi

Effect of Time = (Average of Level 2 Values)
MINUS
(Average of Level 1 Values)

= (8500) - (9000) = - 500 psi

Effect of Temp-Time Interaction =
(Average of Level 2 Value)
MINUS
(Average of Level 1 Values)

= 9500 - 8000 = 1500 psi

We can conclude from this approximate analysis that — for determining extrusion strength after heat treatment — oven temperature is the most critical variable, that oven time is not very important (at least in the 3 to 5 hour range), and that the interaction between temperature and time is somewhat important.

Another informal method of analyzing the experimental results is by constucting graphs as shown in Figure 19.3. In the first of these graphs, *because the lines have different slopes,* existence of an interaction between time and temperature is revealed. In particular, in this example, the crossing lines show that a longer time is more desirable at the lower temperature, but a shorter time is more desirable at the higher temperature.

That there is interaction between temperature and time is also revealed in the second graph in Figure 19.3. The lines there slope in opposite directions, indicating some interaction effect.

Summary. The preceding example was extremely simple; it involved only two variable factors, and we assigned only two values to each of them. Its purpose was only to introduce the main concepts and methods of design of experiments. We continue in the next section, but add in a little more complexity.

19.4 A More Complex Heat Treating Example

From an engineering standpoint, the results obtained in the extrusion heat treating example above are not completely satisfying. Though the best strength was obtained at 500 F and a 3 hour oven time, there may well be an intermediate time and temperature combination that will be better. (There is.) Moreover, to make the example simple, we left out a third variable factor that might be important: it is the different vendors supplying the raw material for the extrusion process. The composition of the material from different vendors can vary, and so can the composition and granularity of the raw material supplied. Both these factors might well influence the heat treatment results. In this section, therefore, we will add the vendor as a third variable factor, but still keep just two levels for each factor. In a subsequent Section, we will extend the example to include three levels for the factors instead of just two.

Problem Formulation. Reformulating our problem including the vendor as a variable factor (and their two levels), we now have:

Factor A: Oven Temperature (300 and 500 F)
Factor B: Oven Time (3 and 5 hours)
Factor C: Supplier (Smith and Jones)

Now we face again the problem of selecting an array. Remember that column 3 measures the interaction of columns 1 and 2. If we use the L4 array and assign the supplier variable to column 3, then the effect that we will get in column 3 will be a combination of the effect of the suppliers *and* the effect of the interaction between temperature and time. We could tolerate this overlap if we were sure that the interaction is small, but we don't know for sure that it is small. In such a case, therefore, we cannot very well use the L4 array.

As noted previously, in Appendix 19A of this chapter, there are a selected number of orthogonal arrays presented taken from "Orthogonal Arrays and Linear Graphs" by G. Taguchi and S. Konishi [12] reproduced by permission of the American Supplier Institute, Dearborn, MI. Looking through these arrays, we obviously want the smallest one that will meet our needs. (The smaller the array, the fewer experiments are needed.) We have decided to stick for now with using two levels of values for the variable factors, so those that are designed for three levels (1, 2, and 3) are not useful. The array to try, then, is L8, as shown in the Appendix 9A and also in Figure 19.4.

Assigning Variables to Columns. To assign variable factors to the columns, we use the process described in the

	1	2	3	4	5	6	7
Run 1	1	1	1	1	1	1	1
Run 2	1	1	1	2	2	2	2
Run 3	1	2	2	1	1	2	2
Run 4	1	2	2	2	2	1	1
Run 5	2	1	2	1	2	1	2
Run 6	2	1	2	2	1	2	1
Run 7	2	2	1	1	2	2	1
Run 8	2	2	1	2	1	1	2

1	2	3	4	5	6	7
(1)	3	2	5	4	7	6
	(2)	1	6	7	4	5
		(3)	7	6	5	4
			(4)	1	2	3
				(5)	3	2
					(6)	1
						(7)

L8
Interactions

Figure 19.4

L8 Array

From "Orthogonal Graphs and Linear Arrays" by G.Taguchi and S. Konishi. Reproduced by permission of the American Supplier Institute, Dearborn, Michigan.

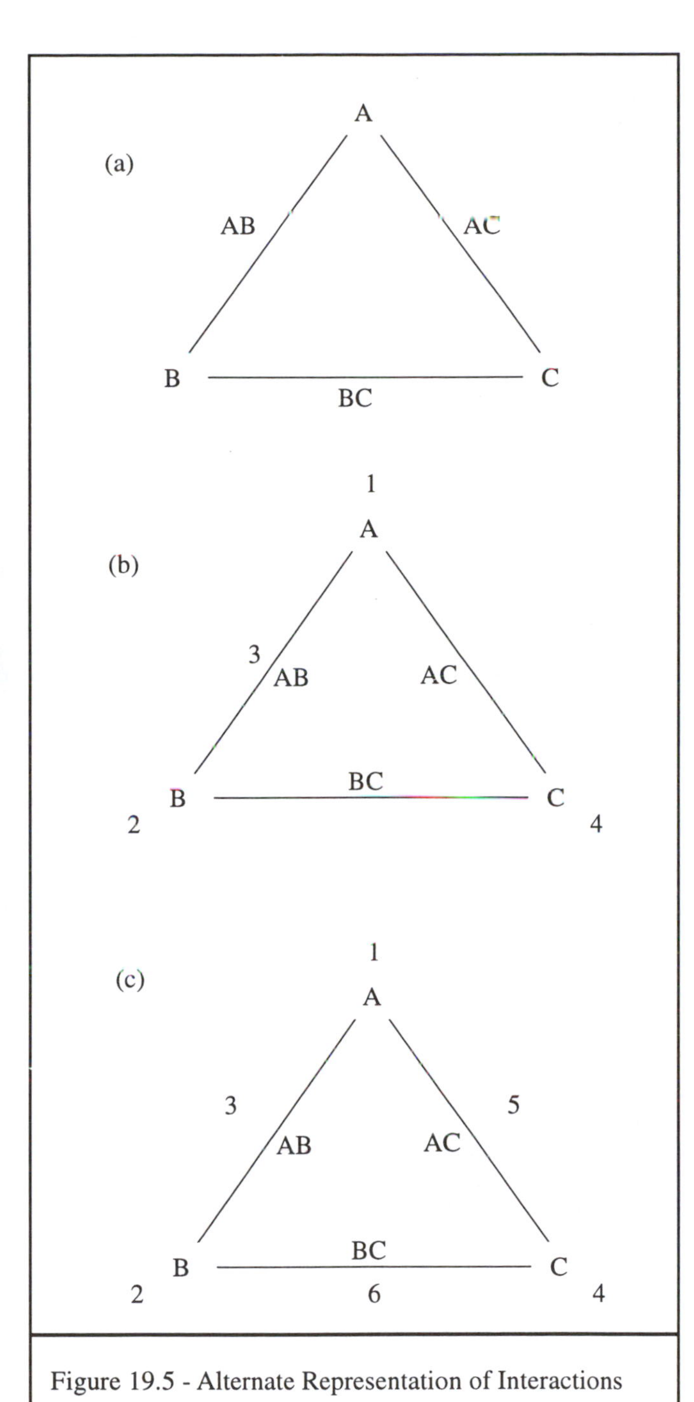

Figure 19.5 - Alternate Representation of Interactions

preceding sub-section. That is, we lay out our variables along a horizontal line, seeing to it that those whose interactions we want to study are next to one another. Where there is an interaction of interest, we connect the variable factors symbols (i.e., A and B in our example) with lines. Otherwise, we just list them. In this case:

```
          (AB)
A --------------------- B ---------------------- C
```

Now we assign factor A to column 1 and factor B to column 2:

```
          (AB)
A -------------------- B -------------------- C
1                      2
```

Next we look in the table of interactions for L8 and find that (same as for L4), the interaction of columns 1 and 2 is in column 3. Thus:

```
          (AB)
A -------------------- B-------------------- C
1         3            2                    4
```

Next we assign factor C to column 4. In this case, where we are not interested in other interactions, we are done with the assignment of variables to columns.

We can, if we wish, simply ignore the rest of the array. It is instructive however, to consider the case where we are interested in other interactions as well. Let's say we are interested in all the interaction effects; that is, in AB, AC, and BC as well as, of course, the individual effects of A, B, and C. Then we could draw diagrams like the ones shown in Figure 19.5 to help us assign variable factors to the columns:

Referring to Figure 19.5, we begin the assignment of variable factors to columns (as always) by assigning column 1 to factor A and column 2 to factor B. Then the interaction table tells us that column 3 will represent the interaction AB.

From the L8 interaction table, we can also now determine that the interaction of columns 2 and 4 is column 6, and the interaction of columns 1 and 4 is column 5.

Column 7 in array L8 is unassigned in the above formulation, but it is not useless. It will reflect a third order interaction of A, B, and C if there is one. Assuming there is not such an interaction, or that it is very weak, we can think of it as a kind of "spy" or "wildcard" column. It will indicate to us the strength of the effect of anything we may have forgotten; thus, its effect should not be large if we have accounted for all the important variables. If column 7 turns out to have a large influence, then we had better look for some factor we have ignored, and reformulate the problem.

Interpreting the Experimental Results. Now we can run the experiments prescribed by the array and again analyze the effects.

Suppose the experimental results are as shown in Figure 19.6 (See page 19-22). The computation of the various effects as the differences of the average results for levels 1 and 2 gives the following:

Effect of A (Temp) = (Avg of Level 2 Values)
MINUS
(Avg of Level 1 Values)
= 12000 -7000 = 5000 psi

COLUMNS

	1	2	3	4
Run 1	1	1	1	1
Run 2	1	2	2	2
Run 3	1	3	3	3
Run 4	2	1	2	3
Run 5	2	2	3	1
Run 6	2	3	1	2
Run 7	3	1	3	2
Run 8	3	2	1	3
Run 9	3	3	2	1

1 —— 3, 4 —— 2

Figure 19.7 - L9 Array

From "Orthogonal Graphs and Linear Arrays" by G.Taguchi and S. Konishi. Reproduced by permission of the American Supplier Institute, Dearborn, Michigan.

Effect of B (Time) = 9250 - 9750 = - 500 psi
Effect of AB = 10250 - 8750 = 1500 psi
Effect of C (Supplier) = 8750 - 10250 = - 1500 psi
Effect of AC = 9750 - 9250 = 500 psi
Effect of BC = 9500 - 9500 = 0
Effect of Column 7 = 9500 - 9500 = 0.

Since the effect of column 7 is zero, we can usually conclude that no crucial variable factors or interactions have been ignored. Also, since the effects of AC (column 5) and BC (column 6) are small, these are apparently not important interactions. Temperature still has the most important influence, and there is still an interaction between temperature and time (column 3) that is of concern, but it is evident that the supplier has a moderate effect also.

19.5 Adding More Complexity

Formulating the Problem. Suppose we now want a still better understanding of the heat treating situation, so we decide to use three levels of values for both temperature and time as follows:

Factor A: Temperature (300, 400, and 500 F)

Factor B: Time (3, 4, and 5 hours)

Factor C: Supplier (Smith and Jones)

To design an experiment now requires that we use one of the arrays that allows for three levels of values for the variable factors. These are L9, L18, and L27 in Appendix C. L9 is appealing because it requires nine rather than 27 experiments, so we try to adapt it first. Figure 19.7 shows L9.

Based on the above results, we are no longer concerned about the strength of temperature-supplier (A-C) and time-supplier (B-C) interactions, so what we would like to have is a plan that gives us the following:

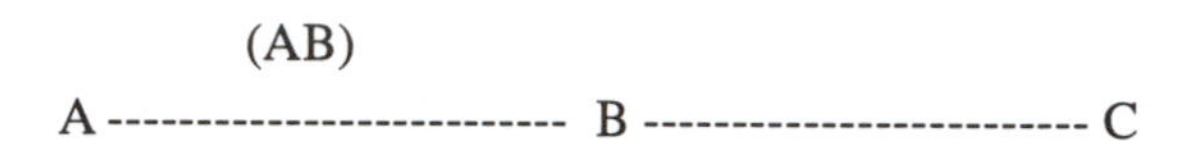

L9 presents two problems when we assign variable factors to the columns. The interaction table for L9 indicates that columns 3 and 4 both involve interactions of columns 1 and 2. That is, if we assign variables to columns in the usual way, we would get:

(AB)
A ------------------------- B ------------------------- C
1 (3, 4) 2

There are no columns available for variable factor, C (supplier). What we can do, however, is assign factor B to column 3, and the supplier to column 2 (and just ignore column 4). Now the plan is this:

A ------------------------- C ------------------------- B
1 2 3

We will still get the interaction AC in columns 3 and 4, but since we know from the previous study that the interaction between A and C is negligible, it will not much influence the effect we find in columns 3 and 4. Thus whatever effect appears in column 3 should be essentially due to oven time.

Though AC and BC interactions will be present in columns 1 and 2, respectively, we have previously determined, as noted above, that these interactions are small. However, a disadvantage of our plan is that we have lost a way to examine the interactions of A and B. We could go to L27 in order to include this, but it is probably better to see first what results from using L9, since L27 requires many

	1 A	2 C	3 B	4	Results psi
Run 1	300 F	Smith	3 hrs	1	7,000
Run 2	300 F	Smith	4 hrs	2	7,500
Run 3	300 F	Jones	5 hrs	3	7,000
Run 4	400 F	Smith	4 hrs	3	20,000
Run 5	400 F	Smith	5 hrs	1	18,000
Run 6	400 F	Jones	3 hrs	2	13,000
Run 7	500 F	Smith	5 hrs	2	12,000
Run 8	500 F	Smith	3 hrs	3	14,000
Run 9	500 F	Jones	4 hrs	1	13,000

Figure 19.8

Experimental Results

more experiments. We will just have to live with this imperfection until we can see the results and determine if going to an L27 array is, in fact, needed.

(It may be noted that the design of the experiments requires the use of qualitative physical reasoning and logic based on pervious tests. Keeping the number of experiments to a minimum is important, and this goal often interferes with establishing a plan that thoroughly and exhaustively, with great certainty, explores all possibilities. Judgment is required. Fortunately, it is usually possible to any detect serious errors after the tests have been run.)

The other problem we have now in this example is that L9 assumes that all the variable factors have three levels. But we have only two suppliers. We handle this by using one of the suppliers twice. That is, we set up the three "levels" for factor C as:

Factor C: Supplier

(Level 1 - Smith

Level 2 - Smith

Level 3 - Jones)

We therefore set up our experimental design as shown in Figure 19.8. The Results of the prescribed set of nine experiments are also shown there.

Interpreting the Results. When there are three levels for the variable factors, we cannot compute the effects as directly and as simply as we did when there were two levels. However, graphs can be used for making an informal analysis, often to great advantage. Figure 19.9 shows graphs made from the information in Figure 19.8.

We can see clearly in Figure 19.9(a) that temperature is crucial, and should be set at about 400 F. We can also see that the time should be about four hours, and that it doesn't matter much which supplier is used if we get the time and temperature right, though slightly better results are obtained from supplier Smith.

19.6 Application to Product and Process Design

19.6.1 Problem Formulation

With the preceding background, we can now move on to apply some of the ideas to the design of products and processes. Here is how a parametric design problem is formulated for solution by Taguchi's method:

1. Define the *design variables* and feasible ranges or, if they are non-numeric, alternatives for their values.

(Note: Sometimes the design variables are referred to as *control factors.*)

2. *Select an array for the design variables or Control Factors.* This will involve deciding whether to use two or three levels, and how thoroughly to avoid possible confusion in the columns due to interactions. The larger the array, the more the experimental cost.

We will refer to the number of runs specified by the selected array as N.

3. *Define the Noise Factors* and feasible ranges or discrete alternatives for their values. The noise factors are factors with the unavoidable variations in the environment or manufacturing or service whose effects on performance we want to minimize in order to create a robust product or process.

It is obvious that the more there are of these variables to be considered the more experiments will be required --

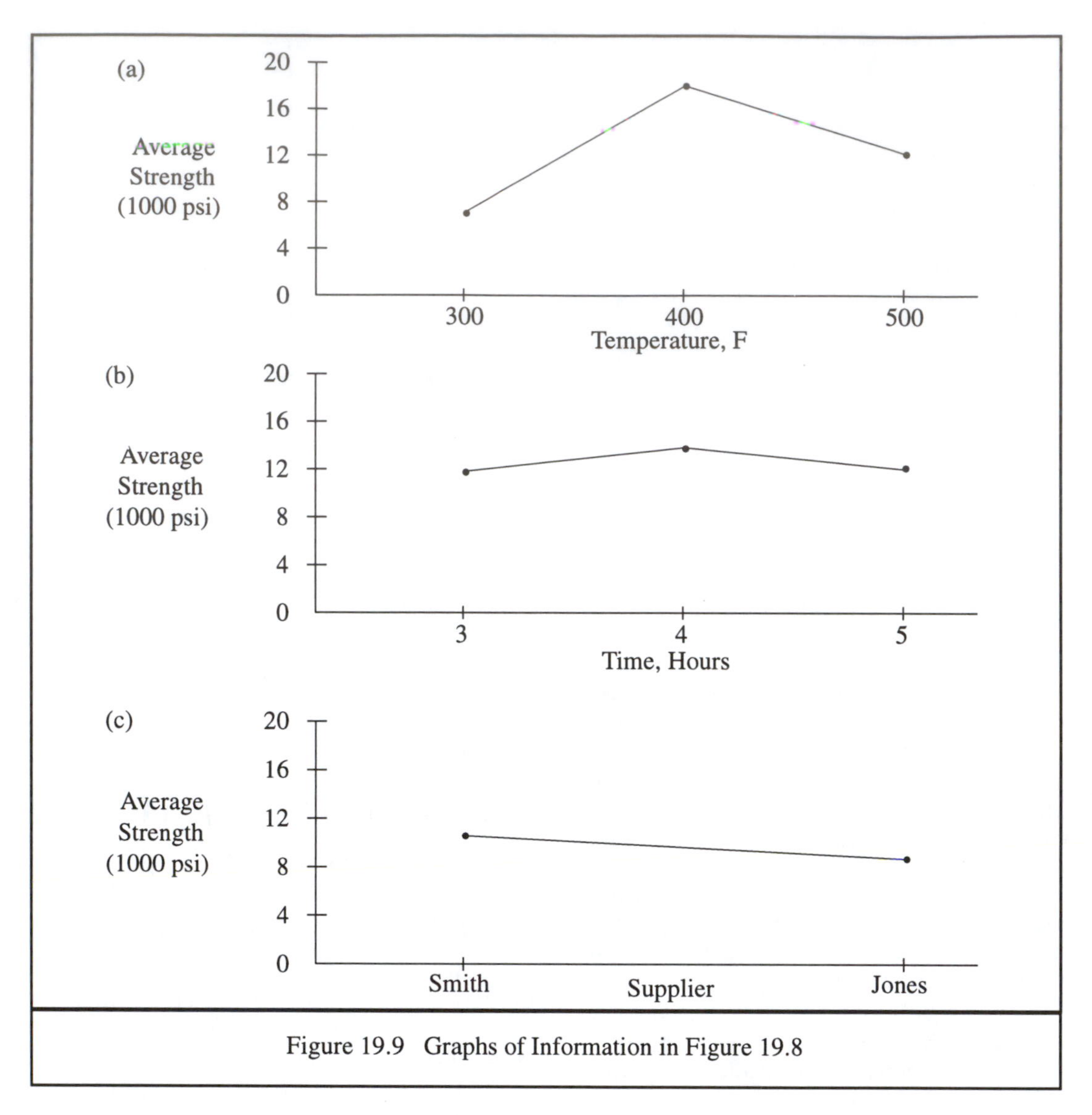

Figure 19.9 Graphs of Information in Figure 19.8

and more experiments means more time and expense. Sometimes two or more factors can be combined into a single combination variable, thus reducing the number of experiments required. (See reference [3], pp 378-382).

4. *Select an array for the Noise Factors* Again, this involves deciding on the number of levels (which need not be the same as in Step 2.)

We will refer to the number of runs specified by the selected array as M.

5.*Specify the performance criterion or Result* desired. Do not include robustness here, just performance.

6. *Arrange the two arrays* selected around an M column by N row matrix as shown generally in Figure 19.10.

Figure 19.11 shows the specific set-up where the Control Factor array is an L9 and the Noise Factor array is an L8. Note that the Noise Factor array is placed on its side along the top of the matrix in the Figures.

This completes the problem formulation. Now the laboratory and/or computation work begins.

19.6.2 Analyzing and Interpreting the Results

7. Perform the N times M experiments or simulations specified by Figure 19.10. For example, Experiment 2-3 involves setting the Control Factors to the values specified in Run 2 of its array, and setting the Noise Factors to the values specified in Run 3 of the its array. (Note: The order in which experiments are performed is important if possible bias is to be eliminated. We don't discuss this issue here.)

8. In the M x N experimental Results matrix, record the results for Result [R(1,1) through R(N,M)] in each experiment. Figure 19.12 shows results in a hypothetical problem used as an illustration here.

9. Now compute and record in the far right column of the Results matrix the value of the "signal-to-noise" ratio. This ratio is a function of the ratio of the mean performance

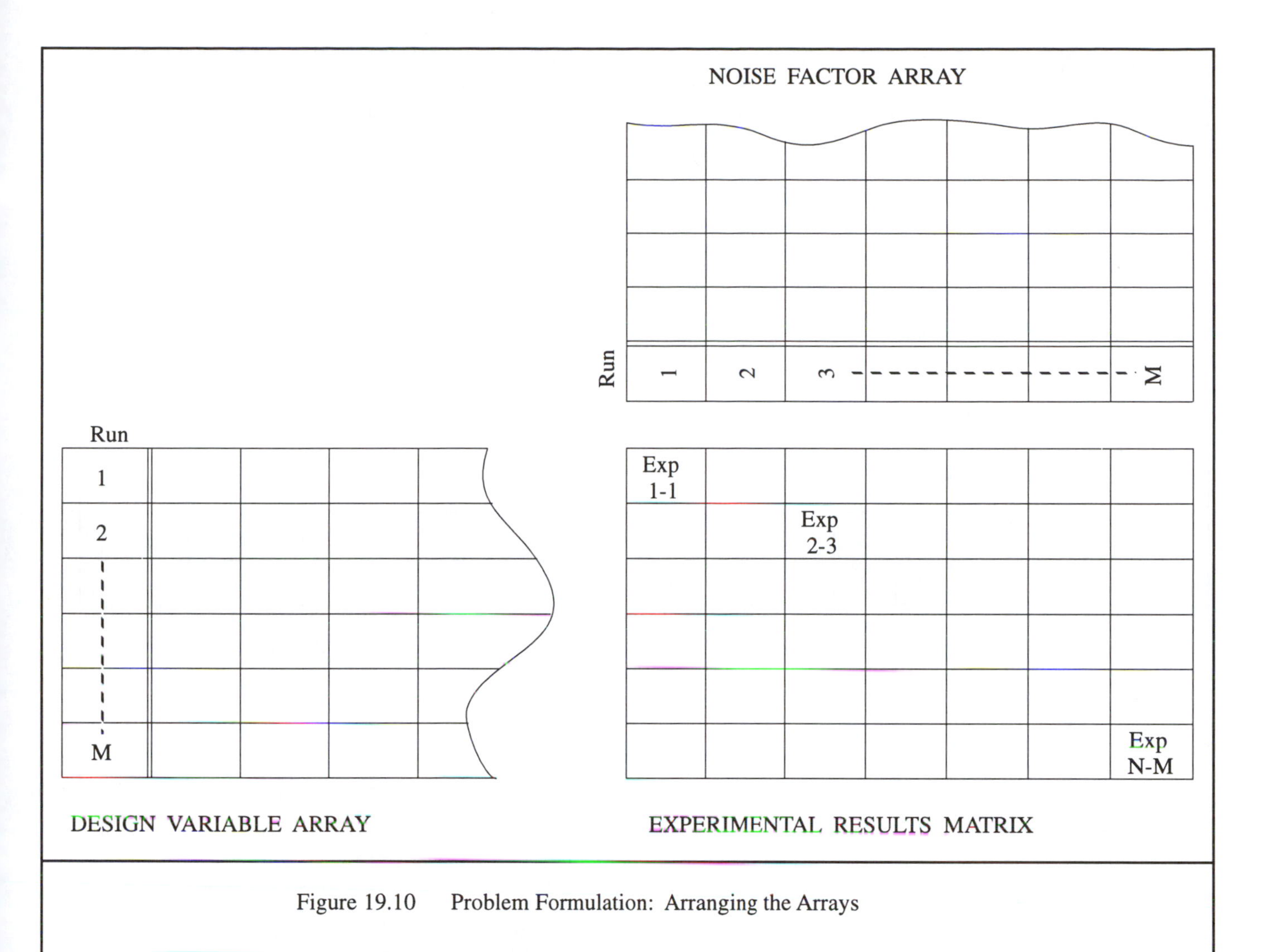

Figure 19.10 Problem Formulation: Arranging the Arrays

value in each row to its standard deviation. In other words, it is a function of the ratio of performance to variation of performance — that is, to robustness.

The method for computing the performance to robustness ratio (or the signal to noise ratio) depends on whether the performance criterion should be (a) as close to some nominal value as possible — like load speed in a v-belt drive system; (b) as small as possible — like cost; or (c) as large as possible — like perhaps the efficiency of a machine. The equations to use in each case are [2]:

(a) When nominal is best:

$$S/N = 10 \cdot \log\left(\frac{y_m^2}{s^2}\right)$$

where the logarithm is to the base 10, and:

$$s^2 = \frac{1}{n-1} \cdot \sum_{i=1}^{n} \langle y_i - y_m \rangle^2$$

y_i = i th value of the Result

y_m = average value of the Result

(b) When smaller is better:

$$S/N = -10 \cdot \log\left(\frac{1}{n} \cdot \sum_{i=1}^{n} y_i^2\right)$$

NOISE FACTOR ARRAY

	1	2	2	1	2	1	1	2
	1	2	2	1	1	2	2	1
	1	2	1	2	2	1	2	1
	1	2	1	2	1	2	1	2
	1	1	2	2	2	2	1	1
	1	1	2	2	1	1	2	2
	1	1	1	1	2	2	2	2
Run Number	1	2	3	4	5	6	7	8

DESIGN VARIABLE ARRAY

Run	A	B	C	D								
1	1	1	1	1								
2	1	2	2	2								
3	1	3	3	3								
4	2	1	2	3								
5	2	2	3	1								
6	2	3	1	2								
7	3	1	3	2								
8	3	2	1	3								
9	3	3	2	1								

Figure 19.11 An L8 Noise Factor Array Used With an L9 Design Variable Array

(c) Larger is Better:

$$S/N = -10 \cdot \log\left(\frac{1}{n} \cdot \sum_{i=1}^{n} \langle 1/y_i \rangle^2\right)$$

The logarithms in the above equations are to the base 10.

Note that, in the Taguchi approach, the signal-to-noise ratio is the function that is maximized to indicate the best balance of performance and robustness. As we noted before, this is an arbitrary measure, but it is reasonable, easy to compute, and obviously works well in practice.

10. Now, using the experimental results, we can determine the best values for each of the design variables. (Note that we need not, and indeed usually will not, set them at the values specified in any one of the N runs specified by the design variable array.) We will show in detail how to do this in the context of an example below.

19.7 V-Belt Drive Example

Readers may want to refer back to Figure 17.18 to review the v-belt drive configuration.

The numbers used in the following example are fictitious, useful only to illustrate the process in a domain of familiarity to readers.

Following the steps outlined above:

1. We suppose that the parameters for a v-belt drive system are to be selected. The design variables are the belt type, the number of belts, the drive pulley diameter, and the center distance between the pulleys. (We omit the load pulley diameter as a design variable because its value is essentially determined by the drive diameter, drive speed, and the desired load speed.) Since there are three common belt types (3v, 5v, and 8v), and since we want to explore a fairly wide range of the other design variables, we will use three levels

(a) NOISE FACTOR ARRAY

		1	2	2	1	2	1	1	2
	BC	1	2	2	1	1	2	2	1
	AC	1	2	1	2	2	1	2	1
Oil	C	1	2	1	2	1	2	1	2
	AB	1	1	2	2	2	2	1	1
Humidity	B	1	1	2	2	1	1	2	2
Temperature	A	1	1	1	1	2	2	2	2
Run Number		1	2	3	4	5	6	7	8

(b) DESIGN VARIABLE ARRAY

Run	A	B	C	D	Results								(S/N)
1	1	1	1	1	100	50	80	45	80	25	70	25	32.3
2	1	2	2	2	150	75	110	70	120	40	110	35	35.8
3	1	3	3	3	175	90	140	80	150	45	130	40	37.0
4	2	1	2	3	125	60	100	55	100	30	80	30	33.9
5	2	2	3	1	200	100	160	90	160	50	140	45	38.0
6	2	3	1	2	225	120	170	110	180	60	160	50	39.2
7	3	1	3	2	150	75	110	65	130	35	110	30	34.9
8	3	2	1	3	175	80	150	70	145	40	125	35	36.0
9	3	3	2	1	150	80	150	70	140	30	120	30	34.5

Figure 19.12 An Example of a Noise Factor and a Design Variable Array Completed for a Hypothetical Problem

for design variables:

	Level 1	Level 2	Level 3
A - Belt Type	3v	5v	8v
B - Number of Belts	1	3	5
C - Drive Pulley Diameter (in)	6	8	10
D - Center Distance (in)	20	25	30

2. The array we select for the design variables is L9. Interactions need not concern us in this array since we are not using it to study effects. All we need to do is get an array that will handle at least four design variables with three levels. See Figure 19.11.

3. The noise factors of concern are assumed to be (1) operating temperature which may vary from about 40 F to about 105 F; (2) the humidity which may vary from about 20% to 165%; and (3) the amount of oil in the operating environment, which may vary from almost none to a 'great deal'.

4. The array we select for the noise factors is L8. It takes only two levels, but that should be sufficient for our purposes. It can be set up to tell us about interactions among the noise factors if we wish to study them. See Figure 19.11.

5. The performance criterion we are interested in is the ratio of the life of the belt system to its cost (initial cost plus belt replacement costs). We will express this in hours per dollar.

6. Figure 19.11 shows the two arrays arranged appropriately.

7, 8. We now imagine that the 72 required experiments are performed, and that the results for our performance criterion (hours of life per dollar) are recorded in the matrix as shown in Figure 19.12.

9. To compute the signal to noise ratio (S/N) for each of the nine designs used in the experiments, we note that our performance factor is a "larger is better" type. Thus we use Equation (c) above. The results of the calculations for S/N are shown in Figure 19.12.

10. Note that design of Run number 6 has the best signal to noise ratio (39.2).

In Run 6, design variable A is level 2, variable B is level 3, C is level 1, and D is level 2. Thus this design is: belt type - 5v, the number of belts - 5, the pulley diameter - 6 inches, and the center distance - 25 inches. This does not necessarily mean, however, that the best settings for all the design variables are those used in Run 6. Remember, we

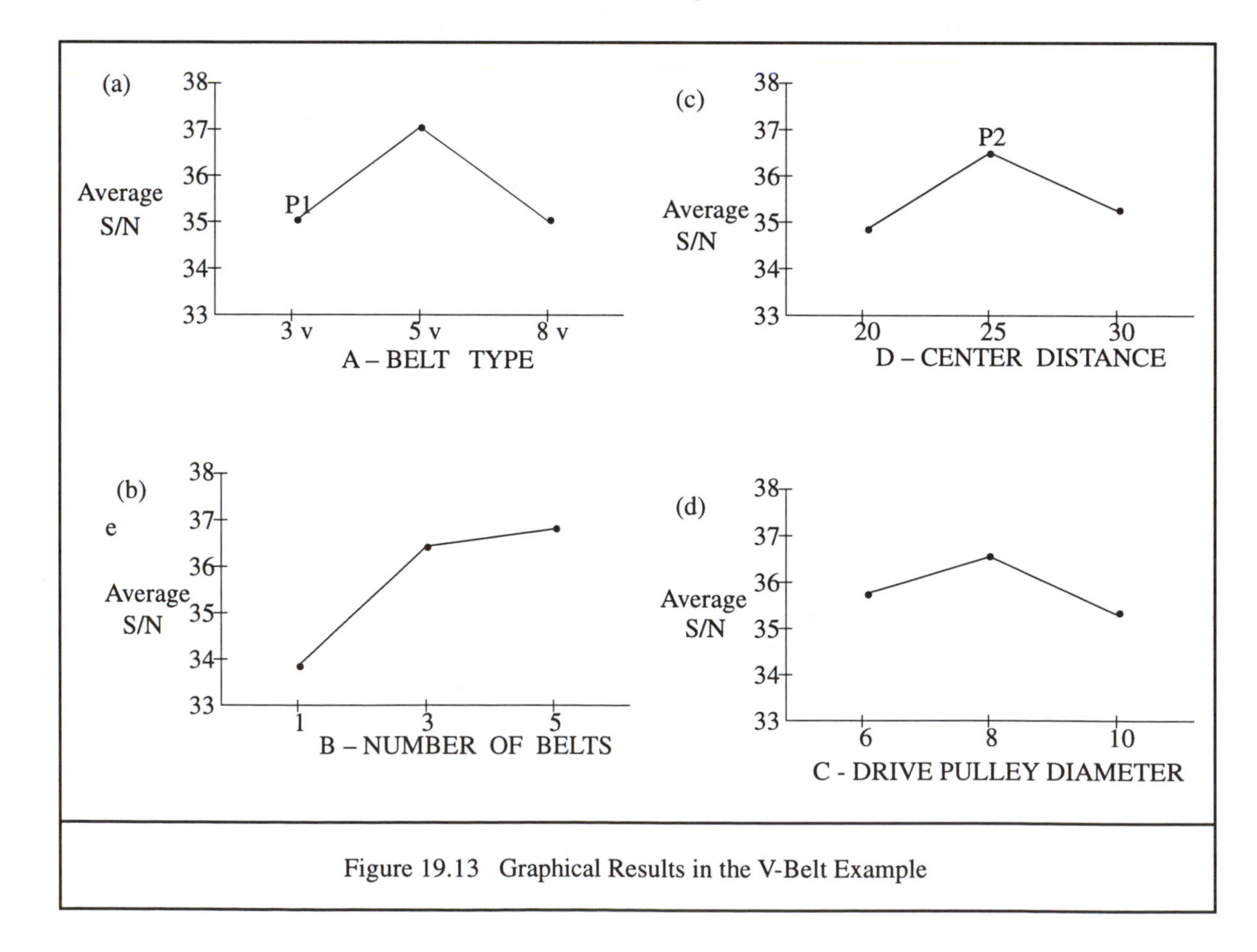

Figure 19.13 Graphical Results in the V-Belt Example

have not tried every possible combination in these experiments. There may well be (and usually are) better combinations, and we can determine them from the results we have.

To do so, it is almost always helpful to make a set of simple graphs in which the S/N ratio is plotted against the values for each design variable separately. Refer to Figure 19.13.

Figure 19.13 (a) shows how belt type influences S/N. The data for the plot are obtained as follows: The belt type is set at level 1 (3v), in the first three rows of experiments. The average of the three S/N results for these runs is the average of 32.3, 35.8, and 37.0, or 35.0. That gives us Point P1 in Figure 19.13 (a).

Students should verify that they understand why and how point P2 in Figure 19.13 (c) is found by averaging 35.8, 39.2, and 34.9 to get 36.6.

By inspecting the set of graphs in Figure 19.13, most engineers will decide that the following design variable settings are the better ones:

Belt Type:	5v
Number of Belts:	3
Pulley Diameter	8 inches
Center Distance	25 inches.

One could argue with the selection for number of belts noting that the S/N is slightly higher for 5 belts than 3 belts. The rationale for the choice made is this: The S/N is only *very* slightly higher, and this is not a precise process. Using three belts will result in lower initial costs, so we tend to favor it.

The design chosen (5v, 3 belts, 8 inch pulley, and 25 inch center distance) is not one of those used in the experiments. We don't really know for certain that it will result in a higher S/N than experimental Run 6. We should, therefore, in a real situation conduct an experiment with the chosen design to confirm that it is indeed better.

It is important to note that the design recommended by this process is a built-in compromise of performance (in this case, expressed as hours of life per dollar of cost) versus its variability of performance (robustness). The 'signal' is a measure of performance; the 'noise' is an inverse measure of robustness. The ratio will be large when the best ratio of relatively high signal to relatively low noise is attained.

Relative Influence of the Factors. We can use the data in Figure 19.12 to estimate the strength of the various noise factors in influencing the performance of the belt system. The process is exactly like we described earlier in this chapter. In Figure 19.14, we have reproduced the noise factor array from Figure 19.12 and put it in its normal orientation. The results shown are those for Run 6. We've used them because they are closest to the optimum design.

Now we can compute the influence of temperature, humidity, etc., on the result. Only absolute values are reported since the sign has no effect on the relative strength of the influences. (Signs can, however, be used in some cases to help choose the better level of a variable to use.)

Effect of Temperature:

[(225 + 120 + 170 + 110) /4] - [(180 + 60 + 160 + 50) /4]

= 44 hours of life/dollar.

Effect of Humidity:	24
Effect of Oil	99
Effect of Temp plus Humidity	9
Effect of Temp plus Oil	16
Effect of Hum plus Oil	14
Other Effects	10

	1 Temp A	2 Hum B	3 AB	4 Oil C	5 AC	6 BC	7 –	Result
Run 1	1	1	1	1	1	1	1	225
Run 2	1	1	1	2	2	2	2	120
Run 3	1	2	2	1	1	2	2	170
Run 4	1	2	2	2	2	1	1	110
Run 5	2	1	2	1	2	1	2	180
Run 6	2	1	2	2	1	2	1	60
Run 7	2	2	1	1	2	2	1	160
Run 8	2	2	1	2	1	1	2	50

Figure 19.14 Results for Design #6 in the V-Belt Example

It is pretty obvious from this analysis that temperature and (especially) oil are the environmental noise factors that most strongly influence the performance of this (remember, imaginary) v-belt system. Humidity is somewhat important. The interaction effects are relatively small.

This kind of information is useful to a designer interested in robustness. Our design (5v belt type, 3 belts, 8 inch pulley, and 25 inch center distance) minimizes the effects of temperature and oil on the performance (measured as hours of life per dollar). It gives us as much robustness as we can get given the temperature and oil variations that are there. However, if there are inexpensive ways to limit the temperature or oil variations, we should do so.

19.8 Tolerances and the Taguchi Loss Function

The above hypothetical example does not present us with a tolerance design issue. The design variables in a v-belt drive system (i.e., belt type, number of belts, etc.) are standard components with tolerances we could not adjust even if we wanted to. In many cases, however, the design variables, whose nominal values are determined in a process just as described above, are made in production processes that do have controllable tolerances. The question then becomes: Are the normal production tolerances suitable for the design, or are tighter tolerances needed?

Tighter tolerances cost money. Thus to answer our question, we must have some way to estimate the costs and benefits of using normal or tighter tolerances. Taguchi approaches this problem by using what he calls a *Loss Function*. Taguchi's Loss Function provides an estimate of the loss that will be incurred as a result of variabilities in production of parts and assemblies. If we then also estimate the cost of providing tighter tolerances, we can make an educated decision about whether, in fact, tighter tolerances are needed.

Consider a case where a design variable (denoted by x) has been assigned a nominal value by the Taguchi parametric design process above. This nominal value balances performance against the effects of noises, but because of variations in the production process, we can only make actual parts with values in a range close to the nominal.

Taguchi defines *Loss* as the long term life cycle cost that results from production of a part or product with a design variable different from the nominal. If a nominal part is produced, the Loss is assumed zero. If other values are produced, the Loss is assumed to increase parabolically as the difference from nominal increases. See Figure 19.15.

By assuming a parabolic loss function (another arbitrary but useful, reasonable assumption by Taguchi), it is possible to obtain an estimate of the cost of producing parts that do not meet the target value. In equation form:

$$\text{Loss} = K\,(T - x)^2$$

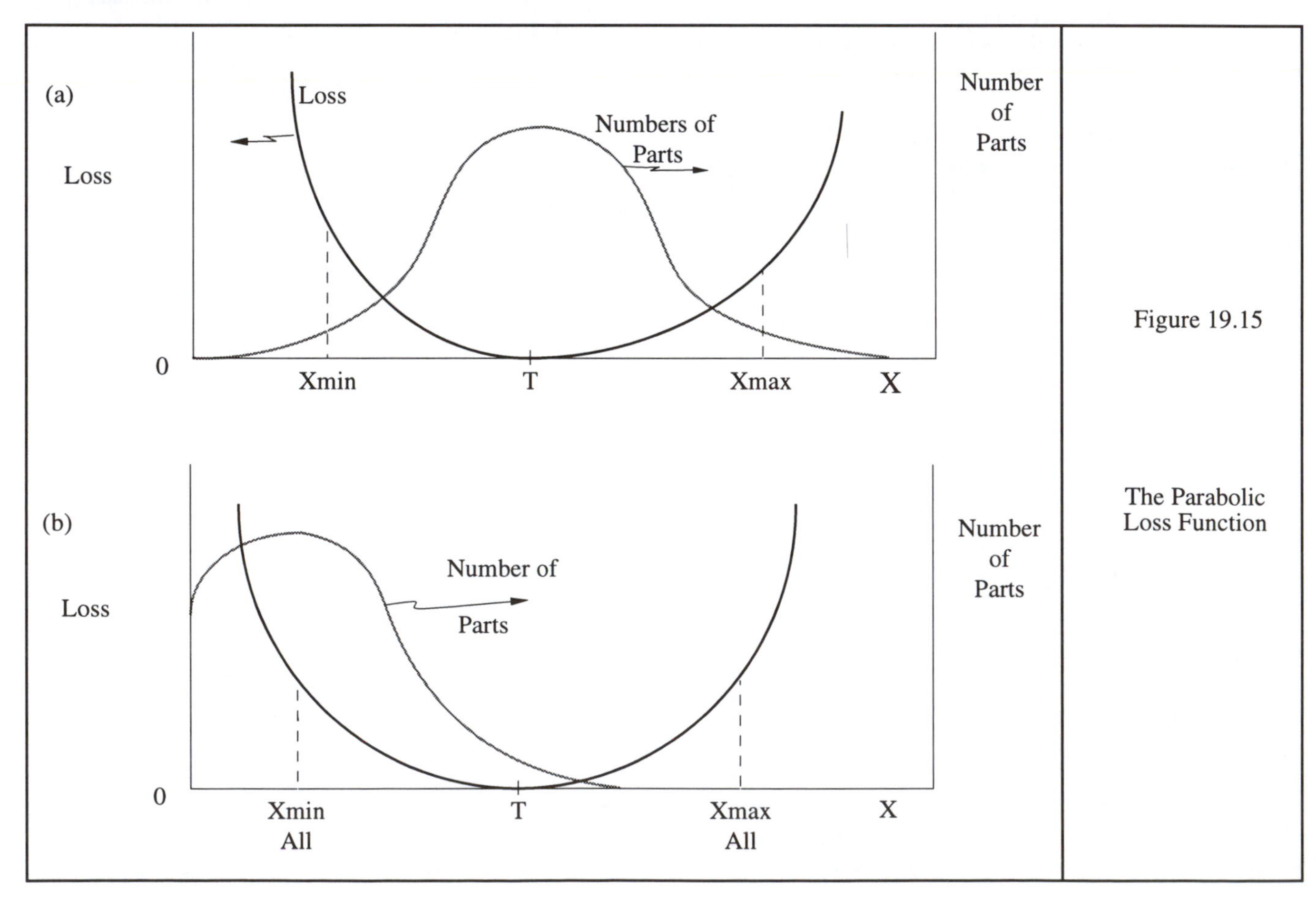

Figure 19.15

The Parabolic Loss Function

where T = Target or Nominal Value of x

K = Constant.

Hopefully, there will be historical data (e.g., on production waste, field repairs, etc.) available to use in the above equation to establish the value for K. If not, estimates must be made. But remember, Loss is to include *all* the life time costs of production waste, repairs, customer dissatisfaction, and so on, caused by variations in the manufacturing process.

Usually the actual values of x for parts or products produced will vary over a range; they will not all be at the target value, x = T. In many firms, it is common to establish an allowed tolerance range and pass all parts between a x-min and a x-max, where x-min and x-max are the smallest and largest values of x considered viable for the part to function satisfactorily. There is nothing wrong with this approach; there need to be limits to sort out parts and products that are clearly unsatisfactory.

However, as illustrated in Figure 19.15, all values of x between x-min and x-max do not produce the same Loss. Those that are close to T produce a small Loss; those out nearer to x-min or x-max produce a larger Loss. Treating all values between x-min and x-max as if they are equally good is incorrect, and does not provide any information about the quality of the set of parts and products being produced.

Though it certainly is not always the case, we can get a much better idea of how quality is affected by variations in x by considering that the values of x actually produced will vary in a normal distribution around T as shown in Figure 19.16 (a). It is beyond our scope here to analyze the situations where the mean of the distribution produced is not T, or where the distribution is not a normal one. Readers are referred to books on statistics and statistical quality control [13]. Notice, however, that the effect on Loss (i.e., quality) will be significant if the mean value of x is considerably off target. Note that in Figure 19.15(a), many parts are produced near the target where the Loss is low. However, in Figure 19.15(b), where there is a distribution of parts with a mean value somewhat less than the target, many parts are produced with a much larger Loss.

the case where the distribution of part dimensions produced is normal with a mean of T, it can be shown that the total Loss is proportional to both K and to the square of the standard deviation (SD) of the distribution of x in the produced parts or products. That is:

$$\text{Total Loss} \quad K \text{ x } (SD)^2$$

There is little designers and manufacturers can do about K (it reflects mostly the responses of customers to poor products, and the costs of repairs and production waste), so the way to control Loss (which is a kind of measure of manufactured quality) is to control the SD of the parts produced. This is a task for both designers and manufacturers. The role of designers, working with manufacturers, is to establish an SD that appropriately trades off Loss and cost; the role of manufacturers is to provide relevant information to designers, and then to produce parts within the established SD.

The above discussion sets the stage and provides the context for tolerance design. The goal of tolerance design is to set tolerances as required to meet a desired quality-cost trade-off. In the next section, we describe how Taguchi recommends doing it.

19.9 Tolerance Design

19.9.1 Introduction

In establishing nominal parameter values by the Taguchi Method as outlined above, the implicit assumption is that standard manufacturing tolerances and materials will be used. If there is any concern at all that these may not be adequate, then the issue must be studied after the nominal values are established from the optimum signal to noise ratio. The way to do the necessary study is to set up another experimental design to determine the effect on performance of anticipated variations in production. In other words, we check to see if in fact the standard tolerances and material property variations are sufficiently good for our purposes.

19.9.2 Example - Tolerance Design

To illustrate how this done, we use another example problem. Let us imagine that a heat treated metal part has been designed to carry a fluctuating load, and that the critical material properties and design variables are:

yield stress, S

modulus of elasticity, E

thickness, t

a crucial inside radius, r.

We further imagine that parametric design has been completed selecting a material with these nominal material properties:

yield stress: S = 24000 psi

modulus of elasticity: E = 10,000 psi

and that the values established for the design variables are:

thickness: t = 0.120 inches

radius: r = 0.020 inches

Now the heat treating and manufacturing processes cannot produce these values exactly, of course, but will produce them only within a range. That is, there are standard tolerances associated with the nominal values determined by parametric design. Suppose that the following ranges are standard:

A yield stress, S:	22000 to 26000 psi
B modulus, E:	9900 to 10100 psi
C thickness, t:	0.115 to 0.125 inches
D radius, r:	0.018 to 0.022 inches

We can now set up an experiment to determine the extent to which the part's performance will be affected by these standard variations. If performance remains satisfactory over these ranges, then nothing need be done. If performance is influenced beyond satisfactory, however, then one or more of the tolerances will have to be tightened — of course, at a cost.

We will want to use a three-valued array for this case, and L9 will do for the four factors. The levels we choose are:

	Level 1	Level 2	Level 3
A	22,000	24,000	26,000
B	9,900	10,000	10,100
C	0.115	0.120	0.125
D	0.018	0.020	0.022

The performance criterion we are concerned about in this example problem is the life of the part under its fluctuating load. Anything less than 2300 hours is to be considered unsatisfactory.

In this example, we would now perform the nine experiments dictated by the L9 array, and determine the life in each experimental set-up. The hypothetical results are shown in Figure 19.16 where we see that, indeed, at least one of the results is less than the required 2300 hours, and two or three other cases are dangerously close. We can conclude that the standard tolerances are not going to be satisfactory. But which one (or ones) to tighten?

The answer to this must be determined in consultation with manufacturing people. First, however, there is more we can learn from the experimental data. The graphs shown in Figure 19.17 are made directly from the data. They reveal clearly that the radius is the most sensitive issue, and that yield stress and thickness have a moderate effect. Apparently the effect of variations in the modulus are negligible. We would thus want to explore the manufacturing costs of tightening the radius tolerance, especially on the low side. If that tolerance could be made 0 to + 0.002, no doubt the performance variations would no longer be a problem. However, should that tighter tolerance turn out to be too expensive, it might be possible to achieve the desired results with small changes in the allowed tolerances in one or more of the radius, yield stress, and/or thickness.

In some cases, it may be necessary to run another set of experiments to verify that the new set of assigned tolerances in fact produces the desired results.

19.10 Reducing the Cost of Experimentation

Designing for robustness is extremely important, but experimentation can be very expensive of both money and time. It is therefore worth the effort to find ways to achieve the desired information with as little experimentation as possible. Qualitative physical reasoning can be used to advantage in selecting noise factors and in selecting arrays and levels that keep experimentation to a minimum. Sometimes several noise factors can be combined into a single factor for experimentation purposes; these are called "combination designs" [3]. And analysis should be used wherever possible to relate the design variables to the result, thus eliminating or at least reducing the need for physical experimentation.

	1 Stress A	2 Modulus B	3 Thick C	4 Radius D	Results Life-Hrs
Run 1	1	1	1	1	2,000
Run 2	1	2	2	2	2,400
Run 3	1	3	3	3	2,800
Run 4	2	1	2	3	2,800
Run 5	2	2	3	1	2,300
Run 6	2	3	1	2	2,400
Run 7	3	1	3	2	2,700
Run 8	3	2	1	3	2,800
Run 9	3	3	2	1	2,300

Figure 19.16

Example: Checking for Acceptability of Standard Tolerances and Material Variations

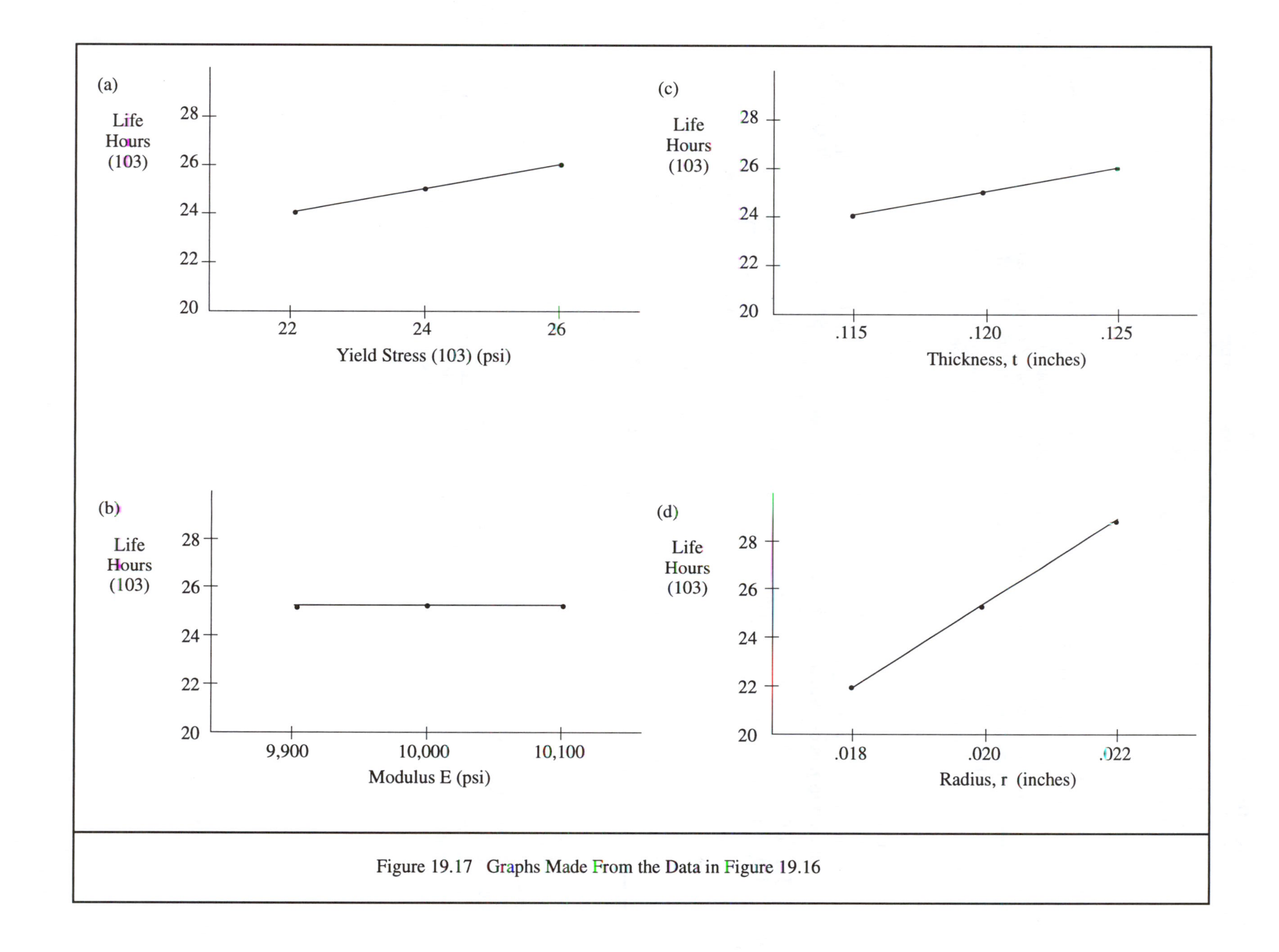

Figure 19.17 Graphs Made From the Data in Figure 19.16

19.11 Summary and Preview

Summary. Robustness in a product is an important element of quality. A robust product is reliable, and performs well over a longer life. Taguchi's Method™ as presented in "how to do it" fashion in this chapter is widely used to help achieve robustness. It is an applied method based on the science of statistics, especially design of experiments. Because experimentation can be expensive of time and money, selecting the right set of noise factors, employing the right array, and interpreting the results correctly are all extremely important in practice. In some cases, designers will be able to apply the methods themselves. In more complex cases, designers should understand when its use is indicated, and be able to work closely with experts in its application.

"You see, of course, if you're not dunce,
How it went to pieces all at once, —
All at once and nothing first, —
Just as bubbles do when they burst.

End of the wonderful one-hoss shay,
Logic is logic. That's all I say."

Oliver Wendell Holmes

The One-Hoss Shay
(The Deacon's Masterpiece)(1878)

Preview. In the next two chapters, we return again to the subject of guided iteration. The goal of these chapters is to show how guided iteration as a general methodology can be employed in the parametric design of standard mechanical components; that is, components usually included within the subject of "machine design" and some common thermal-fluid devices. Instead of each of these component types being designed by its own unique method, we illustrate by example that designers can use guided iteration for them all.

Chapter 20 deals with a sampling of mechanical components (shafts, springs, bolts), and Chapter 21 illustrates guided iteration applied to a sample of thermal-fluid parametric design problems. Our purpose is not to be exhaustive in covering all such components and devices; there are too many of them. Instead, we want to show that guided iteration is a general methodology applicable to parametric design problems of all kinds.

References

[1] Box, G. E. P., Hunter, W.G., and Hunter, J. S., *Statistics for Experimenters*, John Wiley and Sons, New York,1978.

[2] Elliott, J. G. "Statistical Methods and Applications", Allied Signal Corporation, Detroit, MI, 1987.

[3] Peace, G. S. *Taguchi Methods*, Addison-Wesley, Reading, MA, 1993.

[4] Phadke, M. S., *Quality Engineering Using Robust Design*, Prentice Hall, Englewood Cliffs, N.J., 1989.

[5] Ross, P. J., *Taguchi Techniques for Quality Engineering*, McGraw-Hill, New York, 1988.

[6] Taguchi, G. "The Development of Quality Engineering", *The American Supplier Institute Journal*, Vol. 1, No. 1, pp 5-29, 1988.

[7] Taguchi, G, and Clausing, D., "Robust Quality", *Harvard Business Review,* January-February, pp65-75, 1990.

[8] Taguchi, G. and Wu, Y., "Introduction to Off-Line Quality Control", American Supplier Institute, Center for Taguchi Methods, 6 Parkland Blvd., Dearborn, MI., 1980.

[9] Yamada, N. "Let's Learn Quality Engineering By Making Paper Helicopter", Design Productivity International Conference, Honolulu, Hawaii, February, 1991.

[10] Box, G.E. P. and Bisgaard, S., "Statistical Tools for Improving Designs", *Mechanical Engineering*, January, 1988.

[11] Byrne, D. and Taguchi, S., "The Taguchi Approach to Parameter Design" Proceedings ASQE Quality Congress, Anaheim, 1986.

[12] Taguchi, G. and Konishi, S. *Orthogonal Arrays and Linear Graphs*, reproduced by permission of the American Supplier Institute, Dearborn, MI, 1987.

[13] Gulezian, R., *Process Control: Statistical Principles and Tools*, Quality Alert Institute, New York, 1990.

Problems

19.1 Show how you would use the L27 array for solving the example problem described in Section 19.4.

19.2 Aluminum extrusions are often subjected to process called anodizing. In this process, the extrusions are immersed in a sulfuric acid solution and made the anode as direct electrical current is passed through the extrusions. As a result, an aluminum oxide film is created on the surface of the extrusion. This film is hard, has a pleasant matte appearance, and is less likely to tarnish in the weather.

We would like to use Taguchi's method [TM] to establish processing conditions for anodizing. The Result we are interested in is the thickness of the oxide film, which we would like to be as thick as possible, and which varies from about 0.0004 inches to 0.0016 inches. The process variables are: the current density (varies from 10 to 30 amps per square foot), the acid concentration (varies from 10 to 60 per cent), and the solution temperature (varies from 50 to 80 F). The duration of the current flow (varies from 10 to 30 minutes) is also a process variable, but to keep the problem simpler, we will assume here that the oxide thickness is more or less proportional to the time (though this not quite true).

(a) For a two level experiment, which array would you choose, and why?

(b) Assuming you elect to use L8, what variables would you assign to the various columns?

(c) Suppose the experiments are set-up as shown in Figure

	1	2	3	4	5	6	7	
	A %	B F	AB —	C $amps/ft^2$	AC —	BC —	ABC —	Result (inch x 10^3)
Run 1	10	50	1	10	1	1	1	1.0
Run 2	10	50	1	30	2	2	2	1.2
Run 3	10	80	2	10	1	2	2	0.5
Run 4	10	80	2	30	2	1	1	0.7
Run 5	60	50	2	10	2	1	2	0.9
Run 6	60	50	2	30	1	2	1	1.0
Run 7	60	80	1	10	2	2	1	0.4
Run 8	60	80	1	30	1	1	2	0.5

A = Acid Concentration Level, %.
B = Solution Temperatures, F
C = Current Density, $amps/ft^2$.

Figure P19.2 Experimental Results for Problem 19.2

	1	2	3	4	
	A %	B F	AB —	C $amps/ft^2$	Result (inch x 10^3)
Run 1	10	50	1	10	1.0
Run 2	10	65	2	20	1.0
Run 3	10	80	3	30	0.7
Run 4	35	50	2	30	1.2
Run 5	35	65	3	10	0.6
Run 6	35	80	1	20	0.6
Run 7	60	50	3	20	1.1
Run 8	60	65	1	30	0.5
Run 9	60	80	2	10	0.4

Figure P19.3 Experimental Results for Problem 19.3

P19.2, compute the approximate effects of the variables and their interactions. Which variables are the more influential? (Note: The data shown in Figure P19.2 are fictitious.)

(d) Sketch graphs that illustrate the interaction effects.

19.3 (a) Suppose you wished to perform a three level experiment for Problem 19.2. Which array would you choose and why?

(b) Assuming you elect to use L9, what variables would you assign to the columns?

(c) How would you assign variables if L27 were to be used? List the advantages and disadvantages of using L27?

(d) Suppose the experiments are set up as shown in Figure P19.3. Sketch graphs to determine the value for the process variables that will give the best results. (Note: The data shown in Figure P19.3 are fictitious.)

(e) What confirmation experiment(s) would you want to run?

19.4 In the insertion and epoxy assembly of golf club shafts into the club heads, assume the following variables can be controlled by the designer and the assembler:

adhesive type	Available types are A1, A2, and A3
depth of insertion	0.5 inches, 1 inch, 1.5 inches
clearance	0.01 inches, 0.02 inches, 0.03 inches
crimp or no crimp	C or X

"Crimp" here refers to whether or not the inserted portion of the shaft is given a pre-assembly crimp so that it out of round at a section.

The desired result is the impulse force required to pull the club from the shaft.

Uncontrollable variables during assembly include the ambient temperature (50 to 90 F) and the ambient humidity (30 to 80%).

(a) Select and set-up the experimental design to find the best set of design and process variables.

(b) Assuming the experiments are conducted as shown in Figure P19.4, determine the signal to noise ratio for each Run, and make sketches to determine the best set of design variables. (Note: The data shown in Figure P19.4 are fictitious, and the pull-off impulse force is relative to an unspecified reference value.)

(c) What confirmation experiment(s) would you perform?

19.5 Refer to Figure P18.9 and Problem 18.9. We wish to design the instrument now for robustness as well as for performance. Take the control variables to be D_a, D_S, and V. The noise variables are P_S and T_S. The desired result is the sensitivity of the instrument; that is

$$\text{Sensitivity} = \frac{\partial P_1}{\partial x}$$

Assume two levels are to be used for both control variables and noise variables as follows:

	1	2
Diameter, Da	0.02	0.06 in
Diameter, Ds	0.02	0.06 in
Volume, V	1.1	2.2 in3
Supply Pressure, Ps	15	30 psig
Supply Temp, Ts	60	90 F

(a) Select arrays and set up the experimental design.

(b) Assuming the experimental results are as shown in Figure P19.5, determine the best values for the control variables. (Note: The data shown in Figure P19.5 are fictitious.)

(c) Suppose a meter is constructed using the results obtained in part (b) above. Using the simplified analysis proposed in Problem 18.9, what will be the sensitivity of this meter for an x of 0.002 inches? How does this compare with the maximum sensitivity for this x if the supply pressure is 20 psig? How do you explain the difference?

Figure 19.6

Data for "A More Complex Heat Treating Example" Section 19.4

RUN	1 - A	2 - B	3 - AB	4 - C	5 - AC	6 - BC	7 - ABC	Result - PSI
1	300 F	3 hrs	1	Smith	1	1	1	7,000
2	300 F	3 hrs	1	Jones	2	2	2	6,000
3	300 F	5 hrs	2	Smith	1	2	2	8,000
4	300 F	5 hrs	2	Jones	2	1	1	7,000
5	500 F	3 hrs	2	Smith	2	1	2	14,000
6	500 F	3 hrs	2	Jones	1	2	1	12,000
7	500 F	5 hrs	1	Smith	2	2	1	12,000
8	500 F	5 hrs	1	Jones	1	1	2	10,000

Run		1	2	3	4
3		1	2	2	
2	H	30	80	30	80
1	T	50	50	90	90

Run	1	2	3	4					S/N
1	A1	.01	0.5	C	35	30	50	40	
2	A1	.02	1.0	C	18	70	100	90	
3	A1	.03	1.5	X	15	14	25	18	
4	A2	.01	1.0	X	18	25	30	20	
5	A2	.02	1.5	C	60	20	95	65	
6	A2	.03	0.5	C	22	18	70	24	
7	A3	.01	1.5	C	35	32	55	40	
8	A3	.02	0.5	X	23	21	40	28	
9	A3	.03	1.0	C	45	40	70	25	

Figure P19.4

Experimental Results for Problem 19.4

3	–	1	2	2	1
2	Ts	60	90	60	90
1	Ps	15	15	30	30

	1	2	3	4	5	6	7					S/N
	Da	Ds	—	V	—	—	—					
Run 1	.02	.02	1	1.1	1	1	1	1200	880	2300	1600	
Run 2	.02	.02	1	2.2	2	2	2	1170	1200	2400	2300	
Run 3	.02	.06	2	1.1	1	2	2	130	90	135	95	
Run 4	.02	.06	2	2.2	2	1	1	170	173	350	260	
Run 5	.06	.02	2	1.1	2	1	2	3700	2900	5100	4000	
Run 6	.06	.02	2	2.2	1	2	1	4700	3800	9400	7500	
Run 7	.06	.06	1	1.1	2	2	1	400	3200	820	630	
Run 8	.06	.06	1	2.2	1	1	2	520	410	1050	840	

Figure P19.5

Experimental Results for Problem 19.5

Appendix 19A

Orthogonal Arrays

$L_4(2^3)$

Experiment Number	Column 1	Column 2	Column 3
1	1	1	1
2	1	2	2
3	2	1	2
4	2	2	1
	Group 1	Group 2	

Linear Graph for L_4

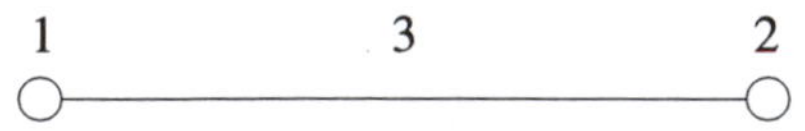

1 3 2

Figure 19A.1 The L_4 Array.

$L_8(2^7)$

Exp. No. \ Col.	1	2	3	4	5	6	7
1	1	1	1	1	1	1	1
2	1	1	1	2	2	2	2
3	1	2	2	1	1	2	2
4	1	2	2	2	2	1	1
5	2	1	2	1	2	1	2
6	2	1	2	2	1	2	1
7	2	2	1	1	2	2	1
8	2	2	1	2	1	1	2
	Group 1	Group 2		Group 3			

Interactions Between Two Columns

Exp. No. \ Col.	1	2	3	4	5	6	7
	(1)	3	2	5	4	7	6
		(2)	1	6	7	4	5
			(3)	7	6	5	4
				(4)	1	2	3
					(5)	3	2
						(6)	1
							(7)

Linear Graph for L_8

(1)

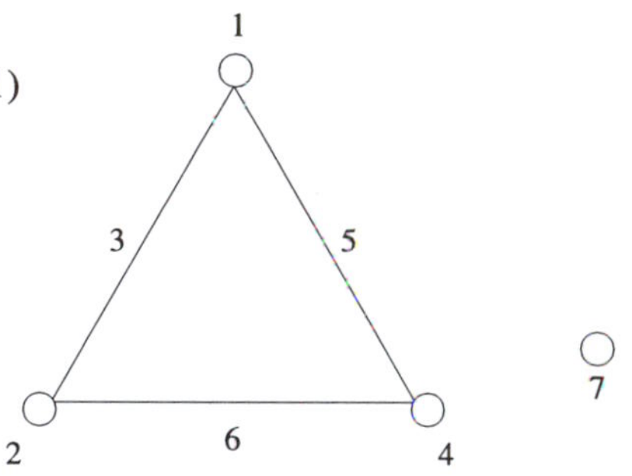

(2)

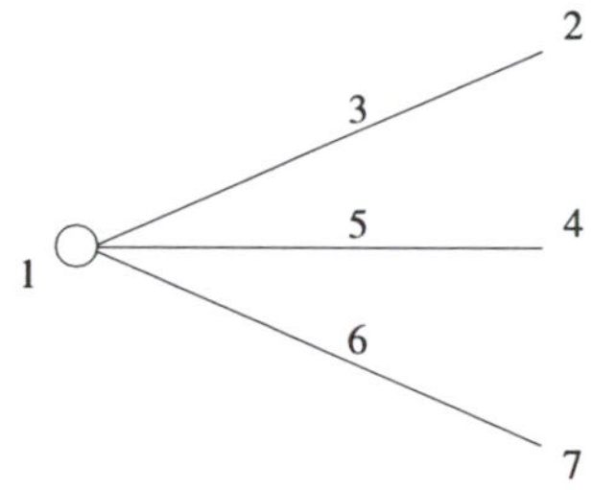

Figure 19A.2 The L_8 Array.

$L_{12}(2^{11})$

Exp. No. \ Col.	1	2	3	4	5	6	7	8	9	10	11
1	1	1	1	1	1	1	1	1	1	1	1
2	1	1	1	1	1	2	2	2	2	2	2
3	1	1	2	2	2	1	1	1	2	2	2
4	1	2	1	2	2	1	2	2	1	1	2
5	1	2	2	1	2	2	1	2	1	2	1
6	1	2	2	2	1	2	2	1	2	1	1
7	2	1	2	2	1	1	2	2	1	2	1
8	2	1	2	1	2	2	2	1	1	1	2
9	2	1	1	2	2	2	1	2	2	1	1
10	2	2	2	1	1	1	1	2	2	1	2
11	2	2	1	2	1	2	1	1	1	2	2
12	2	2	1	1	2	1	2	1	2	2	1
	Group 1	Group 2									

Figure 19A.3 The L_{12} Array.

$L_9(3^4)$

Exp. No. \ Col.	1	2	3	4
1	1	1	1	1
2	1	2	2	2
3	1	3	3	3
4	2	1	2	3
5	2	2	3	1
6	2	3	1	2
7	3	1	3	2
8	3	2	1	3
9	3	3	2	1
	Group 1	Group 2		

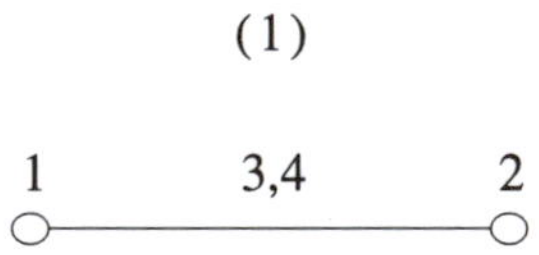

(1)

1 —— 3,4 —— 2

Figure 19A.4 The L_9 Array.

$L_{18}(2^1 \times 3^7)$

Exp. No. \ Col.	1	2	3	4	5	6	7	8
1	1	1	1	1	1	1	1	1
2	1	1	2	2	2	2	2	2
3	1	1	3	3	3	3	3	3
4	1	2	1	1	2	2	3	3
5	1	2	2	2	3	3	1	1
6	1	2	3	3	1	1	2	2
7	1	3	1	2	1	3	2	3
8	1	3	2	3	2	1	3	1
9	1	3	3	1	3	2	1	2
10	2	1	1	3	3	2	2	1
11	2	1	2	1	1	3	3	2
12	2	1	3	2	2	1	1	3
13	2	2	1	2	3	1	3	2
14	2	2	2	3	1	2	1	3
15	2	2	3	1	2	3	2	1
16	2	3	1	3	2	3	1	2
17	2	3	2	1	3	1	2	3
18	2	3	3	2	1	2	3	1
	Group 1	Group 2	Group 3					

Figure 19A.5 The L_{18} Array.

$L_{16}(2^{15})$

Exp. No. \ Col.	1	2	3	4	5	6	7	8	9	10	11	12	13	14	15
1	1	1	1	1	1	1	1	1	1	1	1	1	1	1	1
2	1	1	2	2	2	2	2	2	1	1	2	2	2	2	2
3	1	1	3	3	3	3	3	3	1	1	3	3	3	3	3
4	1	2	1	1	2	2	3	3	1	2	1	1	2	2	3
5	1	2	2	2	3	3	1	1	1	2	2	2	3	3	1
6	1	2	3	3	1	1	2	2	1	2	3	3	1	1	2
7	1	3	1	2	1	3	2	3	1	3	1	2	1	3	2
8	1	3	2	3	2	1	3	1	1	3	2	3	2	1	3
9	1	3	3	1	3	2	1	2	1	3	3	1	3	2	1
10	2	1	1	3	3	2	2	1	2	1	1	3	3	2	2
11	2	1	2	1	1	3	3	2	2	1	2	1	1	3	3
12	2	1	3	2	2	1	1	3	2	1	3	2	2	1	1
13	2	2	1	2	3	1	3	2	2	2	1	2	3	1	3
14	2	2	2	3	1	2	1	3	2	2	2	3	1	2	1
15	2	2	3	1	2	3	2	1	2	2	3	1	2	3	2
16	2	3	1	3	2	3	1	2	2	3	1	3	2	3	1
	Group 1	Group 2		Group 3				Group 4							

Interactions Between Two Columns

Exp. No. \ Col.	1	2	3	4	5	6	7	8	9	10	11	12	13	14	15
	(1)	3	2	5	4	7	6	9	8	11	10	13	12	15	14
		(2)	1	6	7	4	5	10	11	8	9	14	15	12	13
			(3)	7	6	5	4	11	10	9	8	15	14	13	12
				(4)	1	2	3	12	13	14	15	8	9	10	11
					(5)	3	2	13	12	15	14	9	8	11	10
						(6)	1	14	15	12	13	10	11	8	9
							(7)	15	14	13	12	11	10	9	8
								(8)	1	3	2	4	5	6	7
									(9)	3	2	5	4	7	6
										(10)	1	6	7	4	5
											(11)	7	6	5	4
												(12)	1	2	3
													(13)	3	2
														(14)	1

Figure 19A.6 The L_{16} Array.

$L_{27}(3^{13})$

Exp. No. \ Col.	1	2	3	4	5	6	7	8	9	10	11	12	13
1	1	1	1	1	1	1	1	1	1	1	1	1	1
2	1	1	1	1	2	2	2	2	2	2	2	2	2
3	1	1	1	1	3	3	3	3	3	3	3	3	3
4	1	2	2	2	1	1	1	2	2	2	3	3	3
5	1	2	2	2	2	2	2	3	3	3	1	1	1
6	1	2	2	2	3	3	3	1	1	1	2	2	2
7	1	3	3	3	1	1	1	3	3	3	2	2	2
8	1	3	3	3	2	2	2	1	1	1	3	3	3
9	1	3	3	3	3	3	3	2	2	2	1	1	1
10	2	1	2	3	1	2	3	1	2	3	1	2	3
11	2	1	2	3	2	3	1	2	3	1	2	3	1
12	2	1	2	3	3	1	2	3	1	2	3	1	2
13	2	2	3	1	1	2	3	2	3	1	3	1	2
14	2	2	3	1	2	3	1	3	1	2	1	2	3
15	2	2	3	1	3	1	2	1	2	3	2	3	1
16	2	3	1	2	1	2	3	3	1	2	2	3	1
17	2	3	1	2	2	3	1	1	2	3	3	1	2
18	2	3	1	2	3	1	2	2	3	1	1	2	3
19	3	1	3	2	1	3	2	1	3	2	1	3	2
20	3	1	3	2	2	1	3	2	1	3	2	1	3
21	3	1	3	2	3	2	1	3	2	1	3	2	1
22	3	2	1	3	1	3	2	2	1	3	3	2	1
23	3	2	1	3	2	1	3	3	2	1	1	3	2
24	3	2	1	3	3	2	1	1	3	2	2	1	3
25	3	3	2	1	1	3	2	3	2	1	2	1	3
26	3	3	2	1	2	1	3	1	3	2	3	2	1
27	3	3	2	1	3	2	1	2	1	3	1	3	2
	Group 1	Group 2			Group 3								

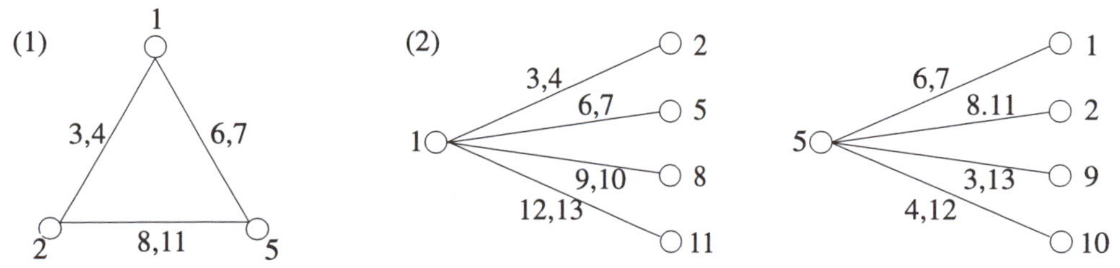

(1)

1
3,4
6,7
2
8,11
5

(2)

1
3,4
2
6,7
5
9,10
8
12,13
11

5
6,7
1
8.11
2
3,13
9
4,12
10

Figure 19A.7 The L_{27} Array.

This page is intentionally blank

Chapter Twenty

Guided Iteration Applied to Common Mechanical Components

"Science is a method that once applied to a problem will produce an answer, and when applied again will produce a comparable answer again, again, and again."

Percy H. Hill [1]
The Science of Engineering Design (1970)

20.1 Introduction

Over the years, specific design procedures have been developed and time-tested for many common mechanical components. We call such procedures *domain-specific* because each is specific to its particular domain, such as to shafts or springs. Many of these domain-specific methods, together with a wealth of related domain information, are well presented in detail in many texts and references. See, for example, [2, 3, 4, 5, 6, 7, 8]. We strongly recommend that readers select one or more of these for their personal reference libraries.

Guided iteration is a more general method, applicable to special purpose parts and components (where no domain methods have been developed) as well as to standard components. In Chapters 14 and 17, we illustrated the guided iteration methodology for parametric design in the domains of beams and v-belt drives. In this chapter, we apply guided iteration to three additional standard components: shafts, springs, and bolts. Then in the next chapter we consider examples from the thermal-fluids area. Our goal in these chapters is to show by examples that guided iteration is applicable to a wide variety of problems.

The guided iteration approach — problem formulation, generation of alternatives, evaluation, and redesign guided by the evaluation — is unchanged from domain to domain. The only differences are in the engineering analysis procedures used in each domain to obtain the numerical values of the solution evaluation parameters (SEPs) for each trial design. As we have stated before, and as the examples in this chapter illustrate, the analysis procedures provide extremely important supporting information for parametric design; without correct analyses, evaluations of trial designs will be based on incorrect values for performance parameters. But as we have also pointed out, *values* are different from *evaluation*. Getting numerical values of the SEPs is a task for analysis; evaluating what the numbers mean to the overall quality of the design (considering all the other evaluation criteria as well), and deciding what changes to make are design tasks.

Since domain-specific design methods are directed at only one domain, they can sometimes be more efficient in their domain than the more general guided iteration methodology. Thus, in practice, when a domain-specific method is known and available, there is no reason not to use it. However, there are too many problem types for anyone to be familiar with all the various domain-specific methods. Moreover, new problems for which there are as yet no domain-specific methods occur in practice regularly. Thus, it is important for students and new design engineers to have a general methodology in their problem solving toolkit.

It is not our purpose with the examples in this chapter to describe completely or in detail specific application information, analysis techniques, and domain-specific design methods so well presented elsewhere. Complete analytical coverage of the example problem domains is sacrificed here in order to focus on how the general design method is employed. Thus in the main body of the chapter, the technical information included with the examples is generally adequate only for the examples themselves.

Nor is it our purpose to present in detail the basic engineering science on which the various analyses required for the examples are based. Most readers of this book will have studied this basic material in previous courses. For those who have not, and for those who wish a review, we present in an Appendix (20A) associated with this chapter a summary of the basic material needed for the analyses in the examples. Many of the analysis equations used for the examples are drawn from this Appendix.

We begin with applications of the use of guided iteration in the domain of shafts.

20.2 Shafts With Static Loads

20.2.1 Pre-Parametric Design Issues: Conceptual Level

Embodiment

The physical embodiment of a shaft is the same as that of a beam: both are long, slender, elastic (though relatively stiff) members. Thus, when the loading is static, there is really no difference between a shaft and a beam; however, when torsional loads are prominent, and certainly when there is rotation, the term *shaft* is commonly used.

Physical Concepts

The physical concepts on which shafts with static torsional loads operate are simple, and much the same as with beams: internal stresses in the material respond to and balance the externally applied loads, transferring them to one or more supports. In Appendix 20A.1.1 yield criteria for materials are discussed. In Appendix 20A.1.2, the physical principles and analysis methods that apply to circular shafts are reviewed in more detail for use in the examples below. These principles and methods are commonly covered in texts on strength of materials and, more completely, in books on machine design as mentioned above.

Shafts that support bending as well as torsional loads develop complex internal stresses that combine tension, compression, and shear. By selecting materials, determining shaft cross sectional configurations, and setting dimensions, designers must insure that the internal stresses do not exceed the strength of the material. In addition, however, special attention must be paid to stress concentrations caused by holes, step changes in diameter, and the like. If loads are variable, or reversing, material fatigue may also become a factor. See Appendix 20A.1.3.

20.2.2 Pre-Parametric Design Issues: Configuration Level

Cross-Sections

Most of the embodiments we call "shafts" are circular, either solid or hollow, though other cross section configurations are occasionally employed. For applications requiring complex thin-walled cross sectional shapes, non-circular thin-walled cross sections of infinite variety can be manufactured by extrusion or rolling, and are used in some applications that involve both bending and torsional loads. The aluminum frames of residential storm windows are a typical example.

We limit our discussion henceforth in this book to circular shafts, solid or hollow. For information about thin-walled complex cross section beams and shafts, see the references cited above.

Hollow circular shafts make more efficient use of materials in supporting torsional loads than solid circular shafts. Per pound, of course, hollow shafts are more expensive — though not much. Nevertheless, unless weight or related vibration problems associated with weight are critical factors, solid circular shafts are almost always selected.

Stress Raising Features

When shafts are designed to include features such as stepped changes in diameter, through holes, grooves, or keyways, then the stresses around these features are raised above the normal levels depending on the specific design and dimensions of the feature. The amount of stress increase is generally accounted for in practice by the use of experimentally determined *stress concentration factors*. In practice, then, the internal stresses are first computed as if the stress concentrating feature were not present, and then multiplied by the value of the stress concentration factor to get an estimate of the actual stress in the local region of concentration.

Once stress raising features have been designed-in at the configuration stage, there is nothing the designer can do to avoid a certain amount of stress increase. The amount of the stress increase can be minimized, perhaps, at the parametric stage by noting how the values of the various stress concentration factors vary with the design details. Fairly precise values for many stress concentration configurations are given in the references noted in Section 20.1. Also, see [9]. The values given below are only to illustrate the approximate ranges of values encountered, and some of the related design concerns.

Step Changes in Diameter

For statically loaded circular shafts in *torsion*, stress concentration factors associated with stepped diameter changes range from about 1.1 to greater than 2.0 depending on (a) the ratio of the diameters (the bigger the diameter change, the greater the stress concentration) and (b) more critically on the internal radius of the fillet at the diameter change. Sharp internal corners, as always, should be avoided where possible.

For statically loaded circular shafts in *bending*, stress concentration factors associated with stepped diameter changes range from about 1.2 to greater than 2.5 depending on ratio of the diameters (the bigger the diameter change, the greater the stress concentration) and more critically on the internal radius of the fillet at the diameter change. Sharp internal corners, as always, should be avoided where possible.

Through Holes

For statically loaded circular shafts with through holes, stress concentration factors for torsion can vary *from* about 2.0 for relatively small holes to over about 2.2 for larger holes.

For statically loaded circular shafts with through holes, stress concentration factors for bending can vary from about 2.0 for relatively large holes to over about 2.5 for smaller holes.

20.2.3 General Considerations in Formulating Solid Circular Statically Loaded Shaft Parametric Design Problems

When the configuration for a shaft has been selected (in this discussion, restricted to circular solid or hollow), then the problem becomes a parametric design problem. Thus the guided iteration problem formulation step requires that we:

- Identify the Problem Definition Parameters (PDPs);
- Specify the Design Variables;
- Specify the Solution Evaluation Parameters (SEPs);
  Including the associated:
  Importance levels,
  Satisfaction curves, and
  Analysis procedures for the SEP values; and
- Establish an initial dependency table.

In the case of a uniform diameter solid circular shafts with static loads, the formulation will usually include at least the specific elements listed below. There may be additional factors in some situations — for example, when the configuration includes a stepped shaft, an additional design variable is the fillet radius at the change which will influence the stress concentration in that region.

Problem Definition Parameters:

- Shaft length, L,
- Loads (torques and transverse loads),
- Locations, directions, magnitudes,
- Supports (location, types (e.g., simple versus built-in)),
- Environmental issues (e.g., those influencing material choice, such as temperatures or corrosive atmospheres),
- Bending deflection limitations (if any),
- Twist deflection limitations (if any),
- Space limitations (if any).

Design Variables:

- Material choice (This gives values for E, G, Yield Stress, cost per pound),
- Shaft diameter, D (For stepped shafts, there will more than one diameter).

Solution Evaluation Parameters (SEPs):

- Tension (or Compression) Failure Ratio (TFR),
- Shear Failure Ratio (SFR),
- Bending deflection,
- Twist deflection,
- Cost or Weight.

The first two solution evaluation parameters above are expressed as ratios of the maximum internal stress (tension, compression, or shear) to the respective yield stresses. The definitions of these terms are shown in the paragraph below on *Analysis Procedures for the SEPs*. The convenience of this formulation is illustrated in the examples.

There can also be other evaluation criteria not considered here. Examples are availability, manufacturability, compatibility, reliability, and life. The operating environment (e.g., temperature, corrosive atmosphere) is also not included.)

Importance Levels and Satisfaction Curves for the SEPs

The Importance levels and satisfaction curves for these parameters will depend on the specific problem. See the example below.

Analysis Procedures for the SEPs

- Tension (or Compression) Failure Ratio = TFR

$$\text{TFR} = \frac{\text{Max Tension or Compression Stress}}{\text{Tensile Yield Stress}} \quad (20.1)$$

- Shear Failure Ratio = SFR

$$\text{SFR} = \frac{\text{Maximum Shear Stress}}{0.5 \ \text{x Tensile Yield Stress}} \quad (20.2)$$

  We have used the maximum shear stress theory here. See Section 20A.1.1.

- Lateral deflection (due to bending loads)
  The analysis procedures to be used will be determined by application of strength of materials principles, and will depend on the load and support locations, and on the geometry of the shaft cross-section.
- Twist deflection (due to torques)
  The analysis procedures to be used will be determined by application of strength of materials principles, and will depend on the load and support locations, and on the geometry of the shaft cross-section.
- Cost or Weight
  Usually the shaft diameter D (or cross sectional area, A) can be used as a surrogate for cost or weight.

20.2.4 Example 20.1 - A Statically Loaded Solid Shaft Design

As an example of designing shafts with static loads using guided iteration, consider the problem defined in Figure 20.1. In this simplest case example, we care only that the shaft not fail, so that the amount of deflection and twisting can be ignored, and the only Solution Evaluation Parameters

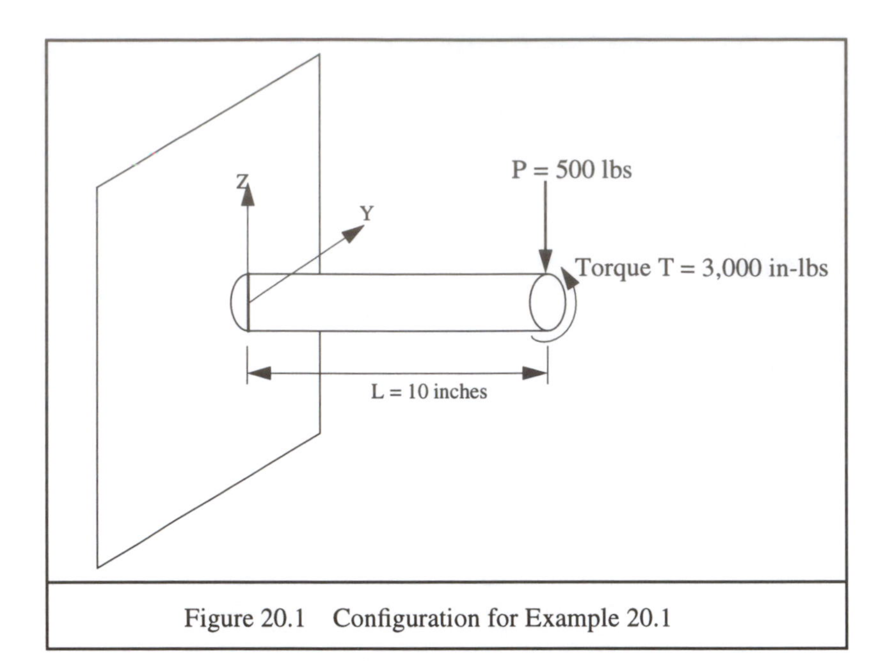

Figure 20.1 Configuration for Example 20.1

(SEPs) are the failure ratios defined by Equations 20.1 and 20.2. Also, assuming there is a material already selected because it is readily available — say cold drawn 1020 steel — then material properties are not active design variables in this case. Achieving low cost in these circumstances simply means finding the smallest safe diameter. With these assumptions, the problem formulation is:

Problem Formulation

Problem Definition Parameters

See configuration in Figure 20.1.

P = 500 lbs

L = 10 inches

T = 3000 inch-lbs

In addition, for the material: E = 30,000,000 psi

G =11,500,000 psi

Yield Stress = 66,000 psi

Design Variable: D

Solution Evaluation Parameters (and their Importance Levels):

- Tension (or Compression) Failure Ratio (TFR) (Very Important)
- Shear Failure Ratio (SFR) (Very Important)

Note: We also want to keep the weight, hence cost, of the shaft as small as possible. To keep this initial example simple, we treat this issue implicitly in this example by trying to find the smallest diameter that will meet the requirements. We could alternatively take the diameter to be a Solution Evaluation Parameter as well as a design variable if we desired.

Satisfaction Curves. See Figure 20.2.

It should be noted that designers have a choice in how these satisfaction curves are constructed. In this case, we have expressed both failure ratio SEPs as hard constraints. We have also elected to introduce a margin of safety by treating values of the tension and shear failure ratios above 0.67 and 0.50, respectively, as unacceptable. This corresponds to traditional "safety factors" of 1.5 and 2.0.

Analysis Procedures (See also Appendix 20A)

The most critical stress state occurs at surface element dA_2 (Figure 20A.3) where:

$$\tau_{x\theta} = \frac{16T}{\pi D^3} = \frac{16 \cdot 3000}{\pi D^3} \text{ psi} \tag{20.3}$$

$$\sigma_x = \frac{32M}{\pi D^3} = \frac{32(500)(10)}{\pi D^3} \text{ psi} \tag{20.4}$$

The failure ratios are (Equations 20.1 and 20.2):

$$\text{TFR} = \frac{\sigma_{max}}{\sigma_y} \tag{20.5}$$

$$\text{SFR} = \frac{\tau_{max}}{\tau_{crit}} = \frac{\tau_{max}}{0.5 \cdot \sigma_y} \tag{20.6}$$

where σ_{max} is the larger of the stresses computed by Equation 20A.12 and τ_{max} is computed from Equation 20A.13.

Dependencies

An initial dependency table is shown in Figure 20.3. From the analysis equations, it is evident that both failure ratios will decrease in proportion to D^3. Thus both dependencies are set at -3.

Initial Design

As always in parametric design problems, we begin the guided iteration process by selecting initial values for the design variables as reasonably as we can. In this case, there is only one design variable (D), so we try (actually, in this case, it is a pure guess) a value of 1.0 inches.

Evaluation

To perform evaluation of this design, we first use the analysis procedures above to compute the values of the SEPs, and then use the satisfaction curves to get the evaluations of individual SEPs. Then we must get an overall evaluation, normally using the Dominic method, though in this simple problem evaluation can easily be done by direct inspection of the satisfaction curves.

From the analysis equations above, we get for DA_2 in Trial #1 (D = 1 inch):

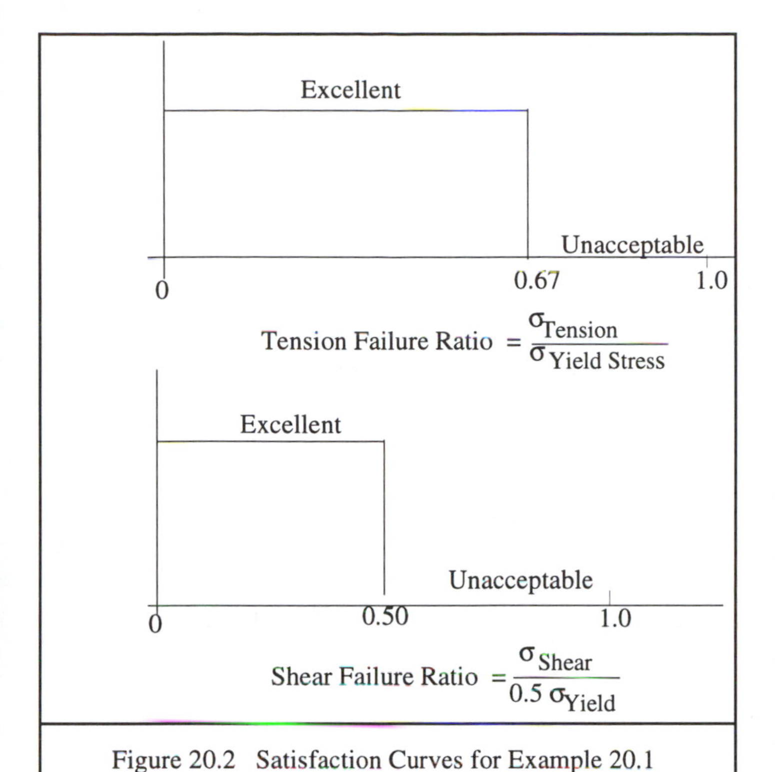

Figure 20.2 Satisfaction Curves for Example 20.1

$$\tau_{x\theta} = \frac{16 \cdot 3000}{\pi \cdot 1^3} = 15,300 \quad \text{psi}$$

$$\sigma_x = \frac{32(500)(50)}{\pi \cdot 1^3} = 50,950 \quad \text{psi}$$

$$\sigma_y = 0$$

$$\sigma_1 = \frac{50,950}{2} + \left[\left(\frac{50,950}{2}\right)^2 + (15300)^2\right]^{\frac{1}{2}}$$

$$= 55,190 \text{ psi}$$

$$\sigma_2 = -4243 \text{ psi}$$

$$\tau_{max} = \frac{\sigma_1 - \sigma_2}{2} = 29,710 = 29,710 \quad \text{psi}$$

$$\text{TFR} = \frac{55,190}{66,000} = 0.836$$

$$\text{SFR} = \frac{29,710}{33,000} = 0.90$$

	X_1 Diameter D
Y_1 Tension Failure Ratio	-3
Y_2 Shear Failure Ratio	-3

Figure 20.3 Initial Dependency Table for Example 20.1

From the satisfaction curves, we see that both Failure Ratio SEPs are in their Unsatisfactory range. Thus, of course, this trial design (i.e., D = 1 inch) is unacceptable.

Guided Redesign

To get an acceptable design, both SEPs must be in their acceptable range. That is, Tension Failure Ratio must be less than 0.67 and the Shear Failure Ratio must be less than 0.50. It appears that the Shear Failure Ratio is farthermost from satisfaction, so we will try to adjust it first. With a dependency of -3, we predict a new value for D in the usual way from:

$$y_j(\text{old}) / y_j(\text{new}) = [\,(x_i(\text{old}) / x_i(\text{new})\,]^{d_{ij}}$$

$$0.900 / 0.500 = [\,1.0 / D(\text{new})\,]^{-3}$$

$$D = 1.22 \text{ inches.}$$

Trial #2 Analysis. The new computational results are:

Maximum internal tension stress = 30,390 psi

Maximum internal shear stress = 16,365 psi

Trial #2 Evaluation:

SEP (Tension Failure Ratio) = 0.46

SEP (Shear Failure Ratio) = 0.496

Now both these SEPs are Satisfactory, and one (the Shear Failure Ratio) is very close to its critical value (0.496 versus 0.50). Since we clearly cannot reduce the shaft diameter any more without causing this SEP to go Unsatisfactory, we have a satisfying solution.

20.2.5 Example 20.2 - A Hollow Shaft With Static Load

For this example, we will use the same loading and supports as shown in Figure 20.1. In this case, how-

"There are moments when everything goes well; don't be frightened, it won't last."

Jules Renard
Journal (c 1900)

ever, we will assume that the configuration is to be hollow; that is, a circular shaft with outer diameter D_2 and inner diameter D_1. In addition, we will assume that the amount of twist is now a factor of concern so that it will be included among the Solution Evaluation Parameters (SEPs). Other assumptions are given in the problem formulation below:

Problem Formulation

Problem Definition Parameters

For loads and supports, see Figure 20.1.

In addition, for the material: E = 30,000,000 psi
G =11,500,000 psi
Yield Stress =66,000 psi

Limiting conditions on the shaft dimensions:

D_2 (max) = 2.0 inches
t (min) = 0.05 inches

Design Variables

D_2 and t, where $t = (D_2 - D_1)/2$.

Note: The change of design variable from D_1 to t is suggested in part by the manufacturing limit imposed below specifically on t. In addition, the change reduces the coupling between the design variables.

Solution Evaluation Parameters (and Assumed Importance Levels)

Tension Failure Ratio (TFR) (Desirable)
Shear Failure Ratio (SFR) (Important)
Area (A) (Desirable)
Area is a surrogate for cost or weight, and hence should be as small as possible.
Angle of Twist (θ) (Very Important)

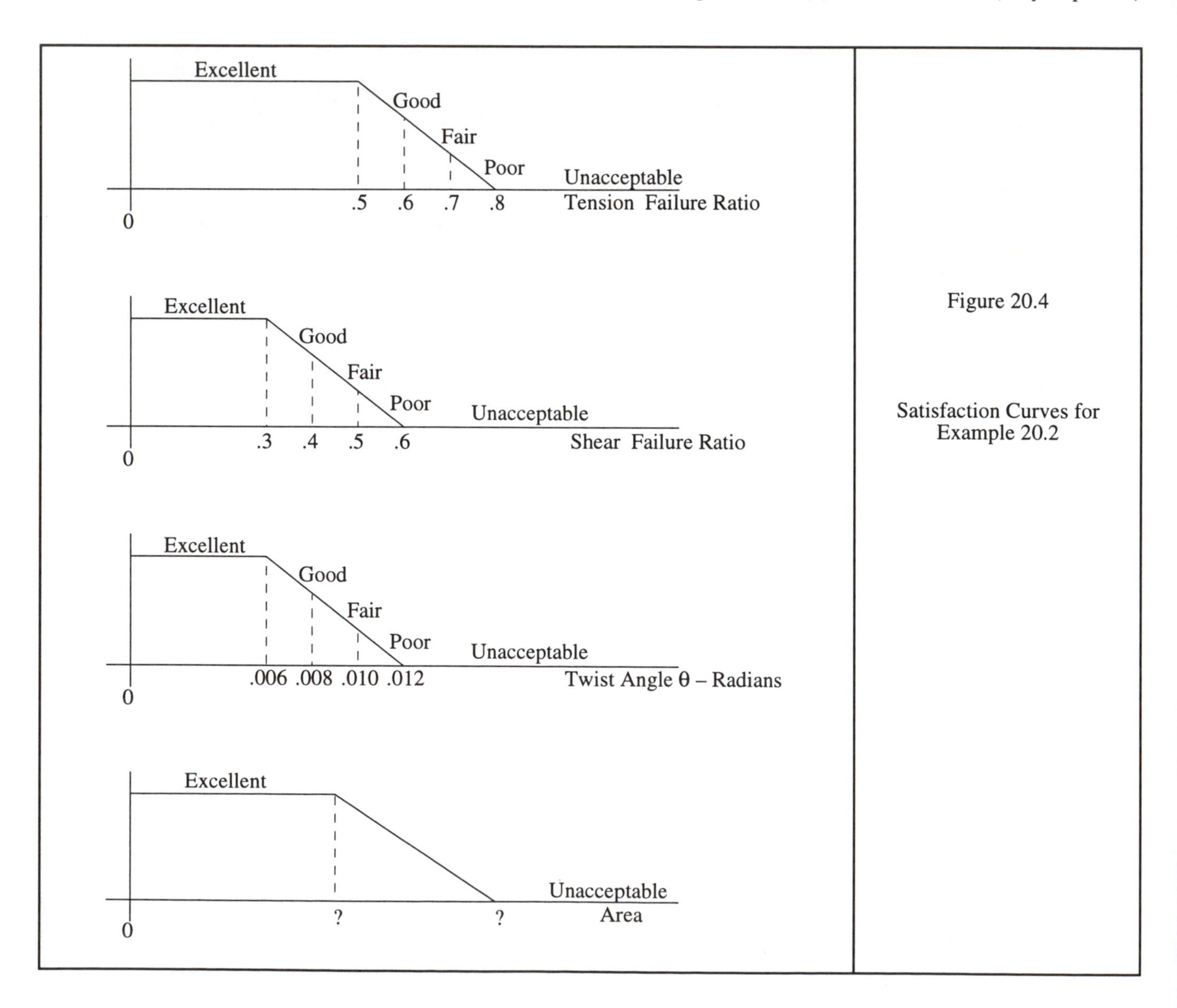

Figure 20.4

Satisfaction Curves for Example 20.2

Satisfaction Curves

See Figure 20.4. For this example, we have made the failure ratio SEPs soft as shown, but have also used the equivalent safety factors of 1.25 and 1.67 to define the boundary of the Unacceptable range.

We cannot yet place numerical values on the SEP Area, and that is okay. After a few iterations, we will be able to do so.

Analysis Procedures

See also Appendix 20A.

The maximum stress state again occurs at surface element dA_2 on the outer diameter where:

$$\tau_{x\theta} = \frac{T \cdot D_2}{2J} \quad (20.7)$$

$$J = \frac{\pi}{32} \cdot \left(D_2^4 - D_1^4\right) \quad (20.8)$$

$$\sigma_x = \frac{MD_2}{2I} \quad (20.9)$$

$$I = \frac{\pi}{64} \cdot \left(D_2^4 - D_1^4\right) \quad (20.10)$$

$$\sigma_{1,2} = \frac{\sigma_x + \sigma_y}{2} \pm \left[\left(\frac{\sigma_x - \sigma_y}{2}\right)^2 + \tau_{x\theta}^2\right]^{\frac{1}{2}} \quad (20.11)$$

$$\tau_{max} = \frac{\sigma_1 - \sigma_2}{2} = \left[\left(\frac{\sigma_x - \sigma_y}{2}\right)^2 + \tau_{x\theta}^2\right]^{\frac{1}{2}} \quad (20.12)$$

$$TFR = \frac{\sigma_{max}}{\sigma_Y} \quad (20.13)$$

$$SFR = \frac{\tau_{max}}{0.5 \cdot \sigma_Y} \quad (20.14)$$

$$\theta = \frac{TL}{GJ} = \frac{(3000)(10)}{11{,}500{,}000\ J} \quad (20.15)$$

$$\text{Area } A = \frac{\pi}{4} \cdot \left(D_2^2 - D_1^2\right) \quad (20.16)$$

Dependencies

An initial dependency table is shown in Figure 20.5. The values shown are obtained by inspection of, and qualitative reasoning about, the various analysis equations above. They are, of course, subject to change as the iteration process proceeds.

It would seem at first glance that the dependency between D_2 and both Failure Ratios would be -3. This is suggested from noting that, for shear stress:

$$\tau = \frac{TD_2}{2J} = \frac{TD_2(32)}{2\pi\left(D_2^4 - D_1^4\right)}$$

$$\tau \propto \frac{D_2}{D_2^4 - D_1^4}$$

However, noting that $D_2 - D_1 = 2t$, and that t is considerably smaller than D_2, we get

$$\tau \propto \frac{D_2}{D_2^4 - \left(D_2 - 2t\right)^4}$$

$$\tau \propto \frac{D_2}{D_2^4 - \left(D_2^4 - \left(4tD_2^3 + \text{---------}\right)\right)}$$

$$\tau \propto \frac{D_2}{t \cdot D_2^3} = \frac{1}{t \cdot D_2^2}$$

	D_2	t
Fail Ratio Tension	-2	-1
Fail Ratio Shear	-2	-1
θ	-3	-2
Area A	+2	+1

Figure 20.5 Initial Dependencies for Example 20.2

Of course, a dependency value of -2 is not precisely correct, but it makes a good place to start. We will see shortly that the actual value is about -2.2. If we had started with -3, no serious problems would have been encountered.

Initial Design

As always in parametric design problems, we begin the guided iteration process by selecting initial values for the design variables as best we can. In this case, we will try D_2 = 1.5 inches and t = 0.05 inches.

Evaluation

To perform evaluation of this design, we first use the analysis procedures to compute the values of the SEPs, and then use the Satisfaction Curves to get the evaluations of individual SEPs. Then we can use the Dominic method to get an overall evaluation. Details of this process are illustrated as we go on with the example.

This problem involves enough repetitive computation that the use of a spreadsheet is of considerable value in reducing computation time (and errors). Figure 20.6 shows a portion of the spreadsheet results developed in this case. Some of the intermediate columns actually used in the spreadsheet have been omitted here to save space. Note that in Trial # 1, both Failure Ratios (for tension and shear) as well as the twist are Unacceptable, so of course the overall evaluation is also Unacceptable.

Guided Redesign

To get an acceptable design, *all* SEPs must be in their acceptable range. In this case, Tension Failure Ratio must be less than 0.8 and the Shear Failure Ratio must be less than 0.6. It appears that the Shear Failure Ratio is the one that is most distant from satisfaction, so we will try to move it into the Acceptable region, say to a new value of 0.50. With a dependency between t and SFR of -1.0, we estimate a required new value for t from:

$$y_j(\text{old}) / y_j(\text{new}) = [\ (x_i(\text{old}) / x_i(\text{new})\ ]^{d_{ij}}$$

$$1.11 / 0.50 = [\ 0.05 / t(\text{new})\ ]^{-1}$$

$$t(\text{new}) = 0.111 \text{ inches}$$

Trial #2 Analysis and Re-Evaluation

With this new value for t, the Tension Failure Ratio has been made Good as shown in Trial 2, Figure 20.6. Also improved, the twist angle is now Poor. However, the Shear Failure Ratio is Poor (see the satisfaction curves in Figure 20.4). We still do not have enough data from only two trial runs to develop a useful Satisfaction Curve for the Area SEP, so the overall evaluation is uncertain.

Guided Redesign

Using the results of the previous trial, we can now compute new dependencies as shown in Figure 20.7. As one example of the calculations involved, here is the solution for the dependency between Shear Failure Ratio and t:

$$y_j(\text{old}) / y_j(\text{new}) = [\ (x_i(\text{old}) / x_i(\text{new})\ ]^{d_{ij}}$$

$$1.11 / 0.564 = [\ 0.05 / 0.111\ ]^{d}$$

$$d = -0.84$$

We show the computation here, but spreadsheet software can be used to compute the new values for the dependencies if desired.

For the next trial design, we choose again to try to bring the Shear Failure Ratio into a better range to a new value of, say, 0.45. This time we will attempt to accomplish our goal by changing design variable D_2. An estimate

Trial No.	D_2	t	I	J	Fail Ratio Tension	Fail Ratio Shear	θ	A	Overall Eval.
1	1.5	0.05	0.060	0.120	1.03 (Unacc.)	1.11 (Unacc.)	0.022 (Unacc.)	0.23 (?)	Unacc.
2	1.5	0.111	.117	.235	0.524 (Good)	0.564 (Poor)	0.011 (Poor)	0.484 (?)	?
3	1.61	0.111			0.448 (Exc.)	0.482 (Fair)	009 (Fair.)	0.522 (?)	?
4	1.61	0.087			.546 (Good.)	.59 (Poor)	.011 (Poor.)	.42 (Good)	Poor
5	1.89	0.087			0.39 (Exc.)	0.42 (Fair)	007 (Exc.)	0.49 (Fair)	Good

Figure 20.6 Summary of Spreadsheet Results for Example 20.2

	D_2	t
Fail Ratio Tension	-2	-0.84
Fail Ratio Shear	-2	-0.84
θ	-3	-0.87
Area A	+2	+0.93

Figure 20.7

Updated Dependency Table After Trial #2 in Example 20.2.

	D_2	t
Fail Ratio Tension	-2.2	-0.84
Fail Ratio Shear	-2.2	-0.84
θ	- 2.8	-0.87
Area A	+1.1	+0.93

Figure 20.8

Updated Dependency Table After Trial 3 in Example 20.2.

of the amount of change required can be found from the dependencies:

$$y_j(\text{old}) / y_j(\text{new}) = [\, (x_i(\text{old}) / x_i(\text{new})\,]^{d_{ij}}$$

$$0.524 / 0.45 = [\, 1.5 / D_2(\text{new})\,]^{-2}$$

$$D_2(\text{new}) = 1.61 \text{ inches}$$

Thus for Trial 3, we use t = 0.111 inches and D_2 = 1.61 inches.

Trial #3 Analysis and Re-Evaluation

The results for Trial 3 are shown in Figure 20.6. We are making progress: now all the SEPs (except Area) are in a better range. However, there is still insufficient data to develop a Satisfaction Curve for the Area. Thus, we will continue attempting to improve the other three SEPs and at the same time gather more data about Area.

Guided Redesign

The new dependency table is shown in Figure 20.8. For another trial, we elect to change t in such a way that the Shear Failure Ratio will be just barely acceptable (at 0.59). This goal is set because it should give some more insight into the smallest feasible area for this problem, and thus help us establish a Satisfaction Curve for Area. Using the dependencies, this calls for a value of t equal to 0.22, so we try that. Here is the computation:

$$y_j(\text{old}) / y_j(\text{new}) = [\, (x_i(\text{old}) / x_i(\text{new})\,]^{d_{ij}}$$

$$0.482 / 0.59 = [\, .111 / t(\text{new})\,]^{-0.84}$$

$$t(\text{new}) = 0.087 \text{ inches}$$

Trial #4 Analysis and Re-Evaluation

The result of Trial 4 is shown in Figure 20.6.

We now can make at least an initial satisfaction curve

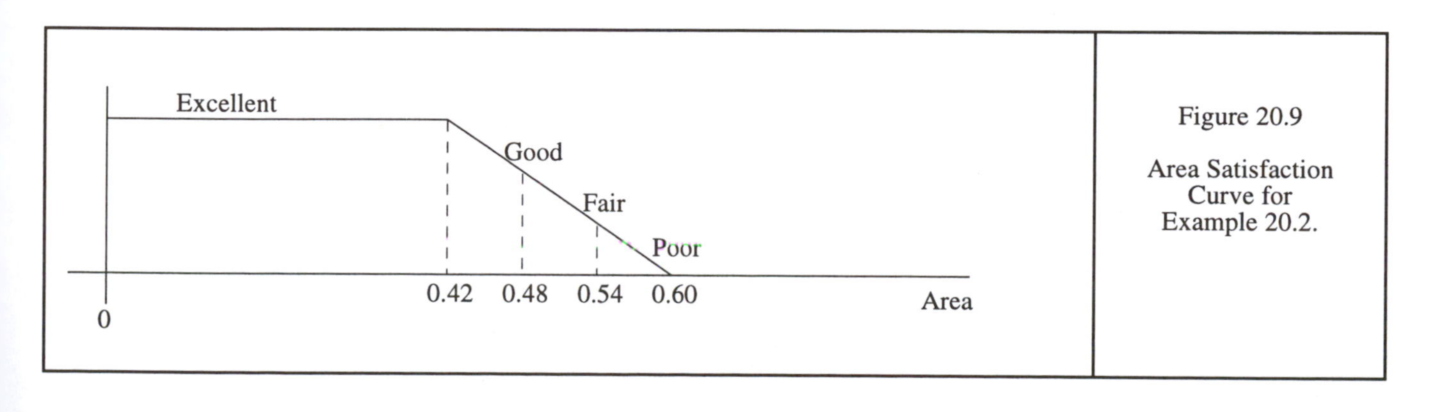

Figure 20.9

Area Satisfaction Curve for Example 20.2.

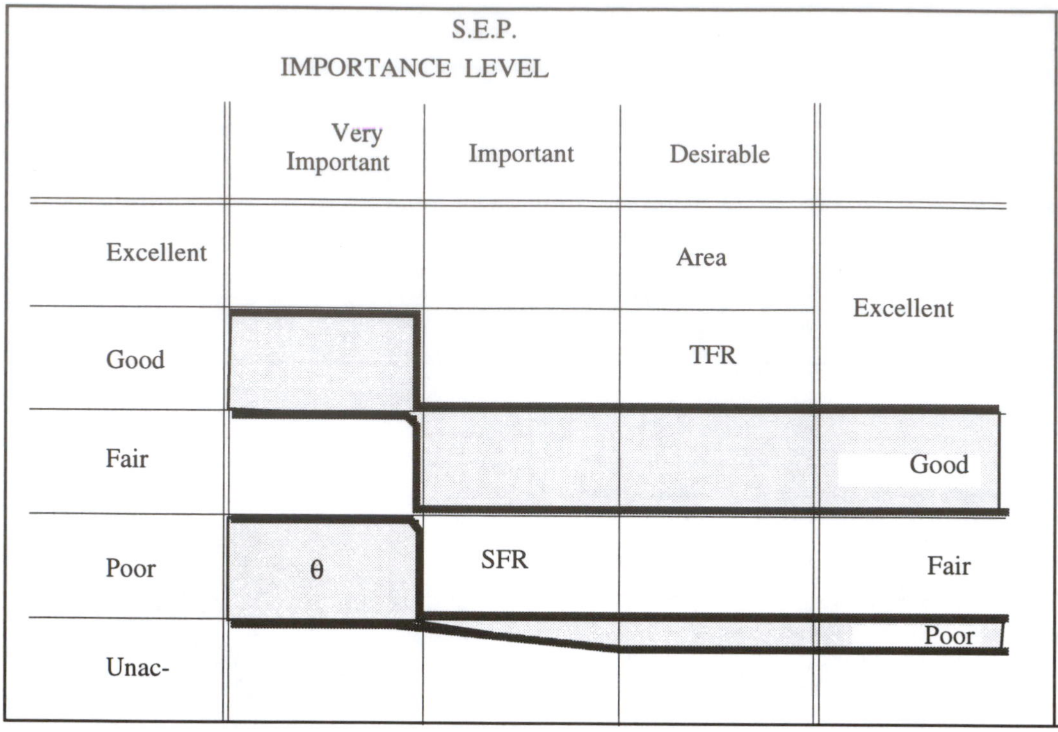

S.E.P. IMPORTANCE LEVEL	Very Important	Important	Desirable	
Excellent			Area	Excellent
Good			TFR	
Fair				Good
Poor	θ	SFR		Fair
Unac-				Poor

Figure 20.10

Dominic Evaluation Table After Trial 4 for Example 20.2

for area. See Figure 20.9. The choice of 0.60 inches as the maximum acceptable value, and of 0.42 as the maximum Excellent value are, of course, judgments made by the designer. But it is clear from inspecting Figure 20.9 that the results of the trials run so far indicate that values for Area when the other SEPs are acceptable will be approximately in this range. If this satisfaction curve does not provide the kind of trade-offs desired, the designer can adjust the area satisfaction curve as more trials are run.

Using the satisfaction curve in Figure 20.9 for area gives an overall evaluation from Trial 4 of Poor. See Figure 20.10. The overall evaluation of Trial 3 can now also be determined; it is also Fair, but for different reasons.

Guided Redesign

For Trial 5, we attempted to get a value for θ of 0.007 (within its Good zone) by changing D_2. The process continues as above, though we will omit additional details here.

It should be noted that there is no assurance that Trial 5 in Figure 20.6 is the best that can be found. Readers with a keen interest in this example may wish to continue the guided iteration process to find a better design.

20.3 Shafts That Transmit Power

Note: An essential reference for design of power transmitting shafts is ANSI B106.1M, 1985 [10].

20.3.1 Pre-Parametric Design Issues: Conceptual Level

Power, Speed, and Torque

It is well known from both mechanics and thermodynamics that the power (P) transmitted by a shaft rotating at constant angular velocity (ω) with torque (T) is given by:

Power = Torque x Angular Velocity.

$$P = T\,\omega \qquad (20.17)$$

If we express power in units of ft-lb/min, torque in ft-lb, and angular velocity in rpm, then this expression becomes:

$$P\,(\text{ft-lb/min}) = T\,(\text{ft-lb})\,(2\,\pi\,N) \qquad (20.18)$$

Fluctuating Stresses and Fatigue

Shafts that transmit power have bending moments as well as torsional loads. Bearings act as support points; the bending loads are caused by the gears or pulleys or turbine or compressor stages that provide for the inputs and outputs of power to and from the shaft at points away from the bearings. As a result, like many statically loaded shafts, power transmitting shafts experience combined bending and torsional stresses.

There is, however, an important difference between static shafts with combined stresses and power transmitting shafts. A power transmitting shaft will have *fluctuating* bending loads, a fact that often makes material fatigue a crucial issue. To visualize the fluctuating loads, suppose the shaft shown in Figure 20.1 is rotating. Then imagine yourself standing and riding around at a point on the surface. When you are at the top during a rotation, the material on which you are standing will be in tension due to the bending load.

When you are at the bottom, the same material will be in compression. Thus there is a complete reversal of the internal bending stresses in the shaft with every rotation. Hence the concern for possible fatigue failure — a concern that is enhanced when there are also stress concentrators present.

When stresses fluctuate, then the concept of *endurance limit* becomes relevant. The basic issues of endurance limit and fatigue are discussed in Appendix 20A.1.3.

20.3.2 Pre-Parametric Design Issues: Configuration Level

The configuration issues in designing power transmitting shafts are essentially the same as for static shafts. Circular shafts are most common; however, square shafts and splines are also used. Hollow circular shafts are helpful where weight is important, and may also help in some cases where vibration is a concern. Deflections may become important at bearing points and at certain other locations. An example is that gears must mesh properly. Where possible, of course, avoiding the use of stress concentrators is obviously important.

Stress Raisers

Values for stress concentration factors (K's) for solid round shafts have been discussed above. The K values are approximately the same for static or fluctuating loads, but with fluctuating loads, the K's are applied only to the fluctuating stress range.

Keyways are often used to transmit force from shafts to gears, pulleys, etc., and they create a stress increase. Stress concentration factors for keyways in shafts are in the vicinity of 1.3. Keyways should be accounted for by computing the bending stresses with a diameter equal to the shaft diameter *minus* the keyway depth.

The equations that account for stress concentration given above apply not only to shafts, but also to plates and flat bars. Values for K in plates and flat bars vary from about 1.1 to over 2.4 depending on the configuration and parametric values. Thus it is important that they be properly included in the analyses for stress. For data on stress concentration factors of many kinds, see [3, 9]. As always, the greater the size change, and the smaller the internal fillet radii, the greater the stress concentration.

20.3.3 An ASME Code Design Procedure

An obsolete ASME Code (B17c-1927) presents a design procedure — a good example of a domain-specific design procedure — that accounts for fatigue as well as the effects of expected shock loading in shafts by defining two empirical factors (C_m and C_i) to be used in the following equations. The form of these equations is clearly inspired by Equations 20.3 and 20.4 above.

$$\tau_{max} = \left[\left(\frac{C_m \cdot \sigma}{2}\right)^2 + \left(C_i \cdot \tau\right)^2\right]^{0.5} \quad (20.19)$$

$$\tau_{max} = \frac{16}{\pi \cdot D^3} \cdot \left[\left(C_m \cdot M\right)^2 + \left(C_i \cdot T\right)^2\right]^{0.5} \quad (20.20)$$

Values for C_m and C_i are listed in the Code. The approximate values for rotating shafts vary from 1.0 to 3.0 as shown here:

	Load Slowly Applied	Load Rapidly Applied	Severe Shock Loading
C_m	1.5	1.5 - 2	2 - 3
C_i	1.0	1.0 - 1.5	1.5 - 3

This procedure is not suitable for a final design, but it does provide a quick way for designers to determine the approximate range for the diameter of a shaft. The current ASME Code is [10].

20.3.4 Problem Formulation

When a shaft is being used to transmit power, there are additional Solution Evaluation Parameters involved (e.g., having to do with fatigue) that are not present in the design of static shafts. Otherwise the formulation is much the same.

Problem Definition Parameters

- Power inputs, outputs, and locations
- Support (i.e. bearing) locations
- frequencies present
- Shaft length
- Geometric limitations (if any)
- Design variables:
- Shaft diameter, D (assuming a solid shaft)
- Shaft material (E, G, Yield Strength, Tensile Strength, Endurance Limit)

Solution Evaluation Parameters

- Tension (or Compression) Failure Ratio, TFR
- Shear Failure Ratio, SFR
- Fatigue Failure Ratio, FFR

Note: Refer to Figure 20A.8 in Appendix 20A. This SEP is the distance AB divided by the distance AC. When it is less than 1.0, the design is in the safe working zone. When AB exceeds AC, the design is in the failure zone. Note that stress concentration must be included in the computation of stress range.

Safety factors are easily included in this formulation. For example, a safety factor of 2 requires that the failure stress ratio be no more than 0.5 instead of 1.0. This will be illustrated in the example below.

- Cost and/or Weight
- Deflection - Lateral
- Lateral deflection limits may have to be specified at

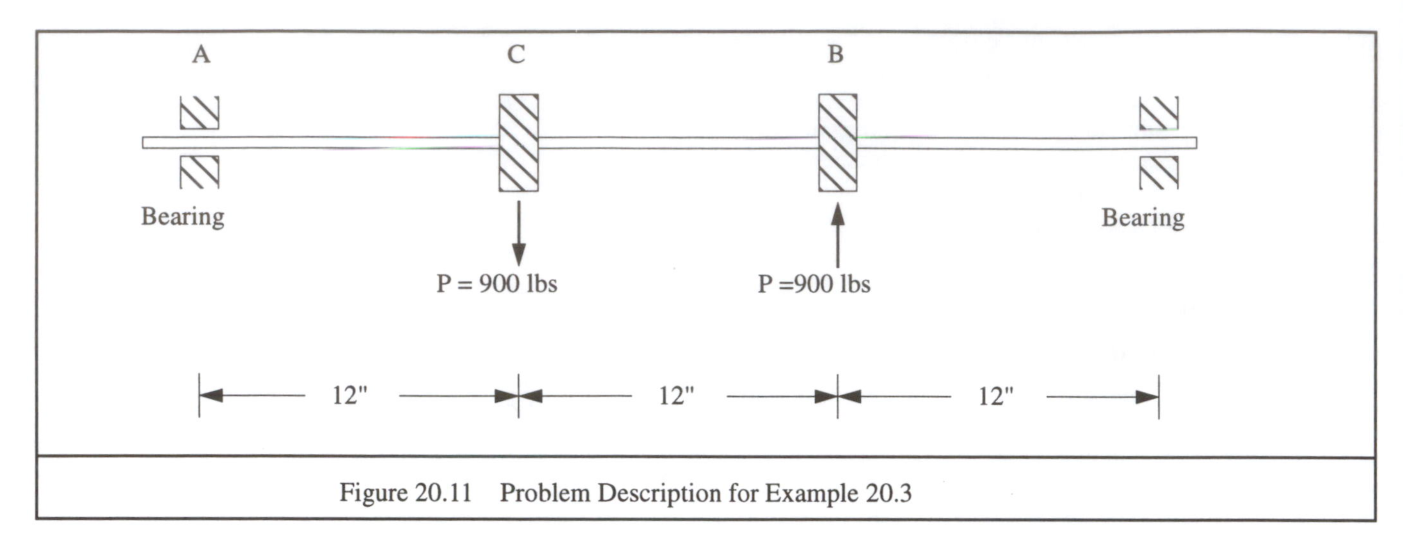

Figure 20.11 Problem Description for Example 20.3

particular point such as at bearing or gear locations.

- Deflection - Angular

  Angular deflection may be important at bearing locations, especially if self-aligning bearings are not to be used.

- Vibration

  A rotating shaft will have a natural frequency of vibration. Clearly this frequency should not coincide closely with the rotation speed or with the natural frequency of connected parts and systems. We do not consider shaft vibration issues in this book.

Analysis Equations Required

- Tension (or Compression) Failure Ratio = TFR

$$TFR = \frac{\text{Max Tension or Compression Stress}}{\text{YieldStress}}$$

- Shear Failure Ratio = SFR

$$SFR = \frac{\text{Max Shear Stress}}{\text{0.5 Yield Stress}}$$

- Fatigue Failure Ratio = FFR (See Eq 20A.20)

$$FFR = \frac{\text{(Stress Range)(Stress Conc Factor)}}{\text{Fatigue Failure Stress}}$$

- Deflections: These depend on the location of loads and supports, and are determined by methods from beam theory normally included in courses and texts on strength of materials.

Dependencies

For a given material and a solid circular shaft:

- Maximum internal stresses vary approximately as D^{-3}.
- Cost and weight (given material) vary approximately as D^2
- Deflections vary approximately as D^{-4}
- Deflection angles vary approximately as D^{-3}

20.3.5 Example 20.3 - A Rotating Shaft Design

As an example of using guided iteration to design rotating shafts, we consider the problem defined in part by Figure 20.11. The torque input at B (and output at C) is 5400 in-lbs. The material is to be hot-rolled 1020 steel whose surface has been machined. The shaft is solid circular. There is a desired limitation on the angle of bending of the shaft at point A, and there are keyways to mount the gears or pulleys. The design task is to decide on the diameter of the shaft.

Problem Formulation

Problem Definition Parameters

See Figure 20.11. Also, for the material:

- Tensile Strength = σ_x = 65,000 psi
- Yield Strength = σ_Y = 43,000 psi
- Endurance Limit = σ_e = 27,000 psi
- Elastic Modulus = E = 30,000,000 psi

Design Variable: D

Solution Evaluation Parameters (and their assumed Importance Levels)

- Shear Stress Failure Ratio (Important)
- Tensile Stress Failure Ratio (Desirable)
- Fatigue Failure Ratio (Very Important)
- Slope of Shaft at A (Important)
- Weight of Shaft, W (Desirable)

Satisfaction Curves for the SEPs

See Figure 20.12. Note that the Tensile and Shear Failure Ratios have been treated as hard constraints with a safety factor of 1/0.7 = 1.4. We cannot yet place numerical values on the Satisfaction Curve for Area (which we use as a surrogate for weight or cost).

Analysis Equations for the SEPs

- Shear Failure Ratio = τ_{max} / τ_{crit} (20.21

where τ_{max} is found from Equation 20.19 and τ_{crit} is found from Equation 20A.1. The maximum stresses in the shaft in this case are located at the gears or pulleys. The magnitude of the bending moment at these points is

- M = (P / 3) 12 = 3600 inch-lbs. (20.22)

- Tensile Stress Failure Ratio = σ_{max} / σ_Y (20.23)

where σ_{max} is the maximum of the stresses computed in Equation 20.20.

- Fatigue Failure Ratio = $K \sigma_r / \sigma_{rF'}$ (20.24)

where K is the stress concentration factor. FFR is found from Equations 20A.19 through 20A.25. See also Figure 20A.8. The stress range, σ_r, in this case of completely reversing stress is just twice the maximum stress, σ_{max}

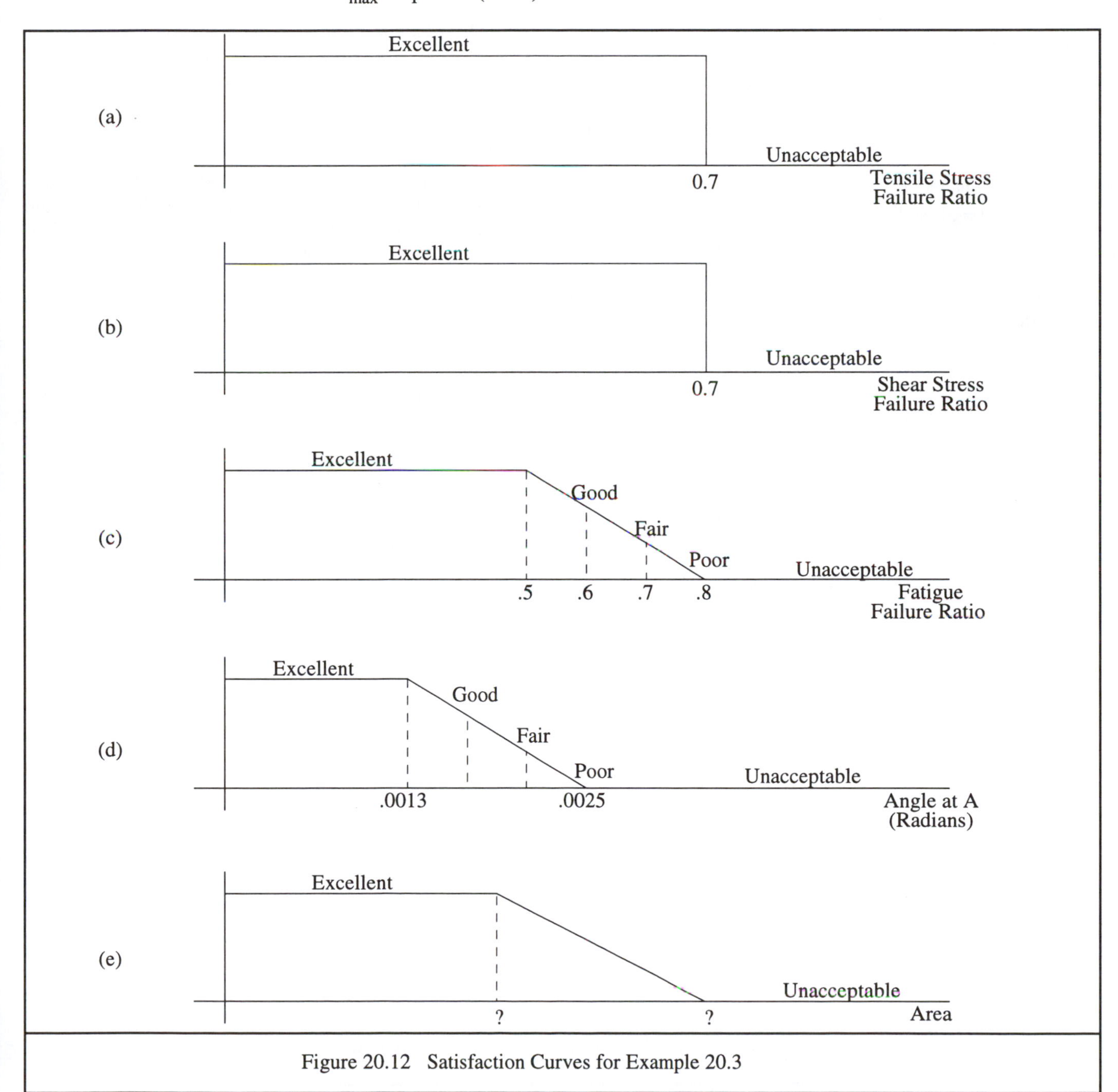

Figure 20.12 Satisfaction Curves for Example 20.3

	D
SFR	-3
TFR	-3
FFR	-3
q	-4
Area	2

Figure 20.13 Initial Dependency Table for Example 20.3

- Slope of Shaft at A = $3 P a^3 / 6 L E I$ (20.25)

where a = 12 inches

L = 3 a = 36 inches

I = $\pi D^4 / 64$

Note: Equation (20.25) for the slope is found from beam theory by superposition. See, for example Figure 1.14 Case 7 in [3].

- Area (which we will use as a surrogate for weight or cost) = A = $\pi D^2 /4$ (20.26)
- Stress Concentration Factor: Based on information found in the literature that present data and discussion of stress concentration factors for keyways, we use a factor in this case of K = 1.6.

Initial Dependencies See Figure 20.13.

Initial Design

We will first try D = 1.0 inches. (At least this choice makes the initial round of computations easy!)

Analysis and Evaluation Results Trial 1

Once again, it is helpful to set up a spreadsheet to deal with the computational details. Figure 20.14 shows a portion of the results from the spreadsheet developed for this example. (A number of intermediate columns are not shown.) Note that the analysis and evaluation results for Trial 1 with D = 1.0 inches are:

Shear Failure Ratio = 1.54 (Unacceptable)
Tension Failure Ratio = 1.20 (Unacceptable)
Fatigue Failure Ratio = 5.46 (Unacceptable
Slope = 0.0147 (Unacceptable)

Redesign

Obviously a 1.0 inch diameter is too small. For a second trial, we elect to try to get the Fatigue Failure Ratio into the acceptable range at about 0.7. Using the initial dependency estimate of -3, we compute a new trial diameter D of:

$$y_j(\text{old}) / y_j(\text{new}) = [\, (x_i(\text{old}) / x_i(\text{new}) \,]^{d_{ij}}$$

$$5.46 / 0.70 = [\, 1.0 / D(\text{new}) \,]^{-3}$$

D(new) = 1.98 inches

We therefore try D = 2.0 inches.

Analysis and Evaluation Results Trials 2 and 3

See Figure 20.14. With the value for D found above, four of the SEPs are in their acceptable range, but we don't yet have a satisfaction range established for area. Moreover,

Trial No.	Diameter D	t_{max}	s_{range}	Shear Fail Ratio	Tension Fail Ratio	Fatigue Fail Ratio	Slope q	Area A	Overall Eval.
1	1	33,070	66,140	1.54 (Unacc.)	1.20 (Unacc.)	5.46 (Unacc.)	0.0147 (Unacc.)	.785	(Unacc.)
2	2	4,130	8,270	0.19 (Exc.)	0.15 (Exc.)	0.51 (Good)	0.0009 (Exc.)	3.14	(Fair)
3	1.5	9,800	19,600	0.46 (Exc.)	0.35 (Exc.)	1.27 (Unacc.)	0.0029 (Unacc.)	1.76 (Good)	(Unacc.)
4	1.8	5,670	11,340	0.26 (Exc.)	0.21 (Exc.)	0.71 (Poor)	0.0014 (Good)	2.54 (Fair)	(Fair)
5	1.81	5,580	11,150	0.26 (Exc.)	0.20 (Exc.)	.69 (Fair)	0.0014 (Good)	2.57 (Fair)	(Fair)
6	1.90	4,820	9,640	0.22 (Exc.)	0.17 (Exc.)	0.60 (Good)	0.0011 (Exc.)	2.83 (Poor)	(Fair)

Figure 20.14 Sample Spreadsheet Results for Example 20.3

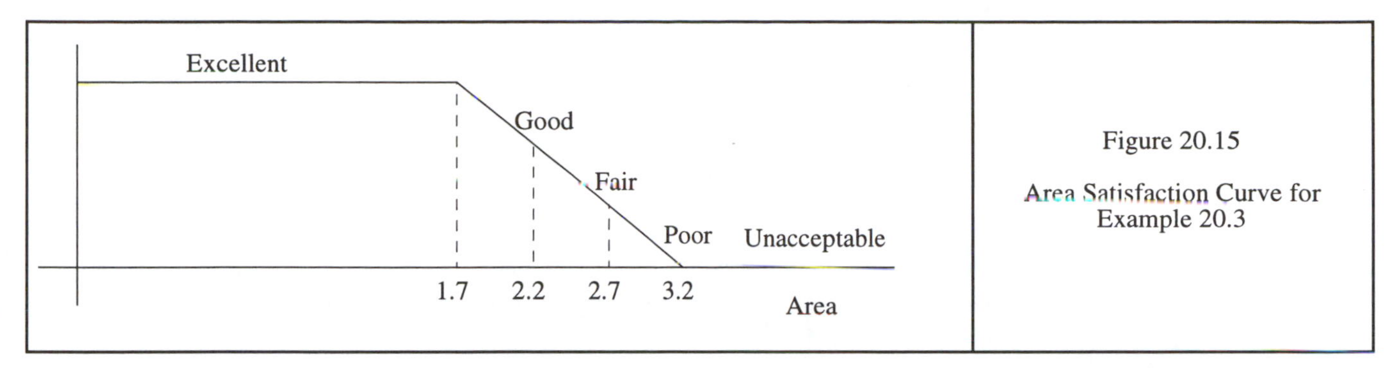

Figure 20.15

Area Satisfaction Curve for Example 20.3

with only two trial runs, it is not really possible to establish reasonable values for the maximum acceptable area and the maximum Excellent range in order to establish the curve.

Just to get some more information about Area, therefore, we try a value of D intermediate to the previous runs: D = 1.5. The results are shown in the third row of Figure 20.14.

With this information, now it is possible to see that an Area above about 3.2 is going to be too high, and that getting an Area below about 1.7 is going to be impossible. Thus we establish the area satisfaction curve shown in Figure 20.15. With this satisfaction curve established, we have the following evaluations of Area for Trials 1 and 2 which can be added to those shown in Figure 20.14.

Trial Number	Area	Area Evaluation
1	.785	Good
2	3.14	Poor

Then we can also get Overall Evaluations for these trials; Figure 20.16 shows that the overall evaluation for Trial 2 is Fair as shown in parentheses in Figure 20.14.

Figure 20.16 shows the Overall Evaluation matrix for Trial 5. Note, for example, the location of the Fatigue Failure Ratio in the matrix. It is a Very Important SEP, and its evaluation is Fair; thus it is placed in the matrix as shown. Area is also evaluated as Fair, but it is Desirable — so it is placed as shown. The overall evaluation is Fair because the Fatigue Failure Ration is below the Fair line in the matrix.

Redesign

For the next iteration (the sixth), we will try to improve the Fatigue Failure Ratio. Note in Figure 20.17 that we can improve the overall evaluation from Fair to Good if we can only improve the Fatigue Failure Ratio itself from Fair to Good.

We can update the initial dependency between D and

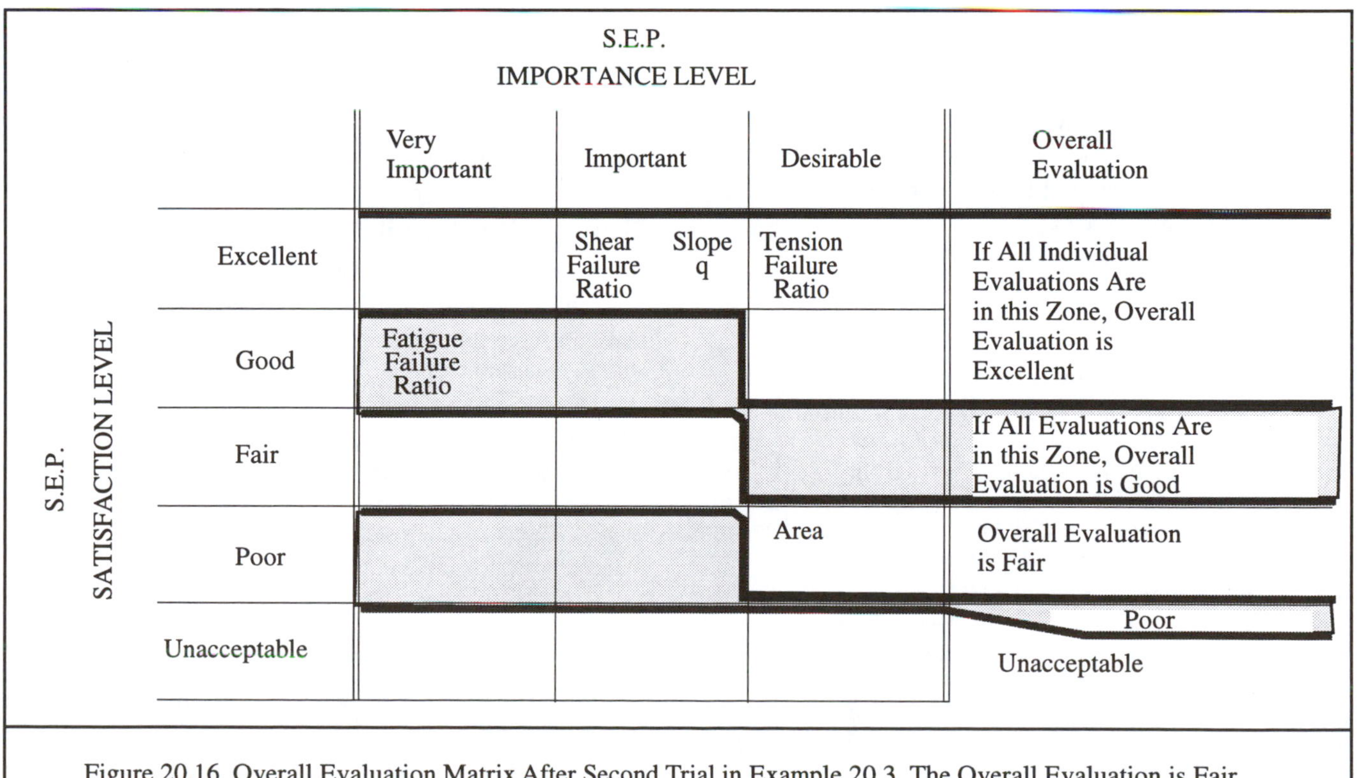

Figure 20.16 Overall Evaluation Matrix After Second Trial in Example 20.3. The Overall Evaluation is Fair.

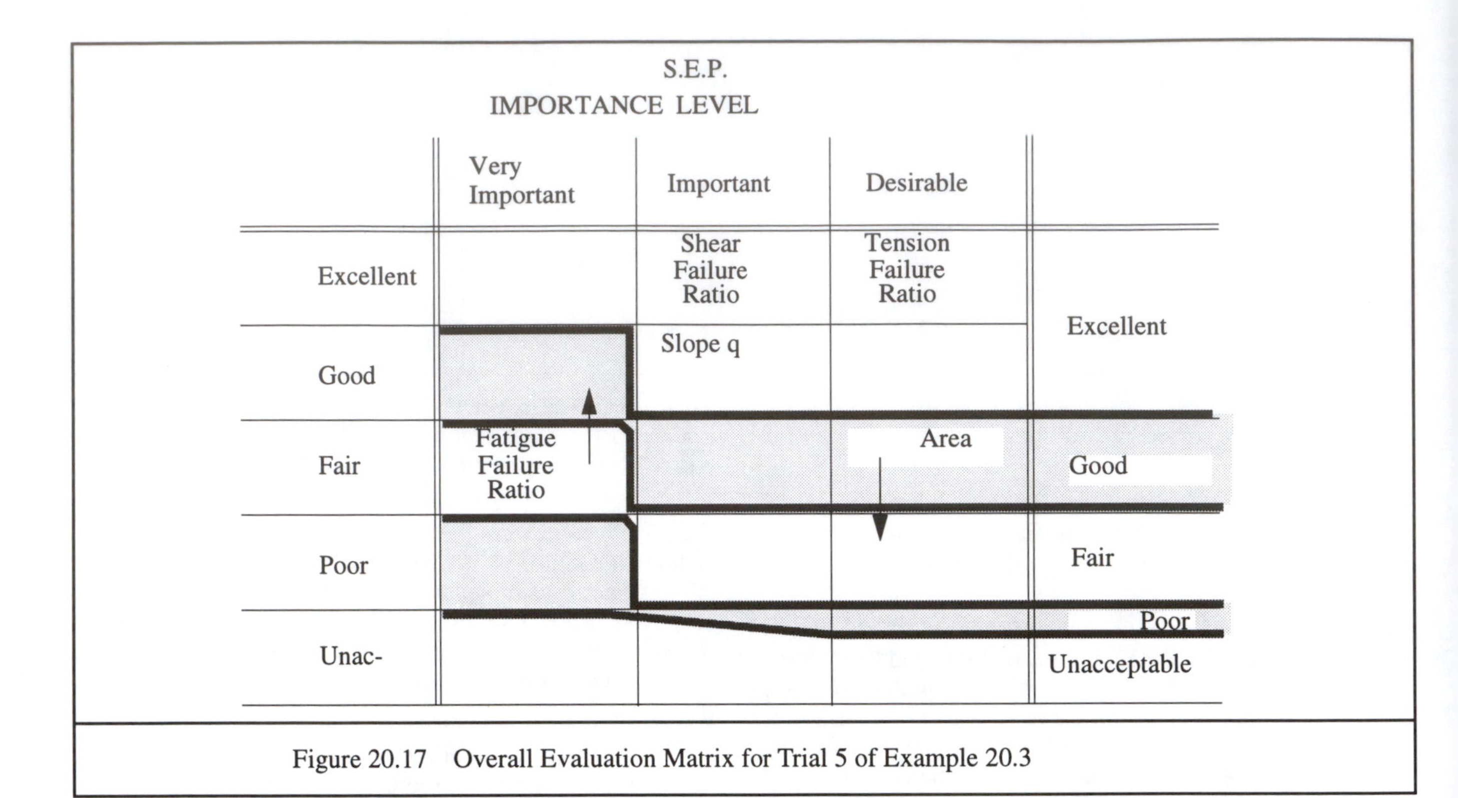

Figure 20.17 Overall Evaluation Matrix for Trial 5 of Example 20.3

FFR using the results of Trials 3 and 4. (We could use Trials 4 and 5 but the change in D was so small that numerical accuracy would be questionable.)

$$y_j(\text{old}) / y_j(\text{new}) = [(x_i(\text{old}) / x_i(\text{new})]^{d_{ij}}$$

$$1.27 / 0.71 = [1.5 / 1.8)]^{d_{ij}}$$

$$d_{ij} = -3.2$$

Thus we use the dependency of -3.2 to try to change the Fatigue Failure Ratio from 0.69 to 0.59.

$$y_j(\text{old}) / y_j(\text{new}) = [(xi(\text{old}) / xi(\text{new})]^{d_{ij}}$$

$$0.69 / 0.59 = [1.81 / D(\text{new})]^{-3.2}$$

$$D(\text{new}) = 1.90 \text{ inches}$$

We therefore try D = 1.9 inches.

Analysis and Evaluation

The analysis and evaluation results for this trial are included in Figure 20.14 We have indeed been able to change the evaluation of Fatigue Failure Ratio from Fair to Good; however, as indicated by the arrows in Figure 20.17, the area evaluation has now dropped to Fair so the overall evaluation has not been improved.

This is about as far as the guided iteration method can take us in this problem. Clearly the best solution is in the vicinity of D = 1.8 to 1.9 or so. Since the exact values used to construct the area satisfaction curve are a bit arbitrary, the designer must now make the final choice of trade-off between area and Fatigue Failure Ratio by considering the numbers in Figure 20.14. With the computations on a spreadsheet, exploring a few more values of D is also an easy option.

20.4 Springs

Note: An essential reference for design of springs is *The Spring Design Handbook* [12].

20.4.1 Pre-Parametric Design Issues: Conceptual Level

The physical concept of a device we call a *spring* is that of a force proportional to its displacement. For a linear spring, the usual case:

$$F = k\ x \tag{20.27}$$

where the constant k is called a *spring constant*, x is the displacement, and F is the force.

The concept applies also to torque and angular displacement. If there is torque proportional to its angular displacement (θ), then the concept is expressed as

$$\text{Torque} = k\ \theta. \tag{20.28}$$

A designed object with such spring behavior stores energy as it is displaced. Assuming no dissipation, the amount of energy stored in the spring in the linear case equals the work done on it, and so is:

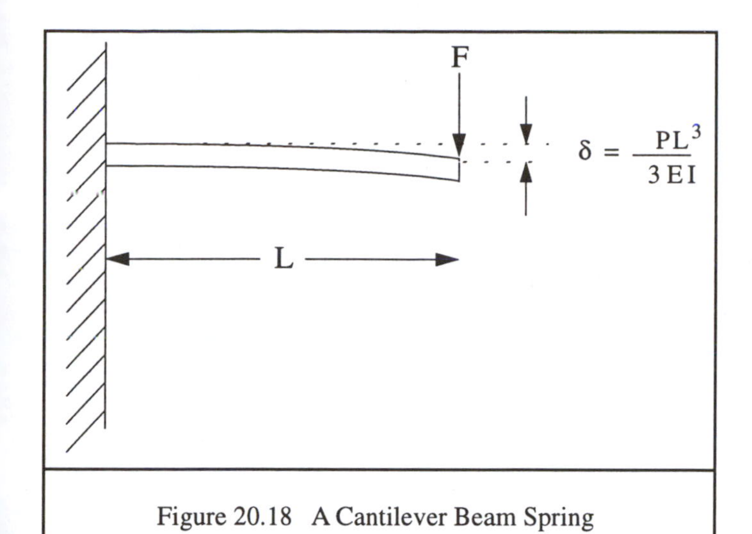

Figure 20.18 A Cantilever Beam Spring

$$\text{Work} = \text{Energy Stored} = \int_{\delta_1}^{\delta_2} F dx = \int_{\delta_1}^{\delta_2} kx dx \qquad (20.29)$$

$$\text{Work} = \frac{1}{2} \cdot k \cdot \left(\delta_2^2 - \delta_1^2\right) \qquad (20.30)$$

This physical concept of "springness" is exhibited by a number of mechanical components in addition to those objects we actually call "springs". Consider, for example, the cantilever beam shown in Figure 20.18. When a load P is applied, it is resisted by a force proportional to P; thus the beam supported and loaded in this way fulfills the concept of a spring. To compute the spring constant, from beam theory:

$$\delta = (PL^3) / (3 E I) \qquad (20.31)$$

Thus for this "spring", the spring constant k is

$$k = P / \delta = (3 E I) / L^3 \qquad (20.32)$$

The energy stored for a beam deflection from 0 to δ is

$$\text{Energy} = (3/2\ E\ I\ \delta) / L^3 \qquad (20.33)$$

There exist useful devices called "constant force springs" that produce a constant force when displaced. They probably should not be called "springs", since their force is not proportional to displacement, but they are.

20.4.2 Pre-Parametric Design Issues: Configuration Level

There are a number of configurations (in addition to beams) that exhibit force proportional to displacement behavior. These include, but are not limited to: helical springs (for compression, tension, or torsion); spiral springs (for torsion); flat springs (which are like very thin beams), washer springs (curved and Belleville types). Figure 20.19 shows some examples of these configuration types. A bar in torsion is also a spring. For a more complete set of spring types, readers are referred to *The Handbook of Spring Design* [12].

When wire is used, as in helical springs, the most common cross sectional configuration of the wire is solid round. However, square or rectangular cross sections are occasionally used.

Materials

For traditional springs like those shown in Figure 20.19 there are many materials available from which designers may choose. The choice is in part determined by the tensile strength of the material, which influences the spring constant, cost, and other performance factors (espe-

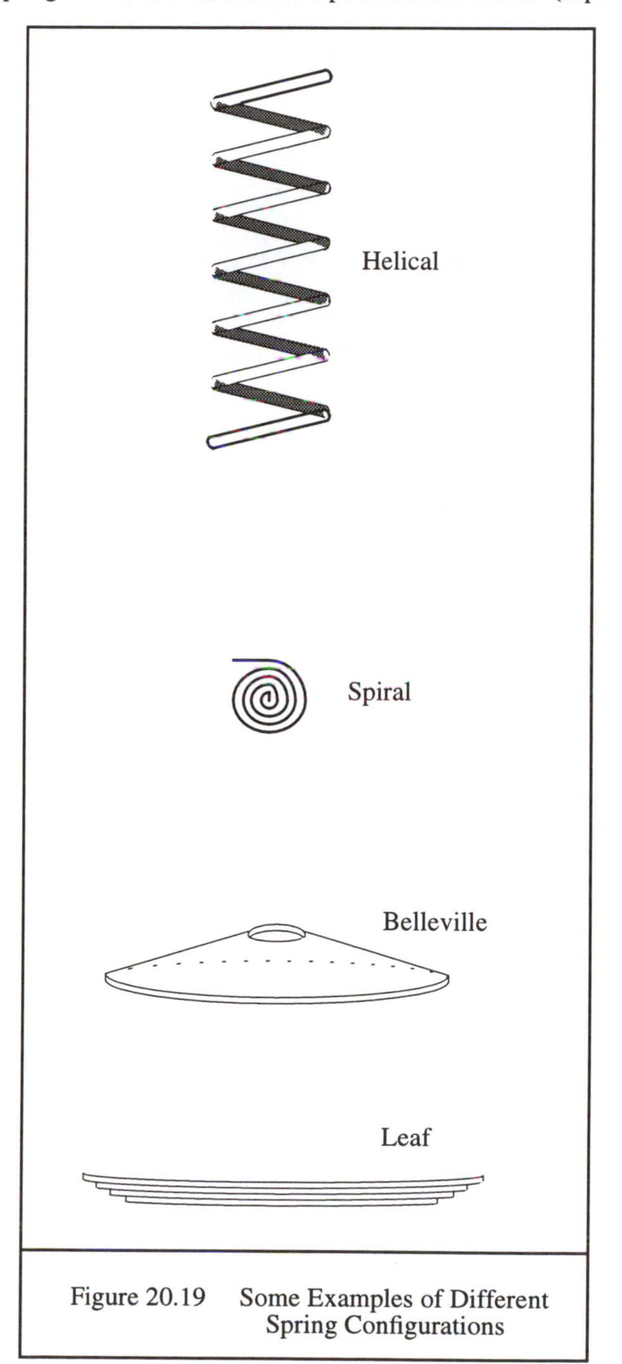

Figure 20.19 Some Examples of Different Spring Configurations

cially fatigue). However, the choice of spring material is also strongly influenced by other (potentially crucial) concerns — especially temperature, corrosion, and possible electrical or thermal conductivity.

For most helical springs used in ordinary applications, round heat treated steel wire is available in three grades: music wire, hard drawn wire, and oil tempered wire. Music wire is the most commonly used because of its more uniform properties. Figures 20A.9 and 20A.10 in Appendix 20A (pages 20-38 and 20-39) give properties of selected wire materials. Please note that the properties given are to be used for wire springs only, not for the materials in other applications.

Scope

We will limit the discussion in this book to the parametric design of helical springs. Readers interested in a full description of spring materials and detailed spring design procedures for all kinds of springs are referred to *The Handbook of Spring Design* [12].

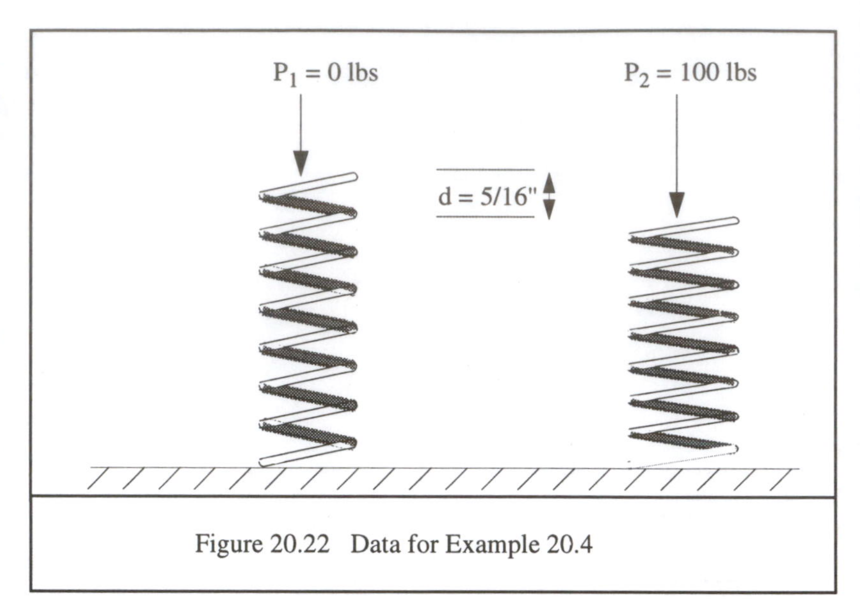

Figure 20.22 Data for Example 20.4

20.4.3 Example 20.4 - Helical Spring

Analysis procedures for helical springs are developed and presented in Appendix 20A.

Refer to Figure 20.22. The problem is to design a helical compression spring that must compress by 5/16 inches after the applied load increases from 0 to 100 pounds. Other conditions to be satisfied are that the maximum allowed diameter of the coil D(max) is 0.75 inches and the maximum allowed length when the spring is unloaded is L(max) = 2 inches. (Such limits on the space available for springs are common, and can often make the spring designers task difficult.)

Problem Formulation

The *Problem Definition Parameters* are:

P_1 = 0 lbs

P_2 = 100 lbs

Desired deflection, = 5/16 inches

Max Diam, D(max) = 0.75 inches

Max Length, L(max) = 2 inches.

Temperature: Normal

Environment: No corrosive elements

End conditions: Ground flat; open

The *design variables* are:

Material

Wire diameter, d

Coil diameter, D

Number of Active Coils, Na

The *Solution Evaluation Parameters* and their assumed importance levels are:

Spring Constant, k [Very Important]

Volume of wire used (This is a surrogate for cost), V [Important]

Shear stress in the wire [Desirable]

Note: If a spring is subject to cyclic loading, then the fatigue failure ratio based on the endurance limit will be a critical SEP as well.)

Satisfaction curves for these SEP's:

Refer to Figure 20. 23(a). The spring constant required by the problem definition parameters is:

$$k = P / \delta_2 = 100 / (5/16) = 128 \text{ lbs/inch.}$$

For shear stress (Figure 20.23(b), we plot the dimensionless parameter, τ_b / τ_y, so that the satisfaction curve need not be changed if the choice of material is changed.

Refer to Figure 20.23(c). We assume here that we have little previous familiarity with spring design, and so have no knowledge to support assigning numbers to V_1 and V_2 in the satisfaction curve for volume. We will show shortly how to get a satisfaction curve for volume in this case.

Initial Design

We will solve this problem by first assuming a trial wire material so that the shear modulus, G, and the ultimate tensile strength of the wire are established. With the material properties known, we can design a spring as an all-numeric problem. Then another trial material can be assumed, and the process repeated until we are certain that a near optimum spring has been found. Of course, all the materials tried will have to meet any environmental requirements (e.g., temperature, corrosive atmosphere, etc.) and will also have to be available.

Since we assume no knowledge of the domain, we make the following design variable guesses for an initial design:

Material: Music wire
Coil Diameter: D = 1.50 inches
Wire Diameter: d = 0.060 inches
Number of Active Coils: Na = 10.
G = 11,500,000 psi (See Equation 20A.2.)

τ_{crit} = 0.45 x (Ultimate Tensile Stress) (20.34)
= 0.45 (296,000) = 133,200 psi.

Note: See Figures 20.A9 and 20.A10 for wire property data

Note: The wire in a helical compression or extension spring is in pure torsion. The spring manufacturer's recommended [12] working shear (yield) stress for music wire of 0.45 times the ultimate stress is based on experimentation and knowledge generalized from their experience.

Analysis and Evaluation

Using the analysis procedures from Appendix 20A.2, the value for the spring index, C, and the values for the SEP's resulting from our initial design are:

Spring Index C = 25
Spring Constant k = 0.552 lb/inch
Shear Stress τ_{max} = 1,812,000 psi
Volume V = 0.160 cubic inches.

Since both k and shear stress evaluate to Unacceptable (See Figure 20.23a), the overall evaluation is Unacceptable also. These results are recorded as Trial 1 in Figure 20.24, which is extracted from the spreadsheet prepared for performing the analyses for this problem.

Initial Dependencies

Before choosing a second trial design, we must have some dependencies to guide our choices. However, with no familiarity of the domain assumed, we cannot make educated estimates in this case. There is a way, however, to generate a set of dependencies by essentially conducting experiments. Here is the procedure:

1. Select a design variable, and make a small change in its value. Do not change the other design variable values.

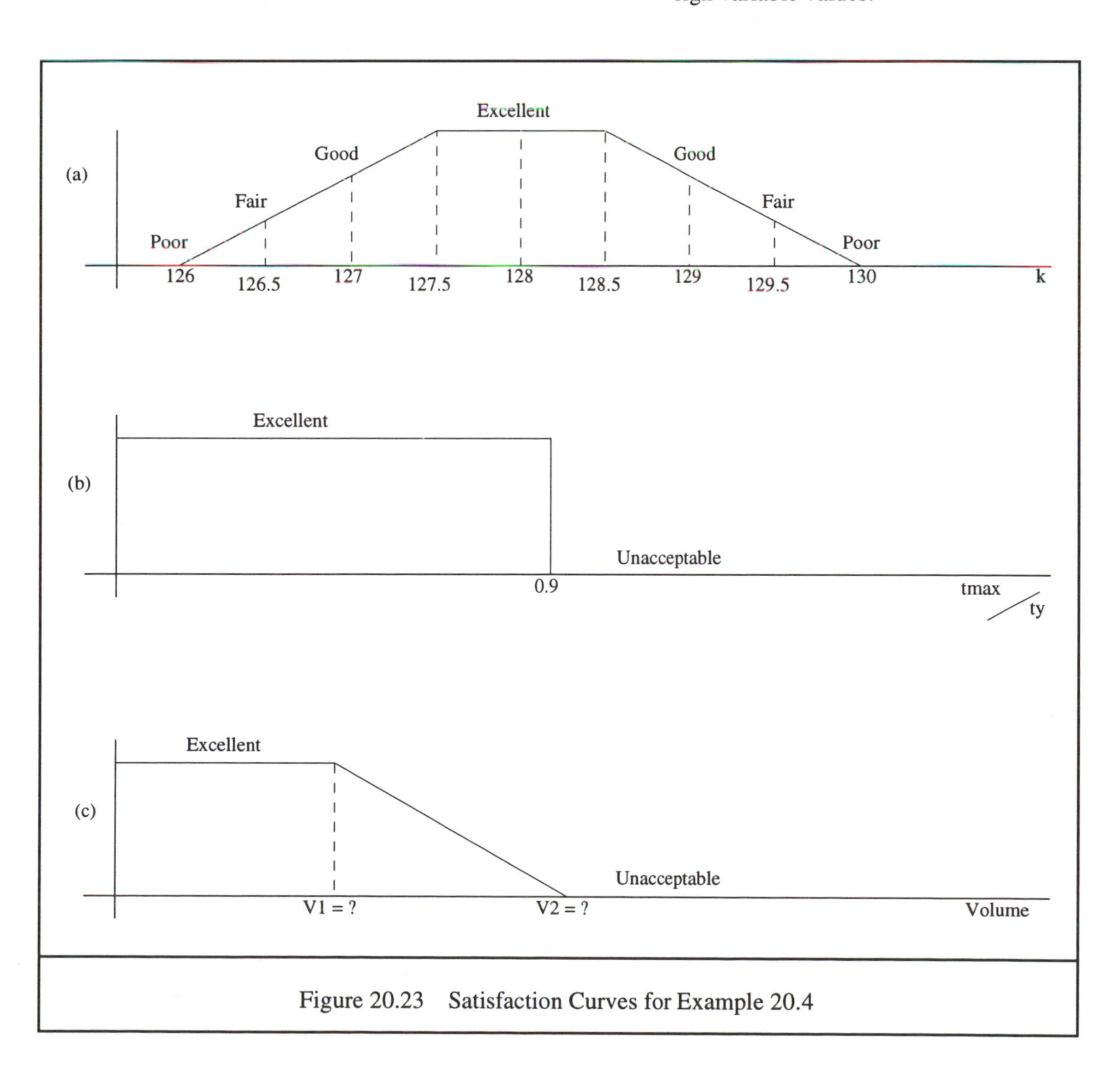

Figure 20.23 Satisfaction Curves for Example 20.4

Trial No.	Coil Diameter D	Wire Diameter d	N_a	t	C	k	τ/τ_y	Volume	Overall Eval.
1	1.50	0.06	10	133,200	25	0.552 (Unacc.)	13.6 (Unacc.)	0.160	(Unacc.)
2	1.7	0.06	10	133,200	28.33	0.379 (Good)	15.37 (Unacc)	0.181	(Unacc/)
3	1.7	0.08	10	126,900	21.25	1.198 (Unacc.)	6.86 (Unacc.)	0.322	(Unacc.)
4	1.7	0.08	11	126,900	21.25	1.090 (Unacc.)	6.86 (Unacc.)	0.349	(Unacc.)
5	1.7	0.13	11	112,950	13.08	7.597 (Unacc.)	1.83 (Unacc.)	0.922	(Unacc.)
6	0.72	0.13	11	112,950	5.54	100.0 (Unacc.)	0.821 (Exc.)	0.390	(Unacc.)
7	0.663	0.13	11	112,950	5.10	128.07 (Exc.)	0.76 (Exc.)	0.359	(Unacc.)
8	0.663	0.10	11	121,950	6.63	44.8 (Unacc.)	1.513 (Unacc.)	0.213 (Poor)	(Unacc.)
9	0.467	0.10	11	121,950	4.67	128.3 (Exc.)	1.104 (Unacc.)	0.150 (Good)	(Unacc.)
10	0.467	0.107	11	120,000	4.36	168.2 (Unacc.)	0.923 (Unacc.)	0.172 (Fair)	(Unacc.)
11	0.511	0.107	11	120,000	4.78	128.4 (Exc.)	0.999 (Unacc.)	0.188 (Fair)	(Unacc.)
12	0.450	0.13	11	112,950	4.26	188.0 (Unacc.)	0.893 (Exc.)	0.165 (Good)	(Unacc.)
13	0.450	0.97	11	122,000	4.64	127.0 (Good)	1.185 (Unacc.)	0.136 (Exc.)	(Unacc.)
⋮	⋮	⋮	⋮	⋮	⋮	⋮	⋮	⋮	⋮
18	0.557	0.114	11	120,000	4.89	127.7 (Exc.)	0.898 (Exc.)	0.232 (Poor)	(Poor)

Figure 20.24 Sample Spreadsheet Results for Example 20.4

	D	d	N_a
k	-3.0	4	-1
t_{max}/t_y	0.98	-3	0
Volume	1.0	2	.84

Figure 20.25 Initial Dependencies for Example 20.4

	D	d	N_a
k	-3.0	4	-1
t_{max}/t_y	0.88	-3	0
Volume	1.0	2	.84

Figure 20.26 Dependencies After Trial 10 for Example 20.4

2. Compute the SEP values for the new design.
3. Compute the dependencies for all the SEP's relative to the changed design variable. Enter them into a dependency table.
4. Go back to step 1 and repeat the process with another design variable until all design variables have been changed and the corresponding dependencies computed. You will now have a complete dependency table.

Implementing this procedure in the spring example, we perform trial runs 2, 3, and 4 shown in Figure 20.24. The dependencies computed from these trials are shown in Figure 20.25. As a sample computation for one of these dependencies, here is how the dependency between D and k is computed:

$$y_j(\text{old}) / y_j(\text{new}) = [(x_i(\text{old}) / x_i(\text{new})]^{d_{ij}}$$

$$0.552 / 0.379 = [1.5 / 1.7]^{d_{kD}}$$

$$d_{kD} = -3.0$$

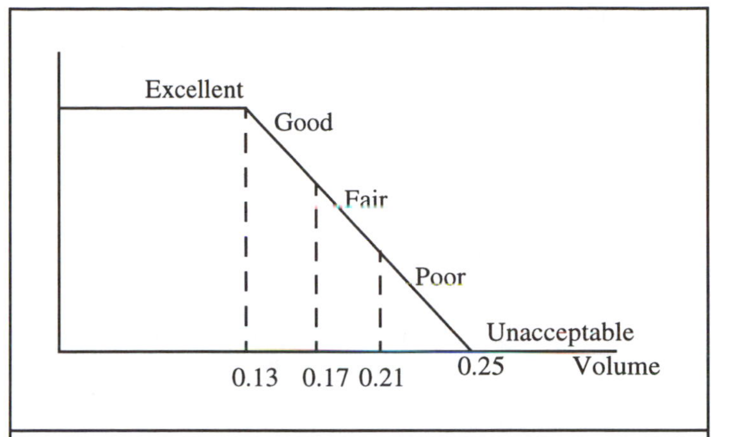

Figure 20.27 Initial Satisfaction Curve for Volume in Example 20.4

Redesign

Based on the results obtained so far, we need very much to increase the spring constant k — by over a hundred fold. The dependency table indicates that this can be done by either increasing wire diameter d or by decreasing coil diameter D.

We choose first to increase d by about as much as is possible. Note that with 13 total coils (Na + Q = 11 + 2) and a 2 inch maximum length, the maximum possible value for d is 0.154 inches. A larger value than this would mean that adjacent coils of the wire will touch. So we choose a slightly smaller new value for d of 0.13 inches. The results using this d are shown as Trial 5 in Figure 20.24.

Trials 6 through 10 shown in Figure 20.24 represent repeated iterations using updated dependencies for guidance. (The dependencies vary only slightly in this domain; see Figure 20.26 for the values after Trial # 10.)

We can't make an overall evaluation for the design of Trial 10 because we have not yet established the values on the satisfaction curve for volume. We can do this now, however, based on the results we have obtained for volume in the trials already made. It is really a matter of designer discretion exactly what values are used to establish the curve.

One way in this case is to note that the best volume achieved so far is in Trial 9. Since Trial 9 is an Excellent design in terms of k, and is not far from an Excellent design in terms of shear stress, it is reasonable to believe this volume (0.150) will be about as good as can be expected. To leave ourselves some room for improvement, however, we place it in the mid portion of the Good zone on the satisfaction curve. The other points are filled in with a view to providing reasonable discriminations as the volume changes during future trials. See Figure 20.27.

Now we can obtain the overall evaluations shown in parentheses Figure 20.24. Continuing and completing this example is now left as an optional exercise for interested readers.

20.5 Bolts

Note: An essential reference for design of bolts is ANSI B1.1, 1974 [13].

20.5.1 Pre-Parametric Design Issues: Conceptual Level

Bolts support loads by means of internal tensile stresses. Thus the basic physical effect in their operation is simple tension. However, there are important complications — mostly in the form of stress raisers and fatigue — that appear when one considers the different types of loading, and the details of the stresses around the head and in the threads where the load is transferred to and from the nut.

The best available guess about what stress concentration factor to use for bolt threads is about 3.85 [3] based on experimental and finite element studies of the region where the load is transferred from the threads of the bolt the those of the nut. We will use 4.0 in the examples in this book. There are various ways to mitigate this stress increase, but still insufficient data to justify use of a lower stress concentration factor. It is likely that bolts are over-designed in most cases, but the cost of bolt over-design is usually small whereas the cost of both failures can be very, very large — including injury or death.

Bolts are sometimes used to clamp gaskets, which in turn provide a seal preventing leakage of a gas or liquid. Note that when the internal pressure in such a case is raised from zero to p, the bolt will extend causing the gap filled by the gasket to increase also. Designers must make certain that with p applied there will still be sufficient compression of the gasket to maintain the seal.

Bolts are generally designed using, not the full tensile strength of the bolt material, but the so called *proof strength* measured on actual bolts. Proof stress is the maximum stress that does not result in a measurable permanent elongation. Generally the proof strength of a bolt is about 90% of the yield strength of the material used. [2].

It is common for bolts to be pulled into considerable tension when they are initially tightened into place; this is called pre-stressing the bolt. The practice reduces the likelihood of separation of the bolted members when forces are applied, and also helps to lower the stress range (see Appendix 20A, Section 20A.3) when fatigue is an issue. The amount of pre-stress is sometimes a design variable. When clamping essentially rigid parts together, the amount of prestress recommended [2] is:

$$T_i = 0.9 \text{ (Proof Stress / Stress Area)} \qquad (20.35)$$

Since the amount of pre-stress is related to the torque used to tighten the bolt (torque can be controlled to some extent with a torque wrench), the torque is also sometimes a design variable. An approximate empirical equation [2] for the torque required to get an initial pre-stress tension of Ti in a bolt of diameter D is given by

$$\text{Torque} = 0.2\ (T_i)(D) \qquad (20.36)$$

Materials and Manufacturing

Bolts can be made of many materials, but are usually made from steel of one type or another. Material properties are shown in Figure 20A.13 in Appendix 20A on page 20-42. See also Reference [14].

Bolts can be manufactured by automatic screw machines that cut the threads from stock large enough to also make the heads (this wastes a lot of material), or by automatic forging machines that waste little or no material. Threads on the forged blanks are formed by grinding or by rolling through dies — a process that moves material from the root of the thread to the top of the thread. Thus the outer diameter of such threads is larger than the forged stock.

20.5.2 Pre-Parametric Design Issues: Configuration Level

There are two levels of configuration issues to be considered: (1) the configuration of the bolt itself (e.g., the types of threads and heads); and (2) the arrangement and locations of bolts in cases where multiple bolts are employed. Bolt thread types are described briefly in Appendix 20A and selected properties listed in Figures 20A.11 and 20A.12 on pages 20-40 and 20-41.

In configuring systems of bolts to support loads, designers must not only insure that there are enough bolts to carry the total load, but also that the alignment of holes, and tolerances on locations of holes, are such that mis-alignment does not result in shear or bending loads on the bolts that were not included among the design considerations. Also, if the distance between bolts is too large, distortions of the parts being held may result in unaccounted for bending loads. For additional information on configuring bolt systems, see especially [2].

20.5.3 Formulating Bolt Design Problems

Bolt problem formulations follow the usual general guided iteration format. There are variations in the specifics, of course, but here are some of the common factors that occur often. There may be others as well in certain cases.

Problem Formulation

Problem Definition Parameters

- The magnitude and nature of the loading, including
- Whether the loading is steady or fluctuating
- Whether or not there are shear loads as well as tensile loads;
- The configuration (e.g., number of bolts and their locations)
- Whether or not a purpose of the bolted joint is to provide a
- seal, say by means of a gasket; and
- The nature and structure of the parts being bolted.
- Environmental conditions (temperatures, corrosive atmosphere, etc.).

There are many possible combinations of these and

other factors — too many to include them all. We will illustrate three problem types in the following examples. Students should note how the general guided iteration formulation is implemented in the examples so that they will be able to make use of it in other specific problems.

Design Variables

- Bolt size (Diameter = D; Stress Area = A)
- Bolt material (Yield Stress, σ_Y; Proof Stress, = σ_{yp})
- Initial bolt tension or tightening torque

Solution Evaluation Parameters

- Tension Failure Ratio = TFR= Stress in bolt /Proof Stress

Satisfaction Curves

See Figure 20.28. A safety factor of 2 is reflected in the satisfaction curve shown, but this is just an example. Also reflected there is a degree of dissatisfaction with over-design.

Analysis Procedures

See also Appendix 20A.3.

Note: The assumption here is that the parts being clamped are rigid.

- Bolt Stress = (Stress Concentration Factor)(Force in Bolt / Stress Area)
- Desired Initial Bolt Tension = T_i = 0.9 (Proof Stress) (Stress Area)
- Desired Initial Torque = (0.2) T_i (D)

Dependencies

Generally the bolt stress will vary inversely with the stress area, or the square of the bolt diameter.

20.5.4. Example 20.5 - Steady Loads, No Sealing, No Shear

We begin with one of the simplest bolt problem types. Refer to Figure 20.29. A cylindrical tank containing a granular material holds a steady weight of W_1 = 10,000 pounds. A flat bottom itself weighing 1000 pounds is bolted on with 10 bolts. The task is to design the bolts.

Problem Formulation

Problem Definition Parameters

W_1 = 10,000 pounds

W_2 = 1000 pounds

Number of bolts = 10 (Course Threads)

Bolt Stress Concentration Factor = 4

Design Variables

Bolt Size (Diameter = D; Stress Area = A)

Bolt Material (Yield Stress σ_Y; Proof Stress = σ_{yp})

Initial Torque, T_i

Solution Evaluation Parameters

Tension Failure Ratio = TFR

Satisfaction Curve

We will use Figure 20.28.

Analysis Procedures

Note: The assumption here is that the parts being clamped are rigid and the stress concentration factor is 4.

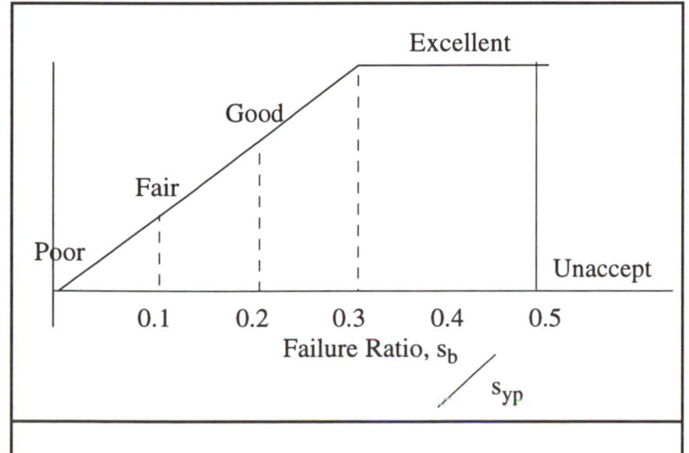

Figure 20.28 Satisfaction Curve for Example 20.5

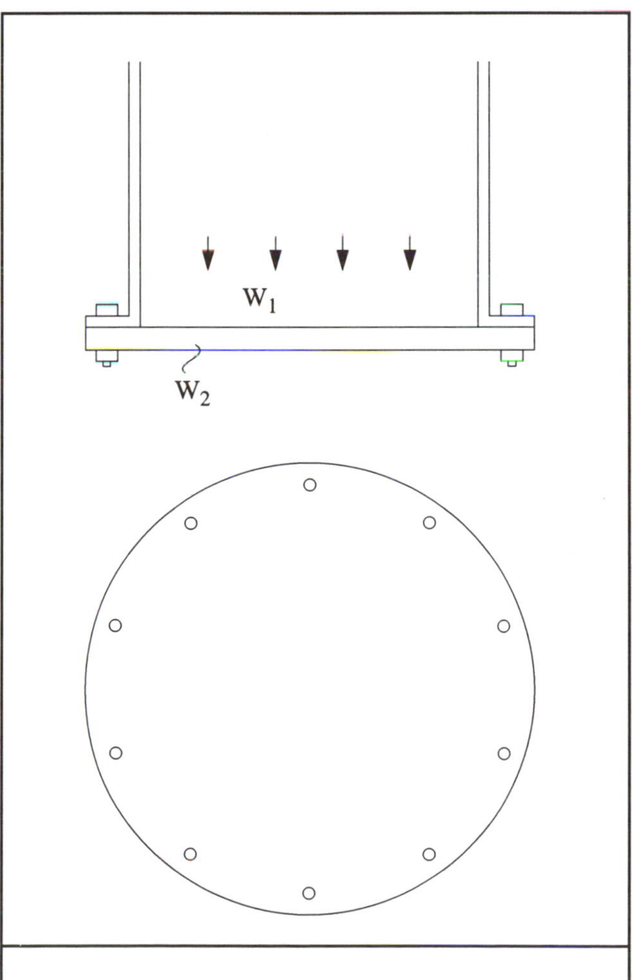

Figure 20.29 Configuration for Example 20.5

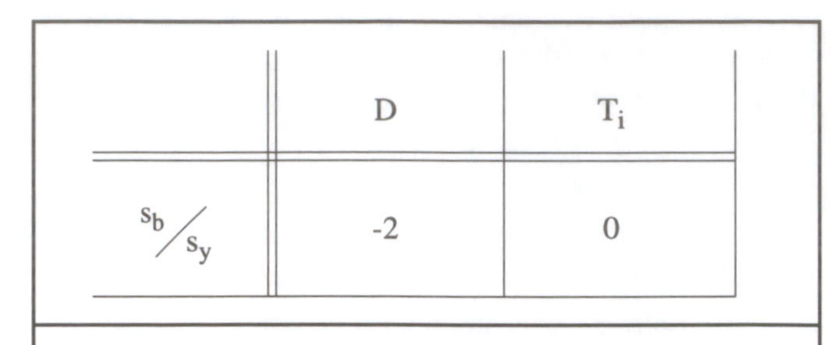

	D	T_i
s_b / s_y	-2	0

Figure 20.30 Initial Dependencies Example 20.5

- Bolt Stress = 4.0 (Force in Bolt / Stress Area)
- Desired Initial Bolt Tension = T_i
- T_i = 0.9 (Proof Stress) (Stress Area)
- Desired Initial Torque = (0.2) T_i (Nominal Diameter)

Dependencies

See Figure 20.30. We reason that bolt stress will vary inversely with the stress area, or the square of the bolt diameter. Hence the stress-diameter dependency is –2. The initial tension is an independent variable which can be set as the designer desires so we show its dependency on diameter as 0.

Initial Design

We must first select a material (a non-numeric design variable), and then determine a bolt size for this material. If the result is not acceptable, another material must be selected, and the process repeated. The initial bolt tension is a sequential design variable that can be determined directly by computation after the material and bolt size have been selected.

For an initial trial, we select a standard material: Grade 5.2. This gives us a Yield Stress of 92,000 psi, and an approximate proof stress of 0.9 (92,000) = 82,800 psi. (See Figure 20A.13.)

As a first trial design with this material choice, we select D = 0.500 inches. The corresponding stress area (see Figure 20A.12)) is 0.1419 inches^2.

Analysis and Evaluation

Bolt Stress = (4) (11000 / 10) / (0.1419) = 31,000 psi.

Tension Failure Ratio = 31,000 / 82,800 = 0.374.

This is rated as Excellent in Figure 20.28. (Obviously, our initial trial was a very lucky guess!)

Redesign

Even though the previous design rated Excellent, we might still try for a smaller bolt. If we try the next smaller size available (7/16"), the resulting bolt stress is 41,390 psi, and the TFR is right on the mark at 0.500. Thus this bolt would do, but we would most likely select the 1/2 inch bolt anyway.

Since the initial trial seems a quite acceptable design, there is no need to consider other materials unless there were other issues to consider.

The initial bolt tension can now be computed from:

T_i = 0.9 (82,800 x 0.1419)= 10, 574 pounds

The approximate initial torque required to achieve this tension is

Torque = 0.2 T_i (0.5) = 1057 inch-pounds.

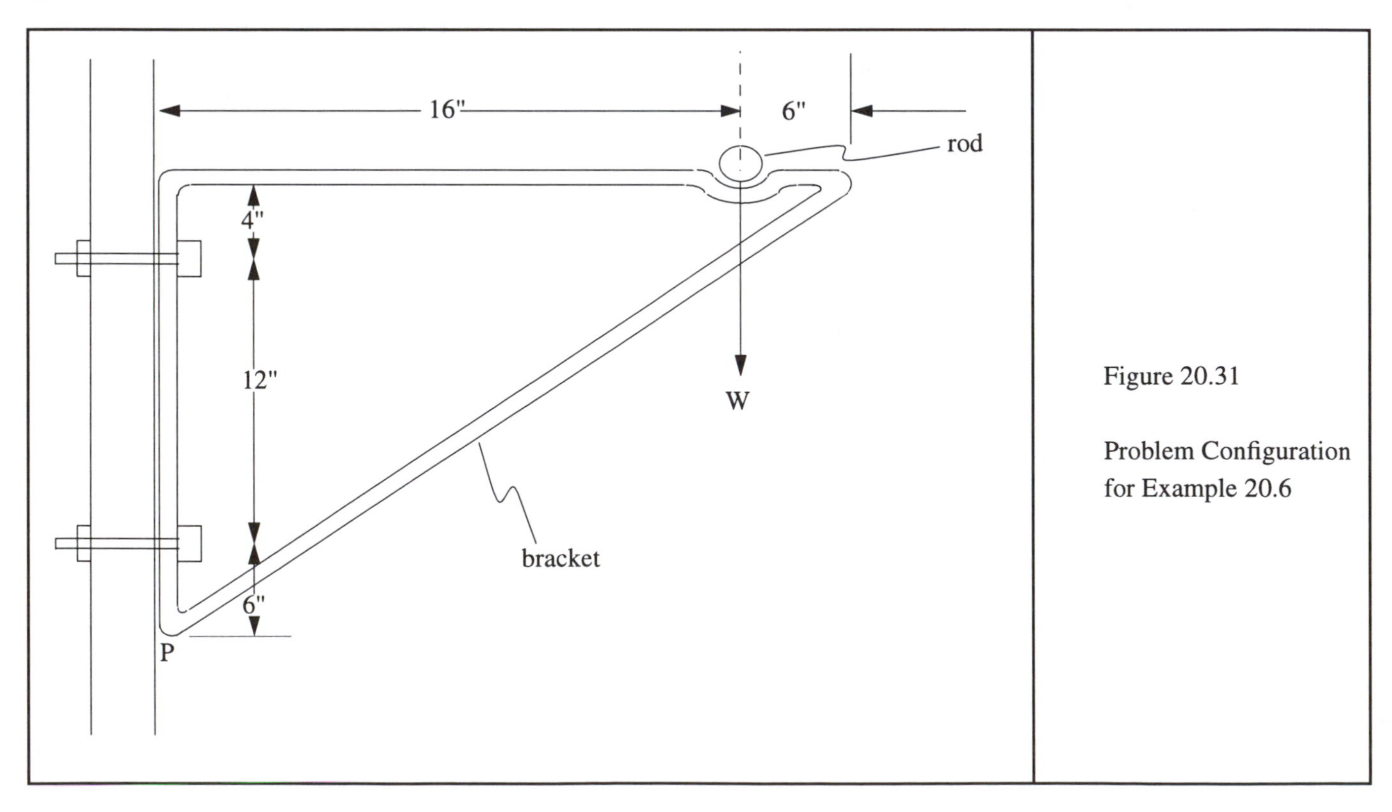

Figure 20.31

Problem Configuration for Example 20.6

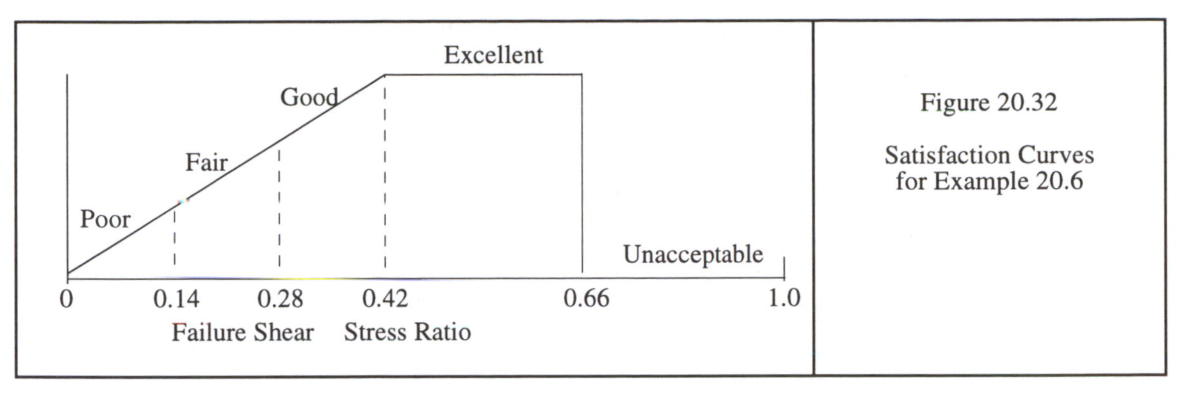

Figure 20.32

Satisfaction Curves for Example 20.6

20.5.5 Example 20.6 - A Bolt With Shear As Well As Tension

Some complexity is added to bolt design when there is shear as well as tensile loading. Consider the case shown in Figure 20.31. A steady weight carrying rod is to be supported by a set of brackets of the type shown in the sketch. (Gusset plates to stiffen the bracket are not shown.)

Problem Formulation

Problem Definition Parameters

See configuration in Figure 20.31.

W (per bracket) = 300 lbs.

Stress Concentration Factor for Bolts = 4

Course Threads

Design Variables

Bolt Size

Bolt Material

Initial Torque

Solution Evaluation Parameters

Shear Failure Ratio (SFR)

= Maximum Shear Stress / 0.5 Proof Stress

Tension Failure Ratio (TFR)

= Max Tensile Stress / Proof Stress

Satisfaction Curves

See Figure 20. 32.

Note: Factors of safety are reflected in the satisfaction curves.

Analysis Procedures

	A	T_i
t_{max}/t_y	-1	0

Figure 20.33 Initial Dependencies Example 20.6

We assume that the tension load must be carried completely by the top bolt. If the bracket were perfectly rigid, the bottom bolt could carry perhaps as much as half of the load of the top bolt, but as an added margin of safety, we will assume a worst case for the top bolt. If this assumption gives an unreasonably large sized bolt, we may have to reconsider the assumption.

The tensile load on the top bolt is found by summing moments around point P in Figure 20.31. The resulting tensile force is 267 lbs.

The shear load (300 lbs) is assumed here to be carried by a single bolt. However, if the bolts holes are well matched, it would be reasonable to assume that sufficient yielding would occur before failure so that the shearing load could be considered shared equally between the two bolts. Again, however, we will assume a worst case, and reconsider later if necessary.

It might also be assumed that the shearing load could be carried by friction between the support wall and the bracket. We will design neglecting this, but then check for an estimate of the friction force later.

The bolt tension stress will be found from

$$\sigma = 4\ (267 / A) \quad (20.37)$$

The bolt shearing stress will be found from

$$\tau = 4\ (300 / A) \quad (20.38)$$

The maximum shearing stress will be found from an analysis of Mohr's circle for this case:

$$\tau_{max} = \left[\left(\frac{\sigma}{2}\right)^2 + \tau^2\right]^{\frac{1}{2}} \quad 20.39)$$

Dependencies

An initial dependency table is shown in Figure 20.33.

Note: We have changed the design variable from D to stress Area, A.

Initial Design

As an initial material, we select Class 8 so that Yield Stress = 130,000 psi and proof stress = 0.9 x 130,000 = 117,000 psi. See Figure 20C.4.

As an initial bolt size, we select 1/4 inch. The stress area (See Figure 20C.1.)) is 0.0318 inches^2.

Analysis and Evaluation

From Equation (20.37):

Bolt Tension = 4 (267 / 0.0318) = 33,585 psi

From Equation (20.38):

Vert Shear Stress = 4 (300 / 0.0318) = 37,736 psi

From Equation (20.39):

Maximum Shear = 41,356 psi

Thus:

Shear Failure Ratio = 41,356 / 0.5 (117,000) = 0.706

And:

Tension Failure Ratio = 33,585 / 117,000 = 0.287

From the satisfaction curve, we see that the TFR is well satisfied (Excellent) but that the value for SFR is Unacceptable.

Redesign

We now want to redesign so that the SFR is in the Excellent zone, which varies from an SFR of 0.42 to 0.66 as shown in Figure 20.32. We decide to try for a value of 0.50, which implies that the shear stress be equal to

Maximum Shear Stress = 0.5 x Proof Stress x SFR

= 0.50 x 117,000 x 0.5 = 29,250 psi.

Then from the dependencies we can compute a new trial stress area:

$$29{,}250 / 41{,}356 = (A(new) / 0.0318)^{-1}$$

Desired $A(new) = 0.045$ inches^2

The next larger standard bolt size is 5/16 inches with a stress area of 0.0524. See Figure 20A.12. In view of all the redundant safety factors that have been included (the safety factor of 4, neglecting the tension bearing contribution of the bottom bolts, assuming the shear is carried by one bolt), we might elect to go with the 1/4 inch bolt. However, there is little added cost in this case to using a larger bolt, and we have also neglected any possible dynamic loads, so we choose to try the 5/16 inch bolt.

Analysis and Evaluation

Bolt Tension = 20,381 psi

Vertical Shear = 22,900 psi

Maximum Shear = 25,065 psi

Failure Stress Ratio = 0.21.

Evaluation: Excellent

We now can determine an appropriate initial torque and estimate the vertical friction available to reduce the shear load on the bolts.

Initial Torque

The initial bolt tension and required torque can now be computed from Equations 76 and 77:

T_i = 0.9 (117,000 x 0.0524)= 5518 pounds

The initial torque = 0.2 T_i 5/16 = 345 inch-pounds.

Vertical Friction Force

The coefficient of friction between the bracket and the supporting wall must be estimated. Assuming (!) no-one has lubricated one of the surfaces, a value of 0.3 is a reasonable estimate. Then the available initial friction force will be:

$F_{initial}$ = 0.3 (5518) = 1665 lbs

There will be some relaxing of the initial bolt tension with time, so that perhaps 80% of this force will be available for carrying the shear load. That is

F = 0.8 (1665) = 1324 lbs.

Though this is a very crude calculation, with a shearing load of only 300 lbs, it certainly appears that the friction load is adequate for the purpose. However, such frictional forces are (in general) a fairly unreliable method of provided for important functions. Beware of some fool with an oil can lubricating the walls, and make sure the bolts *can* carry the shear if necessary.

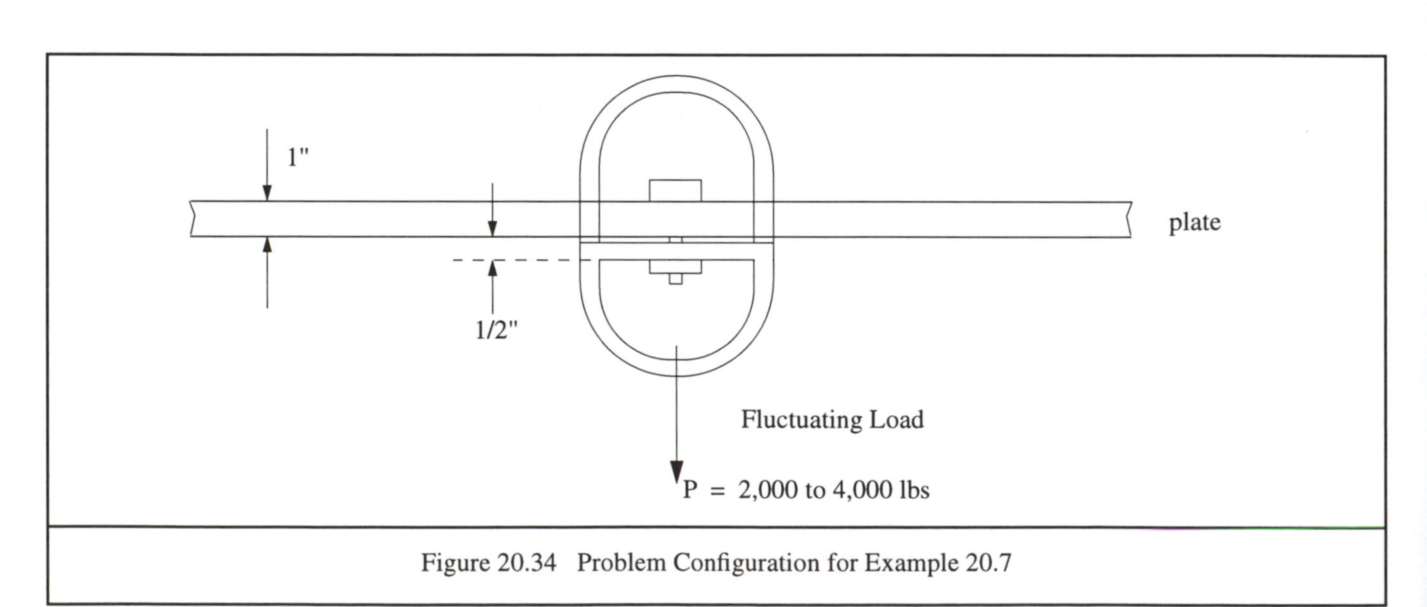

Figure 20.34 Problem Configuration for Example 20.7

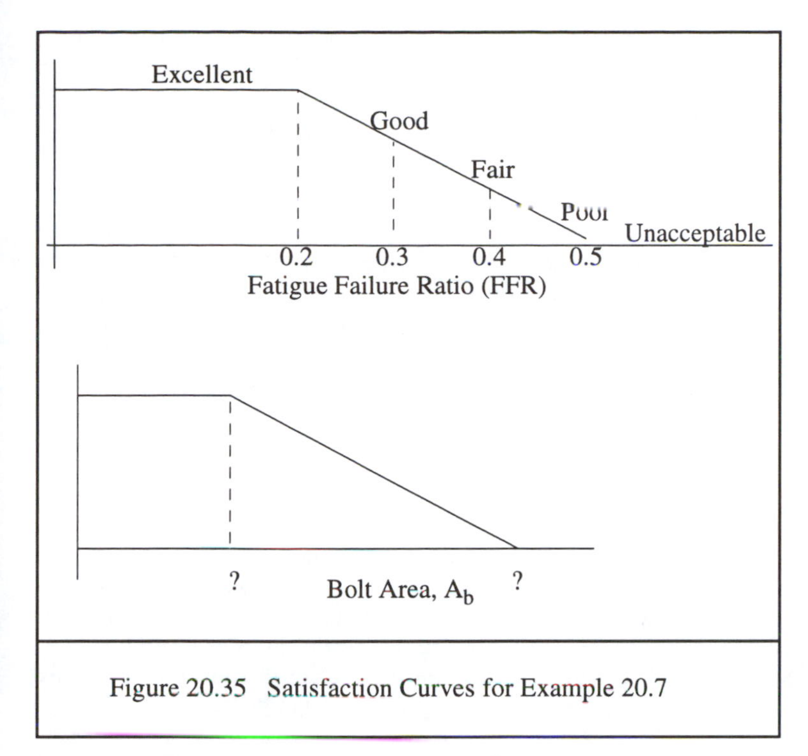

Figure 20.35 Satisfaction Curves for Example 20.7

20.5.6 Example 20.7 - A Bolt With Fluctuating Loads

Refer to Figure 20.34. The lug shown is fastened to the supporting plate by two bolts. Note that the load fluctuates so that bolt fatigue must be considered. The plate and lug are both steel with E = 30,000,000 psi. The problem is to design the bolt.

There is fairly complex analysis required to support evaluation in this problem is presented in Appendix 20A.

Problem Formulation

Problem Definition Parameters

See Figure 20.34.

Design Variables

Bolt Material

Bolt Area, A_b

Required Initial Tension in Bolt, T_i

Solution Evaluation Parameters (and Importance Levels)

Fatigue Failure Ratio, FFR (Very Important)

Tension Stress Failure Ratio, TF (Important)

Bolt Area (A surrogate for weight or cost or convenience; to be as small as practicable) (Desirable)

Satisfaction Curves: See Figure 20.35.

Analysis Procedures (See Appendix 20A.)

Initial Design

For an initial design in this example, we select a Grade 5 bolt material and an initial bolt diameter of 1.0 inch so that:

D = 1.0 inch

A_b = 0.606 inches^2

σ_Y = 120,000 psi

σ_{yp} = 92,000 psi

τ_Y = 0.5 (120,000) = 60,000 psi

Analysis and Evaluation

Figure 20.36 shows analysis results extracted from a spreadsheet developed for this problem.

Based on the analysis results, the evaluations (See Figure 20.35) for the SEPs are:

Fatigue Failure Ratio: Excellent

Area: Indeterminate

We do not have sufficient information yet to establish a satisfaction curve for the area. (We shall see subsequently in this particular example, we really will never need to establish one.)

Redesign

The Fatigue Failure Ratio (FFR) for the initial design is well into the excellent zone — indicat-

Trial No.	Diameter	A_b	D_p	F_{bo}	s_{av}	s_r	s_{rF}	FFR
1	1	0.606	2.0	2,456	7,990	1,312	56,000	0.094 (Exc.)
2	0.625	0.226	1.25	1,920	19,646	3,717	50,180	0.296 (Good)
3	0.75	0.334	1.5	1,991	13,454	2,497	53,270	0.187 (Exc.)
4	0.5625	0.182	1.125	2,356	26,190	4,416	46,900	0.377 (Fair)

Figure 20.36 Selected Spreadsheet Results for Example 20.7

ing the possibility of over-design. Thus we will attempt to use a smaller bolt and get an FFR of about (say) 0.25. To determine the bolt size to use to accomplish this change, we note that the dependency between FFR and bolt area is -1.0, Thus:

$$y_j(old) / y_j(new) = [(x_i(old) / x_i(new)]^{d_{ij}}$$

$$0.094 / 0.25 = [0.606 / A(new)]^{-1}$$

$$A(new) = 0.227 \text{ inches}^2$$

The nearest standard bolt is D = 5/8 inches with an area of 0.226 inches^2.

Analysis and Evaluation

See Trial 2 in Figure 20.36

Fatigue Failure Ratio: Good (but close to Fair zone)

Redesign

It is by now apparent that we can trade-off FFR with bolt area as we wish. Since FFR is a very important, and area only desirable, we elect not to go to smaller bolt than 5/8 inches. However, with the spread sheet set up, it is easy to try as many as we wish. Thus we try both 3/4 inches and 9/16 inches. The results are shown in Figure 20.36. Our selection as the best compromise is D = 3/4 inches.

20.6 Summary and Preview

Summary. This chapter has illustrated the use of the general guided iteration methodology in sample problems from three traditional mechanical component domains: shafts, springs, and bolts. As pointed out in the Introduction to the chapter, the domain information presented here is not complete, and references to more complete information have therefore been included. Our purpose has been to show how the guided iteration design methodology can be used, rather than to develop and present a thorough treatment of the engineering science and analysis procedures from the domains themselves.

The guided iteration design process is, as always, is Formulation, Generation of Alternatives, Evaluation, and Redesign, *guided by* the results of the evaluation. In parametric design, the generation of alternatives is usually sequential beginning with one initial design, so that the first step is simply the generation of an initial design as a place to start.

In parametric design, the key steps in of problem formulation stage are:

1. Establishing the Problem Definition Parameters;
2. Establishing the Design Variables;
3. Establishing the Solution Evaluation Parameters (SEPs), Including:
   Their Importance Levels,
   Their Satisfaction Curves (which relate their numerical values to their evaluation), and
   The analysis procedures by which their values are computed for trial sets of design variables;
4. Establishing an initial dependency table.

For the examples in this chapter, we have placed a number of the supporting analysis procedures — and some of the engineering science from which they are derived — in the chapter appendix to help emphasize that the design process is unaffected by the specific nature of the analysis procedures employed. Stress or deflection analyses, for example, can be done by strength of materials or by finite element methods (or by some third method) but the design process is the same. Analysis is important to design because it is the way numerical values for the SEPs are obtained for trial designs. Incorrect analysis will lead to poor (or worse) designs, and hence designers must know and understand the physical principles and other information on which analyses in their design domains are based. But while analysis supports the parametric design process in this important way, analysis is not design. Parametric design is problem formulation, generating an initial design, evaluation of the values provided by the analyses, and redesign guided by the evaluations.

It should also be remembered that while the engineering analyses are crucial to evaluation of trial designs, evaluation often involves other important criteria not based on engineering science. Manufacturability and cost are almost always involved, for example, and there also may be other evaluation criteria such as aesthetics, reliability, and environmental factors.

Preview. In the next chapter, we illustrate the guided iteration methodology in problems in thermal-fluids domains. Again, since the domains are different, the analysis procedures are different. But the design methodology is the same, and it is the methodology (not the specific domains or problem examples) that is most important to us in this book.

References

[1] Hill, P. H., *The Science of Engineering Design*, Holt Reinhart and Winston, New York, 1970.

[2] Juvinall, R. C. and Marshek, K. M. (1991), *Fundamentals of Machine Component Design*, Wiley, New York, 1991.

[3] Spotts, M. F.), *Design of Machine Elements*, Prentice-Hall, Englewood Cliffs, New Jersey, 1985.

[4] Shigley, J. E. and Mitchell, L. D., Mechanical Engineering Design, McGraw-Hill, 1983.

[5] Walton, J., Engineering Design: From Art to Practice, West Publishing Co., St. Paul, MN, 1991.

[6] Ertas, A. and Jones, J. C., *The Engineering Design Process*, Wiley, New York, 1993.

[7] Edwards, K.S., Jr. and McKee, R. B., *Fundamentals of Mechanical Component Design*, McGraw-Hill, New York, 1991.

[8] Orthwein, W., *Machine Component Design*, West Publishing Co., St. Paul, MN, 1990.

[9] Petersen, R. E. (1974), *Stress Concentration Factors in Design*, Wiley, New York, 1974. (Note: There are also relevant data in two articles by Petersen in *Machine Design*, Volume 23, July, 1951.)

[10] "Design of Transmission Shafting", ANSI B106.1M, ASME, New York, 1985.

[12]*Handbook of Spring Design*, Spring Manufacturers Institute, Inc., Rolling Meadows, IL.

[14] "Unified and American Screw Threads", ASME, New York, 1982.

Suggested Supplementary Reading

Fundamentals of Machine Component Design, Juvinall, R. C. and Marshek, K. M., Wiley, New York, 1991.

Problems

20.1 Perform parametric design the shaft configuration in Figure 20.1 using the following PDPs:

L = 16 inches

P = 1000 lbs

T = 1000 in-lbs

Material = Aluminum 6063-T6

Develop your own satisfaction curves.

20.2 Design the shaft Example 20.2 using the following PDPs:

L = 12 inches

P = 2000 lbs

T = 4000 lbs

D_2 (max) = 4.0 inches

t (min) = 0.15 inches

Use the same material as Example 20.2. Develop your own satisfaction curves.

20.3 Use the old ASME Code (See Section 20.3.3) to design the shaft described in Example 20.3.

20.4 Design the shaft shown in Figure 20.11 but with loads of 1500 lbs and the 12 inch distances changed to 18 inches. Use the same material as in Example 20.3. Develop your own satisfaction curves.

20.5 Design the helical spring shown in Figure 20.22 and described in Example 20.4, except take the PDPs to be:

P_1 = 40 lbs

P_2 = 100 lbs

Desired deflection, = 1/2 inches

Max Diam, D(max) = 1.00 inches

Max Length, L(max) = 3 inches.

Temperature: Normal

Environment: No corrosive elements

End conditions: Ground flat; open

Develop your own satisfaction curves.

20.6 Design the bolts shown in Figure 20.29 and described in Example 20.5 except take the PDPs to be:

W_1 = 20,000 pounds

W_2 = 2000 pounds

Number of bolts = 12 (Coarse Threads)

Bolt Stress Concentration Factor = 4

Develop your own satisfaction curves.

20.7 Design the bolt shown in Figure 20.34 but with the load fluctuating from 1000 to 6000 lbs. Use the same material. The dimensions shown are to be changed from 1 inch to 2 inches, and from 1/2 inches to 3/4 inches. Develop your own satisfaction curves.

20.8 Design a helical spring to support a load of 200 lbs when deflected 4 inches from its unstressed position.

20.9 Curved washers as shown in Figure P20.9 are used as springs where deflections are to be very small. Their spring constant is fairly linear, and they are used for such purposes as eliminating "play" and distributing loads. Suppose that parametric design is needed for a design where the design variables are

t = thickness, inches

D_2 = Outer diameter, inches

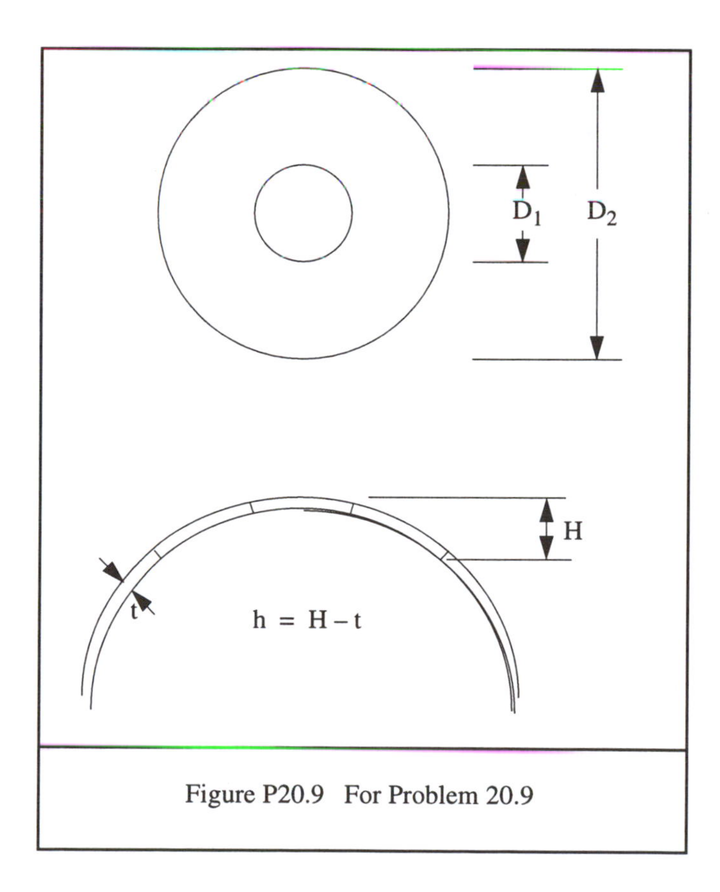

Figure P20.9 For Problem 20.9

D_1 = Inner diameter, inches

and the problem definition parameters are

P = Load = 130 pounds

H = Free overall height = 0.125 inches

E for the spring material = 30,000,000 psi

D_2 maximum = 0.85 inches

t minimum = 0.015 inches

D_1 maximum = 0.30 D_2

f maximum = 0.8 (H - t).

Take the Solution Evaluation Parameters to be the maximum material stress (S), the deflection (f), and the volume of the spring, (V). S is high priority, f is moderate priority, and V is low priority. Analysis equations for values of these SEPs are shown below. The equation for V is approximately proportional to the spring volume. Also, the equations for f and S are approximate only [12], but can be used for preliminary design purposes.

$$f = (\pi \ D_2{}^3) / (4 \ E \ (D_2 - D_1) \ t^3)$$

$$S = (1.5 \ \pi \ D_2) / ((D_2 - D_1) \ t^2)$$

$$V = (\pi/4) \ t \ (D_2{}^2 - D_1{}^2)$$

(a) Develop satisfaction curves for these parameters that reflect the following considerations:

(i) The yield stress of the material is 200,000 psi. A safety factor on the stress should be at least 1.3 but need not be more than 2.

(ii) The deflection should be 0.005 inches give or take about 0.001 inches.

(iii) The volume should be as small as possible.

(b) Develop an initial dependency table.

(c) Using guided iteration, find values for design variables.

Appendix 20A

Engineering Analysis Information for Shafts, Springs, and Bolts

20A.1 Shafts

20A.1.1Material Yield Criteria

Experimentation has indicated that yielding in combined stress situations can be predicted from uniaxial (i.e., tensile) yielding data. In particular, it has been observed that yielding in tension in ductile materials is associated with slippage within the material along oblique surfaces, and is caused primarily by shear stress. This has led to formulation of the maximum shear theory (sometimes called the Tresca theory) which asserts that yielding occurs when the maximum internal shear stress reaches the value it has when yielding occurs in the uniaxial tensile case. We will refer to this value as $\tau_{critical}$ or τ_{crit}. If we sketch a Mohr's circle for a specimen in uniaxial tension See Figure 20A.1(a). At the tensile yield point, it is clear that the critical shear is one-half the yield stress in tension. Thus, the maximum shear stress theory says that yielding occurs when:

Maximum Shear Stress = Tensile Yield Stress / 2

$$\tau_{max} = \sigma_Y/2 \qquad (20A.1)$$

Other theories may also be used (see, for example, [3]), but the above relationship gives reasonable values that are both conservative and easy to use.

It should be noted that the maximum shear stress in a part is, in general, given by

$$\tau_{max} = \frac{\sigma_{max} - \sigma_{min}}{2} \qquad (20A.1a)$$

where σ_{max} and σ_{min} are the maximum and minimum normal stresses in the part. In general, then, the maximum shear theory says that yielding occurs when

$$\sigma_{max} - \sigma_{min} = \sigma_Y \qquad (20A.1b)$$

In three dimensions, the principal stresses (σ_1, σ_2, σ_3) are defined as the normal stresses in planes where the shear stress is zero. In a plane stress situation, when ($\sigma_1 > 0$ and $\sigma_2 < 0$, (See Figure 20A.1(b)), then σ_{max} and σ_{min} correspond to the principal stresses σ_1 and σ_2. Thus, according to the theory, yielding occurs when

$$\sigma_1 - \sigma_2 = \sigma_Y \qquad (20A.1c)$$

However, when σ_1 and σ_2 are both positive ($\sigma_1 > \sigma_2$), then the minimum stress is in the third coordinate direction, and is zero. Thus, yielding occurs in this case when

$$\sigma_1 = \sigma_Y \qquad (20A.1d)$$

Similarly, when σ_1 and σ_2 are both negative ($|\sigma_1| > |\sigma_2|$), the maximum stress is in the third coordinate direction, and is zero. Thus, yielding occurs in this case when

$$\sigma_1 = -\sigma_Y \qquad (20A.1e)$$

Elastic materials in shear deform linearly in response to the torsional loads in a way analogous to the deformations caused by transverse loads — except of course that the deformation caused by shear is twisting, not bending. The amount of the twist, or angular deformation, is dependent upon the torque applied, the geometry of the shaft, and a material property called the *shear modulus* (G) which is analogous to the elastic modulus (E).

It is shown in strength of materials that for an elastic material, the shear modulus is related to the elastic modulus and Poisson's ratio μ as follows:

$$G = E / 2(1 + \mu). \qquad (20A.2)$$

For the vast majority of metals, values for Poisson's ratio are about 0.30. Thus for steels where E = 30,000,000 psi, a typical value for G is about 11,500,000 psi.

20A.1.2 Analysis Procedures to Support Evaluation of Trial Designs of Circular Shafts With Static Loads

Polar Moment of Inertia

When the loads applied to a circular shaft are purely torsional, then the material is in a state of pure shear. The shear stress varies directly in proportion to the distance from the centerline. Thus, for a solid circular shaft with torsional loading only, the shear stress is zero at the center and a max-

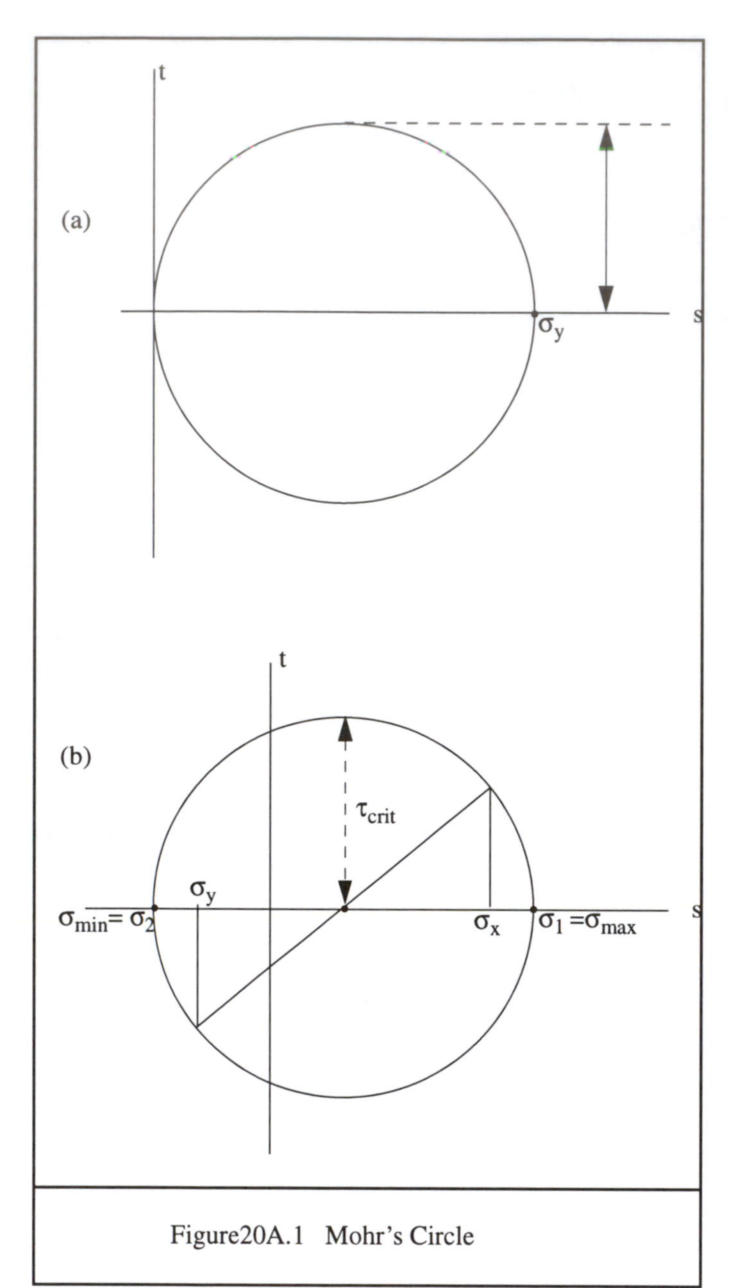

Figure20A.1 Mohr's Circle

imum at the outer radius. This suggests that more efficient use of material is achieved by concentrating material away from the shaft center, leading to the idea of a hollow shaft. This is analogous to the case of beams in pure bending where the desire to concentrate material as far from the neutral axis as possible leads to the I-Beam configuration.

The *polar moment of inertia*, J, about the centroid of a cross section plays a role in torsion analogous to the role played by the moment of inertia, I, in bending. In general, the polar moment of inertia is given by:

$$J = \int r^2 dA \qquad (20A.3)$$

where r is the distance from the centroid of the cross section to an element of cross section area dA. In the case of a solid circular shaft, the polar moment of inertia is:

$$J = \left(\pi \cdot D^4\right)/32 \qquad (20A.4)$$

where D is the diameter of the shaft. For a hollow circular shaft, the expression for J is

$$J = \frac{\pi}{32} \cdot \left(D_2^4 - D_1^4\right) \qquad (20A.5)$$

where D_2 is the outer diameter and D_1 is the inner diameter.

Shearing Stress

The local shearing stress (τ) at radius r in a circular shaft subjected to a torque T is

$$\tau = \frac{Tr}{J} \qquad (20A.6)$$

where r is distance from the centroid. Therefore, the maximum shear stress obviously occurs at the outer radius (R) where r = R (or d = D). For a solid shaft:

$$\tau = \frac{TR}{J} = \frac{16 \cdot T}{\pi \cdot D^3} \qquad (20A.7)$$

The angle of twist (θ) of a circular shaft of length L supporting a torque T is given by

$$\theta = \frac{TL}{JG} \qquad (20A.8)$$

where L is the length of the shaft and G is the modulus of elasticity in shear. See Figure 20A.2.

Torsional and Bending Loads: Combined Stresses

When shafts are loaded with both transverse (i.e.,

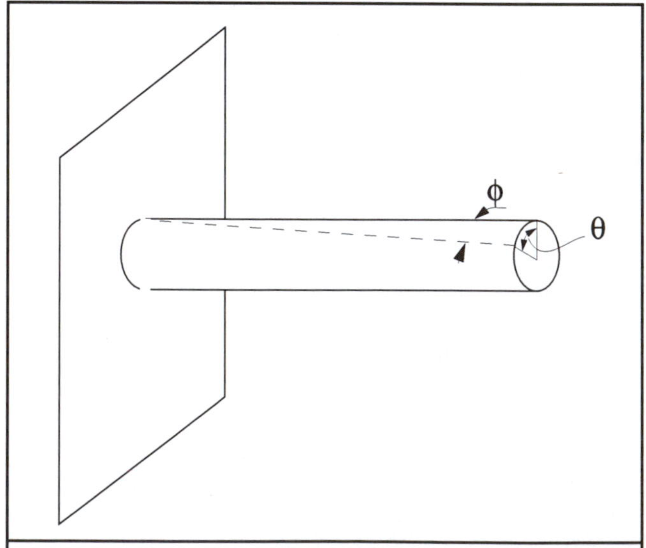

Figure 20A.2 Symbol Definitions for Twist

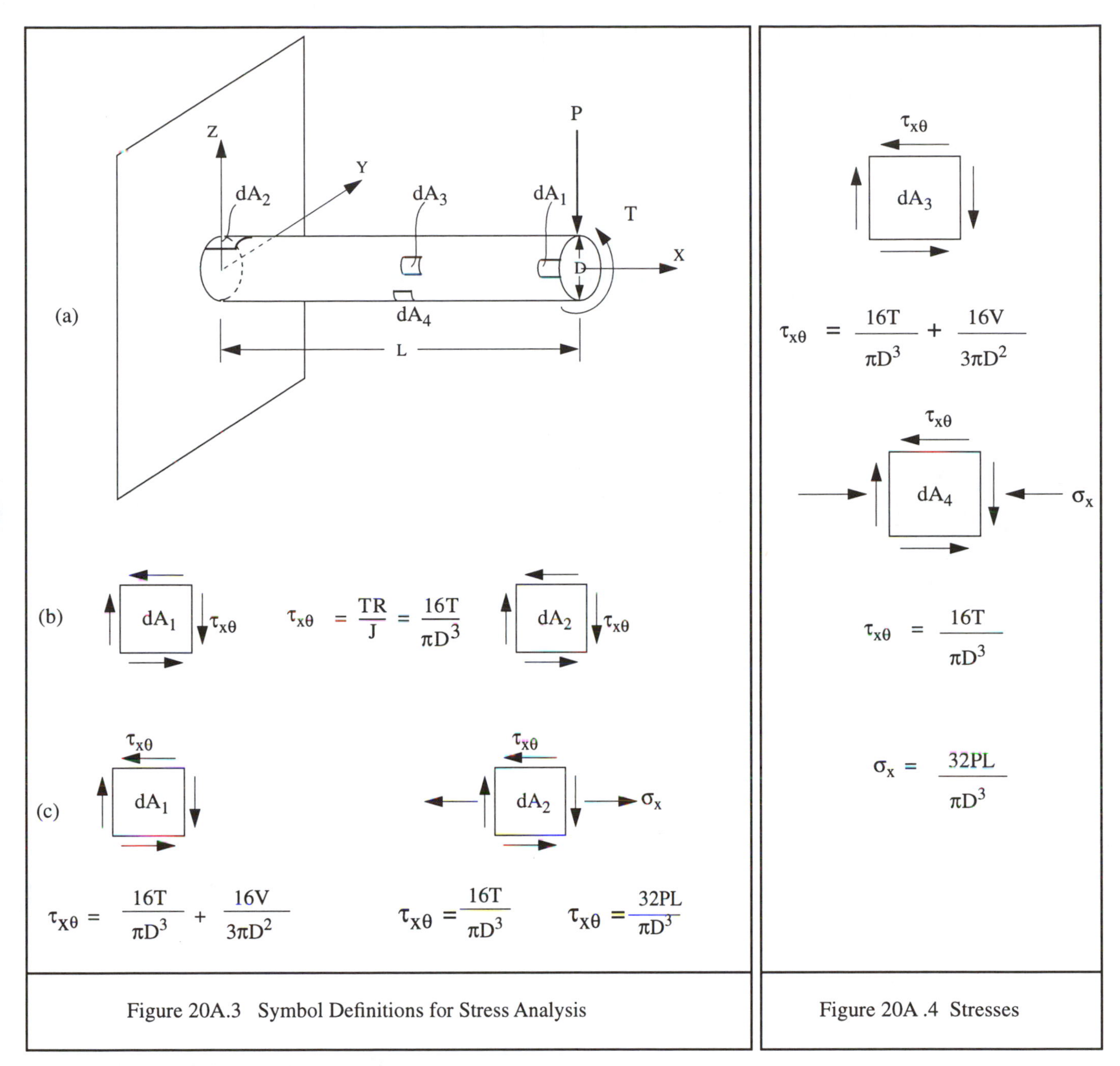

Figure 20A.3 Symbol Definitions for Stress Analysis

Figure 20A .4 Stresses

bending) loads and torsional (i.e., twisting) loads, the internal response of the shaft material is a combined stress. Consider, for example, the solid circular shaft shown in Figure 20A.3(a). Every plane material element on the surface of this shaft (such as, for example, elements dA1 and dA2) will be subjected to shear stresses shown in Figure 20A.3(b). The convention for the notation shown for shear stress is that the first subscript refers to the normal direction from the face, and the second subscript refers to the direction of the shear force.

The magnitude of the shear stress on element DA2 in this case can be computed from Equation 20A.9. In addition, element dA2 will experience a tensile stress due to the bending moment PL:

$$\sigma = \frac{Mc}{I} = \frac{PLD}{2I} = \frac{32 \cdot PL}{\pi \cdot D^3} \tag{20A.9}$$

as shown in Figure 20A.3(c). In contrast to element DA_2, element dA_1 has no bending stresses (since it is at the neutral axis for bending) but it will have a transverse shear stress given by:

$$\tau_T = \frac{4V}{3A} \tag{20A.10}$$

where V is the transverse shear load (500 lbs in the example) and A is the cross sectional area. Thus, there is additional shear on element dA_2 equal to

$$\tau_T = \frac{16 \cdot V}{3\pi \cdot D^2} \tag{20A.11}$$

As a review, and to insure that the principles underlying the above computations are understood, students should verify that the stresses on elements dA_3 and dA_4 are as shown in Figure 20A.4.

When torsional and bending stresses are combined as in the above simple example, the values of the maximum shear and tension (or compression) stresses are not obvious because they occur in directions neither parallel nor perpendicular to the axis of the shaft. Determining the magnitudes of the maximum shearing and tension/compression stress in a plane is then a task that can be accomplished by use of Mohr's circle or by use of the following equations:

$$\sigma_{1,2} = \frac{\sigma_x + \sigma_y}{2} \pm \left[\left(\frac{\sigma_x - \sigma_y}{2}\right)^2 + \tau_{x\theta}^2\right]^{\frac{1}{2}} \tag{20A.12}$$

$$\tau_{max} = \frac{\sigma_1 - \sigma_2}{2} = \left[\left(\frac{\sigma_x - \sigma_y}{2}\right)^2 + \tau_{x\theta}^2\right]^{\frac{1}{2}} \tag{20A.13}$$

where σ_1 and σ_2 ($\sigma_1 > \sigma_2$) are principal stresses. Figure 20A.5 shows the standard Mohr's circle representation for a case when both σ_x and σ_y are positive.

In a simple solid circular shaft, if T is the torque and M is the bending moment, then the tension and compression stresses will be given by:

$$\sigma_x = \frac{32M}{\pi \cdot D^3} \tag{20A.14}$$

The shear stress will be given by:

$$\tau = \frac{16T}{\pi \cdot D^3} \tag{20A.15}$$

Then the maximum shear stress caused by these combined loads will be:

$$\tau_{max} = \left[\left(\frac{16M}{\pi \cdot D^3}\right)^2 + \left(\frac{16T}{\pi \cdot D^3}\right)^2\right]^{\frac{1}{2}} \tag{20A.16}$$

$$\tau_{max} = \frac{16}{\pi \cdot D^3} \cdot \left[M^2 + T^2\right]^{\frac{1}{2}} \tag{20A.17}$$

The maximum tension or compression stress is given by:

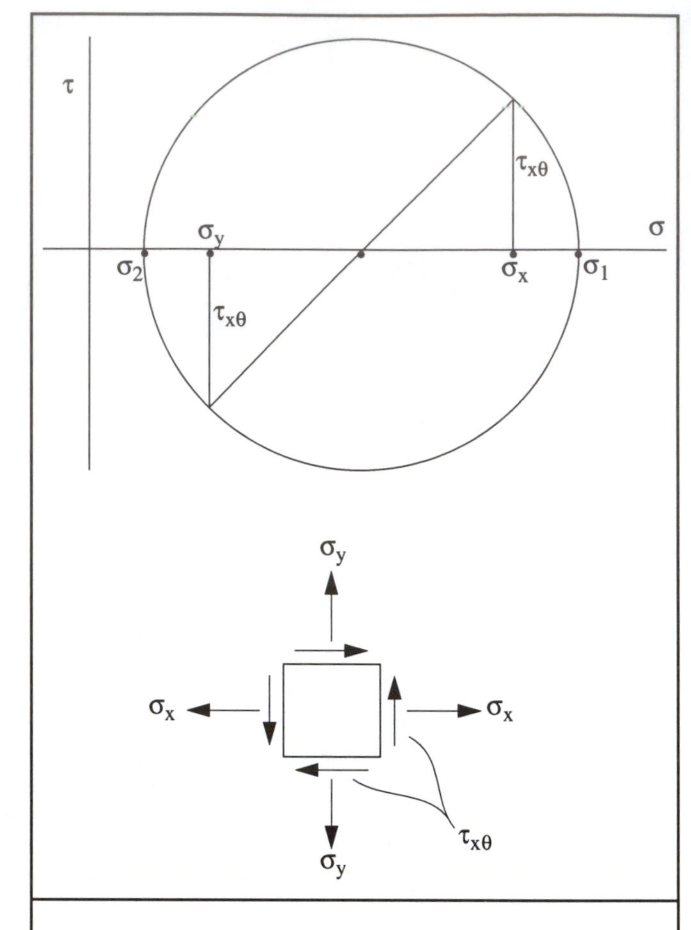

Figure 20A.5 Mohr's Circle Definitions

$$\sigma_{1,2} = \frac{16M}{\pi \cdot D^3} \pm \left(\frac{16}{\pi \cdot D^3} \cdot \left[M^2 + T^2\right]^{\frac{1}{2}}\right) \tag{20A.18}$$

It should be remembered that if there are stress concentrations, these stresses will be increased locally as discussed above.

20A.1.3 Fluctuating Stresses and Fatigue

Fluctuating Stresses and Fatigue

Shafts that transmit power always have bending moments as well as torsional loads. Bearings act as support points; the bending loads are caused by the gears or pulleys or turbine or compressor stages that provide for the inputs and outputs of power to and from the shaft at points away from the bearings. As a result, power transmitting shafts experience combined bending and torsional stresses as do many statically loaded shafts.

There is, however, an important difference between static shafts with combined stresses and power transmitting

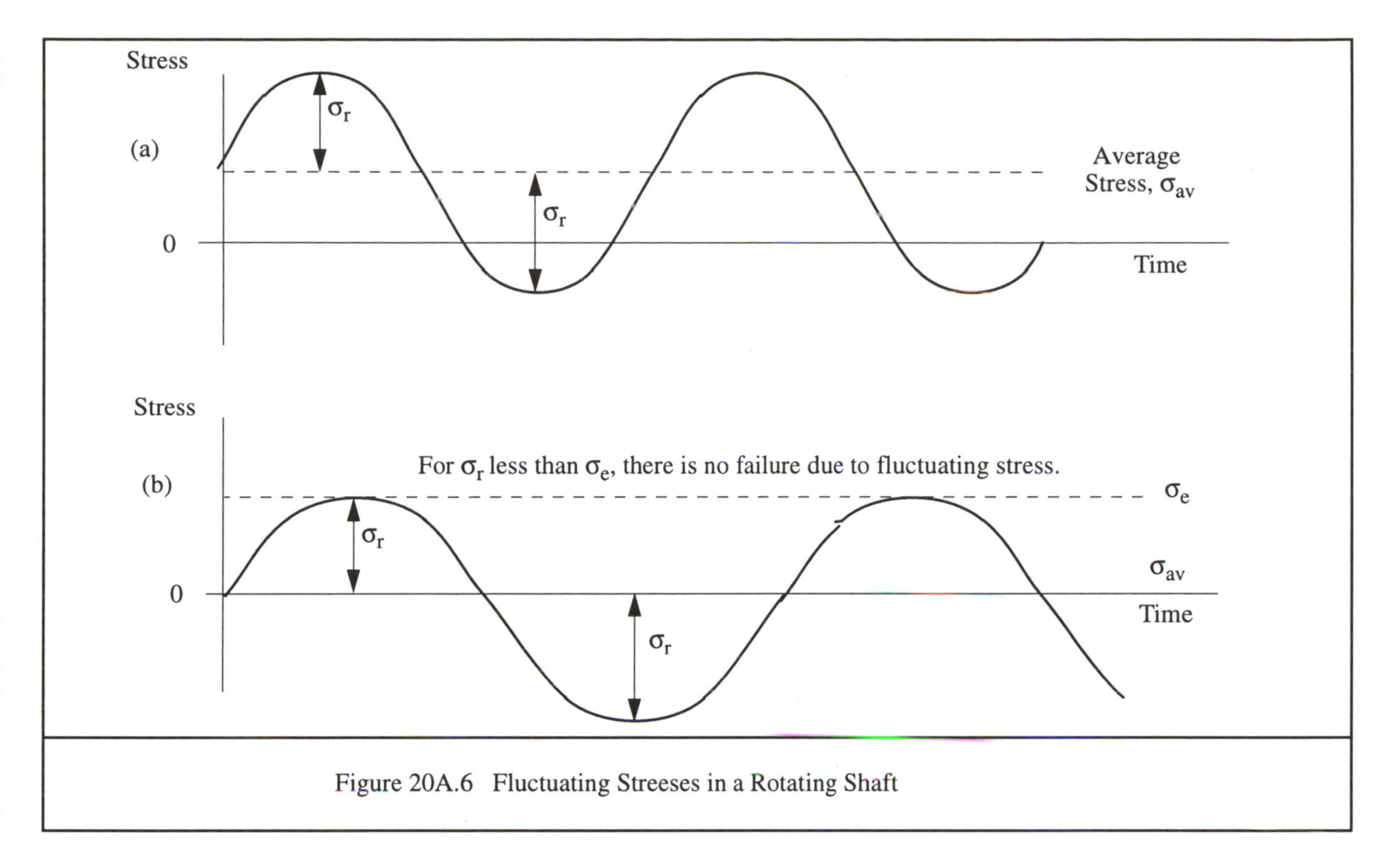

Figure 20A.6 Fluctuating Streeses in a Rotating Shaft

shafts. A power transmitting shaft will have *fluctuating* bending loads, a fact that often makes material fatigue a crucial issue. To visualize the fluctuating loads, suppose the shaft shown in Figure 20A.2 is rotating. Then imagine yourself standing and riding around at a point on the surface. When you are at the top during a rotation, the material on which you are standing will be in tension due to the bending load. When you are at the bottom, the same material will be in compression. Thus there is a complete reversal of the internal bending stresses in the shaft with every rotation. This can be illustrated graphically as in Figure 20A.6(a). Hence the concern for possible fatigue failure — a concern that is enhanced when there are also stress concentrators present.

When stresses fluctuate, then the concept of *endurance limit* becomes relevant. When the *average* fluctuating stress is zero, it has been found that there will be no fatigue failure *if* the fluctuating load range (σ_r) is below a critical value. That critical value is called the endurance limit or endurance stress. See Figure 20A.6(b). For some metals, the endurance limit is related fairly directly to tensile strength as shown in Figure 20A.7 [3].

Experimental studies have led to methods for estimating when the combined effects of average stress σ_{av}, stress range σ_r, and stress concentration K are likely to cause fatigue failure. Refer to Figure 20A.8, which is called a modified Goodman Diagram. See, for example, [2, 3, 8, 15]. In this figure, note that point B is within the safe working zone below the line of fatigue failure. For a part with an average

Surface Condition	Fraction of Ultimate Strength		
	σ_{ult} = 60,000 psi	σ_{ult} = 120,000 psi	σ_{ult} = 180,000 psi
As Forged	0.25	0.18	0.14
Hot Rolled	0.36	0.26	0.19
Machined	0.41	0.37	0.33
Ground	0.45	0.45	0.44

Figure 20A.7

Approximate Endurance Limits for Steels

stress at some point A, the value of AB determines whether or not there is likely to be fatigue failure. If AB is less than AC, then no failure is likely; if AB is greater than AC, then failure is likely. We can define a Fatigue Failure Ratio, then, as the ratio AB / AC. As long as it less than 1.0, fatigue failure is unlikely. We can also use this ratio to define a safety factor; requiring, say, that AB/AC be less than 0.5 is like specifying a safety factor of 2.

In equation form, the line of fatigue failure shown in Figure 20A.8 can be expressed as:

$$\text{When} \quad \sigma_{av} < \frac{\sigma_Y - \sigma_e}{\sigma_T - \sigma_e} \cdot \sigma_T \tag{20A.19}$$

$$\text{Then} \quad FFR = \frac{K\sigma_r}{\sigma_{rF}} \tag{20A.20}$$

Where

$$\sigma_{rF} = \text{Failure Limit} = \left(\frac{\sigma_e}{\sigma_T} \cdot \left(\sigma_T - \sigma_{av}\right)\right) \tag{20A.21}$$

$$\text{And} \quad \sigma_T = \text{Ultimate Tensile Strength} \tag{20A.22}$$

$$\text{When} \quad \sigma_{av} > \frac{\sigma_Y - \sigma_e}{\sigma_T - \sigma_e} \cdot \sigma_T \tag{20A.23}$$

$$\text{Then} \quad FFR = \frac{F\sigma_r}{\sigma_{r\beta}} \tag{20A.24}$$

$$\text{Where} \quad \sigma_{rF'} = \sigma_y - \sigma_{av} \tag{20A.25}$$

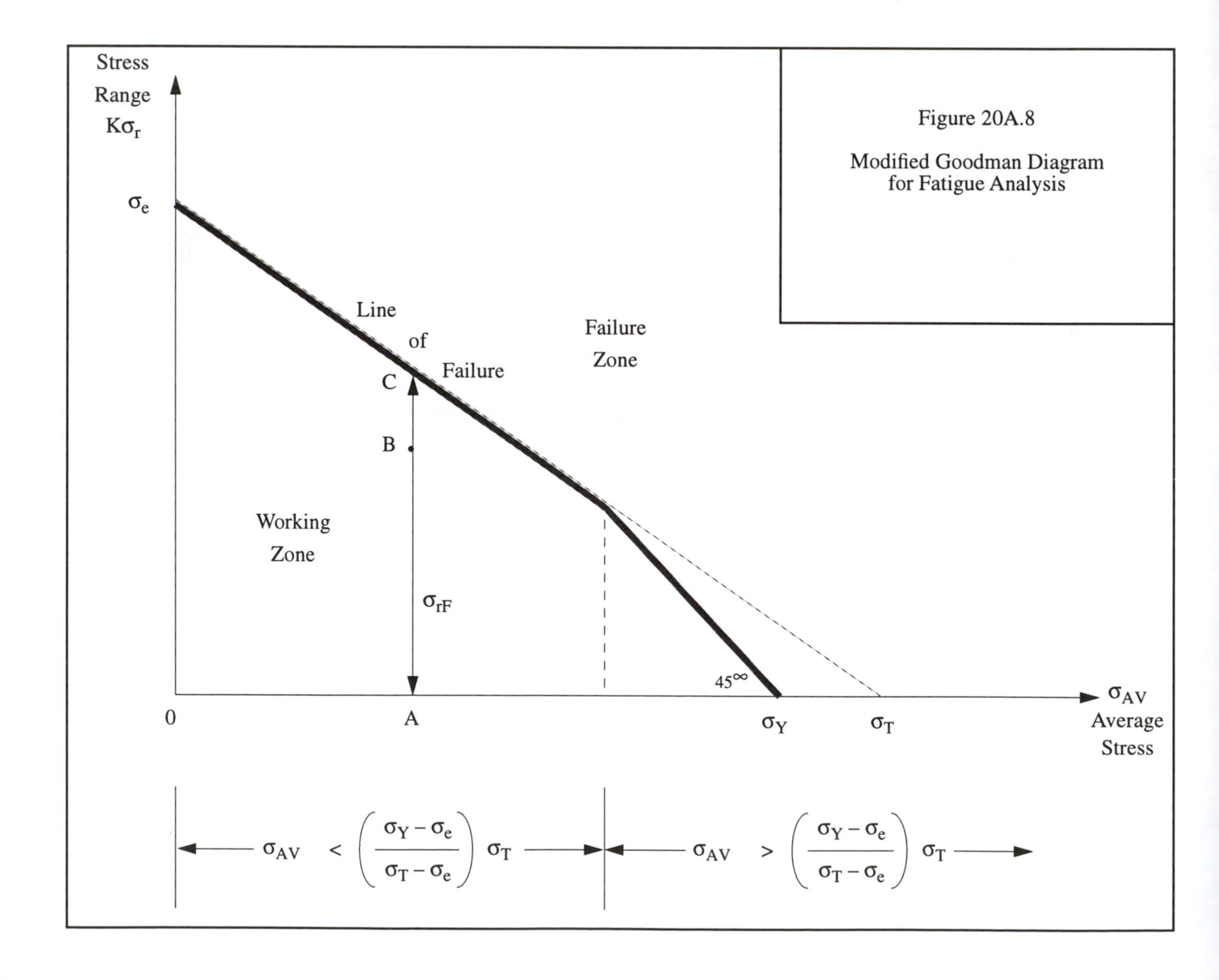

Figure 20A.8

Modified Goodman Diagram for Fatigue Analysis

20A.2 Analysis Procedures for Helical Springs

The analysis equations for springs are presented below.

For shear stress

The torque, T, on the wire is given by:

$$T = 1/2\ P\ D \quad (20A.26)$$

where D is the coil (not the wire) diameter. To visualize this, imagine that the wire of the spring is cut at some point and take one part as a free body. Then the load P acts through the centerline of coil at a distance R = 1/2 D from the wire, which is in torsion to resist this moment.

The maximum shear stress in the wire resulting from this torque is given by

$$\tau_{max} = (T d) / (2 J) \quad (20A.27)$$

where J is the polar moment of inertia of the wire cross section given by

$$J = (\pi d^4) / 32 \quad (20A.28)$$

so

$$\tau = (8 P D) / (\pi d^3) \quad (20A.29)$$

From basic strength of materials considerations, the total shear stress in the wire (τ_{max}) is known to be given by the sum of the shear stress due to torsion (τ) plus the shear stress due to transverse shear (τ_T). That is:

$$\tau_{max} = \tau + \tau_T \quad (20A.30)$$

The maximum transverse shear in the wire is [Spotts 1985]

$$\tau_T = 1.23\ P/A = (1.23\ P)(4 / \pi d^2) \quad (20A.31)$$

The above equations can be combined and reduced to the following:

$$\tau_{max} = K_s [(8 P D) / (\pi d^3) \quad (20A.32)$$

where

$$K_s = [1 + 0.615 / C] \quad (20A.33)$$

and C — called the *spring index* — is given by:

$$C = D / d. \quad (20A.34)$$

For spring constant

It is shown in the literature that the deflection, δ, of a helical spring is given by:

$$\delta = (8 P D^3 N_a) / G d^4 = (8 N_a P C^3) / G d. \quad (20A.35)$$

where Na is the number of active coils in the springs. Usually, a spring will have approximately one coil (depending on exactly how the ends are machined and/or fastened) on each end that is inactive in contributing to the spring's operation. Thus, in this case the total number of coils is

$$N = N_a + Q \quad (20A.36)$$

where Q is the number of inactive coils. Here we will always assume Q = 2. Thus the spring constant, k, is found from:

$$k = P / \delta = (G d) / (8 C^3 N_a) = (G d^4) / 8 D^3 N_a) \quad (20A.37)$$

A Required Relationship

There is a required relationship among N, d, and L(max) since the fully compressed spring surely cannot exceed the maximum length:

$$N d < L(max) \quad (20A.38)$$

For Volume

The length of wire making up the coil is

$$L = \pi D N \quad (20A.39)$$

and so the wire volume is

$$V = (\pi^2 d^2 D N) / 4. \quad (20A.40)$$

20A.3 Bolt Data and Analysis Methods

20A.3.1 Bolt Thread Types, Dimensions, and Materials

Thread Types

Most bolts in the United States today have standardized geometry, and are called *Unified* threads. There are also many other thread configurations (Acme, square, Whitworth, American National pipe threads) as shown in handbooks and machine design references such as [3, 4] and others. Unified threads are the most common, and are also classified as 1A, 2A, and 3A (external threads) and 1B, 2B, and 3B (internal threads) depending on the tolerances held during manufacture. Classes 1A and 1B have the greatest tolerances, and provides the largest clearances when the bolt and nut are assembled. Classes 2A and 2B are the most commonly used.

Data on dimensions of selected common bolts are shown in Figures 20A.11 and 20A.12 [14]. The *pitch diameter* is the diameter of the cylinder at the place where the width (a) of a thread cross section is just equal to the space (b) between the threads. The *stress area* is the area that can be used to carry tensile forces; stress area is very close to the area of a circle with a diameter equal to that of the bolt with the threads shaved off.

Metric (or SI) threads are different enough from the Unified threads that no interchangeability is possible. For dimensions on metric threads, see [14].

Material properties for standard bolts and screws are shown in Figure 20A.13.

20A.3.2 Analysis Procedures for Example 20.6

Refer to Figure 20.34. The spring constant for the bolt is

$$k_b = L_b / A_b\ E_b$$

Wire Size In.	Music Wire 1000 psi	Hard Drawn 1000 psi	Oil Temp. 1000 psi	Wire Size In.	Music Wire 1000 psi	Hard Drawn 1000 psi	Oil Temp. 1000 psi	Wire Size In.	Music Wire 1000 psi	Hard Drawn 1000 psi	Oil Temp. 1000 psi
.008	399	307	315	.046	309	249		.094	274		
.009	393	305	313	.047	309	248	259	.095	274	219	
.010	387	303	311	.048	306	247		.099	274		
.011	382	301	309	.049	306	246		.100	271		
.012	377	299	307	.050	306	245		.101	271		
.013	373	297	305	.051	303	244		.102	270		
.014	369	295	303	.052	303	244		.105	270	216	225
.015	365	293	301	.053	303	243		.106	268		
.016	362	291	300	.054	303	243	253	.109	268		
.017	362	289	298	.055	300	242		.110	267		
.018	356	287	297	.056	300	241		.111	267		
.019	356	285	295	.057	300	240		.112	266		
.020	350	283	293	.058	300	240		.119	266		
.021	350	281		.059	296	239		.120	263	210	220
.022	345	280		.060	296	238		.123	263		
.023	345	278	289	.061	296	237		.124	261		
.024	341	277		.062	296	237	247	.129	261		
.025	341	275	286	.063	293	236		.130	258		
.026	337	274		.064	293	235		.135	258	206	215
.027	337	272		.065	293	235		.139	258		
.028	333	271	283	.066	290			.140	256		
.029	333	267		.067	290	234		.144	256		
.030	330	266		.069	290	233		.145	254		
.031	330	266	280	.070	289			.148	254	203	210
.032	327	265		.071	288			.149	253		
.033	327	264		.072	287	232	241	.150	253		
.034	324	262		.074	287	231		.151	251		
.035	324	261	274	.075	287			.160	251		
.036	321	260		.076	284	230		.161	249		
.037	321	258		.078	284	229		.162	249	200	205
.038	318	257		.079	284			.177	245	195	200
.039	318	256		.080	282	227	235	.192	241	192	195
.040	315	255		.083	282			.207	238	190	190
.041	315	255	266	.084	279			.225	235	186	188
.042	313	254		.085	279	225		.250	230	182	185
.043	313	252		.089	279			.3125		174	183
.044	313	251		.090	276	222		.375		167	180
.045	309	250		.091	276		230	.4375		165	175
				.092	276	220		.500		156	170
				.093	276						

Figure 20A.9 Minimum Tensile Properties of Wire Spring Materials - Ferrous. Reprinted From the *Handbook of Spring Design*, Spring Manufacturers Institute, by permission of the Spring Manufacturers Institute, Rolling Meadows, Illinois. These data should be used only for the design of wire springs.

STAINLESS STEELS

Wire Size In.	Type 302 (kpsi)	Type* 17-7 PH	Wire Size In.	Type 302 (kpsi)	Type* 17-7 PH	Wire Size In.	Type 302 (kpsi)	Type* 17-7 PH
.008	325	345	.033	276		.060	256	
.009	325		.034	275		.061	255	305
.010	320	345	.035	274		.062	255	297
.011	318	340	.036	273		.063	254	
.012	316		.037	272		.065	254	
.013	314		.038	271		.066	250	
.014	312		.039	270		.071	250	297
.015	310	340	.040	270		.072	250	292
.016	308	335	.041	269	320	.075	250	
.017	306		.042	268	310	.076	245	
.018	304		.043	267		.080	245	292
.019	302		.044	266		.092	240	279
.020	300	335	.045	264		.105	232	274
.021	298	330	.046	263		.120	225	272
.022	296		.047	262		.125		272
.023	294		.048	262		.131		260
.024	292		.049	261		.148	210	256
.025	290	330	.051	261	310	.162	205	256
.026	289	325	.052	260	305	.177	195	
.027	287		.055	260		.192		
.028	286		.056	259		.207	185	
.029	284		.057	258		.225	180	
.030	282	325	.058	258		.250	175	
.031	280	320	.059	257		.375	140	
.032	277							

*After aging

CHROME SILICON/CHROME VANADIUM

Wire Size In.	Chrome Silicon	Chrome Vanadium
.020		300
.032	300	290
.041	298	280
.054	292	270
.062	290	265
.080	285	255
.092	280	
.105		245
.120	275	
.135	270	235
.162	265	225
.177	260	
.192	260	220
.218	255	
.250	250	210
.312	245	203
.375	240	200
.437		195
.500		190

COPPER-BASE ALLOYS

Phosphor Bronze (Grade A)	
Wire Size Range — in.	
.007 - .025	145
.026 - .062	135
.063 and over	130
Beryllium Copper (Alloy 25 pretemp)	
.005 - .040	180
.041 and over	170
Spring Brass all sizes	120

NICKEL-BASE ALLOYS

Inconel (Spring Temper)	
Wire Size Range — in.	
up to .057	185
.057 - .114	175
.114 - .318	170
Inconel X Spring Temper 190	After Aging 220

Figure 20A.10

Minimum Tensile Properties of Wire Spring Materials - Non-Ferrous

Reprinted From the *Handbook of Spring Design*, Spring Manufacturers Institute, by permission of the Spring Manufacturers Institute, Rolling Meadows, Illinois.

These data should be used only for the design of wire springs.

Size	Outer Diameter (in.)	Threads per Inch	Tensile Stress Area (in.^{2})
0(0.060)	0.0600	80	0.00180
1(0.073)	0.0730	72	0.00278
2(0.086)	0.0860	64	0.00394
3(0.099)	0.0990	56	0.00523
4(0.112)	0.1120	48	0.00661
5(0.125)	0.1250	44	0.00830
6(0.138)	0.1380	40	0.01015
8(0.164)	0.1640	36	0.01474
10(0.190)	0.1900	32	0.0200
12(0.216)	0.2160	28	0.0258
1/4	0.2500	28	0.0364
5/16	0.3125	24	0.0580
3/8	0.3750	24	0.0878
7/16	0.4375	20	0.1187
1/2	0.5000	20	0.1599
9/16	0.5625	18	0.203
5/8	0.6250	18	0.256
3/4	0.7500	16	0.373
7/8	0.8750	14	0.509
1	1.0000	12	0.663
1 1/8	1.1250	12	0.856
1 1/4	1.2500	12	1.073
1 3/8	1.3750	12	1.315
1 1/2	1.5000	12	1.581

Figure 20A.11 Fine Thread Dimensions UNC
Data from ANSI B1.1-1982 [14], by permission of ASME, New York.

where L_b is the stressed length of the bolt, A_b is the stress area of the bolt, and E_b is the modulus of elasticity of the bolt material. The spring constant for the plate and lug base together is

$$k_p = L_p / A_p \; E_p$$

where Ap is the effective stress carrying cross sectional area of the plate and lug combination and E_p is the modulus of elasticity of the plate material. See Figure 20A.14. Several empirical methods have been proposed for approximating A_b [2, 3]; however, an accurate treatment would require a finite element analysis or considerable experimentation. In this book, we will use the following very simple approximation. For more accuracy (and complexity), readers are referred to the references noted.

$$A_p \cong \frac{\pi D_p^2}{4} - \frac{\pi D_b^2}{4} \qquad (20A.41)$$

$$D_p \cong 1.5 \cdot D_b + \frac{L_b \cdot D_b}{3} \qquad 20A.42)$$

The minimum required initial bolt tension is determined as follows. When the applied load is at its maximum value of 4000 lbs., we do not want the lug to separate from the plate; that is, we want there still to be some amount of compression in the lug and plate. This means that there must also be some tension in the bolt. Assume that the minimum initial tensile force (whatever it is) has been applied by tightening the bolt when there is no load present. Then when a load P is applied to the lug, the bolt increases in length by a value we shall call X. The change in the force in the bolt due to P is given by

$$X = \frac{F_B - F_{B0}}{k_B} = \frac{\Delta F'_B}{k_B} \qquad (20A.43)$$

where F_{B0} is the initial bolt tensile force and F_B is the bolt tensile force with P applied. The thickness of the plate-lug base also changes by the same X, and the change in the force carried is given by

$$X = \frac{F_P - F_{p0}}{k_p} = -\frac{\Delta F'_p}{k_p} \qquad (20A.44)$$

where F_{P0} is the initial compression force in the plate-lug plate combination and F_p is the compressive force after P is applied. Note that as defined ΔF_p is a negative quantity; X is positive. We can find P from a free body diagram of the lug plus the bottom portion of the bolt as shown below:

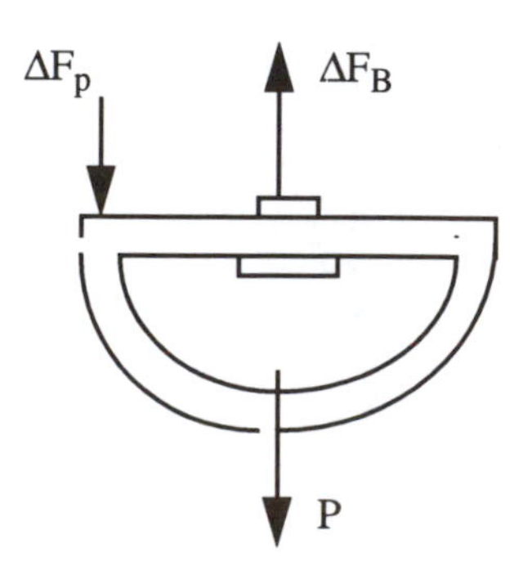

Size	Outer Diameter (in.)	Threads per Inch	Tensile Stress Area (in.^{2})
0(0.060)	0.0600	—	—
1(0.073)	0.0730	64	0.00263
2(0.086)	0.0860	56	0.00370
3(0.099)	0.0990	48	0.00487
4(0.112)	0.1120	40	0.00604
5(0.125)	0.1250	40	0.00796
6(0.138)	0.1380	32	0.00909
8(0.164)	0.1640	32	0.0140
10(0.190)	0.1900	24	0.0175
12(0.216)	0.2160	24	0.0242
1/4	0.2500	20	0.0318
5/16	0.3125	18	0.0524
3/8	0.3750	16	0.0775
7/16	0.4375	14	0.1063
1/2	0.5000	13	0.1419
9/16	0.5625	12	0.182
5/8	0.6250	11	0.226
3/4	0.7500	10	0.334
7/8	0.8750	9	0.462
1	1.0000	8	0.606
1 1/8	1.1250	7	0.763
1 1/4	1.2500	7	0.969
1 3/8	1.3750	6	1.155
1 1/2	1.5000	6	1.405
1 3/4	1.7500	5	1.90
2	2.0000	4 1/2	2.50
2 1/4	2.2500	4 1/2	3.25
2 1/2	2.2500	4	4.00
2 3/4	2.7500	4	4.93
3	3.0000	4	5.97
3 1/4	3.2500	4	7.10
3 1/2	3.5000	4	8.33
3 3/4	3.7500	4	9.66
4	4.0000	4	11.08

Figure 20A.12 Coarse Thread Dimensions UNC
Data from ANSI B1.1 - 1982 [14], by permission of American Society of Engineers, New York.

Grade	Diameter Range (inches)	Yield Strength (psi)	Tensile Strength (psi)
1	0.25 – 1.50	36,000	60,000
5	0.25 – 1.00	92,000	120,000
5	1.00 – 1.50	81,000	105,000
5.2	0.25 – 1.00	92,000	120,000
7	0.25 – 1.50	115,000	133,000
8	0.25 – 1.50	130,000	150,000
8.2	0.25 – 1.50	130,000	150,000

Source: Society of Automotive Engineers (SAE) Standard J429K (1979). Additional data can be found in the SAE Handbooks or in ANSI Standards available from ASME.

Figure 20A.13

Strength of Materials Used in Screws and Bolts

$$P = \Delta F_B - \Delta F_p \qquad (20A.45)$$

The above three equations can be juggled to give:

$$\Delta F_B = P \cdot \left[\frac{k_B}{k_p + k_B}\right] \qquad (20A.46)$$

Since in our problem (with E and L the same for both the bolt and the plate-lug base combination):

$$\frac{k_B}{k_p} = \frac{L/\left(A_B \cdot E\right)}{L/\left(A_p \cdot E\right)} = \frac{A_p}{A_B} \qquad (20A.47)$$

And so Equation (20A.46) becomes:

$$\Delta F_B = P \cdot \left[\frac{A_B}{A_p + A_B}\right]$$

This shows that when a load P is applied to a pre-stressed bolt, the increase in bolt tensile force is less than P.

Minimum Initial Bolt Tension to Prevent Separation

Assume that the bolt clamping the lug and plate has been pre-stressed to a tensile force of F_{bo}. Then as a load P is applied, the above equation shows that the force in the bolt is

$$F_B = F_{BO} + P \cdot \left[\frac{k_B}{k_p + k_B}\right]$$

This equation holds only up to the point of separation of the plate and lug. Beyond this point, the bolt alone must carry the entire load. Since separation is undesirable, it is useful to compute the minimum initial bolt tension $(F_{bo})_{min}$ required to prevent separation when the load P reaches the maximum expected load, P_{max}. Assuming separation just occurs at $P = P_{max}$, then the force in the bolt will also at this point be P_{max}. This establishes the minimum value of F_{bo} since smaller values will have allowed separation at lower values of P. That is:

$$\left(F_{BC}\right)_{min} = P_{max} - \left(P_{min} \cdot \frac{k_B}{k_p + k_B}\right)$$

$$\left(F_{BC}\right)_{min} = P_{max} \cdot \frac{A_B}{A_B + A_p} \qquad (20A.49)$$

Given the uncertainty in the value of A_b, and our desire to be certain that the lug and plate do not separate, we must design for a larger value of initial bolt tension than the minimum. How much more is a matter of designer judgment, like any safety factor. We could treat this factor as a design variable, and tradeoff the cost of a larger bolt with the risk of separation. However, for purposes of this example, we will simply decide that the initial bolt tension to design for should be three times the minimum. That is:

$$F_{BO} \approx 3 \cdot (F_{BO})_{min} \qquad (20A.50)$$

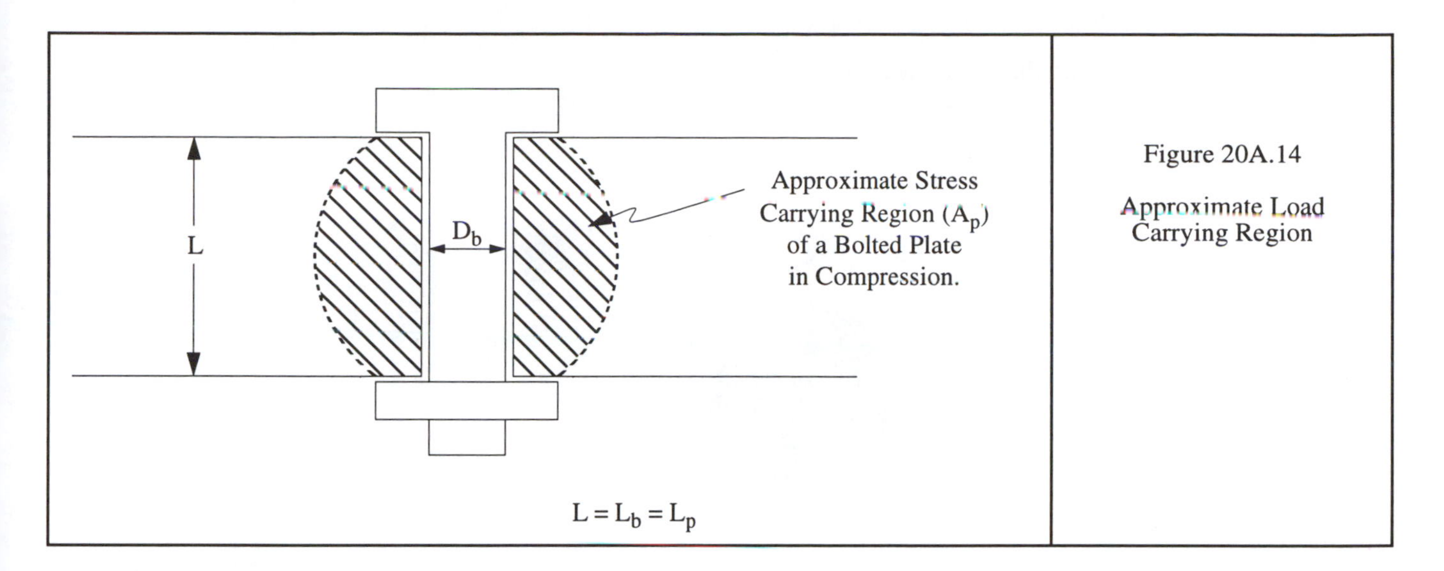

Figure 20A.14

Approximate Load Carrying Region

Stress Range and Average Stress

The stress in the bolt when the load is 4000 pounds is:

$$\sigma_B = \frac{F_{B2}}{A_B} \tag{20A.51}$$

where

$$F_{B2} = F_{B0} + 4000 \cdot \frac{A_B}{A_p + A_B} \tag{20A.52}$$

This is found by combining Equations (20A.46), (20A.47), and (20A.48). Similarly, the stress in the bolt when the load is 2000 pounds is

$$\sigma_{B1} = \frac{F_{B1}}{A_B} \tag{20A.53}$$

where

$$F_{B1} = F_{B0} + 2000 \cdot \frac{A_B}{A_B + A_p} \tag{20A.54}$$

The average stress caused by the fluctuating load is thus

$$\sigma_{av} = \frac{\sigma_{B2} + \sigma_{B1}}{2} \tag{20A.55}$$

The stress range of the fluctuating load is

$$\sigma_r = \frac{\sigma_{B2} - \sigma_{B1}}{2} \tag{20A.56}$$

The Fatigue Failure Ratio is now given by:

$$FFR = \frac{k\sigma_r}{\sigma_{rF}} \tag{20A.57}$$

where σ_{rF} is found from Equation (20A.21) or (20A.25) as appropriate.

This page available for notes, sketches, and calculations.

Chapter Twenty-One

Guided Iteration Applied to Thermal-Fluid Components

"To develop a philosophy of engineering design we must seek out those principles and concepts that are of the greatest generality, consistent with usefulness, and that can lead to a discipline of design."

Morris Asimow [1]
Introduction to Design (1962)

21.1 Introduction

In the preceeding chapter, we illustrated the application of guided iteration to the parametric design of traditional mechanical components. In this chapter, we illustrate its use in examples from the domains of heat transfer and fluid mechanics. We emphasize again that it is the guided iteration method itself that is the subject of interest here, not the domain of the problems used for illustration of the method. What readers should take away with them is the ability to apply the method themselves to future parametic design applications they encounter — regardless of the domain.

We begin with a heat exchanger design problem.

21.2 A Heat Exchanger Design Example

21.2.1 The Problem

The problem is how to take advantage of the availability of a stream of hot waste water (150 gpm at 200 F) in order to raise the temperature of a stream of incoming cold water (200 gpm at 60 F). If the energy available in the hot waste water can be used to help heat the cold water, then a considerable amount of money for fuel can be saved. Of course, there will be cost associated with purchasing and maintaining the heat exchanger, and we must show that the investment will be profitable.

Conceptual Design

The conceptual solution is to employ a water to water heat exchanger. Assume that any contaminants in the hot stream are not such (e.g., corrosive or including solids) that the use of a heat exchanger is precluded.

Other conceptual solutions are possible. For example, the hot waste stream could instead be used as a source of energy for a heat pump. Or a heat pipe might be used.

Configuration Design

There are a huge variety of liquid to liquid heat exchanger configurations possible. See, for example, the types discussed in [2]. We will assume that a shell and tube exchanger has been chosen. In this case, this is a fairly obvious choice, but other configurations are certainly possible. One could also, at least in principle, use a strictly counterflow exchanger.

Parametric Design

As we have indicated in previous chapters, selection of a configuration determines the attributes of a design to which values must be assigned in parametric design. In this case, the attributes (which will be the design variables in this parametric design problem) are as follows. They are illustrated in Figure 21.1:

---

* Reprinted from *Introduction to Design* by Morris Asimow, 1962, with permission of Prentice-Hall, Inc., Englewood Cliffs, NJ.

- Tube Diameter D
- Tube Spacing b
- Number of Shell Passes n
- Number of Tube Passes Nt
- Number of Tubes per Pass Ntp
- Length of Heat Exchanger l

There are the following limitations on these attributes:

Tube Diameter: Only standard sizes are practical. We will, however, work the problem as if D is a continuous variable. This is a reasonable way to begin; then once the approximate value for D at the solution is found, the nearest discrete standard sizes can be tried.

Tube Spacing: This is defined as the center distance between the tubes. Clearly, b must be greater than D. We impose an additional restriction here that b can be no less than 2D in order to allow for reasonable flow through and among the tubes.

In a heat exchanger of this type, the value of b is limited by the pressure drop created for the shell fluid. As b gets small, the pressure drop through the shell increases. At the same time, the velocity of the shell fluid is increased as it flows over the tubes. These effects can be included in the design procedure, but are omitted here in order to keep the space required to present the problem to a reasonable amount.

Number of Shell Passes: A shell and tube heat exchanger can be constructed so that the shell fluid (the hot fluid in this case) flows through and across the tubes several times. Figure 21.1 illustrates a four shell pass exchanger. Usually the number of shell passes is an even number so that the inlet and outlet of the shell fluid are at the same end of the exchanger.

Number of Tube Passes: The number of times that the fluid in the tubes passes through the length of the heat exchanger is the number of tube passes, N_t. In Figure 21.1 that number is 8 since there are two tube passes in each of the four shell passes. The number of tube passes is usually an even multiple of the number of shell passes.

Number of Tubes per Pass: The tube fluid flow usually is divided into a number of tubes; otherwise the flow per tube would be very high resulting in either a very large tube or in very high flow velocities and pressure drops. In Figure 21.1, only two tubes per pass (N_{tp}) are shown, but the number is usually larger.

Note that the total length of tubing in the exchanger is the heat exchanger length (l) multiplied by N_t times N_{tp}.

Heat Exchanger Length: There are standard heat exchanger lengths; for example, six, eight, ten feet, etc. We will treat this discrete variable as a continuous one, but will keep in mind that when a satisfying length has been determined, we will need to explore the effect of adopting a standard length.

There are also material choices, but we will assume here that standard materials will be used.

21.2.2 Formulation of the Problem For Parametric Design

Problem Definition Parameters

The parameters that define the specific problem we will work in this domain are:

Hot water flow rate	150 gpm
Hot water inlet temperature	300 F
Cold water flow rate	200 gpm
Cold water inlet temperature	60 F
Electricity rate	0.10 $/kwhr
Fuel oil rate	7.00 $/Mbtu
Cost of heat exchanger	$50,000 + 200 A_t

where A_t = heat transfer area in square feet.

Annual maintenance cost	$2,000.
Discount rate	5 %
Life of heat exchanger	8 Years
Annual hours of operation	400

It should be noted that the solution method is independent of the values of these parameters.

Design Variables

The design variables are: D, b, n, Nt, Ntp, and l.

Solution Evaluation Parameters

To provide this example with some interesting tradeoffs to consider, we establish the following Solution Evaluation Parameters *(and their importance levels):*

Return on Investment (ROI) (see Chapter 23) *Very Important*

Temperature of cold water leaving after it has been heated (Tc out) *Desirable*

Floor area required for the heat exchanger *Important*

Satisfaction curves for these evaluation parameters are shown in Figure 21.2.

In selecting these parameters, and in constructing their satisfaction curves, we have assumed that, for some reason, floor space available for the heat exchanger is scarce. We also assume that there is a desired range for the heated water temperature.

It may be noted that this problem could be converted to an optimization problem if (1) the return on investment (ROI) is made the sole criterion function, (2) the floor space is is made a constraint, or given a monetary value such as $/ square foot, and (3) the water temperature is constrained to be between a minimum and maximum temperature. This would, however, not enable tradeoffs to be made among the three SEPs we have established.

Analysis Procedures

For each of the evaluation parameters — the ROI, the heated water temperature, and floor area required — there must be analysis procedures available for computing values

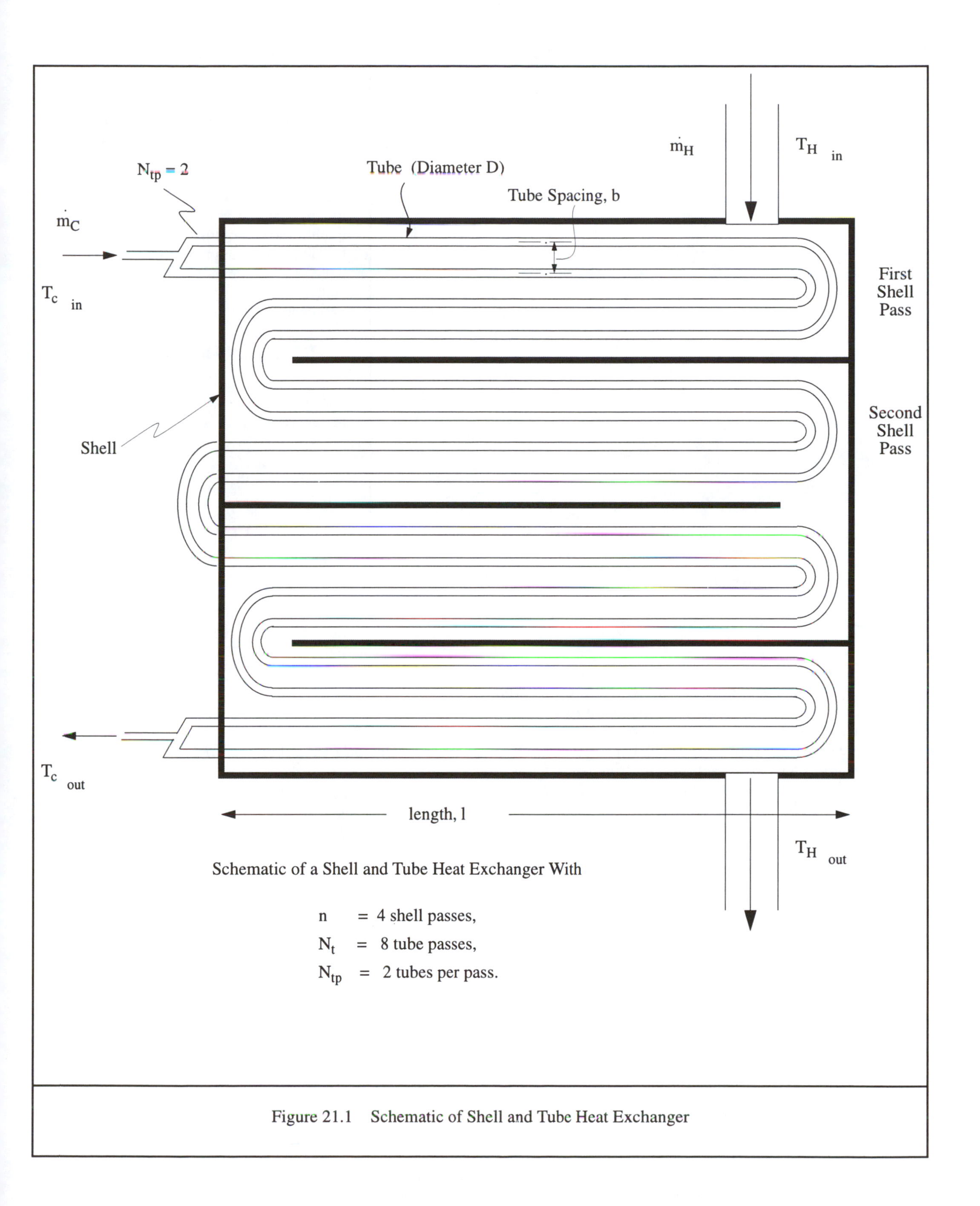

Figure 21.1 Schematic of Shell and Tube Heat Exchanger

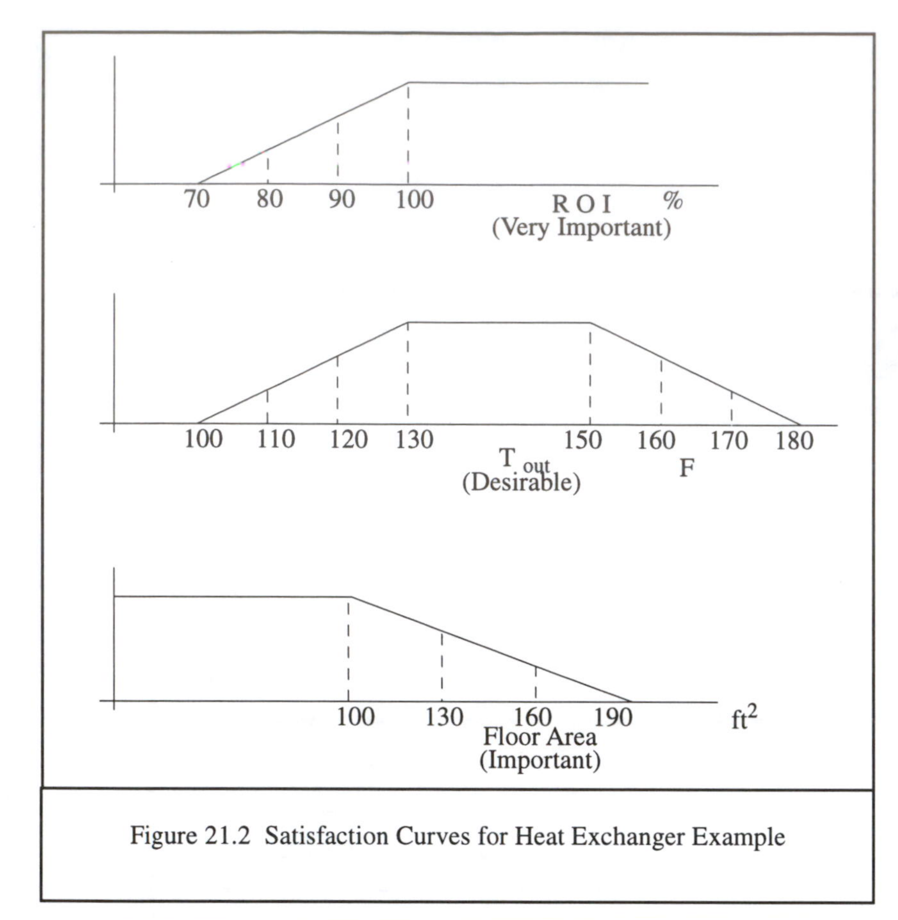

Figure 21.2 Satisfaction Curves for Heat Exchanger Example

Initial Dependency Table

The initial dependency table constructed is shown in Figure 21.3. The reason for the question marks shown is simply that, without experience in heat exchanger design, we would not likely have any idea about how the ROI varies with the design variables. Moreover, the interactions among the design variables are so complex that qualitative reasoning, which can often serve to determine approximate initial dependency values, cannot be used. We will show shortly how to get "experimental" initial values to fill in the table.

21.2.3 Initial Design

The initial design we will use is the one given above to illustrate the results of the analyses. That is:

D = 1 inch,

b = 2 inches,

n = 4 (shell passes),

N_t = 8 (tube passes),

N_{tp} = 4 (tubes per pass),

l = 12 feet.

Using the satisfaction curves shown in Figure 21.2, the performance results of the initial design are:

ROI	88.5%	Fair
Tc out	129 F	Good
Floor Area	79 sq ft	Excellent
Overall Evaluation	*Fair*	(See Figure 21.4)

from a given set of design variables. As noted, this is the place where the analysis procedures learned in engineering science courses serve the design process. In this example, the engineering science involved is heat transfer, and there is also some elementary economic analysis involved. The following equations are extracted primarily from two references [2] and [3]. Procedures required for this example are presented in Appendix 21A at the end of this chapter. They include procedures for computing each of the SEPs: the Return on Investment (ROI), the temperature of the cold water after it has been heated (Tc out), and the required floor area.

It is evident from this evaluation that the best opportunity we have for improving on our initial design is to improve the ROI. However, as noted above, we really have no idea what the ROI-related dependencies are. See Figure 21.3. Therefore, our next step is to develop a complete initial dependency table.

We do this by running the set of trial designs, Runs 2 through 7 shown in Figure 21.5. Note that for Trial 2, we changed only the heat exchanger length. This enables us to

Figure 21.3 Initial Dependencies for Heat Exchanger Example

	D_t	b	n	N_t	N_{ty}	l
ROI	?	?	?	?	?	?
T_c	1	0	1	1	1	1
Area	0.5	0.5	1	0.5	0	1

compute the three length dependencies as shown in Figure 21.5. The dependency between length and ROI, for example, is computed as follows:

$$y_j(old) / y_j(new) = [(x_i(old) / x_i(new)]^{d_{ij}}$$

$$ROI(old) / ROI (new) = [(l(old) / l(new)]^{d_{ij}}$$

$$88 / 78 = (12 / 8)^{d_{ij}}$$

$$d = 0.30$$

In Run 3, we change only the design variable Ntp, and can thus then compute the Ntp dependencies just as we did for length. We proceed in this manner until we have changed each design variable in turn, and computed the associated dependencies. The results are shown in Figures 21.5 and 21.6. Now we have a complete set of dependencies with which to begin our guided redesign efforts.

21.2.4 Guided Redesign

Guided Redesign for Trial 8

The evaluation for Trial 7, the last of our experimental runs to generate dependencies, are shown in Figure 21.5. They are:

ROI	86	Fair
Tc out	142 F	Excellent
Floor Area	200 sq ft	Unacceptable
Overall Evaluation		*Unacceptable*

To improve the floor area into the acceptable range, we note in the dependency table of Figure 21.6 that the design variable b has a fairly large effect on AF, but only very small effects on ROI and Tc out. Thus we elect to make a change in the tube spacing, b. Since b has a limiting value of 2 D (1.5 inches in this case), we try that value.

Results of Trial 8 and Guided Redesign for Trial 9

The results for Trial 8 are shown in Figure 21.5, and summarized as follows:

ROI	88	Fair
Tc out	148F	Excellent
Floor Area	120 sq ft	Good
Overall Evaluation		*Fair*

Now, since ROI is the most important performance parameter, and it has the lowest evaluation rating, we would like to improve it. The dependencies suggest that changing either l or N_{tp} will have the desired effect on ROI. However, l has a bad effect on floor area whereas N_{tp} has no effect on floor area. So we elect to try changing N_{tp}.

By how should we change it? We can estimate a new trial value as follows. Suppose we decide to try for a new ROI = 95. Then:

$$y_j(old) / y_j(new) = [(x_i(old) / x_i(new)]^{d_{ij}}$$

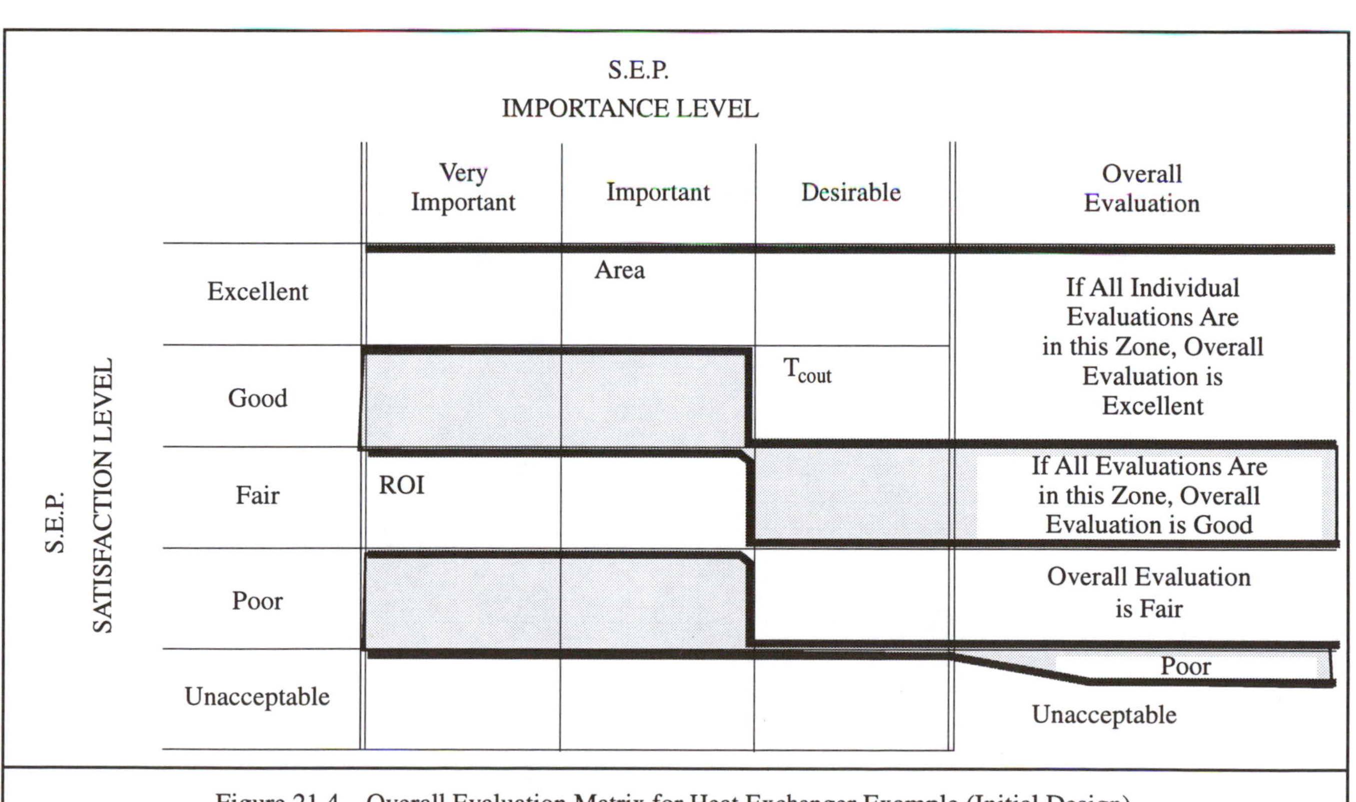

Figure 21.4 Overall Evaluation Matrix for Heat Exchanger Example (Initial Design)

Trial Number	Design Variables						Very Important ROI	Desirable T_c	Important AF	Overall Evaluation
	D_t	b	n	N_t	N_{tp}	l				
1 (6)	1	2	4	8	4	12	88.5 F	129 G	79 E	Fair
2 (7)	1	2	4	8	4	8	78.1 P	118 F	57 E	
3	1	2	4	8	6	8	87 F	126 G	57 E	
4	1	2	4	16	6	8	86 F	143 E	80 E	
5	1	2	8	16	6	8	88 F	144 E	160 F	
6	1	3	8	16	6	8	86 F	143 E	213 U	
7 (12)	0.75	3	8	16	6	8	86 F	142 E	200 U	Unacceptable
8 (13)	0.75	1.5	8	16	6	8	88 F	144 E	120 G	Fair
9 (14)	0.75	1.5	8	16	8	8	88 F	148 E	120 G	Fair
10 (15)	0.75	1.5	8	16	8	10	81 F	152 G	144 F	Fair (but worse)
11 (16)	0.75	1.5	8	1.6	8	6.7	93 G	144 E	104 G	Good!
12 (17)	0.75	1.5	8	1.6	8	8	96 G	137 E	84 E	Good. (Better)
13 (18)	0.65	1.5	8	16	8	8	88 F	136 E	78 E	Fair

Figure 21.5 Summary of Spreadsheet Results for Heat Exchanger Example

$$ROI(old) / ROI(new) = [\, (l(old) / l(new) \,]^{d_{ij}}$$

$$88 / 95 = (6 / N_{tp}(new))^{0.263}$$

$$N_{tp}(new) = 8.03$$

Thus we try a value of $N_{tp} = 8$ for Trial 9.

Results of Trial 9 and Guided Redesign for Trial 10

As shown in Figure 21.5, the results are not as we had hoped. Indeed, there has been practically no change in ROI. The reason is shown in Figure 21.7. Note there that the dependency between ROI and N_{tp} has changed from +0.263 to - 0.002; that is, from a fairly significant positive value to a very tiny negative value.

We might expect something like this in this problem because there are very likely to be optimum values for the design variables with respect to ROI. When the value of a design variable is below its optimum, the dependency will have one sign; when it is above, the dependency will have the opposite sign. Though this is a nuisance while we search for a solution (some of the guidance we receive, as in the last trial, is not helpful), it does not prevent us from finding a solution ultimately. And there is one advantage: As we get close to the optimum solution, we can expect the dependency values for ROI to get very small numerically.

Since Ntp did not work well for us in the previous trial, we decide to go back to our other choice: l. A similar calculation to the one above suggests that we try l = 10 feet for Trial 10.

Results of Trial 10 and Guided Redesign for Trial 11

The results of Trial 10 are shown in Figure 21.5, and they are again disappointing. As shown in the updated dependency table of Figure 21.8, the dependency between

	D_t	b	n	N_t	N_{tp}	l
ROI	.032	-.038	.035	-.012	.263	.307
T_c	.032	.016	.011	.186	.164	.231
Area	.224	.710	1.00	.500	0	.830

Figure 21.6

Dependencies After Run # 7

	D_t	b	n	N_t	N_{tp}	l
ROI	.032	-.048	.025	-.012	-.002	.307
T_c	.032	-.018	.011	.186	.105	.231
AF	.224	.737	1.00	.500	0	.830

Figure 21.7

Dependencies After Run # 9

	D_t	b	n	N_t	N_{tp}	l
ROI	.032	-.048	.012	-.002	-.002	-.389
T_c	.032	-.018	.011	.186	.105	.122
AF	.224	.737	1.00	.500	0	.817

Figure 21.8

Dependencies After Run # 10

	D_t	b	n	N_t	N_{tp}	l
ROI	.032	-.048	.025	-.012	-.002	-.113
T_c	.032	-.018	.011	.186	.105	.171
AF	.224	.737	1.00	.500	0	.743

Figure 21.9

Dependencies After Run # 12

length and ROI also changed sign. However, using the new dependency value for determining a length to try for Trial 11 finally leads to success. We compute a new trial length (6.7 feet) in the usual fashion.

Continuing Results

Note that by continuing on in the usual fashion with guided iteration, we get improved results in Trials 11 and 12 as shown in Figure 21.5. Both these designs have overall evaluations of Good, which may be the best we can in this problem. The dependency table after Trial 12 is shown in Figure 21.9.

It would take some continuing trials to insure that Trial 12 is indeed the best that can be done. With a spreadsheet set up to perform the computations (including computing the updated dependencies), this is not a very difficult or time consuming thing to do. It is, however, consuming of space to report the results in detail. In summary, we were unable to improve on Trial 12.

It should again be pointed out that there is no absolute assurance that the results found by guided iteration are the absolutely best results possible. In the case above, for example, though we explored more alternatives, we still may have missed a better combination. However, based on our own satisfaction curves, we have a Good design; that is all that can be said.

21.3 Preliminary Parametric Design of a Centrifugal Pump Blade

21.3.1 Formulating the Problem

The parametric design problem in this example is the preliminary design of the blade of a centrifugal pump. See Figure 21.10. The configuration of the proposed blade is shown in Figure 21.11. Note the decisions already made: the pump could be other than a centrifugal pump; or the blades could be radial or forward curved.

The goal in this example is to develop a preliminary design of the essential parameters that will determine cost and power required. The analyses will assume frictionless flow, and other simplifying assumptions will also be made, so the design produced will be preliminary, not final. Actual efficiencies of centrifugal pumps are typically over 85% (less for small sizes) so the frictionless assumption is not an awfully bad place to start.

The functional requirement is to deliver Q = 2 cubic feet of water per second at a head (H) of 80 to 90 ft-lb/lbm, though a head between about 70 and 100 ft-lb/lbm would be acceptable.

The Design Variables are:

- Angular velocity of the impeller ω radians/sec
- Outer radius of the impeller R_2 inches
- Outer blade angle β degrees
- Outer width of the blade b_2 inches

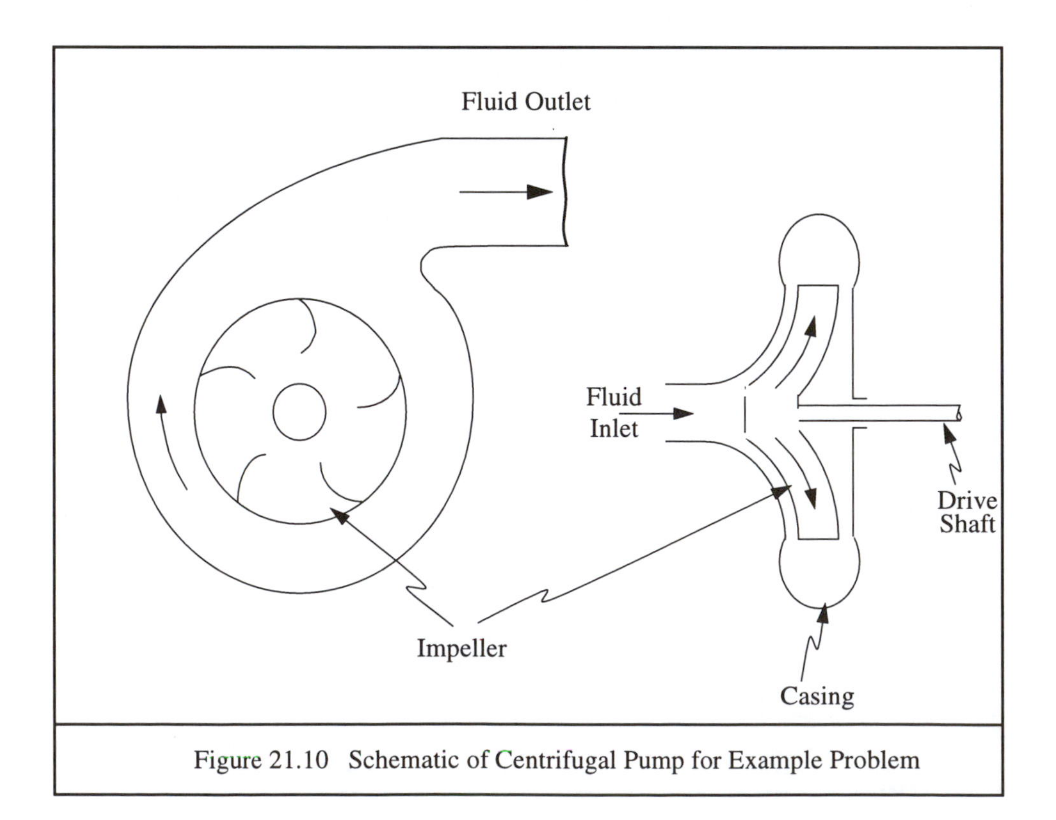

Figure 21.10 Schematic of Centrifugal Pump for Example Problem

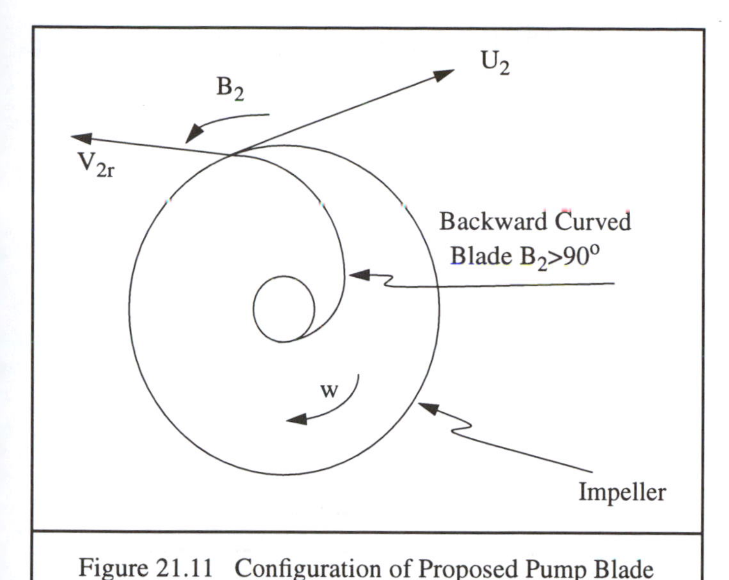

Figure 21.11 Configuration of Proposed Pump Blade

The Solution Evaluation Parameters (SEPs) and their importance levels are shown below. We are using "size" (actually, approximate pump volume) as a surrogate for cost for this preliminary design. Outlet velocity is included because, all other factors being satisfied, a lower fluid outlet velocity is more efficient.

- Head (Very Important) H ft-lb/lbm
- Size (Important) S cubic inches
- Outlet Velocity (Desirable) V_2 ft/second

Satisfaction curves for these SEPs are shown in Figure 21.12. The values used for size and outlet velocity are assumed here to be based on domain knowledge. Remember that these curves can be modified as the design proceeds if needed in order to get the evaluation discriminations desired.

Analysis procedures for computing the values of the SEPs are developed in Appendix 21A.2. See also Figure 21.13.

Initial Dependencies

An initial dependency table developed by reasoning with the above equations is shown in Figure 21.14. In this case, a lot of guessing is required to get these dependencies. It is, however, quite possible to get computed values (using a kind of computational experiment) without great difficulty. Figure 21.15 shows an initial design (that is, a set of trial values for the four design variables) provided as input for a spreadsheet that can compute the corresponding SEPs. Note that for each run after the initial one, only one (underlined in the Figure) design variable value is changed, and that by a relatively small amount. Figure 21.16 shows the SEP values for the five trials.

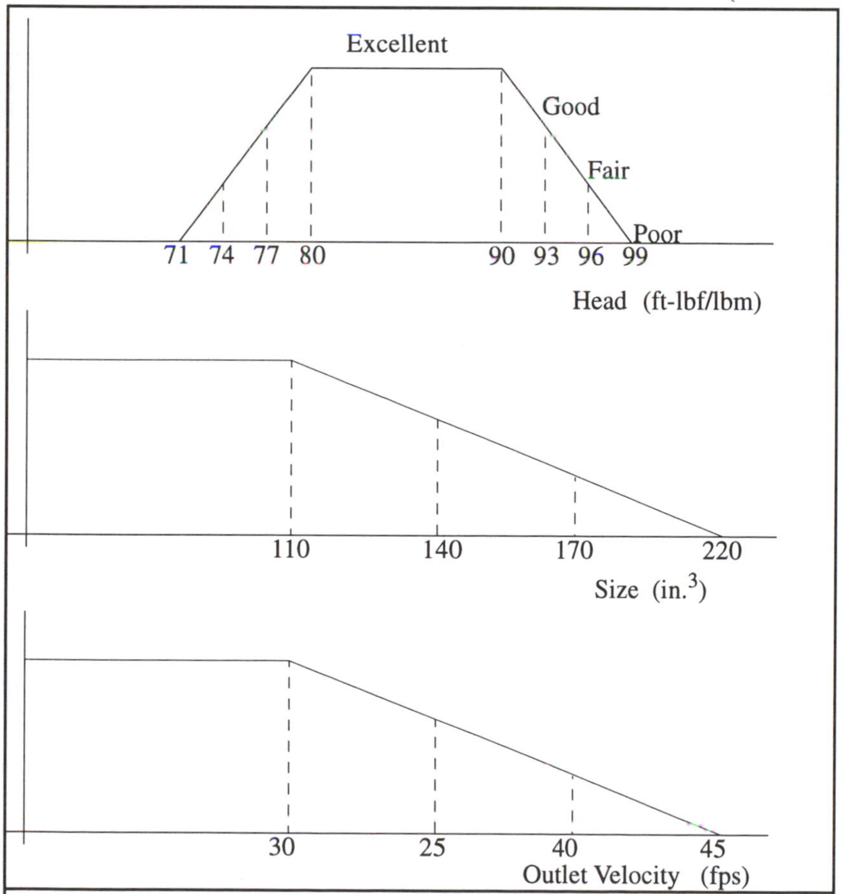

Figure 21.12 Satisfaction Curves for Pump Blade Example

Now we have the data needed to compute all the dependencies. For example, to compute the dependency between, say, b_2 and S, we use:

$$y_j(\text{old}) / y_j(\text{new}) = [(x_i(\text{old}) / x_i(\text{new})]^{d_{ij}}$$

$$301 / 352 = (1.5 / 1.75)^{d_{ij}}$$

from which we get d = 1, which is expected. Other dependencies that are not so obvious can be computed in the same way. For example, the dependency between head H and angle β is found from:

$$(191.9 / 194.1) = (130 / 120)^{d_{ij}}$$

$$d = -0.14$$

Such computations for initial (and subsequent) dependencies need not be done by hand but can be included in the spreadsheet with only a little extra work. The results are shown in Figure 21.17, which becomes a better initial dependency than the one shown in Figure 21.14 where the values were obtained by qualitative reasoning (and a bit of guessing) only.

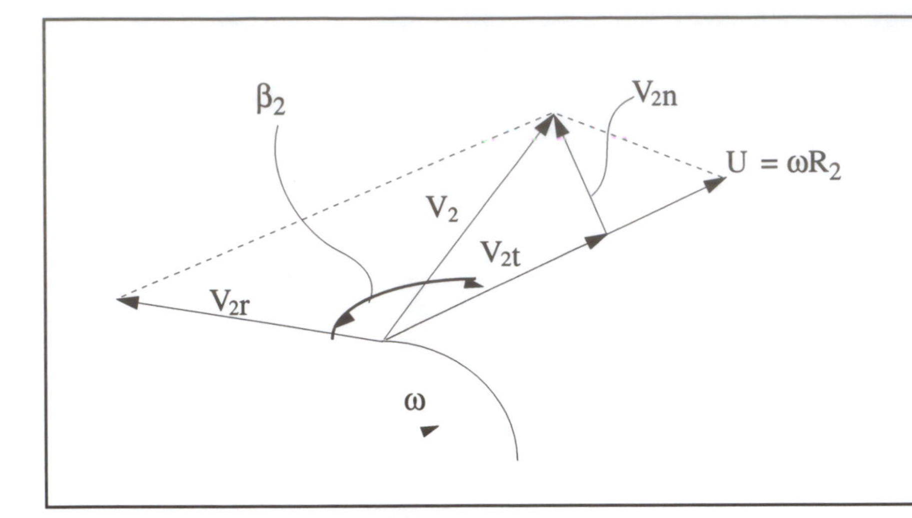

Figure 21.13

Velocity Symbol Definitions for Pump Blade example

	W	R_2	B_2	b_2
H	1.8	2	–0.1	0.5
S	0	2	0	1
V_2	1	1	2	–1

Figure 21.14 Initial Dependencies for Pump Blade example

21.3.2 Evaluation and Redesign

We can now begin the iterative process of evaluation and redesign. The only satisfaction curve with actual numbers at his point is for head, and its evaluation is Unsatisfactory for all of the previous experimental runs. We choose to improve it (that is, lower it into the acceptable range, say to 80) by changing R_2. We note in the dependency table that to lower head will require lowering R_2, and that will also lower S and V_2. Since both these changes will help our design, we proceed.

To determine by how much to lower R_2, we again use:

$$y_j(\text{old}) / y_j(\text{new}) = [\,(x_i(\text{old}) / x_i(\text{new})\,]^{d_{ij}}$$

$$123 \;/\; 80 \;=\; (\;7 \;/\; R_2\text{-new}\;)^{2.05}$$

$$R_2\text{-new} \;=\; 5.7 \text{ inches}$$

Results of trying this new value for R_2 are shown as Trial 6 in Figure 21.18, which is taken from the spreadsheet used for this example.

An updated dependency table is shown in Figure 21.19. Note that the changes are not large, indicating that the initial dependency table generated from Trials 1 through 5 is quite good.

The results of Trial 6 are shown in the overall evalua-

Trial No.	w	R_2	B_2	b_2
1	120	8	130	1.5
2	120	8	130	1.75
3	120	8	120	1.75
4	120	7	120	1.75
5	110	7	120	1.75

Figure 21.15

A Set of Trial Designs to Enable Computation of Initial Dependencies

Trial No.	w	R_2	B_2	b_2	H	S	V_2
1	120	8	130	1.5	190.8	301	94.1
2	120	8	130	1.75	191.9	352	92.0
3	120	8	120	1.75	194.1	352	69.8
4	120	7	120	1.75	147.5	269	58.3
5	110	7	120	1.75	123.6	269	52.5

Figure 21.16 Results of Trial Design to Establish Initial Dependencies

	W	R_2	B_2	b_2
H	2.03	2.05	–0.14	0.04
S	0	2	0	1
V_2	1.21	1.34	3.45	–0.14

Figure 21.17

Computed Initial Dependency Values

tion matrix of Figure 21.20 to be Poor. We can also be guided from this evaluation to see that it is the size that should be improved in order to improve the overall evaluation. From the dependency table in Figure 21.19, we are guided to select width b_2 as the design variable to change in order to improve the size evaluation. Note in the table that changing b_2 has a linear effect on size, but very little effect on either H or S, so it is a good choice.

To estimate a new trial value for b_2, we decide to go for a new value for S of 130. That will make its evaluation Good. We therefore make the usual computation:

$$y_j(\text{old}) / y_j(\text{new}) = [(x_i(\text{old}) / x_i(\text{new})]^{d_{ij}}$$

$$179 / 130 = (1.75 / b_2\text{-new})^1$$

$$b_2\text{-new} = 1.27 \text{ inches}$$

We will omit the remaining details. The process by now should be familiar: We try the new value for the

Trial No.	w	R_2	B_2	b_2	H	S	V_2	Overall Eval.
1	120	8	130	1.5	191(U)	301(U)	94(U)	U
2	120	8	130	1.75	192(U)	352(U)	92(U)	U
3	120	8	120	1.75	194(U)	352(U)	70(U)	U
4	120	7	120	1.75	147(U)	269(U)	58(U)	U
5	110	7	120	1.75	123(U)	269(U)	52(U)	U
6	110	5.7	120	1.75	80(E)	179(P)	38(F)	P
7	110	5.7	120	1.27	79(G)	130(G)	33(E)	G
8	110	5.7	120	1.065	77.7(G)	109(E)	30(E)	G
9	110	5.7	97.1	1.065	83.3(E)	109(E)	50(U)	U
10	110	5.58	97.1	1.065	79.7(G)	104(E)	49(U)	U
11	110	5.58	110	1.065	76.8(F)	104(E)	45(P)	P
12	110	5.7	120	1.27	79(G)	130(G)	33(G)	G
13	111	5.7	120	1.27	80.3(E)	130(G)	34(G)	G

Figure 21.18 Summary Results From Spreadsheet for Pump Blade Example

	W	R_2	B_2	b_2
H	2.03	2.09	–0.14	0.04
S	0	2	0	1
V_2	1.21	1.56	3.45	–0.14

Figure 21.19

Updated Dependency Table After Trial #6 in Pump Blade Example

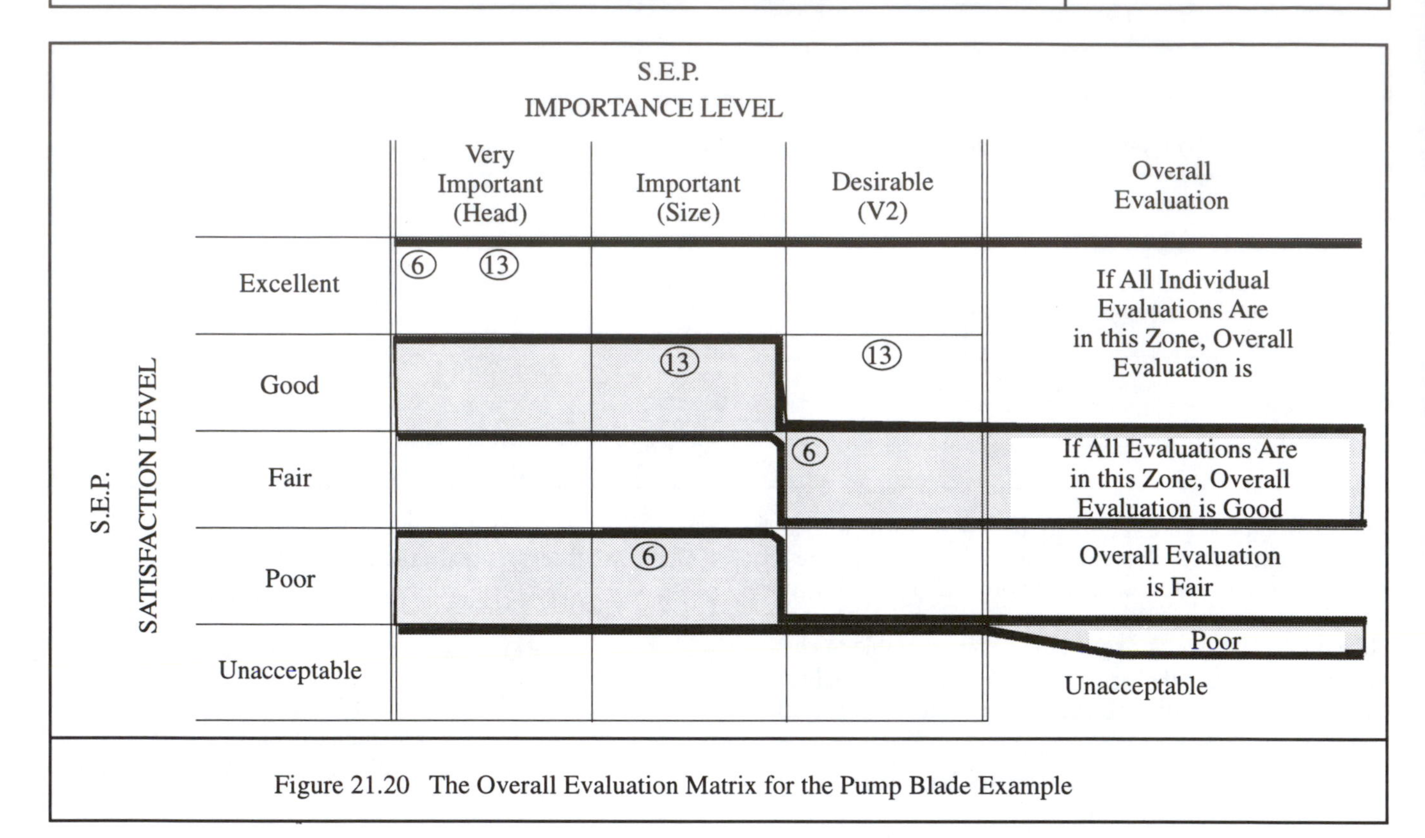

Figure 21.20 The Overall Evaluation Matrix for the Pump Blade Example

selected design variable in a new trial, and perform the evaluation and update the dependency table. Based on the results, we decide the design is satisfactory, or we are guided by the overall evaluation and the updated dependency table to select a design variable to change and by how much. This iterative process continues until we are satisfied.

Figure 21.18 shows the subsequent trials made in this example. Note that several trial designs yield a Good overall evaluation, but that we seem unable to do better. In such as a case, the designer can choose one of these, or refine the satisfaction curves to help discriminate between the competing Good designs.

21.4 Summary and Preview

Summary. This completes our presentation of selected examples using guided iteration to solve parametric design problems in traditional domains. In Chapters 20 and 21, we have employed guided iteration to perform design of shafts, springs, bolts, heat exchangers, and pump impellers. In Chapters 14 and 17, we applied guided iteration to beams and v-belt drives. *In every case, the design method is the same.* Only the analyses vary from problem to problem.

Problem formulation requires:

- Identification of the Problem Definition Parameters
- Identification of the Design Variables
- Identification of the Solution Evaluation Parameters, including
  - Satisfaction Curves
  - Importance Levels
  - Analysis Procedures
- Establishing Initial Dependencies

Then an initial trial design is all that is needed to get the iterative evaluation and redesign process started. The dependencies guide the selection of which design variables to change and the amount by which they should be changed. The computations can be done manually, but the assistance of a computer (especially with a spreadsheet) is very helpful.

Like optimization and Taguchi methods, guided iteration solves only the problem that is formulated. If factors are omitted by the designer, or improperly accounted for, then all the methods will get a correct answer to the wrong problem -- which will be a wrong answer to the real problem. Designers must therefore take care to formulate their parametric design problems correctly. When using guided iteration, it is especially important that the Solution Evaluation Parameters be complete, and that the satisfaction curves reflect the true goals of the design. Then, assuming that the analysis procedures are correct, the method will lead to a satisfactory solution.

Preview. Parametric design methods like guided iteration, optimization, Taguchi Methods™ are very powerful. And there are analysis methods (e.g., finite element methods) and computer tools (e.g., spreadsheets) that can provide the needed information about the expected performance of proposed parametric designs. The methods and supporting tools available at the configuration and conceptual levels are not as detailed or as computationally sophisticated; they rely more on qualitative physical reasoning. But at every stage of the process, designing requires knowledge -- knowledge of facts, physical principles, manufacturing processes, and of methods of many different kinds for applying, integrating, and employing other knowledge effectively. This kind of design process knowledge and how to apply it has been the subject of Parts I through IV of this book.

But there is more to engineering design than conceptual, configuration, and parametric design. As we pointed out in Chapter 1, engineering design takes place in a context of teamwork, a company, a Nation, and a world society. Communications (Chapter 22) are therefore critical to a designer. At least an introductory understanding of how money can influence design is also important (Chapter 23). And there are also legal issues (of patents and liability) and ethical issues that must be considered at all times (Chapter 24). These are the topics in Part V.

References

[1] Asimow, M., *Introduction to Design*, Prentice-Hall, Englewood Cliffs, NJ, 1962.

[2] Mills, A. F. *Heat Transfer*, Irwin Boston, MA 02116, 1992.

[3] McAdams, W. H., *Heat Transmission* McGraw-Hill, New York, 1942.

[4] Olson, R. M., *Essentials of Engineering Fluid Mechanics*, Harper and Row, New York, 1980.

[5] Dixon, J. R., *Design Engineering: Inventiveness, Analysis, and Decision Making*, McGraw-Hill, New York, 1966.

Problems

21.1 A toy rocket can be constructed with a plastic 'rocket' that is partially filled with water and then pumped to a modest pressure with a small hand air pump. See Figure P21.1(a). As a part of the preliminary design of such a rocket, we wish to determine an internal volume for the rocket (V_r) and the volume of water (V_{a1}) that should be added before pumping. The SEPs are the height to which the rocket will fly and the volume. In this case, the rocket volume is both a design variable and a solution evaluation parameter. Satisfaction curves for these SEPs are shown in Figure P21.1(b).

The following equations from a thermodynamic analysis of the rocket after pumping and release leads to the following. (Conservation of momentum, Bernoulli's equation, and ideal gas laws are used along with some assumptions and approximations to simplify the analysis.)These equations are approximate only; they provide only very crude estimates of performance, though perhaps sufficient to develop an initial prototype. See also [5].

- Weight of the rocket:

  $$W_r = 0.27 (V_r^{0.5})$$

  {W_r in ounces; V_r in inches^3}

- Initial velocity of water exiting the nozzle:

  $$S_1 = (2 \; P_1 / d)^{0.5}$$

- Air Pressure when all the water has just exited:

  $$P_2 = P_1 (V_{a1} / V_r)^{1.4}$$

- Final velocity of water exiting the nozzle:

  $$S_2 = (2 \; P_2 / d)^{0.5}$$

- Average velocity of exiting water:

  $$S^* = (S_1 + S_2) / 2$$

- Time required for water to exit:

  $$t = V_{w1} / (S^* A)$$

- Initial velocity of the rocket:

  $$S_r = S^* (W_1 / (W_r + 0.5 W_1))$$

- Height of rocket flight:

  $$h = S_r^2 / 2 g$$

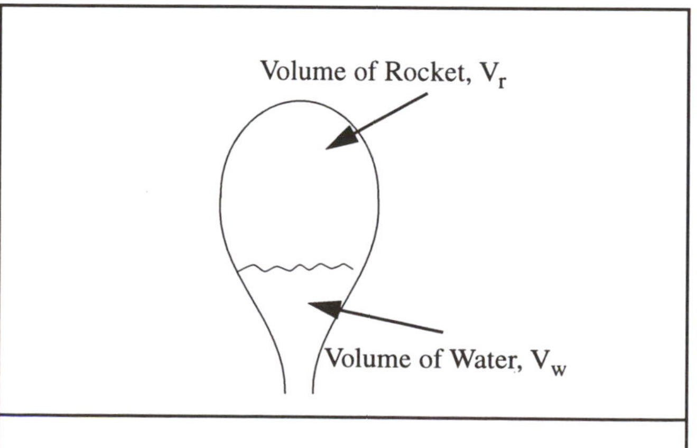

Figure P21.1(a) Toy Rocket Configuration for Problem 21.1

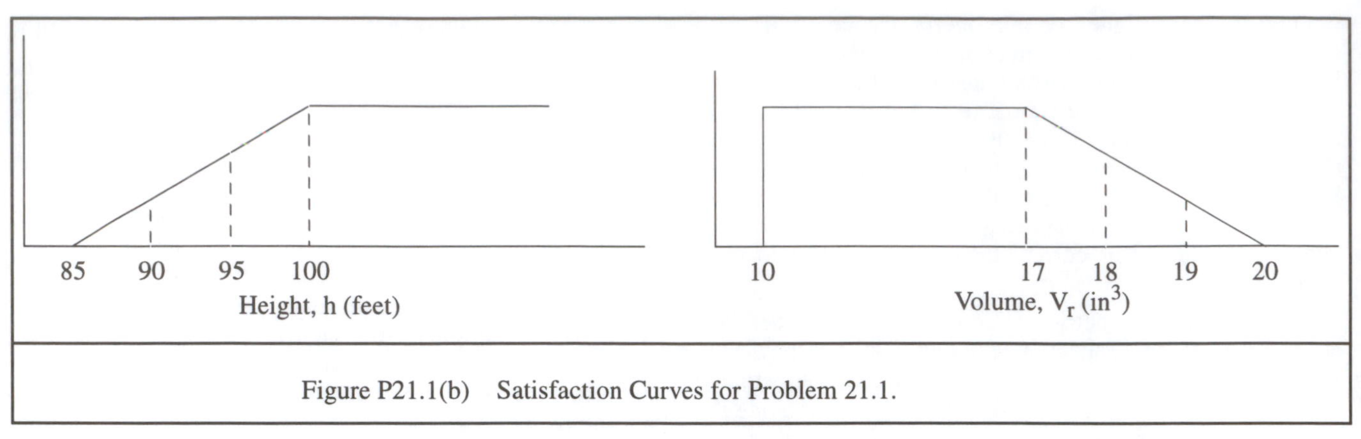

Figure P21.1(b) Satisfaction Curves for Problem 21.1.

where

d = density of the water

V_{w1} = Initial volume of water = V_r - V_{a1}

A = Area of the exit nozzle

W_1 = Initial weight of water

(a) Set up an initial dependency table including estimates of the initial dependencies.

(b) Use guided iteration to find values for the design variables. (Caution: Check units carefully.) Take the maximum feasible initial pressure to be 55 psia.

(c) For your solution, about how high does the rocket fly before all the water has exited? Why isn't the nozzle area a design variable at this stage?

(d) How might the rocket be manufactured?

(e) What safety concerns would you have about this product? How could you protect users, even those who may violate instructions and warnings?

21.2 The functional requirement of a centrifugal pump configured as shown in Figure 21.11 is to deliver Q = 4 cubic feet of water per second at a head (H) of 60 to 70 ft-lb/lbm, though a head between about 60 and 90 ft-lb/lbm would be acceptable. Perform a preliminary parametric design.

Appendix 21A

Analysis Procedures for Heat Exchanger and Pump Examples

21A.1 Analysis Procedures for Heat Exchanger Example

Computing the Heated Water Temperature (Tc out)

(a) The Heat Transfer Coefficients. Given a set of problem definition parameters and a set of trial design variables, this computation is a complex heat transfer analysis problem that must be done in a series of steps. First we compute the overall heat transfer coefficient, U:

$$1 / U = 1 / h_i + 1 / h_o \qquad (21A.1)$$

where hi is the heat transfer coefficient inside the pipes and ho is the outside coefficient. In this equation, we have neglected the thermal resistance of the tubes themselves (a good assumption unless there is corrosion or deposits on or in the tubes). We have also assumed that the inside area of the tubes equals the outside area. From [3], we find, for water:

$$h_i = 150 (1 + 0.011 \, t_{i\,ave}) \, V^{0.8} \, D^{0.2} \qquad (21A.2)$$

where $t_{i\,ave}$ is the average temperature (F) of the water as it passes through the exchanger, and V is the water velocity in the tubes in fps. D is the tube diameter in inches. We assume that the tube is thin-walled. Note that to compute the water velocity in the tubes, we must remember that the total water flow is divided into Ntp pipes. For the outside coefficient, we also find in [3]:

$$h_o = 370 (1 + 0.0067 \, t_{o\,ave}) \, V_{max}^{\;0.6} \, D^{0.4} \qquad (21A.3)$$

where $t_{o\,ave}$ is the average temperature (F) of the water as it passes through the exchanger, and V_{max} is the maximum water velocity in the shell in fps. D is the tube diameter in inches.

To compute V_{max}, we make a simplifying assumption for this example that the bulk velocity of the shell fluid is 3 fps; then as the fluid passes between the parallel tubes and reaches its maximum velocity, that velocity can be computed from:

$$V_{max} = 3\,b / (b - D) \qquad (21A.4)$$

where b is the spacing of the tubes center lines. In order to use these equations, we must assume values for the average temperatures. Then when the results are obtained, we may need to modify our assumptions and repeat the calculation iteratively.

(b) The Heat Exchanger Effectiveness (E). The heat exchanger effectiveness, once we compute it, can be used to compute the temperatures of the water streams as they leave the exchanger:

$$E = Q\,(act) / Q\,(max) \qquad (21A.5)$$

where Q (max) is the maximum possible rate of heat transfer in the exchanger:

$$Q\,(max) = m_h\, c_h\, (T_{h\,in} - T_{c\,in}) \text{ Btu/hr} \qquad (21A.6)$$

and Q (act) is the actual rate of heat transfer:

$$Q\,(act) = m_h\, c_h\, (T_{h\,in} - T_{h\,out})$$

$$= m_c\, c_c\, (T_{c\,out} - T_{c\,in}) \qquad (21A.7)$$

In these equations, c_h and c_c are the specific heats of the two streams (for water, these are both 1 Btu/lb-F) and m_h and m_c are the mass rates of flow of the hot and cold streams, respectively.

To compute the effectiveness, E, we first compute the number of heat transfer units, N_{tu}:

$$N_{tu} = U\, A_t / C_h \qquad (21A.8)$$

where A_t is the total heat transfer area in the exchanger:

$$A_t = N_t\, l\, \pi\, D \qquad (21A.9)$$

and $$C_h = m_h\, c_h \qquad (21A.10)$$

In [2] it is shown that for shell and tube heat exchangers:

$$E = [(A/B)^n - 1] / [(A/B)^n - R] \qquad (21A.11)$$

where $A = 1 - R\,E_1$ and $B = 1 - E_1$ (21A.12)

$$R = C_h / C_c \qquad (21A.13)$$

$$E_1 = 2 / [1 + R + Z\; 1 + \exp(-N_{tu1}\, Z)) / (1 - \exp(-N_{tu1}\, Z))] \qquad (21A.14)$$

$$Z = [1 + R^2]^{0.5} \qquad (21A.15)$$

$$N_{tu1} = N_{tu} / n \qquad (21A.16)$$

n = Number of tube passes (a design variable)

(c) Example Results. As an example of the numbers generated by the above analysis, using a spread sheet for the computations, the following results were obtained. The trial design assumed was:

D = 1 inch

b = 2 inches

n = 4 (shell passes)

N_t = 8 (tube passes)

N_{tp} = 4 (tubes per pass)

l = 12 feet

The computational results are:

Cc = 99600 Btu/hr-F

C_h = 74700 Btu/hr-F

A_t = 100.5 sq ft

T_i ave = 100 F (assumed)

T_o ave = 150 F (assumed)

Velocity inside tubes = 20.33 fps

V max outside flow = 4.5 fps

hi = 3507 Btu/hr-sq ft-F

ho = 1829 Btu/hr-sq ft-F

U = 1202 Btu/hr-sq ft-F

N_tu = 1.617

N_{tu1} = 0.404

R = 0.75

E1 = 0.294

E = 0.660

Q (max) = 10,458,000 Btu/hr

Q (act) = 6,906,275 Btu/hr

T_c out = 129 F

T_h out = 108 F

T_i ave = 94.5F (actual)

T_o ave = 154 F (actual)

Computing the ROI

(a) The Value of the Energy Saved. The problem definition states that oil is available for $7.00 per Mbtu (million Btu). We assume that if the cold water were not heated in the exchanger, it would have to be heated with oil. Thus the value of the energy saved per year is:

M_e = (Q (act)/1000000) (hrs/year of operation) (7.00) $/yr

(b) Cost of Pumping the Water. We are neglecting the cost of pumping the hot water stream. For the cold water stream, the cost of pumping can be computed as follows:

The pressure drop in a tube is:

$$PD = f\,(L'/D)\,\rho\, V^2/2g_o \qquad (21A.17)$$

where f = friction factor = $0.184\, Re^{-0.2}$ (21A.18)

Re = Reynold's Number = $(\rho\ V\ D\ /\ \mu)$ (21A.19)

L' = length of one tube = $1\ N_t$ (ft) (21A.20)

D = Tube Diameter (ft)

V = Velocity inside the tubes (fps)

g_o = 32.2 ft-lbm/lbf-sec^2

ρ = density of water, 62.4 lbm/ft^3

The power required to pump the total flow against this pressure drop is found from

$$P = (1/\rho)\ PD \quad (21A.21)$$

The annual energy used for pumping is

$$E_p = P\ \text{(annual hours of operation)} \quad (21A.22)$$

The annual cost of this energy is

$$M_p = C_e\ E_p \quad (21A.23)$$

(c) Net Annual Savings. The net annual savings are the value of the energy saved minus the cost of pumping and the maintenance cost:

$$M_n = M_e - (M_p + \text{Maintenance Cost}) \quad (21A.24)$$

(d) Initial Cost of the Heat Exchanger. The initial cost of the heat exchanger is given by Mills [1992]:

$$C = C_i + (c_i\ A_t) \quad (21A.25)$$

(e) Economic Analysis. As pointed out in Chapter 23 of this book, there are many ways to compute an ROI. For this example, we will use the following approximation:

ROI =

= Present Value of Ave Ann Savings / Ave Ann Investment

$$ROI = PV_{ave\ sav}\ /\ I_{ave} \quad (21A.26)$$

where I_{ave} = Initial Cost / 2 (21A.27)

and

$PV_{ave\ sav}$ = Present Value of Total Savings / Life of Projec (21A.28)

The annual savings are given above. The present value of these savings over the eight year life of the project are found by applying the following equation from Chapter 23:

$$PV_{ave\ sav} = \text{Annual Savings}\ [((1+i)^n - 1)/\ (i\,(1+i)^n)] \quad (21A.29)$$

(f) Example Results.

Using the same trial design as above, the following results are obtained using these analyses:

Initial Cost of Exchangers	= \$80,144
Reynold's Number	= 230,029
Friction Factor	= 0.0156
L'/D	= 1152
Pressure Drop, PD	= 49.92 psi
Pump kw	= 4.32
Pumping Cost, Annual	= \$3786
Gross Annual Savings	= \$67,681
Net Annual Savings	= \$43,896
Present Value of Savings	= \$283,707
ROI	= 0.88 = 88%

Computing the Floor Area

(a) The Area.

We have developed a simple approximate relationship for the purposes of this example:

$$\text{Floor Area} = AF = (4 + 2\,l)(\,D + b\,)/\ 12\,)\ n\ N_t^{0.5}\ \text{sq ft} \quad (21A.30)$$

(b) Example Result.

For the example above, the result is:

Floor Area, AF = 79 sq ft.

21A.2 Analysis Procedures for Pump Blade Example

For Head (See Figure 21.13):

$$H = 1/g\ (V_{2t} R_2 - V_{1t} R_1)\ \omega \quad (21A.31)$$

We assume for this preliminary design that V_{1t} is small enough to be negligible, so this becomes:

$$H = 1/g\ (V_{2t} R_2)\omega = U_2\ V_{2t}\ /\ g \quad (21A.32)$$

From Figure 21.13, we get:

$$V_{2t} = U_2 - V_{2n}\ \cot(180 - \beta_2) = U - V_{2n}\ \cot\ \beta_2 \quad (21A.33)$$

Also, considering the principle of continuity of mass flow:

$$Q = 2\ \pi R_2\ b_2\ V_{2n} \quad (21A.34)$$

Combining the above three equations, and re-arranging:

$$H = (U_2^2/g) + (Q\ U_2\ \cot\beta_2)/(2\pi\ R_2\ b_2\ g) \quad (21A.35)$$

For Size

We assume that the pump size (external volume) will be approximately proportional to:

$$S = Pi\ R_2^2\ b_2 \quad (21A.36)$$

For Outlet Velocity

$$V_2^2 = V_{2t}^2 + V_{2n}^2 \quad (21A.37)$$

Part V Pandemic Topics

Chapter Twenty-Two

Communications

"It's a question of using the English language in a way that will achieve the greatest strength and least clutter."

William Zinsser [1]*
On Writing Well (5h Edition, 1994)

22.1 Introduction

It is difficult to over-emphasize how important effective communications are to the efficient achievement of quality designs. The communication abilities of the individuals and groups involved are critical, as is the communication system in the organization. As one manager of a complex design activity said: "If I could have just one wish granted, it would be to make our internal communications fewer in number and far more effective."

Communication is also extremely important to personal success. An engineer who can communicate well in person, in written reports, through sketches and diagrams, and in oral presentations — and who can also listen effectively — is far more likely to be given greater responsibility and advancement. There is no substitute for technical competence, but without the ability to communicate well, even the most technically competent engineer will have difficulty advancing in business organizations. An engineer simply cannot fully realize the benefits of technical ability unless he or she is also able to communicate effectively with others.

For example, a very successful former student of ours reported recently that he attributes his rapid rise *primarily* to his ability to communicate effectively. In addition, this engineer believes that his ability to communicate effectively was developed by practicing the ideas in this chapter while working on various engineering projects during his undergraduate years.

In this chapter, we discuss three types of communication: written, oral, and graphical. We point out here some of the major principles, and provide some practical tips and guidelines that will help young engineers get their communication efforts off to a better start. But we strongly urge lots and lots of formal communication practice, coupled with lots and lots of effective criticism.

22.2 Written Communications

22.2.1 Preparation for Writing

There are several types of written communications, including notes, memos, letters, and several types of reports. We will discuss the specific types in later sections; first, we describe some general guidelines that apply to all of them.

Before you write

Question whether writing is the best way to communicate. That is, ask yourself whether this message for this audience is best delivered in writing, in person, over the telephone, or in an oral presentation. Alternatives to writing allow for more interaction and discussion.

You may want to write so there will be a record, or so that your whole message can be heard without interruption. If the message is technical, writing provides receivers with an opportunity to study it and to refer to it later. Some messages need both a written form (e.g., for a record and/or for a permanent reference) and a personal or oral presentation that permits discussion. In this case, however, both the written and the oral messages must be complete in themselves.

Explicitly state for yourself the purpose of the communication. What is it exactly that you want to accomplish? Is there something you want the receiver(s) to do? Is there information you want them to have, or that they need to know?

* From Chapter 1, "The Transition" in *On Writing Well* by William Zinsser, Fifth Edition, Copyright 1994, published by HarperCollins. Permission to quote by Carol Brissie.

Explicitly describe the intended receiver(s) for yourself. Who are they? Why are they interested in getting the message? Do they need it? What symbols (mathematical, verbal, graphical) do they understand and not understand?

Often, there are several types of intended receivers for messages in a company. Each must be described, and the communication constructed and organized accordingly. In a written report, for example, this may call for a brief Abstract for top managers, a main section (or body) for interested engineers, and Appendices with all the details for the colleagues most involved.

List and organize the critical information that is to be communicated. This need not be a formal outline (though it can be), but you need to decide at an early stage, *before* you begin to write, exactly what is the essential content of your message. You must be brief here. For example, consider what such a list might have looked like as we prepared for writing this chapter in this book:

1. Principles and practical guidelines useful to students and young engineers relative to:

- Written communications,
- Oral communications,
- Graphical communication.

Then for each of these major sections, another list, more detailed, is prepared. For example, for this section on written communications in this book, some listed items might be:

A. Preparations for writing:

- Know your purpose,
- Know your readers,
- Know your message.

Select the specific written format that is most appropriate to the audience, your goals, and the message. After you have decided to write, you have your choice of many different writing formats: a handwritten note, a memo, letter, electronic mail, or one of several types of reports. (We will subsequently discuss two types of reports: the Research Report and the Business Technical Report.) There are also notes and records of various kinds that you keep for yourself. You *could* even write a book.

> "If people wish to write in a clear style, let them first be clear in their own thoughts."
>
> J. W. Goethe
> *(circa 1800)*

22.2.2 Standards for Spelling and Grammar

Suppose you got a technical report to review, and it began as follows:

Introduction

A major problem that exists with the production of injection molded plastic parts is acheiving the demensions and shape of the targeted design. Due to the complex numbers of interactive shrinkages that typically are developed from the molding of the part it is virtually impossible for even the most skilled designer to take an emperical approach in predicting what their net effect will be on the final molding. Therefore it is expected that the part cannot be fully evaluated demensionally or mechanically until a mold is actually built and the part produced.

Considering the misspelled words — achieving, dimension (twice), empirical — how would you feel about these authors, and about their technical capability? How do you feel about their use of language? For example, how about those "complex numbers of interactive shrinkages"?

We assume that readers of this book are able to write sentences, and know about such writing basics as the proper structure for sentences and paragraphs. We realize, however, that these basics present problems for some engineering students, so we emphasize again that writing is extremely important to successful engineering practice in any organization. Therefore, if basic writing ability is a problem for *you*, then *you* should immediately and aggressively take whatever steps are needed to correct it.

In all business and technical communications, spelling and grammar must be *perfect*. Yes, *perfect* — because that is the expected professional standard. Anything less will hurt you and the effectiveness of your communications significantly.

The reason that *perfect* is the standard is that, if there are spelling or grammar errors, the people who receive the message will — consciously or unconsciously — infer that you have made other errors as well. They will assume that since you did not care enough to get the spelling and grammar right, then you either (a) don't care much about them, or (b) didn't take care to get the technical content right either, or (c) both. Usually they assume (c). Moreover, many people will stop reading altogether when they begin to encounter more than a rare spelling or grammar error. Of course, an occasional typographical error may get through. No doubt you have found some in this book, though we have tried very hard to get them out. A few such mistakes are inevitable, but the standard to strive for in spelling and grammar is perfection.

If you are not confident of your spelling and grammar, then let someone you know is capable read your messages before they are sent. If you are using a word processor, you can use the spell checker first to save yourself some minor embarrassment. But don't assume that spelling and grammar are the same thing; spelling can be perfect, but grammar may still be bad.

22.2.3 The Importance of Conciseness

In the design of a part or product, as we have pointed out earlier, one good old rule is KISS: Keep It Simple, Stupid. KISS also applies to professional communications, though in a slightly modified form. Here it means: Keep It Short, Stupid. In written communication, conciseness is a primary virtue. But short is not easy to attain. There is a story about the person who received a long, long letter that concludes: "Sorry I didn't have time to write you a short letter".

Usually a written communication should be much shorter than you, the writer, would like it to be. It should probably be even shorter than you think at first it is possible to be. But *any message can be made any length*, and the shorter the better. For example, forced to reduce this section to one short sentence, we could say: Keep written communications short.

The Take-Aways. As a help to achieving conciseness, you may want to try the following exercise. Suppose that an intended reader of one of your communications has just finished reading it, and is then asked by another person passing by in the hall, "What did <he or she> have to say?" Now what, exactly, in just a few brief sentences, would you want your reader to reply in answer to that question?

Your answer is called the *take-away*. The take-away is that brief essential central message you want your readers to *take away* with them (to remember, to hold on to) after getting your communication. Knowing what your desired take-aways are can often help you shorten a communication to include just the take-aways, plus only whatever else is absolutely necessary to support them. The take-aways, then, are what you should emphasize in your message, and everything else can be minimized, if not eliminated.

Reduce the Noise. Another technique for achieving conciseness is to keep the noise level down. By *noise* we mean the frequent communication or inclusion of extraneous or irrelevant information. It is true that you don't always know exactly what all readers will consider useful for their purposes, and you don't want to leave out something useful. But too much noise causes people to tune out, thus missing what you most want to communicate to them. Stick to just the take-aways.

Eliminate Wordiness and Clutter. One of the best ways to achieve concise writing is to eliminate unnecessary words. You can also change fancy words and jargon to plain English. Look again at the third paragraph above that begins with *The Take-Aways.* Here is a rewrite that eliminates some unnecessary words:

The Take-Aways. To achieve conciseness, try the following exercise. Suppose a person who has just read your communication is asked "What did it say?" The short quick answer you hope for is called the take-away.

We have reduced the word count from 71 to 37, without significantly changing the content. This kind of word cutting can almost always be done at least once or twice after a first draft has been written. For example, we can do it again to the above paragraph:

Take-Aways. Try this: Ask someone who has just read your communication "What did it say?" The short quick answer you hope for is the take-away.

Now we have only 27 words, about one-third the original, but essentially the same content. Of course, you may not want the terseness that extreme word reduction can sometimes create. But "too short" is seldom the problem with most writing. The usual problem is excessive length due to unnecessary words; rewriting to eliminate them will improve the result significantly.

22.2.4 The Writing Process

The writing process is, like design, best done by guided iteration; that is, by writing, evaluating, and re-writing until an acceptable result is obtained. At least 90% of writing is re-writing.

The process is this: Write a *complete* first draft, no matter how rough; then re-write, re-write and re-write until you get the result you want. Do *not* try to get any one part fully polished except during a near-final complete re-write.

The first complete, rough draft may be little more than an outline, or just the main ideas jotted down in the order you want them. Then you begin to expand and re-write and re-write and re-write until you have the desired quality. Good writing has generally been re-written and shortened *ten to twenty times or more*.

The process recommended above is not only the best way to get a *quality* result, it is also the *fastest* way to get a quality result. If you are in a hurry, the worst thing to do is to try to get the first paragraph or first section nearly perfect before going on the next paragraph or section. Invariably when you do this, you will find later that you must come back to earlier sections to make major changes or additions. Then some of the time you spent making the first part perfect will have been wasted. The final quality and effectiveness of your writing is determined by the last draft, not by the first draft.

22.2.5 Two Pieces of Practical Advice

1. Always keep a copy of all written technical and business communications, even informal notes and e-mail messages.

2. Be very careful what you write. Sooner or later, the whole world may see it. This is especially true if it is something that you would rather the whole world did not see.

We now turn attention to the specifics of the more common types of written communication required of young design engineers.

22.2.6 Types of Written Communications

Personal Notes and Records.

As will be noted in Chapter 24, design engineers must keep a notebook for patent and liability reasons. However, a notebook is essential for other reasons as well. Memory

> "The horror of that moment," the King went on, "I shall never, *never* forget!"
>
> "You will, though", the Queen said, "if you don't make a memorandum of it."
>
> Lewis Carroll
> *Through the Looking Glass (1872)*

sometimes fails, and when you are busy, memory often fails. Thus a personal daily log of notes and important communications is a personal and professional necessity.

Hardly anyone likes to keep notes; most of us are not used to doing it, and it seems a distraction from pressing work that needs to be done right now. It is therefore important for personal note keeping to include the important items and omit the unimportant ones. In most cases, it will be about right if you devote 5 to 15 minutes a day (on average) to personal notes, and supplement the notes with documents (or references to them) received from others. As noted above, you will also want to keep copies of your own written communications to others.

With your personal notes, you are both sender and receiver. Even so, it is still helpful to think about the purposes of the notes. It is most likely that you will keep them concise to save yourself time, but watch out for too much conciseness here. Like the King in the *Through the Looking Glass*, Your memory a week or a month or a year from now won't be as good as you think. What you fail to include might cost you a lot of time or trouble later.

Informal Hand Written and Electronic Mail Messages

Informal hand written and/or electronic "e-mail" messages are often the most convenient and appropriate medium for communicating in day to day working relationships. Messages can be prepared and sent quickly. You can carry on some discussions via e-mail.

In contrast to longer, more formal media, e-mail messages generally get read and responded to immediately — unless you send too many of them or they are too long. As always, keep the general noise level down if you want to be heard when it is important.

Memos and Letters

Business memos and letters are generally used to express opinions, to make requests, to provide notification of one kind or another, and to provide information that is not highly technical. Memos and letters may provide summaries of the results of technical work, or recommendations based on such results, but usually in this case they will refer to a longer report that contains the complete technical information. When letters or memos accompany technical reports, they are called *covering* or *transmittal* memos or letters.

Memos are generally used inside firms, and letters for outside. However, there are no hard rules about this.

Any business memo or letter longer than one page is highly undesirable. Probably there ought to be law. Certainly two pages is the absolute maximum, and that should be done only rarely. If the memo or letter is longer than one page, then the first and last paragraphs *must* provide the desired take-away information. The reason is that the rest, especially the second page (except perhaps for the last paragraph), is not very likely to get read very carefully — if it is read at all.

Here is an example memo:

Date: March 5, 1993

To: T. S. Jones, Vice-President, Advanced Manufacturing Engineering

From: J. K. Smith, Manager, Manufacturing Engineering

Subject: Systems Manufacturing Testing

We have had double digit growth in productivity throughout the recent period of rapid technological advancement. This growth has continued despite manufacturing system test methods that have not changed for fifteen years. However, as we enter new markets, and our competition increases in the present markets, our development and production processes should now be automated to maximize productivity. Since testing follows a product through its entire life cycle, there is much to be gained from automation of the existing methods. Moreover, some tests can be eliminated.

Preliminary trials conducted on our LS-2 indicate that an automated test system can significantly reduce costs in all our plants by

- *Decreasing the time of highly skilled people doing testing;*
- *Reducing capital and maintenance costs, and*
- *Eliminating redundancy.*

Additional trials with the automated testing system will be conducted during the next quarter, and an updated report issued at the end of the quarter.

Two Types of Technical Reports

A *technical report* describes the results of technical activities such as research, development, analysis, design, experimentation, and related topics including preparing specifications, marketing studies, and the like. In this book, we describe two kinds of technical reports:

- Research Reports,
- Business Technical Reports.

Though we will discuss these forms separately as essentially pure forms, there are often circumstances when a combination of these two formats is the best solution. This is discussed later.

If a technical report includes recommendations for decisions or actions, or if it includes results or conclusions which will be used in the short term as a basis for making or supporting decisions or actions affecting current activities or plans, then we call it a *Business Technical Report*.

If a technical report contains general information, basic data, or research results useful in the longer term — but not immediately and directly relevant to specific current operations — then we call it a *Research Report*. Research Reports do not include recommendations for decisions or actions, except perhaps about continuing research.

The reason for making a distinction between Business Technical Reports and Research Reports is that, because of their different purposes, the two types are best written in quite different formats. We describe two recommended formats below.

There are special customs in some companies about the form of internal technical reports, and if so, readers should follow those customs unless there is general agreement in the company on desired changes. If there are no customs, or possibly as a model for change, we recommend the formats described in the next two sub-sections.

To help make the following discussion more efficient, we will use these definitions:

Data: Experimental data, input values used, survey results, results of previous studies, material properties, machine or product parameters, etc.;

Analyses: Analytical and computational procedures employed, including simulations;

Results: Generalizations of data, and results produced by analyses;

Conclusions: Conclusions of a technical or business nature based on an accumulation of results;

Recommendations: Recommendations for decisions or actions based on the results or conclusions.

Research Reports follow the usual path taught in science and some engineering courses: First the goal of the project is described, next the experiment is described, then the data, analyses, and results, and finally the conclusions and any recommendations are presented.

Business Technical Reports use almost the reverse order; that is, they begin with recommendations, and follow with conclusions, results, analyses and data. This order is used because different people in a business organization read reports for different reasons. Higher managers may want to know only what the recommendations are, and these managers should not have to read through a lot of preliminary detail to find them. If they want to know the reasoning and the details, they can read on, but it should be their option.

We now describe the two types of technical reports separately and in more detail.

22.3 Research Reports

22.3.1 Overview

Most Research Reports are written for internal company use only. However, they are sometimes also prepared for external publication in a trade magazine or research journal. Many companies, to protect patent and/or proprietary information, require that such reports be cleared by the firm's management or legal departments before they are submitted for publication.

Though there are certainly variations possible depending on the content, Research Reports are generally structured more or less as follows:

- Title
- Abstract
- Introduction
- Literature Review
- Main Body
  - Hypothesis and Method of Attack
  - Experiments, Analyses, or Simulations
  - Results
- Discussion of Results
- Summary and Conclusion
- Appendices (if any).

When a research paper is to be submitted for possible publication in a journal, the journal's guidelines regarding format, length, style, and content should be obtained and studied before the report is written. Though there are many similarities among journals on these issues, there are also differences. It is always a good idea to submit papers to journal in the form wanted by the Editor.

Though it is certainly a matter of personal choice, we feel that it is better to avoid use of the first person (that is, "I" or "we") in Research Reports. This necessitates use of the passive voice (e.g., "six tests were conducted" instead of "we conducted six tests") which can produce a rather pedantic style that sometimes gets boring. Whatever choice is made about issues of voice and tense, however, it must be maintained consistently throughout the report. Changing back and forth causes readers a great deal of distress.

22.3.2 Titles

Whether a report is a Business Technical Report or a Research Report, the title is important. This discussion of titles applies to all reports.

A title should indicate the content of the report clearly and concisely. The title provides information needed by a reader to decide whether or not to bother reading the Abstract. Therefore, the title should be accurate. It should not mis-lead readers into thinking that the report is something that it is not. Nor should a title leave readers guessing about the report's actual subject, which happens sometimes when a title is too short and hence too general.

For example, consider this title of a Research Report:

Assembly Analysis of Can Opener.

It provides too little information. There are many types of can openers (e.g., electric or hand operated); which type is being analyzed? Or are all can openers being analyzed? Or maybe it is a study of can openers in general? Is it manual or automatic assembly of the can opener that was studied? A better title would tell the reader more accurately and completely what the report is about. For example:

Analysis and Comparison of the Manual Assembly Times for Three Electric Can Openers.

Here is another short title that is too general:

Comparisons of Robust Design Processes.

A more helpful title would be:

A Comparative Study of Three Robust Design Processes: Optimization, Taguchi methods, and Guided Iteration Using Dominic.

Titles that provide good descriptions of the content of a report can, of course, get too long. There are no hard and fast rules about length, but ten to fifteen words is generally not too long. Twenty words usually is too long.

22.3.3 Abstracts

In a Research Report, the purpose of the Abstract is to enable people to determine whether or not they want to read the rest of the paper. Thus, the Abstract for a Research Report usually contains descriptive as well as summary information. That is, it contains a brief description of *what* was done, as well as a brief summary of the results obtained. When writing an Abstract for your own paper, be sure to write it not as if it is a review of someone else's work, but as a description and/or summary of your work.

Here is a good Abstract for a Research Report:

A design for manufacturing (DFM) analysis of a TOT-50 Swingline stapler is described. The results show that incorporating six simple redesign suggestions can reduce piece part costs by about 25% and manual assembly costs by about 35%.

Notice that the first sentence in this Abstract is descriptive; it says what was done. The second sentence is a summary; it summarizes the results. This Abstract is brief, it describes the project well, and it contains the most important results.

For comparison, the following is an Abstract (taken from a student group report) that is too long and too filled with information that is not important in an Abstract (i.e., noise). The writing is far below professional standards in other ways, too.

The objectives which we used to guide us through the analysis of our part were simple. We analyzed the current design of the part using techniques learned in class. Each component of the part was evaluated for relative die construction cost, relative processing cost, relative material cost, and assembly time. We looked at the analysis and came up with two redesign suggestion.

The first suggestion was to incorporate the cover and the body into one stamped piece. This eliminates a part which reduces the construction, material, processing and assembly cost. Also with one less component to assemble the projected assembly time was reduced. We then did an analysis of the new part to compare with the old.

The second redesign suggestion was to build the cover so that it didn't have any internal undercuts. The cover is a plastic injection molded part and the die construction saw a large increase due to these internal undercuts. We offer a redesign suggestion which eliminate these undercuts and does not change the functionality of the part. The new design for the cover is then re-analyzed to show the advantages of this redesign.

By following this objective we were able to come up with some ideas on how to produce this part less costly. The redesigns are simple and yet they yield some significant savings in cost which can be seen in the analysis presented.

In addition to its excessive length and inclusion of detail inappropriate for an Abstract, the writing can be criticized for at least the following additional faults:

- Research Reports should not be written in the first person. (Business Technical Reports can and usually should be written in the first person.)
- It is incredibly wordy. For example, the first sentence could be reduced to: "Simple objectives guided the analysis." It would be even better, however, to delete the sentence, and just state what the objectives were.
- It is full of grammatical errors like "two redesign suggestion" instead of "two redesign suggestions".
- It contains colloquial phrases like "came up with" that are inappropriate in a Research Report. And how about "the die construction saw....."!
- There are awkward phrases: "By following this objective" and "how to produce this part less costly." Less awkward alternatives might be: "To achieve this objective" and "to produce this part more economically".
- In the third paragraph, there are many switches from first person to passive voice: "We offer......" followed by "The new design is then re-analyzed...."

There is even a kind of technical error: apparently the writer doesn't know about (or doesn't care about) the difference between a part and an assembly. Note in the first sentence that the word *part* is used whereas the analysis was done on an assembly or subassembly that consisted of several parts. Readers can no doubt find other examples of sub-professional writing in the above piece.

The kind of writing illustrated by the above quote is totally unacceptable in the professional work environment, and thus also in engineering courses.

23.3.4 Introductions

The Introduction of a Research Report repeats and expands on an Abstract's descriptive material, explains the motivation for the work, and provides an overview of the organization of the report. It should also provide a general context for the study and its results. An Introduction does not usually contain information about the results.

An effort should be made to make the Introduction short but interesting; if you bore your readers at the beginning, think how they will be dreading the task of reading on.

Suppose the purpose of a research project is to study a company product from an assembly viewpoint. Then the Introduction might be as follows:

Introduction

The current design of the <product name> was developed in 1981, and has been stable for the past eight years. There are four models consisting of a total of eighty-four parts. The product has not been studied or the design revised from an assembly standpoint since its inception. In the interim, new design for assembly (DFA) evaluation methods have been developed, and new assembly technology has also become available.

Competition in the <product name> product line has recently focussed on price reduction, and there is now extreme pressure for cost reduction. Assembly costs currently amount to 37% of the total cost, and therefore present a prime opportunity for savings.

The current design is assembled manually using a progressive line and a two shift operation. The proposed addition of two new models and the continued competitive pressure to reduce costs makes this an ideal time to investigate the possibility of improving the current line or of converting to an automatic line.

Therefore, the goal of this project is to analyze the assembly of the <product name> The specific objectives are:

1. *Evaluate the current design for both manual and automatic assembly;*
2. *Redesign the product for ease of assembly;*
3. *Compare the current design with the proposed new design on the basis of:*

   Assembly cost

   Line balancing efficiency for manual assembly
4. *Identify the tooling required for automatic handling of the parts in the new design via vibratory bowl feeders.*

22.3.5 Literature Review

The purpose of a literature review is to connect the research being reported with previous work done on the same subject, especially by other researchers. Therefore, the related publications are cited and summarized in an organized way. It is shown how the research being reported is similar to, and differs from, the work done by others. Note that the purpose of the Literature Review is not only to report and review on other related research, but also to place the new work in context with that other research.

When there are no, or just a very few, references to work done by others in a research report, then it is either a profound work indeed, or else the authors have not done their homework in searching the literature. One also sometimes sees Literature Reviews that refer primarily or almost exclusively to the work of the same people who are writing the report. Though this may be quite proper occasionally (i.e., when these are the only people working on a subject in a field — a rare situation), probably this too indicates that the Literature Review is incomplete.

A research project should not be undertaken, and certainly a research paper should not be written or submitted for publication, unless the authors are reasonably confident that they are familiar with all the other relevant work previously done. Computer-based literature searches, and searches of the relevant research journals can help develop this confidence.

Here is an example of a good Literature Review. It is taken from a Research Report which describes an alternative approach to the rules-of-thumb based design of metal stampings. The Literature Review is contained under *Related Work* and the references contained in the review are listed under *References*.

Related Work

To assist in the design of easier to produce stampings, several cost estimating and/or group technology (GT) based systems have been developed. These systems were developed to replace the traditional qualitative rules-of-thumb found in handbooks and used by most part designers. Unfortunately, these cost estimating and GT systems require knowledge not generally possessed by designers, and they fail to provide designers with quantitative information concerning difficult to produce features.

For example, sheet metal cost estimators such as the Harig system (Harig, 1977) and the Bradley die estimating method (Bradley, 1980) require die design and process planning knowledge rarely found among product designers. These systems are important advances, but require detailed input, and so are not designed for use at the conceptual stage of design where parametric details have not yet established. Moreover, the systems do not generate redesign suggestions based on part attributes.

The Opitz system (Hohmann et al., 1970), a GT based classification and coding system, uses a nine-digit code to classify sheet metal parts based on similarities in processing. This technique rationalizes design procedures and enables the grouping of parts for production to reduce die costs and lead time. The GT based Salford system (Fogg et al., 1971) uses a six-digit code to classify sheet metal parts based on the industrial sector that they serve, the type of die utilized, and similarities in processing. The objective of the Salford system is to simplify tool design by exploiting similarities in parts. The Hitachi sheet metal CAD/CAM system (Shibata and Kunitomo, 1981) uses a four-digit structure code to access a database of sheet metal configurations, each of which is associated with a graph representation. This improves the retrieval and modification of CAD models that represent sheet metal plates with holes, notches and flanges. None of these systems, however, provides cost data as feedback or assistance in the evaluation of design alternatives from the perspective of manufacturability.

In an effort to provide both quantitative measures of producibility and redesign suggestions, a system that analyzes planar stampings with holes was developed by Schmitz and Desa (1989). This system requires that the design be input to a non-manifold geometric modeler (NOODLES). Relevant part attributes and their parametric values are then extracted from the geometric model and conveyed to a domain mapper that generates a strip layout for a progressive tool. A series of manufacturing cost factors is then computed to provide a measure of producibility, as well as the basis for redesign suggestions. The system is limited to small planar stampings, with holes and notches, produced on progressive dies. It does not account for bent or formed parts, or parts with such features as extruded holes and embossings. The manufacturing cost factors are both quantitative (such as the number of punching tools) and qualitative (such as the need for pressure pads). The unification of these factors into a measure of manufacturability is subjective; it yields a value that is not an objective measure of producibility. Consequently, the importance of redesign suggestions is also dependent on the subjectivity of this measure.

Ulrich and Graham (1990) present a method for synthesizing simple blanked and bent parts used as supports. A solution map is constructed in three-dimensions representing a network of faces that connect two interface surfaces and avoid any intervening obstacles. Candidate configurations are automatically generated and pruned by a limited set of producibility constraints.

The approach adopted in this paper is an attempt to use a part coding and classification approach at the early concept stage of design which communicates process knowledge in terms that are easily understood by product designers. This permits quick evaluation of a broad range of stamped part designs from the perspective of relative part cost, and offers redesign suggestions that can achieve significant percentage savings in cost.

References

Bradley, J., "Die Estimating," Pressworking: Stampings and Dies, Society of Manufacturing Engineers, Dearborn, MI, 1980.

Fogg, B., Jamieson, G.A and Chisholm, A.W.J., "Component Classification as an Aid to the Design of Press Tools," Annals of the C.I.R.P, XVIV, 1971.

Harig, H., "Estimating Stamping Dies," Harig Educational Systems Inc., Scottsdale, AZ, 1977.

Homann, H.W, Guhring, H., and Brankamp, K., "Ein Klassifiizerungsystem fur Stahlbaueinzelteik - Enwicklung und Beschreibung des Systems," Industrie Anziger, Vol. 93, 1970.

Poli, C., and Divgi, J., "Product Design for Economical Injection Molding - Part Shape Analysis for Mold Manufacturability," Proceedings of ANTEC, 1987, pp. 1148-1152.

Poli, C., Dastidar, P., and Mahajan, P., "Design for Stamping - Coding System for Relative Tooling Cost," Mechanical Engineering Dept., Univ. of Massachusetts/Amherst, August, 1990.

Poli, C., Dastidar, P., and Mahajan, P., "Design for Stamping: Part II - Quantification of Part Attributes on the Tooling Cost for Small Parts," Proceedings of the ASME Design Theory and Methodology Conference, Phoenix, 1992.

Sackett, P.J. and Holbrook, A.E.K., "DFA as a Primary Process Decreases Design Deficiencies," Assembly Automation, December 1988.

Schmitz, J.M, and Desa, S., "The Application of a Design for Producibility Methodology to Complex Stamped Parts," Technical Report, General Motors Systems Engineering Center, December. 21, 1989.

Shibata, Y., and Kunitomo, Y., "Sheet Metal CAD/CAM System," Seiki Gakkai, Bulletin of the Japan Society of Precision Engineering, Vol. 15, No. 4, December. 1981.

Ulrich, K.T., and Graham, P.V., "Using Producibility Constraints to Control the Automatic Generation of Sheet Metal Structures, "Proceedings of the ASME Design Theory and Methodology Conference, Chicago, September. 16-19, 1990.

Wozny, Michael J., et al., "A Unified Representation to Support Evaluation of Designs for Manufacturability, "Proceedings of the 18th Annual NSF Grantees Conference on Design and Manufacturing Systems, Atlanta, GA, January 1992.

22.3.6 Main Body

In the main body of a Research Report, there are at least three sub-areas: (a) a section describing the goals, research issues, or hypotheses underlying the research, together with the method used to explore or study them; (b) a description of the experiments, analyses, or simulations performed; and (c) a presentation of the results obtained.

(a) *Goals, Research Issues, Hypotheses and Methods.* Though not all research is done to prove or disprove an hypothesis in the formal scientific sense, there generally are goals and expectations that *something* will be learned about one or more research issues. This first section of the main body is where those goals and issues — and hypotheses if there are any — are stated and discussed.

The methods to be used to prove any hypotheses, achieve the goals, or to study or explore the research issues are also described here. It should be shown how the methods will in fact work to prove or disprove the hypotheses, to achieve the goals, or to elucidate the issues. That is, the methods should be shown to be relevant to the hypotheses, goals, and issues to be studied.

(b) *Experiments, Analyses, or Simulations.* If there are experiments (or the equivalent, such as surveys), then the experimental apparatus and procedures (including survey procedures) are described and discussed. Though Appendices may also be needed, there should ideally be enough detail here for another researcher not only to evaluate what was done, but also to reproduce the experiments if desired.

Analyses and simulations are analogous in this discussion to physical experiments. Assumptions and simplifications employed should be stated. If computer algorithms are constructed, their logic should be explained. Programming details, however, belong only in an Appendix.

(c) *Results.* The results of the experiments and/or analyses and simulations should be presented as completely and as clearly as possible. Actual data can be placed in an Appendix.

As an example of unclear results, consider the following. The writing below is a real example except that the product names have been made fictitious.

"For standard assumptions, the new WhizWidget assembly analysis indicated that the automatic assembly cost was 30% less expensive than the previous design. The competitor, CompWidget, was 12.7% more expensive than the new WhizWidget. Originally the CompWidget was 19.8% less expensive. The new WhizWidget assembly analysis indicated that the manual assembly cost was 42.6% less expensive than the previous WhizWidget design. The CompWidget was 31.9% more expensive to manually assemble than the new WhizWidget. Originally the CompWidget was 21.0% less expensive."

Apparently this writer believes in the following Peter's Law:

"If you can't convince 'em, confuse 'em"

A reader of such confusion can only think, first *"WHAT on earth does that say?"* and then *"How could the study be of any use with such a confused person doing it?"* Indeed, not only is this writing unclear on a first reading, it is even more unclear after careful study! Here is a better presentation of results from a different study:

"The results of the assembly design and cost study on the Wolly Widget are as follows:

1. *Automatic assembly is not practical for the present design.*
2. *For manual assembly of the current design (60 parts):*

   Total manual assembly time: 631 secs/assembly

   Total manual assembly cost: 2.84 $/assembly.
3. *The product can be redesigned to eliminate the following 31 parts: 4 main body screws, 1 charge end #1, 2 charge ends #2, 4 charge ends #3,4 springs, 1 label sticker, 1 light cover, and 1 specification label.*
4. *For manual assembly of the redesigned product (29 parts):*

   Total manual assembly time: 276 sec/assembly,

   Total manual assembly cost: 1.25 $/assembly".

Discussion of Results

In the Discussion of Results section, the results of the study are related back to the goals, issues, and hypotheses. What light do the results shed with respect to them? Also, the significance and implications of the results are described and discussed. Uncertainties and possible errors in the methodology or in its implementation are also described.

Sometimes a Research Report contains ideas or suggestions for continuing or future research work. This is all right, but in a business setting, one should not expect that such recommendations included at the end of a research paper are a substitute for preparing a Business Technical Report to make the recommendations effectively.

As an example of a good discussion, consider the following discussion of results comparing two alternative designs:

Discussion

Tables XX, YY and ZZ show a comparison of the alternative assembly system costs for the original design of the ABC-787 drive and for the alternate version of this drive, the ABC-787B. The analyses assumed that the ratio of faulty parts to acceptable parts is 1 percent, and that both one shift and two shift operations are possible.

Table XX, which contains a parts' list for both the original and alternate designs, shows that for the original version, 21 of the 36 total parts are screws, while the redesigned drive has only 7 screws of 20 total parts. In addition, the rolled point screws utilized in the 787 drive have been replaced by much easier to align "dog point" screws in the 787B version. These design changes, along with the additional changes to facilitate alignment of the various parts, result in a better than 50% reduction in manual assembly time.

Table YY shows that for both designs, the use of manual assembly is less costly than dedicated automatic assembly. For dedicated automatic assembly, however, the capital investment required for the 787B is about half that required for the 787.

Table ZZ shows that while the anticipated production volume for the ABC-787 is insufficient to justify the use of dedicated automatic assembly, use of a product mix may justify the use of programmable automatic assembly. The results shown in Table ZZ also indicate that for the alternate 787B version, even the use of programmable assembly does not appear justified solely on the basis of cost.

Appendices

Details of all kinds (e.g., of experimental apparatus, of analyses, of data, worksheets, etc.) belong in Appendices.

22.4 Business Technical Reports

22.4.1 Overview

The majority of reports written by design engineers in industry are Business Technical Reports. The main purpose of these reports is to provide results and conclusions relevant to short term decisions and actions, and/or to recommend a course of action. Thus the recommendations, conclusions and results, are placed first in the report. That is, when there is an action or decision recommendation, then it is stated concisely in the most prominent position in the report: at the beginning.

After the recommendations, next in importance in a Business Technical Report are the conclusions and logic on which the recommendations are based. Then the results on which the conclusions are based are presented. Data and analyses are included only in Appendices. Note that this is more or less the reverse of the order in a Research Report.

Here is a recommended format for Business Technical Report:

Title

Summary Abstract, 1/4 to 1 page (absolute maximum):

Short summary of recommendations,

Short summary of critical results, conclusions and logic on which recommendations are based,

Short summary of important risks, benefits, and costs associated with the recommendations.

Body of Report, 2 to 4 or 5 pages:

Complete recommendations,

Complete supporting arguments, including important conclusions and results,

Basis for and elaboration on the relevant risks, costs, and benefits,

Associated with the recommendations.

Appendices (As Many as Needed, As long As Needed):

Descriptions and documentation of what was done,

Data,

Analyses,

Methods for generalizing the results from the data and analyses, and

Any research reports whose results are not familiar and that are critical to support conclusions or recommendations.

We now discuss each of these sections in more detail, though not in order. We begin with the Body of the report.

22.4.2 The Body

In the Body of a Business Technical Report the recommendations come first, conclusions next, and then results. Data and analyses, unless there is something remarkable about them, come last, usually in one or more Appendices.

Recommendations

A recommendation is some action you believe the company should take, or some decision that you think should be made. For example, after a study of customer preferences, you might recommend that a more formal marketing study be initiated that would determine the expected added sales if your firm's coffee makers had a customer controlled variable temperature feature. You might also recommend that a parallel engineering design feasibility study be initiated that would determine the technical feasibility and manufacturing costs of producing a suitable variable temperature coffee maker. Or, you might recommend that no changes be made in the design.

Note that recommendations require consideration of business, and perhaps a variety of technical issues, that may not be a direct part of the study being reported. Thus summaries of the benefits, costs, risks, and uncertainties associated with the recommendations are also included. It is very important that risks and costs be included; the last thing you want to do is recommend some action without informing the readers of the associated risks and costs as well as the benefits.

Note that if all you are doing is reporting the results of a study, with no recommendations or conclusions relevant to pending decisions, then the Research Report format is probably the better choice. If, however, you are making recommendations for action (especially actions that cost time or money), then the Business Technical Report format is better because you want those recommendations and the business reasons for them to be up front.

Conclusions

Conclusions are generalizations of results from the reported project. For example, in the coffee temperature control case, a conclusion might be that the best temperature to provide is 165 F. Or we might conclude that an adjustable, variable temperature between 160 and 180 would be the most desirable design goal.

Results

Results are generalizations made from the data and analyses. For example, in a coffee temperature preference study, the results might say, for example, that 20% of 150 people surveyed prefer the temperature to be less than 160 F, that 50% prefer a temperature between 160 and 170 F, and 30% prefer greater than 170 F. The exact form of these summary results will of course depend on the nature of the experiments or analyses done. Note that such results are generalizations of the data, not the actual data. They are, however, objective generalizations, not subjective judgments. The actual data belong only in an Appendix unless there is some important conclusion to be made about the data per se.

Results are included primarily to explain how and why the conclusions were determined. That is, results are selected to show the basis for the conclusions. It may also be necessary to refer to the results, conclusions, or recommendations of other studies to support your recommendations.

Appendices

In a Business Technical Report, there may be any number of Appendices that include the data taken, details of analyses or experiments performed, and whatever other details might be needed by someone at some future date. The Appendices are, in effect, a detailed record of the activities of the project or study. Anyone questioning certain of the results, for example, could find the actual data, computations, and analyses in the Appendices.

22.4.3 Titles

Everything said above about titles for Research Reports applies to titles for Business Technical Reports as well. Of course, in addition, the title of a Business Technical Report should clearly state that there is a recommendation and/or conclusions relevant to short term actions or decisions. For example:

Recommendations for Redesign of the ABC-786 Widget in Order to Reduce Manufacturing Costs

A poor title:

Analysis of the ABC-786 Widget

The first title above tells us exactly what to expect in the report. The second title fails to give us even a clue as to what the report is about. It tells us that it deals with the 786 Widget, but we don't know if the analysis is one concerning manufacturing, functionality, marketing, or something else.

22.4.4 Abstracts

Every Business Technical Report must begin with a short *Summary Abstract*, preferably 1/4 to 1/2 page, and *never* more than one page. The purpose of this Summary Abstract in a Business Technical Report is to provide a short summary of the recommendations, conclusions, and results for readers who do not need or want to know more of the details. The same order (i.e., recommendations first) that is used in the body of a Business Technical Report is also used in its Summary Abstract.

The Summary Abstract of a Business Technical Report contains no descriptive material; that is, it does not describe what was done, or even why. Whatever is necessary for descriptive material to introduce the subject of the report can be included in a brief sentence or two in the covering letter or memo. (But remember KISS.) In other words, the Summary Abstract of a Business Technical Report gives *only* summaries of *only* the most important recommendations, and if needed and appropriate, summaries of the most important conclusions, results, logic, benefits and risks. Whatever descriptions there are of what was done must be somewhere else, mostly in Appendices.

An example of a good Summary Abstract:

The following redesign recommendations are made with regard to the current ABC-786 Widget:

1) Replace the current motor, toggle switch and loose wiring used to connect the switch and motor with the new CW-34 motor which incorporates the motor and toggle switch into a single unit,

2) Incorporate the energy pack into the handle subassembly, and

3) Incorporate the latch button into the energy pack portion of the handle subassembly, rather than assembling the button as a separate part.

These recommendations are made based on the design for manufacturing and assembly analysis recently carried out by the Advanced Design and Manufacturing Group. Their analysis shows that a total savings in assembly time of over 50% can be achieved by incorporating these redesign suggestions. The suggestions can be implemented without the need to redesign any of the special purpose parts which are contained in the 786 Widget.

Here is an example of a poor Abstract:

An analysis is done by decomposing the ABC-786 Widget into its component parts and subassemblies. The motor and its associated wires and switches are not considered here, as the motor is a standard part. The energy pack subassembly includes a snap-fit coupling to the nozzle subassembly and is attached to the handle subassembly with screws.

Each subassembly and part is analyzed for ease of manual assembly and manufacturability. Manual assembly requires 2 min, 13 sec. Relative manufacturing costs are calculated as 20.9 where the reference part is a 1-mm thick flat washer with inner and out diameters of 60 mm and 72 mm respectively.

A savings in assembly time of up to 52.3% can be achieved by:

1. Selecting a standard part motor with a toggle ignition switch and tangle-free wiring.

2. Incorporating the energy pack into the handle subassembly; and

3. Incorporating the latch button into the energy pack portion of the handle subassembly, rather than assembling the button as a separate part.

Notice the difference in the two example Abstracts above. The first starts out, as it should, with concrete recommendations. The second does not even contain any recommendations. The first Abstract does not contain any descriptive material concerning the product. The second contains descriptive material concerning both the product and the procedure used to analyze the product. Finally, the second paragraph in the second Abstract is so unclear that it probably makes the reader suspicious of everything contained in the report.

22.5 Hybrid Technical Reports

Business recommendations are often made on the basis of research results. When this is the case, one can always prepare a Business Technical Report; that format permits the research results to be included as described above. But if the research requires a lengthy explanation, then such a report can become too long. There are also other ways to present both the business recommendations and the research together in a kind of hybrid report format.

One way is to prepare a Research Report and send or deliver it with a covering memo or letter which contains the recommendations. In this case, the reader gets the recommendations first in the short covering document, and then the Abstract of the Research Report will be the next thing read. If that Abstract is well done, then those readers with no need for more details can stop after reading it. This is thus a reasonably efficient format to follow.

Another way is to prepare a Research Report, but use an Abstract that combines the information usually in the abstracts of Research Reports and Business Technical Reports. This way the recommendations and short term conclusions or results are up front like in a Business Technical Report.

Both of these hybrid forms are best used only when the Research Report is fairly short. If the Research Report is long and involved, then it might be necessary to prepare it separately, and then also prepare a Business Technical Report, or covering letter or memo for the recommendations and/or conclusions needed immediately.

Selecting the best format for a report should be done pragmatically: What is the best format for the report given its content, purpose, and the intended audience? Any format will do if it properly meets the purposes of the message and the audience(s). Thus the formats described as above should be modified and adapted as needed to each particular communication situation.

22.6 Oral Reports

22.6.1 Preparation

Oral presentation of both Research Reports and Business Technical reports is commonplace in industry. And Research Reports are also almost always presented in person at technical or trade Conferences. Therefore, the ability to do a presentation well is important. In this section, we provide some guidelines that will help in the preparation and delivery of oral reports.

Begin Preparing Well in Advance. By "well in advance", we do not mean hours, or even days. We mean weeks, or even months in advance! As soon as you know you have an oral presentation to make, that very moment is the time to begin to prepare.

Preparation includes many of the same things that are required in preparing for a written communication. That is:

1. Explicitly state for yourself what is the purpose of the presentation. What is it that you want to accomplish? Is there something you want the listeners(s) to do? Is there information you want them to have, or that they need to know? In other words, understand for yourself *why* you are making the presentation.

2. Explicitly describe the audience for yourself. Who are they? Why are they interested in or wanting to hear your presentation? What words and symbols do they understand and not understand?

Often, there are several classes of receivers in an audience, each with different needs. Each sub-group must then be described, and the presentation constructed and organized accordingly.

3. Decide what the desired take-aways are. In an oral presentation, there can be only a very few, perhaps only one, and no more than two or three.

4. Assuming your presentation will include a set of slides or transparencies (it normally will), then make a tentative title for each.

There can be no more than one slide or transparency for each minute you have been allotted for your presentation. And if you want to allow time for questions and discussion, then you must have fewer than one slide or transparency per minute. Remember, any message can be made any length.

There is no better way to make a lousy impression on an audience than by having a presentation that is too long for the time allowed. Conversely, there hardly is any better way to make a good impression than to have a talk that is short, emphasizes what is important (the take aways), is well organized, and allows time for questions and discussion.

Suppose, for example, that we had an opportunity to make an oral presentation to a group of students concerning the complete content of this chapter in ten minutes. We would want to prepare about eight transparencies, a plan that should allow about two minutes for questions. The two main take-aways we could choose are: (1) the importance of keeping communications short, and (2) the importance of rewriting and rehearsing extensively. The title of the eight slides might then be:

#1. Title Slide.

#2. The Importance of Communications in Engineering.

#3. Preparation: How to keep it short.

#4. The Writing Process: Rewriting to keep it short.

#5. Format of Business Technical Reports.

#6. Format of Research Reports.

#7. Guidelines for Oral Presentations: Rehearsing to keep it short.

#8. Summary (Review the take-aways).

We could supplement such a short talk by handing out an example or two of actual reports that were exemplary.

5. Plan for handouts or pieces of hardware that you can use to supplement your talk. Circulating examples of something real is always a good way to involve an audience in what you are saying.

22.6.2 Guided Iteration

An oral presentation, like a design and like a written communication, is prepared by a process of guided iteration. The steps described above have given you an initial design for your talk. Next you prepare a first draft, try it, evaluate it, try it, evaluate it, etc., improving it repeatedly until you are satisfied with the content and comfortable presenting it.

22.6.3 Planning Slides or Transparencies

The first part of this process of iterative improvement is completing the slides (or transparencies) which initially were given only a title. Slides or transparencies must not be cluttered with words and/or complex drawings. The best slides will have three to five bullets (i.e., main sub-topics) with no more than one very short phrase or sentence attached. If a slide or transparency has more than about forty or fifty words, it is probably too detailed. Shorten it to just the basic ideas and outline. Plan on filling in the elaborative words with your speaking.

The size of letters on your slides and transparencies must be large enough for people in the rear of the meeting room to read easily. The minimum letter size is a 1/4 inch, and 3/8 inches or even 1/2 inches may be necessary for larger rooms. Also make sure that the layout of material on your slides is logical, balanced, and interesting. Leave lots of free space. Though not everyone agrees, color is generally unnecessary, and may even detract from your message.

Never, ever, read what is on your slides to your audience. Not only is this dull and boring, but in addition they can read faster than you can speak. The information on the slides or transparencies is not there for you to read. Its purpose is to provide an outline (only) for both you and audience to follow. Your speaking will explain and elaborate on the brief points made on the slides.

For example, here is a possible way to prepare a slide for Slide #7 listed above:

Guidelines for Oral Presentations

> *Prepare Well in Advance*

+ *Purpose: Analyze Audience and Take Aways*
+ *One Slide per Minute: Prepare Titles First*

> *Use Guided Iteration*

+ *Complete the Slides - 50 Words Each Max.*
+ *Rehearse Out Loud, Again and Again*

An alternative slide for Slide #7, and one that uses too many words is the following:

> *Prepare Well in Advance*

Not hours or days but weeks or months

> *State for yourself the purpose of the presentation.*

Why are you making the presentation?

+ *What you want to accomplish*
+ *What you want listeners to do*
+ *What information you want listeners to have*

> *Analyze the audience*

+ *Who are they?*
+ *Why are should they be interested*

> *Prepare Transparency Titles First*

+ *Keep them Brief*

> *No more than one slide per minute.*

More slides/minute = Talk will be too long

-no time for questions.

-angry (or sleepy) audience

> *Use Guided Iteration*

Draft, evaluate, redo, evaluate, redo,.....

+ *Complete the slides - do not use more than 50 words*
+ *Rehearse out loud*

Here the slide becomes crowded. The audience, instead of listening to you, will simply read the slide. The more detailed ideas on the crowded slide are things you can say to the audience. The slide should be only an outline for the audience and you; they get the details and interesting sidelights from your talk.

22.6.4 Rehearsing

In addition to getting the slides or transparencies right by iteration, the other thing you must do is *rehearse out loud — again and again and again*. You begin rehearsing as soon as you have a rough draft of all the slides. You may need to rehearse out loud dozens of times (not an exaggeration!) over a period of weeks or even months, especially when this whole process is new to you, or if you are nervous about making a particular presentation. Then when the time comes to deliver your presentation, you are so familiar with it, and so confident, that you will do it in a relaxed and polished manner. You will also be able to think clearly about questions when they are asked.

When you feel you are about ready, but long before the actual presentation, practice your presentation before some friends and colleagues. They can give you valuable feedback on how your presentation can be improved.

When you rehearse, you must keep track of the elapsed time. Your actual presentation will usually take longer than when you rehearse. Cut out slides if necessary to keep within time limits. Better to have fewer ideas presented than to exceed your time. Remember, audiences will love you if you use less than your allotted time, and hate you for using more.

22.6.4 The Presentation

Whenever possible, visit the room where your presentation is to take place before the time of your talk. Make sure the projector works, that you know how to control it (e.g., on-off and focus), and that there is a table for your transparencies both before and after you show them. Find the pointer (better yet, bring your own) and think about where you will stand so the audience can see both you and your slides or transparencies.

An oral presentation is a very personal communication. You are face to face with your listeners, sometimes quite close. In small groups in a small room, there is a lot of intimacy. It is always a good idea to begin by saying something that will help you develop a positive and friendly personal relationship with the audience. In planning this, you have to consider the size of the audience, and the nature of the meeting, but it can always be done. At the least, for example, you can thank people for coming, especially considering their busy schedules. If you have been specially invited, you can express your appreciation for the honor of being asked.

A little humor is fine here, but jokes (if used) should be short and clean and funny. Otherwise, forget them. If you can tie some good, short joke or light remark to the local situation, that is good. The purpose of any humor interjected is to make the audience relate more personally and positively to you. If your joke won't do that, omit it. It is enough that you are genuinely pleased to be there, and to have them there; so just say so personally and sincerely and you will be off to a good start.

Never, *ever* read a presentation. Don't read the slides, and don't even read your own notes. By rehearsing sufficiently, and with the slides for your cues, you will know what to say without reading anything. If you absolutely must, it is acceptable (though not especially desirable) to have a few cue cards (e.g., 3 x 5 inches) with sentence-starting phrases on them to help you get started with what you want to say about each of your slides. For example, suppose you were to find in rehearsing the above slide that, for some reason, you have trouble getting started with what is to be said about "Rehearsing Out Loud". Then you might have on a cue card the phrase "It is not enough to rehearse silently.....". Just that much on the card would be enough to get you started on this subject during the actual talk, and you could then go on elaborating from memory.

You must exhibit enthusiasm and energy, and interest in your subject. If you aren't interested, neither will your audience be interested. But after so much rehearsal, how can you still be excited and enthusiastic? If necessary, just turn it On. But mostly, enthusiasm will come naturally. It is your work, you are proud of it, and so you really are interested and excited. Be sure to let it show.

22.7 Graphical Communications

22.7.1 Introduction

A picture, they say, is worth a thousand words. Maybe. It certainly is true that graphical communications are an important and efficient way for design engineers to communicate with each other, with customers, with manufacturing engineers, and with managers. Thus we introduce some of the basic concepts of graphical communications here.

Our scope is limited by space and by the expectation that students and other readers interested in this book will have previously been introduced to the fundamentals of graphical communications. Therefore we provide only the briefest introduction to orthographic projection. We say only a little more about hand sketching, limiting the brief discussion to oblique and isometric forms. Because professional working standards are much higher than most students seem to think they are, we include some examples illustrating the acceptable (and unacceptable) standards for drawings in a working environment.

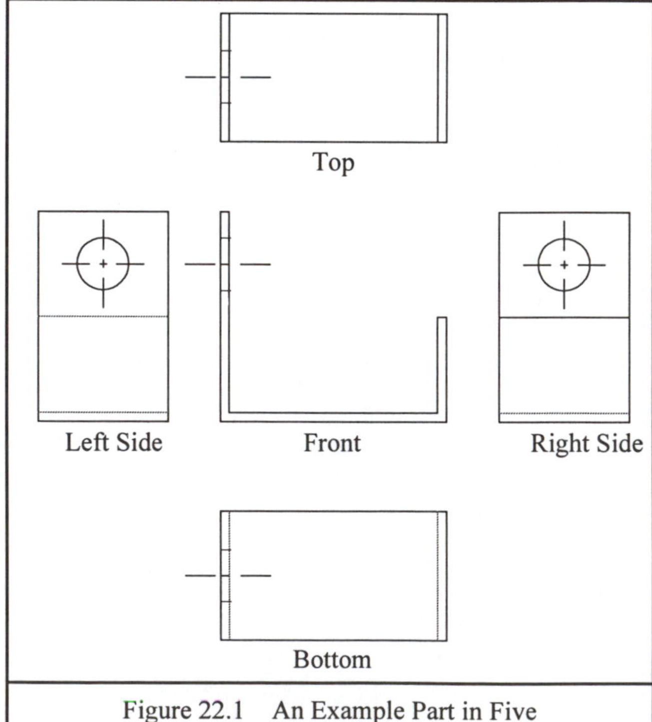

Figure 22.1 An Example Part in Five Orthographic Views

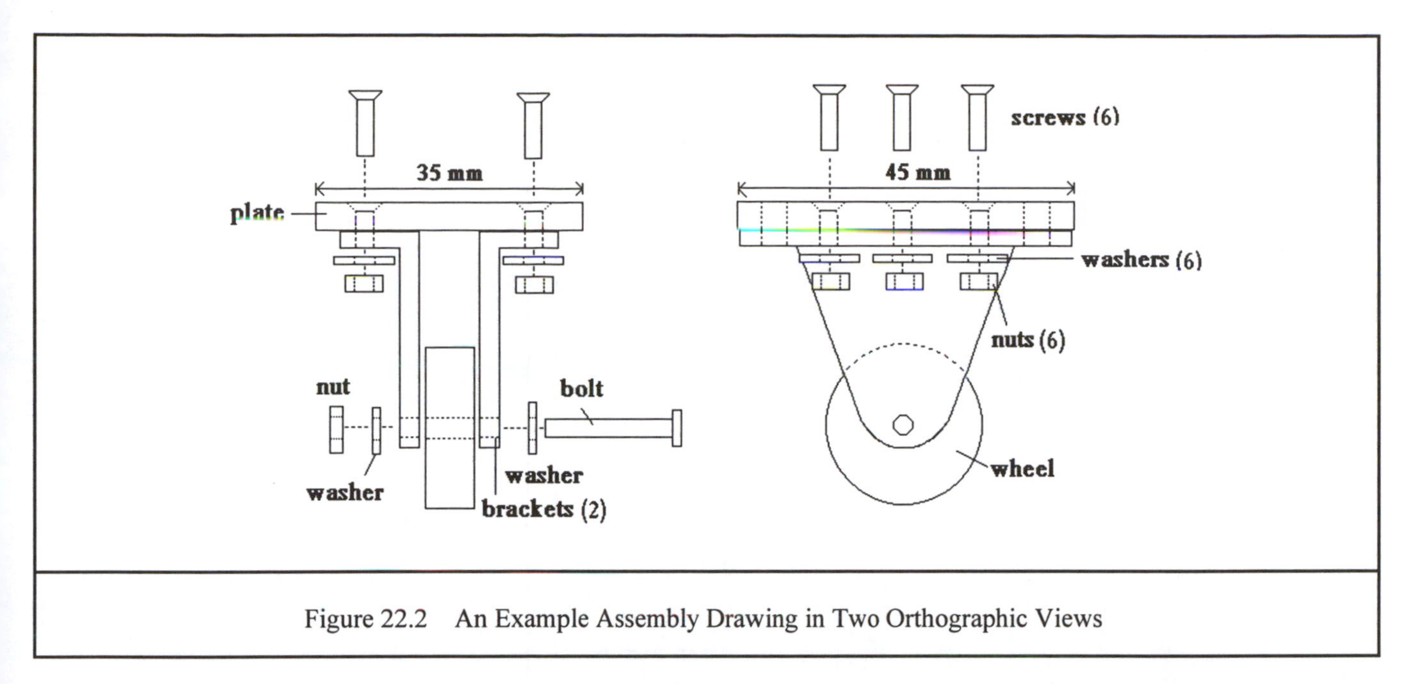

Figure 22.2 An Example Assembly Drawing in Two Orthographic Views

The goal of graphical communications, like all communications, is the effective and efficient transfer of accurate information. Thus the language and grammar used for communication must be correctly done by the sender so it will be correctly interpreted by the receiver. Students and practicing design engineers must understand graphical language and grammar, and use both properly. Proper use includes also preparing drawings that are well laid out and neat, as well as having correct graphical "grammar".

As noted above, we assume that readers of this book are knowledgeable about the rules and conventions of orthographic projection. Nevertheless, the next section provides a brief reminder of the most basic type; readers not thoroughly familiar with orthographic projection should consult one or more of the many books on the subject.

22.7.2 Orthographic Projection

In orthographic projection, an object is imagined to be viewed from an infinite distance along mutually perpendicular (i.e., orthogonal) axes. The resulting views are then drawn to scale and laid out on paper as shown in Figure 22.1. In the figure, five views are shown; a view from the rear of the part, if added, is drawn to left of the left side view. Often it is only necessary for a complete description of a part to show only two or three views.

Figure 22.2 shows an assembly drawing in two orthographic views, and Figure 22.3 shows orthographic views of two parts of a proposed configuration for the food carrying portions of a food scale.

Dimensioning. There are very definite rules and conventions for proper dimensioning which, again, we assume have been or will be learned elsewhere by our readers. Dimensioning is an important subject. Not only does improper dimensioning (like improper spelling and grammar) indicate a serious degree of carelessness, but improper dimensioning may also lead to production of parts not in accordance with designers' expectations. Figure 22.4 shows a set of dimensions added for the part shown in Figure 22.1. Tolerances are not shown.

22.7.3 Pictorial Views

Orthographic projection is the most rigorous way to

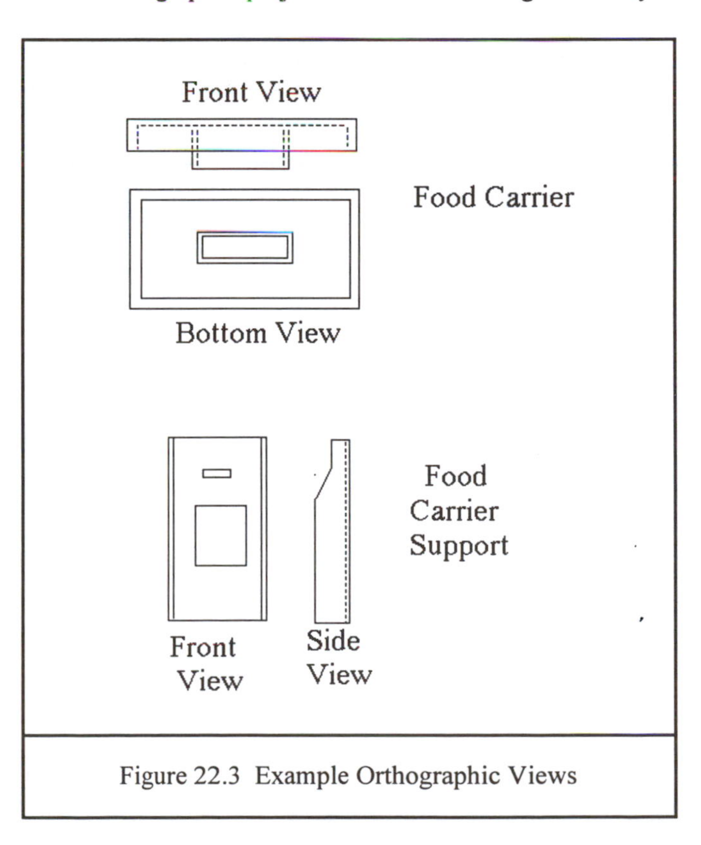

Figure 22.3 Example Orthographic Views

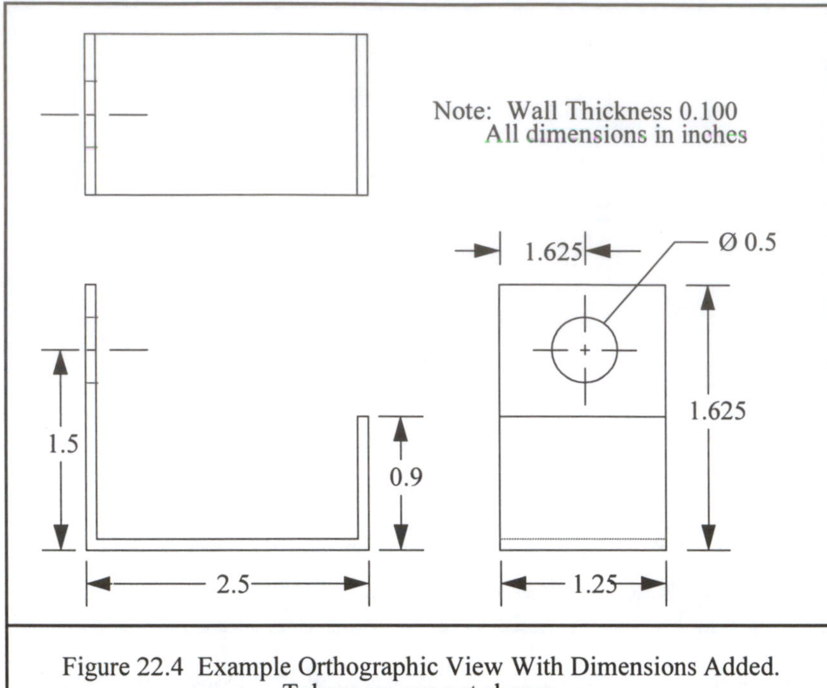

Figure 22.4 Example Orthographic View With Dimensions Added. Tolerances are not shown.

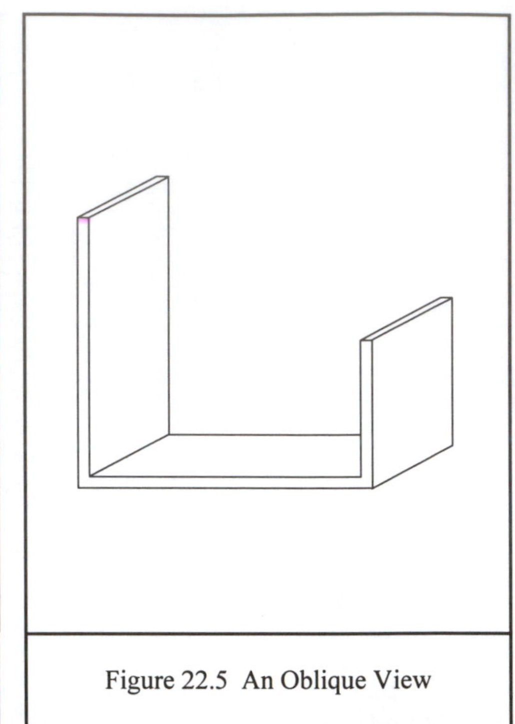

Figure 22.5 An Oblique View

represent an object graphically. The representation can be as complete and detailed as desired. Often, however, it is helpful to the communication process to show a pictorial view that is more easily and quickly visualized. Designing and redesigning at early stages are also often done using pictorial sketches.

There are several types of pictorial views, but we will present only the two most commonly used here. For more information on pictorial views, see [2]. The two we describe are: (a) oblique and (b) isometric. Figure 22.5 shows an oblique view of the object shown in Figure 22.1. Figure 22.6 shows an isometric view of the same object.

Oblique Views. In an oblique view, the orthogonal axes are as shown in Figure 22.7 (a). The front face of the object is drawn in true view to full scale. Thus a one inch solid cube is as shown in Figure 22.8 (a). Note that lengths

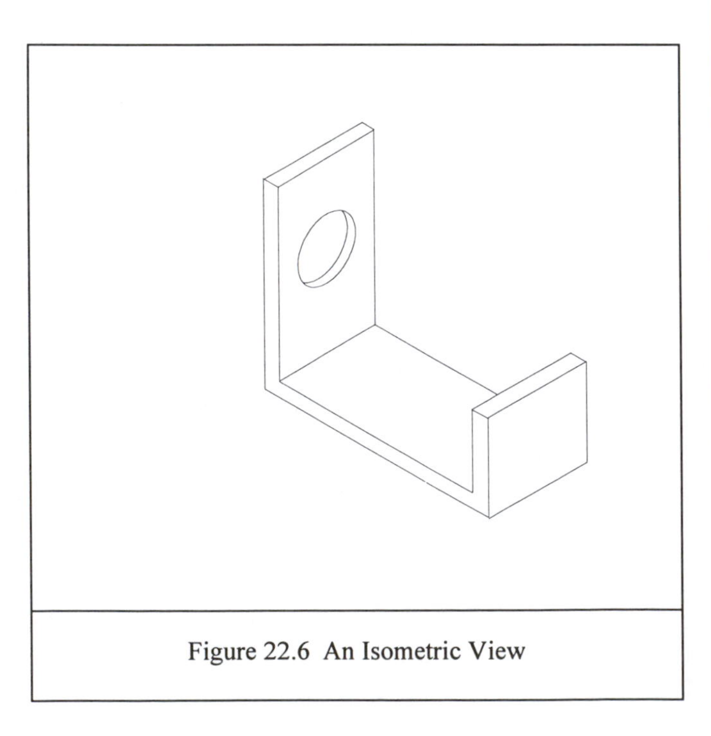

Figure 22.6 An Isometric View

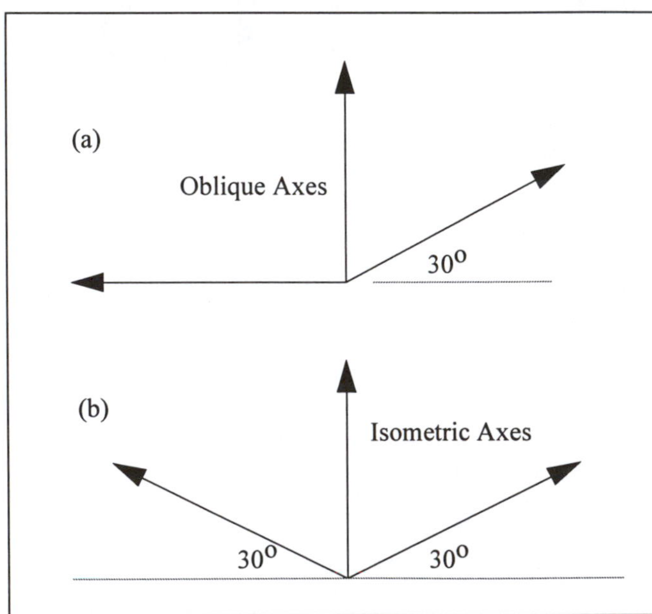

Figure 22.7 Orthogonal Axes for Oblique and Isometric Views

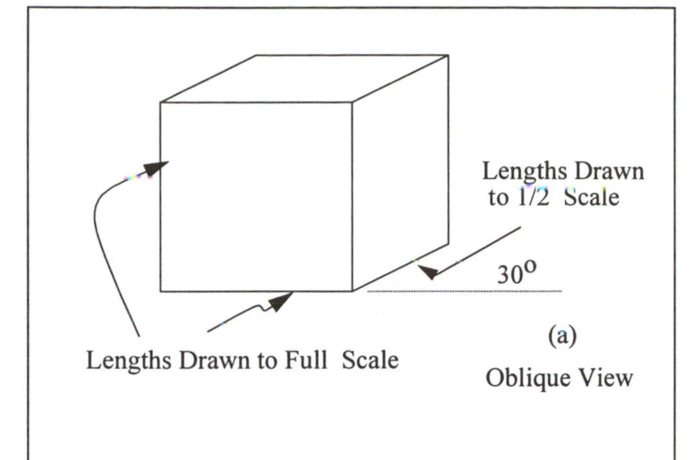

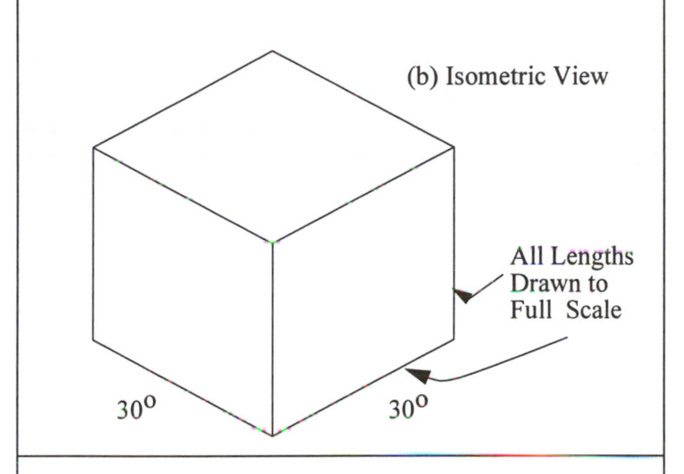

Figure 22.8 Oblique and Isometric Views of a Cube

on the front face are shown at full scale, whereas lengths along the oblique direction (at 30 degrees to the horizontal) are drawn to 1/2 scale. (If the oblique lengths are drawn full scale, the cube would look distorted; try it.)

One advantage to the oblique view is that circles, arcs, and any contours appear in their true shape in the front face of the sketch. In other directions, however, circles appear as ellipses which are a bit more difficult to draw. However, often the object can be oriented in the pictorial so that the circular parts are parallel to the front face.

Isometric Views. In an isometric view the orthogonal axes are shown in Figure 22.7(b). Dimensions along all the axes are drawn to full scale. Thus a one inch solid cube is as shown in Figure 22.8 (b).

Examples of a parts drawn in isometric view are shown in Figure 22.9, 22.10, and 22.11. Note in Figure 22.11 how the cut-away enables visualization of internal details.

22.7.4 Exploded Assembly Views

It is often helpful to develop drawings that illustrate graphically how the parts of an assembly relate spatially to one another. The best way to do this is with a pictorial called an *exploded view* of the assembly. In an exploded view, parts are shown individually but in a location and orientation along their axes of insertion during assembly. As an example, see the isometric exploded view of a three hole paper punch shown in Figure 22.12. Note how the use of dashed center lines helps visualization of how the parts are assembled.

Other examples of isometric exploded assembly drawings are shown in Figures 22.13 and 22.14.

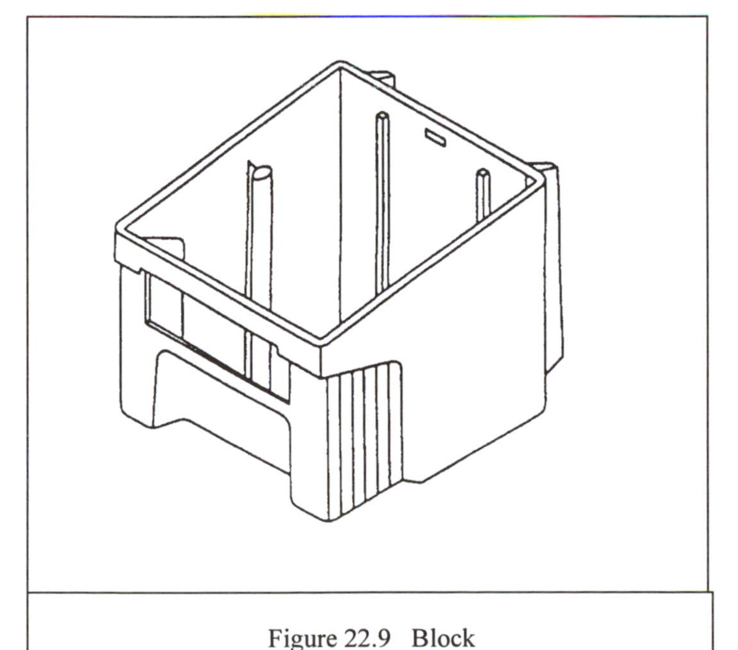

Figure 22.9 Block

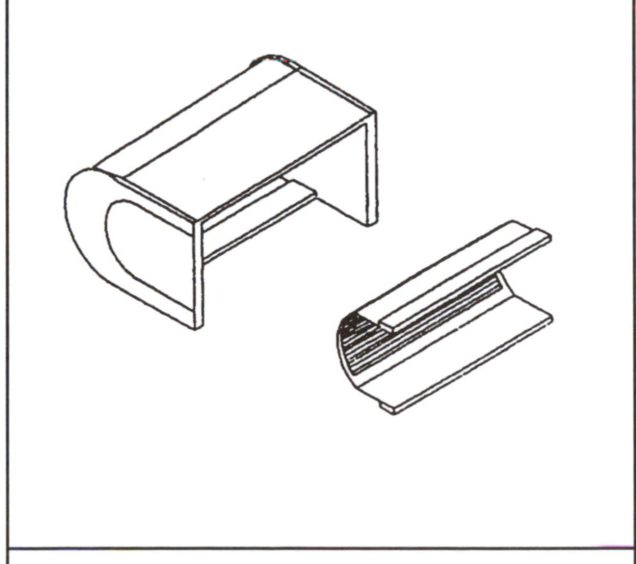

Figure 22.10 Electric Shaver (By L. Longo and W. Hu.)

22.7.5 Tips on Sketching By Hand

One does not have to be an artist to make attractive sketches of designed objects that communicate effectively. There are, however, some tips that will help facilitate learning to sketch well. For a more complete treatment, see [2].

1. Use a soft, sharp pencil, such as an HB.

2. Learn to draw lines of different weights: light, medium, and heavy, at least. This is easy to do with just a little practice.

3. When you begin a sketch, begin with light lines. Don't worry about extraneous lines. Don't erase the light lines unless absolutely necessary; once you begin to make the ones you want medium or heavy, the light lines will probably not be a visual problem.

4. Get a very good eraser so that if you want to erase, you can do it neatly and cleanly.

5. To draw reasonably straight lines, mark the endpoints with dots. Then begin at one dot, and proceed by drawing a series of overlapping, short lines about 1/2 to 1 inch long. Use light weight. As you draw these short lines, keep your eye on the dot where you want to end up. If the line does satisfy you, don't erase it; instead draw a better medium weight line using the same process.

If the line to be drawn is long, you may find it helpful to put in some intermediate dots, starting with one near the middle.

6. To draw circles in true view, first locate the center with a dot. Then place four dots on the desired perimeter of the circle spaced more or less equally. Then place four more dots on the perimeter in between the first four. You now have eight points approximately on the perimeter of the circle. Connect these with short curved lines to make the circle.

7. To draw circles in oblique or isometric views, repeat the above process except that you have to take into account the different scales in oblique drawings, and the sloping axes in both oblique and isometric drawings.

8. For both oblique and isometric sketches, you can purchase and use lined paper (with the 30 degree lines shown) underneath drawing paper that is transparent enough for you to see through.

9. To sketch an object, first sketch in light weight the rectangular box into which the object can fit. This establishes

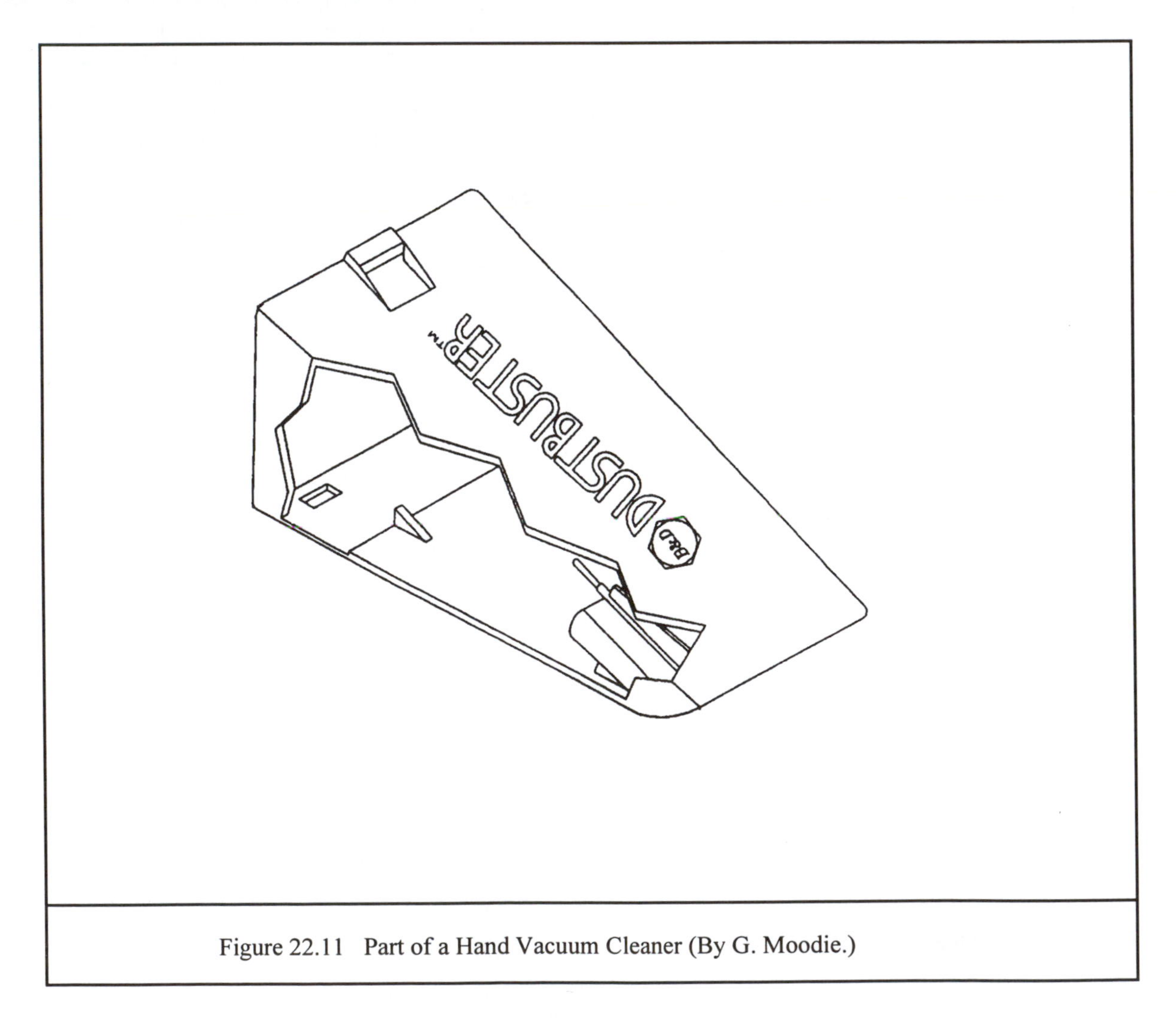

Figure 22.11 Part of a Hand Vacuum Cleaner (By G. Moodie.)

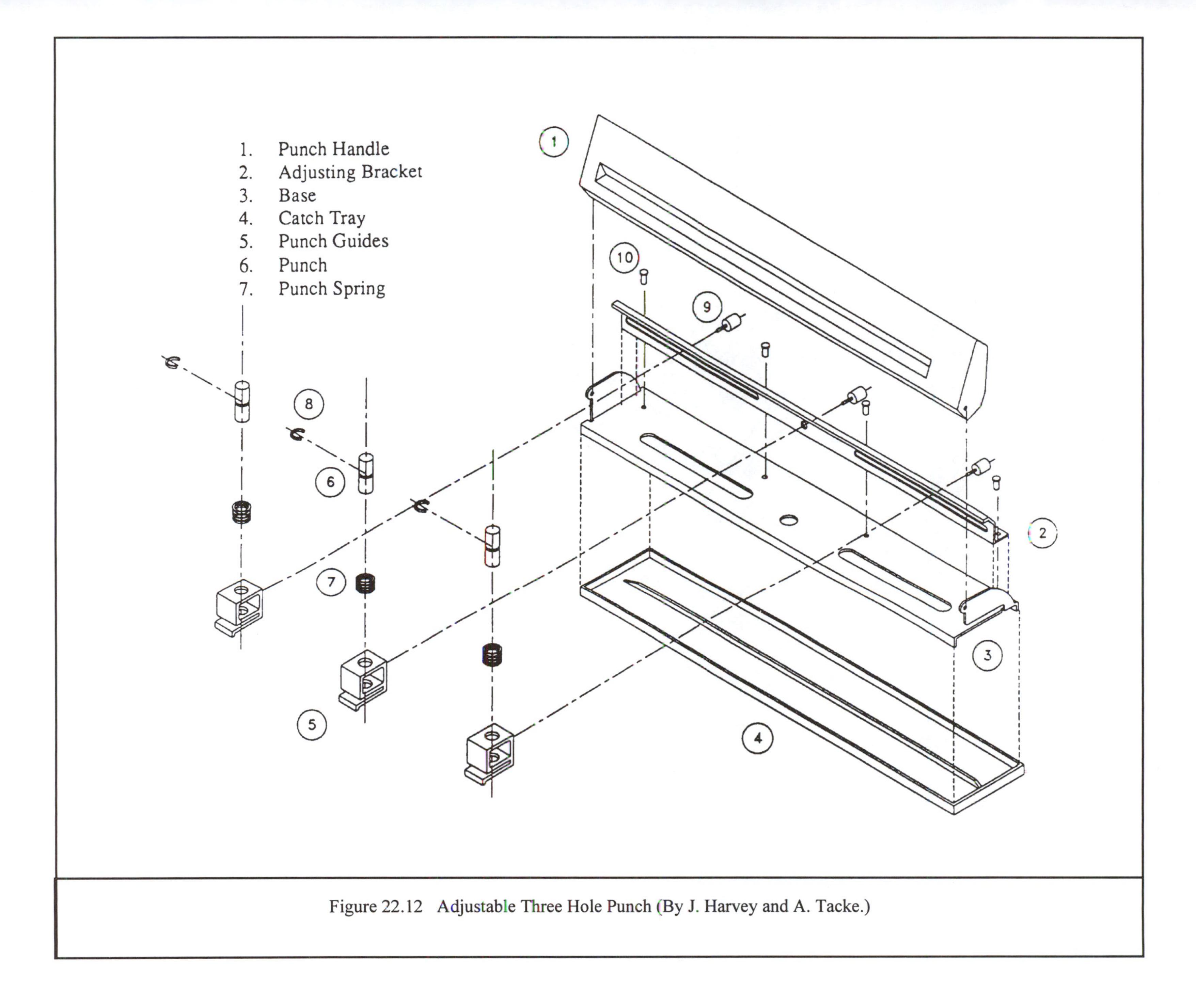

Figure 22.12 Adjustable Three Hole Punch (By J. Harvey and A. Tacke.)

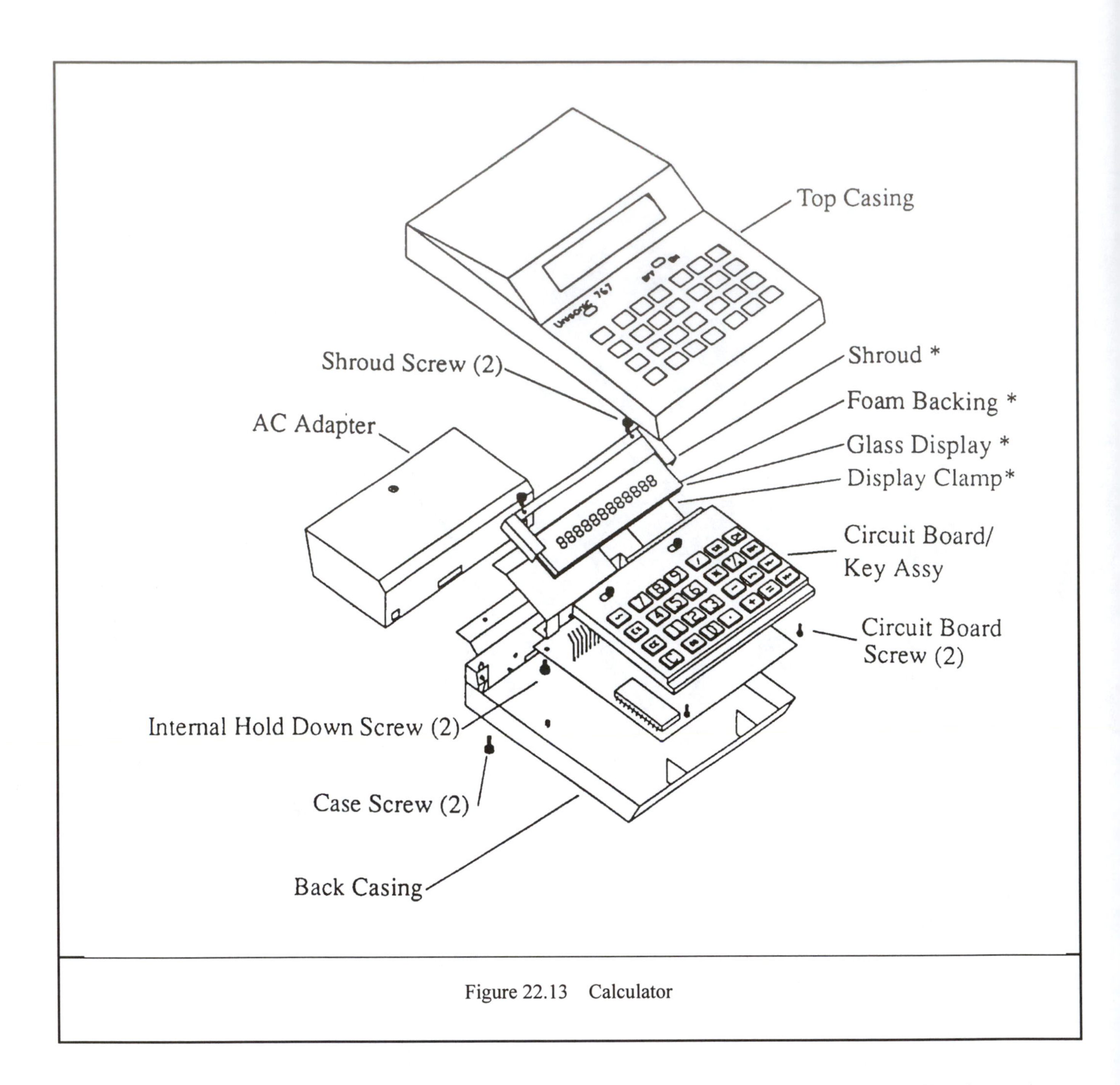

Figure 22.13 Calculator

the outside dimensions of the part. For complex objects, it may help to create a rectangular box for each sub-part.

Of course, if you want you can also make sketches with rulers and templates (say for circles). This is a matter of personal preference. Some will find that the tools somehow interfere with their thought processes or creativity, and so prefer freehand sketching. Others find it easier to use the tools.

Many students are now making drawings with computer graphics packages, which are becoming increasingly more convenient to use for sketching. Again, this is a matter for personal preference.

22.7.6 Standards of Acceptability for Drawings and Sketches

The purpose of drawings and sketches, including those in CAD systems, is effective communication. For effective communication to take place, drawings and sketches must not only express their technical content in proper graphical language, they must also meet an appropriate level of quality in terms of their presentation. A drawing, and certainly a sketch, in an on-going iterative design environment does not have to be finished so as to meet formal drafting standards. But it does have to be neat, be of suitable size, be clear, and communicate without extra effort on the

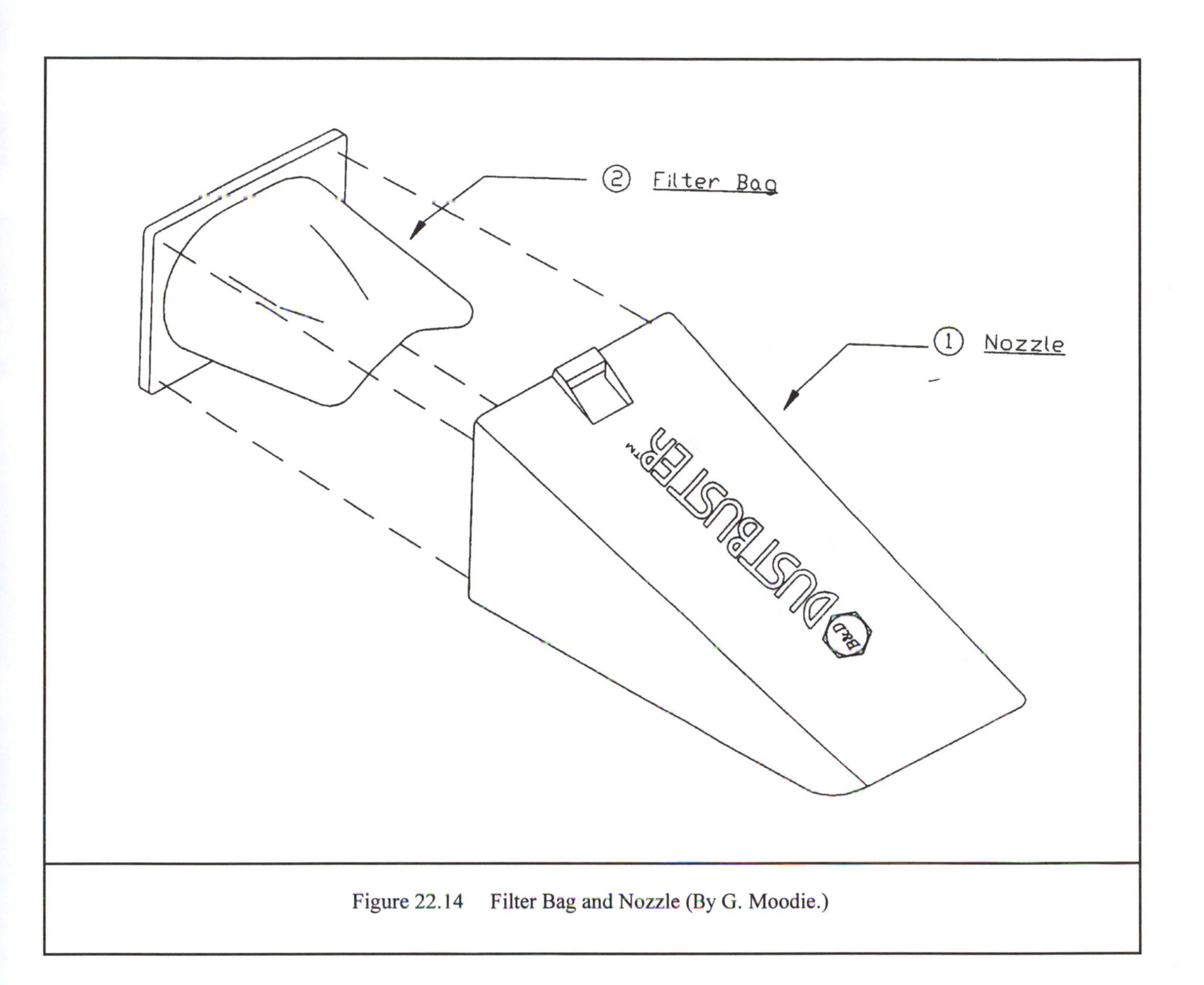

Figure 22.14 Filter Bag and Nozzle (By G. Moodie.)

part of others to overcome sloppiness, carelessness, and important omissions.

For example, lines that are supposed to be in line should look like they are in line. Lines that are supposed to be perpendicular should look like they are perpendicular. Circles should look like circles. Bigger parts should look bigger than the smaller ones. Straight lines should look essentially straight. Lines that are supposed to intersect should in fact intersect. And so on.

There is a tendency for students not familiar with professional standards in these regards to submit drawings and sketches that are far too sloppy and/or fail to communicate properly. The drawings shown thus far in this chapter are all taken from student term project reports and meet at least a minimum standard for such communications. (Your Instructor or boss may disagree; that is fine. Use his or her standards if they are higher than ours, but lower is not acceptable.)

As examples of drawings and sketches that are *not* acceptable, see Figures 22.15 through 22.18. In none of these sketches is it possible to visualize the relationship of the parts. The dashed lines that show the connections are missing. Layouts are off balance. In some, the parts seems just almost randomly placed. Notice that great effort went into preparing these very poor communications. However, the effort was not sufficiently supported by thought, critical evaluation, and re-drawing to make the communications effective.

To end on a happier note, we show Figure 22.19 which again is a student assembly drawing that is up to acceptable standard for a student project report.

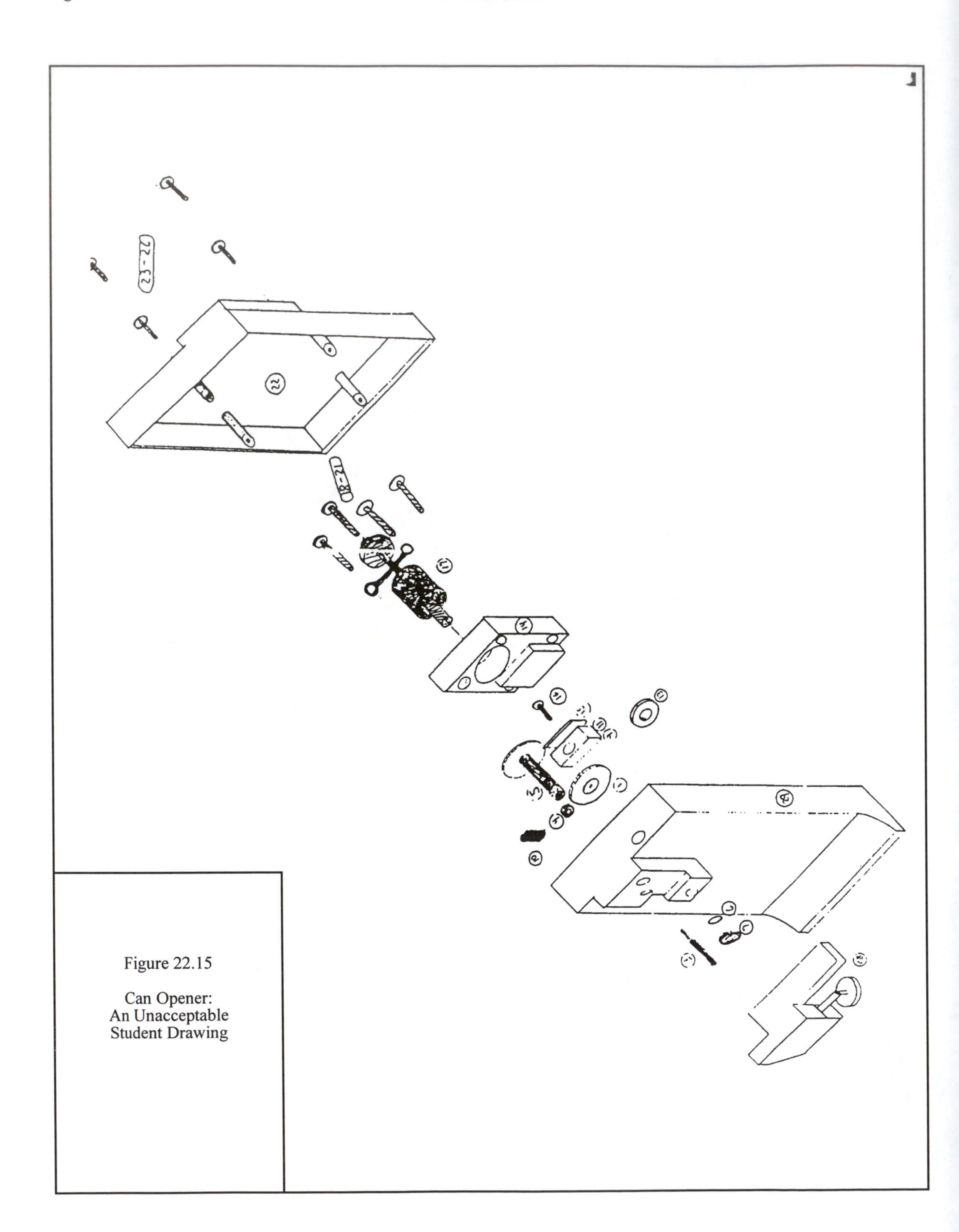

Figure 22.15

Can Opener:
An Unacceptable
Student Drawing

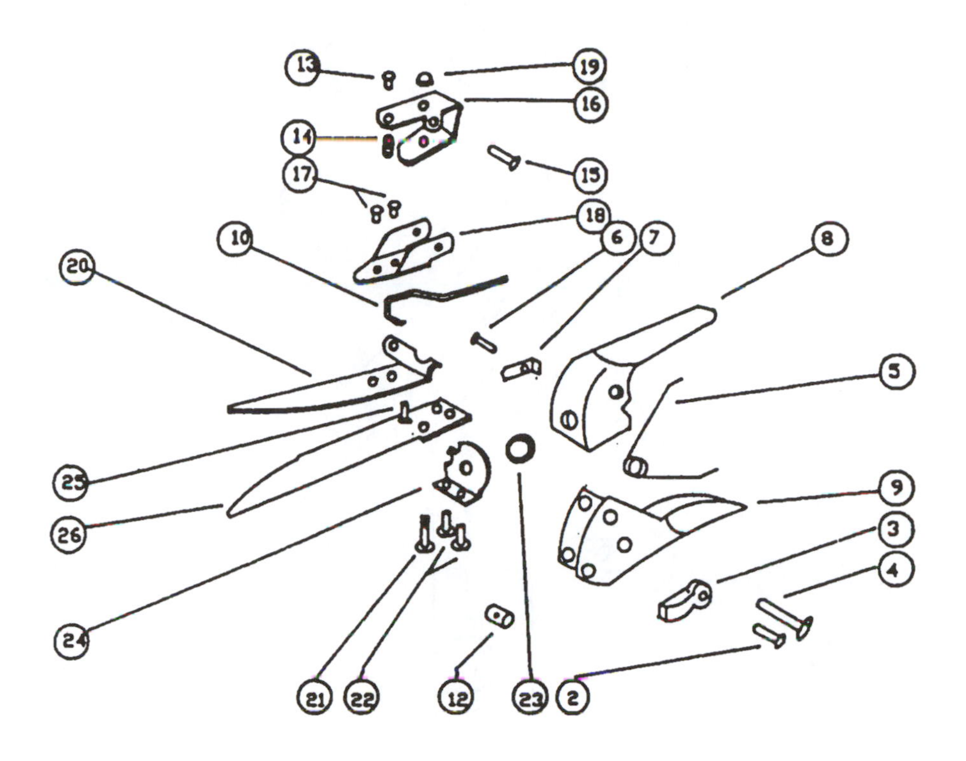

KEY

2. Handle Lock Rivet
3. Handle Lock
4. Pivot Rivet
5. Handle Spring
6. Positioner Lock Rivet
7. Positioner Lock
8. Upper Handle Assembly
9. Lower Handle Assembly
10. Actuator
12. Actuator Nut
13. Spring Assembly Pin
14. Spring Assembly Spring
15. Spring Assembly Rivet
16. Upper Spring Assembly
17. Lower Spring Assembly Rivets
18. Lower Spring Support
19. Blade Pivot Nut
20. Upper Blade Assembly
21. Blade Pivot Bolt
22. Positioner Rivets
23. Positioner Grommet
24. Positioner
25. Bottom Blade Stop
26. Lower Blade

Figure 22.16 Secondary Shears: An Unacceptable Student Drawing

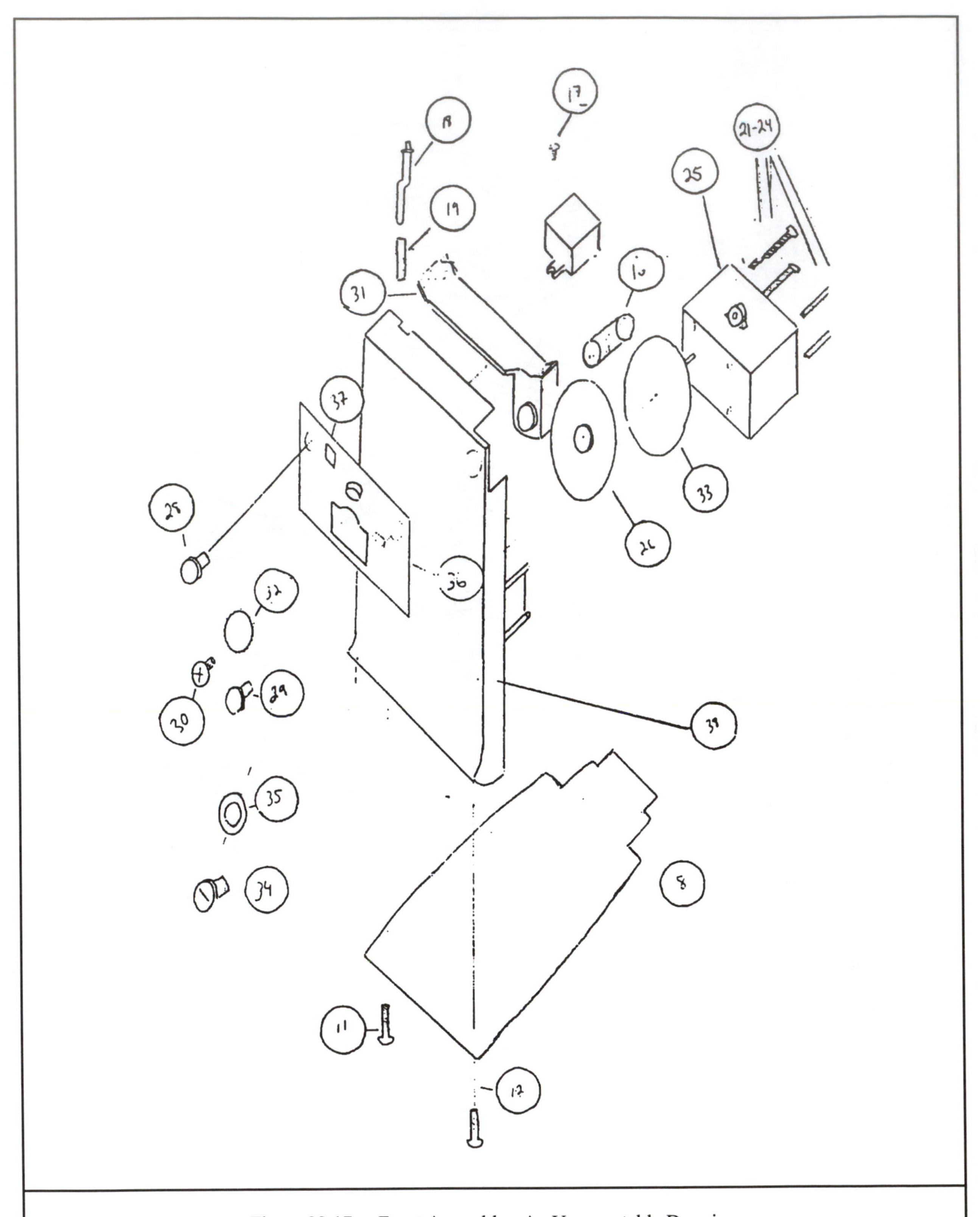

Figure 22.17 Front Assembly: An Unacceptable Drawing

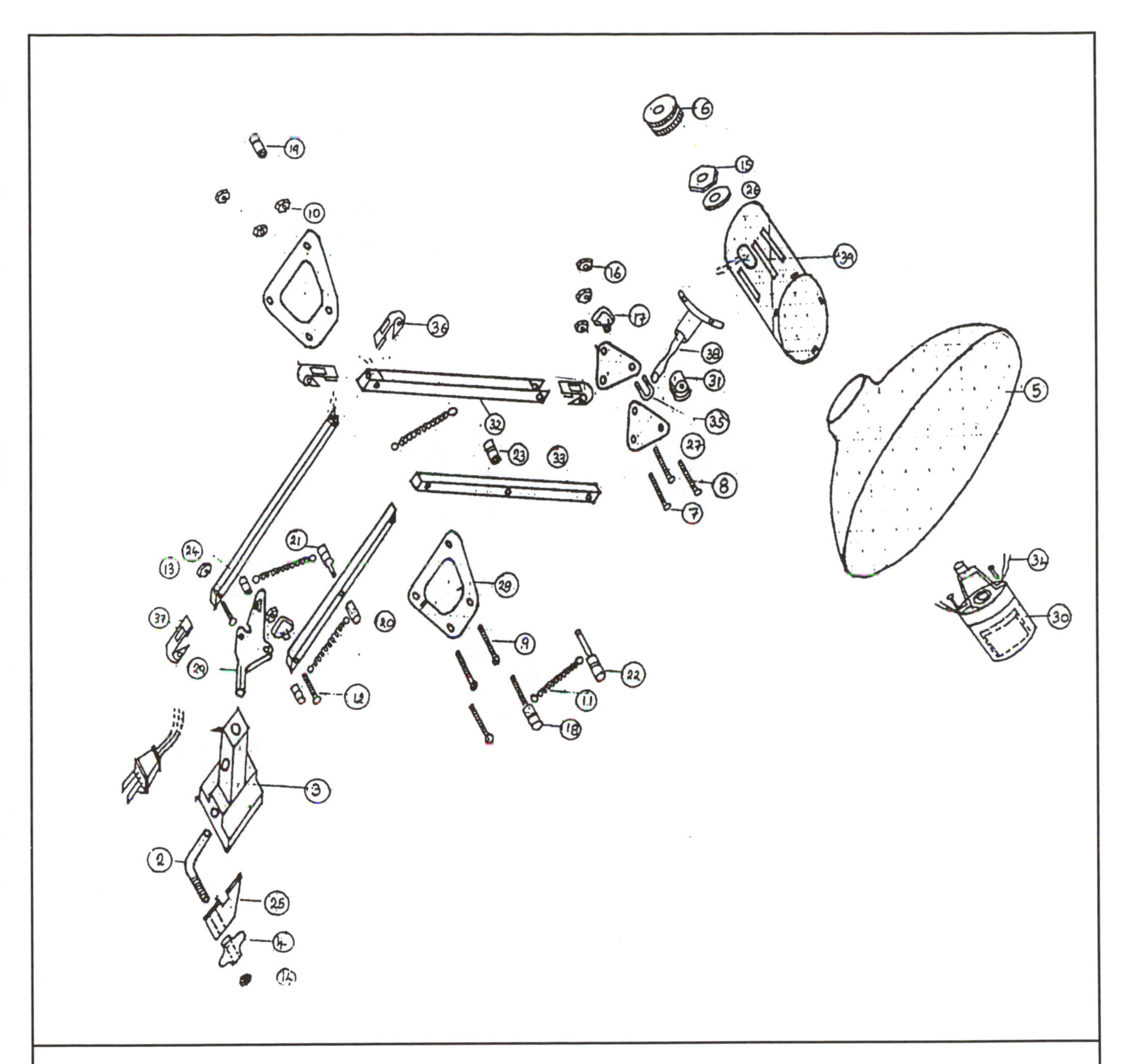

Figure 22.18 Table Lamp: An Unacceptable Student Drawing

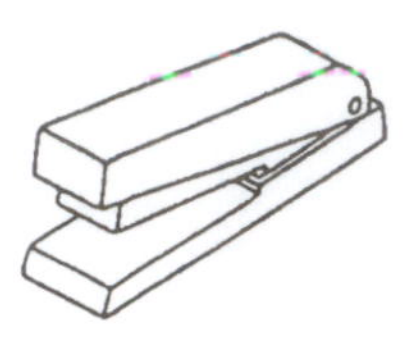

(a) Part 1, The Assembled Stapler

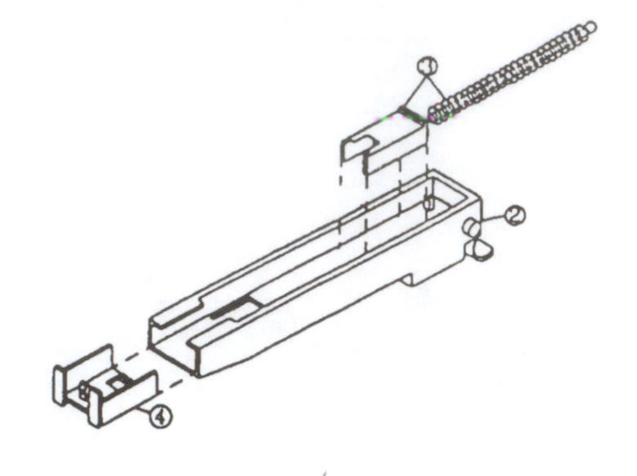

(c) Staple Holder Subassembly Drawing

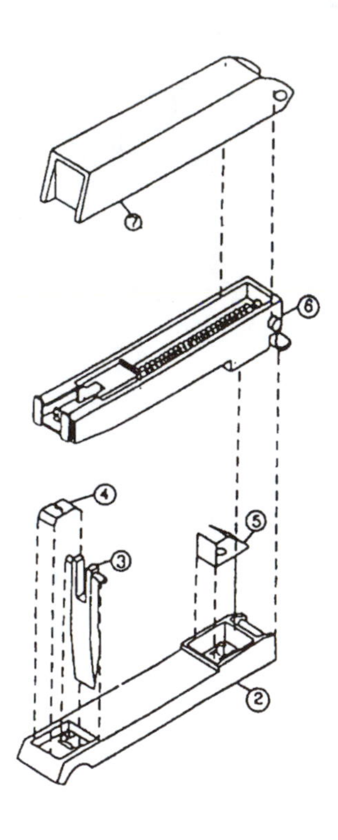

(b) Stapler Assembly Drawing

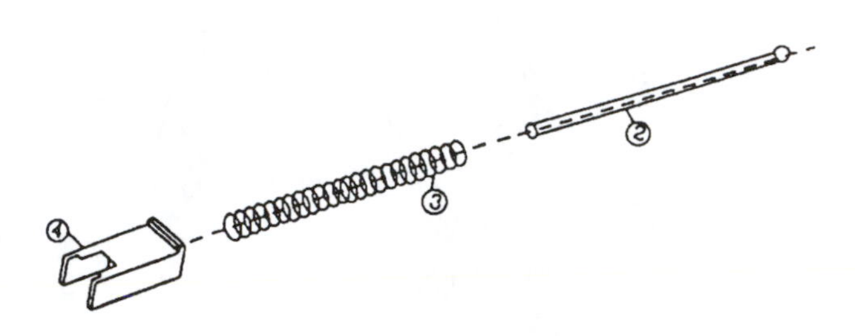

(d) Spring Subassembly Drawing

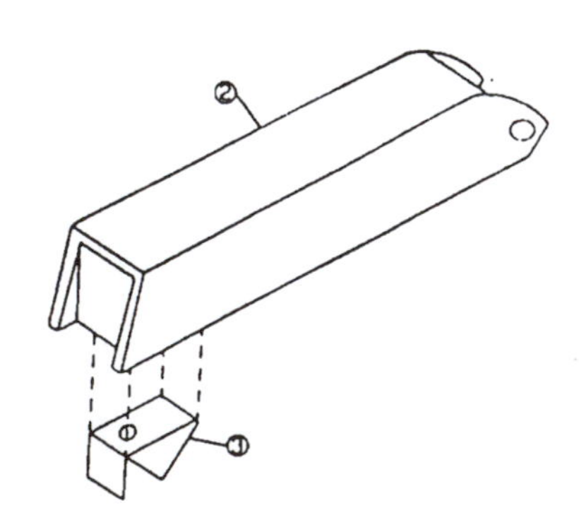

(e) Top Cover Subassembly Drawing

Figure 22.19 Stapler (Drawn by N. Renganath and R. Rajkumar.)

22.8 Summary and Preview

Summary.

- Know the purpose of your communication, and the take-aways;
- Know the receivers - what words and symbols they understand, what their needs are;
- If a written message:
    - + Make an outline and a *complete* rough first draft as a first step,
    - + Rewrite, rewrite, rewrite, rewrite, rewrite......
    - + Spelling and grammar must be perfect (sentences, punctuation, paragraph structure, etc.)
    - + Keep a copy,
    - + Short is critical (KISS: Keep it Short, Stupid.)
- If a Research Report:
    - + Use a descriptive plus summary type Abstract,
    - + Organization: Introduction, Literature Review, Body, Discussion, and Summary and Conclusions,
    - + Appendices.
- If a Business Technical Report:
    - +Use a summary type Abstract,
    - +Focus on the recommendations, conclusions, results, and data (in that order),
    - +Details in Appendices,
    - +Short (except Appendices) is especially critical.
- Make report titles complete and incisively descriptive of a report's contents;
- If an oral report:
    - + One slide or transparency per minute - maximum,
    - + Three or four bullets, and fifty words per slide - maximum,
    - + Never read from a slide or from notes,
    - + Rehearse, rehearse, rehearse, rehearse........
    - + Let your interest and enthusiasm show.
- Do not submit sloppy drawings or sketches. Maintain professional standards.

Preview. In the next chapter, we turn our attention to money.

References

[1] Zinsser, Hans. *On Writing Well*, 5th Edition, HarperCollins, 1994.

[2] Rosenstein, Rathbone and Schneerer, *Communication in Engineering Design*, Prentice Hall, Englewood Cliffs, NJ., 1962.

Supplementary Reading

On Writing Well, 5th Edition, Hans Zinsser, HarperCollins, 1994.

The Elements of Style, W. Strunk, Jr. and E. B. White, MacMillan, 1972.

Communication in Engineering Design, Rosenstein, Rathbone and Schneerer, Prentice Hall, 1962.

The Visual Display of Quantitative Information, Edward R. Tuffte, Graphics Press, Cheshire, Connecticut, 1983.

Problems

22.1 In Chapter 3, Exercise 3.17, you were asked to analyze the assembly shown in Figure P3.17, make redesign suggestions in order to reduce assembly costs, and to estimate the approximate savings in assembly costs. Assume that you are a design engineer working for the ABC Manufacturing Company which produces the assembly in question, and that you are writing a Business Technical Report which is to go to your immediate supervisor.

a) What would you use for a title for this report?
b) Write an Abstract of no more than 200 words for this report.
c) Prepare a Results section for the Main Body of this report.
d) What would you use for a title if this were a Research Report?
e) What would you use for an Abstract if this were a Research Report.

22.2 In Chapter 3, Exercise 3.18, you were asked to analyze the assembly shown in Figure P3.18, make redesign suggestions in order to reduce assembly costs, and to estimate the approximate savings in assembly costs. Assume that you are a design engineer working for the ABC Manufacturing Company which produces the assembly in question, and that you are writing a Business Technical Report which is to go to your immediate supervisor.

a) What would you use for a title for this report?
b) Write an Abstract of no more than 200 words for this report.
c) Prepare a Results section for the Main Body of this report.

22.3 Assume that you work for the ABC Manufacturing Company and that you are involved in a two week training program for new engineering hires. Pretend that as part of this training program YOU have written section 22.7 Graphical Communications as a stand-alone report which is given to the trainees.

a) Write a title for this report which differs from the one used for section 22.7.
b) Write an Abstract of no more than 200 words for this report.
c) Write an Abstract of no more than 100 words for this report.

22.4 Pretend you are the Manager of the Design Group within the ABC Manufacturing Company which is responsible for the design of the floppy disk drive analyzed in Chapter 3, exercise 3.18. Prepare a memo to be sent to A. B. Curtis, Vice-President of Engineering indicating: (a) What the preliminary results of exercise 3.18 indicate, and (b) What you expect to do as a result of these preliminary results. How would your memo differ if instead of the product used in exercise 3.18, your product was the one considered in exercise 3.17?

22.5 In Chapter 3, Figure 3.42, an Assembly Advisor was presented. Pretend you are writing a Research Report discussing the origins and use of this DFA advisor. Prepare the following:

a) A title for this report.
b) An Abstract for this report.
c) A Results section for the Main Body of this report.

22.6 Pretend that *you* are the author of this book. Write a Research Report Abstract for the following Chapters:

a) Chapter 7 - Engineering Conceptual Design: Evaluation and Redesign.
b) Chapter 11 - Evaluation of Part Configurations for Manufacturability: Injection Molding and Die Casting.
c) Chapter 13 - Special Purpose Parts: Evaluation and Redesign.

22.7 Pretend that you are the author of this book. Write a business report Abstract for the following Chapters:

a) Chapter 7 - Engineering Conceptual Design: Evaluation and Redesign.
b) Chapter 11 - Evaluation of Part Configurations for Manufacturability: Injection Molding and Die Casting.
c) Chapter 13 - Special Purpose Parts: Evaluation and Redesign.

22.8 In Chapter 3, Figure 3.14, an Injection Molding DFM Advisor was presented. Pretend you are writing a Research Report discussing the origins and use of this DFM Advisor. Prepare the Results section for the Main Body of this report.

22.9 In Chapter 11, Figure 11.1, a classification system for the basic tool complexity of injection molded parts was presented. Assume that you are preparing a Research Report titled, "Design for Injection Molding - An Analysis of Part Attributes that Impact Die Construction Costs for Injection Molded Parts." Write the Results section for this report indicating what the data presented in Figure 11.1 demonstrates.

22.10 In section 11.11 of Chapter 11, it was shown that if the injection molded part shown in Figure 11.23 was redesigned as shown in Figure 11.25, a 25% reduction in mold construction costs would result. Prepare a Business Technical Report describing these two alternative designs. Be certain to include a recommendation in your report.

22.11 In section 12.8 of Chapter 12, it was shown that a savings in tooling cost could be achieved for the part shown in Figure 12.51, if it were redesigned as shown in Figure 12.52. Prepare a business report describing the three alternative designs and recommend what action should be taken.

22.12 In Chapter 23, Student Exercise 2, you were asked to determine the maximum amount of money that could be spent now to bring about the redesign suggestions you made in Chapter 3. Prepare a Business Technical Report describing the two alternative designs, and make a recommendation as to what action should be taken.

22.13 In section 22.2.2 there is a paragraph used as an example of poor writing. Rewrite it to eliminate the wordiness, and fix the spelling and grammar.

22.14 In Chapter 5 your were introduced to the idea of quality-function deployment (QFD) and the House of Quality as an aid to formulating conceptual design problems. Select a product, preferably one you are involved in designing as part of a semester long project, and prepare an oral presentation describing your use of QFD and the House of Quality.

22.15 In Chapter 12, Figures 12.1 and 12.2, classification systems for shearing, local features, and wipe forming of stamped parts was presented. Prepare an oral presentation in

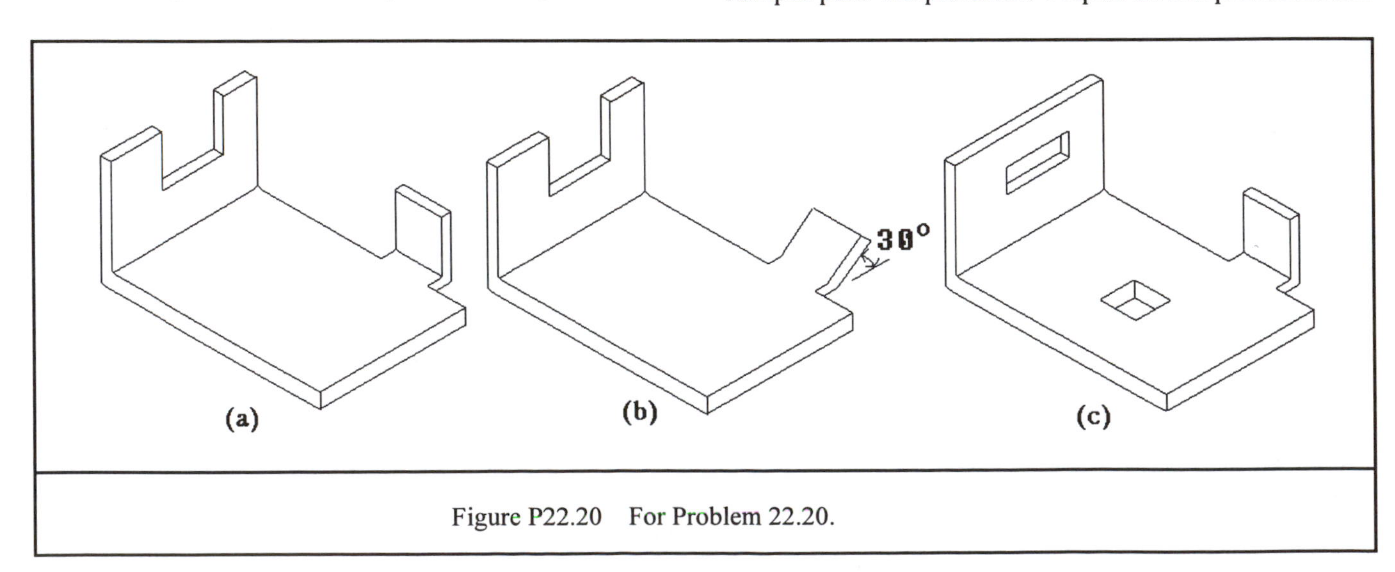

Figure P22.20 For Problem 22.20.

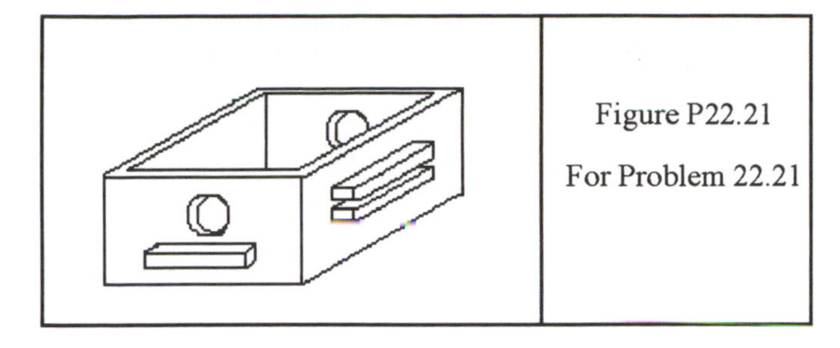

Figure P22.21

For Problem 22.21

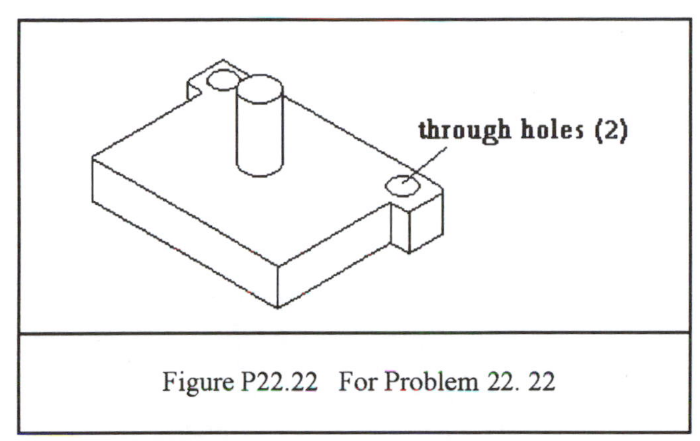

Figure P22.22 For Problem 22. 22

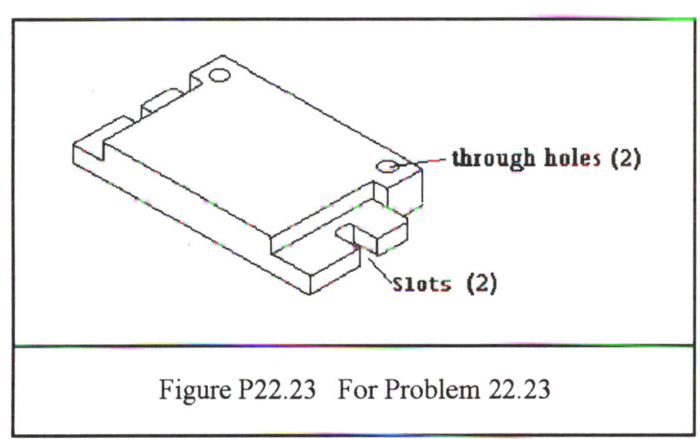

Figure P22.23 For Problem 22.23

which you describe to the audience the effect of part features on the tooling cost of stamped parts.

22.16 Prepare an oral presentation to accompany the Business Technical Report that you prepared for exercise (10) above.

22.17 Prepare an oral presentation to accompany the Business Technical Report that you prepared for exercise (11) above.

22.18 Prepare an oral presentation to accompany the Business Technical Report that you prepared for exercise (12) above.

22.19 In place of the memo you wrote as part of exercise (4), prepare a brief 5 minute oral presentation to be made to Vice-President Curtis.

22.20 Draw the top, front and right side view for the parts shown in Figure P22.20.

22.21 Draw the top, front and right side view for the part shown in Figure P22.21.

22.22 Draw the top, front and right side view for the part shown in Figure P22.22.

22.23 Draw the top, front and right side view for the part shown in Figure P22.23.

22.24 Draw the top, front and right side view for the part shown in Figure P22.24.

22.25 Shown below in Figures P22.25 are orthographic projections for various parts. Are the top, front and right side views shown correct? What changes are required to make them correct?

22.26 An oblique pictorial view of a part is shown in Figure P22.26. Also shown in Figure P22.26 are the top, front and right side views of this part. Are these views correct? What changes are required to make then correct?

22.27 Convert the oblique view shown in Figure P22.26 into an isometric view.

22.28 Convert the oblique view shown in Figure P22.21 into an isometric view.

22.29 Create isometric views of the following parts:

a) The disk carrying case shown in Figure 3.13.

b) Food carrier shown in Figure 22.3 of this chapter.

c) The plate and brackets of Figure 22.2 of this chapter under the assumption that the plate and brackets are produced as a single component via some process such as casting.

22.30 Figure P22.30 shows a part cut by a plane. Create a full sectional view of the part at the location where the plane cuts the part.

22.31 Figure P22.31 shows a part cut by a plane. Create a

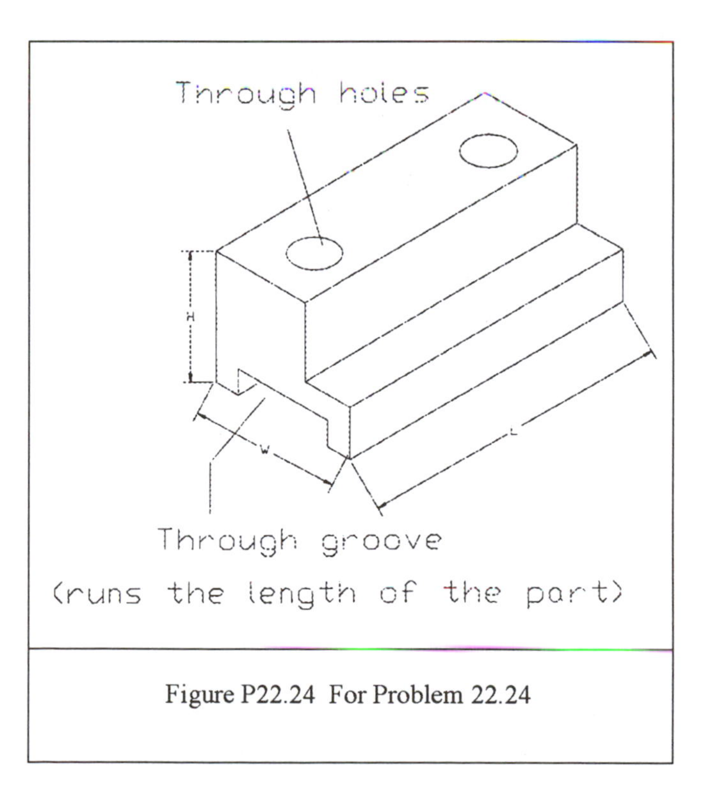

Figure P22.24 For Problem 22.24

full sectional view of the part at the location where the plane cuts the part.

22.32 Figure P22.32 shows a part cut by a plane. Create a full sectional view of the part at the location where the plane cuts the part.

22.33 Figure P22.33 shows a part cut by a plane. Create a full sectional view of the part at the location where the plane cuts the part.

22.34 As part of a student project, the drawing shown in Figure P22.34 was submitted. The drawing was as an exploded assembly drawing of a small hand held hair dryer. What suggestions, if any, would you make to improve this drawing?

22.35 As part of a student project, the drawing shown in Figure P22.35 was submitted. The drawing was as an exploded assembly drawing of a battery-powered pencil sharpener. What suggestions, if any, would you make to improve this drawing?

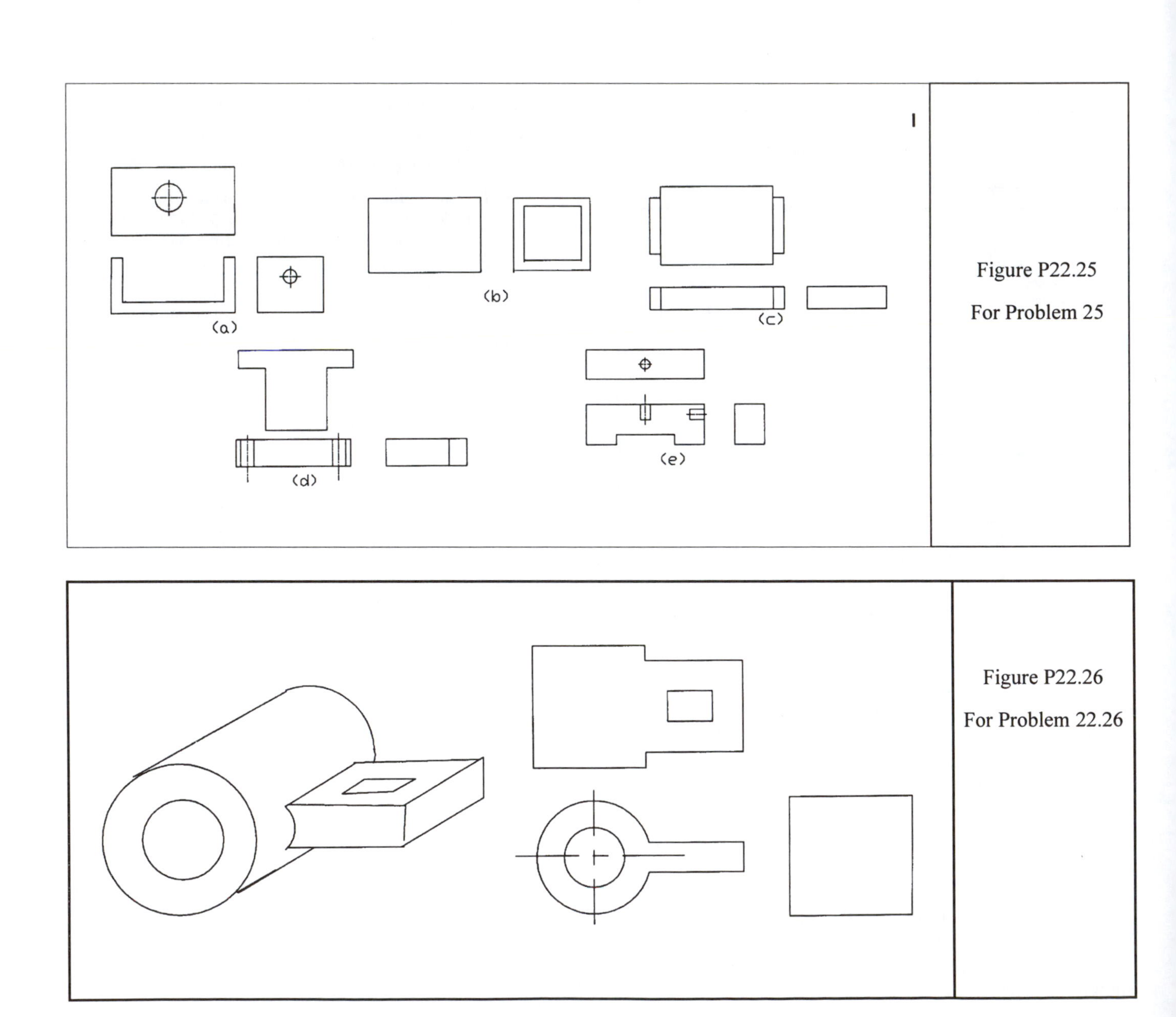

Figure P22.25

For Problem 25

Figure P22.26

For Problem 22.26

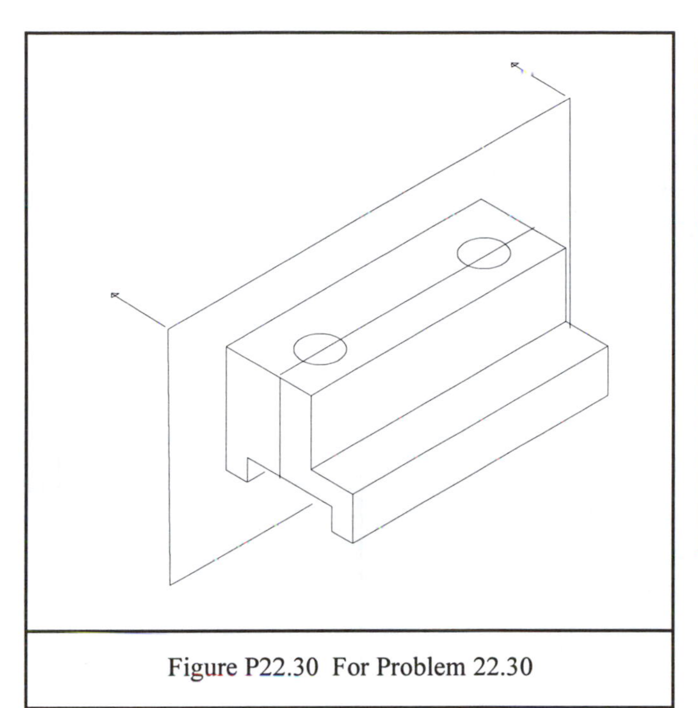

Figure P22.30 For Problem 22.30

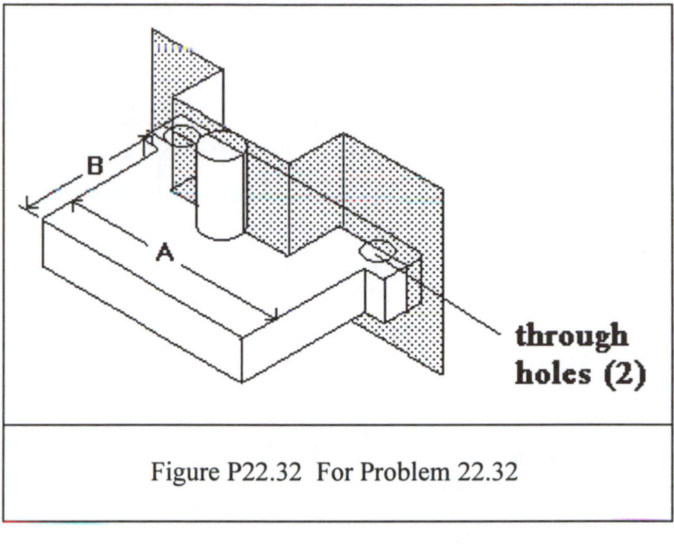

Figure P22.32 For Problem 22.32

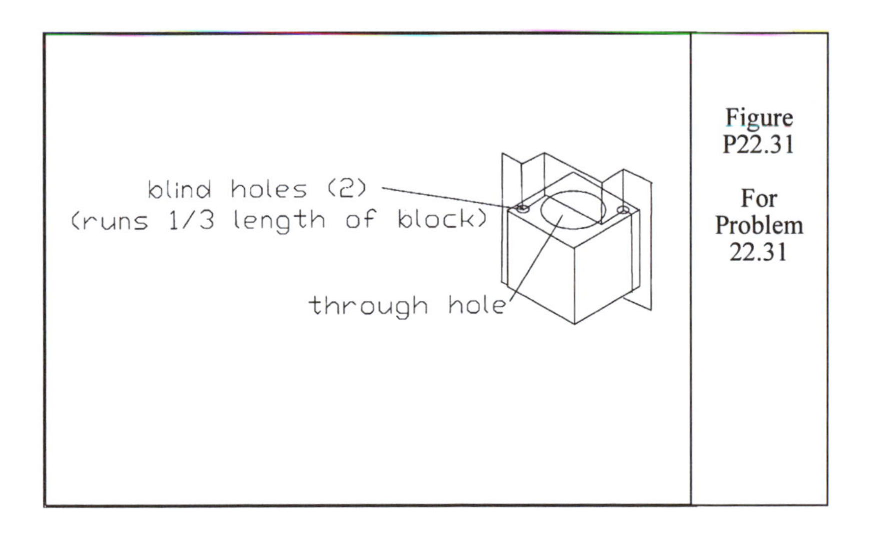

Figure P22.31 For Problem 22.31

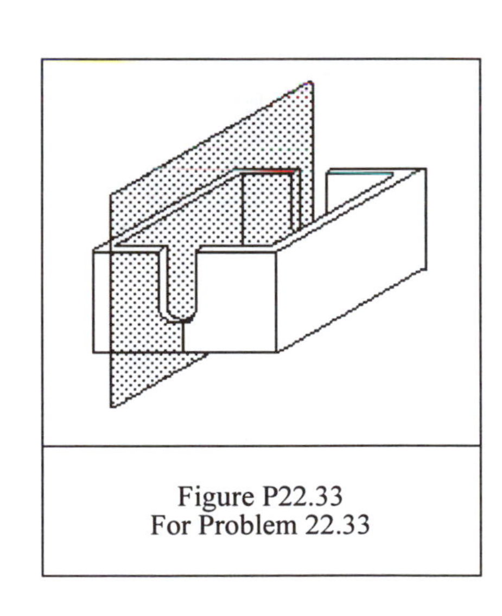

Figure P22.33 For Problem 22.33

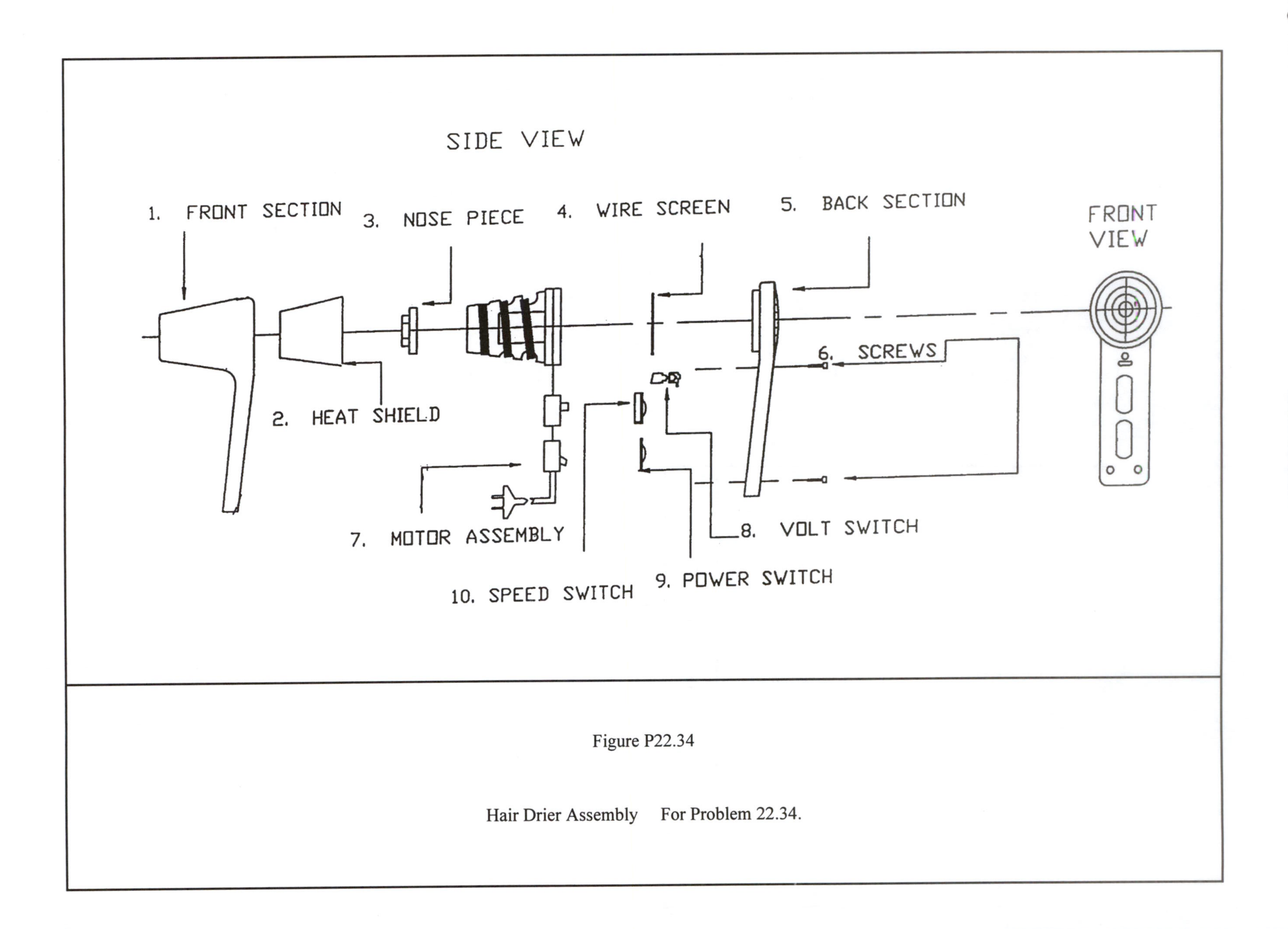

Figure P22.34

Hair Drier Assembly For Problem 22.34.

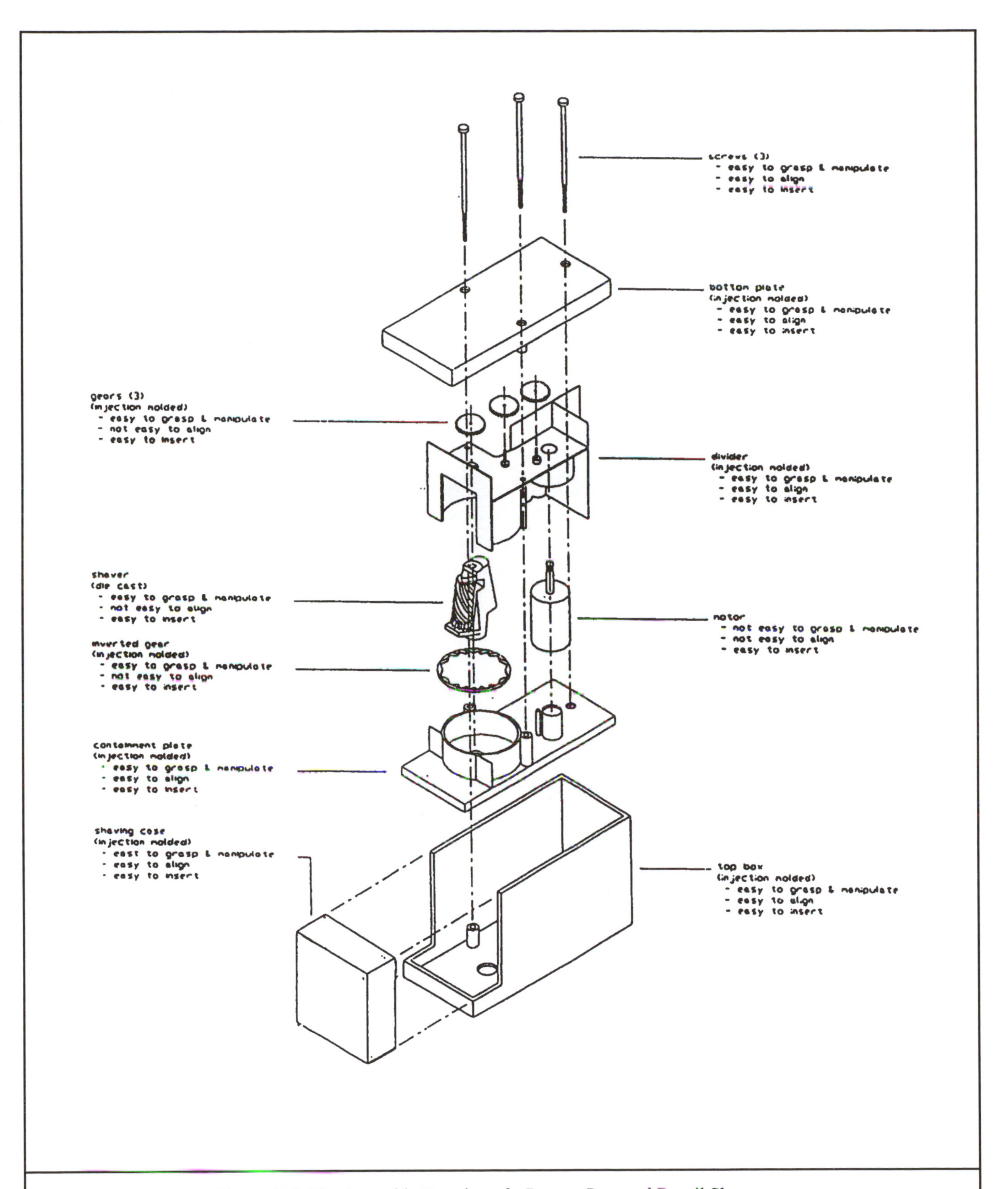

Figure P22.35 Assembly Drawing of a Battery-Powered Pencil Sharpener

This page is intentionally blank

Chapter Twenty-Three

Introduction to Engineering Economics

"You pays your money and you takes your choice"

Punch (1846)

23.1 Introduction

A potential capital investment in designing or manufacturing a new or revised product is always in competition for resources — time, money, talent — with other possible projects. Designers must understand that a firm's available capital need not be spent on new or improved designs or processes. Competition may come from proposals for other design and manufacturing projects, or from energy conservation projects, salary or benefit increases for employees, or just plain money saving. Capital can also be invested in stocks or bonds of various kinds, which is like loaning it to other firms or to the government in return for the interest earned. Thus it is important for engineers and engineering managers to be able to evaluate and justify proposed investments in design and manufacturing projects on economic grounds.

Investing money in design, or in tooling for manufacturing a new or revised product, is an "invest now, profit later — hopefully" kind of situation. The basic question is: Is today's investment justified by the expected future profits? Different projects require different initial investments, and return different profits on different time scales. A method of analysis is needed, therefore, that provides a fair comparison despite such differences. Is it, for example, better to invest $100,000 revising an existing product to improve profits by $40,000 per year for five years, or to invest $150,000 in an energy conservation project that returns $30,000 per year for ten years?

In addition to the purely economic matters, there are always intangibles — non-economic, or at least non-computable considerations — that enter investment decisions. Perhaps the product revision is necessary as a part of the company's strategy for quality or market share. Perhaps energy conservation is considered a social responsibility. Such issues must be considered in making final decisions, but it is still essential to have a clear understanding of the purely economic issues as well.

It should be noted that this chapter is not about economics in a general sense. That is a much broader and more comprehensive subject. Here we present only methods for performing analyses comparing different prospective uses of money. Our purpose is not to attempt to do in one chapter what is the subject of whole books or courses in engineering economics, but only to provide some basic knowledge so that design engineers will understand the issues involved and be better able to work with managers and financial officers. Courses in engineering economics are strongly recommended for engineering designers.

In order to understand the logic of the analyses presented, readers first need to understand some of the basic terms and tools. Thus we begin with explanations of concepts like the "time value" of money, present worth, and capital recovery factor. The notation to be used is as follows:

i = Rate of interest, expressed as a decimal,

n = Number of interest periods, usually years,

P = Present amount of money, *or* the present value of a future amount of money,

S = An amount of money at some future time,

R = An amount of money received or spent periodically, usually annually,

I = An initial amount of money invested in a project.

23.2 The Time Value of Money

For the purposes of this discussion, readers must put the effects of inflation out of their minds. We will come back to the subject of inflation later, but for now we assume that the buying power of a dollar remains constant over time, and hence is not a factor in the analyses performed. We will thus be dealing here strictly with numbers of dollars.

Consider a savings account in a bank. If we deposit today an amount of money P, and the account earns annual interest i, then (if we take nothing out, allowing the interest to accrue) the future amounts of money will be:

After 1 Year: $S_1 = P + iP = P(1 + i)$

After 2 Years: $S_2 = P(1 + i) + iP(1 + i)$

$= P(1 + i)^2$

After 2 Years: $S_3 = P(1 + i)^3$

....

After n Years: $S_n = P(1 + i)^n$

(23.1)

The fact that P has grown by a factor of $(1 + i)^n$ is referred to as the *time value of money.* Because of interest, an amount of money can grow with time. It won't grow if left in a sock under the mattress.

The growth of money earning interest can be quite dramatic, even at modest interest rates. For example, if $1000 is invested at 5% for 14 years, the amount of money available will be nearly doubled:

$$S_{14} = P(1 + 0.05)^{14} = (1000)(1.97993) = \$1979.93$$

This result suggests a useful rule of thumb:

Number of years for an amount to double =

70 / Interest Rate in %

For example, if $12000 is put into an account at 7% interest, it will double to about $24,000 in 70 / 7 = 10 years. The exact amount will be $12000(1 + 0.07)^{10} = \$23,605.82$.

The expression $(1 + i)^n$ is called the Single Payment Compound Amount Factor (SPCAF). Tables exist that list the value of this, and other factors we will define later. Figure 23.1 is a portion from a table for $i = 5\%$ in which values for SPCAF are found in the first column. Note that the value of SPCAF for $i = 5\%$ and $n = 14$ years is 1.97993 — the same as we computed above.

23.3 Present Value

It is often useful to estimate the *present value* of some future amount of money. For example, suppose interest rates are 3% and you have a contract to receive an amount S = $1000 a year from now. What is the present value (P) of that $1000? In other words, how much you would take *now* in place of getting the $1000 a year from now?

To answer this type of question, assume that there is some present amount of money P that can be invested at the available interest rate, i, such that the required amount S is attained at the end of the prescribed period. For example, to get S = $1000 one year from now with an interest rate of 3%, we need a present sum, P, such that:

$$S = P(1 + i) \quad \text{or} \quad P = S / (1 + i)$$

$$P = 1000 / 1.03 = \$970.87$$

Thus we say that the present value of $1000 one year from now at i = .03 is $970.87.

Extending the above reasoning to longer periods (n > 1), the present value equation is:

$$P = S / (1 + i)^n \tag{23.2}$$

$$$$$$$$$$$$

Example

The scrap value of a machine to be sold in four years is estimated to be $7,500. Interest rates are 5%. What is the present value of the scrap?

Solution: $P = S / (1 + i)^n$

$$P = 7500 / (1.05)^4 = \$6170.27$$

$$$$$$$$$$$$

The factor $1 / (1 + i)^n$ is called the Single Payment Present Value Factor (SPPVF). It is simply the reciprocal of the SPCAF. Values for SPPVF for $i = 5\%$ are listed in Figure 23.1 in the second column. Note that the value of SPPVF for four years is 1.2155063. We could have used this to solve the example above:

$$P = 7500 / 1.2155 = \$ 6170.27$$

The concept of present value is often used as a basis for comparing the worth of competing projects.

$$$$$$$$$$$$

Example

The estimated increase in profits from a product revision are as follows:

1st Year	$10,000.
2nd Year	$12,000.
3rd Year	$ 7000.
4th Year	$ 2000.

If the required current investment in design and manufacturing equipment to prepare the revision is $8000, what is the expected present worth of this project? Interest rates are 8%.

Solution:

The Present Value of the Expected Profits are:

1st Year: $P = 10{,}000 / (1 + .08)^1 = 9{,}259.$

2st Year: $P = 12000 / (1 + .08)^2 = 10{,}288.$

3st Year: $P = 7000 / (1 + .08)^3 = 5{,}557.$

4st Year: $P = 2000 / (1 + .08)^4 = 1{,}470.$

$26,574.

5%	SPCAF	SPPAF	SFDF	CRF	USCAF	USPWF
	SINGLE PAYMENT		UNIFORM ANNUAL SERIES			
N Periods	Compound Amount Factor Given P To Find S	Present Worth Factor Given S To Find P	Sinking Fund Factor Given S To Find R	Capital Recovery Fac tor Given P To Find R	Compound Amount Factor Given R To Find S	Present Worth Factor Given R To Find P
	(1 + i)**N	$\frac{1}{(1+i)^N}$	$\frac{i}{(1+i)^N - 1}$	$\frac{i(1+i)^N}{(1+i)^N - 1}$	$\frac{(1+i)^N - 1}{i}$	$\frac{(1+i)^N - 1}{i(1+i)^N}$
1	1.0500000	.9523810	1.0000000	1.0500000	1.0000000	.9523810
2	1.1025000	.9070295	.4878049	.5378049	2.0500000	1.8594104
3	1.1576250	.8638376	.3172085	.3672086	3.1525000	2.7232480
4	1.2155063	.8227025	.2320118	.2820118	4.3101250	3.5459505
5	1.2762816	.7835262	.1809748	.2309748	5.5256313	4.3294767
6	1.3400956	.7462154	.1470175	.1970175	6.8019128	5.0756921
7	1.4071004	.7106813	.1228198	.1720198	8.1420085	5.7863734
8	1.4774554	.6768394	.1047218	.1547218	9.5491089	6.4632128
9	1.5513282	.6446089	.0906901	.1406901	11.0265643	7.1078217
10	1.6288945	.6139133	.0795046	.1295046	12.5776925	7.7217349
11	1.7103394	.5846793	.0703889	.1203889	14.2067872	8.3064142
12	1.7958563	.5568374	.0628254	.1128254	15.9171265	8.8632516
13	1.8856491	.5303214	.0564558	.1064558	17.7129828	9.3935730
14	1.9799316	.5050680	.0510240	.1010240	19.5986320	9.8986409
15	2.0789282	.4810171	.0463423	.0963423	21.5785536	10.3796580
16	2.1828746	.4581115	.0422699	.0922699	23.6574918	10.8377696
17	2.2920183	.4362967	.0386991	.0886991	25.8403664	11.2740662
18	2.4066192	.4155207	.0355482	.0855452	28.1323847	11.6895859
19	2.5269502	.3957340	.0327450	.0827450	30.5390039	12.0853209
20	2.5532977	.3768895	.0302426	.0802426	33.0659341	12.4622103
21	2.7859626	.3589424	.0279961	.0779961	35.7192518	12.8211527
22	2.9252607	.3410499	.0259705	.0759703	38.5052144	13.1630028
23	3.0715238	.3255713	.0241368	.0741368	41.4304751	13.4885739
24	3.2250999	.3100679	.0224709	.0712700	44.5019939	13.7986418
25	3.3863549	.2953028	.0209525	.0709525	47.7270938	14.0939446
26	3.5556727	.2812407	.0195543	.0695643	51.1134538	14.3751253
27	3.7334563	.2678483	.0182919	.0682919	54.6691264	14.6430336
28	3.9201291	.2550936	.0171225	.0671225	58.4025828	14.8981273
29	4.1161356	.2429463	.0160455	.0660455	62.3227119	15.1410736
30	4.3219424	.2313774	.0150514	.0650514	66.4388475	15.3724510
31	4.5380395	.2203595	.0141321	.0641321	70.7607899	15.5928105
32	4.7649415	.2098662	.0132804	.0632804	75.2988294	15.8026767
33	5.0031885	.1998725	.0124900	.0624900	80.0637708	16.0025492
34	5.2533480	.1903548	.0117554	.0617554	85.0669594	16.1929040
35	5.5160154	.1812903	.0110717	.0610717	90.3203074	16.3741943
36	5.7918161	.1726574	.0104345	.0604345	95.8353227	16.5468517
37	6.0814069	.1644356	.0098398	.0598398	101.6281389	16.7112873
38	6.3854773	.1566054	.0092842	.0592842	107.7095458	16.8678927
39	6.7047512	.1491480	.0087646	.0587646	114.0950231	17.0170407
40	7.0399887	.1420457	.0082782	.0582782	120.7997742	17.1590864

Figure 23.1 Tables of Compound Interest Factors for i = 5%

Since the required current investment is $8000., the present net value of the revision is $18,574. This can be compared with the present value of competing projects to help make the investment decision.

$$$$$$$$$$$

23.4 Annuities

An annuity is an amount R that is received regularly, usually in equal installments — in which case it is called a *uniform annuity.* Suppose, for example, that you have won a $1,000,000. lottery that is to pay you R = $33,333.33 per year for 30 years. If you spend this money when you get it, you will have spent a $1,000,000 at the end of thirty years. Or, if you put it under your mattress, you will have a million dollars (and a lumpy mattress) after thirty years. This is the basis for the lottery people saying that you have "won a million dollars".

If, on the other hand, you resist the temptation to spend (or squirrel away) the money, and instead put the R dollars you receive each year into a bank account at interest i, then you will have the following amounts of money in the bank:

After 1 Year: $S_1 = R$

After 2 Years: $S_2 = R(1+i) + R$

After 3 Years: $S_3 = R(1+i)^2 + R(1+i) + R$

After n Years: $S_n =$

$$R(1+i)^{n-1} + R(1+i)^{n-2} + \cdots + R.$$

This type of series expression can be simplified by the following procedure. First, factor out the annual payment, R:

$$S_n = R[(1+i)^{n-1} + (1+i)^{n-2} + \cdots + 1]$$

Next multiply both sides by the factor: $[1 - (1+i)^{-1}]$

This gives:

$$S_n[1 - (1+i)^{-1}] = R[(1+i)^{n-1} - (1+i^{-1})]$$

Now re-arrange in the form $Sn = R$ [Factor]:

$$S_n = R[((1+i)^n - 1)/i] \qquad (23.3)$$

$$$$$$$$$$$

Example

Returning to lottery winnings as an example, suppose the interest rate is 5%. Then after 30 years, a winner who saves the money will have:

$$S_{30} = 33{,}333.33\ [((1+0.05)^{30} - 1)/0.05]$$

$$S_{30} = 33{,}333.33\ [66.439] = \$2{,}214{,}628.$$

Of course, you may be too old to enjoy it!*

$$$$$$$$$$$$

* In this and all lottery examples, we do not consider taxes.

It is important to note that we have not yet computed the present value of those lottery winnings.

The factor $[((1+i)^n - 1)/i]$ in Equation 23.3 is called the Uniform Series Compound Amount Factor (USCAF). It is listed in the fifth column of Figure 23.1. If we use the Table (Figure 23.1) to solve the above lottery problem, we find the value of USCAF for 5% and 30 years is 66.439, which is the same as we computed.

23.5 Sinking Fund

The reciprocal of the USCAF is called the Sinking Fund Deposit Factor (SFDF). It is used to answer the question: What annual payment R is required to accumulate to an amount S in n years at interest i? That is:

$$R = S_n\ [i/((1+i)^n - 1)] \qquad (23.4)$$

$$R = S_n\ (SFDF)$$

The Sinking Fund Factor is found in the third column of Figure 23.1.

$$$$$$$$$$$

Example

Suppose you want to save for your child's first year of college which will cost $10,000 in 18 years. Interest rates are 5%. What must your annual savings be?

Solution: $R = Sn\ [i/((1+i)^n - 1)]$

$$R = S_n\ (SFDF) = 10{,}000\ (0.0355462) = \$355.46$$

That doesn't seem so bad, but you had best get started. And don't forget the remaining years of college and the other children!

$$$$$$$$$$$

23.6 Present Value of an Annuity

The present value of an annuity is an important issue. It can tell us, for example, how much cash those generous lottery folks must *really* spend *now* to pay off the "million dollar" winner who gets R = $33,333.33 per year over 30 years.

Consider the nth year's payment of R dollars. Its present value is

$$P_n = R / (1+i)^n$$

Then the present value of an entire n year annuity is

$P_n =$

$$R/(1+i) + R/(1+i)^2 + R/(1+i)^3 + \cdots + R/(1+i)^n$$

To simplify, again factor out R and multiply both sides by $[1 - (1+i)^{-1}]$ to get (after a bit of juggling)

$$P_n = R \cdot \left[\frac{(1+i)^n - 1}{i \cdot (1+i)^n}\right] \qquad (23.5)$$

The factor in brackets is called the Uniform Series Present Worth Factor (USPWF), and values for a 5% interest rate are found in the last column of Figure 23.1.

$$$$$$$$$$$

Example

As an example, we go back to the lottery — where they are *very* familiar with the USPWF. What is the present value of that $33,333.33 paid to you over 30 years with an interest rate of 5%?

$$P_n = R \cdot \left[\frac{(1+i)^n - 1}{i \cdot (1+i)^n}\right]$$

$$P_{30} = 33,333.33 \cdot \left[\frac{1.05^{30} - 1}{0.05 \cdot 1.05^{30}}\right]$$

$$P_{30} = \$\ 512,415.$$

$$$$$$$$$$$

The example above shows that the lottery needs only come up with about $500,000 now to pay off that so-called million dollar ticket. And that's at 5%. If the average interest they can earn over 30 years is, say, 8% (fairly easy and safe with government bonds, though it depends on interest rates), the present value of that million dollars drops to about $375,000. Thus, lotteries must pay immediately a *much* smaller amount than the million they advertise that winners receive.

There are other applications of the USPWF. Consider the following example.

$$$$$$$$$$$

Example

Suppose you can make a design change that reduces assembly costs by two cents per product produced. 1,000,000 are produced per year, and the product is expected to still have a ten year life. Take interest rates to be 7%. What is the present value of your design change?

Solution: If we think the way the lottery folks would like us to think, we might conclude that the design change is worth $20,000 per year for 10 years, or $200,000. But consider this: How much would you be willing to spend *now* to get that $200,000 return spread out over ten years? We shall see that the present worth of that sum is quite a bit less than $200,000. We can compute it as follows:

$$P_n = R \cdot \left[\frac{(1+i)^n - 1}{i \cdot (1+i)^n}\right]$$

$$P_{10} = 20,000 \cdot \left[\frac{1.07^{10} - 1}{0.07 \cdot 1.07^{10}}\right]$$

$$P_{30} = \$\ 140,470.$$

Thus to break even on this project over the ten year period, you could invest about $140,000 in it now. More than that, and it loses money. But of course, breaking even isn't nearly good enough for most business projects.

$$$$$$$$$$$

Suppose it will cost $35,000 to do the redesign and retooling for the proposed change in the example above. As we mentioned briefly in Chapter 2, most firms will treat these costs of a redesign as an expense. But in reality, that change was an *investment* in a design change that had a handsome return ($140,000 - $35,000 = $105,000).

It should be noted that the assumption made in using USPWF is that the money received each year is reinvested at the interest rate i. This is a reasonable assumption, since something useful will normally be done with the money, even if (in a manufacturing firm) it is not left in the bank.

23.7 Capital Recovery Factor

The Capital Recovery Factor (CRF) is the reciprocal of the USPWF. The CRF is used to answer the question: What annuity R for n years at interest i is required to give a specified present value P?

$$R_n = P \cdot \left[\frac{i \cdot (1+i)^n}{(1+i)^n - 1}\right] \qquad (23.6)$$

$$$$$$$$$$$

Example

Suppose we wish to compute the annual payment required to purchase a house with a present value of $150,000 when the payment period is to be 20 years and the interest rate is 6%. Then

$$R_{20} = 150,000 \cdot \left[\frac{0.06 \cdot 1.06^{20}}{1.06^{20} - 1}\right] = 13,077.$$

We can compute the monthly payment required by noting that the monthly interest rate that is equivalent to an annual rate of 6% is:

$$i\,(\text{monthly}) = 1.06^{1/12} - 1 = 1.004867549 - 1$$
$$= 0.4867549\%$$

That is: $1.004867549^{12} = 1.06$

Therefore, on a monthly basis, the required payment will be:

$R_{240} = 150{,}000\ (USPWF)$

where USPWF =

$$\frac{0.00486755 \cdot 1.0048675^{240}}{1.0048675^{240} - 1} = 0.007073$$

Thus: R240 = $1060.95

$$$$$$$$$$$

23.8 Project Life, Depreciation, Taxes, Maintenance, and Interest Rates

In estimating the economic value of a project, or in comparing competing projects, a number of factors have to be considered including: the expected life the project, depreciation of equipment, taxes on earnings, maintenance costs, and what value to use for interest rate. In this section, we discuss each of these issues briefly.

Project Life. Products have a life expectancy, some short, some long. (Increasing competitiveness generally means shorter life cycles for products.) In developing a product, or in preparing a redesign, there will usually be some estimate over time of the sales and prices that the new or revised product will produce in the marketplace. These data are essential to any analysis of the economic value of the design or design change. The absence of such information contributes to the custom of treating design and redesign (of products and manufacturing facilities) as expenses instead of investments. We will assume in this book that there are marketing studies to support estimates of income from sales of proposed products or product changes.

Equipment. Equipment also has a finite life, and there are several ways of computing it for economic evaluation of projects. For example, the Internal Revenue Service has guidelines for tax purposes. In the past, most equipment has had a 12 year life by these guidelines, though this is subject to change, and there can be shorter or longer periods for certain equipment or in special circumstances. To encourage equipment purchases, the government will (on occasion) allow a shorter tax life on equipment, or provide an "investment tax credit".

Of course, if a piece of manufacturing equipment has a tax life of 12 years, but the product has a market life of only six years, what will become of the equipment after the six years has to be considered. Maybe it will junked, and have some scrap value. Or maybe it can be used in some other project.

Depreciation. Whatever the life of a piece of equipment, it can be depreciated in several ways, again often "guided" by the IRS or by previous company practices. The straight line method is the most common: equipment worth, say, $120,000 that is to have a life of 12 years is assumed to depreciate in value at the rate of $10,000 per year. Other methods may allow for more depreciation early in the life of the project. Whatever method is used for tax purposes should be used for project evaluation also. We will use straight line depreciation in our examples.

Maintenance. Since equipment requires maintenance, this cost must also be included. The same is true if the product itself is going to require maintenance or service by the manufacturer over the years.

Taxes. Company accountants will, of course, know your company's tax rate. It obviously varies with changes in the tax code. We will use a rate of 50% on net earnings, though this is clearly subject to change as tax rates and laws change with economic and political conditions.

Discount Rate. The discussions above have been done in terms of "interest rate" as earned in a bank account or on a loan (e.g., from a bond). But companies seldom put their money in the bank the way individuals may do. What we need for a meaningful value of i is a rate that reflects what the company believes it can earn on available money. We call this rate the *discount rate*, and this term will be used in the rest of the chapter to refer to the rate at which an amount of money might grow if invested.

The term "discount rate" may seem odd, but it makes some sense when one considers that a future amount of money must be *discounted* (by $[1 / (1 + i)^n]$) as described above to find its future value.

The discount rate might be chosen as:

- the current rate being paid on borrowed money (since the company could use the money to pay off a loan), or the rate at which money earns interest if deposited in bank accounts; or

- the current rate of return on a company's external investment, if any, in stocks or bonds (since the company could buy more of these instead of investing in the proposed project); or

- the current company internal rate of return on other projects.

There is not really any way that one of these choices can be considered the most "correct" or appropriate. Companies are free to choose any one they want, or some other one. It can, however, make a tremendous difference to the fortunes of a company what method is used. Methods that use very high discount rates tend to discourage investment in new projects. We shall see soon, however, that there is a method of evaluating projects (called the Discounted Cash Flow Method) that does not require the assumption of an arbitrary discount rate.

23.9 Methods of Evaluating and Comparing Projects on a Economic Basis

There are a number of methods available and in use for evaluating and comparing competing uses of capital investment funds. The ones to be considered here are:

a) Payback Period
   i) Simple Payback (ignores discount rate)
   ii) Present Value Payback
b) Accounting Return on Investment (ROI)
   i) Simple Accounting ROI (ignores discount rate)
   ii) Present Worth Accounting ROI
c) Discounted Cash Flow ROI.

The simple payback period method provides questionable results, but has the advantage of easy of computation. However, with computers available, ease of computation should not by a big issue. Getting the right input information and using a meaningful method are far more important than computation time. Generally, payback methods ignore issues like taxes, depreciation, and even project life. Thus payback periods are recommended only for quick and dirty, first-cut evaluations and comparisons.

A variation on the Payback Period — called Breakeven Date — is being used by some firms. It bases evaluations and comparisons on the expected date that a project will breakeven; that is, when the profits received will equal the money invested. This method places emphasis on shortening the time to market because the quicker the time to market, the shorter will be the time to breakeven. There are a number of ways of defining and computing the "breakeven date", but they are highly company-specific and cannot be discussed here. Discounting future earnings in some way is essential, however.

When methods require the assumption of a "discount rate", the results are obviously somewhat sensitive to the choice made. The Discounted Cash Flow (DCF) method, however, requires no such assumption, and is therefore (we believe) the preferred method.

23.10 Payback Period

Simple Payback. Payback period is the time required for the expected net profits to equal the initial money invested (I). That is, if we let E_m represent the net profits for the project in the mth year, then the Payback Period is the smallest value of m for which

$$\sum_m E_m \geq I \qquad (23.7)$$

A more exact estimate of payback can be computed if months are used for m instead of years.

We can use a simple cash flow analysis to illustrate finding the payback period. Suppose that a project requires a $400,000 initial investment, and the projected net profits after taxes, depreciation, and maintenance are:

1st Year	$100,000.
2nd Year	$150,000.
3rd Year	$150,000.
4th Year	$100,000.

In addition, the equipment involved is expected to have a scrap value of $20,000. In Figure 23.2, a cash flow analysis of this project is shown. In the figure, numbers in parentheses are negative. The "Net Cash Flow" column lists the cash flow for the corresponding year. The "Cumulative Cash Flow" column lists the cash flow for the entire project up to and including the corresponding year.

Year	Investment and Cash Expenses	Net Profits After Taxes, et al.	Net Cash Flow	Cumulative Cash Flow
0	400,000	—	(400,000)	(400,000)
1	—	100,000	100,000	(300,000)
2	—	150,000	150,000	(150,000)
3	—	150,000	150,000	0
4	—	120,000	120,000	120,000
5	—			

Figure 23.2 Computation of a Simple Payback Period

Year	Investment and Expenses	Net Profit	Net Cash Flow	Present Value of Net Cash Flow	Present Value of Cumulative Cash Flow
0	400,000	—	(400,000)	(400,000)	(400,000)
1	—	100,000	100,000	93,458	(306,542)
2	—	150,000	150,000	131,016	(175,526)
3	—	150,000	150,000	122,445	(53,081)
4	—	120,000	120,000	91,547	38,466

Figure 23.3 Calculation of a Present Value Payback Period

Note that this project has a payback period of three years, since after three years the cumulative cash flow becomes zero and the project has "paid back" its initial investment and expenses. The breakeven date is also three years according to this calculation.

Present Value Payback. The usefulness of payback periods in evaluating and comparing projects is enhanced when the present value (or the discounted value) of the predicted net profits is used instead of the amount of the money. If we let PVE_m be the present value of the mth year's (or month's) net profits, then the payback period is the smallest value of m for which

$$\sum_{m} PVE_m \geq I$$

In Figure 23.3 the previous example is re-done using present value. A discount rate of 7% is assumed. Note that the payback period is now more than about 3 1/2 years.

23.11 Accounting Methods for Computing Return on Investment (ROI)

Just as there are different definitions of payback period, there are also different methods of computing a "return on investment" (ROI). All ROI methods involve computing a ratio of some measure of project profits to some measure of the required investment. In this section, we consider two of these methods: the Simple Accounting Method and the Present Worth Accounting Method. There are many other ways that different companies compute ROI.

Simple Accounting Method. The definition of ROI in this method is:

$$\text{SAROI} = \frac{\text{Ave Ann Profits After Tax and Dep}}{\text{Ave Ann Investment}}$$

In using this method, discount rate is not considered. Thus in the example above, the average annual profits are computed as:

Average Annual Profits =

= (1 / 4) ($100,000 + 150,000 + 150,000 + 120,000)

= $130,000.

The investment amount is computed as the average annual capital outlay over the life of the project, but there are several, essentially arbitrary, ways to define "average". One way is to use the total investment made divided by the life of the project. Another way is to assume that the initial investment is recovered at a linear rate throughout the project; thus the average *outstanding* investment is one-half the initial amount. This is clearly a crude assumption, but if used for all projects being compared, it is at least somewhat reasonable. Using this method for our example, we get:

Ave. Ann Investment = (400,000 / 2) = $200,000

With this assumption, then, the Simple Accounting ROI is:

$$\text{SAROI} = 130{,}000 / 200{,}000 = 0.65 \text{ or } 65\%$$

Present Value Accounting ROI. The ROI in this method is denoted as PVROI and is defined as:

$$\text{PVROI} = \frac{\text{Pres Val of Ave Ann Profit After Tax and Dep}}{\text{Ave Ann Investment}}$$

In the above example, this numerator is:

= (1 / 4) (93,458 + 131,016 + 122,445 + 91,547
= $109,616.

Using $200,000 again as the average annual investment results in a PVROI of 54.8%

Using the present value of the profits makes more sense, especially for longer life projects, than ignoring the effects of discount rate. However, neither of these so-called "accounting" methods provides more than a very rough idea of the relative economic value of competing projects. Thus we turn attention to the discounted cash flow method.

23.12 The Discounted Cash Flow (DCF) Method

Both payback methods and accounting methods suffer from the disadvantage that they either ignore the effects of discount rate, or else they assume an arbitrary rate. The discounted cash flow (DCF) method, however, neither ignores the discount rate nor assumes an arbitrary value for it. It also allows a valid comparison of projects with different lifetimes.

Instead of assuming a discount rate, the discounted cash flow method computes the discount rate that makes the net discounted cash flow just equal to zero at the end of project's life. We call the discount rate computed in this way the DCF-ROI. For comparing projects, the one with the largest DCF-ROI is considered the best investment.

The DCF-ROI method is no doubt the best way to compare competing projects, but it also requires more computation. It is a trial and error process. As an example, suppose a project with a life of eight years requires an initial investment of $20,000, and has an expected net income after taxes and depreciation as shown in column 1 in Figure 23.4.

To compute the DCF-ROI we must find the discount rate that makes the net cash flow equal to zero after eight years. We make a first trial guess of a 9% discount rate. The discounted values of the net income are shown in column 2, and the resulting net cash flow values are shown in the column 3. Note that after eight years with this discount rate, the net discounted cash flow is + $9201 — not at all close to zero. Thus we need to try another discount rate.

Since the final cash flow with a 9% discount rate is positive, to get it reduced we need to assume a higher discount rate. We choose 15%, and these results are shown in columns 4 and 5 in Figure 23.4.

Now the final discounted cash flow is + $3001, better but still not close enough to zero. Since the discounted cash flow is still positive, we need an still higher discount rate, so we try 20%. As shown in columns 6 and 7, the final cash flow is now negative, so we try 19% — and so on until the result is as close to zero as we need. A discount rate of 18.85% in this example gives a final cash flow of - $4. Thus we say that the DCF ROI for this project is 18.85%.

Though the computations can be time consuming by hand calculator, it is easy enough to develop an interactive computer program to receive the input information and compute the discounted cash flow for a project. Naturally, this is done when the method is to be used repeatedly.

23.13 More About Taxes and Depreciation

As an example of how taxes and depreciation can be incorporated into a project's evaluation, see Figure 23.5. We assume here a project that has an initial investment in equipment of $20,000. The expected net income before taxes and depreciation is shown in column 1. The life of the equipment is eight years (the same as the life of the project in this case), and it is depreciated in straight line fashion (i.e., in equal amounts each year) over the eight years.

The annual depreciation is $20,000/8 = $2500 so that after eight years, the value of the equipment is zero. For tax purposes, this $2500 is treated as an expense that can be deducted before taxes are computed. Thus the taxable income is as shown in column 3. With a 50% tax rate, the amount of the tax (which is a negative cash flow) is shown in column 4. Then the net income after depreciation and taxes are considered is column 1 minus column 4 as shown in column 5. (It can be noted that the values in column 5 of Figure 23.5 are the same as those in column 1 of Figure 23.4.)

	1	2	3	4	5	6	7	8	9
Year	Net Income after Taxes and Depreciation	9% Discounted Net Income	9% Net Discounted Cash Flow	15% Discounted Net Income	15% Present Value Discounted Cash Flow	20% Discounted Net Income	20% Net Discounted Cash Flow	19% Discounted Net Income	19% Net Discounted Cash Flow
0	(20,000.)	—	(20,000)	—	(20,000)	—	(20,000)	—	(20,000)
1	3,750.	3,440.	(16,560)	3,261.	(16,739)	3,125.	(16,875)	3,151.	(16,849)
2	4,125.	3,472.	(13,088)	3,119.	(13,620)	2,865.	(14,010)	2,913.	(13,936)
3	4,557.	3,519.	(9,569)	2,996.	(10,624)	2,637.	(11,373)	2,704.	(11,232)
4	5,052.	3,560.	(6,009)	2,888.	(7,736)	2,436.	(8,937)	2,519.	(8,713)
5	5,623.	3,655.	(2,354)	2,796.	(4,940)	2,260.	(6,677)	2,356.	(6,357)
6	6,278.	3,743.	1,389.	2,714.	(2,226)	2,102.	(4,575)	2,211.	(4,146)
7	7,033.	3,847.	5,236.	2,644.	418.	1,963.	(2,612)	2,081.	(2,065)
8	7,900.	3,965.	9,201.	2,583.	3,001.	1,837.	(775)	1,964.	(101)

Figure 23.4 A Discounted Cash Flow Computation

	1	2	3	4	5
Year	Net Income Before Taxes and Depreciation	Depreciation	Taxable Income	50% Tax	Net Income After Taxes and Depreciation
0	(20,000.)	—	—	—	(20,000)
1	5,000.	(2,500)	2,500.	(1,250.)	3,750.
2	5,750.	(2,500)	3,250.	(1,625.)	4,125.
3	6,613.	(2,500)	4,113.	(2,056.)	4,557.
4	7,604.	(2,500)	5,104.	(2,552.)	5,052.
5	8,745.	(2,500)	6,245.	(3,122.)	5,623.
6	10,057.	(2,500)	7,557.	(3,779.)	6,278.
7	11,565.	(2,500)	9,065.	(4,532.)	7,033.
8	13,300.	(2,500)	10,800.	(5,400.)	7,900.
					44,318.

Figure 23.5 A Computation Including Taxes and Depreciation

23.14 A Comparative Example

We conclude this chapter with an analysis of two projects using different methods of comparison. See the table below ($ are in thousands). Project A has an initial investment in design and manufacturing facilities of $190,000. Increased productivity as a result of the new design and equipment is expected to produce added revenues of $20,000 the first year, $30,000 the second year, and $40,000 annually thereafter for the ten year life of the project. Maintenance costs are $1000 per year.

Project B has an initial investment of $160,000, and a life of 15 years. Maintenance is expected to be $1000 per year except for major repairs costing $5000 in the fifth and tenth years. Annual earnings are assumed to be $30,000 over the life of the project.

	Project A		Project B	
Yr	Invest-ment	Net Income Before Tax and Deprec	Invest-ment	Net Income Before Tax and Deprec
0	(190)		(160)	
1		19		29
2		29		29
3		39		29
4		39		29
5		39		25
6		39		29
7		39		29
8		39		29
9		39		29
10		39		25
11				29
12				29
13				29
14				29
15				29

For both projects, we will take the discount rate to be 10%, and depreciation is straight line over ten years. The tax rate is 50%.

The investment and net income before taxes and depreciation are as shown in the Table to the left. We have reduced the amount of net income to account for the maintenance costs. Numbers are in thousands.

23.14.1 Comparison by Payback Methods

No Discount Rate. Figures 23.6 and 23.7 show the computation (numbers are thousands in the Figures) of the simple payback period neglecting the effects of depreciation, taxes, and discount rate. For both projects, the simple payback period is about 5.6 years. Thus the projects appear equal on this basis.

With 10% Discount Rate. Figures 23.8 and 23.9 show the computation for the payback periods when the discount rate is included. In this case, Project A has a payback of 8.5 years while Project B has a payback of about 5.7 years. Note how significantly the inclusive of discounting influences the comparative results. This suggests that Project B is the better project.

Year	Cash Outlay	Net Income Before Taxes	Cumulative Cash Flow
0	(190)		(190)
1		19	(171)
2		29	(142)
3		39	(103)
4		39	(64)
5		39	(25)
6		39	14

Figure 23.6 Simple Payback Analysis for Project A. PPs is about 5.6 years. Numbers are in thousands.

23.14.2 Comparison by Accounting Methods

Figures 23.10 and 23.11 show the computations for both the no discount case (see column 5) and the 10% discount rate case (see column 6). Note that Project A has the better ROI using either of these methods.

Year	Cash Outlay	Net Income Before Taxes	Cumulative Cash Flow
0	(160)		(160)
1		29	(131)
2		29	(102)
3		29	(73)
4		29	(44)
5		25	(19)
6		29	10

Figure 23.7 Simple Payback Analysis for Project B. PPs is also about 5.6 years.

23.14.3 Comparison by Discounted Cash Flow ROI

Figure 23.12 shows the Discounted Cash Flow computation for Project A. The first trial value for the discount rate was 10%, resulting in a negative cash flow after ten years of $25,031. Clearly a lower rate is needed to get the cash flow to be zero after ten years. Thus 8% was tried, and then 7%. Note that at 7%, the final cash flow is down to a negative $34, which is very close to zero. Thus we conclude that the DCF-ROI for Project A is 7.0%.

Figure 23.13 shows the Discounted Cash Flow computation for Project B. The result is a DCF-ROI for Project B of 9.8%. Thus, using the Discounted Cash Flow method of comparison, Project B is the better project.

23.14.4 Summary of Example Results for Projects A and B

Figure 23.14 shows a summary of the results for the various methods used above to compare the economic value of Projects A and B. They are equal based on simple payback, but Project B is the better when present value payback is considered. The accounting methods, however, indicate that Project A is preferable. Using Discounted Cash Flow, Project B is the better.

This example has, of course, been intentionally "cooked up" to show how the different methods of comparing projects for their economic value can lead to different conclusions. If these or other economic comparison methods are used to support decisions about whether or not to implement a particular project, or to decide which of several competing projects to implement, it is extremely important which method is used. Firms that require an extremely high ROI or very short payback may fail to invest in projects that enhance longer term market share or competitiveness.

Year	Cash Outlay	Net Income Before Taxes	Discounted Cash Flow	Cumulative Discounted Cash Flow
0	(190,000)	(190,000)	(190,000)	(190,000)
1		19,000	17,272	(172,728)
2		29,000	23,967	(148,761)
3		39,000	29,301	(119,460)
4		39,000	26,637	(92,823)
5		39,000	24,014	(68,607)
6		39,000	22,014	(46,593)
7		39,000	20,013	(26,580)
8		39,000	18,194	(8386)
9		39,000	16,540	8,154

Figure 23.8 Simple Payback With 10% Discount Rate For Project A.

Year	Cash Outlay	Net Income Before Taxes	Discounted Cash Flow	Cumulative Discounted Cash Flow
0	(160,000)	(160,000)	(160,000)	(160,000)
1		29,000	26,364	(133,636)
2		29,000	23,967	(104,669)
3		29,000	21,788	(87,881)
4		29,000	19,807	(36,814)
5		25,000	18,007	(11,953)
6		29,000	16,370	4,417

Figure 23.9 Simple Payback With 10% Discount Rate for Project B

Year	1 Net Income Before Taxes and Depreciation	2 Depreciation	3 Taxable Income	4 Tax	5 Income After Taxes Not Discounted	6 10% Discounted Income After Taxes
0	(190,000)					
1	19,000	19,000	0	0	19,000	17,272
2	29,000	19,000	10,000	5,000	24,000	19,835
3	39,000	19,000	20,000	10,000	29,000	21,788
4	39,000	19,000	20,000	10,000	29,000	19,807
5	39,000	19,000	20,000	10,000	29,000	18,007
6	39,000	19,000	20,000	10,000	29,000	16,370
7	39,000	19,000	20,000	10,000	29,000	14,881
8	39,000	19,000	20,000	10,000	29,000	13,529
9	39,000	19,000	20,000	10,000	29,000	12,299
10	39,000	19,000	20,000	10,000	29,000	11,181

$$(ROI)_{SA} = \frac{275,000/10}{190,000/2} = 0.289 = 28.9\%$$

$$(ROI)_{PWA} = \frac{164,969/10}{190,000/2} = 0.174 = 17.4\%$$

Figure 23.10 Accounting Method ROI's For Project A

Year	1 Net Income Before Taxes and Depreciation	2 Depreciation	3 Taxable Income	4 Tax	5 Income After Taxes Not Discounted	6 10% Discounted Income After Taxes
0	(160,000)					
1	29,000	16,000	13,000	6,500	22,500	20,455
2	29,000	16,000	13,000	6,500	22,500	18,595
3	29,000	16,000	13,000	6,500	22,500	16,905
4	29,000	16,000	13,000	6,500	22,500	15,368
5	25,000	16,000	9,000	4,500	21,500	13,350
6	29,000	16,000	13,000	6,500	22,500	12,701
7	29,000	16,000	13,000	6,500	22,500	11,546
8	29,000	16,000	13,000	6,500	22,500	10,496
9	29,000	16,000	13,000	6,500	22,500	9,542
10	25,000	16,000	9,000	4,500	21,500	8,289
11	29,000		29,000	14,500	14,500	5,082
12	29,000		29,000	14,500	14,500	4,620
13	29,000		29,000	14,500	14,500	4,200
14	29,000		29,000	14,500	14,500	3,818
15	29,000		29,000	14,500	14,500	3,471

$$(\mathrm{ROI})_{SA} = \frac{279,500/15}{160,000/2} = 0.246 = 24.6\%$$

$$(\mathrm{ROI})_{PWA} = \frac{(158,438)/15}{160,000/2} = 0.132 = 13.2\%$$

Figure 23.11 Accounting Method ROI's For Project B

		Discounted Cash Flow		
Year	Net Income After Taxes	At 10% Discount Rate	At 8% Discount Rate	At 7& Discount Rate
0	(190,000)	(190,000)	(190,000)	(190,000)
1	19,000	(172,728)	(172,471)	(172,243)
2	24,000	(152,893)	(151,895)	(151,281)
3	29,000	(131,105)	(128,874)	(127,609)
4	29,000	(111,298)	(107,559)	(105,486)
5	29,000	(93,291)	(87,823)	(84,810)
6	29,000	(76,921)	(69,549)	(65,487)
7	29,000	(62,040)	(52,628)	(47,428*
8	29,000	(48,511)	(36,961)	(30,550)
9	29,000	(36,212)	(22,454)	(14,776)
10	29,000	(25,031)	(9,022)	(34)

Figure 23.12 Discounted Cash Flow Analysis for Project A

23.15 Summary and Preview

> "...in a free enterprise system there can be no prosperity without profit. We want a growing economy, and there can be no growth without investment that is inspired and financed by profit."
>
> John F. Kennedy

Summary. Design engineers must appreciate that business firms have limited available capital and competing needs for that capital, and that the survival and well being of the firm and the employees requires profitability. Thus, decisions must be made with financial considerations carefully considered. This chapter is intended to provide engineers with some of the basic tools and concepts of engineering economics. There is much more to know about economic analysis related to design and engineering work in business firms. Students are thus strongly urges to take a complete course in "engineering economics", or at least to study the book carefully.

There are several unresolved research issues in accounting as it relates to design. As noted earlier in the book, the development of a design is generally treated as an expense, rather than more appropriately as an investment. One difficulty in treating design as an investment is the difficulty in assigning a value to a design. Evan a patented design is difficult to evaluate. With no really good way to assign a value, what happens is that a value of zero is assigned. This is an issue that needs the attention of design and accounting researchers. Another difficulty is in deciding how to depreciate the value of a design over time. Hopefully, these issues will be resolved in the years to come for it is unlikely that management will fully understand the true role of design in product realization so long as it is not properly accounted for in economic terms.

Preview. The next chapter is the final one in the book, but the issues treated are often extremely important. Product liability can never be ignored in a product realization process. Protecting the commercial value of new and unique designs through patents and trademarks is generally the best — possibly only — way to maximize the profits that can result. And engineers, especially design engineers, are continually faced with ethical questions in their professional lives. Each of these topics is discussed in Chapter 24.

		Discounted Cash Flow		
Year	Net Income After Taxes	At 10% Discount Rate	At 9% Discount Rate	At 9.8% Discount Rate
0	(160,000)	(160,000)	(160,000)	(160,000)
1	22,500	(139,546)	(139,358)	(139,508)
2	22,500	(120,951)	(120,420)	(120,845)
3	22,500	(104,046)	(103,046)	(103,848)
4	22,500	(88,678)	(87,106)	(88,368)
5	21,500	(75,328)	(73,132)	(74,896)
6	22,500	(62,627)	(59,716)	(62,056)
7	22,500	(51,081)	(47,408)	(50,362)
8	22,500	(40,585)	(36,116)	(39,712)
9	22,500	(31,043)	(25,750)	(30,012)
10	21,500	(22,754)	(16,674)	(21,570)
11	14,500	(17,672)	(11,055)	(16,386)
12	14,500	(13,052)	(5,900)	(11,664)
13	14,500	(8,852)	(1,170)	(7,363)
14	14,500	(5,034)	3,169	(3,446)
15	14,500	(1,563)	7,150	121

Figure 23.13 Discounted Cash Flow Analysis for Project B

	Project A	Project B
Simple Payback	5.6 years	5.6 years
Present Value Payback	8.5 years	5.7 years*
Simple Accounting ROI	28.9%*	24.6%
Present Worth Accounting ROI	17.4%*	13.2%
Discounted Cash Flow ROI	7%	9.8%*

Figure 23.14

Summary Results Comparing Projects A and B

* Best according to the method used.

Problems

23.1. What is the initial present value of Project A? Of Project B? Would this comparison be a fairly reasonable way to compare projects?

23.2 In Chapter 3, Problem 3.17, you were asked to suggest ways in which the caster assembly shown in Figure P3.17 could be redesigned in order to reduce assembly costs. In addition, you were asked to estimate the approximate savings in assembly time, hence cost, achieved by your redesign.

If the annual production volume for the casters is 500,000, the annual cost for assembly operators is $20,000, and the plant efficiency is 75 percent, determine the maximum amount of money that can be spent now to achieve your redesign suggestions. Assume a product life of 10 years, an annual interest rate of 6 percent, and a single-station manual assembly operation.

Note: The total manual assembly cost for single-station manual assembly, C_{pr}, is given by the following equation:

$$C_{pr} = (W + M)t_{ma}$$

where W is the assembly operator cost in cents/second, M is a rate, in cents/second, to be allocated to each assembly for use of the tooling, and t_{ma} is the manual assembly time in seconds. If we assume an 8 hour day and a 50 week year then the total number of seconds available per year is $7.2x10^6$. If the plant efficiency is 75% (i.e., an assembly operator performs assembly tasks only 75 percent of this time), then the number of seconds available per year becomes $5.4x10^6$. Thus, if the total assembly operator cost is $20,000 per year, then:

$$W = (20{,}000/5.4x10^6) = 3.70\ \$/sec = 0.37\ cents/sec$$

A reasonable value for M is 0.07 cents/sec. Thus, the manual assembly cost per assembly becomes

$$C_{pr} = 0.44t_{ma} \quad \text{cents/assembly}$$

23.3 In Chapter 3, Problem 3.18, you were asked to suggest ways in which the portion of a floppy disk drive assembly shown in Figure P3.18 could be redesigned in order to reduce assembly costs. In addition, you were asked to estimate the approximate savings in assembly time, hence cost, achieved by your redesign.

If the annual production volume for the disk drive is 100,000, the annual cost for assembly operators is $20,000, and the plant efficiency is 75 percent, determine the maximum amount of money that can be spent now to achieve your redesign suggestions. Assume a product life of 5 years, an annual interest rate of 7 percent, and a single-station manual assembly operation.

Note: See the Note provided in Problem 23.2 above.

23.4 You won a lottery that will pay you $10 million dollars over thirty years. They will pay this in thirty annual payments of $250,000 and a final payment in the last year of $2.5 million. What is the present value of your winnings if you take the discount rate to be 5%?

23.5 The design of a new product is projected to earn your company (after expenses, depreciation, and taxes) $200,000 per year for ten years. What is the present value of that design?

23.6 Suppose you want to save money to start your own business ten years from now, and estimate that you will need $50,000 for start-up money. If the available interest rate for your savings is 5%, how much must you save per year?

23.7 You want to purchase a building for your new business. The building cost is $300,000, and you can make a $50,000 down payment. Find the monthly mortgage payments for the following conditions:

a. 15 year loan, 5% interest,

b. 15 year loan, 8% interest,

c. 30 year loan, 5% interest,

d. 30 year loan, 8% interest.

23.8 Compare the following two projects on the basis of

a. Simple Payback,

b. Payback with a 5% discount rate,,

c. Accounting ROI with no discount

d. Accounting ROI with 10% discount rate,

e. Discounted Cash Flow.

Project C has an initial investment in manufacturing facilities of $400,000. Increased productivity as a result of the new design and equipment is expected to produce added revenues of $60,000 the first year, and $100,000 annually thereafter for the eight year life of the project. Maintenance costs are $3000 per year.

Project D has an initial investment of $250,000, and a life of 10 years. Maintenance is expected to be $1000 per year except for major repairs costing $4000 in the sixth year. Annual earnings are assumed to be $70,000 over the life of the project.

For both projects, depreciation of the facilities is straight line over five years. The tax rate is 50%.

Chapter Twenty-Four

Patents, Liability, and Ethics

"Give me the political economist, the sanitary reformer, the engineer; take your saints and virgins, relics and miracles. The spinning-jenny and the railroad, Cunard's liners and electric telegraph, are to me...signs that we are, on some points at least, in harmony with the universe."

Charles Kingsley
Yeast (1848)

24.1 Introduction

For the most part, engineering design is successful, and works in harmony with the universe. Any design not in harmony with the laws of nature surely won't work. And any design that does not serve humankind won't be tolerated for long. But there are also legal and ethical issues in the business and social affairs of people and organizations that do design and manufacturing. This chapter deals with three subjects that relate to the these issues. The subjects, like science and engineering principles, are important in very practical ways to design engineering practice.

The subjects are: patents, liability, and ethics. These topics might also be called: "protecting your design as intellectual property, protecting yourself and your firm from legal liability, and protecting your professional integrity and self esteem".

We deal first with protecting your design as intellectual property.

24.2 Protecting Your Design

24.2.1 Introduction

A primary source of information and views presented in this Section is an excellent booklet entitled *The Basic Tools of Design Protection* by Perry J. Saidman [1]. Readers interested in the subject of design protection are urged to obtain the original, more complete reference.

A design is intellectual property, and therefore may, under certain circumstances, be protected by law from illegal use by others. The types of designs that can be protected include products, packaging, logos, brand names, publications, and buildings or facilities. In this book, we are concerned only with products.

The design of a product can be protected through five legal mechanisms: contract, copyright, trade dress, utility patent, and design patent. (There may at some time in the future be an additional mechanism called *design registration.*) We will now discuss each of the five existing protection mechanisms in turn.

24.2.2 Contracts

A contract is a written or oral agreement between two parties. The parties can be individuals or corporations. Oral agreements, once made, are just as binding as written agreements, but are much more difficult to prove and enforce. This is not usually because one of the parties attempts to "cheat", but because there is so much more likelihood of misunderstanding in an unwritten agreement.

Engineering designers are often asked to sign contracts when they are first employed that say, in effect, every design idea the engineer gets while employed by that company belongs to the company, and therefore can't be used privately or taken along to a subsequent employer. This contract will probably include ideas worked on "after hours" or on the designer's "own time", and will generally have been written carefully by company attorneys. It is sometimes presented to new employees (and consultants) as a *fait accompli* without opportunity for negotiation: "Sign it now as it is, or forget the job." To avoid this frustrating situation, one must find out as a part of the interview process what contracts will be involved in employment. When being interviewed, it is perfectly okay to ask to see copies of any documents or contracts you will be required to sign if finally employed. Then you can study them ahead of time, and if your personal position is strong enough, negotiate more favorable terms before you take the job.

Sometimes these contracts will allow a new employee to list, and hence protect, any ideas previously conceived. It is well to have thought about this earlier, rather than when under pressure to sign immediately.

Another common type of contract is called a "non-disclosure" or "confidentiality agreement". It is often used when a designer or company wishes to reveal a product design or aspects of it to another individual or firm, and still be assured that the other party will not make, use, disclose, or copy the product without permission. An example of a typical contract of this type reprinted by permission from Saidman [1] is shown in Figure 24.1.

Readers will note that the language in the sample contract in Figure 24.1 is quite legalistic, and you are therefore encouraged to employ a competent lawyer whenever a contract is needed or desired. There are generally good reasons for the structure of the language used. The cost of having a contract prepared is modest (say $100 to $200) compared to the protection it affords. A non-disclosure contract will last until the subject product has been made public.

The major advantage to using a contract is that it is simple and inexpensive. Contracts are also flexible; they can say anything that the parties want them to say. Contracts are especially useful during the early stages of invention when marketability and financing are being explored.

There are also disadvantages to contracts. One is, as mentioned above, that non-disclosure contracts become useless once the information is somehow made public; that is, there is no long term protection. Perhaps more important, it is often difficult to define clearly and exactly what is being protected in a short letter form. The implications of ideas are sometimes not apparent until the ideas are explored. Issues can become fuzzy. But non-disclosure contracts are nevertheless a very good way to accomplish temporary protection of a new design idea.

24.2.3 Copyrights

A copyright has only limited usefulness in protecting product designs. It is primarily intended to protect writings, though the interpretation of "writings" can sometimes be extended to include product designs. There are, however, serious problems in making this interpretation. First, a copyright does not protect the idea but only the *expression* of the idea. Thus, a drawing of a product may be copyrighted, but not the novel *idea* used in the product. No-one else could legally use or copy the drawings, but they could use the same idea in a different version of the product. It is therefore difficult to prove that something was actually copied.

Second, in order for a design to be protected by a copyright, the ornamental features of the product must be separable from the functional features. This is very difficult to accomplish, and even more difficult to prove.

Third, even if the interpretation can be made successfully, there is little a copyright owner can do to someone who copies the design. Through the courts, probably at considerable expense, an order to "cease and desist" might be obtained. That could be a moral victory but would rarely have significant economic value.

The advantage to copyrighting is that it is extremely easy to do. Forms (which, believe it or not, are quite simple) are available from the Library of Congress. You need only send two copies of your design — or other "writing" — along with your form and $20 filing fee. In most writing cases, the copyright is then granted without questions. However, in the case of product designs, the application is more likely to be rejected. In these cases, you will need an attorney to help prepare a new application or a rebuttal.

A copyright, once granted, is in effect for your lifetime plus fifty years.

24.2.4 Trade Dress and Trademarks

Trademarks protect their owners from others using the same (or almost the same) names or symbols on the same or similar products.

Trade *dress* is a form of trademark that is becoming increasingly important in design protection. Most people are familiar with trademarks that are distinctive words (e.g., Chevrolet) or symbols (e.g., the Chevrolet logo). Trade *dress* consists of distinctive features or characteristics of a product such as colors, combinations of colors, size, shape, texture, weight, or configuration. Trademark and hence trade dress laws are intended to protect the public from confusion about the source of a product. They also protect their originators and owners from having competitors take business away through purchaser mis-identification of the product.

The law recognizes a spectrum of distinctiveness in trademarks. There are essentially four categories. From the strongest to the weakest, they are: arbitrary or fanciful; suggestive; descriptive; and generic.

Arbitrary or fanciful. A trademark that is not at all related to the product or its use is the strongest kind. A good example is the use of Apple™ and Macintosh™ (a kind of apple) for computers. Another is Peachtree™ for computer software; neither peaches nor trees have any connection with software. Concocted words like Kodak™ also qualify for this category.

An example of arbitrary or fanciful trade *dress* is the orange color of the roofs used by the Howard Johnson™ motel and restaurant chain. That color is neither suggestive or descriptive of the product or services offered, nor is the color itself functional in any way. Another example would be a wine bottle that is one inch high and 12 inches in diameter. (We are indebted to Mr. Perry Saidman for this example.)

YOUR LETTERHEAD

Date

Name of Person
Address of Person

Dear Person:

I wish to consider the possibility of retaining your services for the purpose of assisting in the development of a [ ] device which I have designed.

Naturally, in order to proceed, I will have to provide you with more detailed information in this area. I regard this information as proprietary, and to protect its confidential nature, I request your agreement to the following provisions:

1. For the purpose of this Agreement, "Proprietary Information" shall mean all information furnished to you by me concerning [ ] devices, except:

   a. information which was already known to you (as established by your records) before our initial discussion of the product, or

   b. information which is or becomes part of the public domain (other than from you), or

   c. information which is furnished to you by another who is lawfully in posssession of such information and who lawfully conveys same.

2. You agree you will not disclose such Proprietary Information to any third party for a period of [ ] from today's date, and will not use such Proprietary Information other than on my behalf. You may evaluate such Proprietary Information. Upon the conclusion of your evaluation, or upon my request, you agree to return to me all written matter and samples provided to you by me concerning such Proprietary Information.

3. You agree that you will take all reasonable precautions to safeguard any documents and samples which may be supplied to you or which you may develop or make relating to such Proprietary Information.

4. You agree to assign to me all inventions and designs which are based in any way upon such Proprietary Information and to assist in every way requested by me to obtain Letters Patent for said inventions and designs.

5. Apart from the foregoing, it is mutually understood that you have no obligation to enter into any further agreements with me relating to the instant subject matter, except as in your judgment shall seem advisable.

So that I may have a record of your agreement to the foregoing, please sign and return the enclosed duplicate of this letter to me.

Very truly yours,

Designer

ACCEPTED AND AGREED:

____

Signature of Person

Figure 24.1 Sample Confidentiality Letter of Agreement [1].

Suggestive. A trademark whose distinctiveness is classified as suggestive is the next strongest category. Such a trademark *suggests* something about the nature or qualities of the product, but falls short of being descriptive. An example is Frigidaire™ for the name of a refrigerator. Most automobile names are arbitrary or fanciful (e.g., Buick™, Ford™, Dodge™) but Oldsmobile™ is probably in the suggestive category because the "mobile" part is suggestive of the product's use.

It is very desirable to develop trademarks in one of the two categories above. The reason is that such trademarks, especially those that are arbitrary or fanciful, are deemed "inherently distinctive" and can be registered at any time. Protection begins, however, upon the first use of the product whether or not the trademark has been registered.

Trademarks in the two categories below — descriptive or generic — unlike those in the two categories above, must achieve what is called *secondary meaning* with the public before they provide protection and can be registered. Secondary meaning occurs when the public begins to identify a trademark with a specific product, service, or firm. For example, the orange color of the Howard Johnson™ roof has certainly achieved secondary meaning; so, clearly, have the golden arches of the McDonald's Corporation.

Descriptive. Descriptive trademarks describe in some way the goods or services that they name. For example, the name LampLyter™ is a descriptive trademark for a timer switch to control the on-off cycle of lamps and small appliances. ToastMaster™ is descriptive of toasters.

Generic. The weakest category of trademark is generic. Examples are aspirin, ice box, or The Computer Store™. These probably cannot be strongly protected as trademarks because they are so general.

Protection of the trademark or trade dress that has been designed into a product can happen in two ways: (1) the trademark or trade dress can be registered with the U.S. Patent and Trademark Office (PTO); or (2) by actual use of the trademark or trade dress in the marketplace. In the latter case, no formal registration is required; that is, if the "secondary meaning" develops naturally over time among the public, then the trademark or trade dress is presumed to be in effect. Of course, the vagueness of this condition sets the stage for possible litigation. If your trademark or trade dress is registered, it is more difficult for competitors to argue that they did not know about it.

If a trademark or trade dress is registered, the protection lasts for 20 years and can be re-newed for successive twenty year periods essentially forever so long as the product is in the marketplace. An attorney is necessary for filing a trademark or trade dress application; the fee is likely to be upwards of $1,000. The filing fee is $210.

If a product has been in the marketplace only a short time, or if the trademark or trade dress is not especially distinctive, then it is difficult to prove that "secondary meaning" has been established. Also, it is not easy to prove that some distinctive features (like color, size, shape) are in fact trade dress. If such features claimed as trade dress are "functional", even in part, and are not protected by a patent, then competitors are free to copy them, at least in theory. It adds to the confusion (and hence to the litigation) that different courts may still define and interpret "functional" and "secondary meaning" in slightly different ways.

However, for an individual or firm who believes their trademark or trade dress has been infringed, filing suit is often a reasonable action. In an era of consumer protection, juries are likely to be sympathetic to plaintiffs, especially individuals or small firms. All that must be proved is that confusion was created in the buying public about the source of the product.

Trying to avoid infringing on the trademark or trade dress of a competitor's products is not easy. There are so many trademarks registered that determining if yours will be unique is difficult and may be expensive. The problem is compounded by the fact that there is no central data bank of trademark and trade dress registrations. There are now, however, regional places (often in University libraries, where trademarks can be searched via compact disks.

Avoiding trade dress infringement in your designs is also not easy. You must remember that protection depends on prior use, not on registration. Indeed, most trademarks and trade dress are not registered. Thus to be certain that noone is using a particular trademark is a next to impossible task. Nevertheless, designers are well advised to be as thorough as possible in doing competitive benchmarking on trademark and trade dress issues as well as on functional and patentable features. The goal is to create products that not only avoid competitors' trademarks/trade dress and patents, but also that improve performance and marketability with distinctive designs that can be patented and registered in their own right.

24.2.5 Utility Patents

Utility patents protect the functional features of a product, but only the functional features. Ornamental features and the appearance of products are protected by design patents which are discussed in the next section.

An attorney is required to file for a utility (or a design) patent. The cost is likely to be high, not only to obtain the patent, but also to maintain and enforce it. Be prepared for total fees in the thousands of dollars.

The owner of a utility patent can prevent others from making, using, or selling the claimed invention. The patent protects the functional and structural elements explicitly claimed and allowed in the patent. That is, you must claim something in words — not just show it in a drawing — to have it protected. And the claimed features will be allowed *only* if they are *both* "novel" and "non-obvious" when *everything* else that has *ever* been designed or built before (patented or not) is considered. In the world of patents, that "everything else" is called the "prior art".

Preparing a utility patent application is an important task. It must be done according to the Patent Office rules, and done very thoroughly. Information and drawings —

especially drawings — sufficient for a reasonably knowledgeable person to make and use the product are required. The claims, of course, are especially important as explained above. Figure 24.2 shows a first page of an example utility patent.

It is always strongly advisable to have a utility patent search done as soon as there is an indication of a potentially patentable idea. There are firms that perform patent searches; patent attorneys will have business connections with several. It bears pointing out, however, that there is a huge difference between (a) a patent search to find out whether an idea is sufficiently novel and non-obvious to be patentable, and (b) a patent search to find out if an idea infringes on an existing patent. For one thing, the latter search is *much* more expensive. Just because you can get a patent on an idea does not mean that it will not be found later (when a product is produced and offered for sale) to infringe someone else's patent. Fortunately, this is generally not a problem for an inventor who receives a patent and then sells the rights to a firm that will undertake manufacture and sale. When the patent is sold, it then becomes the responsibility of the producing firm rather than the inventor to worry about infringement.

Almost always, an initial patent application is rejected by the patent examiner, who will assert that the claims are not sufficiently novel and non-obvious considering the "prior art". The examiner's assertion will be supported with a few examples of similar features in other products that have been patented. This places the burden of proof on the applicant to prove to the examiner that the features being claimed are indeed suitably novel and non-obvious. What follows then is a costly and time consuming period of negotiations during which the claims can be amended. The applicant's attorney tries to make the claims as broad as possible while the patent examiner tries to keep them as narrow as possible considering the prior art. This process of negotiation may takes several years and, as we said above, can be very expensive.

When a patent is issued, it lasts for 17 years, though it can be amended with improvements that help keep away predators. You can advertise that there is a "patent pending" as soon as an application is filed; however, you cannot take any legal action to protect your idea until the patent is actually issued. Even then, you may be in for surprises. There are two to three years of patent applications always in the Patent Office pipeline, you have no way of knowing if any of these will invalidate (since they preceded yours) one or more of your claims.

Despite the apparent difficulties, time and expense, a utility patent is the best and most effective way to protect a novel and non-obvious idea from infringement. This is true even for the individual inventor or small firm. Well done patents — emphasis on *well done* — discourage infringement even by wealthy competitors. Moreover,

United States Patent [19] [11] **3,790,773**
Sapper [45] **Feb. 5, 1974**

[54] LAMP WITH AN ARTICULATED SUPPORT
[76] Inventor: Richard Sapper, Loewenstrasse 96-7, Stuttgart, Germany
[22] Filed: Oct. 4, 1971
[21] Appl. No.: 188,709

[52] U.S. Cl. .... 240/81 BD, 240/69, 248/123, 248/280
[51] Int. Cl. F21v 21/26
[58] Field of Search...240/81 BD, 69; 248/123, 280, 248/292

[56] References Cited
UNITED STATES PATENTS

2,878,371 3/1959 Hanlin......................... 240/103 R X
2,621,882 12/1952 Fletcher......................... 248/280 X

FOREIGN PATENTS OR APPLICATIONS

562,676 12/1957 Belgium......................... 248/123
637,550 5/1950 Great Britain......................... 240/81 BD
487,571 12/1953 Italy......................... 248/123

Primary Examiner—Joseph F. Peters, Jr.
Attorney, Agent, or Firm—Karl F. Ross; Herbert Dubno

[57] ABSTRACT

A lamp with an articulated support comprising a stem and a plurality of arms disposed in succession and connected in an articulated manner to each other at intermediate points. A terminal one of the arms remote from the stem supports at the one end a lamp holder and its relative accessories and at the other end a counter weight is arranged in such a manner as to balance the terminal arm and the components supported by it. Each other arm has one end connected in an articulated manner to the next arm in the direction of the terminal arm, and its other end supports a counterweight arranged in such a manner as to balance the said other arm with all the components supported by it.

1 Claim, 3 Drawing Figures

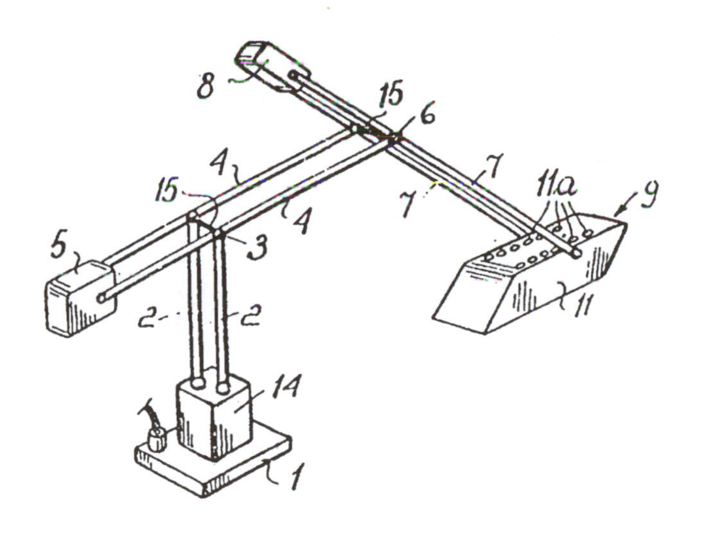

Figure 24.2

Sample First Page of a Utility Patent [1]

juries are prone to be very fair to "little guys" protecting the rights to their inventions.

One very real advantage of a utility patent is that it protects not only the specific embodiment of the ideas shown in the application, but also any functional or structural "equivalents" as well. Thus it really is the essence of the idea that is protected.

Of course, companies do try to protect their new and novel ideas just by keeping them secret, and by beating any would-be competitors to the marketplace. This, of course, is only very temporary protection. As soon as a product is in the marketplace without patent protection, others are free to copy.

There are some guidelines that will help inventors prepare and protect a utility (or a design) patent. We discuss only the more important ones here. Readers are referred to both [1] and [2] for more complete information. The main rules are listed below.

- Your invention must be novel and non-obvious considering the prior art.
- You can't patent an invention that was known or used by others, or described in print, before you invented it (even if they did not apply for a patent).
- You must file an application for a patent within one year of the first public disclosure (including publication of a technical paper) or sale of the product, and it is much better to file prior to any such disclosure. (Note that a non-disclosure contract is a way to protect a patent when discussing it with others, since your disclosure in that case is not a *public* disclosure.)
- If there is a question of two or more parties applying for the same patent, then the patent will be awarded to the party who first conceived the idea, providing that party did not "abandon" the invention in the meantime. Although this situation is not common, it is for this reason designers should keep excellent records of their work on an almost daily basis. Moreover, your drawings and notebooks, to be used in verifying dates and non-abandonment, must be witnessed (i.e., signed and dated) by someone with the technical background to understand them. This is burdensome, but it is essential to protect your invention. Your records will also be of great help to your patent attorney in developing the patent application.

Some famous utility patents that you may be able to find in a local U. S. Patent Office repository are:

- Patent No. 3633 - 1844 - Vulcanization of Rubber - Mr. Goodyear
- Patent No. 6469 - 1849 - Bouying Vessels - Mr. Abe Lincoln
- Patent NO. 608,849 - 1898 - Engine - Mr. Diesel
- Patent No. 223,898 - 1890 - Light Bulb - Mr. Edison
- Patent No. 821,393 - 1906 - Flying Machines - Mrs. Wright.

24.2.6 Design Patents

As noted above, design patents cover the ornamental aspects of a product such as the shape or configuration and/or any surface decoration. Since the functionality of a product is not protected by a design patent, design patents are generally more relevant to what we have called industrial or product design than to engineering design. However, many products involving engineering design have received design patents including, for example, automobile parts, multi-colored toothpaste, toys, and containers. Design patents can be used very effectively to protect designs from infringement.

To receive a design patent, the protected features must be have to do with appearance, not function, be unique, and not be obvious to others who work in the field. Design patents are considerably easier to obtain than utility patents, and they are also easier to enforce. A competitive design that is essentially the same in overall appearance as a result of using the novel features of the protected design will usually be considered to infringe. That is, exact duplication is not required; differences in detail do not matter.

You need an attorney to apply for and complete the process of obtaining a design patent. Pen and ink drawings are required. The cost, by the time the patent is received, will probably be in the two to three thousand dollar range.

Though a utility patent can have as many claims as can be defended, a design patent can make only one claim. This is a serious disadvantage since it means that every unique aspect of a product's design (there may well be several) requires a separate patent. Obviously, this can be expensive of both time and money.

Examples of several interesting design patents from [1] are shown in Figure 24.3.

24.2.7 Foreign Patents

In this era of intense and growing international competition, there is nearly as much concern among U.S. firms for obtaining foreign patents as for U.S. patents. Though U.S. laws allow for filing within one year of public disclosure, most foreign nations do not have that grace period. There is, however, an international treaty (the Paris Convention) that allows filing a design patent six months later, and a utility patent one year later, than the U.S. application without loss of rights. The issue of timing is so important, however, that patents should always be applied for as early as is feasible.

24.2.8 Summary

The table in Figure 24.4 [1] summarizes the basic methods that can be used to protect designs as intellectual property.

24.3 Liability Issues in Product Design

24.3.1 Introduction: The Importance

Note: Nothing in this section (or chapter or book) should be construed as the rendering of legal advice. Readers should always consult an attorney for legal advice.

We have already referred briefly in Chapter 1 to the issue of legal liability for the design of products that cause or contribute to injury (personal or financial) or even death. This is obviously a serious matter, and designers cannot — legally or ethically — take their responsibilities lightly or casually. The situation has become critical in recent years because of the growth of litigiousness in the U.S., the growth of the amounts of monetary awards by consumer-sympathetic juries, and the decline of personal responsibility in the use of products. Indeed, the situation is so extreme that there are calls for reform of the so-called "tort" system in which such suits are tried. Reform is surely needed, but it is not clear exactly what should be done, and it is not likely to occur any time soon. Thus, designers must accept the situation as it is, and protect themselves and their firms as diligently as possible.

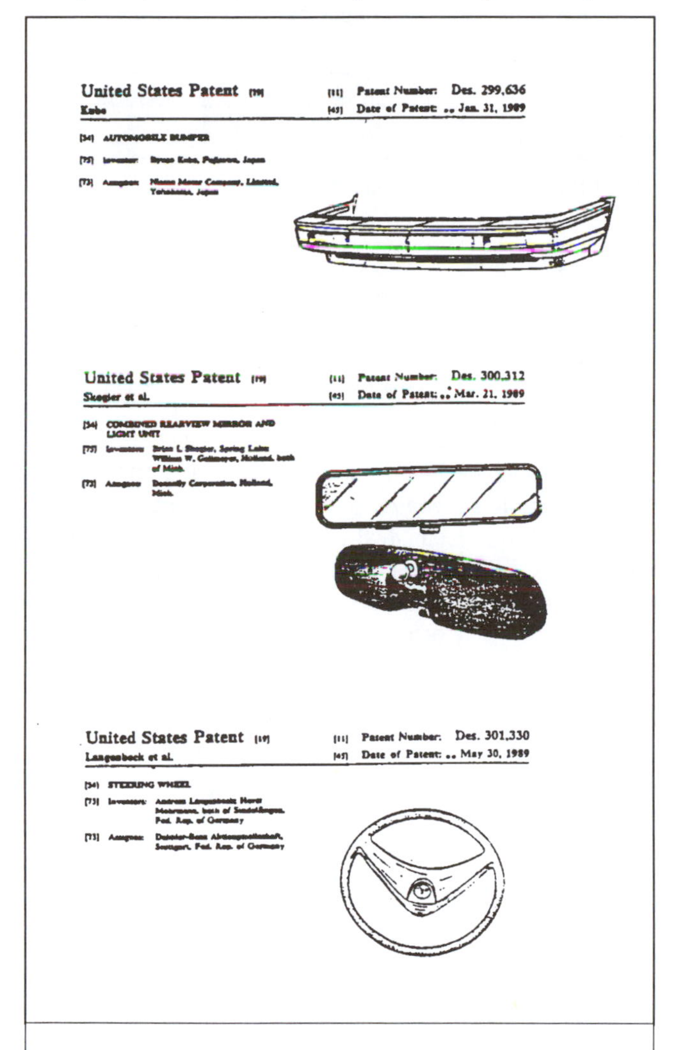

Figure 24.3 Sample Design Patents [1]

Horror stories abound. Here are some examples:

> A small manufacturer of exercise equipment could get product liability insurance of up to $300,000 for a premium of $400,000! (Note: The reason the premium can be higher than the protection is that the insurer has also to consider the costs of legal fees.)

> A safety equipment manufacturer with sales of $500,000 annually found it would have to pay $180,000 for product liability insurance; thus, the firm carries no insurance. One suit will wipe out the company and the employment it provides.

> A machine tool manufacturer was sued successfully for a machine built in 1895 because it does not meet *today's* safety requirements.

> A ladder manufacturer was sued successfully after a farmer fell with the ladder placed on a pile of manure beside the barn.

Clearly, manufacturers and their design engineers must understand the liability laws, and know how to protect themselves — not only by doing proper design but also by attending appropriately to the legal issues.

24.3.2 Trends in Liability Laws

Laws relating to product liability have evolved and changed dramatically over the years. Here is a summary of the major trends:

Let the Buyer Beware (Caveat Emptor). Before the industrial revolution, there were no product liability laws. It was up to purchasers to buy carefully and use products carefully. If there were defective or dangerous products, well, it was not the legal responsibility of designers or manufacturers. Hence the phrase "Caveat emptor", meaning "Let the buyer beware."

Privity. During the industrial revolution, say around 1842, the *law of privity* came into use. This means that liability could occur between those who entered into a contract or direct transaction. Consumers could sue retailers, who could in turn sue wholesalers, and so on. But consumers could not by-pass any step in the chain in order to sue manufacturers directly.

Negligence. Next, around 1916, a precedent was created when an automobile owner was allowed to sue the manufacturer for negligence. At the time, negligence was more difficult to establish than today, but this was nevertheless a significant change. Negligence is discussed in Section 24.3.5 below.

Limited Implied Warranty. Next the *Uniform Commercial Code* was made the law. This Code states that there

Basic Tools Table					
	Contract	Copyright	Trademark Trade Dress	Utility Patent	Design Patent
Protects Appearance	Yes	Yes	Yes	No	Yes
Protects Function	Yes	No	No	Yes	No
Application Required	No	No	No	Yes	Yes
Registration Available	No	Yes	Yes	Yes	Yes
Approximate Expenses	—	$200.	$1,000.	$5,000.-$6,000.	$1250.
Application Pendency Time	—	2-3 months	9-12 months	2-3 years	2-3 years
Expedited Registration Possible	—	Yes (1 day)	No	Yes (6-8 months)	Yes (6-8 months)
Length of Protection	as long as contract says	Life of author plus 50 years	as long as product is in use	17 years	14 years
Major Advantages	Protects design before other types of protection can be obtained; reflects intent of parties	quick; cheap	Broad infringement test (likelihood of confusion); registration not necessary	Patents are in vogue; courts can deal with them; hard to invalidate; can provide broad protection	Can protect essence of designs; injunctive relief readily available
Major Disadvantages	Only protection is with the contracting party; protection expires when product becomes public	"Separability" requirement; must prove "copying"; limited damages	Must prove "secondary meaning" & "nonfunctional"; hard to protect new products	Expensive to get; expensive to enforce	Will not protect functional elements

Figure 24.4 Summary of the Basic Tools of Design Protection [1]

is an *implied* warranty of fitness of products for their purposes and intended uses. Thus, products which proved dangerous when used as intended could be the basis for legal action against manufacturers.

Strict Product Liability. Since about 1962, the law has evolved to what is now called *strict product liability.* This means that manufacturers are "strictly"' liable when a product they cause to be placed on the market proves to have a defect in design or manufacture that causes or contributes to injury. There are only two defenses against liability under this law, and neither is easy to prove:

1. The user knew of the defect, but used the product anyway;
2. The user used the product in a manner in which it was *obviously* not intended. (However, it remains the responsibility of designers and manufacturers to protect users who may use products in ways that, though they are different from their intended purposes, should still be reasonably anticipated.)

The evolution and interpretation of the liability laws continues. The swing of the pendulum from caveat emptor to strict liability has probably gone too far. There is arguable concern that the frequency of suits, the percentage of successful suits, the size of the awards, and the costs of litigation are all discouraging to innovation and the creation or expansion of businesses that are needed for a strong economy. Perhaps there should be more responsibility required of those who use products carelessly or improperly? But designers will always have responsibilities, and they must be understood.

24.3.3 Strange Aspects of the Law

Though the basic tenets of tort law are quite reasonable, there are also some less reasonable aspects. First of all, the interpretation of the law changes rapidly and significantly as cases are tried and verdicts determined. The trend, to this writing, has been towards decreasing responsibility of product users and increasing risk for everyone else, especially manufacturers and designers. For example:

Manufacturers have been found liable even if a user *knowingly* or *deliberately* mis-used the product in a dangerous way, and manufacturers have been found liable for risks that were unknown to science and technology at the time of the product was designed and manufactured. Thus, the pendulum has swung from "Let the buyer beware" to "Let the manufacturer beware".

The law can also be different in different places. There is no uniform Federal code that applies to liability tort cases; instead each of the States has its own laws. Moreover, in most states there is no statute of limitations in product liability cases, and cases have been brought decades after the product was manufactured and sold. The product may even have been modified by the user or intermediate users along the way.

That a product meets all published government standards that exist is *not* a defense. Thus, such standards should be considered by designers only as necessary minimum requirements, not as sufficient requirements for legal purposes.

A manufacturer (or some other party) may be held 100% responsible even though by all logic they are only partly responsible. This is part of what is meant by "strict" liability. Suppose, for example, that something is designed in such a way that it can (say, when being repaired) be reassembled incorrectly. Though there may be clear instructions for assembly, and even a post-assembly test specified, the manufacturer can be held *fully* liable even though the reassembly was done wrong and the test was not performed in accordance with the manufacturer's instructions and training.

24.3.4 Factors in Determining Liability

A number of factors are generally considered in court cases concerning liability. How these play out in relative importance in individual cases is unpredictable (it helps to have the "best" lawyer), but the factors are reasonable. They are:

> The usefulness and desirability of the product. That is, a highly useful product may be forgiven some dangers that would not be acceptable in a product of lesser importance. However, be sure to read on.

> The availability of safer products to perform the same functions. That is, you are more likely to be liable if you design a more dangerous product than already exists for the same purpose.

> The availability of ways (via design or alternative technology) to eliminate or reduce the danger without significantly impairing the product's function or adding to its cost.

> The seriousness of the possible injuries, and their likelihood. Obviously, the more serious or more probable the possible injuries, the more you must protect against them.

> How obvious any dangers are to users, and how common the dangers are known to the segment of the public who uses, or mis-uses, the product? Thus, you must design to avoid the possibility of unexpected happenings during the use, or mis-use, of the product.

> The usefulness to users of adherence to warnings in avoiding dangers. Imagine the effect on a jury of a consumer who followed all the warnings and *still* got injured!

24.3.5 Avoiding Negligence

Negligence is the worst charge that can be leveled at a designer or a manufacturer. In some cases, negligence can even become a criminal instead of a civil (tort) issue. Negligence is, in effect, an error of omission; it is the failure to do something that should reasonably have been done. Following are some examples of non-performance that might be interpreted as negligent, but this is by no means intended to be a complete list of the possibilities:

> Failure to make use of state-of-the-art technology because you did not know about it. You should have taken the trouble to find out.

- Failure to provide adequate warnings.
- Failure to meet published standards. As we have noted above, meeting the standards is not enough, but it is surely negligent not to meet them at all.
- Failure to perform proper analyses.
- Failure to provide reasonable safety features or devices.
- Failure to perform proper tests and/or to interpret the results correctly.
- Failure to consider the possible ways in which a user might mis-use the product.
- Failure to consider the possible consequences of possible material variations, or variations in manufacture, or the effects of wear.
- Failure to consider possible hidden dangers in the product that might surprise users.

Avoiding or defending against charges of negligence involves two things: (1) not being negligent; and (2) keeping the records and documentation that can demonstrate and prove you were not negligent. Without the documentation, you may be found liable if something goes wrong, even if you were not in fact negligent. This means designers must — *must* — be diligent in recording computations, analyses, tests, and logic. Consult the attorney on your product realization team about what you need. For example, it is better to have considered a possible mis-use, but to have made inadequate compensation for it, than not to have considered it at all. Of course, any failures in designing the compensations might also be ruled negligence. But all can be lost without the records to prove what you did not neglect to do.

24.3.6 Two Actions to Take

Consult a Competent Lawyer. The first thing to be said to anyone involved in any way with the design of a product is this: *Consult a competent lawyer* who is very familiar with product liability law and the design implications. This is no place for amateurs. Your lawyer can advise you on the relevant law, on how the law is being interpreted currently in the courts, on cases that will help you determine where you are liable, on the liability possibilities you must consider in the design, on the documentation you should keep, and on the warnings you need to provide to potential customers in the use of the product. We cannot stress this enough: If you are involved in the design of a product, then make very sure that there is a competent product liability lawyer associated with the project.

Be Insured. Designers who work for companies or corporations should make certain at the time of employment that the firm has insurance or their own resources to protect designers in case they are sued as individuals. The reason is that it is increasingly common for consumers to sue not only the firm but also any individuals involved with the product. Actually injured parties will often sue anyone and everyone in sight including, for example, people or firms that sold or distributed the product.

Design engineers who work as consultants in private practice should have professional liability insurance. Though the protection is limited and the cost is high, it is just a fact of life that this insurance is a necessary cost of being in the design business. For information about such insurance, readers may contact the National Society of Professional Engineers or the American Society of Mechanical Engineers. A firm that specializes in professional liability insurance for engineers is Victor O. Schinneerr & Company, Inc., 5028 Wisconsin Avenue, N.W., Washington, D. C. 20016. This firm has an Office of Professional Liability Research that publishes an excellent manual entitled “Guidelines for Improving Practice” that has a wealth of information useful to design professionals related to liability issues.

24.3.7 Where Does the Pendulum Belong?

There is clearly a need for law that protects purchasers and users of products from unexpected, unnecessary, excessively dangerous products. But, as pointed out by the Committee for Economic Development [3], the current system “leads to distortions in production, withdrawal of products that optimally should be marketed, and wasteful litigation costs.” There is strong evidence that the current laws and their interpretation inhibits innovation, development, and marketing of reasonable new and competitive products. Insurance costs and court costs are extraordinarily high for manufacturers, adding to the cost of products. Surely a better balance of responsibility can be achieved.

Those calling for reform have a variety of suggestions. Some of them are listed below. Whether or not there will be reform, and if there is reform, when it will take place, is uncertain. We might, however, consider some of the following changes:

- Extend the concept of fault to include, at least in part, users who knowingly mis-use a product.
- End the use of “strict liability”, and instead apportion responsibility among the involved parties rationally.
- Limit the amounts of awards in a way consistent with other awards of damages.
- Develop a uniform National code for tort liability.
- Allow as a defense that the designer and manufacturer used the best state-of-the-art (science and technology) known at the time of design and manufacture.

We live in a democracy; it is up to the citizens to move the political process.

24.4 Ethical Considerations

24.4.1 Introduction

We have said earlier (in Chapter 1) that engineers, perhaps especially engineers who design, must be aware of the business and social (i.e., human, political, environmental, legal) contexts within which they work. There is also a professional context. The professional engineering societies (for example, the National Society of Professional Engineers

> "There is perhaps no profession besides engineering where probity [honesty, integrity] is so important. Simple honesty toward the end result, toward the object being made, is essential for the engineer....."
>
> Arthur M. Squires [4]*
> *The Tender Ship (1986)*

(NSPE), the American Society of Mechanical Engineers ASME), and most other professional societies for engineers) all have codes or canons of professional ethics to which their members are expected to adhere. The Canons of the ASME are reproduced here in Figure 24.5.

It is assumed that readers will review these Canons now before going on.

The requirements of these Canons make explicit what every thoughtful, responsible human being already knows and feels: Engineers have obligations to society (or the "public welfare") as well as to employers, whether those employers be manufacturing firms, consulting firms, or even their own private practice. And, of course, we know (even if the Canons don't say so) that we also have our own integrity and our own personal welfare, and that of our families, to consider as well. It is not difficult to identify these separate responsibilities; the difficult problem is how to behave when these responsibilities are, or appear to be, in substantial conflict.

It is not possible in this book to more than scratch the surface of how to deal in an ethical manner with these multiple obligations when they conflict, as they occasionally do, in complex ways. We can offer a few principles as guidelines, and references that go into more detail. In some very real cases, there may be no "right" answers, or even "best" answers on which all thoughtful informed persons could agree. But we have to face the issues when they arise, and try our best to do the right thing.

24.4.2 Some Brief General Guidelines to Ethical Behavior

Be honest. It is *never* ethical to be dishonest. Moreover, strict honesty will prevent many ethical problems (and personal disasters as well) from ever developing.

Withholding information that another party needs or would wish to know is as dishonest as is outright lying. Said another way: honesty requires being open. In court, for example, you must swear to tell not only the truth, but "the *whole* truth."

Though honesty is clearly the best policy in the long run, it is not always the easiest course in the short run. Thus, people tell "white lies" or withhold the truth, at least the whole truth, for a variety of supposed short term benefits: to avoid embarrassing someone (often themselves), to gain some financial or political advantage, to expedite a project, or to gain some prestige. But dishonesty never achieves these results in the long term; indeed, it almost always ends up achieving exactly the opposite of the desired results.

Here are examples of some ways engineers have been found to be dishonest; don't let them happen to you:

- "Fudging" or withholding experimental data;
- Withholding information about known risks in a design;
- Failing to report second jobs to employers (secret "moonlighting");
- Padding travel expenses;
- Stealing information from the company or customer;
- Shading opinion, say in a court case, to please a paying client;
- Saying, or giving the impression, that something is known or certain that is in fact not known or certain;
- Saying, or giving the impression, that one has knowledge or capability that one does not really possess;
- Misrepresenting previous training or experience;
- Secretly or knowingly infringing a competitor's patent or trade dress;
- Accepting a bribe (or expensive "gift") from a supplier or vendor.

There are many more possible examples, but most people recognize what is honest and what is dishonest.

Work Within Your Competence. One of the most important ethical requirements is to be competent. This is especially true in relation to social responsibilities and product safety issues. The better you are as an engineer, the less likely you are to make a mistake that endangers someone using your product. You are professionally and personally responsible for being, and continuing throughout your career to being, technically competent.

Noone can be fully competent in every area involved in a design. Partly for this reason, we use design teams. But even on teams, we may be asked to work beyond our competence. We must be open and honest about our capabilities and limitations, and resist such pressures.

This is not at all to say that we must avoid tackling anything new. But there is "new" that we can learn to be competent at, and "new"' that is beyond us (at least in the time frame of the current project). It is the latter that we resist responsibility for as soon as it becomes apparent to us that we are beyond our abilities.

Be Fair, Even to the Point of Generosity. This admonition has to do with giving credit where credit is due. In the case of patents and trade dress, as noted above, this is a legal requirement. In publications, reports, and in design, it is also an ethical responsibility. And it will earn you respect, save you great trouble (if not money), and just plain make things

---

* From *The Tender Ship: Government Management of Technological Change,* (a Pro Scienta title), by Arthur M. Squires, Birkhauser Boston, 1986, by permission of Arthur M. Squires, copyright holder.

Code of Ethics of Engineers

The Fundamental Principles

Engineers uphold and advance the integrity, honor, and dignity of the Engineering profession by:

I. using their knowledge and skill for the enhancement of human welfare;

II. being honest and impartial, and serving with fidelity the public, their employers and clients, and

III. striving to increase the competence and prestige of the engineering profession.

The Fundamental Canons

1. Engineers shall hold paramount the safety, health and welfare of the public in the performance of their professional duties.
2. Engineers shall perform services only in the area of their competence.
3. Engineers shall continue their professional development throughout their careers and shall provide opportunities for the professional and ethical development of those engineers under their supervision.
4. Engineers shall act in professional matters for each employer or client as faithful agents or trustees, and shall avoid conflicts of interest or the appearnace of conflicts of interest.
5. Engineers shall build their professional reputation on the merit of their services and shall not compete unfairly with others.
6. Engineers shall associate only with reputable persons or organizations.
7. Engineers shall issue public statements only in an objective and truthful manner.

Figure 24.5 Code of Ethics Prepared by the Board on Professional Practice and Ethics of the American Society of Mechanical Engineers (ASME), New York. Reprinted by permission of the American Society of Mechanical Engineers.

go better. You may also make some friends, but don't count on it! You must be satisfied within yourself that you are doing the right (and safe) thing.

You may, on the other hand, have your own ideas uncredited by others, co-opted, or even outright stolen. This is very frustrating when it happens, but there is little you can do after the fact. Protect yourself from this as much as possible by documenting your work and communications when they occur, and use contracts when appropriate.

Avoid Conflicts of Interest. Conflicts of interest are sometimes difficult, if not impossible, to avoid. They occur when one person is in a position to profit from actions or decisions under his or her influence. Suppose, for example, that a design engineer has a financial or personal stake (maybe via a relative or friend) in one of the vendors competing to supply parts for a products. Clearly, then, the designer may tend to make design or purchase decisions favorable to this vendor which are not necessarily in the best interest of the product.

Such decisions may be unconscious. It is not possible to argue convincingly that fair decisions can be made by human beings when there is some conflict of interest, though some people will argue that they can do so. It is the conflict that must be avoided. One must also do everything possible to avoid the appearance of conflict.

Sometimes conflicts of interest (usually weak ones) are simply impossible or impractical to avoid. In these cases, the conflict must be made known to everyone involved or concerned. Often, then, ways can be found to live with the situation. But keeping a conflict secret is not only unethical but also an almost sure path to trouble.

You do not, for example:

- Own stock in a competitive company, or in a company that does business with your firm if you are in a position to influence that those dealings;
- Accept expensive gifts from vendors or customers;
- Hire your friends or relatives;

Maintain the Integrity of the Design. In *The Tender Ship* [4], author and engineer Arthur Squires argues persuasively for what he calls "the integrity of the design." He notes how intense can be the pressures on designers and design teams to compromise their designs — pressures to hurry, to lower costs, or to copy from competitive products. Under such pressures, it is natural and human to try to accommodate everyone (especially the boss or a client or the sponsor). That pressure is a part of design. But if, in the process of accommodation, you compromise away the integrity of the design in ways that risk failure or endanger the public, then you will have crossed a line into unethical professional behavior. *You must know when to be firm, and be able to be firm, in defense of the design.* You have to be able to say "No" to the boss, client, customer, etc., if necessary to maintain the integrity of the design.

Of course, no product is 100% safe or 100% environmentally benign. Thus, determining when you are compromising too much is sometimes a difficult matter to judge. There is room for disagreement. But the issue must be faced, and discussed. The pressures can be intense; you must be strong. If the design lacks integrity, the product will fail one way or another, and then*you* — not the folks applying pressures — will be personally and professionally responsible.

24.4.5 A Tough Ethical Problem: "Whistle Blowing"

One of the toughest ethical problems for engineers arises from the dual obligation to serve loyally and simultaneously both their employer and the public interest. The vast majority of manufacturing firms try their very best to be responsible and honest corporate citizens. But problems do occasionally arise. For example, what do you do if your superiors in a firm suppress data that might (not surely, but might) indicate a safety problem with a product? You may not know this happening until the product is on the market. Or what do you do if you know your firm is polluting, say, by dumping some bad stuff out at night? These kinds of questions come in all sizes and shapes, and in all degrees of ambiguity concerning both the facts and the legal, ethical, and moral responsibilities. How serious does the issue have to be for you to risk losing your job, or causing serious damage to your company? How long and hard do you work inside your firm before you take stronger action, like "blowing the whistle" in some public way?

The various Canons of the professional engineering societies require that engineers serve *both* their firm's interests *and* the public interest, but say only a little that is not obvious and over-simplified about how to accomplish both these goals when they come clearly or subtly into conflict with one another. Moreover, the support of these societies for an individual engineer who blows the whistle on his own firm is unlikely at the National level, though occasionally a local chapter has defended a whistle blower. One reason is that such cases are hardly ever absolutely clear-cut. There is always the possibility that a whistle-blower is just a disgruntled employee trying to get even for some real or presumed injustice. The bottom line is: It is up to *you* to find the way to meet your multiple, sometimes conflicting obligations to society and company and self.

Obviously, depending on the degree and imminence of the danger, the first step is to do everything you can within your firm to correct the situation. At first, you can do this without jeopardizing your position. In fact, if done well, you may even enhance your position. Be sure your facts and logic are correct, and document your efforts. Learn how to negotiate effectively. Be skillful in dealing with the people involved; for example, try not to threaten their position, but instead show how they can be helped by fixing the problem. Appeal to the best interests of the people and the firm.

If this fails, or takes too long in cases where a danger is imminent, then at some point you will be forced to adopt a more powerful (and dangerous) approach. There are three broad types of actions you can take: (1) quit the firm (and then you still have to decide whether to blow or not to blow the whistle); (2) step up your pressure inside the firm, risking your position as needed; or (3) blow the whistle by notifying an outside organization of the problem. In the latter case, you should notify the appropriate people in the firm *before* you actually blow the whistle. And yes, face it, you might get fired at this point, so see a competent lawyer *first*.

It is always a good idea to consult with others about these problems. Find experienced people whose judgment you respect, who will listen to you and give you their perspective on the problem. In the end, of course, it is your decision, and you will live with the consequences. But you must try your best; the bottom line in such choices is that you must live with yourself.

A more complete discussion of engineering ethics can be found in [5].

24.5 Summary and Preview

Summary. A design is property. If you want to protect your design property from unauthorized (and unpaid for!) use, then there are several means available — contracts, copyrights, trademarks, design patents, and utility patents. This chapter has given you basic information about these protection methods.

Designers are liable for defects in design that result in injury or harm to users or others. You must use the best state of the art, do everything possible to anticipate and guard against injury (even from possible mis-uses), and keep good records of your efforts in order to minimize the possibility of negligence. Always consult a knowledgeable lawyer to make sure you have all that is required under the law.

Ethics are both personal and professional. Problems arise when our obligations to employers, the public, and/or ourselves come into conflict. Honesty and openness are essential, and will reduce the problems. Maintaining the integrity of the design against pressures to compromise too much is essential. Avoiding, or being open about, possible conflicts of interest is required. Seek the help and advice of knowledgeable people you trust.

Preview. Except for the Epilogue that follows, this concludes our story of Engineering Design. Of course, it is only one story, and there are many others that you should know about. And your learning can never cease. New knowledge relevant to design is being created all the time, and new methods for doing design are also being developed. It is up to you to keep up to date.

References

[1] Saidman, Perry. "The Basic Tools of Design Protection", Design Law Group, 1201 Connecticut Avenue, Suite 550, Washington, D.C. 20036, 1990. Substantial materials reprinted by permission of Mr. Perry Saidman, who reserve all rights under domestic and international copyright laws.

[2] Chicago (1967), "Patents at CP", Chicago Pneumatic Tool Company, New York, NY.

[3] Committee for Economic Development (Undated). "Who Should Be Liable? A Guide to Policy for Dealing With Risk. Committee for Economic Development", New York.

[4] Squires, A. M. (1986). *The Tender Ship*, Birkhausen Boston, Inc., Cambridge, MA 02139. Quotation by prmission of Arthur M. Squires.

[5] Martin, M. M. and Schinzinger, R. (1963), *Ethics in Engineering*, McGraw-Hill.

Recommended Reading

The Tender Ship, Birkhauser Boston, Inc., Cambridge, MA 02139. Squires, A. M., 1986.

Problems

24.1 A repairman is seriously injured by a product as he attempted to re-assemble it. He was new on the job, poorly trained, and had ignored the warnings about the proper assembly procedure. Discuss the possible liability of the designer and the manufacturer.

24.2 A one-of-a kind product newly in use essentially "blew up" destroying any evidence about the cause of the malfunction. Fortunately no one was injured, but there is considerable property damage. How could it be established whether the design or construction of the product was at fault?

24.3 We have argued that the pendulum has swung too far in the direction of strict liability for manufacturers. Others, however, do not agree. What are their arguments? What is your position?

24.4 Look up a utility patent (preferably for a relatively simple product) in a nearby patent repository, or find one in another book. Study the drawings and the claims. Can you find a way to design for the same functionality without violating the patent?

24.5 Suppose you know that a group of four students are constantly working together to cheat on exams. What are your conflicting obligations? How do you resolve the ethical issues?

24.6 You work for a small company that is violating environmental laws by discharging toxic fluid into a nearby stream. Your boss refuses even to discuss the issue with you. You probably can stop the discharges by spending some company money without the boss knowing. What do you do?

Epilogue

Now It Is Up To You

"All lovely things will have an ending."

Conrad Aiken
All Lovely Things Will Have an Ending (1914)

This completes our story of Engineering Design. We have described the context -- product realization -- and the kinds of problems encountered -- conceptual, configuration, and parametric design. We have described the related fields -- materials, manufacturing processes, and design for manufacturing. We have described the problems are solved -- guided iteration. As summarized in Table FM.1 (page FM-14), we have described the methods available to implement the steps in guided iteration. And finally, we have introduced some of the issues with which designers must constantly deal -- communicating, money, liability, patents, and ethics.

Words, and even pictures, are inadequate to present all the human drama, excitement, frustration and satisfaction of engineering design. Thus, like any written story, our story is incomplete and subject to different interpretations. We hope, however, that readers have learned enough to begin to add to and improve on the way design is done. Better yet, perhaps some bright young engineer will even write a new story based on a new, improved foundation and structure than we know at present.

Being teachers as well as researchers and learners ourselves, we cannot resist this final opportunity to offer some advice to younger readers: *Keep learning*. As time goes on, new methods will become available that improve on or replace the methods we are able to present today. Therefore read other books, including the new books and trade and research literature as they come out. Especially be sure to learn from your experiences. Remember that "experience without theory teaches nothing", so build a theory and story of engineering design for yourself as you learn.

Consider this end-of-course message from an unknown Professor of Medicine quoted by Charles Kettering in *Design* by Beakley and Chilton, Macmillan, 1974:

".... I have given you the best information available.....But before we part company, I want to caution you that in a few years from now perhaps half of the things I taught you won't be so. Unfortunately, I don't know which half."

Another physician, Dr. Lawrence Weed, has identified four traits of professional practice in any field: Be reliable, thorough, analytically sound, and efficient. We might add: Try to do the right thing. And be careful, too.

Finally, design is fun -- so enjoy it.

Now it is up to you.

"Begin at the beginning and go on until you come to the end; then stop."

Lewis Carrol
Alice's Adventures in Wonderland (1876)

This page is intentionally blank.

End Matter

Table of Symbols

Index

Table of Symbols

English Letters

A - Area; angle; die set length (for a progressive die); $K_{dmo}/(K_{dmo}+K_{dco})$

A_p - Projected area of a part normal to mold closure

a - Dimension, Length; constant

B - Belt Life Parameter

B - Die set width (for a progressive die); $K_{dco}/(K_{dmo}+K_{dco})$

B_m - Width of part normal to direction of mold closure

b - Dimension, length, thickness, width; tube spacing

b - Rib width; boss width

bt - Belt type

C - Equality Constraint Function; Cost

C - Value obtained from Fig. 11.19

C_{al} - Price per unit volume of aluminum

C_{an} - Cost to assemble n parts

C_b - Approximate relative tooling cost for an injection molded or die cast part due to size and basic complexity; total basic relative tool contruction cost for a stamped part

C_{b1}, C_{b2}, C_{b3} - Basic relative tool construction cost for a part due to shearing and local forming, bending and contour forming, respectively

CD - Center Distance between pulleys

$C_d = K_d/K_{do}$ - Tooling cost for a part relative to the tooling cost for the reference part

$C_{dc} = K_{dc}/K_{dco}$ - Tool construction cost for a part relative to the tool construction costs of the reference part - $C_b C_s C_t$

$C_{dcx} = K_{dcx}/K_{dco}$

$C_{dm} = K_{dm}/K_{dmo}$ - Tool material cost for a part relative to the tool material cost of the reference part

$C_{dmx} = K_{dmx}/K_{dmo}$

$C_{dx} = K_{dx}/K_o$

$C_e = K_e/K_{eo}$ - Processing cost for a part relative to the processing cost for the reference part

$C_{ea} = K_{ea}/K_{eao}$

$C_{ep} = K_{ep}/K_{epo}$

$C_{ex} = K_{ex}/K_{eo}$

C_h - Machine hourly rate

C_{ho} - Machine hourly rate ($/hr) for the reference part

C_{hr} - Relative machine hourly rate - machine hourly rate for a part relative to the machine hourly rate of the reference part - C_h/C_{ho}

C_{hra} - Relative machine hourly rate for aluminum die casting machines

C_{hrp} - Relative machine hourly rate for injection molding machines

$C_m = K_m/K_{mo}$; price per unit volume of metal

C_m, C_i - Constants in ASME Code for Shaft Design

$C_{mr} = K_p/K_{po}$ - relative material cost per unit volume

C_p = Price per unit volume of thermoplastic resin

C_r - Total cost of a part relative to the total cost for the reference part

C_s - Multiplier to account for the cavity detail (subsidiary complexity) of injection molded or die cast parts

C_{ss} - Price per unit volume of stainless steel

C_t - For injection molded and die cast parts, a multiplier to account for the tolerance and surface finish requirements for the tooling

DCF - Discounted cash Flow

D_p - Diameter of the drawn feature after drawing

d - Dependency

d, D - Diameter

d_o - Diameter of the drawn feature before drawing
E - Modulus of Elasticity; Energy; Net Profits
F - Estimated total force in tons required to form a stamped part; Force; Belt Life Parameter
F_b - Bending force required (stamping)
F_{dm} - Factor to account for the effect of sheet thickness on die material cost
F_{dr} - Force required for drawing (stamping)
FFR- Fatigure Failure Ratio
F_{fl} - Flanging force required (stamping)
F_{fo} - Force required for radius forming (stamping)
F_{mb} - Factor to account for effect of part material on basic relative tool construction cost due to bending
F_{mc} - factor to account for effect of part material on basic relative tool construction cost due to shearing and local forming
F_{mf} - Factor to account for the effect of part material on the basic relative tool construction cost due to contour forming
F_{out} - Force required to separate the stamped part from the trip
F_p - Machine tonnage required to mold or die cast a part; total press force required to stamp a part
F_{st} - Stripper force required to removed the strip from round the punches
F_{sz} - Total shearing force required to stamp the part (F_{out} + F_{st})
F_t - Factor to account for the effect of sheet thickness on die construction costs
f - Friction factor; Deflection
$f_d = K_{do}/N_o K_o$
$f_e = K_{eo}/K_o$
$f_m = K_{mo}/K_o$
G - Mass Flow Rate; Torsional Modulus of Elasticity
g, g_o - 32.2 ft-lbm/lbf-sec^2
H - Head
H_m - Height of a part in the direction of mold closure
HP - Horsepower
h - Coefficient of Heat Transfer; Rib height; boss height
h, H - Height
I - Moment of Inertia; An initial amount of money invested in a project
i - Rate of interest expressed as a decimal
J - Polar Moment of Inertia
k - Spring Constant; Belt Life Parameter
K - Stress Concentration Factor; Proportionality Constant
K_d - total tooling cost (mold or die cost) for the part
K_d/N - Tooling cost per part
K_{da} - Tooling cost for an aluminum die casting
K_{dc} - Die construction cost for a part
K_{dco} - Die construction cost for the reference part
K_{dcx} - Die construction cost for the single functionally equivalent part to replace the n parts
K_{di} - Total tooling cost for the ith component
K_{dm} - Die material cost for a part
K_{dmo} - Die material cost for the reference part
K_{dmx} - Die material cost for the single functionally equivalent part to replace the n parts
K_{do} - Tooling cost for the reference part
K_{dp} - Tooling cost for a thermoplastic injection molding
K_{ds} - Tooling cost for a stamped part
K_{dx} - Total tooling cost for the single functionally equivalent part to replace the n parts
K_e - Processing cost per part (approximately the product of the machine hourly rate andthe cycle time)
K_{ea} - Processing cost for an aluminum die casting
K_{eao} - Processing cost for the die cast aluminum reference part
K_{ei} - Processing cost for the ith component
K_{eo} - Processing cost per part for the reference part
K_{ep} - Processing cost for a thermoplastic injection molding
K_{ep} - Processing cost for the thermoplastic injection molded reference part
K_{es} - Processing cost for a stamped part
K_{ex} - Processing cost for the single functionally equivalent part to replace the n parts
K_m - Material cost per part
K_{ma} - Material cost for an aluminum die casting
K_{mi} - Material cost for the ith component
K_{mo} - Material cost per part for the reference part

K_{mp} - Material cost for a thermoplastic injection molding

K_{ms} - Material cost for a stamped part

K_{mx} - Material cost for the single functionally equivalent part to replace the n parts

K_o - Manufacturing cost of the reference part

K_p - Material cost per unit volume

K_{po} - Material cost per unit volume for the reference part

K_t - Total production cost of a single part

K_{ta} - Total cost to produce a part as an aluminum die casting

K_{ti} - Total cost of producing the ith component

K_{tp} - Total cost to produce a part as a thermoplastic injection molding

K_{tx} - Total cost of producing the single functionally equivalent part to replace the n parts

ΔK - Cost to produce and assemble n parts minus the cost of producing one functionally equivalent part

ΔK_{ap} - Difference in cost to produce a part as an aluminum die casting and as a thermoplastic injection molding

ΔK_e - Difference in processing costs between the sum of the n individual parts and the single replacement part; difference in processing cost between an injection molded part and a functionally equivalent stamped part

$\Delta K_m = K_{ma} - K_{mp}$

ΔK_{ps} - Difference in cost to injection mold a part and to stamp a functionally equivalent part

L - Inequality Constraint Function

L, B, H - Lengths of the sides of the basic envelope (L>B>H)

L_b - Maximum bend length

L_{db} - Die block width (progressive die)

L_{dl} - Die block length (progressive die)

L_f - Form length of a stamped part

L_m - The length of the part in a direction normal to the direction of mold closure

L_{out} - Peripheral length of an unfolded part

L_s - Peripheral complexity ratio - $L_{out}/2(L_{ul} + L_{uw})$

L_{ul} - Flat envelope length

L_{uw} - Flat envelope width

l, L - Length; Life

M - Moment of Inertia; Number of Belts

M_a - projected area of the mold base normal to the direction of mold closure

M_t - Thickness of the mold base

M_{wf} - Thickness of core plate

M_{ws} - Thickness of the mold's side walls

N - Number of Spring Coils; Rotational Speed; Production volume

N_a - Number of Active Spring Coils

N_a - Number of active stations for progressive die

N_{i1}, N_{i2}, N_{ie} - Number of idle stations required for shearing and local forming, bending and contour forming, respectively.

N_o - Production volume of reference part - N

N_t - Number of Tube Passes

N_{tp} - Number of Tubes Per Pass

N_{tu} - Number of Heat Transfer Units

n - Number of interest periods, usually years; Number of Shell Passes; Number of parts in an assembly

n_c - Number of cavities in a multple cavity mold

P - Present amount of money, *or* the present value of a future amount of money; Pressure; Power; Force; Perimeter

PD - Pressure drop

PDP- Problem Definition Parameter

PV - Present Value

PVROI - Present Value return on investment

Q - Number of Inactive Spring Coils; Belt Life Parameter

Q - Rate of Heat Transfer; Rate of Volume Flow

R - An amount of money received or spent periodically, usually annually; Ratio; Radius of curvature of radius formed elemental plate

R - Ratio; Radius of curvature of a radius formed elemental plate

Re - Reynold's Number

ROI- Return on Investment

r, R - Radius

S - An amount of money at some future time; Stress; Velocity

S/N - Signal to Noise Ration

SAROI - Simple Accounting Return on Investment

SD - Standard Deviation

S_{ds} - Die set area, AB

SEP- Solution Evaluation Paramter

SFR- Shrear Failure Ratio
T - Temperature; Target Value of a Variable; Torque
TFR- Tensile Failure Ratio
Ti - Pre-Stress Tension in a Bolt
t - cycle time (processing time) for a part; sheet thickness for a stamped part; Thickness; Time
tb - Belt Life Parameter
t_b - Basic relative cycle time
t_{cy} - Effective cycle time for a press (stamping)
t_e - Additional relative cycle time due to inserts and internal threads
t_g - Glass transition temperature
t_o - Cycle time (processing time) for the reference part
t_p - A time penalty factor to account for surface requirements and tolerances on the relative cycle time
t_r - $= t/t_o$ - Relative cycle time of a part
U - Criterion Function; Overall Coefficient of Heat Transfer; Velocity
V - Part volume; Volume; Velocity
V_o - Volume of the reference part
W - Weight
w - Wall thickness; Width
x - Design Variable
Y - Belt Life Parameter
y - Solution Evaluation Parameter
Z - Distance Parameter

Greek Letters

θ - Angle of Twist
ω - Angular Velocity
σ - Normal Stress; Tensile strength
μ - Poisson's Ratio
τ - Shear stress; Shear strength
ϕ - Angular Deflection of Shaft
θ - Coordinate Angle; Slope of Shaft
δ - Spring Displacement

Subscripts

1,2 - principal
av - average
b - belt life parameter
c - belt life parameter
c - cold side
e - endurance
h - hot side
i - i^{th} value
j - j^{th} value
i - inside
m - m^{th} parameter
n - n^{th} parameter
o - outside
p - proof
r - range
rF - fatigue limit
rF' - fatigue limit
T - ultimate; transverse
t - total; tangential
y - yield

Superscripts

Superscripts
* - Value of a design variable at an optimum

Index

Note: The slash mark (/) designates a range of pages.